Lecture Notes in Computer Science   7194

Commenced Publication in 1973
Founding and Former Series Editors:
Gerhard Goos, Juris Hartmanis, and Jan van Leeuwen

Ronald Cramer (Ed.)

Theory
of Cryptography

9th Theory of Cryptography Conference, TCC 2012
Taormina, Sicily, Italy, March 19-21, 2012
Proceedings

 Springer

Volume Editor

Ronald Cramer
Cryptology Group, CWI Amsterdam
and
Mathematical Institute, Leiden University
The Netherlands
E-mail: cramer@cwi.nl, cramer@math.leidenuniv.nl

ISSN 0302-9743   e-ISSN 1611-3349
ISBN 978-3-642-28913-2   e-ISBN 978-3-642-28914-9
DOI 10.1007/978-3-642-28914-9
Springer Heidelberg Dordrecht London New York

Library of Congress Control Number: 2012933240

CR Subject Classification (1998): E.3, C.2, D.4.6, K.6.5, G.2, I.1

LNCS Sublibrary: SL 4 – Security and Cryptology

Typesetting: Camera-ready by author, data conversion by Scientific Publishing Services, Chennai, India

Printed on acid-free paper

Springer is part of Springer Science+Business Media (www.springer.com)

Preface

TCC 2012, the 9th Theory of Cryptography Conference, was held in Taormina (Sicily), Italy, during March 19–21, 2012. It was sponsored by the International Association for Cryptologic Research (IACR). The General Chairs were Nelly Fazio and Rosario Gennaro. The Local Arrangements Chair was Dario Catalano.

By the deadline of September 15, 2011, the Program Committee (PC) had received 131 electronic submissions. As usual, the selection process was carried out using a Web-based interface and consisted of three phases. In the review phase each submission was assigned to at least three PC members for independent review (six in the case of a PC member submission). In the discussion phase, reviews pertaining to the same submission were compared and agreement on a common view was reached where possible. Additional reviews were solicited as needed. The perceived relative merits of all submissions were taken into consideration as a basis for the selection phase. By the December 1 deadline for notification of decisions, the PC had selected 36 submissions for (20-min.) presentation at the conference. Reviewer comments for all submissions were sent out to their respective authors soon after. These proceedings contain the revised versions of the 36 selected submissions, as received by January 3, 2012. These revised versions were not subjected to further review by the PC and authors bear full responsibility for contents.

The program also featured two invited (60-min.) talks. Jens Groth and Sergey Yekhanin treated us to excellent surveys of Non-Interactive Zero-Knowledge and Locally Decodable Codes, respectively. A Best Student Paper Award was shared by Nir Bitansky and Omer Paneth for their paper "Point Obfuscation and 3-round Zero-Knowledge" and by Anindya De for his paper "Lower Bounds in Differential Privacy". In addition, a traditional Rump Session was held, consisting of (5-min.) research announcements. It was organized and chaired by Tal Malkin. The organizers had this evening session well catered for with nice drinks and snacks.

I thank the PC members for their hard work, as well as the external reviewers. Oded Goldreich (TCC Steering Committee Chair) and Yuval Ishai (TCC 2011 Program Chair) provided quick and helpful advice upon my request as well as answers to my questions, for which I am grateful. The PC used Shai Halevi's excellent Web Submission and Review software to handle the submissions. Thanks to Maarten Dijkema of CWI's IT Support for running this software at our system and for his unwavering assistance and thanks to Shai for rendering efficient "customer service" to us. Also, thanks to Tal for running the Rump Session. The

VI Preface

organizers wish to express their gratitude to the TCC 2012 sponsors Alcatel-Lucent Bell Labs, IBM Research and Microsoft Research for generous donations that supported local organization in several ways, including student stipends. In turn, I thank Nelly, Rosario and Dario for our pleasant collaboration. Finally, thanks to all authors of submissions to TCC 2012.

January 2012 Ronald Cramer

TCC 2012

9th IACR Theory of Cryptography Conference

Taormina, Sicily, Italy
March 19–21, 2012

Sponsored by the *International Association for Cryptologic Research (IACR)*

General Chairs

Nelly Fazio   City University of New York (CUNY), USA
Rosario Gennaro   IBM Thomas J. Watson Research Center, USA

Local Arrangements Chair

Dario Catalano   Università di Catania, Italy

Program Chair

Ronald Cramer   CWI Amsterdam and Leiden University,
                         The Netherlands

Program Committee

Masayuki Abe   NTT, Japan
Amos Beimel   Ben-Gurion University, Israel
Alexandra Boldyreva   Georgia Tech, USA
Iftach Haitner   Tel Aviv University, Israel
Martin Hirt   ETH Zurich, Switzerland
Dennis Hofheinz   Karlsruhe Institute of Technology, Germany
Jonathan Katz   University of Maryland, USA
Vadim Lyubashevsky   ENS, France
Tal Malkin   Columbia University, USA
Daniele Micciancio   UCSD, USA
Jesper Buus Nielsen   Aarhus University, Denmark
Carles Padró   Nanyang Technological University, Singapore
Mike Rosulek   University of Montana, USA
Amit Sahai   UCLA, USA
Berry Schoenmakers   TU Eindhoven, The Netherlands

Vinod Vaikuntanathan   University of Toronto, Canada
Daniel Wichs   IBM T.J. Watson Research Center, USA
Jürg Wullschleger   Université de Montréal and McGill University,
                                Canada
Moti Yung   Google Research, New York, USA

Steering Committee

Mihir Bellare   UCSD, USA
Ivan Damgård   Aarhus University, Denmark
Oded Goldreich (Chair)   Weizmann Institute of Science, Israel
Shafi Goldwasser   MIT, USA
Shai Halevi   IBM T.J. Watson Research Center, USA
Russell Impagliazzo   UCSD, USA
Ueli Maurer   ETH Zurich, Switzerland
Silvio Micali   MIT, USA
Moni Naor   Weizmann Institute of Science, Israel
Tatsuaki Okamoto   NTT, Japan

External Reviewers

Michel Abdalla   David Cash
Divesh Aggarwal   Nishanth Chandran
Shweta Agrawal   Mahdi Cheraghchi
Joël Alwen   Kai-Min Chung
Elena Andreeva   Sandro Coretti
Kazumaro Aoki   Dana Dachman-Soled
Benny Applebaum   Bernardo David
Claudio Orlandi   Gregory Demay
Seiko Arita   Yi Deng
Gilad Asharov   Alex Dent
Endre Bangerter   Yevgeniy Dodis
Boaz Barak   Leo Ducas
Kfir Barhum   Stefan Dziembowski
Raef Bassily   Nico Döttling
Itay Berman   Pooya Farshim
Gaetan Bisson   Sebastian Faust
Nir Bitansky   Serge Fehr
Niek J. Bouman   Dario Fiore
Elette Boyle   Eiichiro Fujisaki
Zvika Brakerski   Dave Freeman
Jan Camenisch   Jakob Funder
Ran Canetti   David Galindo
Igor Carboni Oliveira   Sanjam Garg
Ignacio Cascudo   Rosario Gennaro

Wesley George
Sergey Gorbunov
Dov Gordon
Vipul Goyal
Matthew Green
Jens Groth
Robbert de Haan
Shai Halevi
Moritz Hardt
David Harris
Carmit Hazay
Brett Hemenway
Javier Herranz
Viet Tung Hoang
Susan Hohenberger
Thomas Holenstein
Perry Hooker
Pavel Hubacek
Yuval Ishai
Abhishek Jain
Yael Tauman Kalai
Eike Kiltz
Vladimir Kolesnikov
Daniel Kraschewski
Hugo Krawczyk
Stephan Krenn
Ralf Küsters
Virendra Kumar
Eyal Kushilevitz
Robin Künzler
Allison Bishop Lewko
Hendrik Lenstra
Xin Li
Peter van Liesdonk
Huijia Rachel Lin
Adriana López-Alt
Christoph Lucas
Hemanta Maji
Takahiro Matsuda
Sigurd Torkel Meldgaard
Josep M. Miret
Petros Mol
Tal Moran
Ryo Nishimaki
Kobbi Nissim

Peter Sebastian Nordholt
Miyako Ohkubo
Tatsuaki Okamoto
Anat Paskin-Cherniavsky
Rafael Pass
Kenny Paterson
Arpita Patra
Serdar Pehlivanoglu
Chris Peikert
Duong Hieu Phan
Krzysztof Pietrzak
Benny Pinkas
Manoj Prabhakaran
Pavel Raykov
Mariana Raykova
Omer Reingold
Oded Regev
Leonid Reyzin
Alon Rosen
Aaron Roth
Guy Rothblum
Yannis Rouselakis
Louis Salvail
Yu Sasaki
Christian Schaffner
Dominique Schröder
Gil Segev
Ronen Shaltiel
Elaine Shi
Nigel Smart
Bart de Smit
Martijn Stam
Damien Stehlé
John Steinberger
Ron Steinfeld
Marc Stevens
Björn Tackmann
Katsuyuki Takashima
Kunal Talwar
Aris Tentes
Stefano Tessaro
Tomas Toft
Marco Tomamichel
Boaz Tsaban
Dominique Unruh

Table of Contents

Dedicated Encryption I

Security Amplification

Resettable and Parallel Zero Knowledge

Dedicated Encryption II

Pseudorandomness II

Computing on Authenticated Data

Jae Hyun Ahn[1], Dan Boneh[2,*], Jan Camenisch[3,**], Susan Hohenberger[1,***],
Abhi Shelat[4,†], and Brent Waters[5,‡]

[1] Johns Hopkins University
{arjuna,susan}@cs.jhu.edu
[2] Stanford University
dabo@cs.stanford.edu
[3] IBM Research – Zurich
jca@zurich.ibm.com
[4] University of Virginia
abhi@cs.virginia.edu
[5] University of Texas at Austin
bwaters@cs.utexas.edu

Abstract. In tandem with recent progress on computing on encrypted data via fully homomorphic encryption, we present a framework for computing on *authenticated* data via the notion of slightly homomorphic signatures, or P-homomorphic signatures. With such signatures, it is possible for a third party to *derive* a signature on the object m' from a signature of m as long as $P(m, m') = 1$ for some predicate P which captures the "authenticatable relationship" between m' and m. Moreover, a derived signature on m' reveals *no extra information* about the parent m.

Our definition is carefully formulated to provide one unified framework for a variety of distinct concepts in this area, including arithmetic, homomorphic, quotable, redactable, transitive signatures and more. It includes being unable to distinguish a derived signature from a fresh one *even when given the original signature*. The inability to link derived

* Supported by NSF, DARPA, and AFOSR. Applying to all authors, the views and conclusions contained in this document are those of the authors and should not be interpreted as representing the official policies, either expressed or implied, of the Defense Advanced Research Projects Agency or the US government.
** This work has been funded by the European Community's Seventh Framework Programme (FP7/2007-2013) under grant agreement no. 216483 (PrimeLife).
*** Supported by the Defense Advanced Research Projects Agency (DARPA) and the Air Force Research Laboratory (AFRL) under contract FA8750-11-2-0211, the Office of Naval Research under contract N00014-11-1-0470, a Microsoft Faculty Fellowship and a Google Faculty Research Award.
† Supported by NSF CNS-0845811 and TC-1018543, Defense Advanced Research Projects Agency (DARPA) and the Air Force Research Laboratory (AFRL) under contract FA8750-11-2-0211, and a Microsoft New Faculty Fellowship.
‡ Supported by NSF CNS-0915361 and CNS-0952692, AFOSR Grant No: FA9550-08-1-0352, DARPA PROCEED, DARPA N11AP20006, Google Faculty Research award, the Alfred P. Sloan Fellowship, Microsoft Faculty Fellowship, and Packard Foundation Fellowship.

R. Cramer (Ed.): TCC 2012, LNCS 7194, pp. 1–20, 2012.
© International Association for Cryptologic Research 2012

signatures to their original sources prevents some practical privacy and linking attacks, which is a challenge not satisfied by most prior works.

Under this strong definition, we then provide generic constructions for all univariate and closed predicates, and specific efficient constructions for a broad class of natural predicates such as quoting, subsets, weighted sums, averages, and Fourier transforms. To our knowledge, these are the first efficient constructions for these predicates (excluding subsets) that provably satisfy this strong security notion.

1   Introduction

In tandem with recent progress on computing *any function* on encrypted data, e.g., [27,46,44], this work explores computing on unencrypted signed data. In the past few years, several independent lines of research touched on this area:

- Quoting/redacting: [45,32,2,36,30,17,16,18] Given Alice's signature on some message m anyone should be able to derive Alice's signature on a subset of m. Quoting typically applies to signed text messages where one wants to derive Alice's signature on a substring of m. Quoting can also apply to signed images where one wants to derive a signature on a subregion of the image (say, a face or an object) and to data structures where one wants to derive a signature of a subset of the data structure such as a sub-tree of a tree.
- Arithmetic: [33,50,22,13,26,12,11,48] Given Alice's signature on vectors $\mathbf{v}_1, \ldots, \mathbf{v}_k \in \mathbb{F}_p^n$ anyone should be able to derive Alice's signature on a vector $\mathbf{v}$ in the linear span of $\mathbf{v}_1, \ldots, \mathbf{v}_k$. Arithmetic on signed data is motivated by applications to secure network coding [25]. We show that these schemes can be used to compute authenticated linear operations such as computing an authenticated weighted sum of signed data and an authenticated Fourier transform. As a practical consequence of this, we show that an untrusted database storing signed data (e.g., employee salaries) can publish an authenticated average of the data without leaking any other information about the stored data. Recent constructions go beyond linear operations and support low degree polynomial computations [11].
- Transitivity: [41,35,6,31,7,43,49,40] Given Alice's signature on edges in a graph G anyone should be able to derive Alice's signature on a pair of vertices (u, v) if and only if there is a path in G from u to v. The derived signature on the pair (u, v) must be indistinguishable from a fresh signature on (u, v) had Alice generated one herself [35]. This requirement ensures that the derived signature on (u, v) reveals no information about the path from u to v used to derive the signature.

In this paper, we put forth a general framework for computing on authenticated data that encompasses these lines of research and much more. While prior definitions mostly contained artifacts specific to the type of malleability they supported and, thus, were hard to compare to one another, we generalize and strengthen these disparate notions into a single definition. This definition can be instantiated with any predicate, and we allow repeated computation on the signatures (e.g., it is possible to quote from a quoted signature.) During our study, we realized that the "privacy" notions offered by many existing definitions are, in our view, insufficient for some practical applications. We therefore require a stronger (and seemingly a significantly more challenging to achieve) property called *context hiding*. Under this definition, we provide two generic solutions for computing signatures on any univariate, closed predicate; however, these generic constructions are not efficient. We also present efficient constructions for three problems: quoting substrings, a subset predicate, and a weighted average over data (which captures weighted sums and Fourier transforms). Our quoting substring construction is novel and significantly more efficient than the generic solutions. It is detailed in Section 4. For the problems of subsets and weighted averages, we show somewhat surprising connections to respective existing solutions in attribute-based encryption and network coding signatures in Section 5.

1.1   Overview

A general framework. Let $\mathcal{M}$ be some message space and let $2^{\mathcal{M}}$ be its powerset. Consider a predicate $P : 2^{\mathcal{M}} \times \mathcal{M} \to \{0,1\}$ mapping a set of messages and a message to a bit. Loosely speaking we say that a signature scheme supports computations with respect to P if the following holds:

Let $M \subset \mathcal{M}$ be a set of messages and let m' be a *derived* message, namely m' satisfies $P(M, m') = 1$. Then there exists an efficient procedure that can derive Alice's signature on m' from Alice's independent signatures on all of the messages in M.

For the quoting application, the predicate P is defined as $P(M, m') = 1$ iff m' is a quote from the set of messages M. Here we focus on quoting from a single message m so that P is false whenever M contains more than one component[1], and thus use the notation $P(m, m')$ as shorthand for $P(\{m\}, m')$. The predicate P for arithmetic computations is defined in the full version [1] and essentially says that $P\big((\mathbf{v}_1, \ldots, \mathbf{v}_k), \mathbf{v} \big)$ is true whenever $\mathbf{v}$ is in the span of $\mathbf{v}_1, \ldots, \mathbf{v}_k$.

We emphasize that signature derivation can be iterative. For example, given a message-signature pair (m, σ) from Alice, Bob can publish a derived message-signature pair (m', σ') for an m' where $P(m, m')$ holds. Charlie, using (m', σ'), may further derive a signature σ'' on m''. In the quoting application, Charlie is quoting from a quote which is perfectly fine.

Security. We give a clean security definition that captures two properties: unforgeability and context hiding. We briefly discuss each in turn and give precise definitions in the next section.

- Unforgeability captures the idea that an attacker may be given various derived signatures (perhaps iteratively derived) on messages of his choice. The

---

[1] We leave it for future work to construct systems for securely quoting from two messages (or possibly more) as defined next.

attacker should be unable to produce a signature on a message that is not derivable from the set of signed messages at his possession. E.g., suppose Alice generates (m, σ) and gives it to Bob who then publishes a derived signature (m', σ'). Then an attacker given (m', σ') should be unable to produce a signature on m or on any other message m'' such that $P(m', m'') = 0$.

- Context hiding captures an important privacy property: a signature should reveal nothing more than the message being signed. In particular, if a signature on m' was derived from a signature on m, an attacker should not learn anything about m other than what can be inferred from m'. This should be true even if the original signature on m is revealed. For example, a signed quote should not reveal anything about the message from which it was quoted, including its length, the position of the quote, whether its parent document is the same as another quote, whether it was derived from a given signed message or generated freshly, etc.

Defining context hiding is an interesting and subtle task. In the next section, we give a definition that captures a very strong privacy requirement. We discuss earlier attempts at defining privacy following our definition in Section 2.3; while many prior works use a similar sounding *intuition* as we give above, most contain a fundamental difference to ours in their *formalization*.

We note that notions such as group or ring signatures [23,5,19,9,42] have considered the problem of hiding the identity of a signer among a set of users. Context hiding ensures privacy for the data rather than the signer. Our goal is to hide the legacy of how a signature was created.

Efficiency. We require that the size of a signature, whether fresh or derived, depend only on the size of the object being signed. This rules out solutions where the signature grows with each derivation.

Generic Approaches. We begin with two generic constructions that can be inefficient. They apply to *closed, univariate* predicates, namely predicates $P(M, m')$ where M contains a single message (P is false when $|M| > 1$) and where if $P(a, b) = P(b, c) = 1$ then $P(a, c) = 1$. The first construction uses any standard signature scheme S where the signing algorithm is deterministic. (One can enforce determinism using PRFs [28].) To sign a message $m \in \mathcal{M}$, one uses S to sign each message m' such that $P(m, m') = 1$. The signature consists of all these signature components. To verify a signature for m, one checks the signature component corresponding to the message m. To derive a signature m' from m, one copies the signature components for all m'' such that $P(m', m'') = 1$. Soundness of the construction follows from the security of the underlying standard scheme S and context hiding from the fact that signing in S is deterministic.

Unfortunately, these signatures may become large consisting up to $|\mathcal{M}|$ signature components — effecting both the signing time and signature size. Our second generic construction alleviates the space burden by using an RSA accumulator. The construction works in a similar brute force fashion where a signature on m is an accumulator value on all m' such that $P(m, m') = 1$. While this produces short signatures, the time component of both verification and derivation are even worse than the first generic approach. Thus, these generic approaches are too expensive for most interesting predicates. We detail these generic approaches and proofs in the full version [1], where we also discuss a generic construction using NIZK.

Our Quoting Construction. We turn to more efficient constructions. First, we set out to construct a signature for quoting *substrings*[2], which although conceptually simple is non-trivial to realize securely. As an efficiency baseline, we note that the brute force generic construction of the quoting predicate would result in n^2 components for a signature on n characters. So any interesting construction must perform more efficiently than this. We prove our construction selectively secure.[3] In addition, we give some potential future directions for achieving adaptive security and removing the use of random oracles.

Our construction uses bilinear groups to link different signature components together securely, but in such a way that the context can be hidden by a re-randomizing step in the derivation algorithm. A signature in our system on a message of length n consists of $n \lg n$ group elements; intuitively organized as $\lg n$ group elements assigned to each character. To derive a new signature on a substring of ℓ characters, one roughly removes the group elements not associated with the new substring and then re-randomizes the remaining part of the signature. This results in a new signature of $\ell \lg \ell$ group elements. The technical challenge consists in simultaneously allowing re-randomization and preserving the "linking" between successive characters. In addition, there is a second option in our derive algorithm that allows for the derivation of a short signature of $\lg \ell$ group elements; however the derive procedure cannot be applied again to this short signature. *Thus, we support quoting from quotes, and also provide a compression option which produces a very short quote, but the price for this is that it cannot be quoted from further.*

Computing Signatures on Subsets and Weighted Averages. Our final two contributions are schemes for deriving signatures on subsets and weighted averages on signatures. Rather than create entirely new systems, we show connections to existing Attribute-Based Encryption schemes and Network Coding Signatures. We sketch those constructions in Section 5 and provide further details in [1].

Other Predicates. One can also imagine predicates P that support more complex operations on signed messages. One natural set of examples are spreadsheet operations such as median, standard deviation, and rounding on signed data (satisfying unforgeability and context hiding). Other examples include graph algorithms such as computing a signature on a perfect matching in a signed bipartite graph.

---

[2] A substring of $x_1 \ldots x_n$ is some $x_i \ldots x_j$ where $i, j \in [1, n]$ and $i \leq j$. We emphasize that we are not considering *subsequences*. Thus, it is *not* possible, in this setting, to extract a signature on "I like fish" from one on "I do not like fish".

[3] Following an analog of [20], selective security for signatures requires the attacker to give the forgery message before seeing the verification key.

2   Definitions

Definition 1 (Derived messages). *Let $\mathcal{M}$ be a message space and let $P :$ $2^{\mathcal{M}} \times \mathcal{M} \rightarrow \{0,1\}$ be a predicate from sets over $\mathcal{M}$ and a message in $\mathcal{M}$ to a bit. We say that a message m' is* **derivable** *from the set $M \subseteq \mathcal{M}$ if $P(M, m') = 1$. We denote by $P^*(M)$ the set of messages derivable from M by repeated derivation. That is, let $P^0(M)$ be the set of messages derivable from M and for $i > 0$ let $P^i(M)$ be the set of messages derivable from $P^{i-1}(M)$. Then $P^*(M) := \cup_{i=0}^{\infty} P^i(M)$.*

We define the closure of P, denoted P^*, as the predicate defined by $P^*(M, m) = 1$ iff $m \in P^*(M)$.

A P-homomorphic signature scheme $\varPi$ for message space $\mathcal{M}$ and predicate P is a triple of PPT algorithms:

KeyGen(1^{λ}): The key generation algorithm outputs a key pair (pk, sk). We treat the secret key sk as a signature on the empty tuple $\varepsilon \in \mathcal{M}^*$. We also assume that pk is embedded in sk.

SignDerive$(pk, (\{\sigma_m\}_{m \in M}, M), m', w)$: The algorithm takes as input the public key, a set of messages $M \subseteq \mathcal{M}$ and corresponding signatures $\{\sigma_m\}_{m \in M}$, a derived message $m' \in \mathcal{M}$, and possibly some auxiliary information w. It produces a new signature σ' or a special symbol $\perp$ to represent failure. For complicated predicates P, the auxiliary information w serves as a witness that $P(M, m') = 1$. To simplify the notation we often drop w as an explicit argument.

As shorthand we write **Sign**$(sk, m) :=$ **SignDerive**$(pk, (sk, \varepsilon), m, \cdot)$ to denote that any message can be derived when the original signature is the signing key. For a set of messages $M = \{m_1, \ldots, m_k\} \subset \mathcal{M}^*$ it is convenient to let **Sign**(sk, M) denote independently signing each of the k messages, namely:

$$\mathbf{Sign}(sk, M) := \big(\, \mathbf{Sign}(sk, m_1), \ldots, \mathbf{Sign}(sk, m_k) \, \big) \, .$$

Verify(pk, m, σ): Given a public key, message, and purported signature σ, the algorithm returns 1 if the signature is valid and 0 otherwise.

We assume that testing $m \in \mathcal{M}$ can be done efficiently, and that **Verify** returns 0 if $m \notin \mathcal{M}$.

Correctness. We require that for all key pairs (sk, pk) generated by **KeyGen**(1^n) and for all $M \in \mathcal{M}^*$ and $m' \in \mathcal{M}$ we have:

– if $P(M, m') = 1$ then **SignDerive**$(pk, (\mathbf{Sign}(sk, M), M), m') \neq \perp$, and

– for all signature tuples $\{\sigma_m\}_{m \in M}$ such that $\sigma' \leftarrow$ **SignDerive**$(pk,$ $(\{\sigma_m\}_{m \in M}, M), m') \neq \perp$, we have **Verify**$(pk, m', \sigma') = 1$.

In particular, correctness implies that a signature generated by **SignDerive** can be used as an input to **SignDerive** so that signatures can be further derived from derived signatures, if allowed by P.

Derivation efficiency. In many cases it is desirable that the size of a derived signature depend only on the size of the derived message. This rules out signatures that expand as one iteratively calls **SignDerive**. All the constructions in this paper are derivation efficient in this sense.

Definition 2 (Derivation-Efficient). *A signature scheme is derivation-efficient if there exists a polynomial p such that for all $(pk, sk) \leftarrow \mathbf{KeyGen}(1^\lambda)$, set $M \subseteq \mathcal{M}^*$, signatures $\{\sigma_m\}_{m \in M} \leftarrow \mathbf{Sign}(sk, M)$ and derived messages m' where $P(M, m') = 1$, we have $|\mathbf{SignDerive}(pk, \{\sigma_m\}_{m \in M}, M, m')| = p(\lambda, |m'|)$.*

2.1 Security: Unforgeability

To define unforgeability, we extend the basic notion of existential unforgeability with respect to adaptive chosen-message attacks [29]. The definition captures the idea that if the attacker is given a set of signed messages (either primary or derived) then the only messages he can sign are derivations of the signed messages he was given. This is defined using a game between a challenger and an adversary $\mathcal{A}$ with respect to scheme Π over message space $\mathcal{M}$.

— Game **Unforg**$(\Pi, \mathcal{A}, \lambda, P)$:
Setup: The challenger runs **KeyGen**(1^λ) to obtain (pk, sk) and sends pk to $\mathcal{A}$. The challenger maintains two sets T and Q that are initially empty.
Queries: Proceeding adaptively, the adversary issues the following queries to the challenger:
  - *Sign*$(m \in \mathcal{M})$: the challenger generates a unique handle h, runs **Sign**$(sk, m) \rightarrow \sigma$ and places (h, m, σ) into a table T. It returns the handle h to the adversary.
  - *SignDerive*$(\mathbf{h} = (h_1, \ldots, h_k), m')$: the oracle retrieves the tuples (h_i, σ_i, m_i) in T for $i = 1, \ldots, k$, returning $\perp$ if any of them do not exist. Let $M :=$ $(m_1, \ldots, m_k)$ and $\{\sigma_m\}_{m \in M} := \{\sigma_1, \ldots, \sigma_k\}$. If $P(M, m')$ holds, then the oracle generates a new unique handle h', runs **SignDerive**$(pk, (\{\sigma_m\}_{m \in M}, M), m') \rightarrow \sigma'$ and places (h', m', σ') into T, and returns h' to the adversary.
  - *Reveal*(h): Returns the signature σ corresponding to handle h, and adds (σ', m') to the set Q.
Output: Eventually, the adversary outputs a pair (σ', m'). The output of the game is 1 (i.e., the adversary wins the game) if:
    - **Verify**$(pk, m', \sigma') = 1$ and,
    - let $M \subseteq \mathcal{M}$ be the set of messages in Q then $P^*(M, m') = 0$ where P^* is the closure of P from Definition 1.
    Else, the output of the game is 0. Define **Forg**$_\mathcal{A}$ as the probability that $\Pr[\mathbf{Unforg}(\Pi, \mathcal{A}, \lambda, P) = 1]$.

Interestingly, for some predicates it may be difficult to test if the adversary won the game. For all the predicates we consider in this paper, this will be quite easy.

Definition 3 (Unforgeability). *A P-homomorphic signature scheme Π is* **unforgeable** *with respect to adaptive chosen-message attacks if for all PPT adversaries $\mathcal{A}$, the function* **Forg**$_\mathcal{A}$ *is negligible in λ.*

A P-homomorphic signature scheme Π is **selective unforgeable** *with respect to adaptive chosen-message attacks if for all PPT adversaries A who begin the above game by announcing the message m' on which they will forge,* **Forg**$_A$ *is negligible in* λ.

Properties of the definition. By taking P to be the equality oracle, namely $P(x,y) = 1$ iff $x = y$, we obtain the standard unforgeability requirement for signatures.

Notice that *Sign* and *SignDerive* queries return handles, but do not return the actual signatures. A system proven secure under this definition adequately rules out the following attack: suppose (m, σ) is a message signature pair and (m', σ') is a message-signature pair derived from it, namely $\sigma' = \mathbf{SignDerive}(pk, \sigma, m, m')$. For example, suppose m' is a quote from m. Then given (m', σ') it should be difficult to produce a signature on m and indeed our definition treats a signature on m as a valid forgery.

The unforgeability game imposes some constraints on P: (1) P must be reflexive, i.e. $P(m,m) = 1$ for all $m \in \mathcal{M}$, (2) P must be monotone, i.e. $P(M, m') \Rightarrow P(M', m')$ where $M \subseteq M'$. It is easy to see that predicates that do not satisfy these requirements cannot be realized under Definition 3.

2.2   Security: Context Hiding (a.k.a., Privacy)

Let M be some set and let m' be a derived message from M (i.e., $P(M, m') = 1$). Context hiding captures the idea that a signature on m' derived from signatures on M should reveal no information about M beyond what is revealed by m'. For example, in the case of quoting, a signature on a quote from m should reveal nothing more about m: not the length of m, not the position of the quote in m, etc. The same should hold even if the attacker is given signatures on multiple quotes from m.

We put forth the following powerful *statistical* definition of context hiding and discuss its implications following the definition. We were most easily able to leverage a statistical definition for our proofs, although we also give an alternative *computational* definition in the full version [1].

Definition 4 (Strong Context Hiding). *Let* $M \subseteq \mathcal{M}^*$ *and* $m' \in \mathcal{M}$ *be messages such that* $P(M, m') = 1$. *Let* $(pk, sk) \leftarrow \mathbf{KeyGen}(1^\lambda)$ *be a key pair. A signature scheme* (**KeyGen, SignDerive, Verify**) *is strongly context hiding (for predicate P) if for all such triples* $((pk, sk), M, m')$, *the following two distributions are statistically close:*

$$\left\{ \left(sk, \{\sigma_m\}_{m \in M} \leftarrow \mathbf{Sign}(sk, M), \ \mathbf{Sign}(sk, m') \right) \right\}_{sk, M, m'}$$

$$\left\{ \left(sk, \{\sigma_m\}_{m \in M} \leftarrow \mathbf{Sign}(sk, M), \ \mathbf{SignDerive}(pk, (\{\sigma_m\}_{m \in M}, M), m') \right) \right\}_{sk, M, m'}$$

The distributions are taken over the coins of **Sign** *and* **SignDerive**. *Without loss of generality, we assume that pk can be computed from sk.*

The definition states that a derived signature on m', from an honestly-generated original signature, is statistically indistinguishable from a fresh signature on m'. This implies that a derived signature on m' is indistinguishable from a signature generated independently of M. Therefore, the derived signature cannot (provably) reveal any information about M beyond what is revealed by m'. By a simple hybrid argument the same holds even if the adversary is given multiple derived signatures from M.

Moreover, Definition 4 requires that a derived signature look like a fresh signature even if the original signature on M is known. Hence, if for example someone quotes from a signed recommendation letter and somehow the original signed recommendation letter becomes public, it would be impossible to link the signed quote to the original signed letter. The same holds even if the signing key sk is leaked.

Thus, Definition 4 captures a broad range of privacy requirements for derived signatures. Earlier work in this area [32,16,18,15] only considered weaker privacy requirements using more complex definitions. The simplicity and breadth of Definition 4 is one of our key contributions.

Definition 4 uses statistical indistinguishability meaning that even an unbounded adversary cannot distinguish derived signatures from newly created ones. In the full version [1], we give a definition using computational indistinguishability which is considerably more complex since the adversary needs to be given signing oracles. In the unbounded case of Definition 4 the adversary can simply recover a secret key sk from the public key and answer its own signature queries which greatly simplifies the definition of context hiding. All the signature schemes in this paper satisfy the statistical Definition 4.

As mentioned above, the context-hiding guarantee applies to all derivations that begin with an honestly-generated signature. One might imagine a scenario where a malicious signer creates a signature that passes the verification algorithm, but contains a "watermark" that allows the signer to detect if other signatures are derived from it. To prevent such attacks from malicious signers, we could alter the definition so that indistinguishability holds for any derivative that results from a signature that passed the verification algorithm.

A simpler approach to proving unforgeability. For systems that are strongly context hiding, unforgeability follows from a simpler game than that of Section 2.1. In particular, it suffices to just give the adversary the ability to obtain top level signatures signed by sk. In the full version [1], we define this simpler unforgeability game and prove equivalence to Definition 3 using strong context hiding.

2.3   Related Work

Early work on quotable signatures [45,32,38,37,30,17,21,15] supports quoting from a single document, but does not achieve the privacy or unforgeability properties we are aiming for. For example, if *simple quoting* of messages is all that is desired, then the following folklore solution would suffice: simply sign the Merkle hash of a document. A quote represents some sub-tree of the Merkle hash; so a quoter could include enough intermediate hash nodes along with the original signature in any quote. A verifier could simply hash the quote, and then build the Merkle hash tree using the computed hash and the intermediate hashes, and compare with the original signature. Notice, however, that every quote in this scheme reveals information about the original source document. In particular, each quote reveals information about *where in the document* it appears. Thus, this simple quoting scheme is not *context hiding* in our sense.

The work whose definition is closest to what we envision is the recent work on redacted signatures of Chang et al. [21] and Brzuska et al. [15] (see also Naccache [39, p. 63] and Boneh-Freeman [12,11][4]). However, there is a subtle, but fundamental difference between their definition and the privacy notion we are aiming for. In our formulation, a quoted signature should be indistinguishable from a fresh signature, even when the distinguisher is given the original signature. (We capture this by an even stronger game where a derived signature is distributed statistically close to a fresh signature.) In contrast, the definitions of [21,15,12,11] do not provide the distinguisher with the original signature. Thus, it may be possible to link a quoted document to its original source (and indeed it is in the constructions of [21,15,12,11]), which can have negative privacy implications. Overcoming such document linkage while maintaining unforgeability is a real technical challenge. This requires moving beyond techniques that use *nonces* to link parts of messages.

Indeed, in most prior constructions, such as [21,15], nonces are used to prevent "mix-and-match" attacks (e.g., forming a "quote" using pieces of two different messages.) Unfortunately, these nonces reveal the history of derivation, since they cannot change during each derivation operation. Arguably, much of the technical difficulty in our current work comes precisely from the effort to meet our definition and hide the lineage. We introduce new techniques in this work which link pieces together using randomness that can be re-randomized in controlled ways.

Another line of work studies computing on authenticated data by holders of secret information. Examples include *sanitizable* signatures [38,2,36,18,16] that allow a proxy to compute signatures on related messages, but requires the proxy to have a secret key, and *incremental* signatures [4], where the signer can efficiently make small edits to his signed data. In contrast, our proposal is more along the lines of homomorphic encryption and Rivest's vision [41], where *anyone* can compute on the authenticated data.

---

[4] As acknowledged in Section 2.2 of Boneh-Freeman [11], our definitional notion is stronger than and predates the "weak context hiding" notion of [11]. Indeed, the fact that [11] uses our framework lends support to its generality, and the fact that they could not achieve our context hiding notion highlights its difficulty. Their "weak" definition, which is equivalent to [15], only ensures privacy when the original signatures remain hidden. In their system, signature derivation is deterministic and therefore once the original signatures become public it is easy to tell where the derived signature came from. Our signatures achieve full context hiding so that derived signatures remain private no matter what information is revealed. This is considerably harder and is not known how to do for the lattice-based signatures in Boneh-Freeman.

3   Preliminaries: Algebraic Settings

Bilinear Groups and the CDH Assumption. Let $\mathbb{G}$ and $\mathbb{G}_T$ be groups of prime order p. A *bilinear map* is an efficient mapping $\mathbf{e} : \mathbb{G} \times \mathbb{G} \rightarrow \mathbb{G}_T$ which is both: (*bilinear*) for all $g \in \mathbb{G}$ and $a, b \leftarrow \mathbb{Z}_p$, $\mathbf{e}(g^a, g^b) = \mathbf{e}(g, g)^{ab}$; and (*non-degenerate*) if g generates $\mathbb{G}$, then $\mathbf{e}(g, g) \neq 1$. We will focus on the Computational Diffie-Hellman assumption in these groups.

Assumption 1 (CDH [24]). *Let g generate a group $\mathbb{G}$ of prime order $p \in \Theta(2^\lambda)$. For all PPT adversaries $\mathcal{A}$, the following probability is negligible in λ:* $\Pr[a, b, \leftarrow \mathbb{Z}_p; z \leftarrow \mathcal{A}(g, g^a, g^b) : z = g^{ab}].$

4   A Powers-of-2 Construction for Quoting Substrings

We now provide our main construction for quoting substrings in a text document. It achieves the best time/space efficiency trade-off to our knowledge for this problem. We will have two different types of signatures called Type I and Type II, where a Type I signature can be quoted down to another Type I or Type II signature. A Type II signature cannot be quoted any further, but will be a shorter signature. The quoting algorithm will allow us to quote anything that is a substring of the original message. We point out that the Type I, II signatures of this system conform to the general framework given in Section 2. In particular, we can view a message M as a pair $(t, m) \in \{0, 1\}, \{0, 1\}^*$. The bit t will identify the message as being Type I or Type II (assume $t = 1$ signifies Type I signatures) and m will be the quoted substring. The predicate

$$P(M = (t, m), M' = (t', m')) = \begin{cases} 1 & \text{if } t = 1 \text{ and } m' \text{ is a substring of } m; \\ 0 & \text{otherwise.} \end{cases}$$

The bit t' will indicate whether the new message is Type I or II (i.e., whether the system can quote further.) We note that this description allows an attacker to distinguish between any Type I signature from any Type II signature since the "type bit" of the messages will be different and thus they will technically be two different messages even if the substring components are equal. For this reason we will only need to prove context hiding between messages of Type I or Type II, but not across types. In general, flipping the bit t will not result in a valid signature of a different type on the same core message, because the format will be wrong; however, moving from a Type I to a Type II on the same core message is not considered a forgery since Type II signatures can be legally derived from Type I.

For presentational clarity, we will split the description of our quoting algorithm into two quoting algorithms for quoting to Type I and to Type II signatures; likewise we will split the description of our verification algorithm into two separate verification algorithms, one for each type of signature. The type of signature used or created (i.e., bit t) will be implicit in the description.

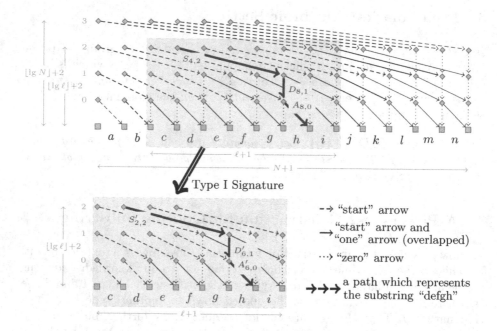

Fig. 1. The top diagram represents a signature on "abcdefghijklmn" with length $N = 14$. Each arrow corresponds to some group elements in the construction. Logically, whenever the elements corresponding to an arrow are included in a quoted signature, the characters underneath this arrow are included in the quoted message. The bold path through the top diagram shows how to construct a Type II signature on "defgh"; it is very short, but cannot be re-quoted. The gray box in this figure shows how to construct a Type I signature on "cdefghi" of length $\ell = 7$; it includes all the arrows in the lower figure and can be re-quoted. A technical challenge is to enforce that following the arrows is the only way to form a valid signature. Details are below.

Notation: We use notation $m_{i,j}$ to denote the substring of m of length j starting at position i.

Intuition: We begin by giving some intuition. We design Type I signatures that allow re-quoting and Type II signatures that cannot be further quoted, but are ultra-short. For an original message of length n, our signature structure should be able to accommodate starting at any position $1 \le i \le n$ and quoting any length $1 \le \ell \le (n - i + 1)$ substring.[5]

To (roughly) see how this works for a message of length n, visualize $(n + 1)$ columns with $(\lfloor \lg n \rfloor + 2)$ rows as in Figure 1. The columns correspond to the characters of the message, so if the 14-character message is "abcdefghijklmn" then there are 15 columns, with a character in between each column. The rows

---

[5] Technically, our predicate $P(m, m')$ will take the quote from the first occurrence of substring m' in m, but for the moment imagine that we allowed quoting from anywhere in m.

correspond to the numbers $\lg n$ down to 0, plus an extra row at the bottom.[6] Each location in the matrix (except along the bottom-most row) contains one or more out-going arrows. We'll establish rules for when these arrows exist and where each arrow ends shortly.

A Type II quote will trace a $(\lg n+1)$-length path on these arrows through this matrix starting in a row (with outgoing arrows) of the column that begins the quote and ending in the lowest row of the first column after the quote ends. The starting row corresponds to the largest power of two less than or equal to the length of the desired quote. E.g., to quote "bcdef", start in row 2 immediately to the left of 'b' (because $2^2 = 4$ is the largest power of two less than 5) and end in row 0 immediately to the right of 'f'. Intuitively, taking an arrow over a character includes it in the quote. A Type II quote on "defgh" is illustrated in Figure 1.

A technical challenge is to make this a $O(\lg n)$-length path rather than a $O(n)$-length path. To do this, the key insight is to view the length of any possible quote as the sum of powers of two and to allow arrows that correspond to covering the quote in pieces of size corresponding to one operand of the sum at a time. Each location (i_c, i_r) in the matrix (except the bottom-most row) contains:

- a *"start" arrow*: an arrow that goes down one row and over 2^{i_r} columns ending in $(i_c + 2^{i_r}, i_r - 1)$, if this end point is in the matrix. This adds all characters from position i_c to $i_c + 2^{i_r} - 1$ to the quoted substring; effectively adding the largest power-of-two-length prefix of the quote characters. This arrow indicates that the quote starts here. These are represented as $S_{i,j}, \widetilde{S_{i,j}}$ pairs in our construction.
- a *"one" arrow*: operate similarly to start arrows and used to include characters after a start arrow includes the quote prefix. These are represented as $A_{i,j}, \widetilde{A_{i,j}}$ pairs in our construction.
- a *"zero" arrow*: an arrow that goes straight down one row ending in $(i_c, i_r - 1)$. This does not add any characters to the quoted substring. These are represented as $D_{i,j}, \widetilde{D_{i,j}}$ pairs in our construction.

A Type II quote always starts with a start arrow and then contains one and zero arrows according to the binary representation of the length of the quote. In our example of original message "abcdefghijklmn", we have 15 columns and 5 rows. We will logically divide our desired substring of "bcdef" (length $5 = 2^2 + 2^0 = 4 + 1$) into its powers-of-two components "bcde" (length $4 = 2^2$) and "f" (length $1 = 2^0$). To form the Type II quote, we start in row 2 (since $4 = 2^2$) of column 2 (to the left of 'b') and take the start arrow ($S_{2,2}$) to row 1 of column 7, take the zero arrow ($D_{7,1}$) to row 0 of column 7, and then take the one arrow ($A_{7,0}$) to the lowest row of column 8. The arrows "pass over" the characters "bcdef". Figure 1 illustrates this for quote "defgh".

For a quote of length ℓ, the elements on this $O(\lg \ell)$-length path of arrows form a very short Type II signature. For Type I signatures, we include all the

---

[6] The lowest row is intentionally not assigned a number. The second lowest row is row 0. We do this so that row i can correspond to a jump of length 2^i.

elements corresponding to all arrows that make connections within the columns corresponding to the quote. We illustrate this in Figure 1. This allows quoting of quotes with a signature size of $O(\ell \lg \ell)$.

It is essential for security that the signature structure and data algorithm enforce that the quoting algorithm be used and not allow an attacker to "splice" together a quote from different parts of the signature. We realize this by adding in random "chaining" variables. In order to cancel these out and get a well formed Type II quote a user must intuitively follow the prescribed procedure (i.e., following the arrows is the only way to form a valid quote.)

The Construction: We now describe our algorithms. While **Sign** is simply a special case of the **SignDerive** algorithm, we will explicitly provide both algorithms here for clarity purposes.

KeyGen(1^λ) : The algorithm selects a bilinear group $\mathbb{G}$ of prime order $p > 2^\lambda$ with generator g. Let L be the maximum message length supported and denote $n = \lfloor \lg(L) \rfloor$. Let $H : \{0,1\}^* \to \mathbb{G}$ and $H_s : \{0,1\}^* \to \mathbb{G}$ be the description of two hash functions that we model as random oracles. Choose random $z_0, \ldots, z_{n-1}, \alpha \in \mathbb{Z}_p$. The secret key is $(z_0, \ldots, z_{n-1}, \alpha)$ and the public key is:

$$PK = (H, H_s, g, g^{z_0}, \ldots, g^{z_{n-1}}, \mathbf{e}(g,g)^\alpha).$$

Sign($sk, M = (t, m) \in \{0,1\} \times \Sigma^{\ell \leq L}$) : If $t = 1$, signatures produced by this algorithm are Type I as described below. If $t = 0$, the Type II signature can be obtained by running this algorithm and then running the Quote-Type II algorithm below to obtain a quote on the entire message. The message space is treated as $\ell \leq L$ symbols from alphabet Σ.

Recall: we use notation $m_{i,j}$ to denote the substring of m of length j starting at position i.

For $i = 3$ to $\ell + 1$ and $j = 0$ to $\lfloor \lg(i-1) - 1 \rfloor$, choose random values $x_{i,j} \in \mathbb{Z}_p$. These will serve as our random "chaining" variables, and they should all "cancel" each other out in our short Type II signatures. By definition, set $x_{i,-1} := 0$ for all $i = 1$ to $\ell + 1$.

A signature is comprised of the following values for $i = 1$ to ℓ and $j = 0$ to $\lfloor \lg(\ell - i + 1) \rfloor$, for randomly chosen values $r_{i,j} \in \mathbb{Z}_p$:

[start arrow: start and include power j]
$$S_{i,j} = g^\alpha g^{-x_{i+2^j, j-1}} H_s(m_{i,2^j})^{r_{i,j}} , \widetilde{S_{i,j}} = g^{r_{i,j}}$$

Together with the following values for $i = 3$ to ℓ and $j = 0$ to $\min(\lfloor \lg(i - 1) - 1 \rfloor, \lfloor \lg(\ell - i + 1) \rfloor)$, for randomly chosen values $r'_{i,j} \in \mathbb{Z}_p$:

[one arrow: include power j and decrease j]
$$A_{i,j} = g^{x_{i,j}} g^{-x_{i+2^j, j-1}} H(m_{i,2^j})^{r'_{i,j}} , \widetilde{A_{i,j}} = g^{r'_{i,j}}$$

Together with the following values for $i = 3$ to $\ell+1$ and $j = 0$ to $\lfloor \lg(i-1)-1 \rfloor$, for randomly chosen values $r''_{i,j} \in \mathbb{Z}_p$:

[zero arrow: decrease j]
$$D_{i,j} = g^{x_{i,j}} g^{-x_{i,j-1}} g^{z_j r''_{i,j}} , \ \widetilde{D_{i,j}} = g^{r''_{i,j}}$$

We provide an example of how to form Type II signatures from this construction shortly. To see why our $A_{i,j}$ and $D_{i,j}$ values start at $i = 3$, note that Type II quotes at position i of length $2^0 = 1$ symbol include only the $S_{i,0}$ value, where the $x_{.,0-1}$ term is 0 by definition. Type II quotes at position i of length $2^1 = 2$ symbols include the $S_{i,1}$ value plus an additional $D_{i+2,0}$ term to cancel out the $x_{i+2,0}$ value (leaving only $x_{i+2,-1} = 0$.) Quotes at position i of length $2^1 + 1 = 3$ symbols include the $S_{i,1}$ value plus an additional $A_{i+2,0}$ term to cancel out the $x_{i+2,0}$ value (leaving only $x_{i+3,-1} = 0$.) Since we index strings from position 1, the first position to include an $A_{i,j}$ or $D_{i,j}$ value is $i + 2 = 3$.

SignDerive$(pk, \sigma, M = (t, m), M' = (t', m'))$: If $P(M, M') = 0$, output $\bot$. Otherwise, if $t' = 1$, output Quote-Type I(PK, σ, m, m'); if $t' = 0$, output Quote-Type II(PK, σ, m, m'), where these algorithms are defined below.

Quote-Type I(pk, σ, m, m') : The quote algorithm takes a Type I signature and produces another Type I signature that maintains the ability to be quoted again. Intuitively, this operation will simply find a substring m' in m, keep only the components associated with this substring and re-randomize them all (both the $x_{i,j}$ and $r_{i,j}$ terms in every component.)
If m' is not a substring of m, then output $\bot$. Otherwise, let $\ell' = |m'|$. Determine the first index k at which substring m' occurs in m. Parse σ as a collection of $S_{i,j}, \widetilde{S_{i,j}}, A_{i,j}, \widetilde{A_{i,j}}, D_{i,j}, \widetilde{D_{i,j}}$ values, exactly as would come from **Sign** with $\ell = |m|$.

First, we choose re-randomization values (to re-randomize the $x_{i,j}$ terms of σ.) For $i = 2$ to $\ell' + 1$ and $j = 0$ to $\lfloor \lg(i-1) - 1 \rfloor$, choose random values $y_{i,j} \in \mathbb{Z}_p$. Set $y_{i,-1} := 0$ for all $i = 1$ to $\ell' + 1$. Later, we will choose $t_{i,j}$ values to re-randomize the $r_{i,j}$ terms of σ.

The quote signature σ' is comprised of the following values:

For $i = 1$ to ℓ' and $j = 0$ to $\lfloor \lg(\ell' - i + 1) \rfloor$, for randomly chosen $t_{i,j} \in \mathbb{Z}_p$:
$$S'_{i,j} = S_{i+k-1,j} \cdot g^{-y_{i+2^j,j-1}} H_s(m_{i+k-1,2^j})^{t_{i,j}}, \ \widetilde{S'_{i,j}} = \widetilde{S_{i+k-1,j}} \cdot g^{t_{i,j}}$$

Together with the following values for $i = 3$ to ℓ' and $j = 0$ to $\min(\lfloor \lg(i-1) - 1 \rfloor, \lfloor \lg(\ell' - i + 1) \rfloor)$, for randomly chosen $t'_{i,j} \in \mathbb{Z}_p$:
$$A'_{i,j} = A_{i+k-1,j} \cdot g^{y_{i,j}} g^{-y_{i+2^j,j-1}} H(m_{i+k-1,2^j})^{t'_{i,j}}, \ \widetilde{A'_{i,j}} = \widetilde{A_{i+k-1,j}} \cdot g^{t'_{i,j}}$$

Together with the following values for $i = 3$ to $\ell' + 1$ and $j = 0$ to $\lfloor \lg(i-1) - 1 \rfloor$, for randomly chosen $t''_{i,j} \in \mathbb{Z}_p$:
$$D'_{i,j} = D_{i+k-1,j} \cdot g^{y_{i,j}} g^{-y_{i,j-1}} g^{z_j t''_{i,j}}, \ \widetilde{D'_{i,j}} = \widetilde{D_{i+k-1,j}} \cdot g^{t''_{i,j}}$$

Quote-Type II(pk, σ, m, m') : The quote algorithm takes a Type I signature and produces a Type II signature. If $P(m, m') \neq 1$, then output $\bot$.

A quote is computed from one start value and logarithmically many subsequent pieces depending on the bits of $|m'|$. All signature pieces must be re-randomized to prevent content-hiding attacks.

Consider the length ℓ' written as a binary string. Let β' be the largest index of $\ell' = |m'|$ that is set to 1, where *we start counting with zero as the least significant bit*. That is, set $\beta' = \lfloor \lg(\ell') \rfloor$. Select random values $v, v_{\beta'-1}, \ldots, v_0 \in \mathbb{Z}_p$. Set the start position as $B := S_{k,\beta'}$ and $k' := k + 2^{\beta'}$. Then, from $j = \beta' - 1$ down to 0, proceed as follows:

- If the jth bit of ℓ' is 1, set $B := B \cdot A_{k',j} \cdot H(m_{k',2^j})^{v_j}$, set $k' := k' + 2^j$, and $Z_j := \widetilde{A_{k',j}} \cdot g^{v_j}$;
- If the jth bit of ℓ' is 0, set $B := B \cdot D_{k',j} \cdot g^{z_j v_j}$ and $Z_j := \widetilde{D_{k',j}} \cdot g^{v_j}$.

To end, re-randomize as $B := B \cdot H_s(m_{k,2^\beta})^v$ and $\widetilde{S} := \widetilde{S_{k,\beta}} \cdot g^v$; output the quote as

$$\sigma' = (B, \widetilde{S}, Z_{\beta-1}, \ldots, Z_0)$$

Verify$(pk, M = (t, m), \sigma)$: If $t = 1$, output Verify-Type I(pk, m, σ). Otherwise, output Verify-Type II(pk, m, σ), where these algorithms are defined immediately below.

Verify–Type I(pk, m, σ) : Parse σ as the set of $S_{i,j}, \widetilde{S_{i,j}}, A_{i,j}, \widetilde{A_{i,j}}, D_{i,j}, \widetilde{D_{i,j}}$. Let $\ell = |m|$.

Let $X_{i,j}$ denote $e(g, g)^{x_{i,j}}$. We can compute these values as follows. The value $X_{i,-1} = 1$, since for all $i = 1$ to $\ell+1$, $x_{i,-1} = 0$. For $i = 3$ to $\ell+1$ and $j = 0$ to $\lfloor \lg(i-1)-1 \rfloor$, we compute $X_{i,j}$ in the following manner: Let $I = i - 2^{j+1}$ and $J = j + 1$. Next, compute $X_{i,j} = \big(e(g,g)^\alpha \cdot e(H_s(m_{I,2^J}), \widetilde{S_{I,J}})\big) / e(S_{I,J}, g)$. The verification accepts if and only if all of the following hold:

- for $i = 3$ to ℓ and $j = 0$ to $\min(\lfloor \lg(i-1)-1 \rfloor, \lfloor \lg(\ell-i+1) \rfloor)$,

$$e(A_{i,j}, g) = X_{i,j}/X_{i+2^j,j-1} \cdot e(H(m_{i,2^j}), \widetilde{A_{i,j}})$$

- and for $i = 3$ to $\ell+1$ and $j = 0$ to $\lfloor \lg(i-1)-1 \rfloor$, $e(D_{i,j}, g) = X_{i,j}/X_{i,j-1} \cdot e(g^{z_j}, \widetilde{D_{i,j}})$.

Verify-Type II(pk, m, σ) : We give the verification algorithm for Type II signatures. Parse σ as $(B, \widetilde{S}, Z_{\beta-1}, \ldots, Z_0)$. Let $\ell = |m|$ and β be the index of the highest bit of ℓ that is set to 1. If σ does not include exactly β Z_i values, reject. Set $C := 1$ and $k = 1$. From $j = \beta - 1$ down to 0, proceed as follows:

- If the jth bit of ℓ is 1, set $C := C \cdot e(H(m_{k,2^j}), Z_j)$ and $k := k + 2^j$;
- If the jth bit of ℓ is 0, set $C := C \cdot e(g^{z_j}, Z_j)$.

Accept if and only if $e(B, g) = e(g, g)^\alpha \cdot e(H_s(m_{1,2^\beta}), \widetilde{S}) \cdot C$.

Theorem 2 (Security under CDH). *If the CDH assumption holds in $\mathbb{G}$, then the above quotable signature scheme is selectively quote unforgeable and context-hiding in the random oracle model.*

In the full version [1], we prove this theorem. We also discuss in detail the efficiency of this construction, how to remove the random oracle, and how to obtain full security.

5   Subsets and Weighted Averages

For the problems of subsets and weighted averages, we show somewhat surprising connections to respective existing solutions in attribute-based encryption and network coding signatures. We sketch these constructions here and provide further details in the full version of this paper [1].

Briefly, our subset construction extends the concept of Naor [10] who observed that every IBE scheme can be transformed into a standard signature scheme by applying the IBE KeyGen algorithm as a signing algorithm. Here we show an analog for known Ciphertext-Policy (CP) ABE schemes. The KeyGen algorithm which generates a key for a set S of attributes can be used as a signing algorithm for the set S. For known CP-ABE systems [8,34,47] it is straightforward to derive a key for a subset S' of S and to re-randomize the signature/key. To verify a signature on S we can apply Naor's signature-from-IBE idea and encrypt a random message X to a policy that is an AND of all the attributes in S and see if the signature can be used as an ABE key to decrypt to X. Signatures for subsets have been previously considered in [31, §6.4], but without context hiding requirements.

Next, we consider a construction for weighted averages, which captures Fourier transforms and weighted sums. This is particularly interesting, because so far we only constructed schemes for *univariate* predicates P. We can now give an example where one computes on multiple signed messages. Let p be a prime, n a positive integer, and $\mathcal{T}$ a set of tags. The message space $\mathcal{M}$ consists of pairs:

$$\mathcal{M} := \mathcal{T} \times \mathbb{F}_p^n$$

Now, define the predicate P as follows: $P(\varepsilon, m) = 1$ for all $m \in \mathcal{M}$ and[7]

$$P\Big(\big((t_1, \mathbf{v}_1), \dots, (t_k, \mathbf{v}_k) \big), (t, \mathbf{v}) \Big) = 1 \iff \begin{cases} t = t_1 = \cdots = t_k, \text{ and} \\ \mathbf{v} \in \mathrm{span}(\mathbf{v}_1, \dots, \mathbf{v}_k) \end{cases}$$

Thus, given signatures on vectors $\mathbf{v}_1, \dots, \mathbf{v}_k$ grouped together by the tag t, anyone can create a signature on a linear combination of these vectors. This can be done iteratively so that given signed linear combinations, new signed linear combinations can be created. Unforgeability means that if the adversary obtains signatures on vectors $\mathbf{v}_1, \dots, \mathbf{v}_k$ for particular tag $t \in \mathcal{T}$ then he cannot create a signature on a vector outside the linear span of $\mathbf{v}_1, \dots, \mathbf{v}_k$.

Signature schemes for this predicate P are presented in [13,12,11,14,3] while schemes over $\mathbb{Z}$ (rather than $\mathbb{F}_p$) are presented in [26]. These schemes were originally designed to secure network coding where context hiding is not needed since there are no privacy requirements for the sender (in fact, the sender is explicitly transmitting all his data to the recipient). The question then is how to construct a system for predicate P above that is both unforgeable and context hiding. Fortunately, we observe that under the CDH assumption, the linearly homomorphic signature scheme, **NCS$_1$**, due to Boneh, Freeman, Katz and Waters [13]

---

[7] Recall, the signature on ϵ is the output the KeyGen algorithm.

is unforgeable and context-hiding in the random oracle model, assuming tags are generated independently at random by the unforgeability challenger when responding to **Sign** queries.

Acknowledgments. We are grateful to the anonymous reviewers for their helpful comments.

References

1. Ahn, J.H., Boneh, D., Camenisch, J., Hohenberger, S., Shelat, A., Waters, B.: Computing on authenticated data. Cryptology ePrint Archive, Report 2011/096 (2011), http://eprint.iacr.org/
2. Ateniese, G., Chou, D.H., de Medeiros, B., Tsudik, G.: Sanitizable Signatures. In: di Vimercati, S.d.C., Syverson, P.F., Gollmann, D. (eds.) ESORICS 2005. LNCS, vol. 3679, pp. 159–177. Springer, Heidelberg (2005)
3. Attrapadung, N., Libert, B.: Homomorphic Network Coding Signatures in the Standard Model. In: Catalano, D., Fazio, N., Gennaro, R., Nicolosi, A. (eds.) PKC 2011. LNCS, vol. 6571, pp. 17–34. Springer, Heidelberg (2011)
4. Bellare, M., Goldreich, O., Goldwasser, S.: Incremental Cryptography: The Case of Hashing and Signing. In: Desmedt, Y.G. (ed.) CRYPTO 1994. LNCS, vol. 839, pp. 216–233. Springer, Heidelberg (1994)
5. Bellare, M., Micciancio, D., Warinschi, B.: Foundations of Group Signatures: Formal Definitions, Simplified Requirements, and a Construction Based on General Assumptions. In: Biham, E. (ed.) EUROCRYPT 2003. LNCS, vol. 2656, pp. 614–629. Springer, Heidelberg (2003)
6. Bellare, M., Neven, G.: Transitive Signatures Based on Factoring and RSA. In: Zheng, Y. (ed.) ASIACRYPT 2002. LNCS, vol. 2501, pp. 397–414. Springer, Heidelberg (2002)
7. Bellare, M., Neven, G.: Transitive signatures: New schemes and proofs. IEEE Transactions on Information Theory 51, 2133–2151 (2005)
8. Bethencourt, J., Sahai, A., Waters, B.: Ciphertext-policy attribute-based encryption. In: IEEE Symposium on Security and Privacy, pp. 321–334 (2007)
9. Boneh, D., Boyen, X., Shacham, H.: Short Group Signatures. In: Franklin, M. (ed.) CRYPTO 2004. LNCS, vol. 3152, pp. 41–55. Springer, Heidelberg (2004)
10. Boneh, D., Franklin, M.K.: Identity-based encryption from the Weil pairing. SIAM J. Comput. 32(3) (2003)
11. Boneh, D., Freeman, D.M.: Homomorphic Signatures for Polynomial Functions. In: Paterson, K.G. (ed.) EUROCRYPT 2011. LNCS, vol. 6632, pp. 149–168. Springer, Heidelberg (2011); Cryptology ePrint Archive, Report 2011/018
12. Boneh, D., Freeman, D.M.: Linearly Homomorphic Signatures over Binary Fields and New Tools for Lattice-Based Signatures. In: Catalano, D., Fazio, N., Gennaro, R., Nicolosi, A. (eds.) PKC 2011. LNCS, vol. 6571, pp. 1–16. Springer, Heidelberg (2011); Cryptology ePrint Archive, Report 2010/453
13. Boneh, D., Freeman, D., Katz, J., Waters, B.: Signing a Linear Subspace: Signature Schemes for Network Coding. In: Jarecki, S., Tsudik, G. (eds.) PKC 2009. LNCS, vol. 5443, pp. 68–87. Springer, Heidelberg (2009)
14. Boneh, D., Hamburg, M.: Generalized Identity Based and Broadcast Encryption Schemes. In: Pieprzyk, J. (ed.) ASIACRYPT 2008. LNCS, vol. 5350, pp. 455–470. Springer, Heidelberg (2008)

15. Brzuska, C., Busch, H., Dagdelen, O., Fischlin, M., Franz, M., Katzenbeisser, S., Manulis, M., Onete, C., Peter, A., Poettering, B., Schröder, D.: Redactable Signatures for Tree-Structured Data: Definitions and Constructions. In: Zhou, J., Yung, M. (eds.) ACNS 2010. LNCS, vol. 6123, pp. 87–104. Springer, Heidelberg (2010)

16. Brzuska, C., Fischlin, M., Freudenreich, T., Lehmann, A., Page, M., Schelbert, J., Schröder, D., Volk, F.: Security of Sanitizable Signatures Revisited. In: Jarecki, S., Tsudik, G. (eds.) PKC 2009. LNCS, vol. 5443, pp. 317–336. Springer, Heidelberg (2009)

17. Brzuska, C., Fischlin, M., Lehmann, A., Schröder, D.: Santizable signatures: How to partially delegate control for authenticated data. In: BIOSIG 2009, pp. 117–128 (2009)

18. Brzuska, C., Fischlin, M., Lehmann, A., Schröder, D.: Unlinkability of Sanitizable Signatures. In: Nguyen, P.Q., Pointcheval, D. (eds.) PKC 2010. LNCS, vol. 6056, pp. 444–461. Springer, Heidelberg (2010)

19. Camenisch, J., Lysyanskaya, A.: Signature Schemes and Anonymous Credentials from Bilinear Maps. In: Franklin, M. (ed.) CRYPTO 2004. LNCS, vol. 3152, pp. 56–72. Springer, Heidelberg (2004)

20. Canetti, R., Halevi, S., Katz, J.: A Forward-Secure Public-Key Encryption Scheme. In: Biham, E. (ed.) EUROCRYPT 2003. LNCS, vol. 2656, pp. 255–271. Springer, Heidelberg (2003)

21. Chang, E.-C., Lim, C.L., Xu, J.: Short Redactable Signatures Using Random Trees. In: Fischlin, M. (ed.) CT-RSA 2009. LNCS, vol. 5473, pp. 133–147. Springer, Heidelberg (2009)

22. Charles, D., Jain, K., Lauter, K.: Signatures for network coding. International Journal of Information and Coding Theory 1(1), 3–14 (2009)

23. Chaum, D., van Heyst, E.: Group Signatures. In: Davies, D.W. (ed.) EUROCRYPT 1991. LNCS, vol. 547, pp. 257–265. Springer, Heidelberg (1991)

24. Diffie, W., Hellman, M.: New directions in cryptography. IEEE Transactions on Information Theory 22, 644–654 (1976)

25. Fragouli, C., Soljanin, E.: Network Coding Fundamentals. Now Publishers Inc., Hanover (2007)

26. Gennaro, R., Katz, J., Krawczyk, H., Rabin, T.: Secure Network Coding over the Integers. In: Nguyen, P.Q., Pointcheval, D. (eds.) PKC 2010. LNCS, vol. 6056, pp. 142–160. Springer, Heidelberg (2010)

27. Gentry, C.: A fully homomorphic encryption scheme. PhD thesis, Stanford University (2009)

28. Goldreich, O., Goldwasser, S., Micali, S.: How to construct random functions (extended abstract). In: FOCS, pp. 464–479 (1984)

29. Goldwasser, S., Micali, S., Rivest, R.L.: A digital signature scheme secure against adaptive chosen-message attacks. SIAM Journal of Computing 17(2), 281–308 (1988)

30. Haber, S., Hatano, Y., Honda, Y., Horne, W., Miyazaki, K., Sander, T., Tezoku, S., Yao, D.: Efficient signature schemes supporting redaction, pseudonymization, and data deidentification. In: ASIACCS 2008, pp. 353–362 (2008)

31. Hevia, A., Micciancio, D.: The Provable Security of Graph-Based One-Time Signatures and Extensions to Algebraic Signature Schemes. In: Zheng, Y. (ed.) ASIACRYPT 2002. LNCS, vol. 2501, pp. 379–396. Springer, Heidelberg (2002)

32. Johnson, R., Molnar, D., Song, D., Wagner, D.: Homomorphic Signature Schemes. In: Preneel, B. (ed.) CT-RSA 2002. LNCS, vol. 2271, pp. 244–262. Springer, Heidelberg (2002)

33. Krohn, M., Freedman, M., Mazieres, D.: On-the-fly verification of rateless erasure codes for efficient content distribution. In: Proc. of IEEE Symposium on Security and Privacy, pp. 226–240 (2004)
34. Lewko, A., Okamoto, T., Sahai, A., Takashima, K., Waters, B.: Fully Secure Functional Encryption: Attribute-Based Encryption and (Hierarchical) Inner Product Encryption. In: Gilbert, H. (ed.) EUROCRYPT 2010. LNCS, vol. 6110, pp. 62–91. Springer, Heidelberg (2010)
35. Micali, S., Rivest, R.L.: Transitive Signature Schemes. In: Preneel, B. (ed.) CT-RSA 2002. LNCS, vol. 2271, pp. 236–243. Springer, Heidelberg (2002)
36. Miyazaki, K., Hanaoka, G., Imai, H.: Digitally signed document sanitizing scheme based on bilinear maps. In: ASIACCS 2006: Proceedings of the 2006 ACM Symposium on Information, Computer and Communications Security, pp. 343–354 (2006)
37. Miyazaki, K., Iwamura, M., Matsumoto, T., Sasaki, R., Yoshiura, H., Tezuka, S., Imai, H.: Digitally signed document sanitizing scheme with disclosure condition control. IEICE Transactions on Fundamentals E88-A(1), 239–246 (2005)
38. Miyazaki, K., Susaki, S., Iwamura, M., Matsumoto, T., Sasaki, R., Yoshiura, H.: Digital document sanitizing problem. IEICE Technical Report, 103(195(ISEC 2003 12-29)), 61–67 (2003)
39. Naccache, D.: Is theoretical cryptography any good in practice? CHES 2010 invited talk (2010), http://www.iacr.org/workshops/ches/ches2010
40. Neven, G.: A simple transitive signature scheme for directed trees. Theoretical Computer Science 396(1-3), 277–282 (2008)
41. Rivest, R.: Two signature schemes. Slides from talk given at Cambridge University (2000), http://people.csail.mit.edu/rivest/Rivest-CambridgeTalk.pdf
42. Rivest, R.L., Shamir, A., Tauman, Y.: How to Leak a Secret: Theory and Applications of Ring Signatures. In: Goldreich, O., Rosenberg, A.L., Selman, A.L. (eds.) Theoretical Computer Science. LNCS, vol. 3895, pp. 164–186. Springer, Heidelberg (2006)
43. Shahandashti, S.F., Salmasizadeh, M., Mohajeri, J.: A Provably Secure Short Transitive Signature Scheme from Bilinear Group Pairs. In: Blundo, C., Cimato, S. (eds.) SCN 2004. LNCS, vol. 3352, pp. 60–76. Springer, Heidelberg (2005)
44. Smart, N.P., Vercauteren, F.: Fully Homomorphic Encryption with Relatively Small Key and Ciphertext Sizes. In: Nguyen, P.Q., Pointcheval, D. (eds.) PKC 2010. LNCS, vol. 6056, pp. 420–443. Springer, Heidelberg (2010)
45. Steinfeld, R., Bull, L., Zheng, Y.: Content Extraction Signatures. In: Kim, K.-c. (ed.) ICISC 2001. LNCS, vol. 2288, pp. 285–304. Springer, Heidelberg (2002)
46. van Dijk, M., Gentry, C., Halevi, S., Vaikuntanathan, V.: Fully Homomorphic Encryption over the Integers. In: Gilbert, H. (ed.) EUROCRYPT 2010. LNCS, vol. 6110, pp. 24–43. Springer, Heidelberg (2010)
47. Waters, B.: Ciphertext-Policy Attribute-Based Encryption: An Expressive, Efficient, and Provably Secure Realization. In: Catalano, D., Fazio, N., Gennaro, R., Nicolosi, A. (eds.) PKC 2011. LNCS, vol. 6571, pp. 53–70. Springer, Heidelberg (2011)
48. Wei, L., Coull, S.E., Reiter, M.K.: Bounded vector signatures and their applications. In: ASIACCS 2011, pp. 277–285 (2011)
49. Yi, X.: Directed Transitive Signature Scheme. In: Abe, M. (ed.) CT-RSA 2007. LNCS, vol. 4377, pp. 129–144. Springer, Heidelberg (2006)
50. Zhao, F., Kalker, T., Médard, M., Han, K.: Signatures for content distribution with network coding. In: Proc. Intl. Symp. Info. Theory (ISIT) (2007)

Identifying Cheaters
without an Honest Majority

Yuval Ishai[1,*], Rafail Ostrovsky[2,**], and Hakan Seyalioglu[3,***]

[1] Department of Computer Science, Technion
[2] Department of Computer Science and Department of Mathematics, UCLA
[3] Department of Mathematics, UCLA

Abstract. Motivated by problems in secure multiparty computation (MPC), we study a natural extension of *identifiable secret sharing* to the case where an arbitrary number of players may be corrupted. An identifiable secret sharing scheme is a secret sharing scheme in which the reconstruction algorithm, after receiving shares from all players, either outputs the correct secret or publicly identifies the set of all cheaters (players who modified their original shares) with overwhelming success probability. This property is impossible to achieve without an honest majority. Instead, we settle for having the reconstruction algorithm inform each *honest* player of the correct set of cheaters. We show that this new notion of secret sharing can be *unconditionally* realized in the presence of arbitrarily many corrupted players. We demonstrate the usefulness of this primitive by presenting several applications to MPC without an honest majority.

- Complete primitives for MPC. We present the first unconditional construction of a complete primitive for fully secure function evaluation whose complexity does not grow with the complexity of the function being evaluated. This can be used for realizing fully secure MPC using *small* and *stateless* tamper-proof hardware. A previous completeness result of Gordon et al. (TCC 2010) required the use of cryptographic signatures.
- Applications to partial fairness. We eliminate the use of cryptography from the online phase of recent protocols for multiparty coin-flipping and MPC with partial fairness (Beimel et al., Crypto 2010 and Crypto 2011). This is a corollary of a more general technique for unconditionally upgrading security against fail-stop adversaries with preprocessing to security against malicious adversaries.

* Work done in part while visiting UCLA. Supported by ERC Starting Grant 259426, ISF grant 1361/10, and BSF grant 2008411.
** Research supported in part by NSF grants 0830803, 09165174, 1065276, 1118126 and 1136174, US-Israel BSF grant 2008411, B. John Garrick Foundation, OKAWA Foundation, IBM, Lockheed-Martin Corporation and the Defense Advanced Research Projects Agency through the U.S. Office of Naval Research under Contract N00014-11-1-0392. The views expressed are those of the author and do not reflect the official policy or position of the Department of Defense or the U.S. Government.
*** Research supported by a NSF Graduate Research Fellowship.

R. Cramer (Ed.): TCC 2012, LNCS 7194, pp. 21–38, 2012.

Finally, we complement our positive results by a negative result on identifying cheaters in unconditionally secure MPC. It is known that MPC without an honest majority can be realized *unconditionally* in the OT-hybrid model, provided that one settles for "security with abort" (Kilian, 1988). That is, the adversary can decide whether to abort the protocol after learning the outputs of corrupted players. We show that such protocols *cannot* be strengthened so that all honest players agree on the identity of a corrupted player in the event that the protocol aborts, even if a broadcast primitive can be used. This is contrasted with the computational setting, in which this stronger notion of security can be realized under standard cryptographic assumptions (Goldreich et al., 1987).

1   Introduction

Consider a scenario in which n mutually distrustful clients wish to distribute a long computation. Instead of directly interacting with each other, they rely on a trusted external *stateless* server. In each invocation, the server receives a share of the current state of computation (and possibly an additional input) from each client, and returns a share of the new state (and possibly an output) to each client. This scenario may apply to distributing sensitive computations using servers in the cloud, where requiring servers to maintain state information between different invocations is undesirable for security reasons.

The question we ask is what form of secret sharing is suitable for distributing the joint state between the clients. Naturally, we do not want to assume that a majority of the clients are honest (this rules out fair [8] or unconditionally secure [6] solutions that use direct interaction between the clients and do not employ the server). Additionally sharing the state fails in protecting the correctness of the computation, allowing each client to change the global state without being detected. A better solution is to use *robust* secret sharing that can *detect* cheating (cf. [29,9] and references therein). When there are three or more clients, this too has the disadvantage that it offers no strong deterrent against cheating: while cheating does not go undetected, it disrupts the computation without identifying a corrupted client. This motivates the use of *identifiable secret sharing*, where a failure of the reconstruction algorithm results in identifying the clients who modified their shares.

Identifiable secret sharing as above can be realized when a majority of the clients are honest [22,20,7,24]. But without an honest majority, there is no way for the server to tell apart cheaters from honest clients. Indeed, $n/2$ cheaters can simulate a consistent sharing of an incorrect secret, which makes it impossible for the server to tell which of the two sets of consistent shares is correct. However, this does not rule out the alternative of allowing the server to *inform each client* (with negligible error probability) which shares have been modified *assuming that this client is honest*. We refer to this as *locally-identifiable secret sharing* (LISS). Note that except with negligible probability, each honest client will agree on which clients are corrupted and should be disqualified. Thus use of LISS to share the state minimizes the incentive to cheat and allows the honest clients in the event of reconstruction failure to *agree* on a strict subset of the clients that includes all honest clients. This subset has the option of restarting the computation on their original inputs, using default values for the inputs of the remaining clients, without losing in this process any of the honest clients.

Settling for *computational* security, LISS can be realized via the use of *digital signatures*: the sharing procedure distributes to all clients *the same* public verification key vk, and gives to each client a signature of its additive share of the secret using the corresponding secret key sk. Reconstruction proceeds by letting each client send to the server its original share, vk, and the signature on the share. The server can then identify definite cheaters as those who supply an inconsistent triplet, and partition the remaining clients according to the value of vk they provide. In fact, such a computationally secure LISS scheme was implicitly used by Gordon et al. [13] in the context of defining a complete primitive for MPC. The possibility of an unconditionally secure construction remained open. This question is motivated not only by the goals of enhancing security and eliminating assumptions, but also by the potential *efficiency* advantages of information-theoretic techniques. This is especially significant in applications (such as those discussed below) where the share generation process is distributed between multiple players.

1.1   Our Results

Constructions. Our main result is an affirmative answer to the above question: we present an unconditional construction of an n-out-of-n LISS scheme whose security holds in the presence of an arbitrary number of corrupted players. More generally, we show how to efficiently transform any secret sharing scheme into one in which the reconstruction function reveals to every honest player of the identity of *all* shares that have been tampered with. In particular, all honest players agree on the same set of cheaters.

We also consider a weaker variant of LISS that we call *unanimously identifiable secret sharing* (UISS) in which only the latter agreement property is required. That is, if reconstruction fails, all honest players should agree on the same (non-empty) set of cheaters. This weaker primitive is easier to construct. (In fact, a construction of UISS is implicit in [25].) In contrast to LISS, however, UISS does not guarantee that *all* cheaters are detected in the event that reconstruction fails.

Applications. We present several applications of the above primitives in the context of MPC without an honest majority. In the following, the term MPC refers to the special case of secure function evaluation, namely MPC of non-reactive (stateless) functionalities. We use poly and neg to represent polynomial and negligible functions, respectively, and κ denote a statistical security parameter. While we mainly consider statistical security, our results are also useful in the domain of computational security.

COMPLETE PRIMITIVES FOR MPC. It is well known that fully secure MPC (with fairness and guaranteed output delivery) is impossible to achieve in general without an honest majority [8]. This naturally raises the question of finding a minimal *complete* primitive that can be used to get around this limitation. Such a primitive is defined by a (stateless) deterministic functionality g mapping n inputs to n outputs, such that any n-party functionality f can be realized using a trusted instance of g initialized between every tuple of players that can supply input to it. The first such results characterized complete boolean primitives for MPC with security against a *passive* adversary [21,19]. In the case of active adversaries, Fitzi et al. [11] presented a complete primitive for fully secure MPC whose computational complexity grows linearly with complexity of f. This left open the question of finding a "simple" complete primitive, whose complexity does not depend on the complexity of f. One such primitive was given by Gordon et al. [13] using digital signatures. We use UISS to get an unconditional variant of this result. In this variant, the complexity of g only grows with the *output length* of f.

Theorem 1. *There is a deterministic, polynomial-time computable functionality g with input and output size $\mathsf{poly}(n, \kappa, \beta)$ such that any n-party function f computed by a circuit of size σ and output length β can be realized with full statistical security (and $2^{-\kappa}$ simulation error) using $\mathsf{poly}(n, \sigma)$ calls to g.*

This result has an interesting interpretation in the context of a recent line of work on basing cryptography on tamper-proof hardware (see [17,15] and references therein). In this line of work, several impossibility results in cryptography (including UC security, unconditional security, software protection and obfuscation) were circumvented by using tamper-proof hardware tokens. These works spent efforts on minimizing the *size* of the tokens, employing *stateless* (rather than stateful) tokens, and minimizing or eliminating cryptographic assumptions. The above result can be viewed as achieving all these goals simultaneously in the context of another major impossibility result: the impossibility of fully secure MPC without an honest majority. It implies that a *small* and *stateless* token, connected via secure channels to the n players, suffices to *unconditionally* realize fully secure MPC. We note that connecting the same token to *all* players is necessary, as implied by the results of Fitzi et al. [11].

We also present other variants of the previous completeness theorem which rely on computational assumptions but still avoid the use of cryptography inside the primitive. These variants have the advantage of requiring only a small number of calls to the primitive (independently of the complexity of f).

APPLICATIONS TO PARTIAL FAIRNESS. A recent line of works studies the extent to which *partial fairness* can be achieved in MPC without an honest majority. Partial fairness can be defined by restricting the simulation error to be small (e.g., inverse polynomial) but not negligible [14]. We show that in partially fair protocols of Beimel et al.[2,1] (extending previous two-party protocols of Moran et al. [23] and Gordon and Katz [14]), the use of a digital signature scheme can be replaced by a unanimously identifiable commitment scheme, a second primitive we define that can be used as a substitute for LISS in certain applications. This yields *unconditional* multiparty protocols for coin-flipping and MPC with partial fairness in the preprocessing model, namely assuming that players have offline access to correlated randomness. We note that trusted preprocessing does not trivialize the problem, because the output needs to be unpredictable in the end of the preprocessing phase. In fact, the negative results on achieving full fairness apply to the preprocessing model as well. The preprocessing model does allow, however, to eliminate the assumptions of secure channels and broadcast, which can be implemented unconditionally in the preprocessing model [27].

The preprocessing phase can be realized either by a trusted offline dealer or via a distributed protocol (possibly employing additional parties for unconditional security). Even if one relies on a computationally secure protocol for distributing the preprocessing phase, the protocols we get have the advantage of making only a *black-box* use of the underlying cryptographic primitives, whereas the original protocols from [2,1] make a non-black-box use of a one-way function.

In the case of coin-flipping, applying our primitive to the offline dealer protocol from [2] implies the following:

Theorem 2. *Assume preprocessing by a trusted off-line dealer. Fix constants n and t such that $t < 2n/3$. Then, for any r, there is an r-round n-party uncon-ditionally secure coin-tossing protocol over point-to-point channels tolerating up to t malicious players with bias $O(1/r)$.*

Our results on MPC with partial fairness are obtained via a general technique for unconditionally upgrading security against fail-stop adversaries to security against malicious adversaries where the messages sent by the players are deter-mined in the preprocessing stage.

A Negative Result. It is known that MPC without an honest majority can be realized *unconditionally* in the OT-hybrid model, provided that one settles for "security with abort" [18,16]. That is, the adversary can decide whether to abort the protocol after learning the outputs of corrupted players but before the honest players receive their output. We show that such protocols *cannot* be strengthened so that all honest players agree on the identity of a corrupted player in the event that the protocol aborts, even if a broadcast primitive and trusted access to an arbitrary pairwise functionality is assumed. This is contrasted with the computational setting, in which this stronger notion of security can be realized under standard cryptographic assumptions [12]. Our negative result strengthens a previous negative result from [11], which shows that pairwise functionalities alone (without broadcast) are not sufficient in general for *fully secure n*-party computation. For lack of space, the details of this result are deferred to the full version.

2   Preliminaries

Our communication model allows for authenticated point to point and broadcast channels unless specified otherwise. While we define our algorithms in terms of finite sets (with fixed input size) and fixed error rate, they can be implemented by uniform algorithms that are polynomial in the bit-length of the inputs, the number of players, and the statistical security parameter κ guaranteeing $\delta = 2^{-\kappa}$ error. The latter is the default convention whenever no value of δ is specified.

We only consider non-adaptive adversaries but our secret sharing definitions and proofs can easily be extended to the adaptive case. We denote the n players by $\mathcal{P} = \{P_1, P_2, \ldots P_n\}$ and will often identify a player with its index. A collection of subsets $\mathbb{A}$ of $\mathcal{P}$ will be called *monotone* if for any $B \in \mathbb{A}$, if $B \subseteq C \subseteq \mathcal{P}$ then, $C \in \mathbb{A}$. We let $[n]$ denote the set $\{1, \ldots, n\}$. We use $x \xleftarrow{\$} X$ to denote a uniform choice of x from a set X.

2.1   Secret Sharing

We briefly describe our notation for standard secret sharing schemes. A secret sharing scheme is defined by a pair of algorithms (Share, Rec), where Share is a randomized algorithm mapping a secret from S to the share space $\prod_{i=1}^{n} S_i$, and Rec is a deterministic reconstruction algorithm mapping the shares of a qualified set of players (along with the identity of this set) to a secret from S. We will refer to S as the secret space and to S_i as the share space of P_i. An *access structure* is a monotone collection of player sets. We say that a secret sharing scheme realizes an access structure $\mathbb{A}$ if sets in $\mathbb{A}$ can reconstruct the secret s and others can learn nothing about it. Throughout this work we define secret sharing schemes to have perfect *correctness* (authorized sets always correctly reconstruct the secret) and perfect *secrecy* (the shares of unauthorized sets reveal no information about the secret). For all additional security guarantees we assume the adversary knows the secret that is being shared; even if the secret is compromised, the adversary should not be able to cause the reconstruction algorithm to behave undesirably (e.g. by outputting an incorrect secret or implicating an honest player of cheating), except with small probability.

As usual, we consider a single adversary who may corrupt one or more players. We distinguish between passive and active corruptions using the following terminology.

Definition 1. (Tampering) *A corrupted player is said to have* tampered *with its share if it provides to the reconstruction algorithm a share different than the one assigned by the distribution algorithm. Such a share is called a* tampered share *and such a player is called a* cheater.

Identifiable Secret Sharing. An identifiable secret sharing scheme is a secret sharing scheme in which the reconstruction algorithm can identify all cheaters in the event that it fails to reconstruct the secret. The above guarantee should hold except with some failure probability δ as long as there are at most t cheaters for an additional parameter t. In our definition we assume that the tampering is done by a single adversary who can observe the shares of a set C of up to t corrupted players and based on this information decide on how to tamper with their shares.

Definition 2. (Identifiable Secret Sharing) *A secret sharing scheme realizing* $\mathbb{A}$ *is* (δ, t)-*identifiable if for any (unbounded) adversary A and any* $s \in S$, *the success probability of A in the following game is at most* δ:

1. $(s_1, s_2, \ldots, s_n) \leftarrow \mathsf{Share}(s)$;
2. A *outputs a set* $C \subset [n]$ *such that* $|C| \leq t$ *and receives* $(s_j)_{j \in C}$;
3. A *outputs* $(B, (s_j')_{j \in C \cap B})$ *where* $B \in \mathbb{A}$;
4. $\mathsf{Out} \leftarrow \mathsf{Rec}(B, (t_j)_{j \in B})$ *where* $t_j = s_j'$ *if* $j \in C$ *and* $t_j = s_j$ *otherwise.*

A *succeeds if for some* $j \in C \cap B$, $s_j' \neq s_j$ *and* $\mathsf{Out} \neq (\perp, \{P_i \in C \cap B : s_i' \neq s_i\})$.

The first work on identifiable secret sharing is due to McEliece and Sarwate [22] who showed that Shamir's k-threshold secret sharing scheme allows perfect identification if $k + 2t$ players of which at most t are cheaters are involved in reconstruction. Several works consider various relaxations of identifiability [28,3,5] which suffice for some applications but not suitable for MPC with a dishonest majority. There is also substantial work on the efficiency of identifiable secret sharing [20,24,7].

Identifiability is not possible with a dishonest majority for a simple reason: If half of the participants are dishonest they can run the sharing algorithm independently among themselves and return as their shares the output of the second run of the algorithm. This strategy makes it impossible for Rec to identify which half of the shares come from the first run of the Share algorithm and which come from the second since they are run independently. This is captured by the following theorem (see full version for proof):

Theorem 3. (No identifiability with a dishonest majority) *For any* t, n, $S, \mathbb{A}$ *with* $t \geq n/2$, $|S| \geq 2$, $\mathbb{A} \neq \emptyset$, *there is no* $(1/4, t)$-*identifiable secret sharing scheme with secret space* S *and access structure* $\mathbb{A}$.

3   Locally Identifiable Secret Sharing

We now give our relaxation of identifiable secret sharing that can be realized when arbitrarily many players may be corrupted. Informally, the guarantee we require is that if the reconstruction fails, the reconstruction algorithm outputs a tuple of players to each player P_i with the guarantee that if P_i is honest, the tuple returned to P_i is precisely the players that tampered with their shares. While this is equivalent to identifiability from the point of view of the honest players, it allows us to circumvent the impossibility result of Theorem 3. Note that we define LISS as being a special type of secret sharing scheme, so the usual correctness and secrecy requirements should hold in addition to the requirements detailed below.

Definition 3. (Lists) *Throughout this paper when we refer to a list* $\mathcal{L}$ *we refer to a subset of the players in the protocol* $(\mathcal{L} \subset \{P_1, P_2, \ldots, P_n\})$.

Definition 4. (LISS) *A secret sharing scheme realizing $\mathbb{A}$ is locally δ-identifiable if it satisfies the following requirements:*

- **Unanimity:** *For any adversary A and $s \in S$, the probability of A's success in the following game is at most δ:*
    1. $(s_1, s_2, \ldots, s_n) \leftarrow \mathsf{Share}(s)$;
    2. A outputs a set $C \subset [n]$ to corrupt and then receives $(s_i : i \in C)$;
    3. A outputs $(B, (s'_j)_{j \in C \cap B})$ such that $B \in \mathbb{A}$ and $B \not\subset C$;
    4. $\mathsf{Out} \leftarrow \mathsf{Rec}(B, (t_j)_{j \in B})$ where $t_j = s'_j$ if $j \in C$ and $t_j = s_j$ otherwise.

    The adversary succeeds unless:
    1. Reconstruction succeeds: $\mathsf{Out} = s$ or,
    2. Each honest player's list is the list of all cheaters: $\mathsf{Out} = (\bot, (\mathcal{L}_j)_{j \in B})$ where for all $j \in B \setminus C$, $\mathcal{L}_j = \{P_i \in C \cap B : s'_j \neq s_j\}$.
- *The scheme has* **Predictable Failures** *(Definition 5).*

We briefly motivate the requirement of Predictable Failures before defining it. The problem to address is that the additional outputs $\mathcal{L}_j$, or even the event of not reconstructing the secret, may leak some information concerning the secret unless a separate guarantee is made. This can cause a problem in applications and therefore we must have a way to simulate the actions of Rec in the case of tampering. Note that this is a new issue not present in identifiable secret sharing: As the Rec function does not simply output a list of tampering players, there are no a-priori guarantees concerning the lists corresponding to dishonest players and therefore we must make requirements on them separately.

Definition 5. (Predictable Failures) *A secret sharing scheme has δ-Predictable Failures if there is an algorithm SRec such that for any adversary A and $s \in S$, the probability of success in the following game is less than δ:*

1. $(s_1, s_2, \ldots, s_n) \leftarrow \mathsf{Share}(s)$;
2. A outputs a set $C \subset [n]$ to corrupt and receives $(s_i)_{i \in C}$;
3. A outputs $(B, (s'_j)_{j \in C \cap B})$ such that $B \in \mathbb{A}$ and $B \not\subset C$;
4. $\mathsf{SOut} \leftarrow \mathsf{SRec}(C, B, (s_i)_{i \in C}, (s'_i)_{i \in C \cap B})$;
5. $\mathsf{Out} \leftarrow \mathsf{Rec}(B, (t_j)_{j \in B})$ where $t_j = s'_j$ if $j \in C$ and $t_j = s_j$ otherwise.

A succeeds unless:

1. SRec correctly predicts success: $\mathsf{SOut} = \textsc{Success}$ and $\mathsf{Out} = s$ or,
2. SRec predicts the output of Rec: $\mathsf{SOut} = \mathsf{Out} \neq s$.

3.1   Our Construction

Let $(\mathsf{Sh}, \mathsf{Rc})$ be a secret sharing scheme realizing access structure $\mathbb{A}$ with $\mathsf{Sh} : S \to \mathbb{F}^n$ where $\mathbb{F}$ is a field. Let $I_{n \times n}$ and $0_{n \times n}$ denote the identity and all zero matrix respectively. We use $\mathbb{F}^{n \times n}$ to denote the set of all $n \times n$ matrices with elements in $\mathbb{F}$ and $GL_n(\mathbb{F})$ to denote the set of all such invertible matrices. For a matrix M we will use $M(i, j)$ to denote the (i, j) entry of M. By default we assume vectors to be column vectors, we will use the notation $\boldsymbol{a}^T$ when referring to a row vector. We use the notation $\mathbb{F}^*$ to denote $\mathbb{F} \setminus \{0\}$.

Share(s):

1. Generate $(t_1, t_2, \ldots, t_n) \leftarrow \mathsf{Sh}(s)$, $u_i, v_i \xleftarrow{\$} \mathbb{F}^*$ for all $i \in [n]$;

2. Define $C_0 \in \mathbb{F}^{n \times n}$ as $\begin{cases} C_0(i, j) = u_i^{j+1} v_j^{i+1} + u_i v_j + 1 & \text{for } i \neq j; \\ C_0(i, i) = t_i & \text{for } i \in [n]. \end{cases}$

3. Define C blockwise as: $\begin{pmatrix} C_0 & I_{n \times n} \\ I_{n \times n} & 0_{n \times n} \end{pmatrix}$;

4. Generate $B \xleftarrow{\$} GL_{2n}(\mathbb{F})$ and define $A = CB^{-1}$;

5. Label row i of A as $\boldsymbol{a}_i^T$ and column j of B as $\boldsymbol{b}_j$;

6. Return $(s_i = (\boldsymbol{a}_i^T, \boldsymbol{b}_i, u_i, v_i))_{i \in [n]}$.

Rec($D, (s_i = (\boldsymbol{a}_i^T, \boldsymbol{b}_i, u_i, v_i))_{i \in D}$) with $D \in \mathbb{A}$, $\boldsymbol{a}_i^T, \boldsymbol{b}_i \in \mathbb{F}^{2n}$, $u_i, v_i \in \mathbb{F}^*$:

1. If for all $i \neq j$, $\boldsymbol{a}_i^T \boldsymbol{b}_j = u_i^{j+1} v_j^{i+1} + u_i v_j + 1$:
   - Set $\boldsymbol{a}_i^T \boldsymbol{b}_i = t_i$ for all $i \in B$;
   - Return $\mathsf{Rc}(D, t_i : i \in D)$.

2. Else, for all $i \in D$ set:

$$\mathcal{L}_i = \{P_j : \boldsymbol{a}_i^T \boldsymbol{b}_j \neq u_i^{j+1} v_j^{i+1} + u_i v_j + 1 \text{ or } \boldsymbol{a}_j^T \boldsymbol{b}_i \neq u_j^{i+1} v_i^{j+1} + u_j v_i + 1\};$$

3. Return $(\perp, (\mathcal{L}_i)_{i \in D}$.

Theorem 4. *If $\delta > n^2(n+1)/(|\mathbb{F}|-1)$, the scheme described above is a δ-LISS scheme realizing $\mathbb{A}$ with secret space S and share space $S_i = \mathbb{F}^{4n+2}$.*

Corollary 1. *Suppose there is a secret sharing scheme which realizes an n-party access structure $\mathbb{A}$ with secret space S and share length β. Then, for any $\delta > 0$ there is a δ-LISS with the same $\mathbb{A}$ and S whose share length is $O(n \log(n/\delta) + n\beta)$.*

Outline of Security. A full proof of security is provided in the full version of this paper but we provide a brief intuition in this section for self containment. We first argue secrecy. Notice that the value t_i is only used in generating the row $\boldsymbol{a}_i^T$, therefore any set of players $E \notin \mathbb{A}$ will have its shares generated using only the t_i values such that $P_i \in E$. The fact that the joint distribution of these t_i values do not depend on the underlying secret (due to the perfect secrecy of $(\mathsf{Sh}, \mathsf{Rc})$) implies secrecy.

Consider now an adversary that is attempting to tamper the share of some P_i (and possibly others) - we will argue that any such attempt will cause the check in Rec between P_i and any honest player P_j to fail with high probability. Assume that the adversary is tampering $\boldsymbol{b}_i \to \boldsymbol{b}_i'$, $v_i \to v_i'$ with one of these values changed (a similar argument will hold if the adversary is tampering a_i^T or u_i). There are then two cases, either b_i' is in $\mathsf{Span}(\{b_i\}_{i\in T})$ where T is the set of corrupted players or it is linearly independent of these values. If b_i' is linearly independent it can be shown that $a_j^T b_i'$ is essentially uniformly distributed over $\mathbb{F}$ conditioned on the view of the adversary even after u_j is fixed and therefore the probability that reconstruction succeeds will be very low (showing this statement is non-trivial).

On the other hand, consider the case where $b_i' \in \mathsf{Span}(\{b_k\}_{k\in T})$. Let $b_i' = \sum_{k\in T} \beta_k b_k$ where $\beta_k \in \mathbb{F}$. Now, the check will succeed only if:

$$a_j^T \sum_{k\in T} \beta_k b_k = u_j^{i+1} v'{}_i^{j+1} + u_j v_i' + 1 \Leftrightarrow$$

$$\sum_{k\in T} \beta_k (u_j^{k+1} v_k^{j+1} + u_j v_k + 1) = u_j^{i+1} v'{}_i^{j+1} + u_j v_i' + 1.$$

Similar to our argument of secrecy, the value u_j is uniformly distributed conditioned on every view of the adversary. Therefore, the check in Rec will succeed only if the above equality is satisfied by a uniformly chosen $u_j \in \mathbb{F}^*$. This will happen rarely unless the polynomials on the left and right of the equality (considered as a polynomial in u_j) are equal. For this equality to hold, we must have $\beta_k = 0$ for all $k \neq i$ since otherwise $v_k \neq 0$ would make the polynomials different. Next notice that we must have $\beta_i = 1$ for the constant terms to match. Finally, the $u_j v_k$ term on the left implies that $v_i' = v_i$. Therefore, unless $v_i' = v_i$ and $b_i' = b_i$ this equality will only occur with low probability, which implies that if P_i tampers with either the v_i or b_i value, it will be detected and placed on P_j's list with high probability for all honest P_j. A similar argument holds if either the a_i^T or u_i value is tampered since the method of generation is equivalent to first generating $A \in GL_{2n}(\mathbb{F})$ and setting $B = A^{-1}C$ since C is always invertible.

Notice that we have actually argued that a dishonest player who modifies his share will be on the list of every honest player and symmetrically that all honest players will be on the list of such a dishonest player with high probability. This implies Predictable Failures since an adversary can easily tell which dishonest players will be on a given dishonest player's list from the shares it has, as well as whether or not the honest players will be on his list depending on whether or not the player modifier his share.

4   Relaxing Local Identifiability

In this section we define a new commitment primitive, *unanimously identifiable commitments* that can be used as a leaner substitute for LISS in certain applications. Additionally, we note that this commitment primitive implies a weaker variant of LISS (called *unanimously identifiable secret sharing*) that can also be used in our applications to MPC.

4.1   Unanimously Identifiable Commitments

A *unanimously identifiable commitment* (UIC) scheme has a single player (called the sender) committed to a value $s \in S$ by having a trusted dealer send commitments c_i to all other players in the protocol and decommitment information d to the sender such that any tampering of the d value will cause all honest players to either reconstruct the original secret or fail reconstruction simultaneously. As with standard commitments, $(c_i)_{i \in [n]}$ should leak no information concerning s.

Definition 6. (Unanimously Identifiable Commitments) *A δ-UIC scheme consists of a randomized algorithm* Offline *and a deterministic algorithm* Decommit *with the following syntax:*

1. **Offline**: $S \to C^n \times D$. *Takes as input a secret* $s \in S$ *outputs* n *commitments* $c_1, c_2, \ldots, c_n$ *and decommitment information* d.
2. **Decommit**: $C \times D \to S \cup \{\bot\}$. *Takes as input* c_i *and the decommitment information* d *and recreates the secret* s *or outputs* $\bot$ *indicating failure.*

Where the algorithms (Offline, Decommit) should satisfy:

- **Completeness.** For any $s \in S$, if $\Pr[\mathsf{Offline}(s) = (c_1, c_2, \ldots, c_n, d)] > 0$ then, $\mathsf{Decommit}(c_i, d) = s$ for any $i \in [n]$.
- **Secrecy.** The values $c_1, c_2, \ldots, c_n$ reveal no information concerning s. Formally, for any $\mathbf{c} = (c_1, c_2, \ldots, c_n)$ and any $s, s' \in S$, the probability that the first n values of $\mathsf{Offline}(s)$ is $\mathbf{c}$ is equal to the probability that the first n values of $\mathsf{Offline}(s')$ is $\mathbf{c}$.

We now present the final requirement placed on this primitive for use in our applications. In the full version of this paper, we include further intuition to the necessity of this condition but omit it here for space restrictions.

There exists simulators W_1, W_2 such that the two guarantees described below hold with probability at least $1 - \delta$ for any A. Consider the following experiment:

1. The adversary, A outputs a set $T \subset [n] \cup \{Q\}$ of players to corrupt;
2. $(c_1, c_2, \ldots, c_n, d) \leftarrow \mathsf{Offline}(s)$;
3. For all $i \in T \cap [n]$ send c_i to the adversary, if $Q \in T$ send d to the adversary;
4. If $Q \notin T$, set $dec = d$; otherwise, dec is output by A.
5. For all $i \in T \cap [n]$, A outputs (c'_i, i), fake commitment information for P_i.

The guarantees around this experiment are as follows:

- **Binding with Agreement on Abort.** $\mathsf{Decommit}(c_i, dec) = s$ for all P_i uncorrupted or $\mathsf{Decommit}(c_i, dec) = \bot$ for all P_i uncorrupted.
- **Simulatable Abort.** Let V be the view of A at the end of 5., then:
  1. If A corrupted Q:
     $W_1(V)$ correctly predicts if $\mathsf{Decommit}(c_i, dec) = \bot$ for all $i \in [n]$.
  2. Otherwise:
     $W_2(V, c'_i)$ correctly predicts if $\mathsf{Decommit}(c'_i, d) = \bot$ for each $i \in T \cap [n]$.

4.2   A Unanimously Identifiable Commitment Scheme

Let $\mathbb{F}$ be a field. We now give a simple construction of a δ-UIC scheme with $S = \mathbb{F}$, $C = \mathbb{F}^{n+2}$ and $D = \mathbb{F}^2$.

Offline(s) :

1. Generate $P(X)$, a random $n+1$ degree polynomial over $\mathbb{F}$ such that $P(0) = s$;
2. For all $i \in [n]$ generate $x_i \xleftarrow{\$} \mathbb{F}$ and let $y_i = P(x_i)$;
3. Set $c_i = (x_i, y_i)$ and $d = P(X)$. Return $((c_i)_{i \in [n]}, d)$.

Decommit($c_i = (x_i, y_i), d = P(X)$ of degree $n + 1$) : If $P(x_i) \neq y_i$ return $\bot$. Else, return $P(0)$.

Theorem 5. *Let* $|\mathbb{F}| > (n+1)^2 \delta^{-1} + 1$. *The scheme described above is a δ-UIC with* $S = \mathbb{F}$, $C = \mathbb{F}^{n+2}$ *and* $D = \mathbb{F}^2$.

Related Concepts. In our applications, we mainly use UIC as a substitute for digital signatures. There are some other unconditional notions that have also been introduced for similar purposes (such pseudosignatures [26], distributed commitments [10] and IC signatures [25]). While the construction itself is not novel (for example, it is used in [25]), the property that all of the honest players accept or reject the same commitment is crucial to our applications and differs from the guarantees placed on the other primitives.

4.3   Unanimously Identifiable Secret Sharing

We note that *unanimously identifiable commitments* actually imply a weaker notion of LISS which we call *unanimously Identifiable Secret Sharing* (UISS). The security requirements for a UISS scheme are identical to the requirements to LISS except that the requirement:

- *Each honest player's list is the list of all cheaters:*

is replaced by the requirement:

- *Each honest player's list is the same subset of corrupted players:*

$$\mathsf{Out} = (\mathcal{L}_j)_{j \in B} \text{ where for all } j, j' \in B \setminus T, \ \mathcal{L}_j \subset T \text{ and } \mathcal{L}_j = \mathcal{L}_{j'}.$$

All other requirements remain unchanged, including the requirement of predictable failures. Implementing UISS for access structure $\mathbb{A}$ using a UIC scheme is straightforward by having each user commit to its share. Note that for most applications, UISS can take the role of LISS, at the cost of not necessarily identifying *all* tampered shares if reconstruction fails.

5   Applications

A secure multiparty computation (MPC) protocol allows a set of players to compute a function evaluated on their individual inputs while revealing no information other than the output of the function. We assume familiarity with (standalone) MPC throughout this section and refer the reader to [4] for formal definitions.

5.1   Model of Computation

By default, we consider static, computationally unbounded adversaries who may corrupt up to t of the n parties ($t = n$ by default). We consider both *active* adversaries, who may arbitrarily control the corrupted players, *passive* adversaries, who can only observe the internal state of corrupted players, and *fail stop* adversaries who behave like passive adversaries except that they make corrupted players stop sending messages. Our network model is synchronous with point-to-point channels and a broadcast channel.

The security of an MPC protocol with respect to an ideal functionality f is defined by comparing a real world execution of the protocol to an *ideal model* execution where a trusted party evaluates f. By default, we refer to *statistical* security, where the statistical advantage of distinguishing between the real world and the ideal model execution is bounded by $2^{-\kappa}$ for a statistical security parameter κ. We will only consider the case of secure function evaluation, in which f is stateless. We will mostly consider *fully secure* MPC in which the ideal model adversary cannot prevent the trusted party from sending the outputs of f to the honest players. Full security cannot be achieved even for simple functionalities such as coin-flipping [8] without an honest majority or other assumptions we will discuss. This impossibility holds even with trusted preprocessing; however, in the latter model the assumptions of secure point-to-point channels and a broadcast primitive are unnecessary as they can be implemented unconditionally [27].

5.2   Complete Primitives for MPC

An n-party functionality g is called a *complete primitive* for n-party MPC if it is possible to securely realize any n-party functionality f in the g-*hybrid model*, namely by using ideal calls to g. Here we consider security against an active adversary who may corrupt an arbitrary number of players.

In prior works, such primitives either depend on the complexity of the function being evaluated [11] or rely on cryptographic assumptions [13]. It remained open to construct an *unconditionally* complete primitive whose complexity is independent of the complexity of the evaluated function f. In the following section, we show how to construct such a primitive. Our contribution can be seen as identifying a cryptographic LISS scheme implicitly present in the construction of Gordon et al. [13] and replacing it with an unconditional construction. In fact, it suffices for this purpose to rely on UISS rather than LISS. For simplicity, we assume that the functionality f being evaluated using g delivers the same output to all players; the general case is handled similarly.

Unconditional Primitive. The first primitive we present is complete for statistically secure MPC and its complexity depends only on the output length of the evaluated functionality f. We give an informal description of the primitive in this version and defer further details to the full version. For expository purposes, we will describe three separate primitives that make up the three modes of operation for the complete primitive.

- FCR_1^1 - Takes as input a bit from a player, runs an n-out-of-n UISS sharing algorithm on this bit and distributes the shares amongst all players.
- FCR_2^1 - Takes as input two n-tuples of shares from the UISS scheme. Internally, the primitive reconstructs the underlying secrets of each, evaluates the NAND of the two secrets, and re-shares the output using the UISS scheme. If reconstruction fails, the functionality will use the lists $\mathcal{L}_j$ output by the UISS scheme to partition the players: If any player is on his own list, the functionality declares this player is disqualified and his input is replaced by a default value by all players. If not, the functionality outputs a partition of the players: P_i and P_j remain in the same partition if $\mathcal{L}_j = \mathcal{L}_i$.
- FCR_3^1 - Takes as input β separate n-tuples of UISS shares, where β is an output length parameter. The functionality either reconstructs each secret and broadcasts all the reconstructed bits, or, if some reconstruction fails, partitions the players as in the previous mode using the first instance of failed reconstruction.

Note that while the first two primitives are randomized, they can be made deterministic by using a standard reduction: the internal randomness can be securely emulated by taking the XOR of shares contributed by the n players.

Using the above primitive, one can securely evaluate any boolean circuit C, which consists of NAND gates and has β output bits, in the following way. The players first use FCR_1^1 to share each of their input bits. After this phase is completed, for each gate in C the players use FCR_2^1 to evaluate shares of the value of each internal value in C. Finally, the players feed the shares of the output values to FCR_3^1 and receive the outputs of C.

Notice that any deviation from the above protocol will result in all honest players identifying the same set of cheaters, and therefore their lists $\mathcal{L}_i$ will be identical. In this case, they are partitioned and the protocol is re-started with default values substituted for the inputs of the corrupted players. Due to the guarantees of the UISS scheme, the partitions can be simulated given only the views of the corrupted players. Defining the three modes of operation as one primitive that can be called on only some partition of the players requires some additional technical steps to fit in with our model of one trusted primitive. In addition to the players declaring which mode they are using, the primitive should also take as input from each player the set of players this player still trusts (as in [13]), we detail this in the full version of this paper.

The above complete primitive yields the following theorem.

Theorem 6. *There is a deterministic, polynomial-time computable functionality g with input and output size $\mathsf{poly}(n, \kappa, \beta)$ such that any functionality f computed by a circuit of size σ and output length β can be realized with full statistical security (and $2^{-\kappa}$ simulation error) using $\mathsf{poly}(n, \sigma)$ calls to g.*

Reducing the Number of Calls. Our second primitive improves on efficiency over the first by requiring fewer calls, but requires a preprocessing phase which is implemented using an MPC with identifiability on aborts (in other words, if the protocol fails then all honest players agree on the identity of a corrupted player.) Settling for computational security, such a protocol can be based on the existence of (two-party) oblivious transfer [12].

The protocol for f begins by the having the players run an MPC protocol as above to compute UISS-shares of the output of f. In case the preliminary MPC protocol fails, all players disqualify the player that caused the abort and restart the protocol by using a default value as the input of disqualified players.

We now describe the second primitive which is used to complete the protocol.

- FCR^2 takes as input an n-tuple of UISS shares for a β-bit secret and reconstructs the secret. In case reconstruction succeeds the primitive returns the reconstructed value to all players. If reconstruction fails, the primitive outputs a partition of the players by the lists output by the UISS scheme as in FCR^1.

The protocol for f proceeds by repeatedly interleaving the preliminary (computational) MPC with calls to FCR^2 until an output value is successfully reconstructed by the latter. Each failure results in the honest players disqualifying at least one corrupted player. As before, in each point of the protocol all honest players agree on the identity of disqualified players.

Theorem 7. *Suppose an oblivious transfer protocol exists with computational security parameter λ. Then there is a deterministic, polynomial-time computable n-party functionality g with input and output size $\mathsf{poly}(n, \beta, \kappa)$ such that any polynomial-time computable f with output size β can be realized with full computational security, up to $\mathsf{neg}(\lambda) + 2^{-\kappa}$ simulation error, using at most n calls to g.*

In the full version we give two variants on the above theorem that eliminate the dependence on the output length at the price of increased complexity of the MPC phase, and reduce the number of calls to 1 at the price of increasing the complexity of the primitive exponentially in n.

5.3 Partial Fairness with Preprocessing

In this section we briefly sketch how the unanimously identifiable commitments (UIC) primitive can be used with the partially fair MPC protocols of Beimel et al. [2,1] to eliminate the assumption of cryptographic signatures in the preprocessing model.

Construction with an Off-Line Dealer. The MPC protocols from [2,1] achieve unconditional security against *fail-stop* adversaries (with a non-negligible error $1/p$) given a trusted preprocessing phase in which a dealer sends some secret information to each player. This information contains the messages each player should send during the protocol, but the choice of which message is sent may depend on the (public) identity of the players that aborted up to this point. To upgrade the security of such a protocol to hold against active adversaries, Beimel et al. rely on digital signatures to ensure that players do not deviate from their designated messages. Our observation is that one could instead rely on the UIC primitive by having the dealer give to the player who should send a message the decommitment information for this message and to all other players the corresponding commitments. Then, if a corrupted player attempts to modify this decommitment information, all honest players will recognize this simultaneously and continue the execution as if this player had aborted.

Note that when considering general MPC in this model (rather than coin-flipping), it may be useful to allow the preprocessing stage to depend on the players' inputs. We refer to such a preprocessing phase as *input dependent preprocessing*. Since we require the outputs of the protocol to be unpredictable in the end of the preprocessing phase,[1] input dependent preprocessing cannot be used to trivially solve the problem by simply delivering the outputs of f to the players.

Theorem 8. *Let $\mathcal{P}$ be an r-round protocol with input dependent preprocessing, which realizes $\mathcal{F}$ with ϵ-security against fail-stop adversaries who can corrupt up to t players. Furthermore, suppose that the online phase of $\mathcal{P}$ has the following structure: in each round, each player sends a subset of the messages it had received in the preprocessing phase, where the identity of this subset can be computed publicly from the pattern of aborts up to this round. Then, there is a protocol $\mathcal{P}'$ with the same features of $\mathcal{P}$ except that it is $(\epsilon + 2^{-\kappa})$-secure against active adversaries.*

In the case of randomized functionalities with no inputs, the above theorem does not require the preprocessing to depend on any inputs. In particular, applying the above theorem to the coin-flipping protocol with preprocessing implicit in the construction from [2], we get the following corollary.

Theorem 9. *Assume preprocessing by a trusted off-line dealer. Fix constants n and t such that $t < 2n/3$. Then, for any r, there is an r-round n-party unconditionally secure coin-flipping protocol over point-to-point channels tolerating up to t malicious players with bias $O(1/r)$.*

In the full version we present a variant of our general UIC-based technique which can make the preprocessing phase independent of the inputs. This variant efficiently applies only when the number of players is constant and the input and output domain of each player is polynomially bounded in the security parameter.

---

[1] More precisely, security in the preprocessing model requires to simulate the adversary's view in the preprocessing phase before invoking the ideal functionality.

Applying this variant to general MPC protocols with $1/p$-security from [1], we obtain the following theorem.

Theorem 10. *Assume preprocessing by a trusted off-line dealer. Let n and t be constants such that $n/2 \le t < 2n/3$ and $\mathcal{F}$ be a deterministic n-party functionality with input domain bounded by a polynomial $d(\kappa)$ for each player. Then, for any polynomial $p(\kappa)$, there is a polynomial-time r-round $1/p$ secure protocol for $\mathcal{F}$ which tolerates up to t corrupt players with $r = pd^{n \cdot 2^t}$.*

Acknowledgements. We would like to acknowledge the helpful comments and suggestions of the anonymous TCC reviewers, and in particular for pointing out the relevance of the notion of IC Signatures from [25].

References

1. Beimel, A., Lindell, Y., Omri, E., Orlov, I.: 1/p-Secure Multiparty Computation without Honest Majority and the Best of Both Worlds. In: Rogaway, P. (ed.) CRYPTO 2011. LNCS, vol. 6841, pp. 277–296. Springer, Heidelberg (2011)
2. Beimel, A., Omri, E., Orlov, I.: Protocols for Multiparty Coin Toss with Dishonest Majority. In: Rabin, T. (ed.) CRYPTO 2010. LNCS, vol. 6223, pp. 538–557. Springer, Heidelberg (2010)
3. Brickell, E.F., Stinson, D.R.: The detection of cheaters in threshold schemes. SIAM J. Discrete Math. 4(4), 502–510 (1991)
4. Canetti, R.: Security and composition of multiparty cryptographic protocols. Journal of Cryptology 13(1), 143–202 (2000)
5. Carpentieri, M.: A perfect threshold secret sharing scheme to identify cheaters. Designs, Codes and Cryptography 5(3), 183–187 (1995)
6. Chor, B., Kushilevitz, E.: A zero-one law for boolean privacy (extended abstract). In: STOC 1989, pp. 62–72 (1989)
7. Choudhury, A.: Simple and asymptotically optimal t-cheater identifiable secret sharing scheme. IACR Cryptology ePrint Archive 2011, 330 (2011)
8. Cleve, R.: Limits on the security of coin flips when half the processors are faulty (extended abstract). In: STOC 1986, pp. 364–369. ACM (1986)
9. Cramer, R., Dodis, Y., Fehr, S., Padró, C., Wichs, D.: Detection of Algebraic Manipulation with Applications to Robust Secret Sharing and Fuzzy Extractors. In: Smart, N.P. (ed.) EUROCRYPT 2008. LNCS, vol. 4965, pp. 471–488. Springer, Heidelberg (2008)
10. Fehr, S., Maurer, U.M.: Linear VSS and Distributed Commitments Based on Secret Sharing and Pairwise Checks. In: Yung, M. (ed.) CRYPTO 2002. LNCS, vol. 2442, pp. 565–580. Springer, Heidelberg (2002)
11. Fitzi, M., Garay, J.A., Maurer, U.M., Ostrovsky, R.: Minimal complete primitives for secure multi-party computation. J. Cryptology 18(1), 37–61 (2005)
12. Goldreich, O., Micali, S., Wigderson, A.: How to play any mental game or a completeness theorem for protocols with honest majority. In: STOC 1987, pp. 218–229 (1987)
13. Gordon, D., Ishai, Y., Moran, T., Ostrovsky, R., Sahai, A.: On Complete Primitives for Fairness. In: Micciancio, D. (ed.) TCC 2010. LNCS, vol. 5978, pp. 91–108. Springer, Heidelberg (2010)

14. Gordon, S.D., Katz, J.: Partial Fairness in Secure Two-Party Computation. In: Gilbert, H. (ed.) EUROCRYPT 2010. LNCS, vol. 6110, pp. 157–176. Springer, Heidelberg (2010)
15. Goyal, V., Ishai, Y., Mahmoody, M., Sahai, A.: Interactive Locking, Zero-Knowledge PCPs, and Unconditional Cryptography. In: Rabin, T. (ed.) CRYPTO 2010. LNCS, vol. 6223, pp. 173–190. Springer, Heidelberg (2010)
16. Ishai, Y., Prabhakaran, M., Sahai, A.: Founding Cryptography on Oblivious Transfer – Efficiently. In: Wagner, D. (ed.) CRYPTO 2008. LNCS, vol. 5157, pp. 572–591. Springer, Heidelberg (2008)
17. Katz, J.: Universally Composable Multi-party Computation Using Tamper-Proof Hardware. In: Naor, M. (ed.) EUROCRYPT 2007. LNCS, vol. 4515, pp. 115–128. Springer, Heidelberg (2007)
18. Kilian, J.: Founding cryptography on oblivious transfer. In: STOC 1988, pp. 20–31. ACM (1988)
19. Kilian, J., Kushilevitz, E., Micali, S., Ostrovsky, R.: Reducibility and completeness in private computations. SIAM J. Comput. 29(4), 1189–1208 (2000)
20. Kurosawa, K., Obana, S., Ogata, W.: t-Cheater Identifiable (k, n) Threshold Secret Sharing Schemes. In: Coppersmith, D. (ed.) CRYPTO 1995. LNCS, vol. 963, pp. 410–423. Springer, Heidelberg (1995)
21. Kushilevitz, E., Micali, S., Ostrovsky, R.: Reducibility and completeness in multiparty private computations. In: FOCS 1994, pp. 478–489 (1994)
22. McEliece, R.J., Sarwate, D.V.: On sharing secrets and Reed-Solomon codes. Commun. ACM 24(9), 583–584 (1981)
23. Moran, T., Naor, M., Segev, G.: An Optimally Fair Coin Toss. In: Reingold, O. (ed.) TCC 2009. LNCS, vol. 5444, pp. 1–18. Springer, Heidelberg (2009)
24. Obana, S., Araki, T.: Almost Optimum Secret Sharing Schemes Secure Against Cheating for Arbitrary Secret Distribution. In: Lai, X., Chen, K. (eds.) ASIACRYPT 2006. LNCS, vol. 4284, pp. 364–379. Springer, Heidelberg (2006)
25. Patra, A., Choudhary, A., Pandu Rangan, C.: Round Efficient Unconditionally Secure Multiparty Computation Protocol. In: Chowdhury, D.R., Rijmen, V., Das, A. (eds.) INDOCRYPT 2008. LNCS, vol. 5365, pp. 185–199. Springer, Heidelberg (2008)
26. Pfitzmann, B., Waidner, M.: Information-theoretic pseudosignatures and byzantine agreement for t= n/3. IBM Research Report RZ 1996, 2882 (1996)
27. Pfitzmann, B., Waidner, M.: Unconditional Byzantine Agreement for any Number of Faulty Processors. In: Finkel, A., Jantzen, M. (eds.) STACS 1992. LNCS, vol. 577, pp. 337–350. Springer, Heidelberg (1992)
28. Rabin, T., Ben-Or, M.: Verifiable secret sharing and multiparty protocols with honest majority (extended abstract). In: STOC 1989, pp. 73–85 (1989)
29. Rogaway, P., Bellare, M.: Robust computational secret sharing and a unified account of classical secret-sharing goals. In: CCS 2007, pp. 172–184 (2007)

On the Security of the "Free-XOR" Technique*

Seung Geol Choi, Jonathan Katz, Ranjit Kumaresan, and Hong-Sheng Zhou**

Dept. of Computer Science, University of Maryland, College Park, MD 20742, USA
{sgchoi,jkatz,ranjit,hszhou}@cs.umd.edu

Abstract. Yao's *garbled-circuit approach* enables constant-round secure two-party computation of any function. In Yao's original construction, each gate in the circuit requires the parties to perform a constant number of encryptions/decryptions and to send/receive a constant number of ciphertexts. Kolesnikov and Schneider (ICALP 2008) proposed an improvement that allows XOR gates to be evaluated "for free," incurring no cryptographic operations and zero communication. Their "free-XOR" technique has proven very popular, and has been shown to improve performance of garbled-circuit protocols by up to a factor of 4.

Kolesnikov and Schneider proved security of their approach in the random oracle model, and claimed that (an unspecified variant of) correlation robustness suffices; this claim has been repeated in subsequent work, and similar ideas have since been used in other contexts. We show that the free-XOR technique *cannot* be proven secure based on correlation robustness alone; somewhat surprisingly, some form of *circular security* is also required. We propose an appropriate definition of security for hash functions capturing the necessary requirements, and prove security of the free-XOR approach when instantiated with any hash function satisfying our definition.

Our results do not impact the security of the free-XOR technique in practice, or imply an error in the free-XOR work, but instead pin down the assumptions needed to prove security.

1 Introduction

Generic protocols for secure two-party computation have been known for over 25 years [35,13]. (By "generic" we mean that the protocol is constructed by starting with a boolean or arithmetic circuit for the function of interest.) For most of that time, generic secure two-party computation was viewed as being only of theoretical interest; much effort was instead devoted to developing more efficient, "tailored" protocols for specific functions of interest.

In recent years, however, a number of works have shown that generic protocols for secure two-party computation may be much more attractive than previously thought. This line of work was initiated by Fairplay [29], which gave an implementation of Yao's garbled-circuit protocol [35] secure in the semi-honest setting.

---

* This work was supported in part by NSF awards #0447075 and #1111599, and by DARPA.
** Supported by an NSF CI postdoctoral fellowship.

R. Cramer (Ed.): TCC 2012, LNCS 7194, pp. 39–53, 2012.

Subsequent works showed improvements in the scalability, efficiency, and usability of garbled circuits [17,24,19,20], extended the garbled-circuit technique to give implementations in the malicious setting [28,33,34], and explored alternatives to the garbled-circuit approach [23,17,11,31].

As secure computation moves from theory to practice, even small improvements can have a significant effect. (Three factor-of-2 improvements can reduce the time from, say, 1 minute to under 8 seconds.) Indeed, several such improvements have been proposed for the garbled-circuit approach: e.g., the *point-and-permute* technique [29] that reduces the circuit evaluator's work (per gate) from four decryptions to one, or *garbled-row reduction* [30,33] that reduces the number of ciphertexts transmitted per garbled gate from four to three.

It is in this spirit that Kolesnikov and Schneider introduced their very influential "free-XOR" approach [26] for improving the efficiency of garbled-circuit constructions. (The free-XOR optimization is compatible with both the point-and-permute technique and garbled-row reduction.) Yao's original construction requires a garbled gate for each boolean gate in the circuit of the function being computed. The free-XOR technique allows XOR gates in the underlying circuit to be evaluated "for free," without the need to construct a corresponding garbled gate. (We defer the technical details to Section 2.2.) XOR gates in the underlying circuit therefore incur no communication cost or cryptographic operations. Because of this, as documented in [26,25,33], it is worth investing the effort to minimize the number of non-XOR gates in the underlying circuit (even if the total number of gates is increased); this results in roughly a 40% overall efficiency improvement for "typical" circuits [33]. For some circuits (e.g., basic arithmetic operations, universal circuits) a factor-of-4 improvement is observed [26,25]. Nowadays, all implementations of garbled-circuit protocols use the free-XOR idea to improve performance [33,17,34,24,19,20].

1.1 Security of the Free-XOR Technique?

Given the popularity of the free-XOR technique, it is natural to ask what are the necessary assumptions based on which it can be proven secure.[1] The free-XOR approach relies on a cryptographic hash function H. Kolesnikov and Schneider [26] gave a proof of security for the free-XOR technique when H is modeled as a random oracle, and claimed that (a variant of) *correlation robustness* [21,14] would be sufficient; this claim has been repeated in several subsequent works [33,3,7]. (Informally, correlation robustness implies that $H(k_1 \oplus R), \ldots, H(k_t \oplus R)$ are all pseudorandom, even given $k_1, \ldots, k_t$, when R is chosen at random. In the context of the free-XOR technique we must consider hash functions taking two inputs. Formal definitions are given in Section 2.3.) Correlation robustness is a relatively mild assumption, and has the advantage

---

[1] It may be interesting to recall here that XOR gates are also "free" when using the GMW approach to secure two-party computation [13]. In that setting, no additional assumptions are needed.

relative to the random-oracle model of being (potentially) falsifiable. Moreover, correlation robustness is already required by existing protocols for oblivious-transfer extension [21], which are used in current efficient implementations of secure two-party computation.

Our Results. It is unclear exactly what variant of correlation robustness is needed to prove security of the free-XOR approach, and Kolesnikov and Schneider (as well as subsequent researchers relying on their result) have left this question unanswered. We show here that the natural variant of correlation robustness (for hash functions taking two inputs instead of one) is *not* sufficient. We describe where the obvious attempt to prove security fails, and moreover show an explicit counterexample (in the random-oracle model) of a correlation-robust hash function H for which the free-XOR approach is demonstrably insecure.

We observe that the difficulty is due to a previously unnoticed *circularity* in the free-XOR construction: in essence, the issue is that $H(k_1 \oplus R)$ is used to encrypt both k_2 and $k_2 \oplus R$. (The actual issue is more involved, and depends on the details of the free-XOR approach; see Section 3.) We thus define a notion of *circular* correlation robustness, and show that any hash function satisfying this definition can be used to securely instantiate the free-XOR technique. Our definition is falsifiable, and is still weaker than modeling H as a random oracle. Our work can be viewed as following the line of research suggested in [10] whose goal is to formalize, and show usefulness of, various concrete properties satisfied by a random oracle.

Besides the original work of Kolesnikov and Schneider, our results also impact security claims made in two other recent papers. Nielsen and Orlandi [32] use an idea similar to that used in the free-XOR approach to construct a (new) protocol for two-party computation secure against malicious adversaries. They, too, prove security in the random-oracle model but claim that correlation robustness suffices; their construction appears to have the same issues with circularity that the free-XOR technique has. Applebaum et al. [3] define a notion of security against passive related-key attacks for encryption schemes, and claim that encryption schemes satisfying this notion can be used to securely instantiate the free-XOR approach (see [3, Section 1.1.2]). However, their definition of related-key attacks does not take into account any notion of circular security, which appears to be necessary for the free-XOR technique to be sound. We conjecture that our new definition of circular correlation robustness suffices to prove security in each of the above works.

We do not claim that our work has any impact on the security of the free-XOR technique (or the protocols of [32,3]) in practice; in most cases, protocol implementors seem content to assume the random-oracle model anyway. Nevertheless, it is important to understand the precise assumptions needed to prove these protocols secure. We also do not claim any explicit error in the work of Kolesnikov and Schneider [26], as they only say that *some* variant of correlation robustness should suffice. Our work pins down exactly what variant of correlation robustness is necessary.

1.2   Related Work

The notion of correlation robustness was introduced by Ishai et al. [21], and has been used in several other works since then [16,22,32,33]. Applebaum et al. [3] and Goyal et al. [14] further study the notion, explore various definitions, and show connections to security against related-key attacks [5,12,4]. To the best of our knowledge, none of the previous definitions of correlation robustness given in the literature suffice to prove security of the free-XOR technique.

As mentioned above, we define a notion of security for hash functions that blends correlation robustness and circular security. The latter notion, as well as the more general notion of key-dependent-message security, has seen a significant amount of attention recently [6,15,18,2,8,1,9].

1.3   Organization

We review Yao's garbled-circuit construction, and the free-XOR modification of it, in Section 2. In that section we also define a notion of correlation robustness that is syntactically suitable for trying to prove security of the free-XOR approach. In Section 3 we explain where a reductionist proof of security for the free-XOR approach fails when trying to base security on correlation robustness alone. We then demonstrate that *no* proof of security is possible by showing an example of a correlation-robust hash function for which the free-XOR approach is demonstrably insecure. This motivates our definition of a stronger notion of security for hash functions in Section 4, one that we show suffices for proving security of the free-XOR technique.

2   Preliminaries

2.1   Yao's Garbled Circuit Construction

Yao's garbled-circuit approach [35], in combination with any oblivious-transfer protocol, yields a constant-round protocol for two-party computation with security against semi-honest parties. We review only those aspects of the construction needed to understand our results; for further details, we refer to [26,27].

Fix a boolean circuit C known to both parties. (For simplicity, we assume the circuit C outputs a single bit; the protocol can be easily extended to handle multi-bit outputs.) One party, the garbled-circuit generator, prepares a garbled version of the circuit as follows. First, two random keys w_i^0, w_i^1 are associated with each wire i in the circuit; key w_i^0 corresponds to the value '0' on wire i, while w_i^1 corresponds to the value '1'. For each wire i, a random bit π_i is also chosen; key w_i^b is assigned the label $\lambda_i^b = b \oplus \pi_i$. For each gate $g : \{0,1\}^2 \to \{0,1\}$ in the circuit, with input wires i, j and output wire k, the circuit generator constructs a "garbled gate" that will enable the other party to recover $w_k^{g(b_i,b_j)}$ (and its label) from $w_i^{b_i}$ and $w_j^{b_j}$ (and their corresponding labels). The garbled gate consists of the four ciphertexts:

$$\mathsf{Enc}^g_{w_i^{\pi_i}, w_j^{\pi_j}}\left(w_k^{g(\pi_i, \pi_j)} \| g(\pi_i, \pi_j) \oplus \pi_k\right) \tag{1}$$

$$\mathsf{Enc}^g_{w_i^{\pi_i}, w_j^{1 \oplus \pi_j}}\left(w_k^{g(\pi_i, 1 \oplus \pi_j)} \| g(\pi_i, 1 \oplus \pi_j) \oplus \pi_k\right) \tag{2}$$

$$\mathsf{Enc}^g_{w_i^{1 \oplus \pi_i}, w_j^{\pi_j}}\left(w_k^{g(1 \oplus \pi_i, \pi_j)} \| g(1 \oplus \pi_i, \pi_j) \oplus \pi_k\right) \tag{3}$$

$$\mathsf{Enc}^g_{w_i^{1 \oplus \pi_i}, w_j^{1 \oplus \pi_j}}\left(w_k^{g(1 \oplus \pi_i, 1 \oplus \pi_j)} \| g(1 \oplus \pi_i, 1 \oplus \pi_j) \oplus \pi_k\right), \tag{4}$$

in that order. (In the above, we use $\mathsf{Enc}^g_{w,w'}(\cdot)$ to denote encryption under the two keys w and w' that may also depend on the gate number g. The exact details of the encryption will be specified in the next section, but for concreteness the reader can for now think of it as being instantiated by $\mathsf{Enc}^g_{w,w'}(m) = \mathsf{Enc}_w(\mathsf{Enc}_{w'}(m))$ with the gate number being ignored. We use here the point-and-permute technique so that the circuit evaluator only needs to decrypt a single ciphertext per garbled gate.) To evaluate this garbled gate, the circuit evaluator who holds $w_i^{b_i} \| \lambda_i^{b_i}$ and $w_j^{b_j} \| \lambda_j^{b_j}$ uses those keys to decrypt the ciphertext at position $\lambda_i^{b_i}, \lambda_j^{b_j}$ of the above array; this will recover $w_k^{g(b_i, b_j)} \| \lambda_k^{g(b_i, b_j)}$, where $\lambda_k^{g(b_i, b_j)} = g(b_i, b_j) \oplus \pi_k$ as required.

Let $i_1, \ldots, i_\ell$ denote the input wires of the circuit. With garbled gates constructed as above for each gate of the circuit (and transmitted to the circuit evaluator), we see that given keys $w_{i_1}^{b_1}, \ldots, w_{i_\ell}^{b_\ell}$ (and their corresponding labels) for the input wires, the circuit evaluator can inductively compute a key (and its label) for the output wire. The keys for input wires belonging to the circuit generator can simply be transmitted to the circuit evaluator along with the garbled gates; the keys for input wires belonging to the circuit evaluator are obtained by the circuit evaluator using oblivious transfer (OT). If the circuit generator also sends π_o for the output wire o, the circuit evaluator can obtain the correct boolean output of the circuit on the given inputs.

The above thus defines a protocol for two-party computation in the OT-hybrid model. If encryption is instantiated via $\mathsf{Enc}^g_{w,w'}(m) = \mathsf{Enc}_w(\mathsf{Enc}_{w'}(m))$ and Enc is a CPA-secure symmetric-key encryption scheme, the protocol is secure against semi-honest adversaries [35,27].

2.2 The Free-XOR Technique

Kolesnikov and Schneider [26] suggested that instead of choosing the keys w_i^0, w_i^1 for each wire i independently at random, one could instead (1) choose a global random value R, (2) choose w_i^0 uniformly and independently at random for every wire i that is not the output of an XOR gate, and (3) set $w_i^1 = w_i^0 \oplus R$. Each such wire is also assigned a random bit π_i as before. If k is the output wire of an XOR gate with input wires i, j (whose keys have already been defined), then the keys for wire k are set to be $w_k^0 = w_i^0 \oplus w_j^0$ and $w_k^1 = w_k^0 \oplus R$; also, π_k

is set to be $\pi_k = \pi_i \oplus \pi_j$. If keys are chosen this way, then for any XOR gate as above the circuit evaluator holding $w_i^{b_i} \| \lambda_i^{b_i}$ and $w_j^{b_j} \| \lambda_j^{b_j}$ can simply compute $w_k^{b_i \oplus b_j} = w_i^{b_i} \oplus w_j^{b_j}$ and $\lambda_k^{b_i \oplus b_j} = \lambda_i^{b_i} \oplus \lambda_j^{b_j}$; this is correct since $w_i^{b_i} = w_i^0 \oplus b_i R$ (and similarly for $w_j^{b_j}$), where the notation $b_i R$ evaluates to $0^{|R|}$ if $b_i = 0$ or to R otherwise, and thus

$$w_i^{b_i} \oplus w_j^{b_j} = w_i^0 \oplus w_j^0 \oplus (b_i \oplus b_j) R = w_k^0 \oplus (b_i \oplus b_j) R = w_k^{b_i \oplus b_j}$$

and

$$\lambda_i^{b_i} \oplus \lambda_j^{b_j} = (b_i \oplus \pi_i) \oplus (b_j \oplus \pi_j) = (b_i \oplus b_j) \oplus \pi_k = \lambda_k^{b_i \oplus b_j}.$$

Note that by doing so, XOR gates incur no communication and require no cryptographic operations by either party. For the remaining (non-XOR) gates in the circuit, the circuit generator prepares garbled gates as in the previous section.

As previously, the above defines a protocol for secure two-party computation in the OT-hybrid model. Kolesnikov and Schneider suggest to implement encryption using a cryptographic hash function H as follows:

$$\mathsf{Enc}_{w,w'}^g(m) = H(w \| w' \| g) \oplus m.$$

When encryption is instantiated in this way, Kolesnikov and Schneider prove security of their protocol, for semi-honest adversaries, when H is modeled as a random oracle. They also claimed that security would hold if H satisfies some "variant" of correlation robustness. While they did not specify precisely what variant of correlation robustness is needed, a natural approach would be that they require the (joint) pseudorandomness of $H(w \| w' \| g)$, $H(w \oplus R \| w' \| g)$, $H(w \| w' \oplus R \| g)$, $H(w \oplus R \| w' \oplus R \| g)$ for w, w', R chosen at random. We discuss this issue further in the following section.

2.3   Correlation-Robust Hash Functions

As noted at the end of the previous section, Kolesnikov and Schneider claim that some variant of correlation robustness would be sufficient to prove security of the free-XOR construction. Let $H = \{H_n : \{0,1\}^{\ell_{in}(n)} \to \{0,1\}^{\ell_{out}(n)}\}$ be a family of hash functions, where for simplicity we write H instead of H_n when the security parameter n is understood. Correlation robustness was defined by Ishai et al. [21] as follows:

Definition 1. *H is* correlation robust *if for any polynomial $p(\cdot)$ and any non-uniform polynomial-time distinguisher $\mathcal{A}$, the following is negligible in the security parameter n:*

$$\left| \Pr_{w_1,\dots,w_p,R \leftarrow \{0,1\}^{\ell_{in}(n)}} \left[\mathcal{A}(w_1,\dots,w_p, H(w_1 \oplus R),\dots,H(w_p \oplus R)) = 1 \right] \right.$$

$$\left. - \Pr_{w_1,\dots,w_p \leftarrow \{0,1\}^{\ell_{in}(n)}, u_1,\dots,u_p \leftarrow \{0,1\}^{\ell_{out}(n)}} \left[\mathcal{A}(w_1,\dots,w_p, u_1,\dots,u_p) = 1 \right] \right|,$$

where $p = p(n)$.

In the context of the free-XOR technique as defined by Kolesnikov and Schneider, an appropriate definition of correlation robustness needs to at least capture the security requirement (informally) that given any pair of keys $w_i^{b_i}, w_j^{b_j}$ for some garbled gate constructed as in Equations (1)–(4), with $\mathsf{Enc}_{w,w'}^g(m) = H(w\|w'\|g) \oplus m$, it should be possible to decrypt only one row while the others remain hidden. Since the hash function H now takes three inputs, the definition of correlation robustness needs to be modified appropriately. Moreover, for the free-XOR approach it appears necessary to allow w_i to take on arbitrary values[2] rather than being chosen uniformly and independently at random; roughly, this is because in the free-XOR construction we have $w_k^0 = w_i^0 \oplus w_j^0$ when k is the output wire of an XOR gate with input wires i, j, and so w_i^0, w_j^0, w_k^0 are not independent. We capture these requirements in the following definition.

Definition 2. $H : \{0,1\}^{3\ell_{in}(n)} \to \{0,1\}^{\ell_{out}(n)}$ is (weakly) 2-correlation robust if for all polynomials $p(\cdot)$ the distribution ensemble

$$\left\{ \begin{aligned} &R \leftarrow \{0,1\}^{\ell_{in}(n)} : \\ &H(w_1 \oplus R \| w_1' \| 1), H(w_1 \| w_1' \oplus R \| 1), H(w_1 \oplus R \| w_1' \oplus R \| 1), \\ &\qquad\qquad\qquad \vdots \\ &H(w_p \oplus R \| w_p' \| p), H(w_p \| w_p' \oplus R \| p), H(w_p \oplus R \| w_p' \oplus R \| p) \end{aligned} \right\}_{\substack{w_1,\ldots,w_p \in \{0,1\}^{\ell_{in}(n)} \\ w_1',\ldots,w_p' \in \{0,1\}^{\ell_{in}(n)}}}$$

is computationally indistinguishable from the uniform distribution over $\{0,1\}^{3p \cdot \ell_{out}(n)}$. (In both cases, $p = p(n)$.)

Simplified to the case $p = 1$ with w_1, w_1' chosen uniformly and independently (and ignoring the last input to H), the definition requires that the values

$$H(w_1 \oplus R \| w_1'), H(w_1 \| w_1' \oplus R), H(w_1 \oplus R \| w_1' \oplus R)$$

be jointly pseudorandom even given w_1, w_1'. Note that this is equivalent to, say, requiring that

$$H(w_1 \| w_1'), H(w_1 \oplus R \| w_1' \oplus R), H(w_1 \| w_1' \oplus R)$$

be jointly pseudorandom given $w_1 \oplus R, w_1'$, and thus may appear to capture the requirements necessary for proving the free-XOR technique secure.

It will be more convenient to rephrase the above as an oracle-based definition, and this also provides a point of departure for the definition we will propose in Section 4. (In fact, the oracle-based definition we give is stronger than Definition 2 as it allows the adversary to adaptively choose w_i, w_i' based on previous outputs of H. But see footnote 2.) Fixing some H, define oracles $\mathsf{Cor}_R(\cdot, \cdot, \cdot)$ and $\mathsf{Rand}(\cdot, \cdot, \cdot)$ as follows:

---

[2] We will show impossibility of proving security based on correlation robustness alone. Thus, a stronger definition of correlation robustness only strengthens that result.

- $\mathsf{Cor}_R(w, w', g)$: output $H(w\|w'\oplus R\|g)$, $H(w\oplus R\|w'\|g)$, and $H(w\oplus R\|w'\oplus R\|g)$.
- $\mathsf{Rand}(w, w', g)$: if this input was queried before then return the answer given previously. Otherwise choose $u \leftarrow \{0,1\}^{3\cdot\ell_{out}(n)}$ and return u.

We now have the following definition:

Definition 3. H *is 2-correlation robust if for all non-uniform polynomial-time distinguishers $\mathcal{A}$ the following is negligible:*

$$\left| \Pr[R\leftarrow\{0,1\}^{\ell_{in}(n)} : \mathcal{A}^{\mathsf{Cor}_R(\cdot)}(1^n) = 1] - \Pr[\mathcal{A}^{\mathsf{Rand}(\cdot)}(1^n) = 1] \right|.$$

3   Insufficiency of Correlation Robustness

In this section, we show that 2-correlation robustness is not enough to prove security of the free-XOR technique. We start by describing where the natural attempt to prove security (following, e.g., the proof of [27]) fails. We then show a construction (in the random-oracle model) of a hash function H that satisfies Definition 3 but for which the free-XOR approach is demonstrably insecure when instantiated using H.

3.1   Where the Reduction Fails

Consider the case where the circuit consists of just a single AND gate, with input wires 1 and 2 (belonging to the circuit generator and evaluator, respectively) and output wire 3. Say the circuit evaluator has input 0 and so receives key w_2^0; assume for concreteness that the circuit generator has input 0 as well and so the circuit evaluator is also given key w_1^0. (The circuit evaluator will also be given the corresponding labels, but these can be left implicit in what follows.) The garbled gate consists of the values

$$H(w_1^0\|w_2^0\|1) \oplus (w_3^0\|0)$$
$$H(w_1^0\|w_2^0\oplus R\|1) \oplus (w_3^0\|0)$$
$$H(w_1^0\oplus R\|w_2^0\|1) \oplus (w_3^0\|0)$$
$$H(w_1^0\oplus R\|w_2^0\oplus R\|1) \oplus ((w_3^0\oplus R)\|1)$$

in some permuted order, for some random value R unknown to the circuit evaluator. (Recall that $w_i^1 = w_i^0\oplus R$ for all i, by construction, when using the free-XOR approach.) The evaluator will be able to decrypt the first row, above, to learn the output; it should not, however, be able to learn any information about the remaining three rows. (In particular, it should not learn whether the other party had input 0 or 1.) The natural way to try to prove security of the above is to argue that the remaining rows are pseudorandom, by reduction to the 2-correlation robustness of H. In the reduction, we would have an adversary $\mathcal{A}$ given access to an oracle $\mathcal{O}$ that is either Cor_R (for a random R) or Rand. The adversary $\mathcal{A}$ can choose random w_1^0, w_2^0, and then query $\mathcal{O}(w_1^0, w_2^0, 1)$ to obtain three values h_1, h_2, h_3 that are either completely random or equal to $H(w_1^0\|w_2^0\oplus R\|1)$,

$H(w_1^0 \oplus R \| w_2^0 \| 1)$, and $H(w_1^0 \oplus R \| w_2^0 \oplus R \| 1)$. But $\mathcal{A}$ cannot complete the simulation, since it has no way to compute values of the form

$$h_1 \oplus (w_3^0 \| 0), \quad h_2 \oplus (w_3^0 \| 0), \quad h_3 \oplus \left((w_3^0 \oplus R) \| 1\right)$$

(since $\mathcal{A}$ does not know R) as would be necessary to simulate the remaining three rows of the garbled gate in case $\mathcal{O} = \mathsf{Cor}_R$.

We show in the next section that this is not just a failure of this particular proof approach, since we can construct a hash function H that satisfies Definition 3 yet for which the free-XOR methodology is demonstrably insecure when instantiated using H.

3.2 A Counter-Example

For simplicity, we fix a value of the security parameter n. Assume further that the last input to H (i.e., the gate index g) is represented using n bits. We construct a pair of oracles $H : \{0,1\}^{3n} \to \{0,1\}^{n+1}$ and $\mathsf{Break} : \{0,1\}^{6n+3} \to \{0,1\}^n$ such that:

- H satisfies Definition 3, even if the distinguisher $\mathcal{A}$ is given oracle access to both H and Break.
- The free-XOR methodology is demonstrably insecure when instantiated using H, against an adversary given oracle access to both H and Break.

Thus, we rule out a fully black-box reduction of the security of the free-XOR technique to 2-correlation robustness.

Let $H : \{0,1\}^{3n} \to \{0,1\}^{n+1}$ be a random function, and define Break as follows:

$\underline{\mathsf{Break}(w \| w' \| g \| z_1 \| z_2 \| z_3)}$: If there exists $r \in \{0,1\}^n$ such that

$$z_1 = H(w \| w' \oplus r \| g), \quad z_2 = H(w \oplus r \| w' \| g), \text{ and } z_3 = H(w \oplus r \| w' \oplus r \| g) \oplus (r \| 0),$$

then output r (if multiple values of r satisfy the above, take the lexicographically smallest one); otherwise, output $\perp$.

We now prove the above claims.

Lemma 1. *H is 2-correlation robust, even when the distinguisher is given oracle access to both H and* Break.

Proof (Sketch). Fix a polynomial-time distinguisher $\mathcal{A}$ who is given access to $H, \mathsf{Break}, \mathcal{O}$ where either $\mathcal{O} = \mathsf{Cor}_R$ (for random $R \in \{0,1\}^n$) or $\mathcal{O} = \mathsf{Rand}$. Without loss of generality, we assume that $\mathcal{A}$ does not repeat queries to $\mathcal{O}$. When $\mathcal{O} = \mathsf{Rand}$, every query to $\mathcal{O}$ is answered with a string of length $3 \cdot (n+1)$ that is uniform and independent of $\mathcal{A}$'s view. When $\mathcal{O} = \mathsf{Cor}_R$, every query $\mathcal{O}(w, w', g)$ is answered with a string of length $3 \cdot (n+1)$ that is uniform and independent of $\mathcal{A}$'s view unless one of the following is true:

- $\mathcal{A}$ at some point queries $\mathcal{O}(\tilde{w}, \tilde{w}', g)$ with $\tilde{w} \oplus w = R$ or $\tilde{w}' \oplus w' = R$ (or both).
- $\mathcal{A}$ at some point queries $H(\tilde{w} \| \tilde{w}' \| g)$ with $\tilde{w} \oplus w = R$ or $\tilde{w}' \oplus w' = R$ (or both).
- $\mathcal{A}$ at some point queries $\mathsf{Break}(w \| w' \| g \| z_1 \| z_2 \| z_3)$ where it holds that $R \| 0 = z_3 \oplus H(w \oplus R \| w' \oplus R \| g)$.

Since R is chosen uniformly from $\{0, 1\}^n$, the probability that $\mathcal{A}$ makes any queries of the above form is negligible.

Lemma 2. *The free-XOR construction, when instantiated using H, is not secure against a semi-honest adversary (with oracle access to H and Break) who corrupts the circuit evaluator.*

Proof. We show that a semi-honest adversary can recover R with high probability. Since a semi-honest adversary can (legitimately) recover one key per wire by following the protocol, knowledge of R allows the adversary to recover *both* keys for every wire in the circuit; thus, this suffices to show that the construction is insecure.

Assume the first gate in the circuit is an AND gate with input wires 1 and 2 (belonging to the circuit generator and evaluator, respectively) and output wire 3. Say the circuit evaluator has input 0 and so receives key w_2^0; assume for concreteness that the circuit generator has input 0 as well and so the circuit evaluator is also given key w_1^0. With constant probability we have $\pi_1 = \pi_2 = \pi_3 = 0$, and in that case the garbled gate consists of the values

$$c_{00} = H(w_1^0 \| w_2^0 \| 1) \oplus (w_3^0 \| 0)$$
$$c_{01} = H(w_1^0 \| w_2^0 \oplus R \| 1) \oplus (w_3^0 \| 0)$$
$$c_{10} = H(w_1^0 \oplus R \| w_2^0 \| 1) \oplus (w_3^0 \| 0)$$
$$c_{11} = H(w_1^0 \oplus R \| w_2^0 \oplus R \| 1) \oplus \left((w_3^0 \oplus R) \| 1\right).$$

The circuit evaluator can compute w_3^0 from c_{00} (as directed by the protocol). It then computes

$$z_1 = c_{01} \oplus (w_3^0 \| 0)$$
$$z_2 = c_{10} \oplus (w_3^0 \| 0)$$
$$z_3 = c_{11} \oplus (w_3^0 \| 1)$$

and queries $\mathsf{Break}(w_1^0 \| w_2^0 \| 1 \| z_1 \| z_2 \| z_3)$. If the answer is some value $R' \neq \bot$ then with overwhelming probability it holds that $R' = R$. (Correctness of R can also be verified by looking at a second garbled gate with known inputs.)

4 Proving Security of the Free-XOR Approach

The essence of the problem(s) described in the previous section is that there is a previously unnoticed *circularity* in the free-XOR approach, since in the general case both $H(w_1 \| w_2 \| g) \oplus w_3$ and $H(w_1 \oplus R \| w_2 \oplus R \| g) \oplus (w_3 \oplus R)$ are revealed to the adversary. (Recall that R is the hidden secret here.) In this section, we introduce a new security definition that explicitly takes this circularity into account, and show that this definition suffices to prove security of the free-XOR approach.

Fix some function $H : \{0,1\}^{3\ell_{in}(n)} \rightarrow \{0,1\}^{\ell_{out}(n)}$. We define an oracle Circ_R as follows:[3]

- $\mathsf{Circ}_R(w, w', g, b_1, b_2, b_3)$ outputs $H(w \oplus b_1 R \| w' \oplus b_2 R \| g) \oplus b_3 R$.

To see the connection with the previous definition (in the context of correlation robustness), note that oracle $\mathsf{Cor}_R(w, w', g)$ defined previously outputs $\mathsf{Circ}_R(w, w', g, 0, 1, 0)$, $\mathsf{Circ}_R(w, w', g, 1, 0, 0)$, and $\mathsf{Circ}_R(w, w', g, 1, 1, 0)$; i.e., b_3 was fixed to 0 there. The possibility of $b_3 = 1$ is exactly what models circularity involving R.

Corresponding to the above we define an oracle Rand in a way analogous to before:

- $\mathsf{Rand}(w, w', g, b_1, b_2, b_3)$: if this input was queried before then return the answer given previously. Otherwise choose $u \leftarrow \{0,1\}^{\ell_{out}(n)}$ and return u.

In our new definition of security for H, we are going to require that oracles Circ_R (for random R) and Rand be indistinguishable. This cannot possibly be true, however, unless we rule out some trivial queries that can be used to distinguish them. Let $\mathcal{O}$ be the oracle to which a distinguisher is given access, where either $\mathcal{O} = \mathsf{Circ}_R$ or $\mathcal{O} = \mathsf{Rand}$. We must restrict the distinguisher as follows: (1) it is not allowed to make any query of the form $\mathcal{O}(w, w', g, 0, 0, b_3)$ (since it can compute $H(w\|w'\|g)$ on its own) and (2) it is not allowed to query both $\mathcal{O}(w, w', g, b_1, b_2, 0)$ and $\mathcal{O}(w, w', g, b_1, b_2, 1)$ for any values w, w', g, b_1, b_2 (since that would allow it to trivially recover R). We say that any distinguisher respecting these restrictions makes *legal queries*.

With this in place we can now define our notion of circular 2-correlation robustness.

Definition 4. H *is* circular 2-correlation robust *if for any non-uniform polynomial-time distinguisher $\mathcal{A}$ making legal queries to its oracle, the following is negligible:*

$$\left| \Pr[R \leftarrow \{0,1\}^{\ell_{in}(n)} : \mathcal{A}^{\mathsf{Circ}_R(\cdot)}(1^n) = 1] - \Pr[\mathcal{A}^{\mathsf{Rand}(\cdot)}(1^n) = 1] \right|.$$

Next, we show that this notion of security suffices to prove security of the free-XOR approach:

Theorem 1. *Consider the protocol described in Section 2.2 for two-party computation in the OT-hybrid model. If H as used there is circular 2-correlation robust, then the resulting protocol is secure against a semi-honest adversary.*

---

[3] Here, we slightly abuse the notation bR so that it evaluates to $0^{\ell_{out}(n)}$ if $b = 0$ or $R\|0^{\ell_{out}(n)-\ell_{in}(n)}$ otherwise.

Proof. The case where the circuit generator is corrupted is trivial. Therefore, we consider corruption of the circuit evaluator B. We describe a simulator who is given the input of B and the output $z \in \{0, 1\}$ of evaluating the function, and must provide B with a simulated garbled circuit that is indistinguishable from the actual one that would be sent during a real execution of the protocol. The high-level idea is exactly the same as in [27]; the crucial difference is that we reduce to circular 2-correlation robustness of H.

The simulator proceeds as follows:

1. For each wire i in the circuit that is not an output wire of an XOR gate, choose $w_i \leftarrow \{0, 1\}^n$ and $\lambda_i \leftarrow \{0, 1\}$.
2. For each wire k in the circuit that is the output wire of an XOR gate with input wires i, j (for which $w_i, \lambda_i, w_j, \lambda_j$ have already been defined), set $w_k = w_i \oplus w_j$ and $\lambda_k = \lambda_i \oplus \lambda_j$.
3. For each non-XOR gate g in the circuit with input wires i, j and output wire k, output the four ciphertexts c_{00}, c_{01}, c_{10}, and c_{11} as the corresponding garbled gate, where $c_{\lambda_i \lambda_j} = H(w_i \| w_j \| g) \oplus (w_k \| \lambda_k)$ and the remaining three ciphertexts are uniform strings of length $n + 1$.
4. For the output wire o, set $\pi_o = \lambda_o \oplus z$.

Say $i_1, \ldots, i_\ell$ are the input wires of the circuit belonging to the circuit generator, and $j_1, \ldots, j_{\ell'}$ are the input wires belonging to the circuit evaluator. The simulator gives to B the values $w_{j_1}, \ldots, w_{j_{\ell'}}$ (as if they came from the calls to the OT functionality), and the simulated communication that includes (1) the keys $w_{i_1}, \ldots, w_{i_\ell}$, (2) the garbled gate for each non-XOR gate in the circuit, and (3) the value π_o corresponding to the output wire.

We claim that the simulated view is indistinguishable from the real-world execution of the protocol. Assume there is an adversary B who can distinguish the two distributions when the inputs to the parties are x and y, respectively, and the output is z. We show an adversary $\mathcal{A}$ who breaks the circular 2-correlation robustness of H. Given access to an oracle $\mathcal{O}$ (that is either Circ or Rand), adversary $\mathcal{A}$ does as follows:

1. For each wire i in the circuit that is not an output wire of an XOR gate, choose $w_i \leftarrow \{0, 1\}^n$ and $\lambda_i \leftarrow \{0, 1\}$.
2. For each wire k in the circuit that is the output wire of an XOR gate with input wires i, j (for which $w_i, \lambda_i, w_j, \lambda_j$ have already been defined), set $w_k = w_i \oplus w_j$ and $\lambda_k = \lambda_i \oplus \lambda_j$.
3. For each wire i, let $b_i \in \{0, 1\}$ be the actual value on wire i; this can be determined since $\mathcal{A}$ knows the actual input (x, y) to the circuit. Set $w_i^{b_i} = w_i$, $\pi_i = \lambda_i \oplus b_i$, $\lambda_i^0 = \pi_i$, and $\lambda_i^1 = 1 \oplus \pi_i$ (i.e., only $w_i^{1-b_i}$s are left undefined).
4. For each non-XOR gate g in the circuit with input wires i, j and output wire k, output the four ciphertexts c_{00}, c_{01}, c_{10}, and c_{11} as the corresponding garbled gate, where these ciphertexts are constructed as follows:
   - $c_{\lambda_i^{b_i} \lambda_j^{b_j}} = H(w_i^{b_i} \| w_j^{b_j} \| g) \oplus (w_k^{b_k} \| \lambda_k^{b_k})$.

- For $(\lambda_i^{\beta_i}, \lambda_j^{\beta_j}) \in \{0,1\}^2$ with $(\beta_i, \beta_j) \neq (b_i, b_j)$, query

$$h_{\beta_i, \beta_j} = \mathcal{O}(w_i, w_j, g, \beta_i \oplus b_i, \beta_j \oplus b_j, g(\beta_i, \beta_j) \oplus b_k),$$

and set $c_{\lambda_i^{\beta_i} \lambda_j^{\beta_j}} = h_{\beta_i, \beta_j} \oplus (w_k^{b_k} \| \lambda_k^{g(\beta_i, \beta_j)})$.

5. For the output wire o, set $\pi_o = \lambda_o \oplus z$ (where z is the known output of the circuit).

$\mathcal{A}$ gives to B the values $w_{j_1}, \dots, w_{j_{\ell'}}$ (as if they came from the calls to the OT functionality), and (1) the keys $w_{i_1}, \dots, w_{i_\ell}$, (2) the garbled gate for each non-XOR gate in the circuit, and (3) the value π_o corresponding to the output wire. Finally, $\mathcal{A}$ outputs whatever B outputs. It is easy to see that $\mathcal{A}$ makes legal queries to its oracle. Furthermore, it is also easy to see that if $\mathcal{O} = \mathsf{Circ}$ the view of B is identically distributed to its view in the real execution of the protocol on the given inputs, whereas if $\mathcal{O} = \mathsf{Rand}$ the view of B is distributed identically to the output of the simulator described previously. This completes the proof.

5 Conclusion

The free-XOR technique has been extremely influential, and it is currently used in all implementations of the garbled-circuit technique because of the speedup that it gives. It was previously known that this approach is secure in the random-oracle model; it was also claimed that *some* variant of correlation robustness would suffice to prove security, but the exact notion of correlation robustness needed was left unspecified. In this work, we explore this question. We show that the natural variant of correlation robustness (extended to handle hash functions taking several inputs, rather than one input) is not sufficient, and identify a previously unnoticed circularity in the free-XOR construction that causes the difficulty. We are thus motivated to propose a new, stronger notion of correlation robustness, and we prove that this notion suffices.

Several intriguing open questions remain. First, is there some variant of the free-XOR approach that does not rely on any assumptions beyond CPA-secure encryption (which is all that is needed to prove security of classical garbled-circuit protocols in the OT-hybrid world)? Alternately, can our definition of circular 2-correlation robustness be realized from standard cryptographic assumptions?

Acknowledgments. The second author would like to thank Vlad Kolesnikov and Thomas Schneider for their feedback and encouragement.

References

1. Acar, T., Belenkiy, M., Bellare, M., Cash, D.: Cryptographic Agility and Its Relation to Circular Encryption. In: Gilbert, H. (ed.) EUROCRYPT 2010. LNCS, vol. 6110, pp. 403–422. Springer, Heidelberg (2010)

2. Applebaum, B., Cash, D., Peikert, C., Sahai, A.: Fast Cryptographic Primitives and Circular-Secure Encryption Based on Hard Learning Problems. In: Halevi, S. (ed.) CRYPTO 2009. LNCS, vol. 5677, pp. 595–618. Springer, Heidelberg (2009)
3. Applebaum, B., Harnik, D., Ishai, Y.: Semantic security under related-key attacks and applications. In: 2nd Symposium on Innovations in Computer Science (ICS), pp. 45–60. Tsinghua University Press (2011)
4. Bellare, M., Cash, D.: Pseudorandom Functions and Permutations Provably Secure against Related-Key Attacks. In: Rabin, T. (ed.) CRYPTO 2010. LNCS, vol. 6223, pp. 666–684. Springer, Heidelberg (2010)
5. Bellare, M., Kohno, T.: A Theoretical Treatment of Related-Key Attacks: RKA-PRPs, RKA-PRFs, and Applications. In: Biham, E. (ed.) EUROCRYPT 2003. LNCS, vol. 2656, pp. 491–506. Springer, Heidelberg (2003)
6. Black, J., Rogaway, P., Shrimpton, T.: Encryption-Scheme Security in the Presence of Key-Dependent Messages. In: Nyberg, K., Heys, H.M. (eds.) SAC 2002. LNCS, vol. 2595, pp. 62–75. Springer, Heidelberg (2003)
7. Blanton, M., Gasti, P.: Secure and Efficient Protocols for Iris and Fingerprint Identification. In: Atluri, V., Diaz, C. (eds.) ESORICS 2011. LNCS, vol. 6879, pp. 190–209. Springer, Heidelberg (2011)
8. Boneh, D., Halevi, S., Hamburg, M., Ostrovsky, R.: Circular-Secure Encryption from Decision Diffie-Hellman. In: Wagner, D. (ed.) CRYPTO 2008. LNCS, vol. 5157, pp. 108–125. Springer, Heidelberg (2008)
9. Brakerski, Z., Goldwasser, S.: Circular and Leakage Resilient Public-Key Encryption under Subgroup Indistinguishability (or: Quadratic residuosity strikes back). In: Rabin, T. (ed.) CRYPTO 2010. LNCS, vol. 6223, pp. 1–20. Springer, Heidelberg (2010)
10. Canetti, R., Goldreich, O., Halevi, S.: The random oracle methodology, revisited. J. ACM 51(4), 557–594 (2004)
11. Choi, S., Hwang, K.-W., Katz, J., Malkin, T., Rubenstein, D.: Secure multi-party computation of boolean circuits with applications to privacy in on-line marketplaces. In: Topics in Cryptology — Cryptographers' Track (CT-RSA 2012) (to appear, 2012)
12. Goldenberg, D., Liskov, M.: On Related-Secret Pseudorandomness. In: Micciancio, D. (ed.) TCC 2010. LNCS, vol. 5978, pp. 255–272. Springer, Heidelberg (2010)
13. Goldreich, O., Micali, S., Wigderson, A.: How to play any mental game, or a completeness theorem for protocols with honest majority. In: 19th Annual ACM Symposium on Theory of Computing (STOC), pp. 218–229. ACM Press (1987)
14. Goyal, V., O'Neill, A., Rao, V.: Correlated-Input Secure Hash Functions. In: Ishai, Y. (ed.) TCC 2011. LNCS, vol. 6597, pp. 182–200. Springer, Heidelberg (2011)
15. Halevi, S., Krawczyk, H.: Security under key-dependent inputs. In: 14th ACM Conference on Computer and Communications Security (CCS), pp. 466–475. ACM Press (2007)
16. Harnik, D., Ishai, Y., Kushilevitz, E., Nielsen, J.B.: OT-Combiners via Secure Computation. In: Canetti, R. (ed.) TCC 2008. LNCS, vol. 4948, pp. 393–411. Springer, Heidelberg (2008)
17. Henecka, W., Kögl, S., Sadeghi, A.-R., Schneider, T., Wehrenberg, I.: TASTY: Tool for automating secure two-party computations. In: 17th ACM Conf. on Computer and Communications Security (CCS), pp. 451–462. ACM Press (2010)
18. Hofheinz, D., Unruh, D.: Towards Key-Dependent Message Security in the Standard Model. In: Smart, N.P. (ed.) EUROCRYPT 2008. LNCS, vol. 4965, pp. 108–126. Springer, Heidelberg (2008)

19. Huang, Y., Evans, D., Katz, J., Malka, L.: Faster secure two-party computation using garbled circuits. In: Proc. 20th USENIX Security Symposium, pp. 539–553. USENIX Association (2011)
20. Huang, Y., Evans, D., Katz, J.: Private set intersection: Are garbled circuits better than custom protocols? In: Network and Distributed System Security Symposium (NDSS) (to appear, 2012)
21. Ishai, Y., Kilian, J., Nissim, K., Petrank, E.: Extending Oblivious Transfers Efficiently. In: Boneh, D. (ed.) CRYPTO 2003. LNCS, vol. 2729, pp. 145–161. Springer, Heidelberg (2003)
22. Ishai, Y., Prabhakaran, M., Sahai, A.: Founding Cryptography on Oblivious Transfer – Efficiently. In: Wagner, D. (ed.) CRYPTO 2008. LNCS, vol. 5157, pp. 572–591. Springer, Heidelberg (2008)
23. Jakobsen, T.P., Makkes, M.X., Nielsen, J.D.: Efficient Implementation of the Orlandi Protocol. In: Zhou, J., Yung, M. (eds.) ACNS 2010. LNCS, vol. 6123, pp. 255–272. Springer, Heidelberg (2010)
24. Katz, J., Malka, L.: VMCrypt — modular software architecture for scalable secure computation, http://eprint.iacr.org/2010/584
25. Kolesnikov, V., Sadeghi, A.-R., Schneider, T.: Improved Garbled Circuit Building Blocks and Applications to Auctions and Computing Minima. In: Garay, J.A., Miyaji, A., Otsuka, A. (eds.) CANS 2009. LNCS, vol. 5888, pp. 1–20. Springer, Heidelberg (2009)
26. Kolesnikov, V., Schneider, T.: Improved Garbled Circuit: Free XOR Gates and Applications. In: Aceto, L., Damgård, I., Goldberg, L.A., Halldórsson, M.M., Ingólfsdóttir, A., Walukiewicz, I. (eds.) ICALP 2008, Part II. LNCS, vol. 5126, pp. 486–498. Springer, Heidelberg (2008)
27. Lindell, Y., Pinkas, B.: A proof of security of Yao's protocol for two-party computation. Journal of Cryptology 22(2), 161–188 (2009)
28. Lindell, Y., Pinkas, B., Smart, N.P.: Implementing Two-Party Computation Efficiently with Security Against Malicious Adversaries. In: Ostrovsky, R., De Prisco, R., Visconti, I. (eds.) SCN 2008. LNCS, vol. 5229, pp. 2–20. Springer, Heidelberg (2008)
29. Malkhi, D., Nisan, N., Pinkas, B., Sella, Y.: Fairplay — a secure two-party computation system. In: Proc. 13th USENIX Security Symposium, pp. 287–302. USENIX Association (2004)
30. Naor, M., Pinkas, B., Sumner, R.: Privacy preserving auctions and mechanism design. In: Proc. 1st ACM Conf. on Electronic Commerce, pp. 129–139. ACM (1999)
31. Nielsen, J.B., Nordholt, P., Orlandi, C., Burra, S.: A new approach to practical active-secure two-party computation, http://eprint.iacr.org/2011/091
32. Nielsen, J.B., Orlandi, C.: LEGO for Two-Party Secure Computation. In: Reingold, O. (ed.) TCC 2009. LNCS, vol. 5444, pp. 368–386. Springer, Heidelberg (2009)
33. Pinkas, B., Schneider, T., Smart, N.P., Williams, S.C.: Secure Two-Party Computation is Practical. In: Matsui, M. (ed.) ASIACRYPT 2009. LNCS, vol. 5912, pp. 250–267. Springer, Heidelberg (2009)
34. Shelat, A., Shen, C.-H.: Two-Output Secure Computation with Malicious Adversaries. In: Paterson, K.G. (ed.) EUROCRYPT 2011. LNCS, vol. 6632, pp. 386–405. Springer, Heidelberg (2011)
35. Yao, A.C.-C.: How to generate and exchange secrets. In: 27th Annual Symposium on Foundations of Computer Science (FOCS), pp. 162–167. IEEE (1986)

Secure Two-Party Computation with Low Communication*

Ivan Damgård, Sebastian Faust, and Carmit Hazay

Department of Computer Science, Aarhus University
{ivan,sfaust,carmit}@cs.au.dk

Abstract. We propose a 2-party UC-secure protocol that can compute any function securely. The protocol requires only two messages, communication that is poly-logarithmic in the size of the circuit description of the function, and the workload for one of the parties is also only poly-logarithmic in the size of the circuit. This implies, for instance, delegatable computation that requires no expensive off-line phase and remains secure even if the server learns whether the client accepts its results. To achieve this, we define two new notions of extractable hash functions, propose an instantiation based on the knowledge of exponent in an RSA group, and build succinct zero-knowledge arguments in the CRS model.

1 Introduction

In the setting of secure two-party computation, two parties with private inputs wish to jointly compute some function of their inputs while preserving certain security properties like privacy, correctness and more. Despite the stringent requirements of the standard simulation-based security definitions [GL90, Can00], it has been shown that any probabilistic polynomial-time two-party functionality can be computed securely against malicious adversaries [Yao86, GMW87, Gol04]. Following these feasibility results many constructions have been proposed to improve the efficiency of the computation [IPS09, PSSW09, NO09, LP11, IKO+11]. A recent work by Gordon et al. [GKK+11] shows an approach using oblivious RAM, with polylogarithmic amortized workload overhead. The best round complexity is obtained by [IPS08, IKO+11] who show a single round protocol in the non-interactive setting. For a general study of multiparty computation with minimal round complexity, see [KK07, IKP10].

The communication complexity of these constructions depends heavily on the size of the computed circuit. To the best of our knowledge, all works that try to minimize the communication complexity do so for particular tasks of interests such as private information retrieval (PIR) [KO97] or functions captured by

---

* The authors acknowledge support from the Danish National Research Foundation and The National Science Foundation of China (under the grant 61061130540) for the Sino-Danish Center for the Theory of Interactive Computation, within part of this work was performed; and from the CFEM research center, supported by the Danish Strategic Research Council.

R. Cramer (Ed.): TCC 2012, LNCS 7194, pp. 54–74, 2012.

branching programs and random access memory machines [NN01]. In all these constructions, the parties do essentially the same amount of work, namely at least the amount of work needed to evaluate the specified circuit. Such constructions are appropriate for settings in which the parties are equally powerful, and offer no solution for "asymmetric settings" in which one of the devices is strictly (computationally) weaker than the other (e.g., smartcards, mobile devices). In this paper we will be interested in solutions for such asymmetric settings, so we want to minimize the workload for one of the parties.

For semi-honest attacks, fully homomorphic encryption [Gen09, BV11a] can be used to design a simple one round protocol with sublinear communication complexity. Here one party, say P_1, sends its encrypted input to party P_2, who uses the homomorphic property to compute ciphertexts that contain the desired output. These ciphertexts are sent to P_1 who can decrypt and learn the result. Obviously, this solution breaks down under malicious attacks. The obvious solution is to have P_2 give a non-interactive zero-knowledge proof (NIZK) that his response is correct, but this will not solve our problem. Even though such a proof can be made very short [Gro11], P_1 would have to work as hard as P_2 to check the NIZK, and hence the *computational* complexity for both parties would be linear in the circuit description of the function to compute. This does not fit our scenario where we want to minimize the work for one party.

To reach our goal, one needs a protocol by which a prover can give a short zero-knowledge argument for an NP statement, where the verifier only needs to do a small amount of work. More precisely, the amount of work needed for the verifier is polynomial in the security parameter and the size of the statement but only poly-logarithmic in the time needed to check a witness in the standard way. Such proofs or arguments are usually called *succinct*. The history of such protocols starts with the work of Kilian [Kil92] who suggested the idea of having the prover commit to a PCP for the statement in question using a Merkle hash tree, and then have the verifier (obliviously) check selected bits from the PCP. This protocol is succinct and zero-knowledge but requires several rounds and so cannot be used towards our goal of a 2-message protocol. Subsequent work in this direction has concentrated on protocols where only a succinct non-interactive argument (and not zero-knowedge) is required. This is known as a SNARG. Micali [Mic00] suggested one-message solution based on Kilian's protocol and the Fiat-Shamir heuristic. In [ABOR00] Aiello at al. suggested a two-message protocol where the verifier accesses bits of the PCP via a private information retrieval scheme (PIR). In such a scheme a client can retrieve an entry in a database held by a server without the server learning which entry was accessed. It seems intuitively appealing that if the prover does not know which bits of the PCP the verifier is looking at, soundness of the PCP should imply soundness of the overall argument. However, it was shown in [DLN+04] that this intuition is not sound. Di Crescenzo and Lipmaa [CL08] suggested a solution where the prover commits to a PCP using the root of a Merkle tree as in Kilian's protocol, but to prove security, they made a very strong type of extractability assumption implying extraction of an entire PCP from the prover in one go.

Our Contribution. Compared to the work on SNARGs just discussed, our work makes two contributions: first, we show how to achieve simulation based privacy also for the prover, even if the verifier is malicious. We need this since our goal is UC-secure 2-party computation and we must have privacy for both parties, even under malicious attacks. This is the reason we need a set-up assumption allowing parties to give non-interactive zero-knowledge proofs of knowledge of their inputs. Also, to get a zero-knowledge SNARG, we do not use the PCP+PIR approach from earlier work for a general PIR, instead we build a PIR-like scheme based on FHE, allowing the prover to compute NIZKs "inside the ciphertexts". Second, we suggest two notions of "extractable hash function" that are more natural and milder than the assumption of Di Crescenzo and Lipmaa but still allow succinct arguments.

Based on these techniques we present a two-party protocol in the common reference string model that computes any PPT functionality f with UC security against malicious adversaries. Our protocol is the first to additionally achieve the following strong properties: *Polylogarithmic communication complexity* in the size of the circuit C that computes f. *One round complexity*, i.e., a single message in each direction. *Polylogarithmic workload* in the size of the circuit C that computes f, for one of the parties. Our protocol is based on fully homomorphic encryption, non-interactive zero-knowledge proofs and the existence of extractable hash functions. While the first two notions are fairly standard, we explain in more detail the new notions of extractability:

The first extractability assumption (EHF1) considers a collision intractable hash function H mapping into a small subset of a large domain and essentially asserts that the only way to generate an element in $\text{Im}(H)$ is to compute the function on a given input. More precisely, we require that for every adversary outputting a value h there exists an efficient extractor that (given the same randomness) outputs a preimage of h, whenever $h \in \text{Im}(H)$. We propose an instantiation of EHF1-extractable and collision intractable hash functions based on a knowledge of exponent assumption [Dam91] in $\mathbb{Z}_N^*$, for N is an RSA modulus.

The second extractability assumption (EHF2) makes a weaker demand on the hash function H: again we require that for each adversary outputting h, there exists an extractor that tries to find a preimage. This time, however, the extractor is allowed to fail even if $h \in \text{Im}(H)$. The demand, however, is that if the extractor fails, the adversary cannot find a preimage either, even if he continues his computation with fresh randomness and auxiliary data that was not known to the extractor.

It is easy to verify that EHF1 implies EHF2: under EHF1, the extractor only fails if it is *impossible* to find a preimage. The more interesting direction is whether EHF2 implies EHF1. In the concurrent and independent work of Bitansky et al. [BCCT11], they consider a variant of EHF1 where the hash function has a stronger notion of collision intractability, so-called proximity collision resistance. They then show that proximity EHF1 is equivalent to proximity EHF2 and furthermore existence of such functions is equivalent to the existence of non-interactive arguments of knowledge (SNARKs). Whether our EHF2 notion implies EHF1 is an interesting open question.

Note that EHF2 is true in the random oracle model, where we let the random oracle play the role of H. In this case it is easy to see that no matter how the adversary produces a string h, there are only two cases: either h was output by the random oracle or not. In the former case a preimage is easy to extract, in the latter case *no one* can produce a preimage except with negligible probability. So the extractor can safely fail in this case.

Finally, it is interesting to note that EHF2 opens the possibility to use many more candidate hash functions, whereas previously only rather slow functions based on number theoretic assumptions seemed to apply. This is because standard hash functions such as SHA (are thought to) behave similarly to a random oracle, and such a function does not satisfy EHF1. However, using, e.g., the random oracle preserving EMD transform from [BR06], one may get interesting candidates for efficient functions satisfying EHF2.

We wish to warn the reader that extractability assumptions are regarded as controversial by some; on the other hand such assumptions have recently been studied quite intensively [BP04, CL08, Gro10, BCCT11, GLR11]. Moreover, Gentry and Wichs [GW11] have recently shown that SNARGs cannot be shown secure via a black-box reduction to a falsifiable assumption [Nao03]. Even more to the point, as mentioned above, [BCCT11], shown that existence of SNARKs imply existence of extractable hash functions. This suggests that non-standard assumptions such as knowledge of exponent are necessary in this setting and hence our construction is essentially tight. Finally, as we pointed out above, the EHF2 assumption is true in the random oracle model and is implied only by the fact that one must call the oracle to get a valid output. So we only use one of the many "magic properties" that the random oracle model has, and this particular one is in fact satisfied in the standard model, if our assumption holds. Therefore, we believe that the assumption on extractable hash functions should be regarded as much less controversial than using the random oracle model.

Applications. Variants of our construction is useful for various settings. We briefly describe some of these applications here, for further details and additional applications, see the full version of this paper [DFH11].

NON-INTERACTIVE SECURE COMPUTATION. In the *non-interactive* setting a receiver wishes to publish an encryption of its secret input x so that any other sender, holding a secret input y, will be able to obliviously evaluate $f(x, y)$ and reveal it to the receiver. This problem is useful for many web applications in which a server publishes its information and many clients respond back. A recent work by Ishai et al. [IKO+11] presents the first general protocol in this model with only black-box calls to a pseudorandom generator (PRG). In contrast, our protocol makes non black-box use of the fully homomorphic encryption but only requires polylogarithmic communication complexity.

DELEGATABLE COMPUTATION. In this setting, a computationally weak client wishes to outsource its computation to a more powerful server, with the aim that the server performs this computation privately and correctly. An important requirement in this scenario is that the amount of work put by the client in order to verify the correctness of the computation is substantially smaller than running this computation by itself. It is also important that the overall amount of work invested by the server grows linearly with the original computation. Lately, the problem has received a lot of attention; see [AIK10, CKV10, GGP10, BGV11] for just a few examples. Our construction implies delegatable computation and can be simplified here because P_1 (the client) is usually assumed to be honest, and P_2 (the server) does not contribute any input y to the computation. Therefore we do not need a set-up assumption, and in contrast to earlier work, the scheme requires no expensive off-line phase and remains secure even if the server learns whether the client accepts its results.

Concurrent Related Work. In recent concurrent and independent work, Bitanski et al [BCCT11] and Goldwasser et al. [GLR11] both define notions of extractable hash function that are technically slightly different from our EHF1 notion, but similar in spirit. They each propose instantiations different from ours. They then build SNARGs based on this assumption, and [BCCT11] also build SNARGs that are in addition proofs of knowledge (SNARK's), and show the very interesting result that existence of SNARKs are equivalent to two notions of extractable hash function similar to EHF1, respectively EHF2, known as strong and weak proximity extractable hash functions.

Privacy for the prover is not considered in [GLR11]. In [BCCT11] zero-knowledge SNARKs and secure computation based on this is shown in the CRS model. They consider only stand-alone rather than UC security, on the other hand they obtain a protocol whose communication complexity is independent of the parties input. This can also be obtained from our construction using a simple modification based on PCP's of knowledge, but UC security would be lost.

2   Notations and Definitions

In this section, we review standard notations. Due to space constraints, we do not give a definition of secure computation here, the definition and proof can be found in [DFH11]. We denote the security parameter by n and adopt the convention whereby a machine is said to run in polynomial-time if its number of steps is polynomial in its *security parameter*. We use the standard definitions of negligible functions and indistinguishability of families of random variables, these can be found in the full version [DFH11]. For convenience, we use a single security parameter for all our primitives and proofs. For an integer t, we denote by $[t]$ the set $\{1, \ldots, t\}$, and by $\{0,1\}^{<t}$ the set of all binary strings of length at most $t - 1$. If X is a random variable then we write $x \leftarrow X$ for the value that the random variable takes when sampled according to the distribution of X. If A is a probabilistic algorithm running on input z, then we write $x \leftarrow A(z)$ for the output of A when run on input z.

2.1 Public Key Encryption Schemes

We specify the notion of public key encryption scheme. We use the standard notion of semantic security and refer to the full version [DFH11] for a formal definition.

Definition 1 (PKE). *We say that Π_E = (KeyGen, Enc, Dec) is a* public key encryption scheme (PKE) *if* KeyGen, Enc, Dec *are algorithms specified as follows.*

- KeyGen, *given a security parameter n (in unary), outputs keys (pk, sk), where pk is a public key and sk is a secret key. We denote this by $(pk, sk) \leftarrow$ KeyGen(1^n).*
- Enc, *given the public key pk and a plaintext message m, outputs a ciphertext c encrypting m. We denote this by $c \leftarrow$ Enc$_{pk}(m)$; and when emphasizing the randomness R used for encryption, we denote this by $c \leftarrow$ Enc$_{pk}(m; R)$.*
- Dec, *given the secret key sk and a ciphertext c, outputs a plaintext message m s.t.* Dec$_{sk}($Enc$_{pk}(m)) = m$.

2.2 Fully Homomorphic Encryption Schemes

We define fully homomorphic encryption and additional desired properties. We will say that a bit string *pk* is a *well-formed* public key, if it can be generated as output from the KeyGen algorithm on input the security parameter and a set of random coins in the range specified for the key generation algorithm. Similarly, a bit string *c* is a well-formed ciphertext if $c =$ Enc$_{pk}(m; r)$ for message *m* and random coins *r* lies in the range specified for the encryption algorithm.

Definition 2 (FHE). *We say that Π_E = (KeyGen, Enc, Dec, Eval) is a* fully homomorphic encryption scheme (FHE) *if* KeyGen, Enc, Dec *are algorithms specified as in Definition 1 and* Eval *is an algorithm specified as follows.*

- Eval, *given a well-formed public key pk, a boolean circuit C with fan-in of size t and well-formed ciphertexts $c_1, \ldots, c_\ell$ encrypting $m_1, \ldots, m_\ell$ respectively, outputs a ciphertext c such that* Dec$_{sk}(c) = C(m_1, \ldots, m_\ell)$.

We further require the existence of a refresh algorithm Refresh so that for well-formed $pk, c_1, \ldots, c_\ell$, the following distributions are statistically close,

$$\{pk, \mathsf{Refresh}_{pk}(\mathsf{Eval}_{pk}(C, c_1, \ldots, c_\ell))\} \equiv_s \{pk, \mathsf{Refresh}_{pk}(\mathsf{Enc}_{pk}(C(m_1, \ldots, m_\ell)))\}$$

Typically, Refresh would run Eval again on ciphertexts Eval$_{pk}(C, c_1, \ldots, c_\ell)$, an appropriately chosen encryption of zero and an addition gate. The idea is that the randomness for the encryption of zero is chosen large enough to "drown" the randomness coming from the original encryptions. We need that Refresh is correct, in the sense that on input well-formed $pk, c_1, \ldots, c_\ell$ as above, it outputs with probability 1 a ciphertext that decrypts to $C(m_1, \ldots, m_\ell)$. We also require that Π_E is semantically secure. Finally, we note that we require compactness in the sense that the output of Eval is upper bounded by some fixed polynomial regardless of C or the input length.

We note that our requirements on correctness of the Eval and Refresh algorithms are stronger than what is usually assumed by existing schemes in the literature: we want them to generate output of the expected form with probability 1 whenever the input is well-formed, whereas other definitions only require correct behavior on average over the distribution we expect the input to have. We need the stronger requirement because we need Eval and Refresh to behave correctly even on adversarially generated input where we cannot assume a particular distribution. All we can require is a ZK proof that the input is well formed. However, the stronger requirement can be assumed for all FHE schemes we are aware of [Gen09, vDGHV10, BV11a, BV11b]: typically, the key generation and encryption involves choosing randomness according to a (discrete) Gaussian distribution. Using a standard tail inequality, we can assume that randomness with the correct distribution is in some small range except with negligible probability and define well-formed public keys and ciphertexts to be those that can be produced using randomness that is in range. Since the probability of being out of range is negligible, this will not affect the security of honestly generated ciphertext, on the other hand, the guaranteed bound on the randomness will give us room to evaluate and refresh without creating incorrect results.

2.3   Efficient Probabilistic Checkable Proofs (PCP)

A PCP system $\Pi = \langle \mathsf{Prov}_{\mathrm{pcp}}, \mathsf{Ver}_{\mathrm{pcp}} \rangle$ for a language L consists of two PPT algorithms: the prover $\mathsf{Prov}_{\mathrm{pcp}}$ and the verifier $\mathsf{Ver}_{\mathrm{pcp}}$. The prover $\mathsf{Prov}_{\mathrm{pcp}}$ takes as input an instance $x \in L$ and a witness w for x and computes a proof π of length $\ell := \mathsf{poly}(|x|, |w|)$. The verifier $\mathsf{Ver}_{\mathrm{pcp}}$ inputs a potential member x and decides whether $x \in L$ given oracle access to the proof oracle π. In this work, we are interested in PCP systems where the verifier only has non-adaptive access to the proof system. To model this, we define the PCP verifier $\mathsf{Ver}_{\mathrm{pcp}}$ as a tuple of algorithms $(\mathsf{Ver}^1_{\mathrm{pcp}}, \mathsf{Ver}^2_{\mathrm{pcp}})$: the first has no access to the PCP π and uses only $\mathsf{polylog}(|x|)$ bits of randomness to compute $t := O(1)$ positions specifying where to read the PCP. The second machine, $\mathsf{Ver}^2_{\mathrm{pcp}}$, is deterministic and takes as input the bit values of the PCP at these t positions. It outputs whether to accept or reject π. We note that non-adaptivity is required as privacy of our protocol may not hold in case of an adaptive corrupted verifier.

Formally, we require the following two properties to hold:

Definition 3 (PCP). *A probabilistically checkable proof (PCP) system $\langle \mathsf{Prov}_{\mathrm{pcp}}, (\mathsf{Ver}^1_{\mathrm{pcp}}, \mathsf{Ver}^2_{\mathrm{pcp}}) \rangle$ for a language L is a triple of (probabilistic) polynomial-time machines, satisfying*

- *Completeness: If $x \in L$, $\pi \leftarrow \mathsf{Prov}_{\mathrm{pcp}}(x, w)$ and $(q_1, \ldots, q_t) \leftarrow \mathsf{Ver}^1_{\mathrm{pcp}}(x, \ell; r)$ with $q_i \in [\ell]$, then $\Pr[\mathsf{Ver}^2_{\mathrm{pcp}}(x, \pi[q_1], \ldots, \pi[q_t], q_1, \ldots, q_t) = 1] = 1$.*

- Soundness: *If $x \notin L$, then for all π we have*

$$\Pr[(q_1, \ldots, q_t) \leftarrow \mathsf{Ver}^1_{\mathrm{pcp}}(x, |\pi|; r) : \mathsf{Ver}^2_{\mathrm{pcp}}(x, \pi[q_1], \ldots, \pi[q_t], q_1, \ldots, q_t) = 1]$$
$$< \mathsf{negl}(n),$$

for negligible function $\mathsf{negl}(\cdot)$, *probability taken over the verifier's internal coins.*

Notice that standard definitions of PCP systems usually require the soundness error to be smaller than $1/2$. We get a negligible soundness error by amplification.

In this paper, we are interested in PCP's for NP languages such that the verifier accepts or rejects after using only $\mathsf{polylog}(|x|)$ bits of randomness and accessing only $O(1)$ bits of π. Moreover, we are interested in efficient protocols and, hence, require that the (probabilistic) prover runs in $\mathsf{poly}(|x|, |w|)$ time. PCP proof systems with efficient verifiers were introduced in the seminal work of Babai, Fortnow, Levin and Szegedy [BFLS91]. More efficient candidates have for instance been proposed in [PS94, AS98, BSS05, Din07]. Most PCP systems require only a non-adaptive verifier and, hence satisfy our additional property from above.

2.4 Collision Resistant Hashing and Merkle Trees

Let in the following $\{\mathcal{H}_n\}_{n \in \mathbb{N}} = \{H : \{0,1\}^{p(n)} \to \{0,1\}^{p'(n)}\}_n$ be a family of hash functions, where $p(\cdot)$ and $p'(\cdot)$ are polynomials so that $p'(n) \leq p(n)$ for sufficiently large $n \in \mathbb{N}$. For a hash function $H \leftarrow \mathcal{H}_n$ a Merkle hash tree [Mer87] is a data structure that allows to commit to $\ell = 2^d$ messages by a single hash value h such that revealing any message requires only to reveal $O(d)$ hash values. A Merkle hash tree is represented by a binary tree of depth d where the ℓ messages $m_1, \ldots, m_\ell$ are assigned to the leaves of the tree. The values that are assigned to the internal nodes are computed using the underlying hash function H. The single hash value h that commits to the ℓ messages $m_1, \ldots, m_\ell$ is assigned to the root of the tree. To open the commitment to a message m_i, one reveals m_i together with all the values assigned to nodes on the path from the root to m_i, and the values assigned to the siblings of these nodes. We denote the algorithm of committing to ℓ messages $m_1, \ldots, m_\ell$ by $h = \mathsf{Commit}(m_1, \ldots, m_\ell)$ and the opening of m_i by $(m_i, \mathsf{path}(i)) = \mathsf{Open}(h, i)$. Verifying the opening of m_i is carried out by essentially recomputing the entire path bottom-up while comparing the final outcome (i.e., the root) to the value given at the commitment phase. For simplicity, we abuse notation and denote by $\mathsf{path}(i)$ both the values assigned to the nodes in the path from the root to decommitted value m_i, together with the values assigned to their siblings.

The standard security property of a Merkle hash tree is collision resistance. Intuitively, this says that it is infeasible to efficiently find a pair (x_1, x_2) so that $H(x_1) = H(x_2)$, where $H \leftarrow \mathcal{H}_n$ for sufficiently large n. One can show that collision resistance of $\{\mathcal{H}_n\}_{n \in \mathbb{N}}$ carries over to the Merkle hashing. Formally,

Definition 4 (Collision Resistance). *A family of hash functions* $\{\mathcal{H}_n\}_n$ *is* collision resistant *if for all PPT adversaries* $\mathcal{A}$ *there exists a negligible function* negl *such that for sufficiently large* $n \in \mathbb{N}$ *we have* $\Pr[\mathsf{Hash}_{\mathcal{A},\mathcal{H}_n}(n) = 1] \leq \mathsf{negl}(n)$ *where game* $\mathsf{Hash}_{\mathcal{A},\mathcal{H}_n}(n)$ *is defined as follows:*

1. *A hash function* H *is sampled* $H \leftarrow \mathcal{H}_n$.
2. *The adversary* $\mathcal{A}$ *is given* H *and outputs* x, x'.
3. *The output of the game is 1 if and only if* $x \neq x'$ *and* $H(x) = H(x')$.

2.5   Non-interactive Zero-Knowledge Proofs

In the following we repeat the definition of non-interactive zero-knowledge proof.

Definition 5. *A* non-interactive zero-knowledge proof *for a language* L *is a tuple of three PPT algorithms* $\langle \mathsf{CRSGen}, \mathcal{P}, \mathcal{V} \rangle$, *such that the following properties are satisfied:*

Completeness: *For every* $(x, \omega) \in R_\mathrm{L}$ *(for* R_L *the witness relation of* L*)*

$$\Pr[crs \leftarrow \mathsf{CRSGen}(1^n) : \mathcal{V}(\mathsf{crs}, x, \mathcal{P}(\mathsf{crs}, x, \omega)) = 1] = 1.$$

Soundness: *For every PPT algorithm* $\mathcal{A}$ *there exists a negligible function* negl *such that for all* $x \notin \mathrm{L}$

$$\Pr[(x, \pi) \leftarrow \mathcal{A}(\mathsf{crs}), \mathsf{crs} \leftarrow \mathsf{CRSGen}(1^n) : \mathcal{V}(\mathsf{crs}, x, \pi) = 1 \ ] \leq \mathsf{negl}(n).$$

Zero-Knowledge: *there exists a PPT simulator* $S = (S_1, S_2)$ *such that for all* $(x, \omega) \in R_\mathrm{L}$ *the distributions (i)* $\{P(\mathsf{crs}, x, \omega)\}$ *and (ii)* $\{S_2(\mathsf{crs}, x, \mathsf{td})\}$ *are computationally indistinguishable, where in (i)* $\mathsf{crs} \leftarrow \mathsf{CRSGen}(1^n)$ *and in (ii)* $(\mathsf{crs}, \mathsf{td}) \leftarrow S_1(1^n)$.

2.6   Extractable Hash Functions

In this work, we are interested in hash functions that are *extractable* – so-called *extractable hash function (EHF)*. We provide two flavors of extractable hash functions. The first extractability assumption (EHF1) considers a hash function H mapping into a small subset of a large domain and essentially asserts that the only way to generate an element in $\mathrm{Im}(H)$ is to compute the function on a given input. More precisely, we require that for every adversary outputting a value h there exists an efficient extractor that (given the same randomness) outputs a preimage of h, whenever $h \in \mathrm{Im}(H)$. We propose later an instantiation of EHF1-extractable and collision intractable hash functions based on a knowledge of exponent assumption (Damgård [Dam91]) in $\mathbb{Z}_N^*$ where N is an RSA modulus. We continue with the formal assumption. For simplicity, we assume that the algorithms below are keeping their state.

Definition 6 (Extractable hash function 1 (EHF1)). *Let* **A** *and* **E** *be PPT algorithms then consider the following game:*

- **EHF1**$_{\mathbf{A},\mathbf{E},\mathcal{H}_n}(1^n, z)$.

   $H \leftarrow \mathcal{H}_n$
   Repeat until **A** halts:
      $h \leftarrow \mathbf{A}(1^n, H, z; R)$
      $z \leftarrow \mathbf{E}(1^n, H, z, R, h; R')$
      If $h \in \mathrm{Im}(H)$ and $H(z) \neq h$ return 1, else reply **A** with z
   Return 0

for R and R' the randomness used by **A** *and* **E** *respectively. Then the family* $\{\mathcal{H}_n\}_{n\in\mathbb{N}}$ *satisfies the first extractability assumption (EHF1) if for every PPT adversary* **A** *there exists a PPT extractor* **E** *such that for any sufficiently large $n \in \mathbb{N}$ and any auxiliary information $z \in \{0,1\}^*$*

$$\Pr[\mathbf{EHF1}_{\mathbf{A},\mathbf{E},\mathcal{H}_n}(1^n, z) = 1] \leq \mathsf{negl}(n).$$

for a negligible function negl, *the probability is over the randomness of the game.*

In the above definition, we require that it should be feasible to verify that a value h is in the image of H; we call this function $\mathrm{Im}(H)$.

The second extractability assumption (EHF2) makes a weaker demand on the hash function H: as before, we require that for each adversary outputting h, there exists an extractor that tries to find a preimage. This time, however, the extractor is allowed to fail even if $h \in \mathrm{Im}(H)$. Specifically, the demand is that if the extractor fails, the adversary cannot *output* a preimage either. For this definition not to be vacuous, one clearly needs that when the adversary tries to "beat" the extractor, it is given randomness/auxiliary input that is not known to the extractor. Otherwise the extractor could simulate the adversary and output whatever the adversary does. To formalize this, we assume a probabilistic algorithm $\mathcal{G}$ that outputs a pair (ζ, ζ'), sampled from some joint distribution. ζ is given to both the adversary and the extractor, while ζ' is only given to the adversary later when she tries to "beat" the extractor. In our case, ζ is a public key for an encryption scheme and ζ' is its corresponding secret key. Notice that our demand on $\mathcal{G}$ is weak as $\mathcal{G}$ does not depend on the choice of the hash function.

Finally, we note that in [BCCT11] a simpler definition is considered, where the adversary runs an arbitrary algorithm in the last stage of the game and the extractor is required to work for any such algorithm. In particular, it must work for an adversary that knows something not known to the extractor. This is a much stronger demand that may exclude some potential constructions of extractable hash functions.[1]

---

[1] [BCCT11] also considers weaker variants. While the basic idea of EHF2 is a contribution of this paper, the precise formulation was in part inspired by discussions with the authors of [BCCT11].

Definition 7 (Extractable hash function 2 (EHF2)). *Let* $\mathbf{A}$ *and* $\mathbf{E}$ *be PPT algorithms then consider the following game:*

- $\mathbf{EHF2}_{\mathbf{A},\mathcal{G},\mathbf{E},\mathcal{H}_n}(1^n, z)$.

  $i = 0, H \leftarrow \mathcal{H}_n, (\zeta, \zeta') \leftarrow \mathcal{G}(1^n)$

  Repeat until $\mathbf{A}$ halts:

  $\quad i = i + 1$

  $\quad h_i \leftarrow \mathbf{A}(1^n, H, z, \zeta; R)$

  $\quad z_i \leftarrow \mathbf{E}(1^n, H, z, R, h_i, \zeta; R')$

  $(z_1^{\mathbf{A}}, \ldots, z_i^{\mathbf{A}}) \leftarrow \mathbf{A}(1^n, H, z, R, \zeta'; R'')$

  If $\exists\, 1 \leq j \leq i$, s.t. $H(z_j) \neq h_j \wedge H(z_j^{\mathbf{A}}) = h_j$ return 1, else return 0

Then $\{\mathcal{H}_n\}_{n \in \mathbb{N}}$ *satisfies the EHF2 assumption if for every PPT adversary* $\mathbf{A}$ *and any PPT algorithm* $\mathcal{G}$ *there exists a PPT extractor* $\mathbf{E}$ *such that for any sufficiently large* $n \in \mathbb{N}$ *and any auxiliary information* $z \in \{0, 1\}^*$

$$\Pr[\mathbf{EHF2}_{\mathbf{A},\mathcal{G},\mathbf{E},\mathcal{H}_n}(1^n, z) = 1] \leq \mathsf{negl}(n).$$

for a negligible function negl, *the probability is over the randomness of the game.*

When we talk in the following of an extractable hash function, then we mean that it satisfies the property given in Definition 7, i.e., any PPT adversary has a negligible advantage in $\mathbf{EHF2}_{\mathbf{A},\mathbf{G},\mathbf{E},\mathcal{H}_n}$.

Note that EHF2 is true in the random oracle model, where we let the random oracle play the role of H. In this case it is easy to see that no matter how the adversary produces a string h, there are only two cases: either h was output by the random oracle or not. In the former case a preimage is easy to extract, in the latter case *no one* can produce a preimage except with negligible probability. So the extractor can safely fail in this case.

It is easy to verify that EHF1 implies EHF2: under EHF1, the extractor only fails if it is *impossible* to find a preimage.

2.7 The Knowledge of Exponent Assumption

The knowledge of exponent assumption proposed by Damgård [Dam91] was previously used in designing 3-round zero-knowledge proofs [HT98], plaintext-aware encryption [BP04, Den06] and more. It was originally defined with respect to prime order groups; here we consider its variant for composite order groups. Say N is a product of two safe primes $p = 2p' + 1$ and $q = 2q' + 1$. We consider the group of so-called signed quadratic residues $\mathcal{QR}_N^+$. It consists of all numbers in $\mathbb{Z}_N$ with Jacobi symbol 1 in the interval $[0, \ldots, (N-1)/2]$. The product of $a, b \in \mathcal{QR}_N^+$ is defined to be $ab \bmod N$ if $ab \bmod N \leq (N-1)/2$ and $N - ab \bmod N$ otherwise. $\mathcal{QR}_N^+$ is isomorphic to the group of quadratic residues mod N and so has order $p'q'$. Furthermore, it has the nice property that membership in $\mathcal{QR}_N^+$ is easy to check. We let g, g' be generators for $\mathcal{QR}_N^+$ where $g' = g^x$ and x is picked at random from $\mathbb{Z}_{p'q'}^*$. Informally, the assumption says that for any PPT algorithm $\mathbf{A}(N, g, g')$ that outputs h, h' such that $h = g^y$ and

$h' = g^{xy}$ there exists an extractor $\mathbf{E}$ such that $(h, h', y) \leftarrow \mathbf{E}(N, g, g')$ with high probability. We refer the reader to the full version for a formal definition of the knowledge of exponent assumption in the group of signed quadratic residues.

Based on the knowledge of exponent assumption, we can construct an extractable hash function according to Definition 6. Moreover, under the factoring assumption our construction is collision resistant. The public parameters of our family of hash functions are a composite N which is the product of two safe primes $p = 2p' + 1$ and $q = 2q' + 1$ and two generators g, h for $\mathcal{QR}_N^+$. For some concrete N, p, q, g, h, we compute the hash function on some input z as $H(z) = (g^z \bmod N, h^z \bmod N)$. Collision resistance follows from factoring, since for every $z \neq z'$ such that $H(z) = H(z')$ it holds that $p'q'$ divides $z - z'$. Moreover, if one knows x such that $h = g^x \bmod N$, then one can check membership of a pair (a, b) in the image of H by checking whether $a \in \mathcal{QR}_N^+$ and $a^x \bmod N = b$. Finally we note that H is an EHF1, which follows from the knowledge of exponent assumption.

3 Secure Two-Party Computation with Low Communication

Consider two parties P_1 with input x and P_2 with input y, respectively, who wish to jointly compute a function $f(x, y)$. Without loss if generality we only consider single-output functions and assume that only P_1 learns the output $f(x, y)$ (the general case can be easily obtained from this special case [Gol04] but this requires additional communication). We are interested in protocols that allow P_1 and P_2 to securely compute $f(x, y)$ in the presence of malicious adversaries that follow arbitrary behavior. Our proof of security guarantees the strongest notion of simulation based UC security [Can01] in the presence of static malicious adversaries. Moreover, we require that our protocol achieves the following strong properties: *Polylogarithmic communication complexity* in the circuit-size C that computes f. *One round complexity*, i.e., a single message in each direction assuming an appropriate trusted setup. In this work we prove our protocol in the common reference string model. *Polylogarithmic workload for P_1* in the circuit-size C.

We introduce our main construction step-by-step. Our starting point is a standard protocol secure against honest-but-curious adversaries for which party P_1 sends its encrypted input to party P_2, who uses the homomorphic property to compute ciphertexts that contain that the output of the specified circuit when evaluated on P_1's (encrypted) input and his own private input. These ciphertexts are sent to P_1 who can decrypt and learn the result. Obviously, this solution completely breaks down against malicious attacks. So additional cryptographic tools must be used in order to ensure correct behavior. We then use this protocol as a building block in our main construction, adding new tools to protect against an increasingly powerful adversary. Namely, we first show how to prove security in the presence of a corrupted P_2 and then prove simulation based security for both corruption cases. For completeness, we formally describe the standard protocol with security against honest-but-curious adversaries.

3.1   Security against Honest-But-Curious Adversaries

We begin with a standard protocol with security in the face of honest-but-curious adversaries. The main building block here is fully homomorphic encryption $\Pi_{\mathsf{E}} = (\mathsf{KeyGen}, \mathsf{Enc}, \mathsf{Dec}, \mathsf{Eval}, \mathsf{Refresh})$.

Protocol 1 (Honest-but-curious adversaries)

- **Inputs:** *Input x for party P_1 and input y for party P_2. A description of function f for both.*
- **The protocol:**
    1. *$P_1(x)$ generates a key pair $(\mathsf{pk}_{\mathsf{comp}}, \mathsf{sk}_{\mathsf{comp}}) \leftarrow \mathsf{KeyGen}(1^n)$ for a fully homomorphic encryption scheme, computes $e_x = \mathsf{Enc}_{\mathsf{pk}_{\mathsf{comp}}}(x)$ and sends $(\mathsf{pk}_{\mathsf{comp}}, e_x)$ to P_2.*
    2. *$P_2(y)$ computes $d = \mathsf{Eval}_{\mathsf{pk}_{\mathsf{comp}}}(C_f, y, e_x)$ and sends $c = \mathsf{Refresh}_{\mathsf{pk}_{\mathsf{comp}}}(d)$ to P_1.*
    3. *P_1 decrypts c and obtains the result of the computation $f(x, y) = \mathsf{Dec}_{\mathsf{sk}_{\mathsf{comp}}}(c)$.*

Security of P_1 follows by the semantic security of Π_{E}. Similarly, security of P_2 follows from the ability to refresh the ciphertext sent back to P_1 so that it only encrypts the outcome. It is easy to see that the communication complexity is independent of the complexity of the circuit-size C that computes f, and only depends on its inputs and outputs lengths, and the security parameter.

3.2   Security against a Malicious P_1

We extend the above protocol and allow P_1 to be malicious (if corrupted), while P_2 remains honest-but-curious. To this end, we use standard techniques to achieve security in the malicious setting by relying on NIZK proof systems $\langle \mathsf{CRSGen}, \mathcal{P}, \mathcal{V} \rangle$ and an idealized setup. Specifically, we let P_1 send two encryptions encrypted under two different keys (one public key for which P_1 knows the secret key and the other public key is placed in the common reference string), so that the same plaintext is encrypted. This enables the simulator to extract x using the trapdoor of the common reference string. In addition to that, P_1 must prove that its public key, together with the ciphertexts, are well-formed. Note that the statement proved below asserts that each ciphertext is produced from a message and randomness of the expected range, so it is implicitly asserted that these ciphertexts are well-formed. Nevertheless, we still need to prove well-formness of $\mathsf{pk}_{\mathsf{comp}}$. This is essentially immediate when specifying the random coins used to generate it as part of the witness, since all it takes is to verify whether these coins are of the expected range. In order to formalize this proof we define language L as follows.

$$L := \{(e_x, e'_x, \mathsf{pk}_{\mathsf{comp}}, \mathsf{pk}_x) : \exists\ (sk_{\mathsf{comp}}, r_{\mathsf{pk}}, r_x, r'_x, x)\ \text{s.t.}\ e_x = \mathsf{Enc}_{\mathsf{pk}_{\mathsf{comp}}}(x; r_x)$$
$$\wedge\ e'_x = \mathsf{Enc}_{\mathsf{pk}_x}(x; r'_x) \wedge (\mathsf{pk}_{\mathsf{comp}}, sk_{\mathsf{comp}}) \leftarrow \mathsf{KeyGen}(1^n, r_{\mathsf{pk}})$$
$$\wedge\ r_{\mathsf{pk}}\ \text{yields a well formed}\ \mathsf{pk}_{\mathsf{comp}}\}.$$

This proof is utilized in Step 1b of Protocol 2. The complete protocol follows.

Protocol 2 (Malicious P_1)

- **Setup:** *Generate keys* $(\mathsf{pk}_x, \mathsf{sk}_x) \leftarrow \mathsf{KeyGen}(1^n)$. *Set the common reference string* $\mathsf{crs} = (\mathsf{pk}_x, \sigma)$, *where* $\sigma \leftarrow \mathsf{CRSGen}(1^n)$ *is the common reference string used for proving membership in* L.
- **Input:** *Input x for party P_1 and input y for party P_2. A description of function f for both.*
- **The protocol:**
    1. **First message computed by party P_1.**
        (a) **Setup.** *Generate a key pair* $(\mathsf{pk}_{\mathsf{comp}}, \mathsf{sk}_{\mathsf{comp}}) \leftarrow \mathsf{KeyGen}(1^n)$ *for a fully homomorphic encryption scheme and compute* $e_x = \mathsf{Enc}_{\mathsf{pk}_{\mathsf{comp}}}(x)$.
        (b) **Proof of consistency.** *Compute* $e'_x = \mathsf{Enc}_{\mathsf{pk}_x}(x)$ *and a NIZK proof π_{L} proving that* $\mathsf{pk}_{\mathsf{comp}}$ *and e_x are well-formed and that e_x and e'_x encrypt the same plaintext x.*
        (c) **The complete message.** *Send* $(e_x, e'_x, \mathsf{pk}_{\mathsf{comp}}, \mathsf{pk}_x, \pi_{\mathrm{L}})$ *to P_2.*

    2. **Second message computed by party P_2.**
        (a) **Verification of NIZK.** *Upon receiving message* $(e_x, e'_x, \mathsf{pk}_{\mathsf{comp}}, \mathsf{pk}_x, \pi_{\mathrm{L}})$ *from P_1, verify π_{L} by running* $\mathcal{V}((e_x, e'_x, \mathsf{pk}_{\mathsf{comp}}, \mathsf{pk}_x), \pi_{\mathrm{L}})$. *If it outputs 0, then abort.*
        (b) **Circuit evaluation.** *Compute* $d = \mathsf{Eval}_{\mathsf{pk}_{\mathsf{comp}}}(C_f, y, e_x)$ *for C_f a PPT circuit computing f, and refresh the ciphertext to get* $c = \mathsf{Refresh}_{\mathsf{pk}_{\mathsf{comp}}}(d)$.
        (c) **The complete message.** *Send the result c to P_1.*

    3. **The output.** *P_1 decrypts c and obtains the result of the computation* $f(x, y) = \mathsf{Dec}_{\mathsf{sk}_{\mathsf{comp}}}(c)$.

Clearly, if both parties behave honestly P_1 learns the correct output.

Theorem 8 (One-Sided Security). *If $\Pi_{\mathrm{E}} = (\mathsf{KeyGen}, \mathsf{Enc}, \mathsf{Dec}, \mathsf{Eval}, \mathsf{Refresh})$ is semantically secure and $\langle \mathsf{CRSGen}, \mathcal{P}, \mathcal{V} \rangle$ is a non-interactive zero-knowledge proof, Protocol 2 securely evaluates f in the presence of malicious P_1 and honest-but-curious P_2 with constant communication in the circuit-size for f.*

Intuitively, security against malicious P_1 follows from the soundness of proof π_{L}. A simulator $\mathcal{S}_1$ for an adversary corrupting P_1 can be designed by first verifying the proof π_{L}. Next, $\mathcal{S}_1$ extracts the adversary's input x' using the secret key sk_x. $\mathcal{S}_1$ sends x' to the trusted party computing f and receives the outcome. It then encrypts this value and sends it back to the adversary. Security against corrupted P_2 follows from the semantic security property of Π_{E}. Communication complexity depends only on the input/output length of f.

3.3 Security against Malicious Adversaries

In this section we present our full protocol that protects against malicious adversarial attacks. Our protocol uses Protocol 2 as a building block but adds additional tools. This essentially amounts to a SNARG allowing P_1 to verify the correctness of the output issued by P_2. More precisely:

1. We first add a PCP system $\langle \mathsf{Prov}_{\mathsf{pcp}}, (\mathsf{Ver}^1_{\mathsf{pcp}}, \mathsf{Ver}^2_{\mathsf{pcp}}) \rangle$ (cf. Definition 3), used by P_2 for proving membership in the language L_1. Formally, L_1 is defined by

$$L_1 := \{(c, e_x, \mathsf{pk}_{\mathsf{comp}}, e_y, \mathsf{pk}_y, f) : \exists\ (d, r_d, r_y, y)\ \text{s.t.}\ d = \mathsf{Eval}_{\mathsf{pk}_{\mathsf{comp}}}(C_f, y, e_x)$$
$$\wedge\ c = \mathsf{Refresh}_{\mathsf{pk}_{\mathsf{comp}}}(d; r_d) \wedge e_y = \mathsf{Enc}_{\mathsf{pk}_y}(y; r_y)\}.$$

Namely, the PCP shows that if one decrypts c it gets the desired result $f(x, y)$, where x is the plaintext contained in e_x and y is the plaintext in e_y. This proof is utilized in Step 2c of Protocol 3. We recall that the statement proved asserts that e_y is produced from a message and randomness of the expected range so it is implicitly asserted that e_y is well-formed.

   We further let P_2 commit to this proof using a Merkle hash tree instantiated with an extractable collision resistance hash function $H : \{0,1\}^* \rightarrow \{0,1\}^\tau$ (cf. Definition 7). The main problem with this is that hashing the proof does not necessarily conceal it, unless a special hiding property is required form the underlying hash function. We fix that by hashing the committed PCP instead, and then prove that the values embedded within these commitments correspond to a valid proof.

2. Furthermore, since the verifier must not see the queried bits from the proof (due to privacy considerations), we consider an NP statement claiming that if the PCP verifier $\mathsf{Ver}^2_{\mathsf{pcp}}$ is run on $\mathsf{Dec}_{\mathsf{sk}_x}(c_{q_1}), \ldots, \mathsf{Dec}_{\mathsf{sk}_x}(c_{q_t})$, denoting the ciphertexts encrypting $(\Gamma_{q_1}, \ldots, \Gamma_{q_t})$ – the openings for the PCP queries $(q_1, \ldots, q_t)$, then it will accept. That is,

$$L_2 := \Big\{(z_{\mathsf{pcp}}, (q_1, \ldots, q_t), (c_{q_1}, \ldots, c_{q_t})) :$$
$$\exists\ (\Gamma_{q_1}, \gamma_{q_1}, \ldots, \Gamma_{q_t}, \gamma_{q_t}, r_{\mathsf{pk}})\ \text{s.t.}\ \big(\forall i \in [t] : c_{q_i} = \mathsf{Enc}_{\mathsf{pk}_y}(\Gamma_{q_i}; \gamma_{q_i})\big)$$
$$\wedge\ \mathsf{Ver}^2_{\mathsf{pcp}}\big(z_{\mathsf{pcp}}, \Gamma_{q_1}, \ldots, \Gamma_{q_t}, q_1, \ldots, q_t\big) = 1\Big\}$$

for the instance $z_{\mathsf{pcp}} \in L_1$. In our protocol, $(\boldsymbol{q}, c_{q_1}, \ldots, c_{q_t})$ are all encrypted under FHE with respect to public key $\mathsf{pk}_{\mathsf{pro}}$, enabling P_1 to verify this proof. Note that the code of $\mathsf{Ver}^2_{\mathsf{pcp}}$ is independent of the strategy followed by a malicious P_1. Furthermore, notice that the we do not explicitly need to include checks of well-formedness for the ciphertext $c_{q_1}, \ldots, c_{q_t}$ since these are implied by the fact that the ciphertext are possible outputs on proper inputs $\Gamma_{q_i}, \gamma_{q_i}$. This proof is utilized in Step 2f in Protocol 3. Importantly, the number of queries asked by P_1 is polylogarithmic in the PCP size (and hence in the circuit-size that computes f).

The above implies that P_1 has to provide encryptions of the queries $q_1, \ldots, q_t$. In order to ensure correctness of these queries, we add a non-interactive zero-knowledge proof for which P_1 proves that the queries were indeed sampled from the correct range. This is formalized in Step 1c of Protocol 3 below.

An overview of our protocol. We summarize the discussion above. (**1**) At first, P_1 sends its input x encrypted under two distinct public keys together with the encrypted PCP queries and a proof of correct behavior. (2) P_2 then replies with ciphertexts that contain the output of the specified circuit, as generated above. It then produces a PCP for this computation and commits to it using a Merkle tree. Finally, P_2 computes ciphertexts that contain the answers for the PCP queries by opening the corresponding paths in the Merkle tree generated above (note that this step is performed obliviously within the fully homomorphic encryption scheme). P_2 sends the computation of $f(x, y)$ and answers to PCP queries with a non-interactive zero-knowledge proof for correct computations.

Intuitively, the overall communication complexity depends on the number of PCP queries, the answers to these queries and the overhead induced by the non-interactive zero-knowledge proofs. Recall first that PCP systems are sound even after observing only polylogarithmic bits of the proof. Moreover, each answer to such a query requires providing the corresponding path in the hashed Merkle tree of the PCP which includes logarithmic number of elements (in the proof's size). Finally, we utilize zero-knowledge proofs with communication that is polynomial in the size of the witness. All these tools ensure that the overall communication is polylogarithmic in the circuit's size. We are now ready to present our protocol.

Protocol 3 (Malicious adversaries)

- **Setup:** *Generate keys* $(\mathsf{pk}_x, \mathsf{sk}_x) \leftarrow \mathsf{KeyGen}(1^n)$ *and* $(\mathsf{pk}_y, \mathsf{sk}_y) \leftarrow \mathsf{KeyGen}(1^n).$[2]
  Set the common reference string $\mathsf{crs} = (\mathsf{pk}_x, \mathsf{pk}_y, \sigma)$, *where* σ *is a joint common reference string used by* P_1 *for proving membership in* L *and by* P_2 *for proving membership in* L_1 *and* L_2. *Pick an extractable collision-resistant hash function* $H \leftarrow \mathcal{H}_n$ *for* $H : \{0,1\}^{p(n)} \to \{0,1\}^{p'(n)}$.
- **Input:** *Input* x *for party* P_1 *and input* y *for party* P_2. *A description of function* f *for both.*
- **The protocol:**
  1. **First message computed by party** P_1.
     - (a) **Setup.** *Generate key pairs for a fully homomorphic encryption scheme* $(\mathsf{pk}_{\mathsf{comp}}, \mathsf{sk}_{\mathsf{comp}}) \leftarrow \mathsf{KeyGen}(1^n)$ *and* $(\mathsf{pk}_{\mathsf{pro}}, \mathsf{sk}_{\mathsf{pro}}) \leftarrow \mathsf{KeyGen}(1^n)$, *and compute* $e_x = \mathsf{Enc}_{\mathsf{pk}_{\mathsf{comp}}}(x)$.
     - (b) **Proof of consistency.** *Compute* $e'_x = \mathsf{Enc}_{\mathsf{pk}_x}(x)$ *and a NIZK proof* π_{L} *proving that* $\mathsf{pk}_{\mathsf{pro}}, \mathsf{pk}_{\mathsf{comp}}, e_x$ *are well-formed and that* e_x *and* e'_x *encrypt the same plaintext* x.
     - (c) **Queries for PCP.** *Sample* t *positions* $(q_1, \dots, q_t) \leftarrow \mathsf{Ver}^1_{\mathrm{pcp}}(z_{pcp}, \ell)$ *and for each* i *encrypt them as* $b_i = \mathsf{Enc}_{\mathsf{pk}_{\mathsf{pro}}}(q_i)$. *Moreover, for each* i *compute a NIZK proof* π_i *that* q_i *lies in the correct range* $[\ell]$.
     - (d) **The complete message.** *Send* $m_1 := ((e_x, e'_x, \mathsf{pk}_{\mathsf{comp}}, \mathsf{pk}_{\mathsf{pro}}, \pi_{\mathrm{L}}), (b_i, \pi_i)_{i \in [t]})$ *to* P_2.

  2. **Second message computed by party** P_2.
     - (a) **Verification of NIZK's.** *Upon receiving message* m_1 *from* P_1, *verify* π_{L} *by running* $\mathcal{V}((e_x, e'_x, \mathsf{pk}_{\mathsf{comp}}, \mathsf{pk}_x), \pi_{\mathrm{L}})$. *If it outputs 0, then abort.*
     - (b) **Circuit evaluation.** *Compute* $d = \mathsf{Eval}_{\mathsf{pk}_{\mathsf{comp}}}(C_f, y, e_x)$ *and refresh it to get* $c = \mathsf{Refresh}_{\mathsf{pk}_{\mathsf{comp}}}(d; r_d)$. *Also, compute* $e_y = \mathsf{Enc}_{\mathsf{pk}_y}(y; r_y)$.

---

[2] We note that these public keys do not have to be associated with the fully homomorphic encryption scheme. For convenience, we assume that they do in order to avoid overload of parameters.

(c) **Compute PCP.** *Compute a PCP* $\Gamma = \mathsf{Prov}_{\mathsf{pcp}}(z_{\mathsf{pcp}}, \omega_{\mathsf{pcp}})$ *of length* $\ell = \mathsf{poly}(n)$, *where* $\omega_{\mathsf{pcp}} := (d, r_d, r_y, y)$ *forms an NP witness for the instance* $z_{\mathsf{pcp}} := (c, e_x, \mathsf{pk}_{\mathsf{comp}}, e_y, \mathsf{pk}_y, f) \in \mathsf{L}_1$.

(d) **Commit to PCP.** *For* $i \in [\ell]$ *compute ciphertexts* $c_i = \mathsf{Enc}_{\mathsf{pk}_y}(\Gamma_i; \gamma_i)$ *and compute the Merkle hash root using* H, *for* $h = \mathsf{Commit}(c_1, \ldots, c_\ell)$, *where for simplicity we let* ℓ *be a power of 2.*

(e) **Answer PCP queries.** *Compute* $p_{q_i} = \mathsf{Enc}_{\mathsf{pk}_{\mathsf{pro}}}(\mathsf{path}(q_i); \rho_{q_i})$ *for* $i \in [t]$ *by running* $\mathsf{Eval}_{\mathsf{pk}_{\mathsf{pro}}}$ *on input* b_i *(sent by* P_1*) and* $(c_1, \ldots, c_\ell)$ *(computed above), where* $\mathsf{path}(q_i) = \mathsf{Open}(h, i)$.

(f) **Proving correctness.** *Compute an encrypted proof* $c_{\pi_{\mathsf{L}_2}} = \mathsf{Enc}_{\mathsf{pk}_{\mathsf{pro}}}(\pi_{\mathsf{L}_2})$ *for proving that* $(z_{\mathsf{pcp}}, (q_1, \ldots, q_t), (c_{q_1}, \ldots, c_{q_t})) \in \mathsf{L}_2$. *This is done by running* $\mathsf{Eval}_{\mathsf{pk}_{\mathsf{pro}}}$ *on input* $z_{\mathsf{pcp}}, (b_1, \ldots, b_t), (c_1, \ldots, c_\ell), (\gamma_1, \ldots, \gamma_\ell)$.

(g) **The complete message.** *Send* $m_2 := (c, e_y, h, (p_{q_1}, \ldots, p_{q_t}), c_{\pi_{\mathsf{L}_2}})$ *to* P_1. *Notice that* c_{q_i} *is part of* $\mathsf{path}(q_i)$ *which is contained in* p_{q_i}.

3. **Verifying the second message** m_2. P_1 *decrypts* c *and obtains the result of the computation* $f(x, y) = \mathsf{Dec}_{\mathsf{sk}_{\mathsf{comp}}}(c)$. *For each* $i \in [t]$ *it also decrypts* $\mathsf{path}(q_i) = \mathsf{Dec}_{\mathsf{sk}_{\mathsf{pro}}}(p_{q_i})$ *and verifies that* $\mathsf{path}(q_i)$ *is correct with respect to the root* h. *It then uses the leaves* $c_{q_1}, \ldots, c_{q_t}$ *and* $\pi_{\mathsf{L}_2} = \mathsf{Dec}_{\mathsf{sk}_{\mathsf{pro}}}(c_{\pi_{\mathsf{L}_2}})$ *together with the common reference string* σ *and verifies the correctness of* π_{L_2}. *If all these checks succeed, then it outputs* $f(x, y)$, *otherwise it aborts.*

Then we claim the following theorem, the proof can be found in [DFH11].

Theorem 9 (Main). *Assuming that* $\Pi_{\mathbb{E}} = (\mathsf{KeyGen}, \mathsf{Enc}, \mathsf{Dec}, \mathsf{Eval}, \mathsf{Refresh})$ *is semantically secure,* $\langle \mathsf{CRSGen}, \mathcal{P}, \mathcal{V} \rangle$ *is a non-interactive zero-knowledge proof,* $\langle \mathsf{Prov}_{\mathsf{pcp}}, (\mathsf{Ver}^1_{\mathsf{pcp}}, \mathsf{Ver}^2_{\mathsf{pcp}}) \rangle$ *is a PCP system,* $\{\mathcal{H}_n\}_{n \in \mathbb{N}}$ *is collision-resistant and satisfies the EHF2 assumption, Protocol 3 evaluates* f *UC-securely against malicious adversaries with polylogarithmic communication in the circuit-size of* f.

We give a brief overview of our proof. We distinct two corruption cases. Let P_1 be controlled by an adversary $\mathcal{A}$. In this case we face the difficulty of protecting the privacy of P_2, since revealing bits from Γ so that the PCP verifier will be able to validate the proof is insecure. Loosely speaking, privacy follows due to hashing the committed proof rather than the proof itself. Thus, secrecy is obtained from the hiding property of the commitment scheme. Simulating $\mathcal{A}$'s view requires from the simulator to verify the correctness of the message m_1 received from $\mathcal{A}$ as the honest P_2 would. Then it extracts $\mathcal{A}$'s input, forwarding it to the trusted party. Finally, upon receiving from the trusted party $f(x, y)$, it encrypts this value under $\mathsf{pk}_{\mathsf{comp}}$ and sends it back to $\mathcal{A}$. Now, since the simulator does not use the real honest party's input, y, it cannot construct a valid proof Γ and therefore has to build the hash tree on commitments to the zero string. It further simulates the NIZK proof for L_2. Indistinguishability follows due to: (**1**) Zero-knowledge property of the proof system of L_2. (**2**) Semantic security of $\Pi_{\mathbb{E}}$. (**3**) Refresh algorithm of $\Pi_{\mathbb{E}}$ that produces a ciphertext indistinguishable from a ciphertext that encrypts $f(x, y)$ directly (without going through homomorphic evaluation). (**4**) Soundness of the proof system of L.

We now consider the case where P_2 is corrupt. Intuitively, security should follow from semantic security of encryptions under $\mathsf{pk}_{\mathsf{pro}}$, soundness of the PCP and the fact that P_2 is committed to a PCP string via sending the root of the

Merkle tree: by soundness of the PCP, the only way P_2 could cheat would be to look at the encrypted PCP queries and adapt the PCP string it commits to, to the specific queries that are asked. Supposedly, this is not possible by semantic security. The technical difficulty, however, is that to have P_2 help us conclude anything on which queries have been encrypted in a given ciphertext (to make a reduction to semantic security), we would need to see the responses P_2 sends back. Unfortunately, these are encrypted under the same key $\mathsf{pk}_{\mathrm{pro}}$, and if we want to do a reduction to semantic security, we cannot know $\mathsf{sk}_{\mathrm{pro}}$ and so cannot see the responses directly. This is solved by first observing that by the extractability of the hash function, we can extract a Merkle tree $\mathcal{T}$ based on the root of the tree sent by P_2, and hence also a PCP string (we can assume we know sk_y so we can decrypt the commitments containing PCP bits). We then show that the encrypted paths $path(q_i)$ must be contained in $\mathcal{T}$, or else we could break extractability or collision resistance of $\mathcal{H}_n$. So the responses we want to see will be embedded in the tree we can extract. The reduction to semantic security can therefore ask for an encryption of one of two sets of queries q^0 or q^1. It shows the ciphertext to P_2 and extracts a PCP string from the root sent by P_2. Then if q^b leads to accept with the extracted PCP P_1 would also accept in a real execution, so we guess that q^b was the encrypted plaintext.

References

[ABOR00] Aiello, W., Bhatt, S., Ostrovsky, R., Rajagopalan, S.R.: Fast Verification of Any Remote Procedure Call: Short Witness-Indistinguishable One-Round Proofs for NP. In: Welzl, E., Montanari, U., Rolim, J.D.P. (eds.) ICALP 2000. LNCS, vol. 1853, pp. 463–474. Springer, Heidelberg (2000)

[AIK10] Applebaum, B., Ishai, Y., Kushilevitz, E.: From Secrecy to Soundness: Efficient Verification via Secure Computation. In: Abramsky, S., Gavoille, C., Kirchner, C., Meyer auf der Heide, F., Spirakis, P.G. (eds.) ICALP 2010, Part I. LNCS, vol. 6198, pp. 152–163. Springer, Heidelberg (2010)

[AS98] Arora, S., Safra, S.: Probabilistic checking of proofs: A new characterization of np. J. ACM 45(1), 70–122 (1998)

[BCCT11] Bitansky, N., Canetti, R., Chiesaz, A., Tromer, E.: From extractable collision resistance to succinct non- interactive arguments of knowledge, and back again. Cryptology ePrint Archive, Report 2011/443 (2011)

[BFLS91] Babai, L., Fortnow, L., Levin, L.A., Szegedy, M.: Checking computations in polylogarithmic time. In: STOC, pp. 21–31 (1991)

[BGV11] Benabbas, S., Gennaro, R., Vahlis, Y.: Verifiable Delegation of Computation over Large Datasets. In: Rogaway, P. (ed.) CRYPTO 2011. LNCS, vol. 6841, pp. 111–131. Springer, Heidelberg (2011)

[BP04] Bellare, M., Palacio, A.: Towards Plaintext-Aware Public-Key Encryption Without Random Oracles. In: Lee, P.J. (ed.) ASIACRYPT 2004. LNCS, vol. 3329, pp. 48–62. Springer, Heidelberg (2004)

[BR06] Bellare, M., Ristenpart, T.: Multi-Property-Preserving Hash Domain Extension and the EMD Transform. In: Lai, X., Chen, K. (eds.) ASIACRYPT 2006. LNCS, vol. 4284, pp. 299–314. Springer, Heidelberg (2006)

[BSS05]   Ben-Sasson, E., Sudan, M.: Simple pcps with poly-log rate and query complexity. In: STOC, pp. 266–275 (2005)

[BV11a]   Brakerski, Z., Vaikuntanathan, V.: Efficient fully homomorphic encryption from (standard) lwe. In: FOCS (2011)

[BV11b]   Brakerski, Z., Vaikuntanathan, V.: Fully Homomorphic Encryption from Ring-LWE and Security for Key Dependent Messages. In: Rogaway, P. (ed.) CRYPTO 2011. LNCS, vol. 6841, pp. 505–524. Springer, Heidelberg (2011)

[Can00]   Canetti, R.: Security and composition of multiparty cryptographic protocols. J. Cryptology 13(1), 143–202 (2000)

[Can01]   Canetti, R.: Universally composable security: A new paradigm for cryptographic protocols. In: FOCS, pp. 136–145 (2001)

[CKV10]   Chung, K.-M., Kalai, Y., Vadhan, S.: Improved Delegation of Computation Using Fully Homomorphic Encryption. In: Rabin, T. (ed.) CRYPTO 2010. LNCS, vol. 6223, pp. 483–501. Springer, Heidelberg (2010)

[CL08]   Di Crescenzo, G., Lipmaa, H.: Succinct NP Proofs from an Extractability Assumption. In: Beckmann, A., Dimitracopoulos, C., Löwe, B. (eds.) CiE 2008. LNCS, vol. 5028, pp. 175–185. Springer, Heidelberg (2008)

[Cle86]   Cleve, R.: Limits on the security of coin flips when half the processors are faulty (extended abstract). In: STOC, pp. 364–369 (1986)

[Dam91]   Damgård, I.: Towards Practical Public Key Systems Secure against Chosen Ciphertext Attacks. In: Feigenbaum, J. (ed.) CRYPTO 1991. LNCS, vol. 576, pp. 445–456. Springer, Heidelberg (1992)

[Den06]   Dent, A.W.: The Cramer-Shoup Encryption Scheme Is Plaintext Aware in the Standard Model. In: Vaudenay, S. (ed.) EUROCRYPT 2006. LNCS, vol. 4004, pp. 289–307. Springer, Heidelberg (2006)

[DFH11]   Damgård, I., Faust, S., Hazay, C.: Secure two-party computation with low communication. Cryptology ePrint Archive, Report 2011/508 (2011), http://eprint.iacr.org/

[Din07]   Dinur, I.: The pcp theorem by gap amplification. J. ACM 54(3), 12 (2007)

[DLN+04]   Dwork, C., Langberg, M., Naor, M., Nissim, K., Reingold, O.: Succinct np proofs and spooky interactions (2004), http://www.openu.ac.il/home/mikel/papers/spooky.ps

[Gen09]   Gentry, C.: Fully homomorphic encryption using ideal lattices. In: STOC, pp. 169–178 (2009)

[GGP10]   Gennaro, R., Gentry, C., Parno, B.: Non-interactive Verifiable Computing: Outsourcing Computation to Untrusted Workers. In: Rabin, T. (ed.) CRYPTO 2010. LNCS, vol. 6223, pp. 465–482. Springer, Heidelberg (2010)

[GKK+11]   Gordon, D., Katz, J., Kolesnikov, V., Malkin, T., Raykov, M., Vahlis, Y.: Secure computation with sublinear amortized work. Cryptology ePrint Archive, Report 2011/482 (2011)

[GL90]   Goldwasser, S., Levin, L.A.: Fair Computation of General Functions in Presence of Immoral Majority. In: Menezes, A., Vanstone, S.A. (eds.) CRYPTO 1990. LNCS, vol. 537, pp. 77–93. Springer, Heidelberg (1991)

[GLR11]   Goldwasser, S., Lin, H., Rubinstein, A.: Delegation of computation without rejection problem from designated verifier cs-proofs. Cryptology ePrint Archive, Report 2011/456 (2011)

[GMW87]   Goldreich, O., Micali, S., Wigderson, A.: How to play any mental game or a completeness theorem for protocols with honest majority. In: STOC, pp. 218–229 (1987)

[Gol04]   Goldreich, O.: Foundations of Cryptography: Basic Applications. Cambridge University Press (2004)

[Gro10]   Groth, J.: Short Pairing-Based Non-interactive Zero-Knowledge Arguments. In: Abe, M. (ed.) ASIACRYPT 2010. LNCS, vol. 6477, pp. 321–340. Springer, Heidelberg (2010)

[Gro11]   Groth, J.: Minimizing non-interactive zero-knowledge proofs using fully homomorphic encryption. Cryptology ePrint Archive, Report 2011/012 (2011)

[GW11]   Gentry, C., Wichs, D.: Separating succinct non-interactive arguments from all falsifiable assumptions. In: STOC, pp. 99–108 (2011)

[HT98]   Hada, S., Tanaka, T.: On the Existence of 3-Round Zero-Knowledge Protocols. In: Krawczyk, H. (ed.) CRYPTO 1998. LNCS, vol. 1462, pp. 408–423. Springer, Heidelberg (1998)

[IKO+11]   Ishai, Y., Kushilevitz, E., Ostrovsky, R., Prabhakaran, M., Sahai, A.: Efficient Non-interactive Secure Computation. In: Paterson, K.G. (ed.) EUROCRYPT 2011. LNCS, vol. 6632, pp. 406–425. Springer, Heidelberg (2011)

[IKP10]   Ishai, Y., Kushilevitz, E., Paskin, A.: Secure Multiparty Computation with Minimal Interaction. In: Rabin, T. (ed.) CRYPTO 2010. LNCS, vol. 6223, pp. 577–594. Springer, Heidelberg (2010)

[IPS08]   Ishai, Y., Prabhakaran, M., Sahai, A.: Founding Cryptography on Oblivious Transfer – Efficiently. In: Wagner, D. (ed.) CRYPTO 2008. LNCS, vol. 5157, pp. 572–591. Springer, Heidelberg (2008)

[IPS09]   Ishai, Y., Prabhakaran, M., Sahai, A.: Secure Arithmetic Computation with No Honest Majority. In: Reingold, O. (ed.) TCC 2009. LNCS, vol. 5444, pp. 294–314. Springer, Heidelberg (2009)

[Kil92]   Kilian, J.: A note on efficient zero-knowledge proofs and arguments (extended abstract). In: STOC, pp. 723–732 (1992)

[KK07]   Katz, J., Koo, C.-Y.: Round-Efficient Secure Computation in Point-to-Point Networks. In: Naor, M. (ed.) EUROCRYPT 2007. LNCS, vol. 4515, pp. 311–328. Springer, Heidelberg (2007)

[KO97]   Kushilevitz, E., Ostrovsky, R.: Replication is not needed: Single database, computationally-private information retrieval. In: FOCS, pp. 364–373 (1997)

[LP11]   Lindell, Y., Pinkas, B.: Secure Two-Party Computation via Cut-and-Choose Oblivious Transfer. In: Ishai, Y. (ed.) TCC 2011. LNCS, vol. 6597, pp. 329–346. Springer, Heidelberg (2011)

[Mer87]   Merkle, R.C.: A Digital Signature Based on a Conventional Encryption Function. In: Pomerance, C. (ed.) CRYPTO 1987. LNCS, vol. 293, pp. 369–378. Springer, Heidelberg (1988)

[Mic00]   Micali, S.: Computationally sound proofs. SIAM J. Comput. 30(4), 1253–1298 (2000)

[Nao03]   Naor, M.: On Cryptographic Assumptions and Challenges. In: Boneh, D. (ed.) CRYPTO 2003. LNCS, vol. 2729, pp. 96–109. Springer, Heidelberg (2003)

[NN01]   Naor, M., Nissim, K.: Communication preserving protocols for secure function evaluation. In: STOC, pp. 590–599 (2001)

[NO09]   Nielsen, J.B., Orlandi, C.: LEGO for Two-Party Secure Computation. In: Reingold, O. (ed.) TCC 2009. LNCS, vol. 5444, pp. 368–386. Springer, Heidelberg (2009)

[PS94]   Polishchuk, A., Spielman, D.A.: Nearly-linear size holographic proofs. In: STOC, pp. 194–203 (1994)

[PSSW09]   Pinkas, B., Schneider, T., Smart, N.P., Williams, S.C.: Secure Two-Party Computation is Practical. In: Matsui, M. (ed.) ASIACRYPT 2009. LNCS, vol. 5912, pp. 250–267. Springer, Heidelberg (2009)

[vDGHV10]   van Dijk, M., Gentry, C., Halevi, S., Vaikuntanathan, V.: Fully Homomorphic Encryption over the Integers. In: Gilbert, H. (ed.) EUROCRYPT 2010. LNCS, vol. 6110, pp. 24–43. Springer, Heidelberg (2010)

[Yao86]   Yao, A.C.-C.: How to generate and exchange secrets (extended abstract). In: FOCS, pp. 162–167 (1986)

Non-interactive CCA-Secure Threshold Cryptosystems with Adaptive Security: New Framework and Constructions

Benoît Libert[1,*] and Moti Yung[2]

[1] Université catholique de Louvain, ICTEAM Institute, Belgium
[2] Google Inc. and Columbia University, USA

Abstract. In threshold cryptography, private keys are divided into n shares, each one of which is given to a different server in order to avoid single points of failure. In the case of threshold public-key encryption, at least $t \leq n$ servers need to contribute to the decryption process. A threshold primitive is said *robust* if no coalition of t malicious servers can prevent remaining honest servers from successfully completing private key operations. So far, most practical non-interactive threshold cryptosystems, where no interactive conversation is required among decryption servers, were only proved secure against static corruptions. In the adaptive corruption scenario (where the adversary can corrupt servers at any time, based on its complete view), all existing robust threshold encryption schemes that also resist chosen-ciphertext attacks (CCA) till recently require interaction in the decryption phase. A specific method (in composite order groups) for getting rid of interaction was recently suggested, leaving the question of more generic frameworks and constructions with better security and better flexibility (*i.e.*, compatibility with distributed key generation).

This paper describes a general construction of adaptively secure robust non-interactive threshold cryptosystems with chosen-ciphertext security. We define the notion of *all-but-one perfectly sound* threshold hash proof systems that can be seen as (threshold) hash proof systems with publicly verifiable and simulation-sound proofs. We show that this notion generically implies threshold cryptosystems combining the aforementioned properties. Then, we provide efficient instantiations under well-studied assumptions in bilinear groups (e.g., in such groups of prime order). These instantiations have a tighter security proof and are indeed compatible with distributed key generation protocols.

Keywords: Threshold cryptography, adaptive corruptions, public-key encryption, chosen-ciphertext security, non-interactivity, robustness.

* This author acknowledges the Belgian Fund for Scientific Research (F.R.S.-F.N.R.S.) for his "Chargé de Recherches" fellowship and the BCRYPT Interuniversity Attraction Pole.

1   Introduction

Threshold cryptography [22,23,12] avoids single points of failure by splitting keys into $n > 1$ shares which are held by servers in such a way that at least t out of n servers should contribute to private key operations. In (t, n)-threshold cryptosystems, an adversary breaking into up to $t - 1$ servers should not jeopardize the security of the system.

Chosen-ciphertext security [45] (or IND-CCA for short) is widely recognized as the standard security notion for public-key encryption. Securely distributing the decryption procedure of CCA-secure public key schemes has proved to be a challenging task. As discussed in, e.g., [49,25], the difficulty is that decryption servers should return their partial decryption results, called "decryption shares", before knowing whether the incoming ciphertext is valid or not and partial decryptions of ill-formed ciphertexts may leak useful information to the adversary.

The first solution to this problem was put forth by Shoup and Gennaro [49] and it requires the random oracle model [5], notably to render valid ciphertexts publicly recognizable. In the standard model, Canetti and Goldwasser [15] gave a threshold variant of the Cramer-Shoup encryption scheme [16]. Unfortunately, their scheme requires interaction among decryption servers to obtain robustness (*i.e.*, ensure that no coalition of $t - 1$ malicious servers can prevent uncorrupted servers from successfully decrypting) as well as to render invalid ciphertexts harmless. The approach of [15] consists in randomizing the decryption process in such a way that partial decryptions of invalid ciphertexts are uniformly random and thus meaningless to the adversary. To avoid the need to jointly generate randomizers at each decryption, shareholders can alternatively store a large number (*i.e.*, proportional to the expected number of decryptions) of pre-shared secrets, which does not scale well. Cramer, Damgård and Ishai suggested [20] a method to generate randomizers without interaction but it is only efficient for a small number of servers.

Other threshold variants of Cramer-Shoup were suggested [1,40] and Abe notably showed [1] how to achieve optimal resilience (namely, guarantee robustness as long as the adversary corrupts a minority of $t < n/2$ servers) in the Canetti-Goldwasser system [15]. In the last decade, generic constructions of CCA-secure threshold cryptosystems with static security were put forth [24,52].

NON-INTERACTIVE SCHEMES. As an application of the Canetti-Halevi-Katz (CHK) paradigm [18], Boneh, Boyen and Halevi [8] came up with the first fully *non-interactive* robust CCA-secure threshold cryptosystem with a security proof in the standard model: in their scheme, decryption servers can generate their decryption shares *without* any communication with other servers. Their scheme takes advantage of bilinear maps to publicly check the validity of ciphertexts, which considerably simplifies the task of proving security in the threshold setting. In addition, the validity of decryption shares can be verified in the same way, which provides robustness. Similar applications of the CHK methodology to threshold cryptography were studied in [13,36].

Recently, Wee [52] defined a framework allowing to construct non-interactive threshold signatures and (chosen-ciphertext secure) threshold cryptosystems in a static corruption model. He left as an open problem the extension of his framework in the scenario of adaptive corruptions.

ADAPTIVE CORRUPTIONS. Most threshold systems (including [49,15,24,25,8]) have been analyzed in a static corruption model, where the adversary chooses which servers it wants to corrupt *before* the scheme is set up. Unfortunately, adaptive adversaries – who can choose whom to corrupt at any time, as a function of their entire view of the protocol execution – are known (see, e.g., [19]) to be strictly stronger. As discussed in [15], properly dealing with adaptive corruptions often comes at some substantial expense like a lower resilience. For example, the Canetti-Goldwasser system can be proved robust and adaptively secure when the threshold t is sufficiently small (typically, when $t = O(n^{1/2})$) but supporting an optimal number of faulty servers is clearly preferable.

Assuming reliable erasures, Canetti *et al.* [14] devised adaptively secure protocols for the distributed generation of discrete-logarithm-based keys and DSA signatures. Their techniques were re-used later on [3] in proactive [44] RSA signatures. In 1999, Frankel, MacKenzie and Yung [26,27] independently showed different methods to achieve adaptive security in the erasure-enabled setting.

Subsequently, Jarecki and Lysyanskaya [34] eliminated the need for erasures and gave an adaptively secure variant of the Canetti-Goldwasser threshold cryptosystem [15] which appeals to interactive zero-knowledge proofs but is designed to remain secure in concurrent environments. Unfortunately, their scheme requires a fair amount of interaction among decryption servers. Abe and Fehr [2] showed how to dispense with zero-knowledge proofs in the Jarecki-Lysyanskaya construction so as to prove it secure in (a variant of) the universal composability framework but without completely eliminating interaction from the decryption procedure. As in most threshold variants of Cramer-Shoup, hedging against invalid decryption queries requires an interactive (though off-line) randomness generation phase for each ciphertext, unless many pre-shared secrets are stored.

Recently, the authors of this paper showed [39] an adaptively secure variant of the Boneh-Boyen-Halevi construction [8] using groups of composite order and the dual system encryption approach [50,38] that was initially applied to identity-based encryption [48,10]. The scheme of [39] is based on a very specific use of the Lewko-Waters techniques [38], which limits its applicability to composite order groups and makes it hard to combine with existing adaptively secure distributed key generation techniques. Also, the concrete security of this initial scheme is not optimal as its security reduction is related to the number of decryption queries made by the adversary. To solve these problems, we need a new approach and different methods to analyze the security of schemes.

OUR CONTRIBUTION. Motivated by an open question raised by Wee [52] and the limitations of [39], we define a general framework for constructing robust, adaptively secure and fully non-interactive threshold cryptosystems with chosen-ciphertext security. Our goal is to have simple and practical client/server protocols, as advocated in [49][Section 2.5], and even avoid the off-line interactive randomness generation stage which is usually needed in threshold versions of Cramer-Shoup.

To this end, we also appeal to hash proof systems (HPS) [17] and take advantage of the property that, in security reductions using the techniques of [16,17], the simulator knows the private keys, which is convenient to answer adaptive corruption queries. Indeed, when the reduction has to reveal the internal state of dynamically-corrupted servers, it is not bound to a particular set of available shares since it knows them all. At the same time, we depart from [15] in that the validity of ciphertexts is made publicly verifiable – which eliminates the need to randomize the decryption operation – using non-interactive proofs satisfying some form of simulation-soundness [46]: in the security reduction, the simulator should be able to generate a proof for a possibly false statement but the adversary should be unable to do it on its own, even after having seen a fake proof.

To this end, we define the notion of *all-but-one perfectly sound threshold hash proof systems* that can be seen as (threshold) hash proof systems [17] with *publicly* verifiable proofs (as opposed to designed-verifier proofs used in traditional HPS [17]). More precisely, each proof is associated with a tag, in the same way as ciphertexts are associated with tags in [41,36]. Real public parameters are indistinguishable from alternative parameters that are generated in an *all-but-one* mode, which is only used in the security analysis. In the latter mode, non-interactive proofs are perfectly sound on all tags, except for a single specific tag where some trapdoor makes it possible to simulate proofs for false statements. While our primitive bears similarities with Wee's extractable hash proof systems [51,52] (where hash proof systems are also associated with tags), it is different in that no extractability property is required and proofs are always used as proofs of membership.

Using all-but-one perfectly sound threshold hash proof systems, we generically construct adaptively secure robust non-interactive threshold cryptosystems with optimal resilience. An additional benefit of this approach is to provide a better concrete security as the security proof requires a constant number of game transitions whereas, in [39], the number of games is proportional to the number of decryption queries.

Then, we show three concrete instantiations using number theoretic assumptions in bilinear groups. The first one uses groups whose order is a product of two primes (whereas three primes are needed in [39]). Our second and third schemes rely on the Groth-Sahai proof systems [31] in their instantiations based on the Decision Linear [9] and symmetric eXternal Diffie-Hellman assumptions [47]. The latter two constructions operate over bilinear groups of prime order, which allows for a significantly better efficiency than composite order groups (as discussed in [28]) and makes them much easier to combine with known adaptively secure discrete-log-based distributed key generation protocols. For example, in the erasure-free setting, the protocols of [34,2] can be used so as to eliminate the need for a trusted dealer at the same time as the reliance on reliable erasures.

2 Background and Definitions

2.1 Definitions for Threshold Public Key Encryption

A non-interactive (t, n)-threshold encryption scheme is a set of algorithms with
these specifications.

Setup(λ, t, n): given a security parameter λ and integers $t, n \in \mathsf{poly}(\lambda)$ (with
   $1 \leq t \leq n$) denoting the number of decryption servers n and the threshold t,
   this algorithm outputs $(PK, \mathbf{VK}, \mathbf{SK})$, where PK is the public key, $\mathbf{SK} =$
   $(SK_1, \ldots, SK_n)$ is a vector of private-key shares and $\mathbf{VK} = (VK_1, \ldots, VK_n)$
   is a vector of verification keys. Decryption server i is given the private key
   share (i, SK_i). For each $i \in \{1, \ldots, n\}$, the verification key VK_i will be used
   to check the validity of decryption shares generated using SK_i.

Encrypt(PK, M): is a randomized algorithm that, given a public key PK and
   a plaintext M, outputs a ciphertext C.

Ciphertext-Verify(PK, C): takes as input a public key PK and a ciphertext
   C. It outputs 1 if C is deemed valid w.r.t. PK and 0 otherwise.

Share-Decrypt(PK, i, SK_i, C): on input of a public key PK, a ciphertext C
   and a private-key share (i, SK_i), this (possibly randomized) algorithm outputs a special symbol $(i, \perp)$ if **Ciphertext-Verify**$(PK, C) = 0$. Otherwise,
   it outputs a decryption share $\mu_i = (i, \hat{\mu}_i)$.

Share-Verify(PK, VK_i, C, μ_i): takes in PK, the verification key VK_i, a ciphertext C and a purported decryption share $\mu_i = (i, \hat{\mu}_i)$. It outputs either
   1 or 0. In the former case, μ_i is said to be a *valid* decryption share. We adopt
   the convention that $(i, \perp)$ is an invalid decryption share.

Combine$(PK, \mathbf{VK}, C, \{\mu_i\}_{i \in S})$: given $PK, \mathbf{VK}, C$ and a subset $S \subset \{1, \ldots, n\}$
   of size $t = |S|$ with decryption shares $\{\mu_i\}_{i \in S}$, this algorithm outputs either
   a plaintext M or $\perp$ if the set contains invalid decryption shares.

CHOSEN-CIPHERTEXT SECURITY. We use a game-based definition of chosen-
ciphertext security which is akin to the one of [49,8] with the difference that the
adversary can adaptively decide which parties it wants to corrupt.

Definition 1. *A non-interactive (t, n)-Threshold Public Key Encryption scheme
is secure against chosen-ciphertext attacks (or IND-CCA2 secure) and adaptive
corruptions if no PPT adversary has non-negligible advantage in this game:*

1. *The challenger runs* **Setup**(λ, t, n) *to obtain PK, a vector of private key
   shares $\mathbf{SK} = (SK_1, \ldots, SK_n)$ and verification keys $\mathbf{VK} = (VK_1, \ldots, VK_n)$.
   It gives PK and $\mathbf{VK}$ to the adversary $\mathcal{A}$ and keeps $\mathbf{SK}$ to itself.*

2 *The adversary $\mathcal{A}$ adaptively makes the following kinds of queries:*
   - *Corruption query: $\mathcal{A}$ chooses $i \in \{1, \ldots, n\}$ and obtains SK_i. No more
     than $t - 1$ private key shares can be obtained by $\mathcal{A}$ in the whole game.*
   - *Decryption query: $\mathcal{A}$ chooses an index $i \in \{1, \ldots, n\}$ and a ciphertext C.
     The challenger replies with $\mu_i = $ **Share-Decrypt**(PK, i, SK_i, C).*

3. *The adversary $\mathcal{A}$ chooses two equal-length messages M_0, M_1 and obtains
   $C^\star = $ **Encrypt**(PK, M_β) for some random bit $\beta \xleftarrow{R} \{0, 1\}$.*

4. $\mathcal{A}$ *makes further queries as in step 2 but is not allowed to make decryption queries on* $C^\star$.
5. $\mathcal{A}$ *outputs a bit* β' *and is deemed successful if* $\beta' = \beta$. *As usual,* $\mathcal{A}$*'s advantage is measured as the distance* $\mathbf{Adv}(\mathcal{A}) = |\Pr[\beta' = \beta] - \frac{1}{2}|$.

CONSISTENCY. A (t, n)-Threshold Encryption scheme provides decryption consistency if no PPT adversary has non-negligible advantage in a three-stage game where stages 1 and 2 are identical to those of Definition 1 with the difference that the adversary $\mathcal{A}$ is allowed to obtain *all* private key shares (alternatively, $\mathcal{A}$ can directly obtain **SK** at the beginning of the game). In stage 3, $\mathcal{A}$ outputs a ciphertext C and two t-sets of decryption shares $\Gamma = \{\mu_1, \ldots, \mu_t\}$ and $\Gamma' = \{\mu'_1, \ldots, \mu'_t\}$. The adversary $\mathcal{A}$ is declared successful if

1. **Ciphertext-Verify**$(PK, C) = 1$.
2. Γ and Γ' only consist of valid decryption shares.
3. **Combine**$(PK, \mathbf{VK}, C, \Gamma) \neq$ **Combine**$(PK, \mathbf{VK}, C, \Gamma')$.

We note that condition 1 prevents an adversary from trivially winning by outputting an invalid ciphertext, for which distinct sets of key shares may give different results. This definition of consistency is identical to the one of [49,8] with the difference that $\mathcal{A}$ can adaptively corrupt servers.

2.2 Hardness Assumptions in Composite Order Groups

In one occasion, we appeal to groups $(\mathbb{G}, \mathbb{G}_T)$ of composite order $N = p_1 p_2$, where p_1 and p_2 are primes, with a bilinear map $e : \mathbb{G} \times \mathbb{G} \to \mathbb{G}_T$ (*i.e.*, for which $e(g^a, h^b) = e(g, h)^{ab}$ for any $g, h \in \mathbb{G}$ and $a, b \in \mathbb{Z}_N$). In the notations hereafter, for each $i \in \{1, 2\}$, $\mathbb{G}_{p_i}$ stands for the subgroup of order p_i in $\mathbb{G}$.

Definition 2 ([11]). *In a group* $\mathbb{G}$ *of composite order* N, *the* **Subgroup Decision** *(SD) problem is given* $(g \in \mathbb{G}_{p_1}, h \in \mathbb{G})$ *and* η, *to decide whether* $\eta \in \mathbb{G}_{p_1}$ *or* $\eta \in_R \mathbb{G}$. *The* **Subgroup Decision** *assumption states that, for any PPT distinguisher* $\mathcal{D}$, *the SD problem is infeasible.*

2.3 Assumptions in Prime Order Groups

We also use bilinear maps $e : \mathbb{G} \times \hat{\mathbb{G}} \to \mathbb{G}_T$ over groups of prime order p. We will work in symmetric pairing configurations, where $\mathbb{G} = \hat{\mathbb{G}}$, and sometimes in asymmetric configurations, where $\mathbb{G} \neq \hat{\mathbb{G}}$.

In the symmetric setting $(\mathbb{G}, \mathbb{G}_T)$, we rely on the following assumption.

Definition 3 ([9]). *In a group* $\mathbb{G}$ *of prime order* p, *the* **Decision Linear Problem** *(DLIN) is to distinguish the distributions* $(g, g^a, g^b, g^{ac}, g^{bd}, g^{c+d})$ *and* $(g, g^a, g^b, g^{ac}, g^{bd}, g^z)$, *with* $a, b, c, d, z \xleftarrow{R} \mathbb{Z}_p$. *The* **Decision Linear Assumption** *is the intractability of DLIN for any PPT distinguisher* $\mathcal{D}$.

The problem amounts to deciding if vectors $\vec{g_1} = (g^a, 1, g)$, $\vec{g_2} = (1, g^b, g)$ and $\vec{g_3} = (g^{ac}, g^{bd}, g^\delta)$ are linearly dependent (i.e., if $\delta = c + d$) or not.

In *asymmetric* bilinear groups $(\mathbb{G}, \hat{\mathbb{G}}, \mathbb{G}_T)$, we assume the hardness of the Decision Diffie-Hellman (DDH) problem in $\mathbb{G}$ and $\hat{\mathbb{G}}$. This implies the unavailability of efficiently computable isomorphisms between $\hat{\mathbb{G}}$ and $\mathbb{G}$. This assumption is called **Symmetric eXternal Diffie-Hellman** (SXDH) assumption. Given vectors $\vec{u_1} = (g, h)$, $\vec{u_2} = (g^a, h^c)$ in $\mathbb{G}^2$ or $\hat{\mathbb{G}}^2$, the SXDH assumption asserts the infeasibility of deciding whether $\vec{u_1}$ and $\vec{u_2}$ are linearly dependent (i.e., whether $a = c \bmod p$).

3 All-But-One Perfectly Sound Threshold Hash Proof Systems

Let $\mathcal{C}$, $\mathcal{K}$ and $\mathcal{K}'$ be sets and let $\mathcal{V} \subset \mathcal{C}$ be a subset. Let also $\mathcal{R}$ be a space where random coins can be chosen. We mandate that $\mathcal{V}$, $\mathcal{K}$, $\mathcal{K}'$ and $\mathcal{R}$ be of exponential size in λ, where $\lambda \in \mathbb{N}$ is a security parameter. In addition, $\mathcal{C}$, $\mathcal{V}$ and $\mathcal{C} \backslash \mathcal{V}$ should be efficiently samplable and we also require the set $\mathcal{K}$ to form a group for some binary operation, which is denoted by $\odot$ hereafter.

An *all-but-one perfectly sound threshold hash proof system* for $(\mathcal{C}, \mathcal{V}, \mathcal{K}, \mathcal{K}', \mathcal{R})$ is a tuple of algorithms (SetupSound, SetupABO, Sample, Prove, SimProve, Verify, PubEval, SharePrivEval, ShareEvalVerify, Combine) of efficient algorithms with the following specifications.

SetupSound(λ, t, n): given a security parameter $\lambda \in \mathbb{N}$ and integers $t, n \in \mathsf{poly}(\lambda)$, this algorithm outputs a public key pk, a vector of private key shares $(\mathsf{sk}_1, \ldots, \mathsf{sk}_n)$ and verification keys $(\mathsf{vk}_1, \ldots, \mathsf{vk}_n)$.

SetupABO$(\lambda, t, n, \mathsf{tag}^\star)$: takes as input a security parameter $\lambda \in \mathbb{N}$, integers $t, n \in \mathsf{poly}(\lambda)$ and a tag $\mathsf{tag}^\star$. It outputs a public key pk, private key shares $(\mathsf{sk}_1, \ldots, \mathsf{sk}_n)$, the corresponding verification keys $(\mathsf{vk}_1, \ldots, \mathsf{vk}_n)$ as well as a simulation trapdoor τ. It is important that τ be independent of $\{\mathsf{sk}_i\}_{i=1}^n$.

Sample(pk): is a probabilistic algorithm that takes as input a public key pk. It draws random coins $r \xleftarrow{R} \mathcal{R}$ and outputs an element $\Phi \in \mathcal{V}$ along with the random coins r that will serve as a witness explaining Φ as an element of $\mathcal{V}$.

Prove$(\mathsf{pk}, \mathsf{tag}, r, \Phi)$: takes in a public key pk, a tag tag, an element $\Phi \in \mathcal{V}$ and the random coins $r \in \mathcal{R}$ that were used to sample Φ. It generates a non-interactive proof $\pi_{\mathcal{V}}$ that $\Phi \in \mathcal{V}$.

SimProve$(\mathsf{pk}, \tau, \mathsf{tag}, \Phi)$: takes as input a public key pk and a simulation trapdoor τ produced by SetupABO$(\lambda, t, n, \mathsf{tag}^\star)$, a tag tag and an element $\Phi \in \mathcal{C}$. If $\mathsf{tag} \neq \mathsf{tag}^\star$, the algorithm outputs $\bot$. If $\mathsf{tag} = \mathsf{tag}^\star$, the algorithm produces a simulated NIZK proof $\pi_{\mathcal{V}}$ that $\Phi \in \mathcal{V}$.

Verify$(\mathsf{pk}, \mathsf{tag}, \Phi, \pi_{\mathcal{V}})$: takes as input a public key pk, a tag tag, an element $\Phi \in \mathcal{C}$ and a purported proof $\pi_{\mathcal{V}}$. It outputs 1 if and only if $\pi_{\mathcal{V}}$ is deemed as a valid proof that $\Phi \in \mathcal{V} \subset \mathcal{C}$.

PubEval(pk, r, Φ): takes as input a public key pk, an element $\Phi \in \mathcal{V}$ and the random coins $r \in_R \mathcal{R}$ such that $(r, \Phi) \leftarrow \mathsf{Sample}(\mathsf{pk})$. It outputs a value $K \in \mathcal{K}$, which is called *public evaluation* of Φ.

SharePrivEval(pk, sk_i, Φ): is a deterministic algorithm that takes in a public key pk, a private key share sk_i and an element $\Phi \in \mathcal{C}$. It outputs a value $K_i \in \mathcal{K}'$, called *private evaluation share* and a proof π_{K_i} that K_i was evaluated correctly.

ShareEvalVerify(pk, vk_i, Φ, K_i, π_{K_i}): given a public key pk, a verification key vk_i, an element $\Phi \in \mathcal{C}$, a private evaluation share $K_i \in \mathcal{K}'$ and its proof π_{K_i}, this algorithm outputs 1 if π_{K_i} is considered as a valid proof of the correct evaluation of K_i. Otherwise, it outputs 0.

Combine(pk, Φ, $\{(K_i, \pi_{K_i})\}_{i \in S}$): takes as input a public key pk, an element $\Phi \in \mathcal{C}$ and a set of t pairs $\{(K_i, \pi_{K_i})\}_{i \in S}$, where $S \subset \{1, \ldots, n\}$, each one of which consists of a private evaluation share $K_i \in \mathcal{K}'$ and its proof π_{K_i}. If ShareEvalVerify(pk, vk_i, Φ, K_i, π_{K_i}) = 0 for some $i \in S$, it outputs $\bot$. Otherwise, it outputs a value $K \in \mathcal{K}$.

We also define this algorithm which is implied by the above ones but will be convenient to use.

PrivEval(pk, $\{sk_i\}_{i \in S}$, Φ): given a public key pk, a set of private key shares $\{sk_i\}_{i \in S}$ where S is an arbitrary t-subset of $\{1, \ldots, n\}$, and an element $\Phi \in \mathcal{C}$, this algorithm outputs the result of Combine(pk, Φ, $\{(K_i, \pi_{K_i})\}_{i \in S}$) where $(K_i, \pi_{K_i}) \leftarrow$ SharePrivEval(pk, sk_i, Φ) for each $i \in S$.

The following properties are required from these algorithms and the sets $(\mathcal{C}, \mathcal{V}, \mathcal{K}, \mathcal{K}', \mathcal{R})$.

(SETUP INDISTINGUISHABILITY): For any integers (λ, t, n) with $1 \leq t \leq n$ and any tag tag^*, the output of SetupSound(λ, t, n) is computationally indistinguishable from the outputs (pk, $\{sk_i\}_{i=1}^n$, $\{vk_i\}_{i=1}^n$) of SetupABO(λ, t, n, tag^*).

(CORRECTNESS AND PUBLIC EVALUABILITY ON $\mathcal{V}$): For any (pk, $\{sk_i\}_{i=1}^n$, $\{vk_i\}_{i=1}^n$) returned by SetupSound or SetupABO, if $(r, \Phi) \xleftarrow{R}$ Sample(pk) (and thus $\Phi \in \mathcal{V}$), it holds that:

1. For any $i \in \{1, \ldots, n\}$, if $(K_i, \pi_{K_i}) \leftarrow$ SharePrivEval(pk, sk_i, Φ), the private evaluation share $K_i \in \mathcal{K}'$ is *uniquely* determined by (pk, vk_i) and Φ. Moreover, the proof π_{K_i} satisfies ShareEvalVerify(pk, vk_i, Φ, K_i, π_{K_i}) = 1.

2. For any t-subset $S \subset \{1, \ldots, n\}$, combining the corresponding private evaluation shares allows recomputing the public evaluation of Φ: namely, PubEval(pk, r, Φ) = PrivEval(pk, $\{sk_i\}_{i \in S}$, Φ).

(UNIVERSALITY): For any (pk, $\{sk_i\}_{i=1}^n$, $\{vk_i\}_{i=1}^n$) produced by SetupSound or SetupABO and any $\Phi \in \mathcal{C} \backslash \mathcal{V}$, for any subset $\bar{S} \subset \{1, \ldots, n\}$ of size $|\bar{S}| = t - 1$, the statistical distance

$$\Delta[(\text{pk}, \{vk_i\}_{i=1}^n, \{sk_i\}_{i \in \bar{S}}, \Phi, \text{PrivEval}(\text{pk}, \{sk_i\}_{i=1}^t, \Phi)),$$
$$(\text{pk}, \{vk_i\}_{i=1}^n, \{sk_i\}_{i \in \bar{S}}, \Phi, K)],$$

where $K \xleftarrow{R} \mathcal{K}$, should be negligible.

(ALL-BUT-ONE SOUNDNESS): For all integers (λ, t, n) such that $1 \leq t \leq n$, any tag tag* and any outputs $(\mathsf{pk}, \{\mathsf{sk}_i\}_{i=1}^n, \{\mathsf{vk}_i\}_{i=1}^n, \tau)$ of SetupABO$(\lambda, t, n, \mathsf{tag}^\star)$, these conditions are satisfied.

1. For any tag $\neq$ tag*, proofs are always perfectly sound. Namely, if a proof $\pi_\mathcal{V}$ satisfies Verify$(\mathsf{pk}, \mathsf{tag}, \varPhi, \pi_\mathcal{V}) = 1$ for some $\varPhi \in \mathcal{C}$, then it necessarily holds that $\varPhi \in \mathcal{V}$.

2. For any $\varPhi \in \mathcal{C}$, the trapdoor τ allows generating as simulated a proof $\pi_\mathcal{V} \leftarrow$ SimProve$(\mathsf{pk}, \tau, \mathsf{tag}^\star, \varPhi)$ such that Verify$(\mathsf{pk}, \mathsf{tag}^\star, \varPhi, \pi_\mathcal{V}) = 1$ (note that $\pi_\mathcal{V}$ is a proof for a false statement if $\varPhi \in \mathcal{C} \backslash \mathcal{V}$). Moreover, if $\varPhi \in \mathcal{V}$, the simulated proof $\pi_\mathcal{V}$ should be perfectly indistinguishable from a real proof (*i.e.*, that would be generated by Prove using a witness $r \in \mathcal{R}$ of the fact that $\varPhi \in \mathcal{V}$).

(SIMULATABILITY OF SHARE PROOFS): For all (λ, t, n) with $1 \leq t \leq n$, any tag tag*, any outputs $(\mathsf{pk}, \{\mathsf{sk}_i\}_{i=1}^n, \{\mathsf{vk}_i\}_{i=1}^n, \tau)$ of SetupABO$(\lambda, t, n, \mathsf{tag}^\star)$ and any $\varPhi \in \mathcal{C}$, the proofs π_{K_i} obtained as $(K_i, \pi_{K_i}) \leftarrow$ SharePrivEval$(\mathsf{pk}, \mathsf{sk}_i, \varPhi)$ should be simulatable using the trapdoor τ instead of $\{\mathsf{sk}_i\}_{i=1}^n$. Using τ and $(\mathsf{pk}, \{\mathsf{vk}_i\}_{i=1}^n, \varPhi)$, an efficient algorithm $\mathcal{S}$ should be able to produce simulated proofs π_{K_i} that are perfectly indistinguishable from real proofs.

(CONSISTENCY): For all (λ, t, n) with $1 \leq t \leq n$, any output $(\mathsf{pk}, \{(\mathsf{vk}_i, \mathsf{sk}_i)\}_{i=1}^n)$ of SetupSound(λ, t, n), given $(\mathsf{pk}, \{(\mathsf{vk}_i, \mathsf{sk}_i)\}_{i=1}^n)$, it should be computationally infeasible to come up with a triple $(\mathsf{tag}, \varPhi, \pi_\mathcal{V})$ as well as two distinct t-sets $\Gamma = \{(K_{i_1}, \pi_{K_{i_1}}), \ldots, (K_{i_t}, \pi_{K_{i_t}})\}$ and $\Gamma' = \{(K'_{j_1}, \pi'_{K_{j_1}}), \ldots, (K'_{j_t}, \pi'_{K_{j_t}})\}$, with $i_k, j_k \in \{1, \ldots, n\}$ for each $k \in \{1, \ldots, t\}$, such that the following three conditions are satisfied: (i) Verify$(\mathsf{pk}, \mathsf{tag}, \varPhi, \pi_\mathcal{V}) = 1$; (ii) for each $k \in \{1, \ldots, t\}$, it holds that ShareEvalVerify$(\mathsf{pk}, \mathsf{vk}_{i_k}, \varPhi, K_{i_k}, \pi_{K_{i_k}}) = 1$ and ShareEvalVerify$(\mathsf{pk}, \mathsf{vk}_{j_k}, \varPhi, K'_{j_k}, \pi'_{K_{j_k}}) = 1$; (iii) Γ and Γ' result in distinct combinations: Combine$(\mathsf{pk}, \varPhi, \Gamma) \neq$ Combine$(\mathsf{pk}, \varPhi, \Gamma')$.

(SUBSET MEMBERSHIP HARDNESS): membership in $\mathcal{C}$ should be easy to check but membership in $\mathcal{V}$ should not. Moreover, this should hold *even* if τ is given. Namely, for all integers (λ, t, n) such that $1 \leq t \leq n$, any tag tag* and any outputs $(\mathsf{pk}, \{\mathsf{sk}_i\}_{i=1}^n, \{\mathsf{vk}_i\}_{i=1}^n, \tau)$ of SetupABO$(\lambda, t, n, \mathsf{tag}^\star)$, for any PPT distinguisher $\mathcal{D}$, it must hold that:

$$\mathbf{Adv}^{\mathrm{SM}}(\mathcal{D}) = |\Pr[\mathcal{D}(\mathcal{C}, \mathcal{V}, C_1, \tau) = 1 | C_1 \xleftarrow{R} \mathcal{C} \backslash \mathcal{V}]$$
$$- \Pr[\mathcal{D}(\mathcal{C}, \mathcal{V}, C_0, \tau) = 1 | C_0 \xleftarrow{R} \mathcal{V}]| \in \mathsf{negl}(\lambda).$$

In the definition of the subset membership hardness property, the trapdoor τ should not carry any side information helping the distinguisher. For this reason, the latter receives τ as part of its input.

4   Adaptively Secure Robust Non-interactive CCA2-Secure Threshold Cryptosystems from All-But-One Perfectly Sound Threshold Hash Proof Systems

Let us assume sets $(\mathcal{C}, \mathcal{V}, \mathcal{K}, \mathcal{K}', \mathcal{R})$ for which we have an all-but-one perfectly sound threshold hash proof system $\Pi^{\mathsf{ABO\text{-}THPS}} = (\mathsf{SetupSound}, \mathsf{SetupABO}, \mathsf{Sample}, \mathsf{Prove},$ $\mathsf{SimProve}, \mathsf{Verify}, \mathsf{PubEval}, \mathsf{SharePrivEval}, \mathsf{ShareEvalVerify}, \mathsf{Combine})$ that satisfies the conditions specified in Section 3. We assume that messages are in $\mathcal{K}$. The generic construction of CCA2-secure threshold cryptosystem goes as follows.

Keygen(λ, t, n)**:** given integers $\lambda, t, n \in \mathbb{N}$, choose a one-time signature scheme $\Sigma = (\mathsf{Gen}, \mathsf{Sig}, \mathsf{Ver})$, generate $(\mathsf{pk}, \{\mathsf{sk}_i\}_{i=1}^n, \{\mathsf{vk}_i\}_{i=1}^n) \leftarrow \mathsf{SetupSound}(\lambda, t, n)$ and output $(PK, \mathbf{SK}, \mathbf{VK})$, where the vectors of private key shares and verification keys are defined as $\mathbf{SK} = (\mathsf{sk}_1, \ldots, \mathsf{sk}_n)$ and $\mathbf{VK} = (\mathsf{vk}_1, \ldots, \mathsf{vk}_n)$, respectively. The public key is $PK = (\mathsf{pk}, \Sigma)$.

Encrypt(M, PK)**:** to encrypt a message $M \in \mathcal{K}$ using $PK = (\mathsf{pk}, \Sigma)$,

1. Generate a one-time signature key pair $(\mathsf{SSK}, \mathsf{SVK}) \leftarrow \Sigma.\mathsf{Gen}(\lambda)$.
2. Choose $r \overset{R}{\leftarrow} \mathcal{R}$, compute $(r, \Phi) \leftarrow \mathsf{Sample}(\mathsf{pk}, r)$ and blind the message as $C_0 = M \odot \mathsf{PubEval}(\mathsf{pk}, r, \Phi)$.
3. Generate a proof $\pi_{\mathcal{V}} \leftarrow \mathsf{Prove}(\mathsf{pk}, \mathsf{SVK}, r, \Phi)$ that $\Phi \in \mathcal{V}$ with respect to the tag SVK.
4. Output $C = (\mathsf{SVK}, C_0, \Phi, \pi_{\mathcal{V}}, \sigma)$, where $\sigma = \Sigma.\mathsf{Sig}(\mathsf{SSK}, (C_0, \Phi, \pi_{\mathcal{V}}))$.

Ciphertext-Verify(PK, C)**:** parse the ciphertext C as $C = (\mathsf{SVK}, C_0, \Phi, \pi_{\mathcal{V}}, \sigma)$ and PK as (pk, Σ). Return 1 if it holds that $\Sigma.\mathsf{Ver}(\mathsf{SVK}, (C_0, \Phi, \pi_{\mathcal{V}}), \sigma) = 1$ and $\mathsf{Verify}(\mathsf{pk}, \mathsf{SVK}, \Phi, \pi_{\mathcal{V}}) = 1$. Otherwise, return 0.

Share-Decrypt(SK_i, C)**:** given $SK_i = \mathsf{sk}_i$ and $C = (\mathsf{SVK}, C_0, \Phi, \pi_{\mathcal{V}}, \sigma)$, return $(i, \perp)$ if it turns out that **Ciphertext-Verify**$(PK, C) = 0$. Otherwise, compute a pair $(K_i, \pi_{K_i}) \leftarrow \mathsf{SharePrivEval}(\mathsf{pk}, \mathsf{sk}_i, \Phi)$ and return $\mu_i = (i, \hat{\mu}_i)$ where $\hat{\mu}_i = (K_i, \pi_{K_i})$.

Share-Verify$(PK, VK_i, C, (i, \hat{\mu}_i))$**:** parse C as $(\mathsf{SVK}, C_0, \Phi, \pi_{\mathcal{V}}, \sigma)$. If $\hat{\mu}_i = \perp$ or if $\hat{\mu}_i$ cannot be properly parsed as a pair (K_i, π_{K_i}), return 0. Otherwise, return 1 if $\mathsf{ShareEvalVerify}(\mathsf{pk}, \mathsf{vk}_i, \Phi, K_i, \pi_{K_i}) = 1$ and 0 otherwise.

Combine$(PK, \mathbf{VK}, C, \{(i, \hat{\mu}_i)\}_{i \in S})$**:** parse C as $(\mathsf{SVK}, C_0, \Phi, \pi_{\mathcal{V}}, \sigma)$. Return $\perp$ if there exists $i \in S$ such that **Share-Verify**$(PK, C, (i, \hat{\mu}_i)) = 0$ or if **Ciphertext-Verify**$(PK, C) = 0$. Otherwise, compute the combined value $K = \mathsf{Combine}(\mathsf{pk}, \Phi, \{(K_i, \pi_{K_i})\}_{i \in S}) \in \mathcal{K}$, which unveils $M = C_0 \odot K^{-1}$.

We observe that there is no need to bind the one-time verification key SVK to the ciphertext components $(C_0, \Phi, \pi_{\mathcal{V}})$ in any other way than by using it as a tag to compute the non-interactive proof $\pi_{\mathcal{V}}$. Indeed, if the adversary attempts to re-use parts $(C_0^\star, \Phi^\star, \pi_{\mathcal{V}}^\star)$ of the challenge ciphertext and simply replaces the one-time verification key $\mathsf{SVK}^\star$ by a verification key SVK of its own, it will be forced to compute a proof π_V that correspond to the same $\Phi^\star$ as in the challenge phase but under the *new* tag SVK. Our security proof shows that this is infeasible as long as $\Pi^{\text{ABO-THPS}}$ satisfies the properties of setup indistinguishability and all-but-one soundness.

The consistency property of the threshold encryption scheme is trivially implied by that of $\Pi^{\text{ABO-THPS}}$ and we focus on proving its IND-CCA security. In the threshold setting, adaptive security is achieved by taking advantage of the fact that, in security reductions using hash proof systems, the simulator typically knows the private key and can thus answer adaptive queries at will. At the same time, invalid ciphertexts are harmless as they are made publicly recognizable due to the use of non-interactive proofs of validity: as long as these proofs are perfectly sound in all decryption queries, the simulator is guaranteed not to leak too much information about the particular private key it is using.

The main problem to solve is thus to make sure that *only* the simulator can simulate a fake proof in the challenge phase and this is where the all-but-one soundness property is handy.

Theorem 1. *The above threshold cryptosystem is IND-CCA secure against adaptive corruptions assuming that: (i) $\Pi^{\text{ABO-THPS}}$ is an all-but-one perfectly sound hash proof system; (ii) Σ is a strongly unforgeable one-time signature.*

Proof. The proof is given in the full version of the paper.   □

5   Instantiations

5.1   Construction in Groups of Composite Order $N = p_1 p_2$

The construction relies on a hash proof system in a group $\mathbb{G}$ of composite order $N = p_1 p_2$ and it is conceptually close to the one in [33] (notably because it builds on a $\log p_2$-entropic hash proof system, as defined in [37]). The public key includes group elements $(g, X = g^x)$ in the subgroup $\mathbb{G}_{p_1}$ of order p_1 and the sets $\mathcal{C}$ and $\mathcal{V}$ are defined to be $\mathbb{G}$ and $\mathbb{G}_{p_1}$, respectively. The sampling algorithm returns $\Phi = g^r \in \mathbb{G}_{p_1}$ for a random exponent $r \stackrel{R}{\leftarrow} \mathbb{Z}_N$, which allows publicly evaluating $H(X^r) = H(\Phi^x)$ using a pairwise independent hash function $H : \mathbb{G} \to \{0,1\}^\ell$. Since the public key is independent of $x \bmod p_2$, for any $\Phi \in \mathbb{G}$ that has a non-trivial component of order p_2, the "hash value" Φ^x has exactly $\log p_2$ bits of min-entropy and the leftover hash lemma implies that $H(\Phi^x)$ is statistically close to the uniform distribution in $\{0,1\}^\ell$ when ℓ is sufficiently small.

In order to turn the scheme into an all-but-one perfectly sound threshold HPS, we need a mechanism that proves membership in the subgroup $\mathbb{G}_{p_1}$ and guarantees the perfect soundness of proofs of membership for all tags $\text{tag} \in \mathbb{Z}_N$ such that $\text{tag} \neq \text{tag}^\star$. To this end, we use additional public parameters $(u, v) \in \mathbb{G}^2$ and a tag-dependent group element $u^{\text{tag}} \cdot v$ will serve as a common reference string to generate a non-interactive proof that $\Phi \in \mathbb{G}_{p_1}$. Membership in $\mathbb{G}_{p_1}$ can be non-interactively proved using a technique that can be traced back to [30]. The proof consists of a group element $\pi_{\text{SD}} \in \mathbb{G}$ satisfying the equality

$e(\Phi, u^{\mathsf{tag}} \cdot v) = e(g, \pi_{\mathsf{SD}})$, which ensures that $\Phi \in \mathbb{G}_{p_1}$ as long as $u^{\mathsf{tag}} \cdot v$ has a $\mathbb{G}_{p_2}$ component. In the public parameters produced by SetupABO, the value $u^{\mathsf{tag}} \cdot v$ thus has to be in $\mathbb{G} \backslash \mathbb{G}_{p_1}$ for any $\mathsf{tag} \neq \mathsf{tag}^\star$ in such a way that generating fake proofs that $\Phi \in \mathbb{G}_{p_1}$ is impossible. At the same time, $u^{\mathsf{tag}^\star} \cdot v$ should be in $\mathbb{G}_{p_1}$ so that fake proofs can be generated for $\mathsf{tag}^\star$.

SetupSound(λ, t, n): choose a group $\mathbb{G}$ of composite order $N = p_1 p_2$ for large primes $p_i > 2^{l(\lambda)}$ for each $i \in \{1, 2\}$ and for some polynomial $l : \mathbb{N} \to \mathbb{N}$. Then, conduct the following steps

1. Pick $g \xleftarrow{R} \mathbb{G}_{p_1}$, $u, v \xleftarrow{R} \mathbb{G}$, $x \xleftarrow{R} \mathbb{Z}_N$ and set $X = g^x \in \mathbb{G}_{p_1}$.
2. Choose a random polynomial $P[X] \in \mathbb{Z}_N[X]$ of degree $t - 1$ such that $P(0) = x$. For each $i \in \{1, \dots, n\}$, compute $Y_i = g^{P(i)} \in \mathbb{G}_{p_1}$.
3. Select a pairwise independent hash function $H : \mathbb{G} \to \{0, 1\}^\ell$, where $\ell \leq l(\lambda) - 2\lambda$. Note that the range $\mathcal{K} = \{0, 1\}^\ell$ of H forms a group for the bitwise exclusive OR operation $\odot = \oplus$.
4. Define private key shares $(\mathsf{sk}_1, \dots, \mathsf{sk}_n)$ as $\mathsf{sk}_i = P(i) \in \mathbb{Z}_N$ for each $i = 1$ to n. The vector $(\mathsf{vk}_1, \dots, \mathsf{vk}_n)$ is defined as $\mathsf{vk}_i = Y_i \in \mathbb{G}_{p_1}$ for each i and the public key consists of $\mathsf{pk} = ((\mathbb{G}, \mathbb{G}_T), N, g, X, u, v, H)$. In addition, we have $(\mathcal{C}, \mathcal{V}, \mathcal{K}, \mathcal{K}', \mathcal{R}) = (\mathbb{G}, \mathbb{G}_{p_1}, \{0, 1\}^\ell, \mathbb{G}, \mathbb{Z}_N)$.

SetupABO$(\lambda, t, n, \mathsf{tag}^\star)$: is like SetupSound with the difference that, instead of being chosen uniformly in $\mathbb{G}$, v is defined as $v = u^{-\mathsf{tag}^\star} \cdot g^\alpha$ for some random $\alpha \xleftarrow{R} \mathbb{Z}_N$. The algorithm also outputs the simulation trapdoor $\tau = \alpha \in \mathbb{Z}_N$.

Sample(pk): parse the public key pk as $((\mathbb{G}, \mathbb{G}_T), N, g, X, u, v, H)$. Choose $r \xleftarrow{R} \mathbb{Z}_N$, compute $\Phi = g^r \in \mathbb{G}_{p_1}$ and output the pair $(r, \Phi) \in \mathbb{Z}_N \times \mathbb{G}_{p_1}$.

Prove$(\mathsf{pk}, \mathsf{tag}, r, \Phi)$: parse pk as $((\mathbb{G}, \mathbb{G}_T), N, g, X, u, v, H)$ and return $\perp$ if $\Phi \neq g^r$. Otherwise, compute and return $\pi_{\mathsf{SD}} = (u^{\mathsf{tag}} \cdot v)^r$.

SimProve$(\mathsf{pk}, \tau, \mathsf{tag}, \Phi)$: return $\perp$ if $\mathsf{tag} \neq \mathsf{tag}^\star$ or if $\Phi \notin \mathbb{G}$. Otherwise, use the simulation trapdoor $\tau = \alpha \in \mathbb{Z}_N$ to compute and output $\pi_{\mathsf{SD}} = \Phi^\alpha$.

Verify$(\mathsf{pk}, \mathsf{tag}, \Phi, \pi_{\mathsf{SD}})$: return 1 iff $(\Phi, \pi_{\mathsf{SD}}) \in \mathbb{G}^2$ and $e(\Phi, u^{\mathsf{tag}} \cdot v) = e(g, \pi_{\mathsf{SD}})$.

PubEval(pk, r, Φ): on input of the public key $\mathsf{pk} = ((\mathbb{G}, \mathbb{G}_T), N, g, X, u, v, H)$, return $\perp$ if $(r, \Phi) \notin \mathbb{Z}_N \times \mathbb{G}$. Otherwise, output $K = H(X^r) \in \{0, 1\}^\ell$.

SharePrivEval$(\mathsf{pk}, \mathsf{sk}_i, \Phi)$: return $\perp$ if $\Phi \notin \mathbb{G}$. Otherwise, compute and return (K_i, π_{K_i}), where $K_i = \Phi^{\mathsf{sk}_i} = \Phi^{P(i)}$ and $\pi_{K_i} = \varepsilon$ is simply the empty string.

ShareEvalVerify$(\mathsf{pk}, \mathsf{vk}_i, \Phi, K_i, \pi_{K_i})$: if $K_i \notin \mathbb{G}$, $\mathsf{vk}_i \notin \mathbb{G}$ or $\pi_{K_i} \neq \varepsilon$, return 0. Otherwise, return 1 if $e(g, K_i) = e(\Phi, \mathsf{vk}_i)$. In any other situation, return 0 (the proof π_{K_i} is ignored in this instantiation since, given key $\mathsf{vk}_i = Y_i$, the private evaluation share K_i is directly verifiable).

Combine$(\mathsf{pk}, \Phi, \{(K_i, \pi_{K_i})\}_{i \in S})$: return $\perp$ if there exists an index $i \in S$ such that ShareEvalVerify$(\mathsf{pk}, \mathsf{vk}_i, \Phi, K_i, \pi_{K_i}) = 0$. Otherwise, compute and output
$$K = H\left(\prod_{i \in S} K_i^{\Delta_{i,S}(0)}\right) = H(\Phi^x) \in \mathcal{K}.$$

Theorem 2. *The above construction is an all-but-one perfectly sound threshold hash proof system if the SD assumption holds in $\mathbb{G}$.* (The proof is given in the full version of the paper).

When the above all-but-one perfectly sound threshold hash proof system is plugged into the generic construction of Section 4, the resulting threshold cryptosystem bears resemblance with the scheme in [39], which makes use of groups whose order is a product of three primes. However, it is more efficient and its security proof is completely different as the dual system encryption approach [50] is not used here.

5.2   Construction in Prime Order Groups

This section presents an all-but-one threshold hash proof system based on the DLIN assumption in prime order bilinear groups. The public key comprises elements $(g, g_1, g_2, X_1, X_2) \in \mathbb{G}^5$, where $X_1 = g_1^{x_1} \cdot g^z$, $X_2 = g_2^{x_2} \cdot g^z$ and (x_1, x_2, z) are part of the private key. The sets $\mathcal{C}$ and $\mathcal{V} \subset \mathcal{C}$ consist of $\mathcal{C} = \mathbb{G}^3$ and $\mathcal{V} = \{(\Phi_1, \Phi_2, \Phi_3) = (g_1^{\theta_1}, g_2^{\theta_2}, g^{\theta_1 + \theta_2}) \mid \theta_1, \theta_2 \in \mathbb{Z}_p\}$, respectively. For any $\Phi = (\Phi_1, \Phi_2, \Phi_3) \in \mathcal{V}$, the public evaluation algorithm computes $X_1^{\theta_1} \cdot X_2^{\theta_2}$, which can be privately evaluated as $\Phi_1^{x_1} \cdot \Phi_2^{x_2} \cdot \Phi_3^z$.

As in the previous instantiation, we append to elements $\Phi \in \mathcal{V}$ a non-interactive proof of their membership of $\mathcal{V}$ (i.e., a proof that $(g, g_1, g_2, \Phi_1, \Phi_2, \Phi_3)$ is a linear tuple) and, in this case, the proof is obtained using the Groth-Sahai techniques. However, we cannot simply combine them with a DLIN-based hash proof system in the obvious way. The reason is that, using parameters produced by SetupABO and under the special tag tag*, SimProve must be able to compute a fake non-interactive proof of the statement $\Phi \in \mathcal{V}$ for an element $\Phi \notin \mathcal{V}$. At the same time, we should make sure that, for any tag such that tag $\neq$ tag*, it will be impossible to simulate such proofs. To solve this problem, we need a form of one-time simulation soundness [46] which can be possibly obtained from Groth's simulation-sound non-interactive proofs [29] or a more efficient variant suggested by Katz and Vaikuntanathan [35]. However, the specific language that we consider allows for even more efficient constructions: it is actually possible to build on the Groth-Sahai proofs essentially without any loss of efficiency.

The solution is as follows. After having sampled a tuple $\Phi = (\Phi_1, \Phi_2, \Phi_3) \in \mathcal{V}$, the sampler generates his proof using a Groth-Sahai CRS that depends on tag. Algorithm SetupABO produces parameters in the fashion of the all-but-one technique [7]: the tag-based CRS is perfectly WI on the special tag tag* (which allows generating NIZK proofs for this tag) and perfectly sound for any other tag, which makes it impossible to convincingly prove false statements on tags tag $\neq$ tag*. Malkin, Teranishi, Vahlis and Yung [42] used a similar idea of message-dependent CRS in the context of signatures. A difference with [42] is that we do not need to extract witnesses from adversarially-generated proofs and only use them as proofs of membership.

Interestingly, the same technique can be applied to have a more efficient simulation-sound proof of plaintext equality in the Naor-Yung-type [43] cryptosystem in [35][Section 3.2.2]: the proof can be reduced from 60 to 22 group elements and the ciphertext size is decreased by more than 50%.

SetupSound(λ, t, n): Choose a group $\mathbb{G}$ of prime order $p > 2^\lambda$ with generators $g, g_1, g_2, f_1, f_2 \xleftarrow{R} \mathbb{G}$.

1. Choose $x_1, x_2, z \xleftarrow{R} \mathbb{Z}_p$ and set $X_1 = g_1^{x_1} g^z$, $X_2 = g_2^{x_2} g^z$. Define the vectors $\vec{g}_1 = (g_1, 1, g)$ and $\vec{g}_2 = (1, g_2, g)$. Then, pick $\xi_1, \xi_2 \xleftarrow{R} \mathbb{Z}_p$ and define $\vec{g}_3 = \vec{g}_1^{\xi_1} \cdot \vec{g}_2^{\xi_2}$.

2. Choose $\phi_1, \phi_2 \xleftarrow{R} \mathbb{Z}_p$ and define vectors $\vec{f}_1 = (f_1, 1, g)$, $\vec{f}_2 = (1, f_2, g)$ and $\vec{f}_3 = \vec{f}_1^{\phi_1} \cdot \vec{f}_2^{\phi_2} \cdot (1, 1, g)$.

3. Choose random polynomials $P_1[X], P_2[X], P[X] \in \mathbb{Z}_p[X]$ of degree $t - 1$ such that $P_1(0) = x_1$, $P_2(0) = x_2$ and $P(0) = z$. For each $i = 1$ to n, compute $Y_{i,1} = g_1^{P_1(i)} g^{P(i)}$, $Y_{i,2} = g_2^{P_2(i)} g^{P(i)}$.

4. Define shares $\mathbf{SK} = (\mathsf{sk}_1, \ldots, \mathsf{sk}_n)$ as $\mathsf{sk}_i = (P_1(i), P_2(i), P(i)) \in (\mathbb{Z}_p)^3$ for each $i \in \{1, \ldots, n\}$. Verification keys $\mathbf{VK} = (\mathsf{vk}_1, \ldots, \mathsf{vk}_n)$ are defined as $\mathsf{vk}_i = (Y_{i,1}, Y_{i,2}) \in \mathbb{G}^2$ for each $i \in \{1, \ldots, n\}$ and the public key is

$$\mathsf{pk} = \left((\mathbb{G}, \mathbb{G}_T), \; g, \; \vec{g}_1, \; \vec{g}_2, \; \vec{g}_3, \; \vec{f}_1, \; \vec{f}_2, \; \vec{f}_3, \; X_1, \; X_2 \right).$$

As for the sets $(\mathcal{C}, \mathcal{K}, \mathcal{K}', \mathcal{R})$, they are defined as $\mathcal{C} = \mathbb{G}^3$, $\mathcal{K} = \mathcal{K}' = \mathbb{G}$ and $\mathcal{R} = (\mathbb{Z}_p)^2$, respectively. The subset $\mathcal{V} \subset \mathcal{C}$ consists of the language $(\Phi_1, \Phi_2, \Phi_3) \in \mathbb{G}^3$ for which there exists $\theta_1, \theta_2 \in \mathbb{Z}_p$ such that $\Phi_1 = g_1^{\theta_1}$, $\Phi_2 = g_2^{\theta_2}$ and $\Phi_3 = g^{\theta_1 + \theta_2}$.

SetupABO$(\lambda, t, n, \mathsf{tag}^\star)$: is like SetupSound with the following differences.

1. In step 1, $\vec{g}_3$ is set as $\vec{g}_3 = \vec{g}_1^{\xi_1} \cdot \vec{g}_2^{\xi_2} \cdot (1, 1, g)^{-\mathsf{tag}^\star}$ so that $\vec{g}_3 \notin \mathsf{span}(\vec{g}_1, \vec{g}_2)$.

2. In step 2, the vectors $(\vec{f}_1, \vec{f}_2, \vec{f}_3)$ are chosen so as to have $\vec{f}_3 = \vec{f}_1^{\phi_1} \cdot \vec{f}_2^{\phi_2}$.

3. The algorithm also outputs the trapdoor $\tau = (\xi_1, \xi_2, \phi_1, \phi_2) \in (\mathbb{Z}_p)^4$.

Sample(pk): choose $\theta_1, \theta_2 \xleftarrow{R} \mathbb{Z}_p$, compute $\Phi = (\Phi_1, \Phi_2, \Phi_3) = (g_1^{\theta_1}, g_2^{\theta_2}, g^{\theta_1 + \theta_2})$ and output $((\theta_1, \theta_2), \Phi)$.

Prove$(\mathsf{pk}, \mathsf{tag}, (\theta_1, \theta_2), \Phi)$: parse pk as $((\mathbb{G}, \mathbb{G}_T), g, \vec{g}_1, \vec{g}_2, \vec{g}_3, \vec{f}_1, \vec{f}_2, \vec{f}_3, X_1, X_2)$. Parse Φ as (Φ_1, Φ_2, Φ_3). Define[1] $\vec{g}_{\mathsf{tag}} = \vec{g}_3 \cdot (1, 1, g)^{\mathsf{tag}}$ and use $\mathbf{g}_{\mathsf{tag}} = (\vec{g}_1, \vec{g}_2, \vec{g}_{\mathsf{tag}})$ as a Groth-Sahai CRS to generate a NIZK proof that $(g, g_1, g_2, \Phi_1, \Phi_2, \Phi_3)$ is a linear tuple. More precisely, generate commitments $\vec{C}_{\theta_1}, \vec{C}_{\theta_2}$ to exponents $\theta_1, \theta_2 \in \mathbb{Z}_p$ (in other words, compute $\vec{C}_{\theta_i} = \vec{g}_{\mathsf{tag}}^{\theta_i} \cdot \vec{g}_1^{r_i} \cdot \vec{g}_2^{s_i}$ with $r_i, s_i \xleftarrow{R} \mathbb{Z}_p$ for each $i \in \{1, 2\}$) and a proof $\pi_{(\theta_1, \theta_2)}$ that they satisfy

$$\Phi_1 = g_1^{\theta_1}, \qquad \Phi_2 = g_2^{\theta_2}, \qquad \Phi_3 = g^{\theta_1 + \theta_2}. \tag{1}$$

The whole proof π_{LIN} for (1) consists of $\vec{C}_{\theta_1}, \vec{C}_{\theta_2}$ and $\pi_{(\theta_1, \theta_2)}$ (see the full version of the paper for details about the generation of this proof) and requires 12 elements of $\mathbb{G}$.

SimProve$(\mathsf{pk}, \tau, \mathsf{tag}, \Phi)$: parses pk as above, τ as $(\xi_1, \xi_2, \phi_1, \phi_2) \in (\mathbb{Z}_p)^4$ and Φ as $(\Phi_1, \Phi_2, \Phi_3) \in \mathbb{G}^3$. If $\mathsf{tag} \neq \mathsf{tag}^\star$, return $\perp$. Otherwise, the commitments $\vec{C}_{\theta_1}, \vec{C}_{\theta_2}$ and the proof π_{LIN} must be generated for the Groth-Sahai CRS

---

[1] We assume that tags are non-zero. This can be enforced by having Prove and Verify output $\perp$ when $\mathsf{tag} = 0$.

$g_{tag^\star} = (\vec{g_1}, \vec{g_2}, \vec{g}_{tag^\star})$, where $\vec{g}_{tag^\star} = \vec{g_3} \cdot (1, 1, g)^{tag^\star} = \vec{g_1}^{\xi_1} \cdot \vec{g_2}^{\xi_2}$, which is a Groth-Sahai CRS for the witness indistinguishability setting.

1. Using the trapdoor (ξ_1, ξ_2), simulate proofs for multi-exponentiation equations (see the full version of the paper for details as to how such proofs can be simulated). That is, generate $\vec{C}_{\theta_1}, \vec{C}_{\theta_2}$ as commitments to 0 and compute $\pi_{(\theta_1, \theta_2)}$ as a simulated proof that relations (1) hold.

2. Output $\pi_{LIN} = (\vec{C}_{\theta_1}, \vec{C}_{\theta_2}, \pi_{(\theta_1, \theta_2)})$ that consists of perfectly hiding commitments and simulated NIZK proofs which, on the CRS $(\vec{g_1}, \vec{g_2}, \vec{g}_{tag^\star})$, are distributed as real proofs.

Verify$(pk, tag, \Phi, \pi_{LIN})$: parse pk and Φ as above. Also, parse the proof π_{LIN} as $(\vec{C}_{\theta_1}, \vec{C}_{\theta_2}, \pi_{(\theta_1, \theta_2)}) \in \mathbb{G}^{12}$. Then, compute $\vec{g}_{tag} = \vec{g_3} \cdot (1, 1, g)^{tag}$ and use $g_{tag} = (\vec{g_1}, \vec{g_2}, \vec{g}_{tag})$ as a Groth-Sahai CRS to verify π_{LIN}. If the latter is deemed as a valid proof for the relations (1), return 1. Otherwise, return 0.

PubEval$(pk, (\theta_1, \theta_2), \Phi)$: parse pk and Φ as above. Return $\perp$ if $(\Phi_1, \Phi_2, \Phi_3) \neq (g_1^{\theta_1}, g_2^{\theta_2}, g^{\theta_1 + \theta_2})$. Otherwise, compute and return $K = X_1^{\theta_1} \cdot X_2^{\theta_2} \in \mathcal{K}$.

SharePrivEval(pk, sk_i, Φ): parse sk_i as $(P_1(i), P_2(i), P(i)) \in (\mathbb{Z}_p)^3$ and return $\perp$ if $\Phi \notin \mathbb{G}^3$. Otherwise, return (K_i, π_{K_i}), where $K_i = \Phi_1^{P_1(i)} \Phi_2^{P_2(i)} \Phi_3^{P(i)} \in \mathcal{K}'$ and $\pi_{K_i} = (\vec{C}_{P_1}, \vec{C}_{P_2}, \vec{C}_P, \pi'_{K_i}) \in \mathbb{G}^{15}$ is a proof consisting of commitments $\vec{C}_{P_1}, \vec{C}_{P_2}, \vec{C}_P$ to exponents $P_1(i), P_2(i), P(i) \in \mathbb{Z}_p$ and a proof π'_{K_i} that these satisfy the equations

$$K_i = \Phi_1^{P_1(i)} \cdot \Phi_2^{P_2(i)} \cdot \Phi_3^{P(i)}, \qquad Y_{i,1} = g_1^{P_1(i)} g^{P(i)}, \qquad Y_{i,2} = g_2^{P_2(i)} g^{P(i)}. \quad (2)$$

The perfectly binding commitments $\vec{C}_{P_1}, \vec{C}_{P_2}, \vec{C}_P$ and the proof π'_{K_i} are generated using the vectors $\mathbf{f} = (\vec{f_1}, \vec{f_2}, \vec{f_3})$ as a Groth-Sahai CRS (in such a way that $\vec{C}_{P_1} = \vec{f_3}^{P_1(i)} \cdot \vec{f_1}^{r_{P_1}} \cdot \vec{f_2}^{s_{P_1}}$, for some $r_{P_1}, s_{P_1} \xleftarrow{R} \mathbb{Z}_p$, for example).

ShareEvalVerify$(pk, vk_i, \Phi, K_i, \pi_{K_i})$: parse vk_i as $(Y_{i,1}, Y_{i,2}) \in \mathbb{G}^2$ and return $\perp$ if (K_i, π_{K_i}) cannot be parsed as a tuple in $\mathbb{G} \times \mathbb{G}^{15}$. Otherwise, parse π_{K_i} as $\pi_{K_i} = (\vec{C}_{P_1}, \vec{C}_{P_2}, \vec{C}_P, \pi'_{K_i}) \in \mathbb{G}^{15}$ and return 1 if π'_{K_i} is a valid proof for equations (2). In any other situation, return 0.

Combine$(pk, \Phi, \{(K_i, \pi_{K_i})\}_{i \in S})$: return $\perp$ if there is an index $i \in S$ for which ShareEvalVerify$(pk, vk_i, \Phi, K_i, \pi_{K_i}) = 0$. Otherwise, compute

$$K = \prod_{i \in S} K_i^{\Delta_{i,S}(0)} = \Phi_1^{x_1} \cdot \Phi_2^{x_2} \cdot \Phi_3^z \in \mathcal{K}.$$

Theorem 3. *The above construction is an all-but-one perfectly sound threshold hash proof system assuming that the DLIN assumption holds in* $\mathbb{G}$. *(The proof is given in the full version of the paper.)*

The proof π_{LIN} takes 6 group elements whereas $\vec{C}_{\theta_1}, \vec{C}_{\theta_2}$ require 3 group elements each. If the scheme is instantiated using Groth's one-time signature [29] (which relies on the discrete logarithm assumption), SVK and σ demand 3 and 2 group elements, respectively. The whole ciphertext C thus consists of 21 group elements. Concretely, if each element has a representation of 512 bits, at the 128-bit security level, the ciphertext overhead amounts to 10240 bits.

From a computational standpoint, assuming that a multi-exponentiation with two base elements has roughly the same cost as a single-base exponentiation, the sender has to compute 19 exponentiations in $\mathbb{G}$ (we include the cost of generating SVK which incurs three exponentiations in Groth's one-time signature [29]). As for the verifier's workload, the validity of a ciphertext can be checked by computing a product of 12 pairings (which is more efficient than naively evaluating 12 individual pairings) using batch verification techniques as in [6].

In the full version of the paper, we show an even more efficient instantiation based on the Symmetric eXternal Diffie-Hellman assumption in prime order groups: only 6 pairing evaluations suffice to check $\pi_\mathcal{V}$.

Acknowledgements. We thank the anonymous reviewers and Carla Ràfols for useful comments.

References

1. Abe, M.: Robust Distributed Multiplication without Interaction. In: Wiener, M. (ed.) CRYPTO 1999. LNCS, vol. 1666, pp. 130–147. Springer, Heidelberg (1999)
2. Abe, M., Fehr, S.: Adaptively Secure Feldman VSS and Applications to Universally-Composable Threshold Cryptography. In: Franklin, M. (ed.) CRYPTO 2004. LNCS, vol. 3152, pp. 317–334. Springer, Heidelberg (2004)
3. Almansa, J.F., Damgård, I., Nielsen, J.B.: Simplified Threshold RSA with Adaptive and Proactive Security. In: Vaudenay, S. (ed.) EUROCRYPT 2006. LNCS, vol. 4004, pp. 593–611. Springer, Heidelberg (2006)
4. Barreto, P.S.L.M., Naehrig, M.: Pairing-Friendly Elliptic Curves of Prime Order. In: Preneel, B., Tavares, S. (eds.) SAC 2005. LNCS, vol. 3897, pp. 319–331. Springer, Heidelberg (2006)
5. Bellare, M., Rogaway, P.: Random oracles are practical: A paradigm for designing efficient protocols. In: ACM CCS, pp. 62–73 (1993)
6. Blazy, O., Fuchsbauer, G., Izabachène, M., Jambert, A., Sibert, H., Vergnaud, D.: Batch Groth–Sahai. In: Zhou, J., Yung, M. (eds.) ACNS 2010. LNCS, vol. 6123, pp. 218–235. Springer, Heidelberg (2010)
7. Boneh, D., Boyen, X.: Efficient Selective-ID Secure Identity-Based Encryption Without Random Oracles. In: Cachin, C., Camenisch, J.L. (eds.) EUROCRYPT 2004. LNCS, vol. 3027, pp. 223–238. Springer, Heidelberg (2004)
8. Boneh, D., Boyen, X., Halevi, S.: Chosen Ciphertext Secure Public Key Threshold Encryption Without Random Oracles. In: Pointcheval, D. (ed.) CT-RSA 2006. LNCS, vol. 3860, pp. 226–243. Springer, Heidelberg (2006)
9. Boneh, D., Boyen, X., Shacham, H.: Short Group Signatures. In: Franklin, M. (ed.) CRYPTO 2004. LNCS, vol. 3152, pp. 41–55. Springer, Heidelberg (2004)
10. Boneh, D., Franklin, M.: Identity-Based Encryption from the Weil Pairing. SIAM J. of Computing 32(3), 586–615 (2003); In: Kilian, J. (ed.) CRYPTO 2001. LNCS, vol. 2139, pp. 213–229. Springer, Heidelberg (2001)
11. Boneh, D., Goh, E.-J., Nissim, K.: Evaluating 2-DNF Formulas on Ciphertexts. In: Kilian, J. (ed.) TCC 2005. LNCS, vol. 3378, pp. 325–341. Springer, Heidelberg (2005)

12. Boyd, C.: Digital Multisignatures. In: Beker, H.J., Piper, F.C. (eds.) Cryptography and Coding, pp. 241–246. Oxford University Press (1989)
13. Boyen, X., Mei, Q., Waters, B.: Direct Chosen Ciphertext Security from Identity-Based Techniques. In: ACM CCS 2005, pp. 320–329 (2005)
14. Canetti, R., Gennaro, R., Jarecki, S., Krawczyk, H., Rabin, T.: Adaptive Security for Threshold Cryptosystems. In: Wiener, M. (ed.) CRYPTO 1999. LNCS, vol. 1666, pp. 98–115. Springer, Heidelberg (1999)
15. Canetti, R., Goldwasser, S.: An Efficient *Threshold* Public Key Cryptosystem Secure against Adaptive Chosen Ciphertext Attack. In: Stern, J. (ed.) EUROCRYPT 1999. LNCS, vol. 1592, pp. 90–106. Springer, Heidelberg (1999)
16. Cramer, R., Shoup, V.: A Practical Public Key Cryptosystem Provably Secure against Adaptive Chosen Ciphertext Attack. In: Krawczyk, H. (ed.) CRYPTO 1998. LNCS, vol. 1462, pp. 13–25. Springer, Heidelberg (1998)
17. Cramer, R., Shoup, V.: Universal Hash Proofs and a Paradigm for Adaptive Chosen Ciphertext Secure Public-Key Encryption. In: Knudsen, L.R. (ed.) EUROCRYPT 2002. LNCS, vol. 2332, pp. 45–64. Springer, Heidelberg (2002)
18. Canetti, R., Halevi, S., Katz, J.: Chosen-Ciphertext Security from Identity-Based Encryption. In: Cachin, C., Camenisch, J.L. (eds.) EUROCRYPT 2004. LNCS, vol. 3027, pp. 207–222. Springer, Heidelberg (2004)
19. Cramer, R., Damgård, I., Dziembowski, S., Hirt, M., Rabin, T.: Efficient Multiparty Computations Secure against an Adaptive Adversary. In: Stern, J. (ed.) EUROCRYPT 1999. LNCS, vol. 1592, pp. 311–326. Springer, Heidelberg (1999)
20. Cramer, R., Damgård, I., Ishai, Y.: Share Conversion, Pseudorandom Secret-Sharing and Applications to Secure Computation. In: Kilian, J. (ed.) TCC 2005. LNCS, vol. 3378, pp. 342–362. Springer, Heidelberg (2005)
21. Damgård, I.: Towards Practical Public Key Systems Secure against Chosen Ciphertext Attacks. In: Feigenbaum, J. (ed.) CRYPTO 1991. LNCS, vol. 576, pp. 445–456. Springer, Heidelberg (1992)
22. Desmedt, Y.: Society and Group Oriented Cryptography: A New Concept. In: Pomerance, C. (ed.) CRYPTO 1987. LNCS, vol. 293, pp. 120–127. Springer, Heidelberg (1988)
23. Desmedt, Y., Frankel, Y.: Threshold Cryptosystems. In: Brassard, G. (ed.) CRYPTO 1989. LNCS, vol. 435, pp. 307–315. Springer, Heidelberg (1990)
24. Dodis, Y., Katz, J.: Chosen-Ciphertext Security of Multiple Encryption. In: Kilian, J. (ed.) TCC 2005. LNCS, vol. 3378, pp. 188–209. Springer, Heidelberg (2005)
25. Fouque, P.-A., Pointcheval, D.: Threshold Cryptosystems Secure against Chosen-Ciphertext Attacks. In: Boyd, C. (ed.) ASIACRYPT 2001. LNCS, vol. 2248, pp. 351–368. Springer, Heidelberg (2001)
26. Frankel, Y., MacKenzie, P., Yung, M.: Adaptively-Secure Distributed Public-Key Systems. In: Nešetřil, J. (ed.) ESA 1999. LNCS, vol. 1643, pp. 4–27. Springer, Heidelberg (1999)
27. Frankel, Y., MacKenzie, P., Yung, M.: Adaptively-Secure Optimal-Resilience Proactive RSA. In: Lam, K.-Y., Okamoto, E., Xing, C. (eds.) ASIACRYPT 1999. LNCS, vol. 1716, pp. 180–195. Springer, Heidelberg (1999)
28. Freeman, D.M.: Converting Pairing-Based Cryptosystems from Composite-Order Groups to Prime-Order Groups. In: Gilbert, H. (ed.) EUROCRYPT 2010. LNCS, vol. 6110, pp. 44–61. Springer, Heidelberg (2010)
29. Groth, J.: Simulation-Sound NIZK Proofs for a Practical Language and Constant Size Group Signatures. In: Lai, X., Chen, K. (eds.) ASIACRYPT 2006. LNCS, vol. 4284, pp. 444–459. Springer, Heidelberg (2006)

30. Groth, J., Ostrovsky, R., Sahai, A.: Perfect Non-interactive Zero Knowledge for NP. In: Vaudenay, S. (ed.) EUROCRYPT 2006. LNCS, vol. 4004, pp. 339–358. Springer, Heidelberg (2006)

31. Groth, J., Sahai, A.: Efficient Non-interactive Proof Systems for Bilinear Groups. In: Smart, N.P. (ed.) EUROCRYPT 2008. LNCS, vol. 4965, pp. 415–432. Springer, Heidelberg (2008)

32. Håstad, J., Impagliazzo, R., Levin, L., Luby, M.: A pseudorandom generator from any one-way function. SIAM Journal on Computing 28(4), 1364–1396 (1999)

33. Hofheinz, D., Kiltz, E.: The Group of Signed Quadratic Residues and Applications. In: Halevi, S. (ed.) CRYPTO 2009. LNCS, vol. 5677, pp. 637–653. Springer, Heidelberg (2009)

34. Jarecki, S., Lysyanskaya, A.: Adaptively Secure Threshold Cryptography: Introducing Concurrency, Removing Erasures (Extended Abstract). In: Preneel, B. (ed.) EUROCRYPT 2000. LNCS, vol. 1807, pp. 221–242. Springer, Heidelberg (2000)

35. Katz, J., Vaikuntanathan, V.: Round-Optimal Password-Based Authenticated Key Exchange. In: Ishai, Y. (ed.) TCC 2011. LNCS, vol. 6597, pp. 293–310. Springer, Heidelberg (2011)

36. Kiltz, E.: Chosen-Ciphertext Security from Tag-Based Encryption. In: Halevi, S., Rabin, T. (eds.) TCC 2006. LNCS, vol. 3876, pp. 581–600. Springer, Heidelberg (2006)

37. Kiltz, E., Pietrzak, K., Stam, M., Yung, M.: A New Randomness Extraction Paradigm for Hybrid Encryption. In: Joux, A. (ed.) EUROCRYPT 2009. LNCS, vol. 5479, pp. 590–609. Springer, Heidelberg (2009)

38. Lewko, A., Waters, B.: New Techniques for Dual System Encryption and Fully Secure HIBE with Short Ciphertexts. In: Micciancio, D. (ed.) TCC 2010. LNCS, vol. 5978, pp. 455–479. Springer, Heidelberg (2010)

39. Libert, B., Yung, M.: Adaptively Secure Non-interactive Threshold Cryptosystems. In: Aceto, L., Henzinger, M., Sgall, J. (eds.) ICALP 2011, Part II. LNCS, vol. 6756, pp. 588–600. Springer, Heidelberg (2011)

40. MacKenzie, P.: An Efficient Two-Party Public Key Cryptosystem Secure against Adaptive Chosen Ciphertext Attack. In: Desmedt, Y.G. (ed.) PKC 2003. LNCS, vol. 2567, pp. 47–61. Springer, Heidelberg (2002)

41. MacKenzie, P., Reiter, M.K., Yang, K.: Alternatives to Non-malleability: Definitions, Constructions, and Applications. In: Naor, M. (ed.) TCC 2004. LNCS, vol. 2951, pp. 171–190. Springer, Heidelberg (2004)

42. Malkin, T., Teranishi, I., Vahlis, Y., Yung, M.: Signatures Resilient to Continual Leakage on Memory and Computation. In: Ishai, Y. (ed.) TCC 2011. LNCS, vol. 6597, pp. 89–106. Springer, Heidelberg (2011)

43. Naor, M., Yung, M.: Public-key cryptosystems provably secure against chosen ciphertext attacks. In: STOC 1990. ACM Press (1990)

44. Ostrovsky, R., Yung, M.: How to Withstand Mobile Virus Attacks. In: 10th ACM Symp. on Principles of Distributed Computing, PODC 1991 (1991)

45. Rackoff, C., Simon, D.R.: Non-interactive Zero-Knowledge Proof of Knowledge and Chosen Ciphertext Attack. In: Feigenbaum, J. (ed.) CRYPTO 1991. LNCS, vol. 576, pp. 433–444. Springer, Heidelberg (1992)

46. Sahai, A.: Non-Malleable Non-Interactive Zero Knowledge and Adaptive Chosen-Ciphertext Security. In: FOCS 1999, pp. 543–553 (1999)

47. Scott, M.: Authenticated ID-based Key Exchange and remote log-in with simple token and PIN number. Cryptology ePrint Archive: Report 2002/164

48. Shamir, A.: Identity-Based Cryptosystems and Signature Schemes. In: Blakely, G.R., Chaum, D. (eds.) CRYPTO 1984. LNCS, vol. 196, pp. 47–53. Springer, Heidelberg (1985)
49. Shoup, V., Gennaro, R.: Securing Threshold Cryptosystems against Chosen Ciphertext Attack. J. of Cryptology 15(2), 75–96 (2002). In: Nyberg, K. (ed.) EUROCRYPT 1998. LNCS, vol. 1403, pp. 1–16. Springer, Heidelberg (1998)
50. Waters, B.: Dual System Encryption: Realizing Fully Secure IBE and HIBE under Simple Assumptions. In: Halevi, S. (ed.) CRYPTO 2009. LNCS, vol. 5677, pp. 619–636. Springer, Heidelberg (2009)
51. Wee, H.: Efficient Chosen-Ciphertext Security via Extractable Hash Proofs. In: Rabin, T. (ed.) CRYPTO 2010. LNCS, vol. 6223, pp. 314–332. Springer, Heidelberg (2010)
52. Wee, H.: Threshold and Revocation Cryptosystems via Extractable Hash Proofs. In: Paterson, K.G. (ed.) EUROCRYPT 2011. LNCS, vol. 6632, pp. 589–609. Springer, Heidelberg (2011)

Round-Optimal Privacy-Preserving Protocols with Smooth Projective Hash Functions*

Olivier Blazy, David Pointcheval, and Damien Vergnaud

ENS, Paris, France

Abstract. In 2008, Groth and Sahai proposed a powerful suite of techniques for constructing non-interactive zero-knowledge proofs in bilinear groups. Their proof systems have found numerous applications, including group signature schemes, anonymous voting, and anonymous credentials. In this paper, we demonstrate that the notion of *smooth projective hash functions* can be useful to design round-optimal privacy-preserving interactive protocols. We show that this approach is suitable for designing schemes that rely on standard security assumptions in the standard model with a common-reference string and are more efficient than those obtained using the Groth-Sahai methodology. As an illustration of our design principle, we construct an efficient oblivious signature-based envelope scheme and a blind signature scheme, both round-optimal.

1 Introduction

In 2008, Groth and Sahai [22] proposed a way to produce efficient and practical non-interactive zero-knowledge and non-interactive witness-indistinguishable proofs for (algebraic) statements related to groups equipped with a bilinear map. They have been significantly studied in cryptography and used in a wide variety of applications in recent years (*e.g.* group signature schemes [8, 9, 20] or blind signatures [2, 5]). While avoiding expensive NP-reductions, these proof systems still lack in practicality and it is desirable to provide more efficient tools.

Smooth projective hash functions (SPHF) were introduced by Cramer and Shoup [13] for constructing encryption schemes. A projective hashing family is a family of hash functions that can be evaluated in two ways: using the (secret) hashing key, one can compute the function on every point in its domain, whereas using the (public) *projected* key one can only compute the function on a special subset of its domain. Such a family is deemed *smooth* if the value of the hash function on any point outside the special subset is independent of the projected key. If it is hard to distinguish elements of the special subset from non-elements, then this primitive can be seen as special type of zero-knowledge proof system for membership in the special subset. The notion of SPHF has found applications in various contexts in cryptography (*e.g.* [18, 26, 1]). We present some other applications with privacy-preserving primitives that were already inherently interactive.

---

* CNRS – UMR 8548 and INRIA – EPI Cascade, Université Paris Diderot.

R. Cramer (Ed.): TCC 2012, LNCS 7194, pp. 94–111, 2012.
© International Association for Cryptologic Research 2012

Applications: Our two applications are *Oblivious Signature-Based Envelope* [27] and *Blind Signatures* [12].

Oblivious Signature-Based Envelope (OSBE) were introduced in [27]. It can be viewed as a nice way to ease the asymmetrical aspect of several authentication protocols. Alice is a member of an organization and possesses a certificate produced by an authority attesting she is in this organization. Bob wants to send a private message P to members of this organization. However due to the sensitive nature of the organization, Alice does not want to give Bob neither her certificate nor a proof she belongs to the organization. OSBE lets Bob sends an obfuscated version of this message P to Alice, in such a way that Alice will be able to find P if and only if Alice is in the required organization. In the process, Bob cannot decide whether Alice does really belong to the organization. They are part of a growing field of protocols, around *automated trust negotiation*, which also include Secret Handshakes [3], Password-based Authenticated Key-Exchange [19], and Hidden Credentials [10]. Those schemes are all closely related, so due to space constraints, we are going to focus on OSBE (as if you tweak two of them, you can produce any of the other protocols [11]).

Blind signatures were introduced by Chaum [12] for electronic cash in order to prevent the bank from linking a coin to its spender: they allow a user to obtain a signature on a message such that the signer cannot relate the resulting message/signature pair to the execution of the signing protocol. In [15], Fischlin gave a generic construction of round-optimal blind signatures in the common-reference string (CRS) model: the signing protocol consists of one message from the user to the signer and one response by the signer. The first practical instantiation of round-optimal blind signatures in the standard model was proposed in [2] but it relies on non-standard computational assumptions. We proposed, recently only [5], the most efficient realizations of round-optimal blind signatures in the common-reference string model under classical assumptions. But these schemes still use the Groth-Sahai proof systems.

Contributions: Our first contribution is to clarify and increase the security requirements of an OSBE scheme. The main improvement residing in some protection for both the sender and the receiver against the Certification Authority. The OSBE notion echoes directly to the idea of SPHF if we consider the language $\mathcal{L}$ defined by encryption of valid signatures, which is hard to distinguish under the security of the encryption schemes. We show how to build, from a SPHF on this language, an OSBE scheme in the standard model with a CRS. And we prove the security of our construction in regards of the security of the commitment (the ciphertext), the signature and the SPHF scheme. We then show how to build a simple and efficient OSBE scheme relying on a classical assumption, DLin. An asymmetrical version is available in the full version [6]. To build those schemes, we use SPHF in a new way, avoiding the need of costly Groth-Sahai proofs when an interaction is inherently needed in the primitive. Our method does not add any other interaction, and so supplement smoothly those proofs.

To show the efficiency of the method, and the ease of application, we then adapt two Blind Signature schemes proposed in [5]. Our approach fits perfectly and decreases significantly the communicational complexity of the schemes (it is divided by more than three in one construction). Moreover one scheme relies on a weakened security assumptions: the XDH assumption instead of the SXDH assumption and permits to use more bilinear group settings (namely, Type-II and Type-III bilinear groups [16] instead of only Type-III bilinear groups for the construction presented in [5]).

2   Definitions

In this section, we briefly recall the notations and the security notions of the basic primitives we will use in the rest of the paper, and namely public key encryption, signature and smooth projective hash functions (SPHF), using the Gennaro-Lindell [18] extension. More details are available in the full version [6]. In a second part, we recall and enhance the security model of oblivious signature-based envelope protocols [27].

2.1   Notations

Encryption Scheme. A (public-key) encryption scheme is defined by four algorithms: $\mathsf{param} \leftarrow \mathsf{ESetup}(1^k)$, $(\mathsf{ek}, \mathsf{dk}) \leftarrow \mathsf{EKeyGen}(\mathsf{param})$, $c \leftarrow \mathsf{Encrypt}(\mathsf{ek}, m; r)$, and $m \leftarrow \mathsf{Decrypt}(\mathsf{dk}, c)$. We will need the classical notion of IND-CPA security. More precisely, we will use commitment schemes (as in [1]), which should be hiding (indistinguishability) and binding (one opening only), with the additional extractability property. The latter property thus needs an extracting algorithm that corresponds to the decryption algorithm. Hence the notation with encryption schemes.

Signature Scheme. A signature scheme is defined by four algorithms: $\mathsf{param} \leftarrow \mathsf{SSetup}(1^k)$, $(\mathsf{vk}, \mathsf{sk}) \leftarrow \mathsf{SKeyGen}(\mathsf{param})$, $\sigma \leftarrow \mathsf{Sign}(\mathsf{sk}, m; s)$, and $\mathsf{Verif}(\mathsf{vk}, m, \sigma)$. We will need the classical notion of EUF-CMA security.

Smooth Projective Hash Function. An SPHF system [13] on a language $\mathcal{L}$ is defined by five algorithms: $\mathsf{SPHFSetup}(1^k)$ that generates the global parameters, $\mathsf{HashKG}(\mathcal{L}, \mathsf{param})$ that generates a hashing key hk, $\mathsf{ProjKG}(\mathsf{hk}, (\mathcal{L}, \mathsf{param}), W)$ that derives the projection key hp, possibly depending on the word W [18,1]. Then, $\mathsf{Hash}(\mathsf{hk}, (\mathcal{L}, \mathsf{param}), W)$ and $\mathsf{ProjHash}(\mathsf{hp}, (\mathcal{L}, \mathsf{param}), W, w)$ outputs the hash value, either from the hashing key, or from the projection key and the witness. The correctness of the scheme assures that if W is indeed in $\mathcal{L}$ with w as a witness, then the two ways to compute the hash value give the same result. The security of a SPHF is defined through two different notions, the smoothness and the pseudo-randomness properties: The smoothness property guarantees that if $W \notin \mathcal{L}$, then the hash value is statistically random (statistically indistinguishable from a random element). The pseudo-randomness guarantees that even for a word $W \in \mathcal{L}$, but without the knowledge of a witness w, then the hash value is random (computationally indistinguishable from a random element). Abdalla *et al.* [1] explained how to combine SPHF to deal with conjunctions and disjunctions of the languages.

2.2   Oblivious Signature-Based Envelope

We now define an OSBE protocol, where a sender S wants to send a private message $P \in \{0,1\}^\ell$ to a recipient R in possession of a certificate/signature on a message M.

Definition 1 (Oblivious Signature-Based Envelope). *An OSBE scheme is defined by four algorithms* (OSBESetup, OSBEKeyGen, OSBESign, OSBEVerif), *and one interactive protocol* OSBEProtocol$\langle S, R \rangle$:

- OSBESetup(1^k), *where k is the security parameter, generates the global parameters* param;
- OSBEKeyGen(param) *generates the keys* (vk, sk) *of the certification authority;*
- OSBESign(sk, m) *produces a signature σ on the input message m, under the signing key* sk;
- OSBEVerif(vk, m, σ) *checks whether σ is a valid signature on m, w.r.t. the public key* vk; *it outputs 1 if the signature is valid, and 0 otherwise.*
- OSBEProtocol$\langle S(\mathsf{vk}, M, P), R(\mathsf{vk}, M, \sigma) \rangle$ *between the sender S with the private message P, and the recipient R with a certificate σ. If σ is a valid signature under* vk *on the common message M, then R receives P, otherwise it receives nothing. In any case, S does not learn anything.*

Such an OSBE scheme should be (the three last properties are additional —or stronger— security properties from the original definitions [27]):

- *correct*: the protocol actually allows R to learn P, whenever σ is a valid signature on M under vk;
- *oblivious*: the sender should not be able to distinguish whether R uses a valid signature σ on M under vk as input. More precisely, if R_0 knows and uses a valid signature σ and R_1 does not use such a valid signature, the sender cannot distinguish an interaction with R_0 from an interaction with R_1;
- *(weakly) semantically secure*: the recipient learns nothing about S input P if it does not use a valid signature σ on M under vk as input. More precisely, if S_0 owns P_0 and S_1 owns P_1, the recipient that does not use a valid signature cannot distinguish an interaction with S_0 from an interaction with S_1;
- *semantically secure* (denoted sem): the above indistinguishability should hold even if the receiver has seen several interactions $\langle S(\mathsf{vk}, M, P), R(\mathsf{vk}, M, \sigma) \rangle$ with valid signatures, and the same sender's input P;
- *escrow free* (denoted esc): the authority (owner of the signing key sk), playing as the sender or just eavesdropping, is unable to distinguish whether R used a valid signature σ on M under vk as input. This notion supersedes the above *oblivious* property, since this is basically oblivious w.r.t. the authority, without any restriction.
- *semantically secure w.r.t. the authority* (denoted sem*): after the interaction, the authority (owner of the signing key sk) learns nothing about P.

We insist that the escrow-free property (esc) is stronger than the oblivious property, hence we will consider the former only. However, the semantic security

$\mathsf{Exp}_{OSBE,A}^{esc-b}(k)$ [Escrow Free property]
1. param $\leftarrow$ OSBESetup(1^k)
2. vk $\leftarrow A(\text{INIT} : \text{param})$
3. $(M, \sigma) \leftarrow A(\text{FIND} : \text{Send}(\text{vk}, \cdot, \cdot), \text{Rec}^*(\text{vk}, \cdot, \cdot, 0), \text{Exec}^*(\text{vk}, \cdot, \cdot, \cdot))$
4. OSBEProtocol$\langle A, \text{Rec}^*(\text{vk}, M, \sigma, b) \rangle$
5. $b' \leftarrow A(\text{GUESS} : \text{Send}(\text{vk}, \cdot, \cdot), \text{Rec}^*(\text{vk}, \cdot, \cdot, 0), \text{Exec}^*(\text{vk}, \cdot, \cdot, \cdot))$
6. RETURN b'

$\qquad$ $\mathsf{Exp}_{OSBE,A}^{sem^*-b}(k)$ [Semantic security w.r.t. the authority]
$\qquad$ 1. param $\leftarrow$ OSBESetup(1^k)
$\qquad$ 2. vk $\leftarrow A(\text{INIT} : \text{param})$
$\qquad$ 3. $(M, \sigma, P_0, P_1) \leftarrow A(\text{FIND} : \text{Send}(\text{vk}, \cdot, \cdot), \text{Rec}^*(\text{vk}, \cdot, \cdot, 0), \text{Exec}^*(\text{vk}, \cdot, \cdot, \cdot))$
$\qquad$ 4. transcript $\leftarrow$ OSBEProtocol$\langle \text{Send}(\text{vk}, M, P_b), \text{Rec}^*(\text{vk}, M, \sigma, 0) \rangle$
$\qquad$ 5. $b' \leftarrow A(\text{GUESS} : \text{transcript}, \text{Send}(\text{vk}, \cdot, \cdot), \text{Rec}^*(\text{vk}, \cdot, \cdot, 0), \text{Exec}^*(\text{vk}, \cdot, \cdot, \cdot))$
$\qquad$ 6. RETURN b'

$\mathsf{Exp}_{OSBE,A}^{sem-b}(k)$ [Semantic Security]
1. param $\leftarrow$ OSBESetup(1^k)
2. (vk, sk) $\leftarrow$ OSBEKeyGen(param)
3. $(M, P_0, P_1) \leftarrow A(\text{FIND} : \text{vk}, \text{Sign}^*(\text{vk}, \cdot), \text{Send}(\text{vk}, \cdot, \cdot), \text{Rec}(\text{vk}, \cdot, 0), \text{Exec}(\text{vk}, \cdot, \cdot))$
4. OSBEProtocol$\langle \text{Send}(\text{vk}, M, P_b), A \rangle$
5. $b' \leftarrow A(\text{GUESS} : \text{Sign}(\text{vk}, \cdot), \text{Send}(\text{vk}, \cdot, \cdot), \text{Rec}(\text{vk}, \cdot, 0), \text{Exec}(\text{vk}, \cdot, \cdot))$
6. IF $M \in SM$ RETURN 0 ELSE RETURN b'

Fig. 1. Security Games for $OSBE$

w.r.t. the authority (sem*) is independent from the basic semantic security (sem) since in the latter the adversary interacts with the sender whereas in the former the adversary (who generated the signing keys) has only passive access to a challenge transcript.

These security notions can be formalized by the security games presented on Figure 1, where the adversary keeps some internal state between the various calls INIT, FIND and GUESS. They make use of the oracles described below, and the advantages of the adversary are, for all the security notions,

$$\mathsf{Adv}_{OSBE,A}^*(k) = \Pr[\mathsf{Exp}_{OSBE,A}^{*-1}(k) = 1] - \Pr[\mathsf{Exp}_{OSBE,A}^{*-0}(k) = 1]$$

$$\mathsf{Adv}_{OSBE}^*(k, t) = \max_{A \leq t} \mathsf{Adv}_{OSBE,A}^*(k).$$

- Sign(vk, m): This oracle outputs a valid signature on m under the signing key sk associated to vk (where the pair (vk, sk) has been outputted by the OSBEKeyGen algorithm);
- Sign*(vk, m): This oracle first queries Sign(vk, m). It additionally stores the query m to the list SM;
- Send(vk, m, P): This oracle emulates the sender with private input P, and thus may consist of multiple interactions;
- Rec(vk, m, b): This oracle emulates the recipient either with a valid signature σ on m under the verification key vk (obtained from the signing oracle Sign)

if $b = 0$ (as the above $\mathcal{R}_0$), or with a random string if $b = 1$ (as the above $\mathcal{R}_1$). This oracle is available when the signing key has been generated by OSBEKeyGen only;

- $\text{Rec}^*(\text{vk}, m, \sigma, b)$: This oracle does as above, with a valid signature σ provided by the adversary. If $b = 0$, it emulates the recipient playing with σ; if $b = 1$, it emulates the recipient playing with a random string;
- $\text{Exec}(\text{vk}, m, P)$: This oracle outputs the transcript of an honest execution between a sender with private input P and the recipient with a valid signature σ on m under the verification key vk (obtained from the signing oracle Sign). It basically activates the $\text{Send}(\text{vk}, m, P)$ and $\text{Rec}(\text{vk}, m, 0)$ oracles.
- $\text{Exec}^*(\text{vk}, m, \sigma, P)$: This oracle outputs the transcript of an honest execution between a sender with private input P and the recipient with a valid signature σ (provided by the adversary). It basically activates the $\text{Send}(\text{vk}, m, P)$ and $\text{Rec}^*(\text{vk}, m, \sigma, 0)$ oracles.

Remark 2. The OSBE schemes proposed in [27] do not satisfy the semantic security w.r.t. the authority. This is obvious for the generic construction based on identity-based encryption which consists in only one flow of communication (since a scheme that achieves the strong security notions requires at least two flows). This is also true (to a lesser extent) for the RSA-based construction: for any third party, the semantic security relies (in the random oracle model) on the CDH assumption in a 2048-bit RSA group; but for the authority, it can be broken by solving two 1024-bit discrete logarithm problems. This task is much simpler in particular if the authority generates the RSA modulus $N = pq$ dishonestly (*e.g.* with $p - 1$ and $q - 1$ smooth). In order to make the scheme secure in our strong model, one needs (at least) to double the size of the RSA modulus and to make sure that the authority has selected and correctly employed a truly random seed in the generation of the RSA key pair [25].

3 An Efficient OSBE Scheme

In this section, we present a high-level instantiation of OSBE with the previous primitives as black boxes. Thereafter, we provide a specific instantiation with linear ciphertexts. The overall security then relies on the DLin assumption, a quite standard assumption in the standard model. Its efficiency is of the same order of magnitude than the construction based on identity-based encryption [27] (that only achieves weaker security notions) and better than the RSA-based scheme which provides similar security guarantees (in the random oracle model).

3.1 High-Level Instantiation

We assume we have an encryption scheme $\mathcal{E}$, a signature scheme $\mathcal{S}$ and a SPHF system onto a set $\mathbb{G}$. We additionally use a key derivation function KDF to derive a pseudo-random bit-string $K \in \{0, 1\}^\ell$ from a pseudo-random element v in $\mathbb{G}$. One can use the Leftover-Hash Lemma [23], with a random seed defined in param during the global setup, to extract the entropy from v, then followed by a pseudo-random generator to get a long enough bit-string. Many uses of the same seed in the Leftover-Hash-Lemma just leads to a security loss linear in the number of extractions. We describe an oblivious signature-based envelope system $\mathcal{OSBE}$, to send a private message $P \in \{0, 1\}^{\ell}$:

- OSBESetup(1^k), where k is the security parameter:
  - it first generates the global parameters for the signature scheme (using SSetup), the encryption scheme (using ESetup), and the SPHF system (using SPHFSetup);
  - it then generates the public key ek of the encryption scheme (using EKeyGen, while the decryption key will not be used);
  The output param consists of all the individual param and the encryption key ek;
- OSBEKeyGen(param) runs SKeyGen(param) to generate a pair (vk, sk) of verification-signing keys;
- The OSBESign and OSBEVerif algorithms are exactly Sign and Verif from the signature scheme;
- OSBEProtocol⟨$\mathcal{S}$(vk, M, P), $\mathcal{R}$(vk, M, σ)⟩: In the following, $\mathcal{L} = \mathcal{L}(vk, M)$ will describe the language of the ciphertexts under the above encryption key ek of a valid signature of the input message M under the input verification key vk (hence vk and M as inputs, while param contains ek).
  - $\mathcal{R}$ generates and sends $c = $ Encrypt(ek, σ; r);
  - $\mathcal{S}$ computes hk = HashKG($\mathcal{L}$, param), hp = ProjKG(hk, ($\mathcal{L}$, param), c), $v = $ Hash(hk, ($\mathcal{L}$, param), c), and $Q = P \oplus $ KDF(v); $\mathcal{S}$ sends hp, Q to $\mathcal{R}$;
  - $\mathcal{R}$ computes $v' = $ ProjHash(hp, ($\mathcal{L}$, param), c, r) and $P' = Q \oplus $ KDF(v').

3.2 Security Properties

Theorem 3 (Correct). $\mathcal{OSBE}$ *is sound*.

Proof. Under the correctness of the SPHF system, $v' = v$, and thus $P' = (P \oplus$ KDF(v)) $\oplus$ KDF(v') $= P$.

Theorem 4 (Escrow-Free). $\mathcal{OSBE}$ *is escrow-free if the encryption scheme $\mathcal{E}$ is semantically secure:* $\mathsf{Adv}_{\mathcal{OSBE}}^{esc}(k, t) \leq \mathsf{Adv}_{\mathcal{E}}^{ind}(k, t')$ *with* $t' \approx t$.

Proof. Let us assume $\mathcal{A}$ is an adversary against the escrow-free property of our scheme: The malicious adversary $\mathcal{A}$ is able to tell the difference between an interaction with $\mathcal{R}_0$ (who knows and uses a valid signature) and $\mathcal{R}_1$ (who does not use a valid signature), with advantage ε.

We now build an adversary $\mathcal{B}$ against the semantic security of the encryption scheme $\mathcal{E}$:

- $\mathcal{B}$ is first given the parameters for $\mathcal{E}$ and an encryption key ek;
- $\mathcal{B}$ emulates OSBESetup: it runs SSetup and SPHFSetup by itself. For the encryption scheme $\mathcal{E}$, the parameters and the key have already been provided by the challenger of the encryption security game;

- $\mathcal{A}$ provides the verification key vk;
- $\mathcal{B}$ has to simulate all the oracles:
  - Send(vk, M, P), for a message M and a private input P: upon receiving c, one computes hk = HashKG($\mathcal{L}$, param), hp = ProjKG(hk, ($\mathcal{L}$, param), c), $v = $ Hash(hk, ($\mathcal{L}$, param), c), and $Q = P \oplus \mathsf{KDF}(v)$. One sends back (hp, Q);
  - Rec*(vk, M, σ, 0), for a message M and a valid signature σ: $\mathcal{B}$ outputs $c = $ Encrypt(ek, σ; r);
  - Exec*(vk, M, σ, P): one first runs Rec(vk, M, σ, 0) to generate c, that is provided to Send(vk, M, P), to generate (hp, Q).
- At some point, $\mathcal{A}$ outputs a message M and a valid signature σ, and $\mathcal{B}$ has to simulate Rec*(vk, M, σ, b): $\mathcal{B}$ sets $\sigma_0 \leftarrow \sigma$ and sets σ_1 as a random string. It sends (σ_0, σ_1) to the challenger of the semantic security of the encryption scheme and gets back c, an encryption of σ_β, for a random unknown bit β. It outputs c;
- $\mathcal{B}$ provides again access to the above oracles, and $\mathcal{A}$ outputs a bit b', that $\mathcal{B}$ forwards as its guess β' for the β involved in the semantic security game for $\mathcal{E}$.

Note that the above simulation perfectly emulates $\mathsf{Exp}^{\mathsf{esc}-\beta}_{\mathcal{OSBE},\mathcal{A}}(k)$ (since basically b is β, and b' is β'):

$$\varepsilon = \mathsf{Adv}^{\mathsf{esc}}_{\mathcal{OSBE},\mathcal{A}}(k) = \mathsf{Adv}^{\mathsf{ind}}_{\mathcal{E},\mathcal{B}}(k) \le \mathsf{Adv}^{\mathsf{ind}}_{\mathcal{E}}(k, t).$$

Theorem 5 (Semantically Secure). $\mathcal{OSBE}$ *is semantically secure if the signature is unforgeable, the SPHF is smooth and the encryption scheme is semantically secure (and under the pseudo-randomness of the KDF):*

$$\mathsf{Adv}^{\mathsf{sem}}_{\mathcal{OSBE}}(k, t) \le q_U \, \mathsf{Adv}^{\mathsf{ind}}_{\mathcal{E}}(k, t') + 2 \, \mathsf{Succ}^{\mathsf{euf}}_{\mathcal{S}}(k, q_S, t'') + 2 \, \mathsf{Adv}^{\mathsf{smooth}}_{\mathcal{SPHF}}(k) \, \text{with} \, t', t'' \approx t.$$

In the above formula, q_U denotes the number of interactions the adversary has with the sender, and q_S the number of signing queries the adversary asked.

Proof. Let us assume $\mathcal{A}$ is an adversary against the semantic security of our scheme: The malicious adversary $\mathcal{A}$ is able to tell the difference between an interaction with $\mathcal{S}_0$ (who owns P_0) and $\mathcal{S}_1$ (who owns P_1), with advantage ε. We start from this initial security game, and make slight modifications to bound ε.

Game $\mathcal{G}_0$. Let us emulate this security game:

- $\mathcal{B}$ emulates the initialization of the system: it runs OSBESetup by itself, and then OSBEKeyGen to generate (vk, sk);
- $\mathcal{B}$ has to simulate all the oracles:
  - Sign(vk, M) and Sign*(vk, M): it runs the corresponding algorithm by itself;
  - Send(vk, M, P), for a message M and a private input P: upon receiving c, one computes hk = HashKG($\mathcal{L}$, param), hp = ProjKG(hk, ($\mathcal{L}$, param), c), $v = $ Hash(hk, ($\mathcal{L}$, param), c), and $Q = P \oplus \mathsf{KDF}(v)$. One sends back (hp, Q);

- Rec(vk, M, 0), for a message M: $\mathcal{B}$ asks for a valid signature σ on M, computes and outputs $c = \mathsf{Encrypt}(\mathsf{ek}, \sigma; r)$;
- Exec(vk, M, P): one simply first runs Rec(vk, M, 0) to generate c, that is provided to Send(vk, M, P), to generate (hp, Q).
- At some point, $\mathcal{A}$ outputs a message M and two inputs (P_0, P_1) to distinguish the sender, and $\mathcal{B}$ call back the above Send(vk, M, P_b) simulation to interact with $\mathcal{A}$;
- $\mathcal{B}$ provides again access to the above oracles, and $\mathcal{A}$ outputs a bit b'.

In this game, $\mathcal{A}$ has an advantage ε in guessing b:

$$\varepsilon = \Pr_0[b' = 1 | b = 1] - \Pr_{\mathcal{G}_0}[b' = 1 | b = 0] = 2 \times \Pr_{\mathcal{G}_0}[b' = b] - 1.$$

Game $\mathcal{G}_1^\beta$. This game involves the semantic security of the encryption scheme: $\mathcal{B}$ is already provided the parameters and the encryption key ek by the challenger of the semantic security of the encryption scheme, hence the initialization is slightly modified. In addition, $\mathcal{B}$ randomly chooses the bit b, and modifies the Rec oracle simulation:

- Rec(vk, M, 0), for a message M: $\mathcal{B}$ asks for a valid signature σ_0 on M, and sets σ_1 as a random string, computes and outputs $c = \mathsf{Encrypt}(\mathsf{ek}, \sigma_b; r)$.

Since $\mathcal{B}$ knows b, it finally outputs $\beta' = (b' = b)$.

Note that $\mathcal{G}_1^0$ is exactly $\mathcal{G}_0$, and the distance between $\mathcal{G}_1^0$ and $\mathcal{G}_1^1$ relies on the Left-or-Right security of the encryption scheme, which can be shown equivalent to the semantic security, with a lost linear in the number of encryption queries, which is actually the number q_U of interactions with a user (the sender in this case), due to the hybrid argument [4]:

$$\begin{aligned}
q_U \times \mathsf{Adv}_{\mathcal{E}}^{\mathsf{ind}}(k) &\geq \Pr[\beta' = 1 | \beta = 0] - \Pr[\beta' = 1 | \beta = 1] \\
&= \Pr[b' = b | \beta = 0] - \Pr[b' = b | \beta = 1] \\
&= (2 \times \Pr_{\mathcal{G}_1^0}[b' = b] - 1) - (2 \times \Pr_{\mathcal{G}_1^1}[b' = b] - 1)
\end{aligned}$$

As a consequence: $\varepsilon \leq q_U \times \mathsf{Adv}_{\mathcal{E}}^{\mathsf{ind}}(k) + (2 \times \Pr_{\mathcal{G}_1^1}[b' = b] - 1)$.

Game $\mathcal{G}_2$. This game involves the unforgeability of the signature scheme: $\mathcal{B}$ is already provided the parameters and the verification vk for the signature scheme, together with access to the signing oracle (note that all the signing queries Sign* asked by the adversary in the FIND stage, i.e., before the challenge interaction with Send(vk, M, P_b), are stored in $\mathcal{SM}$). The simulator $\mathcal{B}$ generates itself all the other parameters and keys, an namely the encryption key ek, together with the associated decryption key dk. For the Rec oracle simulation, $\mathcal{B}$ keeps the random version (as in $\mathcal{G}_1^1$). In the challenge interaction with Send(vk, M, P_b), one stops the simulation and makes the adversary win if it uses a valid signature on a message $M \notin \mathcal{SM}$:

- Send(vk, M, P_b), during the challenge interaction: upon receiving c, if $M \notin \mathcal{SM}$, it first decrypts c to get the input signature σ. If σ is a valid signature, one stops the game, sets $b' = b$ and outputs b'. If the signature is in not valid, the simulation remains unchanged;
- Rec(vk, $M, 0$), for a message M: $\mathcal{B}$ sets σ as a random string, computes and outputs $c = $ Encrypt(ek, $\sigma; r$).

Because of the abort in the case of a valid signature on a new message, we know that the adversary cannot use such a valid signature in the challenge. So, since M should not be in $\mathcal{SM}$, the signature will be invalid. Actually, the unique difference from the previous game $\mathcal{G}_1^1$ is the abort in case of valid signature on a new message in the challenge phase, which probability is bounded by $\mathsf{Succ}_S^{\mathsf{euf}}(k, q_S)$. Using Shoup's Lemma [29]:

$$\Pr_{\mathcal{G}_1^1}[b' = b] - \Pr_{\mathcal{G}_2}[b' = b] \leq \mathsf{Succ}_S^{\mathsf{euf}}(k, q_S).$$

As a consequence: $\varepsilon \leq q_U \times \mathsf{Adv}_{\mathcal{E}}^{\mathsf{ind}}(k) + 2 \times \mathsf{Succ}_S^{\mathsf{euf}}(k, q_S) + (2 \times \Pr_{\mathcal{G}_2}[b' = b] - 1)$.

Game $\mathcal{G}_3$. The last game involves the smoothness of the SPHF: The unique difference is in the computation of v in Send simulation, in the challenge phase only: $\mathcal{B}$ chooses a random $v \in \mathbb{G}$. Due to the statistical randomness of v in the previous game, in case the signature is not valid (a word that is not in the language), this game is statistically indistinguishable from the previous one:

$$\Pr_{\mathcal{G}_2}[b' = b] - \Pr_{\mathcal{G}_3}[b' = b] \leq \mathsf{Adv}_{SPHF}^{\mathsf{smooth}}(k).$$

Since P_b is now masked by a truly random value, no information leaks on b: $\Pr_{\mathcal{G}_3}[b' = b] = 1/2$.

Theorem 6. *$\mathcal{OSBE}$ is semantically secure w.r.t. the authority if the SPHF is pseudo-random (and under the pseudo-randomness of the KDF):*

$$\mathsf{Adv}_{\mathcal{OSBE}}^{\mathsf{sem}^*}(k, t) \leq 2 \times \mathsf{Adv}_{SPHF}^{\mathsf{pr}}(k, t).$$

Proof. Let us assume $\mathcal{A}$ is an adversary against the semantic security w.r.t. the authority: The malicious adversary $\mathcal{A}$ is able to tell the difference between an eavesdropped interaction with S_0 (who owns P_0) and S_1 (who owns P_1), with advantage ε. We start from this initial security game, and make slight modifications to bound ε.

Game $\mathcal{G}_0$. Let us emulate this security game:

- $\mathcal{B}$ emulates the initialization of the system: it runs OSBESetup by itself;
- $\mathcal{A}$ provides the verification key vk;
- $\mathcal{B}$ has to simulate all the oracles:
  - Send(vk, M, P), for a message M and a private input P: upon receiving c, one computes hk = HashKG($\mathcal{L}$, param), hp = ProjKG(hk, ($\mathcal{L}$, param), c), $v = $ Hash(hk, ($\mathcal{L}$, param), c), and $Q = P \oplus$ KDF(v). One sends back (hp, Q);

- Rec*(vk, $M, \sigma, 0$), for a message M and a valid signature σ: $\mathcal{B}$ outputs $c = \mathsf{Encrypt}(\mathsf{ek}, \sigma; r)$;
- Exec*(vk, M, σ, P): one first runs $\mathsf{Rec}(\mathsf{vk}, M, \sigma, 0)$ to generate c, that is provided to $\mathsf{Send}(\mathsf{vk}, M, P)$, to generate (hp, Q).

- At some point, $\mathcal{A}$ outputs a message M with a valid signature σ, and two inputs (P_0, P_1) to distinguish the sender, and $\mathcal{B}$ call back the above $\mathsf{Send}(\mathsf{vk}, M, P_b)$ and $\mathsf{Rec}^*(\mathsf{vk}, M, \sigma, 0)$ simulations to interact together and output the transcript $(c; \mathsf{hp}, Q)$;
- $\mathcal{B}$ provides again access to the above oracles, and $\mathcal{A}$ outputs a bit b'.

In this game, $\mathcal{A}$ has an advantage ε in guessing b:

$$\varepsilon = \Pr_{\mathcal{G}_0}[b' = 1 | b = 1] - \Pr_{\mathcal{G}_0}[b' = 1 | b = 0] = 2 \times \Pr_{\mathcal{G}_0}[b' = b] - 1.$$

Game $\mathcal{G}_1$. This game involves the pseudo-randomness of the SPHF: The unique difference is in the computation of v in Send simulation of the eavesdropped interaction, and so for the transcript: $\mathcal{B}$ chooses a random $v \in \mathbb{G}$ and computes $Q = P_b \oplus \mathsf{KDF}(v)$. Due to the pseudo-randomness of v in the previous game, since $\mathcal{A}$ does not know the random coins r used to encrypt σ, this game is computationally indistinguishable from the previous one.

$$\Pr_{\mathcal{G}_1}[b' = b] - \Pr_{\mathcal{G}_0}[b' = b] \leq \mathsf{Adv}^{\mathsf{pr}}_{\mathcal{SPHF}}(k, t).$$

Since P_b is now masked by a truly random value v, no information leaks on b: $\Pr_{\mathcal{G}_1}[b' = b] = 1/2$.

3.3   Our Efficient OSBE Instantiation

Our first construction combines the linear encryption scheme [7], the Waters signature scheme [30] and a SPHF on linear ciphertexts [13,28]. It thus relies on classical assumptions: CDH for the unforgeability of signatures and DLin for the semantic security of the encryption scheme. The formal definitions are recalled in the full version [6].

Basic Primitives. Given an encrypted Waters signature from the recipient, the sender is able to compute a projection key, and a hash corresponding to the expected signature, and send to the recipient the projection key and the product between the expected hash and the message P. If the recipient was honest (a correct ciphertext), it is able to compute the hash thanks to the projection key, and so to find P, in the other case it does not learn anything.

We briefly sketch the basic building blocks: linear encryption, Waters signature and the SPHF for linear tuples.

All these primitives work in a pairing-friendly environment $(p, \mathbb{G}, g, \mathbb{G}_T, e)$, where $e: \mathbb{G} \times \mathbb{G} \to \mathbb{G}_T$ is an admissible bilinear map, for two groups $\mathbb{G}$ and $\mathbb{G}_T$, of prime order p, generated by g and $g_t = e(g, g)$ respectively.

Waters Signatures. The public parameters are a generator $h \xleftarrow{\$} \mathbb{G}$ and a vector $\boldsymbol{u} = (u_0, \dots, u_k) \xleftarrow{\$} \mathbb{G}^{k+1}$, which defines the *Waters hash* of a message $M = (M_1, \dots, M_k) \in \{0, 1\}^k$ as $\mathcal{F}(M) = u_0 \prod_{i=1}^{k} u_i^{M_i}$. The public verification key is $\mathsf{vk} = g^z$, which corresponding secret signing key is $\mathsf{sk} = h^z$, for a random $z \xleftarrow{\$} \mathbb{Z}_p$. The signature on a message $M \in \{0, 1\}^k$ is $\sigma = (\sigma_1 = \mathsf{sk} \cdot \mathcal{F}(M)^s, \sigma_2 = g^s)$, for some random $s \xleftarrow{\$} \mathbb{Z}_p$. It can be verified by checking $e(g, \sigma_1) = e(\mathsf{vk}, h) \cdot e(\mathcal{F}(M), \sigma_2)$. This signature scheme is *unforgeable* under the CDH assumption.

Linear Encryption. The secret key dk is a pair of random scalars (y_1, y_2) and the public key is $\mathsf{ek} = (Y_1 = g^{y_1}, Y_2 = g^{y_2})$. One encrypts a message $M \in \mathbb{G}$ as $c = (c_1 = Y_1^{r_1}, c_2 = Y_2^{r_2}, c_3 = g^{r_1 + r_2} \cdot M)$, for random scalars $r_1, r_2 \xleftarrow{\$} \mathbb{Z}_p$. To decrypt, one computes $M = c_3 / (c_1^{1/y_1} c_2^{1/y_2})$. This encryption scheme is *semantically secure* under the DLin assumption.

DLin-*compatible Smooth-Projective Hash Function.* This is actually a weaker variant of [28]. The language $\mathcal{L}$ consists of the linear tuples w.r.t. a basis (u, v, g). For a linear encryption key $\mathsf{ek} = (Y_1, Y_2)$, a ciphertext $C = (c_1, c_2, c_3)$ is an encryption of the message M if $(c_1, c_2, c_3/M)$ is a linear tuple w.r.t. the basis (Y_1, Y_2, g). The language $\mathsf{Lin}(\mathsf{ek}, M)$ consists of these ciphertexts. An SPHF for this language can be:

$$\mathsf{HashKG}(\mathsf{Lin}(\mathsf{ek}, M)) = \mathsf{hk} = (x_1, x_2, x_3) \xleftarrow{\$} \mathbb{Z}_p^3$$
$$\mathsf{Hash}(\mathsf{hk}; \mathsf{Lin}(\mathsf{ek}, M), C) = c_1^{x_1} c_2^{x_2} (c_3/M)^{x_3}$$
$$\mathsf{ProjKG}(\mathsf{hk}; \mathsf{Lin}(\mathsf{ek}, M), C) = \mathsf{hp} = (Y_1^{x_1} g^{x_3}, Y_2^{x_2} g^{x_3})$$
$$\mathsf{ProjHash}(\mathsf{hp}; \mathsf{Lin}(\mathsf{ek}, M), C; r) = \mathsf{hp}_1^{r_1} \mathsf{hp}_2^{r_2}$$

This function is defined for linear tuples in $\mathbb{G}$, but it could work in any group, since it does not make use of pairings. And namely, we use it below in $\mathbb{G}_T$.

Smooth-Projective Hash Function for Linear Encryption of Valid Waters Signatures. We will consider a slightly more complex language: the ciphertexts under ek of a valid signature of M under vk. A given ciphertext $C = (c_1, c_2, c_3, \sigma_2)$ contains a valid signature of M if and only if (c_1, c_2, c_3) actually encrypts σ_1 such that (σ_1, σ_2) is a valid Waters signature on M. The latter means

$$(C_1 = e(c_1, g), C_2 = e(c_2, g), C_3 = e(c_3, g)/(e(h, \mathsf{vk}) \cdot e(\mathcal{F}(M), \sigma_2))$$

is a linear tuple in basis $(U = e(Y_1, g), V = e(Y_2, g), g_t = e(g, g))$ in $\mathbb{G}_T$. Since the basis consists of 3 elements of the form $e(\cdot, g)$, the projected key can be compacted in $\mathbb{G}$. We thus consider the language $\mathsf{WLin}(\mathsf{ek}, \mathsf{vk}, M)$ that contains these quadruples $(c_1, c_2, c_3, \sigma_2)$, and its SPHF:

$$\mathsf{HashKG}(\mathsf{WLin}(\mathsf{ek}, \mathsf{vk}, M)) = \mathsf{hk} = (x_1, x_2, x_3) \xleftarrow{\$} \mathbb{Z}_p^3$$
$$\mathsf{Hash}(\mathsf{hk}; \mathsf{WLin}(\mathsf{ek}, \mathsf{vk}, M), C) =$$
$$e(c_1, g)^{x_1} e(c_2, g)^{x_2} (e(c_3, g)/(e(h, \mathsf{vk}) e(\mathcal{F}(M), \sigma_2)))^{x_3}$$
$$\mathsf{ProjKG}(\mathsf{hk}; \mathsf{WLin}(\mathsf{ek}, \mathsf{vk}, M), C) = \mathsf{hp} = (\mathsf{ek}_1^{x_1} g^{x_3}, \mathsf{ek}_2^{x_2} g^{x_3})$$
$$\mathsf{ProjHash}(\mathsf{hp}; \mathsf{WLin}(\mathsf{ek}, \mathsf{vk}, M), C; r) = e(\mathsf{hp}_1^{r_1} \mathsf{hp}_2^{r_2}, g)$$

Instantiation. We now define our OSBE protocol, where a sender $\mathcal{S}$ wants to send a private message $P \in \{0,1\}^{\ell}$ to a recipient $\mathcal{R}$ in possession of a Waters signature on a message M.

- OSBESetup(1^k), where k is the security parameter, defines a pairing-friendly environment $(p, \mathbb{G}, g, \mathbb{G}_T, e)$, the public parameters $h \xleftarrow{\$} \mathbb{G}$, an encryption key ek $= (Y_1 = g^{y_1}, Y_2 = g^{y_2})$, where $(y_1, y_2) \xleftarrow{\$} \mathbb{Z}_p^2$, and $\boldsymbol{u} = (u_0, \ldots, u_k) \xleftarrow{\$} \mathbb{G}^{k+1}$ for the Waters signature. All these elements constitute the string param;
- OSBEKeyGen(param), the authority generates a pair of keys (vk $= g^z$, sk $= h^z$) for a random scalar $z \xleftarrow{\$} \mathbb{Z}_p$;
- OSBESign(sk, M) produces a signature $\sigma = (h^z \mathcal{F}(M)^s, g^s)$;
- OSBEVerif(vk, M, σ) checks if $e(\sigma_1, g) = e(\sigma_2, \mathcal{F}(M)) \cdot e(h, \text{vk})$.
- OSBEProtocol$\langle \mathcal{S}(\text{vk}, M, P), \mathcal{R}(\text{vk}, M, \sigma) \rangle$ runs as follows:
  - $\mathcal{R}$ chooses random $r_1, r_2 \xleftarrow{\$} \mathbb{Z}_p$ and sends a linear encryption of σ: $C = (c_1 = \text{ek}_1^{r_1}, c_2 = \text{ek}_2^{r_2}, c_3 = g^{r_1+r_2} \cdot \sigma_1, \sigma_2)$
  - $\mathcal{S}$ chooses random $x_1, x_2, x_3 \xleftarrow{\$} \mathbb{Z}_p^3$ and computes:
    * HashKG(WLin(ek, vk, M)) = hk = (x_1, x_2, x_3);
    * Hash(hk; WLin(ek, vk, M), C) = v = $e(c_1, g)^{x_1} e(c_2, g)^{x_2} (e(c_3, g)/(e(h, \text{vk})e(\mathcal{F}(M), \sigma_2)))^{x_3}$;
    * ProjKG(hk; WLin(ek, vk, M), C) = hp = $(\text{ek}_1^{x_1} g^{x_3}, \text{ek}_2^{x_2} g^{x_3})$.
  - $\mathcal{S}$ then sends (hp, $Q = P \oplus \text{KDF}(v)$) to $\mathcal{R}$;
  - $\mathcal{R}$ computes $v' = e(\text{hp}_1^{r_1} \text{hp}_2^{r_2}, g)$ and $P' = Q \oplus \text{KDF}(v')$.

An asymmetric instantiation can be found in the full version [6].

3.4 Security and Efficiency

We now provide a security analysis of this scheme. This instantiation differs, from the high-level instantiation presented before, in the ciphertext C of the signature $\sigma = (\sigma_1, \sigma_2)$. The second half of the signature indeed remains in clear. It thus does not guarantee the semantic security on the signature used in the ciphertext. However, granted Waters signature randomizability, one can re-randomize the signature each time, and thus provide a totally new σ_2: it does not leak any information about the original signature. The first part of the ciphertext (c_1, c_2, c_3) does not leak any additional information under the DLin assumption. As a consequence, the global ciphertext guarantees the semantic security of the original signature if a new re-randomized signature is encrypted each time. We can now apply the high-level construction security, and all the assumptions hold under the DLin one:

Theorem 7. *Our OSBE scheme is secure (i.e., escrow-free, semantically secure, and semantically secure w.r.t. the authority) under the* DLin *assumption (and the pseudo-random generator in the* KDF*).*

Our proposed scheme needs one communication for $\mathcal{R}$ and one for $\mathcal{S}$, so it is round-optimal. Communication also consists of few elements, $\mathcal{R}$ sends 4 group elements, and $\mathcal{S}$ answers with 2 group elements only and an ℓ-bit string for the masked $P \in \{0,1\}^\ell$. As explained in Remark 2, this has to be compared with the RSA-based scheme from [27] which requires 2 elements in RSA groups (with double-length modulus). For a 128-bit security level, using standard Type-I bilinear groups implementation [16], we obtain a 62.5% improvement[1] in communication complexity over the RSA-based scheme proposed in the original paper [27].

While reducing the communication cost of the scheme, we have improved its security and it now fits the proposed applications. In [27], such schemes were proposed for applications where someone wants to transmit a confidential information to an agent belonging to a specific agency. However the agent does not want to give away his signature. As they do not consider eavesdropping and replay in their semantic security nothing prevents an adversary to replay a part of a previous interaction to impersonate a CIA agent (to recall their example). In practice, an additional secure communication channel, such as with SSL, was required in their security model, hence increasing the communication cost: our protocol is secure by itself.

4 An Efficient Blind Signature

4.1 Definitions

A more formal definition of blind signatures is provided in the full version [6], but we briefly recall it in this section: A blind signature scheme $\mathcal{BS}$ is defined by a setup algorithm $\mathsf{BSSetup}(1^k)$ that generates the global parameters param, and key generation algorithm $\mathsf{BSKeyGen}(\mathsf{param})$ that outputs a pair $(\mathsf{vk}, \mathsf{sk})$, and interactive protocol $\mathsf{BSProtocol}\langle\mathcal{S}(\mathsf{sk}), \mathcal{U}(\mathsf{vk}, m)\rangle$ which provides $\mathcal{U}$ with a signature on m, and a verification algorithm $\mathsf{BSVerif}(\mathsf{vk}, m, \sigma)$ that checks its validity. The security of a blind signature scheme is defined through the unforgeability and blindness properties: An adversary against the unforgeability tries to generate $q_s + 1$ valid message-signature pairs after at most q_s complete interactions with the honest signer; The blindness condition states that a malicious signer should be unable to decide which of two messages m_0, m_1 has been signed first in two executions with an honest user.

4.2 Our Instantiation

We now present a new way to obtain a blind signature scheme in the standard model under classical assumptions with a common-reference string. This is an improvement over [5]. We are going to use the same building blocks as before, so linear encryption, Waters signatures and a SPHF on linear ciphertexts. More elaborated languages will be required, but just conjunctions and disjunctions of

---

[1] The improvement is even more important for the scheme described in the full version where the size drops down to 3/16-th.

classical languages, as done in [1] (see the full version [6]), hence the efficient construction. Our blind signature scheme is defined by:

- BSSetup(1^k), generates a pairing-friendly system $(p, \mathbb{G}, g, \mathbb{G}_T, e)$ and an encryption key ek $= (u, v, g) \in \mathbb{G}^3$. It also chooses at random $h \in \mathbb{G}$ and generators $\boldsymbol{u} = (u_i)_{i \in [\![1,\ell]\!]} \in \mathbb{G}^\ell$ for the Waters function. It outputs the global parameters param $= (p, \mathbb{G}, g, \mathbb{G}_T, e, \text{ek}, h, \boldsymbol{u})$;
- BSKeyGen(param) picks at random a secret key sk $= x$ and computes the verification key vk $= g^x$;
- BSProtocol$\langle \mathcal{S}(\text{sk}), \mathcal{U}(\text{vk}, m) \rangle$ runs as follows, where $\mathcal{U}$ wants to get a signature on M
    - $\mathcal{U}$ computes the bit-per-bit encryption of M by encrypting each $u_i^{M_i}$ in b_i, $\forall i \in [\![1, \ell]\!], b_i = \text{Encrypt}(\text{ek}, u_i^{M_i}; (r_{i,1}, r_{i,2})) = (u^{r_{i,1}}, v^{r_{i,2}}, g^{r_{i,1}+r_{i,2}} u_i^{M_i})$. Then writing $r_1 = \sum r_{i,1}$ and $r_2 = \sum r_{i,2}$, he computes the encryption c of vk$^{r_1+r_2}$ with $\text{Encrypt}(\text{ek}, \text{vk}^{r_1+r_2}; (s_1, s_2)) = (u^{s_1}, v^{s_2}, g^{s_1+s_2}\text{vk}^{r_1+r_2})$. $\mathcal{U}$ then sends $(c, (b_i))$;
    - On input of these ciphertexts, the algorithm $\mathcal{S}$ computes the corresponding SPHF, considering the language $\mathcal{L}$ of valid ciphertexts. This is the conjunction of several languages :
        1. One checking that each b_i encrypts a bit in basis u_i: in BLin(ek, u_i);
        2. One considering $(d_1, d_2, c_1, c_2, c_3)$, that checks if (c_1, c_2, c_3) encrypts an element d_3 such that (d_1, d_2, d_3) is a linear tuple in basis (u, v, vk): in ELin(ek, vk), where $d_1 = \prod_i b_{i,1}$ and $d_2 = \prod_i b_{i,2}$.
    - $\mathcal{S}$ computes the corresponding Hash-value v, extracts $K = \text{KDF}(v) \in \mathbb{Z}_p$, generates the blinded signature $(\sigma_1'' = h^x \delta^s, \sigma_2' = g^s)$, where $\delta = u_0 \prod_i b_{i,3} = \mathcal{F}(M) g^{r_1+r_2}$, and sends $(\text{hp}, Q = \sigma_1'' \times g^K, \sigma_2')$;
    - Upon receiving $(\text{hp}, Q, \sigma_2')$, using its witnesses and hp, $\mathcal{U}$ computes the ProjHash-value v', extracts $K' = \text{KDF}(v')$ and unmasks $\sigma_1'' = Q \times g^{-K'}$. Thanks to the knowledge of r_1 and r_2, it can compute $\sigma_1' = \sigma_1'' \times (\sigma_2')^{-r_1-r_2}$. Note·that if $v' = v$, then $\sigma_1' = h^x \mathcal{F}(M)^s$, which together with $\sigma_2' = g^s$ is a valid Waters signature on M. It can thereafter re-randomize the final signature $\sigma = (\sigma_1' \cdot \mathcal{F}(M)^{s'}, \sigma_2' \cdot g^{s'})$.
- BSVerif(vk, M, σ), checks whether $e(\sigma_1, g) = e(h, \text{vk}) \cdot e(\mathcal{F}(M), \sigma_2)$.

The idea is to remove any kind of proof of knowledge in the protocol, which was the main concern in [5], and use instead a SPHF. This way, we obtain a protocol where the user first sends $3\ell + 6$ group elements for the ciphertext, and receives back $5\ell + 4$ elements for the projection key and 2 group elements for the blinded signature. So $8\ell + 12$ group elements are used in total. This has to be compared to $9\ell + 24$ in [5]. We both reduce the linear and the constant parts in the number of group elements involved while relying on the same hypotheses. And the final result is still a standard Waters signature.

Remark 8. In [17], Garg *el al.* proposed the first round-optimal blind signature scheme in the standard model, without CRS. In order to remove the CRS, their scheme makes use of ZAPs [14] and is quite inefficient. Moreover, its security relies on a stronger assumption (namely, sub-exponential hardness of one-to-one

one-way functions). A natural idea is to replace the CRS in our scheme with Groth-Ostrovsky-Sahai ZAP [21] based on the DLin assumption. This change would only double the communication complexity, but we do not know how to prove the security of the resulting scheme[2]. It remains a tantalizing open problem to design an efficient round-optimal blind signature in the standard model without CRS.

4.3 Security

In blind signatures, one expects two kinds of security properties:

- *blindness*, preventing the signer to be able to recognize which message was signed during a specific interaction. Due to Waters re-randomizability and linear encryption, this property is guaranteed in our scheme under the DLin assumption;
- *unforgeability*, guaranteeing the user will not be able to output more signed messages than the number of actual interactions. In this scheme, granted the extractability of the encryption (the simulator can know the decryption key) one can show that the user cannot provide a signature on a message different from the ones it asked to be blindly signed. Hence, the unforgeability relies on the Waters unforgeability, that is the CDH assumption.

Theorem 9. *Our blind signature scheme is blind[3] under the* DLin *assumption (and the pseudo-randomness of the* KDF*) and unforgeable under the* CDH *assumption.*

A full proof can be found in the full version [6].

Acknowledgments. This work was supported by the French ANR-07-TCOM-013-04 PACE Project, by the European Commission through the ICT Program under Contract ICT-2007-216676 ECRYPT II.

References

1. Abdalla, M., Chevalier, C., Pointcheval, D.: Smooth Projective Hashing for Conditionally Extractable Commitments. In: Halevi, S. (ed.) CRYPTO 2009. LNCS, vol. 5677, pp. 671–689. Springer, Heidelberg (2009)
2. Abe, M., Fuchsbauer, G., Groth, J., Haralambiev, K., Ohkubo, M.: Structure-Preserving Signatures and Commitments to Group Elements. In: Rabin, T. (ed.) CRYPTO 2010. LNCS, vol. 6223, pp. 209–236. Springer, Heidelberg (2010)

---

[2] Indeed, opening the commitment scheme in the ZAP and forging a signature relies on the same computational assumption, which makes it impossible to apply the complexity leveraging argument from [17].

[3] Our scheme satisfies the *a posteriori blindness* security notion – introduced in [24] – that models exactly what the signer sees in the real world. As mentioned in [24], it formalizes the security desired for most applications of blind signatures (*e.g.* e-cash or e-voting).

3. Balfanz, D., Durfee, G., Shankar, N., Smetters, D., Staddon, J., Wong, H.-C.: Secret handshakes from pairing-based key agreements. In: IEEE Symposium on Security and Privacy, pp. 180–196 (2003)
4. Bellare, M., Desai, A., Jokipii, E., Rogaway, P.: A concrete security treatment of symmetric encryption. In: 38th Annual Symposium on Foundations of Computer Science, pp. 394–403. IEEE Computer Society Press (October 1997)
5. Blazy, O., Fuchsbauer, G., Pointcheval, D., Vergnaud, D.: Signatures on Randomizable Ciphertexts. In: Catalano, D., Fazio, N., Gennaro, R., Nicolosi, A. (eds.) PKC 2011. LNCS, vol. 6571, pp. 403–422. Springer, Heidelberg (2011)
6. Blazy, O., Pointcheval, D., Vergnaud, D.: Round-Optimal Privacy-Preserving Protocols with Smooth Projective Hash Functions. In: Cramer, R. (ed.) TCC 2012. LNCS, vol. 7194, pp. 94–110. Springer, Heidelberg (2012)
7. Boneh, D., Boyen, X., Shacham, H.: Short Group Signatures. In: Franklin, M. (ed.) CRYPTO 2004. LNCS, vol. 3152, pp. 41–55. Springer, Heidelberg (2004)
8. Boyen, X., Waters, B.: Compact Group Signatures Without Random Oracles. In: Vaudenay, S. (ed.) EUROCRYPT 2006. LNCS, vol. 4004, pp. 427–444. Springer, Heidelberg (2006)
9. Boyen, X., Waters, B.: Full-Domain Subgroup Hiding and Constant-Size Group Signatures. In: Okamoto, T., Wang, X. (eds.) PKC 2007. LNCS, vol. 4450, pp. 1–15. Springer, Heidelberg (2007)
10. Bradshaw, R.W., Holt, J.E., Seamons, K.E.: Concealing complex policies with hidden credentials. In: Atluri, V., Pfitzmann, B., McDaniel, P. (eds.) ACM CCS 2004: 11th Conference on Computer and Communications Security, pp. 146–157. ACM Press (October 2004)
11. Castelluccia, C., Jarecki, S., Tsudik, G.: Secret Handshakes from CA-Oblivious Encryption. In: Lee, P.J. (ed.) ASIACRYPT 2004. LNCS, vol. 3329, pp. 293–307. Springer, Heidelberg (2004)
12. Chaum, D.: Blind signatures for untraceable payments. In: Chaum, D., Rivest, R.L., Sherman, A.T. (eds.) Advances in Cryptology – CRYPTO 1982, pp. 199–203. Plenum Press, New York (1983)
13. Cramer, R., Shoup, V.: Universal Hash Proofs and a Paradigm for Adaptive Chosen Ciphertext Secure Public-Key Encryption. In: Knudsen, L.R. (ed.) EUROCRYPT 2002. LNCS, vol. 2332, pp. 45–64. Springer, Heidelberg (2002)
14. Dwork, C., Naor, M.: Zaps and their applications. SIAM J. Comput. 36(6), 1513–1543 (2007)
15. Fischlin, M.: Round-Optimal Composable Blind Signatures in the Common Reference String Model. In: Dwork, C. (ed.) CRYPTO 2006. LNCS, vol. 4117, pp. 60–77. Springer, Heidelberg (2006)
16. Galbraith, S.D., Paterson, K.G., Smart, N.P.: Pairings for cryptographers. Discrete Applied Mathematics 156(16), 3113–3121 (2008)
17. Garg, S., Rao, V., Sahai, A., Schröder, D., Unruh, D.: Round Optimal Blind Signatures. In: Rogaway, P. (ed.) CRYPTO 2011. LNCS, vol. 6841, pp. 630–648. Springer, Heidelberg (2011)
18. Gennaro, R., Lindell, Y.: A Framework for Password-Based Authenticated Key Exchange. In: Biham, E. (ed.) EUROCRYPT 2003. LNCS, vol. 2656, pp. 524–543. Springer, Heidelberg (2003), http://eprint.iacr.org/2003/032.ps.gz
19. Gennaro, R., Lindell, Y.: A framework for password-based authenticated key exchange. ACM Transactions on Information and System Security 9(2), 181–234 (2006)

20. Groth, J.: Fully Anonymous Group Signatures Without Random Oracles. In: Kurosawa, K. (ed.) ASIACRYPT 2007. LNCS, vol. 4833, pp. 164–180. Springer, Heidelberg (2007)
21. Groth, J., Ostrovsky, R., Sahai, A.: Non-interactive Zaps and New Techniques for NIZK. In: Dwork, C. (ed.) CRYPTO 2006. LNCS, vol. 4117, pp. 97–111. Springer, Heidelberg (2006)
22. Groth, J., Sahai, A.: Efficient Non-interactive Proof Systems for Bilinear Groups. In: Smart, N.P. (ed.) EUROCRYPT 2008. LNCS, vol. 4965, pp. 415–432. Springer, Heidelberg (2008)
23. Håstad, J., Impagliazzo, R., Levin, L.A., Luby, M.: A pseudorandom generator from any one-way function. SIAM Journal on Computing 28(4), 1364–1396 (1999)
24. Hazay, C., Katz, J., Koo, C.-Y., Lindell, Y.: Concurrently-Secure Blind Signatures Without Random Oracles or Setup Assumptions. In: Vadhan, S.P. (ed.) TCC 2007. LNCS, vol. 4392, pp. 323–341. Springer, Heidelberg (2007)
25. Juels, A., Guajardo, J.: RSA Key Generation with Verifiable Randomness. In: Naccache, D., Paillier, P. (eds.) PKC 2002. LNCS, vol. 2274, pp. 357–374. Springer, Heidelberg (2002)
26. Kalai, Y.T.: Smooth Projective Hashing and Two-Message Oblivious Transfer. In: Cramer, R. (ed.) EUROCRYPT 2005. LNCS, vol. 3494, pp. 78–95. Springer, Heidelberg (2005)
27. Li, N., Du, W., Boneh, D.: Oblivious signature-based envelope. In: 22nd ACM Symposium Annual on Principles of Distributed Computing, pp. 182–189. ACM Press (July 2003)
28. Shacham, H.: A Cramer-Shoup encryption scheme from the Linear Assumption and from progressively weaker Linear variants. Cryptology ePrint Archive, Report 2007/074 (February 2007), http://eprint.iacr.org/
29. Shoup, V.: OAEP reconsidered. Journal of Cryptology 15(4), 223–249 (2002)
30. Waters, B.: Efficient Identity-Based Encryption Without Random Oracles. In: Cramer, R. (ed.) EUROCRYPT 2005. LNCS, vol. 3494, pp. 114–127. Springer, Heidelberg (2005)

On the Instantiability of Hash-and-Sign RSA Signatures

Yevgeniy Dodis[1], Iftach Haitner[2,*], and Aris Tentes[1]

[1] Department of Computer Science, New York University
{dodis,tentes}@cs.nyu.edu
[2] School of Computer Science, Tel Aviv University
iftachh@cs.tau.ac.il

Abstract. The hash-and-sign RSA signature is one of the most elegant and well known signatures schemes, extensively used in a wide variety of cryptographic applications. Unfortunately, the only existing analysis of this popular signature scheme is in the random oracle model, where the resulting idealized signature is known as the RSA *Full Domain Hash* signature scheme (RSA-FDH). In fact, prior work has shown several "uninstantiability" results for various abstractions of RSA-FDH, where the RSA function was replaced by a family of trapdoor random permutations, or the hash function instantiating the random oracle could not be keyed. These abstractions, however, do not allow the reduction and the hash function instantiation to use the algebraic properties of RSA function, such as the multiplicative group structure of $\mathbb{Z}_n^*$. In contrast, the multiplicative property of the RSA function is critically used in many standard model analyses of various RSA-based schemes.

Motivated by closing this gap, we consider the setting where the RSA function representation is generic (i.e., black-box) *but multiplicative*, whereas the hash function itself is in the standard model, and can be keyed and exploit the multiplicative properties of the RSA function. This setting abstracts all known techniques for designing provably secure RSA-based signatures in the standard model, and aims to address the main limitations of prior uninstantiability results. Unfortunately, we show that it is still impossible to reduce the security of RSA-FDH to any natural assumption even in our model. Thus, our result suggests that in order to prove the security of a given instantiation of RSA-FDH, one should use a non-black box security proof, or use specific properties of the RSA group that are not captured by its multiplicative structure alone. We complement our negative result with a positive result, showing that the RSA-FDH signatures *can* be proven secure under the *standard* RSA assumption, provided that the number of signing queries is *a-priori bounded*.

Keywords: RSA Signature, Hash-and-Sign, Full Domain Hash, Random Oracle Heuristic, Generic Groups, Black-Box Reductions.

---

* Research supported by Check Point Institute for Information Security, and ISF grant 1076/11.

R. Cramer (Ed.): TCC 2012, LNCS 7194, pp. 112–132, 2012.

1   Introduction

Bellare and Rogaway, [3], introduced the *random oracle* (RO) model, as a "paradigm for designing efficient protocols". When following this paradigm, one first builds a provably secure scheme assuming that an access to a *random* function is given, and (possibly) assuming some "standard" hardness assumption (e.g., factoring is hard). Then it instantiates the scheme by replacing the random function with some *concrete* "hash function" (e.g., SHA-1). The intuition underlying this paradigm is that a successful attack on the resulting scheme should indicate (unexpected) weaknesses of the hash function used. This paradigm (also known as the random oracle heuristic) has led to several highly efficient and widely used in practice constructions, such as the RSA *Full Domain Hash* signature scheme (RSA-FDH) [3] and RSA *Optimal Asymmetric Encryption Padding* scheme (RSA-OAEP) [4]. Typically, however, little is known about the provable security of such popular schemes in the *standard model*. In particular, it is unknown whether we can reduce their security to some "natural" assumption.

In this work we revisit this question once again, focusing, in particular, on the instantiability of the RSA hash-and-sign signatures. The RSA signature [31] is one of the most elegant and well known signatures schemes. It is extensively used in a wide variety of applications, and serves as the basis of several existing standards such as PKCS #1 [32]. In its "textbook" form, the signature σ of the message m is simply $\sigma = m^d \bmod n$, which can be verified by checking if $\sigma^e \equiv m \bmod n$, where e is the public RSA exponent, and $d = e^{-1} \bmod \phi(n)$. Of course, the textbook variant is completely insecure, as any σ is a valid signature of some message $m = \sigma^e \bmod n$. The traditional fix, known as RSA *hash-and-sign* signature, is to hash the message m before signing it using some "appropriate" hash function h (i.e., $\sigma = h(m)^d \bmod n$). The key question is how to instantiate this function h?

Bellare and Rogaway, [3], showed that in the random oracle model, where h is modeled as a truly random function (freely available to all the parties including the adversary), the resulting RSA hash-and-sign signature (which they called RSA *Full Domain Hash*, for short, RSA-FDH) is secure assuming that the (standard) RSA assumption holds. When considering an actual instantiation of h, though, a moment's reflection shows that all known security notions for hash functions, such as collision-resistance or pseudorandomness, do not appear to help. In fact, even more "esoteric" notions, such as perfect one-way hash functions or verifiable random functions [5], are not sufficient either. On the other hand, no significant attacks on RSA-FDH signatures are known when h is instantiated using popular "cryptographic hash functions", such as SHA-1. This gave rise to the following important question, which is the main focus of this paper.

Is there an instantiation of RSA-FDH *signature scheme (namely, of the hash function h) that can be proven secure under a natural assumption in the standard model?*

Of course, for any concrete hash function, one can "reduce" the security of RSA-FDH signatures to that of RSA-FDH signatures, which is not very useful. So it is important that the assumption used to argue the security of the scheme should be considerably simpler than the chosen message attack on RSA signatures. The best case scenario would be a reduction to the one-wayness of the RSA function (i.e., the standard "RSA assumption"), which is indeed what happened in the idealistic RO model. Unfortunately, we seem to be very far from this goal. In fact, several works, which we survey next, showed various arguments suggesting that no such reduction is likely to exist.

Existing Impossibility Results. It is well known that in the general case the random oracle heuristic is false. Specifically, there exist schemes secure in the random oracle model that cannot be instantiated by any concrete hash function [8,7,26,18,2]. Most counter-examples of this kind, however, are rather artificial, and do not shed much light on the security of concrete schemes used in practice. The work that seems most relevant to the focus of this paper is those of [12] and [27] described below (whereas other related work is discussed in Section 1.3).

Dodis et.al., [12], considered a generalization of RSA-FDH signatures, known as (general) *Full Domain Hash* (FDH) signatures. In such signatures, the signer has access to an arbitrary trapdoor permutation f, and sets $\sigma = f^{-1}(h(m))$.[1] The main result of [12] rules out proving the security of an instantiation FDH, by reducing it to the one-wayness of f (or more generally, to any assumption on f that is satisfied by a random trapdoor permutation). Their result, however, does not capture reductions that use additional assumptions about f. In particular, it seems likely that if a proof of security of some instantiation of RSA-FDH does exist, then it would use the *algebraic properties* of the RSA function. To demonstrate this point, we present (see Section 1.1) an instantiation of RSA-FDH under the standard RSA assumption, that is secure as long as the number of signing queries is a-priori bounded.[2] Our reduction is black box, and critically uses the algebraic properties of $\mathbb{Z}_n^*$. (Indeed, [12] showed that even *one-time* security of *general* FDH signatures cannot be black-box reduced to the one-wayness of the trapdoor permutation.) In addition, the "RSA-based" signatures [16,10,22], which can be proven secure in the standard model (but, alas, no longer have the simple syntax of the RSA signature), critically use the algebraic properties of the RSA function. Finally, even in the random oracle model, tighter security bounds are sometimes achieved using the algebraic properties of RSA (cf., [9], as compared to the generic proofs from trapdoor permutations [3,13]).

More recently, Paillier, [27], looked at the question of instantiating RSA-FDH using a *fixed* hash function (as opposed to a keyed family), and showed that no such instantiation can be black-box reduced to the traditional RSA assumption,

---

[1] As in the case of RSA-FDH signatures, FDH signatures are known to be secure when the hash function is modeled as a truly random function [3].

[2] With a different motivation, the same result was independently obtained by [21].

assuming the so called "RSA non-malleability" assumption. Informally, this assumption states that calling the RSA inverter on arbitrary "permitted" inputs $(n', e') \neq (n, e)$ does not help in breaking the instance (n, e). We remark that, as observed by Paillier in [27], this assumption is false for various reasonable interpretations of "permitted" tuples (n', e'). More significantly, although the restriction to a fixed hash function h is consistent with the existing use in practice, from a theoretical perspective this assumption is somewhat restrictive. For example, while the result of [27] rules out proving even *one-time* security of RSA-FDH, our positive result (see Section 1.1) circumvents this impossibility result by using a keyed hash family.

1.1   Our Results

Our main result is a new negative result regarding the instantiability of RSA-FDH, which addresses some of the limitations of the previous negative results of [12,27]. To motivate this result, we start by describing our already mentioned positive result.

Theorem 1 (Informal). *Under the standard* RSA *assumption, for every polynomial t there exists an instantiation of* RSA-FDH *that is existentially unforgeable against $t(k)$ signing queries (where k is the security parameter). Furthermore, the reduction treats the group $\mathbb{Z}_n^*$ and the potential adversary in a black-box way.*

The claimed construction is fully described in the full version [], but here we highlight some of its features. First, the result on works for bounded values of t, since the constructed hash function description length, is polynomial (quadratic) in the number of signing queries. Second, our construction uses a keyed family of hash functions (which is needed to overcome the impossibility result of [27]). Third, the hash function depends on the RSA modulus n and critically uses the multiplicative structure of the RSA function (which is needed to overcome one of the impossibility result of [12]). Finally, our reduction does not use any other properties of the RSA function besides its multiplicative homomorphism over $\mathbb{Z}_n^*$. Formally, this means that the reduction works given only oracle access to the multiplication and the inversion operations of $\mathbb{Z}_n^*$.

We now turn to our main, negative result, which can be informally stated as follows:

Theorem 2 (Informal). *It is impossible to reduce the security of an instantiation of* RSA-FDH *to a "natural" assumption (and in particular to the hardness of* RSA*), provided that (1) the reduction treats the potential adversary in a black-box way; (2) the public exponent e used by the scheme is prime with non-negligible probability; (3) the instantiation only "uses the multiplicative properties of $\mathbb{Z}_n^*$", and should "relativize" to any group isomorphic to $\mathbb{Z}_n^*$.*

We now explain this result in more detail. First, our result holds even if the hash function h is allowed to be keyed, and, moreover, to depend on the RSA modulus n (which was used in our positive result). More significantly, we allow both the hash function and the hypothetical security reduction R to use the multiplicative structure of $\mathbb{Z}_n^*$. Finally, we not only rule out reductions to the standard RSA assumption, but also to other non-interactive "RSA-type" assumptions, such as the "strong RSA assumption".

However, our result also has three limitations, (1)-(3). First, and least important, is the assumption that the reduction must treat the adversary in a black-box way. This limitation is met by most existing reductions, and also quite standard in most black-box impossibility results. Technically, it means that the reduction should work given oracle access to any (even inefficient) attacker breaking the security of RSA-FDH. Second, and more significant, is the fact that our current proof relies on the fact that the instantiation will use a prime exponent e (at least with non-negligible probability). Although this limitation appears to be an odd artifact of our specific proof technique, and also seems to be met by most known RSA instantiations, it does leave a possibility for a secure RSA-FDH instantiation always using some composite exponent e. Finally, and most significantly, we assume that the reduction "treats the multiplicative RSA group $\mathbb{Z}_n^*$ in a black-box manner". This is formalized (see Section 2) using the notion of *generic groups* [33,25,23]. Informally, though, it means that nothing is assumed about a group element, apart from what was revealed through the performed group operations (i.e., multiplication, inverse and equality check). In particular, an algorithm that treats $\mathbb{Z}_n^*$ in a black-box way should perform equally well given oracle access to any group isomorphic to $\mathbb{Z}_n^*$ (without knowing the isomorphism).

With this intuition in mind, we can interpret Theorem 2 as an indication that in order to prove the security of a given instantiation of RSA-FDH, one should use a non-black box security proof, or use properties of the RSA group, that are not captured by the generic group abstraction. To the best of our knowledge, *all* known positive results on building "RSA-type" signatures — including our new positive result in Theorem 1, the standard model constructions of [16,10,22], and the random-oracle based analysis of [3,9] — treat $\mathbb{Z}_n^*$ as a black-box, and only use its multiplicative structure. Thus, although still restrictive, our result rules out all known techniques for proving the security of RSA-based signatures, which was not the case for the previous results of [12,27]. Still, the restriction of the reduction to only use the multiplicative structure of $\mathbb{Z}_n^*$ is quite significant, which raises the question if this restriction could be relaxed.

Removing Generic Groups? Unfortunately, removing (or even relaxing) the above mentioned restriction appears to be very challenging. Intuitively, with our current techniques (see more below) we must be able to construct an algorithm Forger which, given any (family of) hash function(s) h, should be able to (1) break the RSA-FDH instantiation using this h, and, yet, (2) do so by only forging the signature which the reduction R must already "know" (so that Forger never helps R compute something which R does not know to begin with, potentially helping R to break some hardness assumption). In particular, satisfying conflicting properties (1) and (2) seems to require some kind of "reverse-engineering" (or "de-obfuscation") techniques on h which seem to be completely beyond our current capabilities, without placing any restriction on the reductions we allow.

Indeed, the introduction of the generic group model was precisely the step which (a) allowed our forger to "reverse engineer" the given hash function h (so as to provably satisfy properties (1)-(2) above), and, yet, (b) allowed the reduction to use the algebraic properties of $\mathbb{Z}_n^*$.

1.2 Our Technique

On a very high level, our proof follows the approach of [12] used to prove that there exists no fully black-box reduction from (general) FDH signature schemes to the one-wayness of random functions. [12] defined an oracle Forger relative to which no FDH signature scheme is secure, yet Forger does not help inverting a random function. In more detail, on input $(h, \{\sigma_i\}_{i \in [t]})$, Forger checks that (1) $\{\sigma_i\}$ are valid signatures for the messages $1, \ldots, t$ (i.e., $f(\sigma_i) = h(i)$ for every $i \in [t]$, where f is the random function), (2) the evaluation of $h(1), \ldots, h(t)$ does not query f on any element of $\{\sigma_i\}$, and (3) t is at least equal to $|h|$ – the description size of h. If positive, Forger returns the signature of 0 (i.e., $f^{-1}(h(0))$).

It is clear that Forger can be used to break the existential security of any FDH scheme: the attacker uses Sign, the signer of the scheme, to compute $\{\sigma_i\}_{i \in [t]}$ for some $t \geq |h|$, and then calls Forger on $(h, \{\sigma_i\})$, where we assume without loss of generality that condition (2) above holds with respect to this query (otherwise, faking a signature *without* Forger is easy). On the other hand, [12] showed that an efficient algorithm (with oracle access to f, but not to Sign) cannot provide all these signatures. Thus, Forger is useless in these settings, and in particular a black-box reduction (i.e., algorithm) cannot make use of Forger for inverting a random function, proving the main result of [12].

Intuitively, Forger is useless for an algorithm with no access to Sign, for the following reason. Fix some efficient oracle-aided algorithm R and let $\{0,1\}^n$ be the domain of the random function f. Since a random function is one way, the only elements that R can invert are those elements it previously received as answers to its f-queries. Hence (since f is random), R only knows how to invert *random* elements inside $\{0,1\}^n$. Since it takes at least t bits to describe t random elements in $\{0,1\}^n$ (actually, it takes tn bits) and since the evaluation of $h(1), \ldots, h(t)$ does not query f on elements inside $\{\sigma_i\}_{i \in [t]}$, there must exist $h(i) \in \{h(1), \ldots, h(t)\}$ that R does not know how to invert, and thus cannot provide a valid signature for the message i.

Moving to our setting, we focus for concreteness on fully black-box reductions from RSA-FDH to the hardness of RSA (i.e., such reductions use the multiplicative RSA group $\mathbb{Z}_n^*$ and the adversary in a black-box way). The blackboxness in the RSA group tells us that such a reduction should work with respect to any group isomorphic to $\mathbb{Z}_n^*$. In particular, it should work well with respect to the group $\pi(\mathbb{Z}_n^*)$, obtained by renaming the elements of $\mathbb{Z}_n^*$ according to a random permutation π over $\mathbb{Z}_n^*$ (i.e., $a \cdot b$ is defined as $\pi(\pi^{-1}(a) \cdot \pi^{-1}(b) \bmod n)$).

Given the above understanding, the first attempt would be to define Forger analogously to that of [12]. Namely, on input $(n, e, h, \{\sigma_i\}_{i \in [t]})$, Forger checks that (1) $\sigma_i^e \equiv h(i)$ for every $i \in [t]$, (2) the evaluation of $h(1), \ldots, h(t)$ does not compute σ_i for some $i \in [t]$, and (3) $t \geq |h|$. If positive, Forger returns the signature of 0 (i.e., $h(0)^d$, for $d = e^{-1} \bmod \phi(n)$, where all group operations are over the group $\pi(\mathbb{Z}_n^*)$).

We would like to argue that if π is chosen at random, then the only way to make a non-aborting query to Forger is via using Sign, the signer of the scheme. It would then follow that Forger is useless for an algorithm R that has no access to Sign (and in particular to a black-box reduction). It turns out, however, that in our settings such R *can* make non aborting calls to Forger. The issue is that unlike in the setting of [12], R can make use of the algebraic structure of $\mathbb{Z}_n^*$ to construct a non-aborting query to Forger. For instance, R can compute $\{j^e\}_{j\in[\ell]}$, and assuming some reasonable mapping M from $[t = \ell^2]$ to $\{j \cdot k\}_{j,k\in[\ell]}$, let $h(i) = M(i)^e \bmod n$ and $\sigma_i = M(i)$. Since the evaluation of $h(1), \ldots, h(t)$ does not query an element of $\{\sigma_i\}_{i\in[t]}$), it follows that $(n, e, h, \{\sigma_i\}_{i\in[t]})$ is a non-aborting query.[3] Alternatively, if R can break the RSA assumption over $\pi(\mathbb{Z}_n^*)$ (say, if it knows the factorization of n), then it can set $h(i) = i$ and compute $\sigma_i = h(i)^d$ (using the factorization of n to compute d).

Fortunately, we manage to prove that a non-aborting query of R is either "degenerated" (as in the first example) or indicates that R knows the factorization of n. To handle the first case, we change Forger to identify and abort on degenerated queries. Where we also show that it is easy to forge a signature with respect to a degenerated h (i.e., h that is part of a degenerated query), even *without* the help of Forger. Namely, we show that there is no secure RSA-FDH scheme relative to the modified Forger. We then show that with respect to this modified Forger, one can efficiently extract the factorization of n from an algorithm that produces a non-aborting query. It follows that for any efficient algorithm R with oracle access to Forger, there exists an efficient algorithm, with no access to Forger, that emulates R^Forger well. In other words, we prove that Forger is useless for the class of efficient algorithms with no oracle access to Sign.

Proving the above intuition is the main challenge of this work, and we achieve that using a novel adaptation of the Gennaro-Trevisan, [17], short description paradigm, described below, to the generic groups realm.[4]

The Gennaro-Trevisan [17] Short Description Paradigm and Its Adaption to Generic Groups. Loosely, [17] shows that an efficient algorithm that inverts a random function too well, can be used to give a too short description for a random function (and thus cannot exist). This elegant approach has turned to be an extremely powerful approach for proving impossibility results in the random functions realm, which typically imply black-box impossibility results for one-way functions/permutations based constructions. While the Gennaro-Trevisan paradigm (from now on, the GT paradigm) has several extensions (e.g., [15,35,19,20,30]), all are given in the random functions realm.

---

[3] Note that to describe h it suffices to describe the set $\{j^e\}_{j\in[\ell]}$. Thus $|h| \in O(\ell \log n)$, which is smaller than t for large enough ℓ.

[4] A side benefit of this proof technique, is an alternative proof to the equivalence of RSA and factoring over generic groups, firstly proven by Aggarwal and Maurer, [1] (the latter, however, also proves it over "generic rings").

We would like to apply a similar approach for arguing that an algorithm that makes a non-aborting query to Forger, can be either used to factor n, or to "compress" the random permutation π (which defines the group $\pi(\mathbb{Z}_n^*)$). Since compressing π is impossible, it follows that a non-aborting query of such an algorithm can be used to factor n. Hence, such queries can be answered efficiently, yielding the existence of an efficient emulator (without access to Forger) for any efficient algorithm.[5]

Extending the GT paradigm to our settings involves many complications. The main part of the GT paradigm is using the (hypothetical) attacker to reconstruct a random function using (too) short advice. This reconstruction involves emulating the attacker, where the key point is to do this without "wasting information": any bit used to emulate, should give a bit of information about the (random) function. Doing the latter is quite easy for random functions; the answer to any query of the attacker gives the same amount of information about the function (i.e., the info that it maps the query input to the provided output). The only subtlety is that there are repeated queries (which are clearly wasteful), but handling such queries is easy: simply keep track of the query history on the emulation.

In our setting, however, things get much more complicated. To begin with, there might be non-repeating queries whose answers yield very little information about the random group $\pi(\mathbb{Z}_n^*)$ (and therefore about π). For instance, for some n's there are only four possible answers for the query $a^{\phi(n)/4}$ over $\pi(\mathbb{Z}_n^*)$. Thus, roughly speaking, the answer for this query contains only two bits of information about π. More generally, it appears that one can create much more intricate examples; e.g., when the answer to the query follows a very complicated distribution, based on the answers given so far.

An even more challenging task is proving the dichotomy that a non-aborting query can either be used to (efficiently) factor n, or implies a (too) short description of π. Handling the above challenges requires an intimate understanding of the algebraic structure of the group $\mathbb{Z}_n^*$, in particular of the set of solutions for linear equations over this group, and critically uses the fact that factoring is solvable in sub-exponential time [11,34].

1.3 Other Related Work

We briefly mention other known results concerning the uninstantiability of popular signature and encryption schemes that can be proven secure in the random oracle model. Paillier and Vergnaud, [28], showed that many popular discrete log based signatures (including *ElGamal, DSA* and *Schnorr*) *cannot* be reduced to the discrete log assumption in the standard model, using the so called "algebraic" reductions. (Similar results also hold for related GQ signatures under the RSA assumption.) Although technically incomparable to our "generic group" modeling, conceptually such reductions are related to our assumption that the

---

[5] In addition, since non-aborting queries are *easy* to generate assuming that RSA is easy over $\pi(\mathbb{Z}_n^*)$, the above would immediately yield that RSA is equivalent to factoring over (random) $\pi(\mathbb{Z}_n^*)$, and thus over generic groups.

reduction can only use the multiplicative structure of a given group. Indeed, in both cases the "meta-reduction" can eventually figure out the multiplicative relations used be the reduction R in its queries to the attacker. The main difference applies in the way the reduction can prepare its queries to the attacker. While the generic group modeling allows the reduction R to use some "hidden values" related to the assumption that R is trying to break, "algebraic" reduction do not allow this flexibility. Thus, much of the technical difficulties in the generic group modeling (e.g., extracting the hidden representations computed by the reduction "on the side") are somewhat trivialized when restricted to "algebraic" reductions. Additionally, the results of [28] are specific to reductions from a concrete assumption (e.g., discrete log), and are conditional on another assumption (e.g., "one-more" discrete log). In contrast, our results are unconditional and rule out all starting assumptions, but only in the generic group model.

Finally, in the realm of factoring/RSA-based CCA encryption, Paillier and Villar, [29] and Brown et.al., [6], showed uninstantiability results analogous to already-mentioned RSA signature result of [27].

Paper Organization

In Section 2 we formally define RSA-FDH and its security in the generic group model and the type of reductions we rule out. Our main result, regarding the impossibility of existentially unforgeable RSA-FDH against unbounded number of signing queries, is proven in Section 3. However, the proof of our main technical lemma using the GT short description paradigm is omitted and can be found in the full version [14].

2   RSA-FDH in the Generic Group Model

In the following we first formally define what we mean by generic group model, then extend the standard definitions of RSA-FDH to this model and finally define weakly black-box proofs of security.

2.1   The Generic Group Model

There are different ways to interpret what it means to "treat the multiplicative RSA group $\mathbb{Z}_n^*$ in a black-box way" (see Theorem 2). In the *generic algorithm model* due to Maurer, [23], "generic" algorithms do not have a direct access to the group elements, but rather to a "black box" containing each element. The only operations allowed with these boxes, are the group operations (inverse and multiplication) and comparing two boxes for equality. The formulation we have chosen here, which we simply call the generic group model, is somewhat less abstract. An algorithm in our model has an oracle access to a group isomorphic to

$\mathbb{Z}_n^*$ (specifically, the group resulting by renaming the elements of $\mathbb{Z}_n^*$ according to some random permutation), through which it can perform the group operations. Unlike the generic algorithm model, however, in our model algorithms we do have access to the representation of the group elements and can manipulate them.

Since any algorithm that "works well" in the generic algorithm model (e.g., breaks the RSA assumption) implies an algorithm that works equally well in our model with respect to *any* group isomorphic to $\mathbb{Z}_n^*$, an impossibility result in our model implies a similar result in the model of Maurer, [23]. Namely, our model can be viewed as a model for proving impossibility results in the generic algorithm model.

We formally define our model as follows: for $n \in \mathbb{N}$, let $\Pi_{\phi(n)}$ be the set of all permutations from $\mathbb{Z}_n^*$ to $\mathbb{Z}_n^*$. For $\pi \in \Pi_{\phi(n)}$, we denote with $\pi(\mathbb{Z}_n^*)$ the group induced by the group $\mathbb{Z}_n^*$ where each element of $\mathbb{Z}_n^*$ is renamed according to π. More specifically, the group operations over $\pi(\mathbb{Z}_n^*)$ are defined as follows: the inverse of $a \in \mathbb{Z}_n^*$ is $\pi((\pi^{-1}(a))^{-1} \bmod n)$ and the (group) product of $a, b \in \pi(\mathbb{Z}_n^*)$ is $\pi(\pi^{-1}(a) \cdot \pi^{-1}(b) \bmod n)$. By $\Pi(\mathbb{Z}_n^*)$ we denote the multiset of all groups $\pi(\mathbb{Z}_n^*)$, where $\mathcal{G} = \{G = \{G_n \colon G_n \in \Pi(\mathbb{Z}_n^*)\}_{n \in \mathbb{N}}\}$ (i.e., $\mathcal{G}$ consists of sets of groups, where each set contains a group of $\Pi(\mathbb{Z}_n^*)$ for every $n \in \mathbb{N}$).

Abusing notation, we view $G \in \mathcal{G}$ as an oracle that given as input $n \in \mathbb{N}$ and one [resp., two elements] of G_n (i.e., of $\mathbb{Z}_n^*$), returns the group inverse [resp., the group product] of the element (if the oracle G is given as input an element outside G_n, lt returns $\bot$), and let $G_n(\cdot) = G(n, \cdot)$. Given a sequence of group operations (e.g., $a \cdot b^{-1}$), we sometimes add the term $[G_n]$, to indicate that the operations are done with respect to the group G_n. In the following, abusing notation again, we will write $G \leftarrow \mathcal{G}$, where this sampling is not well defined because $\mathcal{G}$ is an infinite set. However, we can assume lazy sampling, namely for every query which contains a new n, G_n is sampled uniformly at random from $\Pi(\mathbb{Z}_n^*)$ (which is a finite set).

2.2 RSA-FDH Signature Schemes in the Generic Group Model

RSA-FDH signature schemes over $G \in \mathcal{G}$ is defined as follows:

Definition 1 (RSA-FDH signature scheme in the generic group model). *An* RSA-FDH *signature scheme* $\Sigma^{\mathcal{G}}$ *in the generic group model, consists of the following triplet of oracle-aided* PPT *'s* (KeyGen, Sign, Verify):

- *Given oracle access to $G \in \mathcal{G}$ and input 1^k, KeyGen^G outputs a "public key" (n, e, h), where $n \in \mathbb{N}$ is a product of two primes, $e \in \mathbb{Z}_{\phi(n)}^*$ and h is a (hash) function, represented as an oracle-aided circuit mapping values into $\mathbb{Z}_n^*$, and a "secret key" $d = e^{-1} \bmod \phi(n)$.*
- *Given oracle access to $G \in \mathcal{G}$, input $n \in \mathbb{N}$, $d \in \mathbb{Z}_{\phi(n)}^*$, a circuit h mapping values into $\mathbb{Z}_n^*$ and a "message" m in the domain of h, Sign^G outputs the "signature" $h^G(m)^d$ $[G_n]$.*

– Given oracle access to $G \in \mathcal{G}$, input $n \in \mathbb{N}$, $e \in \mathbb{Z}^*_{\phi(n)}$, a circuit h mapping values into $\mathbb{Z}^*_n$, a "message" m in the domain of h and $\sigma \in \mathbb{Z}^*_n$, Verify^G outputs one iff $\sigma^e \equiv h^G(m)$ $[G_n]$.

For $G \in \mathcal{G}$, we let Σ^G be the instantiation of $\Sigma^{\mathcal{G}}$ with G.

Security Definition. The following definition realizes the security of bounded and unbounded existential unforgeability under chosen message attack of an RSA-FDH signature in the generic group model, analogously to that of the standard model.

Definition 2 (security of RSA-FDH signature in the generic group model). An oracle-aided algorithm F breaks the security of an RSA-FDH signature scheme $\Sigma^{\mathcal{G}} = (\mathsf{KeyGen}, \mathsf{Sign}, \mathsf{Verify})$, if

$$\mathsf{Pr}_{G \leftarrow \mathcal{G}, (sk,pk) \leftarrow \mathsf{KeyGen}^G(1^k)}[(m, \sigma) \leftarrow F^{G, \mathsf{Sign}^G(sk, pk, \cdot)}(pk):$$
$$\mathsf{Verify}^G(\sigma, m, pk) = 1 \wedge \mathsf{Sign}\ was\ not\ queried\ on\ m] > \mathrm{neg}(k)$$

A signature scheme $\Sigma^{\mathcal{G}}$ is EU-CMA-secure, if no (oracle-aided) PPT breaks its security, where $\Sigma^{\mathcal{G}}$ is t-EU-CMA-secure, if no PPT breaks its security when restricted to query Sign at most $t(k)$ times.

2.3   Weakly Black-Box Proofs of Security

Since we would like to rule out an EU-CMA-secure scheme, we ask the security proof of the scheme to be realized via a "black-box reduction" (as discussed in the introduction, we have very little chance to rule out a general proof of security). On the other hand, we consider a very weak form of such a reduction (which strengthens our main impossibility result).

Definition 3 (weakly black-box proof of security of RSA-FDH). An RSA-FDH signature scheme $\Sigma^{\mathcal{G}} = (\mathsf{KeyGen}, \mathsf{Sign}, \mathsf{Verify})$ in the generic group model has a weakly black-box proof of security based on an assumption X, if there exists an oracle-aided PPT R such that if X is true, then the following holds: let F be a (possibly unbounded) adversary that breaks the security of $\Sigma^{\mathcal{G}}$ (see Definition 2), then for any PPT Emul there exists a polynomial-length distribution ensemble $\mathcal{D} = \{D_k\}_{k \in \mathbb{N}}$ such that

$$\mathrm{SD}\left((x, R^{G, F^G}(1^k, x)), (x, \mathsf{Emul}^G(1^k, x))\right)_{G \leftarrow \mathcal{G}, x \leftarrow D_k} > \mathrm{neg}(k).[6]$$

Remark 1 (A black-box proof implies a weakly black-box proof). Assuming that X is true, the above intuitively asks that a security breach of $\Sigma^{\mathcal{G}}$ implies that a

---

[6] Note that F is an adversary which expects oracle access to Sign and R can control the responses of these queries of F. The same does not hold for the queries of F to G.

(slightly) non-trivial task can be performed. Specifically, an efficient oracle-aided algorithm can use a breaker of the scheme (in a black-box way) to sample some unsamplable distribution. Note that this is a very modest demand and indeed, it is implied by most black-box proofs of security one can think of.

Consider for instance a proof of security R that black-box reduces the security of a scheme $\Sigma^{\mathcal{G}}$ to an assumption X, say to the hardness of factoring. It follows that given any adversary F to $\Sigma^{\mathcal{G}}$, the algorithm $R^{G,F^{G}}$ factors integers too well. Assume without loss of generality that $R^{G,F^{G}}(x)$, if succeeds, outputs the factorization of the integer x, let D_k be the distribution that outputs an integer $x = pq$, for two randomly chosen k-bits prime, and consider the distribution $\xi_k = (x, R^{G,F^{G}}(1^k, x))_{G \leftarrow \mathcal{G}, x \leftarrow D_k}$ it induces. Now if factoring is hard, then there is no efficient Emul such that $(x, \mathsf{Emul}^{G}(1^k, x))_{G \leftarrow \mathcal{G}, x \leftarrow D_k}$ is (even computationally) close to ξ_k. Namely, there is no weakly black-box proof of security for $\Sigma^{\mathcal{G}}$ based on factoring.

Now if factoring is hard, then there is no efficient Emul such that $(x, \mathsf{Emul}^{G}(1^k, x))_{G \leftarrow \mathcal{G}, x \leftarrow D_k}$ is (even computational) close to ξ_k. Namely, there is no weakly black-box proof of security for $\Sigma^{\mathcal{G}}$ based on factoring.[7]

3 There Exists No RSA-FDH with a Weakly Black-Box Proof

In this section we prove the main result of this paper.

Theorem 3 (Theorem 2, restated). *Let* $\Sigma^{\mathcal{G}} = (\mathsf{KeyGen}, \mathsf{Sign}, \mathsf{Verify})$ *be an* RSA-FDH *signature scheme in the generic group model in which* $\Pr_{G \leftarrow \mathcal{G}, (n,e,h) \leftarrow \mathsf{KeyGen}^{G}(1^k)}[e \in \mathrm{P}] > \mathrm{neg}(k)$. *If* $\Sigma^{\mathcal{G}}$ *has a weakly black-box proof of security based on (an assumption)* X, *then* X *is false.*

The proof of Theorem 3 immediately follows from the next lemma:

Lemma 1. *Let* $\Sigma^{\mathcal{G}}$ *be as in Theorem 3, then there exist a family of oracles* $\mathsf{Forger} = \{\mathsf{Forger}_G\}_{G \in \mathcal{G}}$ *and oracle-aided* PPT*'s* F *and* Emul, *such that the following hold:*

1. *For every* $G \in \mathcal{G}$, F^{G, Forger_G} *breaks the security of* $\Sigma^{\mathcal{G}}$.
2. *For any oracle-aided* PPT A *and polynomial-length distribution ensemble* $\mathcal{D} = \{D_k\}_{k \in \mathbb{N}}$:

$$\mathrm{SD}\left((x, A^{G, \mathsf{Forger}_G}(1^k, x)), (x, \mathsf{Emul}^{G}(1^k, x, \mathrm{desc}(A)))\right)_{G \leftarrow \mathcal{G}, x \leftarrow D_k} = \mathrm{neg}(k),$$

where $\mathrm{desc}(A)$ *denotes the description of the Turing Machine* A.

Before proving Lemma 1, let us first use it for proving Theorem 3.

---

[7] Note that there nothing specific to the hardness of factoring in the above discussion, but rather it seems to be generic to "any" hardness assumption (e.g., strong RSA).

Proof (of Theorem 3). Let $\Sigma^{\mathcal{G}}$ be an RSA-FDH scheme with $\mathsf{Pr}_{G\leftarrow\mathcal{G},(n,e,h)\leftarrow\mathsf{KeyGen}^G(1^k)}[e \in \mathsf{P}] > \mathrm{neg}(k)$. Assume that $\Sigma^{\mathcal{G}}$ has a weakly black-box proof of security based on (an assumption) X and let R be the algorithm guaranteed by this proof. Let Emul be the algorithm guaranteed by Lemma 1 with respect to $\Sigma^{\mathcal{G}}$. Lemma 1 yields that

$$\mathrm{SD}\left((x, \tilde{R}^{G,\mathsf{Forger}_G}(1^k, x)), (x, \mathsf{Emul}^G(1^k, x, \mathrm{desc}(\tilde{R})))\right)_{G\leftarrow\mathcal{G}, x\leftarrow D_k} = \mathrm{neg}(k)$$

for any polynomial-length distribution ensemble $\mathcal{D} = \{D_k\}$, where $\tilde{R}^{G,\mathsf{Forger}_G}(\cdot) = R^{G,F^{\mathsf{Forger}_G}}(\cdot)$. Letting $\tilde{F}^G(\cdot) = F^{G,\mathsf{Forger}_G}(\cdot)$ and $\mathsf{Emul}_R^G(\cdot) = \mathsf{Emul}^G(\cdot, \mathrm{desc}(\tilde{R}))$, it follows that

$$\mathrm{SD}\left((x, R^{G,\tilde{F}^G}(1^k, x)), (x, \mathsf{Emul}_R^G(1^k, x))\right)_{G\leftarrow\mathcal{G}, x\leftarrow D_k} = \mathrm{neg}(k)$$

for any polynomial-length distribution ensemble $\mathcal{D}$, yielding that X is false.

The rest of this section is devoted for proving Lemma 1. We find it more convenient, however, to prove a variant of Lemma 1 in which the emulator should work for any (polynomial-size) family of circuits. Namely, we prove the following lemma (in the following statement we only focus on the part that changed comparing to the original statement):

Lemma 2 (non uniform variant of Lemma 1)

2. *The following holds for any (no input) polynomial-size family of oracle-aided circuits $\{C_k\}_{k\in\mathbb{N}}$:*

$$\mathrm{SD}\left(C_k^{G,\mathsf{Forger}_G}, \mathsf{Emul}^G(1^k, \mathrm{desc}(C_k))\right)_{G\leftarrow\mathcal{G}} = \mathrm{neg}(k),$$

where $C_k^{G,\mathsf{Forger}_G}$ denotes the output of C_k given access to G and Forger_G, and $\mathrm{desc}(C_k)$ denotes the description of C_k.

It is easy to see that the non-uniform lemma above yields the uniform Lemma 1. In Section 3.1 we define the family of oracles Forger and the efficient algorithm F that uses Forger to break any RSA-FDH scheme, in Section 3.3 we define the emulator Emul, where in Section 3.4 we put things together to prove Lemma 2.

3.1 The Forger

Recall (see Section 1.2) that Forger has to abort on "degenerated queries" — essentially those queries that are easy to produce over any group in $\Pi(\mathbb{Z}_n^*)$. To determine whether a query $(n, e, h, \{\sigma_i\}_{i\in[t]})$ is degenerated, we measure the complexity of the values $\{h(i)\}_{i\in[t]}$,[8] as a function of the group queries done through

---

[8] We actually mean $\{h^G(i)\}_{i\in[t]}$, but for notational convenience we will sometimes omit the superscript G from h.

their evaluations. Since the actual representation of these values is meaningless, we only focus on their representation as functions of the "hardwired terms" — the values used in the evaluation of $\{h(i)\}$ that first appear as an input to a group oracle call. Note that any group element used in the evaluation of $\{h(i)\}$, can be expressed using (only) these hardwired terms. To formally carry the above discussion, we describe the evaluation of $\{h(i)\}$ as a computation over the following group.

Definition 4 (The group Symb**).** *The elements of* Symb *are equivalence classes over the set of all finite strings* "$u_1^{a_1}, \cdots, u_k^{a_k}$", *where the u_i's are in* $\mathbb{N}$ *and the a_i's are in* $\mathbb{Z}$. *The strings* $c =$ "$u_1^{a_1} \cdot \ldots \cdot u_k^{a_k}$" *and* $c' =$ "$u_1'^{a_1'} \cdot \ldots \cdot u_{k'}'^{a_{k'}'}$" *are in the same equivalence class, if for every* $w \in \mathbb{N}$ *it holds that* $\sum_{i \in [k]:\, u_i = w} a_i = \sum_{i \in [k']:\, u_i' = w} a_i'$. *We identify a group element of* Symb, *with any string of its equivalence class. The unit element of* Symb *is the class identified by the empty string* ε *(or by* "$2^1 \cdot 2^{-1}$" *etc), where* $c \cdot c'$ *is the equivalence class identified by the string* "$c \cdot c'$" *and finally* c^{-1} *is the class identified by the string* "$u_1^{-a_1} \cdot \ldots \cdot u_k^{-a_k}$".

We naturally identify an element "$u_1^{a_1} \cdot \ldots \cdot u_k^{a_k}$" $\in$ Symb with an element of a given group V that contains $\{u_i\}_{i \in [k]}$, by identifying it with the result of the sequence of operations it induces over V (i.e., "$u_1 \cdot u_2^{-1}$" with respect to $V = \mathbb{Z}_n^*$, is identified with $u_1 \cdot u_2^{-1} \bmod n$). To avoid confusion over which group a sequence of operations is taken, we typically suffix the sequence with the term $[V]$, indicating that it is done over the group V. It is clear that for any two strings u and u' that identify the same element of Symb (i.e., belong to the same equivalence class), it holds that $u \equiv u'$ $[V]$ for any Abelian group V containing u and u'.

Next we use the above terminology to syntactically describe the computation of an oracle-aided circuit C, where we start by defining the hardwired terms determined by C's computation. To simplify notations, we assume that a circuit evaluates its gates one-by-one, and that its description determines this evaluation order.

Definition 5 (hardwired terms). *Let C be an oracle-aided circuit, $G \in \mathcal{G}$ and $n \in \mathbb{N}$. The* **terms** *of C with respect to G_n, denoted* $\text{Terms}_{C,G,n}$, *are those values that appear either as input or as the answers to non-bottom queries of C to G_n (i.e., G_n returns a non-bottom value). The* **hardwired terms** *of C with respect to G_n, denoted* $\text{HardWired}_{C,G,n}$ *are those element inside* $\text{Terms}_{C,G,n}$ *that first appear as inputs to non-bottom queries to G_n. Finally, the* **answer terms** *are those terms that appear as answers to non-bottom queries (might intersect* $\text{HardWired}_{C,G,n}$). *We assume that the elements of each of the above sets are ordered according to the evaluation order.*

We next use the syntax of the group Symb, to present any term as an expression of the hardwired terms.

Definition 6 (canonical form). *Let C, G and n be as in Definition 5. The* canonical form *of $u \in \text{Terms}_{C,G,n}$ with respect to (C, G, n), denoted* $\text{Can}_{C,G,n}(u)$, *is recursively defined as follows:*

- if $u \in \text{HardWired}_{C,G,n}$, let $\text{Can}_{C,G,n}(u)$ be the element "u^1" $\in \text{Symb}$.
- If u first appears as an output of a query $G_n(u', u'')$, let $\text{Can}_{C,G,n}(u) = \text{Can}_{C,G,n}(u') \cdot \text{Can}_{C,G,n}(u'')$ [Symb].
- Similarly, if u first appears as an output of $G_n(u')$, we let $\text{Can}_{C,G,n}(u) = \text{Can}_{C,G,n}(u')^{-1}$ [Symb].

Let $\{v_i\}_{i \in [\ell]} = \text{HardWired}_{C,G,n}$. Note that the canonical form of any $u \in \text{Terms}_{C,G,n}$ with respect to (C, G, n), can be *uniquely* written as $\prod_{i \in [\ell]} v_i^{a_i}$ [Symb], where a_i might be non zero, only if the hardwired term v_i appears before u does (in the evaluation order of C^G). Finally, the canonical forms of a set of terms, with respect to (C, G, n), is compactly represented using the following matrix.

Definition 7 (canonical-form matrix). *Let C, G and n be as in Definition 5, let $\{v_i\}_{i \in [\ell]} = \text{HardWired}_{C,G,n}$ and let $\mathcal{W} = \{u_i\}_{i \in [t]} \subseteq \text{Terms}_{C,G,n}$. The matrix $M^{G,n,C}(\mathcal{W}) \in \mathbb{Z}_{t \times \ell}$ is defined as $\{a_{ij}\}_{i \in [t], j \in [\ell]}$, assuming that $\text{Can}_{C,G,n}(u_i) = \prod_{j \in [\ell]} v_j^{a_{ij}}$ [Symb] for every $i \in [t]$.*

We actually care for the rank of the canonical-form matrix of the terms output by a circuit C, which shows if there exists an output term which can be expressed as a product of powers of the other output terms. This would imply that if we know the e-th roots of the latter then we can compute the e-th root of the former. Jumping forward, we will exploit this property of the canonical-form matrix to see if a query is degenerated.

We are finally ready to define Forger_G.

Algorithm 4 (Forger_G)
Input: $q = (n, e, h, \{\sigma_i\}_{i \in [t]})$, where n, e and $\{\sigma_i\}_{i \in [t]}$ are integers, and h is an oracle-aided circuit.
Operation:

1. If $e \notin \text{P}$, $|h| (= |\text{desc}(h)|) > t$ or for some $i \in [t]$ $h^G(i) \notin \mathbb{Z}_n^*$ or $h^G(i) \neq \sigma_i^e$ $[G_n]$, return $\perp$.
2. Let $M = M^{G,n,H}(\{h(i)\}_{i \in [t]})$ according to Definition 7, where H is the oracle-aided circuit that first evaluates $h^G(1), \ldots, h^G(t)$ and then queries G_n on the answers (say asking for their inverses).
   If $\text{rank}_e M < t$, return $\perp$.
3. Return $(h^G(0))^d$ $[G_n]$, where $d = e^{-1} \mod \phi(n)$.

That is, Forger_G first checks that $\{\sigma_i\}_{i \in [t]}$ are valid signatures for the messages $\{1, \ldots, t\}$ (with respect to G and the public key (n, e, h)) and that forging a signature for this public key is not easy (reflected by $\text{rank}_e M = t$). If satisfied, Forger_G forges a signature for 0.

Below we describe the PPT F that uses Forger_G for breaking the security of Σ^G.

3.2   The Breaker F

The strategy of the algorithm F that uses Forger for breaking the security of Σ^G is simple: on input (n, e, h) it would like to use Forger on $(n, e, h, \{\sigma_i = \text{Sign}^G(n, e, i)\}_{i \in [t]})$ to forge the signature of 0. It might be the case, however, that Forger returns bottom on such input. Hence, F first checks by himself (without using Sign or Forger) whether Forger will return bottom on this input. If positive, it uses a straightforward approach (see below) for forging a message $k \in [t]$, without using Forger at all.

Algorithm 5 (F)
Input: $pk = (n, e, h)$
Oracles: $G \in \mathcal{G}_n$, $\text{Sign}^G(sk, pk, \cdot)$ and Forger_G.
Operation:

1. *Let $t = |h|$ and let $M = M^{G,n,H}(\{h^G(i)\}_{i \in [t]})$ according to Definition 7, where H is as in Algorithm 4 (with respect to this h and t).*
2. *If $\text{rank}_e(M) = t$, return $\text{Forger}_G(n, e, h, \{\text{Sign}^G(sk, pk, i)\}_{i \in [t]})$. Otherwise,*
   (a) *Using Gaussian Elimination find $k \in [t]$ and a set $\{\lambda_i \in [e]\}_{i \in [t] \setminus \{k\}}$, such that for every $j \in [\ell]$ it holds that $M_{kj} \equiv \sum_{i \in [t] \setminus \{k\}} \lambda_i \cdot M_{ij} \mod e$.*
   (b) *Let $\gamma = \prod_{j \in [\ell]} v_j^{(M_{kj} - \sum_{i \in [t] \setminus \{k\}} \lambda_i \cdot M_{ij})/e} \; [G_n]$, where $\{v_i\}_{i \in [\ell]} = \text{HardWired}_{H,G,n}$ (see Definition 5).*
   (c) *For every $i \in [t] \setminus \{k\}$, let $\sigma_i = \text{Sign}^G(sk, pk, i) \; (\equiv h^G(i)^d \; [G_n])$.*
   (d) *Return $\sigma_k = \gamma \cdot \prod_{i \in [t] \setminus \{k\}} \sigma_i^{\lambda_i} \; [G_n]$.*

The following lemma is immediate, but its proof is omitted and can be found in the full version of this paper [14].

Lemma 3. *For every $G \in \mathcal{G}$, F^{G, Forger_G} breaks the security of Σ^G.*

3.3   The Emulator

Our task is to emulate a family of circuits $\{C_k\}$ with oracle access to $G \in \mathcal{G}$ and Forger_G, using only oracle access to G. We assume without loss of generality that $|C_k| \geq k$ (otherwise we emulate a padded version of this family) and omit k from the input parameter list of the emulator. We also assume without loss of generality that before calling Forger_G on input $(n, e, h, \{\sigma_i\}_{i \in [t]})$, C_k first query G on $\{\sigma_i\}$ (otherwise, we will emulate the circuit C'_k that does so).

Given a circuit C, $\text{Emul}^G(\text{desc}(C))$ emulates the execution of a circuit C^{G, Forger_G} by forwarding the G-calls to G, and answering the Forger_G-calls using the following method: let $q = (n, e, h, \{\sigma_i\}_{i \in [t]})$ be a query that C makes to Forger_G, Emul first checks whether Forger_G returns bottom on this call (which it can do efficiently), and if positive returns bottom to C as well. Otherwise, Emul uses the query q and the description of C to factor n, and then uses this factorization to answer the query efficiently.

The interesting question is how can Emul use such a pair (C, q) to factor n efficiently? Let H and $M^H = M^{G,n,H}(\{h(i)\}_{i \in [t]})$ as computed by $\mathsf{Forger}_G(q)$, and let $M^{(H;C)} = M^{G,n,(H;C)}(\{\sigma_i\}_{i \in [t]}) \in \mathbb{Z}_{t \times \ell'}$, where the circuit $(H; C)$ first evaluates H and then C.[9] Namely, M^H represents the canonical form of $\{h(i)\}_{i \in [t]}$ induced by the (stand alone) computation of H, where $M^{(H;C)}$ represents the canonical form of the "signatures" $\{\sigma_i\}_{i \in [t]}$ induced by the computation of $(H; C)$. Since $(H; C)$ first starts by computing H, it follows that every hardwired term $u \in \mathrm{HardWired}_{H,G,n} \cap \mathrm{HardWired}_{(H;C),G,n}$ has the same index with respect to both ordered sets $\mathrm{HardWired}_{H,G,n}$ and $\mathrm{HardWired}_{(H;C),G,n}$. Hence, the promise that $\sigma_i^e \equiv h(i)$ $[G_n]$ for every $i \in [t]$, yields the following with respect to $\{v_i\}_{i \in [\ell']} = \mathrm{HardWired}_{(H;C),G,n}$:

$$\prod_{j \in [\ell]} v_j^{M_{ij}^H} \equiv \prod_{j \in [\ell']} (v_j^{M_{ij}^{(H;C)}})^e \quad [G_n],$$

for every $i \in [t]$. Since G_n is selected at random, (at least intuitively) C could have satisfied the above equations only if they hold regardless of the choice of G_n. Namely, it is the case that

$$\sum_{j \in [\ell]} M_{ij}^H \equiv e \cdot \sum_{j \in [\ell']} M_{ij}^{(H;C)} \quad \mathrm{mod}\ \phi(n) \tag{1}$$

for every $i \in [t]$. On the other hand, the assumption that $\mathsf{Forger}_G(q) \neq \perp$ yields that $\mathrm{rank}_e M^H = t$. Therefore, Equation (1) is "far" from being satisfied modulo e. In our proof we show how to use this inconsistency to find a multiple of $\phi(n)$, and thus to factor n.

The following description of Emul realizes the above discussion. We start by recalling the following known factoring algorithms. The first one is useful for small n's (for which the above discussion does not hold), and the second one factors arbitrary larger n, given a multiple of $\phi(n)$ as an advice.

Theorem 6 (factoring small numbers, [11,34]). *There exists a procedure* Sef *that on input $n \in \mathbb{N}$, runs in time $2^{O(\sqrt{\log n \log \log n})}$ and factors n with constant probability.*

Lemma 4 (factoring using multiple of $\phi(n)$). *We say that $z = (z_1, z_2) \in \mathbb{Z} \times \mathbb{N}$ is a factoring advice for $n \in \mathbb{N}$, if $z_1^{\lceil \log n \rceil} \cdot \prod_{p \in \mathrm{P}:\ p < z_2} p^{\lceil \log n \rceil}$ is a non-zero multiple of $\phi(n)$.*

There exists a procedure Factor *that on input (n, z_1, z_2), runs in time* $\mathrm{poly}(z_2) \cdot \mathrm{poly}(\log |nz_1|)$, *and factors n with constant probability, assuming that $z = (z_1, z_2)$ is a factoring advice for n.*

Proof. We use the following known algorithm due to Miller, [24].

Theorem 7 (Miller's algorithm [24,34]). *There exists a procedure that on input $n \in \mathbb{N}$ and $\mu \in \mathbb{Z}$, runs in time $\mathrm{poly}(\log |n\mu|)$, and if μ is a non-zero multiple of $\phi(n)$, it factors n with constant probability.*

---

[9] Recall that we allow circuits to have a predetermined evaluating order.

By definition $\mu = z_1^{\lceil \log n \rceil} \cdot \prod_{p \in P \,:\, p < z_2} p^{\lceil \log n \rceil}$ is a non-zero multiple of $\phi(n)$. Thus, Miller's algorithm on input (n, μ), runs in time $\text{poly}(\log |n\mu|) = \text{poly}(z_2 \cdot \log |nz_1|)$ and factors n with constant probability. Finally, note that μ is easily computable in time $\text{poly}(z_2, \log n)$.

We are now finally ready to define Emul.

Algorithm 8 (Emul)
Input: The description of an oracle-aided circuit C.
Oracle: $G \in \mathcal{G}$.
Operation:
Emulate C^G while on every query $q = (n, e, h, \{\sigma_i\}_{i \in [t]})$ to Forger_G, return the following value to C:

1. *If Forger_G would return $\bot$ on q, return $\bot$ as well (and continue to the next query). Else,*
2. *Try to factor n by doing the following for $|C|$ times:*
   If $n \leq |C|^{\frac{\log |C|}{\log \log |C|}}$, execute $\mathsf{Sef}(n)$.
   Otherwise, execute $\mathsf{Factor}(n, \det(Q_{C,G,q}), |C|^4)$, where $Q_{C,G,q}$ is according to Definition 8.
3. *If factoring of n is successful, return $h^G(0)^d$ $[G_n]$, where $d = e^{-1} \bmod \phi(n)$.*
   Otherwise, abort.

The matrix $Q_{C,G,q}$ is defined as follows:

Definition 8 (query matrix). *Let C be an oracle-aided circuit, $G \in \mathcal{G}$ and let $q = (n, e, h, \{\sigma_i\}_{i \in [t]})$ be the query asked by C^{G, Forger_G} to Forger_G. The matrix $Q_{C,G,q} \in \mathbb{Z}_{t \times t}$ is defined as follows:*

1. *If $\mathsf{Forger}_G(q) = \bot$, set $Q_{C,G,q} = 0_{t \times t}$.*
   Otherwise:
2. *Let $M^H = M^{G,n,H}(\{h(i)\}_{i \in [t]})$ according to Definition 7, where H is as in Algorithm 4 with respect to this h and t. (Since $\mathsf{Forger}_G(q) \neq \bot$, the matrix M^H is well defined and of rank t.)*
3. *Let $\mathcal{I} \subseteq [\ell]$ be the first subset of size t (from hereafter we assume some arbitrary order on such sets) with $\text{rank}_e(M_{\mathcal{I}}^H) = t$.[10]*
4. *Let $M^{(H;C)} \in \mathbb{Z}_{t \times \ell'}$ be the matrix $M^{G,n,(H;C)}(\{\sigma_i\}_{i \in [t]})$ according to Definition 7, where $(H; C)$ is the circuit that first evaluates H and then evaluates C.*
5. *Set $Q_{C,G,q} = M_{\mathcal{I}}^H - e \cdot M_{\mathcal{I}}^{(H;C)}$.*

Note that in the code of Emul if Sef is called, and thus n is small, then it runs in time $\text{poly}(|C|)$. In addition, the running time of Factor, if called, is also in $\text{poly}(|C|)$. Thus, Emul runs in polynomial time.

---

[10] Remember that $M_{\mathcal{I}}^H \in \mathbb{Z}_{t \times t}$ is the restriction of M^H to the columns in $\mathcal{I}$.

Moreover, it is clear that the only case where the output of $\mathsf{Emul}^G(\mathrm{desc}(C))$ differs from the output of C^G is when the former aborts. This means that for some query of C to Forger, the latter would not return $\bot$, but either (1) Sef failed, or (2) z was a factoring advice but Factor failed, or (3) z was not a factoring advice for n. As the first two cases happen with negligible probability (by Theorem 6 and Lemma 4), we only have to prove that the latter happens with negligible probability.

This is formally done in the following lemma, whose proof (done via the "short description paradigm") can be found in the full version of this paper [14].

Lemma 5. *A query* $q = (n, \cdot)$ *to* Forger *made by* $C^{G \in \mathcal{G}, \mathsf{Forger}_G}$ *is unexpected, if*

- $\mathsf{Forger}_G(q) \neq \bot,$
- $n > |C|^{\frac{\log |C|}{\log \log |C|}},$ *and*
- $(\det(Q_{C,G,q}), |C|^4)$ *is not a factoring advice for* n, *where* $Q_{C,G,q}$ *is according to Definition 8.*

The following holds for any oracle-aided circuit C:

$$\Pr_{G \leftarrow \mathcal{G}}[C^{G, \mathsf{Forger}_G} \text{ asks } \mathsf{Forger} \text{ an unexpected query}] \leq \delta(|C|),$$

where $\delta(|C|) = 2^{-\log^2 |C|}$.

3.4   Putting It Together

Proof (of Lemma 2). Lemma 3 yields that F^{G, Forger_G} breaks the security of Σ^G with respect to every $G \in \mathcal{G}$, so it is left to prove that $\mathsf{Emul}^G(C_k)$ emulates $C_k^{G, \mathsf{Forger}_G}$ well.

Recall that $|C_k| \in \mathrm{poly}(k)$, and that we assume without loss of generality that $|C_k| \geq k$. Theorem 6 and Lemma 4 yield that $\mathsf{Emul}(C_k)$ answers all "expected" queries of C_k to Forger with probability $1 - |C_k| \cdot 2^{-\Omega(k)} = 1 - \mathrm{neg}(k)$, where Lemma 5 yields that C_k asks unexpected queries with only negligible probability over the choice of $G \in \mathcal{G}$. Hence, with save but negligible probability, $\mathsf{Emul}^G(C_k)$ emulates $C_k^{G, \mathsf{Forger}_G}$ correctly.

Acknowledgments. We thank Nir Bitansky, Thomas Holenstein and Ilya Mironov for very helpful conversations.

References

1. Aggarwal, D., Maurer, U.: Breaking RSA Generically Is Equivalent to Factoring. In: Joux, A. (ed.) EUROCRYPT 2009. LNCS, vol. 5479, pp. 36–53. Springer, Heidelberg (2009)
2. Bellare, M., Boldyreva, A., Palacio, A.: An Uninstantiable Random-Oracle-Model Scheme for a Hybrid-Encryption Problem. In: Cachin, C., Camenisch, J. (eds.) EUROCRYPT 2004. LNCS, vol. 3027, pp. 171–188. Springer, Heidelberg (2004)

3. Bellare, M., Rogaway, P.: Random oracles are practical: A paradigm for designing efficient protocols. In: ACM Conference on Computer and Communications Security, pp. 62–73 (1993)
4. Bellare, M., Rogaway, P.: Optimal Asymmetric Encryption. In: De Santis, A. (ed.) EUROCRYPT 1994. LNCS, vol. 950, pp. 92–111. Springer, Heidelberg (1995)
5. Boldyreva, A., Fischlin, M.: Analysis of Random Oracle Instantiation Scenarios for OAEP and Other Practical Schemes. In: Shoup, V. (ed.) CRYPTO 2005. LNCS, vol. 3621, pp. 412–429. Springer, Heidelberg (2005)
6. Brown, J., González Nieto, J.M., Boyd, C.: Efficient CCA-Secure Public-Key Encryption Schemes from RSA-Related Assumptions. In: Barua, R., Lange, T. (eds.) INDOCRYPT 2006. LNCS, vol. 4329, pp. 176–190. Springer, Heidelberg (2006)
7. Canetti, R., Goldreich, O., Halevi, S.: On the Random-Oracle Methodology as Applied to Length-Restricted Signature Schemes. In: Naor, M. (ed.) TCC 2004. LNCS, vol. 2951, pp. 40–57. Springer, Heidelberg (2004)
8. Canetti, R., Goldreich, O., Halevi, S.: The random oracle methodology, revisited. JACM: Journal of the ACM, 51 (2004)
9. Coron, J.-S.: On the Exact Security of Full Domain Hash. In: Bellare, M. (ed.) CRYPTO 2000. LNCS, vol. 1880, pp. 229–235. Springer, Heidelberg (2000)
10. Cramer, R., Shoup, V.: Signature schemes based on the strong rsa assumption. ACM Trans. Inf. Syst. Secur. 3(3), 161–185 (2000)
11. Dixon, J.D.: Asymptotically fast factorization of integers. Mathematics of Computation 36, 255–260 (1981)
12. Dodis, Y., Oliveira, R., Pietrzak, K.: On the Generic Insecurity of the Full Domain Hash. In: Shoup, V. (ed.) CRYPTO 2005. LNCS, vol. 3621, pp. 449–466. Springer, Heidelberg (2005)
13. Dodis, Y., Reyzin, L.: On the Power of Claw-Free Permutations. In: Cimato, S., Galdi, C., Persiano, G. (eds.) SCN 2002. LNCS, vol. 2576, pp. 55–73. Springer, Heidelberg (2003)
14. Dodis, Y., Haitner, I., Tentes, A.: On the instantiability of hash-and-sign rsa signatures. ePrint, http://eprint.iacr.org/2011/087
15. Gennaro, R., Gertner, Y., Katz, J., Trevisan, L.: Bounds on the efficiency of generic cryptographic constructions. SIAM Journal on Computing 35(1), 217–246 (2005)
16. Gennaro, R., Halevi, S., Rabin, T.: Secure Hash-and-Sign Signatures without the Random Oracle. In: Stern, J. (ed.) EUROCRYPT 1999. LNCS, vol. 1592, pp. 123–139. Springer, Heidelberg (1999)
17. Gennaro, R., Trevisan, L.: Lower bounds on the efficiency of generic cryptographic constructions. In: Proceedings of the 41st Annual Symposium on Foundations of Computer Science, pp. 305–313. IEEE Computer Society (2000)
18. Goldwasser, S., Tauman-Kalai, Y.: On the (in)security of the fiat-shamir paradigm. In: Proceedings of the 44th Annual Symposium on Foundations of Computer Science (FOCS), pp. 102–113. IEEE Computer Society (2003)
19. Haitner, I., Hoch, J.J., Reingold, O., Segev, G.: Finding collisions in interactive protocols – A tight lower bound on the round complexity of statistically-hiding commitments. In: Proceedings of the 48th Annual Symposium on Foundations of Computer Science (FOCS), pp. 669–679. IEEE Computer Society (2007)
20. Haitner, I., Holenstein, T.: On the (Im)Possibility of Key Dependent Encryption. In: Reingold, O. (ed.) TCC 2009. LNCS, vol. 5444, pp. 202–219. Springer, Heidelberg (2009)
21. Hofheinz, D., Jager, T., Kiltz, E.: Short Signatures From Weaker Assumptions. In: Lee, D.H., Wang, X. (eds.) ASIACRYPT 2011. LNCS, vol. 7073, pp. 647–666. Springer, Heidelberg (2011)

22. Hohenberger, S., Waters, B.: Short and Stateless Signatures from the RSA Assumption. In: Halevi, S. (ed.) CRYPTO 2009. LNCS, vol. 5677, pp. 654–670. Springer, Heidelberg (2009)
23. Maurer, U.M.: Abstract models of computation in cryptography. In: IMA Int. Conf., pp. 1–12 (2005)
24. Miller, G.L.: Riemann's hypothesis and tests for primality. Journal of Computer and System Sciences 13(3), 300–317 (1976)
25. Nechaev, V.I.: Complexity of a determinate algorithm for the discrete logarithm. MATHNASUSSR: Mathematical Notes of the Academy of Sciences of the USSR, 55 (1994)
26. Nielsen, J.B.: Separating Random Oracle Proofs from Complexity Theoretic Proofs: The Non-committing Encryption Case. In: Yung, M. (ed.) CRYPTO 2002. LNCS, vol. 2442, pp. 111–126. Springer, Heidelberg (2002)
27. Paillier, P.: Impossibility Proofs for RSA Signatures in the Standard Model. In: Abe, M. (ed.) CT-RSA 2007. LNCS, vol. 4377, pp. 31–48. Springer, Heidelberg (2006)
28. Paillier, P., Vergnaud, D.: Discrete-Log-Based Signatures May Not Be Equivalent to Discrete Log. In: Roy, B. (ed.) ASIACRYPT 2005. LNCS, vol. 3788, pp. 1–20. Springer, Heidelberg (2005)
29. Paillier, P., Villar, J.L.: Trading One-Wayness Against Chosen-Ciphertext Security in Factoring-Based Encryption. In: Lai, X., Chen, K. (eds.) ASIACRYPT 2006. LNCS, vol. 4284, pp. 252–266. Springer, Heidelberg (2006)
30. Pietrzak, K.: Compression from Collisions, or Why CRHF Combiners Have a Long Output. In: Wagner, D. (ed.) CRYPTO 2008. LNCS, vol. 5157, pp. 413–432. Springer, Heidelberg (2008)
31. Rivest, R.L., Shamir, A., Adelman, L.: A method for obtaining digital signatures and public-key cryptosystems. Communications of the ACM 21(2), 120–126 (1978)
32. RSA Laboratories, Redwood City, California. PKCS #1: RSA Encryption Standard (November 1993)
33. Shoup, V.: Lower Bounds for Discrete Logarithms and Related Problems. In: Fumy, W. (ed.) EUROCRYPT 1997. LNCS, vol. 1233, pp. 256–266. Springer, Heidelberg (1997)
34. Shoup, V.: Computational Introduction to Number Theory and Algebra. Cambridge University Press (2005)
35. Wee, H.: One-Way Permutations, Interactive Hashing and Statistically Hiding Commitments. In: Vadhan, S.P. (ed.) TCC 2007. LNCS, vol. 4392, pp. 419–433. Springer, Heidelberg (2007)

Beyond the Limitation of Prime-Order Bilinear Groups, and Round Optimal Blind Signatures

Jae Hong Seo[1] and Jung Hee Cheon[2]

[1] National Institute of Information and Communications Technology, Tokyo, Japan
jaehong@nict.go.jp
[2] ISaC & Dep. of Mathematical Sciences, Seoul National University, Seoul, Korea
jhcheon@snu.ac.kr

Abstract. At Eurocrypt 2010, Freeman proposed a transformation from pairing-based schemes in composite-order bilinear groups to equivalent ones in prime-order bilinear groups. His transformation can be applied to pairing-based cryptosystems exploiting only one of two properties of composite-order bilinear groups: cancelling and projecting. At Asiacrypt 2010, Meiklejohn, Shacham, and Freeman showed that prime-order bilinear groups according to Freeman's construction cannot have two properties simultaneously except negligible probability and, as an instance of implausible conversion, proposed a (partially) blind signature scheme whose security proof exploits both the cancelling and projecting properties of composite-order bilinear groups.

In this paper, we invalidate their evidence by presenting a security proof of the prime-order version of their blind signature scheme. Our security proof follows a different strategy and exploits only the projecting property. Instead of the cancelling property, a new property, that we call *translating*, on prime-order bilinear groups plays an important role in the security proof, whose existence was not known in composite-order bilinear groups. With this proof, we obtain a 2-move (i.e., round optimal) (partially) blind signature scheme (without random oracle) based on the decisional linear assumption in the common reference string model, which is of independent interest.

As the second contribution of this paper, we construct prime-order bilinear groups that possess both the cancelling and projecting properties at the same time by considering more general base groups. That is, we take a rank n $\mathbb{Z}_p$-submodule of $\mathbb{Z}_p^{n^2}$, instead of $\mathbb{Z}_p^n$, to be a base group G, and consider the projections into its rank 1 submodules. We show that the subgroup decision assumption on this base group G holds in the generic bilinear group model for $n = 2$, and provide an efficient membership-checking algorithm to G, which was trivial in the previous setting. Consequently, it is still open whether there exists a cryptosystem on composite-order bilinear groups that cannot be constructed on prime-order bilinear groups.

1 Introduction

Since Boneh, Goh, and Nissim [10] introduced composite-order bilinear groups in 2005, they have been used to solve many challenging problems in cryptography.

R. Cramer (Ed.): TCC 2012, LNCS 7194, pp. 133–150, 2012.
© International Association for Cryptologic Research 2012

Cryptographic systems using composite-order bilinear groups mostly utilize one of two properties, called *cancelling* and *projecting*, which Freeman [17] identified. (Though Freeman named two properties recently, these properties were already used before.) The security of almost all crypto systems using composite-order bilinear groups is based on the subgroup decision assumption, introduced by Boneh, Goh, and Nissim [10], or its variants.

Recently, some literature has aimed at constructing mathematical structures using prime-order bilinear groups with properties similar to (or richer than) composite-order bilinear groups [33,24,17,19]. In particular, Freeman [17] proposed two product groups of prime-order bilinear groups with separately defined bilinear maps. He showed that two proposed product groups satisfy the subgroup decision assumption (in the sense that given g, it is infeasible to determine whether g is in a subgroup or the whole product group), and each product group with a bilinear map satisfies *cancelling* and *projecting*, respectively. One direct benefit of this approach is efficiency improvements of group operations and pairing computations. Loosely speaking, in bilinear groups of composite order, the group order N must be infeasible to factor so that group operations and pairing computations are less efficient than those of bilinear groups of prime order for the same security level. See [17,19] for detailed efficiency comparison between composite-order groups and prime-order groups.

On the other hand, Meiklejohn, Shacham, and Freeman [30] gave a negative result, that is, an evidence of the limitation of constructing in some class of bilinear groups with both the *cancelling* and *projecting* properties, which is constructed on prime-order bilinear groups. To impart meaning to their result, they also proposed a round optimal blind signature scheme in composite-order bilinear groups whose security proof exploits both the *cancelling* and *projecting* properties of the composite-order bilinear group.[1] Their round optimal blind signature scheme is of independent interest since it is the first practical scheme of this type based on static assumptions (not based on q-type assumptions) in the common reference string model. They left two open questions: (1) whether the instantiation in prime-order groups of their round optimal blind signature scheme is provably secure or insecure, and (2) whether their limitation result can be applied to a wider class of bilinear groups constructed from prime-order groups.

In this paper, we answer both questions. We propose a (partially) blind signature scheme in a prime-order bilinear group setting. The proposed scheme can be considered as an adapted version of the scheme in [30] to the prime-order group setting. However, we prove the one-more unforgeability of the proposed scheme by using a completely different strategy from [30]. Our proof does not require the *cancelling* property, and instead we use another property, that we call *translating*, on prime order groups. Informally, the *translating* property is that given $g_1, g_1^a \in G_1, g_2 \in G_2$, where G_1 and G_2 are distinct subgroups of G, there exists a map $\mathcal{T}$ outputting g_2^a. The *translating* property is used, in

---

[1] The scheme in [30] itself does not use *cancelling* and *projecting*. Only the proof of security uses both *cancelling* and *projecting* properties. Thus, the authors do not rule out the existence of different proof strategy.

an essential way, to prove the one-more unforgeability of the proposed scheme. With this proof, we obtain a round optimal (partially) blind signature scheme (without relying on the random oracle heuristic) based on the decisional linear assumption in the common reference string model, which is of independent interest. Our blind signature scheme is more efficient than [30]. For example, our scheme has a shorter signature size (six elements in the prime-order group vs. two elements in the composite-order group). Moreover, the security of our blind signature scheme does not rely on the factoring assumption. (The blindness of the signature scheme in [30] based on the subgroup hiding assumption, which requires that the factorization of group order N is infeasible.)

As the second contribution, we show that there exists a more general class of bilinear groups than Meiklejohn, Shacham, and Freeman considered, and some of theses can be both *cancelling* and *projecting*. That is, we take a rank n $\mathbb{Z}_p$-submodule of $\mathbb{Z}_p^{n^2}$, instead of $\mathbb{Z}_p^n$, to be a base group G, and consider the projections into its rank 1 submodules. In this case, we should carefully consider group membership tests of a subgroup. We provide an efficient membership-checking algorithm to G, which was trivial in the previous setting, and we show that the subgroup decision assumption on this base group G holds in the generic bilinear group model for $n = 2$. Consequently, it is still open as to whether there exists a cryptosystem on composite-order bilinear groups that cannot be constructed on prime-order bilinear groups.

We note that although we construct a structure satisfying both *cancelling* and *projecting*, our construction can not be applied directly to the scheme in [30] to transform it to prime-order setting. The proof of [30] uses a property of composite-order group such that two subgroups' order are relatively prime, and our construction does not support such property so that we could not apply our construction to the round optimal blind signature scheme in [30].

Related Work: Blind Signatures. Since Chaum [11,12] introduced the concept of blind signatures in 1982, it has been studied extensively [6,1,7,8,16,28,31,25,5,18,4,2,21,30,3,20] because of its numerous applications, such as electronic voting [13] and electronic cash [14]. Blind signatures are interactive protocols between a user and a signer. In blind signatures, informally, the user can obtain a signature (signed by the signer) on a message (chosen by the user) without revealing the message to the signer that is signed during the protocol; that is, the signer learns nothing about the message after finishing the protocol.

In particular, round optimal (i.e., 2-move) blind signature schemes have received attention since the round complexity is an important measurement of efficiency in the computer network, and round optimal blind signature schemes directly imply that they are concurrently secure. In the random oracle model, there are elegant round optimal blind signatures by Chaum [12] and Boldyreva [8]. Without relying on the random oracle heuristic, there is an approach using general NIZKs for NP, and its security depends on the assumption that a common reference string exists [16,5]. Very recently, Garg et al. proposed the first round optimal blind signature in the standard model (without random oracle and a setup assumption such as a common reference string) [20]. These approaches without random oracle, however, are not as efficient as an approach, in which we are interested, using a bilinear map [9,10].

In recent years several efficient round optimal blind signatures [18,4,2,30,3] have been proposed in the common reference string model, using a bilinear map, by combining signature schemes with efficient NIWI proofs [23,22,24]. These approaches using a bilinear map either rely on q-type dynamic assumptions [18,4,2,3] or working on the composite-order group [30]. Though there is an analysis of a family of q-type dynamic assumptions by Cheon [15], the security of q-type assumptions still remains obscure. (q-type assumptions used in the above schemes hold in the generic group model [35] and these can be strong evidence for believing such assumptions. However, we believe that as the next step, constructing schemes without relying on such strong assumptions is an encouraging research approach.) In [30], a round optimal blind signature scheme based on static assumptions (not on q-type assumptions) using composite-order groups is proposed.

2   Notations and Definitions

Throughout this paper, we use notation $\oplus$ for the internal direct product: for an abelian group G, we write $G = G_1 \oplus G_2$ when G_1 and G_2 are subgroups of G and $G_1 \cap G_2 = \{1_G\}$ for the identity 1_G of G. In this case, every element g in G can be uniquely written by $g = g_1 \cdot g_2$ for some $g_1 \in G_1$ and $g_2 \in G_2$, where $\cdot$ is a group operation in G, and will be omitted sometimes. We use notation $x \overset{\$}{\leftarrow} A$. If A is a group $\mathbb{G}$, then it means that an element x is randomly chosen from $\mathbb{G}$, and if A is an algorithm, then it means that A outputs x. $[i, j]$ denotes a set of integers $\{i, \cdots, j\}$. We denote an abelian group generated by $g_1, \cdots, g_n$ by $\langle g_1, \cdots, g_n \rangle$.

We give formal definitions of bilinear group generators, and properties and cryptographic assumptions defined on the bilinear group.

Definition 1. *We say that $\mathcal{G}(\cdot, \cdot)$ is a* bilinear group generator *if it takes as input a security parameter λ and a positive integer $n \geq 1$, and it outputs a tuple $(G, G_i, H, H_i, G_t, e, \sigma \mid i \in [1, n]) \overset{\$}{\leftarrow} \mathcal{G}(\lambda, n)$, where G, H, G_t are finite abelian groups, G_i and H_i are cyclic subgroups of G and H of same order, respectively, such that $G = \oplus_{i \in [1,n]} G_i$ and $H = \oplus_{i \in [1,n]} H_i$, and $e : G \times H \to G_t$ is a non-degenerate bilinear map, that is, it satisfies*

> *Bilinearity:* $e(g_1 g_2, h_1 h_2) = e(g_1, h_1) e(g_1, h_2) e(g_2, h_1) e(g_2, h_2)$
> *for $g_1, g_2 \in G$ and $h_1, h_2 \in H$,*
> *Non-degeneracy: for $g \in G$, if $e(g, h) = 1$ for any $h \in H$, then $g = 1$,*
> *for $h \in H$, if $e(g, h) = 1$ for any $g \in G$, then $h = 1$,*

and σ is additional information for group membership-check. Moreover, we assume that group operations, random samplings, and membership-checks in each group, and computation of e can be efficiently performed (i.e. polynomial-time in λ).

We do not exclude the case that $G = H$. When $G = H$, we say that $\mathcal{G}$ is a symmetric bilinear group generator.

Definition 2. *We say that an algorithm $\mathcal{G}_1$ is a bilinear group generator of prime order if $\mathcal{G}_1(\lambda) = \mathcal{G}(\lambda, 1)$, and $\mathcal{G}_1$ outputs groups G, G_1, H, H_1, G_t of prime order p and a map e. Then, $G = G_1$, $H = H_1$. We denote the three distinct groups G, H, G_t by $\mathbb{G}, \mathbb{H}, \mathbb{G}_t$, respectively, and a bilinear map e by $\hat{e}$.*

Now, we provide definitions of two properties, called *cancelling* and *projecting*, which are introduced by Freeman [17].

Definition 3. *A bilinear group generator $\mathcal{G}$ is* cancelling *if $e(g_i, h_j) = 1_t$ whenever $g_i \in G_i$, $h_j \in H_j$, and $i \neq j$, where 1_t is the identity of G_t.*

Definition 4. *A bilinear group generator $\mathcal{G}$ is* projecting *if there exist subgroups $G' \subset G$, $H' \subset H$, and $G'_t \subset G_t$, and non-trivial[2] homomorphisms $\pi : G \to G$, $\bar{\pi} : H \to H$, and $\pi_t : G_t \to G_t$ such that*

1. $G' \subset \ker(\pi)$, $H' \subset \ker(\bar{\pi})$, and $G'_t \subset \ker(\pi_t)$.
2. $\pi_t(e(g, h)) = e(\pi(g), \bar{\pi}(h))$ for $\forall g \in G$ and $\forall h \in H$.

If $\mathcal{G}$ is a symmetric bilinear group generator, that is, $G = H$, then set $G' = H'$ and $\pi = \bar{\pi}$.

To prove the security of the proposed blind signature scheme, we need two widely-known assumptions, the Computational Diffie-Hellman assumption, and k-Linear assumption which is introduced by Hofheinz and Kiltz and Shacham [26,34], in the bilinear group setting.

Definition 5. *Let $\mathcal{G}_1$ be a bilinear group generator of prime order. We define the advantage of an algorithm $\mathcal{A}$ in solving Computational Diffie-Hellman (CDH) problem in G, denoted by $Adv_{\mathcal{A}, \mathcal{G}_1}^{CDHP_G}$, is to be*

$$\Pr\left[\mathcal{A}(G, H, G_t, e, g, g^a, g^b) \to g^{ab} : (G, H, G_t, e) \xleftarrow{\$} \mathcal{G}_1, g \xleftarrow{\$} G, a, b, \xleftarrow{\$} \mathbb{Z}_p\right].$$

We say that $\mathcal{G}$ satisfies Computational Diffie-Hellman (CDH) assumption in G if for any PPT algorithm $\mathcal{A}$, $Adv_{\mathcal{A}, \mathcal{G}_1}^{CDHP_G}$ is a negligible function of λ.

Definition 6. *Let $\mathcal{G}_1$ be a bilinear group generator of prime order and $k \geq 1$. We define the advantage of an algorithm $\mathcal{A}$ in solving the k-Linear problem in G, denoted by $Adv_{\mathcal{A}, \mathcal{G}_1}^{k\text{-}Lin_G}$, is to be*

$$\left| \Pr\left[\mathcal{A}(G, H, G_t, e, g, u_i, u_i^{a_i}, g^b, h \text{ for } i \in [1, k]) \to 1 : \right.\right.$$
$$\left. (G, H, G_t, e) \xleftarrow{\$} \mathcal{G}_1, g, u_i \xleftarrow{\$} G, h \xleftarrow{\$} H, a_i \xleftarrow{\$} \mathbb{Z}_p \text{ for } i \in [1, k], b \xleftarrow{\$} \mathbb{Z}_p\right]$$
$$- \Pr\left[\mathcal{A}(G, H, G_t, e, g, u_i, u_i^{a_i}, g^b, h \text{ for } i \in [1, k]) \to 1 : \right.$$
$$\left.\left. (G, H, G_t, e) \xleftarrow{\$} \mathcal{G}_1, g, u_i \xleftarrow{\$} G, h \xleftarrow{\$} H, a_i \xleftarrow{\$} \mathbb{Z}_p \text{ for } i \in [1, k], b = \sum_{i \in [1, k]} a_i\right] \right|.$$

---

[2] The non-triviality does not appear in the original definition [17]. Without this, however, every bilinear group can be *projecting* by using the trivial homomorphisms.

Then, we say that $\mathcal{G}$ satisfies the k-Linear assumption in G if for any PPT algorithm $\mathcal{A}$, the advantage of $\mathcal{A}$ $Adv_{\mathcal{A},\mathcal{G}_1}^{k\text{-}Lin_G}$ is a negligible function of λ.

We can analogously define the CDH assumption and the k-Linear assumption in H. The *1-Linear assumption* in G is the *DDH assumption* in G and the *2-Linear assumption* in G is the *decisional linear assumption* in G.

Next, we provide the definition of the *subgroup decision assumption*, adapted from [17] to fit our purpose.

Definition 7. *Let $\mathcal{G}$ be a bilinear group generator. We define the advantage of an algorithm $\mathcal{A}$ in solving the (n, k)-subgroup decision problem on the left, denoted by $Adv_{\mathcal{A},\mathcal{G}}^{SDA_L}$, is to be*

$$\Big| \Pr\Big[\mathcal{A}(G, G', H, H', G_t, e, \sigma, g) \to 1 :$$
$$(G, G_i, H, H_i, G_t, e, \sigma) \xleftarrow{\$} \mathcal{G}(\lambda, n), G' := \oplus_{i \in [1,k]} G_i, H' := \oplus_{i \in [1,k]} H_i, g \xleftarrow{\$} G \Big]$$
$$- \Pr\Big[\mathcal{A}(G, G', H, H', G_t, e, \sigma, g') \to 1 :$$
$$(G, G_i, H, H_i, G_t, e, \sigma) \xleftarrow{\$} \mathcal{G}(\lambda, n), G' := \oplus_{i \in [1,k]} G_i, H' := \oplus_{i \in [1,k]}, g' \xleftarrow{\$} G' \Big] \Big|.$$

We say that $\mathcal{G}$ satisfies the (n, k)-subgroup decision assumption on the left if for any PPT algorithm $\mathcal{A}$, its advantage $Adv_{\mathcal{A},\mathcal{G}}^{SDA_L}$ is a negligible function in λ.

We analogously define the (n, k)-*subgroup decision assumption on the right*.

Definition 8. *We say that a bilinear group generator $\mathcal{G}(\cdot, \cdot)$ satisfies the (n, k)-subgroup decision assumption if $\mathcal{G}(\cdot, n)$ satisfies both the (n, k)-subgroup decision assumptions on the left and on the right.*

We will often omit (n, k) term, if it is clear in the context.

3   Round-Optimal Blind Signature in Prime-Order Group

3.1   Symmetric Bilinear Group with Projecting Pairing

We construct a symmetric bilinear group generator with the *projecting* property. (The symmetric bilinear groups mean that $G = H$, and $G_i = H_i$ in our definition of bilinear groups.) We borrow some notations from Freeman's paper [17]. Let $\mathbb{G}$ be a group, $\mathfrak{g}, \mathfrak{g}_1, \cdots, \mathfrak{g}_n$ be elements in $\mathbb{G}$, $\vec{\alpha} = (a_1, \cdots, a_n)$ be a vector in $\mathbb{Z}_p^n$, and $M = (m_{ij})$ be an $n \times n$ matrix. We denote $\mathfrak{g}^{\vec{\alpha}} := (\mathfrak{g}^{a_1}, \cdots, \mathfrak{g}^{a_n}) \in \mathbb{G}^n$ and $(\mathfrak{g}_1, \cdots, \mathfrak{g}_n)^M := (\prod_{i \in [1,n]} \mathfrak{g}_i^{m_{i1}}, \cdots, \prod_{i \in [1,n]} \mathfrak{g}_i^{m_{in}})$. We can see that $(\mathfrak{g}^{\vec{\alpha}})^M = \mathfrak{g}^{(\vec{\alpha}M)}$. We newly define some notations useful to explain product groups. Let $G = \oplus_{i \in [1,n]} G_i$ and $H = \oplus_{j \in [1,n]} H_j$, where G_i and H_j are cyclic groups of same order. Let $e(G_i, H_j)$ be a set $\{e(g_i, h_j) | g_i \in G_i, h_j \in H_j\}$; hence $e(G_i, H_j)$ is a cyclic group since G_i and H_j are cyclic groups. In particular, when G_i and H_j have prime order p, $e(G_i, H_j)$ is a cyclic group of order p or 1.

Now, we construct a symmetric bilinear group generator $\mathcal{G}_{SP}(\lambda, 3)$, which is a generalization of Groth and Sahai's instantiation based on the decisional linear assumption [24], and is also a symmetric version of Freeman's asymmetric bilinear group generator with the *projecting* property [17].

1. $\mathcal{G}_1(\lambda) \xrightarrow{\$} (p, \mathbb{G}, \mathbb{G}_t, \hat{e})$.
2. Set $G = \mathbb{G}^3$, $G_t = \mathbb{G}_t^9$.
3. Choose linearly independent vectors $\vec{x}_1, \vec{x}_2, \vec{x}_3 \in \mathbb{Z}_p^3$, and set $G_1 = \langle \mathfrak{g}^{\vec{x}_1} \rangle$, $G_2 = \langle \mathfrak{g}^{\vec{x}_2} \rangle$ and $G_3 = \langle \mathfrak{g}^{\vec{x}_3} \rangle$. Then, $G = G_1 \oplus G_2 \oplus G_3$.
4. Define a map $e : G \times G \to G_t$ by

$$= e((\mathfrak{g}_1, \mathfrak{g}_2, \mathfrak{g}_3), (\mathfrak{h}_1, \mathfrak{h}_2, \mathfrak{h}_3))$$

$$\Big(\hat{e}(\mathfrak{g}_1, \mathfrak{h}_1)^{1/2}, \hat{e}(\mathfrak{g}_1, \mathfrak{h}_2)^{1/2}, \hat{e}(\mathfrak{g}_1, \mathfrak{h}_3)^{1/2}, \hat{e}(\mathfrak{g}_2, \mathfrak{h}_1)^{1/2}, \hat{e}(\mathfrak{g}_2, \mathfrak{h}_2)^{1/2}, \hat{e}(\mathfrak{g}_2, \mathfrak{h}_3)^{1/2},$$

$$\hat{e}(\mathfrak{g}_3, \mathfrak{h}_1)^{1/2}, \hat{e}(\mathfrak{g}_3, \mathfrak{h}_2)^{1/2}, \hat{e}(\mathfrak{g}_3, \mathfrak{h}_3)^{1/2} \Big)$$

$$\cdot \Big(\hat{e}(\mathfrak{g}_1, \mathfrak{h}_1)^{1/2}, \hat{e}(\mathfrak{g}_2, \mathfrak{h}_1)^{1/2}, \hat{e}(\mathfrak{g}_3, \mathfrak{h}_1)^{1/2}, \hat{e}(\mathfrak{g}_1, \mathfrak{h}_2)^{1/2}, \hat{e}(\mathfrak{g}_2, \mathfrak{h}_2)^{1/2}, \hat{e}(\mathfrak{g}_3, \mathfrak{h}_2)^{1/2},$$

$$\hat{e}(\mathfrak{g}_1, \mathfrak{h}_3)^{1/2}, \hat{e}(\mathfrak{g}_2, \mathfrak{h}_3)^{1/2}, \hat{e}(\mathfrak{g}_3, \mathfrak{h}_3)^{1/2} \Big).$$

Then, $e(\mathfrak{g}^{\vec{x}}, \mathfrak{g}^{\vec{y}}) = \hat{e}(\mathfrak{g}, \mathfrak{g})^{1/2(\vec{x} \otimes \vec{y}) + 1/2(\vec{y} \otimes \vec{x})}$, where $\otimes$ is a tensor product (Kronecker product) of two 3-dimensions vectors.
5. For $i \in [1, 3]$, define maps $\pi_i : G \to G$ and $\pi_{t,i} : G_t \to G_t$ by

$$\pi_i(g) = g^{M^{-1} U_i M} \quad \text{and} \quad \pi_{t,i}(g_t) = g_t^{(M^{-1} U_i M) \otimes (M^{-1} U_i M)}, \quad \text{respectively,}$$

where M is a 3×3 matrix having $\vec{x}_i$ as its i-th row, U_i is a 3×3 matrix with 1 in the (i, i) entry and zeroes elsewhere, and $\otimes$ is a tensor product of matrices: For $\ell_1 \times \ell_2$ matrix $A = (a_{i,j})$ and $\ell_3 \times \ell_4$ matrix $B = (b_{i,j})$, $A \otimes B$ is a $\ell_1 \ell_3 \times \ell_2 \ell_4$ matrix whose (i, j)-th block is equal to $a_{i,j} B$, where we consider $A \otimes B$ as $\ell_1 \times \ell_2$ blocks. Then, π_i is a projection such that for $g_1 \in G_1, g_2 \in G_2, g_3 \in G_3$, $\pi_i(g_1 g_2 g_3)$ is equal to g_i.
6. Output $(p, G, G_1, G_2, G_3, G_t, e, \pi_1, \pi_2, \pi_3, \pi_{t,1}, \pi_{t,2}, \pi_{t,3})$.

We provide a useful lemma to understand the structure of the image of e.

Lemma 1. *The image of e generated by $\mathcal{G}_{SP}$ is equal to $\oplus_{1 \leq i \leq j \leq 3} e(G_i, G_j)$, and each $e(G_i, G_j)$'s order is p.*

We provide the proof of Lemma 1 in the full version of this paper. Non-degeneracy of e is directly coming from the lemma 1. (That is, $e(g, h) \neq 1_t$ for any non-identity elements $g, h \in G$. If not, the image is not equal to $\oplus_{1 \leq i \leq j \leq 3} e(G_i, G_j)$.) The bilinear property of e can be easily checked from the bilinear property of the tensor product. Further, $\mathcal{G}_{SP}$ satisfies the *projecting* property: Let $G' = G_2 \oplus G_3$, $G'_t = \oplus_{2 \leq i \leq j \leq 3} e(G_i, G_j)$, $\pi = \pi_1$, and $\pi_t = \pi_{t,1}$, where G', G'_t, π, and π_t are defined in the definition 4. Then, $G' \subset \ker(\pi)$ and $G'_t \subset \ker(\pi_t)$, and e, π, π_t satisfy the following commutative property.

$$\pi_t(e(\mathfrak{g}^{\vec{x}}, \mathfrak{g}^{\vec{y}})) = e(\pi(\mathfrak{g}^{\vec{x}}), \pi(\mathfrak{g}^{\vec{y}})).$$

We can check this commutative property as follows:

$$\pi_t(e(\mathfrak{g}^{\vec{x}}, \mathfrak{g}^{\vec{y}}))$$
$$= \pi_{t,1}(e(\mathfrak{g}^{\vec{x}}, \mathfrak{g}^{\vec{y}}))$$
$$= \pi_{t,1}(\hat{e}(\mathfrak{g}, \mathfrak{g})^{1/2(\vec{x}\otimes\vec{y})+1/2(\vec{y}\otimes\vec{x})})$$
$$= (\hat{e}(\mathfrak{g}, \mathfrak{g})^{1/2(\vec{x}\otimes\vec{y})+1/2(\vec{y}\otimes\vec{x})})^{(M^{-1}U_iM)\otimes(M^{-1}U_iM)}$$
$$= \hat{e}(\mathfrak{g}, \mathfrak{g})^{1/2(\vec{x}\otimes\vec{y})((M^{-1}U_iM)\otimes(M^{-1}U_iM))+1/2(\vec{y}\otimes\vec{x})((M^{-1}U_iM)\otimes(M^{-1}U_iM))}$$
$$= \hat{e}(\mathfrak{g}, \mathfrak{g})^{1/2(\vec{x}M^{-1}U_iM)\otimes(\vec{y}M^{-1}U_iM)+1/2(\vec{y}M^{-1}U_iM)\otimes(\vec{x}M^{-1}U_iM)}$$
$$= e(\mathfrak{g}^{(\vec{x}M^{-1}U_iM)}, \mathfrak{g}^{(\vec{y}M^{-1}U_iM)})$$
$$= e((\mathfrak{g}^{\vec{x}})^{M^{-1}U_iM}, (\mathfrak{g}^{\vec{y}})^{M^{-1}U_iM})$$
$$= e(\pi_1(\mathfrak{g}^{\vec{x}}), \pi_1(\mathfrak{g}^{\vec{y}})) = e(\pi(\mathfrak{g}^{\vec{x}}), \pi(\mathfrak{g}^{\vec{y}})).$$

The fifth equality comes from the property of the tensor product such as $(A \otimes B)(C \otimes D) = (AC) \otimes (BD)$, where A and B are matrices having ℓ columns and C and D are matrices having ℓ rows for some ℓ. (We can consider a vector as a matrix having one row.)

In contrast to the composite order bilinear group, our product group of prime order group has an additional property, we name *translating* and define as follow.

Definition 9. *A bilinear group generator $\mathcal{G}$ is (i,j)-translating if there exists efficiently computable (that is, polynomial time in λ) maps $\mathcal{T}_{i,j} : G_i^2 \times G_j \to G_j$ defined by $(g_i, g_i^a, g_j) \mapsto g_j^a$ and $\bar{\mathcal{T}}_{i,j} : H_i^2 \times H_j \to H_j$ defined by $(h_i, h_i^a, h_j) \mapsto h_j^a$ for an integer $a \in \mathbb{Z}$. If $\mathcal{G}$ is a symmetric bilinear group generator, then set $\bar{\mathcal{T}}_{i,j} = \mathcal{T}_{i,j}$.*

We show that the above $\mathcal{G}_{SP}$ construction satisfies *translating* property.

Theorem 1. $\mathcal{G}_{SP}(\lambda, 3)$ *satisfies* translating *property for all $i, j \in [1, 3]$.*

Proof. We first construct $\mathcal{T}_{3,1}$. Given g_3^a and a 3×3 matrix M defined as in the description of $\mathcal{G}_{SP}$, we can compute g_1^a without knowing a as follows:

$$(g_3^a)^{M^{-1}} = ((\mathfrak{g}^{\vec{x}_3})^a)^{M^{-1}} = (\mathfrak{g}^{a\vec{e}_3 M})^{M^{-1}} = \mathfrak{g}^{a\vec{e}_3} = (1, 1, \mathfrak{g}^a),$$

$$(\mathfrak{g}^a, 1, 1)^M = (\mathfrak{g}^{a\vec{e}_1})^M = \mathfrak{g}^{a\vec{x}_1} = g_1^a,$$

where $\vec{e}_i$ is the canonical i-th vector in $\mathbb{Z}_p^3$, for example, $\vec{e}_1 = (1, 0, 0)$. We can construct other $\mathcal{T}_{i,j}$ analogously. $\square$

Moreover, $\mathcal{G}_{SP}$ satisfies $(3, 2)$-subgroup decision assumption when the underlying group generator $\mathcal{G}_1$ satisfies the decisional linear assumption.

Lemma 2. *If $\mathcal{G}_1$ satisfies the decisional linear assumption, then $\mathcal{G}_{SP}$ satisfies the $(3, 2)$-subgroup decision assumption.*

We relegate the proof of Lemma 2 in the full version of this paper.

Remark 1. Note that $\mathcal{G}_{SP}$ does not satisfy the *cancelling* property since $e(G_i, G_j)$ is not equal to $\{1_t\}$ for $i \neq j$ (Lemma 1).

3.2 Construction

The abstract of our scheme looks very similar to the Meiklejohn et al.'s construction in the composite order bilinear group [30]. We slightly changed the Meiklejohn et al.'s construction to adapt in the prime order bilinear group setting.

(Partially) blind signature schemes in the common reference model consist of five (interactive) algorithms: Setup, KeyGen, User, Signer, and Verify. We provide the formal definition of (partially) blind signature schemes, and concurrently security, in the full version of this paper. We follow the security definition of [30], which is slightly stronger than [6], by allowing the adversary to choose the public key in the *blindness* definition. As a definition of the blind signature, [30] is modified from [27]; (1) it strengthens the *blindness* game to allow the adversary to generate the public key, and (2) it weakens the *one-more unforgeability* game to require that the messages (instead of pairs of message and signature) must all be distinct.[3]

The proposed partially blind signature scheme for a message space $\mathcal{M} = \{0,1\}^m$ is as follows.[4]:

- Setup(λ): $\mathcal{G}_{SP}(\lambda, 3) \xrightarrow{\$} (p, G, G_1, G_2, G_3, G_t, e, \pi_i, \pi_{t,i})$. Choose $g, u', u_1, \cdots, u_m, v_1 \cdots, v_m \xleftarrow{\$} G$, $h_1 \xleftarrow{\$} G_1$ and $h_2 \xleftarrow{\$} G_2$. Define

$$CRS = (p, G, G_t, e, g, u', u_1, \cdots, u_m, v_1, \cdots, v_m, h_1, h_2).$$

- KeyGen(CRS): Choose $g' \xleftarrow{\$} G$. Set $A = e(g, g')$. The public key is $PK = \{A\}$, and the secret key is $SK = \{g'\}$.
- User($CRS, PK, info, Msg$): Let $info$ be an m_0 bits string and Msg be an $m - m_0$ bit string. We write $info$ bitwise as $b_0 \cdots b_{m_0}$ and Msg as $b_{m_0+1} \cdots b_m$. For $i \in [m_0 + 1, m]$, pick random integers $t_{i,1}, t_{i,2}, s_{i,1}, s_{i,2}, r_i, r'_i \xleftarrow{\$} \mathbb{Z}_p$, and compute

$$c_i = (u_i)^{b_i} h_1^{t_{i,1}} h_2^{t_{i,2}}, \quad d_i = (v_i)^{b_i} h_1^{s_{i,1}} h_2^{s_{i,2}},$$
$$\theta_{i,1} = u_i^{b_i s_{i,1}} (v_i^{b_i-1} h_1^{s_{i,1}} h_2^{s_{i,2}})^{t_{i,1}} h_2^{r_i}, \quad \theta_{i,2} = u_i^{b_i s_{i,2}} (v_i^{b_i-1} h_1^{s_{i,1}} h_2^{s_{i,2}})^{t_{i,2}} h_1^{-r_i},$$
$$\theta_{i,3} = u_i^{(b_i-1)s_{i,1}} (v_i^{b_i} h_1^{s_{i,1}} h_2^{s_{i,2}})^{t_{i,1}} h_2^{r'_i}, \quad \theta_{i,4} = u_i^{(b_i-1)s_{i,2}} (v_i^{b_i} h_1^{s_{i,1}} h_2^{s_{i,2}})^{t_{i,2}} h_1^{-r'_i}.$$

Let $\overrightarrow{\theta}_i = (\theta_{i,1}, \cdots, \theta_{i,4})$, and send $req = \{(c_i, d_i, \overrightarrow{\theta}_i)\}_{i \in [m_0+1,m]}$ to the signer and save $state = \{(t_{i,1}, t_{i,2})\}_{i \in [m_0+1,m]}$.

- Signer($CRS, SK, info, req$): Write $req = \{(c_i, d_i, \overrightarrow{\theta}_i)\}_{i \in [m_0+1,m]}$ and $info = b_1 \cdots b_{m_0}$. For each $i \in [m_0 + 1, m]$, verify c_i is a commitment of 0 or 1 by checking that

$$e(c_i, d_i v_i^{-1}) \overset{?}{=} e(h_1, \theta_{i,1}) e(h_2, \theta_{i,2}) \text{ and } e(c_i u_i^{-1}, d_i) \overset{?}{=} e(h_1, \theta_{i,3}) e(h_2, \theta_{i,4}).$$

---

[3] This weakened definition is necessary if the output signature can be re-randomized. [30]'s partially blind signature and ours are in the case.

[4] For large message spaces, we can use a collision resistance hash function first.

If for some i the above equation does not hold, abort the protocol and output $\perp$. Otherwise, compute

$$c = \left(u' \prod_{i \in [1,m_0]} u_i^{b_i} \right) \left(\prod_{i \in [m_0+1,m]} c_i \right),$$

choose a random integer $r \overset{\$}{\leftarrow} \mathbb{Z}_p$, compute

$$K_1 = g'c^r, \quad K_2 = g^{-r}, \quad K_{3,1} = h_1^{-r}, \quad K_{3,2} = h_2^{-r},$$

send $(K_1, K_2, K_{3,1}, K_{3,2})$ to the user, and output $success$ and $info$.

- User$(state, (K_1, K_2, K_{3,1}, K_{3,2}))$: Write $state = \{(t_{i,1}, t_{i,2})\}_{i \in [m_0+1,m]}$. Check that

$$e(K_{3,1}, g) \overset{?}{=} e(K_2, h_1) \text{ and } e(K_{3,2}, g) \overset{?}{=} e(K_2, h_2).$$

If one of two above equations is fail to hold, then abort the protocol and output $\perp$. Otherwise, unblind the signature by computing

$$S_1 = K_1 \cdot \left(\prod_{i \in [m_0+1,m]} K_{3,1}^{t_{i,1}} K_{3,2}^{t_{i,2}} \right) \text{ and } S_2 = K_2.$$

Check the validity of the signature (S_1, S_2) by running Verify. If it outputs $accept$, then go to the next step. Otherwise, abort the protocol and output $\perp$. Finally re-randomize the signature by picking a random $s \overset{\$}{\leftarrow} \mathbb{Z}_p$ and computing

$$S_1' = S_1 \cdot (u' \prod_{i \in [1,m]} u_i^{b_i})^s \text{ and } S_2' = S_2 \cdot g^{-s}.$$

Output the signature $sig = (S_1', S_2'), info$, and $success$.

- Verify$(CRS, PK, info, Msg, sig)$: Write $PK = \{A\}$, $info = b_1 \cdots b_{m_0}$, $Msg = b_{m_0} \cdots b_m$, and $sig = (S_1, S_2)$. Check that

$$e(S_1, g) \cdot e(S_2, u' \prod_{i \in [1,m]} u_i^{b_i}) \overset{?}{=} A.$$

If the above equality holds, then output $accept$. Otherwise, output $fail$.

In the first procedure of the user, c_i and d_i are GS-commitment to b_i, and $\vec{\theta}_i$ is GS-proof that b_i satisfies the equation $b_i(b_i - 1) = 0$ so that $b_i = 0$ or $b_i = 1$. More precisely, when b_i and b_i' are openings of c_i and d_i, respectively, $\vec{\theta}_i$ is a proof that $b_i(b_i' - 1) = 0$ and $(b_i' - 1)b_i = 0$. Then, $(b_i = 0$ or $b_i' = 1) \bigwedge (b_i = 1$ or $b_i' = 0)$ so that $b_i = b_i' = 0$ or $b_i = b_i' = 1$. We provide three theorems to prove the security of the proposed (partially) blind signature scheme.

Theorem 2. *The above blind signature is correct.*

Theorem 3. *If $\mathcal{G}_1$ satisfies the decisional linear assumption, then the above blind signature satisfies blindness.*

The proof of Theorem 2 and 3 are similar to the previous ones [30]. We provide the proof in the full version of this paper.

Theorem 4. *If $\mathcal{G}_1$ satisfies the the CDH assumption, then the above blind signature is one-more unforgeable.*

Due to space constraints, we leave the proof of Theorem 4 to the full version of this paper. Instead, we briefly explain our idea to prove the one-more unforgeability, and the reason why we cannot apply the Meiklejohn et al. proof strategy to the proposed scheme. At the end of the interaction, the user obtains a Waters-signature, which is existentially unforgeable based on the CDH assumption. If the user obtains only a Waters signature, then the proposed scheme is, loosely speaking, also one-more unforgeable. However, the user obtains not only a Waters signature (of the form $g'(u \prod_{i\in[1,m]} u_i^{b_i})^r$ and g^{-r} for message $b_1 \cdots b_m$), but also some additional information, that is, it eventually gets

$$g'(u \prod_{i\in[1,m]} u_i^{b_i})^r (\prod_{i\in[m_0+1,m]} h_1^{t_{i,1}} h_2^{t_{i,2}})^r, \quad g^{-r}, \quad h_1^{-r}, \text{ and } h_2^{-r}$$

for some (unknown and uniformly distributed) $r \in \mathbb{Z}_p$, and $t_{i,1}$, $t_{i,2}$, and b_i chosen by itself. Therefore, we should show that h_1^{-r}, h_2^{-r}, and $(\prod_{i\in[m_0,m]} h_1^{t_{i,1}} h_2^{t_{i,2}})^r$ will not be helpful for the user to break the one-more unforgeability. In [30], a pairing e satisfies the *cancelling* property, and orders of subgroups are relatively prime so that each part contained in each subgroup in a signature scheme is independent. [30] essentially utilized this independence. If, in our scheme, the $G_1 \oplus G_2$ part and G_3 part were independent, the user could not obtain any additional information about the part in G_3 from the above information. (Since all information other than a Waters signature, which the user gets at the end of the protocol, is related to h_1 and h_2, which are elements in $G_1 \oplus G_2$, this information will not be helpful for forging the Waters signature in the G_3 part.) Hence, the one-more unforgeability of the scheme can be reduced to the existential unforgeability of the Waters signature (in G_3 in the case of our scheme). However, we cannot apply this Meiklejohn et al. proof strategy to our scheme since our bilinear map e does not have the *cancelling* property and each subgroup has the same order p. Instead, we prove the one-more unforgeability using a completely different strategy. Our simulation basically follows the simulation for the existential unforgeability of the Waters signature, and at the same time simulates directly additional information h_1^{-r}, h_2^{-r}, and $(\prod_{i\in[m_0+1,m]} h_1^{t_{i,1}} h_2^{t_{i,2}})^r$. It seems hard to simulate $(\prod_{i\in[m_0+1,m]} h_1^{t_{i,1}} h_2^{t_{i,2}})^r$ since $t_{i,1}$ and $t_{i,2}$ are chosen by the user and r is usually not known to the simulator during the simulation. (r is usually of the form $Ra + S$ for some unknown a and constants R and S, where a is given by the form g^a.) We circumvent this obstacle by using the *projecting* property and the *translating* property mentioned in section 3.1. To simulate this additional information, the simulator first extracts the message, that is, recovers $b_1 \cdots b_m$ by computing $log_{\pi_1(u_i)} \pi_1(c_i) = b_i$, and second computes $\pi_j(c_i/u_i^{b_i}) = h_j^{t_{i,j}}$ and

$$\text{if } b_i = 0, \begin{cases} \pi_3(\theta_{i,1}^{-1}) = \pi_3(v_i)^{t_{i,1}} \\ \pi_3(\theta_{i,2}^{-1}) = \pi_3(v_i)^{t_{i,2}}, \end{cases} \qquad \text{if } b_i = 1, \begin{cases} \pi_3(\theta_{i,3}) = \pi_3(v_i)^{t_{i,1}} \\ \pi_3(\theta_{i,4}) = \pi_3(v_i)^{t_{i,2}}. \end{cases}$$

Though $\pi_3(v_i)^{t_{i,j}}$ is contained in G_3, we can change it to be of the form $h_j^{at_{i,j}}$ for some unknown a by using the *translating* property mentioned in section 3.1 when v_i contains a in the exponent. The simulator can generate $(\prod_{i\in[m_0+1,m]} h_1^{t_{i,1}} h_2^{t_{i,2}})^r$ by using $h_j^{t_{i,j}}$ and $h_j^{at_{i,j}}$.

Remark 2. The decisional linear assumption implies the CDH assumption. (The decisional linear assumption implies the computational linear assumption, and the computational linear assumption implies the CDH assumption. Reductions are quite straightforward.)

Remark 3. In the user's first procedure, the GS-commitment and proof appear to have redundant parts. It would be more natural to change them to

$$c_i = (u_i)^{b_i} h_1^{t_{i,1}} h_2^{t_{i,2}}, \theta_{i,1} = (u_i^{2b_i-1} h_1^{t_{i,1}} h_2^{t_{i,2}})^{t_{i,1}} h_2^{r_i}, \theta_{i,2} = (u_i^{2b_i-1} h_1^{t_{i,1}} h_2^{t_{i,2}})^{t_{i,2}} h_1^{-r_i},$$

and it can be verified by $e(c_i, c_i u_i^{-1}) \overset{?}{=} e(h_1, \theta_{i,1})e(h_2, \theta_{i,2})$. This commitment and proof is GS commitment and proof for $b_i \in \{0,1\}$. However, we note that in this case, we could not prove the one-more unforgeability based on the CDH assumption. We only proved the one-more unforgeability based on the decisional linear assumption and augmented CDH assumption. (Augmented CDH assumption roughly says that given $\mathfrak{g}, \mathfrak{g}^a, \mathfrak{g}^b, \mathfrak{g}^{a^2}$, it is infeasible to compute $\mathfrak{g}^{ab}$.) To avoid requiring $\mathfrak{g}^{a^2}$, in the simulation, that is, to prove the one-more unforgeability based on the CDH assumption, we modified the commitment and the proof to the current form.

4   Bilinear Group: Both Cancelling and Projecting

4.1   Interpreting Limitation Result in [30]

In [30], the authors consider the cases that the bilinear group generator $\mathcal{G}(\lambda, n)$ is defined as follows:

1. $(p, \mathbb{G}, \mathbb{H}, \mathbb{G}_t, \hat{e}) \xleftarrow{\$} \mathcal{G}_1(\lambda)$
2. $G = \mathbb{G}^n$, $H = \mathbb{G}^n$, and $G_t = \mathbb{G}_t^m$ for some positive integer m.
3. a bilinear map $e : G \times G \to G_t$ is defined by

$$e((\mathfrak{g}_1, \cdots, \mathfrak{g}_n), (\mathfrak{h}_1, \cdots, \mathfrak{h}_n)) = (\cdots, e((\mathfrak{g}_1, \cdots, \mathfrak{g}_n), (\mathfrak{h}_1, \cdots, \mathfrak{h}_n))^{(\ell)}, \cdots)$$
$$= (\cdots, \prod_{i,j\in[1,n]} \hat{e}(\mathfrak{g}_i, \mathfrak{h}_j)^{e_{ij}^{(\ell)}}, \cdots),$$

where $e_{ij}^{(\ell)} \in \mathbb{Z}_p$ for all $i, j \in [1, n]$ and $\ell \in [1, m]$.

The authors showed that e can be both the *cancelling* and *projecting* only with negligible probability when e is defined as the above. In the above $\mathcal{G}$ construction, to generate a rank n $\mathbb{Z}_p$-module, G is defined as $\mathbb{G}^n$. In the proof for the limitation result ([30, Proposition 6.4 and Theorem 6.5]), the authors used, in an essential way, the fact that a rank n $\mathbb{Z}_p$-module is of the form $\mathbb{G}^n$.

We can, however, also define, in a different way, a rank n $\mathbb{Z}_p$-module G. First generate a rank $n'(> n)$ $\mathbb{Z}_p$-module $\tilde{G}$, and then define G as a rank n $\mathbb{Z}_p$-submodule of $\tilde{G}$. For example, define $\tilde{G} = \mathbb{G}^4$ and

$$G = \langle(\mathfrak{g}^{a_1}, \mathfrak{g}^{b_1}, \mathfrak{g}^{c_1}, \mathfrak{g}^{d_1}), (\mathfrak{g}^{a_2}, \mathfrak{g}^{b_2}, \mathfrak{g}^{c_2}, \mathfrak{g}^{d_2}), (\mathfrak{g}^{a_3}, \mathfrak{g}^{b_3}, \mathfrak{g}^{c_3}, \mathfrak{g}^{d_3})\rangle,$$

where $\{(a_i, b_i, c_i, d_i)\}_{i \in [1,3]}$ is a set of linearly independent vectors in $\mathbb{Z}_p^4$. Then, G is a rank 3 $\mathbb{Z}_p$-submodule of a rank 4 $\mathbb{Z}_p$-module $\tilde{G}$. This example is not included in the case of the above $\mathcal{G}$ construction. In this example, we should argue about the membership check of G since any group should be easy to check for its membership to be used for cryptographic applications. If there is no additional information, the membership check of G is infeasible since it is equivalent to the decisional 3-linear problem. However, we should not rule out this case when some additional information for membership check is given. Our construction is exactly such a case.

4.2   Our Construction

First, we give an instructive intuition of our construction. To construct a bilinear group generator with *projecting*, we should consider the order of image of a bilinear map, which should be larger than prime p.[5] We start from a bilinear group generator with the *cancelling* property [17]. We consider n different bilinear group generators (of rank n) with *cancelling* property. Let $G^{(i)} = \oplus_{j \in [1,n]} G_{ij}$ (rank n $\mathbb{Z}_p$-module), $H^{(i)} = \oplus_{j \in [1,n]} H_{ij}$ (rank n $\mathbb{Z}_p$-module) and $\bar{e}_i$ (bilinear map) be the output of i-th bilinear group generator. Let $G_{ij} = \langle g_{ij} \rangle$ that is a rank 1 $\mathbb{Z}_p$-submodule of a rank n $\mathbb{Z}_p$-module. Let G_j be $\langle(g_{1j}, \cdots, g_{nj})\rangle$, which is a rank 1 $\mathbb{Z}_p$-submodule of a rank n^2 $\mathbb{Z}_p$-module (n direct product of n $\mathbb{Z}_p$-modules). Define H_j similarly, and define $G = \oplus_{j \in [1,n]} G_j$ and $H = \oplus_{j \in [1,n]} H_j$. We define a map e by using bilinear maps $\bar{e}_i$ defined over each $G^{(i)} \times H^{(i)}$ as follows:

$$e((g_1, \cdots, g_n), (h_1, \cdots, h_n)) = (\bar{e}_1(g_1, h_1), \cdots, \bar{e}_n(g_n, h_n)),$$

where $g_i \in G^{(i)}$ and $h_i \in H^{(i)}$. This construction also satisfies the *cancelling* property. If we can control the basis of the image of e so that the order of image is not prime p, then we may obtain the *projecting* property.

For vectors $\Gamma = (\overrightarrow{\alpha}_1, \cdots, \overrightarrow{\alpha}_n) = (\alpha_{11}, \cdots, \alpha_{nn})$ and $\Lambda = (\overrightarrow{\beta}_1, \cdots, \overrightarrow{\beta}_n) = (\beta_{11}, \cdots, \beta_{nn}) \in \mathbb{Z}_p^{n^2}$, and a group element $\mathfrak{g} \in \mathbb{G}$, we define a notation $\Gamma \circ \Lambda := (\overrightarrow{\alpha}_1 \cdot \overrightarrow{\beta}_1, \cdots, \overrightarrow{\alpha}_n \cdot \overrightarrow{\beta}_n) \in \mathbb{Z}_p^n$, where $\overrightarrow{\alpha}_j$'s and $\overrightarrow{\beta}_j$'s are vectors in $\mathbb{Z}_p^n$, and $\overrightarrow{\alpha}_j \cdot \overrightarrow{\beta}_j = \sum_{\ell \in [1,n]} \alpha_{j\ell} \beta_{j\ell}$. Now, we describe our construction $\mathcal{G}_{CP}$.

1. Take a security parameter and a positive integer n as inputs, run $\mathcal{G}_1$, and obtain $(p, \mathbb{G}, \mathbb{H}, \mathbb{G}_t, \hat{e})$.
2. Choose generators $\mathfrak{g}$ and $\mathfrak{h}$ at random from $\mathbb{G}$ and $\mathbb{H}$, respectively.

---

[5] If the image of a bilinear map is prime p, it cannot satisfy *projecting* property [30].

3. Choose $X_1, \cdots, X_n$ and D from $GL_n(\mathbb{Z}_p)$ at random. Define $D_i \in Mat_n(\mathbb{Z}_p)$ be a diagonal matrix having D's i-th column vector as its diagonal. Define Y_i by $D_i(X_i^{-1})^t$.

4. Let $\overrightarrow{\psi}_{ij}$ be the i-th row of X_j and $\overrightarrow{\phi}_{ij}$ be the i-th row of Y_j. Let $\Psi_i = (\overrightarrow{\psi}_{i1}, \cdots, \overrightarrow{\psi}_{in})$ and $\Phi_i = (\overrightarrow{\phi}_{i1}, \cdots, \overrightarrow{\phi}_{in})$. Then, define G_i by a cyclic subgroup in $\mathbb{G}^{n^2}$ generated by $\langle \mathfrak{g}^{\Psi_i} \rangle$, and define H_i by a cyclic group in $\mathbb{H}^{n^2}$ generated by $\langle \mathfrak{h}^{\Phi_i} \rangle$.

5. Define G and H by the internal direct product of G_i's and H_i's, respectively. That is, $G = \oplus_{i\in[1,n]} G_i \subset \mathbb{G}^{n^2}$, and $H = \oplus_{i\in[1,n]} H_i \subset \mathbb{H}^{n^2}$. Define G_t by $\mathbb{G}_t^n$.

6. Define a map $e : G \times H \to G_t$ as follows:

$$e(\mathfrak{g}^\Gamma, \mathfrak{h}^\Lambda) := (\prod_{\ell\in[1,n]} \hat{e}(\mathfrak{g}^{\alpha_{1\ell}}, \mathfrak{h}^{\beta_{1\ell}}), \cdots, \prod_{\ell\in[1,n]} \hat{e}(\mathfrak{g}^{\alpha_{n\ell}}, \mathfrak{h}^{\beta_{n\ell}})) = \hat{e}(\mathfrak{g}, \mathfrak{h})^{\Gamma \circ \Lambda},$$

for any $\Gamma = (\alpha_{11}, \cdots, \alpha_{nn})$ and $\Lambda = (\beta_{11}, \cdots, \beta_{nn})$.

7. Take a basis of $\langle \Psi_1, \cdots, \Psi_n \rangle^\perp$ at random, say $\{\hat{\Psi}_1, \cdots, \hat{\Psi}_{n^2-n}\}$, and take a basis of $\langle \Phi_1, \cdots, \Phi_n \rangle^\perp$ at random, say $\{\hat{\Phi}_1, \cdots, \hat{\Phi}_{n^2-n}\}$, where the notation $\langle \Gamma_1, \cdots, \Gamma_n \rangle^\perp$ means a set of all orthogonal vectors to $\langle \Gamma_1, \cdots, \Gamma_n \rangle$. Define

$$\sigma := (\hat{e}, \{\mathfrak{h}^{\hat{\Psi}_1}, \cdots, \mathfrak{h}^{\hat{\Psi}_{n^2-n}}\}, \{\mathfrak{g}^{\hat{\Phi}_1}, \cdots, \mathfrak{g}^{\hat{\Phi}_{n^2-n}}\}).$$

8. Output $(G, G_1, \cdots, G_n, H, H_1, \cdots, H_n, G_t, e, \sigma)$.

In the description of $\mathcal{G}_{CP}$ each G_i and H_i is defined to be rank 1, as $\mathbb{Z}_p$-submodules of $\mathbb{G}^{n^2}$, and for $i \neq j$, $G_i \cap G_j = H_i \cap H_j = \{1_{\mathbb{G}^{n^2}}\}$, where $1_{\mathbb{G}^{n^2}}$ is the identity of $\mathbb{G}^{n^2}$. Therefore, in the step 5, $G = \oplus_{i\in[1,n]} G_i$ and $H = \oplus_{i\in[1,n]} H_i$ are well-defined and rank n $\mathbb{Z}_p$-submodules of $\mathbb{G}^{n^2}$.

4.3   Cancelling, Projecting, and Translating

It is straightforward to check that e is a non-degenerate bilinear map. We show that e satisfies *cancelling*, *projecting* and *translating*.

Theorem 5. *Let* $(G = \oplus_{i\in[1,n]} G_i, G_i, H = \oplus_{i\in[1,n]} H_i, H_i, G_t, e, \sigma)$ *be the output of the above* $\mathcal{G}_{CP}$. *Then, e is both* cancelling *and* projecting.

Proof. Let $X_1, \cdots, X_n, Y_1, \cdots, Y_n$ and D be generated in the step 3 of Section 4.2. These satisfy the following three conditions.

(1) X_ℓ and Y_ℓ are in $GL_n(\mathbb{Z}_p)$ for $\ell \in [1, n]$.
(2) For $\ell \in [1, n]$ each $X_\ell \cdot Y_\ell^\top$ is a diagonal matrix with a diagonal $\mathbf{d}_\ell$.
(3) $D = (\mathbf{d}_1 \cdots \mathbf{d}_n)$, that is, the i-th column vector of D is $\mathbf{d}_i$.

From the condition (1) we can see that Ψ_i's are linearly independent and Φ_i's are linearly independent and so $G = \oplus_{i\in[1,n]} G_i$ and $H = \oplus_{i\in[1,n]} H_i$ are well-defined. The condition (2) guarantees that e is a *cancelling* bilinear map: For $i \neq j$,

$\Psi_i \circ \Phi_j := (\overrightarrow{\psi}_{i1} \cdot \overrightarrow{\phi}_{j1}, \cdots, \overrightarrow{\psi}_{in} \cdot \overrightarrow{\phi}_{jn}) = 0$ and so $e(\mathfrak{g}^{\Psi_i}, \mathfrak{h}^{\Phi_j}) = e(\mathfrak{g}, \mathfrak{h})^{\Psi_i \circ \Phi_j} = (1_{\mathbb{G}_t}, \cdots, 1_{\mathbb{G}_t})$ is equal to the identity of the product group $(\mathbb{G}_t)^n$. The third condition (3) implies that $\{\Psi_i \circ \Phi_i\}_{i \in [1,n]}$ is a set of linearly independent vectors in $\mathbb{Z}_p^n$; hence, any pair of groups $e(G_i, H_i) = \langle e(\mathfrak{g}, \mathfrak{h})^{\Psi_i \circ \Phi_i} \rangle = \langle (\mathfrak{g}, \mathfrak{h})^{(d_{i1}, \cdots, d_{in})} \rangle$ has no common element except the identity so that $Im(e) = \oplus_{i \in [1,n]} e(G_i, H_i) = G_t$. We can consider natural projections $\pi_i : G \to G_i$, $\bar{\pi}_i : H \to H_i$, and $\pi_{t,i} : G_t \to e(G_i, H_i)$. We can construct these projections, in a similar way as the construction of the projections in the subsection 3.1. We leave the details to the full version of this paper. Let $G' = \oplus_{[2,n]} G_i$, $H' = \oplus_{[2,n]} H_j$, $G'_t = e(G', H')$, $\pi = \pi_i$, $\bar{\pi} = \bar{\pi}_i$, and $\pi_t = \pi_{t,i}$. Then, e satisfies the definition 4. $\qquad\square$

Theorem 6. $\mathcal{G}_{CP}(\lambda, n)$ *satisfies* translating *property for all* $i, j \in [1, n]$.

Proof. We will construct $\mathcal{T}_{3,1}$. We can construct other $\mathcal{T}_{i,j}$ and $\bar{\mathcal{T}}_{i,j}$ similarly. Given g_3, g_3^a and $n \times n$ matrices X_i defined as in the description of $\mathcal{G}_{CP}$, we can compute g_1^a without knowing a as follows:

Parse g_3^a as $(\mathfrak{g}^{\Psi_3})^a = ((\mathfrak{g}^{\overrightarrow{\psi}_{31}})^a, \cdots, (\mathfrak{g}^{\overrightarrow{\psi}_{3n}})^a)$, and compute

$$\text{for } j \in [1, n], \; ((\mathfrak{g}^{\overrightarrow{\psi}_{3j}})^a)^{X_j^{-1}} = (\mathfrak{g}^{a \overrightarrow{e}_3 X_j})^{X_j^{-1}} = \mathfrak{g}^{a \overrightarrow{e}_3} = (1, 1, \mathfrak{g}^a, \cdots, 1),$$

$$(\mathfrak{g}^a, 1, \cdots, 1)^{X_j} = (\mathfrak{g}^{a \overrightarrow{e}_1})^{X_j} = \mathfrak{g}^{a \overrightarrow{\psi}_{1j}},$$

$$\text{then } (\mathfrak{g}^{a \overrightarrow{\psi}_{11}}, \cdots, \mathfrak{g}^{a \overrightarrow{\psi}_{1n}}) = (\mathfrak{g}^{\Psi_1})^a = g_1^a.$$

where $\overrightarrow{e}_i$ is the canonical i-th vector in $\mathbb{Z}_p^n$, for example, $\overrightarrow{e}_1 = (1, 0, 0, \cdots, 0)$.
$\qquad\square$

We show that anyone knowing σ can test membership of elements in G and H (membership test for G_t is trivial) in the full version. Finally, we should show that $\mathcal{G}$ satisfies the subgroup decision assumption, but it is not easy to prove that $\mathcal{G}$ satisfies the subgroup decision for any n. Instead, in the full version we give a proof that, for $n = 2$, $\mathcal{G}$ satisfies the $(2,1)$-subgroup decision assumption in the generic bilinear group model [35] (that is, we assume that the adversary should access the oracles for group operations of $\mathbb{G}$, $\mathbb{H}$, $\mathbb{G}_t$ and pairing computations for $\hat{e}$, where $\mathcal{G}_1 \to (p, \mathbb{G}, \mathbb{H}, \mathbb{G}_t, \hat{e})$). Though we give a proof for the case $n = 2$, we are positive that $\mathcal{G}_{CP}$ satisfies the subgroup decision assumption for $n > 2$. For $n > 2$, there are several variables, particularly in σ, we should consider for the subgroup decision assumption, so these make it hard to prove for the case $n > 2$, even in the generic bilinear group model.[6]

5 Conclusions and Further Work

In this paper, we answered two open questions left by Meiklejohn, Shacham, and Freeman. First, we showed that the security of the Meiklejohn et al.'s (partial)

---

[6] All variables in σ is public, so to show that $\mathcal{G}_{CP}$ satisfies the subgroup decision assumption, the simulator should simulate σ in the proof.

blind signature can be proved in the prime-order bilinear group setting.[7] Second, we showed that there exist bilinear group generators that are both *cancelling* and *projecting* in the prime-order bilinear group setting.

The proof of the Meiklejohn-Shacham-Freeman blind signature scheme, and the Lewko-Waters identity-based encryption scheme [29] essentially use the fact that orders of subgroups are relatively prime as well as the projecting and/or cancelling properties. For each scheme, the adapted version in prime-order bilinear groups is proposed, with a different security proof strategy, in this paper and [29], respectively. It would be interesting to find a general procedure to transform such schemes using relatively prime orders in composite-order groups to schemes in prime-order groups.

We proposed a new mathematical framework with both *cancelling* and *projecting* in a prime-order bilinear group setting, and gave the proof that the $(2, 1)$ subgroup decision assumption holds in the generic bilinear group model when $n = 2$. This research leaves many interesting open problems. We ask if the subgroup decision assumption holds when $n > 2$, and if the subgroup decision assumption can be reduced to the simple assumption such as the (decisional) k-linear assumption. We did not find good cryptographic applications of this framework. It would be interesting to design cryptographic schemes based on the proposed framework. We expect that this research will provide other directions for our primitive question: whether there exists a cryptosystem on composite-order bilinear groups that cannot be constructed on prime-order bilinear groups.

Acknowledgements. The first author is grateful to MinJae Seo for his useful comments on an early draft of this paper. We are grateful to anonymous reviewers in TCC 2012 for their valuable comments. The second author was supported by the National Research Foundation of Korea (NRF) grant funded by the Korea government (MEST) (No. 20110018345).

References

1. Abe, M.: A Secure Three-Move Blind Signature Scheme for Polynomially Many Signatures. In: Pfitzmann, B. (ed.) EUROCRYPT 2001. LNCS, vol. 2045, pp. 136–151. Springer, Heidelberg (2001)
2. Abe, M., Fuchsbauer, G., Groth, J., Haralambiev, K., Ohkubo, M.: Structure-Preserving Signatures and Commitments to Group Elements. In: Rabin, T. (ed.) CRYPTO 2010. LNCS, vol. 6223, pp. 209–236. Springer, Heidelberg (2010)
3. Abe, M., Groth, J., Haralambiev, K., Ohkubo, M.: Optimal Structure-Preserving Signatures in Asymmetric Bilinear Groups. In: Rogaway, P. (ed.) CRYPTO 2011. LNCS, vol. 6841, pp. 649–666. Springer, Heidelberg (2011)

---

[7] We modified their scheme slightly to prove its security under the CDH assumption. We remark that, however, the security of the direct instantiation of their scheme in the prime-order bilinear group can also be proven secure under the decisional linear assumption and the augmented CDH assumption, which is stronger than the CDH assumption.

4. Abe, M., Haralambiev, K., Ohkubo, M.: Signing on elements in bilinear groups for modular protocol design. Cryptology ePrint Archive, Report 2010/133 (2010), http://eprint.iacr.org/2010/133
5. Abe, M., Ohkubo, M.: A Framework for Universally Composable Non-committing Blind Signatures. In: Matsui, M. (ed.) ASIACRYPT 2009. LNCS, vol. 5912, pp. 435–450. Springer, Heidelberg (2009)
6. Abe, M., Okamoto, T.: Provably Secure Partially Blind Signatures. In: Bellare, M. (ed.) CRYPTO 2000. LNCS, vol. 1880, pp. 271–286. Springer, Heidelberg (2000)
7. Bellare, M., Namprempre, C., Pointcheval, D., Semanko, M.: The one-more-rsa-inversion problems and the security of chaum's blind signature scheme. Journal of Cryptology 16, 185–215 (2003)
8. Boldyreva, A.: Threshold Signatures, Multisignatures and Blind Signatures Based on the Gap-Diffie-Hellman-Group Signature Scheme. In: Desmedt, Y.G. (ed.) PKC 2003. LNCS, vol. 2567, pp. 31–46. Springer, Heidelberg (2002)
9. Boneh, D., Franklin, M.: Identity-Based Encryption from the Weil Pairing. In: Kilian, J. (ed.) CRYPTO 2001. LNCS, vol. 2139, pp. 213–229. Springer, Heidelberg (2001)
10. Boneh, D., Goh, E.-J., Nissim, K.: Evaluating 2-DNF Formulas on Ciphertexts. In: Kilian, J. (ed.) TCC 2005. LNCS, vol. 3378, pp. 325–341. Springer, Heidelberg (2005)
11. Chaum, D.: Blind signatures for untraceable payments. In: CRYPTO, pp. 199–203 (1982)
12. Chaum, D.: Blind signature system. In: CRYPTO 1983 (1983)
13. Chaum, D.: Elections with Unconditionally-Secret Ballots and Disruption Equivalent to Breaking RSA. In: Günther, C.G. (ed.) EUROCRYPT 1988. LNCS, vol. 330, pp. 177–182. Springer, Heidelberg (1988)
14. Chaum, D., Fiat, A., Naor, M.: Untraceable Electronic Cash. In: Goldwasser, S. (ed.) CRYPTO 1988. LNCS, vol. 403, pp. 319–327. Springer, Heidelberg (1990)
15. Cheon, J.H.: Discrete logarithm problems with auxiliary inputs. Journal of Cryptology 23, 457–476 (2010)
16. Fischlin, M.: Round-Optimal Composable Blind Signatures in the Common Reference String Model. In: Dwork, C. (ed.) CRYPTO 2006. LNCS, vol. 4117, pp. 60–77. Springer, Heidelberg (2006)
17. Freeman, D.M.: Converting Pairing-Based Cryptosystems from Composite-Order Groups to Prime-Order Groups. In: Gilbert, H. (ed.) EUROCRYPT 2010. LNCS, vol. 6110, pp. 44–61. Springer, Heidelberg (2010)
18. Fuchsbauer, G.: Automorphic signatures in bilinear groups and an application to round-optimal blind signatures. Cryptology ePrint Archive, Report 2009/320 (2009), http://eprint.iacr.org/2009/320
19. Garg, S., Kumarasubramanian, A., Sahai, A., Waters, B.: Building efficient fully collusion-resilient traitor tracing and revocation schemes. In: ACM Conference on Computer and Communications Security, pp. 121–130. ACM (2010)
20. Garg, S., Rao, V., Sahai, A., Schröder, D., Unruh, D.: Round Optimal Blind Signatures. In: Rogaway, P. (ed.) CRYPTO 2011. LNCS, vol. 6841, pp. 630–648. Springer, Heidelberg (2011)
21. Ghadafi, E., Smart, N.: Efficient two-move blind signatures in the common reference string model. Cryptology ePrimt Archive, Report 2010/568 (2010), http://eprint.iacr.org/2010/568
22. Groth, J., Ostrovsky, R., Sahai, A.: Non-interactive Zaps and New Techniques for NIZK. In: Dwork, C. (ed.) CRYPTO 2006. LNCS, vol. 4117, pp. 97–111. Springer, Heidelberg (2006)

23. Groth, J., Ostrovsky, R., Sahai, A.: Perfect Non-interactive Zero Knowledge for NP. In: Vaudenay, S. (ed.) EUROCRYPT 2006. LNCS, vol. 4004, pp. 339–358. Springer, Heidelberg (2006)
24. Groth, J., Sahai, A.: Efficient Non-interactive Proof Systems for Bilinear Groups. In: Smart, N.P. (ed.) EUROCRYPT 2008. LNCS, vol. 4965, pp. 415–432. Springer, Heidelberg (2008)
25. Hazay, C., Katz, J., Koo, C.-Y., Lindell, Y.: Concurrently-Secure Blind Signatures Without Random Oracles or Setup Assumptions. In: Vadhan, S.P. (ed.) TCC 2007. LNCS, vol. 4392, pp. 323–341. Springer, Heidelberg (2007)
26. Hofheinz, D., Kiltz, E.: Secure Hybrid Encryption from Weakened Key Encapsulation. In: Menezes, A. (ed.) CRYPTO 2007. LNCS, vol. 4622, pp. 553–571. Springer, Heidelberg (2007)
27. Juels, A., Luby, M., Ostrovsky, R.: Security of Blind Digital Signatures. In: Kaliski Jr., B.S. (ed.) CRYPTO 1997. LNCS, vol. 1294, pp. 150–164. Springer, Heidelberg (1997)
28. Kiayias, A., Zhou, H.-S.: Concurrent Blind Signatures Without Random Oracles. In: De Prisco, R., Yung, M. (eds.) SCN 2006. LNCS, vol. 4116, pp. 49–62. Springer, Heidelberg (2006)
29. Lewko, A., Waters, B.: New Techniques for Dual System Encryption and Fully Secure HIBE with Short Ciphertexts. In: Micciancio, D. (ed.) TCC 2010. LNCS, vol. 5978, pp. 455–479. Springer, Heidelberg (2010)
30. Meiklejohn, S., Shacham, H., Freeman, D.M.: Limitations on Transformations from Composite-Order to Prime-Order Groups: The Case of Round-Optimal Blind Signatures. In: Abe, M. (ed.) ASIACRYPT 2010. LNCS, vol. 6477, pp. 519–538. Springer, Heidelberg (2010)
31. Okamoto, T.: Efficient Blind and Partially Blind Signatures Without Random Oracles. In: Halevi, S., Rabin, T. (eds.) TCC 2006. LNCS, vol. 3876, pp. 80–99. Springer, Heidelberg (2006)
32. Okamoto, T., Takashima, K.: Homomorphic Encryption and Signatures from Vector Decomposition. In: Galbraith, S.D., Paterson, K.G. (eds.) Pairing 2008. LNCS, vol. 5209, pp. 57–74. Springer, Heidelberg (2008)
33. Okamoto, T., Takashima, K.: Homomorphic Encryption and Signatures from Vector Decomposition. In: Galbraith, S.D., Paterson, K.G. (eds.) Pairing 2008. LNCS, vol. 5209, pp. 57–74. Springer, Heidelberg (2008)
34. Shacham, H.: A cramer-shoup encryption scheme from the linear assumption and from progressively weaker linear variants. Cryptology ePrimt Archive, Report 2007/074 (2007), http://eprint.iacr.org/2007/074
35. Shoup, V.: Lower Bounds for Discrete Logarithms and Related Problems. In: Fumy, W. (ed.) EUROCRYPT 1997. LNCS, vol. 1233, pp. 256–266. Springer, Heidelberg (1997)

On Efficient Zero-Knowledge PCPs

Yuval Ishai[1,*], Mohammad Mahmoody[2,**], and Amit Sahai[3]

[1] Technion, Israel
yuvali@cs.technion.edu
[2] Cornell, USA
mohammad@cs.cornell.edu
[3] UCLA, USA
sahai@cs.ucla.edu

Abstract. We revisit the question of *Zero-Knowledge PCPs*, studied by Kilian, Petrank, and Tardos (STOC '97). A ZK-PCP is defined similarly to a standard PCP, except that the view of any (possibly malicious) verifier can be efficiently simulated up to a small statistical distance. Kilian et al. obtained a ZK-PCP for **NEXP** in which the proof oracle is in $\mathbf{EXP^{NP}}$. They also obtained a ZK-PCP for **NP** in which the proof oracle is computable in polynomial-time, but this ZK-PCP is only zero-knowledge against *bounded-query* verifiers who make at most an *a priori fixed* polynomial number of queries. The existence of ZK-PCPs for **NP** with efficient oracles and arbitrary polynomial-time malicious verifiers was left open. This question is motivated by the recent line of work on cryptography using tamper-proof hardware tokens: an efficient ZK-PCP (for any language) is *equivalent* to a statistical zero-knowledge proof using only a single stateless token sent to the verifier.

We obtain the following results regarding efficient ZK-PCPs:

Negative Result on Efficient ZK-PCPs. Assuming that the polynomial time hierarchy does not collapse, we settle the above question in the negative for ZK-PCPs in which the verifier is *nonadaptive* (i.e. the queries only depend on the input and secret randomness but not on the PCP answers).

Simplifying Bounded-Query ZK-PCPs. The bounded-query zero-knowledge PCP of Kilian et al. starts from a *weakly-sound* bounded-query ZK-PCP of Dwork et al. (CRYPTO '92) and amplifies its soundness by introducing and constructing a new primitive called *locking scheme* — an unconditional oracle-based analogue of a commitment scheme. We simplify the ZK-PCP of Kilian et al. by presenting an elementary new construction of locking schemes. Our locking scheme is purely combinatorial.

Black-Box Sublinear ZK Arguments via ZK-PCPs. Kilian used PCPs to construct sublinear-communication zero-knowledge arguments for **NP** which make a *non-black-box* use of collision-resistant hash functions (STOC '92). We show that ZK-PCPs can be used to get black-box variants of this result with improved round complexity,

* Research done in part while visiting UCLA.
** Research done in part while visiting UCLA.

R. Cramer (Ed.): TCC 2012, LNCS 7194, pp. 151–168, 2012.

as well as an *unconditional* zero-knowledge variant of Micali's non-interactive CS Proofs (FOCS '94) in the Random Oracle Model.

Keywords: Zero-Knowledge, Probabilistically Checkable Proofs, Arthur Merlin Games, Tamper-Proof Tokens, Sublinear Arguments.

1   Introduction

The seminal work of Goldwasser, Micali, and Rackoff [30] changed the classical notion of a mathematical proof by incorporating randomness and interaction. This change was initially motivated by the intriguing possibility of zero knowledge proofs – proofs that carry no extra knowledge other than being convincing. The result of Goldreich, Micali, and Wigderson [27] showed that any **NP** statement can be proved in a zero-knowledge (ZK) manner, making ZK proofs a central tool for cryptographic protocol design; this was later extended by Ben-Or et al. [8] to any language in **PSPACE**. All these fundamental results, however, relied on the assumption that one-way functions exist. Ostrovsky and Wigderson [46] showed that (similar) computational assumptions are indeed inherent for non-trivial zero-knowledge.

Motivated by the goal of achieving *unconditionally* secure zero-knowledge proofs for **NP**, Ben-Or, Goldwasser, Kilian and Wigderson [9] introduced the model of multi-prover interactive proofs (MIP) and presented a perfect ZK protocol for any statement that is provable in the MIP model. Shortly after, Babai, Fortnow, and Lund [6] showed that in fact any language in **NEXP** can be proved in the MIP model. Fortnow, Rompel, and Sipser [23] studied the MIP model further and observed that as a proof system it is equivalent to another model in which an *oracle* encodes a probabilistically checkable proof (PCP) which is queried by an efficient randomized verifier. (The PCP oracle is often identified with the proof string defined by its truth-table, in which case the output domain of the oracle is referred to as the *PCP alphabet*.) The difference between a prover and a PCP oracle is that a prover can keep an internal state, and hence its answer to a given question can depend on other questions. Therefore, soundness against a PCP oracle is potentially easier to achieve than soundness against a malicious prover. This line of work culminated in the celebrated PCP theorem [4,3].

Zero-Knowledge PCPs. In this work we study *zero-knowledge proofs* in the PCP model. A zero-knowledge PCP (ZK-PCP) is defined similarly to a standard PCP, except that the view of any (possibly malicious) verifier can be efficiently simulated up to a small statistical distance. It is instructive to note that zero-knowledge PCPs are incomparable to traditional ZK proofs: since the PCP model makes the prover less powerful, achieving soundness may become easier whereas achieving zero-knowledge may become harder.

The original ZK protocol of [27] for **NP** implicitly relies on *honest-verifier* zero-knowledge PCP for the **NP**-complete problem of 3-coloring of graphs. In this PCP the prover takes any 3-coloring of the input graph, randomly permutes the 3 colors, and writes down the colors as the PCP string. The verifier chooses a random edge, reads the colors of the vertices of that edge, and accepts iff the colors are different. This ZK-PCP has two disadvantages: (1) it is only zero-knowledge against *honest verifiers* (a malicious verifier can learn whether the colors of two non-adjacent nodes are identical), and (2) the soundness error is very large: $1 - 1/m$ where m is the number of edges. Dwork et al. [19],[1] relying on the PCP theorem [3,4], improved the ZK-PCP implicit in [27] in both directions. Their construction implies a ZK-PCP for **NP** of polynomial length and with a constant alphabet size such that: (1) the PCP is zero-knowledge against verifiers who ask *any* pair of queries (but not more), and (2) the soundness error is constant. However, the soundness error of this ZK-PCP could not be easily reduced further while maintaining ZK against malicious verifiers. Furthermore, it could not be made zero-knowledge against arbitrary polynomial-time verifiers, simply because it has polynomial length and a malicious verifier could read the entire proof string.

Kilian, Petrank, and Tardos [40] were the first to explicitly study the power of ZK-PCPs with malicious verifiers. Their work shows how to get around the above limitations, resulting in two kinds of ZK-PCPs with security against malicious verifiers. For the case of languages in **NP**, [40] obtain a PCP of polynomial length over a binary alphabet which is zero-knowledge with negligible soundness error against malicious verifiers who are limited to ask only up to any *fixed* polynomial $p(|x|)$ number of queries, whereas the honest verifier only asks polylog($|x|$) queries to verify the PCP. (The length of the PCP string can be polynomially larger than $p(|x|)$.) We call such PCPs *bounded-query* ZK. For the case of languages in **NEXP**, a scaled up version of this construction yields a ZK-PCP in which honest verifiers are efficient (i.e. run in poly($|x|$) time), but soundness holds against *arbitrary* polynomial time verifiers. However, the PCP oracle in this case cannot be computed in polynomial time even for languages in **NP**. (By "computable in polynomial time" we mean that the oracle outputs a polynomial-time computable function of its secret randomness, the input x, the **NP**-witness, and the verifier's query.) This is inherent to the approach of [40], as it requires the entropy of the PCP oracle to be bigger than the number of queries made by a malicious verifier.

The above state of affairs leaves open the following natural question.

Main Question: *Are there efficiently computable PCPs for **NP** which are statistically zero-knowledge against any polynomial-time verifier?*

An additional motivation to study the question above comes from the recent line of work on cryptography in an extended model of interaction with "tamper-proof hardware tokens" [38,44,14,29,34,41,33]. This model allows the parties to generate and exchange tamper-proof hardware tokens which are simply circuits (with or without internal state) that are accessible only as a black-box. Indeed, an efficient ZK-PCP for **NP** is equivalent to a statistical zero-knowledge proof for **NP** in this model where the only message sent to the verifier is a single *stateless* token. The stateless nature of the PCP oracle (inside the token) would make such a protocol secure against "resetting attacks" [13]. With this motivation in mind, we revisit the feasibility question of efficient ZK-PCPs for **NP**.

---

[1] This formulation of the result of [19] is due to [40].

2   Our Results

Our main theorem provides a negative answer to the main question above for the case of *nonadaptive* (honest) verifiers whose queries can only depend on their randomness and the input x but not on the prover's answers (so all the queries can be prepared and asked in one round). This theorem may be viewed as supporting the conjecture that efficient ZK-PCPs for **NP** do not exist.

In the setting of bounded-query ZK-PCPs, we revisit the construction of [40] and simplify it considerably. Our contribution is to present a simple combinatorial construction of a "locking schemes" which was the main tool developed in [40] and used in *both* of their constructions for **NP** and **NEXP**.

Finally, motivated by a line of work on the power of black-box constructions in cryptography, we show that efficient bounded-query ZK-PCPs can be used to make the sublinear-communication zero-knowledge argument construction of Kilian [39] *black-box*. Kilian's construction assumes the existence of a collision-resistant hash function, but it uses the hash function in a non-black-box way. We also obtain constant-round variants of this result and an unconditional non-interactive variant in the Random Oracle Model. In the following we describe our results more formally and put them in the proper context

2.1   Efficient Nonadaptive ZK-PCPs

We prove the following negative result about the existence of ZK-PCPs for **NP**.

Theorem 1 (Main Theorem). *If there exists an efficiently computable PCP for* **NP** *with a nonadaptive honest verifier, constant soundness error, and zero-knowledge against arbitrary polynomial-time verifiers, then the polynomial-time hierarchy collapses.*

What we prove is actually more general than the statement of Theorem 1. Namely, we show that any language with an efficient ZK-PCP of polynomial Shannon entropy (see Remark 4) and a nonadaptive verifier is in **coAM**, and Theorem 1 follows by the result of [12]. Also, we only require the zero-knowledge to hold also against nonadaptive verifiers (of arbitrary polynomial time).[2]

We emphasize that even though the zero-knowledge property of ZK-PCPs is defined in a statistical fashion, our main theorem above does *not* follow from the classical result of Fortnow, Aiello, and Håstad [1,22] who proved that **SZK** $\subseteq$ **AM** $\cap$ **coAM**. The reason is that although achieving zero-knowledge in the PCP model is harder, achieving soundness in this model is potentially *easier*.[3] Therefore the languages which posses efficient ZK-PCPs (as far as we know) are not necessarily included in **SZK**. Also recall that if one does not require the

---

[2] The requirement that the honest verifier be nonadaptive is a restriction to our Theorem 1, but only requiring the zero-knowledge to hold against nonadaptive verifiers makes our result stronger.

[3] The latter comparison manifests itself in the following characterizations: it holds that $\mathbf{PCP}(\mathrm{poly}, \mathrm{poly}) = \mathbf{MIP} = \mathbf{NEXP}$ while $\mathbf{IP} = \mathbf{PSPACE} \subseteq \mathbf{EXP}$.

PCP oracle to be efficiently computable, by the result of [40] all of the languages in **NEXP** (including **NP**) *do* have (statistical) ZK-PCPs.

Using Theorem 1 itself, we can extend Theorem 1 to the case of adaptive (honest) verifiers, as long as the total length of the prover's answers returned in an honest PCP verification is $O(\log n)$ bits (see Corollary 7).

Ideas and Tools. At a high level the proof of Theorem 1 uses ideas from many previous influential works [26,1,20,11] and tools from old and new results in the context of constant-round proofs [31,28,36]. The main challenges are in how to force an untrusted prover to extract a PCP oracle from the simulator and run the honest verifier against this PCP. The soundness of this protocol follows from the soundness of the original PCP. To get the completeness, we need to extract this PCP in a way that it is "close" to an actual accepting PCP, and this is where we use efficiency of the PCP and its bounded entropy. Section 3 is dedicated to describing the main result formally and the main ideas behind it. See the full version of the paper for a formal description of our **AM** protocol.

Motivation and Related Work. A recent line of work in cryptography [38,44,14,29,34,41,33] studies the possibility of obtaining secure protocols in an extended model of interaction in which the parties are allowed to exchange more than just classical bits: the parties are allowed to locally construct a (stateful or stateless) circuit, put it inside a tamper-proof token, and send it to another party. The receiver of a token (in this model) is allowed only to use it as a black-box. Namely, she is only allowed to give inputs to the token and receive the output. (If the token is stateful, asking the same query twice might lead to different answers.) Designing protocols in this model is made challenging by the fact that a receiver of a token has no guarantee that the token is indeed well formed. The work of Goyal et al. [34] showed that any two-party functionality (e.g. zero-knowledge proof) can be carried out securely in this model without relying on computational assumptions. Unfortunately the solution of [34] uses *stateful* tokens, which makes it vulnerable to "resetting attacks". Namely, there is no security guarantee if a malicious party receiving a token can reset it to its initial state, say, by cutting off its power.

In another line of research, Kalai and Raz [37] introduced the Interactive PCP (IPCP) model which is a hybrid between the two-prover and the PCP models. In the IPCP model the verifier interacts with a prover and a PCP oracle. Note that when the prover and the PCP oracle are efficiently computable, the IPCP model becomes a special case of the tamper-proof token model in which the prover sends a *stateless* token (computing the PCP) to the verifier.

Although Kalai and Raz [37] introduced the IPCP model for the purpose of optimizing the PCP length at the cost of small amount of interaction with the prover, Goyal, Ishai, Mahmoody, and Sahai [33] showed that the IPCP model is also interesting for cryptographic purposes in the context of achieving unconditional security in the tamper-proof token model. It was shown in [33] that unconditional (statistical) ZK proofs for **NP** exist in the IPCP model, and moreover the prover and the PCP oracle can be implemented efficiently given a witness

w for $x \in L$. The verifier in the protocol of [33] exchanges only four messages with the prover. A main question left open in [33] was whether there exists any protocol that avoids such interaction between the verifier and the prover altogether (i.e. the verifier only interacts with the PCP oracle). It is easy to see that the latter question is equivalent to our main question above! Namely, any positive answer to our main question implies a proof system in which all the communication between the prover and the verifier consists of a single *stateless* token sent to the verifier which hides the circuit computing the PCP oracle and can convince the verifier about the truth of the input statement in a ZK manner.

Therefore, if efficient ZK-PCPs for **NP** exist, they would lead (without any computational assumptions) to "noninteractive" statistical zero-knowledge proofs for **NP** using tamper-proof hardware with the extra feature of being resistant against resetting attacks, since the used token (which computes the PCP oracle) is stateless.

2.2   Simplifying Bounded-Query ZK-PCPs

Our second contribution is a simplification of the ZK-PCP construction of Kilian et al. [40]. The construction of [40] starts from the weakly-sound bounded-query ZK-PCP of [19] and compiles it into a PCP which is zero-knowledge against malicious verifiers of bounded query complexity. The weakly-sound PCP of [19] is zero-knowledge against any k (possibly adaptive) queries, but suffers from the soundness error $1 - 1/\operatorname{poly}(k)$. The main tool introduced and employed in the compiler of [40] is called a "locking scheme", which is an analogue of a commitment scheme in the PCP model. In a locking scheme a sender holds a secret w and randomly encodes it into an oracle σ_w that can be accessed by the receiver R (denoted as R^{σ_w}). The efficient receiver should not be able to learn any information about w through its oracle access to σ_w. On the other hand, the sender can later send a key to the receiver to decommit the value w. The protocol should guarantee that the sender is not able to change his mind about the value w after constructing the oracle σ_w.[4]

Kilian et al. [40] gave an elegant way of using locking schemes to convert a ZK-PCP with $1 - 1/\operatorname{poly}(k)$ soundness error into a standard ZK-PCP of constant or even negligible error. Unfortunately, the locking scheme of [40] which forms the main technical ingredient of their ZK-PCP constructions is quite complicated to describe and analyze (pages 6 to 12 there) and uses ad-hoc algebraic techniques.

Motivation. Most applications of ZK-PCPs considered in this work either require the stronger unbounded variant (see Section 2.1) or alternatively can rely on an honest-verifier variant (see Section 2.3), which is easier to realize. However, efficient bounded-query ZK-PCPs with security against *malicious* verifiers can also be motivated by natural application scenarios. For instance, one can consider

---

[4] In other words, a locking scheme can be thought of as a commitment scheme with statistical security guarantees and minimal interaction such that during its commitment phase the sender sends only a single tamper-proof token (containing the oracle σ_w) to the receiver.

the goal of distributing an **NP**-witness among many servers in a way that simultaneously supports a very efficient verification (corresponding to the work of the honest verifier) and secrecy in the presence of a large number of colluding servers (corresponding to the query bound of a malicious verifier). One can also consider a "time-lock zero-knowledge proof" in which a stateless hardware token contains an embedded witness which can be very quickly validated but requires a lot of time to extract. Another motivation behind our simpler locking schemes comes from the line of work aiming at simplifying PCP constructions and making them combinatorial. The main algebraic and technical components in the final PCP construction of Kilian et al. [40] are **(1)** the PCP theorem of [3,4] (which comes in through the construction of [19]) and **(2)** the locking scheme of [40]. The first (more important) component was considerably simplified by Dinur, and here we give a simplified version of the second component. (For a more extensive survey of this line of research see [42] and the references therein.)

In the full version of this paper, we formally present and analyze a simple combinatorial construction of a locking scheme which can be viewed as a noninteractive implementation of Naor's commitment scheme [45] in the PCP model. In the following we describe the main idea.

Technique. We start by reviewing Naor's commitment scheme. In this commitment scheme, the parties have access to a pseudorandom generator $f : \{0,1\}^n \mapsto \{0,1\}^{3n}$ and the protocol works as follows:

The receiver chooses a random "shift" $r \xleftarrow{\$} \{0,1\}^{3n}$ and sends it to the sender. The sender, who holds a secret input bit b, chooses a random seed $s \xleftarrow{\$} \{0,1\}^n$ and sends $f(s) + b \cdot r = t$ to the receiver (the addition and multiplication are componentwise over the binary field). In the decommitment phase the sender simply sends (b, s) to the receiver, and the receiver makes sure that $f(s) + b \cdot r = t$ holds to accept the decommitted value.

The binding property holds because the support set of f is of size at most $|f(\{0,1\}^n)| \leq 2^n$, and a random shift $r \xleftarrow{\$} \{0,1\}^{3n}$ with overwhelming probability of at least $1 - 2^n \cdot 2^n \cdot 2^{-3n} = 1 - 2^{-n}$ will have the property that $f(\{0,1\}^n) \cap (f(\{0,1\}^n) + r) = \varnothing$. Thus for such "good" r, by sending t to the receiver the sender will be bound to at most one possible value of b (regardless of the structure of the function f).

On the other hand, the hiding property of the scheme reduces in a *black-box* way to the pseudorandomness of $f(\mathbf{U}_n)$. Namely, if an efficient receiver $\widehat{R}$ can distinguish between $f(s) + r$ and $f(s) + r \cdot b$, another efficient algorithm D who uses $\widehat{R}$ internally is able to distinguish $f(\mathbf{U}_n)$ from a random value $\mathbf{U}_{3n}$. Thus it holds that if the function f is random, the scheme will be *statistically* hiding against receivers who ask at most poly(n) oracle queries to f. The reason is that a random function f mapping $\{0,1\}^n$ to random values in $\{0,1\}^{3n}$ is statistically indistinguishable from a truly random function as long as the distinguisher is bound to ask at most $2^{o(n)}$ queries to f.

The above observation about the hiding property of Naor's commitment scheme means that if, in the second round of the commitment phase, the sender chooses f to be a truly random function and sends $f(s) + b \cdot r$ to the receiver as well as (providing oracle access to) $f(\cdot)$, then we get a secure (inefficient) commitment scheme in the *interactive* PCP model without relying on any computational assumption.[5] In our construction of locking schemes we show how to eliminate the first initial message r of the receiver and emulate the role of this shift r by a few more queries asked by the receiver and more structure in the locking oracle.

2.3   Black-Box Sublinear ZK Arguments

Kilian [39], relying on the PCP construction of [5],[6] proved that assuming the existence of exponentially-hard collision-resistant hash functions (CRH) and 2-message statistically-hiding commitments, one can construct a (6-message) statistical ZK argument for **NP** with polylog(n) communication complexity (where n is the input length). Later on, Damgård et al. [17] showed that 2-message statistically-hiding commitments can be obtained from any CRH, which made the existence of exponentially hard CRH sufficient for the construction of Kilian. Micali [43] showed how to make Kilian's protocol noninteractive in the *random oracle model*. The above constructions make a non-black-box use of the underlying collision-resistant hash function.

Our third contribution is to obtain *black-box* constructions of sublinear ZK arguments for **NP** by using bounded-query efficient ZK-PCPs for **NP**. Namely, we observe that the bounded-query ZK-PCP of [19] can be employed to get an alternative to the ZK argument of Kilian [39] for **NP** which uses the underlying CRH function as a black box. (Our protocols are in fact *fully* black-box [49], in the sense that the security reduction makes a black-box use of the adversary, and have black-box simulators.)

Theorem 2 (Black-Box Sublinear ZK Arguments). *Let $\mathcal{H}$ be any family of collision-resistant hash functions. Using $\mathcal{H}$ only as a black-box, one can construct a constant-round ZK argument system for **NP** with negligible soundness error and communication complexity sublinear in the witness size. Furthermore:*

- *For the case of an honest verifier, the zero knowledge is statistical, the round complexity is 4 messages, and the protocol is public coin.*
- *For the case of malicious-verifier zero knowledge, the round complexity is 5 messages, and the proof of security requires that the family of CRH be secure against non-uniform adversaries.*
- *If the family of CRH is secure against adversaries running in time $2^{n^{\Omega(1)}}$, then the communication complexity can be made polylogarithmic in the witness size for both honest verifier and malicious verifier settings.*
- *In the random oracle model, there exists an unconditionally secure non-interactive statistical zero knowledge argument system for **NP** with negligible soundness error and polylogarithmic communication complexity.*

We prove Theorem 2 in the full version; below we describe the main ideas.

---

[5] Note that the random oracle $f(\cdot)$ is *not* efficiently computable. The work of [33] presents an *efficient* construction of unconditionally secure commitments in the IPCP model.

[6] The more advanced PCP constructions of [3,4] were not known at that time.

Motivation and Related Work. Our black-box construction of Theorem 2 is motivated by the recent line of work on studying the power of black-box cryptographic constructions vs. that of non-black-box ones (e.g. [24,18,35,15,16,48,50,32]). The goal in this line of work is to understand whether the non-black-box application of an underlying primitive $\mathcal{P}$ which is used in a construction of another (perhaps more complicated) primitive $\mathcal{Q}$ is *necessary* or a black-box construction exists as well. The reason behind studying this question is that the black-box constructions are generally much more efficient (since the source of the non-black-box-ness usually is an extremely inefficient Cook-Levin reduction to an **NP**-complete language). Moreover, black-box constructions are capable of also incorporating any *physical* implementations of the employed primitive $\mathcal{P}$ in the implementation of $\mathcal{Q}$.

Technique. Kilian's argument system, when only required to be sound (and not ZK), has only four messages and uses the hash function as a black-box. The first three messages can easily made ZK, and it is only the last message from the prover which potentially carries some knowledge. In this last message, the prover reveals some portions of the PCP. To retain the zero-knowledge property, Kilian substitutes the last message (of his 4-message protocol) by a zero-knowledge sub-protocol through which the prover convinces the verifier that he could have revealed the correct portion of the PCP in a way that would cause the verifier to accept. The latter zero-knowledge sub-protocol makes non-black box use of the code of the hash function used in the protocol. Thus, our goal is to remove the zero-knowledge sub-protocol performed at the end.[7]

In order to make Kilian's 6-message ZK argument black-box, we need to know more details about its first 3 rounds. The first message is simply the description of the hash function sent to the prover. Then by using the given hash function and applying a Merkel tree to the PCP the prover hashes down the PCP into a short string which is sent to the verifier as a commitment to whole PCP. With some care, one can make the hash value carry negligible information about the PCP. The third message (from the verifier) consists of the indices of symbols which the PCP verifier chooses to read from the PCP. The prover, in the 4th message reveals the answers to the PCP queries by revealing the relevant paths of the Merkel tree to the verifier. The committed hash value of the PCP (the second message) together with the collision-resistance property of the hash function prevent the prover from changing his mind about the PCP that he committed to in the second message. Thus the soundness of the PCP implies the soundness of the argument system. To keep the last message of this protocol zero-knowledge, as we said, Kilian's prover will *not* simply reveal the relevant preimages, but instead would prove in a zero-knowledge manner, that he knows a set of preimages that would make the PCP verifier accept.

---

[7] Barak and Goldreich [7] also employ Kilian's approach to get a 4-message universal argument without zero-knowledge. Similarly to Kilian's protocol, to make their protocol zero-knowledge (or just witness indistinguishable) [7] use the hash function in a non-black-box way.

Our main intuitive observation is that if instead of using the PCP of [3,4] one feeds (a direct product version of) the the *bounded-query* ZK-PCP of [19] to the construction of Kilian, then the prover can safely reveal the relevant preimages in the last step of the basic 4-message argument of Kilian and this will not hurt the zero-knowledge property. The key point is that although the employed PCP is zero-knowledge only against bounded-query PCP verifiers, since we are in the prover/verifier setting, the prover can control how many queries of the PCP are read by the verifier, and therefore the bounded-query ZK property of the used PCP will suffice for the argument system to be zero-knowledge. Because our construction is black box, an unconditional result in the random oracle model follows immediately. Since this construction based on collision-resistant hash functions is black-box, it immediately implies an unconditional construction of sublinear ZK arguments in the random oracle model. Using the transformation of [21,43] one can eliminate the interaction using the random oracle and obtain an *unconditional* construction of sublinear ZK arguments for **NP** in the random oracle model. To obtain the result for malicious verifiers (and negligible soundness error), we apply a variant of the Goldreich-Kahan [25] where both prover and verifier use statistically hiding commitments. See the full version of the paper for a formal description of the protocol and its analysis.

Using NIZK? A possible alternative way to get a ZK argument (without using ZK-PCPs) is to use noninteractive zero-knowledge (NIZK) proofs for **NP** [10].[8] To do so, the prover and the verifier should perform a coin-tossing protocol along with the first 3 messages of the basic variant of Kilian's argument system, and this will allow the prover to be able to send a noninteractive zero-knowledge message to the verifier in his last message which proves to the verifier that the prover knows the right preimages of the hash function. This approach benefits from having only 4 messages exchanged, but it still uses the code of the hash function in a non-black-box way, and moreover, one needs to assume the existence of NIZK proofs for NP (*in addition* to the assumption that exponentially-hard collision-resistant hash functions exist).

3   On Nonadaptive Efficient ZK-PCPs

In this section we give a formal statement of Theorem 1 and more details about the intuition behind its proof. See the full version for a complete proof.

Definition 3. *In a probabilistically checkable proof (PCP) $\Pi = (P, V)$ for a language L, the prover $P = \{\pi_x\}$ is an (ensemble) of distributions over proof oracles, V is an efficient verifier accessing a proof $\pi_x \xleftarrow{\$} \boldsymbol{\pi}_x$, and the following properties hold.*

- **Completeness:** *For every $x \in L$, it holds that $\Pr_{\pi \xleftarrow{\$} \boldsymbol{\pi}_x}[V^\pi(x) = 1] \geq 2/3$.*
- **Soundness:** *If $x \notin L$, then for every oracle $\hat{\pi}$ we have $\Pr[V^{\hat{\pi}}(x) = 0] \geq 2/3$.*

---

[8] This variant was pointed out to us by Rafael Pass [47].

The verifier V is nonadaptive *if the queries it asks only depend on its own private randomness and the input x. (A nonadaptive verifier can prepare all of its oracle queries in advance and ask them in one "round".) For the case where $L \in \mathbf{NP}$, a PCP Π is called* efficient *if there is an $\mathbf{NP}$-relation $R_L(x, w)$ associated with L with the following efficiency property. Given any input x and witness w such that $(x, w) \in R_L$, one can efficiently sample a circuit computing a PCP oracle $\pi_x \xleftarrow{\$} \pi_x$.*[9]

Remark 4 (The Entropy of PCPs). For an input $x \in L$, the entropy of the PCP oracle π_x is defined similarly to the entropy of any random variable. Note that for a fixed input $x \in L$ (and witness w for $x \in L$, if the PCP is efficient), the distribution of π_x is determined by the prover's private randomness. Since there are at most $2^{\mathrm{poly}(k)}$ circuits of size k, any PCP oracle computable by circuits of size at most $k = \mathrm{poly}(n)$ (regardless of whether these circuits are generated efficiently or not) has entropy at most $\log(2^{\mathrm{poly}(k)}) \leq \mathrm{poly}(k) \leq \mathrm{poly}(n)$, simply because any finite random variable $\mathbf{x}$ has Shannon entropy at most $H(\mathbf{x}) \leq \log |\mathrm{Supp}(\mathbf{x})|$.

Definition 5. *Let $\Pi = (\{\pi_x\}, V)$ be a PCP for the language L. Π is called (statistical)* zero-knowledge *(ZK) if for every malicious $\mathrm{poly}(n)$-time verifier $\widehat{V}$, there is an efficient simulator Sim which runs in (expected) $\mathrm{poly}(n)$-time and for a sequence of inputs $x \in L$ the output of $\mathrm{Sim}(x)$ is $\mathrm{neg}(|x|)$-close to $\mathrm{View}\langle \pi_x, \widehat{V} \rangle(x)$.*[10] *A simulator Sim is called* straight-line *if it uses $\widehat{V}$ only as a black-box and moreover it just outputs the result of a single interaction with $\widehat{V}$. Namely, the simulator Sim interacts with $\widehat{V}$ without knowing its secret randomness $r_{\widehat{V}}$, and its output is distributed statistically close to the view of $\widehat{V}^{\pi_x}$.*

Theorem 1 directly follows from Remark 4 and Theorem 6 below.

Theorem 6. *Let $\Pi = (\{\pi_x\}, V)$ be a ZK-PCP for a language L with a nonadaptive verifier V. If (for every fixed input x) the PCP oracle $\{\pi_x\}$ has entropy at most $\mathrm{poly}(|x|)$, then $L \in \mathbf{AM} \cap \mathbf{coAM}$. Moreover $L \in \mathbf{BPP}$ if the simulator is straight-line.*[11]

Corollary 7. *Let $\Pi = (\{\pi_x\}, V)$ be a ZK-PCP for a language L with oracle entropy at most $\mathrm{poly}(n)$, and suppose the total length of the PCP answers returned to the verifier during a single verification is at most $O(\log n)$ bits, then (regardless of the adaptivity of the verifier), it holds that $L \in \mathbf{AM} \cap \mathbf{coAM}$. (Also $L \in \mathbf{BPP}$ if the simulator is straight-line.)*

---

[9] More formally, in that case we shall index the oracle distributions $\{\pi_{x,w}\}$ by both the input and the witness. Then the completeness should hold for all $x \in L$ when the prover uses *any* witness w that $x \in L$.

[10] In the case of efficient ZK-PCPs, the zero-knowledge property should hold regardless of which witness w (for $x \in L$) is used by the prover to generate the oracle.

[11] Bounded-query ZK-PCPs of [40] and its predecessors [27,19] all have straight-line simulators.

Note that in Corollary 7 there is no bound on the length of the *queries* of the verifier, and particularly it can be applied to cases that the number of queries of V is $O(\log n)$ and the PCP answers (alphabet) are of constant size while the length of the PCP is exponential $2^{\text{poly}(n)}$ (which makes the length of the queries of the verifier at least poly(n)).

Proof (Proof of Corollary 7). Since the total length of oracle answers is $O(\log n)$ bits, we can modify the verifier V into another equivalent verifier V' as follows: the new verifier V' tries to ask a superset of the queries that V would ask, but V' asks its queries in a nonadaptive way. In particular V' enumerates all the possible answers that V might get from the oracle, continues the verification in each case, and prepares all the possible V queries at the beginning. There are at most $2^{O(\log n)} = \text{poly}(n)$ many possibilities caused by different PCP answers in a verification, thus there will be at most poly(n) many queries asked by V'. After getting the answers, V' can emulate V internally and decide as V would. The completeness, soundness, and zero-knowledge of V' are inherited from those of V by definition.

3.1   Main Ideas and Framework

Here we describe the main ideas behind the proof of Theorem 6. Our **AM** protocols for $\overline{L}$ and L follow the same general framework. (The **AM** protocol for $\overline{L}$ is the more interesting case, since it implies the collapse of the hierarchy in case L is **NP**.)

First we show that if a bounded-entropy ZK-PCP for L has a straight-line simulator, then L (and $\overline{L}$) can be decided by an efficient **BPP** algorithm D_L. At a very high level, this step uses ideas from [26] by looking at a particular malicious verifier (in our case a repeated version of the honest verifier) and using its interaction with the straight-line simulator to decide the language. Since the key ideas already appear in the case of straight-line simulation, in Section 3.2 below we start by only describing this basic case.

Beyond Straight-Line Simulation. For the case of general (statistical) simulation, we show how to emulate the efficient algorithm D_L above with the help of an untrusted prover. In particular, we first show how to emulate D_L with the help of some advice α_x sampled from a specific distribution[12], and then we will show how to get this advice α_x from an (untrusted) prover through a constant round protocol GetAdv. The latter protocols are implemented following similar frameworks introduced by Feigenbaum and Fortnow [20] (and extended in the followup works of [11,2]) in the context of studying the possibility of worst-case to average-case reductions for **NP**. Our protocol, however, is more complicated and uses recent and old sampling protocols from [31,28,36].

---

[12] Here we are using the term "advice" in a nonstandard way, because the advice distribution α_x depends on the input x (rather than only depending on the input length $|x|$).

3.2   The Case of Straight-Line Simulation

In this section we present the **BPP** algorithm for L assuming that the ZK-PCP has a perfect straight-line simulator. This special case already captures the main ideas, and we refer the reader to the full version for the general case.

Since the PCP verifier V is assumed to be nonadaptive, we can assume w.l.o.g. that V permutes its queries $a_1, \ldots, a_q$ randomly before querying the oracle.

The Intuition. The general framework is to use the simulator Sim to find a "good enough" *oracle* φ and run a fresh instance of the verifier V against this oracle. This way, the correctness of our algorithm to decide membership in L follows from the soundness of the original PCP system. The challenge is to sample the oracle φ in a way that makes the verifier accept in case the input x is in L. Suppose we run the simulator over the "mildly malicious" verifier who only repeats several (independent) executions of the verifier: $(V^1, \ldots, V^k)$. Then, in case $x \in L$, the simulated transcript of all of these executions $(V^1, \ldots, V^k)$ will be accepted. To define the oracle φ, relying on the straight-line nature of the simulator, we can fix any simulated partial transcript for $(V^1, \ldots, V^i)$ (for $i \in [k]$) and ask Sim to answer any new query *only conditioned* on the simulated transcript of $(V^1, \ldots, V^i)$. (Even though φ is a randomized oracle, its randomness can be fixed independently of the final verification that is executed over φ.) The main intuition is that since the entropy of the simulated transcript for $(V^1, \ldots, V^k)$ is bounded, for most of $i \in [k]$ the simulated transcript of V^i has very small entropy, and relying on the non-adaptivity of V, all of its queries could be thought of as the "first query", and this way the oracle φ (defined above) behaves very close to the actual "oracle" of the simulated transcript of V^i which leads to an accept. The formal argument follows.

Notation. Let $V^{[k]}$ be an execution of k independent copies of the PCP verifier V. By V^i we refer to the i-th execution of V in $V^{[k]}$ (i.e. $V^{[i]} = (V^1, \ldots, V^i)$). $V^{[k]}$ is a potentially malicious verifier whose view $\mathsf{View}\langle \pi_x, V^{[k]} \rangle$ is assumed to be perfectly simulated by the straight-line simulator Sim (when given access to $V^{[k]}$). The view $\mathsf{View}\langle \pi_x, V^{[k]} \rangle$ is composed of k random seeds $r^1, \ldots, r^k$ for V and k transcripts $\tau^1, \ldots, \tau^k$ such that each $\tau^i = (a_1^i, b_1^i, \ldots, a_q^i, b_q^i)$ is a partial transcript where $\{a_1^i, \ldots, a_q^i\}$ are the queries asked by V using the randomness r^i and $b_j^i = \pi_x(a_j^i)$ is (supposedly) a corresponding returned oracle answer. We will only use the fact that Sim simulates $(\tau^1, \ldots, \tau^k)$ correctly and will ignore the fact that this is simulated jointly with random seeds $(r^1, \ldots, r^k)$. Also since we will use Sim only over $V^{[k]}$ and some input x, for simplicity in the following we will use Sim to denote $\mathsf{Sim}(V^{[k]}, x)$. Also, let $m = \mathrm{poly}(n) \geq H(\pi_y)$ be the upper bound on the PCP entropy for every $y \in L \cap \{0,1\}^n$, and let $\epsilon = 1/\mathrm{poly}(n)$ be a parameter controlling the error of the **BPP** algorithm D_L. The formal description of the algorithm D_L is as follows.

Construction 8. BPP Algorithm D_L. *Set $k = m \cdot (\frac{3q}{\epsilon})^2$ where q is the query complexity of V and ϵ is the error parameter.*

1. *Randomly choose $i \overset{\$}{\leftarrow} [k]$, and use* Sim *to generate* $(\tau^1, \ldots, \tau^{i-1})$ *as prefix of* View$\langle \pi_x, V^{[i-1]} \rangle$.

2. *Choose a fresh randomness r^i for the verifier V and generate the queries* $a_1^i, \ldots, a_q^i$ *using r^i.*

3. *Using the simulator* Sim *answer each of the queries a_j^i as follows to get the answer b_j. We extend the execution of the straight-line simulator* Sim *assuming that a_j^i is the first query of V^i conditioned on $(\tau^1, \ldots, \tau^{i-1})$ being generated already for $(V^1, \ldots, V^{i-1})$.*

4. *Finally output whatever V decides over the view* $(r^i, a_1^i, b_1, \ldots, a_q^i, b_q)$.

Lemma 9. *If Π has soundness $1 - \delta_s$, then D_L will reject every $x \notin L$ with probability $\geq 1 - \delta_s$, and if Π has completeness $1 - \delta_c$, then D_L will accept every $x \in L$ with probability $\geq 1 - (\delta_c + \epsilon)$.*

Proof (Proof of Lemma 9). We study the cases $x \in \overline{L}$ and $x \in L$ separately.

When $x \in \overline{L}$. The final verification of the algorithm of Construction 8 is run against a *randomized* oracle, but this oracle can be sampled and fixed independently of the randomness of the verifier, thus the soundness of the PCP implies the soundness of D_L. More formally, define the randomized oracle $\varphi^i = (\pi_x \mid \tau^1, \ldots, \tau^{i-1})$ according to the distribution of the PCP oracle π_x conditioned on the view of $V^{[i-1]}$. Define the oracle $\widehat{\varphi}^i$ as a randomized oracle that for every new query a it samples a fresh instance of the oracle $\varphi \overset{\$}{\leftarrow} \varphi^i$ and then answers a using φ. Based on Construction 8 D_L is indeed running the verifier V against an instance of the oracle $\widehat{\varphi} \leftarrow \widehat{\varphi}^i$ and outputs $V^{\widehat{\varphi}}(x)$. Thus, since $x \notin L$, by the soundness of V, with probability at least $1 - \delta_s$ it holds that $V^{\widehat{\varphi}}(x) = 0$. Note that if instead of asking all of the queries of the verifier "as the first query" we simply ask the simulator to simulate the whole view, the answers might *not* be chosen according to any fixed oracle *independently* of the randomness of V, and V might accept even though $x \in \overline{L}$.

When $x \in L$. Informally speaking, the verifier accepts in this case for the following two reasons: **(1)** If we sample the view of the final verification simply as the view of V^i as an extension of $V^{[i-1]}$ all sampled by the simulator Sim (i.e. using the oracle φ^i rather than $\widehat{\varphi}^i$), then it will be an accepted view by the definition of the simulator, moreover **(2)** since the verifier is nonadaptive and permutes its answers, any of its queries can be thought of as the first query. More formally, consider the following two mental experiments:

1. Sample $(\tau^1, \ldots, \tau^{i-1})$ and $\varphi \overset{\$}{\leftarrow} \varphi^i$ (as defined above) and sample $a_1^i, \ldots, a_q^i$ (by sampling r^i). Then execute q versions of the verifier V as follows. In the j'th execution ask the queries from φ in this order: $(a_j^i, \ldots, a_q^i, a_1^i, \ldots, a_{j-1}^i)$ and receive the answers $(b_j^i, \ldots, b_q^i, b_1^i, \ldots, b_{j-1}^i)$.

2. Do the same as above, but here in the j'th execution first sample a *fresh* oracle $\varphi_j \overset{\$}{\leftarrow} \varphi^i$ and then ask the queries in the order $(a^i_j, a^i_{j+1}, \ldots, a^i_q, a^i_1, \ldots, a^i_{j-1})$ to get the answers $(c^j_1, \ldots, c^j_q)$.

Claim. Let $\alpha = m/k$. Then for every $j \in [q]$, it holds that $\Pr[b^i_j = c^j_1] \geq 1 - 3\sqrt{\alpha}$.

Now we prove Claim 3.2. A crucial point is that the queries of V are already permuted randomly, and therefore rotations inside each execution will still produce a random execution of V (although these random executions are correlated). Therefore by symmetry, it would suffice to prove Claim 3.2 only for the first execution of the two experiments. Since $H(\pi_x) \leq m$ and that a^i_j's are sampled independently of π_x, therefore:

$$m \geq H(\pi_x) \geq \sum_{i \in [k]} \sum_{j \in [q]} H(b^i_j \mid a^1_1, b^1_1, \ldots, a^i_j) \geq \sum_{i \in [k]} H(b^i_1 \mid \tau^1, \ldots, \tau^{i-1}, a^i_1).$$

By averaging over i and using the definition of the conditional entropy it holds that:
$\mathbb{E}_{i \overset{\$}{\leftarrow} [k], \tau^1, \ldots, \tau^{i-1}, a^i_1} H(b^i_1 \mid \tau^1, \ldots, \tau^{i-1}, a^i_1) \leq m/k = \alpha$. By another averaging argument, with probability at least $1 - \sqrt{\alpha}$ over sampling and fixing $(i \overset{\$}{\leftarrow} [k], \tau^1, \ldots, \tau^{i-1}, a^i_1)$, it would hold that $H(b^i_1 \mid \tau^1, \ldots, \tau^{i-1}, a^i_1) \leq \sqrt{\alpha}$. We use the following lemma to bound the collision probability when the Shannon entropy is small.

Lemma 10. *For every finite random variable $\mathbf{x}$ it holds that* $\Pr_{x_1 \overset{\$}{\leftarrow} \mathbf{x}, x_2 \overset{\$}{\leftarrow} \mathbf{x}}[x_1 = x_2] \geq 1 - 1.45 H(\mathbf{x})$.

Proof. Let $C = \Pr_{x_1 \overset{\$}{\leftarrow} \mathbf{x}, x_2 \overset{\$}{\leftarrow} \mathbf{x}}[x_1 = x_2]$ be the collision probability of $\mathbf{x}$, let $p_i = \Pr[\mathbf{x} = i]$, and let $H = H(\mathbf{x})$. By Jensen's inequality: $\sum_i p_i \log p_i \leq \log \sum_i p^2_i$ it holds that $H \geq \log 1/C$ (where $\log 1/C$ is also known as the Renyi entropy). Therefore using $e^{-x} \geq 1 - x$ we conclude that: $C \geq 2^{-H} = e^{(-\ln 2) \cdot H} \geq 1 - (\ln 2) \cdot H > 1 - 1.45 H$.

By Lemma 10, the bounded entropy of $H(b^i_1 \mid \tau^1, \ldots, \tau^{i-1}, a^i_1) \leq \sqrt{\alpha}$ implies that its collision probability is at least $1 - 2\sqrt{\alpha}$ and since c^1_1 and b^i_1 are both sampled from $(b^i_1 \mid \tau^1, \ldots, \tau^{i-1}, a^i_1)$, we have $\Pr[c^1_1 = b^i_1] \geq 1 - 2\sqrt{\alpha}$. Claim 3.2 now follows by a union bound.

Claim 3.2 implies that the sampled $(r^i, a^i_1, b_1, \ldots, a^i_q, b_q)$ in the algorithm D_L (which is the same as using the first query/answer pairs of executions in the second experiment) will also lead to accepting with probability at least $1 - \delta_c - 3q\sqrt{\alpha} = 1 - (\delta_c + \epsilon)$.

Acknowledgement. We thank Vipul Goyal for collaboration at an early stage of this work. We would also like to thank Kai-Min Chung and Rafael Pass for very insightful discussions and the anonymous reviewers for their valuable comments. Yuval Ishai was supported by ERC Starting Grant 259426, ISF grant

1361/10, and BSF grant 2008411. Mohammad Mahmoody was supported in part by NSF Award CCF-0746990, AFOSR Award FA9550-10-1-0093, and DARPA and AFRL under contract FA8750-11-2- 0211. Amit Sahai was supported in part by a DARPA/ONR PROCEED award, NSF grants 1136174, 1118096, 1065276, 0916574 and 0830803, a Xerox Faculty Research Award, a Google Faculty Research Award, an equipment grant from Intel, and an Okawa Foundation Research Grant. This material is based upon work supported by the Defense Advanced Research Projects Agency through the U.S. Office of Naval Research under Contract N00014-11-1-0389. The views and conclusions contained in this document are those of the authors and should not be interpreted as representing the official policies, either expressed or implied, of the Defense Advanced Research Projects Agency, Department of Defense, or the US government.

References

1. Aiello, W., Håstad, J.: Statistical zero-knowledge languages can be recognized in two rounds. Journal of Computer and System Sciences 42(3), 327–345 (1991); Preliminary version in FOCS 1987
2. Akavia, A., Goldreich, O., Goldwasser, S., Moshkovitz, D.: On basing one-way functions on np-hardness. In: Proceedings of the 38th Annual ACM Symposium on Theory of Computing (STOC), pp. 701–710 (2006)
3. Arora, S., Lund, C., Motwani, R., Sudan, M., Szegedy, M.: Proof verification and the hardness of approximation problems. Journal of the ACM 45(3), 501–555 (1998); Preliminary version in FOCS 1992
4. Arora, S., Safra, S.: Probabilistic checking of proofs: a new characterization of NP. Journal of the ACM 45(1), 70–122 (1998); Preliminary version in FOCS 1992
5. Babai, Fortnow, Levin, Szegedy: Checking computations in polylogarithmic time. In: STOC: ACM Symposium on Theory of Computing (STOC) (1991)
6. Babai, L., Fortnow, L., Lund, C.: Non-deterministic exponential time has two-prover interactive protocols. In: FOCS, pp. 16–25 (1990)
7. Barak, B., Goldreich, O.: Universal arguments and their applications, pp. 194–203 (2002)
8. Ben-Or, M., Goldreich, O., Goldwasser, S., Håstad, J., Kilian, J., Micali, S., Rogaway, P.: Everything Provable Is Provable in Zero-Knowledge. In: Goldwasser, S. (ed.) CRYPTO 1988. LNCS, vol. 403, pp. 37–56. Springer, Heidelberg (1990)
9. Ben-Or, M., Goldwasser, S., Kilian, J., Wigderson, A.: Multi-prover interactive proofs: How to remove intractability assumptions. In: STOC, pp. 113–131 (1988)
10. Blum, M., Feldman, P., Micali, S.: Non-interactive zero-knowledge and its applications (extended abstract). In: Proceedings of the 20th Annual ACM Symposium on Theory of Computing (STOC), pp. 103–112 (1988)
11. Bogdanov, A., Trevisan, L.: On worst-case to average-case reductions for np problems. SIAM Journal on Computing 36(4), 1119–1159 (2006)
12. Boppana, R.B., Håstad, J., Zachos, S.: Does co-NP have short interactive proofs? Information Processing Letters 25, 127–132 (1987)
13. Canetti, R., Goldreich, O., Goldwasser, S., Micali, S.: Resettable zero-knowledge (extended abstract). In: STOC, pp. 235–244 (2000)
14. Chandran, N., Goyal, V., Sahai, A.: New Constructions for UC Secure Computation Using Tamper-Proof Hardware. In: Smart, N.P. (ed.) EUROCRYPT 2008. LNCS, vol. 4965, pp. 545–562. Springer, Heidelberg (2008)

15. Choi, S.G., Dachman-Soled, D., Malkin, T., Wee, H.: Black-Box Construction of a Non-malleable Encryption Scheme from Any Semantically Secure One. In: Canetti, R. (ed.) TCC 2008. LNCS, vol. 4948, pp. 427–444. Springer, Heidelberg (2008)

16. Choi, S.G., Dachman-Soled, D., Malkin, T., Wee, H.: Simple, Black-Box Constructions of Adaptively Secure Protocols. In: Reingold, O. (ed.) TCC 2009. LNCS, vol. 5444, pp. 387–402. Springer, Heidelberg (2009)

17. Damgard, Pedersen, Pfitzmann: On the existence of statistically hiding bit commitment schemes and fail-stop signatures. Journal of Cryptology 10 (1997)

18. Damgård, I., Ishai, Y.: Constant-Round Multiparty Computation Using a Black-Box Pseudorandom Generator. In: Shoup, V. (ed.) CRYPTO 2005. LNCS, vol. 3621, pp. 378–394. Springer, Heidelberg (2005)

19. Dwork, C., Feige, U., Kilian, J., Naor, M., Safra, M.: Low Communication 2-Prover Zero-Knowledge Proofs for NP. In: Brickell, E.F. (ed.) CRYPTO 1992. LNCS, vol. 740, pp. 215–227. Springer, Heidelberg (1993)

20. Feigenbaum, J., Fortnow, L.: Random-self-reducibility of complete sets. SIAM Journal on Computing 22(5), 994–1005 (1993)

21. Fiat, A., Shamir, A.: How to Prove Yourself: Practical Solutions to Identification and Signature Problems. In: Odlyzko, A.M. (ed.) CRYPTO 1986. LNCS, vol. 263, pp. 186–194. Springer, Heidelberg (1987)

22. Fortnow, L.: The complexity of perfect zero-knowledge. Advances in Computing Research: Randomness and Computation 5, 327–343 (1989)

23. Fortnow, L., Rompel, J., Sipser, M.: On the power of multi-prover interactive protocols. Theoretical Computer Science 134(2), 545–557 (1994)

24. Gertner, Y., Kannan, S., Malkin, T., Reingold, O., Viswanathan, M.: The relationship between public key encryption and oblivious transfer. In: Proceedings of the 41st Annual IEEE Symposium on Foundations of Computer Science (2000)

25. Goldreich, O., Kahan, A.: How to construct constant-round zero-knowledge proof systems for NP. Journal of Cryptology 9(3), 167–190 (1996)

26. Goldreich, O., Krawczyk, H.: On the Composition of Zero-Knowledge Proof Systems. SIAM Journal on Computing 25(1), 169–192 (1996); In: Paterson, M. (ed.) ICALP 1990. LNCS, vol. 443, pp. 268–282. Springer, Heidelberg (1990);

27. Goldreich, O., Micali, S., Wigderson, A.: Proofs that yield nothing but their validity or all languages in NP have zero-knowledge proof systems. Journal of the ACM 38(1), 691–729 (1991); Preliminary version in FOCS 1986

28. Goldreich, O., Vadhan, S., Wigderson, A.: On Interactive Proofs with a Laconic Prover. In: Yu, Y., Spirakis, P.G., van Leeuwen, J. (eds.) ICALP 2001. LNCS, vol. 2076, pp. 334–345. Springer, Heidelberg (2001)

29. Goldwasser, S., Kalai, Y.T., Rothblum, G.N.: One-Time Programs. In: Wagner, D. (ed.) CRYPTO 2008. LNCS, vol. 5157, pp. 39–56. Springer, Heidelberg (2008)

30. Goldwasser, S., Micali, S., Rackoff, C.: The knowledge complexity of interactive proof systems. SIAM Journal on Computing 18(1), 186–208 (1989); Preliminary version in STOC 1985

31. Goldwasser, S., Sipser, M.: Private coins versus public coins in interactive proof systems. Advances in Computing Research: Randomness and Computation 5, 73–90 (1989)

32. Goyal, V.: Constant round non-malleable protocols using one way functions. In: Fortnow, L., Vadhan, S.P. (eds.) STOC, pp. 695–704. ACM (2011)

33. Goyal, V., Ishai, Y., Mahmoody, M., Sahai, A.: Interactive Locking, Zero-Knowledge PCPs, and Unconditional Cryptography. In: Rabin, T. (ed.) CRYPTO 2010. LNCS, vol. 6223, pp. 173–190. Springer, Heidelberg (2010)

34. Goyal, V., Ishai, Y., Sahai, A., Venkatesan, R., Wadia, A.: Founding Cryptography on Tamper-Proof Hardware Tokens. In: Micciancio, D. (ed.) TCC 2010. LNCS, vol. 5978, pp. 308–326. Springer, Heidelberg (2010)
35. Haitner, I., Ishai, Y., Kushilevitz, E., Lindell, Y., Petrank, E.: Black-box constructions of protocols for secure computation. SIAM J. Comput. 40(2), 225–266 (2011)
36. Haitner, I., Mahmoody, M., Xiao, D.: A new sampling protocol and applications to basing cryptographic primitives on the hardness of NP. In: IEEE Conference on Computational Complexity, pp. 76–87. IEEE Computer Society (2010)
37. Kalai, Y.T., Raz, R.: Interactive PCP. In: Aceto, L., Damgård, I., Goldberg, L.A., Halldórsson, M.M., Ingólfsdóttir, A., Walukiewicz, I. (eds.) ICALP 2008, Part II. LNCS, vol. 5126, pp. 536–547. Springer, Heidelberg (2008)
38. Katz, J.: Universally Composable Multi-party Computation Using Tamper-Proof Hardware. In: Naor, M. (ed.) EUROCRYPT 2007. LNCS, vol. 4515, pp. 115–128. Springer, Heidelberg (2007)
39. Kilian, J.: A note on efficient zero-knowledge proofs and arguments (extended abstract). In: Proceedings of the 24th Annual ACM Symposium on Theory of Computing (STOC), pp. 723–732 (1992)
40. Kilian, J., Petrank, E., Tardos, G.: Probabilistically checkable proofs with zero knowledge. In: STOC: ACM Symposium on Theory of Computing, STOC (1997)
41. Kolesnikov, V.: Truly Efficient String Oblivious Transfer Using Resettable Tamper-Proof Tokens. In: Micciancio, D. (ed.) TCC 2010. LNCS, vol. 5978, pp. 327–342. Springer, Heidelberg (2010)
42. Meir, O.: Combinatorial PCPs with efficient verifiers. In: FOCS, pp. 463–471. IEEE Computer Society (2009)
43. Micali, S.: Computationally sound proofs. SIAM Journal on Computing 30(4), 1253–1298 (2000); Preliminary version in FOCS 1994
44. Moran, T., Segev, G.: David and Goliath Commitments: UC Computation for Asymmetric Parties Using Tamper-Proof Hardware. In: Smart, N.P. (ed.) EUROCRYPT 2008. LNCS, vol. 4965, pp. 527–544. Springer, Heidelberg (2008)
45. Naor, M.: Bit Commitment Using Pseudo-Randomness. Journal of Cryptology 4(2), 151–158 (1991); In: Brassard, G. (ed.) CRYPTO 1989. LNCS, vol. 435, pp. 128–136. Springer, Heidelberg (1990)
46. Ostrovsky, R., Wigderson, A.: One-way functions are essential for non-trivial zero-knowledge. In: Proceedings of the 2nd Israel Symposium on Theory of Computing Systems, pp. 3–17. IEEE Computer Society (1993)
47. Pass, R.: Personal communication
48. Pass, R., Wee, H.: Black-Box Constructions of Two-Party Protocols from One-Way Functions. In: Reingold, O. (ed.) TCC 2009. LNCS, vol. 5444, pp. 403–418. Springer, Heidelberg (2009)
49. Reingold, O., Trevisan, L., Vadhan, S.P.: Notions of Reducibility between Cryptographic Primitives. In: Naor, M. (ed.) TCC 2004. LNCS, vol. 2951, pp. 1–20. Springer, Heidelberg (2004)
50. Wee, H.: Black-box, round-efficient secure computation via non-malleability amplification. In: FOCS, pp. 531–540. IEEE Computer Society (2010)

Progression-Free Sets and Sublinear Pairing-Based Non-Interactive Zero-Knowledge Arguments

Helger Lipmaa

Institute of Computer Science, University of Tartu, Estonia

Abstract. In 2010, Groth constructed the only previously known sublinear-communication NIZK circuit satisfiability argument in the common reference string model. We optimize Groth's argument by, in particular, reducing both the CRS length and the prover's computational complexity from quadratic to quasilinear in the circuit size. We also use a (presumably) weaker security assumption, and have tighter security reductions. Our main contribution is to show that the complexity of Groth's basic arguments is dominated by the quadratic number of monomials in certain polynomials. We collapse the number of monomials to quasilinear by using a recent construction of progression-free sets.

Keywords: Additive combinatorics, bilinear pairings, circuit satisfiability, non-interactive zero-knowledge, progression-free sets.

1 Introduction

By using a zero-knowledge proof, a prover can convince a verifier that some statement is true without leaking any side information. Due to the wide applications of zero-knowledge, it is of utmost importance to construct efficient zero-knowledge proofs. *Non-interactive zero-knowledge* (NIZK) proofs can be generated once can be verified many times by different verifiers and are thus useful in applications like e-voting.

NIZK proofs (or arguments, that is, computationally sound proofs) cannot be constructed in the plain model (that is, without random oracles or any trusted setup assumptions). Blum, Feldman and Micali showed in [4] how to construct NIZK proofs in the common reference string (CRS) model. During the last years, a substantial amount of research has been done towards constructing efficient NIZK proofs (and arguments). Since the communication complexity and the verifier's computational complexity are arguably more important than the prover's computational complexity (again, an NIZK proof/argument is generated once but can be verified many times), a special effort has been made to minimize these two parameters.

One related research direction is to construct efficient NIZK proofs for **NP**-complete languages. Given an efficient NIZK proof for a **NP**-complete language, one can hope to construct NIZK proofs of similar complexity for the whole **NP** either by reduction or implicitly or explicitly using the developed techniques. In some NIZK proofs for the **NP**-complete problem circuit satisfiability (Circuit-SAT), see Tbl. 1, the communication complexity is sublinear in the circuit size. Micali [22] proposed polylogarithmic-communication NIZK *arguments* for all **NP**-languages, but they are based on the PCP theorem (making them computationally unattractive) and on the random oracle model.

R. Cramer (Ed.): TCC 2012, LNCS 7194, pp. 169–189, 2012.

Table 1. Comparison of NIZK Circuit-SAT arguments with (worst-case) sublinear argument size. $|C|$ is the size of circuit, G corresponds to 1 group element and $A/M/E/P$ corresponds to 1 addition/multiplication/exponentiation/pairing

	CRS length	Argument length	Prover comp.	Verifier comp.
	Random-oracle based arguments			
[14]	$O(|C|^{\frac{1}{2}})G$	$O(|C|^{\frac{1}{2}})G$	$O(|C|)M$	$O(|C|)M$
	Knowledge-assumption based arguments from [15]			
$m = 1$	$\Theta(|C|^2)G$	$42G$	$\Theta(|C|^2)E$	$\Theta(|C|)M + \Theta(1)P$
$m = n^{\frac{1}{3}}$	$\Theta(|C|^{\frac{2}{3}})G$	$\Theta(|C|^{\frac{2}{3}})G$	$\Theta(|C|^{\frac{4}{3}})E$	$\Theta(|C|)M + \Theta(|C|^{\frac{2}{3}})P$
	Knowledge-assumption based arguments from the current paper			
$m = 1$	$|C|^{1+o(1)}G$	$39G$	$\Theta(|C|^2)A + |C|^{1+o(1)}E$	$(8|C| + 8)M + 62P$
$m = n^{\frac{1}{3}}$	$|C|^{\frac{1}{3}+o(1)}G$	$\Theta(|C|^{\frac{2}{3}})G$	$\Theta(|C|^{\frac{4}{3}})A + |C|^{1+o(1)}E$	$\Theta(|C|)M + \Theta(|C|^{\frac{2}{3}})P$
$m = n^{\frac{1}{2}}$	$|C|^{\frac{1}{2}+o(1)}G$	$\Theta(|C|^{\frac{1}{2}})G$	$\Theta(|C|^{\frac{1}{2}})A + |C|^{1+o(1)}E$	$\Theta(|C|)M + \Theta(|C|^{\frac{1}{2}})P$

Another NIZK argument for Circuit-SAT, proposed by Groth in 2009 [14], is also based on the random oracle model. It is well-known that some functionalities are secure in the random oracle model and insecure in the plain model. As a safeguard, it is important to design efficient NIZK proofs and arguments that do not rely on the random oracles. Given a fully-homomorphic cryptosystem [10], one can construct efficient NIZK *proofs* for all **NP**-languages in communication that is linear to the witness size [16]. However, since the witness size can be linear in the circuit size, in the worst case the corresponding NIZK proofs are not sublinear.

In 2010, Groth [15] proposed the first (worst-case) sublinear-communication NIZK Circuit-SAT argument in the CRS model. First, he constructed two basic arguments for Hadamard product (the prover knows how to open commitments A, B and C to three tuples $\boldsymbol{a}$, $\boldsymbol{b}$ and $\boldsymbol{c}$ of dimension n, such that $a_i b_i = c_i$ for $i \in [n]$) and permutation (the prover knows how to open commitments A and B to two tuples $\boldsymbol{a}$ and $\boldsymbol{b}$ of dimension n, such that $a_{\varrho(i)} = b_i$ for $i \in [n]$). Groth's Circuit-SAT argument can then be seen as a program in a program language that has two primitive instructions, for Hadamard product and permutation. Some of the public permutations depend on the circuit, while the secret input tuples of the basic arguments depend on the values, assigned to the input and output wires of all gates according to a satisfying assignment. The basic arguments then show that this wire assignment is internally consistent and corresponds indeed to an satisfying input assignment. For example, Groth used one permutation argument to verify that all input wires of all gates have been assigned the same values as the corresponding output values of their predecessor gates.

In the basic variant of Groth's pairing-based Circuit-SAT argument, see Tbl. 1, the argument has $\Theta(1)$ group elements, but on the other hand the CRS has $\Theta(|C|)^2$ group elements, and the prover's computational complexity is dominated by $\Theta(|C|^2)$ bilinear-group exponentiations. A balanced version of Groth's argument has the CRS and argument of $\Theta(|C|^{2/3})$ group elements and prover's computational complexity dominated by $\Theta(|C|^{4/3})$ exponentiations. (See [15] for more details on balancing. Basically, one applies basic arguments on length-m inputs, $m < n$, n/m times in parallel.)

We propose a new Circuit-SAT argument (see Sect. 3 for a description of the new techniques, and subsequent sections for the actual argument) that is strongly related to Groth's argument, but improves upon every step. We first propose more efficient basic arguments. We then use them to construct a (slightly shorter) new Circuit-SAT argument. In the basic variant, while the argument is again $\Theta(1)$ group elements, it is one commitment and one Hadamard product argument shorter. Moreover, in Groth's argument, every commitment consisted of 3 group elements while every basic argument consisted of 2 group elements. In the new argument, most of the commitments consist of 2 group elements. Thus, we saved 3 group elements, reducing the argument size from 42 to 39 group elements, even taking into account that the new permutation argument has higher communication complexity (12 instead of 5 group elements) than that of [15].

A balanced version of the new argument achieves the combined CRS and argument of $\Theta(|C|^{1/2+o(1)})$ group elements. In the full version, we describe a zap for Circuit-SAT that has communication complexity of $|C|^{1/2+o(1)}$ group elements, while Groth's zap from [15] has the communication complexity of $\Theta(|C|^{2/3})$ group elements. We also use much more efficient asymmetric pairings instead of symmetric ones, a (presumably) weaker security assumption (Power Symmetric Discrete Logarithm instead of Power Computational Diffie-Hellman), and have more precise security reductions. The basic version of the new Circuit-SAT argument is more communication-efficient than any prior-art random-oracle based NIZK argument, and it also has a smaller prover's computational complexity than [22].

Our main contribution is to note that the complexity of Groth's basic arguments is correlated to the number of monomials of a certain polynomial. In [15], this polynomial has $\Theta(n^2)$ monomials, where $n = 2|C| + 1$. We show that one can "collapse" the $\Theta(n^2)$ monomials to $\Theta(N)$ monomials, where N is such that $[N]$ has a progression-free subset (that is, a subset that does not contain arithmetic progressions of length 3) of odd integers of cardinality n. By a recent breakthrough of Elkin [9], $N = O(n \cdot 2^{2\sqrt{2(2+\log_2 n)}}) = n^{1+o(1)}$. See Sect. 3 for further elaboration on our techniques.

Thus, one can build an argument of $\Theta(1)$ group elements for every language in **NP**, by reducing the task at hand to a Circuit-SAT instance. Obviously, one can often design more efficient tailor-made protocols, see [21,7] for some follow-up work. In particular, [7] used our basic arguments to construct a non-interactive range proof with communication of $\Theta(1)$ group elements, while [21] used our techniques to design a new basic argument to construct a non-interactive shuffle. (See [6] for a previous use of additive combinatorics in the construction of zero-knowledge proofs.)

Due to the lack of space, several proofs have been deferred to the full version [20].

2 Preliminaries

Let $[n] = \{1, 2, \ldots, n\}$. Let S_n be the set of permutations from $[n]$ to $[n]$. Let $a = (a_1, \ldots, a_n)$. Let $a \circ b$ denote the Hadamard (entry-wise) product of a and b, that is, if $c = a \circ b$, then $c_i = a_i b_i$ for $i \in [n]$. If $y = h^x$, then $\log_h y := x$. Let κ be the security parameter. If $0 < \lambda_1 < \cdots < \lambda_i < \cdots < \lambda_n = \text{poly}(\kappa)$. then $\Lambda = (\lambda_1, \ldots, \lambda_n) \subset \mathbb{Z}$ is an (n, κ)-*nice tuple*. We abbreviate probabilistic polynomial-time as PPT. If Λ_1 and Λ_2 are subsets of some additive group ($\mathbb{Z}$ or $\mathbb{Z}_p$ in this paper), then $\Lambda_1 + \Lambda_2 = \{\lambda_1 + \lambda_2 :$

$\lambda_1 \in \Lambda_1 \wedge \lambda_2 \in \Lambda_2\}$ is their *sum set* and $\Lambda_1 - \Lambda_2 = \{\lambda_1 - \lambda_2 : \lambda_1 \in \Lambda_1 \wedge \lambda_2 \in \Lambda_2\}$ is their *difference set* [25]. If Λ is a set, then $k\Lambda = \{\lambda_1 + \cdots + \lambda_k : \lambda_i \in \Lambda\}$ is an *iterated sumset*, $k \cdot \Lambda = \{k\lambda : \lambda \in \Lambda\}$ is a *dilation* of Λ, and $2\hat{\ }\Lambda = \{\lambda_1 + \lambda_2 : \lambda_1 \in \lambda \wedge \lambda_2 \in \Lambda \wedge \lambda_1 \neq \lambda_2\} \subseteq \Lambda + \Lambda$ is a *restricted sumset*. (See [25].)

Let $\mathcal{G}_{\mathsf{bp}}(1^\kappa)$ be a bilinear group generator that outputs a description of a bilinear group gk $:= (p, \mathbb{G}_1, \mathbb{G}_2, \mathbb{G}_T, \hat{e}) \leftarrow \mathcal{G}_{\mathsf{bp}}(1^\kappa)$, such that p is a κ-bit prime, $\mathbb{G}_1$, $\mathbb{G}_2$ and $\mathbb{G}_T$ are multiplicative cyclic groups of order p, $\hat{e} : \mathbb{G}_1 \times \mathbb{G}_2 \to \mathbb{G}_T$ is a bilinear map (pairing) such that $\forall a, b \in \mathbb{Z}$ and $g_t \in \mathbb{G}_t$, $\hat{e}(g_1^a, g_2^b) = \hat{e}(g_1, g_2)^{ab}$. If g_t generates $\mathbb{G}_t$ for $t \in \{1, 2\}$, then $\hat{e}(g_1, g_2)$ generates $\mathbb{G}_T$. Deciding the membership in $\mathbb{G}_1$, $\mathbb{G}_2$ and $\mathbb{G}_T$, group operations, the pairing $\hat{e}$, and sampling the generators are efficient, and the descriptions of the groups and group elements are $O(\kappa)$ bit long each. Well-chosen asymmetric pairings (with no efficient isomorphism between $\mathbb{G}_1$ and $\mathbb{G}_2$) are much more efficient than symmetric pairings (where $\mathbb{G}_1 = \mathbb{G}_2$). For $\kappa = 128$, the current recommendation is to use an optimal (asymmetric) Ate pairing [18] over a subclass of Barreto-Naehrig curves [2]. In that case, at security level of $\kappa = 128$, an element of $\mathbb{G}_1/\mathbb{G}_2/\mathbb{G}_T$ can be represented in respectively 512/256/3072 bits.

A (tuple) commitment scheme $(\mathcal{G}_{\mathsf{com}}, \mathcal{C}om)$ in a bilinear group consists of two PPT algorithms: a randomized CRS generation algorithm $\mathcal{G}_{\mathsf{com}}$, and a randomized commitment algorithm $\mathcal{C}om$. Here, $\mathcal{G}_{\mathsf{com}}^t(1^\kappa, n)$, $t \in \{1, 2\}$, produces a CRS ck_t, and $\mathcal{C}om^t(\mathsf{ck}_t; \boldsymbol{a}; r)$, with $\boldsymbol{a} = (a_1, \ldots, a_n)$, outputs a commitment value A in $\mathbb{G}_t$ (or in $\mathbb{G}_t^b$ for some $b > 1$). We open $\mathcal{C}om^t(\mathsf{ck}_t; \boldsymbol{a}; r)$ by outputting $\boldsymbol{a}$ and r.

A commitment scheme $(\mathcal{G}_{\mathsf{com}}, \mathcal{C}om)$ is *computationally binding in group* $\mathbb{G}_t$, if for every non-uniform PPT adversary $\mathcal{A}$ and positive integer $n = \mathrm{poly}(\kappa)$, the probability

$$\Pr \begin{bmatrix} \mathsf{ck}_t \leftarrow \mathcal{G}_{\mathsf{com}}^t(1^\kappa, n), (\boldsymbol{a_1}, r_1, \boldsymbol{a_2}, r_2) \leftarrow \mathcal{A}(\mathsf{ck}_t) : \\ (\boldsymbol{a_1}, r_1) \neq (\boldsymbol{a_2}, r_2) \wedge \mathcal{C}om^t(\mathsf{ck}_t; \boldsymbol{a_1}; r_1) = \mathcal{C}om^t(\mathsf{ck}_t; \boldsymbol{a_2}; r_2) \end{bmatrix}$$

is negligible in κ. A commitment scheme $(\mathcal{G}_{\mathsf{com}}, \mathcal{C}om)$ is *perfectly hiding in group* $\mathbb{G}_t$, if for any positive integer $n = \mathrm{poly}(\kappa)$ and $\mathsf{ck}_t \in \mathcal{G}_{\mathsf{com}}^t(1^\kappa, n)$ and any two messages $\boldsymbol{a_1}, \boldsymbol{a_2}$, the distributions $\mathcal{C}om^t(\mathsf{ck}_t; \boldsymbol{a_1}; \cdot)$ and $\mathcal{C}om^t(\mathsf{ck}_t; \boldsymbol{a_2}; \cdot)$ are equal.

A trapdoor commitment scheme has three additional efficient algorithms: (a) A trapdoor CRS generation algorithm inputs t, n and 1^κ, and outputs a CRS ck^* (that has the same distribution as $\mathcal{G}_{\mathsf{com}}^t(1^\kappa, n)$) and a trapdoor td, (b) a randomized trapdoor commitment that takes ck^* and a randomizer r as inputs and outputs the value $\mathcal{C}om^t(\mathsf{ck}^*; 0; r)$, and (c) a trapdoor opening algorithm that takes ck^*, td, $\boldsymbol{a}$ and r as an input and outputs an r' such that $\mathcal{C}om^t(\mathsf{ck}^*; 0; r) = \mathcal{C}om^t(\mathsf{ck}^*; \boldsymbol{a}; r')$.

Let $\mathcal{R} = \{(C, w)\}$ be an efficiently computable binary relation such that $|w| = \mathrm{poly}(|C|)$. Here, C is a statement, and w is a witness. Let $\mathcal{L} = \{C : \exists w, (C, w) \in \mathcal{R}\}$ be an **NP**-language. Let n be some fixed input length $n = |C|$. For fixed n, we have a relation $\mathcal{R}_n$ and a language $\mathcal{L}_n$. A *non-interactive argument* for $\mathcal{R}$ consists of the following PPT algorithms: a common reference string (CRS) generator $\mathcal{G}_{\mathsf{crs}}$, a prover $\mathcal{P}$, and a verifier $\mathcal{V}$. For $\mathsf{crs} \leftarrow \mathcal{G}_{\mathsf{crs}}(1^\kappa, n)$, $\mathcal{P}(\mathsf{crs}; C, w)$ produces an argument ψ. The verifier $\mathcal{V}(\mathsf{crs}; C, \psi)$ outputs either 1 (accept) or 0 (reject).

A non-interactive argument $(\mathcal{G}_{\mathsf{crs}}, \mathcal{P}, \mathcal{V})$ is *perfectly complete*, if $\forall n = \mathrm{poly}(\kappa)$,

$$\Pr[\mathsf{crs} \leftarrow \mathcal{G}_{\mathsf{crs}}(1^\kappa, n), (C, w) \leftarrow \mathcal{R}_n : \mathcal{V}(\mathsf{crs}; C, \mathcal{P}(\mathsf{crs}; C, w)) = 1] = 1 .$$

A non-interactive argument $(\mathcal{G}_{crs}, \mathcal{P}, \mathcal{V})$ is *(adaptively) computationally sound*, if for all non-uniform PPT adversaries $\mathcal{A}$ and all $n = \text{poly}(\kappa)$, the probability

$$\Pr[\text{crs} \leftarrow \mathcal{G}_{crs}(1^\kappa, n), (C, \psi) \leftarrow \mathcal{A}(\text{crs}) : C \notin \mathcal{L} \wedge \mathcal{V}(\text{crs}; C, \psi) = 1]$$

is negligible in κ. The soundness is adaptive, that is, the adversary sees the CRS before producing the statement C. A non-interactive argument $(\mathcal{G}_{crs}, \mathcal{P}, \mathcal{V})$ is *perfectly witness-indistinguishable*, if for all $n = \text{poly}(\kappa)$, if $\text{crs} \in \mathcal{G}_{crs}(1^\kappa, n)$ and $((C, w_0), (C, w_1)) \in \mathcal{R}_n^2$, then the distributions $\mathcal{P}(\text{crs}; C, w_0)$ and $\mathcal{P}(\text{crs}; C, w_1)$ are equal.

A non-interactive argument $(\mathcal{G}_{crs}, \mathcal{P}, \mathcal{V})$ is *perfectly zero-knowledge*, if there exists a PPT simulator $\mathcal{S} = (\mathcal{S}_1, \mathcal{S}_2)$, such that for all stateful non-uniform PPT adversaries $\mathcal{A}$ and $n = \text{poly}(\kappa)$ (with td being the *simulation trapdoor*),

$$\Pr\begin{bmatrix} \text{crs} \leftarrow \mathcal{G}_{crs}(1^\kappa, n), \\ (C, w) \leftarrow \mathcal{A}(\text{crs}), \\ \psi \leftarrow \mathcal{P}(\text{crs}; C, w) : \\ (C, w) \in \mathcal{R}_n \wedge \mathcal{A}(\psi) = 1 \end{bmatrix} = \Pr\begin{bmatrix} (\text{crs}; \text{td}) \leftarrow \mathcal{S}_1(1^\kappa, n), \\ (C, w) \leftarrow \mathcal{A}(\text{crs}), \\ \psi \leftarrow \mathcal{S}_2(\text{crs}; C, \text{td}) : \\ (C, w) \in \mathcal{R}_n \wedge \mathcal{A}(\psi) = 1 \end{bmatrix}.$$

3 Our Techniques

We will first give a more precise overview of Groth's Hadamard product and permutation arguments [15], followed by a short description of our own main contribution. For the sake of simplicity, we will make several simplifications (like the use of symmetric pairings) during this discussion.

Groth uses an additively homomorphic tuple commitment scheme that allows one to commit to a long tuple, while the commitment itself is short. The best known such commitment scheme is the extended Pedersen commitment scheme in a multiplicative cyclic group of order p and a generator g, where the commitment of a tuple $\boldsymbol{a} = (a_1, \ldots, a_n)$ with randomness r_a is equal to $\mathcal{C}om(\boldsymbol{a}; r_a) := g^{r_a} \cdot \prod g_i^{a_i}$. Here, one usually chooses n random secrets $x_i \leftarrow \mathbb{Z}_p$, and then sets $g_i \leftarrow g^{x_i}$. Following [12], Groth [15] chooses a single random secret $x \leftarrow \mathbb{Z}_p$ and then sets $g_i \leftarrow g^{x^i}$. In this case, the commitment

$$\mathcal{C}om(\boldsymbol{a}; r_a) := g^{r_a} \cdot \prod_{i=1}^{n} g_i^{a_i} = g^{r_a + \sum_{i=1}^{n} a_i x^i}$$

can be seen as a lifted polynomial $r_a + \sum_{i=1}^{n} a_i x^i$ in x, that the committer (who does not know x) computes from n given values $g_i = g^{x^i}$. The first obvious benefit of this commitment scheme is that it has a shorter secret (1 element instead of n elements).

Groth's Hadamard product argument, where the prover aims to convince the verifier that the opening of $C = \mathcal{C}om(\boldsymbol{c}; r_c)$ is equal to the Hadamard product of the openings of $A = \mathcal{C}om(\boldsymbol{a}; r_a)$ and $B = \mathcal{C}om(\boldsymbol{b}; r_b)$ (that is, $a_i b_i \equiv c_i \pmod{p}$ for $i \in [n]$), is constructed as follows. Let $A = g^{r_a} \cdot \prod_{i=1}^{n} g_i^{a_i}$ be a commitment of $\boldsymbol{a}$ and $B = g^{r_b} \cdot \prod_{i=1}^{n} g_i^{b_i}$ be a commitment of $\boldsymbol{b}$ by using the generator tuple $(g_1, \ldots, g_n)$. Let $C = g^{r_c} \cdot \prod_{i=1}^{n} g_{i(n+1)}^{c_i}$ be a commitment of $\boldsymbol{b}$ and $D = \prod_{i=1}^{n} g_{i(n+1)}$ be a commitment of $\boldsymbol{1} = (1, \ldots, 1)$ by using a different generator tuple $(g_{n+1}, \ldots, g_{n(n+1)})$.

Groth's Hadamard product argument is based around the verification equation

$$\hat{e}(A, B) = \hat{e}(C, D) \cdot \hat{e}(\psi, g) \tag{1}$$

that (analogously to the Groth-Sahai proofs [17], though the latter only considers the much simpler case $n = 1$) can be seen as a mapping of the required equality $a \circ b = c \circ 1$ to another algebraic domain, with ψ compensating for the use of a randomized commitment scheme. One gets that $\hat{e}(A, B)/\hat{e}(C, D)$ is equal to $\hat{e}(g, g)^{F(x)}$, where $F(x) = (r_a + \sum_{i=1}^{n} a_i x^i) \cdot (r_b + \sum_{i=1}^{n} b_i x^{i(n+1)}) - (r_c + \sum_{i=1}^{n} c_i x^i) \cdot (\sum_{i=1}^{n} x^{i(n+1)})$ is the sum of two formal polynomials in x, $F(x) = F_{con}(x) + F_\psi(x)$, where $F_{con}(x) = \sum_{i=1}^{n}(a_i b_i - c_i)x^{i(n+2)}$ is a *constraint polynomial*, spanned by the powers of x from $\Lambda_{con} = \{i(n+2) : i \in [n]\}$, and

$$F_\psi(x) = r_a r_b + r_b \sum_{i=1}^{n} a_i x^i + \sum_{i=1}^{n}(r_a b_i - r_c)x^{i(n+1)} + \sum_{i=1}^{n} \sum_{\substack{j=1 \\ j \neq i}}^{n}(a_i b_j - c_i)x^{i+j(n+1)}$$

is an *argument polynomial*, spanned by the powers of x from $\Lambda_\psi = \{0\} \cup [n] \cup \{i(n+1) : i \in [n]\} \cup \{i + j(n+1) : i, j \in [n] \wedge i \neq j\}$. One coefficient of $F_{con}(x)$ corresponds to one constraint $a_i b_i = c_i$ that the honest prover has to satisfy, and is 0 if this constraint is true. Thus, all coefficients of F_{con} are equal to 0 iff the prover is honest.

By using homomorphic properties of the commitment scheme, the prover constructs the argument $\psi = g^{F_\psi(x)}$ as $\psi = g^{r_a r_b} \cdot \ldots \cdot \prod_{i=1}^{n} \prod_{j=1:i \neq j}^{n} g_{i+j(n+1)}^{a_i b_j - c_i}$. This can be done, since the prover — who knows how to open the commitments but does not know the secret x — knows all coefficients $r_a r_b, \ldots, a_i b_j - c_i$. He also knows the generators $g, \ldots, g_{i+j(n+1)}$ if the $\Theta(n^2)$ generators g_ℓ, for $\ell \in \Lambda_\psi$, are included to the CRS. Thus, the CRS has $\Theta(n^2)$ group elements and the computational complexity of the prover is $\Theta(n^2)$ bilinear-group exponentiations. On the other hand, the verifier's computational complexity is $\Theta(1)$ pairings, since she only has to check Eq. (1).

For the soundness, one needs that when $a_i b_i \neq c_i$ for some $i \in [n]$, then a satisfying ψ cannot be computed from the elements g^{x^ℓ} that are in the CRS; otherwise, a dishonest prover would be able to compute a satisfying argument. This means that for $i \in [n]$, $g^{x^{i(n+2)}}$ should not belong to the CRS. To be certain that this is true, one needs

(a) that g^{x^ℓ} is in the CRS for values $\ell \in \Lambda_\psi$ but if $\ell \in \Lambda_{con}$, then g^{x^ℓ} does not belong to the CRS (elements from $2 \cdot \Lambda \setminus \hat{\Lambda}$ are allowed),

(b) an appropriate security assumption that states that computing g^{F_ψ} for $F_\psi = \sum_{\ell \in \Lambda_\psi} \mu_\ell x^\ell$ is only possible if one knows all values g^{x^ℓ} for $\ell \in \Lambda_\psi$, and

(c) that $\Lambda_{con} \cap \Lambda_\psi = \emptyset$. (This is also a prerequisite for (a).)

One can guarantee (a) by the choice of the CRS. But also (c) is clearly true, since Λ_{con} and Λ_ψ do not intersect.

To finish off the whole argument, one has to define an appropriate security assumption for (b). Since constructing sublinear NIZK arguments is known to be impossible under standard assumptions (see Sect. 2), one of the underlying assumptions is a knowledge assumption (PKE assumption, as in [15], see Sect. 5). The whole argument will become (slightly!) more complex since all commitments and arguments also have to include a knowledge component.

Groth's permutation argument is based on a very similar idea and has basically the same complexities. The only major difference is that if the permutation is a part of the prover's statement, then the verifier also has to perform $\Theta(n)$ bilinear-group multiplications. Since Groth's Circuit-SAT argument consists of a very small (< 10) number of Hadamard product and permutation arguments, then it just inherits the complexities of the basic arguments, as also seen from Tbl. 1, where, in the basic variation, $|C| = n$ and thus the CRS has $\Theta(|C|^2)$ group elements, the argument length is 42 group elements, the prover's computational complexity is $\Theta(|C|^2)$ exponentiations, and the prover's computational complexity is dominated by $\Theta(|C|)$ bilinear-group multiplications.

Groth's Circuit-SAT argument has several sub-optimal properties that are all inherited from the basic arguments. While it has succinct communication and efficient verification, its CRS of $\Theta(|C|^2)$ group elements and prover's computation of $\Theta(|C|^2)$ exponentiations (in the basic variant) seriously limit applicability. Recall that here $n = 2|C| + 1$. A smaller problem is the use of different generators $(g_1, \ldots, g_n)$ and $(g_{n+1}, \ldots, g_{n(n+1)})$ while committing to different elements.

We note that F_{con} has n monomials (1 per every constraint $a_i b_i = c_i$ that a honest prover must satisfy). On the other hand, F_ψ has $\Theta(n^2)$ distinct — since $i_1 + j_1(n+1) \neq i_2 + j_2(n+1)$ if $i_1, j_1, i_2, j_2 \in [n]$ and $(i_1, j_1) \neq (i_2, j_2)$ — monomials. The number of those monomials is the only reason why the CRS has $\Theta(n^2)$ group elements and the prover has to perform $\Theta(n^2)$ bilinear-group exponentiations.

We now show how to collapse many of the unnecessary monomials into one, so that the full argument still remains secure, obtaining a polynomial $F_\psi(x)$ that has only $n^{1+o(1)}$ monomials. First, we generalize the underlying commitment scheme. We still choose a single $x \leftarrow \mathbb{Z}_p$ and set $g_i \leftarrow g^{x^i}$, but we allow the indexes of n generators $(g_{\lambda_1}, \ldots, g_{\lambda_n})$, that are used to commit, to actually depend on the concrete argument — with the main purpose to be able to obtain as small Λ_ψ as possible, while still guaranteeing that $F_{\mathsf{con}} = 0$ iff the prover is honest, and that $\Lambda_{\mathsf{con}} \cap \Lambda_\psi = \emptyset$. Assume that $\Lambda = (\lambda_1, \ldots, \lambda_n)$ is an (n, κ)-nice tuple of integers, so $\lambda_n = \max_i \lambda_i$. Thus,

$$\mathcal{C}om(\boldsymbol{a}; r_a) := g^{r_a} \prod_{i=1}^{n} g_{\lambda_i}^{a_i} = g^{r_a + \sum_{i=1}^{n} a_i x^{\lambda_i}} .$$

The polynomial $r_a + \sum_{i=1}^{n} a_i x^{\lambda_i}$ has degree (up to) λ_n, but it only has (up to) $n + 1$ non-zero monomials. We now start again with the verification equation Eq. (1), but this time we assume that all A, B, C and D have been committed by using the same set of generators $(g_{\lambda_1}, \ldots, g_{\lambda_n})$. Since $F(x) = (r_a + \sum_{i=1}^{n} a_i x^{\lambda_i})(r_b + \sum_{i=1}^{n} b_i x^{\lambda_i}) - (r_c + \sum_{i=1}^{n} c_i x^{\lambda_i})(\sum_{i=1}^{n} x^{\lambda_i})$, we get that $F(x) = F_{\mathsf{con}}(x) + F_\psi(x)$, where

$$F_{\mathsf{con}}(x) = \sum_{i=1}^{n} (a_i b_i - c_i) x^{2\lambda_i} , \tag{2}$$

$$F_\psi(x) = r_a r_b + \sum_{i=1}^{n} (r_a b_i + r_b a_i - r_c) x^{\lambda_i} + \sum_{i=1}^{n} \sum_{\substack{j=1 \\ j \neq i}}^{n} (a_i b_j - c_i) x^{\lambda_i + \lambda_j} . \tag{3}$$

Here, the powers corresponding to nonzero coefficients belong either to the set $\Lambda_{\text{con}} = 2 \cdot \Lambda := \{2\lambda_i : i \in [n]\}$ or to the set $\Lambda_\psi = \hat{\Lambda} := \{0\} \cup \Lambda \cup 2\hat{\Lambda}$, where $2\hat{\Lambda} := \{\lambda_i + \lambda_j : i, j \in [n] \wedge i \neq j\}$.

If the prover is honest (that is, $a_i b_i - c_i = 0$ for all i), then the coefficients $a_i b_i - c_i$ corresponding to the powers in the set $2 \cdot \Lambda$ are equal to 0. Therefore, an honest prover can compute the argument $\psi = g^{F_\psi(x)}$ as $g^{\sum_{\ell \in \hat{\Lambda}} \mu_\ell x^\ell} = \prod_{\ell \in \hat{\Lambda}} (g^{x^\ell})^{\mu_\ell}$, where the coefficients μ_ℓ are known to the prover. This means that all elements $g^{x^\ell}, \ell \in \hat{\Lambda}$, have to belong to the CRS, and thus the CRS contains at least $|\hat{\Lambda}| < 2\lambda_n$ group elements. Recall that in [15], one had to specify $\Theta(n^2)$ elements in the CRS.

For the soundness, we again need (a–c), as in the case of Groth's argument, to be true. One can again guarantee (a) by the choice of the CRS, and one has to define a reasonable security assumption (PKE assumption) for (b). Finally, achieving (c) is also relatively easy. Namely, one can guarantee that $0 \notin 2 \cdot \Lambda$ and $\Lambda \cap 2 \cdot \Lambda = \emptyset$ by choosing Λ to be a set of odd[1] integers. It is almost as easy to guarantee that $2 \cdot \Lambda \cap 2\hat{\Lambda} = \emptyset$ as soon as one rewrites this condition as $2\lambda_k \neq \lambda_i + \lambda_j$ for $i \neq j$, and notices that this is equivalent to requiring that no 3 elements of Λ are in an arithmetic progression. That is, Λ is a progression-free set [25]. Thus, it is sufficient to assume that Λ is a progression-free set of odd integers. .

Recall that the CRS length (and the prover's computational complexity) depend on $|\hat{\Lambda}|$ and thus it is beneficial to have as small $|\hat{\Lambda}| < 2\lambda_n$ possible. This can be guaranteed by upper bounding λ_n, that is, by finding as small λ_n as possible such that $[\lambda_n]$ contains a progression-free subset of odd integers of cardinality n. To bound λ_n, we show in Sect. 4 (following a recent breakthrough of Elkin [9]) that any range $[N] = \{1, \ldots, N\}$ contains a progression-free set of odd integers of size $n = \Theta(N(\log_2 N)^{1/4}/2^{2\sqrt{2\log_2 N}}) = N^{1-o(1)}$, and thus one can assume that $\lambda_n = n^{1+o(1)}$. (One can obtain $\lambda_n = O(n \cdot 2^{2\sqrt{2(2+\log_2 n)}})$ by inverting a weaker version of Elkin's result.) In the full version, we give another proof of this result that, while based on Green and Wolf's exposition [13] of [9], provides more details and is slightly sharper. In particular, Elkin's progression-free set is efficiently constructible.

Groth's permutation argument uses similar ideas for a different choice of A, B, C, and D, and thus also for a different set Λ_ψ. Unfortunately, if we use it with the new generalized commitment scheme (that is, with general Λ), we obtain the guarantee $a_{\varrho(i)} = b_i$ only if Λ is a part of the Moser-de Bruijn sequence [23]. But then $\lambda_n = \Theta(n^2)$ and one ends up with a CRS of $\Theta(n^2)$ group elements. We use the following idea to get the same guarantees when Λ is an arbitrary progression-free set of odd integers. We show that if Λ is a progression-free set of odd integers, then Groth's permutation argument guarantees that $a_{\varrho(i)} = T_\Lambda(i, \varrho) \cdot b_i$, where $T_\Lambda(i, \varrho) \geq 1$ is an easily computable and public integer. We use this result to show that for some separately committed tuple $\boldsymbol{a}^*$, $a^*_{\varrho(i)} = T_\Lambda(i, \varrho) \cdot b_i$ for $i \in [n]$. We then employ an additional product argument to show that $a^*_i = T_\Lambda(\varrho^{-1}(i), \varrho) \cdot a_i$ for $i \in [n]$. Thus, $a_{\varrho(i)} = b_i$ for $i \in [n]$.

We obtain basic arguments that only use $\Theta(\lambda_n) = n^{1+o(1)}$ generators $\{g^{x^\ell} : \ell \in \hat{\Lambda}\}$. This means that the CRS has $n^{1+o(1)}$ group elements and not $\Theta(n^2)$ as in [15]. In both

---

[1] Oddity is not strictly required. For $\Lambda \cap 2 \cdot \Lambda = \emptyset$ to hold, one can take $\Lambda := \{(2i + 1)2^{2j} : i, j \geq 0\}$, see OEIS sequence A003159. Dealing with odd integers is however almost as good.

basic arguments, the prover has to compute ψ (which takes $\Theta(n^2)$ scalar multiplications or additions in $\mathbb{Z}_p$ and $n^{1+o(1)}$ bilinear-group exponentiations). As in [15], the prover's computation can be optimized even further by using efficient multi-exponentiation algorithms. The verifier has to only perform $\Theta(1)$ bilinear pairings. In the case of the permutation argument, she also has to compute $\Theta(n)$ bilinear-group multiplications, though the multiplications can be done offline if the permutation is fixed. Thus, the new basic arguments are considerably more efficient than Groth's.

The soundness of the new product argument is based on two assumptions, a computational assumption ($\hat{\Lambda}$-PSDL, see Sect. 5) and a knowledge assumption (Λ-PKE, see Sect. 5). Groth [15] used $[an^2]$-PKE (for a constant a) and $[an^2]$-CPDH (which is a presumably stronger assumption than PSDL). Since Λ, Λ_ψ are small subsets of $[an^2]$, then our assumptions can be expected to be somewhat weaker in general. Finally, the security reduction in the proof of the product argument takes time $\Theta(t(\lambda_n))$ in our case and $\Theta(t(an^2))$ in Groth's case, where $t(m)$ is the time to factor a degree-m polynomial.

4 Progression-Free Sets

A set of positive integers $\Lambda = \{\lambda_1, \ldots, \lambda_n\}$ is *progression-free* [25], if no three elements of Λ are in an arithmetic progression, that is, $\lambda_i + \lambda_j = 2\lambda_k$ only if $i = j = k$, or equivalently, $2\hat{}\Lambda \cap 2 \cdot \Lambda = \emptyset$.

Let $r_3(N)$ denote the cardinality of the largest progression-free set that belongs to $[N]$. For any $N > 1$, the set of integers in $[N]$ that have no ternary digit equal to 2 is progression-free. If $N = 3^k$, then there are $2^N - 1$ such integers, and thus $r_3(N) = \Omega(N^{\log_3 2}) = \Omega(N^{0.63})$. Clearly, this set can be efficiently constructed. As shown by Behrend in 1946 [3], this idea can be generalized to non-ternary bases, with $r_3(N) = \Omega(N/(2^{2\sqrt{2\log_2 N}} \cdot \log_2^{1/4} N))$. Behrend's result was improved in a recent breakthrough by Elkin [9], who showed that $r_3(N) = \Omega(N \cdot \log_2^{1/4} N/2^{2\sqrt{2\log_2 N}})$. We have included a proof of Elkin's result in the full version. Our proof is closely based on [13] but it has a sharper constant inside Ω. Moreover, our proof is much more detailed than that given in [13]. While both constructions employ the pigeonhole principle, Elkin's methodology can be used to compute his progression-free set in quasi-linear time $N \cdot 2^{O(\sqrt{\log N})}$, see [9]. On the other hand, Bourgain [5] showed that $r_3(N) = O(N \cdot (\log N/\log\log N)^{1/2})$, and recently Sanders [24] showed that $r_3(N) = O(N \cdot (\log\log N)^5/\log N)$. Thus, according to Behrend and Elkin, the minimal N such that $r_3(N) = n$ is $N = n^{1+o(1)}$, while according to Sanders, $N = \omega(n)$.

We need the progression-free subset to also consist of odd integers. For this, one can take Elkin's set $\Lambda = \{\lambda_1, \ldots, \lambda_n\} \subset [N]$, and then use the set $2 \cdot \Lambda + 1 = \{2\lambda_1 + 1, \ldots, 2\lambda_n + 1\}$. Clearly, if $\Lambda \in [n^{1+o(1)}]$ then also $2 \cdot \Lambda + 1 \in [n^{1+o(1)}]$.

Theorem 1. *Let $r_3^{odd}(N)$ be the size of the largest progression-free set in $[N]$ that only consists of odd integers. For any n, there exists $N = n^{1+o(1)}$, such that $r_3^{odd}(N) = n$.*

5 Cryptographic Tools

In this section, we generalize the PKE assumption from [15] and then define two new cryptographic assumptions, PDL and PSDL, and prove that PSDL is secure in the generic group model. After that, we proceed to describe a generalization of Groth's knowledge commitment scheme from [15] and prove that it is computationally binding under the PDL assumption. Groth proved in [15] that his commitment scheme is computationally binding under the (potentially stronger) CPDH assumption.

Λ-Power (Symmetric) Discrete Logarithm Assumption. Let Λ be an (n, κ)-nice tuple for some $n = \mathrm{poly}(\kappa)$. We say that a bilinear group generator $\mathcal{G}_{\mathsf{bp}}$ is (n, κ)-*PDL secure in group* $\mathbb{G}_t$ for $t \in \{1, 2\}$, if for any non-uniform PPT adversary $\mathcal{A}$, $\Pr[\mathsf{gk} := (p, \mathbb{G}_1, \mathbb{G}_2, \mathbb{G}_T, \hat{e}) \leftarrow \mathcal{G}_{\mathsf{bp}}(1^\kappa), g_t \leftarrow \mathbb{G}_t \setminus \{1\}, x \leftarrow \mathbb{Z}_p : \mathcal{A}(\mathsf{gk}; (g_t^{x^\ell})_{\ell \in \{0\} \cup \Lambda}) = x]$ is negligible in κ. Similarly, we say that a bilinear group generator $\mathcal{G}_{\mathsf{bp}}$ is Λ-*PSDL secure*, if for any non-uniform PPT adversary $\mathcal{A}$,

$$\Pr \left[\begin{array}{l} \mathsf{gk} := (p, \mathbb{G}_1, \mathbb{G}_2, \mathbb{G}_T, \hat{e}) \leftarrow \mathcal{G}_{\mathsf{bp}}(1^\kappa), g_1 \leftarrow \mathbb{G}_1 \setminus \{1\}, \\ g_2 \leftarrow \mathbb{G}_2 \setminus \{1\}, x \leftarrow \mathbb{Z}_p : \mathcal{A}(\mathsf{gk}; (g_1^{x^\ell}, g_2^{x^\ell})_{\ell \in \{0\} \cup \Lambda}) = x \end{array} \right]$$

is negligible in κ. A version of P(S)DL assumption in a non pairing-based group was defined in [12]. Cheon showed in [8] that if n is a prime divisor of $p - 1$ or $p + 1$, then the $[n]$-PDL assumption can be broken by a generic adversary in $O((\sqrt{p/n} + \sqrt{n}) \log p)$ group operations. Clearly, if the Λ-PSDL assumption is hard, then the Λ-PDL assumption is hard in both $\mathbb{G}_1$ and $\mathbb{G}_2$. Moreover, if the bilinear group generator is CPDH secure, then it is also P(S)DL secure. Therefore, by the results of [15], P(S)DL holds in the generic group model.

Theorem 2. *The Λ-PSDL assumption holds in the generic group model for any (n, κ)-nice tuple Λ given that $n = \mathrm{poly}(\kappa)$. Any successful generic adversary for Λ-PSDL requires time $\Omega(\sqrt{p/\lambda_n})$ where λ_n is the largest element of Λ.*

Λ-Power Knowledge of Exponent Assumption (Λ-PKE). Abe and Fehr showed in [1] that no statistically zero-knowledge non-interactive argument for an **NP**-complete language can have a "direct black-box" security reduction to a standard cryptographic assumption unless **NP** $\subseteq$ **P**/poly. (See also [11].) In fact, the soundness of NIZK arguments (for example, of an argument that a perfectly hiding commitment scheme commits to 0) is often unfalsifiable by itself. Similarly to Groth [15], we will base our NIZK argument for circuit satisfiability on Λ-PKE, an explicit knowledge assumption. This assumption was proposed by Groth [15] (though only for $\Lambda = [n]$).

Let $t \in \{1, 2\}$. For two algorithms $\mathcal{A}$ and $X_\mathcal{A}$, we write $(y; z) \leftarrow (\mathcal{A} \| X_\mathcal{A})(x)$ if $\mathcal{A}$ on input x outputs y, and $X_\mathcal{A}$ on the same input (including the random tape of $\mathcal{A}$) outputs z. Let Λ be an (n, κ)-nice tuple for some $n = \mathrm{poly}(\kappa)$. The bilinear group generator $\mathcal{G}_{\mathsf{bp}}$ is Λ-*PKE secure in group* $\mathbb{G}_t$ if for any non-uniform PPT adversary $\mathcal{A}$ there exists a non-uniform PPT extractor $X_\mathcal{A}$, such that

$$\Pr \left[\begin{array}{l} \mathsf{gk} := (p, \mathbb{G}_1, \mathbb{G}_2, \mathbb{G}_T, \hat{e}) \leftarrow \mathcal{G}_{\mathsf{bp}}(1^\kappa), g_t \leftarrow \mathbb{G}_t \setminus \{1\}, (\hat{\alpha}, x) \leftarrow \mathbb{Z}_p^2, \\ \mathsf{crs} \leftarrow (\mathsf{gk}; (g_t^{x^\ell}, g_t^{\hat{\alpha} x^\ell})_{\ell \in \{0\} \cup \Lambda}), (c, \hat{c}; r, (a_\ell)_{\ell \in \Lambda}) \leftarrow (\mathcal{A} \| X_\mathcal{A})(\mathsf{crs}) : \\ \hat{c} = c^{\hat{\alpha}} \wedge c \neq g_t^r \cdot \prod_{\ell \in \Lambda} g_t^{a_\ell x^\ell} \end{array} \right]$$

is negligible in κ. That is, if $\mathcal{A}$ (given access to crs that for a random $\hat{\alpha}$ contains both $g_t^{x^\ell}$ and $g_t^{\hat{\alpha}x^\ell}$ iff $\ell \in \{0\} \cup \Lambda$) can produce c and $\hat{c}$ such that $\hat{c} = c^{\hat{\alpha}}$, then $X_{\mathcal{A}}$ (given access to crs and to the random coins of $\mathcal{A}$) can produce a tuple $(r, (a_\ell)_{\ell \in \Lambda})$ such that $c = g_t^r \cdot \prod_{\ell \in \Lambda} g_t^{a_\ell x^\ell}$. Groth [15] proved that the $[n]$-PKE assumption holds in the generic group model; his proof can be straightforwardly modified to the general case.

New Commitment Scheme. We use the following variant of the *knowledge commitment scheme* from [15] with a generalized choice of generators, defined as follows:

CRS generation: Let Λ be an (n, κ)-nice tuple with $n = \text{poly}(\kappa)$. Define $\lambda_0 = 0$. Given a bilinear group generator $\mathcal{G}_{\text{bp}}$, set $\text{gk} := (p, \mathbb{G}_1, \mathbb{G}_2, \mathbb{G}_T, \hat{e}) \leftarrow \mathcal{G}_{\text{bp}}(1^\kappa)$. Let $g_1 \leftarrow \mathbb{G}_1 \setminus \{1\}$, $g_2 \leftarrow \mathbb{G}_2 \setminus \{1\}$, and $\hat{\alpha}, x \leftarrow \mathbb{Z}_p$. Let $t \in \{1, 2\}$. The CRS is $\text{ck}_t \leftarrow (\text{gk}; (g_{t, \lambda_i}, \hat{g}_{t, \lambda_i})_{i \in \{0, \dots, n\}})$, where $g_{t\ell} = g_t^{x^\ell}$ and $\hat{g}_{t\ell} = g_t^{\hat{\alpha}x^\ell}$.

Commitment: To commit to $a = (a_1, \dots, a_n) \in \mathbb{Z}_p^n$, the committing party chooses a random $r \leftarrow \mathbb{Z}_p$, and defines

$$\mathcal{C}om^t(\text{ck}_t; a; r) := (g_t^r \cdot \prod_{i=1}^n g_{t, \lambda_i}^{a_i}, \hat{g}_t^r \cdot \prod_{i=1}^n \hat{g}_{t, \lambda_i}^{a_i}) .$$

Importantly, we allow Λ to depend on the concrete application. Let $t = 1$. Fix a commitment key ck_1 that in particular specifies $g_2, \hat{g}_2 \in \mathbb{G}_2$. A commitment $(A, \hat{A}) \in \mathbb{G}_1^2$ is *valid* if $\hat{e}(A, \hat{g}_2) = \hat{e}(\hat{A}, g_2)$. The case $t = 2$ is dual.

Theorem 3. *Let $t \in \{1, 2\}$. The knowledge commitment scheme is perfectly hiding in $\mathbb{G}_t$, and computationally binding in $\mathbb{G}_t$ under the Λ-PDL assumption in $\mathbb{G}_t$. If the Λ-PKE assumption holds in $\mathbb{G}_t$, then for any non-uniform PPT $\mathcal{A}$ that outputs some valid knowledge commitments, there exists a non-uniform PPT extractor $X_{\mathcal{A}}$ that, given the input of $\mathcal{A}$ together with $\mathcal{A}$'s random coins, extracts the contents of these commitments.*

In the case of all security reductions in this paper, the tightness of the security reduction depends on the value λ_n. Clearly, the knowledge commitment scheme is also trapdoor, with the trapdoor being $\text{td} = x$: after trapdoor-committing $A \leftarrow \mathcal{C}om^t(\text{ck}; 0; r) = g_t^r$ for $r \leftarrow \mathbb{Z}_p$, the committer can open it to $(a; r - \sum_{i=1}^n a_i x^{\lambda_i})$ for any a.

6 New Hadamard Product Argument

Assume that $(\mathcal{G}_{\text{com}}, \mathcal{C}om)$ is the knowledge commitment scheme. In an *Hadamard product argument* (in group $\mathbb{G}_1$, the case of $\mathbb{G}_2$ is dual), the prover aims to convince the verifier that given commitments A, B and C, he can open them as $A = \mathcal{C}om^1(\text{ck}; a; r_a)$, $B = \mathcal{C}om^1(\text{ck}; b; r_b)$, and $C = \mathcal{C}om^1(\text{ck}; c; r_c)$, s.t. $c_j = a_j b_j$ for $j \in [n]$. Groth constructed an Hadamard product argument [15] with communication of 5 group elements, verifier's computation $\Theta(n)$, prover's computation of $\Theta(n^2)$ exponentiations and the CRS of $\Theta(n^2)$ group elements. We present a more efficient argument in Prot. 1. Intuitively, the discrete logarithm on basis $h = \hat{e}(g_1, g_2)$ of $\hat{e}(A, B_2)/\hat{e}(C, D) = \hat{e}(g_1, \psi)$ is a degree-n formal polynomial in X, which is spanned by $\{X^\ell\}_{\ell \in 2 \cdot \Lambda \cup \hat{\Lambda}}$, where

$$\hat{\Lambda} := \{0\} \cup \Lambda \cup 2\hat{\ }\Lambda . \tag{4}$$

We need that $2 \cdot \Lambda$ and $\hat{\Lambda}$ do not intersect. The next lemma is straightforward to prove.

System parameters: Let $n = \text{poly}(\kappa)$. Let $\Lambda = \{\lambda_i : i \in [n]\}$ be a progression-free set of odd integers, such that $\lambda_{i+1} > \lambda_i > 0$. Denote $\lambda_0 := 0$. Let $\hat{\Lambda}$ be as in Eq. (4).

CRS generation $\mathcal{G}_{\text{crs}}(1^\kappa)$: Let $\text{gk} := (p, \mathbb{G}_1, \mathbb{G}_2, \mathbb{G}_T, \hat{e}) \leftarrow \mathcal{G}_{\text{bp}}(1^\kappa)$. Let $\hat{\alpha}, x \leftarrow \mathbb{Z}_p$. Let $g_1 \leftarrow \mathbb{G}_1 \setminus \{1\}$ and $g_2 \leftarrow \mathbb{G}_2 \setminus \{1\}$. Denote $g_{t\ell} \leftarrow g_t^{x^\ell}$ and $\hat{g}_{t\ell} \leftarrow g_t^{\hat{\alpha}x^\ell}$ for $t \in \{1,2\}$ and $\ell \in \{0\} \cup \hat{\Lambda}$. Let $D \leftarrow \prod_{i=1}^{n} g_{2,\lambda_i}$. The CRS is $\text{crs} \leftarrow (\text{gk}; (g_{1\ell}, \hat{g}_{1\ell})_{\ell \in \{0\} \cup \Lambda}, (g_{2\ell}, \hat{g}_{2\ell})_{\ell \in \hat{\Lambda}}, D)$. Let $\hat{\text{ck}}_1 \leftarrow (\text{gk}; (g_{1\ell}, \hat{g}_{1\ell})_{\ell \in \{0\} \cup \Lambda})$.

Common inputs: $(A, \hat{A}, B, \hat{B}, B_2, C, \hat{C})$, where $(A, \hat{A}) \leftarrow \mathcal{C}om^1(\hat{\text{ck}}_1; \boldsymbol{a}; r_a)$, $(B, \hat{B}) \leftarrow \mathcal{C}om^1(\hat{\text{ck}}_1; \boldsymbol{b}; r_b)$, $B_2 \leftarrow g_2^{r_b} \cdot \prod_{i=1}^{n} g_{2,\lambda_i}^{b_i}$, $(C, \hat{C}) \leftarrow \mathcal{C}om^1(\hat{\text{ck}}_1; \boldsymbol{c}; r_c)$, s.t. $a_i b_i = c_i$ for $i \in [n]$.

Argument generation $\mathcal{P}_\times(\text{crs}; (A, \hat{A}, B, \hat{B}, B_2, C, \hat{C}), (\boldsymbol{a}, r_a, \boldsymbol{b}, r_b, \boldsymbol{c}, r_c))$: Let $I_1(\ell) := \{(i,j) : i,j \in [n] \land j \neq i \land \lambda_i + \lambda_j = \ell\}$. For $\ell \in 2^\frown\Lambda$, the prover sets $\mu_\ell \leftarrow \sum_{(i,j) \in I_1(\ell)} (a_i b_j - c_i)$. He sets $\psi \leftarrow g_2^{r_a r_b} \cdot \prod_{i=1}^{n} g_{2,\lambda_i}^{r_a b_i + r_b a_i - r_c} \cdot \prod_{\ell \in 2^\frown\Lambda} g_{2\ell}^{\mu_\ell}$, and $\hat{\psi} \leftarrow \hat{g}_2^{r_a r_b} \cdot \prod_{i=1}^{n} \hat{g}_{2,\lambda_i}^{r_a b_i + r_b a_i - r_c} \cdot \prod_{\ell \in 2^\frown\Lambda} \hat{g}_{2\ell}^{\mu_\ell}$. He sends $\psi^\times \leftarrow (\psi, \hat{\psi}) \in \mathbb{G}_2^2$ to the verifier as the argument.

Verification $\mathcal{V}_\times(\text{crs}; (A, \hat{A}, B, \hat{B}, B_2, C, \hat{C}), \psi^\times)$: accept iff $\hat{e}(A, B_2)/\hat{e}(C, D) = \hat{e}(g_1, \psi)$ and $\hat{e}(g_1, \hat{\psi}) = \hat{e}(\hat{g}_1, \psi)$.

Protocol 1: New Hadamard product argument $[\![(A, \hat{A})]\!] \circ [\![(B, \hat{B}, B_2)]\!] = [\![(C, \hat{C})]\!]$

Lemma 1. *1) If Λ is a progression-free set of odd integers, then $2 \cdot \Lambda \cap \hat{\Lambda} = \emptyset$. 2) If $2 \cdot \Lambda \cap \hat{\Lambda} = \emptyset$, then Λ is a progression-free set.*

Moreover, since $\hat{\Lambda} \in \{0, \ldots, 2\lambda_n\}$, then by Thm. 1,

Lemma 2. *For any value n there exists a choice of Λ such that $|\hat{\Lambda}| = n^{1+o(1)}$.*

We are now ready to state the security of the new Hadamard product argument for the knowledge commitment scheme. The (knowledge) commitments are $(A, \hat{A})$, $(B, \hat{B})$ and $(C, \hat{C})$. For efficiency reasons, we include another element B_2 to the Hadamard product language. We denote the argument in Prot. 1 by $[\![(A, \hat{A})]\!] \circ [\![(B, \hat{B}, B_2)]\!] = [\![(C, \hat{C})]\!]$. Since $(C, \hat{C})$ is always a commitment of $(a_1 b_1, \ldots, a_n b_n)$ for *some* value of r_c, we cannot claim that Prot. 1 is computationally sound (even under a knowledge assumption). Instead, analogously to [15], we prove a somewhat weaker version of soundness that is however sufficient to achieve soundness of the Circuit-SAT argument. Note that the last statement of the theorem basically says that no efficient adversary can output an input to the Hadamard product argument together with an accepting argument and openings to all commitments and all other pairs of type $(y, \hat{y})$ that are present in the argument, such that $a_i b_i \neq c_i$ for some $i \in [n]$. Intuitively, the theorem statement includes f'_ℓ only for $\ell \in \hat{\Lambda}$ (resp., a_ℓ for $\ell \in \Lambda$ together with r) since $\hat{g}_{2\ell}$ (resp., $\hat{g}_{1\ell}$) belongs to the CRS only for $\ell \in \hat{\Lambda}$ (resp., $\ell \in \{0\} \cup \Lambda$).

Theorem 4. *Prot. 1 is perfectly complete and perfectly witness-indistinguishable. If $\mathcal{G}_{\text{bp}}$ is $\hat{\Lambda}$-PSDL secure, then a non-uniform PPT adversary has negligible chance of outputting $\text{inp}^\times \leftarrow (A, \hat{A}, B, \hat{B}, B_2, C, \hat{C})$ and an accepting argument $\psi^\times \leftarrow (\psi, \hat{\psi})$ together with a witness $w^\times \leftarrow (\boldsymbol{a}, r_a, \boldsymbol{b}, r_b, \boldsymbol{c}, r_c, (f'_\ell)_{\ell \in \hat{\Lambda}})$, s.t. $(A, \hat{A}) = \mathcal{C}om^1(\hat{\text{ck}}_1; \boldsymbol{a}; r_a)$, $(B, \hat{B}) = \mathcal{C}om^1(\hat{\text{ck}}_1; \boldsymbol{b}; r_b)$, $B_2 = g_2^{r_b} \cdot \prod_{i=1}^{n} g_{2,\lambda_i}^{b_i}$, $(C, \hat{C}) = \mathcal{C}om^1(\hat{\text{ck}}_1; \boldsymbol{c}; r_c)$, $(\psi, \hat{\psi}) = (g_2^{\sum_{\ell \in \hat{\Lambda}} f'_\ell x^\ell}, \hat{g}_2^{\sum_{\ell \in \hat{\Lambda}} f'_\ell x^\ell})$, and for some $i \in [n]$, $a_i b_i \neq c_i$.*

The commitment scheme is defined as in Sect. 5 with respect to the set Λ. The following proof will make the intuition of Sect. 3 more formal. Note that the tightness of the reduction depends on the time it takes to factor a degree $(2\lambda_n + 1)$-polynomial.

Proof. Let $h \leftarrow \hat{e}(g_1, g_2)$ and $F(x) \leftarrow \log_h(\hat{e}(A, B_2)/\hat{e}(C, D))$ like in Sect. 3. WITNESS-INDISTINGUISHABILITY: since the argument $\psi^\times = (\psi, \hat{\psi})$ that satisfies the verification equations is unique, all witnesses result in the same argument, and therefore the Hadamard product argument is witness-indistinguishable.

PERFECT COMPLETENESS. Assume that the prover is honest. The second verification is straightforward. For the first one, due to discussion in Sect. 3, $F(x) = F_{\mathsf{con}}(x) + F_\psi(x)$, where $F_{\mathsf{con}}(x)$ and $F_\psi(x)$ are as defined by Eq. (2) and Eq. (3). Consider x to be a formal variable, then $F(X)$ is a formal polynomial of X. This formal polynomial is spanned by $\{X^\ell\}_{\ell \in 2 \cdot \Lambda \cup \hat{\Lambda}}$. If the prover is honest, then $c_i = a_i \cdot b_i$ for $i \in [n]$, and thus $F(X) = F_\psi(X)$ is spanned by $\{X^\ell\}_{\ell \in \hat{\Lambda}}$. Denoting $\psi \leftarrow g_2^{r_a r_b} \cdot \prod_{i=1}^n g_{2,\lambda_i}^{r_a b_i + r_b a_i - r_c} \cdot \prod_{i=1}^n \prod_{j=1: j \neq i}^n g_{2,\lambda_i + \lambda_j}^{a_i b_j - c_i} = g_2^{r_a r_b} \cdot \prod_{i=1}^n g_{2,\lambda_i}^{r_a b_i + r_b a_i - r_c} \cdot \prod_{\ell \in 2^\frown \Lambda} g_{2\ell}^{\mu_\ell}$, we see that clearly $e(g_1, \psi) = h$. Thus, the first verification succeeds.

WEAKER VERSION OF SOUNDNESS. Assume that $\mathcal{A}$ is an adversary that can break the last statement of the theorem. We construct an adversary $\mathcal{A}'$ against the $\hat{\Lambda}$-PSDL assumption. Let $\mathsf{gk} \leftarrow \mathcal{G}_{\mathsf{bp}}(1^\kappa)$, $x \leftarrow \mathbb{Z}_p$, $g_1 \leftarrow \mathbb{G}_1 \setminus \{1\}$, and $g_2 \leftarrow \mathbb{G}_2 \setminus \{1\}$. The adversary $\mathcal{A}'$ receives $\mathsf{crs} \leftarrow (\mathsf{gk}; (g_1^{x^\ell}, g_2^{x^\ell})_{\ell \in \hat{\Lambda}})$ as her input, and her task is to output x. She sets $\hat{\alpha} \leftarrow \mathbb{Z}_p$, $\mathsf{crs}' \leftarrow (\mathsf{gk}; (g_1^{x^\ell}, g_1^{\hat{\alpha} x^\ell})_{\ell \in \{0\} \cup \Lambda}, (g_2^{x^\ell}, g_2^{\hat{\alpha} x^\ell})_{\ell \in \hat{\Lambda}}, \prod_{i=1}^n g_2^{x^{\lambda_i}})$, and then sends crs' to $\mathcal{A}$. Clearly, crs' has the same distribution as $\mathcal{G}_{\mathsf{crs}}(1^\kappa)$. Both $\mathcal{A}$ and $\mathcal{A}'$ set $\mathsf{ck}_t \leftarrow (\mathsf{gk}; (g_t^{x^\ell}, g_t^{\hat{\alpha} x^\ell})_{\ell \in \{0\} \cup \Lambda})$ for $t \in \{1, 2\}$. Assume that $\mathcal{A}$ returns $(inp^\times, w^\times, \psi^\times)$ such that the conditions in the theorem statement hold, and $\mathcal{V}(\mathsf{crs}'; inp^\times, \psi^\times)$ accepts. Here, $inp^\times = (A, \hat{A}, B, \hat{B}, B_2, C, \hat{C})$ and $w^\times = (\boldsymbol{a}, r_a, \boldsymbol{b}, r_b, \boldsymbol{c}, r_c, (f'_\ell)_{\ell \in \hat{\Lambda}})$.

If $\mathcal{A}$ is successful, $(A, \hat{A}) = \mathcal{C}om^1(\widehat{\mathsf{ck}}_1; \boldsymbol{a}; r_a)$, $(B, \hat{B}) = \mathcal{C}om^1(\widehat{\mathsf{ck}}_1; \boldsymbol{b}; r_b)$, $B_2 = g_2^{r_b} \cdot \prod_{i=1}^n g_{2,\lambda_i}^{b_i}$, $(C, \hat{C}) = \mathcal{C}om^1(\widehat{\mathsf{ck}}_1; \boldsymbol{c}; r_c)$, and for some $i \in [n]$, $c_i \neq a_i b_i$. Since $2 \cdot \Lambda \cap \hat{\Lambda} = \emptyset$, $\mathcal{A}$ has thus expressed $F(X)$ as a polynomial $f(X)$ where at least for some $\ell \in 2 \cdot \Lambda$, X^ℓ has a non-zero coefficient $a_i b_i - c_i$.

On the other hand, $\mathcal{A}$ also outputs $(f'_\ell)_{\ell \in \hat{\Lambda}}$, s.t. $F(x) = \log_{g_2} \psi = f'(x)$, where all non-zero coefficients of $f'(X) := \sum_{\ell \in \hat{\Lambda}} f'_\ell X^\ell$ correspond to X^ℓ for some $\ell \in \hat{\Lambda}$. Since Λ is a progression-free set of odd integers and all elements of $2 \cdot \Lambda$ are distinct, then by Lem. 1, $\ell \notin 2 \cdot \Lambda$. Thus, all coefficients of $f'(X)$ corresponding to any X^ℓ, $\ell \in 2 \cdot \Lambda$, are equal to 0. Thus $f(X) = \sum_{\ell \in \hat{\Lambda} \cup (2 \cdot \Lambda)} f_\ell X^\ell$ and $f'(X) = \sum_{\ell \in \hat{\Lambda}} f'_\ell X^\ell$ are different polynomials with $f(x) = f'(x) = F(x)$. Thus, $\mathcal{A}'$ has succeeded in creating a non-zero polynomial $d(X) = f(X) - f'(X)$, such that $d(x) = \sum_{\ell \in \hat{\Lambda} \cup (2 \cdot \Lambda)} d_\ell x^\ell = 0$.

Next, $\mathcal{A}'$ uses an efficient polynomial factorization algorithm [19] in $\mathbb{Z}_p[X]$ to efficiently compute all $< 2\lambda_n + 1$ roots of $d(X)$. For some root y, $g_1^{x^\ell} = g_1^{y^\ell}$. The adversary $\mathcal{A}'$ sets $x \leftarrow y$, thus violating the $\hat{\Lambda}$-PSDL assumption. $\qquad \square$

The Hadamard product argument is not perfectly zero-knowledge. The problem is that the simulator knows $\mathsf{td} = (\hat{\alpha}, x)$, but given td and the common input she will not be able to generate $\psi^\times$. E.g., she has to compute $\psi = g_2^{r_a r_b} \cdot \prod_{i=1}^n g_{2,\lambda_i}^{r_a b_i + r_b a_i - r_c} \cdot \prod_{i=1}^n \prod_{j=1}^n g_{2,\lambda_i + \lambda_j}^{a_i b_j - c_i}$ based on the input, $\hat{\alpha}$ and x, but without knowing the witness.

This seems to be impossible. Technically, the problem is that due to the knowledge of the trapdoor, the simulator can, knowing one opening (a, r), produce an opening (a', r') to any other a'. However, here she does not know any openings. Similarly, the permutation argument of Sect. 7 is not zero-knowledge. On the other hand, in the final circuit satisfiability argument of Sect. 8, the simulator creates all commitments by herself and can thus properly simulate the argument. By the same reason, the subarguments of [15] are not zero-knowledge but the final argument (for circuit satisfiability) is.

Let Λ be as described in Thm. 1. The communication (argument size) of Prot. 1 is 2 elements from $\mathbb{G}_2$. The prover's computational complexity is $\Theta(n^2)$ scalar multiplications in $\mathbb{Z}_p$ and $n^{1+o(1)}$ exponentiations in $\mathbb{G}_2$. The verifier's computational complexity is dominated by 5 bilinear pairings and 1 bilinear-group multiplication. The CRS consists of $n^{1+o(1)}$ group elements, with the verifier's part of the CRS consisting of only the bilinear group description plus 5 group elements.

In the Circuit-SAT argument, all a_i, b_i and c_i are Boolean, and thus all $n^{1+o(1)}$ values μ_ℓ can be computed in $n(n-1) = \Theta(n^2)$ scalar additions (the server also needs to use other operations like comparisons $j \neq i$, but they can be eliminated by using loop unrolling, and λ_i and λ_j can be computed by using table lookups), as follows:

1. For $\ell \in 2\hat{}\Lambda$ do: $\mu_\ell \leftarrow 0$
2. For $i = 1$ to n do:
   - If $a_i = 0$ then for $j = 1$ to n do: if $j \neq i$ then $\mu_{\lambda_i + \lambda_j} \leftarrow \mu_{\lambda_i + \lambda_j} - c_i$
   - Else for $j = 1$ to n do: if $j \neq i$ then $\mu_{\lambda_i + \lambda_j} \leftarrow \mu_{\lambda_i + \lambda_j} + b_j - c_i$

7   New Permutation Argument

In a *permutation argument*, the prover aims to convince the verifier that for given permutation $\varrho \in S_n$ and two commitments A and B, he knows how to open them as $A = \mathcal{C}om^1(\mathsf{ck}; a; r_a)$ and $B = \mathcal{C}om^1(\mathsf{ck}; b; r_b)$, such that $b_j = a_{\varrho(j)}$ for $j \in [n]$. We assume that ϱ is a part of the statement. In [15], Groth constructed a permutation argument, where the prover's computation is $\Theta(n^2)$ exponentiations and the CRS has $\Theta(n^2)$ group elements. We now propose a new argument with the CRS of $n^{1+o(1)}$ group elements. We also improve the prover's concrete computation, and the argument is based on a (probably) weaker assumption.

The new permutation argument $\varrho([[(A, \tilde{A})]]) = [[(B, \tilde{B})]]$, see Prot. 2, uses (almost) the same high-level ideas as the Hadamard product argument from Sect. 6. However, the situation is more complicated. Consider the verification equation $\hat{e}(g_1, \psi^\varrho) = \hat{e}(A, g_2^{\sum_{i=1}^n x^{\lambda_i}}) / \hat{e}(B, g_2^{\sum_{i=1}^n x^{2\lambda_{\varrho_i} - \lambda_i}})$ from [15]. Letting $h = \hat{e}(g_1, g_2)$, $F_\varrho(x) := \log_{g_2} \psi^\varrho = \sum_i (a_{\varrho(i)} - b_i) x^{2\lambda_{\varrho(i)}} + r_a \sum_i x^{\lambda_i} - r_b \sum_i x^{2\lambda_{\varrho(i)} - \lambda_i} + \sum_i a_{\varrho(i)} \cdot \sum_{j \neq i} x^{\lambda_{\varrho(i)} + \lambda_{\varrho(j)}} - \sum_i b_i \cdot \sum_{j \neq i}^n x^{\lambda_i + 2\lambda_{\varrho(j)} - \lambda_j}$. Following Sect. 6, we require that $\tilde{\Lambda} = \Lambda \cup \{2\lambda_k - \lambda_i\} \cup 2\hat{}\Lambda \cup \{\lambda_i + 2\lambda_k - \lambda_j : i \neq j\}$ and $2 \cdot \Lambda$ do not intersect. Since ϱ is a part of the statement, we replaced $\varrho(i)$ and $\varrho(j)$ with a new element k.

Assume that Λ is a progression-free set of odd integers. Since Λ consists of odd integers, $(\Lambda \cup \{2\lambda_k - \lambda_i\}) \cap 2 \cdot \Lambda = \emptyset$. Since Λ is a progression-free set, $2\hat{}\Lambda \cap 2 \cdot \Lambda = \emptyset$. However, we also need that $2\lambda_{k^*} \neq 2\lambda_k + \lambda_i - \lambda_j$ for $i \neq j$. That is, one can uniquely represent any non-negative integer a as $a = 2\lambda_{k^*} + \lambda_j$. (It is only required that any non-negative integer a has at most one representation as $a = 2\lambda_{k^*} + \lambda_j$. See the full version.) The unique sequence $\Lambda = (\lambda_i)_{i \in \mathbb{Z}^+}$ (the *Moser-de Bruijn sequence* [23]) that satisfies this property is the sequence of all non-negative integers that have only 0 or 1 as their radix-4 digits. Since $\lambda_n = \Theta(n^2)$, this sequence is not good enough.

Fortunately, we can overcome this problem as follows. For $i \in [n]$ and a permutation ϱ, let $T_\Lambda(i, \varrho) := |\{j \in [n] : 2\lambda_{\varrho(i)} + \lambda_j = 2\lambda_{\varrho(j)} + \lambda_i\}|$. Note that $1 \leq T_\Lambda(i, \varrho) \leq n$, and that for fixed Λ and ϱ, the whole tuple $\boldsymbol{T_\Lambda(\varrho)} := (T_\Lambda(1, \varrho), \ldots, T_\Lambda(n, \varrho))$ can be computed in $\Theta(n)$ simple arithmetic operations. We can then rewrite $F_\varrho(x)$ as

$$F_\varrho(x) = \sum_{i=1}^{n}(a_{\varrho(i)} - T_\Lambda(i, \varrho) \cdot b_i)x^{2\lambda_{\varrho(i)}} + r_a \sum_{i=1}^{n} x^{\lambda_i} - r_b \sum_{i=1}^{n} x^{2\lambda_{\varrho(i)} - \lambda_i} +$$

$$\sum_{i=1}^{n} a_{\varrho(i)} \sum_{\substack{j=1 \\ j \neq i}}^{n} x^{\lambda_{\varrho(i)} + \lambda_{\varrho(j)}} - \sum_{i=1}^{n} b_i \sum_{\substack{j=1 \\ j \neq i \\ 2\lambda_{\varrho(i)} + \lambda_j \neq \lambda_i + 2\lambda_{\varrho(j)}}}^{n} x^{\lambda_i + 2\lambda_{\varrho(j)} - \lambda_j}, \quad (5)$$

with $\tilde{\Lambda}$ being redefined as

$$\tilde{\Lambda} = \Lambda \cup \{2\lambda_k - \lambda_i\} \cup 2\widehat{\ } \Lambda \cup (\{\lambda_i + 2\lambda_k - \lambda_j : i \neq j\} \setminus 2 \cdot \Lambda). \quad (6)$$

Since $\tilde{\Lambda} \cap 2 \cdot \Lambda = \emptyset$, $\hat{e}(A, D)/\hat{e}(B, E_\varrho) = \hat{e}(g_1, \psi^\varrho)$ convinces the verifier that $a_{\varrho(i)} = T_\Lambda(i, \varrho) \cdot b_i$ for $i \in [n]$. To finish the permutation argument, we let $(A^*, \hat{A}^*)$ to be a commitment to $(a_1^*, \ldots, a_n^*) := (T_\Lambda(\varrho^{-1}(1), \varrho) \cdot a_1, \ldots, T_\Lambda(\varrho^{-1}(n), \varrho) \cdot a_n)$, use an Hadamard product argument to show that $a_i^* = T_\Lambda(\varrho^{-1}(i), \varrho) \cdot a_i$ (and thus $a_{\varrho(i)}^* = T_\Lambda(i, \varrho) \cdot a_{\varrho(i)}$) for $i \in [n]$, and an argument as described above in this section to show that $a_{\varrho(i)}^* = T_\Lambda(i, \varrho) \cdot b_i$ for $i \in [n]$. Therefore, $a_{\varrho(i)} = b_i$ for $i \in [n]$.

Clearly $\hat{\Lambda} \cup \tilde{\Lambda} = \{0\} \cup \tilde{\Lambda}$. Since $\tilde{\Lambda} \subset \{-\lambda_n + 1, \ldots, 3\lambda_n\}$, then by Thm. 1

Lemma 3. *For any n there exists a choice of Λ such that $|\tilde{\Lambda}| = n^{1+o(1)}$.*

We are now ready to state the security of the new permutation argument. The (weaker version of) soundness of this argument is based on exactly the same ideas as that of the Hadamard product argument.

Theorem 5. *Prot. 2 is perfectly complete and perfectly witness-indistinguishable. If $\mathcal{G}_{bp}$ is $\tilde{\Lambda}$-PSDL secure, then a non-uniform PPT adversary has negligible chance of outputting $inp^{perm} \leftarrow (A, \tilde{A}, B, \hat{B}, \tilde{B}, \varrho)$ and an accepting $\psi^{perm} \leftarrow (A^*, \hat{A}^*, \psi^\times, \hat{\psi}^\times, \psi^\varrho, \hat{\psi}^\varrho)$ together with a witness $w^{perm} \leftarrow (\boldsymbol{a}, r_a, \boldsymbol{b}, r_b, \boldsymbol{a}^*, r_{a^*}, (f'_{(\times,\ell)})_{\ell \in \hat{\Lambda}}, (f'_{(\varrho,\ell)})_{\ell \in \tilde{\Lambda}})$, s.t. $(A, \tilde{A}) = Com^1(\tilde{ck}_1; \boldsymbol{a}; r_a)$, $(B, \hat{B}) = Com^1(\tilde{ck}_1; \boldsymbol{b}; r_b)$, $(B, \tilde{B}) = Com^1(\tilde{ck}_1; \boldsymbol{b}; r_b)$, $(A^*, \hat{A}^*) = Com^1(\tilde{ck}_1; \boldsymbol{a}^*; r_{a^*})$, $(\psi^\times, \hat{\psi}^\times) = (g_2^{\sum_{\ell \in \hat{\Lambda}} f'_{(\times,\ell)}}, \hat{g}_2^{\sum_{\ell \in \hat{\Lambda}} f'_{(\times,\ell)}})$, $(\psi^\varrho, \hat{\psi}^\varrho) = (g_2^{\sum_{\ell \in \tilde{\Lambda}} f'_{(\varrho,\ell)}}, \tilde{g}_2^{\sum_{\ell \in \tilde{\Lambda}} f'_{(\varrho,\ell)}})$, $a_i^* = T_\Lambda(\varrho^{-1}(i), \varrho) \cdot a_i$ (for $i \in [n]$), and for some $i \in [n]$, $a_{\varrho(i)} \neq b_i$.*

System parameters: Same as in Prot. 1, but let $\tilde{\Lambda}$ be as in Eq. (6).

CRS generation $\mathcal{G}_{\mathsf{crs}}(1^\kappa)$: Let $\mathsf{gk} := (p, \mathbb{G}_1, \mathbb{G}_2, \mathbb{G}_T, \hat{e}) \leftarrow \mathcal{G}_{\mathsf{bp}}(1^\kappa)$. Let $\hat{\alpha}, \tilde{\alpha}, x \leftarrow \mathbb{Z}_p$. Let $g_1 \leftarrow \mathbb{G}_1 \setminus \{1\}$ and $g_2 \leftarrow \mathbb{G}_2 \setminus \{1\}$. Let $\hat{g}_t \leftarrow \hat{g}_t^{\hat{\alpha}}$ and $\tilde{g}_t \leftarrow \tilde{g}_t^{\tilde{\alpha}}$ for $t \in \{1, 2\}$. Denote $g_{t\ell} \leftarrow g_t^{x^\ell}$, $\hat{g}_{t\ell} \leftarrow \hat{g}_t^{x^\ell}$, and $\tilde{g}_{t\ell} \leftarrow \tilde{g}_t^{x^\ell}$ for $t \in \{1, 2\}$ and $\ell \in \{0\} \cup \tilde{\Lambda}$. Let $(D, \tilde{D}) \leftarrow (\prod_{i=1}^n g_{2,\lambda_i}, \prod_{i=1}^n \tilde{g}_{2,\lambda_i})$. The CRS is

$$\mathsf{crs} \leftarrow (\mathsf{gk}; (g_{1\ell}, \hat{g}_{1\ell}, \tilde{g}_{1\ell})_{\ell \in \{0\} \cup \Lambda}, (g_{2\ell})_{\ell \in \{0\} \cup \tilde{\Lambda}}, (\hat{g}_{2\ell})_{\ell \in \hat{\Lambda}}, (\tilde{g}_{2\ell})_{\ell \in \tilde{\Lambda}}, D, \tilde{D}) \ .$$

Let $\hat{\mathsf{ck}}_1 \leftarrow (\mathsf{gk}; (g_{1\ell}, \hat{g}_{1\ell}, \tilde{g}_{1\ell})_{\ell \in \{0\} \cup \Lambda})$, $\check{\mathsf{ck}}_1 \leftarrow (\mathsf{gk}; (g_{1\ell}, \tilde{g}_{1\ell})_{\ell \in \{0\} \cup \Lambda})$.

Common inputs: $(A, \tilde{A}, B, \hat{B}, \tilde{B}, \varrho)$, where $\varrho \in S_n$, $(A, \tilde{A}) \leftarrow \mathcal{C}om^1(\check{\mathsf{ck}}_1; \boldsymbol{a}; r_a)$, $(B, \hat{B}) \leftarrow \mathcal{C}om^1(\hat{\mathsf{ck}}_1; \boldsymbol{b}; r_b)$, and $(B, \tilde{B}) \leftarrow \mathcal{C}om^1(\check{\mathsf{ck}}_1; \boldsymbol{b}; r_b)$, s.t. $b_j = a_{\varrho(j)}$ for $j \in [n]$.

Argument generation $\mathcal{P}_{\mathsf{perm}}(\mathsf{crs}; (A, \tilde{A}, B, \hat{B}, \tilde{B}, \varrho), (\boldsymbol{a}, r_a, \boldsymbol{b}, r_b))$:

1. Let $(T^*, \hat{T}^*, T_2^*) \leftarrow (\prod_{i=1}^n g_{1,\lambda_i}^{T_\Lambda(\varrho^{-1}(i), \varrho)}, \prod_{i=1}^n \hat{g}_{1,\lambda_i}^{T_\Lambda(\varrho^{-1}(i), \varrho)}, \prod_{i=1}^n g_{2,\lambda_i}^{T_\Lambda(\varrho^{-1}(i), \varrho)})$.

2. Let $r_{a^*} \leftarrow \mathbb{Z}_p$, $(A^*, \hat{A}^*) \leftarrow \mathcal{C}om_1(\hat{\mathsf{ck}}_1; T_\Lambda(\varrho^{-1}(1), \varrho) \cdot a_1, \ldots, T_\Lambda(\varrho^{-1}(n), \varrho) \cdot a_n; r_{a^*})$. Create an argument $\psi^\times$ for $[\![(A, \hat{A})]\!] \circ [\![(T^*, \hat{T}^*, T_2^*)]\!] = [\![(A^*, \hat{A}^*)]\!]$.

3. Let $\tilde{\Lambda}'_\varrho := 2^\frown\Lambda \cup (\{2\lambda_{\varrho(j)} + \lambda_i - \lambda_j : i, j \in [n] \wedge i \neq j\} \setminus 2 \cdot \Lambda) \subset \{-\lambda_n + 1, \ldots, 3\lambda_n\}$.

4. For $\ell \in \tilde{\Lambda}'_\varrho$, $I_1(\ell)$ as in Prot. 1, and $I_2(\ell) := \{(i, j) : i, j \in [n] \wedge j \neq i \wedge 2\lambda_{\varrho(i)} + \lambda_j \neq \lambda_i + 2\lambda_{\varrho(j)} \wedge 2\lambda_{\varrho(j)} + \lambda_i - \lambda_j = \ell\}$, set

$$\mu_{\varrho, \ell} \leftarrow \sum_{(i,j) \in I_1(\ell)} a_i^* - \sum_{(i,j) \in I_2(\ell)} b_i \ .$$

5. Let $(E_\varrho, \tilde{E}_\varrho) \leftarrow (\prod_{i=1}^n g_{2, 2\lambda_{\varrho(i)} - \lambda_i}, \prod_{i=1}^n \tilde{g}_{2, 2\lambda_{\varrho(i)} - \lambda_i})$.

6. Let $\psi^\varrho \leftarrow D^{r_a^*} \cdot E_\varrho^{-r_b} \cdot \prod_{\ell \in \tilde{\Lambda}'_\varrho} g_{2\ell}^{\mu_{\varrho, \ell}}$, $\tilde{\psi}^\varrho \leftarrow \tilde{D}^{r_a^*} \cdot \tilde{E}_\varrho^{-r_b} \cdot \prod_{\ell \in \tilde{\Lambda}'_\varrho} \tilde{g}_{2\ell}^{\mu_{\varrho, \ell}}$,

Send $\psi^{\mathsf{perm}} \leftarrow (A^*, \hat{A}^*, \psi^\times, \psi^\varrho, \tilde{\psi}^\varrho) \in \mathbb{G}_1^2 \times \mathbb{G}_2^4$ to the verifier as the argument.

Verification $\mathcal{V}_{\mathsf{perm}}(\mathsf{crs}; (A, \tilde{A}, B, \hat{B}, \tilde{B}, \varrho), \psi^{\mathsf{perm}})$: Let E_ϱ and $(T^*, \hat{T}^*, T_2^*)$ be computed as in $\mathcal{P}_{\mathsf{perm}}$. If $\psi^\times$ verifies, $\hat{e}(A^*, D)/\hat{e}(B, E_\varrho) = \hat{e}(g_1, \psi^\varrho)$, $\hat{e}(A^*, \hat{g}_2) = \hat{e}(\hat{A}^*, g_2)$, and $\hat{e}(g_1, \tilde{\psi}^\varrho) = \hat{e}(\tilde{g}_1, \psi^\varrho)$, then $\mathcal{V}_{\mathsf{perm}}$ accepts. Otherwise, $\mathcal{V}_{\mathsf{perm}}$ rejects.

Protocol 2: New permutation argument $\varrho([\![(A, \tilde{A})]\!]) = [\![(B, \tilde{B})]\!]$

Proof. Denote $h \leftarrow \hat{e}(g_1, g_2)$ and $F_\varrho(x) := \log_h(\hat{e}(A^*, D)/\hat{e}(B, E_\varrho))$. WITNESS-INDISTINGUISHABILITY: since argument ψ^{perm} that satisfies the verification equations is unique, all witnesses result in the same argument, and therefore the permutation argument is witness-indistinguishable.

PERFECT COMPLETENESS. Completeness of $\psi^\times$ follows from the completeness of the Hadamard product argument. The third and the fourth verifications are straightforward. For the verification $\hat{e}(A^*, D)/\hat{e}(B, E_\varrho) = \hat{e}(g_1, \psi^\varrho)$, consider $F_\varrho(x)$ in Eq. (5). Consider X as a formal variable, then the right-hand side (and thus also $F_\varrho(X)$) is a formal polynomial of X, spanned by $\{X^\ell\}_{\ell \in 2 \cdot \Lambda \cup \tilde{\Lambda}}$. If the prover is honest, then $b_i = a_{\varrho(i)}$ for $i \in [n]$, and thus $F_\varrho(X)$ is spanned by $\{X^\ell\}_{\ell \in \tilde{\Lambda}}$. Defining $\psi^\varrho \leftarrow (\prod_{i=1}^n g_{2,\lambda_i})^{r_{a^*}} \cdot (\prod_{i=1}^n g_{2, 2\lambda_{\varrho(i)} - \lambda_i})^{-r_b} \cdot \prod_{i=1}^n (\prod_{j=1: j \neq i}^n g_{2, \lambda_i + \lambda_j})^{a_i^*} \cdot \prod_{i=1}^n (\prod_{j \in I_2^*(i, \ell)} g_{2, \lambda_i + 2\lambda_{\varrho(j)} - \lambda_j})^{-b_i} = D^{r_{a^*}} \cdot E_\varrho^{-r_b} \cdot \prod_{\ell \in \tilde{\Lambda}'_\varrho} g_{2\ell}^{\mu_{\varrho, \ell}}$, where $I_2^*(i, \ell) := \{j \in [n] : j \neq i \wedge 2\lambda_{\varrho(i)} + \lambda_j \neq \lambda_i + 2\lambda_{\varrho(j)}\}$, we see that the second verification holds.

WEAKER VERSION OF SOUNDNESS. Assume that $\mathcal{A}$ is an adversary that can break the last statement of the theorem. We construct an adversary $\mathcal{A}'$ against the $\tilde{\Lambda}$-PSDL assumption. Let gk $\leftarrow \mathcal{G}_{\mathsf{bp}}(1^\kappa)$, $x \leftarrow \mathbb{Z}_p$, $g_1 \leftarrow \mathbb{G}_1 \setminus \{1\}$, and $g_2 \leftarrow \mathbb{G}_2 \setminus \{1\}$. The adversary $\mathcal{A}'$ receives crs $\leftarrow (\mathsf{gk}; (g_1^{x^\ell}, g_2^{x^\ell})_{\ell \in \{0\} \cup \tilde{\Lambda}})$ as her input, and her task is to output x. She sets $\hat{\alpha} \leftarrow \mathbb{Z}_p$, $\tilde{\alpha} \leftarrow \mathbb{Z}_p$, and crs$' \leftarrow (\mathsf{gk};$ $(g_1^{x^\ell}, g_1^{\hat{\alpha} x^\ell}, g_1^{\tilde{\alpha} x^\ell})_{\ell \in \{0\} \cup \Lambda}, (g_2^{x^\ell})_{\ell \in \{0\} \cup \tilde{\Lambda}}, (g_2^{\hat{\alpha} x^\ell})_{\ell \in \hat{\Lambda}}, (g_2^{\tilde{\alpha} x^\ell})_{\ell \in \tilde{\Lambda}}, \prod_{i=1}^n g_2^{\lambda_i}, \prod_{i=1}^n \tilde{g}_2^{x^{\lambda_i}})$, and forwards crs$'$ to $\mathcal{A}$. Clearly, crs$'$ has the same distribution as $\mathcal{G}_{\mathsf{crs}}(1^\kappa)$. Both parties also set $\hat{\mathsf{ck}}_1 \leftarrow (\mathsf{gk}; (g_1^{x^\ell}, g_1^{\hat{\alpha} x^\ell})_{\ell \in \{0\} \cup \Lambda})$ and $\tilde{\mathsf{ck}}_1 \leftarrow (\mathsf{gk}; (g_1^{x^\ell}, g_1^{\tilde{\alpha} x^\ell})_{\ell \in \{0\} \cup \Lambda})$.

Assume that $\mathcal{A}$ returns $(inp^{\mathsf{perm}}, w^{\mathsf{perm}}, \psi^{\mathsf{perm}})$ such that the conditions in the theorem statement hold, and $\mathcal{V}(\mathsf{crs}'; inp^{\mathsf{perm}}, \psi^{\mathsf{perm}})$ accepts. Here, $inp^{\mathsf{perm}} = (A, \tilde{A}, B, \hat{B}, \tilde{B}, \varrho)$ and $w^{\mathsf{perm}} = (\boldsymbol{a}, r_a, \boldsymbol{b}, r_b, \boldsymbol{a}^*, r_{a^*}, (f'_{(\times, \ell)})_{\ell \in \hat{\Lambda}}, (f'_{(\varrho, \ell)})_{\ell \in \tilde{\Lambda}})$.

If $\mathcal{A}$ is successful, $(A, \tilde{A}) = \mathcal{C}om^1(\tilde{\mathsf{ck}}_1; \boldsymbol{a}; r_a)$, $(B, \hat{B}) = \mathcal{C}om^1(\hat{\mathsf{ck}}_1; \boldsymbol{b}; r_b)$, $(B, \tilde{B}) = \mathcal{C}om^1(\tilde{\mathsf{ck}}_1; \boldsymbol{b}; r_b)$, $\psi^\times$ verifies, and for some $i \in [n]$, $a_{\varrho(i)} \neq T_\Lambda(i, \varrho) \cdot b_i$. Since $\psi^\times$ verifies and the Hadamard product argument is (weakly) sound, we have that $(A^*, \hat{A}^*)$ commits to $(T_\Lambda(\varrho^{-1}(1), \varrho) \cdot a_1, \ldots, T_\Lambda(\varrho^{-1}(n), \varrho) \cdot a_n)$. (Otherwise, we have broken the PSDL assumption.) Since $2 \cdot \Lambda \cap \tilde{\Lambda} = \emptyset$, $\mathcal{A}'$ has expressed $F_\varrho(X)$ as a polynomial $f(X)$ where at least for some $\ell \in 2 \cdot \Lambda$, X^ℓ has a non-zero coefficient.

On the other hand, $\mathcal{A}$ also outputs $(f'_{(\varrho, \ell)})_{\ell \in \tilde{\Lambda}}$, s.t. $F_\varrho(x) = \log_{g_2} \psi = f'_\varrho(x)$, where all non-zero coefficients of $f'_\varrho(X) := \sum_{\ell \in \tilde{\Lambda}} f'_{(\varrho, \ell)} X^\ell$ correspond to X^ℓ for some $\ell \in \tilde{\Lambda}$. Since Λ is a progression-free set of odd integers and all elements of $2 \cdot \Lambda$ are distinct, then by the discussion in the beginning of Sect. 7, $\ell \notin 2 \cdot \Lambda$. Thus, all coefficients of $f'_\varrho(X)$ corresponding to any X^ℓ, $\ell \in 2 \cdot \Lambda$, are equal to 0. Thus, $f(X) \cdot X^{\lambda_n} = \sum_{\ell \in \tilde{\Lambda} \cup (2 \cdot \Lambda)} f_\ell X^{\ell + \lambda_n}$ and $f'_\varrho(X) = \sum_{\ell \in \tilde{\Lambda}} f'_{(\varrho, \ell)} X^{\ell + \lambda_n}$ are different polynomials with $f(x) = f'_\varrho(x) = F_\varrho(x)$. Thus, $\mathcal{A}'$ has succeeded in creating a nonzero polynomial $d_\varrho(X) = f(X) \cdot X^{\lambda_n} - f'_\varrho(X)$, such that $d_\varrho(x) = \sum_{\ell \in \tilde{\Lambda}} d_\ell x^\ell = 0$.

Next, $\mathcal{A}'$ can use an efficient polynomial factorization algorithm [19] in $\mathbb{Z}_p[X]$ to efficiently compute all $\leq 4\lambda_n + 1$ roots of $d_\varrho(X)$. For some root y, $g_1^{x^\ell} = g_1^{y^\ell}$. The adversary $\mathcal{A}'$ sets $x \leftarrow y$, thus violating the $\tilde{\Lambda}$-PSDL assumption. $\square$

Let Λ be as described in Thm. 1. The CRS consists of $n^{1+o(1)}$ group elements. The argument size of Prot. 2 is 2 elements from $\mathbb{G}_1$ and 4 elements from $\mathbb{G}_2$. The prover's computational complexity is dominated by $\Theta(n^2)$ scalar additions in $\mathbb{Z}_p$ and by $n^{1+o(1)}$ exponentiations in $\mathbb{G}_2$. The verifier's computational complexity is dominated by 12 bilinear pairings and $4n - 2$ bilinear-group multiplications.

8   New NIZK Argument for Circuit Satisfiability

In a *NIZK argument for circuit satisfiability* (Circuit-SAT, well-known to be an NP-complete language), the prover and the verifier share a circuit C. The prover aims to prove in non-interactive zero-knowledge that she knows an assignment of input values that makes the circuit output 1. As in [15], the Circuit-SAT argument will use the Hadamard product argument, the permutation argument and a trivial argument for element-wise sum of two tuples — in our case, all operating in parallel on $(2|C| + 1)$-dimensional tuples, where $|C|$ is the circuit size. Those three arguments can be seen

---

System parameters: Define Λ and $\hat{\Lambda}$ as in Prot. 1 and $\tilde{\Lambda}$ as in Prot. 2, but in all cases with n replaced by $2|C| + 1$. Permutation swap.

CRS generation $\mathcal{G}_{\mathsf{crs}}(1^\kappa)$: Let all other variables (including the secret ones) be defined as in the CRS generation of Prot. 2, but let $\mathsf{crs}^{\mathsf{perm}}$ be the CRS of Prot. 2. In addition, let $(\hat{D}, D_2) \leftarrow (\prod_{i=1}^{n} \hat{g}_{1,\lambda_i}, \prod_{i=1}^{n} g_{2,\lambda_i})$. The CRS is $\mathsf{crs} \leftarrow (\mathsf{crs}^{\mathsf{perm}}, \hat{D}, D_2)$. Let $\mathsf{ck}_1 \leftarrow (\mathsf{gk}; (g_{1\ell}, \hat{g}_{1\ell}, \tilde{g}_{1\ell})_{\ell \in \{0\} \cup \Lambda})$.

Common inputs: A satisfiable circuit C, and permutations τ and ζ generated based on C, such that $(\boldsymbol{L}, \boldsymbol{R}, R_{n+1}, \boldsymbol{U}, \boldsymbol{X}, X_{n+1})$ is a "satisfying assignment".

Argument generation $\mathcal{P}(\mathsf{crs}; C, (\boldsymbol{L}, \boldsymbol{R}, R_{n+1}, \boldsymbol{U}, \boldsymbol{X}))$: Denote $\boldsymbol{Y} := (Y_1, \ldots, Y_n)$ for $Y \in \{L, R, U, X\}$. The prover does the following:

1. Set $r_1, \ldots, r_4 \leftarrow \mathbb{Z}_p$, and then compute $(\mathsf{lr}, \widehat{\mathsf{lr}}, \widetilde{\mathsf{lr}}) \leftarrow \mathcal{C}om^1(\mathsf{ck}_1; \boldsymbol{L}, \boldsymbol{R}, R_{n+1}; r_1)$, $\mathsf{lr}_2 \leftarrow g_2^{r_1} \cdot \prod_{i=1}^{n} g_{2,\lambda_i}^{L_i} \cdot \prod_{i=1}^{n+1} g_{2,\lambda_{i+n}}^{R_i}$, $(\mathsf{rl}, \widetilde{\mathsf{rl}}) \leftarrow \mathcal{C}om^1(\widetilde{\mathsf{ck}}_1; \boldsymbol{R}, \boldsymbol{L}, R_{n+1}; r_1)$, $(\mathsf{rz}, \widehat{\mathsf{rz}}) \leftarrow \mathcal{C}om^1(\widehat{\mathsf{ck}}_1; \boldsymbol{R}, 0, \ldots, 0, 0; r_2)$, $(\mathsf{uz}, \widehat{\mathsf{uz}}) \leftarrow \mathcal{C}om^1(\widehat{\mathsf{ck}}_1; \boldsymbol{U}, 0, \ldots, 0, 0; r_3)$, $(\mathsf{ux}, \widehat{\mathsf{ux}}, \widetilde{\mathsf{ux}}) \leftarrow \mathcal{C}om^1(\mathsf{ck}_1; \boldsymbol{U}, \boldsymbol{X}, X_{n+1}; r_4)$.

2. Create an argument ψ_1 for $[\![(\mathsf{lr}, \widehat{\mathsf{lr}})]\!] \circ [\![(\mathsf{lr}, \widehat{\mathsf{lr}}, \mathsf{lr}_2)]\!] = [\![(\mathsf{lr}, \widehat{\mathsf{lr}})]\!]$, ψ_2 for $\mathsf{swap}([\![(\mathsf{rl}, \widetilde{\mathsf{rl}})]\!]) = [\![(\mathsf{lr}, \widehat{\mathsf{lr}}, \widetilde{\mathsf{lr}})]\!]$, ψ_3 for $[\![(\mathsf{rl}, \widetilde{\mathsf{rl}})]\!] \circ [\![(D, \hat{D}, D_2)]\!] = [\![(\mathsf{rz}, \widehat{\mathsf{rz}})]\!]$, ψ_4 for $[\![(\mathsf{ux}, \widehat{\mathsf{ux}})]\!] \circ [\![(D, \hat{D}, D_2)/(g_{1,\lambda_n}, \hat{g}_{1,\lambda_n}, g_{1,\lambda_n})]\!] = [\![(\mathsf{uz}, \widehat{\mathsf{uz}})/(g_{1,\lambda_n}, \hat{g}_{1,\lambda_n}, g_{1,\lambda_n})]\!]$, ψ_5 for $[\![(\mathsf{rz}, \widehat{\mathsf{rz}})]\!] \circ [\![(\mathsf{lr}, \widehat{\mathsf{lr}}, \mathsf{lr}_2)]\!] = [\![(D, \hat{D})]\!] \cdot (\mathsf{uz}^{-1}, \widehat{\mathsf{uz}}^{-1})]\!]$, ψ_6 for $\tau([\![(\mathsf{lr}, \widehat{\mathsf{lr}})]\!]) = [\![(\mathsf{lr}, \widehat{\mathsf{lr}}, \widetilde{\mathsf{lr}})]\!]$, and ψ_7 for $\zeta^{-1}([\![(\mathsf{ux}, \widetilde{\mathsf{ux}})]\!]) = [\![(\mathsf{lr}, \widehat{\mathsf{lr}}, \widetilde{\mathsf{lr}})]\!]$.

3. Send $\psi \leftarrow (\mathsf{lr}, \widehat{\mathsf{lr}}, \widetilde{\mathsf{lr}}, \mathsf{lr}_2, \mathsf{rl}, \widetilde{\mathsf{rl}}, \mathsf{rz}, \widehat{\mathsf{rz}}, \mathsf{uz}, \widehat{\mathsf{uz}}, \mathsf{ux}, \widehat{\mathsf{ux}}, \widetilde{\mathsf{ux}}, \psi_1, \ldots, \psi_7)$ to the verifier.

Verification $\mathcal{V}(\mathsf{crs}; C, \psi)$: The verifier does the following:
- For $A \in \{\mathsf{lr}, \mathsf{rz}, \mathsf{uz}, \mathsf{ux}\}$ check that $\hat{e}(\hat{A}, g_2) = \hat{e}(A, \hat{g}_2)$.
- Check that $\hat{e}(g_1, \mathsf{lr}_2) = \hat{e}(\mathsf{lr}, g_2)$.
- For $A \in \{\mathsf{lr}, \mathsf{rl}, \mathsf{ux}\}$ check that $\hat{e}(\tilde{A}, g_2) = \hat{e}(A, \tilde{g}_2)$.
- Verify all 7 arguments $\psi_1, \ldots, \psi_7$ with corresponding inputs.

Protocol 3: New NIZK argument for Circuit-SAT

---

as basic operations in an NIZK "programming language" for all languages in **NP**. We show that a small constant number of such basic operations is sufficient for Circuit-SAT. The full argument then contains additional cryptographic sugar: a precise definition of the used CRS, computational/communication optimizations, etc.

The first task is to express the underlying argument as a parallel composition of some addition, permutation and Hadamard product arguments. These arguments may include intermediate variables (that will be committed to by the prover) and constants (that can be online committed to by both of the parties separately). When choosing the arguments, one has to keep in mind that we work in an asymmetric setting. This may mean that for some of the inputs to the circuit satisfiability argument, one has to commit to them both in $\mathbb{G}_1$ and $\mathbb{G}_2$ (and the verifier has to check that this is done correctly).

The CRS is basically the CRS of the permutation argument. The total argument consists of commitments to intermediate variables and of all arguments in the program of this "programming language". Finally, the verifier has to check that all commitments are internally consistent, and then verify all used arguments.

Let us now turn to the concrete case of circuit satisfiability. For the sake of simplicity, assume that the circuit C is only composed of NAND gates. Let C have n gates. Assume that the output gate of the circuit is n, and U_n is the output of the circuit. For every gate

$j \in [n]$ of C, let the input wires of its jth gate be L_j and R_j, and let U_j be one of its output wires. We also define an extra value $R_{n+1} = 1$. We let X_j be other "output" wires that correspond to some L_k or R_k that were not already covered by U_k (that is, inputs to the circuit, or duplicates of output wires). That is, $(U_1, \ldots, U_n, X_1, \ldots, X_{n+1})$ is chosen so that for some permutation ζ, (U, X, X_{n+1}) is a ζ-permutation of (L, R, R_{n+1}), where $Y = (Y_1, \ldots, Y_n)$ for $Y \in \{L, R, U, X\}$.

More precisely, the prover and the verifier share the following three permutations, the first two of which completely describe the circuit C. First, $\tau \in S_{2n+1}$ is a permutation, such that for any values $L_{i_1}, \ldots, L_{i_s}, R_{j_1}, \ldots, R_{j_t}$ that correspond to the same wire, τ contains a cycle $i_1 \rightarrow i_2 \rightarrow \cdots \rightarrow i_s \rightarrow j_1 + n \rightarrow \cdots \rightarrow j_t + n \rightarrow i_1$. For unique wires i, $\tau(i) = i$. Second, $\zeta \in S_{2n+1}$ is a permutation that for every input wire (either L_i or R_{i-n}), outputs an index $j \leftarrow \zeta(i)$, such that the output wire U_j or X_{j-n} is equal to that input wire. Third, swap $\in S_{2n+1}$ is a permutation, with swap$(i) = i + n$ and swap$(i + n) = i$ for $i \in [n]$, and swap$(2n + 1) = 2n + 1$. Note that swap $=$ swap^{-1}.

The argument is given by Prot. 3. In every subargument used in Prot. 3, the prover and the verifier use a substring of crs as the CRS. The corresponding substrings are easy to compute, and in what follows, we do not mention this issue. Instead of computing two different commitments $\mathcal{C}om^t(\widehat{\mathsf{ck}}_t; a; r) = (g_t^r \cdot \prod g_{t,\lambda_i}^{a_i}, \hat{g}_t^r \cdot \prod \hat{g}_{t,\lambda_i}^{a_i})$ and $\mathcal{C}om^t(\widetilde{\mathsf{ck}}_t; a; r) = (g_t^r \cdot \prod g_{t,\lambda_i}^{a_i}, \tilde{g}_t^r \cdot \prod \tilde{g}_{t,\lambda_i}^{a_i})$, we sometimes compute a composed commitment $\mathcal{C}om^t(\mathsf{ck}_t; a; r) = (g_t^r \cdot \prod g_{t,\lambda_i}^{a_i}, \hat{g}_t^r \prod \hat{g}_{t,\lambda_i}^{a_i}, \tilde{g}_t^r \cdot \prod \tilde{g}_{t,\lambda_i}^{a_i})$. We assume that the same value $\hat{\alpha}$ is used when creating product arguments and permutation arguments.

Theorem 6. *Let $\mathcal{G}_{\mathsf{bp}}$ be $\tilde{\Lambda}$-PSDL secure, and Λ-PKE secure in both $\mathbb{G}_1$ and $\mathbb{G}_2$. Then Prot. 3 is a perfectly complete, computationally adaptively sound and perfectly zero-knowledge non-interactive Circuit-SAT argument.*

Proof. PERFECT COMPLETENESS: follows from the perfect completeness of the Hadamard product and permutation arguments.

ADAPTIVE COMPUTATIONAL SOUNDNESS: Let $\mathcal{A}$ be a non-uniform PPT adversary that creates a circuit C and an accepting NIZK argument ψ. By the Λ-PKE assumption, there exists a non-uniform PPT extractor $X_{\mathcal{A}}$ that, running on the same input and seeing $\mathcal{A}$'s random tape, extracts all openings. From the (weaker version of) soundness of the product and permutation arguments and by the $\tilde{\Lambda}$-PSDL assumption, it follows that the corresponding relations are satisfied between the opened values. Moreover, by the $\tilde{\Lambda}$-PSDL assumption, the opened values belong to corresponding sets $\hat{\Lambda}$ and $\tilde{\Lambda}$. Let (L, R, R_{n+1}) be the opening of $(\mathsf{lr}, \widehat{\mathsf{lr}})$, where $L = (L_1, \ldots, L_n)$ and $R = (R_1, \ldots, R_n)$, and let $(U_1, \ldots, U_n, X_1, \ldots, X_n, X_{n+1})$ be the opening of $(\mathsf{ux}, \widehat{\mathsf{ux}})$. We now analyze the effect of every subargument in Prot. 3.

The successful verification of $\hat{e}(g_1, \mathsf{lr}_2) = \hat{e}(\mathsf{lr}, g_2)$ shows that lr_2 is correctly formed. The first argument ψ_1 shows that $L_i, R_i \in \{0, 1\}$. The second argument ψ_2 shows that $(\mathsf{rl}, \tilde{\mathsf{rl}})$ commits to (R, L, R_{n+1}). The third argument ψ_3 shows that $(\mathsf{rz}, \hat{\mathsf{rz}})$ commits to $(R, 0, \ldots, 0, 0)$ and is thus consistent with the opening of $(\mathsf{lr}, \widehat{\mathsf{lr}})$. The fourth argument ψ_4 shows that $(\mathsf{uz}, \hat{\mathsf{uz}})$ commits to $(U_1, \ldots, U_{n-1}, U'_n, 0, \ldots, 0, 0)$ for some U'_n. It also shows that $U_n \cdot 0 = U'_n - 1$, and thus $U'_n = 1$. (The value of U_n is not important to get soundness, since it is not used in any other argument.)

The fifth argument shows ψ_5 that the NAND gates are followed. That is, $\neg(L_i \wedge R_i) = U_i$ for $i \in [n-1]$. It also shows that the circuit outputs 1. Really, since $(\text{uz}, \widehat{\text{uz}})$ commits to $(U_1, \ldots, U_{n-1}, U'_n = 1, 0, \ldots, 0, 0)$, then $(D, \hat{D}) \cdot (\text{uz}^{-1}, \widehat{\text{uz}}^{-1})$ commits to $(1 - U_1, \ldots, 1 - U_{n-1}, 1 - 1 = 0, 0, \ldots, 0, 0)$. Thus, the Hadamard product argument verifies only if $L_i \cdot R_i = 1 - U_i$ for $i \in [n-1]$, and $L_n \cdot R_n = 0$, that is, $\neg(L_n \wedge R_n) = 1$.

The sixth argument ψ_6 shows that if $i_1, \ldots, i_s, j_1 + n, \ldots, j_t + n$ correspond to the same wire, then $L_{i_1} = \cdots = L_{i_s} = R_{j_1} = \cdots = R_{j_t}$, that is, the values are internally consistent with the wires. The seventh argument ψ_7 shows that the "input wires" and "output" wires are consistent.

PERFECT ZERO-KNOWLEDGE: we construct the next simulator $\mathcal{S} = (\mathcal{S}_1, \mathcal{S}_2)$. The simulator $\mathcal{S}_1(1^\kappa, n)$ creates a correctly formed CRS together with a simulation trapdoor $\text{td} = (\hat{\alpha}, \tilde{\alpha}, x) \in \mathbb{Z}_p^3$. The adversary then outputs a statement C (a circuit) together with a witness (a satisfying assignment) w. The simulator $\mathcal{S}_2(\text{crs}; C, \text{td})$ creates $(\text{lr}, \widehat{\text{lr}}, \widetilde{\text{lr}}, \text{lr}_2)$, $(\text{rl}, \widetilde{\text{rl}})$, $(\text{rz}, \widehat{\text{rz}})$, $(\text{uz}, \widehat{\text{uz}})$ and $(\text{ux}, \widehat{\text{ux}})$ as commitments to $(0, \ldots, 0)$. Due to the knowledge of trapdoor td, the simulator can simulate all product and permutation arguments. More precisely, he uses $L_i = R_i = U_i = U'_n = 1$ to simulate all product and permutation arguments, except in the case of ψ_5 where he uses $U_i = U'_n = 0$ instead. (Obviously, $(\text{rz}, \widehat{\text{rz}})$ and $(\text{uz}, \widehat{\text{uz}})$ commit to consistent tuples.)

To show that this argument ψ'' simulates the real argument ψ, note that ψ is perfectly indistinguishable from the simulated NIZK argument ψ' where one makes trapdoor commitments but opens them to *real* witnesses L_i, R_i when making product and permutation arguments. On the other hand, also ψ' and ψ'' are perfectly indistinguishable, and thus so are ψ and ψ''.   □

Let Λ be chosen as in Thm. 1. The CRS consists of $|C|^{1+o(1)}$ group elements. The communication (argument length) of the argument in Prot. 3 is 18 elements from $\mathbb{G}_1$ and 21 elements from $\mathbb{G}_2$. The prover's computational complexity is dominated by $\Theta(|C|^2)$ simple arithmetical operations in $\mathbb{Z}_p$ and $|C|^{1+o(1)}$ exponentiations in $\mathbb{G}$. The verifier's computational complexity is dominated by 72 bilinear pairings and $8|C| + 8$ bilinear-group multiplications.

Moreover, the CRS depends on $\hat{\Lambda} \cup \tilde{\Lambda}$. Since 0 may or may not belong to $\tilde{\Lambda}$ (this depends on the choice of Λ) and $\Lambda \cup 2 \widehat{} \Lambda \subseteq \tilde{\Lambda}$, $\hat{\Lambda} \cup \tilde{\Lambda} = \{0\} \cup \tilde{\Lambda}$. Recalling that elements of $\mathbb{G}_1$ can be represented by 512 bits and elements of $\mathbb{G}_2$ can be represented by 256 bits, the communication (argument length) is $18 \cdot 512 + 21 \cdot 256 = 14\,592$ bits.

Acknowledgments. We would like to thank Jens Groth, Amit Sahai, Naomi Benger, Bingsheng Zhang, Rafik Chaabouni, and anonymous reviewers for comments. The author was supported by Estonian Science Foundation, grant #9303, and European Union through the European Regional Development Fund.

References

1. Abe, M., Fehr, S.: Perfect NIZK with Adaptive Soundness. In: Vadhan, S.P. (ed.) TCC 2007. LNCS, vol. 4392, pp. 118–136. Springer, Heidelberg (2007)
2. Barreto, P.S.L.M., Naehrig, M.: Pairing-Friendly Elliptic Curves of Prime Order. In: Preneel, B., Tavares, S. (eds.) SAC 2005. LNCS, vol. 3897, pp. 319–331. Springer, Heidelberg (2006)

3. Behrend, F.A.: On the Sets of Integers Which Contain No Three in Arithmetic Progression. Proceedings of the National Academy of Sciences 32(12), 331–332 (1946)
4. Blum, M., Feldman, P., Micali, S.: Non-Interactive Zero-Knowledge and Its Applications. In: STOC 1988, pp. 103–112. ACM Press (1988)
5. Bourgain, J.: On Triples in Arithmetic Progression. Geom. Funct. Anal. 9(5), 968–984 (1998)
6. Chaabouni, R., Lipmaa, H., Shelat, A.: Additive Combinatorics and Discrete Logarithm Based Range Protocols. In: Steinfeld, R., Hawkes, P. (eds.) ACISP 2010. LNCS, vol. 6168, pp. 336–351. Springer, Heidelberg (2010)
7. Chaabouni, R., Lipmaa, H., Zhang, B.: A Non-Interactive Range Proof with Constant Communication. In: Keromytis, A. (ed.) FC 2012. LNCS, Springer, Heidelberg (2012)
8. Cheon, J.H.: Security Analysis of the Strong Diffie-Hellman Problem. In: Vaudenay, S. (ed.) EUROCRYPT 2006. LNCS, vol. 4004, pp. 1–11. Springer, Heidelberg (2006)
9. Elkin, M.: An Improved Construction of Progression-Free Sets. Israeli J. Math. 184, 93–128 (2011)
10. Gentry, C.: Fully Homomorphic Encryption Using Ideal Lattices. In: Mitzenmacher, M. (ed.) STOC 2009, pp. 169–178. ACM Press (2009)
11. Gentry, C., Wichs, D.: Separating Succinct Non-Interactive Arguments from All Falsifiable Assumptions. In: Vadhan, S. (ed.) STOC 2011, pp. 99–108. ACM Press (2011)
12. Golle, P., Jarecki, S., Mironov, I.: Cryptographic Primitives Enforcing Communication and Storage Complexity. In: Blaze, M. (ed.) FC 2002. LNCS, vol. 2357, pp. 120–135. Springer, Heidelberg (2003)
13. Green, B., Wolf, J.: A Note on Elkin's Improvement of Behrend's Construction. In: Chudnovsky, D., Chudnovsky, G. (eds.) Additive Number Theory, pp. 141–144. Springer, New York (2010)
14. Groth, J.: Linear Algebra with Sub-linear Zero-Knowledge Arguments. In: Halevi, S. (ed.) CRYPTO 2009. LNCS, vol. 5677, pp. 192–208. Springer, Heidelberg (2009)
15. Groth, J.: Short Pairing-Based Non-interactive Zero-Knowledge Arguments. In: Abe, M. (ed.) ASIACRYPT 2010. LNCS, vol. 6477, pp. 321–340. Springer, Heidelberg (2010)
16. Groth, J.: Minimizing Non-interactive Zero-Knowledge Proofs Using Fully Homomorphic Encryption. Tech. Rep. 2011/012, IACR (2011)
17. Groth, J., Sahai, A.: Efficient Non-interactive Proof Systems for Bilinear Groups. In: Smart, N.P. (ed.) EUROCRYPT 2008. LNCS, vol. 4965, pp. 415–432. Springer, Heidelberg (2008)
18. Hess, F., Smart, N.P., Vercauteren, F.: The Eta Pairing Revisited. IEEE Trans. Inf. Theory 52(10), 4595–4602 (2006)
19. van Hoeij, M., Novocin, A.: Gradual Sub-lattice Reduction and a New Complexity for Factoring Polynomials. In: López-Ortiz, A. (ed.) LATIN 2010. LNCS, vol. 6034, pp. 539–553. Springer, Heidelberg (2010)
20. Lipmaa, H.: Progression-Free Sets and Sublinear Pairing-Based Non-Interactive Zero-Knowledge Arguments. Tech. Rep. 2011/009, IACR (2011)
21. Lipmaa, H., Zhang, B.: A More Efficient Computationally Sound Non-Interactive Zero-Knowledge Shuffle Argument. Tech. Rep. 2011/394, IACR (2011)
22. Micali, S.: CS Proofs. In: Goldwasser, S. (ed.) FOCS 1994, pp. 436–453. IEEE (1994)
23. Moser, L.: An Application of Generating Series. Mathematics Magazine 35(1), 37–38 (1962)
24. Sanders, T.: On Roth's Theorem on Progressions. Ann. Math. 174(1), 619–636 (2011)
25. Tao, T., Vu, V.: Additive Combinatorics. Cambridge Studies in Advanced Mathematics. Cambridge University Press (2006)

Point Obfuscation and 3-Round Zero-Knowledge*

Nir Bitansky and Omer Paneth

Tel Aviv University, Boston University

Abstract. We construct 3-round proofs and arguments with negligible soundness error satisfying two relaxed notions of zero-knowledge (ZK): *weak ZK* and *witness hiding* (WH). At the heart of our constructions lie new techniques based on *point obfuscation with auxiliary input* (AIPO).

It is known that such protocols cannot be proven secure using black-box reductions (or simulation). Our constructions circumvent these lower bounds, utilizing AIPO (and extensions) as the "non-black-box component" in the security reduction.

1 Introduction

Interactive proofs and arguments [GMR85, BCC88] are fundamental notions in the theory of computation. In cryptography, these are typically used to prove NP-statements and the proof is required to maintain the prover's privacy. Different notions of privacy were considered, the most comprehensive one being zero-knowledge (ZK). ZK protocols allow proving an assertion without revealing anything but its validity. That is, the information learned by the verifier from the interaction can be simulated only from the (valid) statement itself.

Since ZK was introduced [GMR85], questions regarding the round complexity of ZK protocols were studied extensively. While it is known that 2-round ZK protocols (with auxiliary input) do not exist for languages outside BPP [GO94], a classical open question is whether there exist 3-round ZK protocols for NP with negligible soundness error. The difficulty of this problem is expressed by the lower bound of [GK96]: there do not exist 3-round black-box ZK (BBZK) protocols with negligible soundness for languages outside BPP. Namely, to prove that a 3-round protocol is ZK, one must demonstrate a simulator that uses the verifier in a non-black-box way.

The work of [Bar01] shows that using non-black-box simulation it is possible to go beyond existing black-box bounds. However, so far we do not know how to use similar techniques to obtain 3-round ZK protocols. Nevertheless, 3-round ZK protocols have been constructed based on non-standard "knowledge assumptions". [HT98, BP04] show a 3-round ZK argument based on the *knowledge of exponent assumption* (KEA) and variants of it. A different "knowledge assumption" was used to show the existence of 3-round ZK proofs for NP [LM01]. (See further discussion in Section 1.2.)

---

* This research was funded by the Check Point Institute for Information Security, by Marie Curie grant PIRG03-GA-2008-230640, and ISF grant 0603805843.

R. Cramer (Ed.): TCC 2012, LNCS 7194, pp. 189–207, 2012.

In light of the difficulties in achieving 3-round ZK, it is natural to examine relaxations of ZK that might enable the construction of such protocols. We discuss several previously studied relaxations.

Witness indistinguishability (WI). A protocol is WI [FS90] if any two proofs for the same statement that use two different witnesses are indistinguishable. [FS90] show that, while the parallel repetition of basic (3-round) ZK protocols is not BBZK, it is WI. Furthermore, the soundness error decreases exponentially in the number of repetitions. However, WI protocols do not always guarantee witness secrecy; in particular, for statements with a unique NP-witness, WI is meaningless. Nevertheless, [FS90] show how to use WI to achieve other notions of secrecy such as ZK and *witness-hiding*.

Witness hiding (WH). Roughly speaking, a protocol is WH [FS90] with respect to a distribution $\mathcal{D}$ on an NP-language $\mathcal{L}$ if no verifier can extract a witness from its interaction with the honest prover on a common instance $x \leftarrow \mathcal{D}$. For WH to be meaningful, it should be restricted to *hard distributions*; namely, distributions $\mathcal{D}$ for which poly-size circuits cannot find a witness $w \in \mathcal{R}_{\mathcal{L}}(x)$ for instances $x \leftarrow \mathcal{D}$. WH is in a sense a "minimal" notion of privacy; indeed, leaking the entire witness does not leave much room for imagination.

[FS90] present a 3-round protocol with negligible soundness error that is only WH with respect to a specific type of (hard) distributions on languages, where every instance has two witnesses. In contrast, extending the lower bounds of [GK96], the work of [HRS09] show that, for distributions with unique witnesses, 3-round WH cannot be "black-box reduced" to any "standard cryptographic assumption" (e.g., existence of OWFs), given natural limitations on the reduction.

In this work, we are interested in protocols that are WH with respect to all hard distributions (including the unique witness case). We remark that constructing WH protocols for restricted classes of distributions, where a lower bound on their hardness is apriori known, is a relatively easy task (and is not ruled out by [HRS09]). Indeed, using super-polynomial black-box reductions, it is possible to obtain 3-round WH protocols with respect to super-polynomial hard distributions. (For example, $f(n) = \omega(\log n)$ parallel repetitions of any 3-round ZK protocol with constant soundness error is WH with respect to distributions that are hard for $2^{f(n)}$-size adversaries.) Typical cryptographic scenarios, however, do call for secrecy with respect to general languages/distributions where no apriori super-poly hardness bound is known at the protocol's design time. Here, efficient reductions requiring non-black-box techniques are needed.

Weak zero-knowledge (WZK). The standard notion of ZK requires that for any (potentially adversarial) verifier there exist a simulator that simulates its view in an interaction with the honest prover. The simulated view should be indistinguishable from the real one by any (efficient) distinguisher. The notion of WZK [DNRS99] relaxes ZK by changing the order of quantifiers. Specifically, it allows the ZK simulator to depend on the particular distinguisher in question.

While ZK is often used as a sub-protocol in larger systems, WZK is not always suitable for this purpose due to its weaker simulation guarantee. In particular, WZK is not known to be closed under sequential repetition. Nevertheless, WZK is useful in settings where the verifier tries to learn a specific type of information and we can present a distinguisher that can test whether the verifier succeeded in learning it. Examples include verifiers that try to lean a specific predicate of the witness, or any function of the witness that is efficiently verifiable. In particular, WZK implies WH (by considering a distinguisher that tests if the verifier's view contains a valid witness). We note that, for black-box simulation, WZK and (standard) ZK coincide; hence, by [GK96], a 3-round protocol with negligible soundness error cannot even be shown to be WZK with black-box simulation.

To sum up the above discussion, 3-round arguments with negligible soundness error that are ZK, WH or WZK cannot be constructed using black-box techniques. (From this point on, we only consider proofs/arguments with negligible soundness error). In light of the existing non-black-box constructions, it is interesting to investigate which techniques and assumptions could suffice for constructing such protocols. Another interesting related question is understanding whether the relaxed notions of WH and WZK require simpler techniques than for full-fledged ZK; indeed, all existing WH constructions are based on the stronger notion of ZK as a building block. The question of finding "more direct" constructions of WH was already raised by [FS90]. This work sheds new light on both questions, introducing techniques based on *point obfuscation* (PO). We next briefly review the concept of PO.

Point obfuscation and extensions. Informally, an obfuscator is a randomized algorithm $\mathcal{O}$ that gets as input a program C (given by a circuit) and outputs a new program $\mathcal{O}(C)$ that has the same functionality as the original one, but does not leak any additional information on C [BGI+01]. A stronger variant is *obfuscation with auxiliary input*, in which $\mathcal{O}(C)$ does not leak any information even given a related auxiliary input z_C [GK05].

In this work, we consider obfuscation of *point circuits* and their extensions. A point circuit I_s outputs 1 on s and $\perp$ on all other inputs. A *multibit point circuit* $I_{s \to t}$ outputs t on s and $\perp$ otherwise. We also consider a new extension of point circuits which we call *circular point circuits* - these are circuits $I_{s \leftrightarrows t}$ which output t on input s, s on input t, and $\perp$ otherwise. Obfuscators for multibit point circuits are called *Digital Lockers* (DL). We introduce the new notion of *circular digital lockers* (CDL) that are obfuscators for circular point circuits. Point circuits and their extensions are among the very few functionalities for which obfuscators have been shown (albeit, typically, under rather strong hardness assumptions.) So far, however, POs have found only a handful of applications in cryptographic theory, mostly to strong forms of encryption [Can97, Wee05, CD08, CKVW10, BC10].

1.1 Our Contribution

We construct 3-round WH and WZK protocols based on two different variants of point obfuscation:

- 3-round negligible soundness WH proofs for NP, given auxiliary input point obfuscators that satisfy a relatively mild distributive security requirement. The protocol is WH with respect to general hard distributions (including the unique witness case).
- 3-round WZK arguments for NP, given auxiliary input digital lockers that satisfy a worst-case simulation security requirement.

We next give an overview of our constructions, followed by a discussion on the nature of our obfuscation assumptions and how they relate to previous assumptions used for 3-round ZK protocols.

3-round witness-hiding. The high level idea behind our WH protocol is as follows. Given an NP statement $x \in \mathcal{L}$, have the verifier $\mathcal{V}$ construct a modified NP verification circuit $\mathsf{Ver}^y_{\mathcal{L},x}$ that on a valid witness $w \in \mathcal{R}_{\mathcal{L}}(x)$ outputs a secret random point y and outputs $\perp$ otherwise. $\mathcal{V}$ then "garbles" this circuit using Yao's technique and both parties execute a 2-message oblivious-transfer protocol, at the end of which the prover $\mathcal{P}$ possesses the garbled circuit and the corresponding labels for the witness w. Next, $\mathcal{P}$ evaluates the circuit (on w) and obtains the point y. (This is essentially a *conditional disclosure of secrets* protocol, as termed by [GIKM00, AIR01], where $\mathcal{P}$ learns the output y only if it inputs a valid witness.) In the third message, $\mathcal{P}$ sends back to $\mathcal{V}$ a point obfuscation of y. $\mathcal{V}$ accepts only after verifying it got a valid obfuscation of y.

Informally, soundness follows from the secrecy of the garbled circuit that prevents a dishonest prover from obtaining the random y in case there is no valid witness. In fact, we show that our protocol is a *proof of knowledge*.

The witness-hiding property is based on the security of the underlying obfuscator. To exemplify, consider a version of the protocol where $\mathcal{P}$ sends back y in the clear. Following is an attack on this simple version of the protocol. Consider a cheating verifier $\mathcal{V}^*$ that, instead of garbling $\mathsf{Ver}^y_{\mathcal{L},x}$, garbles the identity circuit. $\mathcal{P}$ now evaluates the garbled circuit on w and obtains the point $y = w$. If $\mathcal{P}$ was to simply send back y in the clear, $\mathcal{V}^*$ would have learned w and the protocol would be completely insecure. Instead, $\mathcal{P}$ sends back an obfuscation $\mathcal{O}(y)$. The security of the obfuscator $\mathcal{O}$ should then assure that $\mathcal{V}^*$ cannot obtain w, unless "it was already known" to $\mathcal{V}^*$ in advance.

The security reduction and required obfuscation assumptions. As we have seen, the WH guarantee of our protocol depends on the security of the underlying point obfuscator $\mathcal{O}$. We now discuss the properties of the obfuscation used to show WH. Concretely, our underlying obfuscator should satisfy a distributional indistinguishability requirement with respect to points and related auxiliary information that are jointly sampled from an *unpredictable distribution*. We say that a distribution ensemble $\mathcal{D} = \{(Z_n, Y_n)\}_{n \in \mathbb{N}}$ on pairs of strings is unpredictable (UPD) if poly-size circuits cannot predict (with noticeable chance) the point Y_n, given the potentially related auxiliary input Z_n. We say that $\mathcal{O}$ is a distributional auxiliary input point obfuscator (AIPO) if, for any UPD $\mathcal{D} = \{(Z_n, Y_n)\}$, no poly-size circuit family can distinguish, given Z_n, an obfuscation of $\mathcal{O}(Y_n)$ from an obfuscation of a random point $\mathcal{O}(U_n)$.

In our setting, Z_n represents the common input x and the prover's first message (during the OT protocol). Y_n is the obfuscated point (returned by the honest prover). That is, Z_n is explicitly known to the verifier, while Y_n is obfuscated. A malicious $\mathcal{V}^*$ might choose its (garbled) circuit to output illegitimate information on the witness (i.e., information it could not predict on its own only from Z_n); the obfuscation, however, should prevent it from doing so.

3-round weak zero-knowledge. The WH protocol described above is not ZK - it enables a cheating verifier $\mathcal{V}^*$ to learn arbitrary predicates of the witness. For example, to learn w_1, the first bit of w, $\mathcal{V}^*$ can maliciously choose its garbled circuit to map w to one of two arbitrary points y_0, y_1 according to w_1. In this case, the honest prover sends an obfuscation $\mathcal{O}(y_{w_1})$, and $\mathcal{V}^*$ learns w_1 by simply running the obfuscation on each of the two points y_0, y_1. (This can be generalized to any function $f(w)$ where $|f(w)| = O(\log n)$, using a poly-size set $\{y_i\}$).

Towards making the protocol ZK, we try to cope with the above attack by requiring that the verifier "proves" it "fully knows" the secret point y (rather than just a poly-size set containing y). To achieve this without adding rounds, we ask that the verifier itself includes an obfuscation of y in its message. The prover then checks the obfuscation's consistency with the point extracted from the circuit evaluation. In case of inconsistency, the prover aborts. This modification, however, still does not prevent the above attack. The verifier $\mathcal{V}^*$ can learn w_1 by sending an obfuscation of the string y_0 and observing whether the prover aborts. Moreover, the protocol may no longer be sound since a cheating prover might use the verifier's obfuscation to create an obfuscation of the same point y without "knowing" y.

We resolve these issues as follows: (a) to regain soundness, we use an obfuscation scheme with non-malleability properties, based on an obfuscated circular point circuit. (b) to achieve WZK, we require that, instead of a plain point obfuscation, the verifier sends an obfuscated multibit point circuit that on the secret input y outputs the coins used by the verifier to garble the circuit. Now, the prover can verify that the garbled circuit is indeed $\mathsf{Ver}^y_{\mathcal{L},x}$ (for some y).

In order to show that the protocol is WZK, we use stronger notions of obfuscation. Since WZK requires worst-case simulation (i.e., simulation for any x), we require that our obfuscators also satisfy a worst-case simulation guarantee (rather than the weaker distributive definition used for WH). To simulate any verifier $\mathcal{V}^*$, our simulator must make use of the obfuscation simulator for $\mathcal{V}^*$. However, an obfuscation simulator for general adversaries with long output could not exist (see [BGI+01]); in fact, known constructions of PO only address simulation of adversaries with a single output bit. To overcome this, we use the fact that the WZK simulator is given a specific distinguisher $\mathcal{D}$, and the simulated verifier view should only needs to fool this specific $\mathcal{D}$. We show how to use an obfuscation simulator for the binary adversary $\mathcal{D}(\mathcal{V}^*)$, which is the composition of the distinguisher and the verifier, in order to construct a WZK simulator. Indeed, this limitation on simulating adversaries with long output is the reason we do not achieve full-fledged ZK.

1.2   Reflections on the Use of Point Obfuscation

The results of [GK96, HRS09] imply that our 3-round protocols cannot be shown secure using reductions that only make black-box use of the adversary. This is not surprising: indeed, neither auxiliary input nor standard point obfuscators can be shown to be secure using black-box reductions [Wee05]. Hence, our use of obfuscation inherently implies that the verifier is not used as a black-box.

To demonstrate the non-black-box nature of POs, we briefly review the techniques used in existing constructions [Can97, Wee05]. We can view POs as a special case of AIPOs, where the auxiliary input Z_n is empty. In this case, Y_n is unpredictable if it is *well-spread* (i.e., has super-logarithmic min-entropy) and the security requirement is that $\mathcal{O}(Y_n) \approx_c \mathcal{O}(U_n)$ for any well-spread Y_n.

The hardness assumptions made in [Can97, Wee05] are shown to imply that the strategy of any distinguisher essentially consists of a poly-size set of "distinguishing elements". That is, only obfuscations of points within this set are distinguishable from an obfuscation of a random point. However, these elements cannot be extracted using black-box access to the adversary. Hence, they are given to the reduction (or simulator) as non-uniform advice.

These techniques allow achieving the stronger worst-case simulation definition, thus showing that the distributive and worst-case definitions are in fact equivalent in the case of no auxiliary input. When considering auxiliary input, we can no longer apply these techniques. Indeed, the set of distinguishing elements can now depend on the auxiliary input in an arbitrary way. That is, no short advice suffices for the reduction to go through. In general, we do not know whether the distributive AIPO definition implies the worst-case simulation definition in the auxiliary input case (the converse still holds).

Concrete constructions. There exist very few constructions that were shown to be secure with respect to auxiliary input. [GK05] show that any point obfuscator is also secure with respect to auxiliary input that is chosen independently of the obfuscated point. [DKL09] suggest a construction that, under a variant of the LWE assumption, satisfies a restricted definition where the distribution $\mathcal{D}$ is "highly unpredictable". Both results are insufficient for our needs.

In this work, we consider two concrete constructions of AIPOs based on two different assumptions. The first AIPO, known as the (r, r^x) obfuscator, was suggested by Canetti [Can97] based on a strong variant of DDH. Informally, the assumption states that there exists an ensemble of prime order groups $\mathcal{G} = \{\mathbb{G}_n : |\mathbb{G}_n| = p_n\}$ such that for any unpredictable distribution $\mathcal{D} = (Z_n, Y_n)$ with support $\{0, 1\}^{\text{poly}(n)} \times \mathbb{Z}_{p_n}$: $(z, r, r^y) \approx_c (z, r, r^u)$, where $(z, y) \leftarrow (Z_n, Y_n), u \xleftarrow{U} \mathbb{Z}_{p_n}$ and r is a random generator of $\mathbb{G}_n$ [1].

For the second construction, we suggest a new assumption that is stated in terms of uninvertibility rather than indistinguishability. The assumption strengthens the assumption made by Wee [Wee05] to account for auxiliary inputs. Roughly, to construct (non auxiliary input) POs, Wee assumes a strong one-way

---

[1] Both [Can97, DKL09], make use of a slightly different formulation for the distributional AIPO requirement. Their formulation is essentially equivalent to ours.

permutation f that is "uninvertible" with respect to all well-spread distributions. A natural extension of the latter to the auxiliary input setting is to assume that the permutation is hard to invert, even given side information Z on the pre-image Y, from which Y cannot be predicted. An additional fact used by Wee is that permutations inherently preserve (information-theoretic) entropy; in particular, if Y is well-spread, so is $f(Y)$. In the (computational) auxiliary input setting, this might not be true; namely, it might be that Y is unpredictable from Z, while $f(Y)$ is predictable from Z. One possible way to deal with this issue is to assume a trapdoor permutation family (with the above strong uninvertibility). Further details can be found in the full version of this paper [BP11].

We remark that both the assumptions we consider (or any assumption that states that a specific obfuscation candidate is an AIPO satisfying either a the worst-case or the distributive security definition) are considered to be non-standard. In particular, any such assumption is non-falsifiable in the terms of Naor [Nao03].

Comparison with previous work on 3-round ZK. As already mentioned, it is known how to construct 3-round ZK arguments and proofs using non-falsifiable "knowledge assumptions," such as the knowledge of exponent assumption (KEA) [HT98, BP04], the POK assumption [LM01], or the existence of "extractable perfect one-way functions" (EPOWF)[CD09].

The KEA assumption [Dam91], essentially asserts that any algorithm that produces a DDH tuple, must "know" the corresponding exponents. Upon the formulation of KEA, [Dam91] raised a more general question regarding the existence of "sparse range one-way functions", such that any algorithm that can sample an element within the function's image, must also "know" a primage (KEA indeed yields such a OWF). The EPOWF primitive of [CD09] formalizes this generalization. All in all, all the above assumptions essentially fall under the abstract notion of EPOWF. (Indeed, [CD09] show that either one of the KEA or the POK assumptions imply the EPOWF primitive, when combined with a hardness assumption such as DDH.)

In this work, we show how to circumvent the black-box impossibility results for 3-round WZK and WH based on a *different* set of primitives; namely, (variants of) point obfuscation with auxiliary input. Currently, we do not know of any formal relation between the AIPO and EPOWF primitives, beyond the relation established in this work (through 3-round ZK). Formalizing such a relation is an interesting question on its own (going beyond the scope of 3-round ZK).

On the efficiency of the construction. Basing our constructions on (Yao-based) secure function evaluation results in efficient protocols with a practical implementation (similarly to [IKOS07]). By working directly with the verification circuit $\mathsf{Ver}_{\mathcal{L}}$, we avoid the overhead of *Karp reductions*, existing in most ZK protocols. Specifically, we can achieve communication complexity $O(ns)$, where n is the security parameter and s is the size of $\mathsf{Ver}_{\mathcal{L}}$. This is not optimal as there exist ZK arguments with polylog communication complexity [Kil92]. However, these require using PCPs, making them less practical.

Finally, we consider the techniques in use. Unlike previous works, our work demonstrates a direct WH construction that is not based on a ZK protocol. We then strengthen it to a limited form of ZK. Our WH to WZK transformation is specifically tailored for our construction. An interesting open question is whether a general transformation of this type exists.

Organization. In Section 2 we present the main definitions and tools used in this work. In Section 3 and Section 4 we introduce our WH and WZK protocols. For lack of space many of the details and proof are omitted and can be found in the full version of this paper [BP11].

2   Definitions and Tools

2.1   Weak Zero-Knowledge and Witness Hiding

In this work, we discuss two relaxations of ZK which are formalized next.

Weak zero-knowledge. In ZK, we require that the view of any verifier $\mathcal{V}^*$, in an interaction with the honest prover $\mathcal{P}$, can be simulated by an efficient simulator $\mathcal{S}$. The simulated view should be indistinguishable from the view of $\mathcal{V}^*$ for any poly-size distinguisher. In weak ZK (WZK), the simulator is only required to output a view that is indistinguishable from that of $\mathcal{V}^*$ for a specific distinguisher. This is modeled by supplying the simulator with the distinguisher circuit as additional auxiliary input.

Definition 2.1 (Weak zero-knowledge). *An argument system $(\mathcal{P}, \mathcal{V})$ is WZK if for every PPT verifier $\mathcal{V}^*$ there exist a PPT simulator $\mathcal{S}$ such that for every poly-size circuit family of distinguishers $\mathcal{D} = \{D_n\}_{n \in \mathbb{N}}$ and any $x \in \mathcal{L} \cap \{0,1\}^n$, $w \in \mathcal{R}_\mathcal{L}(x)$, $z \in \{0,1\}^{\mathrm{poly}(n)}$ it holds that:*

$$|\Pr[D_n((\mathcal{P}(w), \mathcal{V}^*(z))(x)) = 1] - \Pr[D_n(\mathcal{S}(D_n, x, z)) = 1]| \leq \mathrm{negl}(n) .$$

Witness-hiding. A protocol is WH if the verifier cannot fully learn a witness from its interaction with $\mathcal{P}$. This requirement is restricted to instances and witnesses (x, w) sampled from "hard distributions".

Definition 2.2 (Hard distribution). *Let $\mathcal{D} = \{D_n\}_{n \in \mathbb{N}}$ be an efficiently samplable distribution ensemble on $\mathcal{R}_\mathcal{L}$, i.e., the support of D_n is $\mathsf{Supp}(D_n) = \{(x, w) : x \in \mathcal{L} \cap \{0,1\}^n, w \in \mathcal{R}_\mathcal{L}(x)\}$. We say that $\mathcal{D}$ is hard if for any poly-size circuit family $\{C_n\}$ and sufficiently large n it holds that:*

$$\Pr_{(x,w) \overset{D_n}{\leftarrow} \mathcal{R}_\mathcal{L}} [C_n(x) \in \mathcal{R}_\mathcal{L}(x)] \leq \mathrm{negl}(n) .$$

Definition 2.3 ($\mathcal{D}$-witness-hiding). *An argument system $(\mathcal{P}, \mathcal{V})$ for an NP language $\mathcal{L}$ is WH with respect to a hard distribution $\mathcal{D} = \{D_n\}_{n \in \mathbb{N}}$, if for any poly-size verifier $\mathcal{V}^*$ and all large enough $n \in \mathbb{N}$:*

$$\Pr_{(x,w) \leftarrow D_n} [(\mathcal{P}(w), \mathcal{V}^*)(x) \in \mathcal{R}_{\mathcal{L}}(x)] \leq \mathrm{negl}(n) \ .$$

We say that $(\mathcal{P}, \mathcal{V})$ is WH if it is $\mathcal{D}$-WH for every hard distribution $\mathcal{D}$.

As discussed in the introduction, in this work we will be interested in WH protocols (with respect to a every hard distribution), and not with protocols that are WH with respect to a specific hard distribution.

2.2   2-Message Delegation

A central tool used in our constructions is a 2-message delegation protocol in which the prover and verifier jointly evaluate the NP verification circuit of the language on the common instance and the prover's witness. We use this primitive (following the formulation in [IP07]) to abstract the use of the Yao's garbled circuit construction.

A 2-message delegation protocol is executed by parties (A, B) where A has an input x, and B has as input a function f (given by a boolean circuit). The protocol should allow A to obtain $f(x)$ using two messages: $A \rightarrow B \rightarrow A$, without compromising the input secrecy of either party. We additionally require that, given B's message and secret randomness, one can reconstruct f. The protocol is defined by a tuple of algorithms $(\mathsf{Gen}, \mathsf{Enc}, \mathsf{Eval}, \mathsf{Dec}, \mathsf{Open})$ and proceeds as follows:

A: Obtains a key $sk \leftarrow \mathsf{Gen}(1^n)$, computes an encryption of its input $c \leftarrow \mathsf{Enc}(sk, x)$, and sends c.

B: Computes an encrypted output $\hat{c} \leftarrow \mathsf{Eval}(c, f)$ using randomness r, and sends back $\hat{c}$.

A: Outputs $y = \mathsf{Dec}(sk, \hat{c})$.

We briefly describe the security properties required from 2-message delegation schemes in this work:

- **Correctness:** When both parties are honest A outputs $f(x)$.
- **Input Hiding:** An adversarial B cannot learn A's input x (in the semantic security sense).
- **Function Hiding:** An adversarial A learns nothing about B's input f, other than the value of $f(x)$ (security in this case is simulation based).
- **Function Binding:** In a later stage, B can reveal its input function f by exhibiting its random coins. We require that for any message sent by B, it can reveal at most one function. While function-binding is not required in common formulations of delegation protocols, we show that a Yao-based construction (when instantiated with natural forms encryption) has this property.

In the full version of this paper [BP11], we provide a formal definition of secure 2-message delegation and describe a concrete instantiation based on *Yao's garbled circuit* technique and 2-message OT. We also define an information-theoretic version of this primitive, which we use in order achieve a WH protocol with unconditional soundness (i.e., a proof).

2.3   Point Obfuscation with Auxiliary Input

We start by recalling the standard definition for circuit obfuscation with auxiliary input. The definition is a worst-case definition, in the sense that simulation must hold for any circuit in the family and any related auxiliary input.

Definition 2.4 (Worst-case obfuscator with auxiliary input [BGI+01, GK05]). *A PPT $\mathcal{O}$ is an obfuscator with auxiliary input for an ensemble $\mathcal{C} = \{\mathcal{C}_n\}_{n\in\mathbb{N}}$ of families of poly-size circuits if it satisfies:*

- **Functionality.** *For any $n \in \mathbb{N}$, $C \in \mathcal{C}_n$, $\mathcal{O}(C)$ is a circuit that computes the same function as C.*
- **Polynomial slowdown.** *For any $n \in \mathbb{N}$, $C \in \mathcal{C}_n$, $|\mathcal{O}(C)| \leq \text{poly}(|C|)$.*
- **Virtual black box.** *For any PPT adversary $\mathcal{A}$ there is a PPT simulator $\mathcal{S}$ such that for all sufficiently large $n \in \mathbb{N}, C \in \mathcal{C}_n$ and $z \in \{0,1\}^{\text{poly}(n)}$:*

$$\left| \Pr[\mathcal{A}(z, \mathcal{O}(C)) = 1] - \Pr[\mathcal{S}^C(z, 1^{|C|}) = 1] \right| \leq \text{negl}(n) \ ,$$

where the probability is taken over the coins of $\mathcal{A}, \mathcal{S}$ and $\mathcal{O}$.

An obfuscator $\mathcal{O}$ is recognizable if given a program C and an alleged obfuscation of C, $\tilde{C}$, it is easy to verify that C and $\tilde{C}$ compute the same function.

- **Recognizability.** *There exist a polynomial time recognition algorithm $\mathbb{V}$ such that for any $C \in \mathcal{C}_n$:*
  - *$\Pr_{\mathcal{O}}[\mathbb{V}(C, \mathcal{O}(C)) = 1] = 1$*
  - *For any $\tilde{C} \in \{0,1\}^{\text{poly}(n)}$ if $\mathbb{V}(C, \tilde{C}) = 1$ then $\tilde{C}$ and C compute the same function.*

Point obfuscation. We consider obfuscation of *point circuits* and their extensions. A point circuit I_s outputs 1 on string s and $\bot$ on all other inputs.

Definition 2.5 (Worst-Case auxiliary-input point obfuscation (AIPO)). *A PPT algorithm $\mathcal{O}$ is a worst-case AIPO if it is a recognizable obfuscator (according to Definition 2.4) for the circuit ensemble: $\mathcal{C} = \{\mathcal{C}_n = \{I_s | s \in \{0,1\}^n\}\}_{n\in\mathbb{N}}$.*

Remark 2.1. The notion of recognizable obfuscation was not explicitly defined in previous works. We only consider this property in the context of point obfuscation. While, in general, point obfuscators are not required to be recognizable, previously constructed obfuscators [Can97, Wee05] are trivially recognizable. This is due to the fact that they use public randomness, i.e., the randomness used by the obfuscator appears in the clear as part of the obfuscated circuit. The recognition algorithm, given a program and its obfuscation, can simply rerun the obfuscation algorithm with the public randomness and compare the result to the obfuscation in hand.

We next present a weaker distributional definition for point obfuscation with auxiliary input that previously appeared in [Can97] (in a slightly different formulation). We first give a preliminary definition of unpredictable distributions (generalizing Definition 2.2) and then present the obfuscation definition.

Definition 2.6 (Unpredictable distribution). *A distribution ensemble* $\mathcal{D} = \{D_n = (Z_n, Y_n)\}_{n \in \mathbb{N}}$, *on pairs of strings is unpredictable if no poly-size circuit family can predict* Y_n *from* Z_n. *That is, for every poly-size circuit family* $\{C_n\}_{n \in \mathbb{N}}$ *and for all large enough* n:

$$\Pr_{(z,y) \leftarrow D_n} [C_n(z) = y] \leq \operatorname{negl}(n) \ .$$

Definition 2.7 (Auxiliary input point obfuscation for unpredictable distributions (AIPO)). *A PPT algorithm* $\mathcal{O}$ *is a point obfuscator for unpredictable distributions if it satisfies the functionality and polynomial slowdown requirements as in Definition 2.4, and the following secrecy property. For any unpredictable distribution* $\mathcal{D} = \{D_n = (Z_n, Y_n)\}$ *over* $\{0,1\}^{\operatorname{poly}(n)} \times \{0,1\}^n$ *it holds that:*

$$\{z, \mathcal{O}(y) : (z, y) \leftarrow D_n\}_{n \in \mathbb{N}} \approx_c \left\{ z, \mathcal{O}(u) : z \leftarrow Z_n, u \xleftarrow{U} \{0,1\}^n \right\}_{n \in \mathbb{N}} \ .$$

Remark 2.2. Using this definition in our WH construction, we can settle for a slightly relaxed definition with *bounded auxiliary input*; namely $|Y_n| = \omega(|Z_n|)$. We do not know if such a bounded form of auxiliary-input indeed weakens the requirement. However, it does seem to withstand certain "diagonalization attacks" that can be performed for the non-restrictive (under certain obfuscation assumptions).

2.4   Digital Lockers and Circular Digital Lockers

We also consider obfuscation of several extensions of point circuits. Specifically, *multibit point circuits* and *circular point circuits*. A multibit point circuit $I_{s \to t}$ outputs t on s and $\perp$ otherwise. A circular Point circuit $I_{s \leftrightarrows t}$ outputs t on input s, s on input t, and $\perp$ otherwise. Obfuscators satisfying the worst-case AIPO definition (Definition 2.5) for multibit point circuits and circular point circuits are called *digital lockers* (DLs) and *circular digital lockers* (CDLs).

Definition 2.8 (Digital locker (DL)). *A PPT algorithm is a DL if it is a recognizable obfuscator (according to Definition 2.4) for the circuit ensemble:* $\mathcal{C} = \{\mathcal{C}_n = \{I_{s \to t} | s, t \in \{0,1\}^n\}\}_{n \in \mathbb{N}}$.

Definition 2.9 (Circular digital locker (CDL)). *A PPT algorithm is a CDL if it a recognizable obfuscator (according to Definition 2.4) for the circuit ensemble:* $\mathcal{C} = \{\mathcal{C}_n = \{I_{s \leftrightarrows t} | s, t \in \{0,1\}^n\}\}_{n \in \mathbb{N}}$.

Remark 2.3. We note that the "security under circularity" feature is inherently provided by the strong obfuscation guarantees, was already considered in previous work for constructing strong encryption schemes which withstand *key dependent messages* and *related keys attacks* [CKVW10, BC10].

While AIPOs are sufficient for our WH protocol, our WZK protocol requires DLs and CDLs. In the full version of this paper [BP11], we describe how DLs and CDLs can be constructed based on a worst-case AIPOs that satisfy an additional property of *composability*.

3   3-Round WH

Overview of the protocol. As a warmup, consider first the following **unsound** protocol: to prove an NP statement $x \in \mathcal{L}$, the prover $\mathcal{P}$ and verifier $\mathcal{V}$ first engage in a 2-message delegation protocol where $\mathcal{P}$'s (secret) input is the witness w and $\mathcal{V}$'s input function is the NP verification circuit $\mathsf{Ver}_{\mathcal{L},x}$. $\mathcal{P}$ obtains the result $\mathsf{Ver}_{\mathcal{L},x}(w)$ and sends it to $\mathcal{V}$. This is unsound since a cheating prover can always send "1" as its last message.

To make the protocol sound, we augment it as follows. Let $\mathsf{Ver}_{\mathcal{L},x}^{y}$ be a circuit that outputs y on valid witnesses and $\perp$ otherwise. Now, $\mathcal{V}$ will choose a secret string $y \in_R \{0,1\}^n$ and use the circuit $\mathsf{Ver}_{\mathcal{L},x}^{y}$ as its secret input in the delegation protocol. In order to convince $\mathcal{V}$ of the statement, $\mathcal{P}$ should send back y. Indeed, in case $x \notin \mathcal{L}$ we have $\mathsf{Ver}_{\mathcal{L},x}^{y} \equiv \perp$, and hence, the "function hiding" property of the delegation protocol assures that $\mathcal{P}$ does not learn the random y.

However, this protocol is not witness hiding. Indeed, a cheating verifier can try to obtain w by maliciously choosing its input function. For instance, choosing the function to be the Identity results in the prover sending back w.

A natural approach towards fixing the latter problem would be to have the verifier "prove" it behaved honestly, without revealing its secret. In other words, it should give a round-efficient witness-hiding proof, which is what we set out to do to begin with. Thus, we take a different approach. We note that an honest verifier that "knows" y should only be able to verify that the prover "knows" it as well; hence, it suffices to have the prover send a *point obfuscation* of y, instead of sending y in the clear. The security of the obfuscation would then guarantee that any information that the verifier learns on w could also be learned (with noticeable probability) without the obfuscation.

The protocol. Let $\mathsf{DEL} = (\mathsf{Gen}, \mathsf{Enc}, \mathsf{Eval}, \mathsf{Dec}, \mathsf{Open})$ be a secure 2-message delegation protocol and let $\mathcal{O}$ be a point obfuscator for unpredictable distributions (AIPO) with recognition algorithm $\mathbb{V}$. The protocol is given by Figure 1.

Theorem 3.1. *Let* DEL *be a secure 2-message delegation protocol, and let* $\mathcal{O}$ *be an AIPO. Protocol 1 is a WH interactive argument.*

We briefly overview the proof of Theorem 3.1. The full proof as well as an extension from an argument to a proof can be found in the full version of this paper [BP11].

Soundness. The soundness of Protocol 1 follows from the function hiding of the underlying delegation scheme DEL and the recognizability of the point obfuscator. Indeed, in case there is no valid witness the verifier's message reveals

---

Common Input: $x \in \mathcal{L}$. **Auxiliary Input to** $\mathcal{P}$: $w \in \mathcal{R}_\mathcal{L}(x)$.

1. $\mathcal{P}$: Obtains $sk \leftarrow \mathsf{Gen}(1^n)$ and sends $c = \mathsf{Enc}(sk, w)$.
2. $\mathcal{V}$: Samples $y \xleftarrow{U} \{0,1\}^n$, obtains $\hat{c} \leftarrow \mathsf{Eval}(c, \mathsf{Ver}^y_{\mathcal{L},x})$ and sends $\hat{c}$.
3. $\mathcal{P}$: Decrypts $\tilde{y} = \mathsf{Dec}(sk, \hat{c})$, computes a point obfuscation $\mathcal{O}(\tilde{y})$ and sends it.
4. $\mathcal{V}$: Accepts iff $\mathbb{V}(I_y, \mathcal{O}(\tilde{y})) = 1$, i.e., $\mathcal{O}(\tilde{y})$ computes the same function as I_y.

Fig. 1. Protocol 1, 3-round Witness Hiding

no information regarding the verifier's secret random point y. Specifically, the prover's view can be simulated independently of y. Since the obfuscation is recognizable, in order to fool the verifier, the prover must send a point circuit computing I_y and can only succeed with negligible probability.

Proof of Knowledge. In fact, we can show that our WH protocol satisfies a stronger soundness property, namely it is a *proof of knowledge*. For this purpose, we use a similar idea to the one in the "knowledge attack" that shows why the protocol is not ZK (described in the introduction). In order to extract a witness, we essentially apply this attack repeatedly "against" the prover, revealing the witness' bits one by one. Our extractor only makes black-box use of the prover and extracts the witness using rewinding.

Witness hiding. The WH property is based on the input hiding of the delegation scheme DEL and the indistinguishability with respect to unpredictable distributions guarantee of the AIPO $\mathcal{O}$. Concretely, we show how any $\mathcal{V}^*$ that manages to extract a witness w from its interaction with $\mathcal{P}$ can be used to break the input hiding property of DEL. The reduction samples (x, w) from the hard distribution and submits $c_0 = w, c_1 = 1^{|w|}$ to the challenger. Upon receiving a challenge $c = \mathsf{Enc}(sk, c_b)$ it simulates $\mathcal{V}^*(x)$ with c as the first message. $\mathcal{V}^*$ then generates its own message $\hat{c}$, and it is left to simulate the last obfuscation message. To do so, we treat two cases, corresponding to whether the secret point y (induced by $\mathcal{V}^*$'s choice of input circuit to DEL) is (a) unpredictable from (x, c) or (b) is predictable by some poly size predictor Π. Intuitively, the first corresponds to a verifier that chooses its input circuit maliciously to gain information on w. The second, corresponds to a verifier that chooses its circuit honestly. To simulate the obfuscation in the second case, we apply the prediction circuit to compute $y \leftarrow \Pi(x, c)$ and feed $\mathcal{V}^*$ with $\mathcal{O}(y)$. In the case that y is unpredictable, we feed $\mathcal{V}^*$ an obfuscation $\mathcal{O}(u)$ of a random point u. Finally, when $\mathcal{V}^*$ outputs $\tilde{w}$, we check whether it is a valid witness, and if so answer the challenger with $b = 0$. Otherwise, we guess b at random. Indeed, by the indistinguishability guarantee of the AIPO, in case $b = 0$ (i.e., the simulation is done with an encryption of w) the simulated $\mathcal{V}^*$ will manage to extract a witness with noticeable probability (related to the the prediction probability of Π and the success probability of $\mathcal{V}^*$ in a true interaction). In case that $b = 1$, the reduction is unlikely to produce a valid witness since its view is completely independent of w and the underlying distribution is hard. We stress that the reduction is, indeed, not black box in $\mathcal{V}^*$; in particular, it applies the predictor Π implied by the AIPO guarantee, which is not black-box in $\mathcal{V}^*$.

On restricted auxiliary input. In our WH protocol we require the AIPO distributional guarantee to hold with respect to any unpredictable distribution. However, we can in fact settle for less. Specifically, the auxiliary input distribution in Protocol 1 is essentially restricted to a very "benign" form; namely, the first delegation message (ciphertext) and the hard instance x; in particular, the auxiliary input is of fixed polynomial size and can be made much shorter than the obfuscated random point.

Why isn't Protocol 1 ZK? Protocol 1 is not ZK and in fact enables a cheating verifier $\mathcal{V}^*$ to learn arbitrary predicates on the witness. Specifically, $\mathcal{V}^*$ can deviate from the protocol by maliciously selecting its input circuit C for the delegation protocol as follows. Let $B : \{0,1\}^* \to \{0,1\}^t$ be a polynomial time computable function with $t = O(\log(n))$ output bits. To learn $B(w)$, $\mathcal{V}^*$ fixes an arbitrary set of strings $Y = \{y_j\}_{j \in \{0,1\}^t}$ and sets its input circuit $C = C_B$ to map the witness w to $y_{B(w)}$. Indeed, given an obfuscation of $C_B(w)$, $\mathcal{V}^*$ can simply run the obfuscation on all points in $\{y_j\}$ and learn $B(w)$. In the following section we explain how to transform Protocol 1 to a WZK protocol.

4   3-Round WZK

Overview of the protocol. To make Protocol 1 WZK, we try to cope with verifiers executing the "malicious circuit choice attack" described in the previous section. As explained in the introduction, this involves two main modifications:

1. We require that the verifier's message also includes a *digital locker* $\mathsf{DL}(I_{y \to r_\mathcal{V}})$, which on the secret input y "unlocks" the secret coins $r_\mathcal{V}$ used by the verifier in the delegation protocol. Upon receiving this message, the honest prover $\mathcal{P}$ applies Dec as in the previous protocol, obtains y, and then retrieves the coins $r_\mathcal{V}$. Now $\mathcal{P}$ can apply the Open algorithm of the delegation to verify that the input circuit of $\mathcal{V}^*$ was honestly chosen (to be $\mathsf{Ver}^y_{\mathcal{L},x}$). In case it was not, $\mathcal{P}$ returns a *circular digital locker* (CDL) (Definition 2.9) of a randomly selected circular point circuit.

2. The prover is required to send back an obfuscation of y (as in the previous protocol). However, to maintain soundness we should prevent a malicious prover from using (or mauling) the verifier's message $\mathsf{DL}(I_{y \to r_\mathcal{V}})$ to get the required obfuscation. For this purpose, we apply a "non-malleable obfuscation scheme", implemented as follows.[2] In its first message, the prover

---

[2] We only consider a very restricted form of non-malleability where the adversary tries to copy an obfuscation of the same point. A more general notion of non-mailable obfuscation can be found in [CV08].

commits to a random $r \in \{0,1\}^n$ (by sending the image of r under some injective OWF f). Then in the last message, it sends a *circular digital locker* $\mathsf{CDL}(I_{y \leftrightarrows r})$ that "binds" r and the secret point y. The honest verifier then runs the CDL on y, retrieves r and uses the CDL recognition algorithm to validate the CDL.

We now fully describe the protocol and then turn to analyze it.

The protocol. Let $\mathsf{DEL} = (\mathsf{Gen}, \mathsf{Enc}, \mathsf{Eval}, \mathsf{Dec}, \mathsf{Open})$ be a secure 2-message delegation scheme. Let $\mathsf{DL}, \mathsf{CDL}$ be a digital locker and a circular digital locker. Let $\mathbb{V}$ be the recognition algorithm for the CDL. Let f be an *injective one way function*. The protocol is presented in Figure 2.

Common Input: $x \in \mathcal{L}$. **Auxiliary Input to** $\mathcal{P}$: $w \in \mathcal{R}_\mathcal{L}(x)$.

1. $\mathcal{P}$: Obtains $sk \leftarrow \mathsf{Gen}(1^n)$ and $c \leftarrow \mathsf{Enc}(sk, w)$, samples $r \xleftarrow{U} \{0,1\}^n$, sends c and $f(r)$.

2. $\mathcal{V}$: Samples $y \xleftarrow{U} \{0,1\}^n$, obtains $\hat{c} \leftarrow \mathsf{Eval}(c, \mathsf{Ver}^y_{\mathcal{L},x})$ using random coins $r_\mathcal{V}$, sends $\hat{c}$ and $\mathsf{DL}_\mathcal{V} = \mathsf{DL}(I_{y \to r_\mathcal{V}})$.

3. $\mathcal{P}$: Decrypts $\tilde{y} = \mathsf{Dec}(sk, \hat{c})$, obtains $\tilde{r}_\mathcal{V} = \mathsf{DL}_\mathcal{V}(\tilde{y})$, verifies that $\mathbb{V}(I_{\tilde{y} \to r_\mathcal{V}}, \mathsf{DL}_\mathcal{V}) = 1$ and $\mathsf{Open}(\hat{c}, \tilde{r}_\mathcal{V}) = \mathsf{Ver}^{\tilde{y}}_{\mathcal{L},x}$.

   If so, sends back $\mathsf{CDL}_\mathcal{P} = \mathsf{CDL}(I_{\tilde{y} \leftrightarrows r})$. Otherwise, samples $u \xleftarrow{U} \{0,1\}^n$ and sends back $\mathsf{CDL}_\mathcal{P} = \mathsf{CDL}(I_{u \leftrightarrows u})$.

4. $\mathcal{V}$: Obtains $\tilde{r} = \mathsf{CDL}_\mathcal{P}(y)$, accepts iff $f(\tilde{r}) = f(r)$ and $\mathbb{V}(I_{y \leftrightarrows \tilde{r}}, \mathsf{CDL}_\mathcal{P}) = 1$.

Fig. 2. Protocol 2, 3-round WZK

Theorem 4.1. *Let* DEL *be a 2-message delegation protocol, let* DL *be a digital locker and* CDL *a circular digital locker, and let* f *be an injective one way function, then Protocol 2 is a WZK argument.*

We briefly overview the proof of Theorem 3.1. The full proof can be found in the full version of this paper [BP11].

Soundness. Soundness is shown in two stages. First, we argue that given $\mathcal{V}$'s message $(\hat{c}, \mathsf{DL}_\mathcal{V})$, it is hard to recover the underlying secret point y. I.e, no poly-size circuit family can recover y, except with negligible chance. Indeed, the auxiliary input obfuscation guarantee implies that if y can be recovered from $\mathsf{DL}_\mathcal{V}$ and the related auxiliary information $\hat{c}$, it can also be recovered solely from $\hat{c}$. However, since $x \notin \mathcal{L}$ and DEL is function hiding, y cannot be recovered from $\hat{c}$ (similarly to the WH protocol).

Second, we show that any cheating prover $\mathcal{P}^*$ can be used to recover y from $\mathcal{V}$'s message. Assume WLOG that $\mathcal{P}^*$ is deterministic, and note that, in its first message, $\mathcal{P}^*$ sends some (fixed) $f(r)$. Since f is injective, $\mathcal{P}^*$ is in fact "committed" to the corresponding fixed r. We can then feed $\mathcal{P}^*$ with $\mathcal{V}$'s message and get back $\mathsf{CDL}_\mathcal{P}$. Noting that whenever $\mathcal{P}^*$ convinces $\mathcal{V}$, $\mathsf{CDL}_\mathcal{P}(r) = y$, we can run $\mathsf{CDL}_\mathcal{P}$ on r (given as non-uniform advice) and obtain y with noticeable probability.

Weak zero-knowledge. We present a WZK simulator that, given an adversary $\mathcal{V}^*$ and a distinguisher $\mathcal{D}$, simulates the view of V^* with respect to $\mathcal{D}$. Let $\mathcal{V}_\mathcal{D}^*$ be the composition of $\mathcal{D}$ with $\mathcal{V}^*$. $\mathcal{V}_\mathcal{D}^*$ outputs a bit after receiving $\mathsf{CDL}_\mathcal{P} = \mathsf{CDL}(I_{y \leftrightarrows r})$ as the last message. In particular, there exist a PPT $\mathcal{S}_{\mathsf{CDL}}$ that simulates $\mathcal{V}_\mathcal{D}^*$'s output given oracle access to $I_{y \leftrightarrows r}$ and auxiliary input $\mathsf{ai} = (z, x, c, f(r))$, representing the rest of $\mathcal{V}_\mathcal{D}^*$'s view.

The WZK simulator $\mathcal{S}$ will simulate ai on its own, and utilize $\mathcal{S}_{\mathsf{CDL}}$ to simulate $\mathsf{CDL}_\mathcal{P}$ as the last message. To simulate ai, $\mathcal{S}$ samples r and computes $f(r)$. c is simulated by generation a random key $sk \leftarrow \mathsf{Gen}(1^n)$ and computing $c = \mathsf{Enc}(sk, 0^{|w|})$ (instead of w as in a true interaction). The input hiding of DEL implies that the simulated ai is indistinguishable from the true ai. We explain how $\mathcal{S}_{\mathsf{CDL}}$ is used to simulate the last obfuscation message. $\mathcal{S}$ first obtains the verifier's message $(\mathsf{DL}_{\mathcal{V}^*}, \hat{c})$. It then runs $\mathcal{S}_{\mathsf{CDL}}$ with the simulated ai, monitoring all its oracle queries. We treat two separate cases: (a) $\mathcal{S}_{\mathsf{CDL}}$ makes a query y which unlocks $\mathsf{DL}_{\mathcal{V}^*}$; (b) $\mathcal{S}_{\mathsf{CDL}}$ never makes such a query, in which case we always answer its queries with $\bot$.

The first case corresponds to a verifier that "knows" the secret point y. In this case, our simulator can perfectly simulate the behavior of $\mathcal{P}$. That is, it can "open" $\hat{c}$ to check its validity and consistency with $\mathsf{DL}_{\mathcal{V}^*}$, and send back the corresponding CDL.

The second case corresponds to a cheating $\mathcal{V}^*$ that either produces an invalid message or somehow produces a valid message but without actually "knowing" the secret y. In this case, the simulator will always return a "dummy obfuscation". This simulates the behavior of the honest prover $\mathcal{P}$. Indeed, if $\mathcal{V}^*$'s message is invalid, the prover also produces a "dummy obfuscation". If $\mathcal{V}^*$ does not "know" y, it can not distinguish $\mathcal{P}$'s message from a "dummy obfuscation".

The simulator. Let $\mathcal{V}^*$ be any verifier, and let $\mathcal{D}$ be the distinguisher circuit. Denote by $\mathcal{V}_1^*(z, x, c, f(r))$ the algorithm that runs $\mathcal{V}^*(z, x)$, feeds it with $(c, f(r))$ as the first message, and outputs $\mathcal{V}^*$'s message. Denote by $\mathcal{V}_2^*(x, z, c, f(r), \mathsf{CDL}_\mathcal{P})$ the algorithm that runs $\mathcal{V}^*(x, z)$, feeds it with $(c, f(r))$ as a first message, with $\mathsf{CDL}_\mathcal{P}$ as a second message, and returns $\mathcal{V}^*$'s output. Also, denote by $\mathcal{V}_\mathcal{D}^*(x, z, c, f(r), \mathsf{CDL}_\mathcal{P})$ the algorithm that runs $\mathcal{V}_2^*(x, z, c, f(r), \mathsf{CDL}_\mathcal{P})$, applies the circuit $\mathcal{D}$ on the output of $\mathcal{V}_2^*$ and returns the output bit of $\mathcal{D}$.

Let $\mathcal{S}_{\mathcal{V}^*, \mathcal{D}}(x, z, c, f(r))$ be the PPT obfuscation simulator of $\mathcal{V}_\mathcal{D}^*$ as specified by Definition 2.4. Also let $\ell(n)$ be the length of a witness for instances of length n. The description of the simulator is given by Algorithm 1.

Algorithm 1. Simulator $\mathcal{S}$

Input: $x \in \mathcal{L}, z \in \{0,1\}^*$

1: Set $\tilde{y} = \perp$.
2: Sample $r, u \xleftarrow{U} \{0,1\}^n$.
3: Obtain $sk \leftarrow \mathsf{Gen}(1^n)$.
4: Compute $c \leftarrow \mathsf{Enc}(sk, 1^{\ell(|x|)})$
5: Compute $(\hat{c}, \mathsf{DL}_\mathcal{V}) = \mathcal{V}_1^*(x, z, c, f(r))$.
6: Emulate $\mathcal{S}_{\mathcal{V}^*,\mathcal{D}}(x, z, c, f(r))$.
7: **for** each oracle query Q made by $\mathcal{S}_{\mathcal{V}^*,\mathcal{D}}$ **do**
8:     **if** $\mathsf{DL}_\mathcal{V}(Q) = \perp$ **then**
9:         Answer $\mathcal{S}$'s query with $\perp$ and continue the emulation.
10:    **else**
11:        Set $\tilde{r}_\mathcal{V} = \mathsf{DL}_\mathcal{V}(Q)$
12:        **if** $\mathbb{V}(I_{Q \to r_\mathcal{V}}, \mathsf{DL}_\mathcal{V}) = 1$ **then**
13:            Set $\tilde{y} = Q$
14:        **end if**
15:        End the emulation of $\mathcal{S}_{\mathcal{V}^*,\mathcal{D}}$.
16:    **end if**
17: **end for**
18: **if** $\tilde{y} = \perp$ **or** $\mathsf{Open}(\hat{c}, \tilde{r}_\mathcal{V}) \neq \mathsf{Ver}_{\mathcal{L},x}^{\tilde{y}}$ **then**
19:    **return** $\mathcal{V}_2^*(x, z, c, f(r), \mathsf{CDL}(I_{u \leftrightarrows u}))$.
20: **else**
21:    **return** $\mathcal{V}_2^*(x, z, c, f(r), \mathsf{CDL}(I_{\tilde{y} \leftrightarrows r}))$.
22: **end if**

Acknowledgements. We thank Amit Sahai for introducing us to the problem of 3-round witness hiding. We thank Ran Canetti and Yuval Ishai for valuable discussions. In particular, we thank Yuval for the idea of how to transform the witness-hiding protocol from an argument to a proof and for bringing to our attention previous applications of conditional disclosure of secrets.

References

[AIR01]   Aiello, W., Ishai, Y., Reingold, O.: Priced Oblivious Transfer: How to Sell Digital Goods. In: Pfitzmann, B. (ed.) EUROCRYPT 2001. LNCS, vol. 2045, pp. 119–135. Springer, Heidelberg (2001)

[Bar01]   Barak, B.: How to go beyond the black-box simulation barrier. In: FOCS, pp. 106–115 (2001)

[BC10]   Bitansky, N., Canetti, R.: On Strong Simulation and Composable Point Obfuscation. In: Rabin, T. (ed.) CRYPTO 2010. LNCS, vol. 6223, pp. 520–537. Springer, Heidelberg (2010)

[BCC88]   Brassard, G., Chaum, D., Crépeau, C.: Minimum disclosure proofs of knowledge. J. Comput. Syst. Sci. 37(2), 156–189 (1988)

[BGI+01]   Barak, B., Goldreich, O., Impagliazzo, R., Rudich, S., Sahai, A., Vadhan, S.P., Yang, K.: On the (Im)possibility of Obfuscating Programs. In: Kilian, J. (ed.) CRYPTO 2001. LNCS, vol. 2139, pp. 1–18. Springer, Heidelberg (2001)

[BP04]   Bellare, M., Palacio, A.: The Knowledge-of-Exponent Assumptions and 3-Round Zero-Knowledge Protocols. In: Franklin, M. (ed.) CRYPTO 2004. LNCS, vol. 3152, pp. 273–289. Springer, Heidelberg (2004)

[BP11]   Bitansky, N., Paneth, O.: Point obfuscation and 3-round zero-knowledge, Cryptology ePrint Archive, Report 2011/493 (2011), http://eprint.iacr.org/

[Can97]   Canetti, R.: Towards Realizing Random Oracles: Hash Functions that Hide All Partial Information. In: Kaliski Jr., B.S. (ed.) CRYPTO 1997. LNCS, vol. 1294, pp. 455–469. Springer, Heidelberg (1997)

[CD08]   Canetti, R., Dakdouk, R.R.: Obfuscating Point Functions with Multibit Output. In: Smart, N.P. (ed.) EUROCRYPT 2008. LNCS, vol. 4965, pp. 489–508. Springer, Heidelberg (2008)

[CD09]   Canetti, R., Dakdouk, R.R.: Towards a Theory of Extractable Functions. In: Reingold, O. (ed.) TCC 2009. LNCS, vol. 5444, pp. 595–613. Springer, Heidelberg (2009)

[CKVW10] Canetti, R., Tauman Kalai, Y., Varia, M., Wichs, D.: On Symmetric Encryption and Point Obfuscation. In: Micciancio, D. (ed.) TCC 2010. LNCS, vol. 5978, pp. 52–71. Springer, Heidelberg (2010)

[CV08]   Canetti, R., Varia, M.: Non-malleable obfuscation, Cryptology ePrint Archive, Report 2008/495 (2008), http://eprint.iacr.org/

[Dam91]   Damgård, I.: Towards Practical Public Key Systems Secure against Chosen Ciphertext Attacks. In: Feigenbaum, J. (ed.) CRYPTO 1991. LNCS, vol. 576, pp. 445–456. Springer, Heidelberg (1992)

[DKL09]   Dodis, Y., Kalai, Y.T., Lovett, S.: On cryptography with auxiliary input. In: STOC, pp. 621–630 (2009)

[DNRS99] Dwork, C., Naor, M., Reingold, O., Stockmeyer, L.J.: Magic functions. In: FOCS, pp. 523–534 (1999)

[FS90]   Feige, U., Shamir, A.: Witness indistinguishable and witness hiding protocols. In: STOC, pp. 416–426 (1990)

[GIKM00] Gertner, Y., Ishai, Y., Kushilevitz, E., Malkin, T.: Protecting data privacy in private information retrieval schemes. In: JCSS, pp. 151–160. ACM Press (2000)

[GK96]   Goldreich, O., Krawczyk, H.: On the composition of zero-knowledge proof systems. SIAM J. Comput. 25(1), 169–192 (1996)

[GK05]   Goldwasser, S., Kalai, Y.T.: On the impossibility of obfuscation with auxiliary input. In: FOCS, pp. 553–562 (2005)

[GMR85]   Goldwasser, S., Micali, S., Rackoff, C.: The knowledge complexity of interactive proof-systems (extended abstract). In: STOC, pp. 291–304 (1985)

[GO94]   Goldreich, O., Oren, Y.: Definitions and properties of zero-knowledge proof systems. J. Cryptology 7(1), 1–32 (1994)

[HRS09]   Haitner, I., Rosen, A., Shaltiel, R.: On the (Im)Possibility of Arthur-Merlin Witness Hiding Protocols. In: Reingold, O. (ed.) TCC 2009. LNCS, vol. 5444, pp. 220–237. Springer, Heidelberg (2009)

[HT98]   Hada, S., Tanaka, T.: On the Existence of 3-Round Zero-Knowledge Protocols. In: Krawczyk, H. (ed.) CRYPTO 1998. LNCS, vol. 1462, pp. 408–423. Springer, Heidelberg (1998)

[IKOS07] Ishai, Y., Kushilevitz, E., Ostrovsky, R., Sahai, A.: Zero-knowledge from secure multiparty computation. In: STOC, pp. 21–30 (2007)

[IP07]   Ishai, Y., Paskin, A.: Evaluating Branching Programs on Encrypted Data.
            In: Vadhan, S.P. (ed.) TCC 2007. LNCS, vol. 4392, pp. 575–594. Springer,
            Heidelberg (2007)
[Kil92]   Kilian, J.: A note on efficient zero-knowledge proofs and arguments (ex-
            tended abstract). In: STOC, pp. 723–732 (1992)
[LM01]   Lepinski, M., Micali, S.: On the existence of 3-round zero-knowledge proof
            systems. Tech. report, MIT LCS (2001)
[Nao03]   Naor, M.: On Cryptographic Assumptions and Challenges. In: Boneh, D.
            (ed.) CRYPTO 2003. LNCS, vol. 2729, pp. 96–109. Springer, Heidelberg
            (2003)
[Wee05]   Wee, H.: On obfuscating point functions. In: STOC, pp. 523–532 (2005)

Confidentiality and Integrity:
A Constructive Perspective

Ueli Maurer, Andreas Rüedlinger, and Björn Tackmann

Department of Computer Science, ETH Zürich, Switzerland
{maurer,bjoernt}@inf.ethz.ch,
andreas.rueedlinger@gmail.com

Abstract. Traditional security definitions in the context of secure communication specify properties of cryptographic schemes. For symmetric encryption schemes, these properties are intended to capture the protection of the confidentiality or the integrity of the encrypted messages. A vast variety of such definitions has emerged in the literature and, despite the efforts of previous work, the relations and interplay of many of these notions (which are a priori not composable) are unexplored. Also, the exact guarantees implied by the properties are hard to understand.

In constructive cryptography, notions such as confidentiality and integrity appear as attributes of channels, i.e., the communication itself. This makes the guarantees achieved by cryptographic schemes explicit, and leads to security definitions that are composable.

In this work, we follow the approach of constructive cryptography, questioning the justification for the existing (game-based) security definitions. In particular, we compare these definitions with related constructive notions and find that some are too weak, such as INT-PTXT, or artificially strong, such as INT-CTXT. Others appear unsuitable for symmetric encryption, such as IND-CCA.

Keywords: confidentiality, integrity, constructive cryptography.

1   Introduction

Symmetric encryption protects the confidentiality of messages transmitted between two parties that share a secret key. Intuitively, this means that the encrypted message (the ciphertext) transmitted from the sender A to the receiver B does not leak information about the contents of the message (other than, for example, its length). In contrast, encryption generally does not protect integrity: If the ciphertext is modified during transmission, the message obtained by decrypting might differ from the original message.

For some applications of encryption schemes, bare confidentiality is not sufficient. In his analysis of the Authenticate-then-Encrypt (AtE) transformation, Krawczyk [18] constructs an encryption scheme that guarantees confidentiality, but if one uses it to encrypt authenticated plaintexts, the combined scheme does not guarantee both confidentiality and integrity. The vulnerabilities can

R. Cramer (Ed.): TCC 2012, LNCS 7194, pp. 209–229, 2012.

either be seen as a breach of confidentiality [18] or as a breach of integrity, see Sect. 4.4. Natural candidates, such as the cipher block chaining mode (CBC) or stream ciphers, are not vulnerable; they provide weak but sufficient integrity guarantees [25].

In this paper, we use the approach of constructive cryptography [21,22] for a systematic treatment of security notions for symmetric encryption schemes. This approach leads to security definitions that capture the exact conditions that the schemes have to satisfy to achieve certain guarantees for the message transmission. In particular, these definitions are composable, which is instrumental for the soundness of a modular protocol design. We then show how different types of confidentiality and integrity are captured and compare these notions with several security definitions from the literature. This shows that some of the previous definitions are either too weak or artificially strict (which is in general undesired as it may lead to disregarding efficient schemes that are indeed sufficient).

1.1 Game-Based Security Definitions

Most widely-used security definitions for cryptographic schemes in the context of secure communication are game-based. The main concept of these definitions is an interaction of two (hypothetical) entities: The challenger and the attacker. During this interaction, the attacker issues certain "oracle queries" to the challenger; these queries model the use of the scheme in applications. The game also specifies a goal for the attacker, which often corresponds to forging a message or distinguishing encryptions of different messages. The infeasibility of achieving this goal is supposed to capture the guarantees required from the scheme.

Unfortunately, the oracle queries and winning conditions of games encode the use and guarantees only implicitly, and the exact guarantees are often hard to understand. In particular, such security definitions are generally not composable, and subtle details often have a significant impact on the resulting guarantees: Examples where slight slackness in the oracle queries rendered the guarantees of games too weak are discussed in Sect. 4.

1.2 Constructive Cryptography

The foundational idea of constructive cryptography [21,22] is to specify both the setup assumptions and the guarantees of protocols explicitly as resources, and to consider a protocol as a transformation of such resources. Here, a *resource* is a shared functionality accessed by several parties (similar to the ideal functionalities in frameworks such as [2,8]). *Real resources* are assumed functionalities needed for executing protocols (such as a network) and *ideal resources* describe the guaranteed functionalities the parties want to achieve. The way a party accesses a resource is described by the *interface* provided by the resource to this party; the resource provides one interface per party.

A *converter* systems formalizes the actions that a party performs locally, for example when it uses a cryptographic scheme. A converter has two interfaces: The *inner* interface is attached to an interface of the resource, and the *outer* interface is used by the party instead of the original interface of the resource. In particular, the composition of the resource and the converter is again a resource with one interface for each party, which is depicted in Fig 1 for the case of symmetric encryption.

A *protocol* is a tuple (in our context just a pair) of converters, there is one such system for each (honest) party. The goal of a protocol is to *construct* a specified ideal resource from available real resources, where the meaning of "construct" is made precise in Sect. 2.3. The constructed ideal resources can again serve as real resources for other protocols.

1.3   Secure Communication

The resources considered in this work are communication functionalities with different types of security guarantees, and the goal of a cryptographic protocol is to construct a functionality with stronger guarantees from one (or more) with weaker guarantees. As the setting for communication security is described by two (honest) entities that communicate in a potentially hostile environment, we consider resources with three interfaces: One interface labeled A for the sender,[1] one labeled B for the receiver, and a third one that is labeled E and captures potential adversarial access. A resource of this type is called a *channel* (from A to B), and its security properties are described by the capabilities provided at the E-interface. The basic types of channels are (informally) described in the following table, using the notation of [24].

$\longrightarrow$ An *insecure channel* leaks the complete messages at the E-interface, and allows at the E-interface to delete, change, or inject messages.

$\bullet\!\!\longrightarrow$ An *authenticated channel* leaks the complete messages. The E-interface only allows to forward or to delete messages.

$\longrightarrow\!\!\bullet$ A *confidential channel* only leaks the length of the messages, but allows to delete, change, or inject messages.

$\bullet\!\!\longrightarrow\!\!\bullet$ A *secure channel* only leaks the length of the messages and only allows to forward or to delete messages.

The intuitive interpretation of the symbol "$\bullet$" is that the capabilities at the marked (sender's or receiver's) side of the channel are provided exclusively to that party. Consequently, if one side is not marked, the adversary might also be able to send or receive messages. A *shared secret key* is a system $\bullet\!\!=\!\!\bullet$ that outputs the same random value at the A- and B-interfaces, and does not interact at the adversarial interface. This system models the key that is required by (symmetric) schemes; it could be generated in a key agreement protocol.

Security mechanisms such as encryption or MAC schemes are protocols that transform one type of channel (and possibly a shared secret key) into a "more secure" type of channel. In Fig. 1, the protocol (enc, dec) uses as resources a channel $\longrightarrow$ and a key $\bullet\!\!=\!\!\bullet$. The converter enc is attached with its inner interface to the A-interfaces of $\longrightarrow$ and $\bullet\!\!=\!\!\bullet$ (dec is attached to the B-interfaces), and

---

[1] Bidirectional communication also involves the analogous setting with opposite roles.

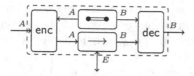

Fig. 1. Encryption protocol (enc, dec) applied to the channel $\longrightarrow$ and the key $\bullet\!=\!\!\bullet$

the outer interfaces of enc and dec are the interfaces of the constructed (dashed) system, which is again a channel. For more examples, we refer to [22,25].

1.4   Related Work

The major part of research on (symmetric) encryption schemes has been pursued in game-based security models. The nowadays "standard" confidentiality notions IND-CPA and IND-CCA are derived from [14] and have been translated to the setting of symmetric encryption schemes by [3]. Further variants of these notions are introduced and compared in [16]. Several types of integrity guarantees have been considered: Notions of non-malleability have been translated in [5] from the respective public-key notions [12]. Further standard notions are INT-CTXT and INT-PTXT (integrity of ciphertext and integrity of plaintext, respectively) introduced and analyzed in [5], their relation is further examined in [28]. Also, various types of unforgeability notions appear in the literature [13,17,18].

The security requirements for schemes used to protect communication over insecure networks is often specified as a combination of properties for confidentiality and integrity, where the standard combination is IND-CPA and INT-CTXT [5,7,17]; combinations with weaker types of integrity properties appear in [5,9,13,17,27]. A single game-based notion for authenticated encryption appeared in [31,34]. A different approach is taken in the definition of [9]: While confidentiality is similar to IND-CPA, authenticity is simulation-based; equivalent fully game-based notions appear in [27]. Fully simulation-based definitions of secure communication have been provided in [29] for Reactive Simulatability and in [10] in the UC framework.

1.5   Outline

We analyze confidentiality and integrity notions for (symmetric) encryption schemes using the paradigm of constructive cryptography. Sect. 2 introduces the notation and the general model, and Sect. 3 shows how different types of confidentiality and integrity guarantees are captured. In Sect. 4, we compare various existing game-based security definitions to the notions in our model.

2   Preliminaries

We use the concept of abstract systems [22,23] to formulate our results. At the highest level of abstraction, a system is an object with interfaces via which it interacts with its environment and with other systems. Every two systems can be composed by connecting one interface of each system, and the composed object is again a system. Also, every two different systems are mutually independent.

2.1   Notation

We consider two distinct types of systems, resources and converters, and we describe topologies of these systems using the notation from [23]. Resources, with three interfaces labeled by A, B, and E, are denoted either by special symbols or by upper case boldface letters. Converters, with one *inner* and one *outer* interface, are denoted either by small Greek letters or by special identifiers such as enc or dec; the set of all converters is denoted as Σ.

The composition of a resource $\mathbf{R}$ and a converter ϕ is written as $\phi^I \mathbf{R}$, where the label $I \in \{A, B, E\}$ means that the inner interface of ϕ is attached to the I-interface of the resource $\mathbf{R}$. Note that the composed system is again a resource that exposes the outer interface of ϕ as the I-interface together with the other interfaces of $\mathbf{R}$. A protocol is a pair of converters, one for each honest party, and applying the protocol (ϕ_1, ϕ_2) to the resource $\mathbf{R}$ is defined as $\phi_1^A \phi_2^B \mathbf{R}$—attaching the converters to the A- and B-interfaces of the resource.

If two resources $\mathbf{R}$ and $\mathbf{S}$ are used in parallel, this is denoted as $\mathbf{R}\|\mathbf{S}$ and is again a resource with the same set of interfaces; each of these interfaces A, B, or E of $\mathbf{R}\|\mathbf{S}$ allows to access the corresponding interfaces of both sub-systems $\mathbf{R}$ and $\mathbf{S}$. The sequential composition of converters is denoted by $\psi \circ \phi$, and is defined by $(\psi \circ \phi)^I \mathbf{R} = \psi^I(\phi^I \mathbf{R})$ for all resources $\mathbf{R}$. The parallel composition $\psi\|\phi$ of converters is defined by $(\psi\|\phi)^I(\mathbf{R}\|\mathbf{S}) = (\psi^I \mathbf{R})\|(\phi^I \mathbf{S})$ for all $\mathbf{R}$ and $\mathbf{S}$. The term id refers to the "identity converter" that forwards all inputs and outputs.

In general, for bit-strings $x = x_1 \cdots x_n \in \{0,1\}^n$ and $l \leq n$, we denote by $x|_l$ the sub-string $x|_l = x_1 \cdots x_l$. We extend the operation "$\oplus$" to bit-strings by defining, for $x = x_1 \cdots x_n$ and $x' = x'_1 \cdots x'_n$, the ith bit of $x \oplus x'$ to be $x_i \oplus x'_i$.

2.2   Discrete Systems

In the analysis of protocols, we model all systems as (probabilistic) *discrete systems* that communicate by passing messages, where the term "discrete" refers to the value spaces of the messages as well as the time. The behavior of discrete systems is formalized by random systems [20], i.e., conditional distributions of the outputs of the system (as random variables) given all previous inputs and outputs. Each input or output is associated to a specific interface.

Discrete systems are an instance of the abstract systems concept described above. The composition of two discrete systems (such as connecting a resource and a converter via interfaces) is a discrete system whose behavior is defined via an interaction of the two sub-systems: A message that is input to the system is processed by the sub-system corresponding to the (external) interface where the message was input, and, if the sub-system provides output at the (internal) connected interface, this value is processed by the other sub-system. Once one of the two sub-systems outputs a message at an external interface, this becomes the output of the composed system. The parallel composition of two resources is defined asynchronously: Each input at an interface A, B, or E explicitly specifies one of the sub-systems, and this sub-system is invoked with the input.

A *distinguisher* $\mathbf{D}$ is a system that connects to all interfaces A, B, and E of a resource $\mathbf{U}$ and outputs (at a separate interface) a single bit, here called W. The complete interaction of $\mathbf{D}$ and $\mathbf{U}$ defines a random experiment, and the probability that the bit W is 1 is written as $\mathsf{P}^{\mathbf{DU}}(W = 1)$. The *distinguishing advantage* of $\mathbf{D}$ for $\mathbf{U}$ and $\mathbf{V}$ measures how much the output of $\mathbf{D}$ differs when it is connected to either $\mathbf{U}$ or $\mathbf{V}$. Intuitively, if no (efficient) distinguisher differentiates between the two systems, they can be used interchangeably in any environment (as otherwise the environment serves as a distinguisher).

Definition 1 (Distinguishing advantage). *The* distinguishing advantage of a distinguisher $\mathbf{D}$ for the systems $\mathbf{U}$ and $\mathbf{V}$ *is defined as*

$$\Delta^{\mathbf{D}}(\mathbf{U}, \mathbf{V}) := \left| \mathsf{P}^{\mathbf{DU}}(W = 1) - \mathsf{P}^{\mathbf{DV}}(W = 1) \right|,$$

where W is the special output of $\mathbf{D}$. *The advantage for a set $\mathcal{D}$ of distinguishers is defined as* $\Delta^{\mathcal{D}}(\mathbf{U}, \mathbf{V}) := \sup_{\mathbf{D} \in \mathcal{D}} \Delta^{\mathbf{D}}(\mathbf{U}, \mathbf{V})$.

2.3 The Simulation-Based Security Definition

The paradigm of constructive cryptography is derived from [23] and follows the ideal world/real world approach similar to [8,29]: The "real world" describes the protocol execution with two honest parties and an adversary, and is defined by the composition of the two converters of the protocol (π_1, π_2) with the real resource $\mathbf{R}$. In the "ideal world", the ideal resource $\mathbf{S}$ specifying the security goals is composed with a simulator σ connected to the E-interface. The purpose of σ is to adapt the E-interface of $\mathbf{S}$ such that it resembles the corresponding interface of $\pi_1^A \pi_2^B \mathbf{R}$. (As the adversary can emulate the behavior of σ, using $\sigma^E \mathbf{S}$ instead of $\mathbf{S}$ can only restrict the adversary's power, so using $\sigma^E \mathbf{S}$ and hence $\pi_1^A \pi_2^B \mathbf{R}$ instead of $\mathbf{S}$ is safe.)

To exclude trivial protocols, we require that if no adversary is present, the protocol must implement the specified functionality. In the definition, we use the special converter "$\perp$" that, when attached to a certain interface of a system, blocks this interface for the distinguisher.

Definition 2 (Secure construction). *The protocol π constructs $\mathbf{S}$ from the resource $\mathbf{R}$ within ε and with respect to the set $\mathcal{D}$ of distinguishers if*

$$\exists \sigma \in \Sigma : \quad \Delta^{\mathcal{D}}(\pi_1^A \pi_2^B \mathbf{R}, \sigma^E \mathbf{S}) \le \varepsilon \qquad \text{and} \qquad \Delta^{\mathcal{D}}(\pi_1^A \pi_2^B \perp^E \mathbf{R}, \perp^E \mathbf{S}) \le \varepsilon.$$

An important property of Definition 2 is its composability. Intuitively, if a resource $\mathbf{S}$ is used in the construction of a larger system, then the composability implies that $\mathbf{S}$ can be replaced by a construction $\pi_1^A \pi_2^B \mathbf{R}$ without affecting the security of the composed system. Theorem 1, taken from [25], shows that security and availability are preserved under sequential and parallel composition.

Theorem 1 (Composition for the 3-party setting). *Let* **R, S, T,** *and* **U** *be resources, and let* $\pi = (\pi_1, \pi_2)$ *and* $\psi = (\psi_1, \psi_2)$ *be protocols such that* π *constructs* **S** *from the resource* **R** *within* ε_1 *and* ψ *constructs* **T** *from* **S** *within* ε_2. *If the considered class of distinguishers is closed under composition with converters, that is* $\mathcal{D} \circ \Sigma \subseteq \mathcal{D}$, *then* $(\psi_1 \circ \pi_1, \psi_2 \circ \pi_2)$ *constructs* **T** *from* **R** *within* $\varepsilon_1 + \varepsilon_2$, $(\pi_1 \| \mathsf{id}, \pi_2 \| \mathsf{id})$ *constructs* **S**$\|$**U** *from* **R**$\|$**U** *within* ε_1 *and* $(\mathsf{id} \| \pi_1, \mathsf{id} \| \pi_2)$ *constructs* **U**$\|$**S** *from* **U**$\|$**R** *within* ε_1.

In asymptotic statements, a system **S** implicitly refers to a family of systems $\{\mathbf{S}_k\}_{k \in \mathbb{N}}$, and the distinguishing advantage is a real-valued function in the parameter k: For each k, one considers the distinguishing advantage where, for all involved systems, one takes the element described by this k. Efficiency notions for sets of systems and a negligibility notion for the distinguishing advantage can be chosen such that they are closed under composition. Examples are the sets of systems with a polynomial bound on the number of queries and/or the run-time, together with the standard notion of negligibility.

2.4   Resources and Protocols as Discrete Systems

This section details the resources and protocols considered in the setting of secure communication.

Channels. Let $\mathcal{M}$ be a discrete set, we usually consider $\mathcal{M} := \{0,1\}^*$. A *channel with message space* $\mathcal{M}$ is a resource that takes at the A-interface inputs from the set $\mathcal{M}$ and provides at the B-interface outputs from $\bar{\mathcal{M}} := \mathcal{M} \cup \{\boxtimes\}$, where the element $\boxtimes$ is interpreted as indicating a transmission error. A *single-use* channel allows for exactly one input at the A-interface and one output at the B-interface, a *multiple-use* channel allows for several (arbitrarily interleaved) such interactions. The possible interactions at the E-interface describe the security properties of the channel. For the insecure channel $\longrightarrow$, every input $m \in \mathcal{M}$ at the A-interface provokes the output m at the E-interface, and every input $m' \in \mathcal{M}$ at the E-interface leads to the output m' at the B-interface. The E-interfaces of the "more secure" types of channels are detailed in Sect. 3.

Keys. Let $\mathcal{K}$ be a discrete set, usually $\mathcal{K} := \{0,1\}^k$ for some $k \in \mathbb{N}$. A key with key space $\mathcal{K}$ is a resource that draws a key $\kappa \in \mathcal{K}$ uniformly at random and outputs it to both A and B. The E-interface does not provide any output.

Encryption Protocols. An *encryption protocol* with key space $\mathcal{K}$, message space $\mathcal{M}$, and ciphertext space $\mathcal{C}$ is a pair (enc, dec) of converters. These converters connect with their inner interfaces to a shared secret key with key space $\mathcal{K}$ and to a channel with message space $\mathcal{M}' \supseteq \mathcal{C}$. The resulting resource is a channel with message space $\mathcal{M}$.

As an example, we describe the one-time pad encryption for bit-strings with length at most n. The key space in this setting is $\mathcal{K} = \{0,1\}^n$, and the message space of the assumed channel is (in general at least) the set of strings of length at most n bits, $\mathcal{M}' = \mathcal{C} = \bigcup_{l \leq n} \{0,1\}^l$.

Example 1 (The one-time pad). The encryption converter otp-enc (generically called enc in Fig. 1) obtains as input the n-bit key κ at the inner interface and a message m with $|m| \leq n$ at the outer interface. The message transmitted via the channel is $c = m \oplus \kappa|_{|m|}$. The decryption converter otp-dec obtains the key κ and the ciphertext c' at its inner interface. It computes $m' = c' \oplus \kappa|_{|c'|}$ and outputs the message m' at the outer interface.

Fig. 1 shows the setting in which the encryption and decryption converters are attached to the resources, the channel $\longrightarrow$ and the key $\bullet\!\!-\!\!\bullet$, with their inner interfaces. Both the A-interface and the B-interface of the combined (dashed) system are of the same type as for the original channel: The A-interface allows to input messages from $\mathcal{M} = \mathcal{C}$, and the B-interface outputs messages from the same set. Hence, the complete system is again a channel with message space $\mathcal{M}$ (but differs at the E-interface).

The scheme extends to multiple, say t, messages as follows. Consider a key with key space $\{0,1\}^{tn}$, and encrypt/decrypt the ith message with the bits $(i-1)n + 1$ through $(i-1)n + |m_i|$. ◆

2.5 Formalizing Games

In game-based definitions, we formalize both the adversary and the game (or challenger) as systems, which are connected via their interfaces as described in Sect. 2.2. The game allows the adversary to make certain "oracle queries" via this interface. Whether or not the game is won is signaled by a special (monotone) output bit of **G** (this can be considered as an additional interface) that is initially 0 but switches to 1 as soon as the winning condition is fulfilled. For a game **G** and an adversary **A**, we define the *game-winning probability* after q steps as

$$\Gamma_q^{\mathbf{A}}(\mathbf{G}) := \mathsf{P}^{\mathbf{AG}}(W_q = 1).$$

For an adversary that halts after (at most) q steps, we write $\Gamma^{\mathbf{A}}(\mathbf{G}) := \Gamma_q^{\mathbf{A}}(\mathbf{G})$. As winning the game with a certain probability might be trivial (such as when the goal is to guess a secret bit), one usually considers the *advantage* of **A**, that is, the (absolute) difference between **A**'s probability of winning **G** and the probability for "trivial" strategies.

If a security property of a scheme is defined by the adversary's inability to win a game **G**, then we say that the scheme is ε-secure with respect to that property and a class[2] $\mathcal{D}$ of adversaries if the advantage for $\mathcal{D}$ in winning **G** is bounded by ε.

3   Notions of Confidentiality and Integrity

The security of communication channels corresponds to restrictions on the capabilities provided at the E-interface, which can be characterized according to two aspects: the amount of information leaked about transmitted messages, and

---

[2] We will often use the same class $\mathcal{D}$ for both adversaries and distinguishers.

the potential influence on messages delivered to the receiver. Consequently, a confidentiality guarantee bounds the amount of information that is leaked, and an integrity guarantee restricts the adversarial influence on delivered messages.

3.1   Confidentiality

A channel is perfectly confidential if no information about the transmitted plaintext message is leaked at the E-interface. We also consider weaker types of confidentiality where the "amount of leakage" is non-trivial but bounded; the (remaining) guarantee is described by a function on the transmitted messages.

Definition 3 (Leakage specification). *For some (discrete) set $\mathcal{S}$, a leakage specification is a family of functions $\mathcal{L} = \{\ell_i : \mathcal{M}^i \to \mathcal{S}\}_{i \geq 1}$.*

Functions ℓ_i on vectors of messages allow to capture, for example, channels that leak whether the same message is sent twice (as in deterministic encryption).

Definition 4 (Confidential channels). *For $\mathcal{L} = \{\ell_i : \mathcal{M}^i \to \mathcal{S}\}_{i \geq 1}$, let $\bullet \xrightarrow{\mathcal{L}} \bullet$ be the channel that, given inputs $m_1, \ldots, m_i$ at the A-interface, outputs the value $\ell_i(m_1, \ldots, m_i)$ at the E-interface (and only allows forwarding or deleting messages). A channel $\mathbf{C}$ is $\mathcal{L}$-confidential if there exists a simulator σ such that*

$$\Delta^{\mathcal{D}}(\perp^B \mathbf{C}, \perp^B \sigma^E(\bullet \xrightarrow{\mathcal{L}} \bullet)) = 0, \qquad \text{and} \qquad \Delta^{\mathcal{D}}(\perp^E \mathbf{C}, \perp^E(\bullet \xrightarrow{\mathcal{L}} \bullet)) = 0,$$

where $\mathcal{D}$ is the set of all distinguishers. If $\mathcal{M} \subseteq \{0,1\}^*$ and the leakage is restricted to $\ell_i : (m_1, \ldots, m_i) \mapsto |m_i|$ for all i, the channel is simply called confidential.

The condition of being $\mathcal{L}$-confidential is merely a restriction on the information leaked at the E-interface; there is no guarantee on the potential influence of the adversary on the delivered messages. In the security condition, this absence of guarantees is expressed by attaching the converter $\perp$ to the B-interface, which hides all messages from the distinguisher.

The goal of an encryption protocol is to construct a confidential channel from one that is not confidential. In particular, the one-time pad encryption achieves confidentiality in this sense.

Example 2 (Confidentiality achieved by the one-time pad). The ciphertext generated by the one-time pad encryption for the message $m \in \mathcal{M} = \bigcup_{l \leq n} \{0,1\}^l$ is an $|m|$-bit string of independent and uniformly distributed random bits. The information leaked to the adversary is exactly the length $|m|$ of the message: There is a simulator that, given the length $|m|$, generates a ciphertext that has exactly the same distribution as the "real" ciphertext for the message m.

This means that the leakage is described by $|\cdot| : \mathcal{M} \to \{1, \ldots, n\}$ (for multiple messages, ℓ_i maps $(m_1, \ldots, m_i)$ to $|m_i|$). The channel that is constructed by the one-time pad from the insecure channel is described in Examples 3 and 4.   ♦

3.2   Integrity

Encryption schemes in general do not protect the integrity of messages: If the adversary replaces the transmitted ciphertext c for a message $m \in \mathcal{M}$ by a ciphertext $c' \neq c$, the receiver will potentially obtain a different message $m' \in \mathcal{M}$ during the decryption. For the adversary (oblivious of m), replacing c by c' corresponds to selecting a transformation $F : \mathcal{M} \to \mathcal{M}$ that describes, for every *potentially* transmitted message $\tilde{m}$, which message $\tilde{m}' = F(\tilde{m})$ the receiver would obtain, given that the original message was $\tilde{m}$.

Example 3 (XOR-Malleability of the one-time pad). For the one-time pad encryption, the adversary can replace the transmitted ciphertext c by an arbitrary ciphertext c'. Assume that $c = m \oplus \kappa$ and $|c| = |c'|$, then this means that the receiver will compute $m' = c' \oplus \kappa = c' \oplus c \oplus m$. Hence, replacing c by c' corresponds to selecting the function $m \mapsto m \oplus (c \oplus c')$. ◆

In general, the distribution of each output at the B-interface depends on the previous inputs and outputs at all interfaces of the channel. But then, conditioned on the complete interaction at the E-interface—the adversary's knowledge— the channel "transforms" all inputs at the A- and all previous outputs at the B-interface into the next output at the B-interface; the interaction at the E- interface can be seen as a choice of a particular such *plaintext transformation*.

Definition 5 (Plaintext transformation). *Let $\mathcal{M}$ be a discrete set. A* plaintext transformation F on $\mathcal{M}$ *is a (probabilistic) transformation* $\mathcal{M}^* \times \mathcal{M}^* \to \bar{\mathcal{M}}$.

The arguments of the plaintext transformation are the sequence of messages transmitted by the sender, and the sequence of messages previously delivered to the receiver; the result is the next message delivered to the receiver. The set of all plaintext transformations available to the adversary formalizes the potential adversarial influence on the delivered messages. Of course, the less such transformations are available to the adversary, the stronger are the integrity guarantees of the channel. This is captured by the concept of *integrity specifications*.

Definition 6 (Integrity specification). *An* integrity specification *is a family* $\mathcal{F} := \{\mathcal{F}_q\}_{q \in \mathbb{N}}$ *of random variables with* $\mathcal{F}_q \subseteq \bar{\mathcal{F}}$, *where* $\bar{\mathcal{F}}$ *is a set of plaintext transformations.*

The random variables $\mathcal{F}_q \subseteq \bar{\mathcal{F}}$ formalize that, depending on the state of the channel, only a subset of the transformations might actually be accessible: After the qth query to the channel, the adversary may choose a transformation from the set $\mathcal{F}_q$ (note that this choice corresponds to replacing the transmitted ciphertext in the "real world"). The generality of this definition is indeed necessary to describe the malleability of certain encryption schemes, such as CBC mode [25]. There, the availability of certain transformations depends on the randomness used during the encryption, so $\mathcal{F}_q \neq \bar{\mathcal{F}}$.

Example 4 (XOR-malleability). Let m, m', c, and c' be as in Example 3. If we set $\delta := c \oplus c'$, the adversary's choice to replace c by $c' = c \oplus \delta$ can be interpreted as selecting the XOR-mask δ for the transmitted message. More generally, the plaintext transformations F_{i,j,δ_j} after i inputs at the A-interface and j inputs at the E interfaces, with $\delta_j \in \bigcup_{l \leq n}\{0,1\}^l$, are described as follows:

- $i < j$: the output is a uniformly random $|\delta_j|$-bit string,
- $i \geq j$ and $|\delta_j| \leq |m_j|$: the output is $m_j|_{|\delta_j|} \oplus \delta_j$,
- $i \geq j$ and $|\delta_j| > |m_j|$: the output is $m_j \oplus \delta_j$ followed by $|\delta_j| - |m_j|$ uniformly random bits.

The transformations available after i inputs at the A- and j inputs at the B-interface are, for each $\delta \in \bigcup_{l \leq n}\{0,1\}^n$, the transformations $F_{i,j,\delta}$.   ◆

The set $\mathcal{F}_q$ of transformations available after the qth query must be (implicitly of explicitly) known to the adversary; abstractly, a description of the set $\mathcal{F}_q$ is output to the adversary by the channel. Of course, for a confidential channel, this description must not leak any information beyond the information specified by the leakage. In the following definition, we refer to the number of queries at the A- and E-interfaces by i and j, respectively, and use $q := i + j$.

Definition 7 (Malleable confidential channel). *Let $\mathcal{L}$ be a leakage specification and $\mathcal{F}$ be an integrity specification such that the distribution of each $\mathcal{F}_q$ depends (only) on the leakage $\ell_s(m^s)$ for $1 \leq s \leq i$ of the messages $m_1, \ldots, m_i$, the previous sets $\mathcal{F}_1, \ldots, \mathcal{F}_{q-1}$, and the selected transformations $F_1, \ldots, F_j$. An $\mathcal{F}$-malleable $\mathcal{L}$-confidential channel $\xrightarrow{\mathcal{L},\mathcal{F}}$ (in the following only $\longrightarrow$ if $\mathcal{L}$ and $\mathcal{F}$ are clear) is an $\mathcal{L}$-confidential channel with malleability described by $\mathcal{F}$.*

On receiving m_i at the A-interface, $\longrightarrow$ outputs $\ell_i(m^i)$ and a description of $\mathcal{F}_q$ at the E-interface. Upon receiving a description of $F \in \mathcal{F}_q$ at the E-interface, $\longrightarrow$ evaluates the transformation F on the plaintexts and outputs the result at the B-interface. If the $\perp$-converter is attached to the E-interface, $\longrightarrow$ immediately forwards each input m_i from the A- to the B-interface.

As an example, we describe the XOR-malleable confidential channel and sketch the proof that the one-time pad constructs this channel from an insecure channel and a shared secret key.[3]

Example 5 (The XOR-malleable channel). The channel $\longrightarrow\!\oplus\!\longrightarrow$ behaves as follows. Upon the ith input $m_i \in \mathcal{M}$ at the A-interface, leak the length $|m_i|$ at the E-interface. Upon the jth input $\delta_j \in \{0,1\}^n$ at the E-interface (after i inputs at the A-interface), output $m'_j := F_{i,j,\delta}(m)$ at the B-interface.

We use the following simulator σ to prove that the one-time pad indeed constructs $\longrightarrow\!\oplus\!\longrightarrow$:

- Upon a message $l_i \in \{1, \ldots, n\}$ at the inner interface (i.e., from $\longrightarrow\!\oplus\!\longrightarrow$), output a uniformly random l_i-bit string $\tilde{c}_i$ as the transmitted ciphertext at the outer interface.

---

[3] For simplicity, we only consider the case $i > j$. For the general case, cf. [25, Sect. 6.1].

- Upon a message $\tilde{c}'_j$ at the outer interface,
  - if $j > i$, input $\delta_j = 0^{|m_j|}$ at —⊕→•,
  - if $j \leq i$ and $|\tilde{c}'_j| \geq |\tilde{c}_j|$, input $\delta_j = \tilde{c}_j|_{|\tilde{c}'_j|} \oplus \tilde{c}'_j$ at —⊕→•,
  - else, input $\delta_j = (\tilde{c}_j \oplus \tilde{c}'_j)|0^{|\tilde{c}'_j|-|\tilde{c}_j|}$ at —⊕→•.

The simulator σ is perfect, i.e., $\Delta^{\mathbf{D}}(\text{otp-enc}^A \text{otp-dec}^B(\longrightarrow \parallel \bullet\!\!=\!\!\bullet), \sigma^E(-⊕→•)) = 0$ for all distinguishers $\mathbf{D}$:

- On input the ith message m_i at the A-interface, in both cases a $|m_i|$-bit uniformly random string is output at the E-interface (generated either by otp-enc using the key or by σ).
- On input the jth message c'_j at the E-interface, the message output at the B interface also has the same distribution in both cases (by construction of σ; this is a simple check for each of the cases).   ◆

Consequently, the one-time pad constructs from the resources $\bullet\!\!=\!\!\bullet$ and $\longrightarrow$ the channel —⊕→•. This channel is confidential according to Definition 4, the simulator assumed in the definition is trivial (both $\bullet\!\!-\!\!\!\twoheadrightarrow$ and —⊕→• leak exactly the length of the message).

4   Relation to Game-Based Security Definitions

In game-based security definitions for encryption schemes, the attacker has access to oracles for encrypting plaintext messages and decrypting or checking the correctness of ciphertexts, sometimes with additional constraints on the number or order of queries. The attacker's goal is either to generate a ciphertext that satisfies a certain condition, or to distinguish two cases in which it is provided with different sets of oracles. For many of these notions, it is not clear which guarantees the proven schemes provide when the ciphertexts are transmitted over a certain type of network.

In contrast, a constructive security statement makes these guarantees explicit: The confidentiality and integrity guarantees appear as the leakage functions and plaintext transformations of the constructed channel. In this section, we analyze the semantics of game-based notions from the literature by proving the (in)equivalence with corresponding constructive notions.

4.1   Goals and Attack Models

Security properties defined using games are often characterized by a *goal* and an *attack model*. The goal is essentially specified by the winning condition (the monotone output switches to 1), and the attack model is characterized by the "oracle queries" the adversary has at its disposal.

The attack model roughly corresponds to adversarial access to the "real resources" used by the protocol in constructive security statements. The more capabilities the game provides, the weaker the security modeled by the real resources, and the stronger the requirements for the protocol. Roughly, the idea of a chosen plaintext attack corresponds to the real resource being an authenticated channel, and a chosen ciphertext attack corresponds to the real resource being an insecure channel. The goal of a game corresponds to the attributes of the constructed resource. For instance, the IND-type of games are often connected with confidentiality, whereas NM (non-malleability) and INT (integrity) are integrity guarantees.

4.2   Indistinguishability of Ciphertexts

The standard security notions for confidentiality are IND-CPA and IND-CCA, i.e., indistinguishability (of ciphertexts) under chosen-plaintext and chosen-ciphertext attack, respectively. Several variants appear in the literature; in all variants, a bit $b \in \{0,1\}$ is chosen uniformly at random, and, depending on the variant, the adversary has access to one of the following settings of oracles:

- multiple queries at a "real-or-random" oracle where, in each query, the adversary inputs a plaintext m_0, the game chooses m_1 with $|m_0| = |m_1|$ uniformly at random, and returns an encryption of m_b;
- multiple queries at a "left-or-right" oracle where the adversary inputs two messages m_0 and m_1 with $|m_0| = |m_1|$ and obtains an encryption of m_b;
- multiple queries at an "encryption" oracle where, on input m, the adversary obtains an encryption of m, as well as one "real-or-random" query;
- multiple "encryption" queries and one "left-or-right" query.

Finally, the adversary has to guess the bit b (with probability non-negligibly different from $1/2$). It turns out that, for any encryption scheme, the advantages that can be achieved in the above games are related by a factor that is either a constant or linear in the number of queries [3].

IND-CPA. The term IND-CPA usually refers to a game where the adversary has access to the oracles described in one of the four settings above. While these settings correspond to assuming that the ciphertexts are transmitted via authenticated channels (and cannot be changed during the transmission), in several practical protocols such as SSL/TLS, the ciphertexts can actually be changed during the transmission. Indeed, as confidentiality in the sense of Definition 4 is defined by restricting only the adversarial interface (the output at the receiver's interface is ignored), one may hope that IND-CPA security will still imply this weak form of confidentiality in this setting. The following example shows that this is not the case.

Consider an encryption scheme where a certain ciphertext $\bar{c} \in \mathcal{C}$ is never used, and append in the encryption to each ciphertexts a perfectly hiding commitment on the plaintext. In particular, expand the secret key using a PRG, use the first part as key for the encryption and the remainder as randomness in the commitment. Also, modify the decryption to output the initial secret key if it receives the special ciphertext $\bar{c}$. As the decryption algorithm does not appear in the IND-CPA game and the erroneous decryption does not hurt correctness (as $\bar{c}$ is never used), the modified scheme is IND-CPA secure. However, for any confidential channel, it is easy to construct a distinguisher that differentiates between the real and the ideal setting (input a message $m \in \mathcal{M}$ at A, input $\bar{c}$ at E, interpret the output at B as the secret key, expand by the PRG, and decrypt the output at E. If this decrypts to m and the decommitment was correct then say 0, otherwise say 1).

IND-CCA. In the IND-CCA game, the adversary is, in addition to one type of oracles of the IND-CPA game, given access to a decryption oracle where it can query ciphertexts that are different from those he obtained from the encryption oracle.[4] While IND-CCA is considered the standard notion for confidentiality in settings where the adversary can modify ciphertexts, it differs considerably from the notion implied by Definition 4. In particular:

1. IND-CCA is artificially strict: A scheme that allows "obvious" modifications of ciphertexts (e.g., appending bits that are ignored) is considered insecure.
2. The definition of IND-CCA already implies strong integrity guarantees.
3. These integrity guarantees seem artificial for symmetric encryption.

These issues are explained further in the following paragraphs.

Replayable CCA. Several authors [1,11,18,19,33] have noticed that IND-CCA is artificially strict in the sense that the decryption oracle will decrypt any ciphertext except for the exact challenge ciphertext. Schemes that allow for "obvious" ciphertext modifications are not IND-CCA secure, the typical separating example being an (otherwise IND-CCA secure) encryption scheme where the encryption always appends a single bit to the ciphertext, and this bit is ignored during decryption. While this modification does not hurt the security guarantees in any meaningful way, the resulting scheme is not IND-CCA secure.

In [11], several variants of "replayable" CCA security are analyzed.[5] In these games, not only the exact challenge ciphertext is disallowed in decryption queries, but also "related" ciphertexts. Intuitively, this means that encryption schemes may allow certain modifications to ciphertexts that do not change the result of the decryption. In more detail, the notions considered in [11] are:

- IND-RCCA, or "replayable CCA": any ciphertext that decrypts to one of the plaintexts issued to the encryption oracle is disallowed;
- IND-sd-RCCA, or "secretly detectable RCCA": intuitively, the receiver can detect whether an adversarially generated ciphertext was generated as a "modification" of an honestly generated one, or whether it is "independent" of all honestly generated ones, these "modified" ciphertexts are disallowed;
- IND-pd-RCCA, or "publicly detectable RCCA": the above distinction can be done publicly, i.e., without knowledge of the secret key.

---

[4] The reason for the latter restriction is that if the adversary were allowed to decrypt the challenge, winning the game would become trivial.

[5] Their original notions regard public-key schemes, but the extensions to symmetric schemes are also described.

The exact formalization is technically involved; for details, we refer to [11].

With respect to achieving secure communication, the guarantees provided by IND-CCA and IND-sd-RCCA secure schemes are indeed equivalent, which can be formalized via bisimulation. Intuitively, the simulator for the IND-sd-RCCA scheme can use the assumed detectability to decide whether a given ciphertext should be considered a replay.

Strong Integrity. An IND-sd-RCCA secure encryption scheme achieves a strong notion of integrity: The remaining malleability is described by the integrity specification $\mathcal{F}_{\mathrm{NM}}$ with the set $\{f_{\bar{m}} : \mathcal{M} \to \mathcal{M}, m \to \bar{m}\}_{\bar{m} \in \mathcal{M}}$ of transformations, where NM refers to "non-malleable." The proof of the following theorem is deferred to the full version of this paper.

Theorem 2 (Informal). *Let* (enc, dec) *be a symmetric encryption protocol. If the protocol is ε-IND-sd-RCCA secure, then it constructs an $\mathcal{F}_{\mathrm{NM}}$-malleable confidential channel from an insecure channel and a secret key within ε.*

Conversely, if the protocol constructs an $\mathcal{F}_{\mathrm{NM}}$-malleable confidential channel from an insecure channel and a secret key within ε (for distinguishers that issue at most q queries and with a special type of simulator) then it is $(q^2 + 1)\varepsilon$-IND-sd-RCCA secure (with respect to the class of adversaries that issue at most q queries). For large message spaces, the special type of simulator is general.[6]

Unnatural Malleability. IND-CCA is not a natural security requirement for symmetric encryption: The adversary may generate valid ciphertexts for arbitrary plaintexts (but only independently of honestly sent messages). Realistic symmetric encryption schemes are either malleable (such as the one-time pad or CBC) or, if they are non-malleable, they will actually already implement the fully secure channel (such as authenticated encryption). Here, it becomes apparent that IND-CCA has evolved as a notion for public-key schemes, where the adversary knows the encryption key and can encrypt arbitrary messages.

4.3   Specific Variants of Integrity

Games that are used to characterize integrity properties express impossibilities (for the adversary) to generate ciphertexts that satisfy certain conditions. In constructive cryptography, integrity guarantees are expressed explicitly by specifying the set of transformations that model the capabilities of the adversary. The correspondence between these two paradigms is as follows: A scheme is secure according to a game if and only if it implements a channel that allows no transformations that contradict the game; the potential probability in winning the game translates into a distinguishing advantage in the constructive security statement.

---

[6] If the distinction between "modified" and "independent" ciphertexts can be performed without the key, then the condition on the size of the message space is not needed. If we assume that the distinction is perfect, the factor $q^2 + 1$ reduces to 1.

NM-CCA. The notion of non-malleable encryption has been introduced in [12] in the context on public-key schemes. Intuitively, no attacker (even given honestly generated ciphertexts) may be able to generate a ciphertext whose decryption relates to "honestly encrypted" messages in a meaningful way. NM-CCA is equivalent to IND-CCA [12]; this extends to the RCCA notions [11]. Consequently, these notions also correspond to $\mathcal{F}_{NM}$-malleable communication.

INT-CTXT. Integrity of ciphertexts has been introduced in [5,6] and formalizes that the adversary cannot produce *any* fresh valid ciphertext. In more detail, an encryption scheme is said to achieve INT-CTXT security if no adversary with access to an encryption oracle can generate a valid ciphertext that is different from all ciphertexts obtained from the oracle. Here, "valid" means that the decryption outputs a message (not an error symbol). Note that existential unforgeability [17] and ciphertext unforgeability [18] are similar: The differences are, for example, that the definition from [5,6] allows multiple queries to the challenge oracle, whereas [17] allows only one.

A symmetric encryption protocol that achieves confidentiality and is additionally INT-CTXT secure constructs a fully secure channel from an insecure channel. Yet, INT-CTXT, as IND-CCA, is artificially strict concerning modifications of ciphertexts. We describe a relaxation of INT-CTXT which is constructed analogously to IND-sd-RCCA. In particular, we also require the existence of a secretly (i.e., given the secret key) computable relation, called $\equiv_\kappa$, on $\mathcal{C}$ with the same properties as for IND-sd-RCCA; this relation formalizes the receiver's ability to distinguish "modified" and "independent" ciphertexts generated by the adversary.

We define INT-sd-CTXT security by changing the INT-CTXT game as follows: The adversary wins only if $\text{dec}(\kappa, c') \neq \boxtimes$ and $\forall r \leq i : c' \not\equiv_\kappa c_r$ for all honestly generated c_r. Note that we also have to change the output of the oracle in the case that $c'_j \equiv_\kappa c_r$ holds (for some r) to be m_r. The proof of the following theorem is deferred to the full version of this paper.

Theorem 3 (Informal). *Let* (enc, dec) *be a symmetric encryption protocol that constructs a confidential channel from an insecure channel and a secret key within* ε_1. *If the protocol is* ε_2-*INT-sd-CTXT secure, then it constructs a secure channel from an insecure channel and a secret key within* $\varepsilon_1 + \varepsilon_2$. *Conversely, if the protocol constructs the secure channel within* ε *for distinguishers in* $\mathcal{D}_q$, *then it is* $(q^2 + 2)\varepsilon$-*INT-sd-CTXT secure with respect to* $\mathcal{D}_q$.[7]

INT-PTXT. Integrity of plaintexts has been defined in [5,6] and is weaker than INT-sd-CTXT. The adversary is also given access to an encryption oracle, but to win the game, it has to fabricate a ciphertext that decrypts to a plaintext that has not been queried at the encryption oracle before. This notion is weaker than INT-sd-CTXT in the sense that the adversary may still be able to generate a ciphertext that decrypts to plaintext that was queried at the encryption oracle but cannot be detected to be a modification of one particular honestly generated

---

[7] The factor $q^2 + 2$ appears for the same technical reasons as for IND-sd-RCCA.

ciphertext (even if all ciphertexts are delivered). This weakens the guarantees in two aspects: First, the adversary can replay messages undetectably, and second, the adversary may fabricate messages that decrypt to any one of the previous messages with some probability *that may even depend on the plaintexts*. Consequently, if the adversary is able to determine which of the original plaintexts has been received, he will potentially obtain information about some transmitted plaintext.

An integrity specification is *value-preserving* if all transformations $F_\alpha :$ $\mathcal{M}^* \times \mathcal{M}^* \to \bar{\mathcal{M}}$ have the property that the output message is either one of the input messages or $\boxtimes$, but any one of these may appear with some probability (which may even depend on the plaintexts). The proof of the following theorem is deferred to the full version of this paper.

Theorem 4. *Let* (enc, dec) *be a symmetric encryption protocol that constructs a confidential channel from an insecure channel and a secret key within* ε_1. *If the protocol is* ε_2-*INT-PTXT secure, then it constructs an* $\mathcal{F}_{VP}$-*malleable confidential channel within* $\varepsilon_1 + \varepsilon_2$, *with* $\mathcal{F}_{VP}$ *being value-preserving. Conversely, if the protocol constructs an* $\mathcal{F}_{VP}$-*malleable confidential channel within* ε_1 *such that* $\mathcal{F}_{VP}$ *is value-preserving, then it is* ε_1-*INT-PTXT secure.*

Namprempre [27] introduces a related but stricter notion called SINT-PTXT, which prohibits replaying messages arbitrarily. There, the adversary also wins the game if it generates ciphertexts such that the decryption outputs any plaintext *more often* than it was queried at the encryption before. Consequently, SINT-PTXT corresponds to a channel with this bounded type of replay.

Fixing the definition from [5,6]. In the original game, the output of the verification oracle is one bit indicating whether the decrypted plaintext is valid. This renders the notion too weak: If (via a higher-level protocol), the adversary learns *which* of the valid plaintexts has been obtained by decrypting (this probability may depend on secret values), this is not captured. Hence, this notion cannot guarantee composability. A slight modification to the game fixes this issue: The verification oracle returns the decrypted message (instead of the single bit). The following (artificial) encryption scheme exemplifies the weakness.

Example 6. Consider a scheme (enc, dec) secure according to the stricter notion. Change the decryption such that for (n, c_0, c_1) with $\mathsf{dec}_\kappa(c_b) \neq \bot$, $b \in \{0, 1\}$, the output is $\mathsf{dec}_\kappa(c_{\kappa_n})$ (with κ_n the nth bit of κ). ◆

The change does not affect the security with respect to the notion of [5,6]: The output of the oracle on (n, c_0, c_1) can be easily computed from the output on c_0 and c_1. In contrast, in the strengthened game, such queries reveal the secret key.

Plaintext Uncertainty. This notion from [13] attempts to capture that the adversary cannot "control" the result of a forgery. While the description is rather informal, it captures that the decrypted message contains a certain amount of entropy (for each message, the probability that this message is obtained by decrypting is small). While this is hard to achieve at least for multiple decryptions—the only entropy in the (otherwise deterministic) decryption is "fresh" key material—the computational (pseudo-entropy) version might prove useful in applications.

The corresponding integrity specification is the set of transformations that have at least a certain min-entropy, meaning that for each input m and transformation F, the min-entropy of the random variable $F(m)$ is larger than some bound. Computational indistinguishability from such a channel means that the output at the receiver's interface has a certain pseudo-entropy.

Known-Plaintext Forgery. This notion from [13] is intended to capture that the adversary providing a forged ciphertext *can* predict the changes to the transmitted message. The (informal) description in [13] states that the adversary could have computed the outcome with overwhelming probability (this can be formalized by means of an extractor). In the language of integrity specifications, this means that all transformations in $\mathcal{F}$ are deterministic (and efficiently computable). Properties of this type can indeed be helpful, as can be seen in the proof of the soundness of Authenticate-then-Encrypt in [25].

4.4 Combining Notions of Confidentiality and Integrity

Traditionally, security requirements for schemes for protecting communication are expressed as a combination of separate properties for confidentiality and integrity [5,7,9,13,17,27]. Such a combination, however, does not necessarily achieve the expected guarantees.

We revisit an example from [18] (modified in [25]): The composition of a tailor-made encryption scheme with a strongly unforgeable MAC. Briefly, the encryption first encodes each bit of the plaintext as two bits, such that the probability whether flipping one of these two bits has an effect depends on the original value (i.e., $0 \mapsto 00, 01$, or $10; 1 \mapsto 11$), and encrypts this expanded string using a one-time pad. Hence, if one encrypts an authenticated message, the probability that flipping a ciphertext bit changes the contained message—and the MAC verification fails—with a probability that depends on the original plaintext value. The resulting scheme achieves both confidentiality (by the one-time pad) and integrity (in the sense of INT-PTXT, by the unforgeability of the MAC), but the different success probabilities for the MAC verification leak information about the message, which is often described as a breach of confidentiality [18].

The described scheme implements a confidential $\mathcal{F}_{\text{VP}}$-malleable channel, where $\mathcal{F}_{\text{VP}}$ is value-preserving as described in Sect. 4.3: The weakness of this scheme is not a deficiency of confidentiality, but it only achieves a weak notion of integrity. Note that, in terms of integrity, INT-PTXT is equivalent to WUF-CMA[8], which is sufficient to construct an authenticated channel (where the adversary can only forward or delete messages) from an insecure channel. Indeed, for channels that are not confidential, the integrity guarantees specified by $\mathcal{F}_{\text{VP}}$ are equivalent

---

[8] Weak unforgeability: Given an oracle for generating tags, it is infeasible for the adversary to generate a tag for a message that has not been queried at the oracle.

to those of an authenticated channel: A simulator that knows the plaintext messages can sample according to distributions that depend on these messages. This equivalence does not hold if the considered channels are confidential.

4.5   A Critique of Game-Based Security Notions

Starting from [14], the major part of research on the security of encryption schemes has been pursued in game-based models. There, however, it is often not immediately clear which assumptions and guarantees are encoded by the oracle queries and winning conditions of games. For instance, which of the a priori different types of IND-CPA security described in Sect. 4.2 captures confidentiality "best" (and why)? This lack of semantics abets the prevalence of security notions that do not capture the security requirements *exactly* (see Sect. 4.2 and 4.3).

A further issue with game-based notions is that seemingly innocent changes may have a significant impact on the security guarantees. The security notion *indistinguishability from random bits* was introduced in [30] and is similar to IND-CPA. Yet, instead of an encryption of a random message, the game returns a uniformly random string of appropriate length. The way this length is chosen, however, is crucial: In the original definition, this is determined by a function of the length of the queried message. If this choice is changed (as done, for example, in [15]) to the length of an encryption of the queried message, this allows to leak information about the plaintext via the length of the ciphertext! A further example is the weakness of the INT-PTXT notion described in Sect. 4.3.

Moreover, several attack models in the definitions described in the literature seem inappropriate for practical applications. One example is IND-CCA1[9], where the receiver stops decrypting adversarially generated ciphertexts after the first message has been sent honestly. Also, certain terms such as NM-CPA are actually misleading: An attack exploiting the malleability of an encryption scheme is necessarily mounted by injecting or replacing ciphertexts. A more appropriate correspondence for this type of notion is a CCA attack on a single-use channel.

5   Conclusion

We have defined and analyzed confidentiality and integrity notions for symmetric encryption schemes using the paradigm of constructive cryptography. The resulting security definitions are composable and have clear semantics: The guarantees of a cryptographic protocol appear explicitly in the description of the constructed resource. We have shown how existing game-based notions can be translated into guarantees in this setting, which makes their semantics explicit. Additionally, this analysis has uncovered a weakness in the notion INT-PTXT, and it has shown that INT-CTXT and IND-CCA are artificially strict.

---

[9] In the CCA1 game, the adversary looses access to the decryption oracle after the first call to the challenge oracle. This corresponds to the situation where the receiver only decrypts messages until the first message has been generated by the sender.

Acknowledgments. We thank Kenny Paterson for fruitful discussions and the anonymous reviewers for their very helpful comments and suggestions. The work was supported by the Swiss National Science Foundation (SNF), project no. 200020-132794.

References

1. An, J.H., Dodis, Y., Rabin, T.: On the Security of Joint Signature and Encryption. In: Knudsen, L.R. (ed.) EUROCRYPT 2002. LNCS, vol. 2332, pp. 83–107. Springer, Heidelberg (2002)
2. Backes, M., Pfitzmann, B., Waidner, M.: The Reactive Simulatability (RSIM) Framework for Asynchronous Systems. Information and Computation 205(12), 1685–1720 (2007)
3. Bellare, M., Desai, A., Jokipii, E., Rogaway, P.: A Concrete Security Treatment of Symmetric Encryption. In: Proceedings of the 38th Symposium on Foundations of Computer Science, pp. 394–403. IEEE (1997)
4. Bellare, M., Desai, A., Pointcheval, D., Rogaway, P.: Relations among Notions of Security for Public-Key Encryption Schemes. In: Krawczyk, H. (ed.) CRYPTO 1998. LNCS, vol. 1462, pp. 26–45. Springer, Heidelberg (1998)
5. Bellare, M., Namprempre, C.: Authenticated Encryption: Relations among Notions and Analysis of the Generic Composition Paradigm. In: Okamoto, T. (ed.) ASIACRYPT 2000. LNCS, vol. 1976, pp. 531–545. Springer, Heidelberg (2000)
6. Bellare, M., Namprempre, C.: Authenticated Encryption: Relations among Notions and Analysis of the Generic Composition Paradigm. Journal of Cryptology 21(4), 469–491 (2008)
7. Bellare, M., Rogaway, P.: Encode-Then-Encipher Encryption: How to Exploit Nonces or Redundancy in Plaintexts for Efficient Cryptography. In: Okamoto, T. (ed.) ASIACRYPT 2000. LNCS, vol. 1976, pp. 317–330. Springer, Heidelberg (2000)
8. Canetti, R.: Universally Composable Security: A New Paradigm for Cryptographic Protocols. In: Proceedings of the 42nd IEEE Symposium on Foundations of Computer Science, pp. 136–145. IEEE (2001)
9. Canetti, R., Krawczyk, H.: Analysis of Key-Exchange Protocols and Their Use for Building Secure Channels. In: Pfitzmann, B. (ed.) EUROCRYPT 2001. LNCS, vol. 2045, pp. 453–474. Springer, Heidelberg (2001)
10. Canetti, R., Krawczyk, H.: Universally Composable Notions of Key Exchange and Secure Channels. In: Knudsen, L.R. (ed.) EUROCRYPT 2002. LNCS, vol. 2332, pp. 337–351. Springer, Heidelberg (2002)
11. Canetti, R., Krawczyk, H., Nielsen, J.B.: Relaxing Chosen-Ciphertext Security. In: Boneh, D. (ed.) CRYPTO 2003. LNCS, vol. 2729, pp. 565–582. Springer, Heidelberg (2003)
12. Dolev, D., Dwork, C., Naor, M.: Non-Malleable Cryptography. SIAM Journal on Computing 30(2), 391–437 (2000)
13. Gligor, V.D., Donescu, P., Katz, J.: On Message Integrity in Symmetric Encryption (February 2002)
14. Goldwasser, S., Micali, S.: Probabilistic Encryption. Journal of Computer and System Sciences 28(2), 270–299 (1984)
15. Iwata, T.: New Blockcipher Modes of Operation with Beyond the Birthday Bound Security. In: Robshaw, M. (ed.) FSE 2006. LNCS, vol. 4047, pp. 310–327. Springer, Heidelberg (2006)

16. Katz, J., Yung, M.: Complete Characterization of Security Notions for Probabilistic Private-Key Encryption. In: Proceedings of the Thirty-Second Annual ACM Symposium on Theory of Computing, pp. 245–254. ACM (2000)

17. Katz, J., Yung, M.: Unforgeable Encryption and Chosen Ciphertext Secure Modes of Operation. In: Schneier, B. (ed.) FSE 2000. LNCS, vol. 1978, pp. 284–299. Springer, Heidelberg (2001)

18. Krawczyk, H.: The Order of Encryption and Authentication for Protecting Communications (or: How Secure Is SSL?). In: Kilian, J. (ed.) CRYPTO 2001. LNCS, vol. 2139, pp. 310–331. Springer, Heidelberg (2001)

19. Krohn, M.: On the Definitions of Cryptographic Security: Chosen-Ciphertext Attack Revisited. Senior Thesis. Harvard University (1999)

20. Maurer, U.: Indistinguishability of Random Systems. In: Knudsen, L.R. (ed.) EUROCRYPT 2002. LNCS, vol. 2332, pp. 110–132. Springer, Heidelberg (2002)

21. Maurer, U.: Constructive Cryptography – A Primer. In: Sion, R. (ed.) FC 2010. LNCS, vol. 6052, p. 1. Springer, Heidelberg (2010)

22. Maurer, U.: Constructive Cryptography – A New Paradigm for Security Definitions and Proofs. In: Mödersheim, S., Palamidessi, C. (eds.) TOSCA 2011. LNCS, vol. 6993, pp. 33–56. Springer, Heidelberg (2012)

23. Maurer, U., Renner, R.: Abstract Cryptography. In: Innovations in Computer Science. Tsinghua University Press (2011)

24. Maurer, U., Schmid, P.: A Calculus for Security Bootstrapping in Distributed Systems. Journal of Computer Security 4(1), 55–80 (1996)

25. Maurer, U., Tackmann, B.: On the Soundness of Authenticate-then-Encrypt: Formalizing the Malleability of Symmetric Encryption. In: ACM Conference on Computer and Communications Security. ACM (2010)

26. Micali, S., Rogaway, P.: Secure Computation. In: Feigenbaum, J. (ed.) CRYPTO 1991. LNCS, vol. 576, pp. 392–404. Springer, Heidelberg (1992)

27. Namprempre, C.: Secure Channels Based on Authenticated Encryption Schemes: A Simple Characterization. In: Zheng, Y. (ed.) ASIACRYPT 2002. LNCS, vol. 2501, pp. 515–532. Springer, Heidelberg (2002)

28. Paterson, K.G., Ristenpart, T., Shrimpton, T.: Tag Size Does Matter: Attacks and Proofs for the TLS Record Protocol. In: Lee, D.H. (ed.) ASIACRYPT 2011. LNCS, vol. 7073, pp. 372–389. Springer, Heidelberg (2011)

29. Pfitzmann, B., Waidner, M.: A Model for Asynchronous Reactive Systems and its Application to Secure Message Transmission. In: Proceedings of the 2001 IEEE Symposium on Security and Privacy, pp. 184–200. IEEE (2001)

30. Rogaway, P., Bellare, M., Black, J.: OCB: A Block-Cipher Mode of Operation for Efficient Symmetric Encryption. ACM Transactions on Information and System Security (TISSEC) 6(3), 365–403 (2003)

31. Rogaway, P., Shrimpton, T.: A Provable-Security Treatment of the Key-Wrap Problem. In: Vaudenay, S. (ed.) EUROCRYPT 2006. LNCS, vol. 4004, pp. 373–390. Springer, Heidelberg (2006)

32. Rüedlinger, A.: Restricted Types of Malleability in Encryption Schemes. Master's thesis, ETH Zürich (2011)

33. Shoup, V.: A Proposal for an ISO Standard for Public Key Encryption. Cryptology ePrint Archive, Report 2001/112 (2001)

34. Shrimpton, T.: A Characterization of Authenticated-Encryption as a form of Chosen-Ciphertext Security. Cryptology ePrint Archive, Report 2004/272 (2004)

Leakage-Resilient Circuits
without Computational Assumptions*

Stefan Dziembowski[1,**] and Sebastian Faust[2,***]

[1] University of Warsaw and Sapienza University of Rome
[2] Aarhus University

Abstract. Physical cryptographic devices inadvertently leak information through numerous side-channels. Such leakage is exploited by so-called side-channel attacks, which often allow for a complete security breache. A recent trend in cryptography is to propose formal models to incorporate leakage into the model and to construct schemes that are provably secure within them.

We design a *general* compiler that transforms *any* cryptographic scheme, e.g., a block-cipher, into a functionally equivalent scheme which is resilient to any *continual* leakage provided that the following three requirements are satisfied: (i) in each observation the leakage is bounded, (ii) different parts of the computation leak independently, and (iii) the randomness that is used for certain operations comes from a simple (non-uniform) distribution. In contrast to earlier work on leakage resilient circuit compilers, which relied on computational assumptions, our results are purely *information-theoretic*. In particular, we do not make use of public key encryption, which was required in all previous works.

1 Introduction

Leakage resilient cryptography attempts to incorporate side-channel information leakage into standard cryptographic models and to design new cryptographic schemes that provably withstand such leakages under reasonable physical assumptions. The "holy grail" in leakage-resilient cryptography is a *generic* method to provably protect *any* cryptographic computation against a broad, well-defined and realistic class of side-channel leakages. This fundamental question has first

---

* This work was supported by the *WELCOME/2010-4/2* grant founded within the framework of the *EU Innovative Economy (National Cohesion Strategy) Operational Programme*.
** The European Research Council has provided financial support to the first author of this paper under the European Community's Seventh Framework Programme (FP7/2007-2013) / ERC grant agreement no CNTM-207908.
*** Sebastian Faust acknowledges support from the Danish National Research Foundation and The National Science Foundation of China (under the grant 61061130540) for the Sino-Danish Center for the Theory of Interactive Computation, within part of this work was performed; and from the CFEM research center, supported by the Danish Strategic Research Council.

R. Cramer (Ed.): TCC 2012, LNCS 7194, pp. 230–247, 2012.

been studied in the work of Ishai et al. [ISW03] who initiated the concept of *leakage resilient circuit compilers*. A circuit compiler takes a description of a (Boolean) circuit Γ as input and outputs a transformed (Boolean) circuit Π^Γ with the same functionality, but with resilience to certain well-defined classes of leakage. The authors consider a very specific type of leakage, namely, an adversary who learns the values of up to $n \in \mathbb{N}$ internal wires in each execution of Π^Γ. Security is proven by a simulation based argument. More precisely, it is shown that any (computationally unbounded) adversary that learns the value of up to n internal wires in each execution of Π^Γ has only a negligible advantage over an adversary that only views the inputs/outputs of the original circuit Γ.

The result of Ishai et al. shows security for a very restricted class of leakages, namely, security is proven only against the *specific attack* of learning the values of n wires. The question that motivates our work is whether, analogously to [ISW03], we can protect any computation against the much broader class of *polynomial-time computable leakages*. This question has been answered affirmatively in the recent feasibility results of Juma and Vahlis [JV10] and Goldwasser and Rothblum [GR10] by making additionally use of the prominent "only computation leaks information" assumption [MR04]. The security of both compilers, however, relies on heavy cryptographic machinery by using public key encryption to "encrypt" the secret state and the whole computation of Γ.[1]

At first sight, it may look natural to rely on some form of cryptographic encryption, if we want to achieve security against any polynomial-time computable leakage function. For instance, it is necessary to "encrypt" the secret state of Γ, as already a single bit of information leaking about the original secret state makes simulation-based security impossible. Perhaps surprisingly, in this paper we show that cryptographically secure encryption schemes are not necessary to construct leakage resilient circuit compilers for *polynomial-time computable leakages*. More precisely, we show that even an *unbounded adversary* with continuous leakage access to Π^Γ only gains a negligible advantage over an adversary with only black-box access to Γ.

Similar to earlier work, we make certain restrictions on the leakage. We follow the work of Dziembowski and Pietrzak [DP08], and allow the leakage to be arbitrary as long as the following two restrictions are satisfied:

1. **Bounded leakage:** the amount of leakage in each round is bounded to λ bits (but overall can be arbitrary large).
2. **Independent leakage:** the computation can be structured into sub computations, where each part of the computations leaks independently (we define the term of a "sub computation" below).

Formally, this is modeled by letting the adversary for each observation choose a leakage function f with range $\{0,1\}^\lambda$, and then giving her $f(\tau)$ where τ is all the data that has been accessed in the current sub-computation. In addition, we require access to a source of correlated randomness generated in a *leak-free* way

---

[1] More precisely, Juma and Vahlis require fully homomorphic encryption, while Goldwasser and Rothblum use a variant of the BHHO encryption scheme.

– e.g., computed by a simple leak free component. We provide more details on our hardware assumptions below.

On Independent Leakages. Variants of the assumption that different parts of the computation leak independently have been used in several works [DP08, Pie09, KP10, GR10, GR10, JV10]. In its weakest form, the assumption says that the state is divided into two parts that leak independently. This type of assumption is used, e.g., in the work on leakage resilient stream ciphers [DP08, Pie09]. Several stronger flavors have been used in the literature. For instance, in the circuit compiler of Goldwasser and Rothblum [GR10] the computation is structured into $O(s)$ sub-computations, where s is the size of the original circuit. Of course, in practice leakage is a global phenomenon and assumptions that require a large number of independent computations is a strong assumption on the hardware. We would like to emphasize, however, hat many relevant global leakage functions can be computed from independent leakages. This is not only true for the prominent Hamming weight leakage, but more generally, for *any affine* leakage function.

On the Relation between Leakage Granularity and the Amount of Leakage. We show a relation between the granularity level of the independent leakage assumption and the amount of leakage that can be tolerated per observation. More precisely, in our basic setting we assume that the computation is structured into $2s$ parts that leak independently, where s is the number of gates in Γ (this is comparable to the model of [GR10]). Here, the amount of leakage can increase linearly with the size of the circuit. Alternatively, we may settle for weaker independency assumptions. That is, in the best case we may require only *two sub-components* that leak independently. Of course this comes at a price: the amount of leakage that is tolerated is *independent* of the circuit's size. We notice that we can tolerate more leakage if we assume some strong form of *memory erasures* between sub-computations (cf. Section 6 for the details).

On Leak-Free Components. Leak-free components are used by recent leakage resilient circuit compilers [GKR08, FRR+10, JV10, GR10]. A leak-free component leaks from its outputs, but the leakage is oblivious to its internals. In this work, we use the leak-free component, $\mathcal{O}$, that was recently introduced by Dziembowski and Faust [DF11]. This component outputs two random vectors $A, B \leftarrow \mathbb{F}^n$ (with $\mathbb{F}$ being a finite field and n being a statistical security parameter) such that their inner product is 0, i.e., $\sum_i A_i \cdot B_i = 0$. As discussed in [DF11], $\mathcal{O}$ exhibits several properties that are beneficial for implementation. We refer the reader to [DF11] for a more thorough discussion on the properties of $\mathcal{O}$.

1.1   Our Contributions

We propose a general transformation (also called the "compiler") that takes any circuit Γ computing over finite fields $\mathbb{F}$ and transforms it into Π^Γ in such a way that (1) the circuit Π^Γ computes the same function as Γ, and (2) any (computationally unbounded) adversary that obtains continuous leakage from

Π^Γ gains only negligible advantage over an adversary with only black-box access to Γ. We emphasize that in contrast to earlier works in similar leakage models [GR10, JV10], we do not use public key encryption to achieve leakage resilience. This makes our results significantly more efficient.

Our construction is secure in the continuous leakage setting with adaptive queries. That is, we assume that the circuit Γ can be initialized (during a trusted step-up phase) with some secret *state*, and is then queried by an adversary S on adaptively chosen inputs $X^1, \ldots, X^\ell$. For each i let $Y^i := \Gamma(X^i, state)$ be the outcome of the ith query. To define security, we consider an adversary $\mathcal{A}$ that attacks Π^Γ and gets the same information (i.e., pairs $(X^1, Y^1), \ldots, (X^\ell, Y^\ell)$ for X^i's chosen by him) *plus* the leakage from each computation. Informally, the security definition requires that for every such (computationally unbounded) adversary $\mathcal{A}$, there exists S with only black-box access to Γ that produces the same output as $\mathcal{A}$. The formal definition is given in Section 5.3. For simplicity, in the formal model we consider only the case where the adversary is allowed to observe the computation once. For readers familiar with the work on leakage resilient circuits [ISW03, FRR⁺10] this is the case of stateless circuits. We briefly discuss how to extend our result to the continuous leakage setting in Section 6.

We emphasize that the running time of our simulator S is polynomial in the running time of $\mathcal{A}$. This is necessary to protect circuits Γ, which hide the secret key only computationally – which is the case for most prominent cryptographic schemes. This is in contrast to the recent work of Dziembowski and Faust [DF11] that consider efficient transformations for cryptographic schemes which hide the secret key information theoretically (e.g., Okamoto signatures or Cramer-Shoup encryption).

1.2   Comparison to Related Work

An extension of the circuit compiler of Ishai et al. [ISW03] (mentioned above) was proposed by Faust et al. [FRR⁺10]. The authors use similar techniques as [ISW03] based on secret sharing but give a significantly improved security analysis considering computationally weak (e.g., AC0) and noisy leakages. Similar to our work, the results of [ISW03, FRR⁺10] work in the information theoretic setting. The leak-free components that are used in earlier works are similar in spirit to the component used in our work. In [FRR⁺10], the leak-free component outputs an n-bit string with parity 0, while in the works of Juma and Vahlis [JV10] and Goldwasser and Rothblum [GR10] it outputs ciphertexts that encrypt 0 using the underlying public-key encryption scheme. Except for the work of Juma and Vahlis all leakage resilient circuit compilers (including ours) require at least one leak-free component for each gate in the original circuit Γ.

We finally remark that our results do not imply the recent results of Dziembowski and Faust [DF11]. More precisely, although we use the same trusted source $\mathcal{O}$ as [DF11], the schemes of [DF11] cannot be obtained by using our circuit compiler. The reason for this are twofold: first, the protocols of [DF11] only use the leak-free component *for the refreshing* of the secret key, while our protocols need to use $\mathcal{O}$ for each gate of the original circuit. Second, their implementation of standard cryptographic schemes are significantly more efficient: while we work on the gate level and blow-up the circuit's size by $O(n^4)$, Dziembowski and Faust directly exploit homomorphic properties of cryptographic schemes and increase the size only by a factor of $O(n)$. Unfortunately, however, these techniques are limited only to certain schemes such as the Okamoto identification and the Cramer-Shoup encryption.

2   Preliminaries

For a set $\mathcal{S}$ we denote by $X \leftarrow \mathcal{S}$ the process of drawing X uniformly from $\mathcal{S}$. A vector V is a row vector, and we denote by V^T its transposition. We let $\mathbb{F}$ be a finite field and for $m, n \in \mathbb{N}$, let $\mathbb{F}^{m \times n}$ denote the set of $m \times n$-matrices over $\mathbb{F}$. For a matrix $M \in \mathbb{F}^{m \times n}$ and an m bit vector $V \in \mathbb{F}^m$ we denote by $V \cdot M$ the n-element vector that results from matrix multiplication of V and M. For a natural number n let $(0)^n = (0, \ldots, 0)$. We use $V[i]$ to denote the ith element of a vector V and $V[i, \ldots, j]$ to denote the elements $i, i + 1, \ldots, j$ of V. For two vectors $V \in \mathbb{F}^m, W \in \mathbb{F}^n$ we denote by $V \| W$ its concatenation and by $V \otimes W$ we will mean a vector in $\mathbb{F}^{m \cdot n}$ defined as

$$V \otimes W := (V_1 W_1, \ldots, V_1 W_m, V_2 W_1, \ldots, V_2 W_m, \quad \ldots \quad , V_n W_1, \ldots, V_n W_m). \quad (1)$$

Finally, let $\langle V, W \rangle$ denote the inner product of V and W. We will use the fact that the inner product is linear, i.e. $\langle a \cdot V + V', W \rangle = a \cdot \langle V, W \rangle + \langle V', W \rangle$.

The "$\stackrel{d}{=}$" symbol denotes the equality of two distributions. For two random variables X_0, X_1 over $\mathcal{X}$ we define the statistical distance between X and Y as $\Delta(X; Y) = \sum_{x \in \mathcal{X}} 1/2 |\Pr[X_0 = x] - \Pr[X_1 = x]|$.

2.1   Leakage Model

To formally model leakage, we follow Dziembowski and Faust [DF11] and only recall some important details here. We model independent leakage from memory parts in form of a *leakage game*, where the adversary can *adaptively* learn information from the memory parts. More precisely, for some $c, \ell, \lambda \in \mathbb{N}$ let $M_1, \ldots, M_\ell \in \{0, 1\}^c$ denote the contents of the memory parts, then we define a λ-*leakage game* played between an adaptive adversary $\mathcal{A}$, called a λ-*limited leakage adversary*, and a *leakage oracle* $\Omega(M_1, \ldots, M_\ell)$ as follows. For some $m \in \mathbb{N}$, the adversary $\mathcal{A}$ can adaptively issue a sequence $\{(x_i, f_i)\}_{i=1}^m$ of requests to the oracle $\Omega(M_1, \ldots, M_\ell)$, where $x_i \in \{1, \ldots, \ell\}$ and $f_i : \{0, 1\}^c \to \{0, 1\}^{\lambda_i}$ with $\lambda_i \leq \lambda$. To each such a query the oracle replies with $f_i(M_{x_i})$ and we say that in this case the adversary $\mathcal{A}$ *retrieved the value* $f_i(M_{x_i})$ *from* M_{x_i}. The only restriction is that in total the adversary does not retrieve more than λ bits from each memory part. In the following, let $(\mathcal{A} \rightleftarrows (M_1, \ldots, M_\ell))$ be the output of $\mathcal{A}$ at the end of this game. Without loss of generality, we assume that $(\mathcal{A} \rightleftarrows (M_1, \ldots, M_\ell)) := (f_1(M_{x_1}), \ldots, f_m(M_{x_m}))$.

LEAKAGE FROM COMPUTATION. We model the computation that is carried out on a device as a ℓ-party protocol $\Pi = (P_1, \ldots, P_\ell)$, which is executed between the parties $(P_1, \ldots, P_\ell)$ and an adversary is allowed to obtain partial information (the leakage) from the internal state of the players. Initially, some parties may hold inputs, and we denote by S_i the input of P_i. The execution of Π with initial inputs $S_1, \ldots, S_\ell$, denoted by $\Pi(S_1, \ldots, S_\ell)$, is structured into sub-computations. In each sub-computation one player is active and sends messages to the other players. These messages can depend on his input (i.e., his initial state), his local randomness, and the messages that he received in earlier rounds. At the end of the protocol's execution, the players $P_1, \ldots, P_\ell$ output values $S'_1, \ldots, S'_\ell$, resp. (some of these values may be empty). For each player P_i, we denote the local randomness that is used by P_i during the execution of Π and all the messages that are received *or* sent (including the messages from the user of the protocol) by view_i. We assume that after the protocol terminates, the adversary $\mathcal{A}$ plays a λ-leakage game against the leakage oracle $\Omega(\text{view}_i, \ldots, \text{view}_\ell)$. We will use the following convention in order to simplify the exposition: while describing a protocol we will explicitly describe the view of each player, sometimes omitting redundant variables. For instance, if the view contains variables X, Y, Z, such that always $Z = X \oplus Y$, then we will omit Z, as it can be calculated by the leakage function from X and Y.

2.2 Leakage-Resilient Storage

Davi et al. [DDV10] recently introduced the notion of leakage-resilient storage (LRS) $\Phi = (\text{Encode}, \text{Decode})$. An LRS allows to store a secret in an "encoded form" such that even given leakage from the encoding no adversary learns information about the encoded values. One of the constructions that the authors propose uses two source extractors and can be shown to be secure in the independent leakage model. More precisely, an LRS for the independent leakage model is defined for message space $\mathcal{M}$ and encoding space $\mathcal{L} \times \mathcal{R}$ as follows:

- Encode : $\mathcal{M} \to \mathcal{L} \times \mathcal{R}$ is a probabilistic, efficiently computable function and
- Decode : $\mathcal{L} \times \mathcal{R} \to \mathcal{M}$ is a deterministic, efficiently computable function such that for every $S \in \mathcal{M}$ we have $\text{Decode}(\text{Encode}(S)) = S$.

An LRS Φ is said to be (λ, ϵ)-secure, if for any $S, S' \in \mathcal{M}$ and any λ-limited adversary $\mathcal{A}$, we have $\Delta(\mathcal{A} \rightleftarrows (L, R); \mathcal{A} \rightleftarrows (L', R')) \le \epsilon$, where $(L, R) \leftarrow \text{Encode}(S)$ and $(L', R') \leftarrow \text{Encode}(S')$, for any two secrets $S, S' \in \mathcal{M}$. In this paper, we consider a leakage-resilient storage scheme $\Phi_\mathbb{F}^n$ that allows to efficiently store elements from $\mathcal{M} = \mathbb{F}$. It is a variant of a scheme proposed in [DF11] and based on the inner-product extractor. For some security parameter $n \in \mathbb{N}$, $\Phi_\mathbb{F}^n := (\text{Encode}_\mathbb{F}^n, \text{Decode}_\mathbb{F}^n)$ is defined as follows:

- $\text{Encode}_\mathbb{F}^n(S)$:
    1. Sample $(L[2, \ldots, n], R[2, \ldots, n]) \leftarrow (\mathbb{F}^{n-1})^2$.
    2. Set $L[1] \leftarrow \mathbb{F} \setminus \{0\}$ and $R[1] := L[1]^{-1} \cdot (S - \langle (L[2, \ldots, n], R[2, \ldots, n]) \rangle)$ Output (L, R).
- $\text{Decode}_\mathbb{F}^n(L, R)$: Output $\langle L, R \rangle$.

The property that $L[1] \neq 0$ will be useful in the "generalized multiplication" protocol (cf. Section 4.2). It is easy to see that $\Phi_{\mathbb{F}}^n$ is correct, i.e.:

$$\mathsf{Decode}_{\mathbb{F}}^n(\mathsf{Encode}_{\mathbb{F}}^n(S)) = S.$$

Security is shown in the following lemma whose proof appears in the full version of this paper.

Lemma 1. *Let $n \in \mathbb{N}$ and let $\mathbb{F}$ such that $|\mathbb{F}| = \Omega(n)$. For any $1/2 > \delta > 0, \gamma > 0$ the LRS $\Phi_{\mathbb{F}}^n$ as defined above is (λ, ϵ)-secure, with $\lambda = (1/2 - \delta)n\log|\mathbb{F}| - \log\gamma^{-1} - 1$ and $\epsilon = 2m(|\mathbb{F}|^{3/2-n\delta} + |\mathbb{F}|\gamma)$.*

We instantiate Lemma 1 with concrete parameters in the next corollary.

Corollary 1. *Suppose $|\mathbb{F}| = \Omega(n)$. Then, LRS $\Phi_{\mathbb{F}}^n$ is $(0.49 \cdot \log_2 |\mathbb{F}^n| - 1, negl(n))$-secure, for some negligible function negl.*

3 An Informal Description of the Protocol

In this section we describe informally our circuit compiler that is based on the LRS scheme $\Phi_{\mathbb{F}}^n$. Our starting point is the result of [DF11] where a protocol $\mathsf{Refresh}_{\mathbb{F}}^n$ is proposed to refresh secrets encoded with $\Phi_{\mathbb{F}}^n$. $\mathsf{Refresh}_{\mathbb{F}}^n$ is run between two parties P_L and P_R, which initially hold L and R in $\mathbb{F}^n$. At the end of the protocol, P_L holds L' and P_R holds R' such that $\langle L, R \rangle = \langle L', R' \rangle$. The protocol can be repeated continuously to refresh the encoding and satisfies the following security requirement: even given continuous leakage independently from the parties P_L and P_R no adversary can learn the encoded secret $\langle L, R \rangle$.

In order to create a general circuit compiler in the independent leakage model, all we need is to perform in a leakage-resilient way arithmetic operations on the encoded secrets using the LRS $\Phi_{\mathbb{F}}^n$. This is similar to the methods used in the MPC literature: first, the secret is secret-shared between the parties (in our case: "encoded"), and then the operations are performed "gate-by-gate" in a secure way. At the end the outputs of the computation are reconstructed in the following way: one of the players, P_L, say, sends his share L' of the output to P_R and P_R computes $\mathsf{Decode}_{\mathbb{F}}^n(L, R)$. We us a similar approach in this paper.

To illustrate this approach, consider the simple case of a circuit that multiplies a constant α with a secret S encoded as (L, R). If L is held by P_L and R is held by P_R, then one of the players, P_L, say, multiplies his vector by α (as $\langle \alpha \cdot L, R \rangle = \alpha \cdot \langle L, R \rangle$). Also, addition of a constant c to S is simple: the player P_L sends $x = L[1]$ to P_R (for simplicity assume that $L[1] \neq 0$), and then P_R sets $R' = R + (x^{-1} \cdot c, 0, \ldots, 0)$ and P_L sets $L' = L$. We notice that (L', R') was computed from (L, R) just by sending one field element from P_L to P_R, and in particular it did not involve computing $\langle L, R \rangle$. We call this protocol $\mathsf{AddConst}_{\mathbb{F}}^n(\alpha, (L, R))$.

The only ingredient that is missing for computing arbitrary functionalities is a protocol for leakage-resilient multiplication of two encoded secrets. The construction of such a protocol is the main contribution of this paper (for technical reasons, we construct in Section 4.2 a protocol for a slightly more general functionality, which we call "generalized multiplication"). Suppose we have two secrets $S^0 \in \mathbb{F}$ and $S^1 \in \mathbb{F}$ encoded as (L^0, R^0) and (L^1, R^1), respectively. Suppose further that player P_L holds (L^0, L^1) and player P_R holds (R^0, R^1). Their goal is to compute $L'', R'' \in \mathbb{F}^n$ in a leakage-resilient way such that $\langle L'', R'' \rangle = S$ and L'' is held by P_L, while R'' is held by P_R. Our first observation is that $\langle L^0 \otimes L^1, R^0 \otimes R^1 \rangle = \langle L^0, R^0 \rangle \cdot \langle L^1, R^1 \rangle = S^0 \cdot S^1$, which follows from simple linear algebra. Hence, $(L^0 \otimes L^1, R^0 \otimes R^1)$ encodes the secret $S^0 \cdot S^1$ in the $\Phi_{\mathbb{F}}^{n^2}$ scheme. Note that this protocol, so far, is non-interactive so it is clearly secure. The disadvantage of this protocol is that the length of the encoding grows exponentially with the depth of Γ. Therefore, we need a method of reducing the length of this encoding. This can be done in the following way. First, the players refresh the $(L^0 \otimes L^1, R^0 \otimes R^1)$ encoding with the $\mathsf{Refresh}_{\mathbb{F}}^{n^2}$ protocol. Let $(L', R') \in \mathbb{F}^{n^2} \times \mathbb{F}^{n^2}$ be the result of this refreshing. Then, the players reconstruct *in clear* the secret encoded by the final $n(n-1)$ elements of L' and R'. More precisely, the player P_L sends $L'[n+1, \ldots, n^2]$ to P_R, and P_R computes $d = \langle L'[n+1, \ldots, n^2], R'[n+1, \ldots, n^2] \rangle$. We now clearly have that $S^0 \cdot S^1 = \langle L', R' \rangle = \langle L'[1, \ldots, n], R'[1, \ldots, n] \rangle + d$. Hence, $(L'[1, \ldots, n], R'[1, \ldots, n])$ encodes $S^0 \cdot S^1$ minus d. Since d can be published by P_R we can now use the protocol $\mathsf{AddConst}_{\mathbb{F}}^n(d, (L'[1, \ldots, n], R'[1, \ldots, n]))$, and add a constant d to $(L'[1, \ldots, n], R'[1, \ldots, n])$. The output (L'', R'') of the protocol is the result of this operation. Observe that the use of the refreshing protocol is crucial, as $(L^0 \otimes L^1)[n+1, \ldots, n^2]$ gives almost complete information about L^0 and L^1.

4   The Ingredients

In this section, we describe the two main ingredients of our compiler construction: the "refreshing" protocol for $\Phi_{\mathbb{F}}^n$ (cf. Section 4.1) and the "generalized multiplication" protocol (cf. Section 4.2). The latter protocol will use the former as a sub-routine. In the full version of this paper, we show that these two components satisfy a simple security property called reconstructibility. This notion was introduced recently in [FRR+10] and essentially says that the view of the parties in a protocol can be efficiently reconstructed from just knowing the encoded inputs and outputs. For our setting, we modify this notion and define reconstruction as a protocol run between players P_L and P_R, where the efficiency criteria of the reconstructor is the amount of information exchanged between the parties. For instance, for the generalized multiplication the reconstructor protocol is run between P_L with input (L^0, L^1, L'') and P_R with input (R^0, R^1, R'') and computes view_L and view_R with only one field element of communication.

4.1   Leakage-Resilient Refreshing of LRS

In this section, we propose a simple variant of the refreshing protocol proposed in [DF11] (cf. Section 3) for the LRS $\Phi_{\mathbb{F}}^n$. As described in the introduction, we assume that the players have access to a leak-free component that samples uniformly at random pairs of orthogonal vectors. Technically, we will assume that we have an oracle $\mathcal{O}'$ that samples a uniformly random vector $((A, \tilde{A}), (B, \tilde{B})) \in (\mathbb{F}^n)^4$, subject to the constraint that the following three conditions hold:

1. $\langle A, B \rangle + \langle \tilde{A}, \tilde{B} \rangle = 0$,
2. $A \neq (0)^n$, and
3. $\tilde{B} \neq (0)^n$.

Note that this oracle is different from the oracle $\mathcal{O}$ described in the introduction (and used earlier in [DF11]) that simply samples pairs (A, B) of orthogonal vectors. It is easy to see, however, that this "new" oracle $\mathcal{O}'$ can be "simulated" by the players that have access to $\mathcal{O}$ that samples pairs (C, D) of orthogonal vectors of length $2n$ each. First, observe that $C \in \mathbb{F}^{2n}$ can be interpreted as a pair $(A, \tilde{A}) \in (\mathbb{F}^n)^2$ (where $A||\tilde{A} = C$), and in the same way $D \in \mathbb{F}^{2n}$ can be interpreted as a pair $(B, \tilde{B}) \in (\mathbb{F}^n)^2$ (where $B||\tilde{B} = D$). By the basic properties of the inner product we get that $\langle A, B \rangle + \langle \tilde{A}, \tilde{B} \rangle = \langle C, D \rangle = 0$. Hence, Condition 1 is satisfied. Conditions 2 and 3 can simply verified by players P_L and P_R respectively. If one these conditions is not met, then the players sample a fresh (C, D) from $\mathcal{O}$. Obviously, this happens with a negligible probability $2 \cdot 2^{-n|\mathbb{F}|}$ only, so it has almost no impact on the efficiency of the protocol.

The reason for introducing Conditions 2 and 3 is to make the exposition simpler as it avoids dealing with the events that happen with negligible probability (cf. the caption of Figure 1). The reason for having Condition 1 is slightly more subtle and will be explained below.

The refreshing scheme is presented in Figure 1. The main idea behind this protocol is as follows (for this high-level overview ignore Step 4, as it anyway influences the execution only with negligible probability). Denote $\alpha := \langle A, B \rangle (= -\langle \tilde{A}, \tilde{B} \rangle)$. The Steps 2 and 3 are needed to refresh the share of P_R. This is done by generating, with the "help" of (A, B) (coming from $\mathcal{O}'$) a vector X such that

$$\langle L, X \rangle = \alpha. \tag{2}$$

Eq. (2) comes from simple linear algebra: $\langle L, X \rangle = \langle L, B \cdot M^T \rangle = \langle L \cdot M, B \rangle = \langle A, B \rangle = \alpha$. Then, vector X is added to the share of P_R by setting (in Step 3) $R' := R + X$. Hence we get $\langle L, R' \rangle = \langle L, R \rangle + \langle L, X \rangle = \langle L, R \rangle + \alpha$. Symmetrically, in Steps 5 and 6 the players refresh the share of P_L, by first generating Y such that $\langle Y, R \rangle = -\alpha$, and then setting $L' = L + Y$. By similar reasoning as before, we get $\langle L', R' \rangle = \langle L, R' \rangle - \alpha$, which, in turn is equal to $\langle L, R \rangle$. Hence, the refreshing is correct.

The security proof of this refreshing scheme appears in the full version of this paper. The key property that is used there is that X is generated "obliviously" from P_L, and Y is generated "obviously" from P_R. In other words: P_L gets no information on X except that $\langle L, X \rangle = -\langle Y, R \rangle$, and a symmetric fact holds for P_R. For more intuition behind this protocol the reader may consult [DF11] (Sect. 3), where a similar refreshing scheme is constructed. The main difference is that the protocol presented here refreshes the shares "completely", i.e. the new encoding (L', R') is completely independent from (L, R) (except that is encodes the same secret), while in [DF11] this was not the case. More precisely, in the refreshing of [DF11] $A, \tilde{A}, B$, and $\tilde{B}$ were such that $\langle A, B \rangle = \langle \tilde{A}, \tilde{B} \rangle = 0$, which implied that in particular $\langle L, R' - R \rangle$ and $\langle L' - L, R' \rangle$ were equal to 0 (and hence (L', R') was not independent from (L, R)). In our protocol it is not the case since $\langle A, B \rangle = \alpha$ and $\langle \tilde{A}, \tilde{B} \rangle = -\alpha$ (where α is random) and hence $\langle L, R' - R \rangle$ and $\langle L' - L, R \rangle$ are random. This "independence" of encodings after refreshing is a very useful property for showing security of composition of larger circuits.

Protocol $(L', R') \leftarrow \mathsf{Refresh}_{\mathbb{F}}^{n}((L, R))$:

Input (L, R): $L \in (\mathbb{F} \setminus \{0\}) \times \mathbb{F}^{n-1}$ is given to P_{L} and $R \in \mathbb{F}^{n}$ is given to P_{R}.

1. Let $(A, \tilde{A}, B, \tilde{B}) \leftarrow \mathcal{O}'$ and give $(A, \tilde{A})$ to P_{L} and $(B, \tilde{B})$ to P_{R}.

Refreshing the share of P_{R}:

2. Player P_{L} generates a random non-singular matrix $M \in \mathbb{F}^{n \times n}$ such that $L \cdot M = A$ and sends it to P_{R}.
3. Player P_{R} sets $X := B \cdot M^{T}$ and $R' := R + X$.

Refreshing the share of P_{L}:

4. If $R' = (0, \ldots, 0)$ then P_{R} sends a message $\mu = $ "**zero**" to P_{L}. Player P_{L} sets $L' \leftarrow (\mathbb{F} \setminus \{0\}) \times \mathbb{F}^{n-1}$. The players output (L', R') and finish this round of refreshing. Otherwise the player P_{R} sends a message $\mu = $ "**nonzero**" to P_{L} and they execute the following:
5. Player P_{R} generates a random non-singular matrix $\tilde{M} \in \mathbb{F}^{n \times n}$ such that $\tilde{M} \cdot R' = \tilde{B}$ and sends it to P_{L}.
6. Player P_{L} sets $Y := \tilde{A} \cdot \tilde{M}^{T}$ and $L' := L + Y$.
7. If $L'[1] = 0$ then restart the procedure of refreshing the share of P_{L}, i.e. go to Step 4.

Output: The players output (L', R').
Views: The view $\mathsf{view}_{\mathsf{L}}$ of player P_{L} is $(L, A, M, \tilde{A}, \tilde{M}, \mu)$ and the view $\mathsf{view}_{\mathsf{R}}$ of player P_{R} is $(R, B, M, \tilde{B}, \tilde{M}, \mu)$.

Fig. 1. Protocol $\mathsf{Refresh}_{\mathbb{F}}^{n}$. Oracle $\mathcal{O}'$ samples random vectors $(A, \tilde{A}, B, \tilde{B}) \in (\mathbb{F}^{n})^{4}$ such that (1) $\langle A, B \rangle = -\langle \tilde{A}, \tilde{B} \rangle$ and (2) $A \neq (0)^{n}$, and (3) $\tilde{B} \neq (0)^{n}$. Note that the conditions (2) and (3) are needed as otherwise it might be impossible to find matrices M and $\tilde{M}$ in Steps 2 and 5, respectively. It is easy to see that $L[1]$ has a uniform distribution over $\mathbb{F}$, and hence restarting part of the protocol in Step 7 happens with probability $|F|^{-1}$. Therefore if $\mathbb{F}$ is large then this probability is negligible. In Sect. 6 we show how to change our protocol so that the probability of restarting is negligible even if $|\mathbb{F}|$ is small (e.g. constant).

4.2   Leakage-Resilient Computation of Generalized Multiplication

We now present a leakage-resilient protocol for computing a "generalized multiplication" function $f(S^{0}, S^{1}, c) = c - S^{0} \cdot S^{1}$, where the values $S^{0} \in \mathbb{F}$ and $S^{1} \in \mathbb{F}$ are encoded by an LRS $\Phi_{\mathbb{F}}^{n} = (\mathsf{Encode}_{\mathbb{F}}^{n}, \mathsf{Decode}_{\mathbb{F}}^{n})$ (let (L^{0}, R^{0}) and

(L^1, R^1) be the respective encodings), and $c \in \mathbb{F}$ is a constant. The result $f(S^0, S^1, c)$ of the computation is encoded by (L'', R''). This construction has already been discussed informally in Section 3. The formal description appears in Figure 2. It uses the $\mathsf{Refresh}_{\mathbb{F}}^{n^2}$ protocol as a sub-routine, and hence also relies on the special free oracle $\mathcal{O}'$. It is easy to see that this protocol is correct. More formally, for any inputs $L^0, R^0, L^1, R^1 \in \mathbb{F}^n$ and $c \in \mathbb{F}$ we have that $\mathsf{Decode}_{\mathbb{F}}^n(L'', R'') = c - \mathsf{Decode}_{\mathbb{F}}^n(L^0, R^0) \cdot \mathsf{Decode}_{\mathbb{F}}^n(L^1, R^1)$, where $(L'', R'') \leftarrow \mathsf{Mult}_{\mathbb{F}}^n((L^0, R^0), (L^1, R^1), c)$. The security properties of this protocol are defined and proven in the full version of this paper, where we show that the multiplication protocol is reconstructible with low communication between the parties P_{L} and P_{R}.

Protocol $(L'', R'') \leftarrow \mathsf{Mult}_{\mathbb{F}}^n((L^0, R^0), (L^1, R^1), c)$:

Input (L, R): $L^0, L^1 \in (\mathbb{F} \setminus \{0\}) \times \mathbb{F}^{n-1}$ are given to P_{L} and $R^0, R^1 \in \mathbb{F}^n$ are given to P_{R}. The field element $c \in \mathbb{F}$ is given to both players.

1. The players P_{L} and P_{R} run the $\mathsf{Refresh}_{\mathbb{F}}^{n^2}(L^0 \otimes L^1, R^0 \otimes R^1)$ protocol. Let L' and R' be their respective outputs, and let $\mathsf{view}_{\mathsf{L}}'$ and $\mathsf{view}_{\mathsf{R}}'$ be their respective views.
2. Player P_{L} sends $x := L'[1]$ and the last $n(n-1)$ bits of L' (i.e. the vector $L'[n+1, \ldots, n^2]$) to P_{R}. Player P_{R} computes $d := \langle L'[n+1, \ldots, n^2], R'[n+1, \ldots, n^2] \rangle$ and sets $R'' := -R'[1, \ldots, n] + (x^{-1}(c - d), 0, \ldots, 0)$.
3. Player P_{L} sets $L'' := L'[1, \ldots, n]$.

Output: The players output (L'', R'').
Views: The view $\mathsf{view}_{\mathsf{L}}$ of player P_{L} is $(L^0, L^1, L', L'', c, \mathsf{view}_{\mathsf{L}}')$ and the view $\mathsf{view}_{\mathsf{R}}$ of player P_{R} is $(R^0, R^1, R', R'', c, d, x, L'[n+1, \ldots, n^2], \mathsf{view}_{\mathsf{R}}')$.

Fig. 2. Protocol $\mathsf{Mult}_{\mathbb{F}}^n$. Note that computing x^{-1} is possible since in our LRS the first bit of L is never equal to 0. This is actually precisely the reason why this restriction was introduced.

5   The Compiler

5.1   Arithmetic Circuits

Before describing our general circuit compiler, we must define how to model arithmetic circuits over finite fields $\mathbb{F}$ as these are used to describe the original circuits. To keep the exposition simple, we consider circuits consisting only of 4 types of gates. The first two types are: the *public-input gates* that will be used by the user, or the adversary, to provide the input X to the circuit, and the *private-input gates* that will be used to provide the secret input *state* (e.g., the cryptographic key) to the scheme. The third type of a gate is the *multiplication gate* (a, b, c). This gate takes as input the values $A \in \mathbb{F}$ and $B \in \mathbb{F}$ of two other gates (indicated by a and b, resp.) and a constant $c \in \mathbb{F}$, and produces a result $c - AB$. Note that in particular the "negated and" function over bits can be expressed by such a gate, as $\overline{A \wedge B} = 1 - AB$, for $A, B \in \{0, 1\}$. Finally, we also have the output gates. Each output gate takes as input a value from of a gate of a previous type and outputs it. Since it is well-known that a NAND gate is complete the above suffices to describe any functionality. Formally, a *circuit over a field* $\mathbb{F}$ is a sequence $\Gamma = (\gamma_1, \ldots, \gamma_t)$, where each γ_i is called a *gate*. The set of gates is divided into the following groups.

public-input gates: $\gamma_1, \ldots, \gamma_m$ — each such a gate is equal to a special symbol pub and takes the inputs provided by the user.

private-input gates: $\gamma_{m+1}, \ldots, \gamma_{m+k}$ — each such a gate is equal to a special symbol priv and represents the memory containing the secret state,

multiplication gates: $\gamma_{m+k+1}, \ldots, \gamma_{t-u}$ — each such a gate γ_i ($i \in [m + k + 1, t - u]$) has a form (a, b, c), where $a, b \in \{1, \ldots, i - 1\}$ and $c \in \mathbb{F}$. We say that the outputs of the gates γ_a and γ_b are *inputs for the gate* γ_i,

output gates: $\gamma_{t-u+1}, \ldots, \gamma_t$ — each such a gate γ_i is equal to some j, where $j \in \{1, \ldots, t - u\}$. We say that γ_j is an *input for the gate* γ_i.

For technical reasons, we also assume that the circuit's fan-out is at most 2, more precisely: each γ_i is an input for at most 2 other gates. This can be clearly done without loss of generality. The *computation* $\mathsf{Comp}(\Gamma, X, state)$ of such a circuit on input $(X, state) = ((x^1, \ldots, x^m), (s^1, \ldots, s^k))$ is a sequence $(\xi^1, \ldots, \xi^t)$ of values on the outputs of circuit gates (one may think of this as the output wires of the gates), defined by the following procedure:

- For $i = 1$ to t do:
  1. if $\gamma_i = \mathsf{pub}$ ("public-input gate") then set $\xi^i := x^i$,
  2. if $\gamma_i = \mathsf{priv}$ ("private-input gate") then set $\xi^i := s^{i-m}$,
  3. if $\gamma_i = (a, b, c)$ ("multiplication gate") then set $\xi^i = c - \xi^a \xi^b$.
  4. if $\gamma_i = j$ ("output gate") then set $\xi^i = \xi^j$,

The *output of the computation* is equal to $(\xi^{t-u+1}, \ldots, \xi^t)$ and will be denoted by $\Gamma(X, state)$.

5.2 Protocols Computing Circuits

Recall the definition of a protocol from Sect. 2.1. In this section we consider a special type of such protocols, that we call *LRS-protocols*. Each such a protocol Π_Φ is parameterized by an LRS $\Phi = (\mathsf{Encode} : \mathcal{M} \to \mathcal{L} \times \mathcal{R}, \mathsf{Decode} : \mathcal{L} \times \mathcal{R} \to \mathcal{M})$ (we will say that Π *works over* Φ). It consists of $2t$ parties $\mathcal{P} = \{P_\mathsf{L}^1, \ldots, P_\mathsf{L}^t, P_\mathsf{R}^1, \ldots, P_\mathsf{R}^t\}$. The parties are divided into following groups:

"public-input parties": $P_\mathsf{L}^1, \ldots, P_\mathsf{L}^m, P_\mathsf{R}^1, \ldots, P_\mathsf{R}^m$ — each P_L^i takes no input and each P_R^i takes as input $x^i \in \mathbb{F}$,

"private-input parties": $P_\mathsf{L}^{m+1}, \ldots, P_\mathsf{L}^{m+k}, P_\mathsf{R}^{m+1}, \ldots, P_\mathsf{R}^{m+k}$ — each P_L^i takes as input $L^i \in \mathcal{L}$, and each P_R^i takes as input $R^i \in \mathcal{R}$,

"multiplication parties": $P_\mathsf{L}^{m+k+1}, \ldots, P_\mathsf{L}^{t-u}, P_\mathsf{R}^{m+k+1}, \ldots, P_\mathsf{R}^{t-u}$ — they have no inputs or outputs,

"output parties": $P_\mathsf{L}^{t-u+1}, \ldots, P_\mathsf{L}^t, R^{t-u+1}, \ldots, P_\mathsf{R}^t$ — each P_R^i produces an output $y^i \in \mathcal{M}$, and the P_L^i's produce no output.

The LRS-protocols will be analyzed only under the assumption that for $i = k+1, \ldots, m$ we have that $(L^i, R^i) \leftarrow \mathsf{Encode}(z^i)$ for some x^i. More precisely for $X = (x^1, \ldots, x^m) \in \mathbb{F}^m$ and $state = (s^1, \ldots, s^k) \in \mathbb{F}^k$ consider the following experiment.

Experiment $\mathsf{ExpExec}(\Pi_\Phi, X, state)$:

1. For each $i = 1, \ldots, m$ give x^i to P_R^i.
2. For each $i = 1, \ldots, k$ sample $(L^{m+i}, R^{m+i}) \leftarrow \Phi(s^i)$. Give L^{m+i} to P_L^{m+i} and R^{m+i} to P_R^{m+i}.
3. Run the protocol Π_Φ with the inputs for the players as described in the previous steps.
4. For $i = 1, \ldots, t$ let $\mathsf{view}_\mathsf{L}^i$ be the view of P_L^i, and let $\mathsf{view}_\mathsf{R}^i$ be the view of P_R^i in the above execution.
   Denote $\mathsf{View}(\Pi_\Phi, (X, state)) := ((\mathsf{view}_\mathsf{L}^1, \mathsf{view}_\mathsf{R}^1), \ldots, (\mathsf{view}_\mathsf{L}^t, \mathsf{view}_\mathsf{R}^t))$.
5. Let $\mathsf{Exec}(\Pi_\Phi, (X, state))$ be the vector containing the outputs of the parties $P_\mathsf{R}^{t-u+1}, \ldots, P_\mathsf{R}^t$ in the above execution.

5.3   The Security Definition

We now present the main security definition of this paper. As mentioned in the introduction, in this definition we consider only the non-adaptive security. In Sect. 6 we show how this definition can be extended to adaptive settings. Let Γ be a circuit with m public-input gates, k private-input gates and u output gates. Let Π_Φ be an LRS-protocol with $2m$ public-input parties, $2k$ private-input parties and $2u$ output parties. We say that *the Π_Φ protocol (λ, ϵ)-securely computes Γ* if:

- Π_Φ *computes* Γ i.e.: for every $(X, state) \in \mathbb{F}^k \times \mathbb{F}^m$ we have that

$$\mathsf{Exec}(\Pi_\Phi, (X, state)) = \Gamma(X, state),$$

  and
- for every λ-limited adversary $\mathcal{A}$ there exists a *simulator* $\mathcal{S}$, running in time polynomial in the running time of $\mathcal{A}$, that for every $(X, state) \in \mathbb{F}^k \times \mathbb{F}^m$, on input $(X, \Gamma(X, state))$ produces a variable $\mathcal{S}(X, \Gamma(X, state))$ such that

$$\Delta((\mathcal{S}(X, \Gamma(X, state)) \; ; \; (\mathcal{A} \rightleftarrows \mathsf{View}(\Pi_\Phi, (X, \Gamma(X, state))))) \leq \epsilon. \qquad (3)$$

Note that *state* is not given directly to the simulator. The only variables that he gets are: the public input X and the output $Y = \Gamma(X, state)$. Therefore, intuitively, the only information that he gets about *state* comes from (X, Y).

5.4   The Construction

We are now ready to present our construction of the circuit compiler. Our compiler takes an arithmetic circuit Γ and a parameter $n \in \mathbb{N}$ and produces an LRS protocol $\Pi_{\Phi_{\mathbb{F}}^n}^{\Gamma}$ over $\Phi_{\mathbb{F}}^n$. To simplify the notation we will write Π_n^{Γ} instead of $\Pi_{\Phi_{\mathbb{F}}^n}^{\Gamma}$. The protocol Π_n^{Γ} is depicted on Fig. 3.

Protocol $(z^{t-u+1}, \dots, z^t) \leftarrow (\Pi_n^{\Gamma}(x^1, \dots, x^m, (L^1, R^1), \dots, (L^k, R^k)))$:

Input $(x^1, \dots, x^m, (L^1, R^1), \dots, (L^k, R^k))$:   Give each $x^i \in \mathbb{F}$ to P_R^i, each $L^i \in \mathbb{F}^n$ to P_L^{m+i} and each $R^i \in \mathbb{F}^n$ to P_R^{m+i}.

1. For $i = 1, \dots, m$ player P_R^i computes $(L^i, R^i) \leftarrow \mathsf{Encode}_{\mathbb{F}}^n(x^i)$ and sends L^i to P_L^i. The view view_L^i of P_L^i is L^i and the view view_R^i of P_R^i is (L^i, R^i).
2. For $i = m+1, \dots, m+k$ the view view_L^i of P_L^i is L^i and the view view_R^i of P_R^i is R^i.
3. For $i = m+k+1, \dots, t-u$ let (a, b, c) be such that $\gamma_i = (a, b, c)$
   (a) Player P_L^a sends L^a to P_L^i.
   (b) Player P_R^a sends R^a to P_R^i.
   (c) Player P_L^b sends L^b to P_L^i.
   (d) Player P_R^b sends R^b to P_R^i.
   (e) Players P_L^i and P_L^i execute the $\mathsf{Mult}^n((L^a, R^a), (L^b, R^b), c)$ protocol. Let L^i and R^i be the respective outputs of the players at the end of this protocol, and let view_L^i and view_R^i be their respective views.
4. For $i = t-u+1, \dots, t$ let j be such that $\gamma_i = j$.
   (a) Player P_L^j sends L^j to P_L^i.
   (b) Player P_R^j sends R^j to P_R^i.
   (c) The players P_L^j and P_R^i execute the $\mathsf{Refresh}^n(L^j, R^j)$ protocol. Let L^i and R^i be the respective outputs of the players at the end of this protocol, and let view_L^i and view_R^i be their respective views.
   (d) Player P_L^i sends L^i to P_R^i. Player P_R^i computes $z^j := \mathsf{Decode}_{\mathbb{F}}^n(L^i, R^i)$ and outputs it. The v^i of P_L^i is view_L^i and the view view_R^i of P_R^i is (view_R^i, L^i).

Fig. 3. The Π_n^{Γ} protocol

We now have the following theorem. Its proof is based on the *hybrid argument* and appears in the full version of this paper.

Theorem 1. *Assume that for some n the LRS $(\mathsf{Encode}_{\mathbb{F}}^n, \mathsf{Decode}_{\mathbb{F}}^n)$ is (λ, ϵ)-secure for some λ and ϵ. Then for any Γ the Π_n^{Γ} protocol $(\lambda/3 - \log_2 |\mathbb{F}|, t\epsilon)$-securely computes Γ.*

The following is an example of the application of Thm. 1 for a concrete LRS.

Corollary 2. *Suppose $|\mathbb{F}| = \Omega(n)$. Then for any Γ the Π_n^{Γ} protocol $(0.16 \cdot \log_2 |\mathbb{F}^n| - 1 - \log_2 |\mathbb{F}|, negl(n))$-securely computes Γ, for some negligible n.*

6 Extensions

The model in Sect. 5 was intentionally kept simple in order to make the proof as easy as possible, and to satisfy the page limit. In this section we present several generalizations and extensions of this model. The formal security definitions and proofs will be presented in the extended version of this paper.

ADAPTIVE SECURITY. Most of the cryptographic security definitions assume that the adversary is *adaptive*, meaning that he can interact with the cryptographic device in rounds, and his queries in the ith round may depend on the answers that he got in rounds $1, \ldots, i-1$. Our model from Sect. 5 obviously does not cover this scenario. We now briefly argue how to extend the model and the protocol to cover also the adaptive security. In the adaptive model one assumes that the circuit Γ is initialized with some secret *state* $\in \mathbb{F}^k$ and it can be queried adaptively on several inputs $X^1, \ldots, X^\ell$ (where ℓ is the number of rounds). To each such a query the circuit responds with $Y^i := \Gamma(X^i, state)$. The input X^i is placed on the "private input gates" at the beginning of each round, and the output Y^i appears on the "output gates".

The protocol Π^Γ that "computes Γ" consists of $2t$ parties, whose role is exactly like in the protocol in Sect. 5. In particular: the "private input parties" are initialized with an encoding of *state*, the "public input parties" in the ith round take X^i as input, and the output Y^i is produced by the "output parties". After the end of each round the memory of all the parties (except the "private-input parties" that hold the encoding of *state*) gets erased. The adversary $\mathcal{A}$ can adaptively choose the X^i's and leak at most λ bits from each party in *each round* of the computation of Π^Γ on input X^i. The security definition assumes that for each round the simulator $\mathcal{S}$ gets a pairs $\{(X^i, Y^i)\}_{i=1}^\ell$ and his goal is to produce the output that is statistically close to the output of $\mathcal{A}$.

The implementation of Π^Γ is similar to the implementation of Π^Γ from Sect. 5. In particular, the protocols for the parties in a single round are the same as before. The only change is that, since *state* does not change between the rounds, the "private input parties" need to refresh the encodings that they hold. This can be done easily with the $\mathsf{Refresh}_\mathbb{F}^n$ protocol from Sect. 4.1: each pair $(P_\mathsf{L}^i, P_\mathsf{R}^i)$ of "private input parties" applies, at the end of each round, the refreshing protocol to their encoding $(L \cdot R^i)$, setting $(L^i, R^i) := \mathsf{Refresh}_\mathbb{F}^n(L^i, R^i)$. The security proof goes along the same lines as the proof of Thm. 1. It will be provided in the extended version of this paper.

MORE GENERAL CIRCUITS. The circuits that we consider in Sect. 5 have a very restricted form in order to make the proof of Thm. 1 as simple as possible. We now argue how some of these restrictions can be avoided. First, observe that we can consider circuits with fan-out $q > 2$. The only price to pay is that the leakage bound in the statement of Thm. 1 changes from "$\lambda/3 - |\mathbb{F}|$" to "$\lambda/(q+1) - |\mathbb{F}|$". This is because now each (L^i, R^i) is given to at most $q+1$ parties (not just 3 parties as before).

For some applications it may also be useful to have a separate procedure for adding values in a leakage resilient way. First, observe that adding a publicly-known constant c to an encoded secret can be done easily, as depicted on Fig. 4 (protocol $\mathsf{AddConst}_{\mathbb{F}}^n$). In fact, this protocol has already been described in Sect. 3 used (implicitly) in protocol $\mathsf{Mult}_{\mathbb{F}}^n$ (cf. Fig. 2, Step 2). The protocol computing the sum of two encoded secrets is presented on Fig. 4. Correctness of this protocols is a simple calculation. Because of the lack of space we the formal pro of their security properties is moved to the full version of this paper.

Protocol $(L', R') \leftarrow \mathsf{AddConst}_{\mathbb{F}}^n((L, R), c)$:

Input (L, R): $L \in (\mathbb{F} \setminus \{0\}) \times \mathbb{F}^{n-1}$ is given to P_L and $c \in \mathbb{F}$ is given to both players.

1. Player P_L sends $x := L[1]$ to P_R.
2. Player P_R computes $\tilde{R} := R + (x^{-1} \cdot c, 0, \ldots, 0)$
3. The players execute the $\mathsf{Refresh}(L, \tilde{R})$ procedure. Let (L', R') be the result.

Output: The players output (L', R').

Protocol $(L', R') \leftarrow \mathsf{Add}_{\mathbb{F}}^n((L^0, R^0), (L^1, R^1))$:

Input (L, R): $L^0, L^1 \in (\mathbb{F} \setminus \{0\}) \times \mathbb{F}^{n-1}$ are given to P_L and $R^0, R^1 \in \mathbb{F}^n$ are given to P_R.

1. Player P_L sets $A := L^0$ and $C := L^1 - L^0$.
2. Player P_L sets $B := R^0 + R^1$ and $D := R^1$.
   Note that $\langle A, B \rangle + \langle C, D \rangle = \langle L^0, R^0 \rangle + \langle L^1, R^1 \rangle$.
3. Refresh (C, D) by $(C', D') \leftarrow \mathsf{Refresh}_{\mathbb{F}}^n(C, D)$.
4. Compute $c := \mathsf{Decode}_{\mathbb{F}}^n(C', D')$.
   Note that this does not reveal any information about the inputs of the protocol, as (C', D') were "refreshed".
5. Set $(L', R') \leftarrow \mathsf{AddConst}_{\mathbb{F}}^n((A, B), c)$

Output: The players output (L', R').

Fig. 4. Protocols $\mathsf{AddConst}_{\mathbb{F}}^n$ and $\mathsf{Add}_{\mathbb{F}}^n$

DEALING WITH SMALL FIELDS. A natural field over which one could use our compiler is Z_2. The problem here is that we assumed that in our encoding we have $L[1] \neq 0$, and in the refreshing protocol, if this condition is not met, then part of the protocol is restarted (cf. Fig. 1). Of course if $\mathbb{F}$ is small then this restarting can happen with a high probability. To avoid this problem one could change the underlying encoding scheme and require that some prefix of L of length $a = \omega(\log_{|\mathbb{F}|}(n))$ (instead of just $L[1]$) is not equal to $(0)^a$. In this way the probability of restarting is at most $|\mathbb{F}|^{-a}$ and hence it is negligible in n. The other change that is also needed in this case is that in Step 2 of the $\mathsf{Mult}_{\mathbb{F}}^n$ protocol the player P_L needs to send $L[1, \ldots, a]$ (instead of $L[1]$) to P_R. The price to pay for it is that the "$- |\mathbb{F}|$" term in the leakage bound needs to be replaced by 2^a.

SMALLER NUMBER OF PARTIES. Recall that the number of parties in the protocol Π^Γ corresponds to the number of independent memory parts in the real implementation of the scheme. In our model this number is linear $(2t)$ in the number t of the gates of Γ. This can be reduced in the following way. First, observe that some parties can be "reused" if we look at the computation of Γ as a procedure that evaluates Γ gate-by-gate (cf. Sect. 5.1). More precisely: if a given gate γ^i is not used anymore as an input to other gates, then the memory of the party P^i that corresponds to γ^i can be erased and P^i can be "assigned" to some other gate. Hence, we can reduce the number of parties to $2t'$, where t' is the *width* of Γ. Here, by the "width" of a circuit we mean the minimal number of gates that needs to be kept in memory in order to compute Γ.

Observe also that we can actually decrease the number of memory parts even to two (call these parts: $\mathcal{L}$ and $\mathcal{R}$), by placing all P_L^i's on $\mathcal{L}$, and all P_R^i's on $\mathcal{R}$. This, however, comes at a price: the leakage bound of $\mathcal{L}$ and $\mathcal{R}$ still needs to be a constant fraction of $|n|$, and hence it is a $\frac{c}{t'} \cdot |\mathcal{L}|$ (where c is a constant and t' is the width of Γ), and the fraction $\frac{c}{t'}$ gets very small for large t'. Hence it is mostly of a theoretical interest.

Acknowledgments. The authors wish to thank Marcin Andrychowicz for pointing out some errors in an earlier version of this paper.

References

[DDV10]   Davì, F., Dziembowski, S., Venturi, D.: Leakage-resilient storage. In: Garay, J.A., De Prisco, R. (eds.) SCN 2010. LNCS, vol. 6280, pp. 121–137. Springer, Heidelberg (2010)

[DF11]   Dziembowski, S., Faust, S.: Leakage-Resilient Cryptography from the Inner-Product Extractor. In: Lee, D.H. (ed.) ASIACRYPT 2011. LNCS, vol. 7073, pp. 702–721. Springer, Heidelberg (2011), http://eprint.iacr.org/

[DP08]   Dziembowski, S., Pietrzak, K.: Leakage-resilient cryptography. In: FOCS 2008: Proceedings of the 49th Annual IEEE Symposium on Foundations of Computer Science. IEEE Computer Society, Washington, DC, USA (2008)

[FRR+10] Faust, S., Rabin, T., Reyzin, L., Tromer, E., Vaikuntanathan, V.: Protecting Circuits from Leakage: the Computationally-Bounded and Noisy Cases. In: Gilbert, H. (ed.) EUROCRYPT 2010. LNCS, vol. 6110, pp. 135–156. Springer, Heidelberg (2010)

[GKR08]   Goldwasser, S., Kalai, Y.T., Rothblum, G.N.: One-Time Programs. In: Wagner, D. (ed.) CRYPTO 2008. LNCS, vol. 5157, pp. 39–56. Springer, Heidelberg (2008)

[GR10]   Goldwasser, S., Rothblum, G.N.: Securing Computation against Continuous Leakage. In: Rabin, T. (ed.) CRYPTO 2010. LNCS, vol. 6223, pp. 59–79. Springer, Heidelberg (2010)

[ISW03]   Ishai, Y., Sahai, A., Wagner, D.: Private Circuits: Securing Hardware against Probing Attacks. In: Boneh, D. (ed.) CRYPTO 2003. LNCS, vol. 2729, pp. 463–481. Springer, Heidelberg (2003)

[JV10]   Juma, A., Vahlis, Y.: Protecting Cryptographic Keys against Continual Leakage. In: Rabin, T. (ed.) CRYPTO 2010. LNCS, vol. 6223, pp. 41–58. Springer, Heidelberg (2010)

[KP10]   Kiltz, E., Pietrzak, K.: Leakage Resilient ElGamal Encryption. In: Abe, M. (ed.) ASIACRYPT 2010. LNCS, vol. 6477, pp. 595–612. Springer, Heidelberg (2010)

[MR04]   Micali, S., Reyzin, L.: Physically Observable Cryptography (Extended Abstract). In: Naor, M. (ed.) TCC 2004. LNCS, vol. 2951, pp. 278–296. Springer, Heidelberg (2004)

[Pie09]   Pietrzak, K.: A Leakage-Resilient Mode of Operation. In: Joux, A. (ed.) EUROCRYPT 2009. LNCS, vol. 5479, pp. 462–482. Springer, Heidelberg (2009)

A Parallel Repetition Theorem for Leakage Resilience

Zvika Brakerski[1] and Yael Tauman Kalai[2]

[1] Stanford University
zvika@stanford.edu
[2] Microsoft Research
yael@microsoft.com

Abstract. A leakage resilient encryption scheme is one which stays secure even against an attacker that obtains a bounded amount of side information on the secret key (say λ bits of "leakage"). A fundamental question is whether parallel repetition amplifies leakage resilience. Namely, if we secret share our message, and encrypt the shares under two independent keys, will the resulting scheme be resilient to 2λ bits of leakage?

Surprisingly, Lewko and Waters (FOCS 2010) showed that this is false. They gave an example of a public-key encryption scheme that is (CPA) resilient to λ bits of leakage, and yet its 2-repetition is not resilient to even $(1 + \epsilon)\lambda$ bits of leakage. In their counter-example, the repeated schemes share secretly generated public parameters.

In this work, we show that under a reasonable strengthening of the definition of leakage resilience (one that captures known proof techniques for achieving non-trivial leakage resilience), parallel repetition *does* in fact amplify leakage (for CPA security). In particular, if fresh public parameters are used for each copy of the Lewko-Waters scheme, then their negative result does not hold, and leakage is amplified by parallel repetition.

More generally, given t schemes that are resilient to $\lambda_1, \ldots, \lambda_t$ bits of leakage, respectfully, we show that their direct product is resilient to $\sum(\lambda_i - 1)$ bits. We present our amplification theorem in a general framework that applies other cryptographic primitives as well.

1 Introduction

In recent years, motivated by a large variety of real-world physical attacks, there has been a major effort by the cryptographic community to construct schemes that are resilient to leakage from the secret keys. This successful line of work gave rise to many constructions of leakage-resilient cryptographic primitives, including stream ciphers [11, 19], signature schemes [15, 12], symmetric and public-key encryption schemes [1, 18, 10, 9], as well as more complicated primitives.

A natural question to ask is: Does parallel repetition amplify leakage? More concretely, suppose we are given a public-key encryption scheme $\mathcal{E}$ that remains secure even if λ bits about the secret key are leaked. Is it possible to amplify the

R. Cramer (Ed.): TCC 2012, LNCS 7194, pp. 248–265, 2012.

leakage-resilience to $t\lambda$ by taking t copies of $\mathcal{E}$, and encrypting a message m by secret sharing it, and encrypting the ith share using $\mathcal{E}_i$ (we denote the resulting scheme by $\mathcal{E}^t$)? Using an appropriate definition of parallel repetition, a similar question can be asked for signatures.

Alwen, Dodis, and Wichs [3] and Alwen, Dodis, Naor, Segev, Walfish and Wichs [2] were able to amplify leakage resilience for particular schemes, using the specific properties of these schemes. They raised the fundamental question of whether leakage resilience can *always* be amplified by parallel repetition. They predicted that such a result will be hard or even impossible to prove under the known definitions.

Recently, Lewko and Waters [16] gave a striking negative result, giving an example of a public-key encryption scheme that is resilient to λ bits of leakage but whose 2 repetition is not resilient to even $(1 + \epsilon)\lambda$ bits. This was followed by a work of Jain and Pietrzak [14] who presented a signature scheme where increasing the number of repetitions does not improve the leakage resilience at all. We elaborate on these negative results (and on how they go hand-in-hand with our positive results) in Section 1.2.

1.1 Our Results

We give positive results, by proving direct product theorems for leakage resilience. In particular, we show that parallel repetition does amplify the leakage resilience (almost) as expected.

The leakage model we consider is based on the *"noisy* leakage" model of Naor and Segev [18].[1] In this model, "legal" leakage functions are poly-size circuits that reduce the min-entropy of the secret key by at most λ. A scheme is said to be λ-*leakage resilient* if every PPT adversary, that asks for a "legal" leakage function, breaks the scheme with only negligible probability.

In this work, we consider a slightly relaxed leakage model. Instead of requiring the leakage function to *always* reduce the min-entropy of sk by at most λ, we require that it should be hard to break the scheme on those leakage values that do reduce the min-entropy by at most λ. In other words, we consider a *point-wise* definition: We say that a scheme is *point-wise λ-leakage resilient* if for any PPT adversary, that asks for a poly-size leakage function L, the probability that both the leakage value $y \leftarrow L(pk, sk)$ reduces the min-entropy of sk by at most λ, and that $\mathcal{A}(pk, y)$ breaks the scheme, is negligible.

We believe that this leakage model is of independent interest, as it captures our "intent" better: As long as the secret key is left with enough min-entropy, the scheme is secure. Moreover, we note that all known constructions that are λ-leakage resilient are also point-wise λ-leakage resilient (including [18, 15, 9, 5]). We elaborate on this in Section 4.

At first it may seem that point-wise leakage is equivalent to noisy leakage. However, the difficulty is that it may be hard to determine whether a leakage

---

[1] While "entropic leakage" may be a more suitable name for this model, we stick with the terminology of [18] for historic reasons.

value $y \leftarrow L(pk, sk)$ indeed reduces the min-entropy of sk by at most λ. If this was efficiently determined, then indeed we would have a reduction between the two models.

For technical reasons (see Section 1.3), we need to further relax our leakage model for our results to go through. We consider two (incomparable) relaxations.

First Relaxation: Almost λ-Leakage. In the first relaxation, instead of requiring that sk has high min-entropy (given pk, y), we require that it is statistically close to a random variable with high min-entropy. A scheme that is secure in this model is said to be *point-wise almost λ-leakage resilient.* We can prove a direct product theorem of any *constant* number of repetitions under this definition.

Theorem 1. *Let $c \in \mathbb{N}$ be a constant, and for every $i \in [c]$, let $\mathcal{E}_i$ be a point-wise almost λ_i-leakage-resilient public-key encryption scheme. Then, $\mathcal{E}_1 \times \ldots \times \mathcal{E}_c$ is point-wise almost λ-leakage-resilient, where $\lambda = \sum_{i=1}^{c}(\lambda_i - 1)$.*

We refer the reader to Section 1.3 and Section 5 for more details.

Second Relaxation: Leakage with Small Advice. In the second relaxation, we give the adversary an additional logarithmic (in the security parameter) number of bits of (possibly hard to compute) advice (quite surprisingly, we were unable to reduce this model to the point-wise λ-leakage model). A scheme that is secure in this model is said to be *point-wise λ-leakage resilient with logarithmic advice.* We can prove a direct product theorem of any *polynomial* number of repetitions under this definition.

We note that it is not clear what it means to have t different leakage resilient schemes when t is super constant, since there is a different number of schemes for each value of the security parameter. While one can come up with a proper definition (involving a generation algorithm that, for every value of the security parameter, gets i and implements $\mathcal{E}_i$), for the sake of clarity, we choose to state the theorem below only for parallel repetition of the same scheme.

Theorem 2. *Let $t = t(k)$ be a polynomial in the security parameter. Let $\mathcal{E}$ be a public-key encryption scheme that is point-wise λ-leakage resilient with logarithmic advice. Then $\mathcal{E}^t$ is point-wise $t(\lambda - 1)$-leakage resilient with logarithmic advice.*

We refer the reader to Section 1.3 for an overview of the proof, and to Section 6 for more details.

The Relation Between our Models. Interestingly, we are not able to show that our relaxations are equivalent to one another, nor to show that they are implied by (plain) point-wise leakage resilience. This is surprising since in the bounded leakage model,[2] a negligible change in the secret-key distribution, or adding a logarithmic number of hard to compute bits, does not change the model.

---

[2] Where the leakage function's output is required to be bounded by λ bits, as opposed to our requirement that the secret key has high residual entropy.

In a nutshell, the reason that this does not carry to our models, is that having high min-entropy is not an efficiently verifiable condition, and that statistical indistinguishability does not preserve min-entropy.

We are able to show, however, that point-wise λ-leakage resilience implies λ-bounded leakage resilience (for the same value of λ), and thus in particular, our relaxed models also imply bounded leakage resilience. We note that proving the above is somewhat nontrivial since we do not want to suffer a degradation in λ. We refer the reader to Section 3.3 for a formal presentation.

Our Models and Current Proof Techniques. We show that for essentially all known schemes that are resilient to non-trivial leakage (i.e. super-logarithmic in the hardness of the underlying problem), amplification of leakage resilience via parallel repetition works. Specifically, this includes the Lewko-Waters counterexample, if the public parameters are chosen independently for each copy of the scheme. In order to do this, we identify a proof template that is used in all leakage resilience proofs, and show that this template is strong enough to prove point-wise leakage resilience, as well as our relaxed notions. See Section 4 for the full details.

The Lewko-Waters counterexample uses its public parameters in a very particular way that makes the argument not go through (see below).

1.2   Prior Work

As we claimed above, all known leakage resilient schemes are proved using the same proof template, and remain secure under our leakage models. This implies that parallel repetition should amplify security for all known schemes, which does not seem to coincide with the negative results of [16, 14]. We explain this alleged discrepancy below.

The Lewko-Waters Counterexample. Lewko and Waters [16] construct a public key encryption scheme that is resilient to non-trivial length-bounded leakage, and prove that parallel repetition does not amplify its leakage resilience. However, the copies of their encryption scheme share public parameters: They are all using the same bilinear group. Their scheme, like all other schemes we are aware of, is (computationally indistinguishable from) point-wise leakage resilient and our theorems imply that parallel repetition does amplify its resilience to leakage. This is true so long as the public parameters are generated anew for each copy of the scheme: In our proof, we need to be able to sample key pairs for the scheme in question. Lewko and Waters use the public parameters in an extremely pathological (and clever!) way: The public parameters enable to generate keys for their actual scheme, but not for the computationally indistinguishable scheme where leakage resilience is actually proven. However, if we consider the generation of public parameters as a part of the key generation process, then new key pairs can always be generated, and parallel repetition works.

The Jain-Pietrzak Counterexample. Jain and Pietrzak [14] give a negative result for signature schemes. They take any secure signature scheme and change it so that if the message to be signed belongs to a set H, then the signature algorithm simply outputs the entire secret key. The set H is computationally hard to hit (given only the public key), and thus the scheme remains secure. It follows that the scheme remains secure also given leakage of length $O(\log k)$, where k is the security parameter (more generally, if the underlying problem is 2^λ hard, then the scheme is resilient to $\sim \lambda$ bits of leakage).

They prove that parallel repetition fails, by proving that if the scheme is repeated t times, for some large enough t, then the leakage can in fact give enough information to find a message m that belongs to all the sets H_i, and thus break security completely. They start with a result that relies on common public parameters: a common (seeded) hash function. Then, they suggest to remove this public parameter by replacing the seeded hash function with an explicit hash function, such as SHA256. However, this explicit hash function is also, in some sense, a joint *non-uniform* public parameter.

This counterexample heavily relies on the "help" of the signing oracle when breaking the repeated scheme. The paper also presents a construction of a CCA encryption scheme, where they use the decryption oracle to break the parallel repetition system.

In general, signature schemes are not covered by our amplification theorems. Our theorems (and proofs) only cover public key primitives where the challenger in the security game does not need to know the secret key (beyond providing the adversary with the leakage value). Our results do extend to schemes such as signature schemes or CCA encryption schemes, if they have the property that the challenger (i.e., the signing oracle or the decryption oracle) can be efficiently simulated given only the public key (or given very little information about the secret key), in a way that is computationally indistinguishable even given the leakage. For example, the signature scheme of Katz and Vaikuntanathan [15] has this property, and thus its leakage resilience is amplified by parallel repetition. Whether our techniques can be applied to other leakage resilient signature schemes (e.g. [4, 17, 13]) is an interesting question that we leave for further research.

1.3   Overview of Our Techniques

In what follows we give a high-level overview of our proofs. For the sake of simplicity, we focus on the case of two-fold parallel repetition. Let $\mathcal{E}$ be any λ-leakage resilient encryption scheme. Our goal is to prove that the scheme $\mathcal{E}^2$ is 2λ-leakage resilient. For technical reasons, in our actual proof, we manage to show that $\mathcal{E}^2$ is $(2\lambda - 1)$-leakage resilient (in both our leakage models).

Our proof is by reduction: Suppose there exists an adversary $\mathcal{B}$ for the parallel repetition scheme $\mathcal{E}^2$ that leaks $L(pk_1, pk_2, sk_1, sk_2)$, where L reduces the min-entropy of (sk_1, sk_2) by at most $2\lambda - 1$. We construct an adversary $\mathcal{A}$, that uses $\mathcal{B}$ to break security of $\mathcal{E}$, and uses a leakage function L' that reduces the min-entropy of the secret key by at most λ.

Intuitively (and, as we will show, falsely), it does not seem too hard to show such a reduction. It only makes sense that when the pair (sk_1, sk_2) looses 2λ bits of entropy, then at least one of the secret keys sk_1, sk_2 "loses" at most λ bits (otherwise the total loss should be more than 2λ). Therefore the adversary $\mathcal{A}$ can sample a key pair by itself and "plant" it either as (pk_1, sk_1) or as (pk_2, sk_2) (at random). Namely, $\mathcal{A}$ will sample a random $i \in \{1, 2\}$, and uniformly sample (pk_i, sk_i), the key pair of the scheme we actually wish to attack will play the role of (pk_{3-i}, sk_{3-i}). Upon receiving a leakage function $L(\cdot)$ from $\mathcal{B}$, the adversary $\mathcal{A}$ will plug the known (pk_i, sk_i) into the function and thus obtain L' to be sent to the challenger. Upon receiving a response from the challenger, it is forwarded back to $\mathcal{B}$, which can then break security with noticeable probability. Notice that $\mathcal{B}$'s view in the game is identical to its view in the repeated game against $\mathcal{E}^2$, and thus it still breaks the security with the same probability. The only worry is whether the function L' only reduces the key entropy by the allowed amount, which is unfortunately not the case. Assume that L leaks some 2λ bits on the bit-wise XOR $sk_1 \oplus sk_2$. Then when plugging in a known sk_i, the resulting L' still leaks 2λ bits on sk_{3-i}.

To solve this problem, we must prevent $\mathcal{A}$ from knowing sk_i. This is achieved by having the key pair (pk_i, sk_i) sampled by the leakage function L', rather than by $\mathcal{A}$. Namely, $L'(pk, sk)$ is now defined as follows: First, sample (pk_i, sk_i) and set $(pk_{3-i}, sk_{3-i}) = (pk, sk)$. Then run $y \leftarrow L(pk_1, pk_2, sk_1, sk_2)$ to obtain the leakage value. Lastly, output (y, pk_1, pk_2). Given the output of L', the adversary $\mathcal{A}$ can forward the value y to $\mathcal{B}$, that uses it to break the scheme, all without ever being exposed to the value of sk_i.

This seems to give $\mathcal{A}$ the least amount of information possible, so we should hope that now we can prove that the entropy of sk is reduced by at most λ. However, again, this is not true. Suppose that with probability $1/2$, the leakage function L outputs 2λ bits about sk_1 and with probability $1/2$ it outputs 2λ bits about sk_2. In this case, L indeed reduces the min-entropy of (sk_1, sk_2) by 2λ, and yet for every $i \in \{1, 2\}$ the leakage function $L'(pk, sk)$ reduces the min-entropy of sk by essentially 2λ as well, and thus is not a valid leakage function for the one shot game.

This abnormality results, to some extent, from using min-entropy (as opposed to Shannon entropy) as our entropy measure: If $L'(pk, sk)$ outputs both $y = L(pk_1, pk_2, sk_1, sk_2)$ and sk_{3-i}, then it would indeed leak at most λ bits on sk (with probability $1/2$). The fact that we have *less* information, namely sk_i is not known, might actually *decrease* the min-entropy of the key.

We arrive at a conflict: On one hand, knowing sk_i is a problem, but on the other, not knowing it seems to also be a problem. We show that revealing sk_i only in some cases, enables to prove parallel repetition. We use a simple lemma (Lemma A.1), which essentially shows how to "split-up" the joint min-entropy of two random variables. More precisely, it says that there is a subset S of all possible secret keys sk_1, such that for every $sk_1 \in S$, the the random variable $sk_2|sk_1$ has high min-entropy. Moreover, given the additional bit of information that $sk_1 \notin S$, causes sk_1 to have high min-entropy (which decreases as the size of S shrinks).

We proceed by a specific analysis for each of our two relaxed models. For explanatory reasons, we first discuss leakage with advice (our second relaxation) and then go back to the almost leakage resilience model (our first relaxation).

Point-Wise λ-Leakage with Advice. In this model, the adversary $\mathcal{A}$ will leak $L'(pk, sk)$, which is a randomized leakage function, defined by choosing a random $\tau \in \{1, 2\}$, setting $(sk_\tau, pk_\tau) = (sk, pk)$, choosing a new fresh key pair (sk_i, pk_i), where $i = 3 - \tau$, and outputting $L(pk_1, pk_2, sk_1, sk_2)$. In addition, it will use one bit of advice which is whether $sk_i \in S$. If so, the leakage function $L'(pk, sk)$ outputs sk_i in addition to $L(pk_1, pk_2, sk_1, sk_2)$, and otherwise it outputs only $L(pk_1, pk_2, sk_1, sk_2)$. Now we can prove that indeed, for many pairs (pk, sk), the leakage $L'(pk, sk)$ leaks at most λ bits about sk (and $\mathcal{B}$ breaks $\mathcal{E}^2$ on the corresponding keys).

Note that the leakage function L' sometimes leaks more than it should. Namely, in some cases the value $y \leftarrow L'(pk, sk)$ reduces the min-entropy of sk by λ; but in other cases it reduces the min-entropy of sk by more than λ,[3] and in these cases it is an invalid leakage function. For this reason, we need to consider the *point-wise* λ-leakage definition. In addition, note that L' used only one bit of additional advice. Therefore when going from $\mathcal{E}$ to $\mathcal{E}^t$ the reduction uses $\log t$ bits of advice.

Point-Wise Almost λ-Leakage. In this model, the idea of the reduction is the following: The adversary $\mathcal{A}$ will leak $L'(sk, pk)$, which is a randomized leakage function, defined by choosing a random $\tau \in \{1, 2\}$, setting $(sk_\tau, pk_\tau) = (sk, pk)$, choosing a new fresh key pair (sk_i, pk_i), where $i = 3 - \tau$, and outputting $L(pk_1, pk_2, sk_1, sk_2)$, and in addition with probability $1/2$ outputting sk_i.

As in the model with advice, the leakage function L' might leak more than λ bits about sk, and thus we use the point-wise definition. In the analysis, we distinguish between the case that the set S is noticeable and the case that it is negligible. In the former, with non-negligible probability the leakage function L' will sample $sk_i \in S$ and will output it. In this case the leakage function is legal. If the set S is negligible, we claim the distribution of the secret key sk_τ is statistically close to the distribution of sk_τ conditioned on the event that $sk_i \notin S$ (as this event happens only with negligible probability). Therefore, if L' did not output the secret key sk_i, the secret key sk_τ is statistically close to a distribution with high enough min-entropy. Due to this analysis, we need to relax our leakage model *almost* λ-leakage resilient.

Since the analysis in this model is asymptotic, we are not able to extend it beyond a constant number of repetitions. See discussion in Section 5.

1.4 Paper Organization

We define our generalized notion of public-key primitives in Section 2, where we also define parallel repetition and leakage attacks on such primitives. Our model

---

[3] This happens when the set S is very small, yet $sk_\tau \in S$.

of point-wise leakage resilience is presented in Section 3. In Section 4 we explain why all known leakage resilient schemes are also point-wise leakage-resilient.

Our parallel repetition theorems for a constant number of repetitions and for a polynomial number of repetitions are presented in Sections 5 and 6, respectively. In Section 7 we discuss what our theorems imply for schemes that are only computationally indistinguishable from being secure in our model. Appendix A contains the min-entropy splitting lemma that is used for all our proofs.

Due to space limitations, some proofs are omitted from this extended abstract and can be found in the full version [6].

2 Public-Key Primitives, Parallel-Repetition, Leakage Attacks

In this section we give a definition of a *public key primitive* which generalizes one-way relations and public-key encryption under chosen plaintext attack (CPA). We then show how to define parallel repetition with respect to public-key primitives in a way that, again, generalizes the intuitive notions of parallel repetition for either one-way relations or public-key encryption.

2.1 A Unified Framework for Public-Key Primitives

We use the following formalization that generalizes both one-way relations and public-key encryption.

Definition 2.1 (public-key primitive). *A* public-key primitive $\mathcal{E} = (G, V)$ *is a pair of* PPT *algorithms such that*

- *The* key generator G *generates a pair of secret and public keys:* $(sk, pk) \leftarrow G(1^k)$.
- *The* verifier V *is an oracle machine such that* $V^{\mathcal{O}(pk)}(pk)$ *either accepts or rejects.*

Definition 2.2 (secure public-key primitive). *A public-key primitive* $\mathcal{E} = (G, V)$ *is secure if for any* PPT *oracle* break, *it holds that*

$$\Pr_{(sk,pk) \leftarrow G(1^k)} [V^{\mathsf{break}(pk)}(pk)] = \mathrm{negl}(k) .$$

To be concrete, for one-way relations, the breaker needs to send a candidate secret key sk (= inversion of the public key), and the verifier runs the relation's verification procedure. To see why public key encryption can be stated in these terms, requires some work. The reason it is not immediate is that typically, we would consider the interaction between the verifier and the breaker, to be the following: The verifier gives the breaker a challenge ciphertext $\mathsf{Enc}_{pk}(b)$, and he accepts if the breaker succeeds in guessing b. However, the breaker can clearly cause the verifier to accept with probability $1/2$, where we need to ensure that the breaker succeeds only with negligible probability. This technical annoyance can be fixed by considering the game where the verifier sends $\mathrm{poly}(k)$ challenge ciphertexts to the breaker, each encrypting a random bit. The breaker succeeds if it succeeded in guessing significantly more than $1/2$ of the bits encrypted. The formal definition and precise analysis are much more cumbersome. The proof appears in the full version [6].

Note that our verifier (which corresponds to the challenger in "security game" based definitions) only gets the public key as input and not the secret key. If the secret key was also given, then all public-key encryption schemes, signature schemes, and one-way relations, would trivially fit into this framework. However, in this work, we only consider primitives where the verifier V does not use the secret key sk to verify, but uses only the public key pk. An example of such a primitive is public-key encryption (under CPA). However, signature schemes or CCA secure encryption schemes do not fall into this category, since for these primitives the verifier in the definition above does need to know the secret key sk in order to simulate the signing oracle, in the case of signature schemes, and to simulate the decryption oracle, in the case of CCA encryption schemes.

2.2 Parallel Repetition

Definition 2.3 (t-parallel repetition). *For any public-key primitive $\mathcal{E} = (G, V)$ and any $t \in \mathbb{N}$, its t-parallel repetition, denoted $\mathcal{E}^t = (G^t, V^t)$, is in itself a public-key primitive defined as follows*

- *The key generator $(sk^t, pk^t) \leftarrow G^t(1^k)$ generates $(sk_i, pk_i) \leftarrow G(1^k)$ for all $i \in [t]$ and outputs $sk^t \triangleq (sk_1, \ldots, sk_t)$, $pk^t \triangleq (pk_1, \ldots, pk_t)$.*
- *The verifier $(V^t)^{\mathcal{O}(pk^t)}(pk^t)$, runs $V^{\mathcal{O}(pk^t, i)}(pk_i)$ for all $i \in [t]$, and accepts if and only if they all accept.*

A direct product of t schemes $\mathcal{E}_1 \times \cdots \times \mathcal{E}_t$ is defined similarly.

While it is straightforward that our definition captures the notion of parallel repetition for one-way relations (where the goal is to find legal pre-images for all input public-keys), let us be a little more explicit about how the above captures parallel repetition for public-key encryption.

Lemma 2.4. *Let $\mathcal{E} = (G, V)$ be a public-key primitive that represents a public-key encryption scheme and let $t \in \mathbb{N}$. Then there exists a public key encryption scheme that is represented by $\mathcal{E}^t$.*

Moreover, this scheme is obtained by secret sharing the message into t shares and encrypting share i with pk_i. To decrypt, decrypt all shares and restore the message.

The proof is straightforward and is omitted.

2.3 Leakage Attacks

In this section, we generalize the notion of leakage attacks to our public-key primitive framework. Note that we do not define what it means for a scheme to be secure, only present a model for an attack.

Definition 2.5 (leakage attack). *We consider adversaries of the form* $\mathcal{A} =$ *(*leak$_{\mathcal{A}}$, break$_{\mathcal{A}}$*), where* leak$_{\mathcal{A}}$*,* break$_{\mathcal{A}}$ *are (possibly randomized) functions. We refer to* leak$_{\mathcal{A}}$ *as the* leakage function *and to* break$_{\mathcal{A}}$ *as the* breaker.

A leakage attack of an adversary $\mathcal{A} = ($leak$_{\mathcal{A}}$, break$_{\mathcal{A}}$*) on a public-key primitive* $\mathcal{E} = (G, V)$ *(with security parameter k) is the following process.*

- **Initialize:** *Generate a key pair* $(sk, pk) \overset{\$}{\leftarrow} G(1^k)$.
- **Leak:** *Apply the leakage function on the key pair to obtain the* leakage value $y \leftarrow$ leak$_{\mathcal{A}}(pk, sk)$.
- **Break:** $\mathcal{A}$ *succeeds if* $V^{\text{break}(pk, y)}(pk)$ *accepts.*

3 Point-Wise Leakage Resilience

In this work, we consider "noisy leakage" functions, which are only allowed to reduce the (average) min-entropy of the secret key by a bounded amount. However, we relax the min-entropy restriction, and consider a *point-wise* definition, where we require that the specific leakage value is legal (as opposed to requiring that the leakage function is *always* legal).

We define our new model below. Then, in Sections 3.1, 3.2, we present two relaxed versions of point-wise leakage resilience that we need in order to prove our parallel repetition theorems. Finally, in Section 3.3 we show that all of these notions are strictly stronger than the old bounded-leakage model of [1]. Namely, security w.r.t. to our definitions imply, as a special case, security w.r.t. bounded leakage.

Definition 3.1 (point-wise λ-leakage). *Let* $\mathcal{E} = (G, V)$ *be a public key primitive. A possibly randomized leakage function* leak *is* λ*-leaky at point* (pk, y)*, where* pk *is a public key and* y *is a leakage value (in the image of* leak*), if*

$$\mathbf{H}_{\infty}(S_{pk,y}) \geq \mathbf{H}_{\infty}(S_{pk}) - \lambda \; ,$$

where S_{pk} *is the distribution of secret keys conditioned on the public key being* pk*, and* $S_{pk,y}$ *is the distribution of secret keys conditioned on both the public key being* pk *and on* leak$(pk, sk) = y$.

Definition 3.2 (point-wise λ-leakage resilience). *A public-key primitive* $\mathcal{E} = (G, V)$ *is* point-wise λ-leakage-resilient *if for any* PPT *adversary* $\mathcal{A}$*, where* $\mathcal{A} = ($leak$_{\mathcal{A}}$, break$_{\mathcal{A}}$*), it holds that*

$$\text{Adv}_{\mathcal{E}, \lambda}[\mathcal{A}] \triangleq \Pr\left[(\text{leak}_{\mathcal{A}} \; is \; \lambda\text{-}leaky \; at \; (pk, y)) \wedge (\mathcal{A}(pk, y) \; succeeds)\right] = \text{negl}(k) \; ,$$

where the probability is taken over $(sk, pk) \leftarrow G(1^k)$*, over the random coin tosses of* $\mathcal{A} = ($leak$_{\mathcal{A}}$, break$_{\mathcal{A}}$*), and over the random coin tosses of the verifier in the verification game.*

In order to obtain our direct product theorems for leakage resilience, we relax the point-wise leakage resilience definition in two (incomparable) ways.

3.1 First Relaxation: Almost Leakage Resilience

In this relaxation, instead of requiring that sk has high min-entropy conditioned on pk and $y = \mathsf{leak}(pk, sk)$, we require that the distribution of sk (conditioned on pk, y) is statistically close to one that has high min-entropy.

Definition 3.3 (close to λ-leaky). *A leakage function* leak *is* μ-close to λ-leaky at point (pk, y) *if there exists a distribution* $\tilde{S}_{pk,y}$ *that is* μ-close to $S_{pk,y}$ *and*

$$\mathbf{H}_{\infty}(\tilde{S}_{pk,y}) \geq \mathbf{H}_{\infty}(S_{pk}) - \lambda .$$

Definition 3.4 (resilience to almost λ-leakage). $\mathcal{E} = (G, V)$ *is point-wise almost λ-leakage-resilient if for any* PPT *adversary* $\mathcal{A} = (\mathsf{leak}_{\mathcal{A}}, \mathsf{break}_{\mathcal{A}})$ *and for any negligible function* μ, *it holds that*

$$\mathrm{Adv}_{\mathcal{E},\lambda,\mu}[\mathcal{A}] \triangleq \Pr\left[(\mathsf{leak}_{\mathcal{A}} \text{ is } \mu\text{-close to } \lambda\text{-leaky at } (pk, y)) \wedge (\mathcal{A}(pk, y) \text{ succeeds})\right]$$
$$= \mathrm{negl}(k) .$$

where the probability is taken over $(sk, pk) \leftarrow G(1^k)$, *over the random coin tosses of* $\mathcal{A} = (\mathsf{leak}_{\mathcal{A}}, \mathsf{break}_{\mathcal{A}})$, *and over the random coin tosses of the verifier in the verification game.*

Under this definition we obtain a direct-product theorem for constant number of repetitions.

3.2 Second Relaxation: Leakage Resilience with Advice

To obtain a direct-product theorem for a super-constant number of repetitions, we use a slightly different (and incomparable) model, where we do not allow statistical closeness, but rather allow the attacker to get a logarithmic number of bits of (possibly inefficient) advice.

Definition 3.5 (PPT$_{-a}$). *We say that a function* f *is* PPT$_{-a}$ *computable if the function* f_{-a}, *defined below, is* PPT *computable. The function* f_{-a} *is identical to* f, *except that the last* a *bits of its output are truncated.*

We say that an adversary $\mathcal{A} = (\mathsf{leak}_{\mathcal{A}}, \mathsf{break}_{\mathcal{A}})$ *is a* PPT$_{-a}$ *adversary if* $\mathsf{leak}_{\mathcal{A}}$ *is* PPT$_{-a}$ *computable and* $\mathsf{break}_{\mathcal{A}}$ *is* PPT *computable.*

Definition 3.6 (point-wise λ-leakage with advice). *A public-key primitive* $\mathcal{E} = (G, V)$ *is resilient to point-wise λ-leakage and logarithmic advice if for any* PPT$_{-O(\log k)}$ *adversary* $\mathcal{A} = (\mathsf{leak}_{\mathcal{A}}, \mathsf{break}_{\mathcal{A}})$ *it holds that*

$$\mathrm{Adv}_{\mathcal{E},\lambda}[\mathcal{A}] \triangleq \Pr\left[(\mathsf{leak}_{\mathcal{A}} \text{ is } \lambda\text{-leaky at } (pk, y)) \wedge (\mathcal{A}(pk, y) \text{ succeeds})\right] = \mathrm{negl}(k) ,$$

where the probability is taken over $(sk, pk) \leftarrow G(1^k)$, *over the random coin tosses of* $\mathcal{A} = (\mathsf{leak}_{\mathcal{A}}, \mathsf{break}_{\mathcal{A}})$, *and over the random coin tosses of the verifier in the verification game.*

3.3   Relation to Bounded Leakage

To conclude, we prove that point-wise λ-leakage resilience implies the basic form of λ-bounded leakage. A proof sketch appears in the full version [6].

Definition 3.7 ([1]). *A public-key primitive* $\mathcal{E} = (G, V)$ *is* λ-bounded leakage resilient *if any* PPT *adversary* $\mathcal{A} = (\text{leak}_{\mathcal{A}}, \text{break}_{\mathcal{A}})$ *for which the output of* $\text{leak}_{\mathcal{A}}$ *is at most* λ *bits, succeeds with negligible probability.*

Lemma 3.8. *If* $\mathcal{E} = (G, V)$ *is point-wise* λ-leakage resilient then it is also λ-bounded leakage resilient.

We note that point-wise almost λ-leakage resilience, and λ-leakage resilience with logarithmic advice, are stronger notions of security (they give the adversary more power) and thus the above immediately applies to these notions as well.

4   Why Known Schemes Are Point-Wise Leakage Resilient

In this section, we show that leakage resilience is amplified by parallel repetition for, essentially, all known schemes that are resilient to non-trivial (i.e. super-logarithmic) leakage. To show this, we sketch a proof template that is shared among all (non trivial) leakage resilient results, and we show that this proof template proves security also w.r.t. our leakage models (the point-wise almost λ-leakage model, and the point-wise λ-leakage with logarithmic advice model).

The Proof Template. The proof template for proving leakage resilience is very simple, and works in two hybrid steps. Recall that the adversary first gets a pair $(pk, y = L(pk, sk))$, where L is a poly-size leakage function chosen by $\mathcal{A}$. Then it chooses messages m_0, m_1 and gets a challenge ciphertext $c_b \leftarrow \text{Enc}_{pk}(m_b)$. The adversary wins if it guesses the bit b correctly.

The first step in the template is to replace the challenge c_b with an "illegally" generated ciphertext c_b^*, such that $(sk, pk, c_b) \stackrel{c}{\approx} (sk, pk, c_b^*)$ (and it is efficient to generate c_b^* given sk, pk, b). Due to computational indistinguishability, the adversary's success probability should remain unchanged. We note that there is no entropy involved in this part, only a requirement that L is efficiently computable.

The second step is completely information theoretic: It is proven that if the distribution of the secret key conditioned on pk, y, which we denote by $S_{pk,y}$, has sufficient min-entropy, then c_b^* carries no information on b (or, more precisely, that conditioned on the view of the adversary, b is statistically close to uniform). Therefore, no adversary can guess its value with non-negligible advantage.

Point-Wise Leakage Resilience. The above proof template also proves point-wise leakage resilience. The second step of the hybrid works in a point-wise manner and therefore we only need to worry about the first step. In the first step, clearly computational indistinguishability still holds, but proving that the point-wise advantage remains unchanged is a bit harder, since we cannot efficiently check the point-wise advantage. Nevertheless, we argue that if the advantage of $\mathcal{A}$ is non-negligible, then it drops by a factor of at most two. Such a claim is sufficient for the next level of the template.

To see why this is the case, consider an adversary $\mathcal{A}$ that has non-negligible point-wise advantage ϵ when given (pk, y, c_b), but less than $\epsilon/2$ when given (pk, y, c_b^*). Recall that the advantage measures the probability of both $\mathcal{A}$ succeeding (in the verification game) and pk, y being point-wise λ-leaky. It follows that with non-negligible probability over pk, y, the conditional success probability of $\mathcal{A}$, conditioned on pk, y, drops by at least $\epsilon/4$ (otherwise the advantage, which measures over a subset of the pk, y, couldn't have dropped).

A distinguisher $\mathcal{B}(sk, pk, c_b/c_b^*)$ is defined as follows: First, compute the leakage $y := L(sk, pk)$. Then generate many samples of c_b/c_b^* and use them to evaluate the success probability of $\mathcal{A}$ conditioned on pk, y in the two cases. If indeed pk, y are such that the success probability drops, use $\mathcal{A}$ to distinguish between the two cases. If no noticeable change in the success probability was noticed, then output a random guess. Putting it all together, we get a polynomial distinguisher between (pk, y, c_b) and (pk, y, c_b^*), in contradiction to the hardness assumption.

We note that this is true even if y is not fully known to the distinguisher: say $O(\log k)$ bits of y are not known, the distinguisher can still try all options and check if for either one the success probability changes by $\epsilon/4$.

Our Relaxed Models of Point-Wise Leakage Resilience. Our first relaxation, of allowing the secret key to be statistically close to λ-leakage resilient, only effects the second step of the template. We can still argue that b is statistically close to uniform by adding another hybrid where the conditional distribution $S_{pk,y}$ is replaced with a statistically indistinguishable $\tilde{S}_{pk,y}$ that has high min-entropy.

Our second relaxation, of allowing logarithmic advice, goes into the first step (this is the only step where we care about the complexity of L). As we explained above, our argument works even if a logarithmic part of the leakage value is not known. Therefore we will use only the efficient part of the leakage function and computational indistinguishability will still hold.

Computationally Indistinguishable Schemes. For some schemes, such as [1, 16], leakage resilient is proven by showing that they are computationally indistinguishable from another scheme which, in turn, is proven leakage resilient using the template. We show in Section 7 that this still implies that parallel repetition amplifies leakage.

5   Direct-Product Theorem for a Constant Number of Repetitions

In this section, we prove a direct-product theorem for a constant number of repetitions, w.r.t. point-wise almost leakage-resilience as defined in Section 3.1.

Theorem 5.1. *Let $c \in \mathbb{N}$ be a constant, and for every $i \in [c]$, let $\mathcal{E}_i = (G_i, V_i)$ be a point-wise almost λ_i-leakage-resilient public-key primitive. Then, $\mathcal{E}_1 \times \ldots \times \mathcal{E}_c$ is point-wise almost λ-leakage-resilient, where $\lambda = \sum_{i=1}^{c} (\lambda_i - 1)$.*

It suffices to prove this theorem for $c = 2$, and apply it successively. In order to simplify notation, we prove it for the case of parallel repetition, where $\mathcal{E}_1 = \mathcal{E}_2$, the proof extends readily to the case of direct product.

Lemma 5.2. *Let $\mathcal{E} = (G, V)$ be a point-wise almost λ-leakage-resilient public-key primitive. Then $\mathcal{E}^2$ is point-wise almost $(2\lambda - 1)$-leakage-resilient.*

Before we present the outline of the proof, let us make a few remarks.

1. Note that there is a loss of one bit in the amplification. Namely, we go from λ to $(2\lambda - 1)$ instead of just 2λ. While some loss in the parameters is implied by our techniques, more detailed analysis can show that the composed scheme is in fact $(2\lambda - \delta)$-leakage resilient for any $\delta(k) = 1/\text{poly}(k)$. Thus the loss incurred is less than a single bit. As our result is qualitative in nature, we chose not to overload with the additional complication.

2. While at first glance one could imagine that Theorem 5.1 should extend beyond constant c, we were unable to prove such an argument. The reason is that super-constant repetition gives a different scheme for each value of the security parameter. This means that we cannot use Theorem 5.1 as black-box. More importantly, our proof techniques rely on the asymptotic behavior of the scheme so we were not able to even change the proof to apply for a super-constant number of repetitions.

   A result for the more general case of any polynomial number of repetitions is presented, in the slightly different and incomparable "advice" model, in Section 6.

   Finally, we remark that known negative results for security of parallel repetition are already effective for a constant number of repetitions. Thus our result contrasts them even for this case.

Proof overview of Lemma 5.2. We consider an adversary $\mathcal{B}$ that succeeds in the parallel repetition game, and construct an adversary $\mathcal{A}$ that succeeds in the single instance game. The straightforward proof strategy would be to "plant" the "real" key pair, that is given as input to $\mathcal{A}$, as one of the key pairs that are input to $\mathcal{B}$, and sample the other pair uniformly.[4] In such case, the input to $\mathcal{B}$ is distributed identically to the parallel repetition case and indeed $\mathcal{B}$ will succeed with noticeable probability. However, we may no longer be able to claim that our leakage leaves sufficient entropy in the secret key. We are guaranteed by the functionality of $\mathcal{B}$ that the key pair (sk_1, sk_2) is left with sufficient min-entropy but it is still possible that neither sk_1 nor sk_2 have *any* min-entropy by themselves.

To solve the above we use Lemma A.1, which essentially says how to split-up the joint entropy of two random variables. Specifically it says that either sk_1 or $sk_2|sk_1$ will have sufficient min-entropy, depending on whether sk_1 belongs to a

---

[4] We note that even this step is impossible when relying on "secretly generated" public parameters as in the scheme presented in [16] (or rather, the scheme that is computationally indistinguishable to theirs and actually has entropic leakage resilient).

hard-to-recognize set R, and conditioned on the knowledge of whether $sk_1 \in R$. Namely, either $sk_1|\mathbb{1}_{sk_1 \in R}$ or $sk_2|(sk_1, \mathbb{1}_{sk_1 \in R})$ have high min-entropy. If we could compute the bit $\mathbb{1}_{sk_1 \in R}$, we would have been done (and indeed if we are allowed one bit of inefficient leakage, an easier proof follows, see Section 6). Since this is impossible, we turn to case analysis:

Obviously, if $\Pr[sk_1 \in R] = \mathrm{negl}(k)$, then we can always guess that $\mathbb{1}_{sk_1 \in R} = 0$ and be right almost always. This implies that in such case $sk_2|sk_1$ is statistically indistinguishable from having high min-entropy, as we wanted.

For the second case, if $\Pr[sk_1 \in R] \geq 1/\mathrm{poly}(k)$, then $sk_2|(sk_1, \mathbb{1}_{sk_1 \in R})$ will have high min-entropy for a noticeable part of the time. To complete the analysis here, we notice that

$$\mathbf{H}_\infty(sk_2|(sk_1, \mathbb{1}_{sk_1 \in R})) = \mathbf{H}_\infty(sk_2|sk_1).$$

This is because R is a well defined set and thus $\mathbb{1}_{sk_1 \in R}$ is a deterministic (though hard to compute) function of sk_1. It follows that $sk_2|sk_1$ will have high min-entropy for a noticeable fraction of the time, which completes the proof.

For the formal proof, see the full version [6].

6 Direct-Product Theorem for Polynomially Many Repetitions

In this section we present a direct product theorem that applies to any polynomial number of repetitions. This theorem is relative to the advice model defined in Section 3.2. For the sake of simplicity, we will assume that the number of repetitions is a power of 2, although the same techniques can be used for any number.

Theorem 6.1. *Let $\mathcal{E} = (G, V)$ be a public-key primitive that is resilient to point-wise λ-leakage and logarithmic advice. Let $t = t(k)$ be a polynomially bounded function of the security parameter such that $t(k)$ is always a power of 2. Then $\mathcal{E}^t$ is resilient to point-wise $t(\lambda - 1)$-leakage and logarithmic adivce.*

Towards proving the theorem, we present the following lemma, which is a parameterized special case of the above theorem, and will imply the theorem by successive applications.

Lemma 6.2. *For any public-key primitive $\mathcal{E} = (G, V)$ and any PPT_{-a} adversary $\mathcal{B} = (\mathsf{leak}_\mathcal{B}, \mathsf{break}_\mathcal{B})$ for $\mathcal{E}^2$, there exists a $\mathrm{PPT}_{-(a+1)}$ adversary $\mathcal{A} = (\mathsf{leak}_\mathcal{A}, \mathsf{break}_\mathcal{A})$ for $\mathcal{E}$, such that for all k,*

$$\mathrm{Adv}_{\mathcal{E},\lambda}[\mathcal{A}] \geq (1/4) \cdot \mathrm{Adv}_{\mathcal{E}^2,(2\lambda-1)}[\mathcal{B}] .$$

The theorem immediately follows by applying the lemma $\log t$ times. See proofs in the full version [6].

7   Leakage from Computationally Indistinguishable Schemes

Our definition of point-wise leakage resilience is based on the residual min-entropy of the secret key, conditioned on the leakage value. In the literature, starting with [18], this is referred to as "resilience to *noisy* leakage". It is self evident that schemes where the public key is an injective function of the secret key cannot be proven leakage resilient in this respect. This is because even leaking the secret key in its entirety, which obviously breaks security, does not reduce its min entropy conditioned on the public key (the conditional min-entropy is 0 to begin with, and it stays 0 after the leakage). We do know, however, of such injective public-key encryption schemes that are proven to be leakage resilient with respect to the weaker notion of "length bounded leakage". There, the restriction on the leakage function is that it has bounded length. Notable examples are the scheme of [1] and the scheme of [16] (which was introduced as a counterexample for parallel repetition of length-bounded leakage resilience, see Section 1.2). While at first glance it may seem that our result is completely powerless with regards to such schemes, we show in this section that for all known schemes, and specifically for the schemes of [1, 16], our theorem in fact *does* imply parallel repetition.

The key observation upon revisiting the proofs of security of [1, 16], is that in both cases, the proof is by presenting a second scheme in which the key distribution is computationally indistinguishable from the original scheme (but may have undesired features such as worse efficiency of key generation), and proving that this second scheme is resilient to leakage of bounded length. This implies that the original scheme is resilient to bounded leakage as well (since otherwise one can distinguish the key generation processes). The second scheme, in these two cases, is in fact resilient to noisy leakage. Furthermore, the second scheme in the two cases adheres to our notion of point-wise leakage resilience.

In light of the above, we put forth the following corollary of Theorems 5.1 and 6.1.

Corollary 7.1. *Let $\mathcal{E} = (G, V)$ be a public-key primitive, and let G' be such that $G(1^k) \stackrel{c}{\approx} G'(1^k)$. Then:*

1. *If $\mathcal{E}' = (G', V)$ is point-wise almost λ-leakage resilient, then $\mathcal{E}^t$ is $t \cdot (\lambda - 1)$-bounded leakage resilient for any constant $t \in \mathbb{N}$.*
2. *If $\mathcal{E}' = (G', V)$ is point-wise λ-leakage resilient with logarithmic advice, then $\mathcal{E}^t$ is $t \cdot (\lambda - 1)$-bounded leakage resilient for any polynomial $t = t(k)$.*

Proof. The proof of the two parts is almost identical: We use either Theorem 5.1 or Theorem 6.1 to show that $(\mathcal{E}')^t$ is point-wise almost $t \cdot (\lambda - 1)$-leakage resilient, or, respectively, leakage resilient with logarithmic advice. By Lemma 3.8, this means that $(\mathcal{E}')^t$ is $t \cdot (\lambda - 1)$-bounded leakage resilient.

264   Z. Brakerski and Y.T. Kalai

By a hybrid argument $(G')^t(1^k) \stackrel{c}{\approx} G^t(1^k)$.[5] Therefore, it must be that $\mathcal{E}^t$ is also $t \cdot (\lambda - 1)$-bounded leakage resilient (otherwise there is a distinguisher between the key generators). This completes the proof.

Using Corollary 7.1, we can show that t-parallel repetition of the schemes of [1, 16] indeed amplifies their leakage resilience.

References

[1] Akavia, A., Goldwasser, S., Vaikuntanathan, V.: Simultaneous Hardcore Bits and Cryptography against Memory Attacks. In: Reingold, O. (ed.) TCC 2009. LNCS, vol. 5444, pp. 474–495. Springer, Heidelberg (2009)
[2] Alwen, J., Dodis, Y., Naor, M., Segev, G., Walfish, S., Wichs, D.: Public-Key Encryption in the Bounded-Retrieval Model. In: Gilbert, H. (ed.) EUROCRYPT 2010. LNCS, vol. 6110, pp. 113–134. Springer, Heidelberg (2010)
[3] Alwen, J., Dodis, Y., Wichs, D.: Leakage-Resilient Public-Key Cryptography in the Bounded-Retrieval Model. In: Halevi, S. (ed.) CRYPTO 2009. LNCS, vol. 5677, pp. 36–54. Springer, Heidelberg (2009)
[4] Boyle, E., Segev, G., Wichs, D.: Fully Leakage-Resilient Signatures. In: Paterson, K.G. (ed.) EUROCRYPT 2011. LNCS, vol. 6632, pp. 89–108. Springer, Heidelberg (2011)
[5] Brakerski, Z., Goldwasser, S.: Circular and Leakage Resilient Public-Key Encryption under Subgroup Indistinguishability (or: Quadratic Residuosity Strikes Back). In: Rabin, T. (ed.) CRYPTO 2010. LNCS, vol. 6223, pp. 1–20. Springer, Heidelberg (2010)
[6] Brakerski, Z., Kalai, Y.T.: A parallel repetition theorem for leakage resilience. Cryptology ePrint Archive, Report 2011/250 (2011), http://eprint.iacr.org/
[7] Brakerski, Z., Segev, G.: Personal communication (2009)
[8] Damgård, I.B., Fehr, S., Renner, R., Salvail, L., Schaffner, C.: A Tight High-Order Entropic Quantum Uncertainty Relation with Applications. In: Menezes, A. (ed.) CRYPTO 2007. LNCS, vol. 4622, pp. 360–378. Springer, Heidelberg (2007)
[9] Dodis, Y., Goldwasser, S., Tauman Kalai, Y., Peikert, C., Vaikuntanathan, V.: Public-Key Encryption Schemes with Auxiliary Inputs. In: Micciancio, D. (ed.) TCC 2010. LNCS, vol. 5978, pp. 361–381. Springer, Heidelberg (2010)
[10] Dodis, Y., Kalai, Y.T., Lovett, S.: On cryptography with auxiliary input. In: Mitzenmacher, M. (ed.) STOC, pp. 621–630. ACM (2009)
[11] Dziembowski, S., Pietrzak, K.: Leakage-resilient cryptography. In: FOCS, pp. 293–302. IEEE Computer Society (2008)
[12] Faust, S., Kiltz, E., Pietrzak, K., Rothblum, G.N.: Leakage-Resilient Signatures. In: Micciancio, D. (ed.) TCC 2010. LNCS, vol. 5978, pp. 343–360. Springer, Heidelberg (2010)
[13] Garg, S., Jain, A., Sahai, A.: Leakage-Resilient Zero Knowledge. In: Rogaway, P. (ed.) CRYPTO 2011. LNCS, vol. 6841, pp. 297–315. Springer, Heidelberg (2011)
[14] Jain, A., Pietrzak, K.: Parallel Repetition for Leakage Resilience Amplification Revisited. In: Ishai, Y. (ed.) TCC 2011. LNCS, vol. 6597, pp. 58–69. Springer, Heidelberg (2011)

---

[5] It is important to note that this will not be necessarily true if either G or G' relies on secret parameters. This relates to our discussion of the result of [16] in Section 1.2.

[15] Katz, J., Vaikuntanathan, V.: Signature Schemes with Bounded Leakage Resilience. In: Matsui, M. (ed.) ASIACRYPT 2009. LNCS, vol. 5912, pp. 703–720. Springer, Heidelberg (2009)

[16] Lewko, A.B., Waters, B.: On the insecurity of parallel repetition for leakage resilience. In: Trevisan, L. (ed.) FOCS, pp. 521–530. IEEE Computer Society (2010)

[17] Malkin, T., Teranishi, I., Vahlis, Y., Yung, M.: Signatures Resilient to Continual Leakage on Memory and Computation. In: Ishai, Y. (ed.) TCC 2011. LNCS, vol. 6597, pp. 89–106. Springer, Heidelberg (2011)

[18] Naor, M., Segev, G.: Public-Key Cryptosystems Resilient to Key Leakage. In: Halevi, S. (ed.) CRYPTO 2009. LNCS, vol. 5677, pp. 18–35. Springer, Heidelberg (2009)

[19] Pietrzak, K.: A Leakage-Resilient Mode of Operation. In: Joux, A. (ed.) EUROCRYPT 2009. LNCS, vol. 5479, pp. 462–482. Springer, Heidelberg (2009)

[20] Wullschleger, J.: Oblivious-Transfer Amplification. In: Naor, M. (ed.) EUROCRYPT 2007. LNCS, vol. 4515, pp. 555–572. Springer, Heidelberg (2007)

A   How to Split Min-Entropy

We present a lemma that shows that the joint min-entropy of two random variables can be split between them under some condition. Variants of this lemma appeared in previous works (e.g. [8, 20]), this formulation is from [7].

Lemma A.1 (min-entropy split). *Let X, Y be such that $\mathbf{H}_\infty(X, Y) \geq a + b$, for $a, b > 0$. Then there exists a set R_X, which is a subset of the support of X such that both:*

1. *For all $x \in R_X$, it holds that $\mathbf{H}_\infty(Y | X = x) \geq b$.*
2. *$\mathbf{H}_\infty(X | X \notin R_X) \geq a - \log(1/\epsilon)$, where $\epsilon \triangleq \Pr[X \notin R_X]$.*

Proof. Define
$$R_X \triangleq \{x : \Pr[X = x] \geq 2^{-a}\} .$$
Then for all $x \in R_X$ and for all y, it holds that $\Pr[Y = y | X = x] \leq 2^{-b}$, and thus $\mathbf{H}_\infty(Y | X = x) \geq b$. In addition, $\mathbf{H}_\infty(X | X \notin R_X) \geq a + \log \Pr[X \notin R_X]$, i.e. $\mathbf{H}_\infty(X | X \notin R_X) \geq a - \log(1/\epsilon)$.

Leakage-Tolerant Interactive Protocols*

Nir Bitansky[1,2], Ran Canetti[1,2], and Shai Halevi[3]

[1] Tel Aviv University
[2] Boston University
[3] IBM T.J. Watson Research Center

Abstract. We put forth a framework for expressing security requirements from interactive protocols in the presence of arbitrary leakage.
The framework allows capturing different levels of leakage-tolerance of
protocols, namely the preservation (or degradation) of security, under
coordinated attacks that include various forms of leakage from the secret
states of participating components. The framework extends the universally composable (UC) security framework. We also prove a variant of
the UC theorem that enables modular design and analysis of protocols
even in face of general, non-modular leakage.

We then construct leakage-tolerant protocols for basic tasks, such
as secure message transmission, message authentication, commitment,
oblivious transfer and zero-knowledge. A central component in several of
our constructions is the observation that resilience to adaptive party corruptions (in some strong sense) implies leakage-tolerance in an essentially
optimal way.

1 Introduction

Traditionally, cryptographic protocols are studied in a model where participants
have a secret state that is assumed to be completely inaccessible by the adversary.
In this model, the adversary can only influence the system via anticipated interfaces (such as, the communication among parties). These interfaces are crossed
only when the adversary manages to fully corrupt a party, thus gaining access
to its entire inner state.

In reality, an intermediate setting often emerges, when the adversary manages
to gain some partial information on the secret state of uncorrupted parties. This
information, termed *leakage*, can be obtained by a variety of side channels attacks
that bypass the usual interfaces and are often undetectable. Known examples
include: timing, power, EM-emission, and cache attacks (see [Sta09] for a survey).

The threat of leakage gained much attention in the past few years, giving
rise to an impressive array of *leakage-resilient* schemes for basic cryptographic
tasks such as encryption and signatures, as well as general non-interactive circuits (e.g., [DP08, AGV09, ADW09, DKL09, Pie09, NS09, ADN+10, BKKV10,

---

* Supported by the Check Point Institute for Information Security. The first two authors are also supported by Marie Curie grant PIRG03-GA-2008-230640, and ISF
grant 0603805843.

R. Cramer (Ed.): TCC 2012, LNCS 7194, pp. 266–284, 2012.

DHLAW10b, DHLAW10a, BSW11]). Most of the work concentrates on preserving, in the presence of leakage, the same functionality and security guarantees that the original primitives guarantee in a leak-free setting. Such strong leakage-resilience is typically guaranteed only when the leakage is restricted in some ways. Examples include: assuming bounded amounts of leakage, assuming that leakage only occurs in specific times (e.g., prior to encryption), or assuming that leakage is limited to specific parts of the state, such as the active parts in the *only computation leaks* model [MR04].

However, in many cases maintaining the same level of security as in a leak-free setting may be too costly, or even outright impossible. To exemplify this, consider the task of *secure message transmission* (SMT), where a sender wishes to transmit a (secret) message m to a receiver, so that the contents of m remain completely hidden from any adversary witnessing the communication. In the leak-free setting, the problem is easily solved using standard semantically secure encryption; however, in the presence of leakage, this is no longer the case. In fact, semantic security is not achievable at all: an adversary that can get even one bit of arbitrary leakage, from either party, can certainly learn any bit of the message, since this bit must reside in the party's leaky memory at some point.

Nevertheless, this inherent difficulty does not imply that we should give up on security altogether, but rather that we should somehow meaningfully relax the security requirements from protocols in the presence of leakage. Concretely, in the above example, we would like to design schemes in which one-bit of leakage on the message does not compromise the security of the entire message. More generally, we would like to establish a framework that will allow to express and analyze security of general cryptographic tasks in the presence of general (non-restricted) leakage, where the level of security may gracefully degrade according to the amount of leakage (that might develop over time). A first step in this direction was taken by Halevi and Lin [HL11] in the context of encryption.

Another intriguing question is what are the composability properties of resilience to leakage. Can one combine two or more schemes and deduce leakage-resilience of the combined system based only on the leakage-resilience properties of the individual schemes? If so, constructs with various levels of leakage-resilience may be composed to obtain new systems that enjoy improved such resilience properties. Some specific examples where this is the case have been recently exhibited [BCG⁺11, BGK11, GJS11]. What can we say in general?

1.1   Our Contribution

We propose a new approach for defining leakage-resilience, or rather *leakage-tolerance*, properties of cryptographic protocols. The approach is based on the ideal model paradigm and, specifically, on the UC framework. The approach allows formulating relaxed security properties of protocols in face of leakage and, in particular, allows specifying how the security of protocols degrades with leakage. It also allows specifying leakage-tolerant variants of interactive, multi-party protocols for general cryptographic tasks. In this context, the new modeling also captures attacks that combine leakage with other "network based" attacks such as controlling the communication and corrupting parties. In addition:

- We prove a general security-preserving composition theorem with respect to the proposed notion. This allows constructing and analyzing protocols in a modular way *while preserving leakage-tolerance properties.* This is a powerful tool, given the inherently modularity-breaking nature of leakage attacks.
- We describe a methodology for constructing leakage-tolerant protocols in this framework. Essentially, we show that any protocol that is secure against adaptive party corruptions (in some strong sense) is already leakage-tolerant.
- Using the above methodology and other techniques, we construct composable leakage-tolerant protocols for secure channels, commitment, zero-knowledge, and honest-but-curious oblivious transfer. (commitment and zero-knowledge are realized in the common reference string model.)

Below we describe these contributions in more detail.

Leakage-Tolerant Security within the Ideal Model Paradigm. Following the ideal model paradigm, we define security by requiring that the protocol π at hand provides the same security properties as in an "ideal world" where processing is done by a trusted party running some functionality $\mathcal{F}$. Specifically, in the UC framework, a protocol π UC-realizes a functionality $\mathcal{F}$ if for any adversary $\mathcal{A}$ there exists a simulator $\mathcal{S}$ such that no environment $\mathcal{Z}$ can tell whether it is interacting with $\mathcal{A}$ and π or with $\mathcal{S}$ and $\mathcal{F}$.

We consider a "real world" where the adversary can get leakage on the state of any party at any time. As we argued above, such attacks may unavoidably degrade the security properties of the protocols at hand and to account for this degradation we also allow leakage from the trusted party in the ideal world. Specifically, the functionality $\mathcal{F}$ defines the "ideal local state" for each party and the party's behavior (and degradation in security) after leakage. (Typically, we will be interested in functionalities where the ideal local state includes the party's inputs and outputs, but weaker functionalities that allow joint leakage on the inputs of several parties can also be considered.) When $\mathcal{A}$ performs a leakage measurement L on the state of some party in the real protocol π, the simulator $\mathcal{S}$ is entitled to a leakage measurement L' on the ideal local state of that party in the ideal protocol. We allow the simulator to choose any function L', so long that its output length is the same as that of L.

For example, we allow our leaky SMT functionality to leak bits from message that it sends and require that a real world attacker that gets ℓ bits of leakage from the state of the implementation can be simulated by a simulator that learns only ℓ bits about the message. Our model also allows the functionality to react to leakage, in order to handle situations where security is only maintained as long as not too much leakage occurred. (For example, an authenticated channels functionality may allow forgeries once the attacker gets more bits of leakage than the security parameter, but not before that.)

Leakage vs. Adaptive Corruptions for Secure Channels. Consider trying to realize leaky SMT in our model using standard encryption; namely, the receiver sends its public key to the sender, who sends back an encryption of m. In the ideal world, the simulator does not witness any communication (and has no information about the message), so it can simulate the cipher by encrypting say the all-zero string, which should be indistinguishable from an encryption of m. However, after seeing the ciphertext the adversary $\mathcal{A}$ can ask for a leakage query specifying (say) the entire secret decryption key and the first bit of m. Although the simulator can now ask for many bits of leakage on m, it can no longer modify the ciphertext that it sent before and therefore cannot maintain a consistent simulation.

A similar problem arises in the well studied setting of *adaptive corruption* (with non-erasing parties), where the adversary can adaptively corrupt parties throughout the protocol and learn their entire state. Also there, the simulator needs to first generate some messages (e.g., the ciphertext) without knowing the inputs of the parties (e.g., the message m), and later it learns the inputs and has to come up with an internal state that explains the previously-generated messages in terms of these inputs. Indeed, it turns out that techniques for handling adaptive corruptions can be used to get leakage-tolerance.

In fact, the problem of secure leaky channels can be solved simply by plugging in *non-committing encryption* (NCE) [CFGN96], which was developed for adaptively secure communication. Recall that an NCE scheme allows generating a "fake" equivocal ciphertext $\tilde{c}$ that can later be "opened" as an encryption of any string of a predefined length ℓ. Namely, $\tilde{c}$ is generated together with a poly-size equivocation circuit E, such that, given any message $m \in \{0,1\}^{\ell}$, $E(m)$ generates randomness $(\tilde{r}_S^m, \tilde{r}_R^m)$, for both the sender and the receiver, that "explains" $\tilde{c}$ as an encryption of m.

To obtain leakage-tolerant secure message transmission, we can simply encrypt the message using an NCE scheme. The simulator can now generate the fake ciphertext $\tilde{c}$ with the associated equivocation circuit E and can then translate any c-dependent leakage function on the entire state (plaintext and randomness) into a leakage function on the plaintext only, which can be queried to the leaky SMT functionality. When leakage on $P \in \{S, R\}$ occurs, the simulator $\mathcal{S}$ translates the leakage function $L(m, r_P)$ into $L'(m) = L(m, E(m)) = L(m, \tilde{r}_P^m)$. Indeed, this idea was used in [BCG+11] in the context of a specific protocol.

The General Case. The above example can be made general. Specifically, we show that, with some limitations, any protocol that realizes a functionality $\mathcal{F}$ under adaptive corruptions also realizes a leaky variant $\mathcal{F}^{+lk}$ under leakage. The "leaky variant" is a natural adaptation that allows leakage on the state of $\mathcal{F}$, just like the leaky SMT allow leakage on the transmitted message. This variant, denoted $\mathcal{F}^{+lk}$, is identical to $\mathcal{F}$ except that $\mathcal{F}^{+lk}$ allows the simulator to apply arbitrary leakage functions to the ideal local state (which is the same as the state defined in a semi-honest corruption). When such leakage occurs the environment is reported on the identity of the leaking party and the number of bits leaked. (This makes sure that the simulator can only leak the same number of bits as in the protocol execution.) After such a leakage event, $\mathcal{F}^{+lk}$ behaves in the same way that $\mathcal{F}$ behaves after a semi-honest corruption of that party. That is, if $\mathcal{F}$ modifies its overall behavior following the corruption of a party, then $\mathcal{F}^{+lk}$

modifies its behavior in the same way. (In the applications considered in this work, we will consider functionalities that do not change their behavior after semi-honest corruptions, see Section 5.)

A limitation of this result is that it only holds when the given proof of security uses a restricted type of simulators, namely ones that work "obliviously" of the state that they learn when corrupting a player. We call such simulators *corruption-oblivious*. We have:

Theorem 1.1 (informal). *If protocol π realizes $\mathcal{F}$ under adaptive corruptions (either semi-honest or Byzantine) with a corruption-oblivious simulator, then it also realizes $\mathcal{F}^{+lk}$ under arbitrary leakage (and the same type of corruptions).*

Composable Leakage-Tolerance. An important property of ideal model based notions of security is that they enable modularity, since the guarantees that they provide are preserved even under (universal) composition of protocols. That is, if a protocol π realizes an ideal functionality $\mathcal{F}$, the security properties of $\mathcal{F}$ carry over to any environment where π is used.

To achieve such modularity, common models of composable security rely crucially on viewing different sub-modules of a large system as autonomous small systems, each with its own local state and well-defined interfaces to the rest of the system. Unfortunately, extending this "modular security" paradigm to the leaky world is problematic: real world leakage is inherently non-modular, in that the adversary can obtain leakage from the joint state of an entire physical device and is not bound by our modular separation to logical modules of the software running on the device. In fact, it is not even clear how to *express* joint leakage from the state of different modules within standard models, let alone how to argue about preservation of security properties.

We extend the UC security framework [Can01] to allow expressing leakage attacks from physical devices that span multiple logical modules. We first allow the protocol analyzer to delineate sets of "jointly leakable modules" (roughly corresponding to physical machines). Then, we introduce a new entity, called an **aggregator**, that has access to the internal states of all the modules in each set.

To get leakage from the joint state of the modules in a set P, the adversary sends the leakage function L to the aggregator, who applies L to the combined state and returns the result to the adversary. The same mechanism is used to obtain leakage from ideal functionalities, except that here the ideal functionality $\mathcal{F}$ hands the aggregator some "ideal local state" that $\mathcal{F}$ associated with the set P. We stress again that our model considers a strong adversary that obtains leakage information in a non-modular way from multiple subroutines that reside on a common device, this makes positive results in this model quite strong.

Having extended the model of protocol execution to capture leakage attacks, we would like to re-assert the composability property described above, i.e., to re-prove the UC composition theorem from [Can01] in our setting. However, that theorem was only proved for systems that behave in a "modular way", and the proof no longer holds in the presence of our modularity-breaking aggregator.

Still, we manage to salvage much of the spirit of the UC theorem, as follows. We formulate a more stringent variant of UC security by putting some technical restrictions on the simulator and then re-assert the UC theorem with respect to this varaint. Similarly to the case of corruption-oblivious simulators, here too we require that the simulator S handles leakage queries "obliviously".

Roughly, S has a "query-independent" way of translating, via a state-translation function, real world leakage queries $L(\text{state}_\pi)$ to ideal world leakage queries $L'(\text{state}_\mathcal{F})$. Furthermore, it ignores the leakage-results in the rest of the simulation. We call such simulators *leakage-oblivious* and show:

Theorem 1.2 (UC-composition with leakage, informal). *Let $\rho^\mathcal{F}$ be a protocol that invokes $\mathcal{F}$ as a sub-routine. Let π be a protocol that UC-emulates $\mathcal{F}$ with a leakage-oblivious simulator. Then the composed protocol $\rho^{\pi/\mathcal{F}}$ (where each call to $\mathcal{F}$ is replaced with a call for π) UC-emulates $\rho^\mathcal{F}$ in face of leakage. Furthermore, it does so with a leakage-oblivious simulator.*

Theorem 1.2 provides a powerful tool in the design of leakage-resilient protocols. In particular, we later use it to (a) combine any leakage-resilient protocol that assumes authenticated communication with a leakage-resilient authentication protocol into a leakage-resilient protocol over unauthenticated channels, and (b) to combine any leakage-resilient zero-knowledge protocol that assumes ideal commitment with leakage-resilient commitment protocols to obtain a composite leakage-resilient zero-knowledge protocol.

Leakage-Tolerant Protocols. We construct leakage-tolerant protocols for a number of basic cryptographic tasks. We first observe that the general result regarding the leakage-tolerance of adaptively secure protocols (Theorem 1.1) in fact guarantees UC security *with leakage-oblivious simulators*. We then observe that existing adaptively secure protocols for secure channels, UC commitment and UC semi-honest oblivious transfer already have corruption-oblivious simulators; hence, we immediately get:

- Assume authenticated communication. Then, any non-committing encryption scheme UC-realizes $\mathcal{F}_{\text{SMT}}^{+\text{lk}}$ in the presence of arbitrary leakage using a leakage-oblivious simulator.
- In the CRS model, the UC commitment protocols of Canetti and Fischlin [CF01] and Canetti, Lindell, Ostrovsky and Sahai [CLOS02], UC-realize $\mathcal{F}_{\text{MCOM}}^{+\text{lk}}$ (the leaky version of the multi-instance commitment functionality) in the presence of arbitrary leakage. Furthermore, they do so with leakage-oblivious simulators.
- Also in the CRS model, the UC (non-interactive) zero-knowledge protocol of Groth, Ostrovsky and Sahai[GOS06] realize $\mathcal{F}_{\text{ZK}}^{+\text{lk}}$ under arbitrary leakage.
- The semi-honest oblivious transfer protocol of [CLOS02] for adaptive corruptions UC-realizes $\mathcal{F}_{\text{OT}}^{+\text{lk}}$ (the leaky version of the ideal oblivious transfer functionality in the presence of arbitrary leakage). Furthermore, it does so with leakage-oblivious simulators.

In this work, we do not consider the generation of a CRS in the presence of leakage; rather, we treat the CRS as an external entity that can be generated in a physically separate location. As in other settings, here too it is interesting to find ways to reduce the setup requirements.

Finally, we note that for certain functionalities $\mathcal{F}$, applying Theorem 1.1 alone may still not give an adequate level of leakage-resilience. Indeed, while the leaky adaptation $\mathcal{F}^{+lk}$ assures graceful degradation of privacy, it may not account for correctness (or soundness) aspects in the face of leakage. In such cases, we may need to further strengthen $\mathcal{F}^{+lk}$. One such example is message authentication. Indeed, $\mathcal{F}_{AUTH}^{+lk}$ gives essentially no security guarantees: as soon as even a single bit of information is leaked from the sender, $\mathcal{F}_{AUTH}^{+lk}$ behaves as if the sender is fully corrupted, in which case forgery of messages is allowed. We thus first formulate a variant of $\mathcal{F}_{AUTH}$ that guarantees authenticity as long as the number of bits leaked is less than some threshold B. We then realize this functionality, denoted $\mathcal{F}_{AUTH}^{+B}$, assuming an initial k-bit shared secret key between the parties and as long as at most $B = O(k)$ bits leak *between each two consecutive message transmissions*. Furthermore, we do this with a leakage-oblivious simulator. The techniques used to realize $\mathcal{F}_{AUTH}^{+B}$ include information-theoretic leakage-resilient message authentication codes, as well as NCE schemes.

We note that the techniques here borrow strongly from the techniques used in [BCG+11] for the related goal of authentication within the context of obfuscation with leaky hardware. That work, however, analyzed these tools in an ad-hoc manner, and the results there apply only to that specific context.

In contrast, using the above UC theorem with leakage, we can combine the above authentication protocol with any protocol that assumes ideally authenticated communication to obtain composite leakage-tolerant protocols that withstand unauthenticated communication.

Finally, we address the task of obtaining zero-knowledge from ideal leaky commitment $\mathcal{F}_{MCOM}^{+lk}$ (the adaptive NIZK protocol of [GOS06] is obtained from specific number-theoretic assumptions on bilinear groups). At first it may seem that, as in the case of commitment, existing protocols for UC-realizing the ideal zero-knowledge functionality, $\mathcal{F}_{ZK}$, would work also in the case of leakage. However, this turns out not to be the case. In particular, while the protocol of [CF01] for UC-realizing $\mathcal{F}_{ZK:R}$, for some relation R, given $\mathcal{F}_{MCOM}$ is indeed secure against adaptive corruptions, the simulator turns out not to be corruption-oblivious and Theorem 1.1 does not apply.

Instead, we settle for UC-realizing, in the presence of leakage, a weaker variant of $\mathcal{F}_{ZK:R}^{+lk}$. This weaker variant permits violation of the soundness requirements if too many bits were leaked from the verifier. We denote this weaker version by $\mathcal{F}_{ZK:R}^{+B}$, where B is the leakage threshold for the verifier. We show how to UC-realize $\mathcal{F}_{ZK:R}^{+B}$ for $B = k - \omega(\log k)$ (where k is the security parameter), given access to $\mathcal{F}_{MCOM}^{+lk}$. Using the (leaky) universal composition theorem and the protocol for realizing $\mathcal{F}_{MCOM}$ (mentioned above), we obtain a protocol for UC-realizing $\mathcal{F}_{ZK:R}^{+B}$ in the CRS model.

Concurrent Work. Garg, Jain, and Sahai [GJS11] also investigate zero-knowledge in the presence of leakage, albeit not in the UC setting. Instead, they consider a stand-alone definition with a rewinding simulator (where a CRS is not needed).

Some of the difficulties that emerge in standard 3-round zero-knowledge protocols, as well as the suggestion to overcome them using the Goldriech-Kahn paradigm, were communicated to us by Amit Sahai.

Damgård, Hazay, and Arpita [DHP11] consider leakage-resilient two-party protocols. Their definition of security, which is also ideal-model based, accounts also for noisy leakage (namely leakage that might not be length-restricted, but is somewhat entropy preserving). They achieve leakage-resilience (or tolerance) for NC_1 functions in a setting where one party is statically and passively corrupted and the other party is leaky. The result, however, only applies in the "only computation leaks" (OCL) model of [MR04] (and with some extra technical limitations). They also prove a security preserving composition theorem, but their modeling considers only separate leakage from each module (rather than overall leakage as considered here). They also construct a leakage-tolerant OT protocol for sufficiently entropic inputs distributions, but only in the OCL model and under a relatively strong hardness assumption; in terms of communication, however, their protocol is more efficient than ours. Finally, we remark that the setting where one party is statically passively corrupted can be seen as a special case of a weak leakage-tolerance model, where the ideal world simulator is allowed to jointly leak from all the parties. See further discussion in Section 5.3.

2   Modeling Leakage in the UC Framework

This section defines the new model of UC security with leakage. Here we provide a high-level overview, the full details can be found in the full version of this work [BCH11]. Recall that the basic UC framework considers realization of an "ideal specification" $\mathcal{F}$ by a "real implementation" π. (Formally both $\mathcal{F}$ and π are just protocols, we call them by different names to guide the intuition.) The realization requirement is that for any "real world attacker" $\mathcal{A}$ against the implementation π there exists another adversary $\mathcal{S}$ (called a simulator) against the specification $\mathcal{F}$, such that an "environment" $\mathcal{Z}$ that interacts with $\mathcal{S}, \mathcal{F}$ has essentially the same view as in an interaction with $\mathcal{A}, \pi$.

The basic UC execution model lets the environment $\mathcal{Z}$ determine the inputs to the parties running the protocol and see the outputs generated by these parties and also allows free communication between the environment and the adversary. The adversary, typically, has full control over the communication between parties and the ability to "corrupt" parties in various ways. Corruption is modeled as just another interface available to the adversary, where it can send a message "you are corrupted" to any party. (In the case of standard passive corruption, the party responds to this message by handing its entire internal state to the adversary. To model Byzantine corruption, the party also changes the program that it is running from then on.)

A crucial aspect of the UC framework is its modularity, where programs can call subroutines, and these subroutines are treated as separate entities that can be analyzed separately for security properties. Importantly, local randomness and secrets that are used by a subroutine should typically not be visible to the calling routine or to other components in the system.

A useful technicality in the UC framework, is that it is sufficient to prove security only with respect to the dummy real world adversary $\mathcal{D}$. This is the adversary that simply reports all the information it receives to the environment and follows all the instructions of the environment regarding sending messages to parties and ideal functionalities. Relying on the fact that any adversary can be emulated by the environment itself, it is easy to show that simulation of the dummy adversary $\mathcal{D}$ implies simulation for any adversary.

Leaky UC. A natural approach to modeling leakage within the UC framework is to view it as a weak form of corruption, where the adversary gets some information about the internal state of the leaky party but perhaps not all of it. Also, leakage resembles "semi-honest" corruption more than "malicious", in that leaky parties keep following the same protocol and do not change their behavior following a leakage event. Thus we could provide yet another interface to the adversary where it can send a "leak L" message to a party (where L is some function) and have that party reply with $L(s)$ where s is its internal state.

The Leakage Aggregator. A serious shortcoming of the modeling approach in the previous paragraphs is that it only lets the adversary obtain leakage on individual processes (or subroutines). In contrast, real life leakage usually provides information that depends on the entire state of a physical device, including all the processes that are currently running on it. To account for this inherently non-modular property of real life leakage, we introduce to the model a new "global entity" that we call the leakage aggregator. The aggregator $\mathcal{G}$ can access the entire internal state of all the components in the system. A leakage query specifies a leakage function L and a set of processes $P = \{p_1, \ldots, p_t\}$. This query is forwarded to the aggregator, who evaluates $L(s_1, \ldots, s_t)$ and returns the result to the adversary. Some important technicalities regarding the working of $\mathcal{G}$ are the following:

- A convention should be set for how to specify the sets of processes and ensure that this is a "legitimate set" for joint leakage. We assume that processes are tagged with "party identifiers" pid (roughly corresponding to physical machines), and joint leakage is allowed only from a set of processes that all have the same pid.
- As done for corruptions, here too the identity of the leaky processes and the amount of leakage needs to be reported to the environment. This forces the simulator, in the ideal world, to use the same amount of leakage from the same processes as in the real world.
- Since ideal functionalities represent idealized constructs that do not necessarily run on physical devices, they are often associated with more than one pid. Thus care should be taken when deciding how an ideal functionality reacts to leakage queries w.r.t. one of its pid's. (For example, the secure-channels functionality runs on behalf of both the sender and the receiver, and would typically react differently to sender-leakage than to receiver-leakage queries.) We let the ideal functionality itself decide how to reply when $\mathcal{G}$ asks it for the state corresponding to any of its pid's. (This is the same convention as

used for corruption, where the functionality gets to decide what to reveal to the adversary when one of its pid's is corrupted.) Typically, the "state" associated with a certain pid will be just the inputs that were received from that pid and the outputs it receives.

- To allow functionalities to react to leakage situations, we have $\mathcal{G}$, upon accessing the state of a module, report to that module the output size of the leakage function L. Typically, "real world implementations" ignore this report (since we assume that real world leakage is undetectable), but "ideal functionalities" may use it to change their behavior (e.g., reduce the security guarantee if too much leakage occurred).

With these conventions in place, a leakage operation is handled as follows: first, the adversary sends a query (leak, L, pid) to $\mathcal{G}$, where L is the leakage function and pid is the leaking party ID. Then, $\mathcal{G}$ obtains state$_{pid}$, the total state of party pid, applies L to state$_{pid}$, and returns the result to $\mathcal{A}$. Finally, $\mathcal{G}$ reports the output length of the function L to all the processes whose state is included in state$_{pid}$ and reports pid and the output length to the environment.

We note that the security guarantee provided by this model may be weaker than one could desire, as the number of leaked bits is reported to *each one* of the processes (or functionalities). This means that when a domain leaks ℓ bits, each one of its components behaves as if the ℓ bits leaked entirely from this component. While this is a relatively weak leakage-resilience guarantee, it seems unavoidable in any general model with modularity-breaking leakage.

Leakage-Oblivious Simulation. Following the approach of basic UC security, the definition of protocol emulation requires that for any adversary $\mathcal{A}$ that attacks the implementation π there exists a simulator S that attacks the specification $\mathcal{F}$ so that no environment can distinguish between an interaction with $\mathcal{A}$ and π, and an interaction with S and $\mathcal{F}$. In particular, S must provide an overall transformation from one interaction scenario to the other, including among others the leakage queries made by $\mathcal{A}$ to the parties (via the aggregator). As noted above, an equivalent requirement considers, instead of any adversary $\mathcal{A}$, only the dummy adversary, $\mathcal{D}$, that merely passes messages between the environment and the protocol's parties.

This natural requirement, however, has (seemingly inherent) difficulties when considering composition of protocols. In particular, we were not able to prove a general composition theorem in this model (see details in Section 3). Consequently, we consider a more restricted notion of protocol emulation, which we term emulation with leakage-oblivious simulators.

To simplify the exposition, we describe here leakage-oblivious simulation only with respect to the dummy adversary $\mathcal{D}$. A leakage-oblivious simulator S for the dummy adversary has a special form: specifically, S has a separate subroutine $\tilde{S}$ for handling leakage. When S receives from the environment a request to apply a leakage function L to a set P of processes, $\tilde{S}$ is invoked to produce a "state translation" function T. This function is meant to transform the internal state of P in the specification $\mathcal{F}$ into "the actual state" in the implementation π. Once T

is produced, the aggregator is given the composed leakage function $L \circ T$. Finally, when the leakage-result is returned, it is forwarded directly to the environment and S returns to its state prior to the leakage event.

The subroutine $\tilde{S}$ should operate **independently of the leakage function** L, its only input is the state of S (prior to the leakage query) and a party identifier pid. Also, the leakage operation has **no side effects on** S. That is, following the leakage event S return to the state that it had before that event.

3   Universal Composition of Leaky Protocols

We now state the universal composition theorem for leaky protocols and leakage-oblivious simulators (as defined in Section 2). Let π be an implementation and $\mathcal{F}$ be a specification. (As mentioned earlier, formally these are just two protocols, and the different names are meant only to help the intuition.) Also let $\rho = \rho[\pi]$ be a protocol that includes subroutine calls to π. Below we denote by ρ^{π} the system where the subroutine calls to π are actually processed by π and by $\rho^{\mathcal{F}/\pi}$ the system where these subroutine calls are processed by $\mathcal{F}$.

The UC theorem [Can01] states that if π UC-realizes $\mathcal{F}$, then ρ^{π} UC-realizes $\rho^{\mathcal{F}/\pi}$; however, that theorem does not hold in the presence of the modularity-breaking aggregator $\mathcal{G}$. The proof of the UC-theorem in [Can01] relies on all the processes being "modular"; namely, a process can only interact with its caller and its subroutines (and the adversary).[1]

As we have seen, modularity is incompatible with the definition of leaky protocols; indeed, all processes are required to interact with the aggregator, which is neither their caller nor their subroutine (nor an adversarial entity). Still, if π realizes $\mathcal{F}$ with a leakage-oblivious simulator, we can recover the same result. Below we call a protocol "modular up to leakage" if it only interacts with its caller, its subroutines, the adversary, and the aggregator.

Theorem 3.1 (UC-composition with leakage). *Let $\rho, \pi, \mathcal{F}$ be protocols as above, all modular up to leakage, such that π UC-emulates $\mathcal{F}$ with a leakage-oblivious simulator. Then ρ^{π} UC-emulates $\rho^{\mathcal{F}/\pi}$. Furthermore, it does so with a leakage-oblivious simulator.*

Proof Overview. The proof follows the outline of the proof of the basic UC theorem; here, we focus on the required adjustments due the leakage. For sake of simplicity, in this overview we assume that ρ invokes only a single instance of the sub-protocol π.

Recall that we need to construct a leakage-oblivious simulator S_{ρ} such that no environment can tell whether it is interacting with ρ^{π} and the dummy adversary $\mathcal{D}$, or with $\rho^{\mathcal{F}/\pi}$ and S_{ρ}. The construction of S_{ρ} is naturally based on the leakage-oblivious simulator S_{π} as guaranteed by the premise. That is, S_{ρ} runs a copy of S_{π}; as in the basic UC theorem, the interaction between Z and the parties is separated into two parts. The interaction with π is dealt with by S_{π}, which

---

[1] Such protocols are also called "subroutine respecting".

generates messages for the corresponding sub-parties and handles incoming messages from these parties. The effect of the environment on rest of the system, is handled by direct interaction with the external parties running ρ.

Leakage queries are handled by way of a subroutine $\tilde{S}_\rho$ that generates a state translation T_ρ, as needed for leakage-oblivious simulation. Recall that the leakage function L that S_ρ receives from the environment was designed to be applied to a "real protocol state" in ρ^π (and since ρ runs a single copy of π then this state is of the form $(\text{state}_\rho, \text{state}_\pi)$). The simulator S_ρ, on the other hand, can only ask the aggregator for leakage on the state of $\rho^{\mathcal{F}}$, which is of the form $(\text{state}_\rho, \text{state}_{\mathcal{F}})$. To bridge this gap, $\tilde{S}_\rho$ runs the "state translation subroutine" $\tilde{S}_\pi$. (This can be done since S_ρ has the entire current state of S_π.) Once $\tilde{S}_\pi$ produces a state translation function T_π, $\tilde{S}_\rho$ generates its own state translation function $T_\rho(\text{state}_\rho, \text{state}_{\mathcal{F}}) = (\text{state}_\rho, T_\pi(\text{state}_{\mathcal{F}}))$ and sends to the aggregator a leakage function L', where

$$L'(\text{state}_\rho, \text{state}_{\mathcal{F}}) = L(T_\rho(\text{state}_\rho, \text{state}_{\mathcal{F}})) = L(\text{state}_\rho, T_\pi(\text{state}_{\mathcal{F}})) .$$

Observe that already at this stage we rely crucially on S_π being leakage-oblivious: if S_π was expecting to see a leakage function $L_\pi(\text{state}_\pi)$ before producing the translation, then we could not use it (since S_ρ does not know the state state_ρ, and therefore cannot write the description of the induced function $L_{\text{state}_\rho}(\text{state}_\pi) = L(\text{state}_\rho, \text{state}_\pi)$). Once the aggregator returns an answer, S_ρ passes it to the environment and returns to its previous state (including the previous state of the sub-simulator S_π).

It is clear from the description that S_ρ is leakage-oblivious. The validity of S_ρ is shown by reduction to the validity of S_π. That is, given an environment $\mathcal{Z}_\rho$ that distinguishes an execution of $(\rho^\pi, \mathcal{D})$ from an execution of $(\rho^{\mathcal{F}/\pi}, S_\rho)$, we construct an environment $\mathcal{Z}_\pi$ that distinguishes an execution of $(\pi, \mathcal{D})$ from an execution of $(\mathcal{F}, S_\pi)$. The environment $\mathcal{Z}_\pi$ simulates an execution of $(\mathcal{Z}_\rho, \rho)$ "in its head", except that all messages corresponding to π are forwarded to the external execution. Indeed, leakage queries aside, we have: (a) if the external execution consists of S_π and $\mathcal{F}$, then the entire (composed) execution amounts to running $\mathcal{Z}_\rho$ with S_ρ and $\rho^{\mathcal{F}}$; (b) if the external execution consists of $\mathcal{D}$ and π, then the entire (composed) execution amounts to running $\mathcal{Z}_\rho$ with $\mathcal{D}$ and ρ^π.

Extending this argument to include leakage, the environment $\mathcal{Z}_\pi$ acts as follows. When $\mathcal{Z}_\rho$ produces a leakage query L to be evaluated on $\text{state}_\rho, \text{state}_\pi$, $\mathcal{Z}_\pi$ computes the simulated state state_ρ and computes the restricted leakage function $L_{\text{state}_\rho}(\text{state}_\pi) = L(\text{state}_\rho, \text{state}_\pi)$, which should be evaluated only on state_π. Note that since S_π is leakage-oblivious, the state-translation function that it outputs when run as a subroutine of S_ρ is the same as the state-translation function that it outputs when run with the environment $\mathcal{Z}_\pi$. The rest of the argument remains unchanged.

The actual proof also deals with the case where multiple instances of the subroutine π are invoked and can be found in the full version of this work [BCH11].

4   From Adaptive Security to Leakage-Tolerance

Recall that the adversary in the UC framework can adaptively corrupt parties during protocol execution, thereby learning their entire internal state. If the corruption is passive (semi-honest), the party keeps following the same program as it did before the corruption, and if it is Byzantine (malicious), then the adversary also gains control of the program that the party runs from now on.

As already pointed out, leakage can be thought of as a form of corruption, where the adversary gains partial information on the inner state of a party. The converse is also true, passive corruption can be viewed simply as leaking the entire internal state. The challenges in simulation are also similar: for both corruption and leakage the simulator must translate some "ideal state" that it gets from the functionality into a "real state" that it can show the environment, and do it in a way that is consistent with the transcript so far. Below we formalize this similarity, showing that "in principle" a protocol that realizes some functionality $\mathcal{F}$ in the presence of passive adaptive corruptions also realizes it in the presence of leakage. There are considerable restrictions, however. Most importantly, the implication holds only for corruption-oblivious simulators (see below). Also, $\mathcal{F}$ must be adapted to handle leakage queries, and we prove the implication for a particular (natural) way of doing this adaptation.

Adapting Functionalities to Leakage. Let $\mathcal{F}$ be functionality that was designed for a leakage-free model with corruptions. This means that $\mathcal{F}$ already has some mechanism to reply to messages from the adversary about corruptions of players. We now need to adapt it by explaining how it reacts to leakage queries from the aggregator $\mathcal{G}$. The adaptation is natural: whenever $\mathcal{G}$ asks for the state of party pid for the purpose of leakage, the functionality replies with exactly the same thing that it would have given the adversary if pid was passively corrupted at this time. Then, once $\mathcal{G}$ reports the number of leakage bits, the functionality forwards this number on the I/O lines of party pid.[2] Thereafter, the functionality behaves just as if party pid was passively corrupted. We denote the resulting functionality by $\mathcal{F}^{+lk}$. We stress that if $\mathcal{F}$ was designed to react differently to passive and Byzantine corruptions, then it uses the passive corruption mode to handle leakage.

Note the implication of viewing leakage as corruption: in principle, reaction to leakage could be gradual - a functionality $\mathcal{F}$ can change its behavior proportionally to the amount of leakage, or to have a leakage threshold up to which it does one thing and after which it does another. However, the reaction of $\mathcal{F}$ to (passive) corruption is typically "all or nothing", it is either not affected or it completely "gives up". Using our convention from above, this "all or nothing" reaction is carried over to $\mathcal{F}^{+lk}$. For example, if $\mathcal{F}$ is an authenticated channels functionality, then $\mathcal{F}^{+lk}$ will permit forgery as soon as even a single bit is leaked. On the other hand, if $\mathcal{F}$ is a commitment functionality then leakage events have no effect on the subsequent behavior of $\mathcal{F}^{+lk}$. Although the transformation that

---

[2] This number-reporting action is meant to allow the environment to do its leakage bookkeeping, and for ideal functionalities to be able to react to leakage.

we prove below works for every functionality $\mathcal{F}$, its usefulness depends crucially on the way $\mathcal{F}$ handles passive corruptions.

Corruption-Oblivious Simulators. The intuition for why adaptive corruption implies leakage-tolerance is that if we can simulate the **entire** state of an adaptively corrupted party, then we should also be able to simulate only parts of its state (according to a particular leakage function). The problem with this intuition, however, is that future behavior of the simulator may depend on the entire state learned during corruption, which is not available to the leakage simulator.

We thus restrict our attention to special simulators that are oblivious of the state learned during corruption (similarly to the leakage-oblivious simulators from Section 2). As for leakage, we only define corruption-oblivious simulators for the **dummy adversary** $\mathcal{D}$ (which is sufficient). The simulator $\mathcal{S}$ for $\mathcal{D}$ should have a special subroutine $\tilde{S}$ for handling passive corruptions. When $\mathcal{S}$ receives from the environment a request (**passive corrupt**, **pid**) to passively corrupt a party **pid**, $\mathcal{S}$ invokes $\tilde{S}$ to produce a state translation function T. T is used to transform the "internal state" that $\mathcal{F}$ (or the hybrid-world protocol) returns for party **pid** into a state of the "real world" implementation protocol π for this party. Then, $\mathcal{S}$ sends a **passive corrupt** message to **pid**, obtains the corresponding **state** (from $\mathcal{F}$ or the hybrid-world instance), applies to it the transformation T and returns the result $\mathsf{state}_\pi = T(\mathsf{state})$ to the environment. After the result is forwarded to the environment, $\mathcal{S}$ returns to its state prior to the time it invoked $\tilde{S}$.

Note that since this is passive corruption, then party **pid** can keep evolving its state after the initial corruption, and the environment can ask to see the updated state from time to time. $\mathcal{S}$ handles each such update request as a new passive corruption query, invoking $\tilde{S}$ again to get state-translation function, calling the functionality again, etc. (We note that there is no restriction on the way that $\mathcal{S}$ handles Byzantine corruptions.)

We stress that $\mathcal{S}$ does not make any direct use of the state of the corrupted parties. In particular, the future operation of $\mathcal{S}$, when simulating the messages generated by corrupted parties, is done independently of their secret local states. As seen in subsequent sections, in some cases this turns out to be a strong restriction (see Example 5.1 in Section 5). We are now ready to state the main result of this section. The proof is provided in the full version [BCH11].

Theorem 4.1. *Let π be a protocol that UC-realizes an ideal functionality $\mathcal{F}$ in the presence of passive adaptive corruptions (but no leakage), with a corruption-oblivious simulator. Then π also UC-realizes $\mathcal{F}^{+\mathsf{lk}}$ with a leakage-oblivious simulator in the UC model with leakage.*

Composition of Corruption-Oblivious Simulators. We note that, viewing corruption-oblivious simulators as a special case of leakage-oblivious simulators (for leaking the identity function), the proof of the leaky UC Theorem 3.1 implies that corruption-oblivious simulation is preserved under universal composition:

Corollary 4.1 (of Theorem 3.1). *Let $\rho, \pi, \mathcal{F}$ be protocols that are modular up to leakage, such that π UC-emulates $\mathcal{F}$ with a corruption-oblivious simulator. Then ρ^π UC-emulates $\rho^{\mathcal{F}/\pi}$ with a corruption-oblivious simulator.*

5 Realizing Leaky Adaptations of Basic Interactive Tasks

This section describes the construction of leakage-tolerant protocols for for several interactive tasks. We describe constructions for secure message transmission, (semi-honest) oblivious transfer, commitment and zero-knowledge. These constructions all assume ideal authenticated channels. We then present a construction of leakage-resilient authenticated channels. All of our constructions are composable. We conclude the section with a discussion on the difficulties in obtaining general leakage-tolerant multi party computation.

The bulk of this section is omitted. It can be found in the full version of this work [BCH11]. Here, we only sketch the constructions for the last two tasks.

5.1 Zero-Knowledge from Ideal Leaky Commitments

We adapt the zero-knowledge ideal functionality to tolerate leakage and demonstrate a protocol that realizes the adapted functionality in the presence of leakage. Recall that $\mathcal{F}_{\mathsf{ZK}:R}$, for a relation R, takes from the prover P an input (x, w), and outputs x to the verifier V only if $R(x, w)$ holds. This formulation guarantees to P perfect secrecy of w. It also guarantees perfect soundness to V.

Adapting $\mathcal{F}_{\mathsf{ZK}}$ to leakage, we can ideally hope to realize a functionality with optimal tolerance, such as $\mathcal{F}_{\mathsf{ZK}}^{+\mathsf{lk}}$, which can "gracefully" tolerate arbitrary leakage from the prover, and in addition does not give up on soundness even in face of arbitrary leakage on the verifier. However, we could not manage to realize such a functionality. Instead, we consider an adaptation that can tolerate arbitrary leakage from the prover, but only a bounded amount of leakage from the verifier before soundness breaks. Before presenting our eventual adaptation and implementation, we briefly sketch the difficulties which prevent us from achieving optimal leakage-tolerance.

As shown in [CF01, CLOS02], the parallel repetition of classic 3-round zero-knowledge protocols, such as Blum's Hamiltonian cycle [Blu86], and GMW's 3-coloring [GMW91], UC-realizes the basic (non-leaky) $\mathcal{F}_{\mathsf{ZK}}$, given access to (non-leaky) ideal commitment. Moreover, they do so even in the presence of adaptive corruptions. However, the proofs of security of these protocols do *not* yield corruption-oblivious simulation. Thus, we cannot conclude that these protocols UC-realize $\mathcal{F}_{\mathsf{ZK}}^{+\mathsf{lk}}$ under leakage.

In fact, without any modifications, these protocols seem inherently impossible to simulate in the face of leakage. To demonstrate this, let us recall GMW's 3-coloring protocol. Here, the prover, who possesses a 3-coloring c, chooses a random permutation σ of the three colors and commits to the permuted coloring $\sigma(c)$. The verifier then requires that the prover opens the colors of a random edge and checks that its endpoints are indeed colored differently. Now, consider a (Byzantinely) corrupted verifier V^* that also obtains leakage on the prover's coloring during the protocol. This verifier can leak, for example, the secret permutation σ and then ask the (honest) prover to open the colors $\sigma(c(i)), \sigma(c(j))$ of some random edge (i, j). Finally, it can leak again the true colors $c(i), c(j)$. Simulating such a behavior seems impossible (assuming 3COL $\notin$ BPP). Indeed, once the simulator simulates σ for the first leakage, it essentially becomes committed to it for the rest of the protocol. Then, when it is required to simulate the opening of $\sigma(c(i)), \sigma(c(j))$, it essentially has no information on c, and hence, if it can consistently simulate the second leakage query, then essentially it must "know" a proper coloring of the entire graph.[3] We stress that this inherent difficulty also fails simulators that are not leakage-oblivious (and are thus allowed to depend on both the leakage function and the leakage-result).

To overcome the above problem, we require that at the beginning of the protocol, the verifier commits to all its challenges. This already allows the simulation to go through; now the simulator can first extract the challenge edge (i, j), choose random colors for it $c'(i), c'(j)$, and then have the leakage return a permutation mapping the real $c(i), c(j)$ to $c'(i), c'(j)$. In fact, we show that this adjustment is enough for simulating any malicious verifier.

This adjustment comes, however, at a price: unlike the original protocols, where the verifier was of the public coins type (and had no secret state), now the verifier commits to its challenges, and the secrecy of these challenges is crucial for the protocol's soundness. Hence, we cannot hope that in such a protocol the verifier will be able to withstand arbitrary amounts of leakage; in particular, once the prover leaks all of the verifier's challenge, soundness is doomed.

Consequently, we only realize a weaker adaptation, where the verifier can only tolerate a bounded amount of leakage. (The prover can still tolerate arbitrary leakage.) More specifically, we can tolerate arbitrary leakage on the verifier's randomness so long that a super-logarithmic amount of min-entropy is maintained.

5.2   Authenticated Channels

We construct a protocol for realizing *leaky authenticated channels* with bounded leakage-resilience. More specifically, the protocol UC-realizes an ideal functionality $\mathcal{F}_{\text{Auth}}^{+B}$ that guarantees authenticated communication *as long as the overall leakage between any two transmissions of some messages does not exceed a pre-specified bound B.*

The protocol we present uses two main building blocks: (a) non-committing encryption (NCE) (b) information theoretic c-time message authentication codes (MACs) that are resilient to a constant leakage rate from the secret key. The idea behind the protocol is simple. The parties initially share a (leaky) secret key K_1. Then the protocol proceeds inductively; at each round, a current authentication key K_i is used to authenticate the i-th message, m_i. In addition, a fresh key K_{i+1} is generated and transmitted using non-committing encryption. These transmitted ciphers are also authenticated using K_i. To allow the authentication to go through, we need our underlying leaky MAC scheme to allow authentication of messages that are polynomially longer than the secret key. This is achieved using *universal hashing*. Concretely, the protocol we present tolerates, between each two transmissions, roughly $k/10$ bits of leakage on the

---

[3] This intuition can be made formal; namely, given such a simulator we can construct an algorithm for 3-coloring arbitrary graphs.

k-long secret key. Similar techniques are used for a related goal in [BCG+11]. However, the security analysis there is different than the one here.

The protocol we construct admits leakage-oblivious simulation and is thus composable. We can, therefore, use it as a basic building block supporting any protocol that requires authenticated channels, when ideally authenticated channels are unavailable. We stress, however, that when doing so the leakage-tolerance of the higher-level protocol, naturally degrades to that of the authentication protocol.

5.3   On the Difficulty in Achieving General Leakage-Tolerant MPC

Equipped with Theorem 4.1, we may hope that, similarly to the tasks considered above, general leakage-tolerant multi-party computation (MPC) would also follow from known results on adaptively secure MPC (such as, [CLOS02]). Unfortunately, known results do not admit corruption-oblivious simulation and are in fact far from being leakage-tolerant. We exemplify the relevant difficulties by giving a protocol that is adaptively secure but not leakage-tolerant. Although seemingly contrived, the protocol suffers from the same caveats that fail known adaptively secure protocols from achieving leakage-tolerance.

Example 5.1. Let $\mathcal{F}$ be a standard corruption functionality that takes n-bit inputs from two parties, P_0 and P_1, and outputs nothing. As soon as party P_i provides input x_i, the virtual local state of P_i is set to x_i. Now, consider the following protocol π: first, the parties give their inputs to some trusted party that returns a random b_i to P_i such that $b_0 + b_1 = \langle x_0, x_1 \rangle$ where $\langle , \rangle$ denotes inner-product in $\mathbb{F}_2$. (The inner product can be replaced by any two-source extractor.) Next, the parties output nothing and halt.

It can be seen that π securely realizes $\mathcal{F}$ with respect to adaptive corruptions. This is so since, once the first party P_i is corrupted, the simulator learns x_i and can give x_i to the adversary, plus a random bit instead of b_i. Now, when P_{1-i} is corrupted, the simulator learns x_{1-i} and can determine the bit b_{1-i} so that $b_0 + b_1 = \langle x_0, x_1 \rangle$. However, notice that here the simulator is not corruption-oblivious: the handling of the second corruption depends on the input value x_i of the first corrupted party. Indeed, π does not realize $\mathcal{F}^{+lk}$ with even one bit of leakage from each party: the adversary can ask to leak b_i from P_i and thus learn $\langle x_0, x_1 \rangle$. However, in the ideal model for $\mathcal{F}^{+lk}$, assuming x_0, x_1 are long random strings, the simulator has no hope of learning $\langle x_0, x_1 \rangle$. This is so since in the ideal model, the simulator can only perform one-bit leakage on x_0 and x_1 separately, and hence it can not guess $\langle x_0, x_1 \rangle$ with non-negligible advantage.

Indeed, the same problem would arise in GMW-based protocols, where the value of each wire is secret-shared between the parties in a non leakage-resilient manner as above. This is actually also the case for YAO-based adaptively secure protocols (for NC_1 functions); there also (although not explicitly), the value of each wire is effectively secret-shared between the parties in a non leakage-resilient way.

Weak   (Joint-State)   Leakage-Tolerance   Vs.   Strong   (Separate-State) Leakage-Tolerance. Note that, had we modified $\mathcal{F}^{+lk}$ in the above example so that the virtual local state of each party includes both inputs, the above protocol would UC-realize $\mathcal{F}^{+lk}$ with leakage. More generally, if we settle for a weaker leakage-tolerance guarantee where the ideal world simulator can jointly leak from the inputs and outputs of all parties (and not only separately from the inputs and outputs of each leaking party alone), then leakage-tolerance can already be achieved. In fact, combining our leakage-tolerant OT protocol with an adaptively secure protocol, such as GMW, it is easy to obtain semi-honest MPC for general functions. (We note that this, in particular, concerns the two party setting where one party is statically corrupted considered by [DHP11], which can be seen as a special case of weak leakage-tolerance.)

However, in a setting where real world adversaries are restricted to separate leakage from each party, an ideal process that allows joint leakage from the internal states of the parties is somewhat unsatisfactory. Achieving strong (separate-state) leakage-tolerant MPC in general (without preprocessing or limitations on the number of honest parties) remains an interesting open question.

Acknowledgments. We thank Amit Sahai for telling us about the problems with proving leakage-tolerance of the standard three round zero-knowledge protocols and about the way this problem is solved in [GJS11].

References

[ADN⁺10]   Alwen, J., Dodis, Y., Naor, M., Segev, G., Walfish, S., Wichs, D.: Public-Key Encryption in the Bounded-Retrieval Model. In: Gilbert, H. (ed.) EUROCRYPT 2010. LNCS, vol. 6110, pp. 113–134. Springer, Heidelberg (2010)

[ADW09]   Alwen, J., Dodis, Y., Wichs, D.: Survey: Leakage Resilience and the Bounded Retrieval Model. In: Kurosawa, K. (ed.) ICITS 2009. LNCS, vol. 5973, pp. 1–18. Springer, Heidelberg (2010)

[AGV09]   Akavia, A., Goldwasser, S., Vaikuntanathan, V.: Simultaneous Hardcore Bits and Cryptography against Memory Attacks. In: Reingold, O. (ed.) TCC 2009. LNCS, vol. 5444, pp. 474–495. Springer, Heidelberg (2009)

[BCG⁺11]   Bitansky, N., Canetti, R., Goldwasser, S., Halevi, S., Kalai, Y.T., Rothblum, G.N.: Program Obfuscation with Leaky Hardware. In: Lee, D.H. (ed.) ASIACRYPT 2011. LNCS, vol. 7073, pp. 722–739. Springer, Heidelberg (2011)

[BCH11]   Bitansky, N., Canetti, R., Halevi, S.: Leakage-tolerant interactive protocols (2011), Full version, http://eprint.iacr.org/2011/204

[BGK11]   Boyle, E., Goldwasser, S., Kalai, Y.T.: Leakage-Resilient Coin Tossing. In: Peleg, D. (ed.) DISC 2011. LNCS, vol. 6950, pp. 181–196. Springer, Heidelberg (2011), http://eprint.iacr.org/2011/291

[BKKV10]   Brakerski, Z., Kalai, Y.T., Katz, J., Vaikuntanathan, V.: Overcoming the hole in the bucket: Public-key cryptography resilient to continual memory leakage. In: FOCS, pp. 501–510 (2010)

[Blu86]   Blum, M.: How to prove a theorem so no one else can claim it. International Congress of Mathematicians, pp. 444–451 (1986)

[BSW11]   Boyle, E., Segev, G., Wichs, D.: Fully Leakage-Resilient Signatures. In:
            Paterson, K.G. (ed.) EUROCRYPT 2011. LNCS, vol. 6632, pp. 89–108.
            Springer, Heidelberg (2011)
[Can01]   Canetti, R.: Universally composable security: A new paradigm for cryp-
            tographic protocols. In: FOCS, pp. 136–145 (2001)
[CF01]   Canetti, R., Fischlin, M.: Universally Composable Commitments. In:
            Kilian, J. (ed.) CRYPTO 2001. LNCS, vol. 2139, pp. 19–40. Springer,
            Heidelberg (2001)
[CFGN96]   Canetti, R., Feige, U., Goldreich, O., Naor, M.: Adaptively secure
            multi-party computation. In: STOC, pp. 639–648 (1996)
[CLOS02]   Canetti, R., Lindell, Y., Ostrovsky, R., Sahai, A.: Universally com-
            posable two-party and multi-party secure computation. In: STOC, pp.
            494–503 (2002)
[DHLAW10a] Dodis, Y., Haralambiev, K., López-Alt, A., Wichs, D.: Cryptography
            against continuous memory attacks. In: FOCS, pp. 511–520 (2010)
[DHLAW10b] Dodis, Y., Haralambiev, K., López-Alt, A., Wichs, D.: Efficient Public-
            Key Cryptography in the Presence of Key Leakage. In: Abe, M. (ed.)
            ASIACRYPT 2010. LNCS, vol. 6477, pp. 613–631. Springer, Heidelberg
            (2010)
[DHP11]   Damgård, I., Hazay, C., Patra, A.: Leakage resilient secure two-
            party computation. IACR Cryptology ePrint Archive, 256 (2011),
            http://eprint.iacr.org/2011/256
[DKL09]   Dodis, Y., Kalai, Y.T., Lovett, S.: On cryptography with auxiliary
            input. In: STOC, pp. 621–630 (2009)
[DP08]   Dziembowski, S., Pietrzak, K.: Leakage-resilient cryptography. In:
            FOCS, pp. 293–302. IEEE Computer Society (2008)
[GJS11]   Garg, S., Jain, A., Sahai, A.: Leakage-Resilient Zero Knowledge. In: Ro-
            gaway, P. (ed.) CRYPTO 2011. LNCS, vol. 6841, pp. 297–315. Springer,
            Heidelberg (2011)
[GMW91]   Goldreich, O., Micali, S., Wigderson, A.: Proofs that yield nothing but
            their validity for all languages in np have zero-knowledge proof systems.
            J. ACM 38(3), 691–729 (1991)
[GOS06]   Groth, J., Ostrovsky, R., Sahai, A.: Non-interactive Zaps and New
            Techniques for NIZK. In: Dwork, C. (ed.) CRYPTO 2006. LNCS,
            vol. 4117, pp. 97–111. Springer, Heidelberg (2006)
[HL11]   Halevi, S., Lin, H.: After-the-Fact Leakage in Public-Key Encryption.
            In: Ishai, Y. (ed.) TCC 2011. LNCS, vol. 6597, pp. 107–124. Springer,
            Heidelberg (2011)
[MR04]   Micali, S., Reyzin, L.: Physically Observable Cryptography. In: Naor,
            M. (ed.) TCC 2004. LNCS, vol. 2951, pp. 278–296. Springer, Heidelberg
            (2004)
[NS09]   Naor, M., Segev, G.: Public-Key Cryptosystems Resilient to Key Leak-
            age. In: Halevi, S. (ed.) CRYPTO 2009. LNCS, vol. 5677, pp. 18–35.
            Springer, Heidelberg (2009)
[Pie09]   Pietrzak, K.: A Leakage-Resilient Mode of Operation. In: Joux, A.
            (ed.) EUROCRYPT 2009. LNCS, vol. 5479, pp. 462–482. Springer,
            Heidelberg (2009)
[Sta09]   Standaert, F.-X.: Introduction to side-channel attacks. In: Ver-
            bauwhede, I.M.R. (ed.) Secure Integrated Circuits and Systems, pp.
            27–44. Springer, Heidelberg (2009)

On the Public Indifferentiability and Correlation Intractability of the 6-Round Feistel Construction

Avradip Mandal[1], Jacques Patarin[2], and Yannick Seurin[3]

[1] University of Luxembourg
avradip.mandal@uni.lu
[2] University of Versailles, France
jacques.patarin@uvsq.fr
[3] ANSSI, Paris, France
yannick.seurin@m4x.org

Abstract. We show that the Feistel construction with six rounds and random round functions is *publicly* indifferentiable from a random invertible permutation (a result that is not known to hold for full indifferentiability). Public indifferentiability (*pub-indifferentiability* for short) is a variant of indifferentiability introduced by Yoneyama *et al.* [29] and Dodis *et al.* [12] where the simulator knows all queries made by the distinguisher to the primitive it tries to simulate, and is useful to argue the security of cryptosystems where all the queries to the ideal primitive are public (as *e.g.* in many digital signature schemes). To prove the result, we introduce a new and simpler variant of indifferentiability, that we call sequential indifferentiability (*seq-indifferentiability* for short) and show that this notion is in fact equivalent to pub-indifferentiability for stateless ideal primitives. We then prove that the 6-round Feistel construction is seq-indifferentiable from a random invertible permutation. We also observe that sequential indifferentiability implies correlation intractability, so that the Feistel construction with six rounds and random round functions yields a correlation intractable invertible permutation, a notion we define analogously to correlation intractable functions introduced by Canetti *et al.* [4].

Keywords: indifferentiability, correlation intractability, Feistel construction.

1 Introduction

Indifferentiability. Indifferentiability has been introduced by Maurer *et al.* [22] as a generalization of the concept of indistinguishability for systems using *public* components (*i.e.* components that can be queried by any party including the adversary). This framework has since then gained much popularity, and starting with [7] it has been widely used to analyze hash functions built from a smaller ideal primitive, *e.g.* a fixed input-length (FIL) random compression function

R. Cramer (Ed.): TCC 2012, LNCS 7194, pp. 285–302, 2012.

or an ideal block cipher. Informally, a construction $\mathcal{C}$ using an ideal primitive F (*e.g.* a hash function based on a FIL random compression function) is said to be indifferentiable from another ideal primitive G (*e.g.* a random oracle) if there exists a simulator $\mathcal{S}$ accessing G such that the two systems $(G, \mathcal{S}^G)$ and $(\mathcal{C}^F, F)$ are indistinguishable. Roughly, the goal of the simulator is twofold: it must provide answers that are consistent with G, without deviating too much from the distribution of answers of F. Indifferentiability allows modular proofs of security in idealized models in the sense that if a construction $\mathcal{C}^F$ is indifferentiable from an ideal primitive G, then any cryptosystem proven secure when used with G remains secure when used with the construction $\mathcal{C}^F$.[1] For example, if a cryptosystem is secure in the random oracle model, and some hash function construction H^f based on a FIL random compression function f is indifferentiable from a random oracle, then the cryptosystem is still secure when used with H^f. More interestingly from a theoretical point of view, Coron *et al.* [7] showed that a number of variants of the Merkle-Damgård construction, used with an ideal cipher in Davies-Meyer mode, are indifferentiable from a random oracle. This implies that any functionality that can be securely implemented in the random oracle model can also be securely realized in the ideal cipher model.

The Feistel Construction with Public Round Functions. The Feistel construction turns a function F from n-bit strings to n-bit strings into an (efficiently invertible) permutation on $2n$-bit strings. It is computed as $\Psi^F(L, R) = (R, L \oplus F(R))$. In their seminal paper [18] which triggered a lot of subsequent work [20,23,24,28], Luby and Rackoff showed that three (resp. four) rounds of the Feistel construction, with independent pseudorandom functions in each round, yields a pseudorandom permutation (resp. strong pseudorandom permutation). The core of this result is in fact purely information-theoretic [20], meaning that the Feistel construction with three (resp. four) rounds and random round functions is indistinguishable from a random permutation (resp. an invertible random permutation) by any *computationally unbounded* distinguisher limited to a *polynomial number of oracle queries*. The Luby-Rackoff theorem crucially relies on the secrecy of the round functions. A few papers studied what happens when the round functions are made public. In particular, Ramzan and Reyzin [25] have shown that the Feistel construction with four rounds remains strongly pseudorandom even when the distinguisher has oracle access to the two middle round functions (but not to the first or the fourth round function). Dodis and Puniya [11] have studied various properties of the Feistel construction (unpredictability, pseudorandomness) when all intermediate round values of the Feistel computation are leaked to the adversary and shown that in that case a super-logarithmic number of rounds was necessary and sufficient for the property to be inherited by the Feistel construction from the round functions.

Indifferentiability of the Feistel Construction. As already mentioned, it is possible to securely instantiate a random oracle in the ideal cipher model.

---

[1] It was recently pointed out that this composition theorem only holds for cryptosystems whose security is defined by so called *single-stage games* [26].

A natural question is whether the other direction holds, namely whether there is a construction using a random oracle that securely implements a random invertible permutation.[2] Given its numerous cryptographic properties, the Feistel construction (with public random round functions) appears as an obvious candidate for this task. Again, this question can be rigorously formulated in the indifferentiability framework: namely, is the Feistel construction with sufficiently many rounds, and public random round functions, indifferentiable from a random invertible permutation? Dodis and Puniya [10] considered the problem in the so-called *honest-but-curious* model, where the distinguisher only sees the queries made by the Feistel construction to the random round functions, but is not allowed to make arbitrary queries to the round functions. In this setting, they showed that a super-logarithmic number of rounds is sufficient to securely realize a random invertible permutation. However, since full indifferentiability is not implied in general by indifferentiability in the honest-but-curious model (these two notions are in fact incomparable [9]), they were not able to conclude in the general setting. Coron, Patarin, and Seurin [9] gave a first proof that the Feistel construction with six rounds is indifferentiable from a random invertible permutation. The proof was rather involved, and Künzler [17] later found a distinguishing attack against the simulator given in [9], therefore invalidating the indifferentiability proof.[3] Only recently, Holenstein *et al.* [14] gave a new proof that the Feistel construction with *fourteen* rounds is indifferentiable from a random invertible permutation, which was inspired from a previous proof for ten rounds that appeared in the PhD thesis of Seurin [27] but had some gaps.

Public Indifferentiability. Yoneyama *et al.* [29] and Dodis *et al.* [12] independently realized that indifferentiability was sometimes stronger than needed to argue security of cryptosystems. In particular, when all queries made to the ideal primitive are public (like in many digital signature schemes such as FDH [2], probabilistic FDH [6], PSS [3]..., where all queries to the hash function can be revealed to the attacker without affecting the security), the weaker notion of *public* indifferentiability is sufficient. [29,12] were both concerned with indifferentiability from a random oracle and respectively called this notion *leaky random oracle* and *public-use random oracle*. Public indifferentiability is defined similarly to indifferentiability, but the task of the simulator is made easier by letting it know all queries made by the distinguisher to the ideal primitive G.

Correlation Intractability. Correlation intractability was introduced by Canetti *et al.* [4] as an attempt to capture as many security properties of the random oracle as possible. A family of functions is said to be correlation intractable if for a random function of the family it is hard to find a sequence of inputs that together with their image satisfy a relation that would be hard to satisfy for a

---

[2] Such a construction easily implies a secure ideal cipher by simply prepending the key of the block cipher to the input of each random oracle queries.

[3] We stress that this does not mean that the 6-round Feistel construction is not indifferentiable from a random invertible permutation, but only that no one is able to give a proof at the moment.

uniformly random function (a so-called *evasive* relation). Correlation intractability in particular implies collision resistance, pre-image resistance and many other security properties usually required for cryptographic hash functions. Unfortunately, Canetti *et al.* also showed that in the standard model, no correlation intractable hash function family exists. A consequence of this non-existence result is that there are cryptosystems that are secure in the random oracle model, but insecure when the random oracle is instantiated by any function family. Though correlation intractability was primarily defined in the standard model, it is easily transposable to idealized models. As we will see our result establishes a connection between correlation intractability and public indifferentiability.

Contributions of This Work. We define a new and weaker notion of indifferentiability that we call *sequential* indifferentiability (*seq-indifferentiability* for short). This new definition only restricts *the order* in which the distinguisher can query the two oracles it is granted access to: it can first query the primitive F (or the simulator S), and then the construction C^F (or the ideal primitive G), but not F/S again. We show that when the ideal primitive G is stateless (which is the most usual case), this notion is equivalent to *public* indifferentiability introduced by [12,29] where all queries to the primitive G are public. However the seq-indifferentiability notion has the advantage of being simpler and easier to use in proofs. This simple restriction on the queries of the distinguisher enables to give a relatively simple proof that the 6-round Feistel construction with random round functions is seq-indifferentiable (and hence also publicly indifferentiable) from a random invertible permutation, a result whose analogue for full indifferentiability seems out of reach at the moment. Our result in particular implies that any scheme proven secure in the random invertible permutation model or the ideal cipher model and where all queries to the ideal primitive can be made public without affecting the security (*e.g.* signature schemes like OPSSR [13] and subsequent variants [15,5]) remains secure in the random oracle model when using a 6-round Feistel construction (while the best generic replacement previously to our work was the 14-round Feistel construction [14]).

Though weaker than full indifferentiability, we also show that seq-indifferentiability is still sufficiently strong to imply correlation intractability. In particular, our result shows that the 6-round Feistel construction with random round functions yields a correlation intractable invertible permutation (we note that previous observations [9] already implied that the 5-round Feistel construction fails to provide a correlation intractable invertible permutation). We discuss the implications of this result for chosen-key and known-key attacks on block ciphers [16].

On a slightly different topic, we also analyze the Feistel-like domain extension construction for ideal ciphers proposed by Coron *et al.* [8] and show that in the seq-indifferentiability model one can obtain a security bound beyond the birthday barrier. See the full version of the paper [19].

Open Problems. The most challenging open question is of course whether the 6-round Feistel construction is fully indifferentiable from a random invertible permutation, and if not, what is the minimal number of rounds needed to achieve this property. We hope that our result will constitute a first step towards a finer understanding of this question. In particular, our result implies that if the 6-round Feistel construction is *not* fully indifferentiable from a random invertible permutation, then this cannot be shown by proving that it is not correlation intractable as was done for five rounds. Another interesting problem is to weaken the assumptions on the round functions and see which property would continue to hold: *e.g.* is the 6-round Feistel construction with correlation intractable round functions still a correlation intractable invertible permutation? A related question is whether our result could be a first step towards proposing plausible constructions of (restricted) correlation intractable function families in the standard model, a question left open by [4, Section 5.1].

Organization. In Section 2, we start by giving the definition of sequential indifferentiability and prove that it is equivalent to public indifferentiability for stateless ideal primitives. In Section 3, we prove the main result of this paper, namely that the 6-round Feistel construction is sequentially (and hence publicly) indifferentiable from a random invertible permutation. In Section 4, we apply this result to prove the correlation intractability of the 6-round Feistel construction.

2 Preliminaries

2.1 Notations and Definitions

Notations. $[i..j]$ will denote the set of integers k such that $i \leq k \leq j$. We will use n to denote the security parameter, and in sections dealing with the Feistel construction we will identify n with the input and output length of the round functions. We will write $f \in \texttt{poly}(n)$ to denote a polynomially bounded function and $f \in \texttt{negl}(n)$ to denote a negligible function. When $\mathcal{X}$ is a non-empty finite set, we write $x \leftarrow_{\mathcal{R}} \mathcal{X}$ to mean that a value is sampled uniformly at random from $\mathcal{X}$ and assigned to x. PPT will stand for probabilistic polynomial-time, and ITM for interactive Turing machine.

Ideal Primitives. Given two sets $\texttt{Dom} \subset \{0,1\}^*$ and $\texttt{Rng} \subset \{0,1\}^*$, we denote $\mathcal{F}(\texttt{Dom}, \texttt{Rng})$ the set of all functions from $\texttt{Dom}$ to $\texttt{Rng}$. A primitive $\mathbb{G}$ is a sequence $\mathbb{G} = (\texttt{Dom}_n, \texttt{Rng}_n, \mathbb{G}_n)_{n \in \mathbb{N}}$ where $\mathbb{G}_n \subset \mathcal{F}(\texttt{Dom}_n, \texttt{Rng}_n)$. The ideal primitive $\boldsymbol{G}$ associated with $\mathbb{G}$ is the sequence of random variables $(\boldsymbol{G}_n)_{n \in \mathbb{N}}$ where $\boldsymbol{G}_n$ is uniformly distributed over $\mathbb{G}_n$. We will often adopt the lazy sampling view [1] to describe ideal primitives queried as oracles.

A random function $\boldsymbol{F} = (\boldsymbol{F}_n)_{n \in \mathbb{N}}$ is the ideal primitive associated to the set of all functions from $\{0,1\}^n$ to $\{0,1\}^n$. Queried as an oracle it returns a uniformly random string in $\{0,1\}^n$ if x was never queried, or the same answer as before if x was previously queried.

A random invertible permutation $\boldsymbol{P} = (\boldsymbol{P}_n)_{n \in \mathbb{N}}$ is the ideal primitive associated with the sequence $\mathbb{P} = (\texttt{Dom}_n, \texttt{Rng}_n, \mathbb{P}_n)_{n \in \mathbb{N}}$ where $\texttt{Dom}_n = \{0,1\} \times \{0,1\}^n$, $\texttt{Rng}_n = \{0,1\}^n$, and $\mathbb{P}_n$ is the set of functions P such that $x \mapsto P(0,x)$ is a permutation of $\{0,1\}^n$, and $y \mapsto P(1,y)$ its inverse. Queries of the form $(0,x)$

and $(1, y)$ will be called respectively *forward* and *backward* queries. In the lazy sampling point of view, P_n keeps two lists L_x and L_y of forward and backward queries whose image is already defined together with an invertible mapping from L_x to L_y. Upon receiving a forward query $(0, x)$ such that $x \notin L_x$ it returns an answer y uniformly random over $\{0, 1\}^n \setminus L_y$, and adds x to L_x and y to L_y and updates the mapping (and reciprocally for a backward query $(1, y)$). Later, we will occasionally refer to L_x and L_y as the *history* of the random invertible permutation. An ideal cipher $E = (E_n)$ takes an additional input, the key, of length $\ell(n)$, and for each key $k \in \{0, 1\}^{\ell(n)}$, $E_n(k, \cdot)$ is an independent random invertible permutation over $\{0, 1\}^n$.

A two-sided random function on $\{0, 1\}^n$, denoted R_n, is very similar to a random invertible permutation. It also keeps to lists L_x and L_y together with an invertible mapping from L_x to L_y. However when receiving a forward query $(0, x)$ such that $x \notin L_x$ or a backward query $(1, y)$ such that $y \notin L_y$, it returns a *uniformly random* answer in $\{0, 1\}^n$. In case a collision happens, the previous image or pre-image is removed from L_y or L_x and the mapping is updated accordingly. Note that a two-sided random function is stateful: it may return different answers to the same query (however at any time it defines an invertible mapping from L_x to L_y). A two-sided random function is statistically indistinguishable from a random invertible permutation: the so called PRF/PRP switching lemma [1] establishes[4] that an oracle machine making at most q oracle queries can distinguish P_n from R_n with advantage at most $q^2/2^{n+1}$.

In the following, we omit the subscripts when the domain and the range of an ideal primitive are clear from the context. A *construction* will simply be a Turing machine having oracle access to an ideal primitive and implementing another given primitive. The main construction we will consider in this work is the Feistel construction.

The Feistel Construction. Given a function $F : \{0, 1\}^n \to \{0, 1\}^n$, the basic (1-round) Feistel construction is the permutation on $\{0, 1\}^{2n}$ defined by $\Psi^F(L, R) = (R, L \oplus F(R))$. Its inverse is computed by $(\Psi^F)^{-1}(S, T) = (T \oplus F(S), S)$. (Here L, R, S, and T are n-bit strings). The k-round Feistel construction associated to round functions $(F_1, \ldots, F_k)$ takes inputs $x \in \{0, 1\} \times \{0, 1\}^{2n}$ and is defined by:

$$\Psi_k^{(F_1, \ldots, F_k)}(0, (L, R)) = \Psi^{F_k} \circ \cdots \circ \Psi^{F_1}(L, R)$$

$$\Psi_k^{(F_1, \ldots, F_k)}(1, (S, T)) = \left(\Psi^{F_1}\right)^{-1} \circ \cdots \circ \left(\Psi^{F_k}\right)^{-1}(S, T) \ .$$

Notations used for denoting the intermediate round values for the 6-round Feistel construction are given in Figure 1. In the following, when considering the Feistel construction using k independent random functions, we will simply note $F = (F_1, \ldots, F_k)$ this tuple of functions and $\Psi_k^F = \Psi_k^{(F_1, \ldots, F_k)}$.

---

[4] Strictly speaking, the result is proven in [1] for one-sided functions and permutations, but the proof can be straightforwardly adapted to two-sided primitives.

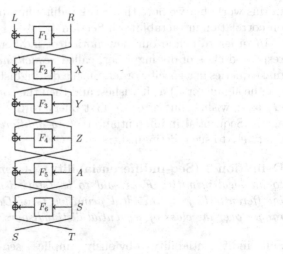

Fig. 1. Notations used for the 6-round Feistel construction

2.2 Sequential Indifferentiability

Indifferentiability was originally formulated within the formalism of *random systems* [21]. We adopt here the simpler formulation using interactive Turing machines as in [7]. We first recall the classical definition of indifferentiability [22]. For this, we slightly change the way one usually measure the cost of queries of a distinguisher (this will make our results simpler to express). Given a distinguisher $\mathcal{D}$, the *total oracle queries cost* of $\mathcal{D}$ is the number of queries received by the oracle $\boldsymbol{F}$ when $\mathcal{D}$ interacts with $(\mathcal{C}^{\boldsymbol{F}}, \boldsymbol{F})$. Hence this is the sum of the number of direct queries of $\mathcal{D}$ to $\boldsymbol{F}$ and the number of queries made by $\mathcal{C}$ to $\boldsymbol{F}$ to answer $\mathcal{D}$'s queries.

Definition 1 ((Statistical, Strong) Indifferentiability). *Let $q, \sigma : \mathbb{N} \to \mathbb{N}$ and $\epsilon : \mathbb{N} \to \mathbb{R}$ be three functions of the security parameter n. A construction $\mathcal{C}$ with oracle access to an ideal primitive $\boldsymbol{F}$ is said to be statistically and strongly (q, σ, ϵ)-indifferentiable from an ideal primitive $\boldsymbol{G}$ if there exists an oracle ITM $\mathcal{S}$ such that for any distinguisher $\mathcal{D}$ of total oracle queries cost at most q, $\mathcal{S}$ makes at most σ oracle queries, and the following holds:*

$$\left| \Pr\left[\mathcal{D}^{\boldsymbol{G}, \mathcal{S}^{\boldsymbol{G}}}(1^n) = 1 \right] - \Pr\left[\mathcal{D}^{\mathcal{C}^{\boldsymbol{F}}, \boldsymbol{F}}(1^n) = 1 \right] \right| \leq \epsilon \ .$$

$\mathcal{C}^{\boldsymbol{F}}$ *is simply said to be statistically and strongly indifferentiable from $\boldsymbol{G}$ if for any $q \in \mathtt{poly}(n)$, the above definition is fulfilled with $\sigma \in \mathtt{poly}(n)$ and $\epsilon \in \mathtt{negl}(n)$.*

Definition 1 does not refer to the running time of $\mathcal{S}$ and $\mathcal{D}$. When only polynomial-time algorithms are considered, indifferentiability is said to be *computational*. Weak indifferentiability is defined as above, but the order of quantifiers for the distinguisher and the simulator are switched (for all distinguisher, there is a simulator...). We will mainly be concerned with statistical strong indifferentiability in this work, but we note that weak indifferentiability is sufficient for our results on correlation intractability in Section 4.

In order to define our new notion of indifferentiability, we will consider a restricted class of distinguisher, called *sequential distinguisher*, which can only make queries in a specific order. Such a distinguisher first queries the primitive F (or the simulator $\mathcal{S}$) as it wishes, and then the construction $\mathcal{C}^F$ (or the primitive G) as it wishes, but after its first query to $\mathcal{C}^F$ or G, it cannot query $\mathcal{S}$ or F again. Sequential indifferentiability (*seq-indifferentiability* for short) is defined relatively to such distinguishers.

Definition 2 (Seq-indifferentiability). *A construction $\mathcal{C}$ with oracle access to an ideal primitive F is said to be (statistically and strongly) (q, σ, ϵ)-seq-indifferentiable from an ideal primitive G if Definition 1 is fulfilled when $\mathcal{D}$ ranges over the class of sequential distinguishers.*

Full indifferentiability obviously implies seq-indifferentiability. Yoneyama *et al.* [29] and Dodis *et al.* [12] have introduced another weakened notion of indifferentiability, where the primitive G is only queried on *public* inputs, that we call here public indifferentiability (*pub-indifferentiability* for short). This can be formalized as follows: given an ideal primitive G, we define the augmented ideal primitive $\overline{G}$ as the primitive exposing two interfaces: the first (regular) one is the same as G, and the second is an interface Reveal that, when queried, returns the ordered sequence of all (regular) queries and corresponding answers made so far by any party to the regular interface. The second interface can only be used by the simulator, not by the distinguisher.

Definition 3 (Pub-indifferentiability). *A construction $\mathcal{C}$ with oracle access to an ideal primitive F is said to be (statistically and strongly) (q, σ, ϵ)-pub-indifferentiable from an ideal primitive G if there exists an oracle ITM $\mathcal{S}$ such that for any distinguisher $\mathcal{D}$ of total oracle queries cost at most q, $\mathcal{S}$ makes at most σ oracle queries, and the following holds:*

$$\left| \Pr\left[\mathcal{D}^{G, \mathcal{S}^{\overline{G}}}(1^n) = 1 \right] - \Pr\left[\mathcal{D}^{\mathcal{C}^F, F}(1^n) = 1 \right] \right| \leq \epsilon .$$

As explained in [12], the composition theorem of [22] still holds with pub-indifferentiability for cryptosystems where all messages queried to G can be inferred from the adversary's query during the security experiment.

Clearly, pub-indifferentiability implies seq-indifferentiability. Indeed, since after its first query to G a sequential distinguisher never queries the simulator again, the interface Reveal is of no use to the simulator. A less trivial result is that seq-indifferentiability implies pub-indifferentiability *for stateless*[5] *ideal primitives* G, thus making seq- and pub-indifferentiability equivalent notions in that case.

---

[5] By stateless we mean that the answer of G to any query only depends on the query and the randomness of G and not on any additional state information. In particular, for fixed randomness, G always returns the same answer to a given query.

Theorem 1. *Let $\mathcal{C}$ be a construction with oracle access to some ideal primitive $\mathbf{F}$. If $\mathcal{C}^{\mathbf{F}}$ is statistically (resp. computationally) strongly $(2q, \sigma, \epsilon)$-seq-indifferentiable from a stateless ideal primitive $\mathbf{G}$, then $\mathcal{C}^{\mathbf{F}}$ is statistically (resp. computationally) strongly $(q, \sigma + q, \epsilon)$-pub-indifferentiable from $\mathbf{G}$.*

Proof. See the full version of the paper [19].   □

Ristenpart[6] observed that the above theorem does not hold (at least in the computational setting) when $\mathbf{G}$ is stateful. This is explained in the full version of the paper [19]. A very simple example enables to separate full indifferentiability from seq/pub-indifferentiability, namely the Merkle-Damgård construction without strengthening using a random compression function: it was proven in [7] that it is not indifferentiable from a random oracle (a consequence of length-extension attacks), and in [12] that it is pub-indifferentiable from a random oracle.

3 Seq-Indifferentiability of the 6-Round Feistel Construction

In this section we prove the main result of this paper which states that the Feistel construction with 6 rounds and random round functions is seq-indifferentiable from a random invertible permutation, and hence also pub-indifferentiable since a random invertible permutation is stateless. Before stating the result, we recall that in [9], it was shown that the Feistel construction with five rounds is not indifferentiable from a random invertible permutation. In fact, the distinguisher they described is sequential, which implies that the 5-round Feistel construction is not even seq-indifferentiable from a random invertible permutation. We recall this attack in the full version of the paper [19].

Theorem 2. *The Feistel construction with six rounds and random round functions is statistically and strongly (q, σ, ϵ)-seq-indifferentiable from a random invertible permutation, where:*

$$\sigma(q) = q^2 \quad and \quad \epsilon(q) = \frac{8q^4}{2^n} + \frac{q^4}{2^{2n}}.$$

The rest of this section is devoted to the proof of Theorem 2. We will consider a sequential distinguisher $\mathcal{D}$ that first issues at most q_f queries to the simulator (or the random functions $\mathbf{F}_i$). These queries will be called F-queries. Then, it issues at most q_p queries to the random permutation $\mathbf{P}$ (or the Feistel construction Ψ_6^F). These queries will be called P-queries. The total oracle queries cost is $q_f + 6q_p$ (for each P-query, the Feistel construction makes 6 F-queries to compute the answer) and is assumed to be less than q.

We start by describing how the simulator $\mathcal{S}$ works. It maintains an history of values for which each round function has been defined (either because this value has been queried by the distinguisher, or because the simulator has set this value

---

[6] Personal communication.

internally). We will note F_i, $i \in [1..6]$ the history of the i-th round function, that is a set of pairs $(U, V) \in \{0,1\}^n \times \{0,1\}^n$, where U is an input to round function F_i and V is the corresponding image (which we denote $F_i(U) = V$). We write $U \in F_i$ to denote that the image of U by F_i is defined in the history. Initially round function values $F_i(U)$ are undefined for all $i \in [1..6]$ and all $U \in \{0,1\}^n$. The images are then modified during the execution of the simulator. $F_i(U) \leftarrow V$ means that the image of U by F_i is set to V and $F_i(U) \leftarrow_{\mathcal{R}} \{0,1\}^n$ means that the image of U by F_i is set uniformly at random in $\{0,1\}^n$. If a round function value is already in the history and a new assignment occurs, the previous value is overwritten (alternatively, we could let the simulator abort in this case, as in [9], but as we will see this happens only with negligible probability so that the exact behavior of the simulator in such a case in unessential). We will note $\mathcal{H} = (F_1, \ldots, F_6)$ the complete history of the six round functions.

When the simulator receives a F-query (i, U) (meaning that the distinguisher asks for the image of U through round function F_i), it calls an internal procedure $\mathtt{Query}(i, U)$. This procedure checks whether the corresponding image is in the history of F_i, in which case it returns this value and stops. Otherwise it sets the image uniformly at random. If $i = 1, 2, 5$, or 6, it does nothing more. If $i = 3$ or 4, the simulator additionally completes all centers $(Y, Z) \in F_3 \times F_4$ newly created so that the corresponding values of (L, R) and (S, T) obtained by evaluating the Feistel construction respectively backward and forward are consistent with the random permutation $\boldsymbol{P}$, meaning that $\boldsymbol{P}(0, (L, R)) = (S, T)$. This is done by calling two internal procedures $\mathtt{CompleteForward}$ (if $i = 4$) or $\mathtt{CompleteBackward}$ (if $i = 3$) which "adapts" two round function values ($F_5(A)$ and $F_6(S)$ for $\mathtt{CompleteForward}$, and $F_1(R)$ and $F_2(X)$ for $\mathtt{CompleteBackward}$) so that the Feistel matches with the random permutation. The pseudo-code for the three procedures is given below. Statements put in boxes in $\mathtt{CompleteForward}$ and $\mathtt{CompleteBackward}$ are replacements for a different system used in the indifferentiability proof and can be ignored for the moment.

There are two points to prove in order to obtain Theorem 2: that the simulator runs in polynomial time, and then that the probabilities that the distinguisher outputs 1 when interacting with $(\boldsymbol{P}, \mathcal{S}^P)$ and $(\Psi_6^F, \boldsymbol{F})$ differ by a negligible quantity ϵ. The following lemma shows that the simulator runs in time polynomial in the number of queries it receives.

Lemma 1. *When the simulator is asked at most q queries, then the size of histories for F_3 and F_4 is at most q, the size of histories for F_1, F_2, F_5 and F_6 is at most $q^2 + q$, the procedures $\mathtt{CompleteForward}$ and $\mathtt{CompleteBackward}$ are called in total at most q^2 times, and the simulator makes at most q^2 queries to the random permutation.*

Proof. Elements are added to the history of F_3 and F_4 only when a corresponding F-query is made to the simulator, so that the size of their history cannot be greater than q. For each pair $(Y, Z) \in F_3 \times F_4$, either $\mathtt{CompleteForward}(Y, Z)$ or $\mathtt{CompleteBackward}(Y, Z)$ is called, at most once, so that in total these procedures are called at most q^2 times. Since the simulator makes one query to the random permutation per execution of $\mathtt{CompleteForward}$ and $\mathtt{CompleteBackward}$

Algorithm 1 Simulator

1: **variable**: round function histories $F_1, \ldots, F_6$

2: **procedure** Query(i, U)
3:    **if** $U \notin F_i$ **then**
4:        $F_i(U) \leftarrow_{\mathcal{R}} \{0,1\}^n$
5:        **if** $i = 3$ **then**
6:            **for all** $Z \in F_4$ **do**
7:                CompleteBackward(U, Z)
8:        **if** $i = 4$ **then**
9:            **for all** $Y \in F_3$ **do**
10:                CompleteForward(Y, U)
11:    **return** $F_i(U)$

12: **procedure** CompleteForward(Y, Z)   22: **procedure** CompleteBackward(Y, Z)
13:    $X := Z \oplus F_3(Y)$ 23: $A := Y \oplus F_4(Z)$
14:    Query($2, X$)   24:   Query($5, A$)
15:    $R := Y \oplus F_2(X)$ 25: $S := Z \oplus F_5(A)$
16:    Query($1, R$)   26:   Query($6, S$)
17:    $L := X \oplus F_1(R)$ 27: $T := A \oplus F_6(S)$
18:    $(S, T) := \boldsymbol{P}(0, (L, R))$ 28: $(L, R) := \boldsymbol{P}(1, (S, T))$
       $\boxed{(S, T) := \boldsymbol{R}(0, (L, R))}$ $\boxed{(L, R) := \boldsymbol{R}(1, (S, T))}$
19:    $A := Y \oplus F_4(Z)$ 29: $X := Z \oplus F_3(Y)$
20:    $F_5(A) \leftarrow Z \oplus S$ 30: $F_2(X) \leftarrow R \oplus Y$
21:    $F_6(S) \leftarrow A \oplus T$ 31: $F_1(R) \leftarrow L \oplus X$

---

this in turns implies that the total number of queries to $\boldsymbol{P}$ is at most q^2. Finally, elements are added to the history of F_1, F_2, F_5 and F_6 either when a query is made to the simulator, or during an execution of CompleteForward or CompleteBackward, so that the size of their history cannot be greater than $q^2 + q$.   □

In order to prove that the two systems $\Sigma_1 = (\boldsymbol{P}, \mathcal{S}^{\boldsymbol{P}})$ and $\Sigma_4 = (\Psi_6^{\boldsymbol{F}}, \boldsymbol{F})$ are indistinguishable, we will use two intermediate systems: $\Sigma_2 = (\Psi_6^{\mathcal{S}^{\boldsymbol{P}}}, \mathcal{S}^{\boldsymbol{P}})$ where the P-queries of $\mathcal{D}$ are answered by the Feistel construction asking round function values to the simulator, which itself interacts with $\boldsymbol{P}$, and $\Sigma_3 = (\Psi_6^{\mathcal{S}^{\boldsymbol{R}}}, \mathcal{S}^{\boldsymbol{R}})$ where the random invertible permutation is replaced by a two-sided random function $\boldsymbol{R}$ (note the corresponding change in procedures CompleteForward and CompleteBackward indicated by a boxed statement). The four systems used in the proof are depicted in Figure 2.

The main part of the analysis is concerned with systems Σ_2 and Σ_3. We will show that unless some bad event happens, the round function values set by the simulator in Σ_2 are consistent with $\boldsymbol{P}$ (which will enable to bound the statistical distance between Σ_1 and Σ_2), and that in Σ_3 they are uniformly random and independent (which will enable to bound the statistical distance between Σ_3 and Σ_4). In systems Σ_2 and Σ_3, the simulator first receives at most q_f queries from

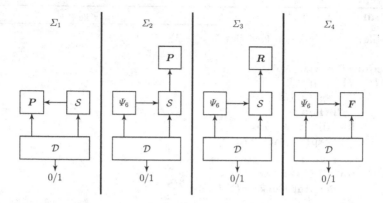

Fig. 2. Systems used in the seq-indifferentiability proof

the distinguisher, and then at most $6q_p$ queries from the Feistel construction (6 for each P-query of the distinguisher). Hence the total number of queries received by the simulator is exactly the total oracle queries cost of $\mathcal{D}$, which is less than q. The statistical distance between answers of systems Σ_2 and Σ_3 is easily bounded.

Lemma 2. *For any distinguisher of total oracle queries cost at most q, the following holds:*

$$\left|\Pr\left[\mathcal{D}^{\Sigma_2}(1^n) = 1\right] - \Pr\left[\mathcal{D}^{\Sigma_3}(1^n) = 1\right]\right| \leq \frac{q^4}{2^{2n+1}} .$$

Proof. Consider the union of $\mathcal{D}$, Ψ_6, and $\mathcal{S}$ as a single distinguisher $\mathcal{D}'$ interacting either with a random invertible permutation or a two-sided random function. Note that $\mathcal{D}'$ makes at most q^2 queries to its oracle (Lemma 1). One can conclude thanks to the PRF/PRP switching lemma [1].   □

Before going further with the proof, we define formally what it means for an input $x \in \{0,1\} \times \{0,1\}^n$ to the Feistel construction to be computable with respect to the history of the simulator.

Definition 4 (Computable input). *Given a simulator history $\mathcal{H}$ and an input $x \in \{0,1\} \times \{0,1\}^{2n}$, the sequence $\rho_{\mathcal{H}}(x) = (\rho_{\mathcal{H}}(x)[i])_{i \in [0..7]}$ is defined as follows:*

- *for a forward input $x = (0,(L,R))$, $\rho_{\mathcal{H}}(x)[0] = L$, $\rho_{\mathcal{H}}(x)[1] = R$, and for $i = 2$ to 7:*

  $$\begin{cases} \text{if } \rho_{\mathcal{H}}(x)[i-1] \in F_{i-1} \text{ then } \rho_{\mathcal{H}}(x)[i] = \rho_{\mathcal{H}}(x)[i-2] \oplus F_{i-1}(\rho_{\mathcal{H}}(x)[i-1]) \\ \text{else } \rho_{\mathcal{H}}(x)[i] = \perp \end{cases}$$

- *for a backward input $x = (1,(S,T))$, $\rho_{\mathcal{H}}(x)[7] = T$, $\rho_{\mathcal{H}}(x)[6] = S$, and for $i = 5$ to 0:*

  $$\begin{cases} \text{if } \rho_{\mathcal{H}}(x)[i+1] \in F_{i+1} \text{ then } \rho_{\mathcal{H}}(x)[i] = \rho_{\mathcal{H}}(x)[i+2] \oplus F_{i+1}(\rho_{\mathcal{H}}(x)[i+1]) \\ \text{else } \rho_{\mathcal{H}}(x)[i] = \perp \end{cases}$$

An input x is said to be computable *with respect to $\mathcal{H}$ iff $\rho_{\mathcal{H}}(x)[i] \neq \perp$ for all $i \in [0..7]$. In that case we note $\Psi_6^{\mathcal{H}}(x) = (\rho_{\mathcal{H}}(x)[6], \rho_{\mathcal{H}}(x)[7])$ if x is a forward input and $\Psi_6^{\mathcal{H}}(x) = (\rho_{\mathcal{H}}(x)[0], \rho_{\mathcal{H}}(x)[1])$ if x is a backward input.*

For a computable input x, we will often use the notation $(L, R, X, Y, Z, A, S, T) = \rho_{\mathcal{H}}(x)$ as depicted on Figure 1.

We now define a bad event that may occur during the execution of the simulator (in Σ_2 or Σ_3) in relation with Lines 20, 21, 30, and 31 of the simulator. We will say that event **Bad** happens if in any execution of CompleteForward or CompleteBackward, the input value whose image is set at Lines 20, 21, 30 or 31 is already in the history of the corresponding round function. This implies that the simulator overwrites a value so that its answers may not be coherent with $\boldsymbol{P}$ or $\boldsymbol{R}$ any more.[7] Reciprocally, if **Bad** does not happen, then the simulator never overwrites any value in its history.

We start with the simple observation that if **Bad** does not happen, then during any execution of CompleteForward or CompleteBackward, the query to $\boldsymbol{P}$ or $\boldsymbol{R}$ made by the simulator is fresh.

Lemma 3. *In system Σ_2, if* **Bad** *does not happen, then in any execution of* CompleteForward *or* CompleteBackward *the query to $\boldsymbol{P}$ made by the simulator is not in the history of $\boldsymbol{P}$. For Σ_3, the corresponding statement holds for $\boldsymbol{R}$.*

Proof. The reasoning is the same for Σ_2 and Σ_3, we use Σ_2 to fix ideas. Consider an execution of CompleteForward(Y, Z). Let $x = (0, (L, R))$ be the query to $\boldsymbol{P}$ made by the simulator, and $(S, T) = \boldsymbol{P}(x)$. If x is already in the history of $\boldsymbol{P}$, it was necessarily added by a previous execution of CompleteForward(Y', Z') or CompleteBackward(Y', Z') (note that the distinguisher does not make any query to $\boldsymbol{P}$ in Σ_2 or to $\boldsymbol{R}$ in Σ_3). But since **Bad** does not happen, round function values are never overwritten so that necessarily $(Y', Z') = (Y, Z)$. This is impossible since by construction the simulator makes at most one call to CompleteForward or CompleteBackward per center $(Y, Z) \in F_3 \times F_4$. □

We are now ready to bound the probability that **Bad** happens in Σ_2 or Σ_3.

Lemma 4. *For any distinguisher of total oracle queries cost at most q, event* **Bad** *happens with probability less than $4q^4/2^n$ in Σ_3 and less than $4q^4/2^n + q^4/2^{2n+1}$ in Σ_2.*

Proof. See the full version of the paper [19]. □

The following lemma says that as long as **Bad** does not happen in Σ_2, the round function values set by the simulator are consistent with $\boldsymbol{P}$.

Lemma 5. *If* **Bad** *does not happen in Σ_2, then for any input $x \in \{0, 1\} \times \{0, 1\}^{2n}$ computable with respect to the final history of the simulator $\mathcal{H}$, $\Psi_6^{\mathcal{H}}(x) = \boldsymbol{P}(x)$.*

---

[7] In previous work on indifferentiability of the Feistel construction [9,27], in such a case the simulator aborted. It does not change much since, as we will prove, this happens only with negligible probability.

Proof. Consider an input $x \in \{0,1\} \times \{0,1\}^{2n}$ computable with respect to the final history $\mathcal{H}$ of the simulator, and let $(L, R, X, Y, Z, A, S, T) = \rho_{\mathcal{H}}(x)$. There was necessarily a call to $\texttt{CompleteForward}(Y, Z)$ or $\texttt{CompleteBackward}(Y, Z)$ during the execution of the simulator. With respect to the history $\mathcal{H}'$ just after the completion of $\texttt{CompleteForward}(Y, Z)$ or $\texttt{CompleteBackward}(Y, Z)$, it is clear that $\Psi_6^{\mathcal{H}'}(x) = P(x)$. Since Bad does not happen the simulator never overwrites a value and the equality remains true until the end of the simulation, hence $\Psi_6^{\mathcal{H}}(x) = P(x)$. $\qquad\square$

A direct consequence of this lemma is that as long as Bad does not happen in Σ_2, the answers of systems Σ_1 and Σ_2 are identically distributed.

Lemma 6. *For any distinguisher of total oracle queries cost at most q, the following holds:*

$$\left| \Pr\left[\mathcal{D}^{\Sigma_1}(1^n) = 1 \right] - \Pr\left[\mathcal{D}^{\Sigma_2}(1^n) = 1 \right] \right| \leq \frac{4q^4}{2^n} + \frac{q^4}{2^{2n+1}} .$$

Proof. Clearly, answers to F-queries of the distinguisher are identically distributed in Σ_1 and Σ_2 since they are answered by $\mathcal{S}^P$ in both systems (may Bad occur or not).[8] Moreover, in Σ_2 any P-query x asked by the distinguisher is computable with respect to the history of the simulator at the time it is answered by Ψ_6, and if Bad does not happen in Σ_2, then according to Lemma 5, $\Psi_6^{\mathcal{H}}(x) = P(x)$ so that answers to P-queries of the distinguisher are also identically distributed in both systems. The result follows from Lemma 4. $\qquad\square$

Lemma 7. *If Bad does not happen in system Σ_3, then the round function values set by the simulator are uniformly random and independent.*

Proof. Since this is clear for round function values set uniformly at random (independently of Bad occurring or not), we only have to examine values that are adapted at Lines 20, 21, 30, and 31 of the simulator. But according to Lemma 3, if Bad does not happen, the query to R made by the distinguisher in any execution of $\texttt{CompleteForward}$ or $\texttt{CompleteBackward}$ is not in the history of R, so that the answer (S, T) or (L, R) is uniformly random. Consequently, round function values set by $F_5(A) \leftarrow Z \oplus S$ and $F_6(S) \leftarrow A \oplus T$ in $\texttt{CompleteForward}$, or $F_2(X) \leftarrow R \oplus Y$ and $F_1(R) \leftarrow L \oplus X$ in $\texttt{CompleteBackward}$ are uniformly random and independent of previous round function values set by the simulator. Since Bad does not happen round function values are not overwritten and the result follows. $\qquad\square$

This lemma finally enables to bound the statistical distance between the answers of Σ_3 and Σ_4.

Lemma 8. *For any distinguisher of total oracle queries cost at most q, the following holds:*

$$\left| \Pr\left[\mathcal{D}^{\Sigma_3}(1^n) = 1 \right] - \Pr\left[\mathcal{D}^{\Sigma_4}(1^n) = 1 \right] \right| \leq \frac{4q^4}{2^n} .$$

---

[8] It is crucial here that the distinguisher is sequential, otherwise the simulation in Σ_2 would be altered by the queries made by Ψ_6.

Proof. If **Bad** does not occur in Σ_3 then answers of $\mathcal{S}^R$ are distributed exactly as answers of $\boldsymbol{F}$ according to Lemma 7. Hence the statistical distance between answers of Σ_3 and Σ_4 is upper bounded by the probability that **Bad** happens in Σ_3, given by Lemma 4. $\qquad\square$

Theorem 2 is now a simple consequence of Lemmata 2, 6, and 8.

Remark 1. The strategy of using the intermediate system Σ_2 is likely to be quite generic for seq-indifferentiability proofs (system Σ_3, on the contrary, is quite specific to the Feistel construction). We believe this could probably make proofs of pub-indifferentiability (*e.g.* [12, Section 7]) much easier, but leave this for future work.

Remark 2. Note that for general distinguishers (not necessarily sequential), the proof would go through exactly as above for Lemmata 2 and 8. The problematic step is clearly going from Σ_1 to Σ_2. To see what could go wrong if the distinguisher can interleave queries to $\boldsymbol{P}$ and $\mathcal{S}$, consider the following simple example. $\mathcal{D}$ first makes a P-query $\boldsymbol{P}(0, (L, R)) = (S, T)$, and then makes the sequence of F-queries $F_1(R)$, $F_2(X)$, $F_6(S)$, $F_5(A)$. In system Σ_1, the simulator returns uniformly answers to the four F-queries and will be unable to adapt F_3 and F_4, whereas in Σ_2 the initial P-query of the distinguisher will trigger six F-queries from Ψ_6 which will lead the simulator to adapt the chain when query $F_4(Y)$ occurs. Making progress towards proving full indifferentiability for six rounds clearly requires to find the right way to deal with these "external" chains without knowing the P-queries of the distinguisher.

4   Applications to Correlation Intractability

Correlation intractability was introduced by Canetti *et al.* in their work on the limits of the random oracle methodology [4]. In the standard model, a function family is said to be correlation intractable if given the description of a random function f of the family, no PPT algorithm can find an input x, or more generally a sequence of inputs $(x_1, \ldots, x_m)$, such that $((x_1, \ldots, x_m), (f(x_1), \ldots, f(x_m)))$ satisfies a relation that would be hard to satisfy for a uniformly random function.

There is no difficulty in extending the definition of correlation intractability to an idealized model: instead of passing the description of the function as input to the algorithm, it is granted access to the ideal primitive used by the construction $\mathcal{C}$. This way one can define a correlation intractable construction (accessing an ideal primitive).

In all the following, we will consider relations over pairs of binary sequences (formally, a subset of $\{0,1\}^* \times \{0,1\}^*$). We assume that the machine $\mathcal{M}$ returns sequences of strings in Dom_n, the domain of the ideal primitive $\boldsymbol{G}_n$ or the construction $\mathcal{C}^{\boldsymbol{F}_n}$.

Definition 5 (Evasive relation). *Let $\boldsymbol{G} = (\boldsymbol{G}_n)$ be an ideal primitive associated to $\mathbb{G} = (\text{Dom}_n, \text{Rng}_n, \mathbb{G}_n)$. A relation $\mathcal{R}$ over pairs of binary sequences is* *said to be evasive with respect to G if for any PPT oracle machine $\mathcal{M}$, there is a negligible function ϵ such that the following holds:*

$$\Pr\left[(x_1,\ldots,x_m) \leftarrow \mathcal{M}^{G_n}(1^n) :\right.$$
$$\left.((x_1,\ldots,x_m),(G_n(x_1),\ldots,G_n(x_m))) \in \mathcal{R}\right] \le \epsilon(n) \ .$$

Definition 6 (Correlation intractable construction). *Let $\mathcal{C}$ be a construction with oracle access to an ideal primitive $F = (F_n)$ and implementing some primitive G. $\mathcal{C}^F$ is said to be (multiple-output) correlation intractable if for any relation $\mathcal{R}$ over pairs of binary sequences evasive with respect to G, and any PPT oracle machine $\mathcal{M}$, there is a negligible function ϵ such that:*

$$\Pr\left[(x_1,\ldots,x_m) \leftarrow \mathcal{M}^{F_n}(1^n) :\right.$$
$$\left.((x_1,\ldots,x_m),(\mathcal{C}^{F_n}(x_1),\ldots,\mathcal{C}^{F_n}(x_m))) \in \mathcal{R}\right] \le \epsilon(n) \ .$$

Weak correlation intractability is defined similarly as above by quantifying only over all polynomial-time recognizable relations (*i.e.* relations $\mathcal{R}$ such that there exists a polynomial-time algorithm that, given $((x_1,\ldots,x_m),(y_1,\ldots,y_m))$, decides whether it belongs to $\mathcal{R}$ or not).

Theorem 3. *Let $\mathcal{C}$ be a construction with oracle access to an ideal primitive $F = (F_n)$ and implementing some primitive G. If $\mathcal{C}^F$ is statistically (resp. computationally) seq-indifferentiable from the ideal primitive G, then $\mathcal{C}^F$ is correlation intractable (resp. weakly correlation intractable).*

Proof. See the full version of the paper [19].   □

A direct consequence of Theorems 2 and 3 is that the 6-round Feistel construction with random round functions is correlation intractable: no polynomial algorithm with oracle access to the round functions can find a sequence of inputs that together with their image by the Feistel satisfy a relation that would be hard to satisfy in the random invertible permutation model. Note that the sole *existence* of correlation intractable invertible permutations in the random oracle model was already implied by the result of Holenstein *et al.* [14] on the full indifferentiability of the 14-round Feistel construction (since full indifferentiability implies seq-indifferentiability and hence correlation intractability), but our results shows that six rounds are sufficient to achieve this property.

Remark 3. According to Theorem 3, sequential indifferentiability implies correlation intractability. However correlation intractability does not necessarily imply sequential indifferentiability. In the full version of the paper [19] we provide a simple counter-example separating the two notions.

Implications for Chosen-Key and Known-Key Attacks on Block Ciphers. Knudsen and Rijmen [16] have introduced so-called known-key attacks on block ciphers. We discuss the implications of our results regarding this attack model in the full version of the paper [19].

References

1. Bellare, M., Ristenpart, T.: Multi-Property-Preserving Hash Domain Extension and the EMD Transform. In: Lai, X., Chen, K. (eds.) ASIACRYPT 2006. LNCS, vol. 4284, pp. 299–314. Springer, Heidelberg (2006)
2. Bellare, M., Rogaway, P.: Random Oracles are Practical: A Paradigm for Designing Efficient Protocols. In: ACM Conference on Computer and Communications Security, pp. 62–73 (1993)
3. Bellare, M., Rogaway, P.: The Exact Security of Digital Signatures - How to Sign with RSA and Rabin. In: Maurer, U.M. (ed.) EUROCRYPT 1996. LNCS, vol. 1070, pp. 399–416. Springer, Heidelberg (1996)
4. Canetti, R., Goldreich, O., Halevi, S.: The Random Oracle Methodology. In: Symposium on Theory of Computing - STOC 1998, pp. 209–218. ACM (1998), revisited (Preliminary Version); Full version, http://arxiv.org/abs/cs.CR/0010019
5. Chevallier-Mames, B., Phan, D.H., Pointcheval, D.: Optimal Asymmetric Encryption and Signature Paddings. In: Ioannidis, J., Keromytis, A.D., Yung, M. (eds.) ACNS 2005. LNCS, vol. 3531, pp. 254–268. Springer, Heidelberg (2005)
6. Coron, J.-S.: Optimal Security Proofs for PSS and Other Signature Schemes. In: Knudsen, L.R. (ed.) EUROCRYPT 2002. LNCS, vol. 2332, pp. 272–287. Springer, Heidelberg (2002)
7. Coron, J.-S., Dodis, Y., Malinaud, C., Puniya, P.: Merkle-Damgård Revisited: How to Construct a Hash Function. In: Shoup, V. (ed.) CRYPTO 2005. LNCS, vol. 3621, pp. 430–448. Springer, Heidelberg (2005)
8. Coron, J.-S., Dodis, Y., Mandal, A., Seurin, Y.: A Domain Extender for the Ideal Cipher. In: Micciancio, D. (ed.) TCC 2010. LNCS, vol. 5978, pp. 273–289. Springer, Heidelberg (2010)
9. Coron, J.-S., Patarin, J., Seurin, Y.: The Random Oracle Model and the Ideal Cipher Model Are Equivalent. In: Wagner, D. (ed.) CRYPTO 2008. LNCS, vol. 5157, pp. 1–20. Springer, Heidelberg (2008)
10. Dodis, Y., Puniya, P.: On the Relation Between the Ideal Cipher and the Random Oracle Models. In: Halevi, S., Rabin, T. (eds.) TCC 2006. LNCS, vol. 3876, pp. 184–206. Springer, Heidelberg (2006)
11. Dodis, Y., Puniya, P.: Feistel Networks Made Public, and Applications. In: Naor, M. (ed.) EUROCRYPT 2007. LNCS, vol. 4515, pp. 534–554. Springer, Heidelberg (2007)
12. Dodis, Y., Ristenpart, T., Shrimpton, T.: Salvaging Merkle-Damgård for Practical Applications. In: Joux, A. (ed.) EUROCRYPT 2009. LNCS, vol. 5479, pp. 371–388. Springer, Heidelberg (2009)
13. Granboulan, L.: Short Signatures in the Random Oracle Model. In: Zheng, Y. (ed.) ASIACRYPT 2002. LNCS, vol. 2501, pp. 364–378. Springer, Heidelberg (2002)
14. Holenstein, T., Künzler, R., Tessaro, S.: The Equivalence of the Random Oracle Model and the Ideal Cipher Model. In: Fortnow, L., Vadhan, S.P. (eds.) Symposium on Theory of Computing - STOC 2011, pp. 89–98. ACM (2011) (revisited)
15. Katz, J., Wang, N.: Efficiency improvements for signature schemes with tight security reductions. In: Jajodia, S., Atluri, V., Jaeger, T. (eds.) ACM Conference on Computer and Communications Security, pp. 155–164. ACM (2003)
16. Knudsen, L.R., Rijmen, V.: Known-Key Distinguishers for Some Block Ciphers. In: Kurosawa, K. (ed.) ASIACRYPT 2007. LNCS, vol. 4833, pp. 315–324. Springer, Heidelberg (2007)

17. Künzler, R.: Are the random oracle and the ideal cipher models equivalent? Master's thesis, ETH Zurich, Switzerland (2009)
18. Luby, M., Rackoff, C.: How to Construct Pseudorandom Permutations from Pseudorandom Functions. SIAM Journal on Computing 17(2), 373–386 (1988)
19. Mandal, A., Patarin, J., Seurin, Y.: On the Public Indifferentiability and Correlation Intractability of the 6-Round Feistel Construction. ePrint Archive Report 2011/496 (2011), http://eprint.iacr.org/2011/496.pdf
20. Maurer, U.M.: A Simplified and Generalized Treatment of Luby-Rackoff Pseudorandom Permutation Generators. In: Rueppel, R.A. (ed.) EUROCRYPT 1992. LNCS, vol. 658, pp. 239–255. Springer, Heidelberg (1993)
21. Maurer, U.M.: Indistinguishability of Random Systems. In: Knudsen, L.R. (ed.) EUROCRYPT 2002. LNCS, vol. 2332, pp. 110–132. Springer, Heidelberg (2002)
22. Maurer, U., Renner, R., Holenstein, C.: Indifferentiability, Impossibility Results on Reductions, and Applications to the Random Oracle Methodology. In: Naor, M. (ed.) TCC 2004. LNCS, vol. 2951, pp. 21–39. Springer, Heidelberg (2004)
23. Naor, M., Reingold, O.: On the Construction of Pseudorandom Permutations: Luby-Rackoff Revisited. Journal of Cryptology 12(1), 29–66 (1999)
24. Patarin, J.: Security of Random Feistel Schemes with 5 or More Rounds. In: Franklin, M. (ed.) CRYPTO 2004. LNCS, vol. 3152, pp. 106–122. Springer, Heidelberg (2004)
25. Ramzan, Z., Reyzin, L.: On the Round Security of Symmetric-Key Cryptographic Primitives. In: Bellare, M. (ed.) CRYPTO 2000. LNCS, vol. 1880, pp. 376–393. Springer, Heidelberg (2000)
26. Ristenpart, T., Shacham, H., Shrimpton, T.: Careful with Composition: Limitations of the Indifferentiability Framework. In: Paterson, K.G. (ed.) EUROCRYPT 2011. LNCS, vol. 6632, pp. 487–506. Springer, Heidelberg (2011)
27. Seurin, Y.: Primitives et protocoles cryptographiques à sécurité prouvée. PhD thesis, Université de Versailles Saint-Quentin-en-Yvelines, France (2009)
28. Vaudenay, S.: Decorrelation: A Theory for Block Cipher Security. Journal of Cryptology 16(4), 249–286 (2003)
29. Yoneyama, K., Miyagawa, S., Ohta, K.: Leaky Random Oracle. IEICE Transactions 92-A(8), 1795–1807 (2009)

Collisions Are Not Incidental: A Compression Function Exploiting Discrete Geometry

Dimitar Jetchev[1], Onur Özen[1], and Martijn Stam[2]

[1] Laboratory for Cryptologic Algorithms, EPFL, Switzerland
[2] Department of Computer Science, University of Bristol, UK

Abstract. We present a new construction of a compression function $H\colon \{0,1\}^{3n} \to \{0,1\}^{2n}$ that uses two parallel calls to an ideal primitive (an ideal blockcipher or a public random function) from $2n$ to n bits. This is similar to the well-known MDC-2 or the recently proposed MJH by Lee and Stam (CT-RSA'11). However, unlike these constructions, we show already in the compression function that an adversary limited (asymptotically in n) to $\mathcal{O}(2^{2n(1-\delta)/3})$ queries (for any $\delta > 0$) has disappearing advantage to find collisions. A key component of our construction is the use of the Szemerédi–Trotter theorem over finite fields to bound the number of full compression function evaluations an adversary can make, in terms of the number of queries to the underlying primitives. Moveover, for the security proof we rely on a new abstraction that refines and strenghtens existing techniques. We believe that this framework elucidates existing proofs and we consider it of independent interest.

1 Introduction

Ever since the initial efforts to turn a blockcipher into a hash function, a major drawback of using blockcipher-based compression functions producing a digest size equal to the block-length is that the digest size is too small to produce a hash function meeting today's security requirements. For example, AES, operating on 128 bits, limits collision resistance to at most 2^{64} operations/queries. As a remedy, *double-length* compression functions and corresponding double-length hash functions have been introduced (e.g. [3–5]): A design that outputs $2n$ bits (while making several calls to a blockcipher with n-bit blocks) could potentially provide collision resistance up to roughly 2^n blockcipher evaluations.

In this work, we are interested in the construction of a provably collision-resistant (beyond $2^{n/2}$ queries) compression function from $3n$ to $2n$ bits making two parallel calls to an ideal primitive from $2n$ to n bits (either a public random function– PuRF–or an ideal blockcipher with n-bit blocks and n-bit keys). Our motivation is a natural one: All existing designs in this class fall short. There is no proof, or the proof is not known to extend to the blockcipher case; (non-trivial) collision resistance is only provided in the iteration; the primitive calls need to be made in sequence; or the number of calls is higher. Yet known impossibility bounds [11, 12, 15] give no reason why such a construction should not be possible.

R. Cramer (Ed.): TCC 2012, LNCS 7194, pp. 303–320, 2012.

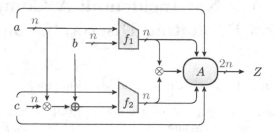

Fig. 1. Our compression function $H^{f_1,f_2}\colon \{0,1\}^{3n} \to \{0,1\}^{2n}$ illustrated (see Construction 1 for the details)

Our Contribution. We provide a construction (see Fig. 1), for which we prove that any adversary limited (asymptotically in n) to $\mathcal{O}(2^{2n(1-\delta)/3})$ queries (for any $\delta > 0$) has disappearing advantage to find collisions. To the best of our knowledge, this is the first design of its kind offering collision resistance beyond $2^{n/2}$ queries. Our construction has two key innovative components (see Fig. 1): a preprocessing function C^{PRE} that transforms the $3n$-bit input into a pair of $2n$-bit strings that are passed as inputs to the two ideal primitive calls; and a postprocessing function C^{POST} that combines the two outputs of the ideal primitives and the $3n$-bit input into the $2n$-bit output of the compression function. Initially, we will concentrate on the PuRF scenario; details for the more complicated ideal-cipher model follow later (Section 6). In either case, we work in the ideal-primitive model (giving separate proofs for each scenario).

A major technical hurdle in the proof of collision resistance is that the standard proof techniques turn out to be insufficient. For concreteness, consider an adversary that adaptively makes three queries trying to create a collision. Customarily, one would upper bound the probability p_i (for $i = 1, 2, 3$) that an adversary causes a collision on the ith query, say with $B_i = 1/4$ each; taking a union bound leads to an overall bound $3/4$. Our first abstraction is a game hop where we allow an adversary to choose its success probability p_i directly, rather than computing it based on which query to some primitive is being made. By requiring $p_i \le B_i$ this leads to the same overall winning bound $3/4$, achieved by a greedy adversary. However, this abstraction allows us to phrase and study different scenarios as well (relevant for our collision resistance proof), for instance one where we only set a global requirement $\sum_i p_i \le 1/2$. Now potentially each of the p_i values could be $1/2$ itself, so using $\sum_i B_i$ would lead to an overall bound of $3/2$ (which is vacuous for a probability), yet intuitively no adversary should be able to do better than $1/2$. A further complication arises when we require the adversary to obtain a success at least twice. While it is easy to deal with non-adaptive adversaries, properly taking care of adaptive adversaries is non-trivial. We provide the abstraction and solutions to the problems just described in Section 3. We believe this framework to be of independent theoretical interest.

The main innovation of our design is the choice for C^{PRE}: the $3n$-bit input is transformed into a pair of an affine line on $\mathbb{F}_{2^n}^2$ and a point on that line.

Hence, any given valid input pair to the underlying ideal primitives corresponds to an incidence between a point and a line in the affine plane $\mathbb{F}_{2^n}^2$ over the finite field $\mathbb{F}_{2^n}$. We then use a classical result of discrete geometry, the Szemerédi–Trotter theorem over finite fields, to bound the number of incidences between a set of q lines and a set of q points on $\mathbb{F}_{2^n}^2$, namely by roughly $q^{3/2}$.

The postprocessing is inspired by the Rogaway–Steinberger construction [10], where a special type of $\mathbb{F}_{2^n}$ linear map is used. However, we add the product of the two primitive-outputs to the inputs to this linear map. This turns out to be crucial for our collision resistance proof. In Section 5 we prove that the best strategy for any collision-finding adversary is (close to) maximizing the number of the aforementioned point-line incidences (in C^{PRE}). Our proof uses the newly developed techniques given in Section 3 to deal with adaptive adversaries.

Putting the pieces together, we achieve the claimed collision resistance of already at the compression function level. We also prove (everywhere) preimage resistance up to $\mathcal{O}(2^{(1-\delta)n})$ queries (for arbitrary $\delta > 0$). From an efficiency perspective, our construction makes two parallel calls to distinct primitives, each with $2n$-bit inputs. The overhead consists of a number of xors (to implement the matrix-multiplication) plus, more significantly, two full ($\mathbb{F}_{2^n}$) finite field multiplications: one during the preprocessing and one during the postprocessing.

2 Preliminaries

Primitive-Based Compression Functions. A *compression function* is a map $H: \{0,1\}^{tn} \to \{0,1\}^{sn}$, where n is an integer (the block-length, which in an asymptotic setting typically takes the role of the security parameter) and $t > s > 0$ are integer parameters. A compression function is *primitive-based* if it is computed by a program making calls to a finite number of specified oracles (primitives). We use superscripts to denote oracle access. For integers c and n, let Func(cn, n) denote the set of all maps $\{0,1\}^{cn} \to \{0,1\}^n$ and let $f \xleftarrow{\$} \text{Func}(cn, n)$ denote that f is sampled uniformly at random from all elements in Func(cn, n). Then we call f a *public random function* (PuRF) and we refer to a compression function making oracle calls to f as *PuRF-based*. For given input W we denote the resulting digest as $H^f(W)$. More generally, when there are r independently sampled primitives $f_1, \ldots, f_r \xleftarrow{\$} \text{Func}(cn, n)$ we write $H^{f_1, \ldots, f_r}(W)$.

Similarly, let Block$((c-1)n, n)$ denote the set of all blockciphers having $(c-1)n$-bit key and operating on n-bit blocks. In other words, Block$((c-1)n, n)$ is the set of all maps $E: \{0,1\}^{(c-1)n} \times \{0,1\}^n \to \{0,1\}^n$, such that for any key $K \in \{0,1\}^{(c-1)n}$, $E(K, \cdot)$ is a permutation on the set $\{0,1\}^n$. (Note that $(c-1)n + n = cn$, so that one can interpret $E \in \text{Func}(cn, n)$ as well.) For a blockcipher E, we denote its inverse by D, so for all $K \in \{0,1\}^{(c-1)n}$ and $X \in \{0,1\}^n$ we have that $D(K, E(K, X)) = X$. When $E \xleftarrow{\$} \text{Block}((c-1)n, n)$ is chosen uniformly at random we call it an *ideal cipher* and refer to a compression function H^E (or more generally, $H^{E_1, \ldots, E_r}$ when there are r independently sampled blockciphers $E_1, \ldots, E_r \xleftarrow{\$} \text{Block}((c-1)n, n)$) as *blockcipher-based*. The definitions and the illustrations below are provided in the PuRF-based setting;

the blockcipher-based case is analogous, where we assume that oracle access to E implicitly implies access to its inverse D as well.

We study *single-layer* compression functions. This means that the oracle calls can be made in parallel and the output of the compression function is computed based on the results of these calls, as well as on the input itself. Formally, let $C_i^{\mathrm{PRE}} \colon \{0,1\}^{tn} \to \{0,1\}^{cn}$ for $i = 1, \ldots, r$, and $C^{\mathrm{POST}} \colon \{0,1\}^{tn} \times (\{0,1\}^n)^r \to \{0,1\}^{sn}$, be pre and postprocessing functions, respectively. Given a tn-bit input W, compute output $Z = H^{f_1,\ldots,f_r}(W)$ as follows: for $i = 1, \ldots, r$, let $x_i \leftarrow C_i^{\mathrm{PRE}}(W)$ and $y_i \leftarrow f_i(x_i)$; the output is then $Z \leftarrow C^{\mathrm{POST}}(W, y_1, \ldots, y_r)$.

Security Notions. An adversary is an algorithm (typically modelled as an interactive Turing machine) that uses its oracle access to the underlying primitives of the compression function in order to 'break' some well-defined property. We will limit ourselves to (everywhere) preimage resistance and collision resistance, and consider information-theoretic adversaries only; our sole resource of interest is the number of queries made to their oracles (adversaries are considered computationally unbounded). Without loss of generality, adversaries are assumed not to repeat queries nor to query an oracle outside of its specified domain.

When, for some $l \in \{1, \ldots, r\}$, an adversary makes an f_l-query obtaining $Y = f_l(X)$, we will append (l, X, Y) to the *query history* $\mathcal{Q}$ (which is initialized empty). For preimage and collision resistance, adversarial success can be determined based on the query history $\mathcal{Q}$ only, which we formalize using the yield set (Definition 1) and which we exploit by dropping the explicit sampling of the primitives f_i and the queries $\mathcal{Q}$ for experiments. We partition $\mathcal{Q}$ in $\mathcal{Q}[1] \ldots \mathcal{Q}[r]$ depending on which of the primitives was called and, although technically elements of $\mathcal{Q}$ are triples, we assume that the context suffices to determine which of the r primitives was used. For $i \leq |\mathcal{Q}|$, we let $\mathcal{Q}_i$ denote the first i elements of $\mathcal{Q}$. Occasionally, we abuse notation by writing $X \in \mathcal{Q}$ or $Y \in \mathcal{Q}$.

Definition 1. *Let $H^{f_1,\ldots,f_r}$ be a primitive-based compression function and let $\mathcal{Q}$ be a set of queries (with answers) to the underlying primitives, then the yield set $\mathrm{yieldset}(\mathcal{Q})$ is the set of all pairs (W, Z) such that $Z = H^{f_1,\ldots,f_r}(W)$ and all queries necessary for the evaluation of the compression function at W are in $\mathcal{Q}$. We refer to the cardinality of $\mathrm{yieldset}(\mathcal{Q})$ as the* yield *and denote it by $\mathrm{yield}(\mathcal{Q})$. Additionally, we define $\mathrm{yield}(q) = \max_{\mathcal{Q}} \mathrm{yield}(\mathcal{Q})$ where $|\mathcal{Q}[i]| \leq q$. (Note that since $\mathcal{Q}$ incorporates the primitives' answers, the maximum implicitly includes a maximization over the choice of the underlying primitives.)*

Definition 2 (Collision resistance). *Let $H^{f_1,\ldots,f_r}$ be a primitive-based compression function. For a given $\mathcal{Q}$ and $Z \in \{0,1\}^{sn}$, define*

$$\mathrm{coll}(\mathcal{Q}) \equiv \exists_{Z, W \neq W'} (W, Z), (W', Z) \in \mathrm{yieldset}(\mathcal{Q}) .$$

The collision-finding advantage of an adversary $\mathcal{A}$ is defined as

$$\mathrm{Adv}_H^{\mathrm{coll}}(\mathcal{A}) = \Pr\left[f_1 \ldots f_r \xleftarrow{\$} \mathrm{Func}(cn, n), \mathcal{Q} \leftarrow \mathcal{A}^{f_1 \ldots f_r} : \mathrm{coll}(\mathcal{Q}) \right] .$$

Similarly, define $\mathsf{Adv}_H^{\mathsf{coll}}(q) = \max_{\mathcal{A}} \mathsf{Adv}_H^{\mathsf{coll}}(\mathcal{A})$, *where the maximum is taken over all adversaries* $\mathcal{A}$ *making at most* q *queries to each of the underlying primitives.*

Definition 3 (Everywhere preimage resistance). *Let* $H^{f_1,\dots,f_r}$ *be a primitive-based compression function. For a given* $\mathcal{Q}$ *and* $Z \in \{0,1\}^{sn}$, *define*

$$\mathsf{epre}_Z(\mathcal{Q}) \equiv \exists_{W'}(W', Z) \in \mathsf{yieldset}(\mathcal{Q}) .$$

The everywhere preimage-finding advantage of an adversary $\mathcal{A}$ *is defined as*

$$\mathsf{Adv}_H^{\mathsf{epre}}(\mathcal{A}) = \max_{Z \in \{0,1\}^{sn}} \left\{ \Pr\left[f_1 \dots f_r \xleftarrow{\$} \mathsf{Func}(cn, n), \mathcal{Q} \leftarrow \mathcal{A}^{f_1 \cdots f_r} : \mathsf{epre}_Z(\mathcal{Q}) \right] \right\} .$$

We also define $\mathsf{Adv}_H^{\mathsf{epre}}(q) = \max_{\mathcal{A}} \mathsf{Adv}_H^{\mathsf{epre}}(\mathcal{A})$, *where the maximum is taken over all adversaries* $\mathcal{A}$ *making at most* q *queries to each of the* r *primitives.*

3 Probabilistic Analysis of Adaptive Adversaries

Most of the security proofs in the literature for compression and hash functions rely on the same principle. Consider the game depicted in Fig. 2, where the adversary has access to some underlying primitive $f()$ and tries to set a predicate **E** that is defined for all collections of query-response pairs. We are primarily interested in monotone predicates **E**, that once set cannot be 'unset' by additional queries. A predicate **E** is monotone if and only if for all $\mathcal{Q} \subseteq \mathcal{Q}'$ it holds that $\mathbf{E}(\mathcal{Q}) \Rightarrow \mathbf{E}(\mathcal{Q}')$. Additionally, we impose non-triviality of the predicate meaning that the predicate is not set from the outset (i.e. $\mathbf{E}(\emptyset) = \mathtt{false}$). For collision resistance, one should read coll (Definition 2) for **E** and for preimage resistance epre_Z (Definition 3). Note that coll and epre_Z are always monotone and that, for our construction, both coll and epre_Z are non-trivial.

Bounding an advantage is then tantamount to bounding $\Pr[\mathbf{E}(\mathcal{Q})]$, where the probabilities are taken over the randomness of f and the coins of $\mathcal{A}$, if any. In the following, we show how we can analyse such events in a stepwise approach to determine useful upper bounds.

There is a distinction between adaptive and non-adaptive adversaries. The latter are required to commit to a fixed set of queries at the very beginning of the game. In the information-theoretic setting, it is customary (and WLOG) to consider deterministic adversaries only. Consequently, maximizing over all q-query (non-adaptive) adversaries becomes equivalent to maximizing over all possible query sets of cardinality q. This considerably simplifies proofs. For instance, when providing a proof in the ideal-cipher model (using a union bound), for a non-adaptive adversary every response can be considered *fully* random, whereas for an adaptive adversary previous queries to the cipher might influence the outcome slightly.

Related work. Maurer [7] (see also Pietrzak [8]) developed a methodology to equate adaptive and non-adaptive adversaries in certain cases. While it is possible to phrase our game from Fig. 2 in their framework, for many of our winning

$\mathsf{Exp}^{\mathbf{E}\text{-ad}}(\mathcal{A})$:
   Let $i \leftarrow 0, \mathcal{Q}_0 \leftarrow \emptyset$
   While $i < q$ do
      $i \leftarrow i + 1$
      $x_i \leftarrow \mathcal{A}(\mathcal{Q}_{i-1})$
      $y_i \leftarrow f(x_i)$
      $\mathcal{Q}_i \leftarrow \mathcal{Q}_{i-1} \cup \{(x_i, y_i)\}$
   Return $\mathbf{E}(\mathcal{Q}_q)$.

$\mathsf{Exp}^{\mathbf{E}\text{-na}}(\mathcal{A})$:
   $(x_1, \dots, x_q) \leftarrow \mathcal{A}()$
   Let $i \leftarrow 0, \mathcal{Q}_0 \leftarrow \emptyset$
   While $i < q$ do
      $i \leftarrow i + 1$
      $y_i \leftarrow f(x_i)$
      $\mathcal{Q}_i \leftarrow \mathcal{Q}_{i-1} \cup \{(x_i, y_i)\}$
   Return $\mathbf{E}(\mathcal{Q}_q)$.

$\mathsf{Exp}^{\mathbf{E},\mathbf{F}}(\mathcal{A})$:
   Let $i \leftarrow 0, \mathcal{Q}_0 \leftarrow \emptyset$
   While $i < q$ do
      $i \leftarrow i + 1$
      $x_i \leftarrow \mathcal{A}(\mathcal{Q}_{i-1})$
      $y_i \leftarrow f(x_i)$
      $\mathcal{Q}_i \leftarrow \mathcal{Q}_{i-1} \cup \{(x_i, y_i)\}$
   Return $\mathbf{E}(\mathcal{Q}_q) \wedge \neg\mathbf{F}(\mathcal{Q}_q)$.

Fig. 2. Standard adaptive ($\mathsf{Exp}^{\mathbf{E}\text{-ad}}(\mathcal{A})$) and non-adaptive ($\mathsf{Exp}^{\mathbf{E}\text{-na}}(\mathcal{A})$) security games for predicate $\mathbf{E}$, as well as the flagged experiment $\mathsf{Exp}^{\mathbf{E},\mathbf{F}}(\mathcal{A})$

predicates adaptive adversaries *do* have an advantage over non-adaptive adversaries. Instead we opt for a more direct approach, where we primarily take our inspiration from existing hash-function security proofs. Henceforth, unless stated otherwise, we will consider adaptive adversaries only (and consequently drop the "ad" suffix in naming experiments and advantages).

The Straightforward Approach. The standard way of dealing with adaptive adversaries, as exemplified for instance by the security proofs [1, 2, 13] for the PGV compression functions [9], is the following. Suppose an adversary makes q queries. These are necessarily made in sequence, so denote with $\mathcal{Q}_i$ the set of query-responses after i queries have been made (where $i \in \{0, \dots, q\}$). The overall winning probability can then be stated as a sum of the probability of winning on the ith step, where these 'stepwise' probabilities are only taken over the choice of y_i. This makes derivation of the overall bound relatively easy (even when taking into account the accompanying maximization).

Proposition 1. *Let $\mathbf{E}$ be a monotone non-trivial predicate. Then the advantage of any (adaptive) adversary $\mathcal{A}$ playing $\mathsf{Exp}^{\mathbf{E}}(\mathcal{A})$ (see Fig. 2) is bounded by*

$$\mathsf{Adv}^{\mathbf{E}}(\mathcal{A}) \le \sum_{i=1}^{q} \max_{\mathcal{Q}_{i-1} \, s.t. \, \neg\mathbf{E}(\mathcal{Q}_{i-1})} \max_{x_i} \Pr\left[\mathbf{E}(\mathcal{Q}_i) \mid \mathcal{Q}_{i-1} \wedge x_i\right] .$$

Using an Auxiliary Flag. Although easy, the standard approach has the disadvantage that for many more involved constructions, the maximum probalities can get too large. This is typically due to the maximum being attained only for relatively obscure values for $\mathcal{Q}_i$, values that themselves are extremely unlikely to occur. To weed out these unwanted cases, the analysis is often enhanced by splitting the monotone predicate into a set of auxiliary events. For some positive integer k, let $\mathbf{E}_1, \dots, \mathbf{E}_k$ be predicates such that (for all $\mathcal{Q}$) $\mathbf{E}(\mathcal{Q}) \Rightarrow \bigvee_i^k \mathbf{E}_i(\mathcal{Q})$, then a union bound implies $\Pr[\mathbf{E}] \le \sum_i^k \Pr[\mathbf{E}_i]$. Several examples of proofs using auxiliary events can be found in the realm of double-length hash functions [6, 14].

The events $\mathbf{E}_i(\mathcal{Q})$ themselves are usually composed as the conjunction of a monotone event and a *negated* monotone event. In the simplest scenario, consider

$\mathsf{Exp}^{\mathbf{B}}(\mathcal{A})$:
   Let $i \leftarrow 0$
    While $i < q$ do
     $i \leftarrow i + 1$
     $p_i \leftarrow \mathcal{A}()$
     if $0 \le p_i \le B_i$ then
      with probability p_i return **true**
   Return **false** .

$\mathsf{Exp}^{\mathbf{B}_\Sigma}(\mathcal{A})$:
   Let $i \leftarrow 0$
    While $i < q$ do
     $i \leftarrow i + 1$
     $p_i \leftarrow \mathcal{A}()$
     if $0 \le \sum_{j=1}^{i} p_j \le B_\Sigma$ then
      with probability p_i return **true**
   Return **false** .

Fig. 3. Game-playing interpretation of the adaptive security game (where $\mathbf{B} = (B_1, \ldots B_q)$) and our refined abstract flagging game.

a second (non-trivial) monotone predicate $\mathbf{F}$. If we define $\mathbf{E}_1 = \mathbf{E} \wedge \neg\mathbf{F}$ and $\mathbf{E}_2 = \mathbf{F}$ then $\mathbf{E} \Rightarrow \mathbf{E}_1 \vee \mathbf{E}_2$ is satisfied. To bound $\Pr[\mathbf{E}_2] = \Pr[\mathbf{F}]$ we can use Proposition 1; for $\Pr[\mathbf{E}_1] = \Pr[\mathbf{E} \wedge \neg\mathbf{F}]$ Proposition 2 shows how the use of the predicate $\mathbf{F}$ effectively allows us to consider a more restricted class of $\mathcal{Q}_i$.

Proposition 2. *Let* $\mathbf{E}$ *be a non-trivial monotone predicate and let* $\mathbf{F}$ *be an arbitrary auxiliary non-trivial monotone predicate. Then the advantage of (adaptive) adversary* $\mathcal{A}$ *setting* $\mathbf{E} \wedge \neg\mathbf{F}$ *is bounded by*

$$\Pr[\mathbf{E} \wedge \neg\mathbf{F}] \le \sum_{i=1}^{q} \Pr[\mathbf{E}(\mathcal{Q}_i) \mid \neg\mathbf{E}(\mathcal{Q}_{i-1}) \wedge \neg\mathbf{F}(\mathcal{Q}_{i-1})] \le \sum_{i=1}^{q} B_i ,$$

where $B_i = \max_{\mathcal{Q}_{i-1} \, s.t. \, \neg\mathbf{E}(\mathcal{Q}_{i-1}) \wedge \neg\mathbf{F}(\mathcal{Q}_{i-1})} \max_{x_i} \Pr[\mathbf{E}(\mathcal{Q}_i) \mid \mathcal{Q}_{i-1} \wedge x_i]$.

An Alternative Interpretation. We now make a far bigger step, removing most of the underlying mechanics of the original game. Instead of letting the adversary output elements x_i and then determining by virtue of y_i whether the adversary wins this round, we directly bound the latter probability. That is, in experiment $\mathsf{Exp}^{\mathbf{B}}(\mathcal{A})$ we let the adversary output a probabilities p_i and imagine that $\mathbf{E}$ is set with probability p_i. To avoid this game becoming vacuous (namely if the adversary would output some $p_i = 1$) we put bounds B_i and B'_i on the adversary's success probability. These bounds correspond to the actual game: they are the highest possible success probabilities any adversary in any run can achieve in round i. These probabilities are reminiscent of the conditional probabilities used in the derivation from the standard approach and indeed we can formalize this relationship. Since in $\mathsf{Exp}^{\mathbf{B}}(\mathcal{A})$ a straightforward application of the union bound leads to an overall upper bound of the winning probability of $\sum_{i=1}^{q} B_i$, we can recover Proposition 2.

Lemma 1. *Consider games* $\mathsf{Exp}^{\mathbf{E},\mathbf{F}}$ *and* $\mathsf{Exp}^{\mathbf{B}}$ *and subject to*

$$B_i = \max_{\mathcal{Q}_{i-1} \ s.t. \ \neg\mathbf{E}(\mathcal{Q}_{i-1}) \wedge \neg\mathbf{F}(\mathcal{Q}_{i-1})} \max_{x_i} \Pr[\mathbf{E}(\mathcal{Q}_i)] \ . \ Then,$$

for all adversaries $\mathcal{A}$ *s.t. there exists an adversary* $\mathcal{A}'$ *s.t.* $\mathsf{Adv}^{\mathbf{E},\mathbf{F}}(\mathcal{A}) \le \mathsf{Adv}^{\mathbf{B}}(\mathcal{A}')$.

3.1   A More Refined Approach

In $\mathsf{Exp}^{\mathbf{B}}(\mathcal{A})$ instead of the guards $0 \leq p_i \leq B_i$ for all i, we could have used $0 \leq \sum_{j=1}^{i} p_j \leq \sum_{j=1}^{i} B_j$ as well. With a seemingly minor modification, this leads to another, different game where instead of bounding step-specific by $\sum_{j=1}^{i} B_j$, we always use the same bound B_Σ, as in the game $\mathsf{Exp}^{\mathbf{B}_\Sigma}(\mathcal{A})$ (Fig. 3).

Proposition 3. *For any adversary $\mathcal{A}$, it holds that* $\mathsf{Adv}^{\mathbf{B}_\Sigma}(\mathcal{A}) \leq B_\Sigma$.

Usage. The $\mathsf{Exp}^{\mathbf{B}_\Sigma}$ game captures a special kind of condition that one can encounter in the $\mathsf{Exp}^{\mathbf{E},\mathbf{F}}$ game. For any given $\mathcal{Q}$, we can *a posteriori* determine the probabilities of success by taking $\mathcal{Q}_{i-1}$ and x_i from $\mathcal{Q}$ and then looking at the probability that a *freshly drawn* y_i causes a $\mathbf{E}$. The overall *a posteriori* probability of some $\mathcal{Q}$ is the sum (over i) of these probabilities. The maximum attainable probability this way determines B_Σ, as formalized in Lemma 2. Of note here is the observation that for certain games the B_Σ obtained here is much smaller than the $\sum_{i=1}^{q} B_i$ one would obtain from application of Lemma 1. Very broadly speaking (and with some abuse of notation), it is the difference between $\max_\mathcal{Q} \{\sum_i p_i(\mathcal{Q})\}$ and $\sum_i \max_\mathcal{Q} \{p_i(\mathcal{Q})\}$.

Lemma 2. *Consider the game* $\mathsf{Exp}^{\mathbf{E},\mathbf{F}}$. *For any given $\mathcal{Q}$, define*

$$p_i(\mathcal{Q}) = \begin{cases} 0, & \text{if } \mathbf{E}(\mathcal{Q}_{i-1}) \vee \mathbf{F}(\mathcal{Q}_{i-1}) \text{ or } |\mathcal{Q}| < i \\ \Pr\left[y_i \leftarrow f(x_i) : \mathbf{E}(\mathcal{Q}_{i-1} \cup \{(x_i, y_i)\}) \mid \mathcal{Q}_{i-1}\right] & \text{otherwise} \end{cases}$$

and let $B_\Sigma = \max_\mathcal{Q} \sum_{i=1}^{q} p_i(\mathcal{Q})$. *Then for all adversaries $\mathcal{A}$ there exists an adversary $\mathcal{A}'$ such that* $\mathsf{Adv}^{\mathbf{E},\mathbf{F}}(\mathcal{A}) \leq \mathsf{Adv}^{\mathbf{B}_\Sigma}(\mathcal{A}')$.

3.2   Counting Successes

In the previous games we considered a predicate $\mathbf{E}(\mathcal{Q})$ that could either be `true` or `false` . In other words, we were interested in at least one success occurring. In some scenarios, counting the number of succcesses is more appropriate. To this end, let `ctr` be a function such that $\mathsf{ctr}(\mathcal{Q}) \in \mathbb{N}$ and $\mathsf{ctr}(\mathcal{Q}_i) - \mathsf{ctr}(\mathcal{Q}_{i-1}) \in \{0,1\}$ for all possible $\mathcal{Q}_i$. For future reference, define the event $\mathsf{hit}(\mathcal{Q}_i) = \mathbf{true}$ iff $\mathsf{ctr}(\mathcal{Q}_i) = \mathsf{ctr}(\mathcal{Q}_{i-1}) + 1$. In the new game $\mathsf{Adv}_\kappa^{\mathbf{B}_\Sigma}(\mathcal{A})$, the predicate $\mathbf{E}(\mathcal{Q})$ is set if and only if $\mathsf{ctr}(\mathcal{Q}_q) > \kappa$.

Proposition 4. *For any* non-adaptive *adversary* $\mathsf{Adv}_\kappa^{\mathbf{B}_\Sigma}(\mathcal{A}) \leq \binom{q}{k+1} \left(\frac{B_\Sigma}{q}\right)^{k+1}$.

Note that for $\kappa = 0$ we retrieve the result of the preceding section given that $\binom{q}{1}\left(\frac{B_\Sigma}{q}\right)^1 = B_\Sigma$ and it might be tempting to think that for larger κ adaptivity can be argued away. This however is not the case, an adaptive adversary does have an increased advantage playing $\mathsf{Adv}_\kappa^{\mathbf{B}_\Sigma,\mathbf{B}'}$ when compared to a non-adaptive one. Nonetheless, we conjecture that the bound just derived is sufficiently loose to apply to adaptive adversaries as well.

Conjecture 1. For any *adaptive* adversary $\mathsf{Adv}_\kappa^{B_\Sigma}(\mathcal{A}) \leq \binom{q}{k+1} \left(\frac{B_\Sigma}{q}\right)^{k+1}$.

Proposition 5. *For any* adaptive *adversary* $\mathsf{Adv}_\kappa^{B_\Sigma}(\mathcal{A}) \leq (B_\Sigma)^{\kappa+1}$.

4   A New Double-Length Compression Function

In this section, we introduce a new compression function (Construction 1, see also Fig. 1) $H \colon \{0,1\}^{3n} \to \{0,1\}^{2n}$ that makes parallel calls to two random functions $f_1, f_2 \colon \{0,1\}^{2n} \to \{0,1\}^n$. For notational convenience, we often write the input $W \in \{0,1\}^{3n}$ as $(a, b, c) \in (\{0,1\}^n)^3$ and identify $\{0,1\}^n$ with $\mathbb{F}_{2^n}$.

Construction 1. Let $f_1, f_2 \colon \{0,1\}^{2n} \to \{0,1\}^n$ be two distinct and independently sampled PuRFs. Define $H^{f_1, f_2} \colon \{0,1\}^{3n} \to \{0,1\}^{2n}$ to be a single-layer compression function using the preprocessing function $C^{\mathrm{PRE}} \colon \mathbb{F}_{2^n}^3 \to (\mathbb{F}_{2^n}^2)^2$ defined by $C^{\mathrm{PRE}} = (C_1^{\mathrm{PRE}}, C_2^{\mathrm{PRE}})$, where

$$C_1^{\mathrm{PRE}}(a, b, c) = (a, b) \quad \text{and} \quad C_2^{\mathrm{PRE}}(a, b, c) = (c, ac + b)$$

and the postprocessing function $C^{\mathrm{POST}} \colon \mathbb{F}_{2^n}^5 \to \mathbb{F}_{2^n}^2$

$$C^{\mathrm{POST}}(a, b, c, y_1, y_2) = A \cdot \begin{pmatrix} a & c & y_1 & y_2 & y_1 y_2 \end{pmatrix}^T, \quad \text{where } A = \begin{pmatrix} \omega_{11} & \omega_{12} & \omega_{13} & \omega_{14} & \omega_{15} \\ \omega_{21} & \omega_{22} & \omega_{23} & \omega_{24} & \omega_{25} \end{pmatrix}$$

is a matrix (over $\mathbb{F}_{2^n}$) satisfying certain non-degeneracy conditions (see Table 1).

Design Rationale. In the security proofs, we abstract as best as we can the properties required of C^{PRE} and C^{POST}. In practice, we recommend using the matrix (cf. Table 1) $A = \begin{pmatrix} 1 & 1 & 0 & 0 & 1 \\ 0 & 0 & 1 & 1 & 0 \end{pmatrix}$. Note that in the context of iterating the compression function, one needs to specify which input blocks represent the message block and which ones represent the state or chaining variable. Our security results are independent of this choice. The choice may, however, significantly affect the efficiency of the design.

Incidence-Based Preprocessing. For a single-layer construction, the preprocessing function C^{PRE} fully determines the relationship between the queries made to the primitive on one hand, and the compression function evaluations this enables on the other. Our search is therefore for a preprocessing function C^{PRE} such that yield(q) does not grow too fast as a function of q. In particular, we are interested in whether we can find a C^{PRE} that has good behaviour for $q < 2^{2n}$ as well. It turns out, we can do well by exploiting results from incidence geometry. We note the following theorem that is a finite field version of a theorem of Szemerédi and Trotter over the reals (see, e.g. [16] for an elementary proof).

Theorem 2. *Let $\mathbb{F}$ be a finite field and P (resp. L) be a set of points (resp. lines) in $\mathbb{F}^2$. Let $I(P, L) = \{(p, \ell) \mid (p, \ell) \in P \times L \text{ and } p \in \ell\}$. Then*

$$|I(P, L)| \leq \min\left(|P||L|^{1/2} + |L|, |L||P|^{1/2} + |P|\right) .$$

Let $(a, b), (c, d) \in \mathbb{F}_{2^n}^2$ denote the query pairs made to f_1 and f_2, respectively. We call a query pair (a, b)–(c, d) *compatible* if and only if $((a, b), (c, d))$ is in the image of C^{PRE}. In addition, a query (a, b) is called (c, d)-compatible or vice versa if the pair (a, b)–(c, d) is compatible. For the preprocessing function C^{PRE} from Construction 1, a pair (a, b)–(c, d) is compatible if and only if $d = ac + b$ is satisfied. Finally, a preprocessing function C^{PRE} satisfies the completion property if and only if (i) (a, b) and c (ii) (c, d) and a uniquely determines a compatible query pair (a, b)–(c, d) for any $a, b, c, d \in \mathbb{F}_{2^n}$.

Proposition 6. *The preprocessing function C^{PRE} from Construction 1 has the completion property and* $\mathrm{yield}(q) \leq q^{3/2} + q$.

Proof. We remark that the completion property can be algebraically verified. To determine the yield, we interpret the (a, b) as the line $y = ax + b$ in $\mathbb{F}_{2^n}^2$ and (c, d) as a point in $\mathbb{F}_{2^n}^2$. This renders bounding the yield an immediate consequence of Theorem 2. To finish the proof, note that the sets $\mathcal{Q}[1]$ and $\mathcal{Q}[2]$ correspond to the lines L and the points P, respectively, and $|I(\mathcal{Q}[2], \mathcal{Q}[1])|$ counts exactly the number of compression function inputs whose mapping can be completely determined by the given queries. Specifying $|\mathcal{Q}[1]| = |\mathcal{Q}[2]| = q$ yields the proposition statement. $\square$

Non-linear Matrix-Style Postprocessing. Our postprocessing is clearly inspired by the use of $\mathbb{F}_{2^n}$-matrices by Rogaway and Steinberger [10], but with the crucial difference that we add the non-linear term $y_1 y_2$. Omitting this non-linear term is fatal for security. For the fully-linear version an adaptive adversary can force its evaluated digest to lie (uniformly) on a *prespecified* set of size 2^n. In contrast, for our construction, the adversary's control is significantly reduced.

Security Claims. We state our security claims for collision and (everywhere) preimage resistance in Theorems 3 and 4, respectively. A sketch of our collision resistance proof is given in Section 5. We refer to the full version for proofs of Theorem 4, Corollaries 1 and 2.

Theorem 3. *Let H^{f_1, f_2} be a single-layer compression function defined by C^{POST} given in Construction 1 where $C^{\mathrm{PRE}} \colon \mathbb{F}_{2^n}^3 \to (\mathbb{F}_{2^n}^2)^2$ is any function that satisfies the completion property. Let $k, \mu, \gamma > 0$ and $\lambda \geq 3$ be integers and let $\kappa = k\lambda + \mu$. Then*

$$\mathsf{Adv}_H^{\mathrm{coll}}(q) \leq \frac{\kappa Y}{2^n} + \frac{q(\gamma^2 + 1)}{2^{n-1}} + \frac{\binom{q}{\gamma}}{2^{(\gamma-1)n-1}} + 2^{2n} \left(\frac{Y}{2^n}\right)^{k+1} + \frac{\binom{q}{\mu}}{2^{(\mu-1)n-1}} + \frac{\binom{q}{\lambda}}{2^{(\lambda-2)n-1}} \ .$$

Corollary 1. *Let H^{f_1, f_2} be the compression function given in Construction 1. For every $\delta > 0$ and $q = 2^{2n(1-\delta)/3}$, one has $\mathsf{Adv}_H^{\mathrm{coll}}(q) = o(1)$ as $n \to \infty$.*

Theorem 4. *Let H^{f_1, f_2} be a single-layer compression function defined by C^{POST} given in Construction 1 and an arbitrary C^{PRE} that satisfies the completion property. For any integer $\kappa > 1$, one has*

$$\mathsf{Adv}_H^{\mathrm{epre}}(q) \leq 2^{n+1} \binom{q}{\kappa} \left(\frac{1}{2^{n-1}}\right)^{\kappa} + \frac{q}{2^{n-1}} + \frac{\kappa q}{2^{n-1}} \ .$$

Corollary 2. *Let H^{f_1,f_2} be the compression function given in Construction 1. Then for all $\delta > 0$ and $q = 2^{n(1-\delta)}$, it holds that $\mathsf{Adv}_H^{\mathsf{epre}}(q) = o(1)$ as $n \to \infty$.*

5 Proof of Collision Resistance (Theorem 3)

5.1 Overall Strategy

Let $\mathcal{A}$ be a collision-finding adversary making at most q queries to each of the public random functions f_1 and f_2 (without loss of generality, we assume that the adversary makes exactly q queries to both). Our goal is to bound $\mathsf{Adv}_H^{\mathsf{coll}}(\mathcal{A})$, in particular $\Pr[\mathsf{coll}(\mathcal{Q})]$, where $\mathcal{Q}$ is adaptively generated by $\mathcal{A}$. We slightly abuse notation and use $\mathcal{Q}$ (and derived symbols such as $\mathcal{Q}_i$) interchangeably as a random variable (when it is the direct result of playing the collision game), or as a dummy variable (e.g. when we want to quantify over all possible instantiations), where the context makes the precise meaning clear. In all cases we can use the global parameter q for the number of f_1 and f_2 queries and $Y = \mathrm{yield}(q)$.

To bound the probability of an adversary finding a collision, we first look at the probability that any specific query completes the collision: fix i and consider the event $\mathsf{coll}(\mathcal{Q}_i) \wedge \neg\mathsf{coll}(\mathcal{Q}_{i-1})$. Here we call query i *fresh* and we say it *causes* a collision. For concreteness, suppose the ith query is an f_1-query (a,b) (the case for an f_2-query (c,d) is analogous), then the first observation is that it adds a new point to the yield set for every (a,b)-compatible pair (c,d) that was already in $\mathcal{Q}_{i-1}$. Now the ith query can cause a collision in two different ways:

Case I. Two compatible and colliding pairs (a,b)–(c,d) and (a',b')–(c',d') are formed with the triple $\{(a',b'),(c,d),(c',d')\} \subseteq \mathcal{Q}_{i-1}$ (where $(a,b) \neq (a',b')$).
Case II. Two distinct compatible and colliding pairs exist with $(a,b) = (a',b')$ and $\{(c,d),(c',d')\} \subseteq \mathcal{Q}_{i-1}$, where $(c,d) \neq (c',d')$.

We associate the events $\mathsf{coll}_I(\mathcal{Q})$ and $\mathsf{coll}_{II}(\mathcal{Q})$ with these two cases; it follows that $\mathsf{coll}(\mathcal{Q}) \equiv (\mathsf{coll}_I(\mathcal{Q}) \vee \mathsf{coll}_{II}(\mathcal{Q}))$. The probability of finding a collision at the ith step depends strongly on the number of compatible queries already in $\mathcal{Q}_{i-1}$; we denote this number by (random variable) n_i. While we know (by design) that $\sum_{i=1}^{2q} n_i \leq \mathrm{yield}(q)$, a straightforward union bound fails to take this into account properly: Because potentially $n_i \approx q$, naive bounding of $\sum_{i=1}^{2q} n_i$ would be quadratic in q (which is typically much larger than $\mathrm{yield}(q)$). Dealing with this in case of non-adaptive adversaries is straightforward (as such an adversary needs to commit to the n_i values in advance), but requires a more careful treatment in the case of adaptive adversaries. To bound the probability of $\mathsf{coll}_I(\mathcal{Q})$, we additionally condition on not having too many collinear output points. For an integer $\kappa > 0$, $\mathsf{bad}_{\mathsf{cl}[\kappa]}(\mathcal{Q})$ is set if and only if $\mathcal{Q}$ leads to more than κ collinear output points in $\mathbb{F}_{2^n}^2$. The reason for collinearity will become evident shortly.

An Overview of the Proof. We start with the observation, for any $\mathcal{Q}$, that

$$\mathsf{coll}(\mathcal{Q}) \equiv (\mathsf{coll}_I(\mathcal{Q}) \vee \mathsf{coll}_{II}(\mathcal{Q})) \equiv (\mathsf{coll}_I(\mathcal{Q}) \wedge \neg\mathsf{coll}_{II}(\mathcal{Q})) \vee \mathsf{coll}_{II}(\mathcal{Q}) ,$$

where the expression $(\mathsf{coll}_I(\mathcal{Q}) \wedge \neg\mathsf{coll}_{II}(\mathcal{Q}))$ is equivalent to

$$\left(\mathsf{coll}_I(\mathcal{Q}) \wedge \neg\mathsf{coll}_{II}(\mathcal{Q}) \wedge \neg\mathsf{bad}_{\mathsf{cl}[\kappa]}(\mathcal{Q})\right) \vee \left(\mathsf{coll}_I(\mathcal{Q}) \wedge \neg\mathsf{coll}_{II}(\mathcal{Q}) \wedge \mathsf{bad}_{\mathsf{cl}[\kappa]}(\mathcal{Q})\right) \ .$$

Using the trivial implications for the above statements, we reach

$$\mathsf{coll}(\mathcal{Q}) \Rightarrow \underbrace{\left(\mathsf{coll}_I(\mathcal{Q}) \wedge \neg\mathsf{bad}_{\mathsf{cl}[\kappa]}(\mathcal{Q})\right)}_{\mathbf{E_1}} \vee \underbrace{\left(\neg\mathsf{coll}_{II}(\mathcal{Q}) \wedge \mathsf{bad}_{\mathsf{cl}[\kappa]}(\mathcal{Q})\right)}_{\mathbf{E_2}} \vee \underbrace{\mathsf{coll}_{II}(\mathcal{Q})}_{\mathbf{E_3}}. \quad (1)$$

The idea of our proof is to find separate upper bounds for the probability of the events $\mathbf{E_i}$ for $i = 1, 2, 3$ and then use the union bound the finalize the proof in Corollary 3 (i.e. $\sum_{i=1}^{3} \Pr[\mathbf{E_i}]$ provides the overall upper bound). An upper bound for $\Pr[\mathbf{E_1}]$ is given in Lemma 7 (corresponding to the term $\kappa Y/2^n$ in Theorem 3). An upper bound for $\Pr[\mathbf{E_3}]$ is established in Lemma 8, which corresponds to the term $q\gamma^2/2^{n-1} + q^\gamma/2^{(\gamma-1)n-1}$ from Theorem 3. Finally, we explain where the bounds for $\Pr[\mathbf{E_2}]$ (i.e. the remaining terms from Theorem 3) come from. We use Proposition 9 to establish an implication that leads to an upper bound for $\Pr[\mathbf{E_2}]$. Moreover, several auxiliary events, which are defined and investigated in Sections 5.2 and 5.4, are required to finalize the bound $\Pr[\mathbf{E_2}]$: The upper bound for the auxiliary events are given in Lemmas 3, 4, 5 and 9.

On the Matrix A Used in C^{POST}. In the following, we consider a general matrix A (see Construction 1) over $\mathbb{F}_{2^n}$ for the proof of Theorem 3. The conditions on the entries of the matrix A required throughout the paper, as well as where they are used, are provided in Table 1. (Note that the probability that a randomly selected matrix A satisfies our criterion is close to one.)

Output Lines. By assumption, we know that an f_1-query (a, b) can only complete a collision using an already present compatible f_2-query (c, d). Let (a, b) be an f_1-query and let (c, d) be a preceding (a, b)-compatible f_2-query with $y_2 = f_2(c, d)$. The output $C^{\text{POST}}(a, b, c, y_1, y_2)$ of the compression function on input (a, b, c) then lies on the line (in $\mathbb{F}_{2^n}^2$)

$$\mathcal{L}_{1:c,d,y_2;a} : \left\{ \underbrace{\begin{pmatrix} a\omega_{11} + c\omega_{12} + y_2\omega_{14} \\ a\omega_{21} + c\omega_{22} + y_2\omega_{24} \end{pmatrix}}_{\text{offset}} + y_1 \underbrace{\begin{pmatrix} \omega_{13} + y_2\omega_{15} \\ \omega_{23} + y_2\omega_{25} \end{pmatrix}}_{\text{slope}} \mid y_1 \in \mathbb{F}_{2^n} \right\}, \quad (2)$$

where we get the actual output point for (a, b, c) by setting $y_1 = f_1(a, b)$. The randomness of f_1 results in a random point on $\mathcal{L}_{1:c,d,y_2;a}$. Note that the line cannot be degenerate (see condition **C1** in Table 1), i.e. it has nonzero slope.

Similarly, let (c, d) be an f_2-query and let (a, b) be a preceding (c, d)-compatible f_1-query. The output of the compression function on (a, b, c) lies on the line:

$$\mathcal{L}_{2:a,b,y_1;c} : \left\{ \begin{pmatrix} a\omega_{11} + c\omega_{12} + y_1\omega_{13} \\ a\omega_{21} + c\omega_{22} + y_1\omega_{23} \end{pmatrix} + y_2 \begin{pmatrix} \omega_{14} + y_1\omega_{15} \\ \omega_{24} + y_1\omega_{25} \end{pmatrix} \mid y_2 \in \mathbb{F}_{2^n} \right\} \ . \quad (3)$$

This time the output point is obtained by setting $y_2 = f_2(c, d)$. Again, the randomness of f_2 results in a random point on $\mathcal{L}_{2:a,b,y_1;c}$. We note that this time

Table 1. Quick recap of the properties of the entries of A (see Construction 1) used in the proof of Theorem 3. (**N**) denotes that the condition is necessary, whereas (**S**) denotes it is sufficient.

The Condition	Where used	Reference
	(Theorem 3)	(Section 5)
C1: $\omega_{13}\omega_{25} \neq \omega_{15}\omega_{23}$	(**S**) Non-degeneracy of $\mathcal{L}_{1:}$-lines	(2)
	(**N**) Non-parallel $\mathcal{P}_{1:}$-partitions	Lemma 4
C2: $\omega_{14}\omega_{25} \neq \omega_{15}\omega_{24}$	(**S**) Non-degeneracy of $\mathcal{L}_{2:}$-lines	(3)
	(**N**) Non-parallel $\mathcal{P}_{2:}$-partitions	Lemma 4
C3: $\omega_{11} \neq 0 \wedge \omega_{21} \neq 0$	(**N**) Non-degeneracy of $\mathcal{P}_{1:}$-partitions	Lemma 3
C4: $\omega_{12} \neq 0 \wedge \omega_{22} \neq 0$	(**N**) Non-degeneracy of $\mathcal{P}_{2:}$-partitions	Lemma 3
C5: $\omega_{15} \neq 0 \wedge \omega_{25} \neq 0$	(**N**) Nonlinearity of C^{POST}	Construction 1

non-degeneracy follows from $\omega_{14}\omega_{25} - \omega_{24}\omega_{15} \neq 0$ (see condition **C2** in Table 1). Now it is easy to see why we do not want too many collinear points: It would ease the collision-finding considerably due to the above output lines.

5.2 Partitions, Bunches and Some Auxiliary Events

Partitions and Bunches. Suppose that an f_2-query (c, d) results in $y_2 = f_2(c, d)$. By the completion property, we obtain, for each $a \in \mathbb{F}_{2^n}$, a unique b such that (a, b) is (c, d)-compatible. Now we recall that if we query $f_1(a, b)$, the resulting yield point lies on the line $\mathcal{L}_{1:c,d,y_2;a}$. From Equation (2) of $\mathcal{L}_{1:c,d,y_2;a}$, it follows that the slope of these lines is fixed (because (c, d) and y_2 are fixed) and independent of a; hence by ranging over all possible $a \in \mathbb{F}_{2^n}$ we achieve a set of (parallel) lines. This is what we call a partition (partitions due to an f_1-query is defined analogously): $\mathcal{P}_{1:c,d,y_2} = \{\mathcal{L}_{1:c,d,y_2;a} \mid a \in \mathbb{F}_{2^n}\}$. The opposite notion to a partition is a bunch: For all preceding and (a, b)-compatible $(c_j, d_j) \in \mathcal{Q}$, for some integer $j \geq 1$, the bunch of interest is the collection of lines (for $y_{2:j} = f_2(c_j, d_j)$)

$$\mathcal{B}_{1:(a,b)}(\mathcal{Q}) = \{\mathcal{L}_{1:c_j,d_j,y_{2:j};a} \mid (c_j, d_j, y_{2:j}) \in \mathcal{Q} \wedge (c_j, d_j) \text{ compatible with } (a, b)\} \ .$$

(We also write $\mathcal{B}_{1:i}$ if the ith query is an f_1-query (a, b).) The answer $y_1 = f_1(a, b)$ specifies a point on each of these lines to be added to the yield set; we refer to this as *realizing* the bunch. For the record, $\mathcal{B}_{2:(c,d)}(\mathcal{Q})$ is defined analogously.

Degenerate Partitions. We have seen that a partition contains a set of parallel lines. If different choices of a lead to different lines, the lines compatible to (c, d) necessarily partition the output plane (justifying our terminology). It is possible however that regardless of the a values, we end up with identical lines (though with a different parametrization). In such a case, a partition collapses to a single line and we speak of a degenerate partition. A degenerate partition causes problems in our proof, because it allows an adversary to create many collinear points (by ranging over a). Let $\mathsf{bad}_{\mathsf{dp}}(\mathcal{Q})$ denote the event that $\mathcal{Q}$ gives rise to a degenerate partition (either via different a or c values).

Lemma 3. *Let $\mathcal{Q}$ be generated adaptively, then* $\Pr\left[\mathsf{bad}_{\mathsf{dp}}(\mathcal{Q})\right] \leq q/2^{n-1}$, *and if* $\omega_{11}\omega_{23} \neq \omega_{13}\omega_{21}, \omega_{12}\omega_{24} \neq \omega_{14}\omega_{22}, \omega_{11}\omega_{25} = \omega_{15}\omega_{21}$ *and* $\omega_{12}\omega_{25} = \omega_{15}\omega_{22}$, *then* $\Pr\left[\mathsf{bad}_{\mathsf{dp}}(\mathcal{Q})\right] = 0$.

Parallel Partitions. We now define another bad event, *parallel partitions*, that can potentially help a collision-finding adversary create collinear points. We have seen that, once answered, a single f_2-query (c, d) determines a well-defined slope for the partition $\mathcal{P}_{1:c,d,y_2}$. If two or more distinct partitions (of the same type) have the same slope, we call the partitions parallel. The number of parallel partitions is tightly related to a standard occupancy problem. Consequently, avoiding parallel partitions altogether is not realistic, yet we can put reasonable bounds on too much parallelism occurring. We define $\mathsf{bad}_{\mathsf{pp}[\mu]}(\mathcal{Q})$ to be the event that $\mathcal{Q}$ results in more than μ parallel partitions (of identical type).

Lemma 4. *Let $\mathcal{Q}$ be generated adaptively. Then, for any integer $\mu > 0$,*

$$\Pr\left[\mathsf{bad}_{\mathsf{pp}[\mu]}(\mathcal{Q})\right] \leq \frac{\binom{q}{\mu}}{2^{(\mu-1)n-1}} .$$

Local Collinearity. Now we discuss another auxiliary event, *local collinearity*, that is used in our collinearity analysis. Suppose an f_1-query results in $y_1 = f_1(a, b)$. We associate with this query-response pair a point $(a, y_1) \in \mathbb{F}_{2^n}^2$. Let $\mathsf{bad}_{\mathsf{lc}[\lambda]}(\mathcal{Q})$ be the event that there exist at least λ pairs of f_1-queries (a_i, b_i) with distinct a_i values, such that the associated points $(a_i, y_{1:i})$ are collinear or, alternatively, that there exist at least λ pairs of f_2-queries (c_i, d_i) with distinct c_i values, such that the points $(c_i, y_{2:i})$ are collinear.

Lemma 5. *Let $\mathcal{Q}$ be generated adaptively. Then, for any integer $\lambda > 0$,*

$$\Pr\left[\mathsf{bad}_{\mathsf{lc}[\lambda]}(\mathcal{Q})\right] \leq \frac{\binom{q}{\lambda}}{2^{(\lambda-2)n-1}} .$$

Target Local Collinearity. For local collinearity, we are interested in any λ associated points being collinear, without worrying about which line they are on. However, in an upcoming case we are only interested in points all lying on a line with a pre-specified slope (the offset of the line is not fixed in advance). Let $\mathsf{bad}_{\mathsf{slc}[\gamma]}(\mathcal{Q})$ be the event that $\mathcal{Q}[1]$ or $\mathcal{Q}[2]$ leads to more than γ associated points collinear with pre-specified, non-vertical slope.

Lemma 6. *Let $\mathcal{Q}$ be generated adaptively. Then, for any integer $\gamma > 0$,*

$$\Pr\left[\mathsf{bad}_{\mathsf{slc}[\gamma]}(\mathcal{Q})\right] \leq \frac{\binom{q}{\gamma}}{2^{(\gamma-1)n-1}} .$$

5.3   Bounding Collisions: Focusing on $\Pr[\mathbf{E_1}]$ and $\Pr[\mathbf{E_3}]$

Lemmas 7 and 8 provide an upper bound for $\Pr[\mathbf{E_1}]$ and $\Pr[\mathbf{E_3}]$, respectively.

Lemma 7. *Let i be a positive integer that satisfies $i \leq q$ and Let $\mathcal{Q}_{i-1}$ be arbitrary query list satisfying $\neg\mathsf{bad}_{\mathsf{cl}[\kappa]}(\mathcal{Q}_{i-1})$ (for some positive integer κ). Then*

$$\Pr[\mathbf{E_1}] = \Pr[\mathsf{coll}_I(\mathcal{Q}) \wedge \neg\mathsf{bad}_{\mathsf{cl}[\kappa]}(\mathcal{Q})] \leq \frac{\kappa Y}{2^n} .$$

Proof (Sketch). We start by noticing that $\Pr[\mathbf{E_1}] \leq \Pr[\mathsf{coll}_I(\mathcal{Q})|\neg\mathsf{bad}_{\mathsf{cl}[\kappa]}(\mathcal{Q})]$. Each of the n_i compatible elements together with the ith query, defines a line such that the random answer to the ith query will determine which point will be added to the yield set. The condition $\neg\mathsf{bad}_{\mathsf{cl}[\kappa]}(\mathcal{Q}_{i-1})$ implies that on each of these lines, there are at most κ previous yield points. Since the underlying primitive is a random function, the answer is fully random and for a given line, one of the previous yield points is hit with probability at most $\kappa/2^n$. A union bound over the n_i lines gives the bound $n_i\kappa/2^n$. To obtain the overall bound, we exploit our refined game $\mathsf{Adv}_\kappa^{\mathsf{B}_\Sigma}(\mathcal{A})$ to determine the $\bigvee$ expression. Here we use the above to determine $B_\Sigma = \kappa Y/2^n$ as $\sum_{i=1}^{2q} n_i \leq Y$ (Proposition 3). □

We now bound the probability of finding an instantaneous collision with a fresh query, first given that $\neg\mathsf{bad}_{\mathsf{slc}[\gamma]}(\mathcal{Q}_{i-1})$ holds. Then we finalize our bound for $\Pr[\mathsf{coll}_{II}(\mathcal{Q})]$ using Proposition 2 along with Lemma 6.

Lemma 8. *Let i be a positive integer that satisfies $i \leq q$ and let $\mathcal{Q}$ be generated adaptively. Then, for any integer $\gamma > 0$,*

$$\Pr[\mathbf{E_3}] = \Pr[\mathsf{coll}_{II}(\mathcal{Q})] \leq \frac{q\gamma^2}{2^{n-1}} + \Pr[\mathsf{bad}_{\mathsf{slc}[\gamma]}(\mathcal{Q})] .$$

5.4 Bounding Overall Collinearity: Bounding $\Pr[\mathbf{E_2}]$

We now bound $\Pr[\mathbf{E_2}]$. The main technical difficulty is to properly separate the randomness of the f_1- and f_2-queries. In order to do this in the adaptive setting, we use a method that we call *bunching*. For a fixed i, suppose that the ith query is an f_1-query (a, b). Recall that for the f_1-query (a, b), the bunch $\mathcal{B}_{1:i}$ consists of the lines $\mathcal{L}_{1:c_j,d_j,y_{2:j};a}$ for the n_i compatible preceding f_2-queries (c_j, d_j) (with $y_{2:j} = f_2(c_j, d_j)$ for $j = 1, \ldots, n_i$). The answer $y_1 = f_1(a, b)$ adds a single point to the yield set for each compatible f_2-query (c_j, d_j). These n_i new points lie on the lines $\mathcal{L}_{1:c_j,d_j,y_{2:j};a}$, thereby realizing the bunch $\mathcal{B}_{1:i}$. We refer to the set of freshly added points inside a bunch as a *constellation* that we denote by

$$\mathcal{C}_{1:i}(\mathcal{Q}) = \{H^{f_1,f_2}(a, b, c_j) \mid (c_j, d_j) \in \mathcal{Q}_{i-1} \wedge (c_j, d_j) \text{ compatible with } (a, b)\} .$$

In order to determine the maximum collinearity within the yield set, we estimate (i) the probability of too much collinearity occurring within a single constellation (Proposition 8) and (ii) the probability of too many constellations being collinear (Lemma 9). Here, a set of constellations is collinear if we can choose a point from each constellation in the set such that all chosen yield points are collinear. If we know that at most λ points are collinear within a single constellation, and at most k constellations are collinear, we can conclude that at most $\kappa = k\lambda$ points are collinear overall. This is formalized in the proposition below, taking into account an additional technicality.

Proposition 7. *Let $k, \lambda, \mu > 0$ be fixed integers, $\kappa = k\lambda + \mu$ and let $\mathsf{bad}_{\mathsf{int}[\lambda]}(\mathcal{Q})$ be the event that there exists a constellation having more than λ collinear points. Define $\mathsf{bad}_{\mathsf{ext}[k]}(\mathcal{Q})$ to be the event that there exists a line ℓ passing through more than k constellations whose bunches do not contain ℓ. Then (for arbitrary $\mathcal{Q}$)*

$$\mathsf{bad}_{\mathsf{cl}[\kappa]}(\mathcal{Q}) \Rightarrow \left(\mathsf{bad}_{\mathsf{int}[\lambda]}(\mathcal{Q}) \vee \mathsf{bad}_{\mathsf{ext}[k]}(\mathcal{Q}) \vee \mathsf{bad}_{\mathsf{pp}[\mu]}(\mathcal{Q})\right) .$$

Proposition 8 is used to decompose the event $\mathsf{bad}_{\mathsf{int}[\lambda]}(\mathcal{Q})$ into two events.

Proposition 8. *For arbitrary $\mathcal{Q}$, if (integer) $\lambda \geq 3$ then*

$$\left(\neg\mathsf{coll}_{II}(\mathcal{Q}) \wedge \mathsf{bad}_{\mathsf{int}[\lambda]}(\mathcal{Q})\right) \Rightarrow \left(\mathsf{bad}_{\mathsf{dp}}(\mathcal{Q}) \vee \mathsf{bad}_{\mathsf{lc}[\lambda]}(\mathcal{Q})\right) .$$

To bound collinearity between constellations, we first consider collinearity with a given line ℓ in the output plane. We are interested in bounding the probability that at least k constellations are incident to ℓ. For a line ℓ, integer k and query history $\mathcal{Q}$, let $\mathsf{bad}_{\ell-\mathsf{hit}[k]}(\mathcal{Q})$ be the following event: there exist at least k constellations whose corresponding bunches do not contain ℓ that are incident to ℓ. Recall that $\mathsf{bad}_{\mathsf{ext}[k]}(\mathcal{Q})$ is the event that there exists a line ℓ passing through more than k constellations whose bunches do not contain ℓ.

Lemma 9. *Let ℓ be given and let $\mathcal{Q}$ be generated adaptively. Then*

$$\Pr\left[\mathsf{bad}_{\ell-\mathsf{hit}[k]}(\mathcal{Q})\right] \leq \left(\frac{Y}{2^n}\right)^{k+1} \quad and \quad \Pr\left[\mathsf{bad}_{\mathsf{ext}[k]}(\mathcal{Q})\right] \leq 2^{2n}\left(\frac{Y}{2^n}\right)^{k+1} .$$

Proof (Sketch). Let $\mathsf{ctr}_{\ell-\mathsf{hit}}(\mathcal{Q})$ be the number of constellations that are incident to ℓ, again restricted to those constellations whose corresponding bunch does not contain ℓ. Clearly, the event $\mathsf{bad}_{\ell-\mathsf{hit}[k]}(\mathcal{Q})$ is equivalent to $\mathsf{ctr}_{\ell-\mathsf{hit}}(\mathcal{Q}) \geq k$. Note that for any i, we have $\mathsf{ctr}_{\ell-\mathsf{hit}}(\mathcal{Q}_i) - \mathsf{ctr}_{\ell-\mathsf{hit}}(\mathcal{Q}_{i-1}) \in \{0, 1\}$ since constellation i can be counted at most once (namely if it is incident to ℓ). Let $\mathsf{hit}_{\ell-\mathsf{hit}}(i)$ be the event that the bunch $\mathcal{B}_i$ upon realization is incident to ℓ. Suppose that $\ell \notin \mathcal{B}_i$ and that $\mathcal{B}_i$ consists of n_i lines (each containing an output point). Since ℓ intersects each line in a bunch in at most one point, we obtain that $\Pr\left[\mathsf{hit}_{\ell-\mathsf{hit}}(i)\right] \leq n_i/2^n$. Due to yield restrictions, $\sum_{i=1}^{2q} n_i \leq Y$. The lemma statement follows from applying Proposition 4 with $B_\Sigma = Y/2^n$. The statement for $\Pr\left[\mathsf{bad}_{\mathsf{ext}[k]}(\mathcal{Q})\right]$ follows from the union bound over all lines ℓ. $\qquad\square$

Proposition 9. *Let $k, \lambda,$ and μ be positive integers with $\lambda \geq 3$ and $\kappa = k\lambda + \mu$. Then, for arbitrary $\mathcal{Q}$,*

$$\Pr[\neg\mathsf{coll}_{II}(\mathcal{Q}) \wedge \mathsf{bad}_{\mathsf{cl}[\kappa]}(\mathcal{Q})] \leq \Pr[\mathbf{F}(\mathcal{Q})] , where$$

$$\Pr[\mathbf{F}(\mathcal{Q})] \leq \Pr[\mathsf{bad}_{\mathsf{ext}[k]}(\mathcal{Q})] + \Pr[\mathsf{bad}_{\mathsf{pp}[\mu]}(\mathcal{Q})] + \Pr[\mathsf{bad}_{\mathsf{lc}[\lambda]}(\mathcal{Q})] + \Pr[\mathsf{bad}_{\mathsf{dp}}(\mathcal{Q})] .$$

Finishing the Proof. The following corollary wraps up what we have discussed so far and finishes the proof of Theorem 3 with the help of earlier obtained bounds (Lemmas 3, 4, 5, 7, 8, and 9).

Corollary 3. *Let $\mathcal{Q}$ be generated adaptively, then for $\mathbf{F}(\mathcal{Q})$ given in Prop. 9*

$$\Pr[\mathsf{coll}(\mathcal{Q})] \leq \Pr[\mathsf{coll}_I(\mathcal{Q}) \wedge \neg\mathsf{bad}_{\mathsf{cl}[\kappa]}(\mathcal{Q})] + \Pr[\mathsf{coll}_{II}(\mathcal{Q})] + \Pr[\mathbf{F}(\mathcal{Q})] .$$

6   Blockcipher-Based Instantiation

A naïve replacement of the underlying PuRFs in Construction 1 with ideal block-ciphers leads to a weaker security due to the availability of the decryption queries (see the full version for the justification). However, adding a layer of "Davies–Meyer" suffices for our purposes. Note that there is no need to change C^{PRE}; the only modification is in C^{POST} (the proofs are given in the full version).

Construction 5. Let $E_1, E_2 \colon \{0,1\}^n \times \{0,1\}^n \to \{0,1\}^n$ be two fixed randomly (and independently) chosen blockciphers. Define a single-layer compression function $H^{E_1,E_2} \colon \{0,1\}^{3n} \to \{0,1\}^{2n}$ by $C^{\text{PRE}} \colon \mathbb{F}_{2^n}^3 \to (\mathbb{F}_{2^n}^2)^2$ from Construction 1 and $C^{\text{POST}} \colon \mathbb{F}_{2^n}^5 \to \mathbb{F}_{2^n}^2$

$$C^{\text{POST}}(a, b, c, y_1, y_2) = A \cdot (a, c, a + y_1, c + y_2, (a + y_1)(c + y_2))^T, \text{ where}$$

A is a matrix satisfying certain non-degeneracy conditions.[1]

Theorem 6. Let H^{E_1,E_2} be given as in Construction 5 where $C^{\text{PRE}} \colon \mathbb{F}_{2^n}^3 \to (\mathbb{F}_{2^n}^2)^2$ is an arbitrary function that satisfies the completion property. Let $k, \mu, \gamma > 0$ and $\lambda \geq 3$ be integers. Then, for $\kappa = k\lambda + \mu$, $\mathsf{Adv}_H^{\text{coll}}(q)$ is upper bounded by

$$\frac{\kappa Y + 2q\gamma^2 + 4q}{2^n - q} + \frac{4\binom{q}{\gamma}}{(2^n - q)^{(\gamma-1)}} + 2^{2n}\left(\frac{Y}{2^n - q}\right)^{k+1} + \frac{4\binom{q}{\mu}}{(2^n - q)^{(\mu-1)}} + \frac{4\binom{q}{\lambda}}{(2^n - q)^{(\lambda-2)}}.$$

Theorem 7. Let H^{E_1,E_2} be given as in Construction 5 where $C^{\text{PRE}} \colon \mathbb{F}_{2^n}^3 \to (\mathbb{F}_{2^n}^2)^2$ is an arbitrary function that satisfies the completion property and let $\kappa > 1$ be an integer. Then

$$\mathsf{Adv}_H^{\text{epre}}(q) \leq 2^{n+2}\binom{q}{\kappa}\left(\frac{2}{2^n - q}\right)^\kappa + \frac{2q}{2^n - q} + \frac{2\kappa q}{2^n - q}.$$

Acknowledgments. This work was initiated while the third author was at EPFL. It has been supported in part by the European Commission through the ICT programme under contract ICT-2007-216676 ECRYPT II. The first two authors are supported by a grant of the Swiss National Science Foundation, 200021-122162. We thank anonymous reviewers of TCC'12 and Tom Shrimpton for their useful comments.

References

1. Black, J., Rogaway, P., Shrimpton, T.: Black-Box Analysis of the Block-Cipher-Based Hash-Function Constructions from PGV. In: Yung, M. (ed.) CRYPTO 2002. LNCS, vol. 2442, pp. 320–335. Springer, Heidelberg (2002)
2. Black, J., Rogaway, P., Shrimpton, T., Stam, M.: An analysis of the blockcipher-based hash functions from PGV. Journal of Cryptology 23(4), 519–545 (2010)

---

[1] We note that we require extra conditions on the entries of A to make our proofs work; yet these are minor and can be easily satisfied, e.g. the proposed matrix A.

3. Brachtl, B., Coppersmith, D., Hyden, M., Matyas Jr., S., Meyer, C., Oseas, J., Pilpel, S., Schilling, M.: Data authentication using modification detection codes based on a public one-way encryption function. U.S. Patent No 4,908,861 (1990)
4. Hirose, S.: Some Plausible Constructions of Double-Block-Length Hash Functions. In: Robshaw, M. (ed.) FSE 2006. LNCS, vol. 4047, pp. 210–225. Springer, Heidelberg (2006)
5. Lai, X., Massey, J.L.: Hash Functions Based on Block Ciphers. In: Rueppel, R.A. (ed.) EUROCRYPT 1992. LNCS, vol. 658, pp. 55–70. Springer, Heidelberg (1993)
6. Lucks, S.: A collision-resistant rate-1 double-block-length hash function. In: Biham, E., Handschuh, H., Lucks, S., Rijmen, V. (eds.) Symmetric Cryptography. No. 07021 in Dagstuhl Seminar Proceedings, Internationales Begegnungs- und Forschungszentrum für Informatik (IBFI), Schloss Dagstuhl, Germany, Dagstuhl, Germany (2007), http://drops.dagstuhl.de/opus/volltexte/2007/1017
7. Maurer, U.M.: Indistinguishability of Random Systems. In: Knudsen, L.R. (ed.) EUROCRYPT 2002. LNCS, vol. 2332, pp. 110–132. Springer, Heidelberg (2002)
8. Pietrzak, K.: Indistinguishability and Composition of Random Systems. Ph.D. thesis, ETH Zurich (2005)
9. Preneel, B., Govaerts, R., Vandewalle, J.: Hash Functions Based on Block Ciphers: A Synthetic Approach. In: Stinson, D.R. (ed.) CRYPTO 1993. LNCS, vol. 773, pp. 368–378. Springer, Heidelberg (1994)
10. Rogaway, P., Steinberger, J.: Constructing cryptographic hash functions from fixed-key blockciphers. In: Wagner (ed.) [17], pp. 433–450
11. Rogaway, P., Steinberger, J.: Security/Efficiency Tradeoffs for Permutation-Based Hashing. In: Smart, N.P. (ed.) EUROCRYPT 2008. LNCS, vol. 4965, pp. 220–236. Springer, Heidelberg (2008)
12. Stam, M.: Beyond uniformity: Better security/efficiency tradeoffs for compression functions. In: Wagner (ed.) [17], pp. 397–412
13. Stam, M.: Blockcipher-Based Hashing Revisited. In: Dunkelman, O. (ed.) FSE 2009. LNCS, vol. 5665, pp. 67–83. Springer, Heidelberg (2009)
14. Steinberger, J.P.: The Collision Intractability of MDC-2 in the Ideal-Cipher Model. In: Naor, M. (ed.) EUROCRYPT 2007. LNCS, vol. 4515, pp. 34–51. Springer, Heidelberg (2007)
15. Steinberger, J.: Stam's Collision Resistance Conjecture. In: Gilbert, H. (ed.) EUROCRYPT 2010. LNCS, vol. 6110, pp. 597–615. Springer, Heidelberg (2010)
16. Tao, T.: The Szemerédi-Trotter theorem and the cell decomposition (2009), http://terrytao.wordpress.com/2009/06/12/the-szemeredi-trotter-theorem-and-the-cell-decomposition/
17. Wagner, D. (ed.): CRYPTO 2008. LNCS, vol. 5157. Springer, Heidelberg (2008)

Lower Bounds in Differential Privacy

Anindya De*

University of California at Berkeley
anindya@cs.berkeley.edu

Abstract. This paper is about private data analysis, in which a trusted curator holding a confidential database responds to real vector-valued queries. A common approach to ensuring privacy for the database elements is to add appropriately generated random noise to the answers, releasing only these *noisy* responses. A line of study initiated in [7] examines the amount of distortion needed to prevent privacy violations of various kinds. The results in the literature vary according to several parameters, including the size of the database, the size of the universe from which data elements are drawn, the "amount" of privacy desired, and for the purposes of the current work, the arity of the query. In this paper we sharpen and unify these bounds. Our foremost result combines the techniques of Hardt and Talwar [11] and McGregor *et al.* [13] to obtain linear lower bounds on distortion when providing differential privacy for a (contrived) class of low-sensitivity queries. (A query has low sensitivity if the data of a single individual has small effect on the answer.) Several structural results follow as immediate corollaries:

- We separate so-called *counting* queries from arbitrary *low-sensitivity* queries, proving the latter requires more noise, or distortion, than does the former;
- We separate $(\varepsilon, 0)$-differential privacy from its well-studied relaxation (ε, δ)-differential privacy, even when $\delta \in 2^{-o(n)}$ is negligible in the size n of the database, proving the latter requires less distortion than the former;
- We demonstrate that (ε, δ)-differential privacy is much weaker than $(\varepsilon, 0)$-differential privacy in terms of mutual information of the transcript of the mechanism with the database, even when $\delta \in 2^{-o(n)}$ is negligible in the size n of the database.

We also simplify the lower bounds on noise for counting queries in [11] and also make them unconditional. Further, we use a characterization of (ϵ, δ) differential privacy from [13] to obtain lower bounds on the distortion needed to ensure (ε, δ)-differential privacy for $\epsilon, \delta > 0$. We next revisit the LP decoding argument of [10] and combine it with a recent result of Rudelson [15] to improve on a result of Kasiviswanathan *et al.* [12] on noise lower bounds for privately releasing ℓ-way marginals.

Keywords: Differential privacy, LP decoding.

* Supported by NSF grant CCF-1017403, CCF-1118083. Most of the work was done while the author was an intern at Microsoft Research, Silicon Valley.

1   Introduction

This is a paper about private data analysis, in which a trusted curator holding a confidential database responds to real vector-valued queries. Specifically, we focus on the practice of ensuring privacy for the database elements by adding appropriately generated random noise to the answers, releasing only these *noisy* responses. A line of study initiated by Dinur and Nissim examines the amount of distortion needed to prevent privacy violations of various kinds [7]. Dinur and Nissim did not have a definition of privacy; rather, they had a notion that has come to be called *blatant non-privacy*; the modest goal, then, was to add enough distortion to avert blatant non-privacy. Since that time, the community has raised the bar by defining (and achieving) powerful and comprehensive notions of privacy [7,9,8], and the goal has been to preserve $(\varepsilon, 0)$-differential privacy and its relaxation, (ε, δ)-differential privacy. A final goal considered herein, *attribute privacy*, has a more complicated description, but may be thought of as preventing blatant non-privacy for a single data attribute [12] in the presence of a certain kind of contingency table query.

The results in the literature vary according to several parameters, including the number n of elements in the database, the size d of the universe from which data elements are drawn, the "amount" and type of privacy desired, and for the purposes of the current work, the arity k of the query. In this paper we strengthen and unify these bounds.

As corollaries of our work, we obtain several "structural" results regarding different types of privacy guarantees:

- We separate so-called *counting* queries from arbitrary *low-sensitivity* queries, proving the latter requires more noise, or distortion, than does the former;
- We separate $(\varepsilon, 0)$-differential privacy from its well-studied relaxation (ε, δ)-differential privacy, even when $\delta \in 2^{-o(n)}$ is negligible in the size n of the database, proving the latter requires less distortion than the former;
- We demonstrate that (ε, δ)-differential privacy is much weaker than $(\varepsilon, 0)$-differential privacy in terms of mutual information of the transcript of the mechanism with the database even when $\delta \in 2^{-o(n)}$ is negligible in the size n of the database.

We also simplify the lower bounds on noise for counting queries in [11] and also make them unconditional removing a technical assumption on the mechanism present in their paper. Next, we use a characterization of (ϵ, δ) differential privacy from [13] to obtain lower bounds on the distortion needed to ensure (ε, δ)-differential privacy for $\epsilon, \delta > 0$. We remark that [12] also obtain quantitatively similar lower bounds on the distortion required to maintain (ϵ, δ) differential privacy for the class of ℓ-way marginals though their proof technique is very different and arguably much more complicated.

After this, we use results of Rudelson [15] and combine it with LP decoding to show that attribute privacy is violated if ℓ-way marginals are released with at least $1 - \eta$ fraction of these marginals are released with $o(\sqrt{n})$ noise for some $\eta > 0$. The results and the technique in [12] required $\eta = 0$ making our results more powerful. Finally, we extend the results of [7] to the case of small universe size achieving stronger lower bounds to prevent blatant non-privacy.

To describe our results even at a high level we must outline the privacy-preserving database model, the notion of *distortion* or *noise* that may be employed in order to preserve privacy, and the meaning of the goals of the adversary: blatant non-privacy, violation of $(\varepsilon, 0)$-differential privacy, violation of (ε, δ)- differential privacy, and attribute non-privacy.

Typically, the curator of a database receives questions to which it responds with potentially noisy answers. There are two possible settings here. One is that the queries are received by the curator one at a time. The other situation is that all the queries are received by the curator at once and it then publishes (noisy) answers to all of them at once. The former is called the interactive setting and the latter is called the non-interactive setting. All our lower bounds are in the non-interactive setting making them applicable to the interactive setting as well.

We now formally describe a database and a query : A database X is an element of $(\mathbb{Z}^+)^d$. Here d is called the universe size and intuitively refers to the number of types of elements present in the database. Also, for a database X, $n = \sum_{i=1}^{d} X_i$ is defined as the size of the database and refers to the number of elements in the database. Note that we are representing databases as histograms. A query (of arity k) is a map $F : (\mathbb{Z}^+)^d \to \mathbb{R}^k$ such that $\forall i \in [k]$, $\forall x, y \in (\mathbb{Z}^+)^d$, $|F(x+y)_i - F(x)_i| \leq 1$ if $\|y\|_1 = 1$. In other words, every coordinate of the map F is 1-Lipschitz. We say F is a counting query if F is a linear map. The meaning of d, k, n throughout the paper shall be the same as above unless mentioned otherwise.

We now formally introduce the definition of mechanism and privacy.

Definition 1. *Let $\mathcal{F}$ be a family of queries such that $\forall F \in \mathcal{F}$, $F : (\mathbb{Z}^+)^d \to \mathbb{R}^k$. Then, a mechanism $M : (\mathbb{Z}^+)^d \times \mathcal{F} \to \mu(\mathbb{R}^k)$ where $\mu(\mathbb{R}^k)$ is simply the set of probability distributions over $\mathbb{R}^k$. On being given a query $F \in \mathcal{F}$ and a database $x \in (\mathbb{Z}^+)^d$, the curator samples z from the probability distribution $M(x, F)$ and returns z.*

We next state the definition of ϵ-differential privacy (introduced by Dwork *et al.* in [9]) and (ϵ, δ)-differential privacy (introduced by Dwork *et al.* in [8]).

Definition 2. *For a family of queries $\mathcal{F}$, a mechanism $M : (\mathbb{Z}^+)^d \times \mathcal{F} \to \mu(\mathbb{R}^k)$ is said to be ϵ-differentially private if for every $x, y \in (\mathbb{Z}^+)^d$ such that $\|x - y\|_1 \leq 1$, every measurable set $S \subseteq \mathbb{R}^k$ and $\forall F \in \mathcal{F}$, the following holds : Let $M(x, F) = M_{x,F}$ and $M(y, F) = M_{y,F}$ and for a probability distribution Γ, let $\Gamma(S)$ denote the probability of set S under Γ. Then,*

$$2^{-\epsilon} \leq \frac{M_{x,F}(S)}{M_{y,F}(S)} \leq 2^{\epsilon}$$

The mechanism is said to be (ϵ, δ)-differentially private if

$$2^{-\epsilon} \cdot M_{y,F}(S) - \delta \leq M_{x,F}(S) \leq 2^{\epsilon} \cdot M_{y,F}(S) + \delta$$

Typically, δ is set to be negligible in n, k.

We remark that we do not define the notion of noise very precisely here as the notion of noise depends on the context. However, in the context of differential privacy, we use the following definition of noise.

Definition 3. *For a family of queries $\mathcal{F}$, a mechanism $M : (\mathbb{Z}^+)^d \times \mathcal{F} \to \mu(\mathbb{R}^k)$ is said to add noise (at most) η if with high probability (say 0.99) over the randomness of M, $\|M(x, F) - F(x)\|_\infty \leq \eta$.*

While differential privacy is a very strong notion of privacy, sometimes one can show that even very modest definitions of privacy get violated. One such notion is that of blatant non-privacy. We say that a mechanism M for answering F over databases of size n and universe size d is blatantly non-private, if there is an attack A such that w.h.p. over the answer y returned by the mechanism M, $A(y)$ differs from the database only at $o(1)$ fraction of the places. Yet another very weak notion of privacy that is interesting to us is that of attribute non-privacy. The formal definition follows :

Definition 4. *For a query $F \in \mathcal{F}$, a mechanism $M : (\{0,1\}^d)^n \times \mathcal{F} \to \mathbb{R}^k$ is said to be attribute non-private if there exists $Y \in (\{0,1\}^{d-1})^n$ and an algorithm A such that for every $x \in \{0,1\}^n$,*

$$\Pr_{z \in M(Y \circ x, F)} [A(z) = x' : \|x - x'\|_1 = o(\|x\|_1)] \geq 1/10$$

where $Y \circ x$ simply denotes the obvious concatenation of Y and x. A need not be computationally efficient and the constant $1/10$ is arbitrary and can be replaced by any positive constant.

We show the following results :

1. Combining techniques from [11] and [13], we obtain tight lower bounds on the noise for arbitrary (non-counting) low-sensitivity queries for any $(\varepsilon, 0)$-differentially private mechanism. Given positive results of Blum, Ligett, and Roth [3], this separates non-counting queries from counting queries, proving that the former require more distortion than the latter for maintaining differential privacy. Also, given the positive results of [8] for arbitrary low-sensitivity queries, this separates (ε, δ)-differential privacy from $(\varepsilon, 0)$-differential privacy, where $\delta = \delta(n, k)$ denotes a function negligible in its argument. We also use this technique to show that the guarantees in terms of information content is drastically weaker for an (ϵ, δ) differentially private protocol as compared to an ϵ-differentially private protocol. Our technique also simplifies the *volume-based* lower bounds on noise for counting queries in [11]. In addition, we also make the lower bounds unconditional. The lower bound in [11] required the mechanism to be defined on "fractional" databases *i.e.*, on $(\mathbb{R}^+)^d$ as opposed to just $(\mathbb{Z}^+)^d$ while we do not have any such restrictions.

2. We give tight lower bounds on noise for ensuring (ε, δ)-differential privacy for $\delta > 0$. This proof relies on a lemma due to [13] showing that (ε, δ)-differentially private mechanisms yield a certain kind of unpredictable source.

On the other hand, any mechanism that is blatantly non-private cannot yield an unpredictable source. Thus, if the noise is insufficient to prevent blatant non-privacy then it cannot provide (ε, δ)-differential privacy. We subsequently use the lower bounds of [7,10] for preventing blatant non-privacy to get lower bounds on the distortion for (ϵ, δ) differential privacy.

3. We revisit the LP decoding attack of Dwork, McSherry, and Talwar [10], observing that any linear query matrix yielding a Euclidean section suffices for the attack. The LP decoding attack succeeds even if a certain constant fraction of the responses have wild noise. Armed with the connection to Euclidean sections, and a recent result of Rudelson [15] bounding from below the least singular value of the Hadamard product of certain i.i.d. matrices, we qualitatively strengthen a lower bound of Kasiviswanathan, Rudelson, Smith, and Ullman [12] on the noise needed to avert attribute non-privacy in ℓ-way marginals release by making the attack resilient to a constant fraction of wild responses.

There is an extension of results of [7] when the size of the universe is smaller than the size of the database which can be found in the full version of this paper [5].

2   Lower Bound by Volume Arguments

We now recall the volume based argument of Hardt and Talwar [11] to show lower bounds on the noise required for ϵ differential privacy.

Theorem 1. *Assume* $x_1, \ldots, x_{2^s} \in (\mathbb{Z}^+)^d$ *such that* $\forall i$, $\|x_i\|_1 \leq n$ *and for* $i \neq j$, $\|x_i - x_j\|_1 \leq \Delta$. *Further, let* $F : (\mathbb{Z}^+)^d \to \mathbb{R}^k$ *such that for any* $i \neq j$, $\|F(x_i) - F(x_j)\|_\infty \geq \eta$. *If* $\Delta \leq (s-1)/\epsilon$, *then any mechanism which is* ϵ-*differentially private for the query* F *on databases of size* n *must add noise* $\eta/2$.

While the line of reasoning in the proof is same as that of [11], we do the proof here as the argument in [11] works only for counting queries *i.e.*, when F is a linear transformation. On the other hand, the statement and proof of our result works for any query F.

Proof. Consider the ℓ_∞ balls of radius $\eta/2$ around each of the $F(x_i)$. By the hypothesis, these balls are disjoint. Now assume, any mechanism M which adds noise $\eta/2$ and consider any x_i. Then, because all the balls are disjoint, we have that there is some $j \neq i$ such that if S is the ℓ_∞ ball of radius $\eta/2$ around $F(x_j)$, then

$$\Pr_{z \in M(x_i, F)} [z \in S] \leq 2^{-s}$$

However, we can also say that because the noise added by the mechanism M is at most η,

$$\Pr_{z \in M(x_j, F)} [z \in S] \geq 1/2$$

Also, because the mechanism M is ϵ-differentially private and $\|x_i - x_j\|_1 \leq \Delta$, then

$$\frac{\Pr_{z \in M(x_i, F)}[z \in S]}{\Pr_{z \in M(x_j, F)}[z \in S]} \geq 2^{-\epsilon \cdot \Delta}$$

This leads to a contradiction if $\Delta \leq (s-1)/\epsilon$ thus proving the assertion.

2.1   Linear Lower Bound for Arbitrary Queries

In this subsection, we prove the following theorem.

Theorem 2. *For any $k, d, n \in \mathbb{N}$ and $1/40 \geq \epsilon > 0$, where $n \geq \min\{k/\epsilon, d/\epsilon\}$, there is a query $F : (\mathbb{Z}^+)^d \to \mathbb{R}^k$ such that any mechanism M which is ϵ-differentially private adds noise $\Omega(\min\{d/\epsilon, k/\epsilon\})$.*

If $\epsilon > 1$, then there is a query $F : (\mathbb{Z}^+)^d \to \mathbb{R}^k$ such that any mechanism M which is ϵ-differentially private adds noise $\Omega(\min\{d/(\epsilon \cdot 2^{5\epsilon}), k/\epsilon\})$ as long as $n \geq \min\{k/\epsilon, d/(\epsilon \cdot 2^{5\epsilon})\}$

Before starting the proof, we make a couple of observations. First of all, note that the statement of the theorem does not give any lower bound for $1 \geq \epsilon > 1/40$. However, any mechanism which is ϵ-differentially private for ϵ in the aforementioned range is also ϵ'-differentially private for $\epsilon' = 10/9$. Hence, the noise lower bounds for ϵ'-differential privacy for $\epsilon' = 10/9$ are also applicable for the range of $1 \geq \epsilon > 1/40$. It is easy to see that up to constant factors, the lower bounds with $\epsilon' = 10/9$ are optimal for ϵ in the aforementioned range.

Secondly, Laplacian mechanism maintains ϵ-differential privacy while adding only $O(k/\epsilon)$ noise. Also, because the databases are of size n, it is enough to add noise $O(n)$ to maintain ϵ-differential privacy for any $\epsilon \geq 0$. Thus, as long as $k = O(d)$, our lower bounds are tight up to constant factors. Next, we do the proof of Theorem 2.

Also, in the subsequent proofs, the databases shall be constructed in clever ways. The full details of these constructions can be found in [5]. We will be referring to the appropriate claims whenever necessary.

Proof. Our proof strategy is to construct a set of databases and a query which meets the conditions stated in the hypothesis of Theorem 1 and then get the desired lower bound on the noise. We first deal with the case when $0 < \epsilon < 1/40$. Let $\ell = \min\{d, k\}$. We can now use Claim A.2 in [5] to construct 2^s databases $x_1, \ldots, x_{2^s}$ (for $s = \ell/400$) such that $x_i \in (\mathbb{Z}^+)^d$ with the property that $\forall i \neq j$, $\|x_i - x_j\|_1 \geq n'/10$ and $\|x_i\|_1 \leq n'$ where $n' = \ell/(1280\epsilon)$ (Application of Claim A.2 uses $d' = \ell/320$). Note that our databases are of size bounded by $n' \leq n$. We now describe a mapping $\mathcal{L} : (\mathbb{Z}^+)^d \to \mathbb{R}^{2^s}$ which is related to a construction in [13]. The mapping is as follows :

- For every x_i, there is a coordinate i in the mapping.
- The ith coordinate of $\mathcal{L}(z)$ is $\max\{n'/30 - \|x_i - z\|_1, 0\}$.

Claim. The map $\mathcal{L}$ is 1-Lipschitz i.e., if $\|z_1 - z_2\|_1 = 1$, then $\|\mathcal{L}(z_1) - \mathcal{L}(z_2)\|_1 \leq 1$.

Proof. We observe that for any z_1, z_2 such that $\|z_1 - z_2\| \leq 1$, if A denotes the set of coordinates where at least one of $\mathcal{L}(z_1)$ or $\mathcal{L}(z_2)$ are non-zero, then A is either empty or is a singleton set. Given this, the statement in the claim is obvious, since the mapping corresponding to any particular coordinate is clearly 1-Lipschitz.

We now describe the queries. Corresponding to any $r \in \{-1, 1\}^{2^s}$, we define $f_r : (\mathbb{Z}^+)^d \to \mathbb{R}$, as

$$f_r(x) = \sum_{i=1}^{d} \mathcal{L}(x)_i \cdot r_i$$

Now, we define a random map $F : (\mathbb{Z}^+)^d \to \mathbb{R}^k$ as follows. Pick $r_1, \ldots, r_k \in \{-1, 1\}^{2^s}$ independently and uniformly at random and define F as follows :

$$F(x) = (f_{r_1}(x), \ldots, f_{r_k}(x))$$

Now consider any $x_h, x_j \in S$ such that $h \neq j$. Because of the way $\mathcal{L}$ is defined, it is clear that for any r_i,

$$\Pr_{r_i}[|f_{r_i}(x_h) - f_{r_i}(x_j)| \geq n'/15] \geq 1/2$$

A basic application of the Chernoff bound implies that

$$\Pr_{r_1, \ldots, r_k}[\text{For at least } 1/10 \text{ of the } r_i\text{'s,} \quad |f_{r_i}(x_h) - f_{r_i}(x_j)| \geq n'/15] \geq 1 - 2^{-k/30}$$

Now, note that the total number of pairs (x_i, x_j) of databases such that $x_i, x_j \in S$ is at most $2^{2s} \leq 2^{\ell/200} \leq 2^{k/200}$. This implies (via a union bound)

$$\Pr_{r_1, \ldots, r_k}[\forall h \neq j, \geq 1/10 \text{ of the } r_i\text{'s,} \quad |f_{r_i}(x_h) - f_{r_i}(x_j)| \geq n'/15] \geq 1 - 2^{-k/40}$$

This implies that we can fix $r_1, \ldots, r_k$ such that the following is true.

$$\forall h \neq j, \quad \text{For at least } 1/10 \text{ of the } r_i\text{'s,} \quad |f_{r_i}(x_h) - f_{r_i}(x_j)| \geq n'/15$$

This implies that for any $x_h \neq x_j \in S$, $\|F(x_h) - F(x_j)\|_\infty \geq n'/15$. In fact, $\|F(x_h) - F(x_j)\|_2 \geq n'\sqrt{k}/150$ which is a much stronger assumption than what we require and is quantitatively similar to the results in [11] where they consider ℓ_2 noise as opposed to ℓ_∞ noise.

We can now apply Theorem 1 by putting $\Delta = 2n'$ and $s = \ell/400 > 3\epsilon n'$ and $\eta = n'/15$ and observe that $\Delta \leq (s-1)/\epsilon$ thus proving the result.

We next deal with the case when $\epsilon > 1$. This part of the proof differs from the case when $\epsilon < 1$ only in the construction of $x_1, \ldots, x_{2^s}$. We also emphasize that had we not insisted on integral databases, our proof would have been identical to the first part. We construct the databases $x_1, \ldots, x_{2^s}$ using combinatorial designs. More precisely, for some sufficiently large constant C, let $\ell = \min\{d/(C \cdot 2^{5\epsilon}), k\}$. We can now use Claim A.3 from [5] to construct 2^s databases $x_1, \ldots, x_{2^s}$ (for $s = \ell/400$) such that $x_i \in (\mathbb{Z}^+)^d$ with the property that $\forall i \neq j, \|x_i - x_j\|_1 \geq$

$n'/10$ and $\|x_i\|_1 \leq n'$ where $n' = \ell/(1280\epsilon)$ (using $d' = \ell/320$ in Claim A.3). Again, we note here that the databases constructed are of size n'.

From this point onwards, we define the map $\mathcal{L}$ and the query F as we did in the proof of Theorem 2 and the proof proceeds identically. In particular, we get a query $F : (\mathbb{Z}^+)^d \to \mathbb{R}^k$ such that for any $i \neq j$, $\|F(x_i) - F(x_j)\|_2 \geq n'\sqrt{k}/150$. As before, we can now apply Theorem 1 by putting $\Delta = 2n'$ and $s = \ell/100 > 3\epsilon n'$ and $\eta = n'/15$ and observe that $\Delta \leq (s-1)/\epsilon$ thus proving the result.

For the subsequent part of this paper, we only consider lower bounds on ϵ-differential privacy for $0 < \epsilon < 1$ as opposed to $\epsilon > 1$. This is because the privacy guarantees one gets becomes unmeaningful when ϵ is large. However, we do remark that the results can be carried in a straightforward way to the regime of $\epsilon > 1$ using combinatorial designs (like we did for Theorem 2).

Consequences of the Linear Lower Bound. We briefly describe the two consequences of the linear lower bound on the noise proven in Theorem 2. The first is separation of counting queries from non-counting queries. While our separation gives quantitatively the same results as long as $d = k^{O(1)}$ and $n = \Theta(k/\epsilon)$, for simplicity, we consider the setting when $k = d$ and $n = k/\epsilon$. In this case, Theorem 2 shows existence of a (non-counting) query such that maintaining ϵ-differential privacy requires noise $\Omega(n)$. On the other hand, [3] had proven that for any counting query with the same setting of parameters, there is a mechanism which adds noise $\tilde{O}(n^{2/3})$ and maintains ϵ-differential privacy. This shows that maintaining ϵ-differential privacy inherently requires more distortion in case of non-counting queries than counting queries.

The next consequence is a separation of (ϵ, δ) differential privacy from $(\epsilon, 0)$ differential privacy for $\delta = 2^{-o(n)}$. We note that Hardt and Talwar [11] had shown such a separation but that was only when $k = O(\log n)$ and $\delta = n^{-O(1)}$. Again, we use the setting of parameters when $k = d$ and $n = k/\epsilon$. The gaussian mechanism of [8] shows that to maintain (ϵ, δ) differential privacy for any k queries, it suffices to add noise $O(\sqrt{k \log(1/\delta)}/\epsilon) = o(n)$. However, Theorem 2 shows that there is a query which requires adding noise $\Omega(n)$ to maintain $(\epsilon, 0)$ differential privacy.

The last consequence of our result is more indirect and is explained next.

2.2   Information Loss in Differentially Private Protocols

In [13], a connection was established between differentially private protocols and the notion of mutual information from information theory. In fact, as [13] was dealing with 2-party protocols, the connection was actually between differentially private protocols and that of information content [1,2] which is a symmetric variant of mutual information useful in 2-party protocols. In that paper, it was shown that the information content (which simplifies to mutual information in our setting) between transcript of a ϵ-differentially private mechanism and the database vector is bounded by $O(\epsilon n)$. Using the construction used in the previous subsection, we show that in case of (ϵ, δ) differentially private protocols (for any $\delta = 2^{-o(n)}$), there is no non-trivial bound on the mutual information between the transcript of the mechanism and the database vector. Thus as far as information theoretic guarantees go, the situation is drastically different for pure differentially private protocols vis-a-vis approximately differentially private protocols. The contents of this subsection are a result of personal communication between the author and Salil Vadhan [6].

We first define the notion of mutual information (can be found in standard information theory textbooks).

Definition 5. *Given two random variables X and Y, their mutual information $I(X;Y)$ is defined as*

$$I(X;Y) = H(X) + H(Y) - H(X,Y) = H(X) - H(X|Y)$$

where $H(X)$ denotes the Shannon entropy of X.

The next claim establishes an upper bound on the mutual information between transcript of a differentially private protocol and the database vector.

Claim. Let $F : (\mathbb{Z}^+)^d \to \mathbb{R}^k$ be a query and $M : (\mathbb{Z}^+)^d \to \mu(\mathbb{R}^k)$ be an ϵ-differentially private protocol for answering F for databases of size n. If X is a distribution over the inputs in $(\mathbb{Z}^+)^d$, then $I(M(X);X) \leq 3\epsilon n$.

Proof. We first note that since the databases are of size bounded by n, hence instead of assuming that μ is a distribution over the inputs $X \in (\mathbb{Z}^+)^d$, we can assume that μ is a distribution over the inputs $X \in [n]^d$ where $[n] = \{0, 1, \ldots, n\}$. Now, we can apply Proposition 7 from [13]. We note that the aforesaid proposition is in terms of information content for 2-party protocols but we observe that we can simply make the second party's input as a constant and get that $I(M(X);X) \leq 3\epsilon n$.

Next, we state the following claim which says that for (ϵ, δ) differentially private protocols, even for an exponentially small δ, the mutual information between the transcript and the input can be as large as $n(1-\eta)$ for any value of $0 < \epsilon, \eta < 1$. In other words, an (ϵ, δ) differentially private protocol does not imply any effective bound on the mutual information between the input and the transcript even as $\epsilon \to 0$ and δ is exponentially small.

Lemma 1. *For $n \in \mathbb{N}$ and $0 < \epsilon, \eta < 1$, there is a constant $C = C(\epsilon, \eta) > 0$ and a distribution X over $(\mathbb{Z}^+)^n$ with a support over databases of size n and a query $F : (\mathbb{Z}^+)^n \to \mathbb{R}^k$ and an (ϵ, δ)-differentially private protocol M for answering F such that $I(X; M(X)) \geq n(1 - 2\eta)$ if $\delta \geq 2^{-C(\epsilon, \eta)n}$.*

Proof. We first construct 2^s vectors in $\{0, 1\}^n$ (for $s = n(1 - \eta)$) with the property that for any x_i, x_j ($i \neq j$), $\|x_i - x_j\|_1 \geq \eta^2 n/8$. It is easy to guarantee the existence of such a set of vectors by a simple application of the probabilistic method. The distribution X is simply the uniform distribution over the set $\{x_1, \ldots, x_{2^s}\}$. By construction, all the databases in X are of size bounded by n.

Next, we define the query $F : (\mathbb{Z}^+)^n \to \mathbb{R}^k$ be defined in the same way as the query F in the proof of Theorem 2. Following, exactly the same calculations, we can show that if we set $k = 80n$, we get a query $F : (\mathbb{Z}^+)^n \to \mathbb{R}^k$ such that for any $i \neq j$, $\|F(x_i) - F(x_j)\|_2 \geq \eta^2 n\sqrt{k}/50$. We now recall the Gaussian mechanism of [8] which maintains (ϵ, δ) differential privacy.

Lemma 2. *[8] Let $F : (\mathbb{Z}^+)^d \to \mathbb{R}^k$ be a query. Let $Y = (Y_1, \ldots, Y_k)$ be a distribution over $\mathbb{R}^k$ such that each Y_i is an i.i.d. $\mathcal{N}(0, \sigma)$ random variable. Here $\sigma^2 = \frac{k \log(1/\delta)}{\epsilon^2}$. Then the mechanism M which for a database x and query F, which samples Y_0 from Y and responds by $F(x) + Y_0$ is an (ϵ, δ) differentially private mechanism.*

Note that for the above mechanism M, and database x, if Z is sampled from $M(x)$, then the distribution of $M(x) - F(x)$ is same as $(Y_1, \ldots, Y_k)$ where each Y_i is an i.i.d. $\mathcal{N}(0, \sigma)$ random variable. Thus,

$$\|M(x) - F(x)\|_2^2 \sim Y_1^2 + \ldots + Y_k^2$$

As the following fact shows, the distribution on the right hand side is concentrated around its mean. The fact is possibly well-known but we could not find a reference and hence we prove it in Appendix C in [5].

Fact 3 . *If $Y_1, \ldots, Y_k$ are i.i.d. $\mathcal{N}(0, \sigma)$ random variables, then,*

$$\Pr_{Y_1, \ldots, Y_k} [Y_1^2 + \ldots + Y_k^2 > 2(1 + \xi) \cdot k \cdot \sigma^2] \leq 2^{-\frac{k\xi}{2}}$$

Using the above fact, we get

$$\Pr\left[\|M(x) - F(x)\|_2^2 > \frac{2(1 + \xi)k^2 \log(1/\delta)}{\epsilon^2}\right] \leq 2^{-\frac{\xi k}{2}}$$

Here the probability is over the randomness of the mechanism. Putting $\xi = 1$ and $\delta = 2^{-C(\epsilon, \eta)n}$ for an appropriate constant $C(\epsilon, \eta)$, we get that

$$\Pr\left[\|M(x) - F(x)\|_2 > \frac{\eta^2 n\sqrt{k}}{200}\right] \leq 2^{-40n}$$

As we know, for any $i \neq j$, $\|F(x_i) - F(x_j)\|_2 \geq \eta^2 n\sqrt{k}/50$. Hence, with probability at least $1 - 2^{-n}$ over the randomness of the mechanism, for any database $x_i \in supp(X)$, if y is sampled from $M(x_i)$,

$$\forall j \neq i \quad \|F(x_j) - y\|_2 > \|F(x_i) - y\|_2$$

Thus, for any x_i, given $M(x_i)$, we can recover x_i with high probability and hence, we can say

$$\Pr_{y \sim M(X)} [H(X|M(X) = y) = 0] > 1 - 2^{-n}$$

This means that

$$H(X|M(X)) \leq 2^{-n}n < 1$$

Recall that $I(X; M(X)) = H(X) - H(X|M(X)) \geq H(X) - 1 = (1 - \eta)n - 1 \geq (1 - 2\eta)n$. This completes the proof of the Lemma 1.

3 Lower Bound on Noise for Counting Queries

In the last section, we proved that to preserve ϵ differential privacy for k queries, one may need to add $\Omega(k/\epsilon)$ noise provided $d, n \gg k$. However, these queries were not counting queries. It is interesting to derive lower bounds on noise required to preserve privacy for counting queries as these are the queries mostly used in practice. While one might initially hope to prove a similar lower bound for counting queries, [3] states that there is a ϵ-differentially private mechanism which adds $\tilde{O}(n^{2/3}/\epsilon)$ noise per query and can answer $O(n)$ counting queries (when $d = n^{O(1)}$).

Still, Hardt and Talwar [11] showed that to answer k counting queries, any mechanism which is ϵ-differentially private must add $\min\{k/\epsilon, \sqrt{k \log(d/k)}/\epsilon\}$ noise (in fact, this is true for k random queries). However, [11] make a technical assumption that the mechanism has a smooth extension which works for "fractional" databases as well. In other words, they require the domain of the mechanism to be $(\mathbb{R}^+)^d$ as opposed to $(\mathbb{Z}^+)^d$. However, it is not clear if this is always true $i.e.$, if given a mechanism which is defined only over true (integral) databases, one can get a mechanism which is defined over "fractional" databases with similar privacy guarantees.

Next, we prove the same result without making any such technical assumptions. Again, our constructions are dependent on combinatorial designs [14]. First, we prove the following simple but useful claim.

Claim. Let $a \in \mathbb{Z}$ and assume $x_1, x_2, \ldots, x_{2^s} \in (\mathbb{Z}^+)^d$ such that $\forall i$, every entry of x_i is either 0 or a. Also, for every $i \neq \ell$, $\|x_i - x_\ell\|_1 \geq \Delta$. Then, for $k \geq 20s$, there is a linear query $F : (\mathbb{Z}^+)^d \to \mathbb{R}^k$ such that for every $i, \ell \in [2^s]$ and $i \neq \ell$, the following holds :

$$\Pr_{j \in [k]} [|F(x_i)_j - F(x_\ell)_j| \geq \Delta'/10] \geq 1/40$$

where $\Delta' = \sqrt{\Delta \cdot a}$.

Proof. Consider any x_i, x_ℓ such that $i \neq \ell$. Note that, z defined as $z = x_i - x_\ell$ is such that all its entries are $0, \pm a$ and also that z has at least Δ/a or more non-zero entries. If we choose $r \in \{-1, 1\}^d$ u.a.r., then note that

$$Y = \sum_{i=1}^d z_i \cdot r_i = \sum_{z_i = \pm a} z_i \cdot r_i$$

Note that the total number of summands is $\ell' \geq \Delta/a$ and hence the distribution of the random variable Y is same as choosing $r' \in \{-1, 1\}^d$ and considering the random variable

$$Y' = a \cdot \left(\sum_{i=1}^{\ell'} r_i' \right)$$

However using Corollary B.2 from [5], we get

$$\Pr\left[|Y'| \geq \frac{\sqrt{\Delta \cdot a}}{10}\right] = \Pr\left[|\sum_{i=1}^{\ell'} r_i'| \geq \frac{\sqrt{\Delta/a}}{10}\right] \geq \frac{9}{10} \tag{1}$$

Now, let us choose $r_1', \ldots, r_k'$ uniformly and independently at random from $\{-1, 1\}^d$ and consider the linear query $F : (\mathbb{Z}^+)^d \to \mathbb{R}^k$ defined as

$$F(x) = \left(\sum_{j=1}^{d} x_j \cdot r_{1j}', \ldots, \sum_{j=1}^{d} x_j \cdot r_{kj}'\right)$$

Set $\Delta' = \sqrt{\Delta \cdot a}$. Now, (1) and an application of Chernoff bound implies that for any x_i, x_ℓ $(i \neq \ell)$

$$\Pr_{r_1', \ldots, r_k'}\left[\Pr_{j \in [k]}[|F(x_i)_j - F(x_\ell)_j| \geq \Delta'/10] \geq 1/40\right] > 1 - 2^{-k/10}$$

We now observe that the total number of pairs (x_i, x_ℓ) $(i \neq \ell)$ is at most $2^{2s} \leq 2^{k/10}$. Applying a union bound, we get that there is some choice of $r_1', \ldots, r_k'$ (and hence a fixed F) such that

$$\Pr_{j \in [k]}[|F(x_i)_j - F(x_\ell)_j| \geq \Delta'/10] \geq 1/40$$

We now prove a lower bound on the noise required to maintain privacy for random counting queries. As we have said before, Hardt and Talwar [11] proved the same result under an additional assumption that the mechanism defined over integral databases can be smoothly extended to fractional databases as well.

Theorem 4. *For every $k, d \in \mathbb{N}$ and $1 > \epsilon > 0$, there is a counting query $F : (\mathbb{Z}^+)^d \to \mathbb{R}^k$ such that any mechanism which maintains ϵ-differential privacy adds noise $\Omega(\min\{k/\epsilon, \sqrt{k \log(d/k)}/\epsilon\})$. The size of the database i.e., $n = O(k/\epsilon)$.*

Proof. The proof strategy is to come up with databases meeting the hypothesis of Claim 3 and use Claim 3 to get a counting query F. We then use Theorem 1 to get a lower bound on the distortion required by any private mechanism to answer F. We consider two cases : $k \leq \log d$ and $k > \log d$.

The first case is trivial : Namely, consider databases $x_1, \ldots, x_{2^{k/20}}$ such that each $x_i = \lfloor (k/80\epsilon) \rfloor \cdot e_i$ where e_i is the standard unit vector in the ith direction. This is possible as there are $d \geq 2^k$ different unit vectors. Note that for any $i \neq \ell$, $\|x_i - x_\ell\|_1 = 2 \cdot \lfloor k/(80\epsilon) \rfloor$. We can now apply Claim 3 and get that there is a linear query $F : (\mathbb{Z}^+)^d \to \mathbb{R}^k$ (using $\Delta = 2 \cdot \lfloor k/(80\epsilon) \rfloor$ and $a = \lfloor k/(80\epsilon) \rfloor$) such that

$$\Pr_{j \in [k]}\left[|F(x_i)_j - F(x_\ell)_j| \geq \frac{\sqrt{2}}{10}\lfloor k/(80\epsilon) \rfloor \geq \frac{k}{800\epsilon}\right] \geq 1/40$$

We see that there are $2^{k/20} = 2^s$ databases which differ by exactly $2 \cdot \lfloor k/(80\epsilon) \rfloor = \Delta$. Note that $\Delta \leq (s-1)/\epsilon$. Hence we can apply Theorem 1 to note that to maintain ϵ-differential privacy, any mechanism needs to add $k/(800\epsilon)$ noise. In fact, we note that the ℓ_2 error of the answer returned by the mechanism needs to be $\Omega(k^{3/2}/\epsilon)$ which is quantitatively the same as the result in [11].

The second case is slightly more complicated. We use Claim A.1 from [5] to construct $x_1, \ldots, x_{2^{k/20}} \in (\mathbb{Z}^+)^d$ with the following properties :

- Every entry of any of the x_i's is either 0 or $a \in \mathbb{Z}$ such that $a \geq \log(d/k)/160\epsilon$.
- $\forall i$, $\|x_i\|_1 \leq k/80\epsilon$ and $\forall i \neq j$, $\|x_i - x_j\|_1 \geq k/160\epsilon$

Again, we can apply Claim 3 and get that there is a linear query $F : (\mathbb{Z}^+)^d \to \mathbb{R}^k$ (using $\Delta \geq k/(160\epsilon)$ and $a \geq (\log(d/k)/160\epsilon)$) such that $\forall i \neq \ell$

$$\Pr_{j \in [k]} \left[|F(x_i)_j - F(x_\ell)_j| \geq \frac{1}{10} \cdot \frac{\sqrt{k \log(d/k)}}{160\epsilon} \right] \geq 1/40$$

Again, we have $2^{k/20}$ databases which differ by at most $k/(40\epsilon)$ and hence we can apply Theorem 1 to get that to maintain ϵ-differential privacy, any mechanism needs to add $\Omega\left(\frac{\sqrt{k \log(d/k)}}{\epsilon} \right)$ noise.

4   Lower Bounds for Approximate Differential Privacy

In this section, we prove lower bounds on the noise required to maintain (ϵ, δ) differential privacy for $\epsilon, \delta > 0$. Our lower bounds are valid for any positive $\delta > 0$ and are in fact tight for a constant ϵ and δ. We note that a quantitatively similar lower bound was proven for the class of ℓ-way marginals by [12] though our proof (for random queries) is arguably much simpler.

In this section, we consider databases which are elements of $\{0,1\}^n$ or in other words we consider the case when the universe size $d = n$ and the databases are allowed to have exactly one element of each type. We note that restricting databases to bit vectors is a well-considered model in literature including [7,10,13] among others.

We prove the following theorem.

Theorem 5. *For any $n \in \mathbb{N}$, $\epsilon > 0$ and $1/20 > \delta > 0$, there exist positive constants α, γ and η such that there is a counting query $F : \{0,1\}^n \to \mathbb{R}^k$ with $k = \alpha n$ such that any mechanism M that satisfies*

$$\Pr_M [\Pr_{i \in [k]} [|M(x,F)_i - F(x)_i| \leq \eta\sqrt{n}] \geq 1/2 + \gamma] \geq 3\sqrt{\delta}$$

is not (ϵ, δ) differentially private. In other words, any mechanism M which with significant probability i.e., $3\sqrt{\delta}$ answers at least $1/2 + \gamma$ fraction of the k queries with at most $\eta\sqrt{n}$ noise, is not (ϵ, δ) differentially private.

An immediate corollary is that there exists a positive constant α and a counting query $F : \{0,1\}^n \to \mathbb{R}^k$ where $k = \alpha n$ such that any mechanism which adds $o(\sqrt{n})$ noise is not (ϵ, δ) differentially private for $\epsilon > 0$ and $\delta < 1/20$.

To do the proof of Theorem 5, we first need to introduce some definitions previously discussed in [13]. We do note that the paper [13] deals with the two-party setting but the relevant definitions and the lemma we use here easily extend to the standard (curator-client) setting of privacy.

Definition 6. *A random variable $Y = (y_1, \ldots, y_{i-1}, y_i, y_{i+1}, \ldots, y_n) \in \{0,1\}^n$ is said to be δ-approximate strongly α-unpredictable bit source (for $\alpha \geq 1$) if with probability $1 - \delta$ over $i \in [n]$*

$$\frac{1}{\alpha} \leq \frac{\Pr[Y_i = 1 | Y_1 = y_1, \ldots, Y_{i-1} = y_{i-1}, Y_{i+1} = y_{i+1}, \ldots, Y_n = y_n]}{\Pr[Y_i = 0 | Y_1 = y_1, \ldots, Y_{i-1} = y_{i-1}, Y_{i+1} = y_{i+1}, \ldots, Y_n = y_n]} \leq \alpha$$

The next lemma (proven in [13] for the two-party setting) roughly says that for any (ϵ, δ) private mechanism, conditioned on the transcript of the mechanism, the distribution of the database is a δ-approximate strong 2^ϵ-unpredictable source. More precisely, we have the following lemma.

Lemma 3. *Let $F : \{0,1\}^n \to \mathbb{R}^k$ be a query and M be a (ϵ, δ)-differentially private mechanism for answering F. Let X be the uniform distribution over $\{0,1\}^n$ and Γ be the probability distribution over the transcripts of $M(x)$ when x is drawn from X. Then for any $\mu > 0$ and $t \leftarrow \Gamma$, the distribution $X|_{\Gamma=t}$ is δ_t approximate strongly $2^{\epsilon+\mu}$-unpredictable sources such that*

$$\mathbb{E}_{t \in \Gamma} [\delta_t] \leq 2\delta \cdot \frac{1 + e^{-\epsilon-\mu}}{1 - e^{-\mu}}.$$

The above lemma trivially follows from Lemma 20 of [13] (full version) and hence we do not prove it here. Before, proving Theorem 5, we need to recall the following theorem from [10] (Theorem 24 in the paper).

Theorem 6. *For any $\gamma > 0$ and any $\nu = \nu(n)$, there is a constant $\alpha = \alpha(\gamma) > 0$ such that for $k = \alpha n$, there is a counting query $F : \{0,1\}^n \to \mathbb{R}^k$ and an algorithm A such that given $\tilde{y}$ which satisfies*

$$\Pr_{i \in [k]}[|\tilde{y}_i - F(x)_i| \leq \nu] \geq \frac{1}{2} + \gamma$$

the output of A on $\tilde{y}$ i.e., $A(\tilde{y}) = x'$ such that $x' \in \{0,1\}^n$ and $\|x - x'\|_1 \leq \frac{4\nu^2}{\gamma^2}$

The following corollary follows immediately from Theorem 6.

Corollary 1. *For any $\delta' > 0$, there are positive constants $\gamma = \gamma(\delta'), \eta = \eta(\delta'), \alpha = \alpha(\delta')$ such that for $k = \alpha n$, there is a counting query $F : \{0,1\}^n \to \mathbb{R}^k$ and an algorithm A such that given $\tilde{y}$ which satisfies*

$$\Pr_{i \in [k]}[|\tilde{y}_i - F(x)_i| \leq \eta\sqrt{n}] \geq \frac{1}{2} + \gamma$$

the output of A on $\tilde{y}$ i.e., $A(\tilde{y}) = x'$ such that $x' \in \{0,1\}^n$ and $\|x - x'\|_1 \leq \delta' n$.

We now prove Theorem 5.

Proof (of Theorem 5).
Let X denote the uniform distribution over $\{0,1\}^n$. First, using Lemma 3, we get that over the randomness of the mechanism M and the choice of $x \in X$, if we sample a transcript t from $M(x, F)$, then for any positive μ, the distribution $X|_{M(x,F)=t}$ is a δ_t-approximate strongly $2^{\epsilon+\mu}$-unpredictable sources where δ_t satisfies

$$\mathop{\mathbb{E}}_{t \in M(x,F)} [\delta_t] \leq 2\delta \cdot \frac{1+e^{-\epsilon-\mu}}{1-e^{-\mu}}.$$

Clearly, we can put $\mu = 10$ and get that the distribution $X|_{M(x,F)=t}$ is a δ_t-approximate strongly $2^{\epsilon+10}$-unpredictable sources where $\mathbb{E}_{t \in M(x,F)} [\delta_t] \leq 3\delta$. By an application of Markov's inequality, we get that with probability $1 - 2\sqrt{\delta}$ over the choice of x and the randomness of the mechanism M, the distribution $X|_{M(x,F)=t}$ is $2\sqrt{\delta}$-approximate strongly $2^{\epsilon+10}$-unpredictable source.

We now apply corollary 1. In particular, we put $\delta' = \sqrt{\delta}$ and get that for some positive γ, η, α (which are functions of δ' and hence δ), there is a counting query $F : \{0,1\}^n \to \mathbb{R}^{\alpha n}$ and an algorithm A such that given $\tilde{y}$ which satisfies

$$\Pr_{i \in [k]} [|\tilde{y}_i - F(x)_i| \leq \eta\sqrt{n}] \geq \frac{1}{2} + \gamma$$

the output of A on $\tilde{y}$ i.e., $A(\tilde{y}) = x'$ such that $x' \in \{0,1\}^n$ and $\|x - x'\|_1 \leq \sqrt{\delta} \cdot n$. Now, consider a mechanism M which satisfies

$$\Pr_M[\Pr_{i \in [k]}[|M(x,F)_i - F(x)_i| \leq \eta\sqrt{n}] \geq 1/2 + \gamma] \geq \beta$$

for $\beta = 3\sqrt{\delta}$. Clearly such a mechanism M is not (ϵ, δ) differentially private because with probability at least $\beta = 3\sqrt{\delta}$, the algorithm A will be able to predict at least $1 - \sqrt{\delta}$ fraction of the positions which contradicts that with probability $1 - 2\sqrt{\delta}$, the distribution $X|_{M(x,F)=t}$ is a $2\sqrt{\delta}$-approximate strongly $2^{\epsilon+10}$-unpredictable source.

5   LP Decoding, Euclidean Sections and Hardness of Releasing ℓ-way Marginals

In this section, we consider attacks on privacy using linear programming. In particular, we use the technique of LP decoding (previously used in [10] in context of privacy) to give attacks which violate even minimal notions of privacy when $1 - \epsilon_0$ (for some $\epsilon_0 > 0$) fraction of the queries are released with insufficient noise. We do this by establishing a connection between Euclidean sections and use of LP decoding in context of privacy which does not seem to have explicitly appeared in the literature before. We remark that the relation between LP decoding and Euclidean spaces is very well known in context of compressed sensing [4]. However, in case of privacy, the adversary is allowed to add small error to say 99% of the entries and arbitrary error to the remaining 1% of the entries. In context of compressed sensing however, the adversary is allowed to add error to only 1% of the entries.

We first describe how to use linear programming in context of privacy. Assume $x \in \mathbb{Z}^{+d}$ is a database and $A : \mathbb{R}^d \to \mathbb{R}^k$ is a linear map which represents a counting query with arity k made on the database x. Further, the right set of answers is given by $y = A \cdot x$. (To make sure that the queries are 1-Lipschitz, all the entries of A come from $[-1, 1]$.) Suppose, $\tilde{y} \in \mathbb{R}^k$ is the answer returned by the mechanism. Then, consider the following optimization problem (which can be written as a linear program) :

$$\text{Minimize } \|y - \tilde{y}\|_1 \text{ subject to } y = A \cdot \tilde{x} \tag{2}$$

The following theorem states the necessary conditions such that the solution to the above linear program, call it $\tilde{x}$, is such that $\|x - \tilde{x}\|_1$ is small. To state the theorem, we will need the definition of a Euclidean section.

Definition 7. $V \subseteq \mathbb{R}^k$ *is said to be a* (δ, d, k) *euclidean section if V is a linear subspace of dimension d and for every $x \in V$, the following holds:*

$$\sqrt{k}\|x\|_2 \geq \|x\|_1 \geq \delta\sqrt{k}\|x\|_2$$

Theorem 7. *Let $A : \mathbb{R}^d \to \mathbb{R}^k$ be a full rank linear map $(k > d)$ and all the singular values of A are at least σ. Further, the range of A (denoted by $\mathcal{L}(A)$) is a (δ, d, k) Euclidean section. Let $F : (\mathbb{Z}^+)^d \to \mathbb{R}^k$ the query corresponding to A. Then, there exists $\gamma = \gamma(\delta)$ such that if*

$$\Pr_{i \in [k]} [|F(x)_i - \tilde{y}_i| \leq \alpha] \geq 1 - \gamma$$

then, any solution $\tilde{x}$ to the linear program (2) satisfies $\|\tilde{x} - x\|_1 \leq O(\alpha\sqrt{kd}/\sigma)$ where the constant inside the $O(\cdot)$ notation depends on δ.

The proof of this theorem can be found in [5]. The specific problem we are interested in is the application of LP decoding to violate attribute privacy when ℓ-way marginals of a contingency table are released. Informally, attribute privacy refers to the situation in a contingency table when all but one of the attributes are public and attacks on privacy amount to revealing the last attribute given the responses to the queries and knowledge of all the other attributes. Releasing the ℓ-way marginals is simply the following : For every subset of size ℓ of the attributes and every configuration of these ℓ-attributes, a count of how many entries in the database have that specific configuration on those ℓ-attributes is released. Due to the lack of space, we refer the reader to [12,5] for the precise definitions of attribute privacy and ℓ-way marginals. We will also need the definition of row products of matrices which can be found in [5]. The next theorem (proven in [5]) shows how if the range of row product of matrices is Euclidean and all the singular values of the row product are large, one can violate attribute privacy when noisy ℓ-way marginals are released.

Lemma 4. *Let $A_1, \ldots, A_{\ell-1} \in \{0,1\}^{d' \times n}$. Let $A = A_1 \circ A_2 \ldots \circ A_{\ell-1}$ (with $d'^{\ell-1} > n$) be their row product. Also, all the singular values of A are at least σ and the range of A i.e., $\mathcal{L}(A)$ is a $(\delta, n, d'^{\ell-1})$ Euclidean section. Then, there exists a constant $\gamma = \gamma(\delta) > 0$ such that any mechanism which answers at least $1 - \gamma$ fraction of the ℓ-way marginals with noise bounded by α is attribute non-private provided $\frac{\alpha \sqrt{d'^{\ell-1} \cdot n}}{\sigma} = o(n)$ or in other words, $\alpha = o(\sqrt{n}\sigma/\sqrt{d'^{\ell-1}})$*

The main technical tool for us is the following theorem of Rudelson [15].

Theorem 8. *[15] Let $q, \ell \in \mathbb{N}$ be constants. Also, let $D \sim \mathbb{R}^{d' \times n}$ be a distribution over matrices such that every entry of the matrix is an independent and unbiased $\{0,1\}$ random variable. Let $A_1, \ldots, A_{\ell-1}$ be i.i.d. copies of random matrices drawn from the distribution D and A be the Hadamard product of $A_1, \ldots, A_{\ell-1}$. Then, provided that $d'^{\ell-1} \gg n \log_{(q)} n$, with probability $1 - o(1)$, the smallest singular value of A denoted by $\sigma_n(A)$ satisfies $\sigma_n(A) = \Omega(\sqrt{d'^{\ell-1}})$ Also, the range of A is a $(n, d'^{\ell-1}, \gamma(q, \ell))$ Euclidean section for some $\gamma(q, \ell) > 0$.*

The above theorem uses the notion of iterated logarithm which is defined as :For $r \in \mathbb{N}$, we define $\log_{(r)} n$ as follows : $\log_{(1)} n = \max\{\log_2 n, 1\}$ and for $r > 1$, $\log_{(r)} n = \log_{(1)} (\log_{(r-1)} n)$. Combining Theorem 8 and Lemma 4, we have the main theorem of this section.

Theorem 9. *Let $q, \ell \in \mathbb{N}$ be constant integers. Then, there exists a constant $\gamma = \gamma(q, \ell) > 0$ such that any mechanism which releases the ℓ-way marginals of a table of size n over d' attributes and $n \leq d'^{\ell-1} \log_{(q)} n$ by adding at most η noise to $1 - \gamma$ fraction of the queries where*

$$\eta = o(\sqrt{n})$$

is attribute non-private. Further, the algorithm which violates attribute privacy is efficient and uses LP decoding.

This improves upon the following result of Kasiviswanathan et al. [12] who could violate attribute privacy only when all the queries were allowed $o(\sqrt{n})$ noise.

Theorem 10. *[12] Let $\ell \in \mathbb{N}$ be a constant and $n, d \in \mathbb{N}$ such that $d'^{\ell-1} \gg n \cdot \log^{2\ell-4} n$. Then, for every mechanism M which releases ℓ-way marginals of a database of size n (and universe $\{0,1\}^{d'}$) such that the noise for every single query is bounded by η where $\eta \ll \frac{\sqrt{n}}{\log^{\ell^2-\ell+1} n}$ is attribute non-private. The attack is an efficient algorithm based on ℓ_2 norm minimization.*

The details of the results in this section can be found in [5].

Acknowledgements. I would like to thank Cynthia Dwork for her contributions to this paper. Even though she declined to co-author, without her contributions, the paper would not have existed. I would also like to thank Salil Vadhan for his kind permission to include the results of subsection 2.2 in this paper. Moritz Hardt and Mark Rudelson answered countlessly many questions.

I also had useful conversations about this work with Ilya Mironov, Elchanan Mossel, Omer Reingold, Adam Smith, Alexandre Stauffer, Kunal Talwar, and Salil Vadhan. I would also like to thank the SODA 2012 and TCC 2012 reviewers for many useful comments including pointing out an error in the earlier proof of Lemma 1.

References

1. Bar-Yossef, Z., Jayram, T.S., Kumar, R., Sivakumar, D.: An information statistics approach to data stream and communication complexity. Journal of Computer and System Sciences 68(4), 702–732 (2004)
2. Barak, B., Braverman, M., Chen, X., Rao, A.: How to compress interactive communication. In: Proceedings of the 42nd ACM Symposium on Theory of Computing, pp. 67–76 (2010)
3. Blum, A., Ligett, K., Roth, A.: A learning theory approach to non-interactive database privacy. In: Proceedings of the 40th ACM Symposium on Theory of Computing, pp. 609–618 (2008)
4. Candès, E.J., Rudelson, M., Tao, T., Vershynin, R.: Error Correction via Linear Programming. In: Proceedings of the 46th IEEE Symposium on Foundations of Computer Science, pp. 295–308 (2005)
5. De, A.: Lower bounds in Differential Privacy, arXiv:1107.2183v1 (2011)
6. De, A., Vadhan, S.: Personal Communication (2010)
7. Dinur, I., Nissim, K.: Revealing information while preserving privacy. Principles of Database Systems, 202–210 (2003)
8. Dwork, C., Kenthapadi, K., McSherry, F., Mironov, I., Naor, M.: Our Data, Ourselves: Privacy Via Distributed Noise Generation. In: Vaudenay, S. (ed.) EUROCRYPT 2006. LNCS, vol. 4004, pp. 486–503. Springer, Heidelberg (2006)
9. Dwork, C., McSherry, F., Nissim, K., Smith, A.: Calibrating Noise to Sensitivity in Private Data Analysis. In: Halevi, S., Rabin, T. (eds.) TCC 2006. LNCS, vol. 3876, pp. 265–284. Springer, Heidelberg (2006)
10. Dwork, C., McSherry, F., Talwar, K.: The price of privacy and the limits of LP decoding. In: Proceedings of the 39th ACM Symposium on Theory of Computing, pp. 85–94 (2007)
11. Hardt, M., Talwar, K.: On the geometry of differential privacy. In: Proceedings of the 42nd ACM Symposium on Theory of Computing, pp. 705–714 (2010)
12. Kasiviswanathan, S.P., Rudelson, M., Smith, A., Ullman, J.: The Price of privately releasing Contingency tables and the spectra of random matrices with correlated rows. In: Proceedings of the 42nd ACM Symposium on Theory of Computing, pp. 775–784 (2010)
13. McGregor, A., Mironov, I., Pitassi, T., Reingold, O., Talwar, K., Vadhan, S.P.: The Limits of Two-Party Differential Privacy. In: Proceedings of the 51st IEEE Symposium on Foundations of Computer Science, pp. 81–90 (2010)
14. Erdős, P., Frankl, P., Füredi, Z.: Families of finite sets in which no set is covered by the union of r others. Israel Journal of Mathematics 51(1,2), 79–89 (1985)
15. Rudelson, M.: Row products of random matrices, arXiv:1102.1947 (2011)

Iterative Constructions and Private Data Release[*]

Anupam Gupta[1,**], Aaron Roth[2,***], and Jonathan Ullman[3,†]

[1] Computer Science Department, Carnegie Mellon University, Pittsburgh, PA, USA
[2] Department of Computer and Information Science, University of Pennsylvania, Philadelphia PA 19104
[3] School of Engineering and Applied Sciences, Harvard University, Cambridge, MA

Abstract. In this paper we study the problem of approximately releasing the *cut function* of a graph while preserving differential privacy, and give new algorithms (and new analyses of existing algorithms) in both the interactive and non-interactive settings.

Our algorithms in the interactive setting are achieved by revisiting the problem of releasing differentially private, approximate answers to a large number of queries on a database. We show that several algorithms for this problem fall into the same basic framework, and are based on the existence of objects which we call *iterative database construction* algorithms. We give a new generic framework in which new (efficient) IDC algorithms give rise to new (efficient) interactive private query release mechanisms. Our modular analysis simplifies and tightens the analysis of previous algorithms, leading to improved bounds. We then give a new IDC algorithm (and therefore a new private, interactive query release mechanism) based on the Frieze/Kannan low-rank matrix decomposition. This new release mechanism gives an improvement on prior work in a range of parameters where the size of the database is comparable to the size of the data universe (such as releasing all cut queries on dense graphs).

We also give a non-interactive algorithm for efficiently releasing private *synthetic data* for graph cuts with error $O(|V|^{1.5})$. Our algorithm is based on randomized response and a non-private implementation of the SDP-based, constant-factor approximation algorithm for cut-norm due to Alon and Naor. Finally, we give a reduction based on the IDC framework showing that an efficient, private algorithm for computing sufficiently accurate rank-1 matrix approximations would lead to an improved efficient algorithm for releasing private synthetic data for graph cuts. We leave finding such an algorithm as our main open problem.

1 Introduction

Consider a graph representing the online communications between a set of individuals; each vertex represents a user, and an edge between two users indicates that they have corresponded by email. It might be useful to allow data analysts to mine this graph for statistical information. However, the graph is also composed of sensitive information,

---

[*] A full version appears at http://arxiv.org/abs/1107.3731
[**] Research was partly supported by NSF awards CCF-0964474 and CCF-1016799.
[***] Work done at Microsoft Research, New England. Email: aaroth@cis.upenn.edu
[†] Supported by NSF grant CNS-0831289. Email: jullman@seas.harvard.edu

R. Cramer (Ed.): TCC 2012, LNCS 7194, pp. 339–356, 2012.

and we cannot release information that reveals much about the existence of specific edges. Thus we would like a way to analyze the structure of this graph while protecting the privacy of individual edges. Specifically we would like to guarantee *differential privacy* [7] (defined in Section 2), which, roughly, requires that our algorithms be randomized, and induce nearly the same distribution over outcomes when given two data sets (e.g. graphs) which differ in only a single point (e.g. an edge).

Table 1. Comparison of accuracy bounds for linear queries. The bounds in the first column are prior to this work, the second column are what we achieve in this work, and the last column are the new bounds instantiated for releasing all cut queries. The bounds listed here are approximate and hide the dependence on certain parameters, such as δ and β. n denotes database size, k denotes the total number of queries answered, and $\mathcal{X}$ represents the data universe. For a graph $G = (V, E)$, $n = n_2 = |E|$, $|\mathcal{X}| = \binom{|V|}{2}$, and for all cut queries, $k = 2^{2|V|}$. Previous efficient results do not achieve non-trivial ($\leq |E|$) error, while all of the new bounds do for sufficiently dense graphs.

	Previous Bounds	This Paper	
		General Bounds	All Cut Queries
Median Mech.[a] [19]	$\dfrac{n^{2/3}(\log k)(\log|\mathcal{X}|)^{1/3}}{\epsilon^{1/3}}$	$\dfrac{n^{1/2}(\log k)^{3/4}(\log|\mathcal{X}|)^{1/4}}{\epsilon^{1/2}}$	$\dfrac{|E|^{1/2}|V|^{3/4}(\log|V|)^{1/4}}{\epsilon^{1/2}}$
Online MW [15]	$\dfrac{n^{1/2}(\log k)(\log|\mathcal{X}|)^{1/4}}{\epsilon}$	$\dfrac{n^{1/2}(\log k)^{1/2}(\log|\mathcal{X}|)^{1/4}}{\epsilon^{1/2}}$	$\dfrac{|E|^{1/2}|V|^{1/2}(\log|V|)^{1/4}}{\epsilon^{1/2}}$
Frieze/Kannan IDC	New in this paper	$\dfrac{n_2^{1/4}(\log k)^{1/2}|\mathcal{X}|^{1/4}}{\epsilon^{1/2}}$ [b]	$\dfrac{|E|^{1/4}|V|}{\epsilon^{1/2}}$
K-Norm Mech.[16]	$\dfrac{\sqrt{k}}{\epsilon}\left(\log\left(\dfrac{|\mathcal{X}|}{k}\right)\right)^{1/2}$ [c]	Not in IDC Framework	Not Applicable

[a] The bounds listed here are for linear queries. The Median Mechanism more generally works for any set of low sensitivity queries $\mathcal{Q}$ that have an α-net of size $N_\alpha(\mathcal{Q})$. We improve the bound from the solution to $\alpha = \dfrac{\log(N_\alpha(\mathcal{Q}))\log^2 k}{\epsilon}$ to the solution to $\alpha = \dfrac{\sqrt{\log N_\alpha(\mathcal{Q})}\log k}{\epsilon}$.
[b] Here we use $n_2 = \|\mathcal{D}\|_2^2$, in contrast to other known IDCs, whose error is in terms of $n = \|\mathcal{D}\|_1$. Note that $n \leq n_2 \leq n^2$.
[c] For $k \leq |\mathcal{X}|/2$. This is an approximate bound on *average* per-query error. All other algorithms listed bound worst-case per-query error.

One natural objective is to provide private access to the *cut function* of this graph. That is, to provide a privacy preserving way for a data analyst to specify any two (of the exponentially many) subsets of individuals, and to discover (up to some error) the number of email correspondences that have passed between these two groups. There are two ways we might try to achieve this goal: We could give an *interactive* solution where we give the analyst private oracle access to the cut function. Here the user can write down any sequence of cut queries and the oracle will respond with private, approximate answers. We may also try for a stronger, *non-interactive* solution, in which we release a private *synthetic dataset*; a new, private graph that approximately preserves the cut function of the original graph.

The case of answering cut queries on a graph is just one instance of the more general problem of query release for exponentially sized families of *linear queries* on a data set. Although this problem has been extensively studied in the differential privacy literature, we observe that no previously known efficient solution is suitable for the case of releasing all cut queries on graphs. In the setting of cut queries on a graph, we use "efficient solution" roughly to mean one in which each query is answered in time $\text{poly}(|V|)$, in the interactive setting, or one in which the whole construction runs in time $\text{poly}(|V|)$, in the non-interactive setting. In this paper we provide both efficient interactive and non-interactive solutions for this problem.

We give a generic framework that converts objects we call *iterative database construction (IDC)* algorithms into private query release mechanisms in both the interactive and non-interactive settings. This framework generalizes the median mechanism [19], the online multiplicative weights mechanism [15], and the offline multiplicative weights mechanism [12, 14]. Our framework gives a simple, modular analysis of all of these mechanisms, which lead to tighter bounds in the interactive setting than those given in [19] and [15]. These improved bounds are crucial to our objective of giving non-trivial approximations to all possible cut queries. We also instantiate this framework with a new IDC algorithm for arbitrary linear queries that is based on the Frieze/Kannan low-rank matrix decomposition [10] and is tailored to releasing cut queries. This algorithm leads to a new online query release mechanism for linear queries that gives a better approximation in settings (such as we would encounter trying to answer all cut queries on a dense graph) where the database size is comparable to the size of the data universe. We summarize our bounds in Table 1.

We also give a new algorithm (building on techniques for constructing private synthetic data in [2, 8]) in the non-interactive setting that efficiently generates private synthetic graphs that approximately preserve the cut function. Finally, we use our IDC framework to show that an efficient, private algorithm for privately computing good rank-1 approximations to matrices would automatically yield efficient private algorithms for releasing synthetic graphs with improved approximation guarantees.

1.1   Our Results and Techniques

Our main conceptual contribution is to define the abstraction of *iterative database construction (IDC)* algorithms (Section 3) and to show that an efficient IDC for any class of queries Q automatically yields an efficient private data release mechanism for Q in both the interactive and non-interactive settings. Informally, IDCs construct a data structure that can be used to answer all the queries in Q by iteratively improving a hypothesis data structure. Moreover, they update the hypothesis when given a query witnessing a significant difference between the hypothesis data structure and the underlying database.

In hindsight, this framework generalizes the median mechanism [19] and the online multiplicative weights mechanism [15]. It also generalizes the offline multiplicative weights mechanism [12, 14]. All of these mechanisms can be seen to use IDCs of the sort we define in this work. (In Appendix A we show how these algorithms fall into the IDC framework.)

Our generalization and abstraction also allows for a simple, modular analysis of mechanisms based on IDCs. Using this analysis, we are able to show improved bounds on the accuracy of both the median mechanism and multiplicative weights mechanism. These improved bounds are significant in our application of using an interactive mechanism to release a large number of cut queries and crucial if we want to answer all cut queries. When answering all cut queries, the previous bounds would not guarantee error that is $\leq |E|$, meaning that the error may be larger than the largest cut in the graph. Of course, we can privately guarantee error $\leq |E|$ simply by releasing the answer 0 for every cut query. Our new analysis shows that these mechanisms are capable of answering all $2^{2|V|}$ cut queries with error $o(|E|)$ on sufficiently dense graphs; e.g., multiplicative weights gives sublinear error for graphs with $|E| = \omega(|V|\sqrt{\log |V|})$.

Although it may seem unrealistic to answer all cut queries using an interactive mechanism, our new analysis allows us to give a best-of-both-worlds guarantee that we can answer each query efficiently with non-trivial accuracy without ever having to "shut off" the algorithm for answering too many queries. In practice it may be preferable to limit the number of queries the interactive mechanism will have to answer, in order to improve the accuracy of the responses. In this case our new bounds still offer significant improvements in accuracy.

We also define a new IDC based on the Frieze/Kannan low-rank matrix decomposition [10], which yields a private interactive mechanism for releasing linear queries. Our new mechanism outperforms previously known techniques when the size of the database is comparable to the size of the data universe, as is the case on a dense graph. The error for the Frieze/Kannan IDC is smaller than that for multiplicative weights for extremely dense graphs, where $|E| = \Omega(|V|^2 / \log |V|)$.

We then consider the problem of efficiently releasing private synthetic data for the class of cut queries. We show that a technique based on randomized response efficiently yields a private data structure (but not a synthetic database) capable of answering any cut query on a graph with $|V|$ vertices up to maximum error $O(|V|^{1.5})$. (Note this error is independent of the density of the graph and the Frieze/Kannan and multiplicative weights IDCs introduce smaller error for sparser graphs.) We then show how to use this data structure to efficiently construct a synthetic database with only a small constant factor blowup in our error. Our algorithm is based on a technique for constructing synthetic data in [2, 8]. Their observation is that, for linear queries, the set of accurate synthetic databases is described by a (large) set of linear constraints. In the case of cut queries, we are able to use a constant-factor approximation to the cut-norm due to Alon and Naor [1] as the separation oracle to find a feasible solution (and thus a synthetic database) efficiently. Finally, we show how the existence of an efficient private algorithm for finding good low-rank approximations to matrices would imply the existence of an improved algorithm for privately releasing synthetic data for cut queries, using our IDC framework.

To summarize the results for cut queries: between the multiplicative weights IDC, the Frieze/Kannan IDC, and randomized response, the best mechanism depends on $|E|$. When $|E|$ is below $O(|V|^2 / \log |V|)$, the multiplicative weights IDC introduces the least error. For $|E|$ lying between $O(|V|^2 / \log |V|)$ and $O(|V|^2)$, the Frieze/Kannan IDC introduces the least error. Both IDC mechanisms have error increasing with $|E|$, finally matching the error for randomized response when $|E| = \Theta(|V|^2)$. When answering k queries, the error for all three mechanisms depends on $\sqrt{\log k}$, so these thresholds are independent of the number of queries.

1.2 Related Work

Differential privacy, introduced in a series of papers [4, 6, 7] in the last decade, has become a standard solution concept for statistical database privacy. The first mechanism for simultaneously releasing the answers to exponentially large classes of statistical queries was given in [5]. They showed that the existence of small nets for a class of queries $\mathcal{Q}$ automatically yields a (computationally inefficient) non-interactive, private algorithm for releasing answers to all the queries in $\mathcal{Q}$ with low error. Subsequent improvements were given by Dwork et al. [8, 9].

Roth and Roughgarden [19] showed that large classes of queries could also be released with low error in the *interactive* setting, in which queries may arrive online, and the mechanism must provide answers before knowing which queries will arrive in the future. Subsequently, Hardt and Rothblum [15] gave improved bounds for the online query release problem based on the multiplicative weights algorithm. In hindsight, both of these algorithms follow the same basic framework, which is to use an IDC.

Gupta et al. [12] gave a *non-interactive* data release mechanism based on the multiplicative weights algorithm and an arbitrary agnostic learner for a class of queries. An instantiation of this algorithm (the *offline* multiplicative weights algorithm) using the generic agnostic learner of Kasiviswanathan et al. [17] (who use the exponential mechanism of [18]) was implemented and experimentally evaluated on the task of releasing small conjunctions to low error on real data by Hardt, Ligett, and McSherry [14]. This algorithm gives bounds comparable to those given in this paper, but it does not work in the interactive setting, and is not computationally efficient for settings in which the number of queries is exponentially larger than the database size (as is the case with graph cuts). We note in Section 7 that this generic algorithm can also be instantiated with any iterative database construction algorithm.

Hardt and Talwar [16] consider the setting where the number of queries is smaller than the universe size and introduced the K-Norm mechanism. Subsequent improvements were given by [3]. When the number of queries and the database size are comparable to the universe size (i.e. $|\mathcal{Q}| = \Omega(|\mathcal{X}|)$, $n \geq \Omega(|\mathcal{X}|/\log|\mathcal{X}|)$), the K-Norm mechanism gives *average error* that is smaller than the worst-case error promised by the online multiplicative weights mechanism. In this range of parameters the Frieze/Kannan IDC and the K-Norm mechanism both improve on the online multiplicative weights, and give roughly the same error. However, the Frieze/Kannan IDC has bounded worse-case error, as opposed to average-case error. In general the two mechanisms are incomparable, as the error of the Frieze/Kannan IDC has bounded worse-case error and applies even when $|Q| > |\mathcal{X}|$, but its error has polynomial, rather than logarithmic dependence on $|\mathcal{X}|$.

The Frieze-Kannan low-rank approximation (or the weak regularity lemma) shows that every matrix can be approximated by a sum of a small number of cut matrices [10, 11], and this fact has many important algorithmic applications. We also use the fact that the proof extends to more general settings, as was noted by [20].

2 Preliminaries

In this paper, we study datasets $\mathcal{D}$ that consist of collections of n elements from some universe $\mathcal{X}$. We can also write $\mathcal{D} \in \mathbb{N}^{|\mathcal{X}|}$ when it is convenient to represent $\mathcal{D}$ as a histogram over $\mathcal{X}$. We say that two databases $\mathcal{D}, \mathcal{D}'$ are adjacent if they differ in only a single element. As histograms, they are adjacent if $\|\mathcal{D} - \mathcal{D}'\|_1 \leq 1$. We will require that our algorithms satisfy *differential privacy*:

Definition 1 (Differential Privacy). *A randomized algorithm* $M : \mathbb{N}^{|\mathcal{X}|} \to R$ *(for any abstract range* R*) satisfies* (ϵ, δ)*-differential privacy if for all adjacent databases* $\mathcal{D}$ *and* $\mathcal{D}'$*, and for all events* $S \subseteq R$*,* $\Pr[M(\mathcal{D}) \in S] \leq \exp(\epsilon) \Pr[M(\mathcal{D}') \in S] + \delta$

We will generally think of ϵ as being a small constant, and δ as being negligibly small – i.e. smaller than any inverse polynomial function of n.

We note that when we will discuss interactive mechanisms, we must view the output of a mechanism as the *transcript* of an interaction between an adaptive adversary who supplies questions about the database based on previous outcomes of the mechanism, and the mechanism itself. For clarity, in this paper we will elide specifics about the model of adaptive private composition. For a detailed treatment of this issue, see [9].

A useful distribution is the *Laplace* distribution.

Definition 2 (The Laplace Distribution). *The Laplace Distribution with mean* 0 *and scale* b *is the distribution with probability density function:* $\mathrm{Lap}(x|b) = \frac{1}{2b} \exp(-\frac{|x|}{b})$. *We will sometimes write* $\mathrm{Lap}(b)$ *to denote the Laplace distribution with scale* b*, and will sometimes abuse notation and write* $\mathrm{Lap}(b)$ *simply to denote a random variable* $X \sim \mathrm{Lap}(b)$.

A fundamental result in data privacy is that perturbing low sensitivity queries with Laplace noise preserves $(\epsilon, 0)$-differential privacy.

Theorem 1 ([7]). *Suppose* $Q : \mathbb{N}^{|\mathcal{X}|} \to \mathbb{R}^k$ *is a function such that for all adjacent databases* $\mathcal{D}$ *and* $\mathcal{D}'$*,* $\|Q(\mathcal{D}) - Q(\mathcal{D}')\|_1 \leq 1$*. Then the procedure which on input* $\mathcal{D}$ *releases* $Q(\mathcal{D}) + (X_1, \dots, X_k)$*, where each* X_i *is an independent draw from a* $\mathrm{Lap}(1/\epsilon)$ *distribution, preserves* $(\epsilon, 0)$*-differential privacy.*

It will be useful to understand how privacy parameters for individual steps of an algorithm compose into privacy guarantees for the entire algorithm. The following useful theorem is due to Dwork, Rothblum, and Vadhan:

Theorem 2 ([9]). *Let* $0 \leq \epsilon \leq 1$ *be a parameter. Let* P, Q *be probability measures supported on a set* S *such that* $\max_{s \in S} |\log (P(s)/Q(s))| \leq \epsilon$. *Then*

$$\mathbb{E}_P [\log (P(s)/Q(s))] \leq 2\epsilon^2.$$

We are interested in privately releasing accurate answers to large collections of queries. Queries are functions $Q : \mathbb{N}^{|\mathcal{X}|} \to \mathbb{R}$, and we denote collections of queries by $\mathcal{Q}$. We write $k = |\mathcal{Q}|$ to denote the cardinality of the set of queries.

A common type of queries are *linear queries*. A linear query Q has a representation as a vector $[0, 1]^{|\mathcal{X}|}$, and can be evaluated on a database by taking the dot product between the query and the histogram representation of the database: $Q(\mathcal{D}) = Q \cdot \mathcal{D}$.

Definition 3 (Accuracy). *Let Q be a set of queries. A mechanism $M : \mathbb{N}^{|\mathcal{X}|} \to \mathcal{R}$ is (α, β)-accurate for Q if there exists a function* Eval $: Q \times \mathcal{R} \to \mathbb{R}$ *s.t. for every database $\mathcal{D} \in \mathbb{N}^{|\mathcal{X}|}$, with probability at least $1 - \beta$ over the coins of M, $M(\mathcal{D})$ outputs $r \in \mathcal{R}$ such that $\max_{Q \in Q} |Q(\mathcal{D}) - \text{Eval}(Q, r)| \leq \alpha$. We will abuse notation and write $Q(r) = \text{Eval}(Q, r)$.*

We say that an algorithm M *releases synthetic data* (as is the case for our new IDC, as well as the multiplicative weights IDC [15]) if $\mathcal{R} = \mathbb{N}^{|\mathcal{X}|}$ In this case, $M(\mathcal{D}) = \widehat{\mathcal{D}} \in \mathbb{N}^{|\mathcal{X}|}$ and $\text{Eval}(\widehat{\mathcal{D}}, Q) = Q(\widehat{\mathcal{D}})$. We say that a synthetic data release algorithm is *efficient* if it runs in time polynomial in $n = \|\mathcal{D}\|_1$, the size of the data set. Note that if $n \ll |\mathcal{X}|$, efficient algorithms will have to input and output concise representations of the dataset (i.e., as collections of items from the universe) instead of using the histogram representation. Nevertheless, it will be convenient to think of datasets as histograms.

We say an algorithm efficiently releases k queries from a class Q in the *interactive* setting if on an arbitrary, adaptively chosen stream of queries $Q_1, \ldots, Q_k$, it outputs answers $a_1, \ldots, a_k$. The algorithm must output each a_i after receiving query Q_i but before receiving Q_{i+1}, and is only allowed poly(n) run time per query. We are typically interested in the case when k can be exponentially large in n. Note that as far as computational efficiency is concerned, releasing synthetic data for a class of queries k is at least as difficult as releasing queries from k in the interactive setting, since we can use the synthetic data to answer queries interactively.

Graphs and Cuts. When we consider datasets that represent graphs $G = (V, E)$, we think of the database as being the edge set $\mathcal{D}_G = E$, and the data-universe being the collection of all possible edges in the complete graph: $|\mathcal{X}| = \binom{|V|}{2}$. That is, we consider the vertex set to be common among all graphs, which differ only in their edge sets. One example we care about is approximating the cut function of a sensitive graph G.

For any real-valued matrix $A \in \mathbb{R}^{m \times m'}$, for $S \subseteq [m]$ and $T \subseteq [m']$, we define $A(S, T) := \sum_{s \in S, t \in T} A_{st}$. The *cut norm of the matrix* A is now defined as $\|A\|_C := \max_{S \subseteq [m], T \subseteq [m']} |A(S, T)|$. A graph G can be represented as its adjacency matrix $A_G \in \{0, 1\}^{|V| \times |V|}$. In this paper, a *cut* in a graph G is defined by any two subsets of vertices $S, T \subseteq V$. We write the value of an S, T cut in G as $G(S, T) := A_G(S, T)$, where A_G is the adjacency matrix of G. Similarly, we extend the definition of *cut norm* to n vertex graphs naturally by defining $\|G\|_C := \|A_G\|_C = \max_{S, T \subseteq V} |G(S, T)|$ and $\|G - H\|_C := \|A_G - A_H\|_C$. The class of cut queries $Q_{\text{Cut}} = \{Q_{S, T} : S, T \subseteq V\}$, where $Q_{S, T}(G) = A_G(S, T)$. Note that cut queries are an example of a class of linear queries, because we can represent them as a vector in which $Q_{S, T}[i, j] = 1$ if $i \in S, j \in T$ and 0 otherwise, and evaluate $Q_{S, T}(G) = \sum_{i, j \in V} Q_{S, T}[i, j] \cdot A_G[i, j]$.

Note that as linear queries, we can write cut queries as the outer product of two vectors: $Q_{S, T} = \chi_S \cdot \chi_T^T$, where $\chi_S, \chi_T \in \{0, 1\}^{|V|}$ are the characteristic vectors of the sets S and T respectively. Let us define a more general class of *rank-1* queries on graphs to be a subset of all linear queries: $Q_{r1} = \{Q \in [0, 1]^{|V| \times |V|}$ such that $Q = u \cdot v^T$ for some vectors $u, v \in [0, 1]^{|V|}\}$. Of course the set of rank-1 queries includes the set of cut queries, and any mechanism that is accurate with respect to rank-1 queries is also accurate with respect to cut queries.

Proofs. Because of space constraints, many of the proofs in this paper have been omitted. The interested reader can see full proofs in the full version of this paper: [13].

3   Iterative Database Constructions

In this section we define the abstraction of *iterative database constructions* that includes our new Frieze/Kannan construction and several existing algorithm [19, 15] as a special case. Roughly, each of these mechanisms works by maintaining a sequence of data structures $\mathcal{D}^{(1)}, \mathcal{D}^{(2)}, \ldots$ that give increasingly good approximations to the input database $\mathcal{D}$ (in a sense that depends on the IDC). Moreover, these mechanisms produce the next data structure in the sequence by considering only one query Q that *distinguishes* the real database in the sense that $Q(\mathcal{D}^{(t)})$ differs significantly from $Q(\mathcal{D})$.

Syntactically, we will consider functions of the form $\mathbf{U} : \mathcal{R}_\mathbf{U} \times \mathcal{Q} \times \mathbb{R} \to \mathcal{R}_\mathbf{U}$. The inputs to $\mathbf{U}$ are a data structure in $\mathcal{R}_\mathbf{U}$, which represents the current data structure $\mathcal{D}^{(t)}$; a query Q, which represents the distinguishing query, and may be restricted to a certain set $\mathcal{Q}$; and also a real number. which estimates $Q(\mathcal{D})$. Formally, we define a *database update sequence* , to capture the sequence of inputs to $\mathbf{U}$ used to generate the database sequence $\mathcal{D}^{(1)}, \mathcal{D}^{(2)}, \ldots$.

Definition 4 (Database Update Sequence). *Let $\mathcal{D} \in \mathbb{N}^{|\mathcal{X}|}$ be any database and let* $\left\{ (\mathcal{D}^{(t)}, Q^{(t)}, \widehat{A}^{(t)}) \right\}_{t=1,\ldots,C} \in (\mathcal{R}_\mathbf{U} \times \mathcal{Q} \times \mathbb{R})^C$ *be a sequence of tuples. We say the sequence is an $(\mathbf{U}, \mathcal{D}, \mathcal{Q}, \alpha, C)$-database update sequence if it satisfies the following properties:*

1. *$\mathcal{D}^{(1)} = \mathcal{D}(\emptyset, \cdot, \cdot)$,*
2. *for every $t = 1, 2, \ldots, C$, $\left| Q^{(t)}(\mathcal{D}) - Q^{(t)}(\mathcal{D}^{(t)}) \right| \geq \alpha$,*
3. *for every $t = 1, 2, \ldots, C$, $\left| Q^{(t)}(\mathcal{D}) - \widehat{A}^{(t)} \right| < \alpha$,*
4. *and for every $t = 1, 2, \ldots, C - 1$, $\mathcal{D}^{(t+1)} = \mathbf{U}(\mathcal{D}^{(t)}, Q^{(t)}, \widehat{A}^{(t)})$.*

We note that for all of the iterative database constructions we consider, the approximate answer $\widehat{A}^{(t)}$ is used only to determine the *sign* of $Q^{(t)}(\mathcal{D}) - Q^{(t)}(\mathcal{D}^{(t)})$, which is the motivation for requiring that $\widehat{A}^{(t)}$ have error smaller than α. The main measure of efficiency we're interested in from an iterative database construction is the maximum number of updates we need to perform before the database $\mathcal{D}^{(t)}$ approximates $\mathcal{D}$ well with respect to the queries in $\mathcal{Q}$. To this end we define an iterative database construction as follows:

Definition 5 (Iterative Database Construction). *Let $\mathbf{U} : \mathcal{R}_\mathbf{U} \times \mathcal{Q} \times \mathbb{R} \to \mathcal{R}_\mathbf{U}$ be an update rule and let $B : \mathbb{R} \to \mathbb{R}$ be a function. We say $\mathbf{U}$ is a $B(\alpha)$-iterative database construction for query class $\mathcal{Q}$ if for every database $\mathcal{D} \in \mathbb{N}^{|\mathcal{X}|}$, every $(\mathbf{U}, \mathcal{D}, \mathcal{Q}, \alpha, C)$-database update sequence satisfies $C \leq B(\alpha)$.*

Note that, by definition, if $\mathbf{U}$ is a $B(\alpha)$-iterative database construction, then given any maximal $(\mathbf{U}, \mathcal{D}, \mathcal{Q}, \alpha, C)$-database update sequence, the final database $\mathcal{D}^{(C)}$ must satisfy $\max_{Q \in \mathcal{Q}} |Q(\mathcal{D}) - Q(\mathcal{D}^{(C)})| \leq \alpha$ or else there would exist another query satisfying property 2 of Definition 4, and thus there would exist a $(\mathbf{U}, \mathcal{D}, \mathcal{Q}, \alpha, C+1)$-database update sequence, contradicting maximality.

4 Query Release from Iterative Database Construction

In this section we describe an interactive algorithm for releasing linear queries using an arbitrary iterative database construction.

Algorithm 1. Online Query Release Mechanism

$\mathcal{M}^{\mathbf{U}}(\mathcal{D}, \epsilon, \delta, \alpha, \beta, k)$:

Input: A database $\mathcal{D} \in \mathbb{N}^{|\mathcal{X}|}$, a parameter $\alpha \in \mathbb{R}$, parameters $\epsilon, \delta, \beta \in [0, 1]$, and the number of queries $k \in \mathbb{N}$. Oracle access to $\mathbf{U}$, a $B = B(\alpha)$-iterative database construction for $\mathcal{Q}$.

Parameters:

$$\sigma = \sigma(\alpha) := \frac{1000\sqrt{B(\alpha)} \cdot \log(4/\delta)}{\epsilon} \qquad T = T(\alpha) := 4\sigma(\alpha) \cdot \log(2k/\beta).$$

Set $\mathcal{D}^{(1)} := \mathbf{U}(\emptyset, \cdot, \cdot)$, $C = 0$.

For: $t = 1, 2, \ldots, k$

1. Receive a query $Q^{(t)} \in \mathcal{Q}$ and compute

$$Z^{(t)} \sim \mathrm{Lap}(\sigma) \qquad A^{(t)} = Q^{(t)}(\mathcal{D}) \qquad \widehat{A}^{(t)} = Q^{(t)}(\mathcal{D}) + Z^{(t)} \qquad \Lambda^{(t)} = Q^{(t)}(\mathcal{D}^{(t)})$$

2. **If:** $|\widehat{A}^{(t)} - \Lambda^{(t)}| \leq T$ **then:** output $\Lambda^{(t)}$ and set $\mathcal{D}^{(t+1)} = \mathcal{D}^{(t)}$.

   Else: output $\widehat{A}^{(t)}$, set $\mathcal{D}^{(t+1)} = \mathbf{U}\left(\mathcal{D}^{(t)}, Q^{(t)}, \widehat{A}^{(t)}\right)$, and set $C = C + 1$.

3. **If:** $C = B(\alpha)$ **then:** terminate.

4.1 Privacy Analysis

Theorem 3. *Algorithm 1 is (ϵ, δ)-differentially private.*

Proof (Proof Sketch). Our privacy analysis follows the approach of [15] straightforwardly. The details appear in the full version of the paper. Intuitively, we will try to classify the answers to the queries by the amount of "information leaked about the database." This classification will lead to a bound on the total amount of information leaked, and a tighter bound can be deduced using Theorem 2.

At a very high level, the argument can be thought of in two steps. The first is to argue that the noise we add has large enough magnitude that the information leaked in the (small number of) "update rounds" is small. This step is simple and follows from the bound on the number of update rounds and the well-known properties of the Laplace distribution. The second step is to argue that the *location* of the update rounds also leaks little information. This second step is more difficult, and requires reasoning carefully about rounds that are "close to update rounds."

More specifically, though still informally we will consider three possible ranges for the value of the *noise* $Z^{(t)}$ in each round $t = 1, 2, \ldots, k..$ Intuitively the three cases are as follows: 1) The noise is sufficiently small that there would never be an update, even if the input database were exchanged with an adjacent one. Here we argue no information is leaked. 2) The noise is sufficiently large that there would always be an update, even if the database were exchanged with an adjacent one. In these rounds there is information leaked, but we also increment C, and thus there cannot be too many of these before terminating. 3) The noise is intermediate, such that we do not do an update and increment C, but might if we switched to an adjacent database. In principle there may be as many as k such rounds, however it will turn out with high probability the number of such rounds is not much bigger than B.

We then complete the proof by applying Theorem 2 to bound the *expected* privacy loss over the course of all the rounds, and apply Azuma's inequality to argue that except with probability δ, the total privacy loss does not exceed ϵ.

4.2 Utility Analysis

Theorem 4. *Let* $\mathcal{D} \in \mathbb{N}^{|\mathcal{X}|}$ *be any database. And* $\mathbf{U}$ *be a* $B(\alpha)$-*iterative database construction for query class* $\mathcal{Q}$. *Then for any* $\beta, \epsilon, \delta > 0$, *Algorithm 1 is* $\left(\frac{5T(\alpha)}{4}, \beta\right)$-*accurate for* $\mathcal{Q}$, *as long as* $T(\alpha) \in [4\alpha/3, 2\alpha]$.

Proof (Proof sketch). Roughly, the argument is as follows: Assume we did not add any noise to the queries. Then we would answer each query with the exactly-correct answer $A^{(t)}$ or with $\Lambda^{(t)}$ so long as $\Lambda^{(t)}$ is sufficiently close to $A^{(t)}$. Essentially, all we do in the proof is show that this intuition remains correct when noise is added.

When adding noise we answer with either $A^{(t)} + Z^{(t)}$ or $\Lambda^{(t)}$, so long as $\Lambda^{(t)}$ is sufficiently close to $A^{(t)} + Z^{(t)}$. It is not hard to argue that $Z^{(t)}$ remains small in every round, and thus the answers in the latter case are not much less accurate than the answers in the former case.

What remains to be shown is that the mechanism does not terminate early due to the condition $C = B$. In order to do this, we show that the sequence of updates forms a database update sequence, and thus cannot be too long if $\mathbf{U}$ is an efficient iterative database construction. In order to do this, we argue that $Z^{(t)}$ is sufficiently small that the condition for performing an update ($|A^{(t)} + Z^{(t)} - \Lambda^{(t)}| \geq T$) is sufficient to ensure that the query is a good distinguisher ($|A^{(t)} - \Lambda^{(t)}| \geq \alpha$).

In order to get the best accuracy parameters, one can just solve for the equation $\alpha = 3T(\alpha)/4$; substituting for $T(\cdot)$, this is the same as solving the following equation for α: $\alpha = \frac{96\sqrt{B(\alpha)}\log(4/\delta)\log(k/\beta)}{\epsilon}$. Using this method we obtain bounds on the error for various IDCs, which are summarized both in Table 1 and in the full version. $\quad\blacksquare$

5 An Iterative Database Construction Based on Frieze/Kannan

In this section we describe and analyze an iterative database construction based on the Frieze/Kannan "cut decomposition" [10]. Although the style of analysis we use was originally applied specifically to cuts in [10], their argument generalizes to arbitrary linear queries. To our knowledge, such a generalization was first observed in [20].

Note that the sum in Algorithm 2 denotes entrywise vector addition.

Theorem 5. *Let* $\mathcal{D} \in \mathbb{N}^{|\mathcal{X}|}$ *be a dataset. For any* $\alpha > 0$, $\mathbf{U}_\alpha^{FK}$ *is a* $B(\alpha)$-*iterative database construction for a class of linear queries* $\mathcal{Q}$, *where* $B(\alpha) = \frac{\|\mathcal{D}\|_1^2 |\mathcal{X}|}{\alpha^2}$.

Algorithm 2. The Frieze/Kannan-based IDC

$\mathbf{U}_\alpha^{FK}(\mathcal{D}, Q, \widehat{A})$:

  If: $\mathcal{D} = \emptyset$ then: output $\mathcal{D}' = \emptyset$

  Else if: $Q(\mathcal{D}) - \widehat{A} > 0$ then: output $\mathcal{D}' = \mathcal{D} - \frac{\alpha}{|\mathcal{X}|} \cdot Q$

  Else if: $Q(\mathcal{D}) - \widehat{A} < 0$ then: output $\mathcal{D}' = \mathcal{D} + \frac{\alpha}{|\mathcal{X}|} \cdot Q$

Proof (Proof sketch). Let $\mathcal{D} \in \mathbb{N}^{|\mathcal{X}|}$ be any database and let $\left\{ (\mathcal{D}^{(t)}, Q^{(t)}, \widehat{A}^{(t)}) \right\}_{t=1,\dots,C}$ be $(\mathbf{U}_\alpha^{FK}, \mathcal{D}, \mathcal{Q}, \alpha, B)$-database update sequence (Definition 4). We want to show that $C \leq \|\mathcal{D}\|_2^2 |\mathcal{X}|/\alpha^2$. Specifically, after $\|\mathcal{D}\|_2^2 |\mathcal{X}|/\alpha^2$ invocations of $\mathbf{U}_\alpha^{FK}$, the database $\mathcal{D}^{(\|\mathcal{D}\|_2^2 |\mathcal{X}|/\alpha^2)}$ is $(\alpha, \mathcal{Q})$-accurate for $\mathcal{D}$, and thus there cannot be a sequence of longer than $\|\mathcal{D}\|_2^2 |\mathcal{X}|/\alpha^2$ queries that satisfy property 2 of Definition 4.

In order to formalize this intuition, we use a potential argument as in [10] to show that for every $t = 1, 2, \dots, B$, $\mathcal{D}^{(t+1)}$ is significantly closer to $\mathcal{D}$ than $\mathcal{D}^{(t)}$. Specifically, our potential function is the L_2^2 norm of the database $\mathcal{D} - \mathcal{D}^{(t)}$, defined as $\|\mathcal{D}\|_2^2 = \sum_{i \in \mathcal{X}} \mathcal{D}(i)^2$. Observe that $\|\mathcal{D} - \mathcal{D}^{(1)}\|_2^2 = \|\mathcal{D}\|_2^2$, and $\|\mathcal{D}\|_2^2 \geq 0$. Thus it will suffices to show, as we do in the full proof, that in every step, the potential decreases by $\alpha^2/|\mathcal{X}|$.

Corollary 1. *Let* $\gamma = O\left(\epsilon^{-1/2} n_2^{1/4} |\mathcal{X}|^{1/4} \sqrt{\log(k/\beta)} \right)$. *Then Algorithm 1, instantiated with* $\mathbf{U}_\gamma^{FK}$ *is* (ϵ, δ)-*differentially private and an* (α, β)-*accurate interactive release mechanism for query set* $\mathcal{Q}$ *with* $\alpha = O\left(\frac{n_2^{1/4} |\mathcal{X}|^{1/4} \sqrt{\log(k/\beta)\log(1/\delta)}}{\sqrt{\epsilon}} \right)$ *where* $n_2 = \|\mathcal{D}\|_2^2$. *Note that for databases that are subsets of the data universe (rather than multisets),* $n_2 = n$.

Remark 1. For the setting in which the database represents a graph and the queryset contains all cut queries, this bounds is $O(|V||E|^{1/4}/\sqrt{\epsilon})$. This improves on the accuracy of the multiplicative weights IDC for dense graphs with $|E| \geq \Omega(|V|^2/\log|V|)$.

6   Results for Synthetic Data

In this section, we consider the more demanding task of efficiently releasing *synthetic data* for the class of cut queries on graphs. Our algorithm is simple, and is based on releasing a noisy histogram. Note that for a graph, $|\mathcal{X}| = \binom{|V|}{2}$, and $\mathcal{D} = E$, so as long as $|E| = \Omega(|V|)$, the universe is at most a polynomial in the database size. (Moreover, it is easy to show that there does not exist any $(\epsilon, 0)$-private mechanism that has error $o(|V|)$, so the only interesting cases are when $|E| = \Omega(|V|)$.)

Consider a database whose elements are drawn from $\mathcal{X}$; we represent this as a vector (histogram) $\mathcal{D} \in \mathbb{N}^{|\mathcal{X}|}$. Let $\widehat{\mathcal{D}} = \mathcal{D} + (Y_1, \dots, Y_{|\mathcal{X}|})$ be a "noisy" database, where each $Y_i \sim \mathrm{Lap}(1/\varepsilon)$ is an independent draw from the Laplace distribution. Note that by Theorem 1, the procedure which on input $\mathcal{D}$ releases the noisy database $\widehat{\mathcal{D}}$ preserves $(\epsilon, 0)$-differential privacy. This follows because the histogram vector can be viewed as simply the evaluation of the identity query $Q : \mathbb{N}^{|\mathcal{X}|} \to \mathbb{N}^{|\mathcal{X}|}$, which can be easily seen to be 1-sensitive. At this stage, we could release $\widehat{\mathcal{D}}$ and be satisfied that we have designed a private algorithm. There are two issues: first, we must analyze the utility guarantees that $\widehat{\mathcal{D}}$ has with respect to our query set $\mathcal{Q}$. Second, $\widehat{\mathcal{D}}$ is not quite synthetic data. It will be a vector with possibly negative entries, and so does not represent a histogram. Interpreted as a graph, it will be a weighted graph with negative edge weights. Such an answer may be insufficient for some applications, so in Section 6.1 we show how to convert such an answer into $[0,1]$ weighted graph with similar accuracy guarantees.

The utility guarantee of this procedure over the collections $\mathcal{Q}$ of linear queries is also not difficult; i.e., each query $Q \in \mathcal{Q}$ is a vector in $[0,1]^{|\mathcal{X}|}$, and on any database $\mathcal{D}$ evaluates to $Q(\mathcal{D}) = \langle Q, \mathcal{D} \rangle$.

Lemma 1. *Suppose that $\mathcal{Q} \subseteq [0,1]^{|\mathcal{X}|}$ is some collection of linear queries. For the case $|\mathcal{Q}| \leq (\beta/2)\, 2^{|\mathcal{X}|/6}$, it holds that with probability at least $1 - \beta$, for every query $Q \in \mathcal{Q}$, $|Q(\mathcal{D}) - Q(\widehat{\mathcal{D}})| \leq \varepsilon^{-1}\sqrt{6|\mathcal{X}|\log(|\mathcal{Q}|/\beta)}$. For general $\mathcal{Q}$, the error bound is $O(\varepsilon^{-1}\sqrt{|\mathcal{X}|\log(|\mathcal{X}|/\beta)\log(|\mathcal{Q}|/\beta)})$.*

The proof of this lemma uses standard moment-generating function techniques and is deferred to the full version.

In summary, the bounds on the error are $\approx \varepsilon^{-1}\sqrt{|\mathcal{X}|\log|\mathcal{Q}|}$, with some correction terms depending on whether the size of the query set is at most $2^{O(|\mathcal{X}|)}$ or larger.

6.1 Randomized Response and Synthetic Data for Cut Queries

For the case of cuts in graph on a vertex set V, the database is a vector in $\{0,1\}^{\binom{|V|}{2}}$, and the noisy database just adds independent $\mathrm{Lap}(1/\varepsilon)$ noise to each bit value. Since the query set $\mathcal{Q}_{cuts}$ has size $2^{2|V|}$, (namely it consists of all (S,T) pairs), we have $|\mathcal{Q}_{cuts}| \ll (\beta/2)2^{|\mathcal{X}|/6}$ for all reasonable β and $|V|$, we can use the randomized response analysis above to get accuracy

$$O\left(\left(\binom{|V|}{2}\log(|\mathcal{Q}_{cuts}|/\beta)\right)^{1/2}/\varepsilon\right) = O((|V|^{3/2} + |V|\log 1/\beta)/\varepsilon)$$

with probability at least $1 - \beta$. In fact, one can give a slightly tighter analysis where the accuracy depends on the size of the sets S, T—by observing that the number of random variables participating in a cut query (S,T) is exactly $|S||T|$, one can show that the accuracy for all cuts is whp $O(\varepsilon^{-1}\sqrt{|V||S||T|})$.

Viewing the noisy database $\widehat{\mathcal{D}}$ as a weighted graph $\widehat{G}$, where the weight of (u,v) is $\mathbf{1}_{(u,v)\in E(G)} + \mathrm{Lap}(1/\varepsilon)$, note that $\widehat{G}$ has negative weight edges and hence cannot be considered synthetic data. We can remedy the situation (using the idea of solving a suitable linear program [2, 8]):

Lemma 2 (Synthetic Data for Cuts). *There is a computationally efficient $(\varepsilon, 0)$-differentially private randomized algorithm that takes a unweighted graph G and outputs a synthetic graph G' such that, with high probability, $\|G - G'\|_C \leq O(|V|^{3/2}/\varepsilon)$—all cuts in G and G' are within $O(|V|^{3/2}/\varepsilon)$ additive error.*

The proof is deferred to the full version, but the idea is straightforward: we write a linear program with exponentially many constraints to solve for a synthetic database, and use an SDP-based approximation algorithm of [1] for the cut-norm problem as an approximate separation oracle to solve the LP.

7   Towards Improving on Randomized Response for Synthetic Data

In this section, we consider one possible avenue towards giving an efficient algorithm for privately generating synthetic data for graph cuts that improves over randomized response. We first show how generically, any efficient Iterative Database Construction algorithm can be used to give an efficient *offline* algorithm for privately releasing synthetic data when paired with an efficient *distinguisher*. The analysis here follows the analysis of [12], who analyzed the corresponding algorithm when instantiated with the multiplicative weights algorithm, rather than a generic Iterative Database Construction algorithm.

We will pair an Iterative Database Construction algorithm for a class of queries $\mathcal{C}$ with a corresponding *distinguisher*.

Definition 6 $((F(\epsilon), \gamma)$-**Private Distinguisher).** *Let Q be a set of queries, let $\gamma \geq 0$ and let $F(\epsilon) : \mathbb{R}^+ \to \mathbb{Z}$ be a function. An algorithm $Distinguish_\epsilon : \mathbb{N}^{|\mathcal{X}|} \times \mathbb{N}^{|\mathcal{X}|} \to Q$ is an $(F(\epsilon), \gamma)$-Private Distinguisher for Q if for every setting of the privacy parameter ϵ, it is ϵ-differentially private with respect to $\mathcal{D}$ and if for every $\mathcal{D}, \mathcal{D}' \in \mathbb{N}^{|\mathcal{X}|}$ it outputs a $Q^* \in Q$ such that $|Q^*(\mathcal{D}) - Q^*(\mathcal{D}')| \geq \max_{Q \in Q} |Q(\mathcal{D}) - Q(\mathcal{D}')| - F(\epsilon)$ with probability at least $1 - \gamma$.*

We present the algorithm in the full version, but the idea is very simple. Rather than waiting for a query to arrive online that induces an update step, we find queries which will induce update steps using the distinguisher. The IDC algorithm will guarantee that there will not be too many update steps, and so an efficient distinguisher will yield an efficient algorithm for releasing synthetic data.

Theorem 6. *There is an (ϵ, δ)-differentially private mechanism for releasing synthetic data such that given an $(F(\epsilon), \gamma)$-private distinguisher and a $B(\alpha)$-IDC, it is (α, β)-accurate for:*

$$\alpha \geq \max \left[\frac{16\sqrt{B(\alpha)\log(1/\delta)} \log(2B(\alpha)/\beta)}{\epsilon}, 2F\left(\frac{\epsilon}{4\sqrt{B(\alpha)\log(1/\delta)}}\right) \right]$$

as long as $\gamma \leq \beta/(2B(\alpha))$

We defer the proof until the full version. Note that the running time of the algorithm is dominated by the running time of the IDC algorithm and of the distinguishing algorithm: efficient IDC algorithms paired with efficient distinguishing algorithms for a class of queries Q automatically correspond to efficient algorithms for privately releasing synthetic data useful for Q. For the class of graph cut queries, both the multiplicative weights IDC and the Frieze/Kannan IDC are computationally efficient. Therefore, one approach to finding a computationally efficient algorithm for releasing synthetic data useful for cut queries is to find an efficient private distinguisher for cut queries.

One curious aspect of this approach is that it might in fact be computationally easier to release a *larger* class of queries than cut queries, even though this is a strictly more difficult task from an information theoretic perspective. For example, solving the distinguishing problem for cut queries on graphs $\mathcal{D}$ and $\mathcal{D}'$ is equivalent to finding a pair of sets (S, T) which witness the cut-norm on the graph $\mathcal{D} - \mathcal{D}'$. On the other hand, solving the distinguishing problem for rank-1 queries (which include cut queries, and are a larger class) is equivalent to finding the best rank-1 approximation to the adjacency matrix $\mathcal{D} - \mathcal{D}'$. The former problem is NP-hard, whereas the latter problem can be quickly solved non-privately using the singular value decomposition.

Corollary 2. *An efficient $(F(\epsilon), \gamma)$-distinguisher for the class of rank-1 queries for $F(\epsilon) = T/\epsilon$ would yield an (α, β)-accurate mechanism for releasing synthetic data for graph cuts (and all rank-1 queries) for any $\beta \geq \Omega(\exp(-\epsilon T))$ and: $\alpha_{MW} = 2\sqrt[4]{2}\epsilon^{-1/2}\sqrt{Tm}\,(\log|V|\log(1/\delta))^{1/4}$ using the multiplicative weights IDC, or: $\alpha_{FK} \geq 2\epsilon^{-1/2}(m\log(1/\delta))^{1/4}\sqrt{|V|T}$ using the Frieze/Kannan IDC*

The proof, deferred to the full version, only requires plugging in the parameters for these two IDC algorithms. We remark that for the class of rank-1 queries, an efficient $(F(\epsilon), \gamma)$-distinguisher with $F(\epsilon) = \tilde{O}\left(\frac{|V|}{\epsilon}\right)$ would be sufficient to yield an efficient algorithm for releasing synthetic data useful for cut queries, with guarantees matching those of the best known algorithms for the interactive case, as listed in Table 1. For graphs for which the size of the edge set $m \leq \Omega(n^2)$, this would yield an improvement over our randomized response mechanism, which is the best mechanism currently for privately releasing synthetic data for graph cuts. We observe that such a distinguisher is information-theoretically possible, and the only question is whether such a private distinguisher exists that is also computationally efficient. To see this, observe that an $O(|V|)$-net for the set of all rank-1 queries can be constructed by considering all pairs of vectors $x, y \in \{0, 1/|V|, 2/|V|, \ldots, 1\}^{|V|}$ and their associated outer-products $x \cdot y^T$. Since there are at most $|V|^{2|V|}$ such pairs, the exponential mechanism serves as an inefficient $F(\epsilon)$ distinguisher for $F(\epsilon) = O(|V|\log|V|/\epsilon)$.

We note that a distinguisher for rank-1 queries must simply give a good rank-1 approximation to the matrix $\mathcal{D} - \mathcal{D}'$, which will always be symmetric in this setting (because both the hypothesis is at every step simply the adjacency matrix for an undirected graph, as of course is the private database), and hence an algorithm for finding accurate rank-1 approximations merely for symmetric matrices would already yield an algorithm for releasing synthetic data for cuts! Unlike classes of queries like conjunctions, for which their are imposing barriers to privately outputting useful synthetic data [21, 12], there are as far as we know no such barriers to improving our randomized-response based results for synthetic data for graph cuts. We leave finding such an algorithm, for privately giving low rank approximations to matrices, as an intriguing open problem.

Acknowledgements. This paper benefited from interactions with many people. We particularly thank Moritz Hardt and Kunal Talwar for extensive, enlightening discussions. In particular, the observation that randomized response leads to a data structure for graph cuts with error $O(|V|^{1.5})$ is due to Kunal Talwar. We thank Salil Vadhan for helpful discussions about the Frieze/Kannan low-rank matrix decomposition, and Frank McSherry and Adam Smith for helpful discussions about algorithms for computing low-rank matrix approximations. We thank Cynthia Dwork for always fruitful conversations.

References

[1] Alon, N., Naor, A.: Approximating the cut-norm via Grothendieck's inequality. SIAM J. Comput. 35(4), 787–803 (2006) (electronic)

[2] Barak, B., Chaudhuri, K., Dwork, C., Kale, S., McSherry, F., Talwar, K.: Privacy, accuracy, and consistency too: a holistic solution to contingency table release. In: PODS, pp. 273–282 (2007)

[3] Bhaskara, A., Krishnaswamy, R., Talwar, K.: Unconditional differentially private mechanisms for linear queries (2011) (manuscript)

[4] Blum, A., Dwork, C., McSherry, F., Nissim, K.: Practical privacy: the SuLQ framework. In: PODS, pp. 128–138 (2005)

[5] Blum, A., Ligett, K., Roth, A.: A learning theory approach to non-interactive database privacy. In: STOC, pp. 609–618 (2008)

[6] Chawla, S., Dwork, C., McSherry, F., Smith, A., Wee, H.: Toward Privacy in Public Databases. In: Kilian, J. (ed.) TCC 2005. LNCS, vol. 3378, pp. 363–385. Springer, Heidelberg (2005)

[7] Dwork, C., McSherry, F., Nissim, K., Smith, A.: Calibrating Noise to Sensitivity in Private Data Analysis. In: Halevi, S., Rabin, T. (eds.) TCC 2006. LNCS, vol. 3876, pp. 265–284. Springer, Heidelberg (2006)

[8] Dwork, C., Naor, M., Reingold, O., Rothblum, G., Vadhan, S.: On the complexity of differentially private data release: efficient algorithms and hardness results. In: STOC, pp. 381–390 (2009)

[9] Dwork, C., Rothblum, G., Vadhan, S.: Boosting and differential privacy. In: FOCS, pp. 51–60 (2010)

[10] Frieze, A., Kannan, R.: Quick approximation to matrices and applications. Combinatorica 19(2), 175–220 (1999)

[11] Frieze, A., Kannan, R.: A simple algorithm for constructing Szemerédi's regularity partition. Electron. J. Combin. 6, Research Paper 17, 7 (1999)

[12] Gupta, A., Hardt, M., Roth, A., Ullman, J.: Privately Releasing Conjunctions and the Statistical Query Barrier. In: STOC. ACM, New York (2011)

[13] Gupta, A., Roth, A., Ullman, J.: Iterative constructions and private data release, Arxiv preprint arXiv:1107.3731 (2011)

[14] Hardt, M., Ligett, K., McSherry, F.: A simple and practical algorithm for differentially private data release, Arxiv preprint arXiv:1012.4763 (2011)

[15] Hardt, M., Rothblum, G.: A multiplicative weights mechanism for privacy-preserving data analysis. In: FOCS, pp. 61–70 (2010)

[16] Hardt, M., Talwar, K.: On the Geometry of Differential Privacy. In: STOC (2010)

[17] Kasiviswanathan, S., Lee, H., Nissim, K., Raskhodnikova, S., Smith, A.: What Can We Learn Privately? In: FOCS, pp. 531–540 (2008)

[18] McSherry, F., Talwar, K.: Mechanism design via differential privacy. In: FOCS (2007)

[19] Roth, A., Roughgarden, T.: Interactive Privacy via the Median Mechanism. In: STOC (2010)

[20] Trevisan, L., Tulsiani, M., Vadhan, S.P.: Regularity, boosting, and efficiently simulating every high-entropy distribution. In: CCC (2009)

[21] Ullman, J., Vadhan, S.: PCPs and the Hardness of Generating Private Synthetic Data. In: Ishai, Y. (ed.) TCC 2011. LNCS, vol. 6597, pp. 400–416. Springer, Heidelberg (2011)

A   Other Iterative Database Construction Algorithms

In this section, we demonstrate how the median mechanism and the multiplicative weights mechanism fit into the IDC framework. These mechanisms apply to general classes of linear queries $\mathcal{Q}$.

A.1   The Median Mechanism

In this section, we show how to use the median database subroutine as an Iterative Database Construction.

Definition 7 (Median Datastructure). *A median datastructure* $\mathbf{D}$ *is a collection of databases* $\mathbf{D} \subset \mathbb{N}^{|\mathcal{X}|}$. *Any query can be evaluated on a median datastructure as follows:*
$Q(\mathbf{D}) = Median(\{Q(\mathcal{D}') : \mathcal{D}' \in \mathbf{D}\})$.

Algorithm 3. The Median Mechanism (MM) Algorithm

$\mathbf{U}_{k,\alpha}^{MM}(\mathbf{D}^t, Q^{(t)}, \widehat{A}^{(t)})$

    If: $\mathbf{D}^t = \emptyset$ **then:** output $\mathbf{D}^0 = \{\mathcal{D} \in \mathbb{N}^{|\mathcal{X}|} : |\mathcal{D}| = n^2 \log k/\alpha^2\}$

    Else if: $Q^{(t)}(\mathbf{D}^t) - \widehat{A}^{(t)} > 0$ **then:** output $\mathbf{D}' = \mathbf{D}' \setminus \{\mathcal{D} \in \mathbf{D} : Q^{(t)}(\mathcal{D}) \geq Q^{(t)}(\mathbf{D})\}$

    Else if: $Q^{(t)}(\mathbf{D}^t) - \widehat{A}^{(t)} < 0$ **then:** output $\mathbf{D}' = \mathbf{D}' \setminus \{\mathcal{D} \in \mathbf{D} : Q^{(t)}(\mathcal{D}) \leq Q^{(t)}(\mathbf{D})\}$

Theorem 7. *The Median Mechanism algorithm is a* $B(\alpha) = n^2 \log |\mathcal{X}| \log k/\alpha^2$ *iterative database construction algorithm for every class of* k *linear queries* $\mathcal{Q}$.

Proof. Let $\mathcal{D} \in \mathbb{N}^{|\mathcal{X}|}$ be any database and consider a $(\mathbf{U}_k^{MM}, \mathbf{D}^*, \mathcal{Q}, \alpha, B)$-database update sequence, $\left\{ (\mathbf{D}^t, Q^{(t)}, \widehat{A}^{(t)}) \right\}_{t=1,\dots,B}$. It will be sufficient if we can show that $B(\alpha) \leq n^2 \log |\mathcal{X}| \log k/\alpha^2$. Specifically, that after $n^2 \log |\mathcal{X}| \log k/\alpha^2$ invocations of $\mathbf{U}_{k,\alpha}^{MM}$, the median datastructure $\mathbf{D}^{n^2 \log |\mathcal{X}| \log k/\alpha^2}$ is $(\alpha, \mathcal{Q})$-accurate for $\mathcal{D}$. The argument is simple. First, we have a simple fact from [5]:

Claim. For any set of k linear queries $\mathcal{Q}$ and any database $\mathcal{D}$ of size n, there is a database $\mathcal{D}'$ of size $|\mathcal{D}'| = n^2 \log k/\alpha^2$ so that $\mathcal{D}'$ is α-accurate for $\mathcal{D}$ with respect to $\mathcal{Q}$.

From this claim, we have that $|\mathbf{D}^t| \geq 1$ for all t, and so can always be used to evaluate queries. On the other hand, each update step eliminates half of the databases in the median datastructure: $|\mathbf{D}^t| = |\mathbf{D}^{t-1}|/2$. This is because the update step eliminates every database either above or below the median with respect to the last query. Initially $|\mathbf{D}^0| = |\mathcal{X}|^{n^2 \log k/\alpha^2}$, and so there can be at most $B(\alpha) \leq \log n^2 |\mathcal{X}| \log k/\alpha^2$ update steps before we would have $|\mathbf{D}^B| < 1$, a contradiction.

A.2   The Multiplicative Weights Mechanism

In this section we show how to use the multiplicative weights subroutine as an Iterative Database Construction. The analysis of the multiplicative weights algorithm is not new, and follows [15]. It will be convenient to think of our databases in this section as probability distributions, i.e. normalized so that $||\mathcal{D}||_1 = 1$. Note that if we are α/n accurate for the normalized database, we are α-accurate for the un-normalized database with respect to any set of linear queries.

Algorithm 4. The Multiplicative Weights (MW) Algorithm

$\mathbf{U}_\alpha^{MW}(\mathcal{D}^t, Q^{(t)}, \widehat{A}^{(t)})$:

  Let $\eta \leftarrow \alpha/(2n)$.
  If: $\mathcal{D}^t = \emptyset$ **then:** output $\mathcal{D}' = \mathcal{D} \in \mathbb{R}^{|\mathcal{X}|}$ such that $D_i^0 = 1/|\mathcal{X}|$ for all i.
  if $\widehat{A}^{(t)} < Q^{(t)}(\mathcal{D}^t)$ **then**
    Let $r_t = Q^{(t)}$
  else
    Let $r_t = 1 - Q^{(t)}$
  end if
  Update: For all $i \in [|\mathcal{X}|]$ Let

$$\widehat{\mathcal{D}}_i^{t+1} = \exp(-\eta r_t(\mathcal{D}_i^t)) \cdot \mathcal{D}_i^t$$

$$\mathcal{D}_i^{t+1} = \frac{\widehat{\mathcal{D}}_i^{t+1}}{\sum_{j=1}^{|\mathcal{X}|} \widehat{\mathcal{D}}_j^{t+1}}$$

  Output $\mathcal{D}^{t+1}$.

Theorem 8. *The Multiplicative Weights algorithm is a $B(\alpha) = 4n^2 \log |\mathcal{X}|/\alpha^2$ iterative database construction algorithm for every class of linear queries $\mathcal{Q}$.*

Proof. Let $\mathcal{D} \in \mathbb{N}^{|\mathcal{X}|}$ be any database and consider a $(\mathbf{U}^{MW}, \mathcal{D}^*, \mathcal{Q}, \alpha, B)$-database update sequence, $\left\{ (\mathcal{D}^{(t)}, Q^{(t)}, \widehat{A}^{(t)}) \right\}_{t=1,\ldots,B}$. It will be sufficient if we can show that $B(\alpha) \leq 4n^2 \log |\mathcal{X}|/\alpha^2$. Specifically, that after $4n^2 \log |\mathcal{X}|/\alpha^2$ invocations of $\mathbf{U}^{MW}$, the database $\mathcal{D}^{(4n^2 \log |\mathcal{X}|/\alpha^2)}$ is $(\alpha, \mathcal{Q})$-accurate for $\mathcal{D}$. First let $\widehat{\mathcal{D}} \in \mathbb{R}^{|\mathcal{X}|}$ be a normalization of the database $\mathcal{D}$: $\widehat{\mathcal{D}}_i = \mathcal{D}_i/||\mathcal{D}||_1$. Note that for any linear query, $Q(\mathcal{D}) = n \cdot Q(\widehat{\mathcal{D}})$. We define:

$$\Psi_t \stackrel{\text{def}}{=} D(\widehat{\mathcal{D}}||D^t) = \sum_{i=1}^{|\mathcal{X}|} \widehat{\mathcal{D}}_i \log\left(\frac{\widehat{\mathcal{D}}_i}{D_i^t}\right)$$

We begin with a simple fact:

Claim ([15]). For all t: $\Psi_t \geq 0$, and $\Psi_0 \leq \log |\mathcal{X}|$.

We will argue that in every step for which $|Q^{(t)}(\mathcal{D}) - Q^{(t)}(\mathcal{D}^t)| \geq \alpha/n$ the potential drops by at least $\alpha^2/4n$. Because the potential begins at $\log |\mathcal{X}|$, and must always be non-negative, we know that there can be at most $B(\alpha) \leq 4n^2 \log |X|/\alpha^2$ steps before the algorithm outputs a database $\mathcal{D}^t$ such that $\max_{Q \in \mathcal{Q}} |Q(\mathcal{D}) - Q(\mathcal{D}^t)| < \alpha/n$, which is exactly the condition that we want.

Lemma 3 ([15])

$$\Psi_t - \Psi_{t+1} \geq \eta \left(r_t(\mathcal{D}^t) - r_t(\mathcal{D}) \right) - \eta^2$$

Proof

$$\Psi_t - \Psi_{t+1} = \sum_{i=1}^{|\mathcal{X}|} \hat{\mathcal{D}}_i \log \left(\frac{D_i^{t+1}}{D_i^t} \right)$$

$$= -\eta r_t(\mathcal{D}) - \log \left(\sum_{i=1}^{|\mathcal{X}|} \exp(-\eta r_t(x_i)) D_i^t \right)$$

$$\geq -\eta r_t(\mathcal{D}) - \log \left(\sum_{i=1}^{|\mathcal{X}|} D_i^t(1 + \eta^2 - \eta r_t(x_i)) \right)$$

$$\geq \eta \left(r_t(\mathcal{D}^t) - r_t(\mathcal{D}) \right) - \eta^2$$

The rest of the proof now follows easily. By the conditions of an iterative database construction algorithm, $|\widehat{A}^{(t)} - Q^{(t)}(\mathcal{D})| \leq \alpha/(2n)$. Hence, for each t such that $|Q^{(t)}(\mathcal{D}) - Q^{(t)}(\mathcal{D}^t)| \geq \alpha/n$, we also have that $Q^{(t)}(\mathcal{D}) > Q^{(t)}(\mathcal{D}_t)$ if and only if $\widehat{A}^{(t)} > Q^{(t)}(\mathcal{D}_t)$. In particular, $r_t = Q^{(t)}$ if $Q^{(t)}(\mathcal{D}^t) - Q^{(t)}(\mathcal{D}) \geq \alpha/n$, and $r_t = 1 - Q^{(t)}$ if $Q^{(t)}(\mathcal{D}) - Q^{(t)}(\mathcal{D}^t) \geq \alpha/n$. Therefore, by Lemma 3 and the fact that $\eta = \alpha/2n$:

$$\Psi_t - \Psi_{t+1} \geq \frac{\alpha}{2n} \left(r_t(\mathcal{D}^t) - r_t(\mathcal{D}) \right) - \frac{\alpha^2}{4n^2} \geq \frac{\alpha}{2n} \left(\frac{\alpha}{n} \right) - \frac{\alpha^2}{4n^2} = \frac{\alpha^2}{4n^2}$$

From Non-adaptive to Adaptive Pseudorandom Functions

Itay Berman and Iftach Haitner*

School of Computer Science, Tel Aviv University
itayberm@post.tau.ac.il, iftachh@cs.tau.ac.il

Abstract. Unlike the standard notion of pseudorandom functions (PRF), a *non-adaptive* PRF is only required to be indistinguishable from random in the eyes of a *non-adaptive* distinguisher (i.e., one that prepares its oracle calls in advance). A recent line of research has studied the possibility of a *direct* construction of adaptive PRFs from non-adaptive ones, where direct means that the constructed adaptive PRF uses only few (ideally, constant number of) calls to the underlying non-adaptive PRF. Unfortunately, this study has only yielded negative results, showing that "natural" such constructions are unlikely to exist (e.g., Myers [EUROCRYPT '04], Pietrzak [CRYPTO '05, EUROCRYPT '06]).

We give an affirmative answer to the above question, presenting a direct construction of adaptive PRFs from non-adaptive ones. Our construction is extremely simple, a composition of the non-adaptive PRF with an appropriate pairwise independent hash function.

1 Introduction

A pseudorandom function family (PRF), introduced by Goldreich, Goldwasser, and Micali [11], cannot be distinguished from a family of *truly* random functions by an efficient distinguisher who is given an oracle access to a random member of the family. PRFs have an extremely important role in cryptography, allowing parties, which share a common secret key, to send secure messages, identify themselves and to authenticate messages [10, 13]. In addition, they have many other applications, essentially in any setting that requires random function provided as black-box [2, 3, 6, 7, 14, 18]. Different PRF constructions are known in the literature, whose security is based on different hardness assumption. Constructions relevant to this work are those based on the existence of pseudorandom generators [11] (and thus on the existence of one-way functions [12]), and on, the so called, synthesizers [17].

In this work we study the question of constructing (adaptive) PRFs from *non-adaptive* PRFs. The latter primitive is a (weaker) variant of the standard PRF we mentioned above, whose security is only guaranteed to hold against non-adaptive distinguishers (i.e., ones that "write" all their queries before the

---

* Research supported by Check Point Institute for Information Security and BSF grant 2010196.

R. Cramer (Ed.): TCC 2012, LNCS 7194, pp. 357–368, 2012.

first oracle call). Since a non-adaptive PRF can be easily cast as a pseudorandom generator or as a synthesizer, [11, 17] tell us how to construct (adaptive) PRF from a non-adaptive one. In both of these constructions, however, the resulting (adaptive) PRF makes $\Theta(n)$ calls to the underlying non-adaptive PRF (where n being the input length of the functions).[1]

A recent line of work has tried to figure out whether more efficient reductions from adaptive to non-adaptive PRF's are likely to exist. In a sequence of works [16, 19, 20, 5], it was shown that several "natural" approaches (e.g., composition or XORing members of the non-adaptive family with itself) are unlikely to work. See more in Section 1.3.

1.1 Our Result

We show that a simple composition of a non-adaptive PRF with an appropriate pairwise independent hash function, yields an adaptive PRF. To state our result more formally, we use the following definitions: a function family $\mathcal{F}$ is $T = T(n)$-adaptive PRF, if no distinguisher of running time at most T, can tell a random member of $\mathcal{F}$ from a random function with advantage larger than $1/T$. The family $\mathcal{F}$ is T-non-adaptive PRF, if the above is only guarantee to hold against non-adaptive distinguishers. Given two function families $\mathcal{F}_1$ and $\mathcal{F}_2$, we let $\mathcal{F}_1 \circ \mathcal{F}_2$ [resp., $\mathcal{F}_1 \oplus \mathcal{F}_2$] be the function family whose members are all pairs $(f, g) \in \mathcal{F}_1 \times \mathcal{F}_2$, and the action $(f, g)(x)$ is defined as $f(g(x))$ [resp., $f(x) \oplus g(x)$]. We prove the following statements (see Section 3 for the formal statements).

Theorem 1 (Informal). *Let $\mathcal{F}$ be a $(p(n) \cdot T(n))$-non-adaptive PRF, where $p \in$ poly is function of the evaluating time of $\mathcal{F}$, and let $\mathcal{H}$ be an efficient pairwise-independent function family mapping strings of length n to $[T(n)]_{\{0,1\}^n}$, where $[T]_{\{0,1\}^n}$ is the first T elements (in lexicographic order) of $\{0,1\}^n$. Then $\mathcal{F} \circ \mathcal{H}$ is a $\left(\sqrt[3]{T(n)}/2 \right)$-adaptive PRF.*

For instance, assuming that $\mathcal{F}$ is a $(p(n) \cdot 2^{cn})$-non-adaptive PRF and that $\mathcal{H}$ maps strings of length n to $[2^{cn}]_{\{0,1\}^n}$, Theorem 1 yields that $\mathcal{F} \circ \mathcal{H}$ is a $\left(2^{\frac{cn}{3}-1} \right)$-adaptive PRF.

Theorem 1 is only useful, however, for polynomial-time computable T's (in this case, the family $\mathcal{H}$ assumed by the theorem exists, see Section 2.2). Unfortunately, in the important case where $\mathcal{F}$ is only assumed to be polynomially secure non-adaptive PRF, no useful polynomial-time computable T is guaranteed to exists.[2]

We suggest two different solutions for handling polynomially secure PRFs. In Section 4 we observe (following Bellare [1]) that a polynomially secure non-adaptive PRF is a T-non-adaptive PRF for some $T \in n^{\omega(1)}$. Since this T can

---

[1] We remark that if one is only interested in *polynomial security* (i.e., no adaptive PPT distinguishes with more than negligible probability), then $w(\log n)$ calls are sufficient (cf., [8, Sec. 3.8.4, Exe. 30]).

[2] Clearly $\mathcal{F}$ is p-non-adaptive PRF for any $p \in$ poly, but applying Theorem 1 with $T \in$ poly, does not yield a polynomially secure adaptive PRF.

be assumed without loss of generality to be a power of two, Theorem 1 yields a non-uniform (uses n-bit advice) polynomially secure adaptive PRF, that makes a single call to the underlying non-adaptive PRF. Our second solution is to use the following "combiner", to construct a (uniform) adaptively secure PRF, which makes $\omega(1)$ parallel calls to the underlying non-adaptive PRF.

Corollary 1 (Informal). *Let $\mathcal{F}$ be a polynomially secure non-adaptive PRF, let $\mathcal{H} = \{\mathcal{H}_n\}_{n \in \mathbb{N}}$ be an efficient pairwise-independent length-preserving function family and let $k(n) \in \omega(1)$ be polynomial-time computable function.*

For $n \in \mathbb{N}$ and $i \in [n]$, let $\widehat{\mathcal{H}_n}^i$ be the function family $\widehat{\mathcal{H}_n}^i = \{\widehat{h} \colon h \in \mathcal{H}\}$, where $\widehat{h}(x) = 0^{n-i} \| h(x)_{1,\ldots,i}$ ('$\|$' stands for string concatenation). Then the ensemble $\{\bigoplus_{i \in [k(n)]} \left(\mathcal{F}_n \circ \widehat{\mathcal{H}_n}^{\lfloor i \cdot \log n \rfloor} \right)\}_{n \in \mathbb{N}}$ is a polynomially secure adaptive PRF.

1.2 Proof Idea

To prove Theorem 1 we first show that $\mathcal{F} \circ \mathcal{H}$ is indistinguishable from $\Pi \circ \mathcal{H}$, where Π being the set of *all* functions from $\{0,1\}^n$ to $\{0,1\}^{\ell(n)}$ (letting $\ell(n)$ be $\mathcal{F}$'s output length), and then conclude the proof by showing that $\Pi \circ \mathcal{H}$ is indistinguishable from Π.

$\mathcal{F} \circ \mathcal{H}$ **Is indistinguishable from** $\Pi \circ \mathcal{H}$**.** Let D be (a possibly adaptive) algorithm of running time $T(n)$, which distinguishes $\mathcal{F} \circ \mathcal{H}$ from $\Pi \circ \mathcal{H}$ with advantage $\varepsilon(n)$. We use D to build a *non-adaptive* distinguisher $\widehat{\mathsf{D}}$ of running time $p(n) \cdot T(n)$, which distinguishes $\mathcal{F}$ from Π with advantage $\varepsilon(n)$. Given an oracle access to a function ϕ, the distinguisher $\widehat{\mathsf{D}}^\phi(1^n)$ first queries ϕ on *all* the elements of $[T(n)]_{\{0,1\}^n}$. Next it chooses at uniform $h \in \mathcal{H}$, and uses the stored answers to its queries, to emulate $\mathsf{D}^{\phi \circ h}(1^n)$.

Since $\widehat{\mathsf{D}}$ runs in time $p(n) \cdot T(n)$, for some large enough $p \in \text{poly}$, makes *non-adaptive* queries, and distinguishes $\mathcal{F}$ from Π with advantage $\varepsilon(n)$, the assumed security of $\mathcal{F}$ yields that $\varepsilon(n) < \frac{1}{p(n) \cdot T(n)}$.

$\Pi \circ \mathcal{H}$ **Is indistinguishable from** Π**.** We prove that $\Pi \circ \mathcal{H}$ is *statistically* indistinguishable from Π. Namely, even an unbounded distinguisher (that makes bounded number of calls) cannot distinguish between the families. The idea of the proof is fairly simple. Let D be an s-query algorithm trying to distinguish between $\Pi \circ \mathcal{H}$ and Π. We first note that the distinguishing advantage of D is bounded by its probability of finding a collision in a random $\phi \in \Pi \circ \mathcal{H}$ (in case no collision occurs, ϕ's output is uniform). We next argue that in order to find a collision in ϕ, the distinguisher D gains nothing from being adaptive. Indeed, assuming that D found no collision until the i'th call, then it has only learned that h does not collide on these first i queries. Therefore, a random (or even a constant) query as the $(i+1)$ call, has the same chance to yield a collision, as any other query has. Hence, we assume without loss of generality that D is non-adaptive, and use the pairwise independence of $\mathcal{H}$ to conclude that D's probability in finding a collision, and thus its distinguishing advantage, is bounded by $s(n)^2/T(n)$.

Combining the above two observations, we conclude that an adaptive distinguisher whose running time is bounded by $\frac{1}{2}\sqrt[3]{T(n)}$, cannot distinguish $\mathcal{F} \circ \mathcal{H}$ from Π (i.e., from a random function) with an advantage better than $\frac{T(n)^{\frac{2}{3}}/4}{T(n)} + \frac{1}{p(n)T(n)} \leq 2/\sqrt[3]{T(n)}$. Namely, $\mathcal{F} \circ \mathcal{H}$ is a $\left(\sqrt[3]{T(n)}/2\right)$-adaptive PRF.

1.3   Related Work

Maurer and Pietrzak [15] were the first to consider the question of building adaptive PRFs from non-adaptive ones. They showed that in the *information theoretic* model, a self composition of a non-adaptive PRF *does* yield an adaptive PRF.[3]

In contrast, the situation in the *computational model* (which we consider here) seems very different: Myers [16] proved that it is impossible to reprove the result of [15] via fully-black-box reductions. Pietrzak [19] showed that under the Decisional Diffie-Hellman (DDH) assumption, composition does not imply adaptive security. Where in [20] he showed that the existence of non-adaptive PRFs whose composition is not adaptively secure, yields that key-agreement protocol exists. Finally, Cho et al. [5] generalized [20] by proving that composition of two non-adaptive PRFs is not adaptively secure, iff (uniform transcript) key agreement protocol exists. We mention that [16, 19, 5], and in a sense also [15], hold also with respect to XORing of the non-adaptive families.

2   Preliminaries

2.1   Notations

All logarithms considered here are in base two. We let '$\|$' denote string concatenation. We use calligraphic letters to denote sets, uppercase for random variables, and lowercase for values. For an integer t, we let $[t] = \{1, \ldots, t\}$, and for a set $\mathcal{S} \subseteq \{0,1\}^*$ with $|\mathcal{S}| \geq t$, we let $[t]_{\mathcal{S}}$ be the first t elements (in increasing lexicographic order) of $\mathcal{S}$. A function $\mu \colon \mathbb{N} \to [0,1]$ is *negligible*, denoted $\mu(n) = \text{neg}(n)$, if $\mu(n) = n^{-\omega(1)}$. We let poly denote the set all polynomials, and let PPT denote the set of probabilistic algorithms (i.e., Turing machines) that run in *strictly* polynomial time.

Given a random variable X, we write $X(x)$ to denote $\Pr[X = x]$, and write $x \leftarrow X$ to indicate that x is selected according to X. Similarly, given a finite set $\mathcal{S}$, we let $s \leftarrow \mathcal{S}$ denote that s is selected according to the uniform distribution on $\mathcal{S}$. The *statistical distance* of two distributions P and Q over a finite set $\mathcal{U}$, denoted as $\text{SD}(P,Q)$, is defined as $\max_{\mathcal{S} \subseteq \mathcal{U}} |P(\mathcal{S}) - Q(\mathcal{S})| = \frac{1}{2}\sum_{u \in \mathcal{U}} |P(u) - Q(u)|$.

---

[3] Specifically, assuming that the non-adaptive PRF is (Q, ε)-non-adaptively secure, no Q-query non-adaptive algorithm distinguishes it from random with advantage larger than ε, then the resulting PRF is $(Q, \varepsilon(1 + \ln \frac{1}{\varepsilon}))$-adaptively secure.

2.2 Ensemble of Function Families

Let $\mathcal{F} = \{\mathcal{F}_n \colon \mathcal{D}_n \mapsto \mathcal{R}_n\}_{n \in \mathbb{N}}$ stands for an ensemble of function families, where each $f \in \mathcal{F}_n$ has domain $\mathcal{D}_n$ and its range contained in $\mathcal{R}_n$. Such ensemble is *length preserving*, if $\mathcal{D}_n = \mathcal{R}_n = \{0,1\}^n$ for every n.

Definition 1 (efficient function family ensembles). *A function family ensemble $\mathcal{F} = \{\mathcal{F}_n\}_{n \in \mathbb{N}}$ is* efficient, *if the following hold:*

Samplable. *$\mathcal{F}$ is samplable in polynomial-time: there exists a* PPT *that given 1^n, outputs (the description of) a uniform element in $\mathcal{F}_n$.*
Efficient. *There exists a polynomial-time algorithm that given $x \in \{0,1\}^n$ and (a description of) $f \in \mathcal{F}_n$, outputs $f(x)$.*

Operating on Function Families

Definition 2 (composition of function families). *Let $\mathcal{F}^1 = \{\mathcal{F}_n^1 \colon \mathcal{D}_n^1 \mapsto \mathcal{R}_n^1\}_{n \in \mathbb{N}}$ and $\mathcal{F}^2 = \{\mathcal{F}_n^2 \colon \mathcal{D}_n^2 \mapsto \mathcal{R}_n^2\}_{n \in \mathbb{N}}$ be two ensembles of function families with $\mathcal{R}_n^1 \subseteq \mathcal{D}_n^2$ for every n. We define the* composition *of $\mathcal{F}^1$ with $\mathcal{F}^2$ as $\mathcal{F}^2 \circ \mathcal{F}^1 = \{\mathcal{F}_n^2 \circ \mathcal{F}_n^1 \colon \mathcal{D}_n^1 \mapsto \mathcal{R}_n^2\}_{n \in \mathbb{N}}$, where $\mathcal{F}_n^2 \circ \mathcal{F}_n^1 = \{(f_2, f_1) \in \mathcal{F}_n^2 \times \mathcal{F}_n^1\}$, and $(f_2, f_1)(x) := f_2(f_1(x))$.*

Definition 3 (XOR of function families). *Let $\mathcal{F}^1 = \{\mathcal{F}_n^1 \colon \mathcal{D}_n^1 \mapsto \mathcal{R}_n^1\}_{n \in \mathbb{N}}$ and $\mathcal{F}^2 = \{\mathcal{F}_n^2 \colon \mathcal{D}_n^2 \mapsto \mathcal{R}_n^2\}_{n \in \mathbb{N}}$ be two ensembles of function families with $\mathcal{R}_n^1, \mathcal{R}_n^2 \subseteq \{0,1\}^{\ell(n)}$ for every n. We define the* XOR *of $\mathcal{F}^1$ with $\mathcal{F}^2$ as $\mathcal{F}^2 \bigoplus \mathcal{F}^1 = \{\mathcal{F}_n^2 \bigoplus \mathcal{F}_n^1 \colon \mathcal{D}_n^1 \cap \mathcal{D}_n^2 \mapsto \{0,1\}^{\ell(n)}\}_{n \in \mathbb{N}}$, where $\mathcal{F}_n^2 \bigoplus \mathcal{F}_n^1 = \{(f_2, f_1) \in \mathcal{F}_n^2 \times \mathcal{F}_n^1\}$, and $(f_2, f_1)(x) := f_2(x) \oplus f_1(x)$.*

Pairwise Independent Hashing

Definition 4 (pairwise independent families). *A function family $\mathcal{H} = \{h \colon \mathcal{D} \mapsto \mathcal{R}\}$ is* pairwise independent *(with respect to $\mathcal{D}$ and $\mathcal{R}$), if*

$$\Pr_{h \leftarrow \mathcal{H}}[h(x_1) = y_1 \wedge h(x_2) = y_2] = \frac{1}{|\mathcal{R}|^2},$$

for every distinct $x_1, x_2 \in \mathcal{D}$ and every $y_1, y_2 \in \mathcal{R}$.

For every $\ell \in$ poly, the existence of efficient pairwise-independent family ensembles mapping strings of length n to strings of length $\ell(n)$ is well known ([4]). In this paper we use efficient pairwise-independent function family ensembles mapping strings of length n to the set $[T(n)]_{\{0,1\}^n}$, where $T(n) \leq 2^n$ and is without loss of generality a power of two.[4] Let $\mathcal{H}$ be an efficient length-preserving, pairwise-independent function family ensemble and assume that $t(n) := \log T(n)$ is polynomial-time computable. Then the function family $\widehat{\mathcal{H}} = \{\widehat{\mathcal{H}}_n = \{h' \colon h \in \mathcal{H}_n, h'(x) = 0^{n-t(n)} \| h(x)_{1,\dots,t(n)}\}\}$, is an efficient pairwise-independent function family ensemble, mapping strings of length n to the set $[T(n)]_{\{0,1\}^n}$.

---

[4] For our applications, see Section 3, we can always consider $T'(n) = 2^{\lfloor \log(T(n)) \rfloor}$, which only causes us a factor of two loss in the resulting security.

Pseudorandom Functions

Definition 5 (pseudorandom functions). *An efficient function family ensemble* $\mathcal{F} = \{\mathcal{F}_n\colon \{0,1\}^n \mapsto \{0,1\}^{\ell(n)}\}_{n\in\mathbb{N}}$ *is a* $(T(n), \varepsilon(n))$*-adaptive PRF, if for every oracle-aided algorithm (distinguisher)* D *of running time* $T(n)$ *and large enough* n*, it holds that*

$$\left|\mathrm{Pr}_{f\leftarrow\mathcal{F}_n}[\mathsf{D}^f(1^n) = 1] - \mathrm{Pr}_{\pi\leftarrow\Pi_n}[\mathsf{D}^\pi(1^n) = 1]\right| \leq \varepsilon(n),$$

where Π_n *is the set of all functions from* $\{0,1\}^n$ *to* $\{0,1\}^{\ell(n)}$*. If we limit* D *above to be non-adaptive (i.e., it has to write all his oracle calls before making the first call), then* $\mathcal{F}$ *is called* $(T(n), \varepsilon(n))$*-non-adaptive PRF.*

The ensemble $\mathcal{F}$ *is a* t*-adaptive PRF, if it is a* $(t, 1/t)$*-adaptive PRF according to the above definition. It is* polynomially secure *adaptive PRF (for short,* adaptive PRF*), if it is a* p*-adaptive PRF for every* $p \in$ poly*. Finally, it is* super-polynomial secure *adaptive PRF, if it* T*-adaptive PRF for some* $T(n) \in n^{\omega(1)}$*. The same conventions are also used for non-adaptive PRFs.*

Clearly, a super-polynomial secure PRF is also polynomially secure. In Section 4 we prove that the converse is also true: a polynomially secure PRF is also super-polynomial secure PRF.

3   Our Construction

In this section we present the main contribution of this paper — a direct construction of an adaptive pseudorandom function family from a non-adaptive one.

Theorem 2 (restatement of Theorem 1). *Let* T *be a polynomial-time computable integer function, let* $\mathcal{H} = \{\mathcal{H}_n\colon \{0,1\}^n \mapsto [T(n)]_{\{0,1\}^n}\}$ *be an efficient pairwise independent function family ensemble, and let* $\mathcal{F} = \{\mathcal{F}_n\colon \{0,1\}^n \mapsto \{0,1\}^{\ell(n)}\}$ *be a* $(p(n) \cdot T(n), \varepsilon(n))$*-non-adaptive PRF, where* $p \in$ poly *is determined by the computation time of* T*,* $\mathcal{F}$ *and* $\mathcal{H}$*. Then* $\mathcal{F} \circ \mathcal{H}$ *is a* $\left(s(n), \varepsilon(n) + \frac{s(n)^2}{T(n)}\right)$*-adaptive PRF for every* $s(n) < T(n)$*.*

Theorem 2 yields the following simpler statement.

Corollary 2. *Let* T*,* p *and* $\mathcal{H}$ *be as in Theorem 2. Assuming* $\mathcal{F}$ *is a* $(p(n)T(n))$*-non-adaptive PRF, then* $\mathcal{F} \circ \mathcal{H}$ *is a* $\left(\sqrt[3]{T(n)}/2\right)$*-adaptive PRF.*

Proof. Applying Theorem 2 with respect to $s(n) = \sqrt[3]{T(n)}/2$ and $\varepsilon(n) = \frac{1}{p(n)T(n)}$, yields that $\mathcal{F} \circ \mathcal{H}$ is a $\left(s(n), \frac{1}{p(n)T(n)} + \frac{s(n)^2}{T(n)}\right)$-adaptive PRF. Since $\frac{1}{p(n)T(n)} < \frac{1}{2s(n)}$ and $\frac{s(n)^2}{T(n)} \leq \frac{1}{2s(n)}$, it follows that $\mathcal{F} \circ \mathcal{H}$ is a $(s, 1/s)$-adaptive PRF. $\qquad\square$

To prove Theorem 2, we use the (non efficient) function family ensemble $\Pi \circ \mathcal{H}$, where $\Pi = \Pi_\ell$ (i.e., the ensemble of all functions from $\{0,1\}^n$ to $\{0,1\}^\ell$), and $\ell = \ell(n)$ is the output length of $\mathcal{F}$. We first show that $\mathcal{F} \circ \mathcal{H}$ is *computationally* indistinguishable from $\Pi \circ \mathcal{H}$, and complete the proof showing that $\Pi \circ \mathcal{H}$ is *statistically* indistinguishable from Π.

3.1 $\mathcal{F} \circ \mathcal{H}$ Is Computationally Indistinguishable From $\Pi \circ \mathcal{H}$

Lemma 1. *Let T, $\mathcal{F}$ and $\mathcal{H}$ be as in Theorem 2. Then for every oracle-aided distinguisher D of running time T, there exists a non-adaptive oracle-aided distinguisher $\widehat{\mathsf{D}}$ of running time $p(n) \cdot T(n)$, for some $p \in$ poly (determined by the computation time of T, $\mathcal{F}$ and $\mathcal{H}$), with*

$$\left| \Pr_{g \leftarrow \mathcal{F}_n}[\widehat{\mathsf{D}}^g(1^n) = 1] - \Pr_{g \leftarrow \Pi_n}[\widehat{\mathsf{D}}^g(1^n) = 1] \right| =$$
$$\left| \Pr_{g \leftarrow \mathcal{F}_n \circ \mathcal{H}_n}[\mathsf{D}^g(1^n) = 1] - \Pr_{g \leftarrow \Pi_n \circ \mathcal{H}_n}[\mathsf{D}^g(1^n) = 1] \right|$$

for every $n \in \mathbb{N}$, where Π_n is the set of all functions from $\{0,1\}^n$ to $\{0,1\}^{\ell(n)}$.

In particular, the pseudorandomness of $\mathcal{F}$ yields that $\mathcal{F} \circ \mathcal{H}$ is computationally indistinguishable from the ensemble $\{\Pi_n \circ \mathcal{H}_n\}_{n \in \mathbb{N}}$ by an adaptive distinguisher of running time T.

Proof. The distinguisher $\widehat{\mathsf{D}}$ is defined as follows:

Algorithm 3 ($\widehat{\mathsf{D}}$)

Input: 1^n.
Oracle: *a function ϕ over $\{0,1\}^n$.*

1. *Compute $\phi(x)$ for every $x \in [T(n)]_{\{0,1\}^n}$.*
2. *Set $g = \phi \circ h$, where h is uniformly chosen in $\mathcal{H}_n$.*
3. *Emulate $\mathsf{D}^g(1^n)$: answer a query x to ϕ made by D with $g(x)$, using the information obtained in Step 1.*

....

Note that $\widehat{\mathsf{D}}$ makes $T(n)$ *non-adaptive* queries to ϕ, and it can be implemented to run in time $p(n)T(n)$, for large enough $p \in$ poly. We conclude the proof by observing that in case ϕ is uniformly drawn from $\mathcal{F}_n$, the emulation of D done in $\widehat{\mathsf{D}}^\phi$ is identical to a random execution of D^g with $g \leftarrow \mathcal{F}_n \circ \mathcal{H}_n$. Similarly, in case ϕ is uniformly drawn from Π_n, the emulation is identical to a random execution of D^π with $\pi \leftarrow \Pi_n$. $\square$

3.2 $\Pi \circ \mathcal{H}$ Is Statistically Indistinguishable From Π

The following lemma is commonly used for proving the security of hash based MACs (cf., [9, Proposition 6.3.6]), yet for completeness we give it a full proof below.

Lemma 2. *Let n, T be integers with $T \leq 2^n$, and let $\mathcal{H}$ be a pairwise-independent function family mapping string of length n to $[T]_{\{0,1\}^n}$. Let D be an (unbounded) s-query oracle-aided algorithm (i.e., making at most s queries), then*

$$\left| \Pr_{g \leftarrow \Pi \circ \mathcal{H}}[\mathsf{D}^g = 1] - \Pr_{\pi \leftarrow \Pi}[\mathsf{D}^\pi = 1] \right| \leq s^2/T,$$

where Π is the set of all functions from $\{0,1\}^n$ to $\{0,1\}^\ell$ (for some $\ell \in \mathbb{N}$).

Proof. We assume for simplicity that D is deterministic (the reduction to the randomized case is standard) and makes exactly s valid (i.e., inside $\{0,1\}^n$) distinct queries, and let $\Omega = (\{0,1\}^\ell)^s$. Consider the following random process:

Algorithm 4

1. *Emulate* D, *while answering the* i*'th query* q_i *with a uniformly chosen* $a_i \in \{0,1\}^\ell$.
   Set $\overline{q} = (q_1, \ldots, q_s)$ *and* $\overline{a} = (a_1, \ldots, a_s)$.
2. *Choose* $h \leftarrow \mathcal{H}$.
3. *Emulate* D *again, while answering the* i*'th query* q'_i *with* $a'_i = a_i$ *(the same* a_i *from Step 1), if* $h(q'_i) \notin \{h(q'_j)\}_{j \in [i-1]}$, *and with* $a'_i = a_j$, *if* $h(q'_i) = h(q'_j)$ *for some* $j \in [i-1]$.
   Set $\overline{q'} = (q'_1, \ldots, q'_s)$ *and* $\overline{a'} = (a'_1, \ldots, a'_s)$.

....

Let $\overline{A}, \overline{Q}, \overline{A'}, \overline{Q'}$ and H be the (jointly distributed) random variables induced by the values of $\overline{q}, \overline{a}, \overline{q'}, \overline{a'}$ and h respectively, in a random execution of the above process. It is not hard to verify that $\overline{A}$ is distributed the same as the oracle answers in a random execution of D^π with $\pi \leftarrow \Pi$, and that $\overline{A'}$ is distributed the same as the oracle answers in a random execution of D^g with $g \leftarrow \Pi \circ \mathcal{H}$. Hence, for proving Lemma 2, it suffices to bound the statistical distance between $\overline{A}$ and $\overline{A'}$.

Let Coll be the event that $H(\overline{Q}_i) = H(\overline{Q}_j)$ for some $i \neq j \in [s]$. Since the queries and answers in both emulations of Algorithm 4 are the same until a collision with respect to H occurs, it follows that

$$\Pr[\overline{A} \neq \overline{A'}] \leq \Pr[\text{Coll}] \tag{1}$$

On the other hand, since H is chosen *after* $\overline{Q}$ is set, the pairwise independent of $\mathcal{H}$ yields that

$$\Pr[\text{Coll}] \leq s^2/T, \tag{2}$$

and therefore $\Pr[\overline{A} \neq \overline{A'}] \leq s^2/T$. It follows that $\Pr[\overline{A} \in C] \leq \Pr[\overline{A'} \in C] + s^2/T$ for every $C \subseteq \Omega$, yielding that $\text{SD}(\overline{A}, \overline{A'}) \leq s^2/T$. □

3.3   Putting It Together

We are now finally ready to prove Theorem 2.

Proof (of Theorem 2). Let D be an oracle-aided algorithm of running time s with $s(n) < T(n)$. Lemma 1 yields that $|\Pr_{g \leftarrow \mathcal{F}_n \circ \mathcal{H}_n}[D^g(1^n) = 1] - \Pr_{g \leftarrow \Pi_n \circ \mathcal{H}_n}[D^g(1^n) = 1]| \leq \varepsilon(n)$ for large enough n, where Lemma 2 yields that $|\Pr_{g \leftarrow \Pi_n \circ \mathcal{H}_n}[D^g(1^n) = 1] - \Pr_{\pi \leftarrow \Pi_n}[D^\pi(1^n) = 1]| \leq s(n)^2/T(n)$ for every $n \in \mathbb{N}$. Hence, the triangle inequality yields that $|\Pr_{g \leftarrow \mathcal{F}_n \circ \mathcal{H}_n}[D^g(1^n) = 1] - \Pr_{\pi \leftarrow \Pi_n}[D^\pi(1^n) = 1]| \leq \varepsilon(n) + s(n)^2/T(n)$ for large enough n, as requested. □

3.4 Handling Polynomial Security

Corollary 2 is only useful when the security of the underlying non-adaptive PRF (i.e., T) is efficiently computable (or when considering non-uniform PRF constructions, see Section 1.1). In this section we show how to handle the important case of polynomially secure non-adaptive PRF. We use the following "combiner".

Definition 6. *Let $\mathcal{H}$ be a function family into $\{0,1\}^n$. For $i \in [n]$, let $\widehat{\mathcal{H}}^i$ be the function family $\widehat{\mathcal{H}}^i = \{\widehat{h}\colon h \in \mathcal{H}\}$, where $\widehat{h}(x) = 0^{n-i}||h(x)_{1,\dots,i}$.*

Corollary 3. *Let $\mathcal{F}$ be a $T(n)$-non-adaptive PRF, let $\mathcal{H}$ be an efficient length-preserving pairwise-independent function family ensemble, and let $\mathcal{I}(n) \subseteq [n]$ be polynomial-time computable (in n) index set. Define the function family ensemble $G = \{G_n\}_{n\in\mathbb{N}}$, where $G_n = \bigoplus_{i\in\mathcal{I}(n)} \left(\mathcal{F}_n \circ \widehat{\mathcal{H}_n}^i\right)$.*

There exists $q \in$ poly such that G is a $\left(\sqrt[3]{2^{t(n)}}/2\right)$-adaptive PRF, for every polynomial-time computable integer function t, with $t(n) \in \mathcal{I}(n)$ and $2^{t(n)} \leq T(n)/q(n)$.

Before proving the corollary, let us first use it for constructing adaptive PRF from non-adaptive polynomially secure one.

Corollary 4 (restatement of Corollary 1). *Let $\mathcal{F}$ be a polynomially secure non-adaptive PRF, let $\mathcal{H}$ be an efficient pairwise-independent length-preserving function family ensemble and let $k(n) \in \omega(1)$ be polynomial-time computable function. Then $G := \{\bigoplus_{i\in[k(n)]} \left(\mathcal{F}_n \circ \widehat{\mathcal{H}_n}^{\lfloor i\cdot\log n\rfloor}\right)\}_{n\in\mathbb{N}}$ is polynomially secure adaptive PRF.*

Proof. Let $\mathcal{I}(n) := \{\lfloor\log n\rfloor, \lfloor 2\cdot\log n\rfloor\dots, \lfloor k(n)\cdot\log n\rfloor\}$. Applying Corollary 3 with respect to $\mathcal{F}$, $\mathcal{H}$, $\mathcal{I}$ and $t(n) = \lfloor c\cdot\log n\rfloor$, where $c \in \mathbb{N}$, yields that G is a $O(\sqrt[3]{n^c})$-adaptive PRF. It follows that G is p-adaptive PRF for every $p \in$ poly. Namely, G is polynomially secure adaptive PRF. $\qquad\square$

Remark 1 (unknown security). Corollary 3 is also useful when the security of $\mathcal{F}$ is "not known" in the construction time. Taking $\mathcal{I}(n) = \{1, 2, 4, \dots, 2^{\lfloor\log n\rfloor}\}$ (resulting in $\log n$ calls to $\mathcal{F}$) and assuming that $\mathcal{F}$ is found to be $T(n)$-non-adaptive PRF for some polynomial-time computable T, the resulting PRF is guaranteed to be $O(\sqrt[6]{T(n)})$-adaptive PRF (neglecting polynomial factors).

Proof (of Corollary 3). It is easy to see that G is efficient, so it is left to argue for its security. Let $q(n) = q'(n)p(n)$, where p is as in the statement of Corollary 2, and $q' \in$ poly to be determined later. Let t be a polynomial-time computable integer function with $t(n) \in \mathcal{I}(n)$ and $2^{t(n)} \leq T(n)/q(n)$. It follows that $\widehat{\mathcal{H}}^t = \{\widehat{\mathcal{H}_n}^{t(n)}\}_{n\in\mathbb{N}}$ is an efficient pairwise-independent function family ensemble, and Corollary 2 yields that $\mathcal{F} \circ \widehat{\mathcal{H}}^t$ is a $\left(\sqrt[3]{q'(n)2^{t(n)}}/2\right)$-adaptive PRF.

Assume towards a contradiction that there exists an oracle-aided distinguisher D that runs in time $T'(n) = \sqrt[3]{2^{t(n)}}/2$ and

$$\left| \Pr_{g \leftarrow G_n}[\mathsf{D}^g(1^n) = 1] - \Pr_{\pi \leftarrow \Pi_n}[\mathsf{D}^\pi(1^n) = 1] \right| > 1/T'(n) \tag{3}$$

for infinitely many n's. We use the following distinguisher for breaking the pseudorandomness of $\mathcal{F} \circ \widehat{\mathcal{H}}^t$:

Algorithm 5 ($\widehat{\mathsf{D}}$)

Input: 1^n.
Oracle: a function ϕ over $\{0,1\}^n$.

1. For every $i \in \mathcal{I}(n) \setminus \{t(n)\}$, choose $g^i \leftarrow \mathcal{F}_n \circ \widehat{\mathcal{H}}_n^i$.
2. Set $g := \phi \oplus \bigoplus_{i \in \mathcal{I}(n) \setminus \{t(n)\}} g^i$.
3. Emulate $\mathsf{D}^g(1^n)$.

. . . .

Note that $\widehat{\mathsf{D}}$ can be implemented to run in time $|\mathcal{I}(n)| \cdot r(n) \cdot T'(n)$ for some $r \in$ poly, which is smaller than $\sqrt[3]{q'(n)2^{t(n)}}/2$ for large enough q'. Also note that in case ϕ is uniformly distributed over Π_n, then g (selected by $\widehat{\mathsf{D}}^\phi(1^n)$) is uniformly distributed in Π_n, where in case ϕ is uniformly distributed in $\mathcal{F}_n \circ \widehat{\mathcal{H}}_n^{t(n)}$, then g is uniformly distributed in G_n. It follows that

$$\left| \Pr_{g \leftarrow (\mathcal{F} \circ \widehat{\mathcal{H}}^t)_n}[\widehat{\mathsf{D}}^g(1^n) = 1] - \Pr_{\pi \leftarrow \Pi_n}[\widehat{\mathsf{D}}^\pi(1^n) = 1] \right| =$$
$$\left| \Pr_{g \leftarrow G_n}[\mathsf{D}^g(1^n) = 1] - \Pr_{\pi \leftarrow \Pi_n}[\mathsf{D}^\pi(1^n) = 1] \right| \tag{4}$$

for every $n \in \mathbb{N}$. In particular, Equation (3) yields that

$$\left| \Pr_{g \leftarrow (\mathcal{F} \circ \widehat{\mathcal{H}}^t)_n}[\widehat{\mathsf{D}}^g(1^n) = 1] - \Pr_{\pi \leftarrow \Pi_n}[\widehat{\mathsf{D}}^\pi(1^n) = 1] \right| > \frac{2}{\sqrt[3]{2^{t(n)}}} > \frac{2}{\sqrt[3]{q'(n)2^{t(n)}}}$$

for infinitely many n's, in contradiction to the pseudorandomness of $\mathcal{F} \circ \widehat{\mathcal{H}}^t$ we proved above.   □

Acknowledgment. We are very grateful to Omer Reingold for very useful discussions, and for challenging the second author with this research question a long while ago.

References

1. Bellare, M.: A note on negligible functions. Journal of Cryptology, 271–284 (2002)
2. Bellare, M., Goldwasser, S.: New Paradigms for Digital Signatures and Message Authentication Based on Non-interactive Zero Knowledge Proofs. In: Brassard, G. (ed.) CRYPTO 1989. LNCS, vol. 435, pp. 194–211. Springer, Heidelberg (1990)

3. Blum, M., Evans, W.S., Gemmell, P., Kannan, S., Naor, M.: Checking the correctness of memories. Algorithmica 12(2/3), 225–244 (1994)
4. Carter, L.J., Wegman, M.N.: Universal classes of hash functions. Journal of Computer and System Sciences, 143–154 (1979)
5. Cho, C., Lee, C.-K., Ostrovsky, R.: Equivalence of Uniform Key Agreement and Composition Insecurity. In: Rabin, T. (ed.) CRYPTO 2010. LNCS, vol. 6223, pp. 447–464. Springer, Heidelberg (2010)
6. Chor, B., Fiat, A., Naor, M., Pinkas, B.: Tracing traitors. IEEE Transactions on Information Theory 46(3), 893–910 (2000)
7. Goldreich, O.: Towards a Theory of Software Protection. In: Odlyzko, A.M. (ed.) CRYPTO 1986. LNCS, vol. 263, pp. 426–439. Springer, Heidelberg (1987)
8. Goldreich, O.: Foundations of Cryptography: Basic Tools. Cambridge University Press (2001)
9. Goldreich, O.: Foundations of Cryptography. Basic Applications, vol. 2. Cambridge University Press (2004)
10. Goldreich, O., Goldwasser, S., Micali, S.: On the Cryptographic Applications of Random Functions. In: Blakely, G.R., Chaum, D. (eds.) CRYPTO 1984. LNCS, vol. 196, pp. 276–288. Springer, Heidelberg (1985)
11. Goldreich, O., Goldwasser, S., Micali, S.: How to construct random functions. Journal of the ACM, 792–807 (1986)
12. Håstad, J., Impagliazzo, R., Levin, L.A., Luby, M.: A pseudorandom generator from any one-way function. SIAM Journal on Computing, 1364–1396 (1999)
13. Luby, M.: Pseudorandomness and cryptographic applications. Princeton computer science notes. Princeton University Press (1996) ISBN 978-0-691-02546-9
14. Luby, M., Rackoff, C.: How to construct pseudorandom permutations from pseudorandom functions. SIAM Journal on Computing
15. Maurer, U.M., Pietrzak, K.: Composition of Random Systems: When Two Weak Make One Strong. In: Naor, M. (ed.) TCC 2004. LNCS, vol. 2951, pp. 410–427. Springer, Heidelberg (2004)
16. Myers, S.: Black-Box Composition Does Not Imply Adaptive Security. In: Cachin, C., Camenisch, J.L. (eds.) EUROCRYPT 2004. LNCS, vol. 3027, pp. 189–206. Springer, Heidelberg (2004)
17. Naor, M., Reingold, O.: Synthesizers and their application to the parallel construction of psuedo-random functions. In: Proceedings of the 36th Annual Symposium on Foundations of Computer Science (FOCS), pp. 170–181 (1995)
18. Ostrovsky, R.: An Efficient Software Protection Scheme. In: Brassard, G. (ed.) CRYPTO 1989. LNCS, vol. 435, pp. 610–611. Springer, Heidelberg (1990)
19. Pietrzak, K.: Composition Does Not Imply Adaptive Security. In: Shoup, V. (ed.) CRYPTO 2005. LNCS, vol. 3621, pp. 55–65. Springer, Heidelberg (2005)
20. Pietrzak, K.: Composition Implies Adaptive Security in Minicrypt. In: Vaudenay, S. (ed.) EUROCRYPT 2006. LNCS, vol. 4004, pp. 328–338. Springer, Heidelberg (2006)

4   From Polynomial to Super-Polynomial Security

The standard security definition for cryptographic primitives is *polynomial security*: any PPT trying to break the primitive has only negligible success probability. Bellare [1] showed that for any polynomially secure primitive there exists a *single* negligible function μ, such that no PPT can break the primitive with probability larger than μ. Here we take his approach a step further, showing that for a polynomially secure primitive there exists a super-polynomial function T, such that no adversary of running time T breaks the primitive with probability larger than $1/T$.

In the following we identify algorithms with their string description. In particular, when considering algorithm A, we mean the algorithm defined by the string A (according to some canonical representation). We prove the following result.

Theorem 6. *Let $v\colon \{0,1\}^* \times \mathbb{N} \mapsto [0,1]$ be a function with the following properties: 1) $v(\mathsf{A}, n) \le 1/p(n)$ for every oracle-aided* PPT *A, $p \in$ poly and large enough n; and 2) if the distributions induced by random executions of $\mathsf{A}^f(x)$ and $\mathsf{B}^f(x)$ are the same for any input $x \in \{0,1\}^n$ and function f (each distribution describes the algorithm's output and oracle queries), then $v(\mathsf{A}, n) = v(\mathsf{B}, n)$.*

Then there exists an integer function $T(n) \in n^{\omega(1)}$ such that following holds: for any algorithm A of running time at most $T(n)$, it holds that $v(\mathsf{A}, n) \le 1/T(n)$ for large enough n.

Remark 2 (Applications). Let f be a polynomially secure OWF (i.e., $\Pr[\mathsf{A}(f(U_n)) \in f^{-1}(f(U_n))] = \mathrm{neg}(n)$ for any PPT A). Applying Theorem 6 with $v(\mathsf{A}, n) := \Pr[\mathsf{A}(f(U_n)) \in f^{-1}(f(U_n))]$ (where if A expects to get an oracle, provide him with the constant function $\phi(x) = 1$), yields that f is super-polynomial secure OWF (i.e., exists $T(n) \in n^{\omega(1)}$ such that $\Pr[\mathsf{A}(f(U_n)) \in f^{-1}(f(U_n))] \le 1/T(n)$ for any algorithm of running time T and large enough n).

Similarly, for a polynomially secure PRF $\mathcal{F} = \{\mathcal{F}_n\}_{n\in\mathbb{N}}$ (see Definition 5), applying Theorem 6 with $v(\mathsf{A}, n) := \left| \Pr_{f \leftarrow \mathcal{F}_n}[A^f(1^n) = 1] - \Pr_{\pi \leftarrow \Pi_n}[A^\pi(1^n) = 1] \right|$, where Π_n is the set of all functions with the same domain/range as $\mathcal{F}_n$, yields that $\mathcal{F}$ is super-polynomial secure PRF.

Proof (of Theorem 6). Given a probabilistic algorithm A and an integer i, let A_i denote the variant of A that on input of length n, halts after n^i steps (hence, A_i is a PPT). Let $\mathcal{S}_i$ be the first i strings in $\{0,1\}^*$, according to some canonical order, viewed as descriptions of i algorithms. Let $\mathcal{I}(n) = \{i \in [n]\colon \forall \mathsf{A} \in \mathcal{S}_i, k \ge n\colon v(\mathsf{A}_i, k) < 1/k^i\} \cup \{1\}$, let $t(n) = \max \mathcal{I}(n)$ and $T(n) = n^{t(n)}$.

Let A be an algorithm of running time $T(n)$, and let i_A be the first integer such that $\mathsf{A} \in \mathcal{S}_{i_\mathsf{A}}$. In Claim 7 we prove that $t(n) \in \omega(1)$, hence it follows that $t(n) > i_\mathsf{A}$ for any large enough n. For any such n, the definition of t guarantees that $v(\mathsf{A}_{t(n)}, n) < 1/n^{t(n)} = 1/T(n)$. Since A is of running time $T(n)$, the second property of v yields that $v(\mathsf{A}, n) = v(\mathsf{A}_{t(n)}, n)$, and therefore $v(\mathsf{A}, n) < 1/T(n)$. $\qquad\square$

Claim 7. It holds that $t(n) \in \omega(1)$.

Proof. Fix $i \in \mathbb{N}$. For each $\mathsf{A} \in \mathcal{S}_i$, let n_A be the first integer such that $v(\mathsf{A}_i, n) \le 1/n^i$ for every $n \ge n_\mathsf{A}$ (note that such n_A exists by the first property of v), and let $n_i = \max\{n_\mathsf{A}\colon \mathsf{A} \in \mathcal{S}_i\}$. It follows that $v(\mathsf{A}_i, n) \le 1/n^i$ for every $n \ge n_i$ and $\mathsf{A} \in \mathcal{S}_i$, and therefore $t(n_i) \ge i$. $\qquad\square$

Hardness Preserving Constructions of Pseudorandom Functions

Abhishek Jain[1,*], Krzysztof Pietrzak[2,**], and Aris Tentes[3,***]

[1] UCLA
abhishek@cs.ucla.edu
[2] IST Austria
pietrzak@ist.ac.at
[3] New York University
tentes@cs.nyu.edu

Abstract. We show a hardness-preserving construction of a PRF from any length doubling PRG which improves upon known constructions whenever we can put a non-trivial upper bound q on the number of queries to the PRF. Our construction requires only $O(\log q)$ invocations to the underlying PRG with each query. In comparison, the number of invocations by the best previous hardness-preserving construction (GGM using Levin's trick) is logarithmic in the *hardness* of the PRG.

For example, starting from an exponentially secure PRG $\{0,1\}^n \mapsto \{0,1\}^{2n}$, we get a PRF which is exponentially secure if queried at most $q = \exp(\sqrt{n})$ times and where each invocation of the PRF requires $\Theta(\sqrt{n})$ queries to the underlying PRG. This is much less than the $\Theta(n)$ required by known constructions.

1 Introduction

In 1984, the notion of *pseudorandom functions* was introduced in the seminal work of Goldreich, Goldwasser and Micali [10]. Informally speaking, a pseudorandom function (PRF) is a keyed function $F : \{0,1\}^n \times \{0,1\}^m \to \{0,1\}^n$, such that no efficient oracle aided adversary can distinguish whether the oracle implements a uniformly random function, or is instantiated with $F(k,.)$ for a random key $k \leftarrow \{0,1\}^n$. PRFs can be used to realize a shared random function, which has found many applications in cryptography [9,7,8,2,16,15,12].

Goldreich et al. [10] gave the first construction of a PRF from any length-doubling pseudorandom generator $G : \{0,1\}^n \to \{0,1\}^{2n}$; this is known as the GGM construction. In this work, we revisit this classical result. Although we will state the security of all constructions considered in a precise quantitative way, it helps to think in asymptotic terms to see the qualitative differences between constructions. In the discussion below, we will therefore think of n as a parameter

---

* Research conducted while at CWI Amsterdam.

** Supported by the European Research Council under the European Union's Seventh Framework Programme (FP7/2007-2013) / ERC Starting Grant (259668-PSPC).

*** Research conducted while at CWI Amsterdam and IST Austria.

R. Cramer (Ed.): TCC 2012, LNCS 7194, pp. 369–382, 2012.

(and assume the PRG G is defined for all input lengths $n \in \mathbb{N}$, and not just say $n = 128$). Moreover, for concreteness we assume that G is exponentially hard, that is, for some constant $c > 0$ and all sufficiently large n, no adversary of size 2^{cn} can distinguish $G(U_n)$ from U_{2n} (where U_n denotes a variable with uniform distribution over $\{0, 1\}^n$) with advantage more than 2^{-cn}. We will also refer to this as "G having cn bits of security".

The GGM construction $\mathsf{GGM_G} : \{0,1\}^n \times \{0,1\}^m \to \{0,1\}^n$ is **hardness preserving**, which means that if the underlying PRG G has cn bits of security, it has $c'n$ bits of security for some $0 < c' < c$. The domain size $\{0,1\}^m$ can be arbitrary, but the **efficiency** of the construction depends crucially on m as every invocation of $\mathsf{GGM_G}$ requires m calls to the underlying PRG G.

Levin [13] proposed a modified construction which improves efficiency for long inputs: first hash the long m-bit input to a short u-bit string using a universal hash function $h : \{0,1\}^m \to \{0,1\}^u$, and only then use the GGM construction on this short u-bit string. The smaller a u we choose, the better the efficiency. If we just want to achieve security against polynomial size adversaries, then a super-logarithmic $u = \omega(\log n)$ will do. But if we care about exponential security and want this construction to be hardness preserving, then we must choose a $u = \Omega(n)$ that is linear in n. Thus, the best hardness-preserving construction of a PRF $\mathsf{F^G}$ from a length-doubling PRG G requires $\Theta(n)$ invocations of G for every query to F (unless the domain $m = o(n)$ is *sublinear*, then we can use the basic GGM construction.) In this work we ask if one can improve upon this construction in terms of efficiency. We believe that in this generality, the answer actually is no, and state this explicitly as a conjecture. But our main result is a new construction which dramatically improves efficiency in many practical settings, namely, whenever we can put a bound on the number of queries the adversary can make.

In the discussion above, we didn't treat the number of queries an adversary can make as a parameter. Of course, the size of the adversary considered is an upper bound on the number of queries, but in many practical settings, the number of outputs an adversary can see is tiny compared to its computational resources.

For example consider an adversary of size 2^{cn} who can make only $q = 2^{\sqrt{n}} \ll 2^{cn}$ queries to the PRF. If the domain of the PRF is small, $m = \Theta(\sqrt{n})$, then using GGM we get a hardness-preserving construction with efficiency $\Theta(\sqrt{n})$ (where efficiency is measured by the number of queries to G per invocation of the PRF.) If we want a larger domain $m = \omega(\sqrt{n})$, then the efficiency drops to $m = \omega(\sqrt{n})$. We can get efficiency $\Theta(n)$ regardless of how large m is by using Levin's trick, but cannot go below that without sacrificing hardness preservation.

In this paper we give a hardness-preserving construction which, for any input length m, achieves efficiency $\Theta(\sqrt{n})$. The construction works also for other settings of the parameters. In particular, for $q = 2^{n^\epsilon}$ (note that above we considered the case $\epsilon = 1/2$) we get a construction with efficiency $\Theta(\log q) = \Theta(n^\epsilon)$. Actually, this is only true for $\epsilon \geq 1/2$; whether there exists a hardness-preserving black-box construction with efficiency $\Theta(\log q)$ for $q = 2^{n^\epsilon}$ where $\epsilon < 1/2$ is an interesting open question.

Other Applications. Although we described our result as an improved reduction from PRFs to PRGs, the main idea is more general. Viewing it differently, ours is a new technique for extending the domain of a PRF. If we apply our technique to PRFs with an input domain of length ℓ bits, Levin's trick would require roughly a domain of size ℓ^2 to achieve a comparable quality of hardness preservation.

This technique can be used to give more efficient constructions in other settings, for example to the work of Naor and Reingold [18] who construct PRFs computable by low depth circuits from so called pseudorandom synthesizer (PRS), which is an object stronger than a PRG, but weaker than a full blown PRF. Very briefly, [18] gives a hardness-preserving construction of a PRF from PRS which can be computed making $\Theta(n)$ queries to the PRS in depth $\Theta(\log n)$ (GGM also makes $\Theta(n)$ queries, but sequentially, i.e. has depth $\Theta(n)$; on the other hand, GGM only needs a PRG, not a PRS as building block). Our domain extension technique can also be used to improve on the Naor-Reingold construction, and improves efficiency from $\Theta(n)$ to $\Theta(\log q) = \Theta(n^\epsilon)$ whenever one can put an upper bound $q = 2^{n^\epsilon}$ ($\epsilon \geq 1/2$) on the number of adversarial queries.

Subsequent to [18], several number-theoretic constructions of PRFs have been proposed, inspired by the PRS based construction and GGM [19,20,14,5,1]. In particular, in [19], Naor and Reingold gave an efficient construction of a PRF from the DDH assumption that requires only n multiplications and one exponentiation instead of the n exponentiations required for GGM or the PRS based construction. This is achieved by exploiting particular properties of the underlying assumptions like the self reducibility of DDH. Our technique does not seem to be directly applicable to improve upon these constructions [19].

The Construction. Before we describe our construction in more detail, it is instructive to see why the universal hash-function $h : \{0,1\}^m \to \{0,1\}^u$ used for Levin's trick must have range $u = \Omega(n)$ to be hardness-preserving. Consider any two queries x_i and x_j made by the adversary. If we have a collision $h(x_i) = h(x_j)$ for the initial hashing, then the outputs $\mathsf{GGM_G}(k, h(x_i)) = \mathsf{GGM_G}(k, h(x_j))$ of the PRF will also collide. To get exponential security, we need this collision probability to be exponentially small. The probability for such a collision depends on the range u and is $\Pr_h[h(x_i) = h(x_j)] = 2^{-u}$. So we must choose $u = \Theta(n)$ to make this term exponentially small.

Similar to Levin's trick, we also use a hash function $h : \{0,1\}^m \to \{0,1\}^t$ to hash the input down to $t = 3 \log q$ bits (Recall that q is an upper bound on the queries to the PRF, so if say $q = 2^{\sqrt{n}}$, then $t = 3\sqrt{n}$.) As discussed earlier, the collision probability with such a short output length will not be exponentially small. However, we can prove something weaker, namely, if h is t-wise independent, then the probability that we have a $t + 1$-wise collision (i.e. any $t + 1$ of the q inputs hash down to the same value.) is exponentially small.

Next, the hashed value $x_i' = h(x_i)$ is used as input to the standard GGM PRF to compute $x_i'' := \mathsf{GGM_G}(k, x_i')$. Note, however, that we can't simply set x_i'' as the output of our PRF because several of the inputs $x_1, \ldots, x_q$ can be mapped by h to the same x', and thus also the same x'', which would not look random at all.

We solve this problem by using $x_i'' = \mathsf{GGM}_\mathsf{G}(k, h(x_i))$ to sample a t-wise independent hash function h_i. The final output $z_i := h_i(x_i)$ is then computed by hashing the original input x_i using this h_i. Note that with very high probability, for every i, at most $t' \leq t$ different $x_{i_1}, \ldots, x_{i_{t'}}$ will map to the same t-wise independent h_i. Thus, the corresponding outputs $h_i(x_{i_1}), \ldots, h_i(x_{i_{t'}})$ will be random.

The invocation of the GGM construction and the sampling of h_i from x_i'' can both be done with $\Theta(t)$ invocations of G, thus we get an overall efficiency of $\Theta(\sqrt{n})$.

2   Preliminaries

Variables, Sets and Sampling. By lowercase letters we denote values and bit strings, by uppercase letters we denote random variables and by uppercase calligraphic letters we denote sets. Specifically, by U_m we denote the random variable which takes values uniformly at random from the set of bit strings of length m and by $\mathcal{R}_{m,n}$ the set of all functions $F : \{0,1\}^m \mapsto \{0,1\}^n$. If $\mathcal{X}$ is a set, then by $\mathcal{X}^t$ we denote the t'th direct product of $\mathcal{X}$, i.e., $(\mathcal{X}_1, \ldots, \mathcal{X}_t)$ of t identical copies of $\mathcal{X}$. If X is a random variable, then by $X^{(t)}$ we denote the random variable which consists of t independent copies of X. By $x \leftarrow X$ we denote the fact that x was chosen according to the random variable X and analogously by $x \leftarrow \mathcal{X}$, that x was chosen uniformly at random from set $\mathcal{X}$.

Computational/Statisical Indistinguishability. For random variables X_0, X_1 distributed over some set $\mathcal{X}$, we write $X_0 \sim X_1$ to denote that they are identically distributed, we write $X_0 \sim_\delta X_1$ to denote that they have statistical distance δ, i.e. $\frac{1}{2} \sum_{x \in \mathcal{X}} |\mathsf{Pr}_{X_0}[x] - \mathsf{Pr}_{X_1}[x]| \leq \delta$, and $X_0 \sim_{(\delta,s)} X_1$ to denote that they are (δ, s) indistinguishable, i.e. for all distinguishers D of size at most $|D| \leq s$ we have $\sum_{x \in \mathcal{X}} |\mathsf{Pr}_{X_0}[D(x) \to 1] - \mathsf{Pr}_{X_1}[D(x) \to 1]| \leq \delta$. In informal discussions we will also use $\sim_s$ to denote statistical closeness (i.e. $\sim_\delta$ for some "small" δ) and $\sim_c$ to denote computational indistinguishability (i.e. $\sim_{(\delta,s)}$ for some "large" s and "small" δ.)

3   Definitions

We will need two information theoretic notions of hash functions, namely, δ-universal and t-wise independent hash functions. Informally, a hash function is t-wise independent if its output is uniform on any t distinct inputs. A function is δ-universal if any two inputs collide with probability at most δ.

Definition 1 (almost universal hash function). *For $\ell, m, n \in \mathbb{Z}$, a function $h : \{0,1\}^\ell \times \{0,1\}^m \to \{0,1\}^n$ is δ-almost universal if for any $x \neq x' \in \{0,1\}^m$*

$$\mathsf{Pr}_{k \leftarrow \{0,1\}^\ell}[h_k(x) = h_k(x')] \leq \delta$$

Universal hash functions were studied in [6,21], who also gave explicit constructions.

Proposition 1. *For any m, n there exists a 2^{-n+1}-universal hash function with key length $\ell = 4(n + \log m)$. Further, no such function can be δ-universal for $\delta < 2^{-n-1}$.*

Definition 2 (t-wise independent hash function family). *For $\ell, m, n, t \in \mathbb{Z}$, a function $h : \{0,1\}^\ell \times \{0,1\}^m \to \{0,1\}^n$ is t-wise independent, if for every t distinct inputs $x_1, \ldots, x_t \in \{0,1\}^m$ and a random key $k \leftarrow \{0,1\}^\ell$ the outputs are uniform, i.e.*

$$h_k(x_1) \| \ldots \| h_k(x_t) \sim U_n^{(t)}$$

Proposition 2. *For any $t, m, n \leq m$ there exits a t-wise independent hash function with key length $\ell = m \cdot t$.*

Remark 1. Note that 2-wise independence implies 2^{-n}-universality. The reason to consider the notion of δ-universality for $\delta > 2^{-n}$ is that it can be achieved with keys of length linear in the output, as opposed to the input.

Definition 3 (PRG[4,22]). *A length-increasing function $\mathsf{G} : \{0,1\}^n \mapsto \{0,1\}^m$ ($m > n$) is a (δ, s)-hard pseudorandom generator if*

$$\mathsf{G}(U_n) \sim_{(\delta,s)} U_m$$

*We say G has σ bits of security if G is $(2^{-\sigma}, 2^\sigma)$-hard. G is **exponentially hard** if it has cn bits of security for some $c > 0$, and G is **sub-exponentially hard** if it has cn^ϵ bits of security for some $c > 0, \epsilon > 0$.*

The following lemma, which follows from a standard hybrid argument, will be useful.

Lemma 1. *If $\mathsf{G} : \{0,1\}^n \mapsto \{0,1\}^m$ is a (δ, s)-hard PRG of size $|\mathsf{G}| = s'$, then for any $q \in \mathbb{N}$*

$$\mathsf{G}(U_n)^{(q)} \sim_{(q \cdot \delta, s - q \cdot s')} U_m^{(q)}$$

Definition 4 (PRF[10]). *A function $\mathsf{F} : \{0,1\}^\ell \times \{0,1\}^m \to \{0,1\}^n$ is a (q, δ, s)-hard pseudorandom function (PRF) if for every oracle aided distinguisher D^* of size $|D^*| \leq s$ making at most q oracle queries*

$$\left| \Pr_{k \leftarrow \{0,1\}^\ell}[D^{\mathsf{F}(k,\cdot)} \to 1] - \Pr_{f \leftarrow \mathcal{R}_{m,n}}[D^{f(\cdot)} \to 1] \right| \leq \delta$$

F *has σ bits of security against q queries if F is $(q, 2^{-\sigma}, 2^\sigma)$ secure.*

If q is not explicitly specified, it is unbounded (the size 2^σ of the distinguisher considered is a trivial upper bound on q.)

3.1 The GGM Construction

Goldreich, Goldwasser and Micali [10] gave the first construction of a PRF from any length doubling PRG. We describe their simple construction below.

For a length-doubling function $G : \{0,1\}^n \to \{0,1\}^{2n}$ and $m \in \mathbb{N}$, let $GGM_G : \{0,1\}^n \times \{0,1\}^m \to \{0,1\}^n$ denote the function

$GGM_G(k, x) = k_x$ where k_x is recursively defined as $k_\varepsilon = k$ and $k_{a\|0}\|k_{a\|1} := G(k_a)$

Proposition 3 ([10]). *If G is a (δ_G, s_G)-hard PRG, then for any $m, q \in \mathbb{N}$, $GGM_G : \{0,1\}^n \times \{0,1\}^m \to \{0,1\}^n$ is a (q, δ, s)-hard PRF where*

$$\delta = m \cdot q \cdot \delta_G \qquad s = s_G - q \cdot m \cdot |G| \tag{1}$$

3.2 Levin's Trick

One invocation of the GGM construction $GGM_G : \{0,1\}^n \times \{0,1\}^m \to \{0,1\}^n$ requires m invocations of the underlying PRG G, so the efficiency of the PRF depends linearly on the input length m. Levin observed that the efficiency can be improved if one first hashes the input using a universal hash function. Using this trick one gets a PRF on long m-bit inputs at the cost of evaluating a PRF on "short" u bit inputs plus the cost of hashing the m-bit string down to u bits.[1]

Proposition 4 (Levin's trick). *Let $h : \{0,1\}^\ell \times \{0,1\}^m \to \{0,1\}^u$ be a δ_h-universal hash function and $F : \{0,1\}^{\ell'} \times \{0,1\}^u \to \{0,1\}^n$ be a (q, δ_F, s)-hard PRF, then the function $F^h : \{0,1\}^{\ell+\ell'} \times \{0,1\}^m \to \{0,1\}^n$ defined as*

$$F^h(k_F\|k_h, x) := F(k_F, h(k_h, x))$$

is a (q, δ, s)-hard PRF where

$$\delta = q^2 \cdot \delta_h + \delta_F \qquad s = s_F - q \cdot |h| \tag{2}$$

3.3 Hardness Preserving and Good Constructions

Definition 5 (Hardness Preserving Construction). *A construction F^* of a PRF from a PRG is **hardness preserving** up to $q = q(n)$ queries, if for every constant $c > 0, \epsilon > 0$ there is a constant $c' > 0$ and $n' \in \mathbb{Z}$ such that for all $n \geq n'$: if G is of polynomial size and has cn^ϵ bits of security, F has $c'n^\epsilon$ bits of security against q queries. It is hardness preserving if it is hardness preserving for any q.*

*If the above holds for every $c' < c$, we say that it is **strongly** hardness preserving.*

---

[1] As universal hash functions are non-cryptographic primitives, hashing is generally much cheaper than evaluating pseudorandom objects.

Proposition 5. *The GGM construction is hardness preserving, more concretely (1) if* G *has* cn^ϵ *bits of security,* GGM_G *has* $c'n^\epsilon$ *bits of security for any* $c' < c/2$ *(2) GGM for* $q = n^{\epsilon'}$*,* $\epsilon' < \epsilon$ *queries is* <u>strongly</u> *hardness preserving.*

Proof. By eq.(1), if G has cn^ϵ bits of security, then the GGM construction has

$$\min\{cn^\epsilon - \log(q) - \log(m), cn^\epsilon - \log|G| - \log(m)\} \tag{3}$$

bits of security. To see (1), we observe that for any $c' < c/2$ eq.(3) is $c'n^\epsilon$ for sufficiently large n as required (using that m and $|G|$ are polynomial in n and $q = 2^{c'n}$.) To see (2) observe that for $\log(q) = n^{\epsilon'}$ where $\epsilon' < \epsilon$, the term eq.(3) is $c'n^\epsilon$ for sufficiently large n and every $c' < c$.

Recall that one invocation of GGM requires m invocations of the underlying PRG, where m must be at least $\lceil \log(q) \rceil$. We conjecture that $\Omega(\log(q))$ invocations are necessary for any hardness preserving construction.

Conjecture 1. Any construction[2] $F^G(.,.) : \{0,1\}^n \times \{0,1\}^m \to \{0,1\}^n$ that preserves hardness for q queries and has a black-box security proof must make $\Omega(\log q)$ invocation to G per invocation of F^G.

In the appendix we give some intuition as why we believe this conjecture holds. We show that the standard black-box security proof technique as used e.g. for GGM will not work for constructions making $o(\log q)$ invocations.

Definition 6 ((Very) Good Construction). *We call a construction as in Definition 5* **good** *for* q *queries, if it is hardness preserving up to* q *queries and each invocation of* F^G *results in* $O(\log q)$ *invocations of* G*. We call it* <u>very</u> *good, if it is even* <u>strongly</u> *hardness-preserving.*

Thus, GGM_G is good as long as the domain m is in $O(\log q)$, but not if we need a large domain $m = \omega(\log q)$. Let's look at Levin's construction.

Proposition 6. *The GGM construction with Levin's trick* GGM_G^h *(with* $h : \{0,1\}^\ell \times \{0,1\}^m \to \{0,1\}^u$ *as in Proposition 4) is hardness preserving if and only if* u *is linear in the security of the underlying PRG (e.g.* u *has to be linear in* n *if* G *is exponentially hard.)*

Proof. For concreteness, we assume G is exponentially hard, the proof is easily adapted to the general case. The number of queries to G per invocation of GGM_G^h is u, where $\{0,1\}^u$ is the range of the δ_h universal hash function. By eq.(2), GGM_G^h has security $\delta = q^2 \cdot \delta_h + \delta_{GGM_G}$. To preserve exponential hardness, δ must be exponentially small. So also $\delta_h \leq \delta$ must be exponentially small. By Proposition 1 $\delta_h > 2^{-u-1}$, thus u must be linear in n.

---

[2] We restrict the key length to n bits. This is not much of a restriction, as one can use G to expand the key. If we allow polynomially sized keys directly, then the conjecture would be wrong for polynomial q as the key could just contain the entire function table.

Summing up, the GGM construction is hardness preserving for any q, but only good if the domain is restricted to $m = O(\log q)$ bits. By using Levin's trick, we can get a hardness preserving construction where $u = \Theta(n)$ (if G is exponentially hard), but this will only be a good construction for q queries if q is also exponentially large.

In a practical setting we often know that a potential adversary will never see more than, say $2^{\sqrt{n}}$ outputs of a PRF F^{G}. If we need a large domain for the PRF, and would like the construction to preserve the exponential hardness of the underlying PRG G, then the best we can do is to use GGM with Levin's trick, which will invoke G a linear in n number of times with every query. Can we do better? If Conjecture 1 is true, then one needs $\Theta(\sqrt{n})$ invocations, which is much better than $\Theta(n)$. The main result in this paper is a construction matching this (conjectured) lower bound.

4   Our Construction

Let $\mathsf{G} : \{0,1\}^n \to \{0,1\}^{2n}$ be a length doubling function. For $e \in \mathbb{N}$, we denote with $\mathsf{G}^e : \{0,1\}^n \to \{0,1\}^{en}$ the function that expands an n bit string to a

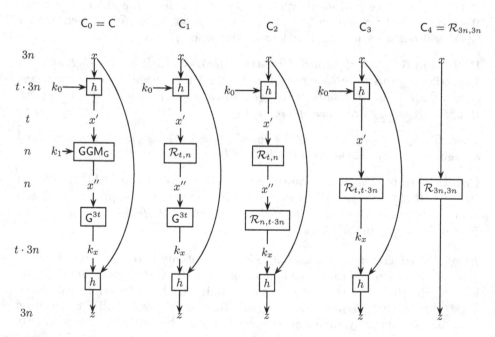

Fig. 1. The leftmost figure illustrates our construction $\mathsf{C}(.,.)$ using key $k = k_0 \| k_1$ on input x. The numbers $3n, t \cdot 3n, \ldots$ on the left indicate the bit-length of the corresponding values $x, k_0, \ldots$. The remaining figures illustrate the games from the proof of Theorem 1. $t = 3\log(q)$ is a parameter which depends on the number of queries q we allow.

en bits string using $e - 1$ invocations of G (this can be done sequentially, or in parallel in depth $\lceil \log e \rceil$.) We will use the following simple lemma which follows by a standard hybrid argument.

Lemma 2. *Let* G *be a* (δ, s)*-hard PRG, then* G^e *is a* $(e \cdot \delta, s - e \cdot |G|)$*-hard PRG.*

Further, our constructions uses a *t*-wise independent (cf. Proposition 2) hash function.

$$h : \{0,1\}^{t \cdot 3n} \times \{0,1\}^{3n} \to \{0,1\}^{3n}$$

Our construction $C^G : \{0,1\}^{t \cdot 3n+n} \times \{0,1\}^{3n} \to \{0,1\}^{3n}$ of a PRF which will be good for large ranges of q, on input a key $k = k_0 \| k_1$ (where $k_0 \in \{0,1\}^{t \cdot 3n}$ and $k_0 \in \{0,1\}^n$) and $x \in \{0,1\}^{3n}$, computes the output as ($X_{|t}$ denotes the t bit prefix of X.)

$$C(k,x) = h(G^{3t}(\mathsf{GGM}_G(k_1, h(k_0, x)_{|t})) , x)$$

Remark 2 (About the domain size and key-length). This construction has a domain of $3n$ bits and key-length $t \cdot 3n + n$. We can use Levin's trick to expand the domain to any m bits, and this will not affect the fact that the construction is good: by eq.(2) we get an additional $q^2 \cdot \delta_h = q^2/2^{3n-1}$ term in the distinguishing advantage, which can be ignored compared to the other terms.

We can also use a short n-bit key (like in plain GGM) and then expand it to a longer $t \cdot 3n + n$ bit key with every invocation (if we use Levin's trick we will need an extra $4(3n + \log m)$ bits.) This also will preserve the fact that the construction is good.

Theorem 1 (Main Theorem). *If* G *is a* (δ_G, s_G)*-hard PRG, then* C^G *is a* (q, δ, s)*-secure PRF where*

$$\delta = 4 \cdot q \cdot t \cdot \delta_G + q^2/2^n + q^t/2^{t^2} \qquad s = s_G - q \cdot |C^G| - q \cdot 3 \cdot t \cdot n \cdot |G| \qquad (4)$$

Before we prove this Theorem, let's see what it implies for some concrete parameters. Assume G is $(\delta_G = 2^{-cn}, s_G = 2^{cn})$-hard and we want security against $q = 2^{\sqrt{n}}$ queries. If we set $t := 3 \log q$ then the construction is very good (cf. Def. 6):

- It makes $7t = 21 \log(q) = O(\log q)$ invocation to G per query to C^G.
- C^G is strongly hardness preserving for $q = 2^{\sqrt{n}}$ queries. By eq.(4) we get

$$\delta < 2^{-cn+2+\sqrt{n}+\log(3\sqrt{n})} \qquad s \geq 2^{cn} - 2^{\sqrt{n}} \cdot |C^G|$$

If $|C^G|$ is polynomial in n (which is the only case we care about), then by the above equation, for every $c' < c$, we have $\delta \leq 2^{-c'n}$ and $s \geq 2^{c'n}$ for sufficiently large n as required by Definition 5.

The above argument works for any $q = 2^{n^\epsilon}$ where $0.5 \leq \epsilon < 1$. It also works if q is unbounded (i.e. $\epsilon = 1$), but we only get normal (and not strong) hardness-preservation. The argument fails for $\epsilon < 0.5$, that is whenever $q = 2^{o(\sqrt{n})}$. Technically, the reason is that the $q^t/2^{t^2}$ term in eq.(4) is not exponentially small (as required for hardness-preservation) when we set $t = O(\log q)$ (as required for a good construction.) It is an interesting open question if an optimal and hardness preserving construction with range $\omega(\log(q))$ exits for any $q = 2^{o\sqrt{n}}$. Summing up, we get the following corollary of Theorem 1

Corollary 1. *For any $0 < \delta \leq 1$ and $\epsilon \in [\delta/2, \delta[$, the construction C^G (setting $t := 3\log(q)$) is very good for $q = 2^{n^\epsilon}$ queries for any G with cn^δ bits of security (for any $c > 0$.) It is good for $\epsilon = \delta$.*

Proof (Proof of Theorem 1). Let D^* be any q-query distinguisher of size s. We denote with C_0 our construction C, with $C_4 = \mathcal{R}_{3n,3n}$ a random function and with C_1, C_2, C_3 intermediate constructions as shown in Figure 1. With $p(i)$ we denote the probability that $D^{C_i(\cdot)}$ outputs 1 (where e.g. in C_0 the probability is over the choice of $k_0 \| k_1$, in C_1 the probability is over the choice of k_0 and $f \leftarrow \mathcal{R}_{t,n}$, etc.)

Note that the advantage of D^* in breaking C is $\delta = |p(0) - p(4)|$, to prove the theorem we will show that

$$|p(0) - p(4)| = \left| \sum_{i=0}^{3} p(i) - p(i+1) \right| \leq \sum_{i=0}^{3} |p(i) - p(i+1)| \leq 4 \cdot q \cdot t \cdot \delta_G + q^2/2^n + q^t/2^{t^2}$$

The last step follows from the four claims below.

Claim. $|p(0) - p(1)| \leq q \cdot t \cdot \delta_G$

Proof (Proof of Claim). Assume $|p(0) - p(1)| > q \cdot t \cdot \delta_G$. We will construct a distinguisher D_1 for GGM_G and $\mathcal{R}_{t,n}$, which is of size $s_G - q \cdot t \cdot |G|$ and has advantage $> q \cdot t \cdot \delta_G$, contradicting Proposition 3. $D_1^{\mathcal{O}(\cdot)}$ chooses a random $k_0 \in \{0,1\}^{3tn}$, and then runs D^* where it answers its oracle queries by simulating C (using k_0), but replacing the GGM_G invocation with its oracle $\mathcal{O}(\cdot)$. In the end D_1 outputs the same as D. If $\mathcal{O}(\cdot) = GGM_G(k_1, \cdot)$ (for some random k_1) then this simulates C_0, and if $\mathcal{O}(\cdot) = \mathcal{R}_{t,n}$ it will simulate C_1. Thus D_1 will distinguish GGM_G and $\mathcal{R}_{t,n}$ with exactly the same advantage $> q \cdot t \cdot \delta_G$ that D has for C_0 and C_1.

Claim. $|p(1) - p(2)| \leq q \cdot 3 \cdot t \cdot \delta_G$

Proof (Proof of Claim). Assume $|p(1) - p(2)| > q \cdot 3 \cdot t \cdot \delta_G$. We will construct a distinguisher D_2 which is of size $s_G - q \cdot 3 \cdot n \cdot t \cdot |G|$ who can distinguish q-tuples of samples of U_{3tn} from $G^{3t}(U_n)$ with advantage $> q \cdot 3 \cdot t \cdot \delta_G$. Using a standard hybrid argument this then gives a distinguisher D_2' who distinguishes a single sample of U_{3tn} from $G^{3t}(U_n)$ with advantage $> q \cdot 3 \cdot t \cdot \delta_G/q = 3 \cdot t \cdot \delta_G$, contradicting Lemma 2.

D_2 on input $v_1, \ldots, v_q \in \{0,1\}^{3tn}$, runs D^* and answers its oracle queries by simulating C_1, but replacing the output of G^{3t} with the v_i's (using a fresh v_i for every query, except if x' appeared in a previous query, then it uses the same v_i as in this previous query. If the v_i's have distribution $G^{3t}(U_n)$ this perfectly simulates C_1, and if they have distribution U_{3tn} this simulates C_2. So D_2 has the same distinguishing advantage as D^* has for C_1 and C_2.

The proofs of the final two claims are completely information theoretic.

Claim. $|p(2) - p(3)| \leq q^2/2^n$

Proof (Proof of Claim). We claim that the distinguishing advantage of any (even computationally unbounded) q-query distinguisher for C_2 and C_3 is $\leq q^2/2^n$.

We get C_3 from C_2 by replacing the two nested uniformly random functions $f_2(.) = \mathcal{R}_{n,t\cdot 3n}(\mathcal{R}_{t,n}(.))$ with a single $f_3(.) = \mathcal{R}_{t,t\cdot 3n}(.)$. As each invocation of C_2 results in exactly one invocation of $f_2(.)$, the distinguishing advantage of the best q-query distinguisher for C_2 from C_3 can be upper bounded by the distinguishing advantage of the best such distinguisher for $f_2(.)$ and $f_3(.)$.

Let $\mathcal{E}$ denote the event that the q distinct queries $x'_1, \ldots, x'_q$ to $f_2(.) = \mathcal{R}_{n,t\cdot 3n}(\mathcal{R}_{t,n}(.))$ do not contain a collision on the inner function, i.e. $\mathcal{R}_{t,n}(x'_i) \neq \mathcal{R}_{t,n}(x'_j)$ for all $x'_i \neq x'_j$. Conditioned on $\mathcal{E}$, the outputs of f_2 and f_3 have the same distribution, namely $U_{t\cdot 3n}^{[q]}$. Using this observation, we can bound (using e.g. Theorem 1.(i) in [17] or the "fundamental Lemma" from [3])[3] the distinguishing advantage of any q-query distinguisher for f_2 and f_3 by the probability that one can make the event $\mathcal{E}$ fail. This is the probability that q uniformly random elements from $\{0,1\}^n$ (i.e. the outputs of the inner function) contain a collision. This probability can be upper bounded as $q^2/2^n$.

Claim. $|p(3) - p(4)| \leq q^t/2^{t^2}$

Proof (Proof of Claim). We claim that the distinguishing advantage of any (even computationally unbounded) q-query distinguisher for C_3 and C_4 is $\leq q^t/2^{t^2}$. We will first prove this only for non-adaptive distinguishers. To show security against adaptive adversaries, security against adaptive adversaries will then follow by a result from [17].

Let $x_1, \ldots, x_q$ denote the q distinct queries non-adaptively chosen by D^*. Let $\mathcal{E}$ denote the event which holds if there is no $t+1$-wise collision after the evaluation of the initial hash function h in C_3. That is, there is no subset $\mathcal{I} \subseteq [q]$ of size $|\mathcal{I}| = t+1$ such that $h(k_0, x_i) = h(k_0, x_j)$ for all $i,j \in \mathcal{I}$. Below we show that conditioned on $\mathcal{E}$, the outputs $y_1, \ldots, y_q$ (where $y_i := C_3(x_i)$) are uniformly random, and thus have the same distribution like the outputs of C_4. Using Theorem 1.(i) in [17] or the "fundamental Lemma" from [3], this means we can upper bound the distinguishing advantage of any non-adaptive distinguisher for C_3 and C_4 by the the probability that the event $\mathcal{E}$ fails to hold. Which means we have a $t+1$ wise collision in q t-wise independent strings over $\{0,1\}^t$. This can be upper bounded as $q^t/2^{t^2}$.[4]

We now show that the outputs of C_3 are uniform conditioned on $\mathcal{E}$. Consider a subset $\mathcal{J} \subseteq [q]$ with $|\mathcal{J}| \leq t$ such that $h(k_0, x_i) = h(k_0, x_j) = a$ for all

---

[3] Informally, the statement we use is the following: given two systems F and G and an event $\mathcal{E}$ defined for F, if F conditioned on $\mathcal{E}$ behaves exactly as G, then distinguishing F from G is at least as hard as making the event $\mathcal{E}$ fail.

[4] The probability that any t particular strings in $\{0,1\}^t$ collide is exactly $(2^{-t})^{t-1} = 2^{-t^2+t}$, we get the claimed bound by taking the union bound over all $q^t/t!$ possible t-element subsets of the q element set.

$i, j \in \mathcal{J}$. The fact that $\mathcal{R}_{t,3tn}(a)$ is a uniformly random together with the fact that h is a t-wise independent hash function implies that the joint distribution of $(h(a, x_i))_{i \in \mathcal{J}}$ follows the uniform distribution. Now let $\mathcal{J}_1, \ldots, \mathcal{J}_{q'}$ be the subsets of $[q]$ of size at most t, such that for $j \in \mathcal{J}_i$ $h(k_0, x_j) = a_i$ and all a_i's are distinct. The fact that $(\mathcal{R}_{t,3nt}(a_i))_{i \in [q']}$ follows the uniform distribution and $\mathcal{J}_1, \ldots, \mathcal{J}_{q'}$ are of size at most t implies that $(h(\mathcal{R}_{t,3nt}(a_i), x_i))_{i \in [q]}$ follows the uniform distribution as well.

So far, we only established the indistinguishability of C_3 and C_4 against non-adaptive distinguishers. We get the same bound for adaptive distinguishers using Theorem 2 from [17], which (for our special case) states that adaptivity does not help if the outputs of the system (C_3 in our case) are uniform conditioned on the event we want to provoke. Very recently [11] found that the precondition stated in[17] is not sufficient, but one additionally requires that the probability of the event failing is independent of the outputs observed so far. Fortunately in our case (and also for all applications in [17]) this stronger precondition is easily seen to be satisfied.

References

1. Banerjee, A., Peikert, C., Rosen, A.: Pseudorandom functions and lattices. In: EUROCRYPT (2012)
2. Bellare, M., Goldwasser, S.: New Paradigms for Digital Signatures and Message Authentication Based on Non-interactive Zero Knowledge Proofs. In: Brassard, G. (ed.) CRYPTO 1989. LNCS, vol. 435, pp. 194–211. Springer, Heidelberg (1990)
3. Bellare, M., Rogaway, P.: The Security of Triple Encryption and a Framework for Code-Based Game-Playing Proofs. In: Vaudenay, S. (ed.) EUROCRYPT 2006. LNCS, vol. 4004, pp. 409–426. Springer, Heidelberg (2006)
4. Blum, M., Micali, S.: How to generate cryptographically strong sequences of pseudo random bits. In: FOCS, pp. 112–117 (1982)
5. Boneh, D., Montgomery, H.W., Raghunathan, A.: Algebraic pseudorandom functions with improved efficiency from the augmented cascade. In: ACM Conference on Computer and Communications Security, pp. 131–140 (2010)
6. Carter, L., Wegman, M.N.: Universal classes of hash functions. J. Comput. Syst. Sci. 18(2), 143–154 (1979)
7. Goldreich, O.: Two Remarks Concerning the Goldwasser-Micali-Rivest Signature Scheme. In: Odlyzko, A.M. (ed.) CRYPTO 1986. LNCS, vol. 263, pp. 104–110. Springer, Heidelberg (1987)
8. Goldreich, O.: Towards a theory of software protection and simulation by oblivious rams. In: STOC, pp. 182–194 (1987)
9. Goldreich, O., Goldwasser, S., Micali, S.: On the Cryptographic Applications of Random Functions. In: Blakely, G.R., Chaum, D. (eds.) CRYPTO 1984. LNCS, vol. 196, pp. 276–288. Springer, Heidelberg (1985)
10. Goldreich, O., Goldwasser, S., Micali, S.: How to construct random functions. J. ACM 33(4), 792–807 (1986)
11. Jetchev, D., Özen, O., Stam, M.: Probabilistic analysis of adaptive adversaries revisited. Manuscript in preparation (2011)
12. Katz, J., Lindell, Y.: Introduction to Modern Cryptography

13. Levin, L.A.: One-way functions and pseudorandom generators. Combinatorica 7(4), 357–363 (1987)
14. Lewko, A.B., Waters, B.: Efficient pseudorandom functions from the decisional linear assumption and weaker variants. In: ACM Conference on Computer and Communications Security, pp. 112–120 (2009)
15. Luby, M.: Pseudorandomness and cryptographic applications. Princeton computer science notes. Princeton University Press, Princeton (1996)
16. Luby, M., Rackoff, C.: A Study of Password Security. In: Pomerance, C. (ed.) CRYPTO 1987. LNCS, vol. 293, pp. 392–397. Springer, Heidelberg (1988)
17. Maurer, U.M.: Indistinguishability of Random Systems. In: Knudsen, L.R. (ed.) EUROCRYPT 2002. LNCS, vol. 2332, pp. 110–132. Springer, Heidelberg (2002)
18. Naor, M., Reingold, O.: Synthesizers and their application to the parallel construction of psuedo-random functions. In: FOCS, pp. 170–181 (1995)
19. Naor, M., Reingold, O.: Number-theoretic constructions of efficient pseudo-random functions. In: 38th Annual Symposium on Foundations of Computer Science, pp. 458–467. IEEE Computer Society Press (October 1997)
20. Naor, M., Reingold, O., Rosen, A.: Pseudo-random functions and factoring (extended abstract). In: STOC, pp. 11–20 (2000)
21. Wegman, M.N., Carter, L.: New hash functions and their use in authentication and set equality. J. Comput. Syst. Sci. 22(3), 265–279 (1981)
22. Yao, A.C.-C.: Theory and applications of trapdoor functions (extended abstract). In: FOCS, pp. 80–91 (1982)

A   Intuition for Conjecture 1

We can think of the GGM construction with domain $\{0,1\}^m$ as a tree, where the outputs are leaves at depth m. More generally, we can think of any construction F^G as a directed loop-free graph, which is separated in layers. Each invocations starts at the root which holds the secret key K, and the computation follows a path, crossing layers, where the path within each layer contains at most one invocation of G.

We can define the "entropy" of a layer, as the amount of randomness leaving the layer (assuming G is a uniformly random function.) In the GGM contsruction, the fist layer has $2n$ bits of randomness, namely $\mathsf{G}(K)$, the ith layer has $2^i n$ bits of randomness. In a construction F^G contradicting the conjecture, there must be a layer which has significantly more than twice as much randomness as the layer before. To see this note that if each layer at most doubles the randomness, we need $\log(q)$ layers to get the qn bits of randomness. And moreover the last layer must have qn bits of randomness, as for a black-box security proof the only source of randomness is G.

Now, if a layer more than doubles its randomness, it must be the case that in this layer, G is invoked on either (1) inputs that are not uniformly random, or (2) the inputs to G are not independent. In a black-box reduction from F^G to G, one considers a series of hybrids $H_1, H_2, \ldots, H_t$, where H_1 is F^G and H_t is a random function. One gets from a hybrid H_i to H_{i+1} by replacing some internal value $Y := \mathsf{G}(X)$ with a uniform U_{2n}. If we have an adversary $\mathcal{A}$ who can distinguish H_i from H_{i+1}, we can use it to tell if a random variable Z has distribution U_{2n} or $\mathsf{G}(U_n)$ by replacing Y with Z in H_i, and using $\mathcal{A}$ to tell if what we get is H_i (which will be the case if $Z = \mathsf{G}(U_n)$) or H_{i+1}.

In the above argument, it is crucial that X has distribution U_n, but if (1) or (2) holds, this will not be the case. It is hard to imagine a black-box technique which works differently than by replacing some internal variables Y with the challenge Z, which as just explained will not work here.

Computational Extractors
and Pseudorandomness[*]

Dana Dachman-Soled[1], Rosario Gennaro[2], Hugo Krawczyk[2], and Tal Malkin[3]

[1] Microsoft Research New England
dadachma@microsoft.com
[2] IBM Research
rosario@us.ibm.com, hugo@ee.technion.ac.il
[3] Columbia University
tal@cs.columbia.edu

Abstract. *Computational extractors* are efficient procedures that map a source of sufficiently high min-entropy to an output that is computationally indistinguishable from uniform. By relaxing the statistical closeness property of traditional randomness extractors one hopes to improve the efficiency and entropy parameters of these extractors, while keeping their utility for cryptographic applications. In this work we investigate computational extractors and consider questions of existence and inherent complexity from the theoretical and practical angles, with particular focus on the relationship to pseudorandomness.

An obvious way to build a computational extractor is via the "extract-then-prg" method: apply a statistical extractor and use its output to seed a PRG. This approach carries with it the entropy cost inherent to implementing statistical extractors, namely, the source entropy needs to be substantially higher than the PRG's seed length. It also requires a PRG and thus relies on one-way functions.

We study the necessity of one-way functions in the construction of computational extractors and determine matching lower and upper bounds on the "black-box efficiency" of generic constructions of computational extractors that use a one-way permutation as an oracle. Under this efficiency measure we prove a direct correspondence between the complexity of computational extractors and that of pseudorandom generators, showing the optimality of the extract-then-prg approach for generic constructions of computational extractors and confirming the intuition that to build a computational extractor via a PRG one needs to make up for the entropy gap intrinsic to statistical extractors.

On the other hand, we show that with stronger cryptographic primitives one can have more entropy- and computationally-efficient constructions. In particular, we show a construction of a very practical computational extractor from any weak PRF without resorting to statistical extractors.

1 Introduction

Randomness extractors (or simply 'extractors') are algorithms that map sources of sufficient min-entropy to outputs that are statistically close to uniform.

---

[*] See [DGKM11] for the full version of this paper.

R. Cramer (Ed.): TCC 2012, LNCS 7194, pp. 383–403, 2012.

Randomness extraction has become a central and ubiquitous notion in complexity theory and theoretical computer science with innumerable applications and surprising connections to other notions. Cryptography, too, has greatly benefited from this notion. Cryptographic applications of randomness extractors range from the construction of pseudorandom generators from one-way functions to the design of cryptographic functionalities from noisy and weak sources (including applications to quantum cryptography) to the more recent advances in areas such as leakage- and exposure-resilient cryptography, circular encryption, lattice-based cryptosystems, and more. Randomness extractors have also found important uses in practical applications, particularly for the construction of key derivation functions. In many of these cryptographic applications, the defining property of randomness extractors, namely, statistical closeness of their output to a uniform distribution, can often be relaxed and replaced with computational indistinguishability. Extractors that provide this relaxed guarantee are called *computational extractors*, and they are the main object studied in this paper.

Let us review informally some basic facts about statistical extractors and the associated parameters n, m, k, δ. A function $Ext : \{0,1\}^n \times \{0,1\}^\ell \rightarrow \{0,1\}^m$ is a $(k, 2^{-\delta})$-*statistical extractor* if for any distribution X on $\{0,1\}^n$ with min-entropy k, the statistical distance between $Ext(X, U_\ell)$ and U_m is at most $2^{-\delta}$, where U_ℓ, U_m denote the uniform distribution over $\{0,1\}^\ell, \{0,1\}^m$, respectively. Note that extractors are randomized via the second argument called a seed or key (in our actual definitions we require the seed to be output, i.e., the so called *strong* extractor). We are interested in extractors where the values k and $2^{-\delta}$ are as small as possible (i.e., we want to minimize the entropy requirement from the source and get as small as possible statistical distance of the output to uniform). It is known how to construct statistical extractors that achieve $\delta = (k+\ell-m)/2$ [NZ96, HILL99]. Radhakrishnan and Ta-Shma [RTS00] show that this bound on δ is optimal, by showing how to build, for every extractor with parameters as above, a source distribution of min-entropy k for which the output of the extractor is $2^{-\delta}$-far from uniform for $\delta = (k + \ell - m)/2$. In the sequel we refer to this as the *RT bound*.

A major motivation to study computational extractors is that they allow us to go beyond the RT bound by replacing statistical closeness to uniform with computational indistinguishability. Indeed, an obvious way to do so is to first use a statistical extractor applied to the source distribution to obtain a short statistically close-to-uniform string and then use this string as a seed to a pseudorandom generator (PRG) to obtain more bits that are indistinguishable from uniform. We will refer to this as the *extract-then-prg* approach.

While the latter is a natural way to build computational extractors, it is not the only one or necessarily the best one, especially when implemented in practical settings. In particular, this approach carries with it the entropy limitations of statistical extractors as set by the RT bound, a serious concern in cases where the entropy of the source is too small to produce (via the statistical extractor) a sufficiently long key for the PRG. For example, consider the use of an extractor to convert a 160-bit elliptic curve Diffie-Hellman value (which by the DDH assumption has 160 bits of computational min-entropy) into a 128-bit seed for an AES-based PRG. Applying a statistical extractor to the DH value only guarantees a poor indistinguishability bound of 2^{-16} (i.e., $\delta = (160 - 128)/2$). If we wanted to preserve, say, 100-bit security we would need $\delta = 100$ bringing the required source entropy to 328 ($= 128 + 2 \cdot 100$).

One way around this problem is to build dedicated computational extractors based on cryptographic functions. Such an approach is taken in [Kra10, DGH+04], where computational extractors are built using *specific* schemes (HMAC and CBC) under assumptions that are specific to these schemes (and directed to the use of these extractors in the context of key derivation functions) including random-oracle type assumptions. On the other hand, the recent results of [BDK+11] show that for some key derivation applications one may relax the entropy requirements dictated by the RT bound (see more discussion on these issues in Section 7).

In this work, we further investigate computational extractors and consider questions of existence and inherent complexity from the theoretical and practical angles, with particular focus on the relationship to pseudorandomness. In particular, we ask how intrinsic is the use of pseudorandomness in constructing computational extractors, to what extent can we build computational extractors without resorting to a statistical extractor, and whether the "entropy penalty" of the extract-then-prg approach is avoidable.

Our Results

On the existence of computational extractors. The most basic question with respect to computational extractors is whether they exist at all and if they do under what (if any) assumption. The trivial answer is affirmative: statistical extractors are also computational. But we are interested in non-trivial computational extractors that output "more bits" than a statistical one. To capture this, we define the notion of *stretch*. For a security parameter p consider an extractor acting on a $k(p)$-entropy source: its stretch σ is the difference between the extractor's output length and its input's min-entropy, i.e., $\sigma(p) = m(p) - k(p) - \ell(p)$. Computational extractors with negative stretches of the form[1] $-\omega(\log p)$ exist unconditionally since a statistical extractor (that matches the RT bound) generates an output that is $2^{-\omega(\log p)}$-close to uniform and therefore is computationally indistinguishable from uniform. Thus, non-trivial computational extractors are those for which the stretch is at least $-O(\log p)$: we call such stretches and their associated extractors *proper*. The fact that proper computational extractors can be built on the basis of one-way functions via the extract-then-prg approach, raises the fundamental question: *Are one-way functions necessary for building proper computational extractors?* One would expect the answer to be "of course they are!". However, we can only provide a partial answer: We can show this to be the case for proper extractors of *positive* stretch. But for stretches in the range between $-O(\log p)$ and 0 the question remains open. Interestingly,

---

[1] $\omega(\cdot)$ stands for any superlinear function (i.e., one that grows faster than any linear function of its argument).

however, we can provide an affirmative answer under the assumption that *the RT bound applies to efficiently samplable distributions*. We refer to this as the *SRT Assumption* (see details in Section 3):

SAMPLABLE RT (SRT) ASSUMPTION. *Let $Ext : \{0,1\}^n \times \{0,1\}^\ell \to \{0,1\}^m$ be a poly-time computable statistical extractor. Then, for $k < n$ there exists a* poly-time samplable *source X of min-entropy k such that the statistical distance between the distributions $Ext(X, U_\ell)$ and U_m is at least $2^{-O(k+\ell-m)}$.*

In other words, the SRT assumption strengthens the RT bound by requiring it to hold even if we restrict our attention to efficiently samplable distributions. On the other hand, it weakens the RT bound by only requiring it to hold for efficient extractors and by reducing the lower bound requirement to $2^{-c \cdot (k+\ell-m)}$ for *any* constant c (in the RT bound, $c = 1/2$). To the best of our knowledge, the validity of this assumption has not been settled. Interestingly, given our results, any resolution of the assumption will have significant consequences. Disproving the assumption would open the door to the possibility of more effective statistical extractors for applications that are only concerned with efficient sources; e.g., it would mean that extractors based on the Leftover Hash Lemma may not be the best in practice (a surprising conclusion that may actually indicate the plausibility that the SRT does hold). And if the SRT assumption does hold, then our work settles affirmatively the question of existential equivalence of proper computational extractors and one-way functions.

Black-box constructions of proper extractors from OWPs. After investigating the relationship between proper extractors and one-way functions, we examine the question of whether we can have black-box constructions of proper extractors from OWPs that are more efficient than going through the extract-then-prg approach. As the measure of efficiency we use "OWP-complexity", namely, the number of invocations to the OWP in a black-box construction, following [GGKT05]. We prove a lower bound on the OWP-complexity of black-box constructions of proper extractors from OWPs. We show that, under the SRT assumption, the OWP-complexity of the extract-then-prg construction is optimal by showing a *tight lower bound* on the number of invocations to the OWP for *any* black-box construction of proper extractors from OWPs. Interestingly, this result confirms the intuition that in order to build a proper computational extractor one needs to make up for the entropy gap intrinsic to the RT bound (as explained above).

The above result applies to *any* black-box construction of a proper extractor that has oracle access to a OWP and it puts no restriction on the security reduction (which efficiently transforms an extractor-attacker into a OWP-attacker). A more restricted form of black-box constructions, known as *fully black-box*, also requires that the reduction between attackers be black-box (i.e., the reduction cannot access the code of the extractor-attacker). Interestingly, we prove a similar bound for fully black-box constructions, but *unconditionally*, i.e. without a need for the SRT assumption. Thus, we trade the more restricted form of black-box reduction for a lower bound that fully dispenses with the SRT assumption.

(For a thorough treatment of the semi- and fully-black-box notions and their meanings and implications please refer to [RTV04].)

Constructions based on Stronger Primitives. Next, we investigate the possibility of avoiding the intrinsic entropy loss in the generic extract-then-prg construction by assuming stronger primitives as the basis for the construction.

Our first result in this direction shows that given an exponentially-hard OWP, one can build a proper computational extractor where the OWP is applied *directly* to the high-entropy source without having to go through an initial extraction phase, hence avoiding the need to compensate for the entropy gap of the extractor. In order to achieve this result we replace the standard extract-then-prg approach with a dual prg-then-extract scheme that exploits the exponential hardness of the OWP to build a PRG that uses as its seed the very input from the high-entropy source.[2]

A practical computational extractor based on wPRF. We show a very simple construction of computational extractors based on *weak pseudorandom functions* (i.e., PRFs whose output is indistinguishable from uniform by adversaries that only see values of the function computed on *random* independent inputs). For this we resort to a lemma by Pietrzak [Pie09] showing that weak PRFs retain some of their security even when the keys are chosen from an imperfect source. More specifically, [Pie09] shows that if the original keys are of length n but they are chosen from a source with min-entropy $k \leq n$ then their security degrades roughly by an (optimal) factor of $2^{-(n-k)}$. This allows us to construct (strong) computational extractors where the source distribution is used to sample a key for the PRF and the extractor's random seed is used as an input to the PRF. This results in a very practical construction of computational extractors that fully dispenses with statistical extractors and perfectly fits the needs of randomness extraction in the context of key derivation functions (KDF) as studied in [Kra10] and as extensively used in real-world applications. In particular, one obtains a very practical KDF for cases where the input to the KDF (the source of key material) is at most of the size of the wPRF key. The security of the scheme solely depends on the security of the underlying (weak) PRF and it implies meaningful security bounds even in constrained cases where the entropy-output gap is small (or even negative). See Section 7 for details.

Relations to work on statistical extractors. While a main theme of our work is the role of pseudorandom generators in the construction of *computational* extractors, it is interesting to point out that pseudorandomness also plays a fundamental role in the development of *statistical* extractors. Starting with the work of Trevisan [Tre01] it has been realized that constructions of "non-cryptographic" pseudorandom generators such as [NW88, IW97] can lead to efficient statistical extractors. The notion of pseudorandomness in these works is usually weaker than the traditional cryptographic notion (that we use in our definition of computational

---

[2] This construction is somewhat reminiscent of the techniques used by Kalai *et al.* in [KLR09] for building two-source or network extractors, though the context and goals of these constructions are different.

extractors), e.g., they allow for super-polynomial (on the seed length) running time or consider more limited adversaries. Also the focus on efficiency in statistical extractors has traditionally been geared towards minimizing the size of the random seed as this determines the utility of these extractors in derandomization applications. See [Sha02] for a survey of results in this area. It would be very interesting to find closer relations between results in the above area and the questions raised by our work. In particular, in spite of the large body of work on statistical extraction, there seems to be little work that investigates statistical extractors against (efficiently) samplable sources. The only paper on the subject that we are aware of is by Trevisan and Vadhan [TV00] who show that if we only care about samplable distributions we can use deterministic extractors; however, this only works as long as the sampler of the source is computationally weaker than the extractor itself. Indeed, [TV00] shows that if we allow the source to depend on the extractor and to have higher computational complexity then deterministic extraction is not possible. In terms of our SRT assumption, what this shows is that the SRT does apply to deterministic extractors (for each such extractor there is a samplable source where the extractor fails). For all we know, the seemingly fundamental question of the entropy bounds that apply to statistical extractors when acting on samplable sources has not been studied. We hope that our work will provide motivation to investigate this question.

2   Proper Computational Extractors

We recall the definitions of statistical extractors, define proper computational extractors and give some of their basic properties. All extractor definitions presented here are stated in an asymptotic setting; in Section 5 we provide definitions in a concrete-complexity framework.

2.1   Preliminaries

Terminology. A *probability ensemble* X is an infinite sequence of probability distributions $\{X_p\}$ indexed by a parameter p. We usually assume that for all p, X_p has support in $\{0,1\}^{n(p)}$ where $n(\cdot)$ is a polynomially bounded function. For any integer t we use the symbol U_t to denote the uniform distribution on $\{0,1\}^t$. The *statistical distance* between two probability ensembles X, Y with common support ensemble $\{0,1\}^{n(p)}$ is defined as the function $\Delta_{X,Y}(p) = \max_{T \subseteq \{0,1\}^{n(p)}} |\Pr[X_p \in T] - \Pr[Y_p \in T]|$. We say that a distribution X has *min-entropy* $k(p)$ if for all x in the support of X_p it holds that $Pr_{X_p}[x] \leq 2^{-k(p)}$. For simplicity, in what follows we assume that the entropies denoted $k(p)$ are positive integers (in case $k(p)$ is not an integer, our results hold by replacing it with $\lceil k(p) \rceil$).

Definition 1. *An* extractor family *(or simply* extractor*) is an infinite family* $E = \{E_p\}$*, indexed by a parameter* p*, of the form* $E_p : \{0,1\}^{n(p)} \times \{0,1\}^{\ell(p)} \to \{0,1\}^{m(p)}$ *where the functions* $n(p), \ell(p), m(p)$*, are all polynomial in* p*. The extractor family* E *is called* $(k(p), \varepsilon(p))$-statistical *if for any probability ensemble*

X with support in $\{0,1\}^{n(p)}$ and min-entropy $k(p)$, it holds that the statistical distance between $\langle U_{\ell(p)}, E_p(X_p, U_{\ell(p)}) \rangle$ and $U_{\ell(p)+m(p)}$ is at most $\varepsilon(p)$.

The probability distribution from which the first input is taken is called the *source* and the second input is the *seed*. This definition of an extractor, requiring the joint distribution of output and seed to be ε statistically-close to uniform, is sometimes referred to in the literature as a *strong* extractor. A weaker flavor of this definition, referred to as a *weak* extractor, is one where one only considers the distance between the output $E_p(X_p, U_{\ell(p)})$ and the uniform distribution $U_{m(p)}$ (without the seed, which may remain hidden). *In this paper, unless otherwise noted, an "extractor" refers to a strong extractor.*

Intuitively, the goal of an extractor is to extract close-to-uniform bits out of a source with sufficiently high min-entropy, using a "short" uniformly random seed. We require that the output is longer than the seed,[3] specifically that $m(p) > \ell(p) + 1$.

Ideally, we'd like to extract all the randomness from the input, getting $m = k + \ell$ truly uniform bits (with $\varepsilon = 0$). However, this is impossible in general. From the results of [RTS00, NZ96, HILL99] we have the following lemma (which holds even for weak extractors) showing a tight relationship between how much of the input entropy $k + \ell$ can be extracted, and the distance ε from uniform.

Lemma 1 (RT Bound [RTS00]). Let E be a $(k(p), \varepsilon(p))$-statistical extractor with parameters $n(p), \ell(p), m(p)$ where $k(p) < n(p) - O(1)^4$ and $\varepsilon(p) < 1/2$. Then $\varepsilon(p) \geq 2^{-\frac{k(p)+\ell(p)-m(p)+O(1)}{2}}$. That is, for every such E there is a probability ensemble X with min-entropy $k(p)$ for which $E_p(X_p, U_{\ell(p)})$ has statistical distance $\min\{\frac{1}{2}, 2^{-\frac{k(p)+\ell(p)-m(p)}{2}}\}$ from $U_{m(p)}$. This bound is tight and achieved, in particular, by statistical extractors implemented via pairwise independent hash functions.

2.2 Proper Computational Extractors and Proper Stretch

We start by defining computational extractors, which differ from statistical ones in that the output is only required to be computationally indistinguishable from uniform rather than statistically close, the extractor itself needs to be efficient, and it is only required to work on efficiently samplable distributions.

Definition 2. *A family E of extractors is called $k(p)$-computational if E_p is polynomial-time computable, and for all efficiently-samplable probability ensembles X with min-entropy $k(p)$, the joint distribution $(U_{\ell(p)}, E_p(X_p, U_{\ell(p)}))$ is computationally indistinguishable from $U_{\ell(p)+m(p)}$.*

In this definition "efficiently samplable" means samplable by a polynomial-time algorithm and "computationally indistinguishable" refers to the regular notion of negligible advantage for all polynomial-time distinguishers. In a non-uniform setting, polynomial-time is be replaced by poly-size circuits.

---

[3] Without this condition, the trivial extractor that outputs its seed works for any source (even with 0 entropy).

[4] The symbol $O(1)$ represents a specific constant calculated in [RTS00].

Discussion. The defined notion corresponds to a strong extractor (see Section 2.1). A *weak* computational extractor is defined similarly but only requiring that the output $E_p(X_p, U_{\ell(p)})$ (without the seed) is indistinguishable from uniform. Although *our lower bounds hold even for weak extractors,* we focus our treatment on strong extractors, because in the computational setting, weak extractors are not very interesting. Indeed, any PRF is, by definition, a weak computational extractor that works for *any* source distribution.

We require the output of the computational extractor to be pseudorandom only when the input is an efficiently samplable distribution. Indeed, for computational uses (where we model feasible computation as polynomial-time) a hard-to-sample distribution is of little interest. In particular, we would not want to disqualify a good computational extractor just because it fails on a hard to compute source. Also, samplable sources allow to use the same seed – as long as it has been chosen at random and independently of the source – with multiple samples (this is crucial in some applications, including key derivation as discussed in Section 7).

At the same time, it is worth noting that we could consider a flavor of our definition where efficient samplability is replaced with oracle access (for the attacker) to an arbitrary distribution. The lower-bound results from Sections 3 and 4 hold for this definition, while the upper bound from Lemma 7 holds as long as the OWP is secure against non-uniform attackers (non-uniformity is necessary to argue that access to a hard-to-compute distribution does not help the attacker break the OWP or other primitives such as a PRG). Finally, we note that for our results on fully black-box reductions from Section 5, we do consider the latter setting, namely, arbitrary distributions to which the attacker gets oracle access.

It is clear that any efficient $(k(p), \varepsilon(p))$-statistical extractor for a negligible $\varepsilon(p)$, is also a $k(p)$-computational extractor. Thus, the upper bound of Lemma 1 implies the following.

Lemma 2. *There exist extractors with parameters $n(p), \ell(p), m(p)$ that are $k(p)$-computational for any $k(p) < n(p) - O(1)$ such that*

$$k(p) = m(p) - \ell(p) + \omega(\log p) \tag{1}$$

Note that the Lemma is unconditional, i.e., computational extractors with parameters as in (1) exist unconditionally. In this sense, *non-trivial* computational extractors are those whose parameters beat (1), and in particular have an output that is (indistinguishable from but) statistically far from uniform. We call such extractors *proper*, defined as follows.

Definition 3. *The stretch $\sigma(p)$ of a $k(p)$-computational extractor with parameters $n(p), \ell(p), m(p)$ is defined as $\sigma(p) = m(p) - k(p) - \ell(p)$. The stretch $\sigma(p)$ is proper if $\sigma(p) \geq -O(\log p)$ (i.e., there exists a constant c such that $\sigma(p) \geq -c \log p$ for all p). A $k(p)$-computational extractor is proper if its stretch is proper.*

Note that the stretch does not only depend on the extractor but also on the input entropy $k(p)$ (though, for simplicity, we sometimes omit the explicit $k(p)$ notation when talking about proper extractors). Since, for simplicity, we have assumed that $k(p)$ is integer (or else we consider $\lceil k(p) \rceil$) then the stretch is integer and can be negative, zero, or positive. Hereafter, when we say "proper extractor" we mean "proper computational extractor."

3 The Equivalence of Proper Extractors and One-Way Functions

Note that statistical extractors have statistical distance from uniform of at least $\varepsilon(p) = 2^{-\frac{k(p)+\ell(p)-m(p)}{2}}$ which is $1/poly(p)$ (hence non-negligible) in the case of proper extractors. Thus, statistical extractors do not immediately yield proper computational extractors.

This raises the question: *Do proper computational extractors exist?* The following Lemma answers this in the affirmative, assuming one-way functions exist.

Lemma 3. *If one-way functions exist then strong proper computational extractors exist too.*

Proof Sketch. Let $E = \{E_p\}$ be a $k(p)$-computational extractor with parameters $n(p), \ell(p), m(p)$ for which equation (1) holds (such an extractor exists for any functions $m(p), k(p)$ as in Lemma 2). Also assume $\omega(\log p) \le p$. Let $\{G_p\}$ be a pseudorandom generator with seed length $m(p)$ and output length $k(p)+\ell(p)$ (assuming OWFs, PRGs exist for some function $m(p)$ and output length $m(p)+p$). Construct extractor E' that first applies E and uses the output to seed the PRG. It is easy to see that E has parameters $n(p), \ell(p), m'(p) = k(p) + \ell(p)$ and its output is indistinguishable from $U_{m'(p)}$. But $m'(p) = k(p) + \ell(p)$, thus E' is proper.

Somewhat surprisingly we can't immediately prove *equivalence* between proper extractors and one-way functions. The opposite direction of Lemma 3 can be easily proven only for proper computational extractors with *positive stretch* as shown in the following Lemma.

Lemma 4. *From any (even weak) computational extractor with positive stretch one can build a pseudorandom generator.*

Proof Sketch. Let E be a $k(p)$-computational extractor with parameters $n(p)$, $\ell(p), m(p)$ and positive stretch $\sigma(p)$, i.e. $m(p) > k(p) + \ell(p)$. We build a PRG G with random seeds of length $s(p) = k(p) + \ell(p)$ and output length $m(p) > s(p)$. G partitions its seed into a $k(p)$-long value x and an $\ell(p)$-long value y, and calls E on (x', y) where x' consists of x padded with $n(p) - k(p)$ zeros. Clearly, the input distribution to E has entropy $k(p)$, hence its output is pseudorandom. Since G outputs more bits than its seed then G is a pseudorandom generator.

The last two lemmas leave the following question: *Does the existence of proper computational extractors, even those with non-positive proper stretch imply the existence of one-way functions? In particular, is this the case for computational extractors of stretch 0?* To provide an affirmative answer we need to resort to an additional assumption about the RT bound.

Samplable RT (SRT) Assumption. *For every polynomial-time computable extractor E with parameters $n(p), \ell(p), m(p)$ and every function k such that $k(p) < n(p) - O(1)$, there exists a poly-time samplable probability ensemble X of min-entropy k such that the statistical distance between the distributions $E_p(X_p, U_{\ell(p)})$ and $U_{m(p)}$ is at least $\min\{\frac{1}{2}, 2^{-O(k(p)+\ell(p)-m(p))}\}$.*

In other words, we are assuming that if we restrict attention to efficiently samplable sources then the RT bound still applies. More accurately, we assume a weaker bound where the RT bound $2^{-\frac{1}{2}(k(p)+\ell(p)-m(p))}$ is replaced with $2^{-c \cdot (k(p)+\ell(p)-m(p))}$ for *any* constant c, possibly much larger than $1/2$. In addition, we assume this to be the case only for efficient extractors[5]. This assumption is not implied by the proof in [RTS00] which builds a source on which the extractor incurs the claimed bound but this source may not be efficiently samplable. Quite interestingly, the question raised by this conjecture does not seem to have been widely researched. Any answer to it, positive or negative, would be of interest. If true it implies the equivalence of proper computational extractors and pseudorandom generators (see Theorem 1). If disproven it would open the possibility of building efficient extractors that beat the RT and Leftover-Hash-Lemma bounds on efficient sources.

Lemma 5. *Under the SRT assumption, the existence of a proper extractor implies the existence of a OWF.*

Proof Sketch. Let E be a proper $k(p)$-computational extractor and let X be a polynomial-time samplable ensemble of min-entropy $k(p)$, then the output of E on X induces a polynomial-time samplable distribution that is statistically far from uniform but computationally indistinguishable. Thus, the pair of distributions $(E_P(X_P, U_{\ell(p)}), U_{m(p)})$ are efficiently samplable, have statistical distance greater than $1/\text{poly}(p)$ for some polynomial and are computationally indistinguishable. Using the results of [Gol90, HILL99], constructing such a pair of distributions is sufficient to construct pseudorandom generators (PRG). This in turn implies the existence of OWF.

From Lemmas 3 and 5 we get:

Theorem 1. *Under the SRT assumption, proper computational extractors exist if and only if one-way functions exist.*

---

[5] It is most likely (using a counting argument) that the conjecture does not hold for super-polynomial extractors, namely, there may be inefficient extractors that beat the RT bound on all efficiently samplable distributions.

4   The Cost of Black-Box Constructions of Proper Extractors from OWPs

In this section we follow the methodology from [GGKT05] for quantifying the cost, as a number of OWP invocations, of (semi) black-box constructions of proper computational extractors from OWPs. We show a lower bound on the number of calls to the OWP that depends on the strength of the OWP and the stretch of the extractor. This result reflects the intuition that in order to build a computational extractor one needs to first make up for the entropy gap intrinsic to the RT bound. Indeed, the result shows that it is not enough to call the OWP just to generate as many bits as the extractor's stretch but one needs to generate $\omega(\log p)$ additional bits to cover for the loss of entropy. Comparing with the corresponding results of [GGKT05] about pseudorandom generators, we see that making up for this entropy gap is the only intrinsic difference between proper extractors and PRGs (under this black-box complexity measure). We also prove that the lower bound is tight.

Remark: Our lower bounds deal with constructions of computational extractors from one-way permutations. However, we note that our results extend to the case of one-way functions since our lower bounds are proven using random permutations which are not efficiently distinguishable from one-way functions. However we do not know if for the case of OWF our bounds are tight (i.e. the currently known constructions based on OWF have a larger number of queries).

In Section 3 we showed that proper extractors are equivalent to one-way functions. Here we formalize a notion of *black-box* constructions for computational extractors: such constructions access a one-way function as an *oracle*, rather than having access to the code of an algorithm computing it.

We start by developing an analogue of the treatment from [GGKT05] to the asymptotic setting of our analysis. For any integers t, n, $t \leq n$, we denote by Π_n the set of all permutations over $\{0,1\}^n$ and by $\Pi_{t,n}$ the set of permutations in Π_n that arbitrarily permute the first t bits of input while leaving the remaining $n - t$ bits fixed.

For a security parameter p denote with $n(p), k(p), \ell(p), m(p)$ and $t(p)$ integer functions that grow polynomially in p. Assume also that $t(p) \leq n(p)$ and $k(p) \leq n(p) - O(1)$ for all p. Consider an infinite family of permutations $\Pi = \{\pi_p\}_{p=1}^{\infty}$ where π_p is chosen in $\Pi_{n(p)}$. We say that Π is $T(p)$-hard if for sufficiently large p, any attacker running in time $T(p)$ succeeds in inverting π_p with probability less than $1/T(p)$. We say that Π is *one-way* if it is $T(p)$-hard for every polynomial $T(\cdot)$.

With Π^* we denote such a family $\Pi^* = \{\pi_p^*\}_{p=1}^{\infty}$ where each permutation π_p^* is chosen at random from the set $\Pi_{t(p),n(p)}$. The following Lemma (based on [IR89]) proves that for any hardness $T(p)$, if we choose $t(p) = 3 \log T(p)$ (and an additional technical condition that $t(p) \geq 6 \log p$), then this family is $T(p)$-hard with probability 1.

Lemma 6. *Let* $t(p) \geq 6 \log p$. *Then with probability 1,* Π^* *constructed as above is* $T(p)$-*hard for* $T(p) = 2^{t(p)/3}$.

Proof. Let A be an adversary that runs time $T(p)$ and attempts to invert Π^*. On expectation, over the choice of π_p^*, A succeeds in inverting with probability $T(p)/2^{t(p)} = 1/T^2(p)$, namely:

$$E_{\pi \sim \Pi_{t(p),n(p)}} \left[\Pr_{x \sim U_{n(p)}} [A(\pi(x)) = x] \right] = 1/T^2(p).$$

Using Markov's inequality we have that the probability over the choice of π_p^* that A inverts successfully with probability better than $T(p) \cdot 1/T^2(p)$ is at most $1/T(p)$:

$$\Pr_{\pi \sim \Pi_{t(p),n(p)}} \left[\Pr_{x \sim U_{n(p)}} [A(\pi(x)) = x] \geq T(p) \cdot 1/T^2(p) \right] \leq 1/T(p). \tag{2}$$

Since by choice of $t(p) \geq 6 \log p$ we have $1/T(p) \leq 1/p^2$ we get that the sum $\sum_{p \to \infty} 1/T(p)$ is finite. The convergence of this sum allows us to apply the Borel-Cantelli Lemma to (2) which implies that with probability 1 over the choice of Π^* the inequality $\Pr_{x \sim U_{n(p)}}[A(\pi_p^*(x)) = x] < 1/T(p)$ (where A is assumed to run time $T(p)$) holds for all but a finite number of p's. In other words, with probability 1 over the choice of Π^*, the resultant family Π^* is $T(p)$-hard.

Definition 4. *An* oracle extractor *construction (from a one-way permutation) is a family of oracle procedures* $\mathcal{E}^{(\cdot)} = \{E_p^{(\cdot)} : \{0,1\}^{n(p)} \times \{0,1\}^{\ell(p)} \to \{0,1\}^{m(p)}\}$ *such that* $E_p^{(\cdot)}$ *expects as an oracle a permutation* $\pi_p \in \Pi_{n(p)}$ *and* $E_p^{(\cdot)}$ *is computable in time polynomial in* p. *We say that* $\mathcal{E}^{(\cdot)}$ *has black-box access to a family* $\Pi = \{\pi_p\}_{p=1}^{\infty}$ *(and denote it as* $\mathcal{E}^{(\Pi)}$ *) if* $E_p^{(\cdot)}$ *uses* $\pi_p \in \Pi$ *as its oracle.*

We say that $\mathcal{E}^{(\cdot)}$ *is a* $k(p)$-computational oracle extractor *if for every one-way family* Π *the family* $\mathcal{E}^{(\Pi)}$ *is a* $k(p)$-computational extractor *according to Definition 2.*

Another way to restate the above definition is that there must be an efficient reduction from distinguishing the output of the extractor from uniform to inverting the permutation family. In other words, any distinguishing adversary can be used to construct an inverter for the permutation family. Note that the above definition formalizes the notion of *semi black-box* construction in which the construction (the extractor) has oracle access to the underlying primitive (the one-way permutation), but no restriction is made on the reduction (in particular, the reduction might be able to access the code of the adversary). The more restricted notion of fully black-box constructions (in which additionally the security reduction only has oracle access to the adversary breaking the construction) will be discussed in Section 5.

We now state the main theorem in this section. It shows that under the SRT assumption, proving a semi-black-box construction of a computational extractor for which $q(p) \cdot t(p) - \sigma(p) = O(\log p)$ is at least as hard as proving that OWFs exist (or, equivalently, proving such a construction is at least as hard as proving that the SRT assumption implies OWF).

Theorem 2. *Let $\mathcal{E}^{(\cdot)}$ be a proper $k(p)$-computational oracle extractor according to Definition 4, which has access to a $T(p)$-hard family where $T(p)$ is super-polynomial. Let $t(p) = 3\log T(p) = \omega(\log p)$. Assuming SRT, if $E_p^{\pi_p}$ has proper stretch $\sigma(p)$ and it calls the oracle π_p a total of $q(p)$ times, then $q(p)\cdot t(p) - \sigma(p) = \omega(\log p)$ or else one-way functions exist. This lower bound on $q(p)$ is tight.*

Proof. Let $\mathcal{E}^{(\cdot)}$ be a proper $k(p)$-computational oracle extractor with parameters $(n(p), \ell(p), m(p))$ and proper stretch $\sigma(p) = m(p) - k(p) - \ell(p)$. By assumption $\mathcal{E}^{\Pi}$ is $k(p)$-computational whenever the oracle Π is implemented with one-way permutation family, i.e. Π is $T(p)$-hard where $T(p)$ is a function growing faster than any polynomial. In particular, by Lemma 6 this is the case (with probability 1) when Π is implemented by the family Π^* with parameter $t(p) = 3\log T(p) = \omega(\log p)$. We will show that if $E_p^{\pi_p}$ calls $\pi_p \in \Pi^*$ a total of $q(p)$ times, we can construct a computational extractor E_p' with parameters $(n(p), \ell'(p) = \ell(p) + q(p)t(p), m(p))$ (and no oracle calls) such that for any distribution X_p with min-entropy $k(p)$, the output distributions $E_p'(X_p, U_{\ell'(p)})$ and $E_p^{\pi_p^*}(X_p, U_{\ell(p)})$ are $q^2(p)/2^{t(p)}$-statistically close, and since the latter distribution is pseudorandom so is the former (here we use the fact that $q(p)/2^{t(p)}$ is negligible since $q(p)$ is polynomial and $2^{t(p)} = T^3(p)$ super-polynomial).

More specifically, we construct $E_p' : \{0,1\}^{n(p)} \times \{0,1\}^{\ell'(p)} \to \{0,1\}^{m(p)}$, where $\ell'(p) = \ell(p) + q(p) \cdot t(p)$, in the following way: Let x, z' denote the input to E_p'. The string x and the first $\ell(p)$ bits of z' are used by E_p' to define the input (x, z) to $E_p^{(\cdot)}$ and the remaining bits of z' are used to select $q(p)$ distinct elements $y_1, \ldots, y_{q(p)} \in \{0,1\}^{t(p)}$. We then define: $E_p'(x, z, y_1, \ldots, y_{q(p)}) \stackrel{def}{=} E_p^{y_1, \ldots, y_{q(p)}}(x, z)$, namely, when $E_p^{(\cdot)}$ presents its i-th query to its oracle, call it w_i, we return as response the string y_i followed by the last $n(p) - t(p)$ bits of w_i.

Note that as long as all the y_i's are different the output distributions $E_p'(X_p, U_{\ell'(p)})$ and $E_p^{\pi_p^*}(X_p, U_{\ell(p)})$ are identical. The probability of a repeated y_i is $q^2(p)/2^{t(p)}$ and therefore the actual statistical distance between these distributions is negligible. In particular, we have that the output from E_p' is indistinguishable from random and therefore E_p' is a $k(p)$-computational extractor which makes no oracle calls. Moreover, its stretch $\sigma'(p)$ equals

$$\sigma'(p) = m(p) - k(p) - \ell'(p) = m(p) - k(p) - \ell(p) - q(p)t(p) = \sigma(p) - q(p)t(p).$$

If, for the sake of contradiction, we assume that $q(p)t(p) \leq \sigma(p) + c\log p$ for some constant c then we would get $\sigma'(p) \geq -c\log p$ meaning that E_p' is a regular (non-oracle) proper computational extractor from which, using Lemma 5 and the SRT assumption, we can construct a one-way function. This proves the theorem (the tightness of the bound on $q(p)$ is proven in Lemma 7 below).

Lemma 7. *The bound of Theorem 2 is tight: For any function $\sigma(p)$, polynomial in p, and any function $W(p)$ that grows as $\omega(\log p)$ there is a black-box construction of a strong proper extractor from OWP that attains stretch $\sigma(p)$ and calls the OWP $q(p)$ times such that $q(p)t(p) \leq \sigma(p) + W(p)$.*

Proof Sketch. We start by noting that there are black-box constructions of pseudorandom generators from OWPs that for any PRG-stretch function $\sigma'(p)$ (defined as the length of the PRG output less the length of PRG seed) call the OWP $\sigma'(p)/t(p)$ times where $t(p)$ is defined as in Theorem 2. This is the case, in particular, for the Blum-Micali construction using Goldreich-Levin hard-core bits. Therefore, to prove the Lemma it suffices to show how to build a proper extractor of stretch $\sigma(p)$ using a PRG of stretch $\sigma(p) + W(p)$ for any $W(p)$ that grow as $\omega(\log p)$.

Let $G = \{G_p\}_p$ be a PRG family, indexed by a parameter p, with seed length $s(p)$ and output size $r(p) = s(p) + \sigma(p) + W(p)$, for a given (polynomial in p) function $\sigma(p)$. Assume G is $(T(p), \varepsilon(p))$-secure (where $\varepsilon(p)$ is negligible in p). Let E be a strong statistical extractor (e.g., based on pairwise independent hash functions) with parameters $n(p), \ell(p), m(p) = s(p) + \ell(p)$ that on input distributions of min-entropy $k(p)$ outputs a distribution that is $2^{-\frac{k(p)-s(p)}{2}}$-close to $U_{m(p)}$. Using both G and E we build a proper computational extractor E' with parameters $n(p), \ell(p), m'(p) = r(p) + \ell(p)$. On input (x, z), E' calls E on (x, z) and uses the $s(p)$-bit output from E as the seed to G to produce an output of bit length $r(p) = s(p) + \sigma(p) + W(p)$. This, plus the $\ell(p)$-bit input salt, are the outputs from E'.

Note that on distributions of min-entropy $k(p) = r(p) - \sigma(p)$, E' has stretch $\sigma(p)$; moreover, we claim that the output from E' is $(T(p), \varepsilon'(p))$-indistinguishable from uniform where $\varepsilon'(p)$ equals $\varepsilon(p)$ plus a negligible term $2^{-W(p)/2} = 2^{-\omega(\log p)}$. Indeed, the only loss of security with respect to G is in the derivation of the seed $z \in \{0,1\}^{s(p)}$ that is chosen from a distribution that is $2^{-(k(p)-s(p))/2} = 2^{-W(p)/2} = 2^{-\omega(\log p)}$-close to $U_{s(p)}$. Thus E' is a proper computational extractor with stretch $\sigma(p)$ built on the basis of a PRG of stretch $\sigma(p) + W(p)$ which, as said, implies the tightness of the bound.

Note. The Blum-Micali construction with a randomized hardcore like Goldreich-Levin [GL], requires extra perfect but non-secret randomness. Hence this auxiliary randomness can be supplied by the extractor's seed and be output as part of the strong extractor's output.

5 Unconditional Fully Black-Box Lower Bound

Next, we pose the question of what can be shown *without assuming SRT*. We show that by restricting our attention to fully black box constructions, not only can we get rid of the SRT but actually can show an *unconditional* lower bound on the number of OWP invocations.

We first show an analogous lower bound to the semi black-box case (though unconditional) in the asymptotic, uniform setting. We then show a tighter concrete-complexity result in the non-uniform setting.

To begin, we review the notion of fully black box construction/reduction.

Definition 5. *A fully black-box reduction from a primitive Q to a primitive P is a pair of oracle PPT Turing machines $(G^{(\cdot)}, S^{(\cdot,\cdot)})$ such that the following two properties hold:*

Correctness: For every implementation f of primitive P, $g = G^f$ implements Q.

Security: For every implementation f of primitive P, and every adversary A, if A breaks G^f (as an implementation of Q) then $S^{A,f}$ breaks f. (Thus, if f is "secure", then so is G^f.)

Notice that in a full black-box reduction, the adversary is only accessed as an oracle. One consequence of this fact is that the adversary does not have to be efficient. We remark that an *implementation* of a primitive is any specific scheme that meets the requirements of that primitive (e.g., an implementation of a public-key encryption scheme provides samplability of key pairs, encryption with the public-key, and decryption with the private key).

5.1 Unconditional Lower Bound in the Asymptotic, Uniform Setting

In this section we show an analogue of the lower bound in Theorem 2 for the fully black-box setting. While the bound on the number of queries is the same as in Theorem 2, this result can be proven unconditionally (i.e., without requiring the SRT and without concluding that a construction that violates the bound implies a proof of the existence of one-way functions). However, Theorem 3 holds only when we consider a slightly modified definition of computational extractors where the output of the extractor is required to be computationally indistinguishable from uniform for *every* input probability ensemble X of min-entropy k. Observe that the construction outlined in Lemma 7 satisfies this stronger notion of security.

Theorem 3. *Let $\mathcal{E}^{(\cdot)}$ be a proper $k(p)$-computational fully black box extractor construction, which has access to a $T(p)$-hard family where $T(p)$ is super-polynomial. Further assume that such extractor remains proper $k(p)$-computational on any $k(p)$-entropy source, including those that are not efficiently samplable. Let $t(p) = 3\log T(p) = \omega(\log p)$. If $E_p^{\pi_p}$ has proper stretch $\sigma(p)$ and it calls the oracle π_p a total of $q(p)$ times, then $q(p) \cdot t(p) - \sigma(p) = \omega(\log p)$.*

Proof. See full version [DGKM11].

Next, we present a stronger version of this result. It will be a tighter concrete (rather than asymptotic) lower bound, for *non-uniform* fully black-box constructions of proper extractors from OWP. In order to do that, we need to revisit definitions and preliminary Lemmas in a concrete, non-uniform context.

5.2 Unconditional Lower Bounds in the Concrete, Non-uniform Setting

We start by adapting the definition of (oracle) computational extractors to the non-uniform and concrete (i.e., non-asymptotic) complexity setting.

We say that a permutation π over $\{0,1\}^n$ is S-hard if no circuit of size $\leq S$ and oracle access to π can invert π with probability better than $1/S$. Additionally, we say that two distributions are (S, ε)-indistinguishable if no circuit of size $\leq S$ can distinguish between them with probability better than ε.

Definition 6. $E : \{0,1\}^n \times \{0,1\}^\ell \to \{0,1\}^m$ *is a* $(k, S, 2^{-\delta})$-*computational extractor* (CompEXT) *if for any distribution X on $\{0,1\}^n$ with $H_\infty(X) \geq k$, we have that $(E(X, U_\ell), U_\ell)$ and $U_{m+\ell}$ are $(S, 2^{-\delta})$-indistinguishable (where indistinguishability holds even for circuits given oracle access to a* Sampler *which samples from distribution X).*

Definition 7. *An* oracle computational extractor *(OCompEXT) construction (from a one-way permutation) is an oracle procedure $E^{(\cdot)} : \{0,1\}^n \times \{0,1\}^\ell \to \{0,1\}^m$ that expects as an oracle a permutation $\pi \in \Pi_n$. We are interested in constructions where $E^{(\cdot)}$ is computable in time polynomial in n.*

We say that $E^{(\cdot)}$ is is an $(k, S_\pi, S_E, 2^{-\delta})$-OCompEXT construction from OWP if for every permutation π that is S_π-hard, E^π is an $(k, S_E, 2^{-\delta})$-secure CompEXT (where indistinguishability holds even for circuits given oracle access to both Sampler *and π).*

Using a standard averaging argument, the existence of a non-uniform attacker that succeeds in inverting a OWP with the help of such an oracle implies the existence of another attacker (of slightly larger size) that inverts the OWP without access to the oracle (just wire-in into the attacker circuit the source samples that maximize the attacker's inverting probability).

We now restate the lower bound of Radhakrishnan and Ta-Shma [RTS00] regarding the efficiency of statistical extractors (which was given in Lemma 1 for the asymptotic, uniform setting).

Lemma 8. *Let $E' : \{0,1\}^n \times \{0,1\}^{\ell'} \to \{0,1\}^m$ be a statistical extractor. Then, for any $k < n - C'$ there exists a distribution X of min-entropy k such that the two distributions $E'(X, U_{\ell'})$ and U_m are statistically $\min\{\frac{1}{2}, 2^{-((k+\ell'-m+C)/2)}\}$-far, where C and C' are universal constants.*

We are now ready to state our main result in this section, namely, a lower bound on the number of queries to the OWP by a fully black box construction of a computational extractor.

Theorem 4. *Let $E^{(\cdot)}$ be a fully black-box construction of a $(k, S_\pi, S_E, 2^{-\delta})$ proper oracle extractor which expects an S_π-hard one-way permutation π over n bits. Assume that $E^{(\cdot)}$ makes $q \leq S_\pi$ queries to its oracle and that $S_E \leq S_\pi$ and $2^{-\delta} \geq 1/S_\pi$. If $E^{(\cdot)}$ has proper stretch σ then $E^{(\cdot)}$ must call the one-way permutation q times, where $q \geq (2\delta + \sigma - C)/(5 \log S_\pi)$ for some constant C.*

Proof. See full version [DGKM11]. □

6   Construction from Exponentially-Hard One-Way Permutations

The results from Sections 4 and 5 indicate the optimality of the "extract-then-prg" approach when all we are interested in is minimizing the number of calls to a OWP in a black-box construction. However, a significant cost of this approach is that in order to use an n-bit OWP we need to start with an input distribution whose entropy is noticeably larger than n so we can apply the extraction part of the construction to it and still get n bits that are close to uniform and serve as input to the OWP. Here we show that one can make up for the entropy gap if the OWP has exponential hardness. In this case, we show a black-box construction based on such a OWP where one applies the OWP directly on the entropy source without an intermediate extractor step. For this we reverse the extract-then-prg approach and use instead an "prg-then-extract" construction where the OWP is applied first to expand (pseudo) entropy and then a statistical extractor is applied on this expanded entropy to generate a close-to-uniform output.[6]

The Construction.   Given an $(S_\pi, 2^{-\delta}/2^{n-k})$-hard OWP, π, we present a construction of a k-entropy strong computational extractor

$$F : \{0,1\}^n \times \{0,1\}^{(2\delta+\sigma)\cdot n+\ell} \to \{0,1\}^{k+(2\delta+\sigma)\cdot n+\ell+\sigma}$$

with proper stretch σ in Figure 1.

On input $(x, z' = (r_0, \ldots, r_{2\delta+\sigma-1}, z))$, where $x \in \{0,1\}^n$, $r_i \in \{0,1\}^n$, $z \in \{0,1\}^\ell$, the extractor F does the following:

Step 1:
  – Compute $(w_1, w_2) =$
    $\left(\left(\pi^{2\delta+\sigma}(x), \langle r_{2\delta+\sigma-1}, \pi^{2\delta+\sigma-1}(x) \rangle \right), \ldots, \langle r_0, x \rangle \right), (r_{2\delta+\sigma-1}, \ldots, r_0) \right)$
Step 2:
  – Let $F' : \{0,1\}^{n+2\delta+\sigma} \times \{0,1\}^\ell \to \{0,1\}^{k+\ell+\sigma}$ be a statistical $(k+2\delta+\sigma, 2^{-\delta})$ strong extractor.
  – Compute $(v, z) = F'(w_1, z)$.
Step 3: F outputs $(v, z, w_2) \in \{0,1\}^{k+(2\delta+\sigma)\cdot n+\ell+\sigma}$.

Fig. 1. Strong Computational Extractor from Exponentially-Hard OWP

The proof of Lemma 10 that F is indeed a strong extractor when π is an exponentially-hard OWP is based on the following lemma showing that exponentially-hard OWP's are "hard to invert" on arbitrary distributions of sufficiently high min-entropy.

Lemma 9. *Let $\pi : \{0,1\}^n \to \{0,1\}^n$ be an (S, ε)-one way permutation and let X be a distribution over $\{0,1\}^n$ of min-entropy k where $k = n - \alpha$. Then for all adversaries A of size at most S it is the case that:*

$$\Pr_{x \sim X}[A(\pi(x)) = x] \le \varepsilon \cdot 2^\alpha.$$

---

[6] [BDK+11] also uses the prg-then-extract approach for constructing an extractor; in their case, however, the prg is used to expand the seed rather than for increasing the computational entropy of the source as in our case.

Lemma 10. *The construction of F from Figure 1 is a black-box construction of a $(k, S_\pi/\mathrm{poly}(n), \mathrm{poly}(n) \cdot 2^{-O(\delta)})$-strong* CompEXT *with proper stretch σ from any $(S_\pi, 2^{-\delta}/2^{n-k})$-OWP π.*

Proof. See full version [DGKM11].

7 Practical Computational Extractors from Weak PRF

In this section we explore a connection between computational extractors and pseudo-random functions. We show a very efficient construction of a strong computational extractor using *any* PRF, and demonstrate its practical utility in the context of key derivation functions. Actually, we do not need the full security of a PRF; it suffices that the PRF is secure against attackers that do not choose inputs to the function but only see pairs of (input, output) where the inputs are chosen uniformly at random. Such PRFs are referred to as *weak PRF (wPRF)* (note that in our application 'weak' is stronger). The proof of our scheme follows directly from recent results by Pietrzak [Pie09] about *leakage-resilient* wPRFs.

Weak PRF. A pseudo-random function family is a family of functions $\mathcal{F} = \{f_a : \{0,1\}^\ell \to \{0,1\}^m\}_{a \in \{0,1\}^n}$ with the property that if a is chosen uniformly at random in $\{0,1\}^n$, then the function f_a is computationally indistinguishable from a random function from $\{0,1\}^\ell$ to $\{0,1\}^m$. More specifically, no efficient algorithm which has oracle access to either f_a or to a random function, can decide which is the case. If the oracle access is restricted to query the function on randomly chosen inputs, one obtains the notion of weak PRF (wPRF). We quantify this notion by saying that $\mathcal{F}$ is a (S, q, ε)-wPRF family if no circuit of size S can distinguish between f_a (for a chosen uniformly at random) and a random function with advantage better than ε when seeing the value of the function on q *random* inputs.

The main contribution in this section is in presenting the following construction of a simple computational extractor from any wPRF and demonstrating its *practical* security.

wPRF-based computational extractor. Let $\mathcal{F} = \{f_a : \{0,1\}^\ell \to \{0,1\}^m\}_{a \in \{0,1\}^n}$ be a wPRF family. We define the extractor $F : \{0,1\}^n \times \{0,1\}^\ell \to \{0,1\}^{m+\ell}$ as $F(a, s) = (f_a(s), s)$.

Theorem 5. *If $\mathcal{F} = \{f_a : \{0,1\}^\ell \to \{0,1\}^m\}_{a \in \{0,1\}^n}$ is (S, q, ε)-weak PRF with $q^2 < \varepsilon 2^{\ell+1}$, then for $k \leq n$ the extractor F defined above is a (k, S', ε') strong (and proper[7]) computational extractor with $\varepsilon' \approx \varepsilon \cdot 2^{n-k}$ and $S' \approx S \cdot \varepsilon'$.*

Proof. See full version [DGKM11].

---

[7] We assume $m \geq n$; if this is not the case in the given family $\mathcal{F}$ we can achieve it using standard range expansion techniques to increase m, possibly at the cost of somewhat strengthening the weak PRF requirement.

7.1   Application to Key Derivation

A main application of a strong computational extractor in cryptography is for key derivation [Kra10]. In this case, the source distribution is some *key material*, derived from some statistical process or a key agreement protocol, that has some significant amount of min-entropy but is not uniformly random as needed to key cryptographic functions. Thus, we need a way to produce a cryptographic key (random or pseudorandom) out of this key material. This is where our computational extractor is useful. One *restriction* is that if we use a wPRF whose key size is n we need to consider sources of key material whose length is at most n. In this case, we simply use the key material (without any processing, except maybe for padding to n bits) as the key to the wPRF and choose as the input to the wPRF a random value of length ℓ. The latter is the seed of the extractor and is assumed that the application provides such random but public "salt" (see [Kra10] for discussions on this issue). Next we show concrete examples of the applicability of this method when the key material is derived from a Diffie-Hellman value (as is common in the settings of key exchange and ElGamal encryption).

We are given (S, q, ε)-wPRF and consider the ratio S/ε as its measure of security (here S is a function of ε and q). Assume that the wPRF has full security, i.e. for a key of size n we have $S/\varepsilon \approx 2^n$. In this case, Theorem 5 guarantees that the extractor F (the KDF in our application) has parameters (S', ε') such that:

$$\varepsilon' \approx \varepsilon \cdot 2^{n-k} \text{ and } S' \approx S \cdot \varepsilon' = 2^n \cdot \varepsilon \cdot \varepsilon' \approx 2^k \varepsilon'^2$$

As a concrete example, consider the case of a wPRF with a 256-bit key and security $S/\varepsilon = 2^{256}$ (this would apply, given current knowledge, to a PRF based on SHA-256, especially that we only consider attacks where the attacker cannot chose any inputs – it only sees the function applied to a set of random values). Assume now that the key, instead of being sampled uniformly at random, follows a distribution with min-entropy $k = 160$; this is the case, for example, when the key material is a Diffie-Hellman value computed over an elliptic curve of size 2^{160} [GKR04]. In this case we have that to distinguish the 256-bit output of the extractor from random with advantage $\varepsilon' \approx 2^{-40}$ we must invest $S' \approx 2^{80}$. If we want to double the advantage ε' we need to invest four times more work (circuit size). For example, to obtain $\varepsilon' = 2^{-20}$ we need to work $S' = 2^{120}$ and for $\varepsilon' = 0.001$ one needs $S' = 2^{140}$. Even if we consider a less-perfect function, say $S/\varepsilon = 2^{200}$ one still gets $S' = 2^{64}$ for $\varepsilon' = 2^{-20}$ and $S' = 2^{84}$ for $\varepsilon' = 0.001$. Note that in all these cases we are outputting more pseudorandom bits (256) than the source entropy (160).

In comparison, if we were applying a *statistical* extractor to the key material of min-entropy 160 to obtain a key of size 256, we could not claim any security at all (this is the case even if we only needed a 160-bit of output, and we would get security of only 2^{-16} if were outputting a 128-bit key). In comparing with statistical extractors another main advantage of our PRF-based computational extractor is the fact that PRFs are already available in practical cryptographic protocols for other uses (including key expansion as often needed in the context of key derivation) and hence do not require of additional mechanisms such as a statistical extractor.

Related Schemes. It is worth noting the *duality* between the above KDF construction and the HKDF scheme from [Kra10]. In our case, the imperfect key material is used to key the (weak) PRF and the seed is used as an input to the KDF. In HKDF these roles are reversed. This gives HKDF the advantage of being appropriate for input distributions of arbitrary length while in our scheme we are limited to the key size. On the other hand, the very non-standard use of a *known value* (the seed) as a key to a PRF in the HKDF scheme, makes the latter much more restricted on the type of PRFs one can use (actually, the known analysis of HKDF is for particular PRFs, mainly HMAC, and under dedicated assumptions). In contrast, our scheme can use any PRF and even any wPRF.

The recent work of Barak et al. [BDK+11] builds a computational extractor in the traditional way, namely, using a statistical extractor to get a close-to-uniform key and using a PRG or PRF to get additional pseudorandom bits as needed. The novelty of that work, however, is that they show that if the output from the statistical extractor (implemented via a suitable hash function) is used as a key to a wPRF and this wPRF is applied to a random point then the best possible distinguishing advantage against the output of this scheme is the wPRF's best distinguishing advantage plus $2^{-(k-m)}$. This is an improvement over the generic analysis using statistical extractors where the latter term would be $2^{-(k-m)/2}$. This relaxes the entropy requirement from the source and is significant in cases as those considered above (e.g. when generating keys from Diffie-Hellman protocols of relative small order). Moreover, depending on the security parameters, the analysis from [BDK+11] can sometimes be used, as in our case, to generate keys that are even larger than the available entropy. The crucial difference with our construction, however, is that [BDK+11] requires the implementation of a statistical extractor (with its corresponding seed) in addition to the wPRF. In contrast, our scheme re-uses the PRF already available in most cryptographic implementations without requiring extra machinery (which may seem a minor issue considering the relative simplicity of statistical extractors but represents a significant barrier for adoption into standardized protocols, particularly those requiring hardware support). On the downside, our scheme is limited to situations where the source of key material produces values that are no longer than the key of the wPRF, while [BDK+11, Kra10] have no such length restrictions.

References

[BDK+11]   Barak, B., Dodis, Y., Krawczyk, H., Pereira, O., Pietrzak, K., Standaert, F.-X., Yu, Y.: Leftover Hash Lemma, Revisited. In: Rogaway, P. (ed.) CRYPTO 2011. LNCS, vol. 6841, pp. 1–20. Springer, Heidelberg (2011)

[DGH+04]   Dodis, Y., Gennaro, R., Håstad, J., Krawczyk, H., Rabin, T.: Randomness Extraction and Key Derivation Using the CBC, Cascade and HMAC Modes. In: Franklin, M. (ed.) CRYPTO 2004. LNCS, vol. 3152, pp. 494–510. Springer, Heidelberg (2004)

[DGKM11]   Dachman-Soled, D., Gennaro, R., Krawczyk, H., Malkin, T.: Computational extractors and pseudorandomness (2011), full version of this paper eprint.iacr.org/2011/708

[GGKT05]   Gennaro, R., Gertner, Y., Katz, J., Trevisan, L.: Bounds on the efficiency of generic cryptographic constructions. SIAM J. Comput. 35(1), 217–246 (2005)

[GKR04]   Gennaro, R., Krawczyk, H., Rabin, T.: Secure Hashed Diffie-Hellman over Non-DDH Groups. In: Cachin, C., Camenisch, J.L. (eds.) EUROCRYPT 2004. LNCS, vol. 3027, pp. 361–381. Springer, Heidelberg (2004)

[Gol90]   Goldreich, O.: A note on computational indistinguishability. Inf. Process. Lett. 34(6), 277–281 (1990)

[HILL99]   Håstad, J., Impagliazzo, R., Levin, L., Luby, M.: A pseudorandom generator from any one-way function. SIAM J. on Computing 28(4) (1999)

[IR89]   Impagliazzo, R., Rudich, S.: Limits on the provable consequences of one-way permutations. In: Proc. 21st Annual ACM Symposium on Theory of Computing (STOC), pp. 44–61 (1989)

[IW97]   Impagliazzo, R., Widgerson, A.: P = bpp unless e has subexponential circuits: derandomizing the xor lemma. In: Proceedings of the Twenty-Ninth Annual Symposium on Theory of Computing, pp. 220–229 (1997)

[KLR09]   Kalai, Y.T., Li, X., Rao, A.: 2-source extractors under computational assumptions and cryptography with defective randomness. In: FOCS, pp. 617–626 (2009)

[Kra10]   Krawczyk, H.: Cryptographic Extraction and Key Derivation: The HKDF Scheme. In: Rabin, T. (ed.) CRYPTO 2010. LNCS, vol. 6223, pp. 631–648. Springer, Heidelberg (2010)

[NW88]   Nisan, N., Wigderson, A.: Hardness vs. randomness. In: Proc. 29th IEEE Symposium on Foundations of Computer Science (FOCS), pp. 2–11. IEEE Computer Society Press (1988)

[NZ96]   Nisan, N., Zuckerman, D.: Randomness is linear in space. J. Comput. Syst. Sci. 52(1), 43–52 (1996)

[Pie09]   Pietrzak, K.: A Leakage-Resilient Mode of Operation. In: Joux, A. (ed.) EUROCRYPT 2009. LNCS, vol. 5479, pp. 462–482. Springer, Heidelberg (2009)

[RTS00]   Radhakrishnan, J., Ta-Shma, A.: Bounds for dispersers, extractors, and depth-two superconcentrators. SIAM J. Discrete Math. 13(1), 2–24 (2000)

[RTV04]   Reingold, O., Trevisan, L., Vadhan, S.P.: Notions of Reducibility between Cryptographic Primitives. In: Naor, M. (ed.) TCC 2004. LNCS, vol. 2951, pp. 1–20. Springer, Heidelberg (2004)

[Sha02]   Shaltiel, R.: Recent developments in explicit constructions of extractors. Bulletin of the EATCS 77, 67–95 (2002)

[Tre01]   Trevisan, L.: Extractors and pseudorandom generators. Journal of the ACM 48(4), 860–879 (2001)

[TV00]   Trevisan, L., Vadhan, S.P.: Extracting randomness from samplable distributions. In: FOCS, pp. 32–42 (2000)

Functional Re-encryption
and Collusion-Resistant Obfuscation

Nishanth Chandran[1,*], Melissa Chase[1], and Vinod Vaikuntanathan[2,**]

[1] Microsoft Research
[2] University of Toronto

Abstract. We introduce a natural cryptographic functionality called *functional re-encryption*. Informally, this functionality, for a public-key encryption scheme and a function F with n possible outputs, transforms ("re-encrypts") an encryption of a message m under an "input public key" pk into an encryption of the same message m under one of the n "output public keys", namely the public key indexed by $F(m)$.

In many settings, one might require that the program implementing the functional re-encryption functionality should reveal nothing about both the input secret key sk as well as the function F. As an example, consider a user Alice who wants her email server to share her incoming mail with one of a set of n recipients according to an access policy specified by her function F, but who wants to keep this access policy private from the server. Furthermore, in this setting, we would ideally obtain an even stronger guarantee: that this information remains hidden even when some of the n recipients may be corrupted.

To formalize these issues, we introduce the notion of *collusion-resistant obfuscation* and define this notion with respect to average-case secure obfuscation (Hohenberger *et al.* - TCC 2007). We then provide a construction of a functional re-encryption scheme for any function F with a polynomial-size domain and show that it satisfies this notion of collusion-resistant obfuscation. We note that collusion-resistant security can be viewed as a special case of dependent auxiliary input security (a setting where virtually no positive results are known), and this notion may be of independent interest.

Finally, we show that collusion-resistant obfuscation of functional re-encryption for a function F gives a way to obfuscate F in the sense of Barak *et al.* (CRYPTO 2001), indicating that this task is impossible for arbitrary (polynomial-time computable) functions F.

1 Introduction

Informally, a program obfuscator is an algorithm that transforms a program into another, functionally equivalent program whose inner workings are "completely unintelligible". Starting from the formalization of program obfuscation in the work of Barak, Goldreich, Impagliazzo, Rudich, Sahai, Vadhan and Yang [3], the problem

---

* Part of this work was done while this author was at UCLA.
** Part of this work was done while this author was at Microsoft Research.

R. Cramer (Ed.): TCC 2012, LNCS 7194, pp. 404–421, 2012.

has received considerable attention in the cryptographic community. A method of obfuscating programs is an exceedingly valuable tool, both in theory and practice.

Despite its potential for far-reaching applications, the area of program obfuscation is wrought with impossibility results. The seminal work of Barak *et al.* [3] demonstrated a class of circuits which cannot be obfuscated even under a weak notion of obfuscation, thereby diminishing the hope of achieving general-purpose obfuscation. Further impossibility results for obfuscation of more natural functionalities were shown in [14,27,18,4]. Positive results for obfuscation, on the other hand, have been largely limited to relatively simple classes of functions such as point functions [7,9,21,27,14,8], proximity testing [12], encrypted permutations [1] and more recently, testing hyperplane membership [10].

In one of the few exceptions to this trend, Hohenberger *et al.* [19] showed how to obfuscate a complex cryptographic functionality called re-encryption [5,2]. Informally, a re-encryption program associated with two public keys transforms an encryption of a message m under the first of these keys to an encryption of the same message m under the second public key. Hohenberger *et al.* (and independently, [18]) also introduce a strong definition of (average-case) secure obfuscation which we will use and build on in this work. Following [19], Hada [17] showed how to securely obfuscate an encrypted signature functionality.

Despite the slow and steady stream of positive results for obfuscation, we have relatively few techniques and paradigms for obfuscation. In particular,

- The key point that enables obfuscation in both [19] and [17] is that they obfuscate functionalities that compute a function "inside a ciphertext". For example, in [19], this is the decryption function and in [17], it is the signature function. Not surprisingly, it has been noted that given a fully homomorphic encryption scheme [22,13], the functionalities of [19,17] can be easily obfuscated. Thus, we would like to *find other paradigms for obfuscating complex functionalities*.
- Both re-encryption and obfuscated signatures can be thought of as access control mechanisms. The catch, though, is that both of them embody an "all-or-nothing" form of access control – for example, in the case of re-encryption, neither the re-encryptor nor the recipient alone can decrypt a ciphertext created by the initiator although together, the two of them can learn the entire contents of the ciphertext. We would like to consider functionalities that capture a *finer grained delegation of access*.
- An issue that is important in both theory and practice is the presence of auxiliary inputs. Most positive results on obfuscation (including [19,17], but also others) do not achieve any form of security against auxiliary inputs that depend on the function being obfuscated. Indeed, this task seems quite hard, as indicated by impossibility results of [14] (for some limited positive results against auxiliary inputs, see [4]). *Can we achieve obfuscation against a large, meaningful class of auxiliary inputs?*

In this work, we make progress on the above lines of inquiry. Firstly, we relax (somewhat) the definition of secure obfuscation in the presence of auxiliary inputs, and introduce the notion of *collusion resistant obfuscation*. Secondly, we show how to obfuscate a natural and complex cryptographic functionality called *functional re-encryption* in a way that satisfies this notion of security. This functionality captures a finer grained delegation of access, and also protects against collusion between various participating parties.

1.1 Collusion Resistant Obfuscation

Consider the following scenario. A department would like to create a login program that will grant access to several users - say, Alice, Bob, and Carol, who have different passwords. The department would like to obfuscate this program and give it to the server that will run it. Now, we would like to guarantee that this obfuscation remains secure even if, for example, Alice were to collude with the server. One can view Alice's password as being specific auxiliary information that an adversary obtains about the program. Note that this is a restricted form of auxiliary information as we do not allow an adversary to learn, say specific bits of Bob or Carol's passwords. In this work, we are interested in the notion of average-case secure obfuscation (as defined by [19,18]) and hence in the above example we assume that all passwords are chosen uniformly at random.

One can generalize the above functionality and obtain a general definition of collusion resistant obfuscation. We would like to obfuscate a function family $\{\mathcal{C}_\lambda\}$ that has the following form. Any $C_{\mathcal{K}} \in \mathcal{C}_\lambda$ is parameterized by a set of "secret" keys $\mathcal{K} = \{k_1, k_2, \cdots, k_\ell\}$ (in addition to any other parameters that the circuit might take) that are chosen at random from some specified distribution. Now, define a subset of keys represented through a set of indices $\mathcal{T} \subseteq [\ell]$. ($[\ell]$ denotes the set $\{1, 2, \cdots, \ell\}$.) We would like to construct an obfuscation of the circuit, denoted by $\mathsf{Obf}(C_{\mathcal{K}})$, so that $\mathsf{Obf}(C_{\mathcal{K}})$ is a "secure obfuscation" of $C_{\mathcal{K}}$ (in the sense of [19]) even against an adversary that knows the set of keys $\{k_i\}_{i \in \mathcal{T}}$.

1.2 Functional Re-encryption

Functional re-encryption is an expressive generalization of re-encryption [5,2]. A functional re-encryption functionality is parameterized by a policy function $F : \mathcal{D} \to [n]$ (i.e, F has domain $\mathcal{D}$ and has n possible outputs) chosen from some class of functions, an input public key pk, and n output public keys. The functionality receives as input a ciphertext of message m with "identity" id under the input public key pk.[1] It decrypts the ciphertext using the secret key sk to get m and id, and then re-encrypts m under the "appropriate" output public key $\widehat{\mathsf{pk}}_{F(\mathsf{id})}$. Following our desiderata from before, one could think of functional re-encryption as a form of fine-grained delegation of access.

To motivate the functional re-encryption functionality, consider the following scenario: Alice wishes to have her e-mail server "route" her incoming mail to one of a set of n recipients. The particular recipient to which the ciphertext should be routed depends on both the contents of the ciphertext – essentially,

---

[1] This is a slight generalization of the description given earlier in the abstract where the function F is applied to the entire message. We choose to view the message as an identity on which the function F is applied, and a separate "payload" for conceptual cleanliness.

the identity id – as well as Alice's access policy encoded by her function F. The e-mail server does this by "re-encrypting" the contents of the ciphertext under the appropriate public key. The minimal requirement from such a system is that the "re-encryption mechanism" hide both the message and Alice's access policy – it should merely provide a means for the server to do the appropriate routing. [2]

One (not particularly appealing) way for Alice to do this would be to give the e-mail server her secret key and her access policy; this lets the server decrypt all incoming messages and determine where to route them. Unfortunately, this "solution" completely fails this minimum requirement. Ideally, Alice would like to "obfuscate" the trivial functional re-encryption program above and give it to the server. We show how to *securely obfuscate* functional re-encryption which, informally speaking, guarantees that any "attack" that the server can carry out given the obfuscated functional re-encryption program, could also be carried out given only oracle access to the functional re-encryption program (which is no power at all!).

Furthermore, in reality we could reasonably expect the server to collude with some of the recipients to learn additional information about messages or about Alice's access policy function F. Clearly, collusion helps the server – he can use a recipient's decryption key together with the re-encryption program to learn the output of F on certain inputs. If we consider the auxiliary input to be the secret keys of the colluding recipients, then our strong notion of collusion-resistant secure obfuscation guarantees that this is the only information that the server could possibly learn by colluding.

Selectively delegating access is also the central theme of a recently introduced notion of predicate encryption [20,25] (which can be viewed as attribute based encryption in which ciphertexts hide their attributes). In fact, (predicate-hiding, public key) predicate encryption schemes can potentially be used to solve Alice's dilemma. This is done by completely ignoring the email server and giving each of the recipients a "little secret key" that is just powerful enough to decrypt the appropriate ciphertexts (dictated by the access policy). Aside from the fact that there are no known public-key predicate hiding encryption schemes (nor even good definitions of them), this solution has two drawbacks – first, there is no way to revoke access from a recipient other than by having Alice choose a fresh key for herself (which could be quite expensive). Second, this solution requires all recipients to be aware of the existence of an access policy, while the solution based on functional re-encryption is completely invisible to the recipients – they continue using their already registered public keys, and they do not even have to know of the existence of the functional re-encryption mechanism.

1.3 Overview of Results and Techniques

Collusion resistant obfuscation. We define the notion of collusion resistant obfuscation which guarantees security against a natural form of auxiliary inputs. This

---

[2] Of course, since the e-mail server does not know who the recipient is, it either sends the resulting ciphertext to all the recipients or publishes it on a bulletin board from which the intended recipient can then access it.

notion of auxiliary input security might be realizable (without random oracles) for many common cryptographic tasks.

Functional Re-encryption. We show, informally:

Theorem 1 (Informal). *Under the Symmetric External Diffie-Hellman assumption there exists an encryption scheme such that for any function $F : \mathcal{D} \to \mathcal{R}$ with polynomial-sized domain $\mathcal{D}$, there is a collusion-resistant average-case secure obfuscation of the functional re-encryption program w.r.t. F. The size of the input ciphertext in the encryption scheme is $O(|\mathcal{D}| \cdot \mathsf{poly}(\lambda))$, and the size of the output ciphertext is $O(\mathsf{poly}(\lambda))$ (i.e., independent of the domain and the range of F).*

We now present the ideas behind our construction at a very high level. One can think of a functional re-encryption program as a program that must achieve two goals - a) it must "hide" the policy function F, and b) it must also "hide" the input secret key (that it uses to decrypt the input ciphertext). These two goals must simultaneously be achieved while maintaining the right functionality. Informally, the main innovation in our work is a technique to hide the policy function - this combined with techniques from [19] allows us to achieve both goals simultaneously. We shall now describe this first technique in more detail.

Let $\mathbb{G}, \mathbb{H}, \mathbb{G}_T$ be groups such that there is a bilinear map $\mathsf{e} : \mathbb{G} \times \mathbb{H} \to \mathbb{G}_T$. Let $\boldsymbol{a}_1, \cdots, \boldsymbol{a}_d \in \mathbb{Z}_q^d$ be vectors that denote elements in the domain $\mathcal{D}$ of function F and let $\hat{a}_1, \cdots, \hat{a}_n \in \mathbb{Z}_q$ denote elements in the range $\mathcal{R}$ of F. Now consider a function OF that maps elements in $\mathbb{G}^d$ to elements in $\mathbb{G}_T$ in the following way. OF is parameterized by random generators $g \in \mathbb{G}$ and $h \in \mathbb{H}$. On input $g^{\boldsymbol{a}_i}$, OF outputs $\mathsf{e}(g, h)^{\hat{a}_{F(i)}}$. We shall now informally sketch how to publish a program that achieves the functionality provided by OF, but at the same time hides F.

The program computes a vector $\boldsymbol{\alpha} \in \mathbb{Z}_q^d$ such that the inner product $\langle \boldsymbol{a}_i, \boldsymbol{\alpha} \rangle = \hat{a}_{F(i)}$ for all i. Note that this is indeed possible as $\boldsymbol{\alpha}$ is a solution to a system of d equations in d variables. The program description simply contains $h^{\boldsymbol{\alpha}}$. (This can be computed given only $h^{\hat{a}_{F(i)}}$ and $\boldsymbol{a}_i$ for all i, so we do not actually need the recipient secret keys $\hat{a}_{F(i)}$.) On input $g^{\boldsymbol{a}_i}$, the program computes and outputs $\prod_{j=1}^{d} \mathsf{e}(g^{a_{ij}}, h^{\alpha_j}) = \mathsf{e}(g, h)^{\langle \boldsymbol{a}_i, \boldsymbol{\alpha} \rangle} = \mathsf{e}(g, h)^{\hat{a}_{F(i)}}$, which is the output as desired.

Unfortunately, this solution does not completely hide the function. Note that if $F(1) = F(2)$ (say), then an adversary can learn this by simply running the above program and checking if the output is the same on both the inputs. To get around this problem, we modify the program in the following way. The program picks random w_i, for all i, and computes two vectors $\boldsymbol{\alpha}, \boldsymbol{\beta} \in \mathbb{Z}_q^d$ such that the inner product $\langle \boldsymbol{a}_i, \boldsymbol{\alpha} \rangle = w_i \hat{a}_{F(i)}$ and $\langle \boldsymbol{a}_i, \boldsymbol{\beta} \rangle = w_i$, for all i (in our actual solution we require the R.H.S of the second equation to be $w_i - 1$ instead of w_i, but we will ignore that for now). The program description now contains $h^{\boldsymbol{\alpha}}, h^{\boldsymbol{\beta}}$. On input $g^{\boldsymbol{a}_i}$, the program computes and outputs $\prod_{j=1}^{d} \mathsf{e}(g^{a_{ij}}, h^{\alpha_j}) = \mathsf{e}(g, h)^{w_i \hat{a}_{F(i)}}$, as well as $\prod_{j=1}^{d} \mathsf{e}(g^{a_{ij}}, h^{\beta_j}) = \mathsf{e}(g, h)^{w_i}$. Now, on two different inputs (of F) that have the same output, the above program outputs elements of the form $(\mathsf{e}(g, h)^{xa}, \mathsf{e}(g, h)^x)$ and $(\mathsf{e}(g, h)^{ya}, \mathsf{e}(g, h)^y)$, for random a, x and y. However, these tuples are indistinguishable from random, even given $\mathsf{e}(g, h)$ and

$e(g, h)^a$, (by DDH) and hence an adversary cannot tell if $F(1) = F(2)$. This construction now ensures that F is completely hidden.

Now, note that if we let $\{g^{a_i}\}$, $1 \le i \le d$ be the input public key and $e(g, h)^{\hat{a}_j}$ be the output public key, then one can potentially use the above construction to build a scheme that converts an encryption of message m under g^{a_i} to one under $e(g, h)^{\hat{a}_{F(i)}}$. This is precisely what we do. Our encryption schemes are ElGamal-like, the input encryption key contains a set of vectors $g^{a_i}, \cdots, g^{a_d}$, and an input encryption of message m with identity i uses the key g^{a_i}. Finally, in order to obtain a secure obfuscation, we apply techniques from [19] to re-randomize the ciphertexts.

Obfuscating Functional Re-encryption for Arbitrary Policy Functions? A natural question raised by our result is whether it is possible to achieve collusion-resistant obfuscation of functional re-encryption for *arbitrary* (polynomial-time computable) policy functions F (in particular, functions F with domains of super-polynomial size). We show that this goal is impossible to achieve. In particular, we show that a collusion-resistant obfuscation with respect to a policy function F already contains within it a [3]-style obfuscation (a so-called "predicate obfuscation") of the policy function F. In some sense, this is not entirely surprising, and corresponds to the intuition that a collusion-resistant obfuscation of functional re-encryption allows computation of the function F [3], and yet hides all internal details of F except the input-output behavior. Together with the impossibility result of [3] for obfuscating general (families of) functions, this shows that there are classes of (polynomial-time computable) policy functions for which it is impossible to construct collusion-resistant secure obfuscation of functional re-encryption. See the full version of the paper [11] for a formal statement and proof of this result. The next question to ask is whether there is any non-trivial policy function (with a domain of super-polynomial size) for which this goal can be achieved. We informally argue that this may require some new innovation on the question of constructing *public-key* predicate encryption schemes which satisfy a strong security notion called *predicate-hiding*. Predicate encryption schemes were defined by Katz, Sahai and Waters [20], following [23,15] (in particular, the predicate-hiding property was defined in the work of Shi, Shen and Waters [25]). Constructions of predicate encryption schemes (even ones that do not achieve predicate-hiding) are known only for simple classes of functions such as inner products [20]. Moreover, in the public-key setting, we do not know how to achieve (any reasonable definition of) *predicate-hiding*, even for simple functions. Since collusion-resistant obfuscation of functional re-encryption seems to have the same flavor in functionality as predicate-hiding public-key predicate encryption, advancements in the class of policy functions that these primitives can handle seem to be correlated.

---

[3] A collusion-resistant obfuscation of functional re-encryption allows computation of the function F since given an output secret key $\widehat{sk}_i$ and the re-encryption program, one can test if $F(\mathsf{id}) = i$ for any id in the domain of F. Simply encrypt a random message with identity id, run it through the re-encryption program and decrypt it using $\widehat{sk}_i$. If this returns the same message that was encrypted, then conclude that $F(\mathsf{id}) = i$.

2 Collusion Resistant Secure Obfuscation

2.1 Average-Case Secure Obfuscation

Throughout this paper, we will implicitly assume that the adversary (as well as simulator) can obtain arbitrary polynomial-size independent auxiliary input z. We remark that our construction is secure even against the presence of such auxiliary information. We now recall the notion of average-case secure obfuscation introduced in [19] below.

Definition 1. *An efficient algorithm* Obf *that takes as input a (probabilistic) circuit C from the family $\{\mathcal{C}_\lambda\}$ and outputs a new (probabilistic) circuit, is an average-case secure obfuscator, if it satisfies the following properties:*

- Preserving functionality: *With overwhelming probability* Obf(C) *behaves "almost identically" to C on all inputs. Formally, there exists a negligible function* neg(λ), *such that for any input length λ and any $C \in \mathcal{C}_\lambda$:*

$$\Pr_{coins\ of\ \mathtt{Obf}}[\exists x \in \{0,1\}^\lambda : C' \leftarrow \mathtt{Obf}(C);\ \mathtt{SD}(C'(x), C(x)) \geq \mathtt{neg}(\lambda)] \leq \mathtt{neg}(\lambda)$$

  where SD$(\mathcal{X}, \mathcal{Y})$ *denotes the statistical distance between two distributions $\mathcal{X}$ and $\mathcal{Y}$.*
- Polynomial slowdown: *There exists a polynomial $p(\lambda)$ such that for sufficiently large input lengths λ, for any $C \in \mathcal{C}_\lambda$, the obfuscator* Obf *only enlarges C by a factor of p. That is, $|\mathtt{Obf}(C)| \leq p(|C|)$.*
- Average-case Virtual Black-Boxness: *There exists an efficient simulator $\mathcal{S}$ and a negligible function* neg(λ), *such that for every efficient distinguisher* D, *and for every input length λ:*

$$|\Pr[C \leftarrow \mathcal{C}_\lambda : \mathsf{D}^C(\mathtt{Obf}(C)) = 1] - \Pr[C \leftarrow \mathcal{C}_\lambda : \mathsf{D}^C(\mathcal{S}^C(1^\lambda)) = 1]| \leq \mathtt{neg}(\lambda)$$

  The probability is over the selection of a random circuit C from $\mathcal{C}_\lambda$, and the coins of the distinguisher, the simulator, the oracle, and the obfuscator. [4]

2.2 Average-Case Secure Obfuscation with Collusion

Consider the case where we would like to obfuscate a function family $\{\mathcal{C}_\lambda\}$ that has the following particular form. Any $C_\mathcal{K} \in \mathcal{C}_\lambda$ is parameterized by a set of "secret" keys $\mathcal{K} = \{k_1, k_2, \cdots, k_\ell\}$ (in addition to any other parameters that the circuit might take) that are chosen at random from some specified distribution. Now, define a (non-adaptively chosen) subset of keys represented through a set of indices $\mathcal{T} \subseteq [\ell]$, where $[\ell]$ denotes the set $\{1, 2, \cdots, \ell\}$. We would like to construct an obfuscation of the circuit, denoted by Obf$(C_\mathcal{K})$, so that Obf$(C_\mathcal{K})$ is a "secure obfuscation" of $C_\mathcal{K}$ (in the sense of [19]) even against an adversary that knows the set of keys $\{k_i\}_{i \in \mathcal{T}}$.

We accomplish this using a definition that is similar in spirit to the notion of obfuscation against *dependent auxiliary inputs* [14]. More precisely, in addition

---

[4] This is the definition in [19] but with a dummy adversary. The authors of that paper note that this is equivalent to the definition they give.

to their usual inputs and oracles, we give both the adversary and the simulator access to a (non-adaptively chosen) subset $\{k_i\}_{i \in \mathcal{T}} \subseteq \mathcal{K}$ of the keys. This can be seen as auxiliary information about the circuit $C_{\mathcal{K}} \leftarrow C_\lambda$. The formal definition of collusion-resistant secure obfuscation is as follows.

Definition 2. *An efficient algorithm* Obf *that takes as input a (probabilistic) circuit and outputs a new (probabilistic) circuit, is a* collusion-resistant (average-case) secure obfuscator *for the family* $\{C_\lambda\}$ *if it satisfies the following properties:*

- *"Preserving functionality" and "Polynomial Slowdown", as in Definition 1.*
- *Average-case Virtual Black-Boxness against Collusion: There exists an efficient simulator* $\mathcal{S}$, *and a negligible function* $\mathsf{neg}(\lambda)$, *such that for every input length* λ, *every efficient distinguisher* D, *and any subset* $\mathcal{T} \subseteq [\ell]$:

$$\Big| \Pr[C_{\mathcal{K}} \leftarrow C_\lambda : \mathsf{D}^{C_{\mathcal{K}}}(\mathtt{Obf}(C_{\mathcal{K}}), \{k_i\}_{i \in \mathcal{T}}) = 1] -$$
$$\Pr[C_{\mathcal{K}} \leftarrow C_\lambda : \mathsf{D}^{C_{\mathcal{K}}}(\mathcal{S}^{C_{\mathcal{K}}}(1^\lambda, \{k_i\}_{i \in \mathcal{T}}), \{k_i\}_{i \in \mathcal{T}}) = 1] \Big| \leq \mathsf{neg}(\lambda)$$

The probability is over the selection of a random *circuit* $C_{\mathcal{K}}$ *from* C_λ, *and the coins of the distinguisher, the simulator, the oracle, and the obfuscator.*

Remarks on the Definition. An even stronger attack model allows the adversary to obtain an obfuscation of a circuit $C_{\mathcal{K}}$ where some of the keys in $\{k_i\}_{i \in \mathcal{T}}$ are adversarially chosen. Furthermore, one could allow the adversary to select the set $\mathcal{T}$ adaptively, after seeing the public keys and/or the obfuscated program. We postpone a full treatment of these issues to future work.

2.3   Securely Obfuscating Functional Re-encryption

We would like to obtain a collusion-resistant average-case obfuscator for the functional re-encryption functionality. A Functional Re-encryption (FR) functionality associated to function $F : \mathcal{D} \to \mathcal{R}$, input public/secret key pair (pk, sk), and output public keys $\widehat{\mathsf{pk}}_1, \ldots, \widehat{\mathsf{pk}}_{|\mathcal{R}|}$ [5] is a functionality that takes as input a ciphertext $c = \mathtt{I\text{-}Enc}(\mathsf{pk}, \mathsf{id}, m)$ and re-encrypts m under the output public key $\widehat{\mathsf{pk}}_{F(\mathsf{id})}$. More precisely, for a given function $F : \mathcal{D} \to \mathcal{R}$, we are interested in the family of circuits $\mathcal{FR}_{F, \mathcal{D}, \mathcal{R}} = \{\mathsf{FR}_{\lambda, F, \mathcal{D}, \mathcal{R}}\}_{\lambda > 0}$ where each circuit $C_{\mathsf{pk}, \mathsf{sk}, \widehat{\mathsf{pk}}_1, \ldots, \widehat{\mathsf{pk}}_{|\mathcal{R}|}} \in \mathsf{FR}_{\lambda, F, \mathcal{D}, \mathcal{R}}$ is a *probabilistic circuit* indexed by a key pair $(\mathsf{pk}, \mathsf{sk}) \leftarrow \mathtt{I\text{-}Gen}(1^\lambda)$, and public keys $(\widehat{\mathsf{pk}}_i, \star) \leftarrow \mathtt{O\text{-}Gen}(1^\lambda)$, and works as follows:

$C_{\mathsf{pk}, \mathsf{sk}, \widehat{\mathsf{pk}}_1, \ldots, \widehat{\mathsf{pk}}_{|\mathcal{R}|}}$, on input c :
    Computes $(\mathsf{id}, m) \leftarrow \mathtt{I\text{-}Dec}(\mathsf{sk}, c)$, and outputs $\widehat{c} \leftarrow \mathtt{O\text{-}Enc}(\widehat{\mathsf{pk}}_{F(\mathsf{id})}, m)$.
    If $\mathtt{I\text{-}Dec}(\mathsf{sk}, c)$ returns $\bot$ then outputs random elements in the format of $\widehat{c}$.
$C_{\mathsf{pk}, \mathsf{sk}, \widehat{\mathsf{pk}}_1, \ldots, \widehat{\mathsf{pk}}_{|\mathcal{R}|}}$, on a special input keys:
    Outputs $\mathsf{pk}, \widehat{\mathsf{pk}}_1, \ldots, \widehat{\mathsf{pk}}_{|\mathcal{R}|}$.

---

[5] Without loss of generality, and for simplicity of notation, throughout the paper we will often assume that the domain $\mathcal{D} = \{1, 2, \ldots, d\}$ and the range $\mathcal{R} = \{1, 2, \ldots, n\}$.

Now, for a class of functions, $\mathcal{F}$, we say that a re-encryption program *securely obfuscates re-encryption for* $\mathcal{F}$, if there exists a simulator $\mathcal{S}$ that satisfies the collusion resistant obfuscation property w.r.t. $\mathcal{FR}_{F,\mathcal{D},\mathcal{R}}$ for all $F : \mathcal{D} \to \mathcal{R} \in \mathcal{F}$.

In other words, all public keys in the circuit $C_{\mathsf{pk},\mathsf{sk},\widehat{\mathsf{pk}}_1,\ldots,\widehat{\mathsf{pk}}_{|\mathcal{R}|}}$ are considered public knowledge; the only pieces of information we are interested in protecting are the input secret key sk and the function F. Also, note that we are interested in guaranteeing security for arbitrarily chosen F, not F chosen at random.

The set of secret keys that will parameterize a functional re-encryption functionality is $\mathcal{K} = \{\widehat{\mathsf{sk}}_1, \cdots, \widehat{\mathsf{sk}}_{|\mathcal{R}|}\}$. The definition of collusion-resistant average-case secure obfuscation guarantees security against an adversary who not only knows the re-encryption program, but also has access to a subset $\{\widehat{\mathsf{sk}}_i\}_{i \in \mathcal{T}} \subseteq \mathcal{K}$ of the output secret keys. This scenario endows the adversary with considerable power and knowledge. For instance,

- The adversary will inevitably be able to decrypt all ciphertexts $c = \mathtt{I\text{-}Enc}(\mathsf{pk}, \mathsf{id}, m)$, where $F(\mathsf{id}) \in \mathcal{T}$, simply by using the re-encryption program to convert the ciphertext c into an encryption of m under the output public key $\widehat{\mathsf{pk}}_{F(\mathsf{id})}$, and then decrypting it using $\widehat{\mathsf{sk}}_{F(\mathsf{id})}$.
- Moreover, the power to selectively decrypt a subset of the input ciphertexts gives the adversary information about the access policy function F itself. For instance, the adversary can determine if $F(\mathsf{id}) = i$ whenever $i \in \mathcal{T}$.
- Finally, we remark that the definition of obfuscation for functional re-encryption by itself does not guarantee the semantic security of the input and output encryption schemes. We define these separately and prove the security of the encryption schemes (even in the presence of the re-encryption program). In more detail, we will require the semantic security of the input encryption scheme, on messages encrypted with an identity id^*, whenever $F(\mathsf{id}^*) \notin \mathcal{T}$, even when the adversary is given access to a re-encryption oracle. We will similarly require that the input ciphertext hides the identity id^*, under which the message is encrypted. The security of the output encryption scheme will be that of standard semantic security. Since we wish to hide everything about the function F, we will also require the output encryption scheme to be key private; i.e., an encryption under public key $\widehat{\mathsf{pk}}_i$ will be indistinguishable from an encryption under public key $\widehat{\mathsf{pk}}_j$. For formal definitions and proofs of these properties, see the full version [11].

3   Preliminaries

We let λ be the security parameter throughout this paper. By $\mathtt{neg}(\lambda)$ we denote some *negligible* function, namely a function μ such that for all $c > 0$ and all sufficiently large λ, $\mu(\lambda) < 1/\lambda^c$. For two distributions $\mathcal{D}_1$ and $\mathcal{D}_2$, $\mathcal{D}_1 \overset{c}{\approx} \mathcal{D}_2$ means that they are computationally indistinguishable (to be precise, this statement holds for *ensembles* of distributions).

We let $[\ell]$ denote the set $\{1, \cdots, \ell\}$. We denote vectors by bold-face letters, e.g., $\boldsymbol{a}$. Let $\mathbb{G}$ be a group of prime order q. For a vector $\boldsymbol{a} = (a_1, a_2, \cdots, a_\ell) \in \mathbb{Z}_q^\ell$

and group element $g \in \mathbb{G}$, we write g^a to mean the vector $(g^{a_1}, g^{a_2}, \cdots, g^{a_\ell})$. For two vectors a and b where a and b are either both in $\mathbb{Z}_q^\ell$ or both in $\mathbb{G}^\ell$, we write ab to denote their component-wise product and a/b to denote their component-wise division. In case $b \in \mathbb{Z}_q^\ell$, we let a^b denote their component-wise exponentiation. For a vector a and scalar x, $xa = ab, a/x = a/b$, and $a^x = a^b$, where $b = (x, x, \cdots, x)$ of dimension ℓ.

Assumptions. We assume the existence of families of groups $\{\mathbb{G}^{(\lambda)}\}_{\lambda > 0}$, $\{\mathbb{H}^{(\lambda)}\}_{\lambda > 0}$ and $\{\mathbb{G}_T^{(\lambda)}\}_{\lambda > 0}$ with prime order $q = q(\lambda)$, endowed with a bilinear map $\mathsf{e}_\lambda : \mathbb{G}^{(\lambda)} \times \mathbb{H}^{(\lambda)} \to \mathbb{G}_T^{(\lambda)}$. When clear from the context, we omit the superscript that refers to the security parameter from all these quantities. The mapping is efficiently computable, and is bilinear – namely, for any generators $g \in \mathbb{G}$ and $h \in \mathbb{H}$, and $a, b \in \mathbb{Z}_q$, $\mathsf{e}(g^a, h^b) = \mathsf{e}(g, h)^{ab}$. We also require the bilinear map to be non-degenerate, in the sense that if $g \in \mathbb{G}, h \in \mathbb{H}$ generate $\mathbb{G}$ and $\mathbb{H}$ respectively, then $\mathsf{e}(g, h) \neq 1$.

We assume the *Symmetric External Diffie-Hellman Assumption* (SXDH)), which says that the decisional Diffie-Hellman (DDH) problem is hard in both of the groups $\mathbb{G}$ and $\mathbb{H}$. That is, when $(q, \mathbb{G}, \mathbb{H}, \mathbb{G}_T, \mathsf{e}) \leftarrow \mathsf{BilinSetup}(1^\lambda); g \leftarrow \mathbb{G}; a, b, c \leftarrow \mathbb{Z}_q$, the following two ensembles are indistinguishable:

$$\{(q, \mathbb{G}, \mathbb{H}, \mathbb{G}_T, \mathsf{e}, g, g^a, g^b, g^{ab})\} \stackrel{c}{\approx} \{(q, \mathbb{G}, \mathbb{H}, \mathbb{G}_T, \mathsf{e}, g, g^a, g^b, g^c)\}$$

and a similar statement when $g \in \mathbb{G}$ is replaced with $h \in \mathbb{H}$. In contrast, the assumption that DDH is hard in one of the two groups $\mathbb{G}$ or $\mathbb{H}$ is simply called the external Diffie-Hellman assumption (XDH). These assumptions were first proposed and used in various works, including [26,6,24,16]. In this work, we use the SXDH assumption.

4   Collusion-Resistant Functional Re-encryption

We are now ready to present our construction of a functional re-encryption scheme from the symmetric external Diffie-Hellman (SXDH) assumption. We first construct our basic encryption schemes in Section 4.1. In Section 4.2, we describe a program that implements the functional re-encryption scheme. Finally, in Section 4.3, we prove that our functional re-encryption program satisfies the notion of collusion-resistant average-case secure obfuscation.

4.1   Construction of the Encryption Schemes

A functional re-encryption scheme transforms a ciphertext under an *input public key* into a ciphertext of the same message under one of many *output public keys*. In our construction, the input and the output ciphertexts have different shapes – namely, the input ciphertext lives in the "source group" $\mathbb{G}$ whereas the output ciphertext lives in the "target group" $\mathbb{G}_T$. We now proceed to describe our input and output encryption schemes which are both variants of the ElGamal encryption scheme.

Parameters. The public parameters for both the input and the output encryption scheme consist of the description of three groups $\mathbb{G}$, $\mathbb{H}$ and $\mathbb{G}_T$ of prime order $q = q(\lambda)$, with a bilinear map $e : \mathbb{G} \times \mathbb{H} \to \mathbb{G}_T$. Also included in the public parameters are two generators – $g \in \mathbb{G}$ and $h \in \mathbb{H}$. Let $\mathcal{M} = \mathcal{M}(\lambda) \subseteq \mathbb{G}$ denote the message space of both the input and output encryption schemes. We assume that $|\mathcal{M}|$ is polynomial in λ. The construction of our output encryption scheme requires this to be the case; however, one can encrypt longer messages by breaking the message into smaller blocks and encrypting the blocks separately.

The Input Encryption Scheme. We first construct the input encryption scheme, which is parameterized by $d = d(\lambda)$ which is an upper bound on the size of the domain of the policy function that we intend to support. We will also use a NIZK proof system; we note that [16] provides an efficient scheme for the type of statements we use, which is perfectly sound and computationally zero-knowledge based on SXDH. We remark that, while the semantic security of the input encryption scheme does not require this NIZK proof, the obfuscation guarantee provided by our construction relies on it; if, for example the adversary were to provide an invalid ciphertext as input to the re-encryption program (e.g. by combining 2 valid ciphertexts with different i's), the program might output some group elements that are distinguishable from random to an adversary that possesses some of the recipient secret keys.

The input encryption scheme is as follows:

1. I-Gen($1^\lambda, 1^d$): Pick random vectors $\boldsymbol{a}_1, \cdots, \boldsymbol{a}_d$ from $\mathbb{Z}_q^d$ that are linearly independent. We also generate crs, a common reference string (abbreviated CRS) for the NIZK proof system. Output pk $= (\text{crs}, g, g^{\boldsymbol{a}_1}, \cdots, g^{\boldsymbol{a}_d})$, and sk $= (\boldsymbol{a}_1, \cdots, \boldsymbol{a}_d)$. We remark that the public key pk can be viewed as being made up of d public keys $\text{pk}_i = (g, g^{\boldsymbol{a}_i})$ of a simpler scheme.

2. I-Enc($\text{pk}, i \in [d], m$): To encrypt a message $m \in \mathcal{M}$, with "identity" $i \in [d]$, choose random exponents r and r' from $\mathbb{Z}_q$, and compute:
   (a) $\mathbf{C} = g^{r\boldsymbol{a}_i}$; $D = g^r m$, and
   (b) $\mathbf{C}' = g^{r'\boldsymbol{a}_i}$; $D' = g^{r'}$
   (c) π, a proof that these values are correctly formed, i.e. that they correspond to one of the vectors $g^{\boldsymbol{a}_i}$ contained in the public key.

   Output the ciphertext $(\mathbf{E}, \mathbf{E}', \pi)$ where $\mathbf{E} = (\mathbf{C}, D)$ and $\mathbf{E}' = (\mathbf{C}', D')$. (Looking ahead, we remark that $\mathbf{E}$ looks like an encryption of message m under pk_i, while $\mathbf{E}'$ looks like an encryption of $1_\mathbb{G}$ under pk_i. $\mathbf{E}'$ is primarily used by the re-encryption program for input re-randomization, and is not required if the encryption scheme is used stand-alone without the functional re-encryption program.)

3. I-Dec($\text{sk}, (\mathbf{E}, \mathbf{E}')$): If any of the components of the ciphertext $\mathbf{E}'$ is $1_\mathbb{G}$ or if the proof π does not verify, output $\perp$.[6] Ignore $\mathbf{E}', \pi$ subsequently, and parse $\mathbf{E}$ as $(\mathbf{C}, D)$. Check that for some $i \in [d]$ and $m \in \mathcal{M}$, $D \cdot (\mathbf{C}^{1/a_i})^{-1} = (m, \cdots, m)$. If yes, output (i, m). Otherwise output $\perp$.

---

[6] This "sanity check" is to ensure the correctness of the re-encryption program. Note that if $(\mathbf{E}, \mathbf{E}')$ is honestly generated, this event happens only with negligible probability.

The Output Encryption scheme. We now describe the output encryption scheme.

1. $\mathsf{O\text{-}Gen}(1^\lambda)$: Pick $\hat{a} \leftarrow \mathbb{Z}_q$. Let $\widehat{\mathsf{pk}} = h^{\hat{a}}$ and $\widehat{\mathsf{sk}} = \hat{a}$.
2. $\mathsf{O\text{-}Enc}(\widehat{\mathsf{pk}}, m)$: To encrypt a message $m \in \mathcal{M} \subset \mathbb{G}$,
   - Choose random number $r \leftarrow \mathbb{Z}_q$.
   - Compute $\widehat{Y} = (h^{\hat{a}})^r$ and $\widehat{W} = h^r$.
   - Output the ciphertext as $[\widehat{F}, \widehat{G}] := [\mathsf{e}(g, \widehat{Y}), \, \mathsf{e}(g, \widehat{W}) \cdot \mathsf{e}(m, h)]$.
3. $\mathsf{O\text{-}Dec}(\widehat{\mathsf{sk}} = \hat{a}, (\widehat{F}, \widehat{G}))$: The decryption algorithm does the following:
   - Compute $\widehat{Q} = \widehat{G} \cdot \widehat{F}^{-1/\hat{a}}$.
   - For each $m \in \mathcal{M}$, test if $\mathsf{e}(m, h) = \widehat{Q}$. If so, output m and halt. (Note that if $\mathsf{e}(m, h)$ are precomputed for all $m \in \mathcal{M}$, then this step can be implemented with a table lookup.)

4.2   Obfuscation for Functional Re-encryption

We now describe our scheme for securely obfuscating the functional re-encryption functionality for the input and output encryption schemes described above.

The Functional Re-encryption Key. The obfuscator gets an input *secret key* sk, the n output public keys $\widehat{\mathsf{pk}}_i$, and the description of a function $F : [d] \to [n]$. It outputs a functional re-encryption key which is a description of a program that takes as input a ciphertext of message $m \in \mathcal{M}$ and identity $i \in [d]$ under public key pk, and outputs a ciphertext of m under $\widehat{\mathsf{pk}}_{F(i)}$.

The obfuscator does the following:

1. Pick $w_i \leftarrow \mathbb{Z}_q$ for all $i \in [d]$ uniformly at random.
2. Solve for $\boldsymbol{\alpha} = (\alpha_1, \ldots, \alpha_d)$ and $\boldsymbol{\beta} = (\beta_1, \ldots, \beta_d)$ such that for all $i \in [d]$:

$$\langle \boldsymbol{a}_i, \boldsymbol{\alpha} \rangle \;=\; w_i \cdot \hat{a}_{F(i)} \qquad \text{and} \qquad \langle \boldsymbol{a}_i, \boldsymbol{\beta} \rangle \;=\; w_i - 1$$

The re-encryption key consists of the tuple $(\mathbf{A}, \mathbf{B})$ where $\mathbf{A} = h^{\boldsymbol{\alpha}}$ and $\mathbf{B} = h^{\boldsymbol{\beta}}$. We remark that computing the re-encryption key does not require knowledge of the output secret keys. To compute $h^{\boldsymbol{\alpha}}$, one can take the output public keys $h^{\hat{a}_1}, \cdots, h^{\hat{a}_d}$, and with the knowledge of the input secret keys $\{\boldsymbol{a}_1, \cdots, \boldsymbol{a}_d\}$ and random values $w_1, \ldots, w_d$, one can solve the set of equations $h^{\langle \boldsymbol{a}_i, \boldsymbol{\alpha} \rangle} = h^{w_i \cdot \hat{a}_{F(i)}}$, for all $i \in [d]$, to obtain $h^{\boldsymbol{\alpha}}$. $h^{\boldsymbol{\beta}}$ can be computed in a similar manner.

The Functional Re-encryption Program. Given the functional re-encryption key $(\mathbf{A}, \mathbf{B})$ and an input ciphertext $(\mathbf{E}, \mathbf{E}')$ where $\mathbf{E} = (\mathbf{C}, D)$ and $\mathbf{E}' = (\mathbf{C}', D')$, the functional re-encryption program performs the following steps:

1. *Sanity Check:* If any of the components of the input ciphertext $\mathbf{E}'$ is $1_\mathbb{G}$ or if the proof π does not verify, output $(\widehat{F}, \widehat{G})$ for random $\widehat{F}, \widehat{G} \in \mathbb{G}_T$. The sanity check is to ensure that the next step – namely, input re-randomization – randomizes the ciphertext $\mathbf{E}$.
2. *Input Re-Randomization:* Pick a random exponent $t \leftarrow \mathbb{Z}_q$ and compute $\widehat{\mathbf{C}} = \mathbf{C}(\mathbf{C}')^t$ and $\widehat{D} = D(D')^t$.
   Note that the random exponent t is used to re-randomize the encryption of $1_\mathbb{G}$, and this re-randomized encryption of $1_\mathbb{G}$ is multiplied with the encryption of m to get a re-randomized encryption of m.

3. *The main Re-encryption step:* Write $\widehat{\mathbf{C}} := (\widehat{C}_1, \ldots, \widehat{C}_d)$, $\mathbf{A} := (A_1, \ldots, A_d)$ and $\mathbf{B} := (B_1, \ldots, B_d)$. Compute

$$\widehat{F} = \prod_{j=1}^{d} \mathsf{e}(\widehat{C}_j, A_j) \qquad \text{and} \qquad \widehat{G} = \prod_{j=1}^{d} \mathsf{e}(\widehat{C}_j, B_j) \cdot \mathsf{e}(\widehat{D}, h)$$

Output the ciphertext $(\widehat{F}, \widehat{G})$.

Preserving functionality. Let the input ciphertext be $(\mathbf{C}, D, \mathbf{C}', D', \pi)$. Given that π verifies, we know these values will be of the form $\mathbf{C} = g^{r a_i}, D = g^r m$ and $\mathbf{C}' = g^{r' a_i}, D = g^{r'}$. (If π does not verify, then both the functionality and the above program will output random group elements.) Let the re-encryption key be $(\mathbf{A}, \mathbf{B})$ where $\mathbf{A} = h^\alpha$ and $\mathbf{B} = h^\beta$.

- First, the input re-randomization step computes $\widehat{\mathbf{C}} = \mathbf{C}(\mathbf{C}')^t = g^{(r+tr')a_i} = g^{\hat{r} a_i}$ and $\widehat{D} = D(D')^t = g^{r+tr'} m = g^{\hat{r}} m$, where we defined $\hat{r} \triangleq r + tr'$.

- Second, the main re-encryption step computes $\widehat{F} = \prod_{j=1}^{d} \mathsf{e}(\widehat{C}_j, A_j) = \mathsf{e}(g, h)^{\hat{r}\langle a_i, \alpha \rangle} = \mathsf{e}(g, h)^{\hat{r} w_i \hat{a}_{F(i)}}$ and

$$\widehat{G} = \prod_{j=1}^{d} \mathsf{e}(\widehat{C}_j, B_j) \cdot \mathsf{e}(\widehat{D}, h)$$

$$= \mathsf{e}(g, h)^{\hat{r}\langle a_i, \beta \rangle} \cdot \mathsf{e}(g^{\hat{r}} m, h) = \mathsf{e}(g, h)^{\hat{r}(w_i - 1)} \cdot \mathsf{e}(g^{\hat{r}}, h) \cdot \mathsf{e}(m, h)$$

$$= \mathsf{e}(g, h)^{\hat{r} w_i} \cdot \mathsf{e}(m, h)$$

- Now the ciphertext looks like $\widehat{F} = \mathsf{e}(g, h^{\hat{a}_{F(i)} \rho})$, $\widehat{G} = \mathsf{e}(g, h^\rho) \cdot \mathsf{e}(m, h)$, where $\rho = \hat{r} w_i$ is uniformly random in $\mathbb{Z}_q$, even given all the randomness in the input ciphertext. The claim about ρ being uniformly random crucially relies on the "sanity check" step in the re-encryption program (in particular, since $r' \neq 0$).

Thus, the final ciphertext is distributed exactly like the output of $\texttt{O-Enc}(\widehat{\mathsf{pk}}_{F(i)}, m)$.

Semantic security of encryption schemes. We show that the input and output encryption schemes are semantically secure (in particular, the input scheme hides both the message and the "identity", and the output scheme is also key-private) under the DDH assumption over different groups, even given the re-encryption program. We present a detailed proof in the full version [11].

Remark. Note that if $d = n = 1$, then our construction (with the removal of certain now unnecessary parts, such as the NIZK proof) reduces to something very similar to that of [19]. Also, note that if the function F were to have larger (super-polynomial) domain, then our solution would satisfy the property of polynomial slowdown only if F were represented as a truth table. If F has large domain but a concise representation, then this property no longer holds.

4.3 Proof of Collusion-Resistant Secure Obfuscation

We show that our construction is a collusion-resistant average-case secure obfuscator for the functional re-encryption functionality. In order to satisfy collusion-resistance, the encryption as well as the obfuscation scheme have to be modified somewhat. The modifications do not affect the functionality or the security of the scheme, and are merely artifacts that seem necessary to show that our functional re-encryption scheme meets the rigorous demands of being a secure obfuscation.

A necessary modification to the encryption and obfuscation schemes. Consider the case where a corrupt recipient that holds secret key $\widehat{\mathsf{sk}}_j$ colludes with the re-encryption program. Now, essentially, this recipient has access to a program that selectively decrypts *input* ciphertexts that are encrypted with an identity i such that $F(i) = j$. However, the simulator only has oracle access to such a program and must yet produce a "fake" re-encryption program, that on input a ciphertext of message m with identity id, outputs a correct ciphertext of m under $\widehat{\mathsf{pk}}_j$. Hence, in order to put the simulator on an equal footing with the adversary we need to give the simulator the power to produce an explicit program which can selectively decrypt input ciphertexts. One way to do this is to cheat and give the simulator the vector $\boldsymbol{a}_i$, for all i such that $F(i) = j$, in our construction, which we will refer to as sk_i. (Note that sk_i is a secret key that allows for the selective decryption of ciphertexts with identity i, but not any other ciphertext.). For ease of exposition, we shall for now assume that the simulator obtains sk_i for all i such that $F(i) \in \mathcal{T}$. However, we would not like to resort to this cheat — we show in the full version [11] how this can be avoided. In other words, we first show the security of our scheme in the modified model where the simulator obtains the sk_i values for all i such that $F(i) \in \mathcal{T}$. Next (in the full version [11]), we show that if the scheme is secure in this model, then it can be easily transformed into a scheme that is secure in the standard model where the simulator, like the real world adversary, only gets $\widehat{\mathsf{sk}}_j$ values for $j \in \mathcal{T}$. We will now focus on proving the former statement. Towards showing that our obfuscation satisfies the collusion-resistant secure obfuscation definition in the model where the simulator obtains the sk_i values for all i such that $F(i) \in \mathcal{T}$, we first construct a simulator.

Simulator. Let $\mathcal{C} \leftarrow \mathsf{FR}_{\lambda,F,d,n}$ be a functional re-encryption circuit for the function $F : [d] \to [n]$, parameterized by the input keys $(\mathsf{pk}, \mathsf{sk})$ and the output keys $(\widehat{\mathsf{pk}}_j, \widehat{\mathsf{sk}}_j)$ for all $j \in [n]$. Let $\mathcal{T} \subseteq [n]$ be a set of corrupted receivers. We construct a simulator $\mathcal{S}$ that gets as input the secret keys $\widehat{\mathsf{sk}}_j$ of all the corrupted receivers (where $j \in \mathcal{T}$) and the secret keys sk_i such that $F(i) \in \mathcal{T}$, and has oracle access to the functionality $\mathcal{C}$.

First, consider the case where none of the receivers is corrupted. Then, the simulator works as follows. Recall that the obfuscated re-encryption program consists of the tuple $(h, h^{\alpha}, h^{\beta})$ where and α and β are solutions to some linear equations involving the input and output secret keys. The simulator, instead, simply picks α and β uniformly at random (with no relation to the input or the output keys). It then runs the adversary on this "junk functional re-encryption program" (along with the secret keys of the corrupted receivers). Under the

SXDH assumption, we manage to show that this is indistinguishable from the obfuscated program that the adversary expects to get (even if the adversary is also given oracle access to the real re-encryption circuit $\mathcal{C}$).

If some of the receivers are corrupted, the simulator cannot choose α and β at random any more. Indeed, since the distinguisher has the corrupted output keys, it can check if the α and β (in the exponent) satisfy the equations involving the corrupted keys, namely $\{\widehat{sk}_j\}_{j \in \mathcal{T}}$. Thus, the simulator has to choose α and β as *uniformly random solutions to a set of equations that involve the corrupted keys.* It turns out that this can be done efficiently since the simulator knows the keys of the corrupted receivers as well.

Without further ado, let us present the simulator $\mathcal{S}^{\mathcal{C}}(1^\lambda, \mathcal{T}, \{\widehat{sk}_i\}_{i \in \mathcal{T}},$ $\{sk_j\}_{j \in F^{-1}(\mathcal{T})})$ that works as follows:

1. Query the oracle $\mathcal{C}$ on input the string "keys" to get all the public keys, including the input public key $pk = (g, g^{a_1}, \cdots, g^{a_d})$; and the output public keys $\widehat{pk}_1 = (h, h^{\hat{a}_1}), \cdots, \widehat{pk}_n = (h, h^{\hat{a}_n})$.
2. Sample random $w_1, \ldots, w_d$ from $\mathbb{Z}_q$. Sample random α, β from $\mathbb{Z}_q^d$ such that

$$\forall i \text{ s.t. } F(i) \in \mathcal{T}: \qquad \langle a_i, \alpha \rangle = w_i \hat{a}_{F(i)} \qquad \text{and} \qquad \langle a_i, \beta \rangle = w_i - 1$$

   Note that this can be done efficiently using the knowledge of the vectors a_i that we obtained in $\{sk_j\}_{j \in F^{-1}(\mathcal{T})}$, as well as the $\hat{a}_{F(i)}$ values which are part of the corrupted secret keys. Compute $\mathbf{A} = h^\alpha$, and $\mathbf{B} = h^\beta$. Output the tuple $(\mathbf{A}, \mathbf{B})$ as the re-encryption key.

We now show that the output of the simulator described above is indistinguishable from an obfuscation of the re-encryption functionality (given in Section 4.2), even to a distinguisher that has the corrupted receivers' secret keys and oracle access to the re-encryption functionality. This proves that the obfuscation scheme we constructed in section 4.2 is a collusion-resistant average-case secure obfuscation satisfying Definition 2. More formally, we show:

Theorem 2. *Under SXDH, for any ppt distinguisher* D *and corrupt set* $\mathcal{T} \subseteq [n]$,

$$\mathsf{D}^{\mathcal{C}} \left[\mathcal{O}bf(\mathcal{C}), \mathcal{T}, \{\widehat{sk}_j\}_{j \in \mathcal{T}}, \{sk_j\}_{j \in F^{-1}(\mathcal{T})} \right] \stackrel{c}{\approx}$$

$$\mathsf{D}^{\mathcal{C}} \left[\mathcal{S}^{\mathcal{C}}(1^\lambda, \mathcal{T}, \{\widehat{sk}_j\}_{j \in \mathcal{T}}, \{sk_j\}_{j \in F^{-1}(\mathcal{T})}), \{\widehat{sk}_j\}_{j \in \mathcal{T}}, \{sk_j\}_{j \in F^{-1}(\mathcal{T})} \right]$$

for obfuscator $\mathcal{O}bf$, *where* $\mathcal{C} \leftarrow \mathsf{FR}_{\lambda, F, d, n}$ *is a uniformly random re-encryption circuit parameterized by* $(pk, sk) \leftarrow \mathit{I\text{-}Gen}(1^\lambda)$ *and* $(\widehat{pk}_i, \widehat{sk}_i) \leftarrow \mathit{O\text{-}Gen}(1^\lambda)$.

From the above theorem, our main theorem (which we stated informally as Theorem 1) follows after making the necessary modifications to the construction outlined earlier. We now describe a sketch of the proof of Theorem 2. For the formal proof, see the full version [11].

Proof. (sketch.) At a high level, the proof will go through the following steps:

- **Step 1:** For simplicity, let us first consider the case when there is no collusion – that is, neither the distinguisher nor the simulator has access to any of the output secret keys. Later, we will point out the necessary modifications to achieve collusion-resistance.

We first show that the re-encryption key is indistinguishable from random group elements to any distinguisher D who is given the public keys for the input and output encryption scheme (but *no oracle access*). In other words, we will show that constructing a re-encryption key $(\mathbf{A}, \mathbf{B})$ where $\mathbf{A} = h^{\alpha}$ and $\mathbf{B} = h^{\beta}$ with α, β being solutions to the equations

$$\langle \mathbf{a}_i, \boldsymbol{\alpha} \rangle = w_i \hat{a}_{F(i)} \qquad \text{and} \qquad \langle \mathbf{a}_i, \boldsymbol{\beta} \rangle = w_i - 1 \qquad \text{for all } i \in [d] \qquad (1)$$

is indistinguishable from constructing a re-encryption key with uniformly random α and β. This follows from two ideas – first, under the DDH assumption in group $\mathbb{H}$, it is hard to distinguish between $(h, h^{\alpha}, h^{\beta})$ where α and β are solutions to Equations 1, from the case where they are solutions to the same set of equations with the right-hand sides replaced by *uniformly random elements* in $\mathbb{Z}_q^*$. [7] Next, we note that choosing α, β as a solution to a set of equations with uniformly random right-hand side is equivalent to simply choosing random α, β. This completes the first step - in the full version [11] we show that this generalizes to the case where $\mathcal{T}$ is non-empty, and the simulator's α, β are chosen as a random solution to the resulting underconstrained set of equations.

- **Step** 2: Next, we will provide our distinguisher D with oracle access to a random oracle that simply returns random group elements of the same format as the output ciphertext of the re-encryption program. (The only exception is that, when it receives a ciphertext encrypted under id such that $F(\mathsf{id}) \in \mathcal{T}$, it honestly performs the re-encryption.) We then show that the re-encryption key is indistinguishable from random group elements to this distinguisher $\mathsf{D}^{\mathcal{RO}}$ as well.

This follows from Step 1 fairly easily once we note that the distinguisher in Step 1 could easily simulate this random oracle itself.

- **Step** 3: Next, we will provide our distinguisher D with oracle access to either the re-encryption oracle or the random oracle, and argue that D will not be able to determine which oracle it is given, even if it is also given the real re-encryption key.

The main intuition behind this proof is that, based on SXDH, we can show that honestly generated outputs ciphertexts are indistinguishable from random tuples. This is fairly easy to see: consider public key $h^{\hat{a}}$, and the following tuple $[\mathsf{e}(g, h^w), \mathsf{e}(g, h^r) \cdot \mathsf{e}(m, h)]$ for random $\hat{a}, r \in \mathbb{Z}_q$. If $w = \hat{a}r$, this is a valid encryption of m, if w is a random element of $\mathbb{Z}_q$, then this is a random tuple from $\mathbb{G}_T \times \mathbb{G}_T$.

A fairly straightforward hybrid argument then shows that a real encryption oracle for public keys $\widehat{\mathsf{pk}}_1, \ldots, \widehat{\mathsf{pk}}_n$ is indistinguishable from a random oracle

---

[7] Note that the right-hand sides of Equation 1 are not random as such – for example, consider the case where $F(1) = F(2) = 1$. Then, the right-hand sides of the four equations corresponding to $i = 1$ and $i = 2$ are $w_1 \hat{a}_1, w_1 - 1, w_2 \hat{a}_1, w_2 - 1$, which are clearly correlated.

which only produces valid ciphertexts for $\widehat{\mathsf{pk}}_i$ with $i \in \mathcal{T}$ (even when the distinguisher is given $\widehat{\mathsf{sk}}_i$ for $i \in \mathcal{T}$).

Now, we note that we can generate a real re-encryption key and perfectly simulate either the real re-encryption oracle or the random re-encryption oracle given only $\widehat{\mathsf{pk}}_1, \ldots, \widehat{\mathsf{pk}}_n$, and either the encryption oracle or the random oracle described above. We conclude that the real re-encryption oracle and random re-encryption oracle are indistinguishable even given the real re-encryption key (and $\widehat{\mathsf{sk}}_i$ for $i \in \mathcal{T}$).

- **Step 4:** Finally, we will again provide our distinguisher D with oracle access to either the re-encryption oracle or the random oracle and argue that it will not be able to determine which oracle it is given, this time when given the simulated re-encryption key instead.

  Again, this follows from Step 3, when we note that the distinguisher in Step 3 could easily ignore the re-encryption key it is given and instead run the simulator to generate a simulated one.

We have argued that the distinguisher has the same behavior given the real re-encryption key and real re-encryption oracle or the real re-encryption key and random oracle (Step 3), that it has the same behavior given the real re-encryption key and random oracle or the simulated re-encryption key and random oracle (Step 2), and that it has the same behavior given the simulated re-encryption key and random oracle or the simulated re-encryption key and real re-encryption oracle (Step 4). Putting everything together, we conclude that the real re-encryption key and simulated re-encryption key are indistinguishable, even given access to the real re-encryption oracle. Thus, we obtain the proof of Theorem 2.

Acknowledgements. We wish to thank Markulf Kohlweiss for suggesting the use of the SXDH assumption which simplified our construction.

References

1. Adida, B., Wikström, D.: How to Shuffle in Public. In: Vadhan, S.P. (ed.) TCC 2007. LNCS, vol. 4392, pp. 555–574. Springer, Heidelberg (2007)
2. Ateniese, G., Fu, K., Green, M., Hohenberger, S.: Improved proxy re-encryption schemes with applications to secure distributed storage. In: NDSS 2005 (2005)
3. Barak, B., Goldreich, O., Impagliazzo, R., Rudich, S., Sahai, A., Vadhan, S., Yang, K.: On the (Im)possibility of Obfuscating Programs. In: Kilian, J. (ed.) CRYPTO 2001. LNCS, vol. 2139, pp. 1–18. Springer, Heidelberg (2001)
4. Bitansky, N., Canetti, R.: On Strong Simulation and Composable Point Obfuscation. In: Rabin, T. (ed.) CRYPTO 2010. LNCS, vol. 6223, pp. 520–537. Springer, Heidelberg (2010)
5. Blaze, M., Bleumer, G., Strauss, M.: Divertible Protocols and Atomic Proxy Cryptography. In: Nyberg, K. (ed.) EUROCRYPT 1998. LNCS, vol. 1403, pp. 127–144. Springer, Heidelberg (1998)
6. Boneh, D., Boyen, X., Shacham, H.: Short Group Signatures. In: Franklin, M. (ed.) CRYPTO 2004. LNCS, vol. 3152, pp. 41–55. Springer, Heidelberg (2004)

7. Canetti, R.: Towards Realizing Random Oracles: Hash Functions that Hide All Partial Information. In: Kaliski Jr., B.S. (ed.) CRYPTO 1997. LNCS, vol. 1294, pp. 455–469. Springer, Heidelberg (1997)
8. Canetti, R., Dakdouk, R.R.: Obfuscating Point Functions with Multibit Output. In: Smart, N.P. (ed.) EUROCRYPT 2008. LNCS, vol. 4965, pp. 489–508. Springer, Heidelberg (2008)
9. Canetti, R., Micciancio, D., Reingold, O.: Perfectly one-way probabilistic hash functions. In: STOC 1998, pp. 131–140 (1998)
10. Canetti, R., Rothblum, G.N., Varia, M.: Obfuscation of Hyperplane Membership. In: Micciancio, D. (ed.) TCC 2010. LNCS, vol. 5978, pp. 72–89. Springer, Heidelberg (2010)
11. Chandran, N., Chase, M., Vaikuntanathan, V.: Collusion Resistant Obfuscation and Functional Re-encryption. IACR Eprint Archive, http://eprint.iacr.org/2011/337
12. Dodis, Y., Smith, A.: Correcting errors without leaking partial information. In: STOC 2005, pp. 654–663 (2005)
13. Gentry, C.: Fully homomorphic encryption using ideal lattices. In: STOC 2009, pp. 169–178 (2009)
14. Goldwasser, S., Kalai, Y.: On the impossibility of obfuscation with auxiliary input. In: FOCS 2005, pp. 553–562 (2005)
15. Goyal, V., Pandey, O., Sahai, A., Waters, B.: Attribute-based encryption for fine-grained access control of encrypted data. In: CCS 2006, pp. 89–98 (2006)
16. Groth, J., Sahai, A.: Efficient Non-interactive Proof Systems for Bilinear Groups. In: Smart, N.P. (ed.) EUROCRYPT 2008. LNCS, vol. 4965, pp. 415–432. Springer, Heidelberg (2008)
17. Hada, S.: Secure Obfuscation for Encrypted Signatures. In: Gilbert, H. (ed.) EUROCRYPT 2010. LNCS, vol. 6110, pp. 92–112. Springer, Heidelberg (2010)
18. Hofheinz, D., Malone-Lee, J., Stam, M.: Obfuscation for Cryptographic Purposes. In: Vadhan, S.P. (ed.) TCC 2007. LNCS, vol. 4392, pp. 214–232. Springer, Heidelberg (2007)
19. Hohenberger, S., Rothblum, G.N., Shelat, A., Vaikuntanathan, V.: Securely Obfuscating Re-encryption. In: Vadhan, S.P. (ed.) TCC 2007. LNCS, vol. 4392, pp. 233–252. Springer, Heidelberg (2007)
20. Katz, J., Sahai, A., Waters, B.: Predicate Encryption Supporting Disjunctions, Polynomial Equations, and Inner Products. In: Smart, N.P. (ed.) EUROCRYPT 2008. LNCS, vol. 4965, pp. 146–162. Springer, Heidelberg (2008)
21. Lynn, B., Prabhakaran, M., Sahai, A.: Positive Results and Techniques for Obfuscation. In: Cachin, C., Camenisch, J.L. (eds.) EUROCRYPT 2004. LNCS, vol. 3027, pp. 20–39. Springer, Heidelberg (2004)
22. Rivest, R., Adleman, L., Dertouzos, M.: On data banks and privacy homomorphisms. In: Foundations of Secure Computation, pp. 169–177. Academic Press (1978)
23. Sahai, A., Waters, B.: Fuzzy Identity-Based Encryption. In: Cramer, R. (ed.) EUROCRYPT 2005. LNCS, vol. 3494, pp. 457–473. Springer, Heidelberg (2005)
24. Scott, M.: Authenticated ID-based key exchange and remote log-in with insecure token and PIN number (2002), http://eprint.iacr.org/2002/164
25. Shen, E., Shi, E., Waters, B.: Predicate Privacy in Encryption Systems. In: Reingold, O. (ed.) TCC 2009. LNCS, vol. 5444, pp. 457–473. Springer, Heidelberg (2009)
26. Verheul, E.: Evidence that xtr is more secure than supersingular elliptic curve Cryptosystems. J. Cryptology 17(4), 277–296 (2004)
27. Wee, H.: On obfuscating point functions. In: STOC 2005, pp. 523–532 (2005)

How to Delegate and Verify in Public: Verifiable Computation from Attribute-Based Encryption

Bryan Parno[1], Mariana Raykova[2,*], and Vinod Vaikuntanathan[3,**]

[1] Microsoft Research
[2] Columbia University
[3] University of Toronto

Abstract. The wide variety of small, computationally weak devices, and the growing number of computationally intensive tasks makes it appealing to delegate computation to data centers. However, outsourcing computation is useful only when the returned result can be trusted, which makes verifiable computation (VC) a must for such scenarios.

In this work we extend the definition of verifiable computation in two important directions: *public delegation* and *public verifiability*, which have important applications in many practical delegation scenarios. Yet, existing VC constructions based on standard cryptographic assumptions fail to achieve these properties.

As the primary contribution of our work, we establish an important (and somewhat surprising) connection between verifiable computation and attribute-based encryption (ABE), a primitive that has been widely studied. Namely, we show how to construct a VC scheme with public delegation and public verifiability from any ABE scheme. The VC scheme verifies any function in the class of functions covered by the permissible ABE policies (currently Boolean formulas). This scheme enjoys a very efficient verification algorithm that depends only on the output size. Efficient delegation, however, requires the ABE encryption algorithm to be cheaper than the original function computation. Strengthening this connection, we show a construction of a *multi-function* verifiable computation scheme from an ABE scheme with outsourced decryption, a primitive defined recently by Green, Hohenberger and Waters (USENIX Security 2011). A multi-function VC scheme allows the verifiable evaluation of multiple functions on *the same preprocessed input*.

In the other direction, we also explore the construction of an ABE scheme from verifiable computation protocols.

---

* Research conducted as part of an internship with Microsoft Research.
** Supported by an NSERC Discovery Grant and by DARPA under Agreement number FA8750-11-2-0225. The U.S. Government is authorized to reproduce and distribute reprints for Governmental purposes notwithstanding any copyright notation thereon. The views and conclusions contained herein are those of the author and should not be interpreted as necessarily representing the official policies or endorsements, either expressed or implied, of DARPA or the U.S. Government.

R. Cramer (Ed.): TCC 2012, LNCS 7194, pp. 422–439, 2012.

1   Introduction

In the modern age of cloud computing and smartphones, asymmetry in computing power seems to be the norm. Computationally weak devices such as smartphones gather information, and when they need to store the voluminous data they collect or perform expensive computations on their data, they outsource the storage and computation to a large and powerful server (a "cloud", in modern parlance). Typically, the clients have a pay-per-use arrangement with the cloud, where the cloud charges the client proportional to the "effort" involved in the computation.

One of the main security issues that arises in this setting is – *how can the clients trust that the cloud performed the computation correctly?* After all, the cloud has the financial incentive to run (occasionally, perhaps) an extremely fast but incorrect computation, freeing up valuable compute time for other transactions. Is there a way to *verifiably outsource* computations, where the client can, without much computational effort, check the correctness of the results provided by the cloud? Furthermore, can this be done without requiring much interaction between the client and the cloud? This is the problem of *non-interactive verifiable computation*, which was considered implicitly in the early work on efficient arguments by Kilian [18] and computationally sound proofs (CS proofs) by Micali [20], and which has been the subject of much attention lately [2–5, 10, 11, 13, 14].

The starting point of this paper is that while the recent solutions consider and solve the bare-bones verifiable computation problem in its simplest form, there are a number of desirable features that they fail to achieve. We consider two such properties – namely, *public delegatability* and *public verifiability*.

Public Delegatability. In a nutshell, public delegatability says that everyone should be able to delegate computations to the cloud. In some protocols [2, 4, 10, 11], a client who wishes to delegate computation of a function F is required to first run an expensive pre-processing phase (wherein her computation is linear in the size of the circuit for F) to generate a (small) secret key SK_F and a (large) evaluation key EK_F. This large initial cost is then amortized over multiple executions of the protocol to compute $F(x_i)$ for different inputs x_i, but the client needs the secret key SK_F in order to initiate each such execution. In other words, *clients can delegate computation to the cloud only if they put in a large initial computational investment*. This makes sense only if the client wishes to run the same computation on many different inputs. Can clients delegate computation without making such a large initial commitment of resources?

As an example of a scenario where this might come in handy, consider a clinic with a doctor and a number of lab assistants, which wishes to delegate the computation of a certain expensive data analysis function F to a cloud service. Although the doctor determines the structure and specifics of F, it is in reality the lab assistants who come up with inputs to the function and perform the delegation. In this scenario, we would like to ask the doctor to run the (expensive) pre-processing phase once and for all, and generate a (small) *public key* PK_F and an evaluation key EK_F. The public key lets anyone, including the lab assistants, delegate the computation of F to the cloud and verify the results. Thus, once the doctor makes the initial investment, any of the lab assistants can delegate computations to the cloud without the slightest involvement of the doctor. Needless to say, the cloud should not be able to cheat even given PK_F and EK_F.

Goldwasser, Kalai and Rothblum [13] present a *publicly delegatable* verifiable computation protocol for functions in the complexity class NC (namely, functions that can be computed by circuits of size $\mathsf{poly}(n)$ and depth $\mathsf{polylog}(n)$); indeed, their protocol is stronger in that it does not even require a pre-processing phase. In contrast, as mentioned above, many of the protocols for verifying general functions [2, 4, 10, 11] are *not publicly delegatable*. In concurrent work, Canetti, Riva, and Rothblum propose a similar notion (though they call it "public verifiability") [9] and construct a protocol, based on collision-resistant hashing and poly-logarithmic PIR, for general circuits C where the client runs in time $\mathsf{poly}(\log(|C|), \mathsf{depth}(C))$; they do not achieve the public verifiability property we define below. Computationally sound (CS) proofs achieve public delegatability; however the known constructions of CS proofs are either in the random oracle model [20], or rely on non-standard "knowledge of exponent"-type assumptions [5, 14]. Indeed, this seems to be an inherent limitation of solutions based on CS proofs since Gentry and Wichs [12] showed recently that CS proofs cannot be based on any falsifiable cryptographic assumption (using a black-box security reduction). Here, we are interested in standard model constructions, based on standard (falsifiable) cryptographic assumptions.

Public Verifiability. In a similar vein, the delegator should be able to produce a (public) "verification key" that enables anyone to check the cloud's work. In the context of the example above, when the lab assistants delegate a computation on input x, they can also produce a verification key VK_x that will let the patients, for example, obtain the answer from the cloud and check its correctness. Neither the lab assistants nor the doctor need to be involved in the verification process. Needless to say, the cloud cannot cheat even if it knows the verification key VK_x.

Papamanthou, Tamassia, and Triandopoulos [23] present a verifiable computation protocol for set operations that allows anyone who receives the result of the set operation to verify its correctness. In concurrent work, Papamanthou, Shi, and Tamassia [22] propose a similar notion, but they achieve it only for multivariate polynomial evaluation and differentiation, and the setup and evaluation run in time exponential in the degree; they do not consider the notion of public delegation. Neither the Goldwasser-Kalai-Rothblum protocol [13] nor any of the later works [2, 4, 10, 11] seem to be publicly verifiable.

Put together, we call a verifiable computation protocol that is both publicly delegatable and publicly verifiable a *public verifiable computation* protocol. We are not aware of any such protocol (for a general class of functions) that is non-interactive and secure in the standard model. Note that we still require the party who performs the initial function preprocessing (the doctor in the example above) to be trusted by those delegating inputs and verifying outputs.

As a bonus, a public verifiable computation protocol is immune to the "rejection problem" that affects several previous constructions [2, 10, 11]. Essentially, the problem is that these protocols do not provide reusable soundness; i.e., a malicious cloud that is able to observe the result of the verification procedure (namely, the accept/reject decision) on polynomially many inputs can eventually break the soundness of the protocol. It is an easy observation that public verifiable computation protocols do not suffer from the rejection problem. Roughly speaking, verification in such protocols depends only on the public key and some (instance-specific) randomness generated by the delegator, and not on any long-term secret state. Thus, obtaining the result of the verification procedure on one instance does not help break the soundness on a different instance.[1]

This paper is concerned with the design of public (non-interactive) verifiable computation protocols.

1.1 Our Results and Techniques

Our main result is a (somewhat surprising) connection between the notions of attribute-based encryption (ABE) and verifiable computation (VC). In a nutshell, we show that a *public* verifiable computation protocol for a class of functions $\mathcal{F}$ can be constructed from any attribute-based encryption scheme for a related class of functions – namely, $\mathcal{F} \cup \overline{\mathcal{F}}$. Recall that attribute-based encryption (ABE) [15, 25] is a rich class of encryption schemes where secret keys $\mathsf{ABE.SK}_F$ are associated with functions F, and can decrypt ciphertexts that encrypt a message m under an "attribute" x if and only if $F(x) = 1$.

For simplicity, we state all our results for the case of Boolean *functions*, namely functions with one-bit output. For functions with many output bits, we simply run independent copies of the verifiable computation protocol for each output bit.

Theorem 1 (Main Theorem, Informal). *Let $\mathcal{F}$ be a class of Boolean functions, and let $\overline{\mathcal{F}} = \{\overline{F} \mid F \in \mathcal{F}\}$ where $\bar{F}$ denotes the complement of the function F. If there is a key-policy ABE scheme for $\mathcal{F} \cup \overline{\mathcal{F}}$, then there is a public verifiable computation protocol for $\mathcal{F}$.*

Some remarks about this theorem are in order.

1. First, our construction is in the pre-processing model, where we aim to outsource the computation of the same function F on polynomially many inputs x_i with the goal of achieving an amortized notion of efficiency. This is the same as the notion considered in [10, 11], and different from the one in [13]. See Definition 1.

2. Secondly, since the motivation for verifiable computation is outsourcing computational effort, efficiency for the client is obviously a key concern. Our protocol will be efficient for the client, as long as computing an ABE encryption (on input a message m and attribute x) takes less time than evaluating the function F on x. We will further address the efficiency issue in the context of concrete instantiations below (as well as in Section 3.2).

---

[1] In fact, this observation applies also to any protocol that is publicly delegatable and not necessarily publicly verifiable.

3. Third, we only need a *weak* form of security for attribute-based encryption which we will refer to as *one-key security*. Roughly speaking, this requires that an adversary, given a single key ABE.SK$_F$ for any function F of its choice, cannot break the semantic security of a ciphertext under any attribute x such that $F(x) = 0$. Much research effort on ABE has been dedicated to achieving the much stronger form of security against collusion, namely when the adversary obtains secret keys for not just one function, but polynomially many functions of its choice. We will not require the strength of these results for our purposes. On the same note, constructing one-key secure ABE schemes is likely to be much easier than full-fledged ABE schemes.

Note on Terminology: Attribute-based Encryption versus Predicate Encryption. In this paper, we consider attribute-based encryption (ABE) schemes to be ones in which each secret key ABE.SK$_F$ is associated with a function F, and can decrypt ciphertexts that encrypt a message m under an "attribute" x if and only if $F(x) = 1$. This formulation is implicit in the early definitions of ABE introduced by Goyal, Pandey, Sahai and Waters [15, 25]. However, their work refers to F as an access structure, and existing ABE instantiations are restricted to functions (or access structures) that can be represented as polynomial-size span programs (a generalization of Boolean formulas) [15, 19, 21]. While such restrictions are not inherent in the definition of ABE, the fully general formulation we use above was first explicitly introduced by Katz, Sahai, and Waters, who dubbed it predicate encryption [17]. Note that we do not require attribute-hiding or policy/function-hiding, properties often associated with predicate encryption schemes (there appears to be some confusion in the literature as to whether attribute-hiding is inherent in the definition of predicate encryption [8, 17, 19], but the original formulation [17] does not seem to require it).

Thus, in a nutshell, our work can be seen as using ABE schemes for general functions, or equivalently, predicate encryption schemes that do not hide the attributes or policy, in order to construct verifiable computation protocols.

Let us now describe an outline of our construction. The core idea of our construction is simple: attribute-based encryption schemes naturally provide a way to "prove" that $F(x) = 1$. Say the server is given the secret key ABE.SK$_F$ for a function F, and a ciphertext that encrypts a *random message* m under the attribute x. The server will succeed in decrypting the ciphertext and recovering m if and only if $F(x) = 1$. If $F(x) = 0$, he fares no better at finding the message than a random guess. The server can then prove that $F(x) = 1$ by returning the decrypted message.

More precisely, this gives an effective way for the server to convince the client that $F(x) = 1$. The pre-processing phase for the function F generates a master public key ABE.MPK for the ABE scheme (which acts as the public key for the verifiable computation protocol) and the secret key ABE.SK$_F$ for the function F (which acts as the evaluation key for the verifiable computation protocol). Given the public key and an input x, the delegator encrypts a random message m under the attribute x and sends it to the server. If $F(x) = 1$, the server manages to decrypt and return m, but otherwise, he returns $\perp$. Now,

– If the client gets back the same message that she encrypted, she is convinced beyond doubt that $F(x) = 1$. This is because, if $F(x)$ were 0, the server could not have found m (except with negligible probability, assuming the message is long enough).

– However, if she receives no answer from the server, it could have been because $F(x) = 0$ and the server is truly unable to decrypt, or because $F(x) = 1$ but the server intentionally refuses to decrypt.

Thus, we have a protocol with one-sided error – if $F(x) = 0$, the server can never cheat, but if $F(x) = 1$, he can.

A verifiable computation protocol with no error can be obtained from this by two independent repetitions of the above protocol – once for the function F and once for its complement $\bar{F}$. A verifiable computation protocol for functions with many output bits can be obtained by repeating the one-bit protocol above for each of the output bits. Intuitively, since the preprocessing phase does not create any secret state, the protocol provides public verifiable computation. Furthermore, the verifier performs as much computation as is required to compute two ABE encryptions.

Perspective: Signatures on Computation. Just as digital signatures authenticate messages, the server's proof in a non-interactive verifiable computation protocol can be viewed as a "signature on computation", namely a way to authenticate that the computation was performed correctly. Moni Naor has observed that identity-based encryption schemes give us digital signature schemes, rather directly [7]. Given our perspective, one way to view our result is as a logical extension of Naor's observation to say that just as IBE schemes give us digital signatures, ABE schemes give us signatures on computation or, in other words, non-interactive verifiable computation schemes.

Instantiations. Instantiating our protocol with existing ABE schemes creates challenges with regard to functionality, security, and efficiency. We discuss this issues briefly below and defer a detailed discussion to Section 3.2.

As mentioned earlier, existing ABE schemes only support span programs or polynomial-size Boolean formulas [15, 19, 21], which restricts us to this class of functions as well. In particular, the more recent ABE schemes, such as that of Ostrovsky, Sahai, and Waters [21], support the class of all (not necessarily monotone) formulas.

Another challenge is that most ABE schemes [15, 21, 25] are proven secure only in a selective-security model. As a result, instantiating the protocol above with such a scheme would inherit this limitation. If we instantiate our protocol with the scheme of Ostrovsky, Sahai, and Waters [21], we achieve a VC protocol for the class of polynomial-size Boolean formulas, which has delegation and verification algorithms whose combined complexity is more efficient than the function evaluation. Essentially, the complexity gain arises because the delegation algorithm is essentially running the ABE encryption algorithm whose complexity is a fixed polynomial in $|x|$, the size of the input to the function, as well as the security parameter. The verification algorithm is very simple, involving just a one-way function computation. The resulting verifiable computation protocol is selectively secure.

Unfortunately, removing the "selective restriction" seems to be a challenge with existing ABE schemes. Although there have recently been constructions of adaptively secure ABE schemes, starting from the work of Lewko et al. [19], all these schemes work for *bounded* polynomial-size Boolean formulas. The up-shot is that the amount of work required to generate an encryption is proportional to the size of the formula, which makes the delegation as expensive as the function evaluation (and thus, completely useless)!

Much work in the ABE literature has been devoted to constructing ABE schemes that are secure against collusion. Namely, the requirement is that even if an adversary obtains secret keys for polynomially many functions, the scheme still retains security (in a precise sense). However, for our constructions, we require much less from the ABE scheme! In particular, we only need the scheme to be secure against adversaries that obtain the secret key for a single function. This points to instantiating our general construction with a *one-key secure* ABE scheme from the work of Sahai and Seyalioglu [24] for the class of bounded polynomial-size *circuits*. Unfortunately, because their scheme only supports bounded-size circuits, it suffers from the same limitation as that of Lewko et al. [19]. However, we can still use their construction to obtain a VC protocol where the *parallel complexity* of the verifier is significantly less than that required to compute the function.

We also note that when we instantiate our VC protocol with existing ABE schemes, the computation done by both the client and the worker is significantly cheaper than in any previous VC scheme, since we avoid the overhead of PCPs and FHE. However, existing ABE schemes restrict us to either formulas or a less attractive notion of parallel efficiency. It remains to be seen whether this efficiency can be retained while expanding the security offered and the class of functions supported. Fortunately, given the amount of interest in and effort devoted to new ABE schemes, we expect further improvements in both the efficiency and security of these schemes. Our result demonstrates that such improvements, as well as improvements in the classes of functions supported, will benefit verifiable computation as well.

1.2   Other Results

Multi-Function Verifiability and ABE with Outsourcing. The definition of verifiable computation focuses on the evaluation of a single function over multiple inputs. In many constructions [4, 10, 11] the evaluated function is embedded in the parameters for the VC scheme that are used for the input processing for the computation. Thus evaluations of multiple functions on the same input would require repeated invocation for the ProbGen algorithm. A notable difference are approaches based on PCPs [5, 13, 14] that may require a single offline stage for input processing and then allow multiple function evaluations. However, such approaches inherently require verification work proportional to the depth of the circuit, which is at least logarithmic in the size of the function and for some functions can be also proportional to the size of the circuit. Further these approaches employ either fully homomorphic encryption or private information retrieval schemes to achieve their security properties.

Using the recently introduced definition of ABE with outsourcing [16], we achieve a multi-function verifiable computation scheme that decouples the evaluated function from the parameters of the scheme necessary for the input preparation. This VC scheme provides separate algorithms for input and function preparation, which subsequently can be combined for multiple evaluations. When instantiated with an existing ABE scheme with outsourcing [16], the verification algorithm for the scheme is very efficient: its complexity is linear in the output size but independent of the input length and the complexity of the computation. Multi-function VC provides significant efficiency improvements whenever multiple functions are evaluated on the same input, since a traditional VC scheme would need to invoke ProbGen for every function.

Attribute-Based Encryption from Verifiable Computation. We also consider the opposite direction of the ABE-VC relation: can we construct an ABE scheme from a VC scheme? We are able to show how to construct an ABE scheme from a *very special* class of VC schemes with a particular structure. Unfortunately, this does not seem to result in any new ABE constructions.

Due to space constraints, we defer the details to the full version of this paper.

2   Definitions

2.1   Public Verifiable Computation

We propose two new properties of verifiable computation schemes, namely

- *Public Delegation*, which allows arbitrary parties to submit inputs for delegation, and
- *Public Verifiability*, which allows arbitrary parties (and not just the delegator) to verify the correctness of the results returned by the worker.

Together, a verifiable computation protocol that satisfies both properties is called a *public verifiable computation* protocol. The following definition captures these two properties.

Definition 1 (Public Verifiable Computation). *A public verifiable computation scheme (with preprocessing) $\mathcal{VC}$ is a four-tuple of polynomial-time algorithms* (KeyGen, ProbGen, Compute, Verify) *which work as follows:*

- $(PK_F, EK_F) \leftarrow$ KeyGen$(F, 1^\lambda)$: *The randomized* key generation *algorithm takes as input a security parameter λ and the function F, and outputs a public key PK_F and an evaluation key EK_F.*
- $(\sigma_x, VK_x) \leftarrow$ ProbGen(PK_F, x): *The randomized* problem generation *algorithm uses the public key PK_F to encode an input x into public values σ_x and VK_x. The value σ_x is given to the worker to compute with, whereas VK_x is made public, and later used for verification.*
- $\sigma_{\text{out}} \leftarrow$ Compute(EK_F, σ_x): *The deterministic* worker *algorithm uses the evaluation key EK_F together with the value σ_x to compute a value σ_{out}.*
- $y \leftarrow$ Verify$(VK_x, \sigma_{\text{out}})$: *The deterministic* verification *algorithm uses the verification key VK_x and the worker's output σ_{out} to compute a string*

$y \in \{0,1\}^* \cup \{\perp\}$. Here, the special symbol $\perp$ signifies that the verification algorithm rejects the worker's answer σ_{out}.

A number of remarks on the definition are in order.

First, in some instantiations, the size of the public key (but not the evaluation key) will be independent of the function F, whereas in others, both the public key and the evaluation key will be as long as the description length of F. For full generality, we refrain from making the length of the public key a part of the syntactic requirement of a verifiable computation protocol, and instead rely on the definition of efficiency to enforce this (see Definition 4 below).

Secondly, our definition can be viewed as a "public-key version" of the earlier VC definition [10, 11]. In the earlier definition, KeyGen produces a secret key that was used as an input to ProbGen and, in turn, ProbGen produces a secret verification value needed for Verify (neither of these can be shared with the worker without losing security). Indeed, the "secret-key" nature of these definitions means that the schemes could be attacked given just oracle access to the verification function (and indeed, there are concrete attacks of this nature against the schemes in [2, 10, 11]). Our definition, in contrast, is stronger in that it allows any party holding the public key PK_F to delegate and verify computation of the function F on any input x, even if the party who originally ran ProbGen is no longer online. This, in turn, automatically protects against attacks that use the verification oracle.

Definition 2 (Correctness). *A verifiable computation protocol $\mathcal{VC}$ is correct for a class of functions $\mathcal{F}$ if for any $F \in \mathcal{F}$, any pair of keys $(PK_F, EK_F) \leftarrow$ KeyGen$(F, 1^\lambda)$, any $x \in$ Domain(F), any $(\sigma_x, VK_x) \leftarrow$ ProbGen(PK_F, x), and any $\sigma_{\mathsf{out}} \leftarrow$ Compute(EK_F, σ_x), the verification algorithm Verify on input VK_x and σ_{out} outputs $y = F(x)$.*

Providing public delegation and verification introduces a new threat model in which the worker knows both the public key PK_F (which allows him to delegate computations) and the verification key VK_x for the challenge input x (which allows him to check whether his answers will pass the verification).

Definition 3 (Security). *Let $\mathcal{VC}$ be a public verifiable computation scheme for a class of functions $\mathcal{F}$, and let $A = (A_1, A_2)$ be any pair of probabilistic polynomial time machines. Consider the experiment $\mathbf{Exp}_A^{PubVerif}[\mathcal{VC}, F, \lambda]$ for any $F \in \mathcal{F}$ below:*

$$Experiment\ \mathbf{Exp}_A^{PubVerif}[\mathcal{VC}, F, \lambda]$$
$$(PK_F, EK_F) \leftarrow \mathsf{KeyGen}(F, 1^\lambda);$$
$$(x^*, \mathsf{state}) \leftarrow A_1(PK_F, EK_F);$$
$$(\sigma_{x^*}, VK_{x^*}) \leftarrow \mathsf{ProbGen}(PK_F, x^*);$$
$$\sigma_{\mathsf{out}}^* \leftarrow A_2(\mathsf{state}, \sigma_{x^*}, VK_{x^*});$$
$$y^* \leftarrow \mathsf{Verify}(VK_{x^*}, \sigma_{\mathsf{out}}^*)$$
$$If\ y^* \neq \perp\ and\ y^* \neq F(x^*),\ output\ '1',\ else\ output\ '0';$$

A public verifiable computation scheme $\mathcal{VC}$ is secure for a class of functions $\mathcal{F}$, if for every function $F \in \mathcal{F}$ and every p.p.t. adversary $A = (A_1, A_2)$:

$$Pr[\mathbf{Exp}_A^{PubVerif}[\mathcal{VC}, F, \lambda] = 1] \leq \mathtt{negl}(\lambda). \tag{1}$$

where negl *denotes a negligible function of its input.*

Later, we will also briefly consider a weaker notion of "selective security" which requires the adversary to declare the challenge input x^* before it sees PK_F.

For verifiable outsourcing of a function to make sense, the client must use "less resources" than what is required to compute the function. "Resources" here could mean the running time, the randomness complexity, space, or the depth of the computation. We retain the earlier efficiency requirements [11] – namely, we require the complexity of ProbGen and Verify combined to be less than that of F. However, for KeyGen, we ask only that the complexity be $\mathtt{poly}(|F|)$. Thus, we employ an *amortized* complexity model, in which the client invests a larger amount of computational work in an "offline" phase in order to obtain efficiency during the "online" phase. We provide two strong definitions of efficiency – one that talks about the running time and a second that talks about computation depth.

Definition 4 (Efficiency). *A verifiable computation protocol $\mathcal{VC}$ is efficient for a class of functions $\mathcal{F}$ that act on $n = n(\lambda)$ bits if there is a polynomial p s.t.:* [2]

- *the running time of* ProbGen *and* Verify *together is at most $p(n, \lambda)$, the rest of the algorithms are probabilistic polynomial-time, and*
- *there exists a function $F \in \mathcal{F}$ whose running time is $\omega(p(n, \lambda))$.* [3]

In a similar vein, $\mathcal{VC}$ is depth-efficient if the computation depth of ProbGen *and* Verify *combined (written as Boolean circuits) is at most $p(n, \lambda)$, whereas there is a function $F \in \mathcal{F}$ whose computation depth is $\omega(p(n, \lambda))$.*

We now define the notion of unbounded circuit families which will be helpful in quantifying the efficiency of our verifiable computation protocols.

Definition 5. *We define a family of circuits $\{C_n\}_{n \in \mathbb{N}}$ to be unbounded if for every polynomial p and all but finitely many n, there is a circuit $C \in C_n$ of size at least $p(n)$. We call the family depth-unbounded if for every polynomial p and all but finitely many n, there is a circuit $C \in C_n$ of depth at least $p(n)$.*

2.2 Key-Policy Attribute-Based Encryption

Introduced by Goyal, Pandey, Sahai and Waters [15], Key-Policy Attribute-Based Encryption (KP-ABE) is a special type of encryption scheme where a Boolean function F is associated with each user's key, and a set of attributes (denoted as a string $x \in \{0, 1\}^n$) with each ciphertext. A key SK_F for a function F will decrypt a ciphertext corresponding to attributes x if and only if $F(x) = 1$.

---

[2] To be completely precise, one has to talk about a family $\mathcal{F} = \{\mathcal{F}_n\}_{n \in \mathbb{N}}$ parameterized by the input length n. We simply speak of $\mathcal{F}$ to implicitly mean $\mathcal{F}_n$ whenever there is no cause for confusion.

[3] This condition is to rule out trivial protocols, e.g., for a class of functions that can be computed in time less than $p(\lambda)$.

KP-ABE can be thought of as a special-case of predicate encryption [17] or functional encryption [8], although we note that a KP-ABE ciphertext need not hide the associated policy or attributes. We will refer to KP-ABE simply as ABE from now on. We state the formal definition below, adapted from [15, 19].

Definition 6 (Attribute-Based Encryption). *An attribute-based encryption scheme $\mathcal{ABE}$ for a class of functions $\mathcal{F} = \{\mathcal{F}_n\}_{n \in \mathbb{N}}$ (where functions in $\mathcal{F}_n$ take n bits as input) is a tuple of algorithms (Setup, Enc, KeyGen, Dec) that work as follows:*

- *$(PK, MSK) \leftarrow \mathsf{Setup}(1^\lambda, 1^n)$: Given a security parameter λ and an index n for the family $\mathcal{F}_n$, output a public key PK and a master secret key MSK.*
- *$C \leftarrow \mathsf{Enc}(PK, M, x)$: Given a public key PK, a message M in the message space MsgSp, and attributes $x \in \{0,1\}^n$, output a ciphertext C.*
- *$SK_F \leftarrow \mathsf{KeyGen}(MSK, F)$: Given a function F and the master secret key MSK, output a decryption key SK_F associated with F.*
- *$\mu \leftarrow \mathsf{Dec}(SK_F, C)$: Given a ciphertext $C \in \mathsf{Enc}(PK, M, x)$ and a secret key SK_F for function F, output a message $\mu \in \mathsf{MsgSp}$ or $\mu = \perp$.*

Definition 7 (ABE Correctness). *Correctness of the ABE scheme requires that for all $(PK, MSK) \leftarrow \mathsf{Setup}(1^\lambda, 1^n)$, all $M \in \mathsf{MsgSp}$, $x \in \{0,1\}^n$, all ciphertexts $C \leftarrow \mathsf{Enc}(PK, M, x)$ and all secret keys $SK_F \leftarrow \mathsf{KeyGen}(MSK, F)$, the decryption algorithm $\mathsf{Dec}(SK_F, C)$ outputs M if $F(x) = 1$ and $\perp$ if $F(x) = 0$. (This definition could be relaxed to hold with high probability over the keys (PK, MSK), which suffices for our purposes).*

We define a natural, yet relaxed, notion of security for ABE schemes which we refer to as "one-key security". Roughly speaking, we require that adversaries who obtain a single secret key SK_F for any function F of their choice and a ciphertext $C \leftarrow \mathsf{Enc}(PK, M, x)$ associated with any attributes x such that $F(x) = 0$ should not be able to violate the semantic security of C. We note that much work in the ABE literature has been devoted to achieving a strong form of security against collusion, where the adversary obtains not just a single secret key, but polynomially many of them for functions of its choice. We do not require such a strong notion for our purposes.

Definition 8 (One-Key Security for ABE). *Let $\mathcal{ABE}$ be a key-policy attribute-based encryption scheme for a class of functions $\mathcal{F} = \{\mathcal{F}_n\}_{n \in \mathbb{N}}$, and let $A = (A_0, A_1, A_2)$ be a three-tuple of probabilistic polynomial-time machines. We define security via the following experiment.*

Experiment $\mathbf{Exp}_A^{ABE}[\mathcal{ABE}, n, \lambda]$
$(PK, MSK) \leftarrow \mathsf{Setup}(1^\lambda, 1^n);$
$(F, \mathsf{state}_1) \leftarrow A_0(PK);$
$SK_F \leftarrow \mathsf{KeyGen}(MSK, F);$
$(M_0, M_1, x^*, \mathsf{state}_2) \leftarrow A_1(\mathsf{state}_1, SK_F);$
$b \leftarrow \{0,1\}; \ C \leftarrow \mathsf{Enc}(PK, M_b, x^*);$
$\hat{b} \leftarrow A_2(\mathsf{state}_2, C);$
If $b = \hat{b}$, output '1', else '0';

The experiment is valid if $M_0, M_1 \in$ MsgSp and $|M_0| = |M_1|$. We define the advantage of the adversary in all valid experiments as

$$Adv_A(\mathcal{ABE}, n, \lambda) = |Pr[b = b'] - 1/2|.$$

We say that $\mathcal{ABE}$ is a one-key secure ABE scheme if $Adv_A(\mathcal{ABE}, n, \lambda) \leq \mathtt{negl}(\lambda)$.

3 Verifiable Computation from ABE

In Section 3.1, we present our main construction and proof, while Section 3.2 contains the various instantiations of our main construction and the concrete verifiable computation protocols that we obtain as a result.

3.1 Main Construction

Theorem 2. *Let $\mathcal{F}$ be a class of Boolean functions (implemented by a family of circuits $\mathcal{C}$), and let $\overline{\mathcal{F}} = \{\overline{F} \mid F \in \mathcal{F}\}$ where $\bar{F}$ denotes the complement of the function F. Let $\mathcal{ABE}$ be an attribute-based encryption scheme that is one-key secure (see Definition 8) for $\mathcal{F} \cup \overline{\mathcal{F}}$, and let g be any one-way function.*

Then, there is a verifiable computation protocol $\mathcal{VC}$ (secure under Definition 3) for $\mathcal{F}$. If the circuit family $\mathcal{C}$ is unbounded (resp. depth-unbounded), then the protocol $\mathcal{VC}$ is efficient (resp. depth-efficient) in the sense of Definition 4.

We first present our verifiable computation protocol.

Let $\mathcal{ABE}$ = (ABE.Setup, ABE.KeyGen, ABE.Enc, ABE.Dec) be an attribute-based encryption scheme for the class of functions $\mathcal{F} \cup \overline{\mathcal{F}}$. Then, the verifiable computation protocol $\mathcal{VC}$ = (VC.KeyGen, ProbGen, Compute, Verify) for $\mathcal{F}$ works as follows.[4] We assume, without loss of generality, that the message space $\mathcal{M}$ of the ABE scheme has size 2^λ.

Key Generation VC.KeyGen: The client, on input a function $F \in \mathcal{F}$ with input length n, runs the ABE setup algorithm twice, to generate two independent key-pairs

$$(\mathsf{msk}_0, \mathsf{mpk}_0) \leftarrow \mathsf{ABE.Setup}(1^n, 1^\lambda) \quad \text{and} \quad (\mathsf{msk}_1, \mathsf{mpk}_1) \leftarrow \mathsf{ABE.Setup}(1^n, 1^\lambda)$$

Generate two secret keys $\mathsf{sk}_{\overline{F}} \leftarrow \mathsf{ABE.KeyGen}(\mathsf{msk}_0, \overline{F})$ (corresponding to $\overline{F}$) and $\mathsf{sk}_F \leftarrow \mathsf{ABE.KeyGen}(\mathsf{msk}_1, F)$ (corresponding to F).

Output the pair $(\mathsf{sk}_{\overline{F}}, \mathsf{sk}_F)$ as the evaluation key and $(\mathsf{mpk}_0, \mathsf{mpk}_1)$ as the public key.

Delegation ProbGen: The client, on input x and the public key PK_F, samples two uniformly random messages $m_0, m_1 \overset{R}{\leftarrow} \mathcal{M}$, computes the ciphertexts

$$\mathsf{CT}_0 \leftarrow \mathsf{ABE.Enc}(\mathsf{mpk}_0, m_0) \quad \text{and} \quad \mathsf{CT}_1 \leftarrow \mathsf{ABE.Enc}(\mathsf{mpk}_1, m_1)$$

Output the message $\sigma_x = (\mathsf{CT}_0, \mathsf{CT}_1)$ (to be sent to the server), and the verification key $VK_x = (g(m_0), g(m_1))$, where g is the one-way function.

---

[4] We denote the VC key generation algorithm as VC.KeyGen in order to avoid confusion with the ABE key generation algorithm.

Computation Compute: The server, on receiving the ciphertexts (CT_0, CT_1) and the evaluation key $EK_F = (sk_{\overline{F}}, sk_F)$ computes

$$\mu_0 \leftarrow \text{ABE.Dec}(sk_{\overline{F}}, CT_0) \quad \text{and} \quad \mu_1 \leftarrow \text{ABE.Dec}(sk_F, CT_1)$$

and send $\sigma_{\text{out}} = (\mu_0, \mu_1)$ to the client.

Verification Verify: On receiving $VK_x = (v_0, v_1)$ and $\sigma_{\text{out}} = (\mu_0, \mu_1)$, output [5]

$$y = \begin{cases} 0 & \text{if } g(\mu_0) = v_0 \text{ and } g(\mu_1) \neq v_1 \\ 1 & \text{if } g(\mu_1) = v_1 \text{ and } g(\mu_0) \neq v_0 \\ \perp & \text{otherwise} \end{cases}$$

Remark 1. Whereas our main construction requires only an ABE scheme, using an attribute-hiding ABE scheme (a notion often associated with predicate encryption schemes [8, 17]) would also give us input privacy, since we encode the function's input in the attribute corresponding to a ciphertext.

Remark 2. To obtain a VC protocol for functions with multi-bit output, we repeat this protocol (including the key generation algorithm) independently for every output bit. To achieve better efficiency, if the ABE scheme supports attribute hiding for a class of functions that includes message authentication codes (MAC), then we can define $F'(x) = MAC_K(F(x))$ and verify F' instead, similar to the constructions suggested by Applebaum, Ishai, and Kushilevitz [2], and Barbosa and Farshim [3].

Remark 3. The construction above requires the verifier to trust the party that ran ProbGen. This can be remedied by having ProbGen produce a non-interactive zero-knowledge proof of correctness [6] of the verification key VK_x. While theoretically efficient, the practicality of this approach depends on the particular ABE scheme and the NP language in question.

Proof of Correctness: The correctness of the VC scheme above follows from:

- If $F(x) = 0$, then $\overline{F}(x) = 1$ and thus, the algorithm Compute outputs $\mu_0 = m_0$ and $\mu_1 = \perp$. The algorithm Verify outputs $y = 0$ since $g(\mu_0) = g(m_0)$ but $g(\mu_1) = \perp \neq g(m_1)$, as expected.
- Similarly, if $F(x) = 1$, then $\overline{F}(x) = 0$ and thus, the algorithm Compute outputs $\mu_1 = m_1$ and $\mu_0 = \perp$. The algorithm Verify outputs $y = 1$ since $g(\mu_1) = g(m_1)$ but $g(\mu_0) = \perp \neq g(m_0)$, as expected. ∎

We now consider the relation between the efficiency of the algorithms for the underlying ABE scheme and the efficiency for the resulting VC scheme. Since the algorithms Compute and Verify can potentially be executed by different parties, we consider their efficiency separately. It is easily seen that:

- The running time of the VC key generation algorithm VC.KeyGen is twice that of ABE.Setup plus ABE.KeyGen.

---

[5] As a convention, we assume that $g(\perp) = \perp$.

- The running time of Compute is twice that of ABE.Dec.
- The running time of ProbGen is twice that of ABE.Enc, and the running time of Verify is the same as that of computing the one-way function.

In short, the combined running times of ProbGen and Verify is polynomial in their input lengths, namely $p(n, \lambda)$, where p is a fixed polynomial, n is the length of the input to the functions, and λ is the security parameter. Assuming that $\mathcal{F}$ is an *unbounded* class of functions (according to Definition 5), it contains functions that take longer than $p(n, \lambda)$ to compute, and thus our VC scheme is efficient in the sense of Definition 4. (Similar considerations apply to depth-efficiency).

We now turn to showing the security of the VC scheme under Definition 3. We show that an attacker against the VC protocol must either break the security of the one-way function g or the one-key security of the ABE scheme.

Proof of Security: Let $A = (A_1, A_2)$ be an adversary against the VC scheme for a function $F \in \mathcal{F}$. We construct an adversary $B = (B_0, B_1, B_2)$ that breaks the one-key security of the ABE, working as follows. (For notational simplicity, given a function F, we let $F_0 = \overline{F}$, and $F_1 = F$.)

1. B_0 first tosses a coin to obtain a bit $b \in \{0, 1\}$. (Informally, the bit b corresponds to B's guess of whether the adversary A will cheat by producing an input x such that $F(x) = 1$ or $F(x) = 0$, respectively.)
   B_0 outputs the function F_b, as well as the bit b as part of the state.

2. B_1 obtains the master public key mpk of the ABE scheme and the secret key sk_{F_b} for the function F_b. Set $\mathsf{mpk}_b = \mathsf{mpk}$.
   Run the ABE setup and key generation algorithms to generate a master public key mpk' and a secret key $\mathsf{sk}_{F_{1-b}}$ for the function F_{1-b} under mpk'. Set $\mathsf{mpk}_{1-b} = \mathsf{mpk}'$.
   Let $(\mathsf{mpk}_0, \mathsf{mpk}_1)$ be the public key for the VC scheme and $(\mathsf{sk}_{F_0}, \mathsf{sk}_{F_1})$ be the evaluation key. Run the algorithm A_1 on input the public and evaluation keys and obtain a challenge input x^* as a result.
   If $F(x^*) = b$, *output a uniformly random bit and stop.* Otherwise, B_1 now chooses two uniformly random messages $M^{(b)}, \rho \leftarrow \mathcal{M}$ and outputs $(M^{(b)}, \rho, x^*)$ together with its internal state.

3. B_2 obtains a ciphertext $C^{(b)}$ (which is an encryption of either $M^{(b)}$ or ρ under the public key mpk_b and attribute x^*).
   B_2 constructs an encryption $C^{(1-b)}$ of a uniformly random message $M^{(1-b)}$ under the public key mpk_{1-b} and attribute x^*.
   Run A_2 on input $\sigma_{x^*} = (C^{(0)}, C^{(1)})$ and $VK_{x^*} = (g(M^{(0)}), g(M^{(1)}))$, where g is the one-way function. As a result, A_2 returns σ_{out}.
   If $\mathsf{Verify}(VK_{x^*}, \sigma_{\mathsf{out}}) = b$, output 0 and stop.

We now claim the algorithms (B_0, B_1, B_2) described above distinguish between the encryption of $M^{(b)}$ and the encryption of ρ in the ABE security game with non-negligible advantage.

We consider two cases.

Case 1: $C^{(b)}$ is an encryption of $M^{(b)}$. In this case, B presents to A a perfect view of the execution of the VC protocol, meaning that A will cheat with probability $1/p(\lambda)$ for some polynomial p.

Cheating means one of two things. Either $F(x^*) = b$ and the adversary produced an inverse of $g(M^{(1-b)})$ (causing the Verify algorithm to output $1 - b$), or $F(x^*) = 1 - b$ and the adversary produced an inverse of $g(M^{(b)})$ (causing the Verify algorithm to output b).

In the former case, B outputs a uniformly random bit, and in the latter case, it outputs 0, the correct guess as to which message was encrypted. Thus, the overall probability that B outputs 0 is $1/2 + 1/p(\lambda)$.

Case 2: $C^{(b)}$ is an encryption of the message ρ. In this case, as above, B outputs a random bit if $F(x^*) = b$. Otherwise, the adversary A has to produce σ_{out} that makes the verifier output b, namely a string σ_{out} such that $g(\sigma_{\text{out}}) = g(M^{(b)})$, while given only $g(M^{(b)})$ (and some other information that is independent of $M^{(b)}$).

This amounts to inverting the one-way function which A can only do with a negligible probability. (Formally, if the adversary wins in this game with non-negligible probability, then we can construct an inverter for the one-way function g).

The bottom line is that the adversary outputs 0 in this case with probability $1/2 + \texttt{negl}(\lambda)$.

This shows that B breaks the one-key security of the ABE scheme with a non-negligible advantage $1/p(\lambda) - \texttt{negl}(\lambda)$. ∎

Remark 4. If we employ an ABE scheme that is selectively secure, then the construction and proof above still go through if we adopt a notion of "selectively-secure" verifiable computation in which the VC adversary commits in advance to the input on which he plans to cheat.

3.2 Instantiations

We describe two different instantiations of our main construction.

Efficient Selectively Secure VC Scheme for Formulas. The first instantiation uses the (selectively secure) ABE scheme of Ostrovsky, Sahai and Waters [21] for the class of (not necessarily monotone) polynomial-size Boolean formulas (which itself is an adaptation of the scheme of Goyal et al. [15] which only supports monotone formulas[6]). This results in a selectively secure public VC scheme for the same class of functions, by invoking Theorem 2. Recall that selective security

---

[6] Goyal et al.'s scheme [15] can also be made to work if we use DeMorgan's law to transform f and $\bar{f}$ into equivalent monotone formulas in which some variables may be negated. We then double the number of variables, so that for each variable v, we have one variable representing v and one representing its negation $\bar{v}$. Given an input x, we choose an attribute such that all of these variables are set correctly.

in the context of verifiable computation means that the adversary has to declare the input on which she cheats at the outset, before she sees the public key and the evaluation key.

The efficiency of the resulting VC scheme for Boolean formulas is as follows: for a boolean formula C, KeyGen runs in time $|C| \cdot \text{poly}(\lambda)$; ProbGen runs in time $|x| \cdot \text{poly}(\lambda)$, where $|x|$ is the length of the input to the formula; Compute runs in time $|C| \cdot \text{poly}(\lambda)$; and Verify runs in time $O(\lambda)$. In other words, the total work for delegation and verification is $|x| \cdot \text{poly}(\lambda)$ which is, in general, more efficient than the work required to evaluate the circuit C. Thus, the scheme is efficient in the sense of Definition 4. The drawback of this instantiation is that it is only selectively secure.

Recently, there have been constructions of fully secure ABE for formulas starting from the work of Lewko et al. [19] which, one might hope, leads to a fully secure VC scheme. Unfortunately, all known constructions of fully secure ABE work for *bounded* classes of functions. For example, in the construction of Lewko et al., once a bound B is fixed, one can design the parameters of the scheme so that it works for any formula of size at most B. Furthermore, implicit in the work of Sahai and Seyalioglu [24] is a construction of an (attribute-hiding, one-key secure) ABE scheme for *bounded* polynomial-size circuits (as opposed to formulas).

These constructions, unfortunately, do not give us efficient VC protocols. The reason is simply this: the encryption algorithm in these schemes run in time polynomial (certainly, at least linear) in B. Translated to a VC protocol using Theorem 2, this results in the worker running for time $\Omega(B)$ which is useless, since given that much time, he could have computed any circuit of size at most B by himself!

Essentially, the VC protocol that emerges from Theorem 2 is non-trivial if the encryption algorithm of the ABE scheme for the function family $\mathcal{F}$ is (in general) more efficient than computing functions in $\mathcal{F}$.

Depth-Efficient Adaptively Secure VC Scheme for Arbitrary Functions. Although the (attribute-hiding, one-key secure) ABE construction of Sahai and Seyalioglu [24] mentioned above does not give us an efficient VC scheme, it does result in a *depth-efficient VC scheme* for the class of polynomial-size circuits. Roughly speaking, the construction is based on Yao's Garbled Circuits, and involves an ABE encryption algorithm that constructs a garbled circuit for the function F in question. Even though this computation takes at least as much time as computing the circuit for F, the key observation is that it can be done in parallel. In short, going through the VC construction in Theorem 2, one can see that both the Compute and Verify algorithms can be implemented in constant depth (for appropriate encryption schemes and one-way functions, e.g., the ones that result from the AIK transformation [1]), which is much faster in parallel than computing F, in general.

Interestingly, the VC protocol thus derived is very similar to the protocol of Applebaum, Ishai and Kushilevitz [2]. We refer the reader to [2, 24] for details.

We believe that this scheme also illuminates an interesting point: unlike other ABE schemes [15, 19, 21], this ABE scheme is only *one-key secure*, which suffices for verifiable computation. This relaxation may point the way towards an ABE-based VC construction that achieves generality, efficiency, and adaptive security.

4   Conclusions and Future Work

In this work, we introduced new notions for verifiable computation: *public delegatability* and *public verifiability*. We demonstrated a somewhat surprising construction of a public verifiable computation protocol from any (one-key secure) attribute-based encryption (ABE) scheme.

Our work leaves open several interesting problems. Perhaps the main open question is the design of *one-key secure* ABE schemes for general, unbounded classes of functions. Is it possible to come up with such a scheme for the class of all polynomial-size circuits (as opposed to circuits with an a-priori bound on the size, as in [24])? Given the enormous research effort in the ABE literature devoted to achieving the strong notion of security against collusion, our work points out that achieving even security against the compromise of a single key is a rather interesting question to investigate!

Acknowledgements. We wish to thank Seny Kamara and David Molnar for joint work in the early stages of this project, and Melissa Chase for her useful comments on this work. Our gratitude also goes to Daniel Wichs and the anonymous TCC reviewers whose comments helped improve the exposition of this paper.

References

1. Applebaum, B., Ishai, Y., Kushilevitz, E.: Cryptography in NC^0. In: Proceedings of the IEEE Symposium on Foundations of Computer Science, FOCS (2004)
2. Applebaum, B., Ishai, Y., Kushilevitz, E.: From Secrecy to Soundness: Efficient Verification via Secure Computation. In: Abramsky, S., Gavoille, C., Kirchner, C., Meyer auf der Heide, F., Spirakis, P.G. (eds.) ICALP 2010, Part I. LNCS, vol. 6198, pp. 152–163. Springer, Heidelberg (2010)
3. Barbosa, M., Farshim, P.: Delegatable homomorphic encryption with applications to secure outsourcing of computation. Cryptology ePrint Archive, Report 2011/215 (2011)
4. Benabbas, S., Gennaro, R., Vahlis, Y.: Verifiable Delegation of Computation over Large Datasets. In: Rogaway, P. (ed.) CRYPTO 2011. LNCS, vol. 6841, pp. 111–131. Springer, Heidelberg (2011)
5. Bitansky, N., Canetti, R., Chiesa, A., Tromer, E.: From extractable collision resistance to succinct non-interactive arguments of knowledge, and back again. Cryptology ePrint Archive, Report 2011/443 (2011)
6. Blum, M., Feldman, P., Micali, S.: Non-interactive zero-knowledge and its applications (extended abstract). In: Proceedings of the ACM Symposium on Theory of Computing, STOC (1988)

7. Boneh, D., Franklin, M.: Identity-based encryption from the Weil pairing. SIAM Journal of Computing 32(3), 586–615 (2003)
8. Boneh, D., Sahai, A., Waters, B.: Functional Encryption: Definitions and Challenges. In: Ishai, Y. (ed.) TCC 2011. LNCS, vol. 6597, pp. 253–273. Springer, Heidelberg (2011)
9. Canetti, R., Riva, B., Rothblum, G.N.: Two 1-round protocols for delegation of computation. Cryptology ePrint Archive, Report 2011/518 (2011)
10. Chung, K.-M., Kalai, Y., Vadhan, S.: Improved Delegation of Computation Using Fully Homomorphic Encryption. In: Rabin, T. (ed.) CRYPTO 2010. LNCS, vol. 6223, pp. 483–501. Springer, Heidelberg (2010)
11. Gennaro, R., Gentry, C., Parno, B.: Non-interactive Verifiable Computing: Outsourcing Computation to Untrusted Workers. In: Rabin, T. (ed.) CRYPTO 2010. LNCS, vol. 6223, pp. 465–482. Springer, Heidelberg (2010)
12. Gentry, C., Wichs, D.: Separating succinct non-interactive arguments from all falsifiable assumptions. In: Proceedings of the ACM Symposium on Theory of Computing, STOC (2011)
13. Goldwasser, S., Kalai, Y.T., Rothblum, G.N.: Delegating computation: interactive proofs for muggles. In: Proceedings of the ACM Symposium on the Theory of Computing, STOC (2008)
14. Goldwasser, S., Lin, H., Rubinstein, A.: Delegation of computation without rejection problem from designated verifier CS-proofs. Cryptology ePrint Archive, Report 2011/456 (2011)
15. Goyal, V., Pandey, O., Sahai, A., Waters, B.: Attribute-based encryption for fine-grained access control of encrypted data. In: Proceedings of the ACM Conference on Computer and Communications Security, CCS (2006)
16. Green, M., Hohenberger, S., Waters, B.: Outsourcing the decryption of ABE ciphertexts. In: Proceedings of the USENIX Security Symposium (2011)
17. Katz, J., Sahai, A., Waters, B.: Predicate Encryption Supporting Disjunctions, Polynomial Equations, and Inner Products. In: Smart, N.P. (ed.) EUROCRYPT 2008. LNCS, vol. 4965, pp. 146–162. Springer, Heidelberg (2008)
18. Kilian, J.: Improved Efficient Arguments (Preliminary Version). In: Coppersmith, D. (ed.) CRYPTO 1995. LNCS, vol. 963, pp. 311–324. Springer, Heidelberg (1995)
19. Lewko, A., Okamoto, T., Sahai, A., Takashima, K., Waters, B.: Fully Secure Functional Encryption: Attribute-Based Encryption and (Hierarchical) Inner Product Encryption. In: Gilbert, H. (ed.) EUROCRYPT 2010. LNCS, vol. 6110, pp. 62–91. Springer, Heidelberg (2010)
20. Micali, S.: CS proofs (extended abstract). In: Proceedings of the IEEE Symposium on Foundations of Computer Science, FOCS (1994)
21. Ostrovsky, R., Sahai, A., Waters, B.: Attribute-based encryption with non-monotonic access structures. In: Proceedings of the ACM Conference on Computer and Communications Security, CCS (2007)
22. Papamanthou, C., Shi, E., Tamassia, R.: Publicly verifiable delegation of computation. Cryptology ePrint Archive, Report 2011/587 (2011)
23. Papamanthou, C., Tamassia, R., Triandopoulos, N.: Optimal Verification of Operations on Dynamic Sets. In: Rogaway, P. (ed.) CRYPTO 2011. LNCS, vol. 6841, pp. 91–110. Springer, Heidelberg (2011)
24. Sahai, A., Seyalioglu, H.: Worry-free encryption: functional encryption with public keys. In: Proceedings of the ACM Conference on Computer and Communications Security, CCS (2010)
25. Sahai, A., Waters, B.: Fuzzy Identity-Based Encryption. In: Cramer, R. (ed.) EUROCRYPT 2005. LNCS, vol. 3494, pp. 457–473. Springer, Heidelberg (2005)

On Black-Box Reductions between Predicate Encryption Schemes

Vipul Goyal[1], Virendra Kumar[2,*], Satya Lokam[1], and Mohammad Mahmoody[3]

[1] Microsoft Research, Bangalore, India
{vipul,satya}@microsoft.com
[2] Georgia Institute of Technology, Atlanta, GA, USA
virendra@cc.gatech.edu
[3] Cornell University, Ithaca, NY, USA
mohammad@cs.cornell.edu

Abstract. We prove that there is no black-box construction of a threshold predicate encryption system from identity-based encryption. Our result signifies nontrivial progress in a line of research suggested by Boneh, Sahai and Waters (TCC '11), where they proposed a study of the relative power of predicate encryption for different functionalities. We rely on and extend the techniques of Boneh et al. (FOCS '08), where they give a black-box separation of identity-based encryption from trapdoor permutations.

In contrast to previous results where only trapdoor permutations were used, our starting point is a more powerful primitive, namely identity-based encryption, which allows planting exponentially many trapdoors in the public-key by only planting a single master public-key of an identity-based encryption system. This makes the combinatorial aspect of our black-box separation result much more challenging. Our work gives the first impossibility result on black-box constructions of any cryptographic primitive from identity-based encryption.

We also study the more general question of constructing predicate encryption for a complexity class $\mathbb{F}$, given predicate encryption for a (potentially less powerful) complexity class $\mathbb{G}$. Toward that end, we rule out certain natural black-box constructions of predicate encryption for $\mathbf{NC}^1$ from predicate encryption for $\mathbf{AC}^0$ assuming a widely believed conjecture in communication complexity.

Keywords: Predicate Encryption, Black-Box Reductions, Identity-based Encryption, Communication Complexity.

1 Introduction

An encryption scheme enables a user to securely share data with other users. Traditional methods based on Secret-Key Cryptography and Public-Key Cryptography consider the scenarios where a user securely shares data with another *fixed* user whose identity (characterized by the possession of *the* decryption-key) it

---

* Part of the work was done while visiting Microsoft Research, India.

R. Cramer (Ed.): TCC 2012, LNCS 7194, pp. 440–457, 2012.

knows in advance. In particular, in these schemes, there is a bijection between the encryption-key and the decryption-key, fixed by the chosen encryption scheme.

As systems and networks grow in complexity, and in particular with the emergence of the cloud computing, the above viewpoint may be too narrow to cover many important applications. Often, a user might want to encrypt data to be shared with a large set of other users based on some common "property", or attribute, they satisfy. Membership in this set may not be known to the encryptor, or may not even be decidable in advance. Furthermore, a user might want to share data selectively so different users are able to decrypt different parts of that data. To cater to these scenarios, the notion of Predicate Encryption (or Attribute-based Encryption) has recently emerged. Predicate encryption was introduced by Sahai and Waters [31], and further developed in the work of Goyal et al. [17]. It has been the subject of several recent works, e.g., [11,19,24,28,10]. Predicate encryption is useful in a wide variety of applications; in particular, for fine-grained access control. It has also been a useful technical tool in solving seemingly unrelated problems, e.g., key escrow[15] and user revocation [5] in Identity-based Encryption (IBE). IBE [32,8,12] can be seen as the most basic form of a predicate encryption, where the predicate corresponds to a point function.

A predicate encryption scheme is defined in terms of a family $\mathbb{F}$ of Boolean functions (predicates) on a universe $\mathbb{A}$ of attributes. Decryption-keys are associated to a predicate $f \in \mathbb{F}$ and ciphertexts are labeled with (or are created based on) an attribute string $a \in \mathbb{A}$. A user with a decryption-key corresponding to f can decrypt a ciphertext labeled with x if and only if $f(x) = 1$. As argued by Boneh et al. [10], the key challenge in the study of predicate encryption (or Functional Encryption in general) is understanding what classes of functionalities $\mathbb{F}$ can be supported. If we could support any polynomial time computable predicate f, then any polynomial-time access control program that acts over a user's credentials could be supported [10].

Unfortunately, the current state of the art is far from being able to support an arbitrary polynomial-time f. Given this, an important direction Boneh et al. [10] suggested was to understand the relative strengths of predicate encryption schemes with respect to the functionalities they can support: When does a scheme for one functionality imply a scheme for another? In the absence of such a reduction, can we prove that predicate encryption for one functionality is inherently harder than for another? A meaningful approach to address this latter question is via *black-box separations* [18]; see [30,27] for a comprehensive survey on the topic. A proof that a cryptographic primitive P_1 cannot be constructed given black-box access to another primitive P_2 (and of course without incurring any additional assumptions) can be viewed as an indication that P_1 is in some sense a stronger primitive than P_2. Hence, to construct P_1 one may have to look for more powerful techniques, or stronger assumptions than for P_2 (or try non-black-box reductions). Thus, studying these questions would help us better understand the extent to which techniques for current predicate encryption systems might or might not be useful in obtaining systems for more general functionalities. The broad goal of this work is to make progress toward answering these questions.

Since a predicate encryption scheme has an associated family $\mathbb{F}$ of Boolean functions, a natural way to classify such schemes is according to the complexity class the corresponding family comes from. For example, we can call a scheme $(\mathbb{A}, \mathbb{F})$ an $\mathbf{AC}^0$-PE scheme, if every member of $\mathbb{F}$ can be computed by a constant-depth polynomial size circuit (an $\mathbf{AC}^0$ circuit) on an attribute string from $\mathbb{A}$. Hence, a concrete approach to compare predicate encryption schemes is to ask questions of the kind: *Given a predicate encryption scheme for predicates in complexity class $\mathbb{G}$, can we construct a scheme for predicates in a (potentially larger) complexity class $\mathbb{F}$ in a black-box way?* For example, it is well-known that the circuit class $\mathbf{NC}^1$ is strictly larger than $\mathbf{AC}^0$. Thus a concrete question is: *Is $\mathbf{NC}^1$-predicate encryption provably harder than $\mathbf{AC}^0$-predicate encryption with respect to black-box reductions?* A second aspect of our work is to try to relate (perhaps conjectured) separations among Boolean function complexity classes to black-box separations among the corresponding predicate encryption schemes.

1.1   Our Results

Our main result is a black-box separation of threshold predicate encryption (TPE) from identity-based encryption (IBE) schemes. To our knowledge, this is the first result on the *impossibility* of constructing a cryptographic primitive from IBE in a blackbox manner. Recall that IBE can be viewed as the most basic form of predicate encryption in which the decryption tests exact equality (in other words, the predicate is a point function). Hence, the first natural step in the study of the above question is whether IBE can be used to construct more general predicate encryption systems. Our results show that IBE cannot be used to construct even a basic system for threshold predicates (introduced by Sahai and Waters [31]). We believe that the question of IBE vs. more advanced predicate encryption systems is of special interest. IBE as a primitive is very well studied [8,12,7,6,34,14], and constructions of IBE are now known based on a variety of hardness assumptions.

Returning to our more general question, we rule out certain "natural" black-box constructions of predicate encryption for the class $\mathbf{NC}^1$ from predicate encryption for the class $\mathbf{AC}^0$, *assuming* a widely believed conjecture in the area of two-party communication complexity. Given black-box access to a predicate encryption scheme for $(\mathbb{B}, \mathbb{G})$, a natural way to construct a predicate encryption scheme for a "larger" system $(\mathbb{A}, \mathbb{F})$ is to use a a *Sharing-Based Construction* as follows. The decryption-key for an $f \in \mathbb{F}$ is simply the set of decryption keys for a set $S(f) = \{g_1, \ldots, g_q\}$ of predicates $g_i \in \mathbb{G}$ from the smaller system. Similarly, for each attribute $a \in \mathbb{A}$, we associate a set $S(a) = \{\alpha_1, \ldots, \alpha_q\}$ of attributes from $\mathbb{B}$. To encrypt a message m under an attribute a for the big system, we generate q *shares* $m_1, \ldots, m_q$ of m and encrypt m_j under the attribute α_j of the small system. The concatenation of these encrypted shares is the ciphertext of m under a. To decrypt, we try to decrypt each m_j using the decryption keys of each $g_i \in S(f)$. The sharing construction ensures that the shares m_j that are successfully decrypted, if any, in this process suffice to recover m. Thus the sharing-based construction is a rather natural and obvious way to build predicate encryption schemes for more complex functionalities from simpler ones. Our result shows that such a sharing-based construction is impossible if $\mathbb{F}$ is a family of functions in $\mathbf{NC}^1$ and $\mathbb{G}$ is any family of functions from $\mathbf{AC}^0$, assuming certain conjectures in communication complexity. It is worth noting that combinatorial arguments about sharing-based constructions form a core component of our main result on (unrestricted) black-box separation of TPE from IBE.

1.2   Techniques

We build upon and extend the techniques of Boneh et al. [9] (and a follow-up work by Katz and Yerukhimovich [20]) which rule out black-box construction of IBE from Trapdoor Permutations (TDP). Along the way, we also simplify several aspects of their proof. Given a black-box construction of TPE from IBE, our proof proceeds by designing an attack on TPE which succeeds with high probability (in fact arbitrarily close to the completeness probability of the purported TPE scheme). Somewhat more formally, we build an oracle $\mathcal{O}$ relative to which a CCA secure IBE exists, but any purported construction of a TPE relative to this oracle is insecure.

Our analysis of the attack roughly consists of a combinatorial part and a cryptographic part. The combinatorial aspect of our analysis is new and completely different from that in [9]. While the cryptographic part is similar in structure to that of [9], we do make several crucial modifications that makes our attack simpler and analysis cleaner.

A Comparison of the Combinatorial Aspects. At the heart of the proof of [9] is a combinatorial argument as follows. An IBE system obtained by a black-box construction from a TDP must embed in its public parameters the public keys of some permutations of the TDP oracle. The adversary's main goal is to collect all the trapdoors corresponding to these permutations. Such trapdoors are embedded in the decryption keys corresponding to identities in the IBE system. The main point is that there are only $q = \mathrm{poly}(\kappa)$ many permutations planted in the public parameters of the IBE, but they must also encode an exponential number of identities. Therefore, if we look at a sufficiently large set of random identities and their secret keys, and encrypt and decrypt a random message under these identities, during at most q of these decryptions we might encounter a "new" trapdoor (which is planted in the public-key to be used during encryption, but was not discovered during other decryptions). It follows, if we choose our identity set S to be of size $k \cdot q$ (and encrypt and decrypt random messages under them), and then choose an identity $id \xleftarrow{\$} S$ at random from those $q \cdot k$ identities, then with probability at least $1 - 1/k$ there is no new (undiscovered) trapdoor left for this identity id. Therefore, whatever is learned during the decryptions of the encryptions of random messages under the identities $S \setminus \{id\}$, is sufficient to decrypt a message encrypted under id without knowing its decryption-key.

This combinatorial argument immediately suggest the following attack. Get decryption-keys for all but a random identity id_* chosen from a large enough random set $S = id_1, \ldots, id_{k \cdot q}$ of identities. Collect the trapdoors learned from the encryptions of random messages under the identities in $S \setminus id_*$, and their decryptions using the corresponding decryption-keys. Try to decrypt the challenge ciphertext C encrypted under the identity id_*.

In our case, we have a related but more difficult question: what if we start with a more powerful primitive like an IBE and want to construct another "target" predicate encryption scheme? Now the intuition behind the combinatorial argument of [9] completely breaks down. The reason is that in our new setting, by planting only one (master) public-key of the IBE scheme in the public-key of the target predicate encryption, the encryption algorithm potentially has access to an *exponential* number of permutations (each indexed by an identity) whose trapdoors can be planted in the decryption-keys. In fact, each decryption-key of the predicate encryption system might have a unique trapdoor (corresponding to a unique identity derived from the description of the predicate). Hence, one can't hope to learn all trapdoors and use them to decrypt the challenge ciphertext. Thus, roughly speaking, by moving from a trapdoor permutation oracle to various forms of PE oracles such as IBE (as the primitive used in the construction), we are allowing the "universe" of trapdoor permutations planted in the public-key and decryption-keys to be exponentially large (rather than some fixed polynomial). The latter difference is the main reason behind the complications in the combinatorial aspect of our problem, because suddenly the regime of positive results becomes much richer, making the job of proving an impossibility result much more challenging.

Our proof relies on the collusion-resistance property of the predicate encryption. The "hope" that an attack exists comes from the following observations:

- The decryption key for each predicate may still consist of only a polynomial number of IBE decryption-keys.
- Each ciphertext is encrypted using a polynomially large set of identities such that a decryption-key for at least one of these identities is required to decrypt the ciphertext. On the other hand, each ciphertext can be decrypted by keys for an exponential number of different predicates (this follows from the property of a threshold encryption scheme). Call such predicates "related".
- This exponentially large set of related predicates must share an IBE decryption-key since they can decrypt a common ciphertext.

Our attack works by requesting sufficient number of decryption-keys for related predicates (which would still be unable to decrypt challenge ciphertext). Since related predicates share IBE decryption-keys, the adversary is able to collect all "useful" IBE decryption-keys. It is not surprising that the above combinatorial arguments sound as though they could already be used to attack sharing based constructions. Indeed, our core combinatorial lemma (Lemma 10) is used to refute any sharing-based construction of a TPE from an IBE (Corollary 11).

A Comparison of the Cryptographic Aspects. As in [9], turning the combinatorial analysis into a full-fledged impossibility result requires non-trivial black-box separation machinery. For this reason, even though the combinatorial argument of [9] is relatively simple, the full proof is quite complicated. The explanation for the complexity of such proofs is that one has to handle *all possible* constructions using a trapdoor permutation oracle (and not just where, for example, a decryption-key simply consists of decryption keys for various identities).

Although the overall structure of our proof is similar to that of [9], there are several differences in the detailed arguments. In fact, we make some crucial modifications which lead to a more direct attack and cleaner analysis. The first major modification is that our attacker "directly" learns the heavy queries (following the paradigm of [2,3]). In [9], the attack proceeds by having steps (such as several encryptions of a random bit under the challenge identity, repeating a few steps several times) whose indirect purpose is to learn the heavy queries. Secondly, since we start with an oracle which roughly provides four functionalities (as opposed to the three functionalities of a trapdoor permutation oracle), we need to modify and adapt the techniques of [9] to the new setting. Apart from these, there are significant differences in the manner we compare the various experiments which we believe makes the analysis cleaner and more general. The details regarding these can be found in Section 5 and in the full version [16] where we have deferred most of the proofs due to space constraints.

2   Preliminaries

Notation. For any probabilistic algorithm A, by $y \leftarrow A(x)$ we denote the process of executing A over the input x while using fresh randomness (which we do not represent explicitly) and getting the output y. By a *partial oracle* we refer to an oracle which is defined only for some of the queries it might be asked. By $[x \mapsto y] \in \mathcal{P}$ we mean that $\mathcal{P}(x) = y$ is defined. For a query x and a partial oracle $\mathcal{P}$, we misuse the notation and denote $x \in \mathcal{P}$ whenever an answer for x is defined in $\mathcal{P}$. By $\text{Supp}(X)$ we refer to the support set of the random variable X. For a random variable S whose values are sets, we call an element ϵ-heavy, if $\Pr[x \in S] \geq \epsilon$. The view of any probabilistic oracle algorithm A, denoted as $\text{View}(A)$ refers to its input, private randomness, and oracle answers (which all together determine the whole execution of A).

Definition 1 (Predicate Encryption). *A predicate encryption scheme* **PE** *for the predicate set* $\mathbb{F}_\kappa$ *and attribute set* $\mathbb{A}_\kappa$ *with completeness* ρ *consists of four probabilistic polynomial time algorithms* $\mathbf{PE} = (\mathbf{G}, \mathbf{K}, \mathbf{E}, \mathbf{D})$ *such that for every predicate* $f \in \mathbb{F}$, *every attribute* $a \in \mathbb{A}$ *such that* $f(a) = 1$, *and every message* M, *if we do the following steps, then with probability at least* ρ *it holds that* $M' = M$: (i). *generate a public-key and a master secret-key* $(\mathsf{PK}, \mathsf{SK}) \leftarrow \mathbf{G}(1^\kappa)$, (ii). *get a decryption-key* $\mathsf{DK}_f \leftarrow \mathbf{K}(\mathsf{SK}, f)$ *for the predicate* $f \in \mathbb{F}$, (iii). *encrypt the message* M *under the attribute* $a \in \mathbb{A}$ *and get* $C \leftarrow \mathbf{E}(\mathsf{PK}, a, M)$, *and finally,* (iv). *decrypt* C *using the decryption-key* DK_f *and get* $M' \leftarrow \mathbf{D}(\mathsf{PK}, \mathsf{DK}_f, C)$.

Definition 2 (Neighbor Sets of Predicates and Attributes). *For every set of predicates* $\mathbb{F}$ *and* $f \in \mathbb{F}$, *and for every set of attributes* $\mathbb{A}$ *and* $a \in \mathbb{A}$ *we define the following terminology:*

- $N(f) = \{a \mid a \in \mathbb{A}, f(a) = 1\}$ *and similarly* $\mathbb{N}(a) = \{f \mid f \in \mathbb{F}, f(a) = 1\}$.
- $\deg(f) = |N(f)|$ *and* $\deg(a) = |N(a)|$.

Since we always work with families of algorithms and sets indexed by a security parameter κ, when it is clear from the context we might omit the index κ.

Definition 3 (Security of Predicate Encryption). *Let* $\mathbf{PE} = (\mathbf{G}, \mathbf{K}, \mathbf{E}, \mathbf{D})$ *be a predicate encryption scheme with the predicate set* $\mathbb{F}$ *and the attribute set* $\mathbb{A}$. $\mathbf{PE}$ *is said to be CPA secure if for any* $\mathbf{PPT}$ *adversary* $\mathbf{Adv}$ *participating in the experiment below, the probability of* $\mathbf{Adv}$ *correctly outputting the bit* b *is at most* $1/2 + \mathrm{neg}(\kappa)$:

1. **Setup:** *Generate the keys* $(\mathsf{PK}, \mathsf{SK}) \leftarrow \mathbf{G}(1^\kappa)$ *and give* PK *to* $\mathbf{Adv}$.
2. **Query Keys:** $\mathbf{Adv}$ *adaptively queries some predicates* $f_i \in \mathbb{F}$ *for* $i = 1, 2, \dots$ *and is given the corresponding decryption-keys* $\mathsf{DK}_i \leftarrow \mathbf{K}(\mathsf{SK}, f_i)$.
3. **Challenge:** $\mathbf{Adv}$ *submits an attribute* $a \in \mathbb{A}$ *and a pair of messages* $M_0 \neq M_1$ *of the same length* $|M_0| = |M_1|$ *conditioned on*

$$f_i(a) = 0 \text{ for every predicate } f_i \text{ whose key } \mathsf{DK}_i \text{ is acquired by } \mathbf{Adv} \qquad (1)$$

   and is given $C \leftarrow \mathbf{E}(\mathsf{PK}, a, M_b)$ *for a randomly selected* $b \xleftarrow{\$} \{0, 1\}$.
4. $\mathbf{Adv}$ *continues to query keys for predicates subject to condition (1) and finally outputs a bit.*

$\mathbf{PE}$ *is said to be CCA secure if for any* $\mathbf{PPT}$ *adversary* $\mathbf{Adv}$ *participating in a modified experiment (explained next), the probability of* $\mathbf{Adv}$ *correctly outputting the bit* b *is at most* $1/2 + \mathrm{neg}(\kappa)$. *The modified experiment proceeds identically as the above experiment, except that after Step 3,* $\mathbf{Adv}$ *is also allowed to adaptively query ciphertexts* C_i *for* $i = 1, 2, \dots$ *encrypted under the attribute* a, *with the condition that* $C_i \neq C$ *for any* i, *and he is given the decrypted message* $M \leftarrow \mathbf{D}(\mathsf{DK}_f, C_i)$, *where* $\mathsf{DK}_f \leftarrow \mathbf{K}(\mathsf{SK}, f)$ *is a decryption-key for a predicate* f *such that* $f(a) = 1$.

Definition 4 (Identity-based Encryption [32]). *An Identity Based Encryption scheme is a predicate encryption scheme where* **(1)** *the predicate and attribute sets are equal* $\mathbb{A} = \mathbb{F} = \{0, 1\}^\kappa$ *(and are called the set of identities), and* **(2)** *for every predicate* $f \in \{0, 1\}^\kappa$ *and every attribute* $a \in \{0, 1\}^\kappa$ *we have that* $f(a) = 1$ *if and only if* $f = a$.

Definition 5 (Threshold Predicate Encryption [31]). *A Threshold Predicate Encryption with threshold* $0 < \tau < 1$ *(or simply a* τ-**TPE**) *is a predicate encryption where both the predicate and the attribute sets are equal to* $\{0, 1\}^\kappa$ *and for any predicate* $f \in \{0, 1\}^\kappa$ *and any attribute* $a \in \{0, 1\}^\kappa$ *we have that* $f(a) = 1$ *if and only if* $\langle f, a \rangle \geq \tau \cdot \kappa$ *where* $\langle f, a \rangle$ *is the inner product of the Boolean vectors* $f = (f_1, \dots, f_\kappa), a = (a_1, \dots, a_\kappa)$ *defined as* $\langle f, a \rangle = \sum_{i \in [\kappa]} a_i \cdot f_i$.

The notion of threshold predicate encryption was defined in [31] and is also known as the *fuzzy* IBE.

3 Sharing-Based Constructions and Impossibility Results

In this section, we describe two intuitive and simple approaches to build a predicated encryption scheme using another predicate encryption scheme as a black-box. It is interesting that the simpler of the two, the OR-based approach turns out to be as powerful as the seemingly more general Sharing-based approach. Even though ruling out constructions using these approaches is a weaker impossibility result than an unrestricted black-box separation (as we will do in Section 5), it seems instructive to refute these natural and general approaches to black-box reductions among predicate encryption schemes. In fact, our proof refuting OR-based constructions of TPE from this section forms the combinatorial core of our subsequent proof of a general black-box separation in Section 5. Moreover, the basic approach to building the attack needed in our proof (as well as that in [9]) of the general black-box separation results seems to benefit by keeping the sharing-based constructions in mind. In Section 4, we investigate a new approach to refute sharing-based constructions using (proved or conjectured) separation results in two-party communication complexity. In particular, we can use conjectures in communication complexity to give evidence that $\mathbf{NC}^1$-predicate encryption is strictly harder than $\mathbf{AC}^0$-predicate encryption.

Definition 6. *Let* $(\mathbb{F}, \mathbb{A})$ *and* $(\mathbb{G}, \mathbb{B})$ *be two pairs of predicate and attribute sets. We call* $S(\cdot)$ *a q-set system for* $(\mathbb{F}, \mathbb{A})$ *using* $(\mathbb{G}, \mathbb{B})$ *if* S *is a mapping defined over* $\mathbb{F} \cup \mathbb{A}$ *such that: (1) For every* $f \in \mathbb{F}$ *it holds that* $S(f) \subset \mathbb{G}$, *and for every* $a \in \mathbb{A}$ *it holds that* $S(a) \subset \mathbb{B}$, *and (2) For every* $x \in \mathbb{F} \cup \mathbb{A}$ *it holds that* $|S(x)| \leq q$.

Definition 7 (OR-based Construction). *We say there is an* OR-*based construction with set-size q for the pair of predicate and attribute sets* $(\mathbb{F} = \{f_1, \ldots\},$ $\mathbb{A} = \{a_1, \ldots\})$ *using another pair* $(\mathbb{G} = \{\varphi_1, \ldots\}, \mathbb{B} = \{\alpha_1, \ldots\})$ *if there exists a q-set system* $S(\cdot)$ *for* $(\mathbb{F}, \mathbb{A})$ *using* $(\mathbb{G}, \mathbb{B})$ *such that: For every* $f \in \mathbb{F}$ *and* $a \in \mathbb{A}$, *if* $S(f) = \{\varphi_1, \ldots, \varphi_{d_f}\}$ *and* $S(a) = \{\alpha_1, \ldots, \alpha_{d_a}\}$, *then* $f(a) = \bigvee_{i \in [d_f], j \in [d_a]} \varphi_i(\alpha_j)$. *We call the* OR-*based construction efficient if the mapping* $S(\cdot)$ *is efficiently computable.*

The encryption under attribute a of an OR-based construction works by encrypting a message M independently under every $\alpha_i \in S(a)$ and concatenating the corresponding ciphertexts. The decryption key for a predicate f is simply the set of keys DK_j for all $j \in [d_f]$, where DK_j is the decryption key for φ_j.

Lemma 8. *Suppose there exists an efficient* OR-*based construction for* $(\mathbb{F}, \mathbb{A})$ *using* $(\mathbb{G}, \mathbb{B})$. *Then a secure predicate encryption scheme* $\mathbf{PE}_1 = (\mathbf{G}_1, \mathbf{K}_1, \mathbf{E}_1, \mathbf{D}_1)$ *for* $(\mathbb{F}, \mathbb{A})$ *with completeness* ρ *can be constructed (in a black-box way) from any secure predicate encryption scheme* $\mathbf{PE}_2 = (\mathbf{G}_2, \mathbf{K}_2, \mathbf{E}_2, \mathbf{D}_2)$ *for* $(\mathbb{G}, \mathbb{B})$ *with completeness* ρ.

Clearly, the OR-based construction of Lemma 8 is not the only way that one can imagine to construct an $\mathbb{F}$-$\mathbf{PE}$ from a $\mathbb{G}$-$\mathbf{PE}$. In fact, as noted also by [20] in the context of using trapdoor permutations, there is a possibility of employing a more complicated "sharing-based" approach that generalizes the OR-based construction. The idea is to use a set system $S(\cdot)$ in a similar way to the OR-based construction, but to encrypt the message M differently: instead of encrypting the message M d_a times, first construct some "shares" $M_1, \ldots, M_{d_a}$ of M, and then encrypt each M_i using α_i. To get the completeness and the security, we need the following two properties.

- *Completeness:* For every $f \in \mathbb{F}$ such that $f(a) = 1$, the set of indices $I_S(a, f) = \{j \mid \exists \varphi \in S(f) \text{ such that } \varphi(\alpha_j) = 1\}$ is rich enough that $\{M_i \mid i \in I_S(a, f)\}$ can be used to reconstruct M.
- *Security:* For every choice of $a_*, f_*, f_1, \ldots, f_k$ for $k = \mathrm{poly}(\kappa)$ such that $f_*(a_*) = 1$ and $f_i(a_*) = 0$ for all $i \in [k]$, it holds that $C_S(a_*, f_*) \not\subseteq \bigcup_{j \in [k]} C_S(a_*, f_j)$, where $C_S(a, f) = \{\alpha_i \mid i \in I_S(a, f)\}$. This is because otherwise the adversary can acquire keys for $f_1, \ldots, f_k$ and use the sub-keys planted in them to decrypt enough of the shares of M_i's and reconstruct M which is encrypted under the attribute a_*.

Despite the fact that the sharing-based approach is more general than the OR-based approach, for the case of polynomial sized sets $q = \mathrm{poly}(\kappa)$, we show that the construction of Lemma 8 is indeed as powerful as any sharing-based approach:

Lemma 9. *There is a sharing based construction for the predicate system* $\mathbb{F}$ *using* $\mathbb{G}$ *if and only if there exists an* OR-*based construction.*

Note that by proving Theorem 19, we shall rule out an OR-based (and hence sharing-based) constructions along the way. A special case of the following combinatorial lemma, Corollary 11, shows that no OR-based (nor sharing-based) construction of τ-**TPE** from **IBE** exists for any constant $0 < \tau < 1$. Moreover, not surprisingly, we will use this lemma in our proof of Theorem 19.

Lemma 10. *Let* $\mathbb{F} = \mathbb{A} = \{0, 1\}^\kappa$ *denote the set of attributes and predicates for* τ-**TPE** *for a constant* $0 < \tau < 1$. *Also suppose that the following sets of size at most* $q = \mathrm{poly}(\kappa)$ *are assigned to* $\mathbb{F}$, $\mathbb{A}$, *and* $\mathbb{F} \times \mathbb{A}$: $S(a)$ *for* $a \in \mathbb{A}$, $S(f)$ *for* $f \in \mathbb{F}$, *and* $S(a, f)$ *for* $(a, f) \in \mathbb{A} \times \mathbb{F}$. *Then, there exists a sampling algorithm* Samp *that, given an input parameter* $\epsilon > 1/\mathrm{poly}(\kappa)$, *outputs* $k + 1 = \mathrm{poly}(\kappa)$ *pairs* $(f_*, a_*), (f_1, a_1), \ldots, (f_k, a_k)$ *such that with probability at least* $1 - \epsilon$ *over the randomness of* Samp *the following holds:*

1. $f_*(a_*) = 1$ *and* $f_i(a_i) = 1$ *for all* $i \in [k]$ *(this part holds with probability 1),*
2. $f_i(a_*) = 0$ *for all* $i \in [k]$,
3. $S(a_*) \cap S(f_*) \cap S(a_*, f_*) \subseteq \bigcup_{i \in [k]} S(a_i, f_i)$.

Moreover, the algorithm Samp *chooses its* $k + 1$ *pairs without the knowledge of the set system* $S(\cdot)$. *Therefore we call* Samp *an* oblivious sampler *against the predicate structure of* τ-**TPE**.

Note that although $\mathbb{F} = \mathbb{A}$, the sets $S(a)$ for $a \in \mathbb{A}$ and $S(f)$ for $f \in \mathbb{F}$ are potentially different even if a and f represent the same string. Intuitively, the set $S(a)$ refers to the set of sub-attributes (or identities in case of using IBE as the black-box primitive) used during an encryption of a random message under the attribute a, the set $S(f)$ refers to the set of decryption-keys planted in the decryption-key of f, and finally $S(a, f)$ refers to the decryption-keys discovered during the decryption of the mentioned random encryption (under the attribute a) using the generated key for f.

Proof. Let $\mathcal{A}$ be the set of vectors in $\{0,1\}^\kappa$ of normalized Hamming weight τ, namely $\mathcal{A} = \{a \mid a = (a_1, \ldots, a_\kappa) \in \{0,1\}^\kappa, \sum_i a_i = \tau \cdot \kappa\}$. Also let $\mathcal{F}$ be the set of vectors in $\{0,1\}^\kappa$ of normalized Hamming weight $\tau' = \tau + \frac{1-\tau}{2}$. Consider a bipartite graph G with nodes $(\mathcal{A}, \mathcal{F})$ and connect $a \in \mathcal{A}$ to $f \in \mathcal{F}$ iff $f(a) = 1$ according to τ-**TPE** (i.e., the indexes of the nonzero components of a is a subset of those of f). We will later use the fact that G is a regular graph (on its $\mathcal{F}$ side). For any vertex x in G let $N(x)$ be the set of neighbors of x in the graph G. The covering-sampler acts as follows: Choose $p = \mathrm{poly}(\kappa)$ and $h = \mathrm{poly}(\kappa)$ to satisfy $q(\frac{1}{p} + \frac{1}{h} + (1 - \frac{1}{h})^p) < \frac{\epsilon}{2}$ (e.g., this can be done by setting $h = \sqrt{p}$ and choosing p large enough). Choose $f_* \overset{\$}{\leftarrow} \mathcal{F}$ at random. Choose $a_*, a_1, \ldots, a_p \overset{\$}{\leftarrow} N(f_*)$ at random *with possible repetition* from the neighbors of f_*. For each $i \in [p]$, choose p random neighbors $f_{i1}, \ldots, f_{ip} \overset{\$}{\leftarrow} N(a_i)$ of a_i (repetition is allowed). Output the $p^2 + 1$ pairs: $(a_*, f_*), (a_i, f_{ij})_{i \in [p], j \in [p]}$.

Now we prove that with probability at least $1 - \epsilon/2 - \mathrm{neg}(\kappa) > 1 - \epsilon$ the output pairs have the properties specified in Lemma 10.

Property (1) holds by construction.

Since $0 < \tau < \tau' < 1$ are constants, using standard probabilistic arguments one can easily show that the probability of f_{ij} being connected to a_* in G (i.e., $f_{ij}(a_*) = 1$) is $\mathrm{neg}(\kappa)$ (given a_*, a_i are random subsets of f_*, a random superset f_{ij} of a_i is exponentially unlikely to pick all the elements of a_*). Thus (2) holds.

The challenging part is to show that (3) holds, i.e., the following: With probability at least $1 - q(\frac{1}{p} + \frac{1}{\sqrt{p}} + (1 - \frac{1}{\sqrt{p}})^p) \geq 1 - \epsilon/2$ it holds that $S(a_*) \cap S(f_*) \cap S(a_*, f_*) \subset \cup_{ij} S(a_i, f_{ij})$. The proof will go through several claims.

In the following let $h = \sqrt{p}$. For an attribute node $a \in \mathcal{A}$ of G, define $H(a)$ to be the set of "heavy" elements that with probability at least $1/h$ are present in $S(a, f)$ for a random neighbor f of a, i.e., $H(a) = \{x : \Pr[x \in S(a, f) \mid f \overset{\$}{\leftarrow} N(a)] > 1/h\}$. Note that $H(a)$ is not necessarily a subset of $S(a)$.

Claim. Define BE_1 to be the bad event "$S(a_*) \cap S(a_*, f_*) \not\subseteq H(a_*)$." Then, $\Pr[\mathsf{BE}_1] \leq q/h$.

Proof. Since G is regular on its $\mathbb{F}$ side, conditioned on a fixed a_* the distribution of f_* is still uniform over $N(a_*)$. Now fix a_* and fix an element $b \in S(a_*)$. If b is not in $H(a_*)$, then over the random choice of $f_* \overset{\$}{\leftarrow} N(a_*)$, it holds that $\Pr[b \in S(a_*, f_*)] \leq 1/h$. The claim follows by a union bound over the q elements in $S(a_*)$.   □

Claim. Define BE_2 to be the bad event "there exists a $b \in S(f_*)$ such that $b \in H(a_*)$ but for every $i \in [p]$, $b \notin H(a_i)$, i.e., $S(f_*) \cap H(a_*) \not\subseteq \cup_i H(a_i)$." Then, $\Pr[\mathsf{BE}_2] \leq q/p$.

Proof. It is enough to bound BE_2 by $1/p$ for a fixed $b \in S(f_*)$ and the claim follows by union bound over the elements of $S(f_*)$. But when $b \in S(f_*)$ is fixed, we can pretend that a_* is chosen at random from the sequence $a_0, \ldots, a_p$ *after* they are chosen and are fixed. In that case BE_2 happens if there is only a unique $j \in \{0, \ldots, p\}$ such that $b \in H(a_j)$ and a_* chooses to be a_j. The latter happens with probability at most $1/(p+1) < 1/p$.   □

Claim. Define $\mathsf{BE_3}$ to be the bad event *"given neither* $\mathsf{BE_1}$ *nor* $\mathsf{BE_2}$ *happens,* $S(a_*) \cap S(f_*) \cap S(a_*, f_*) \not\subseteq \cup_{i,j} S(a_i, f_{ij})$*."* Then, $\Pr[\mathsf{BE_3}] \leq q(1 - 1/h)^p$.

Proof. We assume events $\mathsf{BE_1}$ and $\mathsf{BE_2}$ have not happened and perform the analysis. By $\neg\mathsf{BE_1}$, we have $S(a_*) \cap S(a_*, f_*) \subseteq H(a_*)$. Moreover, since $\neg\mathsf{BE_2}$ holds, any element $b \in S(f_*) \cap H(a_*)$ will be in $H(a_i)$ for at least one $i \in [p]$. Therefore for each $j \in [p]$, $\Pr[b \in S(a_i, f_{ij})] \geq 1/h$ holds by the definition of heavy sets, and thus $b \notin \cup_j S(a_i, f_{ij})$ can hold only with probability at most $(1 - 1/h)^p$. By union bound, the probability that there exists a $b \in S(a_*) \cap S(f_*) \cap S(a_*, f_*)$ such that $b \notin \cup_j S(a_i, f_{ij})$ is bounded by $q(1 - 1/h)^p$. □

From Claims 3, 3, and 3, it follows that (3) fails with probability at most $q(\frac{1}{p} + \frac{1}{h} + (1 - \frac{1}{h})^p) < \frac{\epsilon}{2}$. Therefore, the sampled $[a_*, f_*, \{f_{ij}\}_{i\in[p], j\in[p]}]$ will have the desired properties with probability at least $1 - \text{neg}(\kappa) - \epsilon/2$ which finishes the proof of Lemma 10. □

Using Lemma 10, it is almost straightforward to prove the following.

Corollary 11. *For any constant $0 < \tau < 1$, there is no OR-based (nor sharing-based) construction of τ-TPE schemes from IBE schemes.*

4   The Communication Complexity Approach

In this section, we show an alternative general approach to refute sharing-based constructions of predicate encryption schemes using separation results in two-party communication complexity. In particular, using conjectured separations in communication complexity, we prove the impossibility of a sharing-based construction of $\mathbf{NC^1}$-PE from $\mathbf{AC^0}$-PE, thus making some progress toward the question of separating PE schemes based on the complexity classes the underlying predicates come from. On the other hand, we are currently able to apply this approach only to sharing-based constructions rather than to general black-box constructions.

Let $(\mathbb{A}, \mathbb{F})$ be a predicate encryption scheme. W.l.o.g. we identify $\mathbb{A}$ with $\{0,1\}^\kappa$ and think of $\mathbb{F}$ as a family of functions $\{f_b : \{0,1\}^\kappa \to \{0,1\}\}_{b\in\{0,1\}^\kappa}$, i.e., we assume for simplicity that $|\mathbb{F}| = 2^\kappa$ and its members are also indexed by $b \in \{0,1\}^\kappa$. We may abuse this notation and refer to b itself as a member of $\mathbb{F}$. We can then talk about the communications complexity of $\mathbb{F}$ when $b \in \mathbb{F}$ is given to Bob and $a \in \mathbb{A}$ to Alice. We can represent this communication complexity problem by the $\{0,1\}$-matrix with rows indexed by $\mathbb{A}$ and columns by $\mathbb{F}$. With a little more abuse of notation, we denote this matrix also by $\mathbb{F} = (f_b(a))_{a,b}$ and refer to the communication complexity of $\mathbb{F}$. Recall that the essential resource in communication complexity is the number of bits Alice and Bob need to communicate to determine $f_b(a)$. Various models such as deterministic, randomized (public or private coins), nondeterministic, etc., communication complexity can be defined naturally. For details on such models, we refer to the classic book by Kushilevitz and Nisan [22], the paper by Babai et al. [1], and the surveys by Lokam [26] and Lee and Shraibman [23].

To connect communication complexity to OR-based constructions using IBE, we use the model of *Merlin-Arthur (MA) games* in communication complexity:

Definition 12 (Merlin-Arthur Protocols [21]). *A matrix* $\mathbb{F}$ *is said to have an MA-protocol of complexity* $\ell + c$ *if there exists a c-bit randomized public-coin verification protocol* Π *between Alice and Bob such that*

- $\mathbb{F}(a,b) = 1 \Rightarrow \exists w \in \{0,1\}^{\ell} \; \Pr[\Pi((a,w),(b,w)) = 1] \geq 2/3,$
- $\mathbb{F}(a,b) = 0 \Rightarrow \forall w \in \{0,1\}^{\ell} \; \Pr[\Pi((a,w),(b,w)) = 1] \leq 1/3.$

The MA-complexity of $\mathbb{F}$, *denoted* $\mathsf{MA}(\mathbb{F})$, *is the minimum complexity of an MA protocol for the matrix* $\mathbb{F}$.

With this definition, the well-known fact (see, for example, [22]) that EQUALITY has public coin randomized communication complexity of $O(1)$, and our Definition 7 of OR-construction, the following lemma is easy.

Lemma 13. *Suppose there is an* OR-*based construction of a predicate encryption scheme* $(\mathbb{A}, \mathbb{F})$ *using an IBE scheme* $(\mathbb{B}, \mathbb{G})$. *Then* $\mathsf{MA}(\mathbb{F}) = O(\log \kappa)$.

Using a result due to Klauck [21] that $\mathsf{MA}(\mathsf{DISJOINTNESS}) = \Omega(\sqrt{\kappa})$, we can show.

Theorem 14. *For some constant* $0 < \tau < 1$, *e.g.,* $\tau = 1/3$, *there is no* OR-*based (and hence no sharing-based) construction of a* τ-*TPE scheme from IBE.*

To derive separations among stronger predicate encryption schemes based on sharing constructions, we need to recall definitions of languages and complexity classes in two-party communication complexity, in particular, $\mathbf{PH}^{cc}$ and $\mathbf{PSPACE}^{cc}$.

Complexity classes in two-party communication complexity are defined in terms of languages consisting of pairs of strings (a, b) such that $|a| = |b|$. Denote by $\{0,1\}^{2*}$ the universe $\{(a,b) : a, b \in \{0,1\}^* \text{ and } |a| = |b|\}$. For a language $L \subseteq \{0,1\}^{2*}$, we denote its characteristic function on pairs of strings of length κ by L_κ. The language L_κ is naturally represented as a $2^\kappa \times 2^\kappa$ matrix with $\{0,1\}$ or ± 1 entries.

Definition 15. *Let* $l_1(\kappa), \ldots, l_d(\kappa)$ *be nonnegative integers such that* $l(\kappa) := \sum_{i=1}^{d} l_i(\kappa) \leq (\log \kappa)^c$ *for a fixed constant* $c \geq 0$. *A language* $L \subseteq \{0,1\}^{2*}$ *is in* $\mathbf{\Sigma}_d^{cc}$ *if there exist* $l_1(\kappa), \ldots, l_d(\kappa)$ *as above and Boolean functions* $\varphi, \psi :$ $\{0,1\}^{\kappa + l(\kappa)} \longrightarrow \{0,1\}$ *such that* $(a,b) \in L_\kappa$ *if and only if* $\exists u_1 \forall u_2 \ldots Q_d u_d$ $(\varphi(a, u) \Diamond \psi(b, u))$, *where* $|u_i| = l_i(\kappa), u = u_1 \ldots u_d, Q_d$ *is* $\forall$ *for d even and is* $\exists$ *for d odd, and,* $\Diamond$ *stands for* $\vee$ *if d is even and for* $\wedge$ *if d is odd.*

- *By allowing a bounded number of alternating quantifiers, we get an analog of the polynomial time hierarchy:* $\mathbf{PH}^{cc} = \bigcup_{d \geq 0} \mathbf{\Sigma}_d^{cc}$.
- *By allowing an unbounded, but at most* $\mathrm{polylog}(\kappa)$ *alternating quantifiers, we get an analog of* $\mathbf{PSPACE}$: $\mathbf{PSPACE}^{cc} = \bigcup_{c > 0} \bigcup_{d \leq (\log \kappa)^c} \mathbf{\Sigma}_d^{cc}$.

The following lemma shows a connection between the communication complexity class $\mathbf{PH}^{cc}$ and OR-based constructions using $\mathbf{AC}^0$-predicate encryption.

Lemma 16. *Suppose a predicate encryption scheme* $(\mathbb{A}, \mathbb{F})$ *is obtained by an* OR-*based construction using an* $\mathbf{AC}^0$-*predicate encryption scheme. Then the language given by the sequence of matrices* $\{\mathbb{F}\}_\kappa$ *is in* $\mathbf{PH}^{\mathrm{cc}}$.

Proof. By hypothesis, for a given $f_b \in \mathbb{F}$, we have $\mathbf{AC}^0$ circuits $\varphi_{1b}, \ldots, \varphi_{qb}$ and for a given $a \in \mathbb{A}$, we have $\alpha_{1a}, \ldots, \alpha_{qa}$ such that $f_b(a) = \vee_{i,j} \varphi_{ib}(\alpha_{ja})$. Knowing f_b, Bob can compute the circuit $C_b(z) \equiv \bigvee_{ij} \varphi_{iy}(z_j)$, where $z = (z_1, \ldots, z_q)$, $|z_j| = |\alpha_j|$. Knowing a, Alice can compute $\alpha_a = (\alpha_{1a}, \ldots, \alpha_{qa})$ on which C_b needs to be evaluated. We give a protocol with a bounded number of alternations for $\mathbb{F}$. Let the depth of C_b be d (including the top OR-gate). An existential player will have a move for an OR gate in C_b and a universal player will have a move for an AND gate. Their d moves will describe an accepting path in C_b on α_a. For example, assuming AND and OR gates alternate in successive layers, $\exists w_1 \forall w_2 \cdots Q_d w_d \; \gamma(C_b, w_1, \ldots, w_d)(\alpha_a)$ describes a path in C_b – start with the top OR gate and follow the wire w_1 to the AND gate below and then the wire w_2 from this gate and so on – ending in a gate $\gamma := \gamma(\ldots)$ to witness the claim that $f_b(a) = 1$. Since Bob knows C_b, he can verify the correctness of the path $w_1 w_2 \cdots w_k$ in the circuit and the type of the gate γ given by the path. He then sends the labels of the inputs and the type (AND or OR) of the gate to Alice, who responds with $\gamma(\alpha_a)$. Bob can verify that this will ensure $C_b(\alpha_a) = 1$. On the other hand, if $C_b(\alpha_a) = 0$, then it is easy to see that the existential player will not have a winning strategy to pass verification protocol of Alice and Bob on their inputs a and C_b. It follows that $\mathbb{F}$ has a protocol with at most d alternations and hence $\{\mathbb{F}\}_\kappa \in \mathbf{PH}^{\mathrm{cc}}$.   □

This lemma enables us to show the impossibility of OR-based constructions of predicate encryption schemes using $\mathbf{AC}^0$-predicate encryption. In particular,

Theorem 17. *Suppose* $\mathbf{PH}^{\mathrm{cc}} \neq \mathbf{PSPACE}^{\mathrm{cc}}$. *Then, there is no* OR-*based construction of an* $\mathbf{NC}^1$-*PE scheme from any* $\mathbf{AC}^0$-*PE scheme. In particular, there is an* $\mathbf{NC}^1$-*function family* $\mathbb{F}$ *(derived from so-called Sipser functions [33]) such that* $(\mathbb{A}, \mathbb{F})$ *does not have an* OR-*based construction from any* $\mathbf{AC}^0$-*PE scheme.*

However, it is a longstanding open question in communication complexity to separate $\mathbf{PSPACE}^{\mathrm{cc}}$ from $\mathbf{PH}^{\mathrm{cc}}$. Currently it is known that such a separation holds if certain Boolean matrices can be shown to have high rigidity, a connection explained in [29,25].

Corollary 18. *Suppose Hadamard matrices are as highly rigid as demanded in [29,25]. Then, predicate encryption defined by the parity functions (arising from Inner Product mod 2 matrix) does not have an* OR-*based construction from any* $\mathbf{AC}^0$-*predicate encryption scheme.*

5   Separating TPE from IBE

In this section, we prove that there is no general black-box construction of threshold predicate encryption schemes from identity-based encryption schemes.

Theorem 19. *Let $\kappa \in \mathbb{N}$ be the security parameter. Then, there exists an oracle $\mathcal{O}$ relative to which CCA secure IBE schemes exist, as per Definition 3. However, for any constant $0 < \tau < 1$, there exists a query-efficient (i.e., that makes at most $\mathrm{poly}(\kappa)$ queries to $\mathcal{O}$) adversary* **Adv** *that can break even the CPA security of any τ-TPE scheme relative to $\mathcal{O}$, again as per Definition 3. Moreover,* **Adv** *can be implemented in $\mathrm{poly}(\kappa)$-time if given access to a* **PSPACE** *oracle, and its success probability can be made arbitrarily close to the completeness of the τ-TPE scheme.*

We will first define our random IBE oracle, $\mathcal{O}_{\mathbf{IBE}}$, also denoted by $\mathcal{O}$ for short, (which trivially implies a CCA secure IBE as outlined in Remark 21), and then break any τ-TPE (with a constant τ) relative to this oracle.

Construction 20 (Randomized oracle $\mathcal{O} = (\mathbf{g}, \mathbf{k}, \mathbf{id}, \mathbf{e}, \mathbf{d})$). *By $\mathcal{O}_\lambda$ we refer to the part of $\mathcal{O}$ whose answers are λ bits, and $\mathcal{O}$ is the union of $\mathcal{O}_\lambda$ for all λ.*

- *The* master-key generating oracle $\mathbf{g} : \{0,1\}^\lambda \mapsto \{0,1\}^\lambda$ *is a random permutation that takes as input a secret-key $\mathsf{sk} \in \{0,1\}^\lambda$, and returns a public-key $\mathsf{pk} \in \{0,1\}^\lambda$.*
- *The* decryption-key generating oracle $\mathbf{k} : \{0,1\}^{2\lambda} \mapsto \{0,1\}^\lambda$ *takes as input a secret-key $\mathsf{sk} \in \{0,1\}^\lambda$ and an identity $\alpha \in \{0,1\}^\lambda$, and returns a decryption-key $\mathsf{dk}_\alpha \in \{0,1\}^\lambda$. We require $\mathbf{k}(\mathsf{sk}, \cdot)$ to be a random permutation over $\{0,1\}^\lambda$ for every $\mathsf{sk} \in \{0,1\}^\lambda$.*
- *The* Identity finding oracle $\mathbf{id} : \{0,1\}^{2\lambda} \mapsto \{0,1\}^\lambda$ *takes as input a public-key $\mathsf{pk} \in \{0,1\}^\lambda$ and a decryption-key $\mathsf{dk} \in \{0,1\}^\lambda$, and returns the unique α such that $\mathbf{k}(\mathsf{sk}, \alpha) = \mathsf{dk}$, where $\mathsf{sk} = \mathbf{g}^{-1}(\mathsf{pk})$.*
- *The* encryption oracle $\mathbf{e} : \{0,1\}^{3\lambda} \mapsto \{0,1\}^\lambda$ *takes as input a public-key $\mathsf{pk} \in \{0,1\}^\lambda$, an identity $\alpha \in \{0,1\}^\lambda$ and a message $m \in \{0,1\}^\lambda$, and returns a ciphertext $c \in \{0,1\}^\lambda$. We require $\mathbf{e}(\mathsf{pk}, \alpha, \cdot)$ to be a random permutation over $\{0,1\}^\lambda$ for every $(\mathsf{pk}, \alpha) \in \{0,1\}^{2\lambda}$.*
- *The* decryption oracle $\mathbf{d} : \{0,1\}^{3\lambda} \mapsto \{0,1\}^\lambda$ *takes as input a public-key $\mathsf{pk} \in \{0,1\}^\lambda$, a decryption-key $\mathsf{dk} \in \{0,1\}^\lambda$ and a ciphertext $c \in \{0,1\}^\lambda$, and returns the unique m such that $\mathbf{e}(\mathsf{pk}, \alpha, m) = c$, where $\alpha = \mathbf{id}(\mathsf{pk}, \mathsf{dk})$.*

By an IBE oracle, we refer to an oracle in the support set of $\mathcal{O}$, $\mathrm{Supp}(\mathcal{O})$, and by a partial IBE oracle we refer to a partial oracle that could be extended to an oracle in $\mathrm{Supp}(\mathcal{O})$.

Remark 21 (CCA secure IBE relative to $\mathcal{O}$). To encrypt a bit $b \in \{0,1\}$ under identity α and public-key pk, the encryption algorithm extends b to a λ-bit random string: $m = (b, b_1, \ldots, b_{\lambda-1}), b_i \xleftarrow{\$} \{0,1\}$ and gets the encryption $c = \mathbf{e}(\mathsf{pk}, \alpha, m)$. To decrypt, we decrypt c and output its first bit. By independently encrypting the bits of a message $m = (m_1, \ldots, m_n)$, with $n = \mathrm{poly}(\kappa)$, and using a standard hybrid argument, one can generalize the scheme to arbitrarily long messages. This construction is only CPA secure, where any adversary has advantage at most $2^{-\Theta(\kappa)}$. But, this can easily be transformed in a blackbox manner into a CCA secure construction, without incurring any additional assumptions, using the Fujisaki-Okamoto transform [13] in the random oracle model [4]. We note that even though $\mathcal{O}$ is not exactly a random oracle, for our purposes it suffices to use one of the sub-oracles of $\mathcal{O}$ as a random oracle in the above transform.

Now we present an attack that aims to break any τ-TPE in an $\mathcal{O}$-relativized world by asking only poly(κ) queries to the random IBE oracle $\mathcal{O}$, where κ is the security parameter of the τ-TPE scheme. We prove the query-efficiency and the success probability of our attack in the full version [16]. Similar to the attack of [9], our attack can easily be implemented in poly(κ)-time if $\mathbf{P} = \mathbf{PSPACE}$[1], and the relativizing reductions can be ruled out by adding a $\mathbf{PSPACE}$ oracle to $\mathcal{O}$.

We first note that any black-box construction of τ-TPE schemes from IBE schemes can potentially call the oracle $\mathcal{O}_\lambda$ over different values of λ which are potentially different from the security parameter of the τ-TPE scheme itself. However, similar to [9], we assume that the τ-TPE scheme asks its queries to $\mathcal{O}_\lambda$ only for one value of λ. This assumption is purely to simplify our presentation of the attack and its analysis, and all the arguments below extend to the general case (of asking queries over any parameter $\lambda > \log s$) in a straightforward way.

We also assume that λ is large enough in the sense that $2^\lambda > s$ for an arbitrarily large $s = \text{poly}(\kappa)$ that can be chosen in the description of the attack. The reason for the latter assumption is that the adversary can always ask and learn *all* the oracle queries to $\mathcal{O}$ that are of logarithmic length $O(\lambda) = O(\log \kappa)$, simply because there are at most $2^{O(\lambda)} = \text{poly}(\kappa)$ many queries of this form.[2]

Construction 22 (Adv Attacking the Scheme τ-TPE$^{\mathcal{O}}$). *The parameters are as follows. q: the total number of queries asked by the components of the scheme τ-TPE all together, κ: the security parameter of τ-TPE, $\epsilon = 1/\text{poly}(\kappa)$ and $s = \text{poly}(\kappa)$: input parameter to the adversary $\mathbf{Adv}$, $\lambda \leq \text{poly}(\kappa)$: the parameter which determines the output length of the queries asked by the components of τ-TPE to the oracle $\mathcal{O}$. It is assumed that $2^\lambda > s$ for some $s = \text{poly}(\kappa)$ to be chosen later. Our adversary $\mathbf{Adv}$ executes the following.*

1. **Sampling Predicates and Attributes:** $\mathbf{Adv}$ *executes the sampling algorithm* Samp *of Lemma 10 with the parameter ϵ, over the predicate structure of τ-TPE, to get $k+1$ pairs $(a_*, f_*), \{(a_i, f_i)\}_{i \in [k]}$. Recall that this sampling is done only by knowing the predicate structure of τ-TPE and is independent of the actual implementation of the scheme. It can be done, for example, without the knowledge of* PK.

2. **Receiving the Keys:** $\mathbf{Adv}$ *receives from the challenger: the public-key* PK *and the decryption-keys $\{DK_i\}_{i \in [k]}$, where DK_i is the generated decryption-key for f_i. We also assume that DK_* is generated by the challenger, although $\mathbf{Adv}$ does not receive it. Let V be the view of the algorithms executed by the challenger so far that generated the keys $PK, DK_*, DK_1, \ldots, DK_k$. Let $Q(V)$ be the partial oracle consisting of the queries (and their answers) specified in V. By writing in the bold font $\mathbf{V}$, we refer to V as a random variable.*

3. **Encrypting Random Bits:** *For all $i \in [k]$, $\mathbf{Adv}$ chooses a random bit $d \xleftarrow{\$} \{0,1\}$, computes the encryption $C_i \leftarrow \mathbf{E}(PK, a_i, d)$, and then the decryption*

---

[1] A good "approximation" of the attack can also be implemented assuming $\mathbf{P} = \mathbf{NP}$.

[2] In [9] a scheme that asks such queries is called "degenerate" and is handled similarly.

$D(PK, DK_i, C_i)$. Let $\mathcal{L}_0$ be the partial oracle consisting of the oracle queries (and their answers) that **Adv** observes in this step.

4. **Learning Heavy Queries:** *This step consists of some internal rounds. For $j = 1, 2, \ldots$ do the following. Let $\mathcal{L}_j$ be the partial oracle consisting of the oracle queries (and their answers) that **Adv** has learned about $\mathcal{O}$ till the end of the j'th round[3] of this learning step. Let $\mathbf{V}_j = (\mathbf{V} \mid \mathcal{L}_j, PK, \{DK\}_{i \in [k]})$ be the distribution of the random variable $\mathbf{V}$ (also including the randomness of $\mathcal{O}$) conditioned on the knowledge of $(\mathcal{L}_j, PK, \{DK\}_{i \in [k]})$. For a partial oracle $\mathcal{P}$, let $\overline{\mathcal{P}}$ denote its closure[4]. Now, if there is any query x such that $x \notin \mathcal{L}_j$ but $\Pr[x \in \overline{Q(\mathbf{V}_j)}] \geq \epsilon$, **Adv** asks the lexicographically first such x from the oracle $\mathcal{O}$, sets $\mathcal{L}_{j+1} = \mathcal{L}_j \cup (x, \mathcal{O}(x))$, and goes to round $j+1$. In other words, as long as there is any new query x that is ϵ-heavy to be in the closure of the queries of the view of the key-generations, **Adv** asks such a query x. If no such query exists, **Adv** breaks the loop and goes to the next step.*

  *(Note that the above and the following steps may require a **PSPACE**-complete oracle to be implemented efficiently.)*

5. **Guessing Challenger's View:** *Let $\mathcal{L}$ be the partial oracle consisting of the oracle queries (and their answers) that **Adv** learned in Steps 3 and 4 (i.e., $\mathcal{L} = \mathcal{L}_\ell$, where $\overline{Q(\mathbf{V}_\ell)}$ had no ϵ-heavy queries to be learned). Let $\mathbf{V}_{chal} = (\mathbf{V} \mid \mathcal{L}, PK, \{DK_i\}_{i \in [k]})$, and sample $V' \xleftarrow{\$} \mathbf{V}_{chal}$. Let SK' and DK'_* be in order, the "guessed" values for the secret-key and the decryption-key of f_* determined by the sampled V'. We note that by definition the other keys $PK', \{DK'_i\}_{i \in [k]}$ determined by V' are the same as the ones that **Adv** has received: $PK, \{DK_i\}_{i \in [k]}$.*

6. **Receiving the Challenge and the Final Decryption:** **Adv** *receives $C_* (= \mathbf{E}^{\mathcal{O}}(PK, a_*, b))$ for a random bit $b \in \{0, 1\}$. Then, **Adv** uses the oracle $\mathcal{O}'$ defined below and outputs the decrypted value $b' \leftarrow \mathbf{D}^{\mathcal{O}'}(PK, DK'_*, C_*)$ as his guess about the bit b.*

  The Oracle $\mathcal{O}'$: At the beginning of the decryption of Step 6, the partially defined oracle $\mathcal{O}'$ is equal to $\mathcal{L} \cup Q(V')$, namely the learned queries (and their answers) together with the guessed ones specified in V'. Afterwards, if a new query x is asked: (i) if $x \in \mathcal{O}'$, return $\mathcal{O}'(x)$, otherwise (ii) if $x \in \overline{\mathcal{O}'}$, then return $y = \overline{\mathcal{O}'}(x)$ and add (x, y) to $\mathcal{O}'$, and finally (iii) if $x \notin \overline{\mathcal{O}'}$, ask x from $\mathcal{O}$ and add $(x, \mathcal{O}(x))$ to $\mathcal{O}'$.

This finishes the description of our attack. We prove the query-efficiency and the success probability of our attack in the full version [16].

Acknowledgement. We thank Brent Waters for suggesting to us the problem of separating predicate encryption from IBE and for useful collaborations in

---

[3] Step 3 can be thought of as the 0'th round.

[4] Informally, the closure of a partial oracle is a superset consisting of all the queries (in addition to the partial oracle itself) that are dependent on the queries in the partial oracle (or its closure), e.g. if the partial oracle contains queries $[\mathbf{g}(sk) = pk]$ and $[\mathbf{k}(sk, \alpha) = dk]$ then its closure must also contain the query $[\mathbf{id}(pk, dk) = \alpha]$. Please refer to the full version [16] for a formal definition.

initial stages of this work. We thank Yevgeniy Vahlis for clarifying several doubts in [9] and the anonymous reviewers for their valuable comments. Virendra Kumar was supported in part by Alexandra Boldyreva's NSF CAREER award 0545659 and NSF Cyber Trust award 0831184. Mohammad Mahmoody was supported in part by NSF Award CCF-0746990, AFOSR Award FA9550-10-1-0093, and DARPA and AFRL under contract FA8750-11-2- 0211. The views and conclusions contained in this document are those of the authors and should not be interpreted as representing the official policies, either expressed or implied, of the Defense Advanced Research Projects Agency or the US government.

References

1. Babai, L., Frankl, P., Simon, J.: Complexity classes in communication complexity theory (preliminary version). In: FOCS, pp. 337–347 (1986)
2. Barak, B., Mahmoody-Ghidary, M.: Lower Bounds on Signatures From Symmetric Primitives. In: FOCS, pp. 680–688 (2007)
3. Barak, B., Mahmoody-Ghidary, M.: Merkle Puzzles Are Optimal — An $O(n^2)$-Query Attack on Any Key Exchange from a Random Oracle. In: Halevi, S. (ed.) CRYPTO 2009. LNCS, vol. 5677, pp. 374–390. Springer, Heidelberg (2009)
4. Bellare, M., Rogaway, P.: Random Oracles are Practical: A Paradigm for Designing Efficient Protocols. In: ACM Conference on Computer and Communications Security, pp. 62–73 (1993)
5. Boldyreva, A., Goyal, V., Kumar, V.: Identity-based encryption with efficient revocation. In: ACM Conference on Computer and Communications Security, pp. 417–426 (2008)
6. Boneh, D., Boyen, X.: Efficient Selective-ID Secure Identity-Based Encryption Without Random Oracles. In: Cachin, C., Camenisch, J.L. (eds.) EUROCRYPT 2004. LNCS, vol. 3027, pp. 223–238. Springer, Heidelberg (2004)
7. Boneh, D., Boyen, X.: Secure Identity Based Encryption Without Random Oracles. In: Franklin, M. (ed.) CRYPTO 2004. LNCS, vol. 3152, pp. 443–459. Springer, Heidelberg (2004)
8. Boneh, D., Franklin, M.K.: Identity-Based Encryption from the Weil Pairing. SIAM J. Comput. 32(3), 586–615 (2003)
9. Boneh, D., Papakonstantinou, P.A., Rackoff, C., Vahlis, Y., Waters, B.: On the Impossibility of Basing Identity Based Encryption on Trapdoor Permutations. In: FOCS, pp. 283–292 (2008)
10. Boneh, D., Sahai, A., Waters, B.: Functional Encryption: Definitions and Challenges. In: Ishai, Y. (ed.) TCC 2011. LNCS, vol. 6597, pp. 253–273. Springer, Heidelberg (2011)
11. Boneh, D., Waters, B.: Conjunctive, Subset, and Range Queries on Encrypted Data. In: Vadhan, S.P. (ed.) TCC 2007. LNCS, vol. 4392, pp. 535–554. Springer, Heidelberg (2007)
12. Cocks, C.: An Identity Based Encryption Scheme Based on Quadratic Residues. In: IMA Int. Conf., pp. 360–363 (2001)
13. Fujisaki, E., Okamoto, T.: Secure Integration of Asymmetric and Symmetric Encryption Schemes. In: Wiener, M. (ed.) CRYPTO 1999. LNCS, vol. 1666, pp. 537–554. Springer, Heidelberg (1999)
14. Gentry, C.: Practical Identity-Based Encryption Without Random Oracles. In: Vaudenay, S. (ed.) EUROCRYPT 2006. LNCS, vol. 4004, pp. 445–464. Springer, Heidelberg (2006)

15. Goyal, V.: Reducing Trust in the PKG in Identity Based Cryptosystems. In: Menezes, A. (ed.) CRYPTO 2007. LNCS, vol. 4622, pp. 430–447. Springer, Heidelberg (2007)
16. Goyal, V., Kumar, V., Lokam, S., Mahmoody, M.: On Black-Box Reductions Between Predicate Encryption Schemes. In: Cramer, R. (ed.) TCC 2012. LNCS, vol. 7194, pp. 440–457. Springer, Heidelberg (2012), http://eprint.iacr.org
17. Goyal, V., Pandey, O., Sahai, A., Waters, B.: Attribute-based encryption for fine-grained access control of encrypted data. In: ACM Conference on Computer and Communications Security, pp. 89–98 (2006)
18. Impagliazzo, R., Rudich, S.: Limits on the Provable Consequences of One-Way Permutations. In: STOC, pp. 44–61 (1989)
19. Katz, J., Sahai, A., Waters, B.: Predicate Encryption Supporting Disjunctions, Polynomial Equations, and Inner Products. In: Smart, N.P. (ed.) EUROCRYPT 2008. LNCS, vol. 4965, pp. 146–162. Springer, Heidelberg (2008)
20. Katz, J., Yerukhimovich, A.: On Black-Box Constructions of Predicate Encryption from Trapdoor Permutations. In: Matsui, M. (ed.) ASIACRYPT 2009. LNCS, vol. 5912, pp. 197–213. Springer, Heidelberg (2009)
21. Klauck, H.: Rectangle Size Bounds and Threshold Covers in Communication Complexity. In: IEEE Conference on Computational Complexity, pp. 118–134 (2003)
22. Kushilevitz, E., Nisan, N.: Communication complexity. Cambridge University Press (1997)
23. Lee, T., Shraibman, A.: Lower Bounds in Communication Complexity. Foundations and Trends in Theoretical Computer Science 3(4), 263–398 (2009)
24. Lewko, A.B., Okamoto, T., Sahai, A., Takashima, K., Waters, B.: Fully Secure Functional Encryption: Attribute-Based Encryption and (Hierarchical) Inner Product Encryption. In: Gilbert, H. (ed.) EUROCRYPT 2010. LNCS, vol. 6110, pp. 62–91. Springer, Heidelberg (2010)
25. Lokam, S.V.: Spectral Methods for Matrix Rigidity with Applications to Size-Depth Trade-offs and Communication Complexity. J. Comput. Syst. Sci. 63(3), 449–473 (2001)
26. Lokam, S.V.: Complexity Lower Bounds using Linear Algebra. Foundations and Trends in Theoretical Computer Science 4(1-2), 1–155 (2009)
27. Mahmoody-Ghidary, M., Wigderson, A.: Black Boxes, Incorporated (2009), http://www.cs.cornell.edu/~mohammad/files/papers/BlackBoxes.pdf
28. Okamoto, T., Takashima, K.: Fully Secure Functional Encryption with General Relations from the Decisional Linear Assumption. In: Rabin, T. (ed.) CRYPTO 2010. LNCS, vol. 6223, pp. 191–208. Springer, Heidelberg (2010)
29. Razborov, A.: On Rigid Matrices (1989) (in Russian), http://people.cs.uchicago.edu/~razborov/rigid.pdf
30. Reingold, O., Trevisan, L., Vadhan, S.P.: Notions of Reducibility between Cryptographic Primitives. In: Naor, M. (ed.) TCC 2004. LNCS, vol. 2951, pp. 1–20. Springer, Heidelberg (2004)
31. Sahai, A., Waters, B.: Fuzzy Identity-Based Encryption. In: Cramer, R. (ed.) EUROCRYPT 2005. LNCS, vol. 3494, pp. 457–473. Springer, Heidelberg (2005)
32. Shamir, A.: Identity-Based Cryptosystems and Signature Schemes. In: Blakely, G.R., Chaum, D. (eds.) CRYPTO 1984. LNCS, vol. 196, pp. 47–53. Springer, Heidelberg (1985)
33. Sipser, M.: Borel Sets and Circuit Complexity. In: STOC, pp. 61–69 (1983)
34. Waters, B.: Efficient Identity-Based Encryption Without Random Oracles. In: Cramer, R. (ed.) EUROCRYPT 2005. LNCS, vol. 3494, pp. 114–127. Springer, Heidelberg (2005)

Lossy Functions Do Not Amplify Well

Krzysztof Pietrzak[1], Alon Rosen[2], and Gil Segev[3]

[1] IST Austria
pietrzak@ist.ac.at
[2] Efi Arazi School of Computer Science, IDC Herzliya, Israel
alon.rosen@idc.ac.il
[3] Microsoft Research, Mountain View, CA 94043, USA
gil.segev@microsoft.com

Abstract. We consider the problem of amplifying the "lossiness" of functions. We say that an oracle circuit $C^* : \{0,1\}^m \to \{0,1\}^*$ amplifies relative lossiness from ℓ/n to L/m if for every function $f : \{0,1\}^n \to \{0,1\}^n$ it holds that

1. If f is injective then so is C^f.
2. If f has image size of at most $2^{n-\ell}$, then C^f has image size at most 2^{m-L}.

The question is whether such C^* exists for $L/m \gg \ell/n$. This problem arises naturally in the context of cryptographic "lossy functions," where the relative lossiness is the key parameter.

We show that for every circuit C^* that makes at most t queries to f, the relative lossiness of C^f is at most $L/m \leq \ell/n + O(\log t)/n$. In particular, no black-box method making a polynomial $t = poly(n)$ number of queries can amplify relative lossiness by more than an $O(\log n)/n$ additive term. We show that this is tight by giving a simple construction (cascading with some randomization) that achieves such amplification.

1 Introduction

Lossy trapdoor functions, introduced by Peikert and Waters [14], are a powerful cryptographic primitive. Soon after their introduction, they were found to be useful for realizing new constructions of traditional cryptographic concepts, as well as for demonstrating the feasibility of new ones. Their wide applicability, simple definition, and realizability under a variety of cryptographic assumptions make them a clear candidate for induction into the "pantheon" of cryptographic primitives.

1.1 Lossy Trapdoor Functions

A collection of lossy trapdoor functions consists of two families of functions. Functions in the first family are injective (and can be inverted using a trapdoor), whereas functions in the second are "lossy," meaning that the size of their image is significantly smaller than the size of their domain. The security requirement

R. Cramer (Ed.): TCC 2012, LNCS 7194, pp. 458–475, 2012.

is that the description of a function sampled from the injective family is computationally indistinguishable from the description of a function sampled from the lossy family.

As demonstrated by Peikert and Waters, lossy trapdoor functions imply primitives such as trapdoor functions, collision-resistant hash functions, and oblivious transfer [14]. Amongst "higher level" applications, we can find chosen-ciphertext secure public-key encryption [14], deterministic public-key encryption [4], OAEP-based public-key encryption [10], "hedged" public-key encryption for protecting against bad randomness [2], security against selective opening attacks [3], and non-interactive universally-composable string commitments [13].[1]

1.2 Relative Lossiness

A key parameter in all the applications of lossy trapdoor functions is the amount of lossiness guaranteed in case that a lossy function was sampled. We say that a function $f : \{0,1\}^n \to \{0,1\}^n$ is (n, ℓ)-*lossy* if its image size is at most $2^{n-\ell}$. Intuitively, this means that an application of f on an input $x \in \{0,1\}^n$ loses at least ℓ bits of information, on average, about x. We refer to ℓ as the *absolute* lossiness of the function and to ℓ/n as the *relative* lossiness of the function.

Peikert and Waters [14] showed how to obtain chosen ciphertext secure encryption assuming relative lossiness $\ell/n = \Omega(1)$. This was subsequently improved by Mol and Yilek [12] who, building on work by Rosen and Segev [16], demonstrated how to obtain the same result assuming relative lossiness of only $1/\text{poly}(n)$. One-way functions and similarly trapdoor functions and oblivious transfer, can be constructed assuming relative lossiness of $1/\text{poly}(n)$. Collision resistant hashing requires relative lossiness of at least $1/2 + 1/\text{poly}(n)$. All other known applications of lossy trapdoor functions currently assume relative lossiness that is at least as large as $1 - o(1)$.

Currently, relative lossiness of $1 - o(1)$ seems to be necessary for most "non-traditional" applications of lossy trapdoor functions. While some of the known instantiations are able to guarantee such a high rate of lossiness, some other constructions fall short. Most notably, the lattice-based construction of Peikert and Waters [14], which is the only one based on a worst-case assumption and the only one for which no sub-exponential attack is known, only guarantees relative lossiness of $\Omega(1)$.

High relative lossiness is also relevant for applications that do not necessitate it. This is because the lossiness rate typically has a pronounced effect on the efficiency of the resulting construction. Specifically, higher lossiness rate enables the use of a smaller security parameter, and in many applications also enables the extraction of a larger number of "information theoretic" hard-core bits from the underlying function. This is useful, for example, for efficiently handling long messages.

---

[1] We note that for some of these constructions (e.g., collision-resistant hashing) the existence of a trapdoor is not required.

1.3   Lossiness Amplification

All of the above leads to the question of whether, given a specific construction of lossy trapdoor functions, it is possible to apply an efficient transformation that would result in a construction with significantly higher lossiness. It can be easily seen that parallel evaluation of t independent copies of an (n, ℓ)-lossy function amplifies the absolute lossiness from ℓ to $t\ell$. Specifically, given an (n, ℓ)-lossy function $f : \{0,1\}^n \to \{0,1\}^n$ the function $g : \{0,1\}^{tn} \to \{0,1\}^{tn}$, defined as

$$g(x_1, \ldots, x_t) = (f(x_1), \ldots, f(x_t))$$

is $(tn, t\ell)$-lossy. However, this comes at the cost of blowing up the input size by a factor of t and hence leaves the relative lossiness ℓ/n unchanged. What we are really looking for is a construction of a (m, L)-lossy function $h : \{0,1\}^m \to \{0,1\}^m$ where $L/m \gg \ell/n$. A natural candidate is sequential evaluation (also known as "cascading"), defined as

$$h(x) = \underbrace{f(f(\ldots, f(f(x))\ldots)}_{t \text{ times}}$$

Unfortunately, in general h might not be more lossy than f. In particular, this is the case when f is injective on its own range. One can do a bit better though. By shuffling the outputs in-between every invocation, using randomly chosen $r_1, \ldots, r_t$, one obtains the function

$$h_{r_1, \ldots, r_t}(x) = f(f(\ldots, f(f(x) \oplus r_1) \oplus r_2) \ldots \oplus r_t),$$

for which it is possible to show that, if f is say $(n, 1)$-lossy, then with overwhelming probability over the choice of $r_1, \ldots, r_t$, the function $h_{r_1, \ldots, r_t}$ has relative lossiness of $\Omega(\log t)/n$.

While already not entirely trivial, relative lossiness of $\Omega(\log t)/n$ is a fairly modest improvement over $\Omega(1)/n$, and would certainly not be considered sufficient for most applications. Still, it is not a-priori inconceivable that there exists more sophisticated ways to manipulate f so that the relative lossiness is amplified in a more significant manner. In this paper, we show that an additive gain of $O(\log n)/n$ is actually the best one can hope for, at least with respect to black-box constructions.

1.4   Our Results

We show that no efficient black-box amplification method can additively improve the relative lossiness of a given function f by more than $O(\log n)/n$. To this end, we consider a circuit $C^* : \{0,1\}^m \to \{0,1\}^*$ with oracle access to a function $f : \{0,1\}^n \to \{0,1\}^n$ such that the following hold:

1. If f is injective then so is C^f.
2. If f has image size of at most $2^{n-\ell}$, then C^f has image size at most 2^{m-L}.

Our main result is that, if $\ell < n - \omega(\log n)$, then for every C^* that makes at most t queries to f, the relative lossiness, L/m, of C^f is at most $(\ell + O(\log t))/n$. The impossibility result holds regardless of whether the injective mode of f has a trapdoor, and rules out even probabilistic constructions C^* (i.e., ones which amplify lossiness only with high probability over the choice of some randomness). In Section 2 we provide a high-level overview of our approach, and in Section 3 we formally present our proof. We then show (in Section 4) how to extend the above result to a "full fledged" cryptographic setting, in which one does not simply get black-box access to a single lossy or injective function f. In this setting, lossy functions are defined by a triple of algorithms $\{g_0, g_1, f\}$, where one requires that a function f_k is injective if the key is sampled by $k \leftarrow g_1$, and lossy if the key is sampled by $k \leftarrow g_0$. Moreover, the distributions generated by the injective and lossy key generation algorithms g_0, g_1 must be computationally indistinguishable.

1.5 Relation to the Collision Problem

Closely related to our setting is the *collision problem*, in which one is given black-box access to a function $f : \{0,1\}^n \rightarrow \{0,1\}^n$ and is required to distinguish between the case that f is injective and the case that it is 2^ℓ-to-1. A simple argument shows that any (randomized) classical algorithm that tries to distinguish between the cases must make $\Omega(2^{(n-\ell)/2})$ calls to f. Kutin [11], extending work of Aaronson and Shi [1], proves an analogous bound of $\Omega(2^{(n-\ell)/3})$ in the quantum setting.

Lower bounds on the collision problem can be seen to directly imply a weak version of our results. Specifically, if non-trivial lossiness amplification were possible then one could have applied it, and then invoked known upper bounds for the collision problem (either $O(2^{(n-\ell)/2})$ randomized classical or $O(2^{(n-\ell)/3})$ quantum), resulting in a violation of the corresponding lower bounds. However, this approach will only work if the amplification circuit does not blow up f's input size (specifically, only if $m < n + (L - \ell)$). In contrast, our results also hold with respect to arbitrary input blow-up.

1.6 Related Work

Several instantiations of lossy trapdoor functions guarantee relative lossiness of $1 - o(1)$. Peikert and Waters present constructions based on the Decisional Diffie-Hellman assumption [14]. These are further simplified by Freeman et al, who also present a generalization based on the d-linear assumption [6]. Boldyreva et al. [4], and independently Freeman et al. [6], present a direct construction based on Paillier's Composite Residuosity assumption.

Hemenway and Ostrovsky [7] generalize the approach of Peikert and Waters, and obtain relative lossiness of $1 - o(1)$ from any homomorphic hash proof system (a natural variant of hash proof systems [5]). In turn, this implies a unified construction based on either Decisional Diffie Hellman, Quadratic Residuosity, or Paillier's Composite Residuosity assumptions.

Constructions with relative lossiness $\Omega(1)$ are known based on the hardness of the "learning with errors" problem, which is implied by the worst case hardness of various lattice problems [14]. Kiltz et al. argue that RSA with exponent e satisfies relative lossiness $(\log e)/n$ under the phi-hiding assumption, and that use of multi-prime RSA increases relative lossiness up to $(m \log e)/n$ where m is the number of prime factors of the modulus [10]. Finally, Freeman et al. [6] propose an instantiation based on the Quadratic Residuosity assumption with relative lossiness of $\Omega(1/n)$.

1.7   On Black-Box Separations

The use of black-box separations between cryptographic primitives was pioneered by Impagliazzo and Rudich [9], who proved that there is no black-box construction of a key-exchange protocol from a one-way permutation. Since then, black-box separations have become the standard tool for demonstrating such assertions. We note that our main result is "unconditional", in the sense that it holds regardless of any cryptographic assumption. Our "cryptographic" result, in contrast, is more standard in that it relies on the indistinguishability property of lossy functions (see the work of Reingold et al. [15] for an extensive discussion on black-box separations).

Strictly speaking, it is not clear whether black-box separations should be interpreted as strong impossibility results. Certainly not as long as non-black-box techniques are still conceivable. Nevertheless, since as far as we know any of the primitives could exist unconditionally (cf. [8]), it is currently not clear how else one could have gone about proving cryptographic lower bounds . In addition, most of the known construction and reductions in cryptography are black-box. Knowing that no such technique can be used to establish an implication serves as a good guideline when searching for a solution. Indeed, it would be extremely interesting to see if non-black box techniques are applicable in the context of lossy function amplification.

2   Overview of Our Approach

We say that a function $f : \{0,1\}^n \to \{0,1\}^{n'}$ is (n, ℓ)-lossy if its image $\{f(x) : x \in \{0,1\}^n\}$ has size at most $2^{n-\ell}$. We refer to ℓ as the *absolute lossiness*, and ℓ/n as the *relative lossiness* of f. An (n, ℓ)-lossy function f is *balanced* if $f(x)$ has exactly 2^ℓ preimages for every $x \in \{0,1\}^n$, i.e. $|\{z : f(z) = f(x)\}| = 2^\ell$. We denote with $\mathcal{F}_{n,\ell}$ the set of all balanced (n, ℓ)-lossy functions.

Definition 2.1 (Lossiness amplification). *We say that an oracle circuit $C^* : \{0,1\}^m \to \{0,1\}^{m'}$ amplifies the relative lossiness from ℓ/n to L/m if*

1. *for every injective function f_0 over $\{0,1\}^n$, C^{f_0} is injective.*
2. *for every $f_1 : \{0,1\}^n \to \{0,1\}^n$ with image size $2^{n-\ell}$, the image of C^{f_1} has size at most 2^{m-L}.*

We say C^* weakly amplifies if C^* is probabilistic and the second item above only holds with probability ≥ 0.9 over the choice of C^*'s randomness.

Remark 2.2 (Permutations vs. injective functions). *In order to make our negative result as strong and general as possible, we require the oracle to be length preserving (and thus the injective f_0 is a permutation), whereas the input and output domain of C^* can be arbitrary.*

For concreteness, in this proof sketch we only consider the case $\ell = 1$. We will also assume that $m = nk$ is an integer multiple of n. The basic idea of our proof is to show that for any C^*, property 1. of Definition 2.1 implies that C^{f_1} has very low collision probability if $f_1 \in \mathcal{F}_{n,1}$ is a *randomly* chosen 2-1 function. More concretely, let t denote the number of oracle gates in C^* and assume we could prove that

$$\Pr_{X,Y \in \{0,1\}^m}[C^{f_1}(X) = C^{f_1}(Y)] \leq 2^{-k \cdot n + O(k \log t)} \tag{1}$$

Such a low collision probability implies that C^{f_1} must have a large range and thus cannot be too lossy. In particular, Eq. (1) implies that the absolute lossiness of C^{f_1} is at most $O(k \log t)$, or equivalently, the relative lossiness is $O(k \log t)/kn = O(\log t)/n$, which matches (ignoring the constant hidden in the big-oh) the lossiness of the construction $h_{r_1,\ldots,r_t}$ from Section 1.3. Unfortunately Eq. (1) is not quite true. For example consider a circuit $\tilde{C}^* : \{0,1\}^{kn} \to \{0,1\}^{kn}$ which makes only $t = 2$ queries to its oracle and is defined as

$$\tilde{C}^f(x_1, x_2, \ldots, x_k) \stackrel{\text{def}}{=} \begin{cases} 0^{kn} & \text{if } f(x_1) = f(x_2) \text{ and } x_1 \neq x_2 \\ (x_1, x_2, \ldots, x_k) & \text{otherwise} \end{cases}$$

If $f_0 : \{0,1\}^n \to \{0,1\}^n$ is a permutation, so is $\tilde{C}^{f_0}$ (in fact, it's the identity function), thus property 1. holds. On the other hand, for any $(n,1)$-lossy f_1 we have $f_1(x_1) = f_1(x_2)$ and $x_1 \neq x_2$ with probability 2^{-n} for uniform x_1, x_2. Thus the probability that $\tilde{C}^{f_1}$ outputs 0^{kn} on a random input is also 2^{-n}, which implies

$$\Pr_{X,Y \in \{0,1\}^m}[\tilde{C}^{f_1}(X) = \tilde{C}^{f_1}(Y)] \geq \Pr_{X,Y \in \{0,1\}^m}[\tilde{C}^{f_1}(X) = \tilde{C}^{f_1}(Y) = 0^{2k}]$$
$$\geq 2^{-2n}$$

contradicting Eq. (1) for $k > 2$.

The idea behind the counterexample $\tilde{C}^f$ is to query f on two random inputs and check if f collides on these inputs. If this is the case, C^f "knows" that f is not a permutation and so it must not be a permutation itself as required by property 1, in this case mapping to some fixed output. Although Eq. (1) is wrong, we can prove a slightly weaker statement, where we exclude inputs X where the evaluation of C^f on X involves two invocations of f on inputs $x \neq x'$ where $f(x) = f(x')$ (we will call such bad inputs "burned"). As with high probability, for a random $(n,1)$-lossy f, most inputs are not burned, already this weaker statement implies that C^f has large range.

The cryptographic setting. In a cryptographic setting, one usually does not simply get black-box access to a single lossy or injective function f, but lossy functions are defined by a collection (indexed by a security parameter λ) of triples of algorithms $\{g_0, g_1, f\}_{\{\lambda \in \mathbb{N}\}}$, where one requires that $f(k, \cdot)$ is injective if the key is sampled by $k \leftarrow g_1$, and lossy if the key is sampled by $k \leftarrow g_0$. Moreover the distributions generated by the injective and lossy key generation algorithms g_0, g_1 must be computationally indistinguishable.

In this setting one can potentially do more sophisticated amplification than what is captured by Definition 2.1, e.g. by somehow using the key-generation algorithms g_0, g_1. In Section 4 we prove that black-box lossiness amplification is not possible in this setting either.

In a nutshell, we show that constructions which amplify collections of lossy functions can be classified in two classes depending on whether the lossiness of the construction depends only on the lossiness of the oracle (we call such amplifiers "non-communicating") or if the property of being lossy is somehow encoded into the key. In the first case, the proof goes along the lines of the proof of Theorem 3.1 (in particular, amplifiers as in Definition 2.1 are "non-communicating" as there's not even a key). In the second case, where the construction is "communicating", we show that the output of the key-generation algorithms (of the amplified construction) will not always be indistinguishable. This proof borrows ideas from the work of Impagliazzo and Rudich [9] who show that one cannot construct a key-agreement from one-way permutations. Their proof shows that for any two parties Alice and Bob who can communicate over a public channel and who have access to random oracle $\mathcal{R}$, there exists an adversary Eve who can with high probability make all queries to $\mathcal{R}$ that both, Alice and Bob, made. As a consequence, Alice and Bob cannot use $\mathcal{R}$ to "secretly" communicate. In a similar vein we show that the lossy key-generation algorithm cannot "communicate" the fact that the key it outputs is lossy to the evaluation function or we can catch it, and thus distinguish lossy from injective keys.

3   An Upper Bound on Black-Box Lossiness Amplification

We now state our main theorem, asserting that simple sequential composition is basically the best black-box amplification that can be achieved.

Theorem 3.1 (Impossibility of Black-Box Amplification). *Consider any* $n, \ell, t \in \mathbb{N}$ *where*

$$n \geq \ell + 2 \log t + 2 \tag{2}$$

and any oracle aided circuit $C^* : \{0,1\}^m \rightarrow \{0,1\}^{m'}$ *which makes t oracle queries per invocation, then the following holds: If C^* weakly amplifies relative lossiness from ℓ/n to L/n,[2] then $L \leq \ell+3 \log t+4$. More concretely, for a random $f \in \mathcal{F}_{n,\ell}$, the construction C^f will have relative lossiness less than $(\ell + 3 \log t + 4)/n$ with probability at least $1/2$.*

---

[2] Note that we denote the relative lossiness of C^* by L/n, not L/m like in the previous sections. In particular, the absolute lossiness of C^* is Lm/n (not L).

Remark 3.2. *The bound $n \geq \ell + 2\log t + 2$ is basically tight, as for $n = \ell + 2\log t - O(1)$ one can with constant advantage p distinguish any (n, ℓ)-lossy function from an injective one by simply making t random queries and looking for a collision. The exact value of p depends on the $O(1)$ term, in particular, replacing the $O(1)$ with a sufficiently large constant we get a $p \geq .9$ as required by Definition 2.1. Then $C^f(x)$ which outputs x if no such collision is found, and some fixed value (say $0^{m'}$) otherwise is a weak amplifier as in Definition 2.1.*

Remark 3.3 (Probabilistic C^* vs. random f). *Instead of considering a probabilistic C^* and constructing a particular lossy f such that C^f is not too lossy with high probability over C^*'s randomness (as required by Definition 2.1), we consider a deterministic C^* and show that C^f fails to be lossy with high probability for a randomly chosen f. As f is sampled independently of (the description of) C^*, the latter implies the former.*

Below we formally define what we mean by an input being burned as already outlined.

Definition 3.4 (Burned input). *For $X \in \{0,1\}^m$, we denote with $\mathrm{in}(X)$ and $\mathrm{out}(X)$ the inputs and outputs of the t invocations of f in an evaluation of $C^f(X)$. Consider an input $X \in \{0,1\}^m$ and let $\{x_1, \ldots, x_t\} \leftarrow \mathrm{in}(X)$, we say that X is burned if for some $1 \leq i < j \leq t$, $x_i \neq x_j$ and $f(x_i) = f(x_j)$. $\phi(X)$ denotes the event that X is burned.*

Below is the main technical Lemma which we will use to prove Theorem 3.1 (recall that $m = nk$).

Lemma 3.5. *For a random balanced (n, ℓ)-lossy function f, and two random inputs X, Y, the probability that X, Y are colliding inputs for C^f and at the same time both are not burned can be upper bounded as*

$$\Pr_{\substack{f \in \mathcal{F}_{n,\ell} \\ X,Y \in \{0,1\}^m}} \left[(C^f(X) = C^f(Y)) \wedge \neg\phi(X) \wedge \neg\phi(Y) \right] \leq 2^{-kn + k(3\log t + \ell)} \quad (3)$$

We postpone the proof of this Lemma to Section 3.1. The following simple claim upper bounds the probability (over the choice of $f \in \mathcal{F}_{n,\ell}$) that an input x to C^f is burned

Claim 3.6. *For any $x \in \{0,1\}^m$*

$$\Pr_{f \in \mathcal{F}_{n,\ell}} [\phi(x)] \leq \frac{2^\ell t^2}{2^n} \quad (4)$$

Proof. For $i \in \{1, \ldots, t\}$, the probability that the ith query to f made during the evaluation of $C^f(x)$ provides a collision for f (assuming there's been no collision so far) is at most $\frac{(i-1)(2^\ell - 1)}{2^n - i - 1}$. To see this, note that as f is balanced, there are exactly $(i - 1)(2^\ell - 1)$ possible inputs which will lead to a collision as each of the $(i - 1)$ queries we did so far has $2^\ell - 1$ other preimages. As f is random, the probability the ith query (for which there are $2^n - i - 1$ choices) will hit one of these values is $\frac{(i-1)(2^\ell-1)}{2^n-i-1}$. The claim follows by taking the union bound over all i

$$\Pr_{f \in \mathcal{F}_{n,\ell}}[\phi(x)] \leq \sum_{i=1}^{t} \frac{(i-1)(2^\ell - 1)}{2^n - i - 1} \leq \frac{2^\ell t^2}{2^n}$$

The second step above used $t \leq 2^{n/2}$ which is implied by Eq. (2). ∎

Proof of Theorem 3.1. Consider a C^* as in the statement of the theorem and a random $f \in \mathcal{F}_{n,\ell}$. Let $\Phi \stackrel{\text{def}}{=} \{x \in \{0,1\}^m : \phi(x)\}$ denote the set of inputs which are burned (cf. Definition 3.4) and $\overline{\Phi} = \{0,1\}^m \setminus \Phi$. Using the chain rule, we can state eq.(3) as

$$\Pr_{\substack{f \in \mathcal{F}_{n,\ell} \\ X,Y \in \overline{\Phi}}}[C^f(X) = C^f(Y)] \leq \frac{2^{-kn+k(3\log t+\ell)}}{\Pr_{\substack{f \in \mathcal{F}_{n,\ell} \\ X,Y \in \{0,1\}^m}}[X,Y \in \overline{\Phi}]} \tag{5}$$

Using eq.(4) we can bound the expected size (over the choice of $f \in \mathcal{F}_{n,\ell}$) of Φ as

$$\mathsf{E}[|\Phi|] = |\{0,1\}^n| \cdot \Pr_{\substack{f \in \mathcal{F}_{n,\ell} \\ X \in \{0,1\}^m}}[\phi(X)] \leq 2^n \cdot \frac{2^\ell t^2}{2^n} = 2^\ell t^2$$

Using the Markov inequality and eq.(2), this implies that Φ is not too big, say at most half of the domain $\{0,1\}^m$, with probability $1/2$

$$\Pr_{f \in \mathcal{F}_{n,\ell}}[|\Phi| \geq 2^{n-1}] = \Pr_{f \in \mathcal{F}_{n,\ell}}[|\Phi| \geq 2^{n-1-\ell-2\log t}\mathsf{E}[|\Phi|]]$$

$$\leq 1/2^{n-1-\ell-2\log t}$$

$$\stackrel{(2)}{\leq} 1/2$$

By the above equation, $|\overline{\Phi}| > 2^{n-1}$ with probability $\geq 1/2$ over the choice of f, and for such a "good" f, two random X,Y are in $\overline{\Phi}$ with probability at least $(1/2)^2 = 1/4$. Thus the denominator on the right side of eq.(5) is at least $1/8$, replacing the denominator in eq.(5) with $2^{-3} = 1/8$ we get

$$\Pr_{\substack{f \in \mathcal{F}_{n,\ell} \\ X,Y \in \{0,1\}^m}}[C^f(X) = C^f(Y)] \leq 2^{-kn+k(3\log t+\ell)+3} \tag{6}$$

Again using Markov, this means that for a randomly chosen $f \in \mathcal{F}_{n,\ell}$, with probability at least $1/2$

$$\Pr_{X,Y \in \{0,1\}^m}[C^f(X) = C^f(Y)] \leq 2^{-kn+k(3\log t+\ell)+4} \tag{7}$$

As two values sampled independently from a distribution with support of size u collide with probability at least $1/u$ (this is tight if the distribution is flat), eq.(7) implies that the range of C^f must be at least of size $2^{kn-k(3\log t+\ell)-4}$, thus the relative lossiness (recall that $m = nk$) is $(k\ell+k3\log t+4)/kn \leq (\ell+3\log t+4)/n$. ∎

3.1   Proof of Lemma 3.5

We consider a random experiment denoted Γ where $X_0, Y_0 \in \{0,1\}^m$ and $f \in \mathcal{F}_{n,\ell}$ are chosen at random, and then $C^f(X_0)$ and $C^f(Y_0)$ are evaluated. This evaluations result in $2t$ invocations of f. Let $\{x_1, \ldots, x_t\} \leftarrow \mathsf{in}(X_0)$ and $\{X_1, \ldots, X_t\} \leftarrow \mathsf{out}(X_0)$ denote the inputs and outputs of f in the evaluation of $C^f(X_0)$. Analogously we define values y_i, Y_i occurring in the evaluation of $C^f(Y_0)$. For $I \subseteq \{1, \ldots t\}$, we define an event E_I which holds if for every $i \in I$ (and only for such i), there exits a j such that $y_i \neq x_j$ and $f(y_i) = f(x_j)$ and $y_i \neq y_k$ for all $k < i$ (i.e. we have a fresh, non-trivial collision). Γ defines a "transcript"

$$v_{f,X_0,Y_0} \stackrel{\text{def}}{=} \{X_0, Y_0, x_1, \ldots, x_t, f(x_1), \ldots, f(x_t), y_1, \ldots, y_t, f(y_1), \ldots, f(y_t)\}$$

The values x_i and y_i in the transcript are redundant, i.e., they can be computed from values X_0, Y_0, $f(x_i)$ and $f(y_i)$, and only are added for convenience. For $I \subseteq \{1, \ldots, t\}$ we define V_I as all transcripts where (1) both inputs are not burned (2) we have a collision and (3) E_I holds, i.e.

$$V_I \stackrel{\text{def}}{=} \{v_{f,X_0,Y_0} : \neg\phi(X_0) \wedge \neg\phi(Y_0) \wedge (C^f(X_0) = C^f(Y_0)) \wedge E_I\}$$

V_{col} is the union of all V_I, i.e.

$$V_{col} = \cup_I V_I = \{v_{f,X_0,Y_0} : \neg\phi(X_0) \wedge \neg\phi(Y_0) \wedge C^f(X_0) = C^f(Y_0)\} \quad (8)$$

For a set of transcripts V, we denote with $\Pr_\Gamma[V]$ the probability that the transcript generated by Γ is in V. It is not hard to see[3] that $\Pr_\Gamma[V_\emptyset] \leq 2^{-nk}$, we prove that this bound (up to a factor 2) holds for any V_I.

Lemma 3.7. *For any $I \subseteq \{1, \ldots, t\}$ we have (recall that $m = nk$)*

$$\Pr_\Gamma[V_I] \leq 2^{-nk+1}$$

We postpone the proof of this main technical lemma and first prove how it implies Theorem 3.1. But let us here give some intuition as to why Lemma 3.7 holds. The experiment Γ generates a transcript in V_I if (besides $C^f(X_0) = C^f(Y_0)$ colliding and X_0, Y_0 not being burnt) for every $i \in I$, the ith invocation of f during the evaluation of $C^f(Y_0)$ produces a fresh collision. Now, conditioned on such a collision happening, the probability of actually getting a collision $C^f(X_0) = C^f(Y_0)$ can potentially raise significantly (by something like $2^{n-\ell}$) as this is a rare event, but then, the probability of having such a collision is also around $2^{n-\ell}$, and if this collision does not occur, we definitely will not end up

---

[3] We have $\Pr_\Gamma[V_\emptyset] \leq \Pr_\Gamma[X_0 = Y_0] = 2^{-nk}$. The second step follows as $X_0, Y_0 \in \{0,1\}^{nk}$ are uniformly random. The first step follows as $\neg\phi(X_0), \neg\phi(Y_0)$ and $E_\emptyset$ together imply that there are no collisions in the $2t$ invocations of f, and thus f is "consistent" with being a permutation. But in this case, $C^f(X_0) = C^f(Y_0)$ implies $X_0 = Y_0$.

with a transcript in V_I. These two probabilities even out, and we end up with roughly the same probability for a transcript V_I as we had for $V_\emptyset$.

Before we can prove the theorem we need one more lemma, which bounds the probability of Γ generating a transcript with lots (k or more) collisions.

Lemma 3.8

$$\sum_{I:|I|\geq k} \Pr_\Gamma[V_I] \leq \sum_{I:|I|\geq k} \Pr_\Gamma[E_I] \leq 2^{k(\ell+2\log t-(n-1))} \tag{9}$$

Proof. The first step of Eq. (9) follows as V_I implies E_I. Let E_I^+ denote the event which holds if $E_{I'}$ holds for any $I' \supseteq I$. We have

$$\Pr_\Gamma[E_I^+] \leq \left(\frac{(2^\ell-1)t}{2^n-2t}\right)^{|I|} \leq \left(\frac{2^\ell t}{2^{n-1}}\right)^{|I|} \tag{10}$$

To see this, note that to get E_I^+, in every step $i \in I$, x_i must be fresh, and then $f(y_i)$ must "hit" one of the at most t distinct $f(x_i)$. As f is a random 2^ℓ-1 function evaluated on at most $2t$ inputs, this probability can be upper bounded by $(2^\ell-1)t/(2^n-2t)$ as at most $(2^\ell-1)t$ of the at least 2^n-2t fresh inputs can "hit" as described above. The probability that we have such a "hit" for all $i \in I$ is the $|I|$'th power of this probability. The number of different I where $|I| = k$ can be upper bounded by $2^{k\log t}$, using this and Eq. (10) we get

$$\sum_{I:|I|\geq k} \Pr_\Gamma[E_I] \leq \sum_{I:|I|=k} \Pr_\Gamma[E_I^+]$$

$$\leq 2^{k\log t}\left(\frac{2^{\ell k}t^k}{2^{(n-1)k}}\right)$$

$$= 2^{k(\ell+2\log t-(n-1))}$$

∎

Proof of Lemma 3.5. Lemma 3.5 states that $\Pr_\Gamma[V_{col}] \leq 2^{-kn+O(k\log(t))}$, which we can write as

$$\Pr_\Gamma[V_{col}] \overset{Eq.(8)}{=} \sum_{I:|I|<k} \Pr_\Gamma[V_I] + \sum_{I:|I|\geq k} \Pr_\Gamma[V_I]$$

Using Lemma 3.7 and 3.8 and the fact that there are $\binom{t}{k-1} < t^k$ different I's with $|I| < k$, we get

$$\sum_{I:|I|<k} \Pr_\Gamma[V_I] + \sum_{I:|I|\geq k} \Pr_\Gamma[V_I] \leq t^k \cdot 2^{-nk+1} + 2^{k(\ell+2\log t-(n-1))}$$

$$\leq 2^{1+k(\ell+2\log t-(n-1))}$$

$$< 2^{-nk+k(3\log t+\ell)}$$

∎

Proof of Lemma 3.7. For any I, we consider a new random experiment Γ_I. This experiment will define a distribution $X'_t \in \{0,1\}^m, Y'_t \in \{0,1\}^m \cup \bot$. We'll show that

$$\Pr_{\Gamma_I}[X'_t = Y'_t] \leq 2^{-m} \tag{11}$$

and

$$\Pr_{\Gamma}[V_I] \leq 2 \cdot \Pr_{\Gamma_I}[X'_t = Y'_t] \tag{12}$$

Note that the two equations above imply Lemma 3.7. The experiment Γ_I is defined as follows

1. We sample random $X'_0, Y'_0 \in \{0,1\}^m$ and a random *permutation* g over $\{0,1\}^n$.
2. Let $x'_1, \ldots, x'_t$ be the inputs to g in the evaluation of $C^g(X'_0)$. Let $X'_t \stackrel{\text{def}}{=} C^g(X'_0)$.
3. Now evaluate $C^g(Y'_0)$ in steps (one invocation of g per step), where for any $i \in I$ do the following:
   - if y'_i is "fresh" (that is $y'_i \neq x'_j$ for any $1 \leq j \leq t$ and $y'_i \neq y'_j$ for any $1 \leq j < i$). we change the value of $g(y'_i)$ and set it to some uniformly random value $z_i \in_U \{0,1\}^n$ (note that g is no longer a permutation).
   - If y_i is no fresh set $Y'_t = \bot$ and stop.

We will first prove Eq. (11). Let's consider a new random experiment Γ^*_I which will define outputs $X''_t, Y''_t \in \{0,1\}^m$. This experiment is defined exactly as the experiment Γ_I defining X'_t, Y'_t, but when y'_i is not fresh we nonetheless redefine $g(y'_i)$ to a random z_i (instead of setting $Y'_t = \bot$ and aborting). As the two experiments only differ when $Y'_t = \bot$, but X'_t cannot be $\bot$, we have.

$$\Pr_{\Gamma_I}[X'_t = Y'_t] \leq \Pr_{\Gamma^*_I}[X''_t = Y''_t]$$

Moreover $X''_t = C^g(X''_0)$ is uniformly random (as X''_0 is uniform and C^g is a permutation) and Y''_t is *independent* of X''_t (the reason we consider the experiment Γ^*_I is because in Γ_I we don't have this independence), thus

$$\Pr_{\Gamma^*_I}[Y''_t = X''_t] = 2^{-m}$$

The two equations above imply Eq. (11). Now we show Eq. (12), i.e.

$$\Pr_{\Gamma}[V_I] \leq 2 \cdot \Pr_{\Gamma'}[X'_t = Y'_t] \tag{13}$$

We will show a stronger statement, namely that for every transcript $\hat{v} \in V_I$ we have

$$\Pr_{\Gamma}[\hat{v}] \leq 2 \cdot \Pr_{\Gamma_I}[\hat{v}] \tag{14}$$

This implies (13) as

$$\Pr_{\Gamma}[V_I] = \sum_{\hat{v} \in V_I} \Pr_{\Gamma}[\hat{v}] \leq 2 \cdot \sum_{\hat{v} \in V_I} \Pr_{\Gamma_I}[\hat{v}] = 2 \cdot \Pr_{\Gamma_I}[V_I] \leq 2 \cdot \Pr_{\Gamma_I}[X'_t = Y'_t]$$

We'll use the following notation for the transcript $\hat{v}$ and the transcripts generated by Γ and Γ_I, respectively.

$$\hat{v} \stackrel{\text{def}}{=} \{\hat{X}_0, \hat{Y}_0, \hat{x}_1, \ldots, \hat{x}_t, a_1, \ldots, a_t, \hat{y}_1, \ldots, \hat{y}_t, b_1, \ldots, b_t\}$$

$$v \stackrel{\text{def}}{=} \{X_0, Y_0, x_1, \ldots, x_t, f(x_1), \ldots, f(x_t), y_1, \ldots, y_t, f(y_1), \ldots, f(y_t)\}$$

$$v' \stackrel{\text{def}}{=} \{X_0', Y_0', x_1', \ldots, x_t', g(x_1'), \ldots, g(x_t'), y_1', \ldots, y_t', g(y_1'), \ldots, g(y_t')\}$$

As $\hat{X}_0, \hat{Y}_0, X_0, Y_0, X_0', Y_0'$ are uniformly random, we have

$$\Pr[(\hat{X}_0, \hat{Y}_0) = (X_0, Y_0)] = \Pr[(\hat{X}_0, \hat{Y}_0) = (X_0', Y_0')] = 2^{-2m}$$

Further

$$\Pr_{\Gamma}[(\hat{x}_1, \ldots, \hat{x}_t, a_1, \ldots, a_t) = (x_1, \ldots, x_t, f(x_1), \ldots, f(x_t)) \mid (\hat{X}_0, \hat{Y}_0) = (X_0, Y_0)] \leq$$

$$\Pr_{\Gamma_I}[(\hat{x}_1, \ldots, \hat{x}_t, a_1, \ldots, a_t) = (x_1', \ldots, x_t', g(x_1'), \ldots, g(x_t')) \mid (\hat{X}_0, \hat{Y}_0) = (X_0', Y_0')]$$

Using the chain rule, the above is implied by

$$\prod_{i=1}^{t} \Pr_{\Gamma}[a_i = f(x_i)| \ldots] \leq \prod_{i=1}^{t} \Pr_{\Gamma_I}[a_i = g(x_i')| \ldots] \tag{15}$$

where here and below we use the convention that "$\ldots$" always means that the transcript defined up to this point is consistent with the transcript $\hat{v}$. E.g. on the left side of eq.(15) the "$\ldots$" stands for

$$(\hat{X}_0, \hat{Y}_0) = (X_0, Y_0) \quad , \quad \forall j = 1 \ldots i-1 : f(x_j) = a_j \tag{16}$$

Note that we don't have to explicitly require $\forall j = 1 \ldots i-1 : x_j = \hat{x}_j$ as this is already implied by (16).[4]

For $i = 1, \ldots, 2t$ we will denote with $q_i \leq i$ the number of *distinct* elements that appeared as inputs to f in the first i queries. I.e., for $i \leq t$ $q_i = |\{\hat{x}_1, \ldots, \hat{x}_i\}|$ and for $t < i \leq 2t$, $q_i = |\{\hat{x}_1, \ldots, \hat{x}_t, \hat{y}_1, \ldots, \hat{y}_{i-t}\}|$.

To see that Eq. (15) holds, note that for any i where $\hat{x}_i$ is not fresh (i.e. $x_i = x_j$ for some $j < i$) we have

$$\Pr[a_i = f(x_i)| \ldots] = \Pr[a_i = g(x_i')| \ldots] = 1$$

For i's where x_i is fresh, let q_i denote the number of distinct elements in $\hat{x}_1, \ldots, \hat{x}_{i-1}$. A g is a random permutation and $a_i \neq g(x_j')$ for $j < i$ because $\neg\phi(X_0)$, we have

$$\Pr_{\Gamma_I}[g(x_i') = a_i| \ldots] = \frac{1}{2^n - q_i}$$

---

[4] If the inputs $(\hat{X}_0, \hat{Y}_0) = (X_0, Y_0)$ are identical, and all the oracle queries so far gave the same outputs, also all intermediate values (including the next oracle query) will be the same.

On the other hand

$$\Pr_{\Gamma}[f(x_i) = a_i | \ldots] \leq \frac{1}{2^n - q_i}$$

To see this note that $\Pr_{\Gamma}[f(x_i) = a_i | \ldots]$ is exactly $\frac{1}{2^n - q_i}$ if one additionally conditions on the fact that $f(x_i) \neq a_j$ for all $j < i$. Not conditioning on this event can only decrease the probability as $a_i \neq a_j$ for $j < i$ as $\neg \phi(X_0)$.

Now we come to the second part of the transcript. Here we will show that

$$\Pr_{\Gamma}[(\hat{y}_1, \ldots, \hat{y}_t, a_1, \ldots, a_t) = (y_1, \ldots, y_t, f(y_1), \ldots, f(y_t)) | \ldots] \leq$$
$$2 \cdot \Pr_{\Gamma_I}[(\hat{y}_1, \ldots, \hat{y}_t, a_1, \ldots, a_t) = (y'_1, \ldots, y'_t, g(y'_1), \ldots, g(y'_t)) | \ldots]$$

The proof is almost identical as for the first part, except that now for fresh y'_i we have a slightly smaller probability

$$\Pr_{\Gamma_I}[g(y'_i) = b_i | \ldots] = 2^{-n}$$

that g maps to the right value b_i in the experiment Γ_I, as by definition of Γ_I the output of g is assigned a uniformly random value in this case. Using the fact that $t \leq 2^{-n/4}$ this difference is covered by the extra factor 2.   ∎

4   Extension to Collections of Lossy Functions

By Theorem 3.1 no circuit C^* (of size polynomial in n) can amplify relative lossiness better than sequential composition. That is, if C^f is injective for any permutation $f : \{0,1\}^n \to \{0,1\}^n$, then there exists an (n, ℓ)-lossy f (i.e. it has relative lossiness ℓ/n) such that C^f has relative lossiness only $(\ell + O(\log n))/n$. In fact, a random (n, ℓ)-lossy f will have this property with very high probability. In a cryptographic setting, lossy functions are not given as a single function, but by a collection of triple of algorithms as defined below.

Definition 4.1 (Collection of Lossy Functions). *Let $\lambda \in \mathbb{N}$ denote a security parameter and $n = n(\lambda)$, $n' = n'(\lambda)$, $\ell = \ell(\lambda)$ be functions of λ. A collection of (n, n', ℓ)-lossy function is a sequence (indexed by λ) of functions $\pi = \{g_0, g_1, f\}_{\lambda \in \mathbb{N}}$ where g_0, g_1 are probabilistic key-generation functions, such that*

1. **Evaluation of lossy functions:** *For every function index $\sigma \leftarrow g_0(1^\lambda)$, $f(\sigma, \cdot)$ is a function $f_\sigma : \{0,1\}^n \to \{0,1\}^{n'}$ whose image is of size at most $2^{n-\ell}$.*
2. **Evaluation of injective functions:** *For every function index $\sigma \leftarrow g_1(1^\lambda)$, the function $f(\sigma, \cdot)$ computes an injective function $f_\sigma : \{0,1\}^n \to \{0,1\}^{n'}$.*
3. **Security:** *The ensembles $\{\sigma : \sigma \leftarrow g_0(1^\lambda)\}_{\lambda \in \mathbb{N}}$ and $\{\sigma : \sigma \leftarrow g_1(1^\lambda)\}_{\lambda \in \mathbb{N}}$ are computationally indistinguishable.*

We refer to ℓ as the absolute lossiness *of π, and to ℓ/n as the* relative lossiness *of π.*

Definition 4.2 (Black-Box Amplification of Lossy Collection). *A triple of probabilistic polynomial-time oracle algorithms $\Pi^* = \{G_0^*, G_1^*, F^*\}$ is a black-box amplification for relative lossiness from $\alpha = \alpha(\lambda)$ to $\beta = \beta(\lambda)$ $(\beta > \alpha)$ if for every oracle $\pi = \{g_0, g_1, f\}_{\lambda \in \mathbb{N}}$ that implements a $(n, n, \alpha n)$-lossy collection, Π^π is a $(m, m', \beta m)$-lossy collection (where $m = m(\lambda), m' = m'(\lambda)$).*

Note that if π is efficient (i.e. can be implemented by polynomial time algorithms), so is Π^π. We will prove the following theorem.

Theorem 4.3 (Impossibility of Black-Box Amplification). *Let t, ℓ, n be functions of λ such that $n(\lambda) \leq \ell(\lambda) + 2\log(t(\lambda)) + \omega(\lambda)$. If each of the algorithms in $\Pi^* = \{G_0^*, G_1^*, F^*\}$ makes at most $t = t(\lambda)$ oracle queries per invocation and Π^* amplifies relative lossiness from $\alpha(\lambda) = \ell/n$ to $\beta(\lambda) = L/n$ then $L = \ell + O(\log t)$.*

To save on notation, we will identify the security parameter λ with the domain size n of the lossy-function we try to amplify (which will be given as an oracle).

To prove Theorem 4.3, we will show that for any construction Π^*, if we choose a random (n, ℓ)-lossy $\pi_n = \{g_0, g_1, f\}$ ("random" to be defined in Section 4.1), then with overwhelming probability either the outputs of $G_0^{\pi_n}$ and $G_1^{\pi_n}$ can be distinguished relative to π_n, or for a random lossy key $k \leftarrow G_0^{\pi_n}$, the function $F^{\pi_n}(k, \cdot)$ has very small collision probability and thus cannot be too lossy.

4.1   The Random $\pi = \{g_0, g_1, f\}$

For $n, \ell \in \mathbb{N}$ let $\mathcal{L}_{n,\ell}$ denote the set of triples of functions $g_0, g_1 : \{0,1\}^{n-1} \to \{0,1\}^n$, $f : \{0,1\}^n \to \{0,1\}^n$ where the range of g_0, g_1 covers all of $\{0,1\}^n$ (note this means that the range of g_0 and g_1 are disjoint) and (with $\mathcal{F}_{n,\ell}$ as defined in the first paragraph of Section 2)

$$\forall x \in \{0,1\}^{n-1} : f(g_0(x), \cdot) \in \mathcal{F}_{n,\ell} \quad \text{and} \quad f(g_1(x), \cdot) \in \mathcal{F}_{n,0}$$

Claim 4.4. *For $\ell(n) \leq n - \omega(n)$, let $\pi = \{\pi_n\}_{n \in \mathbb{N}}$ where $\pi_n = \{g_0, g_1, f\}$ is chosen uniformly in $\mathcal{L}_{n,\ell}$ (for every $n \in \mathbb{N}$.) Then with overwhelming probability π is (n, ℓ)-lossy even relative to an EXPTIME-complete oracle.*

4.2   (Non-)Communicating Π^*

Consider a $\Pi^* = \{G_0^*, G_1^*, F^*\}$ as in Definition 4.2. We will classify such Π^* in two classes, depending on whether Π^* is close to being "non-communicating" or not. Intuitively, we say Π^* is non-communicating if the lossiness of Π^π comes entirely from the lossiness of π, that is, if π is not lossy, then also Π^π will not be lossy.

Definition 4.5 ((close to) non-communicating). *Π^* is non-communicating if for every $n \in \mathbb{N}$ and $\pi_n \in \mathcal{L}_{n,0}$ the function computed by $F^\pi(k, \cdot)$ is injective for every $k \leftarrow G_0^\pi(1^n)$. In addition, Π^* is close to being non-communicating if* *for all but finitely many* $n \in \mathbb{N}$, *with probability* $1/2$ *over the choice of a random* $\pi_n \in \mathcal{L}_{n,0}$, *for at least* $1/2$ *of the keys* $k \leftarrow \mathsf{G}_0^{\pi_n}$, *there's a subset* $\mathcal{M}_k \subseteq \{0,1\}^m$ *of size at least* $2^m/2$ *such* $\mathsf{F}^{\pi_n}(k,x)$ *is injective on* $\mathcal{M}_k$ *(i.e. for* $x, x' \in \mathcal{M}_k$, $\mathsf{F}^{\pi_n}(k,x) = \mathsf{F}^{\pi}(k,x')$ *implies* $x = x'$*).*

In order to prove that Theorem 4.3 holds for some particular construction Π^*, we will use a different argument depending on whether Π^* is close to being non-communicating or not. The proof for the first case is almost identical to that of Theorem 3.1, where we rely on the fact that C^* is injective for any injective key k. The proof for the second case relies on the indistinguishability of injective and lossy functions, and requires new ideas. More specifically, in this case we prove the following lemma:

Lemma 4.6. *If* Π^* *(as in the statement of Theorem 4.3) is far (i.e. not close) from being non-communicating, then for infinitely many* $n \in \mathbb{N}$ *the following holds. For a random* $\pi_n \in \mathcal{L}_{n,\ell}$ *the outputs of* $\mathsf{G}_0^{\pi_n}$ *and* $\mathsf{G}_1^{\pi_n}$ *can be distinguished with constant advantage making* $poly(t,n)$ *oracle queries to* π_n *(and one query to an EXPTIME oracle).*

Due to space limitations in the remainder of this section we describe a high-level outline for the proof of Lemma 4.6, and refer the reader to the full version for the formal proof.

Proof outline. For $b \in \{0,1\}$ consider a key $k \leftarrow \mathsf{G}_b^{\pi_n}(R)$ and let $\mathcal{Q}_k$ denote all the queries that $\mathsf{G}_b^{\pi_n}(R)$ made to its oracle π_n during sampling this key using randomness R. Now consider a $(n,0)$-lossy $\hat{\pi}_n \in \mathcal{F}_{n,0}$ which is sampled at random except that we require it to be consistent with the queries in $\mathcal{Q}_k$. As $\hat{\pi}_n$ is consistent with π_n on $\mathcal{Q}_k$, we have $\mathsf{G}_b^{\pi_n}(R) = \mathsf{G}_b^{\hat{\pi}_n}(R) = k$. Thus if

- $b = 1$, then k is a valid injective key relative to $\hat{\pi}_n$ and thus $\mathsf{F}^{\hat{\pi}_n}(k,\cdot)$ has image size 2^m.
- $b = 0$, then k is a valid lossy key relative to $\hat{\pi}_n$. As Π^* is far from being non-communicating, with constant probability $\mathsf{F}^{\hat{\pi}_n}(k,\cdot)$ will have an image size of $\leq 2^{m-1}$ despite the fact that $\hat{\pi}_n$ is not lossy at all.

Using the above two observations, here's a way to distinguish the case $b = 0$ from $b = 1$ (i.e. lossy from injective keys) with constant advantage given $\mathcal{Q}_k$ and access to an EXPTIME oracle: query the oracle on input $k, \mathcal{Q}_k$ and ask for the image size of $\mathsf{F}^{\hat{\pi}_n}(k,\cdot)$ for a $\hat{\pi}_n$ randomly sampled as described above. If the image size is $\leq 2^{m-1}$, guess $b = 0$, guess $b = 1$ otherwise.

Unfortunately we are only given the key k, but not $\mathcal{Q}_k$. What we'll do is consider a random $\bar{\pi}_n$ which is consistent with π_n on a set of inputs/outputs $\mathcal{Q}_{k,q}^{sam}$ to π_n which is sampled by invoking $\mathsf{F}^{\pi_n}(k,\cdot)$ on $q = poly(n,t)$ random inputs (i.e. $\mathcal{Q}_{k,q}^{sam}$ contains all inputs/outputs to π_n made during these q invocations).

We will prove that for such a $\bar{\pi}_n$ the image size of $\mathsf{F}(k,\cdot)^{\bar{\pi}_n}$ is still close to 2^m if $b = 1$, but with constant probability $\ll 2^m$ if $b = 0$, so we can use our EXPTIME oracle to distinguish these cases by sending $k, \mathcal{Q}_{k,q}^{sam}$ (which, unlike

$\mathcal{Q}_k$, we do have) to the EXPTIME oracle asking for the image size of $\mathsf{F}(k, \cdot)^{\bar{\pi}_n}$ when $\bar{\pi}_n \in \mathcal{F}_{n,0}$ is chosen at random but consistent with $\mathcal{Q}_{k,q}^{sam}$.

The reason it is good enough to consider a $\bar{\pi}_n$ that is consistent with $\mathcal{Q}_{k,q}^{sam}$ and not $\mathcal{Q}_k$, is that for sufficiently many samples $q = poly(n,t)$, $\mathcal{Q}_{k,q}^{sam}$ will with high probability contain all "heavy" queries in $\mathcal{Q}_k$, where we say a query is heavy if there's a good probability that $\mathsf{F}^{\bar{\pi}_n}(k, \cdot)$ will make that query if invoked on a random input.

So for most inputs x, $\mathsf{F}^{\bar{\pi}_n}(k, x)$ will not query $\bar{\pi}_n$ on a query which is in $\mathcal{Q}_k$ (i.e. which was made during key-generation), but is not in $\mathcal{Q}_{k,q}^{sam}$. As a consequence, $\mathsf{F}^{\bar{\pi}_n}(k, \cdot)$ "behaves" differently from what we would get by using $\hat{\pi}_n$ (which is consistent with all of $\mathcal{Q}_k$) instead of $\bar{\pi}_n$ only for a small fraction of the inputs. In particular, the image size is close to what we would have gotten by using $\hat{\pi}_n$.

Acknowledgements. We would like to thank Oded Goldreich and Omer Reingold for discussions at an early stage of this project, and Scott Aaronson for clarifications regarding the collision problem.

References

1. Aaronson, S., Shi, Y.: Quantum lower bounds for the collision and the element distinctness problems. Journal of the ACM 51(4), 595–605 (2004)
2. Bellare, M., Brakerski, Z., Naor, M., Ristenpart, T., Segev, G., Shacham, H., Yilek, S.: Hedged Public-Key Encryption: How to Protect against Bad Randomness. In: Matsui, M. (ed.) ASIACRYPT 2009. LNCS, vol. 5912, pp. 232–249. Springer, Heidelberg (2009)
3. Bellare, M., Hofheinz, D., Yilek, S.: Possibility and Impossibility Results for Encryption and Commitment Secure under Selective Opening. In: Joux, A. (ed.) EUROCRYPT 2009. LNCS, vol. 5479, pp. 1–35. Springer, Heidelberg (2009)
4. Boldyreva, A., Fehr, S., O'Neill, A.: On Notions of Security for Deterministic Encryption, and Efficient Constructions without Random Oracles. In: Wagner, D. (ed.) CRYPTO 2008. LNCS, vol. 5157, pp. 335–359. Springer, Heidelberg (2008)
5. Cramer, R., Shoup, V.: Universal Hash Proofs and a Paradigm for Adaptive Chosen Ciphertext Secure Public-Key Encryption. In: Knudsen, L.R. (ed.) EUROCRYPT 2002. LNCS, vol. 2332, pp. 45–64. Springer, Heidelberg (2002)
6. Freeman, D.M., Goldreich, O., Kiltz, E., Rosen, A., Segev, G.: More Constructions of Lossy and Correlation-Secure Trapdoor Functions. In: Nguyen, P.Q., Pointcheval, D. (eds.) PKC 2010. LNCS, vol. 6056, pp. 279–295. Springer, Heidelberg (2010)
7. Hemenway, B., Ostrovsky, R.: Lossy trapdoor functions from smooth homomorphic hash proof systems. Electronic Colloquium on Computational Complexity, Report TR09-127 (2009)
8. Impagliazzo, R.: A personal view of average-case complexity. In: Structure in Complexity Theory Conference, pp. 134–147 (1995)
9. Impagliazzo, R., Rudich, S.: Limits on the provable consequences of one-way permutations. In: Proceedings of the 21st Annual ACM Symposium on Theory of Computing, pp. 44–61 (1989)

10. Kiltz, E., O'Neill, A., Smith, A.: Instantiability of RSA-OAEP under Chosen-Plaintext Attack. In: Rabin, T. (ed.) CRYPTO 2010. LNCS, vol. 6223, pp. 295–313. Springer, Heidelberg (2010)
11. Kutin, S.: Quantum lower bound for the collision problem with small range. Theory of Computing 1(1), 29–36 (2005)
12. Mol, P., Yilek, S.: Chosen-Ciphertext Security from Slightly Lossy Trapdoor Functions. In: Nguyen, P.Q., Pointcheval, D. (eds.) PKC 2010. LNCS, vol. 6056, pp. 296–311. Springer, Heidelberg (2010)
13. Nishimaki, R., Fujisaki, E., Tanaka, K.: Efficient Non-interactive Universally Composable String-Commitment Schemes. In: Pieprzyk, J., Zhang, F. (eds.) ProvSec 2009. LNCS, vol. 5848, pp. 3–18. Springer, Heidelberg (2009)
14. Peikert, C., Waters, B.: Lossy trapdoor functions and their applications. In: Proceedings of the 40th Annual ACM Symposium on Theory of Computing, pp. 187–196 (2008)
15. Reingold, O., Trevisan, L., Vadhan, S.P.: Notions of Reducibility between Cryptographic Primitives. In: Naor, M. (ed.) TCC 2004. LNCS, vol. 2951, pp. 1–20. Springer, Heidelberg (2004)
16. Rosen, A., Segev, G.: Chosen-ciphertext security via correlated products. SIAM Journal on Computing 39(7), 3058–3088 (2010)

Counterexamples to Hardness Amplification beyond Negligible

Yevgeniy Dodis[1], Abhishek Jain[2], Tal Moran[3], and Daniel Wichs[4]

[1] New York University (NYU)
[2] University of California, Los Angeles (UCLA)
[3] Interdisciplinary Center (IDC) Herzliya
[4] IBM Research, T.J. Watson

Abstract. If we have a problem that is mildly hard, can we create a problem that is significantly harder? A natural approach to *hardness amplification* is the "direct product"; instead of asking an attacker to solve a single instance of a problem, we ask the attacker to solve several independently generated ones. Interestingly, proving that the direct product amplifies hardness is often highly non-trivial, and in some cases may be false. For example, it is known that the direct product (i.e. "parallel repetition") of general interactive games may not amplify hardness at all. On the other hand, positive results show that the direct product does amplify hardness for many basic primitives such as one-way functions, weakly-verifiable puzzles, and signatures.

Even when positive direct product theorems are shown to hold for some primitive, the parameters are surprisingly weaker than what we may have expected. For example, if we start with a weak one-way function that no poly-time attacker can break with probability $> \frac{1}{2}$, then the direct product provably amplifies hardness to *some negligible* probability. Naturally, we would expect that we can amplify hardness exponentially, all the way to 2^{-n} probability, or at least to some fixed/known negligible such as $n^{-\log n}$ in the security parameter n, just by taking sufficiently many instances of the weak primitive. Although it is known that such parameters cannot be proven via black-box reductions, they may seem like reasonable conjectures, and, to the best of our knowledge, are widely believed to hold. In fact, a conjecture along these lines was introduced in a survey of Goldreich, Nisan and Wigderson (ECCC '95). In this work, we show that such conjectures are *false* by providing simple but surprising counterexamples. In particular, we construct weakly secure *signatures* and *one-way functions*, for which standard hardness amplification results are known to hold, but for which hardness does not amplify *beyond just negligible*. That is, for *any negligible* function $\varepsilon(n)$, we instantiate these primitives so that the direct product can always be broken with probability $\varepsilon(n)$, *no matter how many copies we take*.

1 Introduction

Hardness amplification is a fundamental cryptographic problem: given a "weakly secure" construction of some cryptographic primitive, can we use it to build a

R. Cramer (Ed.): TCC 2012, LNCS 7194, pp. 476–493, 2012.
© International Association for Cryptologic Research 2012

"strongly secure" construction? The first result in this domain is a classical conversion from weak one-way functions to strong one-way function by Yao [32] (see also [13]). This result starts with a function f which is assumed to be *weakly one-way*, meaning that it can be inverted on at most (say) a *half* of its inputs. It shows that the *direct-product* function $F(x_1, \ldots, x_k) = (f(x_1), \ldots, f(x_k))$, for an appropriately chosen polynomial k, is one-way in the standard sense, meaning that it can be inverted on only a *negligible* fraction of its inputs. The above result is an example of what is called the *direct product theorem*, which, when true, roughly asserts that simultaneously solving many independent repetitions of a mildly hard task yields a much harder "combined task".[1] Since the result of Yao, such direct product theorems have been successfully used to argue security amplification of many other important cryptographic primitives, such as collision-resistant hash functions [8], encryption schemes [12], weakly verifiable puzzles [7,20,22], signatures schemes/MACs [11], commitment schemes [18,9], pseudorandom functions/generators [11,26], block ciphers [24,27,25,30], and various classes of interactive protocols [5,28,19,17].

Direct product theorems are surprisingly non-trivial to prove. In fact, in some settings, such as general interactive protocols [5,29], they are simply false and hardness does not amplify at all, irrespective of the number of repetitions. Even for primitives such as one-way functions, for which we do have "direct product theorems", the parameters of these results are surprisingly weaker than what we may have expected. Let us say that a cryptographic construction is *weakly secure* if no poly-time attacker can break it with probability greater than $\frac{1}{2}$. Known theorems tell us that the direct product of $k = \Theta(n)$ independent instances of a weakly secure construction will become secure in the standard sense, meaning that no poly-time attacker can succeed in breaking security with better than *some negligible probability in the security parameter* n. However, we could naturally expect the direct product of k instances will amplify hardness exponentially, ensuring that no poly-time attacker can break security with more than 2^{-k} probability. Or, we would at least expect that a sufficiently large number of $k = \text{poly}(n)$ repetitions can amplify hardness to some fixed/known negligible probability such as $\varepsilon(n) = 2^{-n^\delta}$ for some constant $\delta > 0$, or even less ambitiously, $\varepsilon(n) = n^{-\log n}$. We call such expected behavior *amplification beyond negligible*.

LIMITATION OF EXISTING PROOFS. One intuitive reason that the positive results are weaker than what we expect is the limitation of our reduction-based proof techniques. In particular, assume we wanted to show that the k-wise direct product amplifies hardness down to some very small probability ε. Then we would need an *efficient reduction* that uses an adversary $\mathcal{A}$ breaking the security of the k-wise direct product with probability ε, to break the security of a single instance with a much larger probability, say one half. Unfortunately, the

---

[1] A related approach to amplifying the hardness of decisional problems is the "XOR Lemma" which roughly asserts the hardness of predicting an XOR of the challenge bits of many independent instances of a decisional problem will amplify. In this work, we will focus of "search" problems such as one-way functions and signatures and therefore only consider amplification via direct product.

reduction cannot get "anything useful" from the attacker $\mathcal{A}$ until it succeeds at least once. And since $\mathcal{A}$ only succeeds with small probability ε, the reduction is forced to run $\mathcal{A}$ at least (and usually much more than) $1/\varepsilon$ times, since otherwise $\mathcal{A}$ might never succeed. In other words, the reduction is only efficient as long as ε is an *inverse polynomial*. This may already be enough to show that the direct product amplifies hardness *to some negligible probability*, since the success probability of $\mathcal{A}$ must be smaller than *every* inverse polynomial ε. But it also tells us that black-box reductions cannot prove any stronger bounds *beyond negligible*, since the reduction would necessarily become inefficient.[2] For example, we cannot even prove that the k-wise direct product of a weak one-way function will amplify hardness to $n^{-\log n}$ security (where n is the security parameter), no matter how many repetitions k we take.

OUR QUESTION. The main goal of this work is to examine whether the limitations of current hardness amplification results are just an artifact our proof technique, or whether they reflect reality. Indeed, we may be tempted to ignore the lack of formal proofs and nevertheless make the seemingly believable conjecture that *hardness does amplify beyond negligible*. In more detail, we may make the following conjecture:

Conjecture (Informal): *For all primitives for which standard direct product theorems hold (e.g., one-way functions, signatures etc.), the k-wise direct product of any weakly secure instantiation will amplify hardness all the way down to some fixed negligible bound $\varepsilon(n)$, such as $\varepsilon(n) = 2^{-\Omega(n)}$, or, less ambitiously, $\varepsilon(n) = n^{-\log n}$, when $k = \mathsf{poly}(n)$ is sufficiently large.*

To the best of our knowledge, such a conjecture is widely believed to hold. The survey of Goldreich et al. [14] explicitly introduced a variant of the above conjecture in the (slightly different) context of the XOR Lemma and termed it a *"dream version"* of hardness amplification which, although seemingly highly reasonable, happens to elude a formal proof.

OUR RESULTS. In this work, we show that, surprisingly, the above conjecture does *not* hold, and give strong counterexamples to the conjectured hardness amplification beyond negligible. We do so in the case of signature schemes and one-way functions for which we have standard direct-product theorems showing that hardness amplifies to negligible [32,11]. Our result for the signature case, explained in Section 3, relies on techniques from the area of *stateless (resettably-secure) multiparty computation* [6,3,10,16,15]. On a high level, we manage to embed an execution of a stateless mutliparty protocol Π into the design of our signature scheme, where Π generates a random instance of a hard relation $\mathcal{R}$, and the signer will output its secret key if the message contains a witness for $\mathcal{R}$. The execution of Π can be driven via carefully designed signing queries. Since Π is secure and $\mathcal{R}$ is hard, the resulting signature scheme is still secure by itself.

---

[2] This "folklore" observation has been attributed to Steven Rudich in [14].

However, our embedding is done in a way so as to allow us to attack the direct product of *many independent* schemes by forcing them to execute a *single correlated* execution of Π resulting in a common instance of the hard relation $\mathcal{R}$. This allows us to break all of the schemes simultaneously by breaking a single instance of $\mathcal{R}$, and thus with some negligible probability $\varepsilon(n)$, which is independent of the number of copies k. Indeed, we can make $\varepsilon(n)$ an arbitrarily large negligible quantity (say, $n^{-\log n}$) by choosing the parameters for the relation $\mathcal{R}$ appropriately.

One may wonder whether such counterexamples are particular to signature schemes. More specifically, our above counterexample seems to crucially rely on the fact that the security game for signatures is highly interactive (allowing us to embed an interactive MPC computation) and that the communication complexity between the challenger and attacker in the security game can be arbitrarily high (allowing us to embed data from all of the independent copies of the scheme into the attack on each individual one). Perhaps hardness still amplifies beyond negligible for simpler problems, such as one-way functions, where the security game is not interactive and has an a-priori bounded communication complexity. Our second result gives strong evidence that this too is unlikely, by giving a counterexmaple for one-way functions. The counterexample relies on a new assumption on a hash functions called *Extended Second Preimage Resistance* (ESPR), which we introduce in this paper. Essentially, this assumption says that given a random challenge x, it is hard to find a bounded-length *Merkle path* that starts at x, along with a collision on it. To break many independent copies of this problem, the attacker takes the independent challenges $x_1, \ldots, x_k$ and builds a *Merkle tree* with them as leaves. If it manages to find a *single* collision at the root of tree (which occurs with some probability independent of k), it will be able to find a witness (a Merkle path starting at x_i with a collision) for *each* of the challenges x_i. So far, this gives us an amplification counterexample for a *hard relation* based on the ESPR problem (which is already interesting), but, with a little more work, we can also convert it into a counterexample for a *one-way function* based on this problem. For the counterexample to go through, we need the ESPR assumption to hold for some fixed hash function (not a family), and so we cannot rely on collision resistance. Nevertheless, we argue that the ESPR assumption for a fixed hash function is quite reasonable and is likely satisfied by existing (fixed) cryptographic hash functions, by showing that it holds in a variant of the random oracle model introduced by Unruh [31], where an attacker gets arbitrary "oracle-dependent auxiliary input". As argued by [31], such model is useful for determining which security properties can be satisfies by a single hash function rather than a family.

Overall, our work gives strong indications that the limitations of our reductionist proofs for the direct product theorems might actually translate to real attacks for some schemes.

RELATED WORK. Interestingly, a large area of related work comes from a seemingly different question of *leakage amplification* [2,1,23,21]. These works ask the following: given a primitive P which is resilient to ℓ bits of leakage on its secret key, is it true that breaking k independent copies of P is resilient to almost $L = \ell k$ bits of leakage? At first sight this seems to be a completely unrelated question. However, there is a nice connection between hardness and leakage-resilience: if a primitive (such as a signature or one-way function) is hard to break with probability ε, then it is resilient to $\log(1/\varepsilon)$ bits of leakage. This means that if some counter-example shows that the leakage bound L does not amplify with k, then neither does the security. Therefore, although this observation was never made, the counterexamples to leakage amplification from [23,21] seem to already imply some counterexample for hardness. Unfortunately, both works concentrate on a modified version of parallel repetition, where some *common public parameters* are reused by all of the instances and, thus, they are *not truly independent*. Indeed, although showing counterexamples for (the harder question of) leakage amplification is still interesting in this scenario, constructing ones for hardness amplification becomes trivial.[3] However, the work of [21] also proposed that a variant of their counterexample for leakage amplification may extend to the setting without common parameters under a highly non-standard assumption about computationally sound (CS) proofs. Indeed, this suggestion led us to re-examine our initial belief that such counterexamples should not exist, and eventually resulted in this work. We also notice that our counterexample for signature schemes (but not one-way functions) can be easily extended to give a counterexample for leakage amplification without common parameters.

2 Hardness Amplification Definitions and Conjectures

In this work, we will consider a non-uniform model of computation. We equate entities such as challengers and attackers with circuit families, or equivalently, Turing Machines with advice. We let n denote the *security parameter*. We say that a function $\varepsilon(n)$ is *negligible* if $\varepsilon(n) = n^{-\omega(1)}$.

We begin by defining a general notion of (single prover) cryptographic games, which captures the security of the vast majority of cryptographic primitives, such as one-way functions, signatures, etc.

Definition 1 (Games). *A game is defined by a probabilistic interactive challenger $\mathcal{C}$. On security parameter n, the challenger $\mathcal{C}(1^n)$ interacts with some attacker $\mathcal{A}(1^n)$ and may output a special symbol* win. *If this occurs, we say that $\mathcal{A}(1^n)$ wins $\mathcal{C}(1^n)$.*

We can also define a *class* $\mathbb{C}$ of cryptographic games $\mathcal{C} \in \mathbb{C}$. For example the factoring problem fixes a particular game with the challenger $\mathcal{C}_{FACTOR}$ that chooses two random n-bit primes p, q, sends $N = p \cdot q$ to $\mathcal{A}$, and outputs win iff it gets back p, q. On the other hand, *one-way functions* can be thought of as a class of games $\mathbb{C}_{OWF}$, where each candidate one-way function f defines

---

[3] E.g., the hard problem could ask to break either the actual instance or the common parameter. While such an example does not necessarily contradict leakage amplification, it clearly violates hardness amplification.

a particular game $C_f \in \mathbb{C}_{OWF}$. So far, this definition of games and classes of games such as one-way function is purely syntactic and we now define what it means for a game to be hard.

Definition 2 (Hardness). *We say that the game C is $(s(n), \varepsilon(n))$-hard if, for all sufficiently large $n \in \mathbb{N}$ and all $\mathcal{A}(1^n)$ of size $s(n)$, we have*

$$\Pr[\mathcal{A}(1^n) \text{ wins } C(1^n)] < \varepsilon(n).$$

We say that the game C is $(\mathsf{poly}, \varepsilon(n))$-hard if it is $(s(n), \varepsilon(n))$-hard for all polynomial $s(n)$. We say that the game C is $(\mathsf{poly}, \mathsf{negl})$-hard if it is $(s(n), 1/p(n))$-hard for all polynomials $s(n), p(n)$.

Definition 3 (Direct Product). *For a cryptographic game C we define the k-wise direct-product game C^k, which initializes k independent copies of C and outputs the* win *symbol if and only if all k copies individually output* win.

Finally, we are ready to formally define what we mean by hardness amplification. Since we focus on negative results, we will distinguish between several broad levels of hardness amplification and ignore exact parameters. For example, we do not pay attention to the number of repetitions k needed to reach a certain level of hardness (an important parameter for positive results), but are more concerned with which levels of hardness are or are not reachable altogether.

Definition 4 (Hardness Amplification). *For a fixed game C, we say that hardness amplifies to $\varepsilon = \varepsilon(n)$ if there exists some polynomial $k = k(n)$ such that C^k is $(\mathsf{poly}, \varepsilon)$-hard. We say that hardness amplifies to negligible if there exists some polynomial $k = k(n)$ such that C^k is $(\mathsf{poly}, \mathsf{negl})$-hard. For a class $\mathbb{C}$ of games, we say that:*

1. *The hardness of a class $\mathbb{C}$ amplifies to negligible if, for every game $C \in \mathbb{C}$ which is $(\mathsf{poly}, \frac{1}{2})$-hard, the hardness of C amplifies to negligible.*
2. *The hardness of a class $\mathbb{C}$ amplifies to $\varepsilon(n)$ if, for every game $C \in \mathbb{C}$ which is $(\mathsf{poly}, \frac{1}{2})$-hard, the hardness of C amplifies to $\varepsilon(n)$.*
3. *The hardness of a class $\mathbb{C}$ amplifies beyond negligible if there exists some global negligible function $\varepsilon(n)$ for the entire class, such that the hardness of $\mathbb{C}$ amplifies to $\varepsilon(n)$.*

Remarks on Definition. The standard "direct product theorems" for classes such as one-way functions/relations and signatures show that the hardness of the corresponding class amplifies *to negligible* (bullet 1). For example, if we take *any* $(\mathsf{poly}, 1/2)$-hard function f, then a sufficiently large direct product f^k will be $(\mathsf{poly}, \mathsf{negl})$-hard.[4] However, what "negligible" security can we actually get? The result does not say and it may depend on the function f that we start with.[5] One

---

[4] The choice of $1/2$ is arbitrary and can be replaced with any constant or even any function bounded-away-from 1. We stick with $1/2$ for concreteness and simplicity.

[5] It also seemingly depends on the exact polynomial size $s(n)$ of the attackers we are trying to protect against. However, using a result of Bellare [4], the dependence on $s(n)$ can always be removed.

could conjecture that there is *some* fixed negligible $\varepsilon(n)$ such that a sufficiently large direct product of *any* weak instantiation will amplify its hardness to $\varepsilon(n)$. This is amplification *beyond negligible* (bullet 3). More ambitiously, we could expect that this negligible $\varepsilon(n)$ is very small such as $\varepsilon(n) = 2^{-n^{\Omega(1)}}$ or even $2^{-\Omega(n)}$. We explicitly state these conjectures below.

Dream Conjecture (Weaker): For any class of cryptographic games $\mathbb{C}$ for which hardness amplifies *to negligible*, it also amplifies *beyond negligible*.

Dream Conjecture (Stronger): For any class of cryptographic games $\mathbb{C}$ for which hardness amplifies *to negligible*, it also amplifies to $\varepsilon(n) = 2^{-n^{\Omega(1)}}$.

Our work gives counterexamples to both conjectures. We give two very different types counterexamples: one for the classes of *signature schemes* (Section 3) and one for the class of *one-way functions* (Section 4). Our counterexamples naturally require that some hard instantiations of these primitives exist to begin with, and our counterexamples for the weaker versions of the dream conjecture will actually require the existence of *exponentially hard* versions of these primitives. In particular, under strong enough assumptions, we will show that for *every* negligible function $\varepsilon(n)$ there is stand-alone scheme which is *already* $(\mathsf{poly}, \mathsf{negl})$-hard, but whose k-wise direct product is *not* $(\mathsf{poly}, \varepsilon(n))$-hard, no matter how large k is.

2.1 Hard and One-Way Relations

As a component of both counterexamples, we will rely on the following definition of hard relations phrased in the framework of cryptographic games:

Definition 5 (Hard Relations). *Let $R \subseteq \bigcup_{n \in \mathbb{N}} \{0,1\}^n \times \{0,1\}^{p(n)}$ be an NP relation consisting of pairs (y, w) with instances y and witnesses w of polynomial size $p(|y|)$. Let $L = \{ y : \exists\, w \text{ s.t. } (y, w) \in R \}$ be the corresponding NP language. Let $y \leftarrow \mathrm{SAML}(1^n)$ be a PPT algorithm that samples values $y \in L$. For a relation $\mathcal{R} = (R, \mathrm{SAML})$, we define the corresponding security game where the challenger $\mathcal{C}(1^n)$ samples $y \leftarrow \mathrm{SAML}(1^n)$ and the adversary wins if it outputs w s.t. $(y, w) \in R$. By default, we consider $(\mathsf{poly}, \mathsf{negl})$-hard relations, but we can also talk about $(s(n), \varepsilon(n))$-hard relations.*

Note that, for hard relations, we only require that there is an efficient algorithm for sampling hard instances y. Often in cryptography we care about a sub-class of hard relations, which we call *one-way relations*, where is it also feasible to efficiently sample a hard instance y *along with* a witness w. We define this below.

Definition 6 (One-Way Relation). *Let R be an NP relation and L be the corresponding language. Let $(y, w) \leftarrow \mathrm{SAMR}(1^n)$ be a PPT algorithm that samples values $(y, w) \in R$, and define $y \leftarrow \mathrm{SAML}(1^n)$ to be a restriction of SAMR to its first output. We say that (R, SAMR) is a one-way relation if (R, SAML) is a hard relation.*

3   Counterexample for Signature Schemes

3.1   Overview

The work of [11] shows that the direct product of any *stateless* signature scheme amplifies hardness *to negligible*. We now show that it does not (in general) amplify hardness beyond negligible. In fact, we will give a transformation from any standard signature scheme (Gen, Sign, Verify) into a new signature scheme (GEN, SIGN, VERIFY) whose hardness does not amplify (via a direct product) beyond negligible. We start by giving an informal description of the transformation to illustrate our main ideas. In order to convey the intuition clearly, we will first consider a simplified case where the signing algorithm SIGN of the (modified) scheme (GEN, SIGN, VERIFY) is *stateful*, and will then discuss how to convert the stateful signing algorithm into one that is *stateless*.[6]

Embedding MPC in Signatures. Let (Gen, Sign, Verify) be any standard signature scheme. Let $\mathcal{F} = \{\mathcal{F}_k\}_{k \in \mathbb{N}}$ be a randomized k-party "ideal functionality" that takes no inputs and generates a random instance y of a hard relation $\mathcal{R} = (R, \text{SAML})$ according to the distribution SAML. Further, let $\Pi = \{\Pi_k\}_{k \in \mathbb{N}}$ be a multi-party computation protocol that securely realizes the functionality $\mathcal{F}$ for any number of parties k. Then, the new signature scheme (GEN, SIGN, VERIFY) works as follows.

Algorithms GEN and VERIFY are identical to Gen and Verify respectively. The signing algorithm SIGN is essentially the same as Sign, except that, on receiving a signing queries of a "special form", SIGN interprets these as "protocol messages" for Π_k and (in addition to generating a signature of them under SIGN) also executes the *next message function* of the protocol and outputs its response as part of the new signature. A special initialization query specifies the number of parties k involved in the protocol and the role P_i in which the signing algorithm should act. The signing algorithm then always acts as the honest party P_i while the user submitting signing queries can essentially play the role of the remaining $k - 1$ parties. When Π_k is completed yielding some output y (interpreted as the instance of a hard relation $\mathcal{R}$) the signing algorithm SIGN will look for a signing query that contains a corresponding witness w, and, if it receives one, will respond to it by simply outputting its entire *secret key* in the signature. The security of the transformed signature (GEN, SIGN, VERIFY) immediately follows from the security of the MPC protocol Π_k against all-but-one corruptions, the hardness of the relation $\mathcal{R}$ and the security of the original signature scheme.

Attacking the Direct-Product. Let us briefly demonstrate an adversary $\mathcal{A}$ for the k-wise direct product. Very roughly, $\mathcal{A}$ carefully chooses his signing queries

---

[6] We note that in the setting of stateful signatures, hardness fails to amplify even *to* negligible since we can embed the counterexamples of [5,29] into the signature scheme. Nevertheless our initial description of our counterexample for the stateful setting will clarify the main new result, which is a counterexample for the stateless setting.

so as to force $\text{SIGN}_1, \ldots, \text{SIGN}_k$ to engage in a *single* execution of the protocol Π_k, where each SIGN_i plays the role of a different party P_i, while $\mathcal{A}$ simply acts as the "communication link" between them. This results in all component schemes SIGN_i generating a common instance y of the hard relation. Finally, $\mathcal{A}$ simply "guesses" a witness w for y at random and, if it succeeds, submits w as a signing query, thereby learns the secret key of *each* component signature scheme thereby breaking all k of them! Note that the probability of guessing w is bounded by some negligible function in n and is *independent* of the number of parallel repetitions k.

Stateful to Stateless. While the above gives us a counterexample for the case where SIGN is a *stateful* algorithm, (as stated above) we are mainly interested in the (standard) case where SIGN is *stateless*. In order to make SIGN a stateless algorithm, we can consider a natural approach where we use a modified version Π'_k of protocol Π_k: each party P_i in Π'_k computes an outgoing message in essentially the same manner as in Π_k, except that it also attaches an *authenticated encryption* of its current protocol state, as well as the previous protocol message. This allows each (stateless) party P_i to "recover" its state from the previous round to compute its protocol message in the next round. Unfortunately, this approach is insufficient, and in fact insecure, since an adversarial user can *reset* the (stateless) signing algorithm at any point and achieve the effect of *rewinding* the honest party (played by the signing algorithm) during the protocol Π_k. To overcome this problem, we leverage techniques from the notion of resettably-secure computation. Specifically, instead of using a standard MPC protocol in the above construction, we use a recent result of Goyal and Maji [15] which constructs an MPC protocol that is secure against reset attacks and works for stateless parties for a large class of functionalities, including "inputless" randomized functionalities (that we will use in this paper).

The above intuitive description hides many details of how the user can actually "drive" the MPC execution between the k signers within the direct-product game where all signers respond to a single common message. We proceed to make this formal in the following section.

3.2   Our Signature Scheme

We now give our transformation from any standard signature scheme into one whose hardness does not amplify beyond negligible. We first establish some notation.

Notation. Let n be the security parameter. Let $(\mathsf{Gen}, \mathsf{Sign}, \mathsf{Verify})$ be any standard signature scheme. Further, let $(\mathcal{R}, \text{SAML})$ be a hard relation as per Definition 5. Let $\{PRF_K : \{0,1\}^{\mathsf{poly}(n)} \to \{0,1\}^{\mathsf{poly}(n)}\}_{K \in \{0,1\}^n\}}$ be a pseudo-random function family.

Stateless MPC. We consider a randomized k-party functionality $\mathcal{F} = \{\mathcal{F}\}_{k \in \mathbb{N}}$ that does not take any inputs; $\mathcal{F}$ simply samples a random pair $y \leftarrow \text{SAML}(1^n)$

and outputs y to all parties. Let $\{\Pi_k\}_{k \in \text{poly}(n)}$ be a family of protocols, where each $\Pi_k = \{P_1, \ldots, P_k\}$ is a k-party MPC protocol for computing the functionality $\mathcal{F}$ in the *public state model*. This model is described formally in the full version, and we only give a quick overview here. Each party P_i is completely described by the next message function NM_i, which takes the following four values as input: (a) a string π_{j-1} that consists of all the messages sent in any round $j - 1$ of the protocol, (b) the *public state* state_i of party P_i, and (c) the secret randomness r_i. On receiving an input of the form $\pi_{j-1}\|\text{state}_i\|r_i$, NM_i outputs P_i's message in round j along with the updated value of state_i. We assume that an attacker corrupts (exactly) $k - 1$ of the parties. In the real-world execution, the attacker can arbitrarily call the next-message function NM_i of the honest party P_i with arbitrarily chosen values of the public state state_i and arbitrary message π_{j-1} (but with an honestly chosen and secret randomness r_i). Nevertheless, the final output of P_i and the view of the attacker can be simulated in the ideal world where the simulator can "reset" the ideal functionality. In our case, that means that the attacker can adaptively choose one of polynomially many honestly chosen instances $y_1, \ldots, y_q$ of the hard relation which P_i will then accept as output.

The Construction. We describe our signature scheme (GEN, SIGN, VERIFY).

GEN(1^n): Compute $(pk, sk) \leftarrow \text{Gen}(1^n)$. Also, sample a random tape $K \leftarrow \{0, 1\}^{\text{poly}(n)}$ and a random identity id $\in \{0, 1\}^n$. Output $PK = (pk, \text{id})$ and $SK = (sk, K, \text{id})$.

SIGN(SK, m): To sign a message m using secret key $SK = (sk, K, \text{id})$, the signer outputs a signature $\sigma = (\sigma^1, \sigma^2)$ where $\sigma^1 \leftarrow \text{Sign}(sk, m)$. Next, if m does not contain the prefix "prot", then simply set $\sigma^2 = \{0\}$. Otherwise, parse $m = (\text{"prot"}\|\text{IM}\|\pi_j\|\text{state}\|w)$, where $\text{IM} = k\|\text{id}_1\|\ldots\|\text{id}_k$ such that $\text{state} = \text{state}_1\|\ldots\|\text{state}_k$, then do the following:

- Let $i \in [k]$ be such that $\text{id} = \text{id}_i$. Compute $r_i = PRF_K(\text{IM})$. Then, apply the next message function NM_i of (stateless) party P_i in protocol Π_k over the string $\pi_j\|\text{state}_i\|r_i$ and set σ^2 to the output value.[7]
- Now, if σ^2 contains the output y of protocol Π_k,[8] then further check whether $(y, w) \in \mathcal{R}$. If the check succeeds, set $\sigma^2 = SK$.

VERIFY(PK, m, σ): Given a signature $\sigma = (\sigma^1, \sigma^2)$ on message m with respect to the public key $PK = (pk, \text{id})$, output 1 iff VERIFY(pk, m, σ^1) = 1.

---

[7] Note that here σ^2 consists of party P_i's protocol message in round $j + 1$, and its updated public state state_i.

[8] Note that this is the case when j is the final round in Π_k. Here we use the property that the last round of Π_k is the *output delivery round*, and that when NM_i is computed over the protocol messages of this round, it outputs the protocol output.

This completes the description of our signature scheme. In the full version, we prove the following theorem showing that the signature scheme satisfies basic signature security.

Theorem 1. *If (Gen, Sign, Verify) is a secure signature scheme, $\{PRF_K\}$ is a PRF family, $\mathcal{R}$ is a hard relation, and Π_k is a stateless MPC protocol for functionality $\mathcal{F}$, then the proposed scheme (GEN, SIGN, VERIFY) is a secure signature scheme.*

3.3   Attack on the Direct Product

Theorem 2. *Let (GEN, SIGN, VERIFY) be the described signature scheme and let $\mathcal{R} = (\text{SAML}, R)$ be the hard relation used in the construction. Assume that for any $y \xleftarrow{\$} \text{SAML}(1^n)$, the size of the corresponding witness w is bounded by $|w| = p(n)$. Then, for any polynomial $k = k(n)$, there is an attack against the k-wise direct product running in time $\text{poly}(n)$ with success probability $\varepsilon(n) = 2^{-p(n)}$.*

We will prove Theorem 2 by constructing an adversary $\mathcal{A}$ that mounts a key-recovery attack on any k-wise direct product of the signature scheme (GEN, SIGN, VERIFY).

k-wise Direct Product. Let (**Gen, Sign, Verify**) denote the k-wise direct product of the signature scheme (GEN, SIGN, VERIFY), described as follows. The algorithm **Gen** runs GEN k-times to generate $(PK_1, SK_1), \ldots, (PK_k, SK_k)$. To sign a message m, **Sign** computes $\sigma_i \leftarrow \text{SIGN}(SK_i, m)$ for every $i \in k$ and outputs $\sigma = (\sigma_1, \ldots, \sigma_k)$. Finally, on input a signature $\sigma = (\sigma_1, \ldots, \sigma_k)$ on message m, **Verify** outputs 1 iff $\forall i \in k$, $\text{VERIFY}(PK_i, m, \sigma_i) = 1$.

Description of $\mathcal{A}$. We now describe the adversary $\mathcal{A}$ for (**Gen, Sign, Verify**). Let $(PK_1, \ldots, PK_k)$ denote the public key that $\mathcal{A}$ receives from the challenger of the signature scheme (**Gen, Sign, Verify**), where each $PK_i = (pk_i, \text{id}_i)$. The adversary $\mathcal{A}$ first sends a signing query m_0 of the form "prot"$\|\text{IM}\|\pi_0\|\text{state}\|w$, where $\text{IM} = k\|\text{id}_1\|\ldots\|\text{id}_k$, and $\pi_0 = \text{state} = w = \{0\}$. Let $\sigma = (\sigma_1, \ldots, \sigma_k)$ be the response it receives, where each $\sigma_i = \sigma_i^1, \sigma_i^2$. $\mathcal{A}$ now parses each σ_i^2 as a first round protocol message π_1^i from party P_i followed by the public state state_i of P_i (at the end of the first round) in protocol Π_k.

$\mathcal{A}$ now prepares a new signing query m_1 of the form "prot"$\|\text{IM}\|\pi_1\|\text{state}\|w$, where IM and w are the same as before, but $\pi_1 = \pi_1^1\|\ldots\|\pi_1^k$, and state $= \text{state}_1\|\ldots\|\text{state}_k$. On receiving the response, $\mathcal{A}$ repeats the same process as above to produce signing queries $m_2, \ldots, m_{t-1}$, where t is the total number of rounds in protocol Π_k. (That is, each signing query $m_2, \ldots, m_{t-1}$ is prepared in the same manner as m_1.)

Finally, let $\sigma = (\sigma_1, \ldots, \sigma_k)$ be the response to the signing query m_{t-1}. $\mathcal{A}$ now parses each σ_i^2 as the round t protocol message π_t^i from party P_i followed by the state state_i of P_i. Now, since the final round (i.e., round t) of protocol Π_k is the *output delivery round*, and further, Π_k satisfies the *publicly computable output* property, $\mathcal{A}$ simply computes the protocol output y from the messages $\pi_t^1, \ldots, \pi_t^k$.

Now, $\mathcal{A}$ guesses a $p(n)$-sized witness $w^* \xleftarrow{\$} \{0,1\}^{p(n)}$ at random and, if $(y, w^*) \in \mathcal{R}(x)$, it now sends the final signing query $m_t =$ "prot"$\|$IM$\|\pi_t\|$state$\|w$, where IM is the same as before, $\pi_t = \pi_t^1\|\ldots\|\pi_t^k$, state $=$ state$_1\|\ldots\|$state$_k$, and $w = w^*$. Thus, $\mathcal{A}$ obtains $SK_1, \ldots, SK_k$ from the challenger and can forge arbitrary signatures for the direct product scheme. It's clear that its success probability is at least $2^{-p(n)}$.

Corollary 1. *Assuming the existence hard relations and a general stateless MPC compilers, the hardness of signature schemes does* not *amplify to any $\varepsilon(n) = 2^{-n^{\Omega(1)}}$. This gives a counterexample to the strong dream conjecture. If we, in addition, assume the existence of $(2^{\Omega(n)}, 2^{-\Omega(n)})$-hard relations with witness size $p(n) = O(n)$, then there exist signature schemes whose hardness does not amplify beyond negligible. This gives a counterexample to the weak dream conjecture.*

Proof. For the first result, assume that the witness size of the relation $\mathcal{R}$ is bounded by $p(n) = O(n^c)$ for some constant c. Given any constant $\delta > 0$, we can simply instantiate the signature scheme (GEN, SIGN, VERIFY) used in our counterexample with the hard relation $\mathcal{R}'$ that uses security parameter $m(n) = n^{\delta/c}$ so that its witness size is $p'(n) = p(m) = O(n^\delta)$. It's clear that $\mathcal{R}'$ is still $(\mathsf{poly}(n), \mathsf{negl}(n))$-secure but, by Theorem 2, the k-wise direct product can be broken in $\mathsf{poly}(n)$ time with probability $\varepsilon(n) = 2^{-O(n^\delta)}$. Therefore security does not amplify 2^{n^δ} for any $\delta > 0$. The second part of the theorem follows in the same way, except that, for *any fixed* negligible function $\delta(n)$ we set $m(n) = -\log(\delta(n))$.

4   Counterexample for One-Way Relations and Functions

In Section 3, we proved that there exist signature schemes whose hardness does not amplify. This already rules out the general conjecture that "for *any* game for which hardness amplifies to negligible, hardness will also amplify to exponential (or at least beyond negligible)". Nevertheless, one might still think that the conjecture hold for more restricted classes of games. Perhaps the simplest such class to consider is one-way functions. Note that, unlike the case for signature schemes, the one-wayness game does not allow interaction and has bounded communication between attacker and challenger. Thus, the general strategy we employed in Section 3 of embedding a multiparty computation inside signature queries, will no longer work. In this section, we propose an alternate strategy for showing that one-way relation hardness does not amplify beyond negligible.

4.1   Our Construction

We begin by giving a counterexample for *hard relations*. We then extend it to counterexamples for one-way relations and and one-way functions. Our constructions are based on a new (non-standard) cryptographic security assumption on hash functions. Let $h : \{0,1\}^{2n} \mapsto \{0,1\}^n$ be a hash function. We define a *Merkle path of length* ℓ to be a tuple of the form

$$p_\ell = (x_0, (b_1, x_1), \ldots, (b_\ell, x_\ell)) \quad : \quad b_i \in \{0,1\}, x_i \in \{0,1\}^n.$$

Intuitively, x_0 could be the *leaf* of some Merkle tree of height ℓ, and the values $x_1, \ldots, x_\ell$ are the siblings along the path from the leaf to the *root*, where the bits b_i indicate whether the sibling x_i is a left or right sibling. However, we can also talk about a path p_ℓ on its own, without thinking of it as part of a larger tree. Formally, if p_ℓ is a Merkle path as above, let $p_{\ell-1}$ be the path with the last component (b_ℓ, x_ℓ) removed. The *value* of a Merkle path p_ℓ as above is defined iteratively via:

$$\bar{h}(p_\ell) = \begin{cases} h(\bar{h}(p_{\ell-1}), x_\ell) & \ell > 0, b_\ell = 1 \\ h(x_\ell, \bar{h}(p_{\ell-1})) & \ell > 0, b_\ell = 0 \\ x_0 & \ell = 0 \end{cases}$$

We call x_0 the *leaf* of the path p_ℓ, and $z = \bar{h}(p_\ell)$ is its *root*. We say that $y = (x_L, x_R) \in \{0,1\}^{2n}$ is the *known preimage of the path* p_ℓ if x_L, x_R are the values under the root, so that either $x_L = x_\ell$, $x_R = \bar{h}(p_{\ell-1})$ if $b_\ell = 0$, or $x_L = \bar{h}(p_{\ell-1})$, $x_R = x_\ell$ if $b_\ell = 1$. Note that this implies $h(y) = \bar{h}(p_\ell)$. We say that $y' \in \{0,1\}^{2n}$ is a *second preimage of the path* p_ℓ if $y' \neq y$ is *not* the known preimage of p_ℓ, and $h(y') = \bar{h}(p_\ell)$. We are now ready to define the *extended second-preimage resistance (ESPR)* assumption. This assumption says that, given a random challenge $x_0 \in \{0,1\}^n$, it is hard to find a (short) path p_ℓ containing x_0 as a leaf, and a second-preimage y' of p_ℓ.

Definition 7 (ESPR). *Let* $h : \{\{0,1\}^{2n} \mapsto \{0,1\}^n\}_{n \in \mathbb{N}}$ *be a poly-time computable hash function. We define the* Extended Second Preimage Resistance *(ESPR) assumption on* h *via the following security game between a challenger and an adversary* $\mathcal{A}(1^n)$:

1. *The challenger chooses* $x_0 \xleftarrow{\$} \{0,1\}^n$ *at random and gives it to* $\mathcal{A}$.
2. $\mathcal{A}$ *wins if it outputs a tuple* (p_ℓ, y'), *where* p_ℓ *is a Merkle path of length* $\ell \leq n$ *containing* x_0 *as a leaf, and* y' *is a second-preimage of* p_ℓ.

Discussion. In the above definition, we want h to be a single *fixed* hash function and not a function family. The notion of ESPR security seems to lie somewhere in between second-preimage resistance (SPR) and collision resistance (CR), implying the former and being implied by the latter.[9] Unfortunately, collision resistance *cannot* be achieved by any fixed hash function (at least w.r.t non-uniform attackers), since the attacker can always know a single hard-coded collision as auxiliary input. Fortunately, there does not appear to be any such trivial non-uniform attack against ESPR security, since the attacker is forced to "incorporate" a random leaf x_0 into the Merkle path on which it finds a collision. Therefore, in this regard, it seems that ESPR security may be closer to SPR

---

[9] A hash function is SPR if, given a uniformly random y, it's hard to find any $y' \neq y$ such that $h(y) = h(y')$. It is CR if it is hard to find any $y \neq y'$ s.t. $h(y) = h(y')$.

security, which can be achieved by a fixed hash function (if one-way functions exist). Indeed, in Section 4.2, we give a heuristic argument that modern (fixed) cryptographic hash functions already satisfy the ESPR property, even against non-uniform attackers. We do so by analyzing ESPR security in a variant of the random-oracle model, where the attacker may observe some "oracle-dependent auxiliary input". This model, proposed by Unruh [31], is intended to capture the properties of hash functions that can be achieved by fixed hash functions, rather than function families.

A Hard Relation from ESPR. Given a hash function h we can define the NP relation R_h with *statements* $x \in \{0,1\}^n$ and *witnesses* $w = (p_\ell, y')$ where p_ℓ is a Merkle path of length $\ell \leq n$ containing x as leaf, and y' is a second-preimage of p_ℓ. The corresponding NP language is defined as $L_h \stackrel{\text{def}}{=} \{x : \exists\, w \text{ s.t. } (x, w) \in R_h\}$. We say that h is *slightly regular*, if for every $z \in \{0,1\}^n$ there exist at least two distinct pre-images $y \neq y'$ such that $h(y) = h(y') = z$. If this is the case, then $L_h = \{0,1\}^*$ is just the language consisting of all bit strings. Now, we can define the distribution $x \leftarrow \text{SAML}(1^n)$ which just samples $x \stackrel{\$}{\leftarrow} \{0,1\}^n$ uniformly at random. It is easy to see that, if h is an $(s(n), \varepsilon(n))$-hard ESPR hash function, then $\mathcal{R}_h = (R_h, \text{SAML})$ is an $(s(n), \varepsilon(n))$-*hard relation.*

Hardness Non-Amplification. We now show our counterexample to the hardness amplification for the hard relation $\mathcal{R}_h$. The main idea is that, given k random and independent challenges $x^{(1)}, \dots, x^{(k)}$, the attacker builds a Merkle tree with the challenges as leaves. Let z be the value at the top of the Merkle tree. Then the attack just guesses some value $y' \in \{0,1\}^{2n}$ at random and, with probability $\geq 2^{-2n}$, y' will be a second-preimage of z (i.e. $h(y') = z$ and y' is distinct from the known preimage y containing the values under the root). Now, for each leaf $x^{(i)}$, let p_ℓ^i be the Merkle path for the leaf $x^{(i)}$. Then the witness $w_i = (y', p_\ell^i)$ is good witness for $x^{(i)}$. So, with probability $\geq 2^{-2n}$ with which the attack correctly guessed y', it breaks *all* k independent instances of the relation $\mathcal{R}_h$, no matter how large k is! By changing the relation $\mathcal{R}_h = (R_h, \text{SAML})$ so that, on security parameter n, the sampling algorithm $\text{SAML}(1^n)$ chooses $x \stackrel{\$}{\leftarrow} \{0,1\}^m$ with $m = m(n)$ being some smaller function of n such as $m(n) = n^\delta$ for a constant $\delta > 0$ or even $m(n) = \log^2(n)$, we can get more dramatic counterexamples where hardness does not amplify beyond $\varepsilon(n) = 2^{-n^\delta}$ or even $\varepsilon(n) = n^{-\log n}$. We now summarize the above discussion with a formal theorem.

Theorem 3. *Let h be a slightly regular, ESPR-secure hash function and let $\mathcal{R}_h = (R_h, \text{SAML})$ be the corresponding (poly, negl)-hard relation. Then, for any polynomial $k = \text{poly}(n)$, the k-wise direct product of $\mathcal{R}_h$ is not (poly, 2^{-2n}) secure. That is, for any polynomial k, there is a poly-time attack against the k-wise direct product of $\mathcal{R}_h$ having success probability 2^{-2n}.*

Proof. We first describe the attack. The attacker gets k independently generated challenges $x^{(1)}, \dots, x^{(k)}$. Let ℓ be the unique value such that $2^{\ell-1} < k \leq 2^\ell$, and let $k^* = 2^\ell$ be the smallest power-of-2 which is larger than k. Let us define additional "dummy values" $x^{(k+1)} = \ldots = x^{(k^*)} := 0^n$. The attack constructs a *Merkle Tree*, which is a full binary tree of height ℓ, whose k^* leaves are associated with the values $x^{(1)}, \ldots, x^{(k^*)}$. The value of any non-leaf node v is defined recursively as $val(v) = h(val(v_L), val(v_R))$ where v_L, v_R are the left and right children of v respectively. For any leaf $v^{(i)}$ associated with the value $x^{(i)}$, let $(v_1 = v^{(i)}, v_2, \ldots, v_\ell, r)$ be the nodes on the path from the leaf v_1 to the root r in the Merkle tree. The Merkle path associated with the value $x^{(i)}$ is then defined by $p_\ell^{(i)} = (x^{(i)}, (x_1, b_1), \ldots, (x_\ell, b_\ell))$ where each x_j is the value associated with the *sibling* of v_j, and $b_j = 0$ if v_j is a right child and 1 otherwise. Note that, if r is the root of the tree and $z = val(r)$ is the value associated with it, then $\bar{h}(p_\ell^{(i)}) = z$ for all paths $p_\ell^{(i)}$ with $i \in \{1, \ldots, k^*\}$. Furthermore let us label the nodes v_L, v_R to be the children of the root r, the values x_L, x_R be the values associated with them, and set $y := (x_L, x_R)$. Then y is the known preimage such that $h(y) = z$, associated with each one of the paths $p_\ell^{(i)}$.

The attack guesses a value $y' \xleftarrow{\$} \{0,1\}^{2n}$ at random and, outputs the k-tuple of witnesses $(w_1, \ldots, w_k)$ where $w_i = (p_\ell^{(i)}, y')$. With probability at least 2^{-2n}, y' is a second-preimage of z with $h(y') = z$ and $y' \neq y$ (since h is slightly regular, such second preimage always exists). If this is the case, then y' is also a second preimage of *every* path $p_\ell^{(i)}$. Therefore, with probability $\geq 2^{-2n}$ the attack finds a witness for each of the k instances and wins the hard relation game for the direct product relation $\mathcal{R}_h^k$.

Corollary 2. *Assuming the existence of a slightly regular* (poly, negl)*-hard ESPR hash functions, the hardness of hard relations does* not *amplify to* $2^{-n^{\Omega(1)}}$, *giving a counterexample to the stronger dream conjecture. If we instead assume the existence of* $(2^{\Omega(n)}, 2^{-\Omega(n)})$*-hard ESPR hash functions, then the hardness of hard relations does* not *amplify beyond negligible, giving a counterexample to the weaker dream conjecture.*

Proof. Let h be the ESPR hash-function. We define a modified relation $\mathcal{R}_h^m = (R_h, \textsc{SamL})$ where the sampling algorithm $\textsc{SamL}(1^n)$ samples an instance $x \xleftarrow{\$} \{0,1\}^m$ where $m = m(n)$ is some function of n. For the first part of the corollary, let $\delta > 0$ be any constant, and set $m(n) = n^\delta/2$. Then $\mathcal{R}_h^m$ is still a (poly, negl)-hard relation. However, by appplying Theorem 3 with m replacing n, we see that for any $k = \mathsf{poly}(m) = \mathsf{poly}(n)$, there is an attack against the k-wise direct product which succeeds with probability $\geq 2^{-2m} = 2^{-n^\delta}$. In other word, for any $\delta > 0$, there is a (poly, negl)-hard relation whose direct product is *not* (poly, 2^{-n^δ})-hard, no matter how large k is. This proces the first part of the corollary. The second part of the corollary works the same way as the first part but, for *any fixed* negligible function $\delta(n)$ we set $m(n) = -\frac{1}{2}\log(\delta(n))$. Assuming that h is a $(2^{\Omega(n)}, 2^{-\Omega(n)})$-hard ESPR hash function, the relation $\mathcal{R}_h^m$ is then still (poly, negl)-hard, but it's direct product is *not* (poly, $\delta(n)$)-hard. This proves the second part of the corollary.

Extension to One-Way Relations. We can get essentially the same results as above for *one-way* relations rather than just *hard* relations. Assume that $\mathcal{R}_{ow} = (R_{ow}, \text{SAMR}_{ow})$ is *any one-way* relation, and $\mathcal{R}_h = (R_h, \text{SAML}_H)$ is the hard relation used in our counterexample. Define the OR relation $\mathcal{R}_{or} = (R_{or}, \text{SAMR}_{or})$ via:

$$R_{or} \overset{\text{def}}{=} \{(y_1, y_2), (w_1, w_2) \; : \; (y_1, w_1) \in R_H \text{ or } (y_2, w_2) \in R_{ow}\}$$

$$\text{SAMR}_{or}(1^n) \; : \; \text{Sample } y_1 \leftarrow \text{SAML}_h(1^n), (y_2, w_2) \leftarrow \text{SAMR}_{ow}(1^n)$$

$$\text{Output: } ((y_1, y_2), (0, w_2)).$$

Then Theorem 3 applies as-is to the one-way relation $\mathcal{R}_{or}$ replacing $\mathcal{R}_H$ and Corollary 2 applies to *one-way relations* as well.

Extension to One-Way Functions. We can also extend the above counterexample to one-way functions. Let $i(n) \geq n$ be a polynomial and $f \; : \; \{\{0,1\}^{i(n)} \to \{0,1\}^n \}_{n \in \mathbb{N}}$ be a *regular one-way function* so that, for $x \overset{\$}{\leftarrow} \{0,1\}^{i(n)}$, the output $f(x)$ is uniformly random over $\{0,1\}^n$. Let $\mathcal{R} = (R, \text{SAML})$ be the hard relation for which we have a counterexample, with witness-size bounded by $u(n)$. We define $F \; : \; (\{0,1\}^{i(n)} \times \{0,1\}^n \times \{0,1\}^{u(n)} \times \{0,1\}^n) \to \{0,1\}^n$ via:

$$F(x, y, w, z) \overset{\text{def}}{=} \begin{cases} y & \text{If } (y,w) \in R \wedge z = 0^n \\ f(x) & \text{Otherwise.} \end{cases}$$

Note that the distribution of $F(x, y, w, z)$ is statistically close to that of $f(x)$ since the probability of $z = 0^n$ is negligible. The preimage of any $y \in \{0,1\}^n$ is either of the form $(\cdot, y, w, \cdot)$ where $(y, w) \in R$ or of the form $(x, \cdot, \cdot)$ where $f(x) = y$, and hence breaking the one-wayness of F is no easier then breaking that of f or breaking the hard relation $\mathcal{R}$. On the other hand, it is possible to break the k-wise direct product of F just by breaking the k-wise direct product of the hard relation $\mathcal{R}$. Therefore, the results of Corollary 2 apply to *one-way relations* as well, if we also assume the existence of a (fixed) regular one-way function f (and an exponentially secure one for the counterexample to the weaker conjecture). In the full version of this work, we also show how to instantiate f using the ESPR function h so as to get the results of Corollary 2 for one-way functions, without needing any additional assumptions.

4.2   Justifying the ESPR Assumption

We now give some justification that ESPR hash functions may exist by showing how to construct them in in a variant of the random-oracle (RO) model. Of course, constructions in the random-oracle model do not seem to offer any meaningful guarantees for showing that the corresponding primitive may be realized by a *fixed* hash function: indeed the RO model immediately implies collision resistance which cannot be realized by a fixed hash function. Rather, the RO model is usually interpreted as implying that the given primitive is likely to be realizable by a *family of hash functions*. Therefore, we will work with a variant of the RO model in which the attacker is initialized with some arbitrary "oracle-dependent auxiliary input". This model was proposed by Unruh [31] with the explicit motivation of capturing properties the can be satisfied by a fixed hash function. For example, the auxiliary input may include some small number of fixed collisions on the RO and therefore collision-resistance is unachievable in this model. By showing that ESPR security *is* achievable, we provide some justification for this assumption.

Let $\mathcal{O} : \{0,1\}^{2n} \mapsto \{0,1\}^n$ be a fixed length random oracle. Following [31], we define "oracle-dependent auxiliary input" of size $p(n)$ as an arbitrary function $z : \{\{0,1\}^{2n} \mapsto \{0,1\}^n\} \mapsto \{0,1\}^{p(n)}$ which can arbitrarily "compresses" the entire oracle $\mathcal{O}$ into $p(n)$ bits of auxiliary information $z(\mathcal{O})$. When considering security games in the oracle-dependent auxiliary input model, we consider attackers $\mathcal{A}^{\mathcal{O}}(z(\mathcal{O}))$ which are initialized with polynomial-sized oracle-dependent auxiliary input $z(\cdot)$. In the full version, we show that the ESPR security game is hard in the random oracle model with auxiliary input.

Theorem 4. *Let $\mathcal{O}$ be modeled as a random oracle, and consider the ESPR game in which h is replaced with $\mathcal{O}$. Then, for any attacker $\mathcal{A}^{\mathcal{O}}(z(\mathcal{O}))$ with polynomial-sized auxiliary input $z(\cdot)$ and making at most polynomially many queries to $\mathcal{O}$, its probability of winning the ESPR game is at most $\varepsilon = 2^{-\Omega(n)}$.*

References

1. Alwen, J., Dodis, Y., Naor, M., Segev, G., Walfish, S., Wichs, D.: Public-Key Encryption in the Bounded-Retrieval Model. In: Gilbert, H. (ed.) EUROCRYPT 2010. LNCS, vol. 6110, pp. 113–134. Springer, Heidelberg (2010)
2. Alwen, J., Dodis, Y., Wichs, D.: Leakage-Resilient Public-Key Cryptography in the Bounded-Retrieval Model. In: Halevi, S. (ed.) CRYPTO 2009. LNCS, vol. 5677, pp. 36–54. Springer, Heidelberg (2009)
3. Barak, B., Goldreich, O., Goldwasser, S., Lindell, Y.: Resettably-sound zero-knowledge and its applications. In: FOCS, pp. 116–125 (2001)
4. Bellare, M.: A note on negligible functions. J. Cryptology 15(4), 271–284 (2002)
5. Bellare, M., Impagliazzo, R., Naor, M.: Does parallel repetition lower the error in computationally sound protocols? In: FOCS, pp. 374–383 (1997)
6. Canetti, R., Goldreich, O., Goldwasser, S., Micali, S.: Resettable zero-knowledge (extended abstract). In: STOC, pp. 235–244 (2000)
7. Canetti, R., Halevi, S., Steiner, M.: Hardness Amplification of Weakly Verifiable Puzzles. In: Kilian, J. (ed.) TCC 2005. LNCS, vol. 3378, pp. 17–33. Springer, Heidelberg (2005)
8. Canetti, R., Rivest, R., Sudan, M., Trevisan, L., Vadhan, S.P., Wee, H.: Amplifying Collision Resistance: A Complexity-Theoretic Treatment. In: Menezes, A. (ed.) CRYPTO 2007. LNCS, vol. 4622, pp. 264–283. Springer, Heidelberg (2007)
9. Chung, K.-M., Liu, F.-H., Lu, C.-J., Yang, B.-Y.: Efficient String-Commitment from Weak Bit-Commitment. In: Abe, M. (ed.) ASIACRYPT 2010. LNCS, vol. 6477, pp. 268–282. Springer, Heidelberg (2010)
10. Deng, Y., Goyal, V., Sahai, A.: Resolving the simultaneous resettability conjecture and a new non-black-box simulation strategy. In: FOCS, pp. 251–260 (2009)
11. Dodis, Y., Impagliazzo, R., Jaiswal, R., Kabanets, V.: Security Amplification for *Interactive* Cryptographic Primitives. In: Reingold, O. (ed.) TCC 2009. LNCS, vol. 5444, pp. 128–145. Springer, Heidelberg (2009)

12. Dwork, C., Naor, M., Reingold, O.: Immunizing Encryption Schemes from Decryption Errors. In: Cachin, C., Camenisch, J.L. (eds.) EUROCRYPT 2004. LNCS, vol. 3027, pp. 342–360. Springer, Heidelberg (2004)

13. Goldreich, O.: Foundations of Cryptography: Basic Tools. Cambridge University Press (2001)

14. Goldreich, O., Nisan, N., Wigderson, A.: On Yao's XOR-Lemma. Electronic Colloquium on Computational Complexity (ECCC) 2(50) (1995)

15. Goyal, V., Maji, H.K.: Stateless cryptographic protocols. In: FOCS (2011)

16. Goyal, V., Sahai, A.: Resettably Secure Computation. In: Joux, A. (ed.) EUROCRYPT 2009. LNCS, vol. 5479, pp. 54–71. Springer, Heidelberg (2009)

17. Haitner, I.: A parallel repetition theorem for any interactive argument. In: FOCS, pp. 241–250 (2009)

18. Halevi, S., Rabin, T.: Degradation and Amplification of Computational Hardness. In: Canetti, R. (ed.) TCC 2008. LNCS, vol. 4948, pp. 626–643. Springer, Heidelberg (2008)

19. Håstad, J., Pass, R., Wikström, D., Pietrzak, K.: An Efficient Parallel Repetition Theorem. In: Micciancio, D. (ed.) TCC 2010. LNCS, vol. 5978, pp. 1–18. Springer, Heidelberg (2010)

20. Impagliazzo, R., Jaiswal, R., Kabanets, V.: Chernoff-Type Direct Product Theorems. Journal of Cryptology (September 2008); In: Menezes, A. (ed.) CRYPTO 2007. LNCS, vol. 4622, pp. 500–516. Springer, Heidelberg (2007)

21. Jain, A., Pietrzak, K.: Parallel Repetition for Leakage Resilience Amplification Revisited. In: Ishai, Y. (ed.) TCC 2011. LNCS, vol. 6597, pp. 58–69. Springer, Heidelberg (2011)

22. Jutla, C.S.: Almost Optimal Bounds for Direct Product Threshold Theorem. In: Micciancio, D. (ed.) TCC 2010. LNCS, vol. 5978, pp. 37–51. Springer, Heidelberg (2010)

23. Lewko, A., Waters, B.: On the insecurity of parallel repetition for leakage resilience. In: FOCS, pp. 521–530 (2010)

24. Luby, M., Rackoff, C.: Pseudo-random permutation generators and cryptographic composition. In: STOC, pp. 356–363 (1986)

25. Maurer, U., Tessaro, S.: Computational Indistinguishability Amplification: Tight Product Theorems for System Composition. In: Halevi, S. (ed.) CRYPTO 2009. LNCS, vol. 5677, pp. 355–373. Springer, Heidelberg (2009)

26. Maurer, U., Tessaro, S.: A Hardcore Lemma for Computational Indistinguishability: Security Amplification for Arbitrarily Weak PRGs with Optimal Stretch. In: Micciancio, D. (ed.) TCC 2010. LNCS, vol. 5978, pp. 237–254. Springer, Heidelberg (2010)

27. Naor, M., Reingold, O.: On the construction of pseudo-random permutations: Luby-Rackoff revisited. J. of Cryptology 12, 29–66 (1999); Preliminary version in: Proc. STOC 1997

28. Pass, R., Venkitasubramaniam, M.: An efficient parallel repetition theorem for Arthur-Merlin games. In: STOC, pp. 420–429 (2007)

29. Pietrzak, K., Wikström, D.: Parallel Repetition of Computationally Sound Protocols Revisited. In: Vadhan, S.P. (ed.) TCC 2007. LNCS, vol. 4392, pp. 86–102. Springer, Heidelberg (2007)

30. Tessaro, S.: Security Amplification for the Cascade of Arbitrarily Weak PRPs: Tight Bounds via the Interactive Hardcore Lemma. In: Ishai, Y. (ed.) TCC 2011. LNCS, vol. 6597, pp. 37–54. Springer, Heidelberg (2011)

31. Unruh, D.: Random Oracles and Auxiliary Input. In: Menezes, A. (ed.) CRYPTO 2007. LNCS, vol. 4622, pp. 205–223. Springer, Heidelberg (2007)

32. Yao, A.C.-C.: Theory and applications of trapdoor functions (extended abstract). In: FOCS, pp. 80–91 (1982)

Resettable Statistical Zero Knowledge

Sanjam Garg[1], Rafail Ostrovsky[1,2], Ivan Visconti[3], and Akshay Wadia[1]

[1] Department of Computer Science, UCLA, USA
{sanjamg,rafail,awadia}@cs.ucla.edu
[2] Department of Mathematics, UCLA, USA
[3] Dipartimento di Informatica, University of Salerno, Italy
visconti@dia.unisa.it

Abstract. Two central notions of Zero Knowledge that provide strong, yet seemingly incomparable security guarantees against malicious verifiers are those of Statistical Zero Knowledge and Resettable Zero Knowledge. The current state of the art includes several feasibility and impossibility results regarding these two notions *separately*. However, the question of achieving Resettable Statistical Zero Knowledge (i.e., Resettable Zero Knowledge and Statistical Zero Knowledge *simultaneously*) for non-trivial languages remained open. In this paper, we show:

- Resettable Statistical Zero Knowledge with unbounded prover: under the assumption that sub-exponentially hard one-way functions exist, $\mathrm{r}\mathcal{SZK} = \mathcal{SZK}$. In other words, every language that admits a Statistical Zero-Knowledge ($\mathcal{SZK}$) proof system also admits a Resettable Statistical Zero-Knowledge ($\mathrm{r}\mathcal{SZK}$) proof system. (Further, the result can be re-stated unconditionally provided there exists a sub-exponentially hard language in $\mathcal{SZK}$). Moreover, under the assumption that (standard) one-way functions exist, all languages L such that the complement of L is random self reducible, admit a $\mathrm{r}\mathcal{SZK}$; in other words: co-$\mathcal{RSR} \subseteq \mathrm{r}\mathcal{SZK}$.
- Resettable Statistical Zero Knowledge with efficient prover: efficient-prover Resettable Statistical Zero-Knowledge proof systems exist for all languages that admit hash proof systems (e.g., QNR, QR, $\mathcal{DDH}$, DCR). Furthermore, for these languages we construct a two-round resettable statistical witness-indistinguishable argument system.

The round complexity of our proof systems is $\tilde{O}(\log \kappa)$, where κ is the security parameter, and all our simulators are *black-box*.

1 Introduction

The notion of a Zero-Knowledge (ZK, for short) Proof System introduced by Goldwasser, Micali and Rackoff [19] is central in Cryptography. Since its introduction, the concept of a ZK proof has been extremely influential and useful for many other notions and applications (e.g., multi-party computation [18], CCA encryption [27]). Moreover, the original definition has been then extended under several variations, trying to capture additional security guarantees. Well

R. Cramer (Ed.): TCC 2012, LNCS 7194, pp. 494–511, 2012.

known examples are the notions of non-malleable ZK [14] introduced by Dolev, Dwork and Naor, which concerns security against man-in-the-middle attacks, of ZK arguments introduced by Brassard, Chaum and Crepeau [4] where soundness is guaranteed only with respect to probabilistic polynomial-time adversarial provers, and of concurrent ZK [16] introduced by Dwork, Naor and Sahai, which concerns security against concurrent malicious verifiers. Another important variant is that of Statistical Zero Knowledge [19,3,33], where it is guaranteed that a transcript of a proof will remain zero knowledge even against computationally unbounded adversaries.

An important model of security against malicious verifiers, known as *Resettable Zero-Knowledge*, was introduced by Canetti, Goldreich, Goldwasser and Micali in [5]. In this setting, the malicious verifier is allowed to *reset* the prover, and make it re-use its randomness for proving new theorems. Indeed, one of the main motivations for studying resettable ZK was to understand the consequences of re-using limited randomness on the zero-knowledge property. In [5], it was shown that *computational* zero-knowledge for all of $\mathcal{NP}$ is possible even in this highly adversarial setting. Although resettable zero knowledge has received considerable attention since its inception (see for example [1,24,13,39,12,8,35]), almost all the work has been focused on the computational setting.

In this work, we continue the line of research on resettable ZK by investigating the question of resettability when the zero-knowledge property is required to be statistical, i.e., Resettable Statistical Zero Knowledge. This model constrains the prover strategy severely: not only should the prover somehow re-use its limited randomness, it must do so in a way that makes the transcript of the proof statistically secure. Known solutions in the setting of computational resettable ZK involve converting prover's bounded randomness to unbounded pseudo-randomness by using pseudo-random functions (PRF). However, this approach fails in our case, as an unbounded adversary can break the PRF and gain critical information, breaking zero knowledge. In this paper, we develop a new technique to handle this problem. Using this technique, we study resettable statistical zero knowledge in the form of following two *distinct* questions.

- Do there exist *efficient-prover* resettable statistical ZK proofs? This question is motivated by practical applications of resettable ZK, for example, in smart cards. If a prover is to be implemented in a small device like a smart card, it is essential that the prover strategy is polynomial-time.
- What languages in $\mathcal{SZK}$ have resettable statistical ZK proofs? The class $\mathcal{SZK}$ is the class of problems which admit statistical zero-knowledge proofs. This question is purely theoretical in nature, and tries to ascertain the difficulty of achieving resettability where statistical zero-knowledge already exists. In this setting we consider prover's which are forced into giving multiple proofs using the same *limited* random coins. This work can be thought of a natural extension of the recent work on Concurrent Statistical Zero-Knowledge ($c\mathcal{SZK}$) [25,30].

1.1   Our Contribution

In this paper we address the above questions and present the following results. We stress that our techniques may be of independent interest.

Resettable Statistical Zero Knowledge with efficient prover. We show the existence of *efficient-prover* resettable statistical ZK proof systems for all languages in $\mathcal{SZK}$ that admit hash proof systems [10] (e.g., Quadratic Non-Residuosity (QNR), Decisional Diffie-Hellman ($\mathcal{DDH}$), Decisional Composite Residuosity (DCR)). Therefore, our techniques show that *efficient-prover* resettable statistical ZK proof systems also exist for non-trivial languages (like $\mathcal{DDH}$) where each instance is associated to more than one witness, where intuitively reset attacks are harder to deal with.[1] Furthermore, using our techniques, for these languages we also construct a two-round resettable statistical witness-indistinguishable argument system.

Resettable Statistical Zero Knowledge with unbounded prover. We show that if a family of sub-exponentially hard one-way functions exists then $r\mathcal{SZK} = \mathcal{SZK}$, i.e., all languages that admit a statistical ZK proof systems also admit a resettable statistical ZK proof system. If there exists an $\mathcal{SZK}$ language L which is (worst-case) sub-exponentially hard for all input length[2] then $r\mathcal{SZK} = \mathcal{SZK}$ without any additional assumptions, as it already implies the existence of sub-exponentially hard one-way functions [29]. Informally, a sub-exponentially hard one-way function is a one-way function that is secure against sub-exponential (2^{κ^ϵ} for some $0 < \epsilon < 1$) size circuits. Moreover, we show that if a family of (standard) one-way functions exists (or, if there are languages which are hard on the average and admit statistical zero-knowledge proofs [29]) then co-$\mathcal{RSR} \subseteq r\mathcal{SZK}$. Our results are achieved through a novel use of instance-dependent (ID, for short) commitment schemes, a new simulation technique, and a coin-tossing protocol that is secure under reset attacks that we build on top of a new ID commitment for all $\mathcal{SZK}$.

Our simulators are *black-box* and the round complexity of all our constructions is $\tilde{O}(\log \kappa)$ which is optimal considering the lower bounds achieved so far for black-box concurrent ZK [6,26].

We stress that since the very introduction in [5] of the notion of resettable ZK, our results are the first in establishing Resettable *Statistical* Zero Knowledge.

---

[1] When there are multiple witnesses that can prove membership of an instance in a language, in a reset attack we allow the adversarial verifier to force the prover to reuse the same randomness for proving the same instance but using a different witness. We therefore achieve a stronger definition of resettability than the one used in previous work.

[2] If there exists a language $L \in \mathcal{SZK}$ such that for infinite sequence of input lengths, the worst-case decision problem for L is sub-exponentially-hard, Ostrovsky showed that there exists a non-uniform sub-exponentially hard one-way functions for that sequence of input length [29].

We finally leave open an interesting question of proving that $\mathcal{SZK} = \text{r}\mathcal{SZK}$ unconditionally or under relaxed complexity-theoretic assumptions and of establishing whether resettable statistical ZK *arguments* are achievable for all $\mathcal{NP}$.

As a final note, we remark upon the complexity of the verifiers in our protocols. Historically, the notion of $\mathcal{SZK}$ was developed with bounded verifiers (and unbounded distinguishers), for example, see [3,37]. Moving in the same direction, we obtain our results in this model, where the verifiers are computationally bounded. In subsequent literature on $\mathcal{SZK}$, the stronger notion of statistical zero-knowledge against *unbounded* verifiers was developed. In this scenario, the notion of resettability seems hard to achieve: unbounded verifiers can compute statistical correlations on the fly by making multiple reset queries to the prover. We leave the question of constructing such protocols or showing impossibility in a setting with unbounded verifiers as an open problem for future work.

1.2   Technical Difficulties and New Techniques

We begin by asking the general question: "Why is the problem of constructing resettable statistical zero-knowledge proof systems hard?" The problem lies in the fact that the prover has limited randomness and can be reset. Therefore, prover's messages are essentially a deterministic function of the verifier's messages, and the verifier can probe this function by resetting the prover and thereby obtaining information that might be useful for an unbounded distinguisher. We highlight the issues by demonstrating why existing techniques fail. The most well studied way of achieving resettable *computational* zero-knowledge proofs [5], is by using a pseudorandom function. In particular, very informally, using this technique the prover applies a pseudorandom function on the common input and the verifier's first messages (this message is called the *determining message*), which fixes all future messages of verifier, and uses the output as its random tape. Now, when the verifier resets and changes its determining message, prover's random tape changes, and thus, intuitively, the verifier does not gain any advantage by resetting the prover. However, for our goal of obtaining resettable *statistical* zero knowledge, this approach is not sufficient. In fact, intuitively, any protocol (as far as we know) in which there exists a message computed using both the witness and the randomness, where the randomness is fixed but the witnesses could change with theorem statements, can not be statistically "secure" in presence of reset attacks. Indeed, an adversarial verifier could interact multiple times with provers that use a fixed randomness but different statements and witnesses. This information can be used by an unbounded distinguisher to establish certain correlations among the values used in different executions, ultimately breaking the statistical ZK property. Because of these restrictions, previously known techniques, which were sufficient for resettable [5,1] and statistical ZK [25,20,21] independently, turn out to be insufficient for achieving both of them simultaneously.

In light of the intuition above resettable statistical ZK for non-trivial languages at first sight might be considered impossible to achieve. But, on the contrary we develop a new technique that overcomes the above problems.

We demonstrate this new technique by first considering a *toy version* of our protocol. The protocol consists of three phases. In the first phase the verifier sends a "special" *instance-dependent non-interactive* (ID, for short) *commitment* of a random string m to the prover. (In this commitment, if the prover is lying and $x \notin L$, then m will be undefined, while if $x \in L$, then m will be unique.) The second phase consists of a *PRS preamble* [32]. Very roughly, in the PRS preamble the verifier commits to random shares of m, which are opened depending on the provers challenges. Finally, the prover is required to send m to the verifier. The prover can obtain m by extracting it from the commitment either efficiently using a witness in case of efficient-prover proofs, or running in exponential time in case of unbounded-prover proofs. We stress that when the theorem being proved is true the message m that can be extracted is unique.

First, the protocol just described has the following property: every message sent by the prover is *public coin*[3] except its last message, which is *uniquely* determined by the first message of the verifier (we use [5] terminology and refer to it as the *determining message*). Most importantly, no message depends on the witness of the prover. It is this property that allows a simulator to generate a transcript that is statically close to the transcript generated in the interaction with a real prover. An honest prover uses a pseudorandom function on the common input and the determining message and uses the output as its random tape. A simulator can sample the messages from the *same* distribution as the real prover. Finally, the simulator will be able to obtain m by using rewinding capabilities, through a variation of a PRS rewinding strategy [32]. The need for the variation arises from the fact that a simulator that uses pseudo-random coins does not gain anything by rewinding (i.e., after a rewind it would re-send the same message). We deal with this problem by having the simulator use pseudorandom coins for some messages while using pure random coins for others. We elaborate on this in § 4. This toy version, described above, illustrates the key ideas that we use in achieving simultaneously both resettable and statistical zero knowledge. To transform our toy version into a full proof system, for even the most basic languages that we consider in this paper, we need an extra instance-dependent primitive. But we defer this discussion to § 3 and § 5.

Second, our protocol also has the property that if the theorem is false then the prover has almost no chance (in the information-theoretic sense) of sending an accepting last message. This follows from the fact that the ID commitment from verifier is statistically hiding. This property guarantees soundness.

Unfortunately, the above ideas are insufficient to prove that $r\mathcal{SZK} = \mathcal{SZK}$. This is because statistically hiding non-interactive ID commitments, introduced by Chailloux, Ciocan, Kerenidis and Vadhan [7] for $\mathcal{SZK}$ are "honest-sender." To force the sender into using purely random coins we need a coin-flipping protocol secure against resetting senders. For this coin-flipping protocol an *ID commitment* scheme which is *computationally* binding with respect to a resetting sender for instances in the language and statistically hiding for instances not in the language, suffices. We will use some techniques introduced by Barak, Goldreich,

---

[3] Looking ahead, we will use a pseudorandom function to generate these messages.

Goldwasser and Lindell in [1] on top of a previous result of Ong and Vadhan [28] for obtaining such an ID commitment scheme.

However the more subtle problem arises in the use of pseudorandom functions. To obtain security against reset attacks, the coin-flipping message played by the receiver of the commitment must be computed by using a pseudorandom function. This again turns out to be insufficient for our analysis since the use of the pseudorandom function does not guarantee that the outcome of the coin-flipping protocol is a uniform string to be used in the honest-sender non-interactive ID commitment scheme. In order to solve this additional problem, we use sub-exponentially hard pseudorandom functions (constructed from sub-exponentially hard one-way functions). These stronger primitives have the additional property that they are secure against sub-exponential size circuits. This technique is referred to as *complexity leveraging*, and has been previously used in various applications (e.g., [5,23,2,11,9,31,38]). However, we stress that in all our constructions, the simulator runs in expected polynomial time, and the above assumptions play a role only inside our security proof.

Before concluding this section, we point out an important difference between our approach and ideas developed by Micciancio, Ong, Sahai and Vadhan in [25], where the authors give *unconditional* constructions of concurrent statistical zero-knowledge proofs for many non-trivial problems. Like their construction we use similar ID commitments but our general approach and overall protocol is different from their approach. In [25], a compiler is constructed that (using ID commitments) provides a generic way to construct statistical zero-knowledge protocols. But, as pointed earlier, such a compiling technique along with standard resettability techniques [5] is not sufficient for us. Therefore, we develop our zero-knowledge protocol from scratch. This is needed because obtaining resettability along with statistical zero knowledge is different and (as pointed earlier) harder than obtaining concurrent statistical zero knowledge. We further note that in fact our techniques imply that $\mathcal{SZK} = \mathrm{c}\mathcal{SZK}$ unconditionally. We refer the reader to the full version [17] for further discussion on this.

Road map. We start by giving some preliminary definitions in § 2. We use three ID primitives in this paper. We elaborate on those in § 3. In § 4 we construct a resettable statistical ZK proof secure against partially honest verifiers. Then in § 5 we remove this limitation for certain classes of languages. In § 6, we construct the proof system that works for all language in $\mathcal{SZK}$.

2   Notation and Tools

We say that a function is *negligible* in the security parameter κ if it is asymptotically smaller than the inverse of any fixed polynomial. Otherwise, the function is said to be *non-negligible* in κ. We say that an event happens with *overwhelming* probability if it happens with a probability $p(\kappa) = 1 - \nu(\kappa)$ where $\nu(\kappa)$ is a negligible function of κ. In this section, we provide an overview of the primitives used in this paper. Formal definitions can be found in the full version [17].

Resettable/Statistical Zero Knowledge. In this paper we consider resettable [5] and statistical [19,3,33] notions of zero-knowledge. The notion of resettability requires that a protocol remains zero-knowledge even if the verifier can reset the prover. The notion of statistical zero knowledge provides security guarantees against unbounded distinguishers. This paper constructs resettable statistical zero-knowledge proof systems. In other words we try to achieve both the resettability and the statistical guarantees simultaneously.

PRS Preamble from [32]. A PRS preamble is a protocol between a committer $\mathcal{C}$ and a receiver $\mathcal{R}$ that consists of two main phases, namely: 1) the commitment phase, and 2) the challenge-response phase.

Let k be a parameter that determines the round-complexity of the protocol. Then, in the commitment phase, very informally, the committer commits to a secret string σ and k^2 pairs of its 2-out-of-2 secret shares. The challenge-response phase consists of k iterations, where in each iteration, very informally, the committer "opens" k shares, one each from k different pairs of secret shares as chosen by the receiver.

The goal of this protocol is to enable the simulator to be able to rewind and extract the "preamble secret" σ with overwhelming probability. In the concurrent setting, rewinding can be difficult since one may rewind to a time step that precedes the start of some other protocol [16]. However, as it has been demonstrated in [32], there is a fixed "time-oblivious" rewinding strategy that the simulator can use to extract the preamble secrets from every concurrent cheating committer, except with negligible probability. Moreover this works as long as $k = \tilde{\Omega}(\log \kappa)$ for some positive ϵ. We refer to this as the PRS rewinding strategy or the PRS simulation strategy. We refer the reader to [32] for more details.

Sub-exponentially hard one-way functions. A sub-exponentially hard one-way function is a one-way function that is hard to invert even by sub-exponential ($2^{\kappa^{\epsilon}}$ for some $1 > \epsilon > 0$) size circuits. They imply the existence of *sub-exponentially hard pseudorandom functions.* We stress that we need this assumption only for proving that $\mathcal{SZK} = r\mathcal{SZK}$.

3 Instance-Dependent Commitments and Proofs

In this section we construct three instance-dependent primitives, that we use in this paper: (1) a non-interactive instance-dependent commitment scheme, (2) an interactive instance-dependent commitment scheme, and finally (3) an instance-dependent argument system.

Non-Interactive Instance-Dependent Commitment Scheme. An important tool that we will re-define, construct and use in our proof systems, is that of "special" *non-interactive instance-dependent* (ID, for short) *commitment schemes.* A commitment scheme allows one party (referred to as the *sender*) to commit to a value while keeping it *hidden*, with the ability to *reveal* the committed value later. Commitments also have the property that once the sender commits to a value, it can not change its mind later. This property is refereed to as the *binding* property. In certain settings, commitment schemes for which these properties are not required to hold simultaneously, suffice. Such schemes are parameterized by a value x and a language L and either the binding or the hiding property holds depending upon the membership of x in L. These schemes are referred to as ID commitment schemes [7]. Typically, the ID commitment schemes that have been considered in the literature require hiding property to hold when $x \in L$ and binding property to hold otherwise. We actually need the reverse properties, i.e., we need hiding property when $x \notin L$ and binding property otherwise.

In particular we consider an ID commitment scheme with further special properties. We require that the commitment scheme be statistically binding for $x \in L$ and statistically hiding otherwise. In other words we want binding and hiding properties to hold against unbounded adversaries. Also we require that our ID commitment scheme be secure against a resetting sender. This always holds when the commitment scheme is *non-interactive*. All the non-interactive ID commitments that we consider are statistically hiding. So to simplify notation we refer to a non-interactive instance-dependent commitment scheme with perfect (binding holds with probability 1) binding and statistical hiding as a *perfect non-interactive ID commitment*. Similarly, we refer to a non-interactive instance-dependent commitment scheme with statistical binding and statistical hiding as a *statistical non-interactive ID commitment*.

Since the commitment is statistically binding, when $x \in L$, the committed value can always (with overwhelming probability) be extracted in exponential time. Extractability instead becomes tricky when the extractor has to run in polynomial time. We will call an ID commitment scheme *efficiently extractable* if when $x \in L$ then there exists an extractor that takes as input a witness for the membership of x in L and the commitment, and outputs in polynomial-time the committed message.

It turns out that perfect non-interactive ID commitment schemes are actually known to exist for all languages in co-$\mathcal{RSR}$ [22,36,34]. co-$\mathcal{RSR}$ is the class of languages such that the complement of each of these languages is random self-reducible. Another class of languages that is amenable to our techniques is the class of languages that are in $\mathcal{SZK}$ and that admit a hash proof system. Observing that these languages imply instance-dependent primitives that are analogous to ID commitments described above, we get efficient-prover resettable statistical ZK proof systems for this interesting class. In particular, for $\mathcal{DDH}$ (the language that consists of all Diffie-Hellman quadruples and that admits two different witnesses for proving the membership of a quadruple to the language), we give a separate ID commitment scheme highlighting how our techniques work with multiple witnesses.

We notice that for the whole $\mathcal{SZK}$ we only know a weak form of statistical non-interactive ID commitment scheme where statistical binding holds with respect to honest senders only. The details have been provided in the full version.

We will denote the extractable perfect (or, statistical) non-interactive ID commitment scheme by $\mathcal{COM}$. The commitment function for a value x and language L will be denoted by $\mathsf{C}_{L,x}$. Also we use the notation $\mathsf{C}_{L,x}(m;r)$ for the function used to generate the commitment to message $m \in \{0,1\}^{\ell_0}$ using random coins $r \in \{0,1\}^{\ell_1}$. The extractability property of these commitments is very important for our constructions.

Interactive ID Commitment Scheme. We use an *interactive ID commitment scheme* $\mathcal{COM}_{L,x} = (S_x, R_x)$, where S_x and R_x are the sender and the receiver respectively, with common input x. This ID commitment scheme is *computationally binding* against a *resetting* sender when the instance x is in the language, and is *statistically hiding* otherwise. Very roughly, we construct such a scheme by using the constant-round public-coin ID commitment scheme of [28]. This scheme has statistical binding and statistical hiding properties. We make it secure under resetting senders by having the receiver determine its messages by applying a pseudo-random function (similarly to Proposition 3.1 in [1]) to the transcript so far. Because of this, the statistical binding property is degraded to computational[4] binding. We stress that unlike the non-interactive ID commitment described earlier, we will not need any extractability from these commitments. We obtain this new ID commitment scheme for all of $\mathcal{SZK}$ under the assumption that one-way functions exist. The details have been provided in the full version.

Instance-Dependent Argument System $\langle \mathsf{PrsSWI}_x, \mathsf{VrsSWI}_x \rangle$. We will need an instance-dependent argument system $\langle \mathsf{PrsSWI}_x, \mathsf{VrsSWI}_x \rangle$ where PrsSWI_x and VrsSWI_x are the prover and the verifier respectively, with common inputs x and a statement ξ. When[5] x is in the language, we want that $\langle \mathsf{PrsSWI}_x, \mathsf{VrsSWI}_x \rangle$ be a resettably sound argument of knowledge for $\mathcal{NP}$. In this case, very roughly, $\langle \mathsf{PrsSWI}_x, \mathsf{VrsSWI}_x \rangle$ has the additional property that the soundness holds even when the prover can reset the verifier. If instead x is not in the language then $\langle \mathsf{PrsSWI}_x, \mathsf{VrsSWI}_x \rangle$ must be statistical witness indistinguishable. We construct this argument system by instantiating Blum's Hamiltonicity protocol with the constant-round public-coin ID commitment scheme of [28]. We make it resettably sound by using a pseudorandom function [1]. Details, definition and constructions are given in the full version.

---

[4] However, looking ahead we note that, computational binding will be sufficient for our applications since the role of the sender will be played by a polynomially bounded party.

[5] In general, in proof systems when an ID commitment is used, it is parameterized by the theorem statement ξ being proven. In our case the ID commitment is actually parameterized by a different value x. Looking ahead, x would be the theorem statement of an interactive proof system that uses the sub-protocol $\langle \mathsf{PrsSWI}_x, \mathsf{VrsSWI}_x \rangle$ to prove the $\mathcal{NP}$ statement ξ.

4   Resettable Partially Honest-Verifier Statistical Zero Knowledge

We aim at constructing a resettable statistical zero-knowledge proof system. We start by building a simpler protocol which is resettable statistical zero knowledge only against a restricted class of adversarial verifiers. In subsequent sections, we build upon this simpler protocol to achieve our general results. The adversarial verifiers that we consider here are restricted to "act honestly" but only in a limited manner. We call such verifiers *partially honest*. As pointed out in § 3, we use a non-interactive ID commitment scheme. Looking ahead, in our protocol this commitment is used by the verifier to commit to certain messages. A partially honest verifier is required to behave honestly when computing the commitment function, using pure randomness to commit to messages. Besides this it can cheat in any other way. We state this restriction more concretely after we have described the protocol.

We begin by construction a *concurrent* statistical zero-knowledge proof system secure against such partially honest adversaries, and then transform it into a resettable statistical zero-knowledge proof system under the same restricted class of adversarial verifiers.

Concurrent Partially Honest-Verifier SZK. We start by informally describing the protocol cpHSZK of Fig. 1. It consists of three phases. The first phase, called the *Determining Message Phase*, consists of the verifier sending a commitment to a string m to the prover. We use the extractable non-interactive ID commitment scheme described earlier. The second phase is roughly a PRS preamble [32] and we refer to it as the *PRS Phase*. Note that some commitments are made in the PRS preamble, but we lump these with the commitment to m, in the Determining Message Phase itself. Finally the prover sends to the verifier the value m. This is referred to as the *Final Message*. An adversarial verifier, denoted by V^*, is called a partially honest verifier if it generates the non-interactive ID commitments of the Determining Message Phase "honestly." This requires that these ID commitments are: (1) "well-formed" and (2) have unique[6] openings (except with negligible probability).

We begin by briefly sketching why cpHSZK is a concurrent statistical zero-knowledge proof system for L. Full details of the proof are in the full version. Completeness follows from binding property of $\mathcal{COM}$: when $x \in L$, the commitments in the Determining Message Phase are statistically binding with unique openings with overwhelming probability. Thus, the prover can extract the unique message m and make the verifier accept in the Final Message Phase. For soundness, note that when $x \notin L$, the commitments in the first phase are statistically hiding. Thus, m committed to in the Determining Message Phase is information theoretically hidden from a cheating prover (also shares received during the

---

[6] It follows from the description in the full version, that a perfectly non-interactive ID commitment always has a unique opening. On the other hand an honest sender statistical non-interactive ID commitment, has a unique opening with overwhelming probability, for honest senders only.

Common Input: $x \in L \cap \{0,1\}^n$, $k = \omega(\log \kappa)$ and $n = \mathsf{poly}(\kappa)$, for a security parameter κ..

Secret Input to P: Witness w such that $(x, w) \in R_L$ (not needed in case of unbounded prover).

1. **Determining Message** $(V \to P)$ V chooses message m randomly from $\{0,1\}^{\ell_0}$, and computes $\alpha = \mathsf{C}_{L,x}(m; \rho_0)$ for some random $\rho_0 \in \{0,1\}^{\ell_1}$. For $1 \leq i \leq k$ and $1 \leq j \leq k$, V randomly chooses $\sigma_{i,j}^0$ and $\sigma_{i,j}^1$ such that $\sigma_{i,j}^0 \oplus \sigma_{i,j}^1 = m$. For each (i,j,b), where $1 \leq i \leq k$, $1 \leq j \leq k$ and $b \in \{0,1\}$, V randomly chooses $\rho_{i,j}^b \in \{0,1\}^{\ell_1}$ computes the commitment $\alpha_{i,j}^b := \mathsf{C}_{L,x}(\sigma_{i,j}^b; \rho_{i,j}^b)$. Finally, V sends all the commitments $\alpha, \alpha_{1,1}^0, \alpha_{1,1}^1, \ldots, \alpha_{k,k}^1$ to the prover.
2. **PRS Phase** $(V \Leftrightarrow P)$ For $1 \leq l \leq k$:
   (a) P sends b_l chosen randomly in $\{0,1\}^k$ to V.
   (b) Let b_l^i be the ith bit of b_l. V sends the openings of $\alpha_{l,1}^{b_l^1}, \ldots, \alpha_{l,k}^{b_l^k}$.
   (c) If the opening sent by the verifier is invalid, then P sends ABORT to verifier, and aborts the protocol.
3. **Final Message** $(P \to V)$ P runs the extractor associated to the ID commitment of the Determining Message Phase. If the extractor aborts then P aborts, else P sends the output of the extractor m' to V, who accepts if $m' = m$.

Fig. 1. Concurrent Partially Honest-Verifier Statistical Zero-Knowledge Proof System: cpHSZK

preamble do not give any information), and therefore, it can convince the verifier only with negligible probability.

To argue zero knowledge, we use the rewinding strategy of [32]. Using the PRS rewinding strategy we can construct a simulator that obliviously rewinds the verifier and is guaranteed (except with negligible probability) to obtain the opening m committed to in the Determining Message Phase, before the end of the PRS Phase for every session (except with negligible probability) initiated by the cheating verifier. Once the cpHSZK simulator knows the message m committed in the Determining Message Phase, it can play it back to the verifier in the Final Phase.

Note that to prove zero knowledge, we crucially use the fact that the verifier is partially honest. First, we need that the commitment sent by the verifier are correctly formed. This is to make sure that the commitments are done in accordance with the specifications for the first message of PRS preamble. Secondly, we need that these commitments have unique openings with overwhelming probability. If, for example, verifier's commitment to m in the Determining Message Phase has two openings, then the simulation would fail. Indeed, an unbounded prover unable to decide which is the right opening, would always abort while the simulator would still extract some message from the PRS Phase and send that to the verifier in the Final Phase. The case of an efficient prover instead would result in extracting a message that could depend on the witness used, while the one obtained by the simulator would not depend on the witness, therefore potentially generating a distinguishable deviation in the transcript.

Resettable Partially Honest-Verifier SZK. We now exploit a key property of cpHSZK and transform it into a resettable statistical zero-knowledge proof system secure against partially honest verifiers. We note that the final message of cpHSZK depends only on the first message of the verifier. In particular, it depends neither on the random tape of the prover, nor on its witness. Also messages of the prover in the PRS phase are just random strings. Thus, very informally, an adversarial verifier can not obtain any advantage by resetting the prover, as after every reset, the verifier will get the same message back in the final round. This is a crucial fact that allows us to achieve resettability.

Common Input: $x \in L \cap \{0,1\}^n, k = \omega(\log \kappa), n = \mathsf{poly}(\kappa)$ for a security parameter κ.

Secret Input to P: Witness w such that $(x, w) \in R_L$ (not needed in case of unbounded prover).

1. **Determining Message** Same as in Fig. 1.
2. **PRS Phase** $(V \Leftrightarrow P)$ P chooses a random seed s, and sets $\omega = f_s(x, \alpha, \alpha^0_{1,1}, \ldots, \alpha^1_{k,k})$. Now P divides ω into k blocks of k-bits each, i.e., $\omega = \omega^1 \circ \ldots \circ \omega^k$. For $1 \le l \le k$,
   (a) P sends ω^l to V.
   (b) Same as Fig. 1 Step 2b.
   (c) Same as in Fig. 1 Step 2c.
3. **Final Message** Same as in Fig. 1.

Fig. 2. Resettable Statistical Partially Honest Verifier Zero-Knowledge Proof System rpHSZK

The transformed protocol, called rpHSZK (Fig. 2), is the same as cpHSZK, except for one difference: in the PRS Phase, instead of sending random challenges in Step 2(a), the prover uses pseudorandom challenges. The prover chooses a random seed s for selecting a function from a PRF family $\{ f_s \}_{s \in \{0,1\}^*}$, and sets ω as the output of $f_s()$ evaluated on the message received during the Determining Message Phase. The prover uses this ω as its random tape for the PRS phase. A modification of the PRS simulation where the simulator uses both pseudorandom and random messages during the preamble, along with other known tricks [5] allows us to prove that this protocol is a resettable statistical zero-knowledge proof system with respect to partially honest verifiers.

5 Resettable Statistical ZK from Perfect Non-interactive ID Commitments

In this section we consider languages that admit perfect non-interactive ID commitments and we construct a resettable statistical ZK proof system which is secure against *all* malicious verifiers.

Let L be a language that admits a perfect non-interactive ID commitment scheme. We extend the proof system rpHSZK for L to handle arbitrary malicious verifiers. The main idea is to enforce "partially honest behavior" on the malicious verifier. We recall that the partially honest restriction on a verifier required that the verifier generates commitments honestly. More specifically, we required that these commitments have unique openings and are correctly constructed. A fully malicious verifier however can deviate and compute commitments that do not have the prescribed form. Therefore, the only concern we have is to make sure that commitments are correctly generated. We enforce this by modifying rpHSZK and adding an extra step to it. This step requires that the verifier proves to the prover that shares constructed in Step 1 (as part of the Determining Message) are correct. If this proof is accepted then the prover can conclude that the first message of the verifier is indeed honestly generated and the malicious verifier is forced into following the desired partially honest behavior. In our protocol the verifier uses an instance-dependent argument system $\langle \mathsf{PrsSWI}_x, \mathsf{VrsSWI}_x \rangle$ such that: when $x \in L$, $\langle \mathsf{PrsSWI}_x, \mathsf{VrsSWI}_x \rangle$ is a resettably sound argument of knowledge, while when $x \notin L$, $\langle \mathsf{PrsSWI}_x, \mathsf{VrsSWI}_x \rangle$ is statistically witness indistinguishable. Since the protocol is resettably sound the malicious verifier can not go ahead with incorrect commitments even when it can reset the prover. For the protocol see Fig. 3.

Sub-protocol: $\langle \mathsf{PrsSWI}_x, \mathsf{VrsSWI}_x \rangle$ is a resettably sound argument of knowledge when $x \in L$ and a statistical witness indistinguishable argument when $x \notin L$.

Common Input: $x \in L \cap \{0,1\}^n, k = \omega(\log \kappa), n = \mathsf{poly}(\kappa)$ for a security parameter κ.

Secret Input to P^a: Witness w such that $(x, w) \in R_L$ (not needed in case of unbounded prover).

1. **Determining Message:** Same as in Fig. 2.
2. **Proof of Consistency:** $(V \Leftrightarrow P)$ V and P run $\langle \mathsf{PrsSWI}_x, \mathsf{VrsSWI}_x \rangle$, where V plays the role of PrsSWI_x, and P plays the role of VrsSWI_x. V proves to P knowledge of $m, \sigma_{i,j}^b, \rho_0, \rho_{i,j}^b$ for $1 \le i, j, \le k, b \in \{0,1\}$, such that:
   (a) $\alpha = \mathsf{C}_{L,x}(m, \rho_0)$, and,
   (b) $\alpha_{i,j}^b = \mathsf{C}_{L,x}(\sigma_{i,j}^b; \rho_{i,j}^b)$ for each $1 \le i, j \le k$ and $b \in \{0,1\}$, and,
   (c) $\sigma_{i,j}^0 \oplus \sigma_{i,j}^1 = m$ for $1 \le i, j \le k$.
3. **PRS Phase:** Same as in Fig. 2.
4. **Final Message:** Same as in Fig. 2.

---

[a] P aborts the protocol in case any proof from the verifiers does not **accept** or some message is not well formed.

Notice that P uses two different seeds for the PRF f (one in Step 2 and the other one in Step 3).

Fig. 3. Resettable Statistical Zero-Knowledge from Perfect Non-Interactive ID Commitments: rSZK

Application to co-$\mathcal{RSR}$ and hash proof systems. Languages in co-$\mathcal{RSR}$ and $\mathcal{DDH}$ have perfect non-interactive ID commitment schemes. Thus, from the discussion above, it follows that these languages have resettable statistical zero-knowledge proofs. For languages in $\mathcal{SZK}$ that admit hash proof systems, a minor modification of our resettable statistical zero-knowledge protocol suffices. The details are provided in the full version [17].

6   Resettable Statistical ZK for All Languages in $\mathcal{SZK}$

In this section we construct the general proof system which is actually resettable statistical zero knowledge for all languages that have a statistical zero knowledge proof. Just like in previous section, we start with a resettable partially honest verifier statistical ZK proof system. But we look at all languages in $\mathcal{SZK}$ and construct a resettable statistical ZK proof system which is secure against *all* malicious verifiers.

Let L be a language that admits an honest sender statistical non-interactive ID commitment scheme $\mathcal{COM}$. We extend the proof system rpHSZK for L to handle arbitrary malicious verifiers. The main idea is to enforce "partially honest behavior" on the malicious verifier. We recall that the partially honest restriction on a verifier required that the verifier uses $\mathcal{COM}$ to generate commitments honestly. More specifically, we required that these commitments are correctly constructed and have unique openings. The first requirement can be handled in a way just like in previous section, i.e. by having the verifier prove to the prover that shares constructed in Step 1 (as part of the Determining Message) are correct. We use the ID argument system $\langle \mathsf{PrsSWI}_x, \mathsf{VrsSWI}_x \rangle$ to achieve this. The problem of uniqueness is more tricky, and we discuss that next.

The difficulty lies in the fact that the statistical non-interactive ID commitment scheme for all languages in $\mathcal{SZK}$ [7], only works with respect to honest senders. Indeed, if the sender chooses the randomness for the commitment uniformly, then, with overwhelming probability, the computed commitment has a unique valid opening. However a malicious sender could focus on a set of negligible size, $\mathcal{B}$, of *bad* random strings r, such that $C_{L,x}(m; r)$ does not have a unique opening. If a malicious verifier (that plays as sender of this commitment scheme) is able to pick random strings from $\mathcal{B}$, then the real interaction and the simulation can be easily distinguished. In the real protocol, the prover tries to invert the commitment α, finds it does not have a unique opening, and aborts. In the simulation, the simulator extracts some message m from the PRS phase, and sends m as the final message. As the simulator is polynomially bounded, it *can not* detect if the commitment has a unique opening or not. To use this commitment scheme, we must somehow ensure that the verifier does not use bad randomness for its commitments. We do this by adding a special coin-flipping subprotocol at the beginning of the protocol. However, because of reset attacks, the coin-flipping subprotocol introduces several technical problems.

We begin by describing our coin-flipping protocol. The coin-flipping protocol requires a commitment scheme such that computational binding holds against resetting senders when $x \in L$ and statistical hiding holds when $x \notin L$. We use the interactive ID commitment scheme $\mathcal{COM}_{L,x} = (S_x, R_x)$. The coin flipping proceeds as follows: first the verifier commits to a random string r_1. Let the transcript of the interactive commitment be c. Then prover applies the sub-exponentially hard PRF $f_s(c)$ and obtains r_2 that is sent to the verifier. The randomness that the verifier will use for the non-interactive ID commitment is $r_1 \oplus r_2$. For technical reasons, the verifier also needs to prove knowledge of r_1 after it has committed to r_1. We use the interactive ID argument system $\langle \mathsf{PrsSWI}_x, \mathsf{VrsSWI}_x \rangle$ for this.

Next we highlight the reasons behind the use of sub-exponentially hard pseudorandom (PRF) functions for our construction. Let α be the statistical non-interactive ID commitment of some message m sent by the verifier. There are two ways in which α might not have a unique opening. In the first case, a malicious V^*, after looking at prover's response r_2, might use an opening of c such that $r_1 \oplus r_2 \in \mathcal{B}$. This however would violate the computational binding of the interactive ID commitment scheme secure against resetting senders used in the coin flipping, thus this event occurs with negligible probability. The second case is more subtle. It might be possible that performing reset attacks, the verifier can study the behavior of the PRF, and then can be able to succeed in obtaining that $r_1 \oplus r_2 \in \mathcal{B}$ with non-negligible probability (even though the polynomial-time V^* does not know the two openings). In this case, we can not construct a polynomial-time adversary that breaks f_s, as we can not efficiently decide if $r \in \mathcal{B}$. This is where we need the sub-exponential hardness of the one-way function and in turn of the PRF. As $|\mathcal{B}|$ is only 2^ℓ while the size of the set of all random strings is 2^L, where $l = o(L)$, we can give the entire set $\mathcal{B}$ as input to the sub-exponential size circuit that aims at breaking the PRF. The circuit can now check if the string r is a bad string or not, by searching through its input. Notice that one can give as input to the circuit the whole $\mathcal{B}$ for each of the polynomial number of statements (since for each x there can be a different $\mathcal{B}$) on which the reset attack is applied. This sub-exponential size circuit has still size $o(L)$ and breaks the PRF which contradicts the sub-exponential hardness of the PRF.

Completeness follows from the fact that when $x \in L$, with overwhelming probability, the commitment α in the determining message will have a unique opening. Thus, the prover will be able to extract the committed message and send it as the final message to the verifier, that will accept.

Statistical resettable zero knowledge property of our protocol also follows the same argument. Indeed, when $x \in L$ even a resetting verifier can not cheat during the proofs in Steps 1(c) and 3. Moreover, the above discussion about the security of the coin-flipping protocol implies that a resetting adversarial verifier is forced into following partially honest behavior when computing the non-interactive ID commitments.

Finally, we look at soundness. Note that when $x \notin L$ non-interactive ID commitments are statistically hiding and the protocol $\langle \mathsf{PrsSWI}_x, \mathsf{VrsSWI}_x \rangle$ is statistical WI in Steps 1(c) and 3. Also note that only a single share is revealed in the PRS phase. From this it follows that the prover's view when verifier commits message m is statistically close to its view when verifier commits to m', where $m \neq m'$. Thus, the probability that it replies with the correct final message is negligible. The complete protocol and proof appear in the full version [17].

7 2-Round Statistical Witness Indistinguishability

In this section, we highlight the applicability of our techniques, and construct a simple two-round resettable statistical witness-indistinguishable argument for languages that have efficiently extractable perfectly binding instance-dependent commitment schemes. As discussed before, this class contains, in particular, all languages that admit hash proof systems. We note that all results in this section hold in the stronger model of statistical zero-knowledge where the verifier is computationally unbounded.

Informally, the two-round WI argument consists of the verifier committing to a randomly chosen message m using the instance-dependent commitment scheme for that language. The prover, using the witness and the efficient extractor, extracts a message m' from the commitment and sends it to the verifier. The verifier accepts if $m = m'$. Intuitively, as long as verifier's commitment is well-formed, this protocol is a perfect WI, as irrespective of the witness and randomness, the prover always extracts the same message (in fact, prover's strategy is deterministic). Thus, the only complication is to ensure that verifier's commitment is well-formed in a round efficient manner. We enforce this by making the verifier provide a non-interactive WI proof (i.e., a one-round ZAP [21,15]) of "well-formedness" in the first round.

For lack of space, details of the protocol and proof of security can be found in the full version of this paper.

Acknowledgments. We thank Giuseppe Persiano for suggesting the open problem of achieving resettable statistical zero knowledge and Omkant Pandey for several discussions on our techniques.

Research supported in part by NSF grants 0830803, 09165174, 1065276, 1118126 and 1136174, US-Israel BSF grant 2008411, B. John Garrick Foundation, OKAWA Foundation Award, IBM Award, Lockheed-Martin Corporation and the Defense Advanced Research Projects Agency through the U.S. Office of Naval Research under Contract N00014-11-1-0392. The views expressed are those of the authors and do not reflect the official policy or position of the Department of Defense or the U.S. Government. The work of the third author was partially done while visiting UCLA and has also been supported in part by the European Commission through the FP7 programme under contract 216676 ECRYPT II, and in part by a research grant of Provincia di Salerno.

References

1. Barak, B., Goldreich, O., Goldwasser, S., Lindell, Y.: Resettably-sound zero-knowledge and its applications. In: FOCS, pp. 116–125 (2001), full version, http://eprint.iacr.org/2001/063

2. Barak, B., Lindell, Y., Vadhan, S.: Lower bounds for non-black-box zero knowledge. In: FOCS 2003, pp. 384–393 (2003)
3. Bellare, M., Micali, S., Ostrovsky, R.: The (true) complexity of statistical zero knowledge. In: STOC, pp. 494–502 (1990)
4. Brassard, G., Chaum, D., Crépeau, C.: Minimum disclosure proofs of knowledge. J. Comput. Syst. Sci. 37(2), 156–189 (1988)
5. Canetti, R., Goldreich, O., Goldwasser, S., Micali, S.: Resettable zero-knowledge (extended abstract). In: STOC, pp. 235–244 (2000)
6. Canetti, R., Kilian, J., Petrank, E., Rosen, A.: Black-box concurrent zero-knowledge requires (almost) logarithmically many rounds. SIAM J. Comput. 32(1), 1–47 (2002)
7. Chailloux, A., Ciocan, D.F., Kerenidis, I., Vadhan, S.P.: Interactive and Noninteractive Zero Knowledge are Equivalent in the Help Model. In: Canetti, R. (ed.) TCC 2008. LNCS, vol. 4948, pp. 501–534. Springer, Heidelberg (2008)
8. Cho, C., Ostrovsky, R., Scafuro, A., Visconti, I.: Simultaneously Resettable Arguments of Knowledge. In: Cramer, R. (ed.) TCC 2012. LNCS, pp. 530–547. Springer, Heidelberg (2012)
9. Cook, J., Etesami, O., Miller, R., Trevisan, L.: Goldreich's One-Way Function Candidate and Myopic Backtracking Algorithms. In: Reingold, O. (ed.) TCC 2009. LNCS, vol. 5444, pp. 521–538. Springer, Heidelberg (2009)
10. Cramer, R., Shoup, V.: Universal Hash Proofs and a Paradigm for Adaptive Chosen Ciphertext Secure Public-Key Encryption. In: Knudsen, L.R. (ed.) EUROCRYPT 2002. LNCS, vol. 2332, pp. 45–64. Springer, Heidelberg (2002)
11. Damgård, I., Fazio, N., Nicolosi, A.: Non-interactive Zero-Knowledge from Homomorphic Encryption. In: Halevi, S., Rabin, T. (eds.) TCC 2006. LNCS, vol. 3876, pp. 41–59. Springer, Heidelberg (2006)
12. Deng, Y., Goyal, V., Sahai, A.: Resolving the simultaneous resettability conjecture and a new non-black-box simulation strategy. In: FOCS (2009)
13. Di Crescenzo, G., Persiano, G., Visconti, I.: Constant-Round Resettable Zero Knowledge with Concurrent Soundness in the Bare Public-Key Model. In: Franklin, M. (ed.) CRYPTO 2004. LNCS, vol. 3152, pp. 237–253. Springer, Heidelberg (2004)
14. Dolev, D., Dwork, C., Naor, M.: Non-Malleable Cryptography. SIAM J. on Computing 30(2), 391–437 (2000)
15. Dwork, C., Naor, M.: Zaps and their applications. In: FOCS, pp. 283–293 (2000)
16. Dwork, C., Naor, M., Sahai, A.: Concurrent zero-knowledge. In: STOC, pp. 409–418 (1998)
17. Garg, S., Ostrovsky, R., Visconti, I., Wadia, A.: Resettable statistical zero knowledge. Cryptology ePrint Archive, Report 2011/457 (2011), http://eprint.iacr.org/
18. Goldreich, O., Micali, S., Wigderson, A.: How to play any mental game - a completeness theorem for protocols with honest majority. In: STOC, pp. 218–229 (1987)
19. Goldwasser, S., Micali, S., Rackoff, C.: The knowledge complexity of interactive proof-systems. SIAM J. on Computing 18(6), 186–208 (1989)
20. Goyal, V., Moriarty, R., Ostrovsky, R., Sahai, A.: Concurrent Statistical Zero-Knowledge Arguments for NP from One Way Functions. In: Kurosawa, K. (ed.) ASIACRYPT 2007. LNCS, vol. 4833, pp. 444–459. Springer, Heidelberg (2007)
21. Groth, J., Ostrovsky, R., Sahai, A.: Non-interactive Zaps and New Techniques for NIZK. In: Dwork, C. (ed.) CRYPTO 2006. LNCS, vol. 4117, pp. 97–111. Springer, Heidelberg (2006)
22. Itoh, T., Ohta, Y., Shizuya, H.: A language-dependent cryptographic primitive. J. Cryptology 10(1), 37–50 (1997)

23. Lindell, Y.: Bounded-concurrent secure two-party computation without setup assumptions. In: STOC, pp. 683–692. ACM (2003)
24. Micali, S., Reyzin, L.: Soundness in the Public-Key Model. In: Kilian, J. (ed.) CRYPTO 2001. LNCS, vol. 2139, pp. 542–565. Springer, Heidelberg (2001)
25. Micciancio, D., Ong, S.J., Sahai, A., Vadhan, S.P.: Concurrent Zero Knowledge Without Complexity Assumptions. In: Halevi, S., Rabin, T. (eds.) TCC 2006. LNCS, vol. 3876, pp. 1–20. Springer, Heidelberg (2006)
26. Micciancio, D., Yilek, S.: The Round-Complexity of Black-Box Zero-Knowledge: A Combinatorial Characterization. In: Canetti, R. (ed.) TCC 2008. LNCS, vol. 4948, pp. 535–552. Springer, Heidelberg (2008)
27. Naor, M., Yung, M.: Public-key cryptosystems provably secure against chosen ciphertext attacks. In: STOC 1990, pp. 427–437 (1990)
28. Ong, S.J., Vadhan, S.P.: An Equivalence Between Zero Knowledge and Commitments. In: Canetti, R. (ed.) TCC 2008. LNCS, vol. 4948, pp. 482–500. Springer, Heidelberg (2008)
29. Ostrovsky, R.: One-way functions, hard on average problems, and statistical zero-knowledge proofs. In: Structure in Complexity Theory Conference, pp. 133–138 (1991)
30. Pass, R., Tseng, W.L.D., Venkitasubramaniam, M.: Concurrent zero knowledge: Simplifications and generalizations. Technical Report (2008), http://hdl.handle.net/1813/10772
31. Pass, R., Wee, H.: Constant-Round Non-malleable Commitments from Subexponential One-Way Functions. In: Gilbert, H. (ed.) EUROCRYPT 2010. LNCS, vol. 6110, pp. 638–655. Springer, Heidelberg (2010)
32. Prabhakaran, M., Rosen, A., Sahai, A.: Concurrent zero knowledge with logarithmic round-complexity. In: FOCS, pp. 366–375 (2002)
33. Sahai, A., Vadhan, S.P.: A complete problem for statistical zero knowledge. J. ACM 50(2), 196–249 (2003)
34. Santis, A.D., Crescenzo, G.D., Persiano, G., Yung, M.: On monotone formula closure of szk. In: FOCS, pp. 454–465 (1994)
35. Scafuro, A., Visconti, I.: On round-optimal zero knowledge in the bare public-key model. In: EUROCRYPT. LNCS. Springer, Heidelberg (2012)
36. Tompa, M., Woll, H.: Random self-reducibility and zero knowledge interactive proofs of possession of information. In: FOCS, pp. 472–482 (1987)
37. Vadhan, S.: A Study of Statistical Zero-Knowledge Proofs. Ph.D. thesis. MIT (1999)
38. Wee, H.: Black-box, round-efficient secure computation via non-malleability amplification. In: FOCS (2010)
39. Yung, M., Zhao, Y.: Generic and Practical Resettable Zero-Knowledge in the Bare Public-Key Model. In: Naor, M. (ed.) EUROCRYPT 2007. LNCS, vol. 4515, pp. 129–147. Springer, Heidelberg (2007)

The Knowledge Tightness of Parallel Zero-Knowledge

Kai-Min Chung*, Rafael Pass**, and Wei-Lung Dustin Tseng

Department of Computer Science, Cornell University, Ithaca, NY, USA
{chung,rafael,wdtseng}@cs.cornell.edu

Abstract. We investigate the concrete security of black-box zero-knowledge protocols when composed in parallel. As our main result, we give essentially tight upper and lower bounds (up to logarithmic factors in the security parameter) on the following measure of security (closely related to knowledge tightness): the number of queries made by black-box simulators when zero-knowledge protocols are composed in parallel. As a function of the number of parallel sessions, k, and the round complexity of the protocol, m, the bound is roughly $k^{1/m}$.

We also construct a modular procedure to amplify simulator-query lower bounds (as above), to generic lower bounds in the black-box concurrent zero-knowledge setting. As a demonstration of our techniques, we give a self-contained proof of the $o(\log n / \log \log n)$ lower bound for the round complexity of black-box concurrent zero-knowledge protocols, first shown by Canetti, Kilian, Petrank and Rosen (STOC 2002). Additionally, we give a new lower bound regarding constant-round black-box concurrent zero-knowledge protocols: the running time of the black-box simulator must be at least $n^{\Omega(\log n)}$.

Keywords: Zero-Knowledge, Knowledge Tightness, Concrete Security, Concurrent Zero-Knowledge Lower Bounds.

1   Introduction

Zero-knowledge interactive proofs, introduced by Goldwasser, Micali and Rackoff [GMR89] are paradoxical constructions allowing one player (called the prover) to convince another player (called the verifier) of the validity of a mathematical statement $x \in L$, while providing no additional knowledge to the verifier. In addition to being an independent construction of interest, zero-knowledge have become an extremely useful tool in construction of numerous cryptographic protocols.

* Chung is supperted by a Simons Foundation Fellowship.
** Pass is supported in part by a Alfred P. Sloan Fellowship, Microsoft New Faculty Fellowship, NSF CAREER Award CCF-0746990, AFOSR YIP Award FA9550-10-1-0093, and DARPA and AFRL under contract FA8750-11-2-0211. The views and conclusions contained in this document are those of the authors and should not be interpreted as representing the official policies, either expressed or implied, of the Defense Advanced Research Projects Agency or the US government.

R. Cramer (Ed.): TCC 2012, LNCS 7194, pp. 512–529, 2012.

A fundamental question regarding zero-knowledge protocols is whether their composition remains zero-knowledge. In theoretical constructions as well as in practice, a zero-knowledge protocol is sometimes composed in parallel (to amplify soundness or to improve efficiency, for example). It is well-known that the definition of zero-knowledge (ZK) is not closed under parallel composition [GK96b]. Nevertheless, we know numerous constructions of constant-round zero-knowledge protocols that are secure when composed in parallel [FS90, GK96a, Gol02]. As a result, the subject of ZK with respect to parallel composition is widely considered closed.

We turn our attention to another fundamental question regarding zero-knowledge: its knowledge tightness. In its original definition, the zero-knowledge property is formalized by requiring that the view of any probabilistic polynomial time (PPT) verifier V in an interaction with a prover can be "indistinguishably reconstructed" by a PPT simulator S that interacts with no one. Since whatever V "sees" in the interaction can be reconstructed by the simulator, the interaction does not yield any knowledge to V that V can already compute by itself. Because the simulator is allowed to be an arbitrary PPT machine, this traditional notion of ZK only guarantees that the *class* of PPT verifiers learn nothing.

To more concretely measure the knowledge gained by a particular verifier, Goldreich, Micali and Wigderson [GMW91] (see also [Gol01]) put forward the notion of *knowledge tightness*: informally, the "tightness" of a simulation is the ratio of the (expected) running-time of the simulator, divided by the (worst-case) running-time of the verifier. Thus, in a knowledge-tight ZK proof, the verifier is expected to gain no more knowledge than what it could have computed in time closely related to its *worst-case* running-time. In addition to theoretical interests, the knowledge tightness of a zero-knowledge protocol is a helpful aid for setting the security parameter in practice. It is easy to check that the original zero-knowledge protocols [GMR89, GMW91, Blu86] all enjoy constant knowledge tightness. The aforementioned protocols secure under parallel composition [FS90, GK96a, Gol02] also enjoy constant knowledge tightness when executed in isolation; however, when composed in parallel, the tightness of these protocols seem increase/loosen linearly (sometimes even quadratically) with respect to the number of parallel sessions (based on the currently known analysis of their simulators)!

Since we do want to execute zero-knowledge protocols in parallel (for instance in the application of secure multi-party computation), a natural question is to ask: how does the knowledge tightness of a protocol vary when we increase the number of parallel repetitions?

1.1   Our Results

In this work we give essentially tight upper and lower bounds to the above question. Our results focus on *black-box* zero-knowledge and "simulator queries", which we explain below.

Informally, a protocol is *black-box* zero-knowledge if there exists a *universal* simulator S, called the *black-box simulator*, such that S generates the view of any adversarial verifier V^* if S is given black-box access to V^*. Essentially all known constructions of zero-knowledge (with the notable exception of [Bar01]) and all practical zero-knowledge protocols are black-box zero-knowledge. Given a black-box simulator S, we focus on bounding the number of black-box queries made by S to a given adversarial verifier V^*; we refer to this as the *simulator-query* complexity. It is easy to see that the number of queries made by a black-box simulator is closely related to knowledge tightness; in fact, for the case of constant round protocols, they are asymptotically equivalent.

We state our main theorems below:

Theorem 1. *Let n be the security parameter. For any $m = m(n)$, there exists a $2m + 7$-round black-box zero-knowledge argument Π for all of NP based on one-way functions, with perfect completeness and negligible soundness error, such that for any polynomially bounded $k = k(n)$, the parallel composition of k-copies of the protocol, Π^k, remains black-box zero-knowledge with simulator-query complexity $O(mk^{1/m} \log^2 n)$.*

The above theorem can be extended to proofs assuming the existence of collision-resistant hash-functions. We complement Theorem 1 with a lower bound:

Theorem 2. *Let n be the security parameter, L be a language, and $m = m(n) \in O\left(\frac{\log n}{\log \log n}\right)$. Suppose Π is a $m(n)$-round black-box zero-knowledge argument for L with perfect completeness and negligible soundness error, and suppose there exist a polynomially bounded $k(n) \geq n$ such that the parallel composition of k-copies of the protocol, Π^k, remains black-box zero-knowledge with simulator-query complexity $O(k^{1/m}/(\log^2 n))$. Then, $L \in$ BPP.*

For protocols with sub-logarithmic number of rounds, Theorem 1 and 2 are tight up to logarithmic factors in the security parameter; essentially, the simulator-query complexity is asymptotically close to $k^{1/m}$ (in most cases, think of k as a low polynomial in n). We mention that one can achieve simulator-query complexity $O(m)$ (independent of k) when $m = \omega(\log n)$.

Briefly, our results show that the concrete security of constant-round black-box zero-knowledge protocols actually decays polynomially in the number of parallel sessions. Fortunately, this decay can be significantly slowed if we consider protocols with more rounds (even if we simply use a large constant m).

1.2   Related Works

While we are unaware of any past work that explicitly studies the knowledge tightness of parallelized zero-knowledge protocols, there are numerous related publications that focus on the composition of zero-knowledge protocols, or on the concrete security of zero-knowledge simulator. Dwork, Naor and Sahai [DNS04] introduces the notion of *concurrent zero-knowledge* protocols; these protocols must stay zero-knowledge even when composed arbitrarily (a strengthening over parallel composition). Micali and Pass [MP06] introduces the notion of *precision*; in a precise zero-knowledge protocol, the running time of the simulator should be closely related to the running time of the adversarial verifier, on a view by view basis[1] (a strengthening over knowledge tightness).

Even with these stronger requirements, Pandey et. al. [PPS+08] is able to construct protocols that are simultaneously precise and (black-box) concurrent zero-knowledge. Note that our results are incomparable with the result of [PPS+08] for many reasons, one of which being that black-box concurrent zero-knowledge protocols require logarithmically many rounds [CKPR01], while our setting is mainly interesting for sub-logarithmic-round protocols. Interestingly, [PPS+08] actually gives a construction of a family of precise concurrent zero-knowledge protocols, with trade-offs between round-complexity and precision, much like our observed trade-off between round-complexity and knowledge tightness for the case of parallelized zero-knowledge.

1.3   Connection to Concurrent Zero-Knowledge

We also present a connection from simulator-query lower bounds for zero-knowledge, to round-complexity lower bounds for concurrent zero-knowledge (cZK). Due to lack of space we postpone the result on concurrent zero-knowledge to the full version. We briefly discuss the ideas as follows.

We start by describing the common framework for all known black-box zero-knowledge lower bounds (e.g., [KPR98, Ros00, CKPR01, BL02, Kat08, HRS09]). Let Π be a protocol for a language L. To show that Π cannot be zero-knowledge unless the language L is trivial (i.e., $L \in \mathsf{BPP}$), we start by constructing a decision procedure for L. Let S be the black-box zero-knowledge simulator of Π, and let V^* be some "hard to simulate" adversarial verifier, and consider the following decision procedure $\mathcal{D}$: on input x, $\mathcal{D}(x)$ accepts if and only if $S^{V^*}(x)$ generates an accepting view of $V^*(x)$. Usually, the completeness of $\mathcal{D}$ follows easily from the zero-knowledge property; to show that $\mathcal{D}$ is sound often requires more work. Our query-complexity lower bounds (Theorem 2) also follow the same framework. That is, we construct some adversarial verifier V^*_{para} that schedules multiple sessions in parallel, and show that for any zero-knowledge simulator S with appropriately bounded query-complexity, if $x \notin L$, then $S^{V^*_{\text{para}}}(x)$ cannot generate an accepting view of $V^*_{\text{para}}(x)$.

Inspired by the work of Canetti, Kilian, Petrank and Rosen [CKPR01], we next present a modular construction of a concurrent adversarial verifier V^*_{conc} whose purpose is to *amplify* query-complexity lower bounds of more basic verifiers. For example, consider V^*_{para}, an adversarial verifier that is restricted to parallel composition. Our modular construction would take V^*_{para} as input, and output an adversarial verifier $V^*_{\text{conc}} = V^*_{\text{conc}}(V^*_{\text{para}})$ that, among other things, nests multiple incarnations of V^*_{para} in a way that takes full advantage of the concurrent scheduling. Under appropriate parameters, our analysis would conclude that for any zero-knowledge simulator S with *polynomially* bounded query-complexity, if $x \notin L$,

---

[1] For example, to achieve precision 2, if the simulator S generates a view of V^* and the running time of V^* on that view is T, then the simulator S must have run in time $2T$.

then $S^{V^*_{\text{conc}}}(x)$ cannot generate an accepting view of $V^*_{\text{conc}}(x)$ (recall again that this is the key step for most zero-knowledge lower bounds).

To demonstrate our framework, we re-prove the result of [CKPR01] — a $o(\log n / \log \log n)$ round-complexity lower bound for black-box concurrent zero-knowledge (the currently best known round-complexity lower bound); we believe the resulting analysis is quite clean. We also give a second lower bound concerning constant-round cZK protocols:

Theorem (Informal). *Let L be a non-trivial language, and let Π be a constant-round black-box concurrent zero-knowledge protocol with a potentially possibly super-polynomial time simulator. Then the simulator must run in time $n^{\Omega(\log n)}$.*

Incidentally, Pass and Venkitasubramaniam [PV08] do construct constant-round black-box concurrent zero-knowledge protocols for all of NP in the model where both the simulator and the adversarial verifier runs in quasi-polynomial time $n^{\text{poly}(\log n)}$.

We also find our modular framework satisfying on a philosophical level: it serves as an framework in which lower bounds for restricted compositions of zero-knowledge (in this example parallel composition) can be transformed into lower bounds for zero-knowledge in the fully concurrent setting. A similar and celebrated example occurs in the work of Goldreich [Gol02], where it is shown that constructions of zero-knowledge protocols secure under parallel composition directly leads to constructions of concurrent zero-knowledge protocols secure in the timing model.

2   Preliminaries

We use $\mathbb{N}$ to denote the natural numbers $\{0, 1, \ldots\}$, $[n]$ to denote the set $\{1, \ldots, n\}$, and $|x|$ to denote the length of a string $x \in \{0,1\}^*$. By $\mathsf{ngl}(n)$, we mean a function negligible in n (i.e., $1/n^{\omega(1)}$). We assume familiarity with indistinguishability.

Interactive Protocols. An interactive protocol Π is a pair of interactive Turing machines, (P, V), where V is probabilistic polynomial time (PPT). P is called the prover, while V is called the verifier. $\langle P, V \rangle (x)$ denotes the random variable (over the randomness of P and V) representing V's output at the end of the interaction on common input x. If additionally V receives auxiliary input z, we write $\langle P(x), V(x, z) \rangle$ to denote V's output. We assume WLOG that Π starts with a verifier message and ends with a prover message, and say Π has k **rounds** if the prover and verifier each sends k messages alternately. A full or partial **transcript** of Π is a sequence of alternating verifier and prover messages, $(v_1, p_1, \ldots)$, where v denotes verifier messages and p denotes prover messages.

We may compose an interactive proof in parallel. Let $\Pi^k = (P^k, V^k)$ be the **parallel composition** of k copies of Π; that is, each prover and verifier message in Π^k is just concatenation of k independent copies of the corresponding message in Π. Upon completion, V^k accepts if and only if all k sessions are accepted by V. We note that an adversarial verifier may choose to abort in one session but not another.

Zero Knowledge Protocols. In the setting of zero knowledge, we consider an adversarial verifier that attempts to "gain knowledge" by interacting with an honest prover. An **adversarial verifier** V^* is a probabilistic polynomial time machine that, on common input x and auxiliary input z, interacts with the prover P. Let $\text{View}^P_{V^*}(x, z)$ be the random variable that denotes the **view** of V^* in an interaction with P (this includes the random coins of V^* and the messages received by V^*).

A **black-box simulator** S is a probabilistic polynomial time machine that is given black-box access to V^* (written as S^{V^*}). Formally, S fixes the random coins r of V^* a priori, and S is allowed to specify a valid partial transcript $\tau = (v_1, p_1, \ldots, p_i)$ of V^*_r, and query V^*_r for the next verifier message v_{i+1}. Here, τ is **valid** if it is consistent with V^*_r, i.e., each verifier message v_j in τ is what V^*_r would have responded given the previous prover messages $p_1, \ldots, p_{j-1}$ and the fixed random tape r. Note that S is allowed to "rewind" V^* by querying V^* with different partial transcripts that shares a common prefix.

Intuitively, an interactive proof is zero-knowledge (ZK) if the view of any adversarial verifier V^* can be generated by a simulator. The formal definition follows.

Definition 3 (Black-Box Zero-Knowledge [GMR89, GO94]). *Let $\Pi = \langle P, V \rangle$ be an interactive proof (or argument) for a language L. Π is **black-box zero-knowledge** if there exists a black-box simulator S such that for every common input x, auxiliary input z and every adversary V^*, $S^{V^*(x,z)}(x)$ runs in time polynomial in $|x|$, and the ensembles $\{\text{View}^P_{V^*}(x, z)\}_{x \in L, z \in \{0,1\}^*}$ and $\{S^{V^*(x,z)}(x)\}_{x \in L, z \in \{0,1\}^*}$ are computationally indistinguishable as a function of $|x|$.*

Other Primitives. In our construction of zero-knowledge arguments we use a few other primitives including Witness-Indistinguishable (WI) Proofs [FS90], Proofs of Knowledge (POK) [FS90, BG02], and Special-Sound (SS) Proofs [CDS94]. Due to lack of space, we refer the readers to the full version of this paper for a more detailed description of these primitives.

3 Construction

We define a zero-knowledge argument PARALLELZK in Section 3.1, and show that it satisfies Theorem 1 in Section 3.2.

3.1 The Protocol

Our ZK argument PARALLELZK (also used in [PV08, PTV10]) is a slight variant of the precise ZK protocol of [MP06], which in turn is a generalization of the Feige-Shamir protocol [FS89]. The protocol for language $L \in$ NP proceeds in three stages, given a security parameter n, a common input statement $x \in \{0,1\}^n$, and a round-parameter m:

Stage Init: The verifier picks two random strings $r_1, r_2 \in \{0,1\}^n$ and sends their images $c_1 = f(r_1)$, $c_2 = f(r_2)$ through a one-way function f to the prover. The verifier then acts as the prover in m parallel instances of a 4-round witness indistinguishable and special sound proof of knowledge (WI and SS-POK) of the NP statement "c_1 or c_2 is in the image set of f" (a witness here would be a pre-image of c_1 or c_2). All but the last two messages of each SS-POK is exchanged in this stage; we denote their partial transcripts by $(\alpha_1, \alpha_2, \ldots, \alpha_m)$.

Stage 1: m rounds of message exchanges occur in Stage 1. In the jth round, the prover sends β_j, a random second last message of the jth SS-POK, and the verifier replies with the last message γ_j of the proof. These m rounds are called *slots*. Slot i is *convincing* if the verifier produces an accepting proof (i.e., the transcript $(\alpha_i, \beta_i, \gamma_i)$ is accepting). If there is ever an *unconvincing* slot, the prover aborts the whole session.

Stage 2: The prover provides a 4-round witness indistinguishable proof of knowledge (WI-POK) of knowledge of the statement "$x \in L$, or one of c_1 or c_2 is in the image set of f".

Completeness and soundness follows directly from the proof of Feige and Shamir [FS89]; in fact, the protocol is an instantiation of theirs. Intuitively, to cheat in the protocol a prover must "know" an inverse to c_1 or c_2 (because Stage 2 is an argument of knowledge), which requires the prover to invert the one-way function f (it is shown in [FS90] that Stage Init and Stage 1 of the protocol cannot aid the prover in inverting f). A formal description of protocol ParallelZK is shown in Figure 1.

Remark 4. *We note that here we use multiple slots to improve the knowledge tightness of parallel zero knowledge, whereas previously, multiple slots was typically used to achieve concurrent zero knowledge and $\omega(\log n)$ slots were considered. In contrast, we show that in the context of parallel zero knowledge, using even constant number of slots improves the knowledge tightness significantly. Indeed, both our simulation technique and its analysis presented in the next section are new, where we rewind each slot to resolve all sessions in parallel (as opposed to previous works that focused on one session at a time).*

3.2   The Simulator

To show that protocol $\Pi = \text{PARALLELZK}$ satisfies Theorem 1, given any polynomially bounded $k = k(n)$, we need to construct a black-box zero-knowledge simulator $S = S_k$ for protocol Π^k (PARALLELZK repeated k times in parallel). On a very high-level, our simulator follows that of Feige and Shamir [FS90]: after fixing the SS-POK prefixes in Stage Init, the simulator rewinds one of the "slots" in Stage 1 (the last two messages of the SS-POKs). If the verifier responds with two convincing slots, the simulator uses the special-soundness property to extract a "fake witness" r such that $f(r) = c_1$ or c_2, and uses this fake witness to simulate Stage 2 of the protocol.

Common Input: an instance x of a language L with witness relation R_L.
Auxiliary Input for Prover: a witness w, such that $(x, w) \in R_L(x)$.
Stage Init:

V uniformly chooses $r_1, r_2 \in \{0, 1\}^n$.

V $\to$ P: $c_1 = f(r_1), c_2 = f(r_2)$.

V $\leftrightarrow$ P: Exchange in parallel (interactively) all but the last two messages $\alpha_1, \ldots, \alpha_m$ of m WI and SS-POKs on common input (c_1, c_2) with respect to the witness relation:

$$R_{L'}(c_1, c_2) = \{r : f(r) = c_1 \text{ or } c_2\}$$

Note that V acts as the prover in these SS-POK's.

Stage 1: For $j = 1$ to m, exchange the ith "slot"

P $\to$ V: The second last message β_i of the ith SS-POK.

V $\to$ P: The last message γ_i of the ith SS-POK.

P aborts if $(\alpha_i, \beta_i, \gamma_i)$ is not a valid SS-POK.

Stage 2:

P $\leftrightarrow$ V: a 4-round computational-WI proof of knowledge from P to V on common input (c_1, c_2, x) with respect to the witness relation:

$$R_{L' \vee L}(c_1, c_2, x) = \{(r, w) : r \in R_{L'}(c_1, c_2) \text{ or } w \in R_L(x)\}$$

Fig. 1. PARALLELZK: a ZK argument for NP with round parameter m

Given an adversarial verifier V^* (for protocol Π^k) and a common input $x \in \{0, 1\}^n$, the simulator $S^{V^*}(x)$ does the following:

1. The simulator S interacts with V^*, following the honest prover strategy, until the end of Stage 1. We call this the *reference simulation*.

2. The simulator S attempts to *resolve* all k parallel sessions in the reference simulation by extracting a fake witness r from the SS-POKs for each non-aborting session; aborted sessions are automatically considered resolved (and no fake witnesses are needed). To do so, S repeats the following step (called a rewinding pass) as many times as necessary, until all sessions are resolved.

3. **A rewinding pass.** For each slot i, the simulator rewinds the reference simulation back to the beginning of slot i, sends V^* a fresh random message β_i', and receives a new reply γ_i' (of course this is done in parallel for all k sessions). Note that for each unresolved session j, S already knowns an accepting transcript $(\alpha_i, \beta_i, \gamma_i)$ of SS-POK from the reference simulation. If session j does not abort during slot i in this rewinding pass, then S learns another accepting transcript $(\alpha_i, \beta_i', \gamma_i')$ of SS-POK. In this case, S can resolve the session j by extracting a fake witness using the special-sound property.

4. S completes the reference simulation using extracted fake witnesses to simulate the Stage 2 proof (only needed in each parallel session that did not abort). S outputs the view of V^* on the reference simulation and this completion.

For simplicity, we assume that for sessions that did not abort in the reference simulation, the extraction of fake witnesses always succeeds whenever S receives an accepting slot in a rewinding pass (i.e., we assume that S never sends the same value for β twice). This assumption can be made without loss of generality by the following modifications of the simulation strategy.

- Let the simulator S performs at most 2^n rewinding passes. If there exist any unsolved sessions j after 2^n rewinding passes, S resolves the session by brute force, i.e., by directly inverting the one-way function f to obtain a fake witness of length n. This modification increases the running time (but not the number of queries) of S by at most a poly(n) factor (multiplicatively), and makes sure that S makes at most poly(2^n) queries to V^*.
- Let the final verifier challenge in the SS-POK have length $|\beta| = n^2$. In this case, the probability of S ever querying V^* with the same value of β twice is poly(2^n) $\cdot 2^{-n^2} = 2^{-\Omega(n^2)}$, definitely negligible in n.

We now show two lemmas regarding S that together show that PARALLELZK is zero-knowledge when composed in parallel.

Lemma 5. *S runs in expected polynomial time, and makes $O(mk^{1/m} \log^2 n)$ queries in expectation.*

Lemma 6. *On common input $x \in L$, the output of S is indistinguishable from the real view of V^*.*

We give a sketch of proof of Lemma 6 first, and then prove Lemma 5 by bounding the *expected number of rewinding passes* before S extracts all necessary fake witnesses.

Proof (Proof Sketch of Lemma 6). The output of S up to the end of Stage 1 (i.e., the reference simulation) is identical to the view of V^*, because S follows the honest prover strategy. The output of S in Stage 2 of the protocol is computationally indistinguishable from the view of V^* because the Stage 2 proof is witness indistinguishable. Formally, this can be shown with a hybrid argument where we incrementally exchange each of the k parallel Stage 2 proofs from using "fake witnesses" r such that $f(r) = c_1$ or c_2 (the simulator strategy), to a real witnesses w for $x \in L$ (the honest prover strategy).

Proof of Lemma 5. We proceed to prove Lemma 5 by bounding the expected number of rewinding passes in an execution of S. Let R be a random variable that denotes the number of rewinding passes. We will show that:

$$\mathbb{E}[R] = \mathbb{E}[\# \text{ rewinding passes }] \leq O(k^{1/m} \cdot \log^2 n).$$

This then implies Lemma 5 because outside of rewinding passes, $S^{V^*}(x)$ makes only $O(m)$ queries to V^* and runs in polynomial time.

Before presenting our analysis for the general case of m slots, we revisit the classical analysis for the case of single slot for intuition.

The case of single slot. The analysis is very simple. For every $j \in [k]$, let R_j denote the number of rewinding passes to resolve session j, and let p_j be the probability that session j does not abort during the single slot. Recall that session j is resolved if it aborts in the reference simulation, and otherwise, the simulator needs to rewind the slot several times until session j does not abort again. Hence, the expected number of rewinding passes to resolve session j is

$$\mathbb{E}[R_j] = (1 - p_j) \cdot 0 + p_j \cdot \frac{1}{p_j} = 1.$$

By linearity of expectation, the expected number of rewinding passes is

$$\mathbb{E}[R] = \sum_j \mathbb{E}[R_j] = k \leq O(k \cdot \log^2 n).$$

We note that the above simple analysis is tight. Consider the case where during the slot, each session aborts independently with probability $(1 - 1/k)$. It is not hard to see that in this case, with constant probability, at least one session does not abort during the slot, and the simulator needs to rewind k times in expectation to resolve the survival session. Therefore, the expected number of rewinding passes is $\Omega(k)$.

In fact, it is instructive to note that the following natural generalization of the above example is essentially the worse-case example for the general case of m slots: during each slot $i \in [m]$, each survival session j aborts independently with probability $(1 - k^{-1/m})$. In this case, each session does not abort during the m slots with probability $(k^{-1/m})^m = 1/k$, and hence with constant probability, at least one session survives after m slots. Resolving the survival session requires $k^{1/m}/m$ rewinding passes in expectation, and hence the expected number of rewinding passes is $\Omega(k^{1/m}/m)$.

We note that although in the above example, each session aborts during each slot independently, in general, the aborting probability of each session at each slot can depends arbitrarily on the history and correlated arbitrarily.

The general case of m slots. To analyze the expected number of rewinding passes, we define the following $[0, 1]$-valued random variables based on the reference simulation generated in Step 1. Let h_i denote the partial transcript of the reference simulation before slot i. For every slot $i \in [m]$ and session $j \in [k]$, we define random variable $p_{i,j}$ as follows.

– If session j is already aborted at the end of slot i, then we define $p_{i,j} \triangleq 1$.
– Otherwise, we define $p_{i,j}$ to be the conditional probability

$$p_{i,j} \triangleq \Pr[\text{ session } j \text{ does not abort during slot } i \mid h_i].$$

For intuition, $p_{i,j}$ is essentially the probability that S can resolve session j by rewinding slot i. Now consider the *best slot* for each session — the slot with the highest $p_{i,j}$ value (this is the slot that S wants to rewind). We record this value as

$$p_j^* = \max_i p_{i,j}$$

Note that for a session j that aborts in the reference simulation, we have $p_j^* = 1$, indicating that sessions j is already resolved and matching the above intuition. Finally, the number of rewinding passes depends heavily on the *worst session* — the session with the worst p_j^* value (the "worst best slot"). We record this value as the *critical probability*:

$$p^* = \min_j p_j^*.$$

To see how the critical probability p^* plays an important role in the expected number of rewinding passes, note that on one hand, S needs roughly $1/p^*$ rewinding passes to resolve the worse-case session; on the other hand, the chance of having a reference simulation with small critical probability (say, $p^* \leq p$) is rare (at most p^m). Therefore, to upper bound $\mathbb{E}[R]$, we define the following events, which partition the probability space according to the critical probability. For every $t \in \mathbb{N}$, let

$$\alpha_t \stackrel{\text{def}}{=} \left(\frac{1}{2^t \cdot k^{1/m}} \right)$$

- Let A_0 be the event that $p^* \geq \alpha_0 = k^{-1/m}$, and for every $t \in \mathbb{N}$, let A_t be the event that

$$\alpha_t \leq p_j^* < \alpha_{t-1}.$$

Similarly for every session $j \in [k]$,

- Let $A_{0,j}$ be the event that $p_j^* \geq \alpha_0 = k^{-1/m}$, and for every $t \in \mathbb{N}$, let $A_{t,j}$ be the event that

$$\alpha_t \leq p_j^* < \alpha_{t-1}.$$

We can now express the expectation of the number of rewinding passes as follows.

$$\mathbb{E}[R] = \sum_{t \geq 0} \Pr[A_t] \cdot \mathbb{E}[R \mid A_t]$$

$$\leq \Pr[A_0] \cdot \mathbb{E}[R \mid A_0] + \sum_{t \geq 1} \left(\sum_{j=1}^{k} \Pr[A_{t,j}] \right) \cdot \mathbb{E}[R \mid A_t],$$

where the last inequality follows by $A_t \subseteq \cup_j A_{t,j}$ (which follows from definition). We proceed to bound each term. For A_0, we use trivial bound $\Pr[A_0] \leq 1$. For general $t \geq 1$ and every $j \in [k]$, we first observe that when $A_{t,j}$ happens, session j does not abort all of its m slots in the reference simulation (since otherwise, $p_j^* = 1$). This happened despite the fact that each slot i in session j in the reference simulation could have only survived (not aborted) with probability $p_{i,j} \leq \alpha_{t-1}$. Thus,

$$\Pr[A_{t,j}] \leq \alpha_{t-1}^m = \left(\frac{1}{2^{t-1} \cdot k^{1/m}} \right)^m = \frac{1}{2^{m(t-1)} \cdot k},$$

and,

$$\sum_{j=1}^{k} \Pr[A_{t,j}] \leq k \cdot \frac{1}{2^{m(t-1)} \cdot k} = \frac{1}{2^{m(t-1)}}.$$

It remains to bound $\mathbb{E}[R \mid A_t]$, which is given in the follow lemma.

Lemma 7. *For every $t \geq 0$, we have*

$$\mathbb{E}[R \mid A_t] \leq O\left(2^t \cdot k^{1/m} \cdot \log^2 n\right).$$

We apply Lemma 7 to upper bound $\mathbb{E}[R]$ first.

$$\mathbb{E}[R] \leq \mathbb{E}[R \mid A_0] + \sum_{t \geq 1} \frac{1}{2^{m(t-1)}} \cdot \mathbb{E}[R \mid A_t]$$

$$\leq O\left(k^{1/m} \cdot \log^2 n\right) + \sum_{t \geq 1} \frac{2^t}{2^{m(t-1)}} \cdot O\left(k^{1/m} \cdot \log^2 n\right)$$

$$\leq O\left(k^{1/m} \cdot \log^2 n\right).$$

This completes the proof of Lemma 5.

Proof (Proof of Lemma 7). The event A_t means that in the reference simulation, for every non-aborting session j, there exists a useful slot $i \in [m]$ such that

$$\Pr[\text{ session } j \text{ is not aborted after slot } i \mid h_i] = p_{i,j} \geq \alpha_t.$$

Therefore, in each rewinding pass, the simulator S may learn an (additional) accepting transcript of SS-POK in session j with probability at least α_t, allowing it to extract a fake witness.

Fix a non-aborting session j, and define

$$q = \left(\frac{10 \cdot \log^2 n}{\alpha_t}\right) = O\left(2^t \cdot k^{1/m} \cdot \log^2 n\right),$$

Because the rewinding passes are independent, we have

$$\Pr[\text{session } j \text{ is resolved after } q \text{ rewinding passes}] = 1 - (1 - \alpha_t)^q \geq 1 - \mathsf{ngl}(n).$$

Since there are at most k survival sessions, by the union bound,

$$\Pr[\text{all sessions are resolved after } q \text{ rewinding passes}] \geq 1 - \mathsf{ngl}(n).$$

In other words, every q rewinding passes can solve all the sessions with probability at least $1 - \mathsf{ngl}(n)$. It follows that

$$\mathbb{E}[R \mid A_t] \leq (1 - \mathsf{ngl}(n)) \cdot q + \mathsf{ngl}(n)\,(1 - \mathsf{ngl}(n)) \cdot 2q + \cdots$$

$$\leq O(q) = O\left(2^t \cdot k^{1/m} \cdot \log^2 n\right).$$

4   Lower Bound

The proof of Theorem 2 follows a well-known framework (e.g., [GK96b, CKPR01]). Let S be a black-box zero-knowledge simulator for $\Pi^k = (P^k, V^k)$ that makes less than $q = O(k^{1/m}/\log^2 n)$ queries, and let V^{k*} be a particular adversarial verifier to be specified later. We define $\mathcal{D}$, a BPP decision procedure for L by combining S and V^{k*}: on input instance x, $\mathcal{D}(x)$ accepts if and only if $S^{V^{k*}}(x)$ outputs an accepting view of V^{k*} (i.e., all k sessions of V^{k*} accept). Using the zero-knowledge property, it is easy to show (see for example [GK96b]) that if the modified protocol $\Pi^{k*} = (P^k, V^{k*})$ is complete for L (based on our choice of V^{k*}), then $\mathcal{D}$ is complete for L as well. The main effort of the proof is to show that $\mathcal{D}$ is sound; this relies both on the choice of V^{k*} and the fact that S makes less than q queries to V^{k*}. We discuss our choice of V^{k*} in Section 4.1, and analyze the soundness of $\mathcal{D}$ in Section 4.2.

4.1   The Random Termination Verifier V^{k*}

In this section, we define a verifier V^{k*} for the parallelized protocol with two goals in mind: the protocol $\Pi^{k*} = (P^k, V^{k*})$ should be complete (so that $\mathcal{D}$ is complete), and V^{k*} should be sound against any rewinding simulator S that makes less than q queries to V^{k*} (so that $\mathcal{D}$ is sound).

Just as [CKPR01], we define V^{k*} to follow the honest verifier strategy V^k with one extra property: random termination.[2] Whenever the prover P^k or the rewinding simulator S makes a query to V^{k*}, V^{k*} determines, with independent and fresh randomness,[3] whether or not to terminate immediately and *accept* with probability $\rho \in [0, 1]$, a parameter to be specified later; this is done *independently* for each of the k parallel sessions (i.e., one session may be terminated while other sessions continue). Due to this independence among parallel sessions, we often treat V^{k*} as k machines, $(V_1^*, \ldots, V_k^*)$, each responsible for making the decision to terminate and generating the verifier messages for one session. Note that the fresh randomness is only used to decide whether to terminate or not; V^{k*} generates protocol messages using its default random tape that is kept the same between rewinds (as expected by following the honest verifier strategy).

Clearly, $\Pi^{k*} = (P^k, V^{k*})$ is still complete. It remains to show that V^{k*} is "sound" against the rewinding S; that is, on input $x \notin L$, $S^{V^{k*}}$ is unlikely to

---

[2] The term "random termination" was first used by Haitner [Hai09], but the random termination verifier we considered already appeared in the earlier work of [CKPR01].

[3] We use a well-known technique (see for example [GK96b, CKPR01]) to generate fresh independent randomness on the fly for each query from the simulator S, despite the fact that S may rewind V^{k*} between queries and force V^{k*} to use the same random tape. Let $\mathcal{H}$ be a family of q-wise independent hash-functions, and let V^{k*} sample one hash-function $h \leftarrow \mathcal{H}$ in the very beginning. Then whenever V^{k*} receives a query (from P^k or S), V^{k*} applies h to the current protocol transcript (the sequence of messages exchanged in the protocol so far) and use the output as a fresh random tape. Since S makes at most q queries to V^{k*}, the output distribution of the hash-function is truly uniformly random.

generate an accepting transcript of V^{k*}. From now on we drop the common input $x \notin L$. Intuitively, by randomly terminating, V^{k*} can better protect its randomness against S's rewinds (when V^{k*} terminates, S learns nothing about V^{k*}'s fixed random tape), thus ensuring soundness. To make this intuition more concrete, suppose for example that S made q queries $\tau_1, \ldots, \tau_q$ to V^{k*}, and without loss of generality outputs the view of V^{k*} on a subset of size m of those queries[4], $T = \{\tau_{i_1}, \ldots, \tau_{i_m}\}$. Further suppose that there exists a parallel session $j \in [k]$ such that V^{k*} does not terminate on the queries in T, but terminates on all remaining queries. Then intuitively, S's rewinding does not help S convince V^{k*} in session j, and the soundness of the original protocol Π should imply that V^{k*} rejects with overwhelming probability in session j (and therefore rejects overall).

The core of our proof is to show that, with high probability, for every subset of size m of queries $T = \{\tau_{i_1}, \ldots, \tau_{i_m}\}$ made by S, there exists a session $j \in [k]$ with *overwhelming probability* such that rewinds are "not helpful" for session j with respect to T in the above manner. We make this possible by setting the termination probability to $\rho = (1 - 1/q)$.

We now state the formal lemmas. Let n be the security parameter and L be a language. Suppose there exists a $m(n) \in O\left(\frac{\log n}{\log \log n}\right)$-round argument $\Pi = (P, V)$ for L with perfect completeness and negligible soundness error. For any polynomially bounded $k(n) \geq n$, let S be a black-box zero-knowledge simulator of the parallelized protocol $\Pi^k = (P^k, V^k)$ that makes at most

$$q = k^{1/m}/(\log^2 n)$$

queries, and let V^{k*} be a random termination verifier of the parallelized protocol with termination probability

$$\rho = \left(1 - \frac{1}{q}\right) = \left(1 - \frac{1}{k^{1/m}} \cdot (\log^2 n)\right).$$

(These parameters passes the following sanity checks: q is polynomially bounded and $q \geq m$ — the simulator queries V^{k*} at least once for each round of the protocol. It is also useful later to know that $\binom{q}{m} \leq q^m \leq k$.) Then:

Lemma 8. *On input $x \in L$, $\mathcal{D}(x)$ accepts with probability 1, i.e., $S^{V^{k*}}(x)$ outputs an accepting view of V^{k*} with probability $1 - \mathsf{ngl}(n)$.*

Lemma 9. *On input $x \notin L$, the probability that $S^{V^{k*}}(x)$ generates an accepting view of V^{k*} is negligible, i.e., $\mathcal{D}$ has negligible soundness error.*

We sketch the proof of Lemma 8 now, and give the proof of Lemma 9 in the next section.

---

[4] Without loss of generality, we may assume that before S outputs a view of V^{k*}, S first queries V^{k*} with the messages in the view (if S hasn't already). This may increase the number of queries by m, and thus weaken the resulting lower bound from q to $q - m$. Nevertheless, this does not change our lower bound since $q = \omega(m)$ in Theorem 2.

Proof (Proof Sketch). Using the zero-knowledge property, the output of S is indistinguishable from the view of V^{k*} in an execution with P^k. Therefore it is enough to show that $\langle P^k, V^{k*} \rangle (x)$ accepts with probability 1. In each parallel session $j \in [k]$, V_j^* accepts by definition if it decides to terminate in some protocol round. Otherwise, V_{j*} is identical to V and would still accept with probability 1 because the original protocol $\Pi = (P, V)$ has perfect completeness.

4.2   Soundness of $\mathcal{D}$

Proof (Proof of Lemma 9). We prove Lemma 9 with a reduction. Suppose for the sake of contradiction that S convinces V^{k*} on some input $x \notin L$ with probability more than $1/p(n)$ for some polynomial p. Using S, we construct a cheating prover P^* for the original protocol $\Pi = (P, V)$ that convinces V with non-negligible probability.

Before we start, assume without loss of generality that S makes exactly q queries, and that before S outputs a view of V^{k*}, S would first query V^{k*} on all previous messages in the view. For technical convenience, we let V^{k*} make a fresh decision to terminate for each query and each session, *even if V^{k*} has already terminated previously in the same session.* I.e., regardless of history or message content, for each query and each parallel session, V^{k*} always terminates independently with probability ρ.

Our P^* is a natural extension of the classic reduction of [GK96b] — P^* guesses a session $j_0 \in [k]$ and m indices $T_0 = \{i_1, \ldots, i_m\} \subseteq [q]$ uniformly at random, and interacts with an outside honest V by internally simulating an interaction of (S, V^{k*}) with V embedded in session j_0, queries $\tau_{i_1}, \ldots, \tau_{i_m}$ of V^{k*}. In comparison, the idea of guessing a random query subset is exactly as in [GK96b]. The difference is that the reduction in [GK96b] is for single session protocols, and in contrast, we reduce from parallel protocols to single session protocols. Hence, our reduction P^* guesses a random session as well.

In more details, P^* runs S and V^{k*} internally. It simulates $k - 1$ sessions of V^{k*} honestly (except $V_{j_0}^*$). When simulating $V_{j_0}^*$ for the ith query where S queries τ_i, P^* first simulates (with fresh randomness) $V_{j_0}^*$'s decision on termination. If $V_{j_0}^*$ decides to terminate but $i \in T_0$ or if $V_{j_0}^*$ does not terminate but $i \notin T_0$, P^* aborts (in both these cases, the termination decision of $V_{j_0}^*$ is incompatible with P^*'s choice of queries to forward). If the forwarded queries (index set T_0) are not "consistent" (e.g., if they query for the same round of the protocol more than once, or the query contains inconsistent transcript), P^* aborts as well. Note that if P^* does not abort, then V^{k*} is perfectly simulated (even in session j_0).

Now consider the following best case scenario. Suppose that at the end of the simulation, S successfully outputs an accepting view of V^{k*}. Moreover, suppose that the accepting view consists exactly of the queries in index set T_0 (this automatically guarantees that the forwarded queries are consistent), and suppose that P^* does not abort (i.e., termination decisions are compatible with the forwarded queries). Then, P^* will have successfully convinced the outside honest V. The rest of the proof is devoted to show that this best case scenario occurs with noticeable probability (roughly $1/(p \cdot k^2)$).

Let $T \subset [q]$ denote an index set $\{i_1, \ldots, i_m\}$ of size m. For an index set $T \subset [q]$ and a session $j \in [k]$, we define $A(T, j)$ to be the event that, on session j, V^{k*} terminates session j on query τ_i iff $i \notin T$. Referring back to our intuition earlier, $A(T, j)$ denotes the event that for session j, S's rewinds are not helpful with respect to the queries indexed by T. If event $A(T, j)$ holds, and S uses the queries indexed by T to form an accepting view of V^{k*}, and P^* guesses both $T_0 = T$ and $j_0 = j$ in the beginning, then P^* will have successfully convinced the outside honest V.

We claim that by the setting of parameters, we have

$$\Pr[\forall T \subset [q], \exists j \in [k] \text{ s.t. } A(T, j)] \geq 1 - \mathsf{ngl}(n) \tag{1}$$

where $\mathsf{ngl}(n)$ denotes a negligible quantity in n. In words, with overwhelming probability, for every possible index set T of size m that S may use to output a view of V^{k*}, there exists a session j such that S's rewinds are not helpful with respect to the queries indexed by T.

Before proving (1), we first use the claim to show that P^* convinces V with noticeable probability. Recall that S outputs an accepting view of V^{k*} with probability $1/p$. By a union bound, we have

$$\Pr[(S \text{ outputs accepting view of } V^{k*}) \wedge (\forall T \subset [q], \exists j \in [k] \text{ s.t. } A(T, j))]$$
$$\geq (1/p) - \mathsf{ngl}(n).$$

Note that when the above event holds, there exist a unique index $\hat{T}$ of m queries used by S to form an accepting view of V^{k*}, and there exists a session $\hat{j} \in [k]$ such that $A(\hat{T}, \hat{j})$ holds. As mentioned earlier, if P^* guesses $j_0 = \hat{j}$ and $T_0 = \hat{T}$ correctly, P^* will have successfully convinced V. Since P^* guesses j and T uniformly at random and independent of the interaction between S and V^{k*}, we have

$$\Pr[P^* \text{ convinces } V]$$
$$\geq \Pr[(S \text{ convinces } V^{k*}) \wedge (\forall T \subset [q], \exists j \in [k] \text{ s.t. } A(T, j))$$
$$\wedge (P^* \text{ guesses } \hat{T} \text{ and } \hat{j} \text{ correctly})]$$
$$\geq \frac{(1/p - \mathsf{ngl}(n))}{k \cdot \binom{q}{m}} \geq \frac{1}{p \cdot k^2},$$

where in the last line we used $\binom{q}{m} \leq q^m \leq k$. This contradicts to the fact that Π has negligible soundness error and completes our analysis.

It remains to show (1). By definition, each session j terminates on each query τ_i with probability exactly ρ, independent from any other session or query. Hence, for any session j and index set T of size m, the probability that event $A(T, j)$ holds is

$$\Pr[A(T, j)] = \rho^{q-m} \cdot (1 - \rho)^m \geq \left(1 - \frac{1}{q}\right)^q \cdot \left(\frac{1}{q}\right)^m \geq \Omega\left(\frac{1}{k} \cdot (\log^{2m} n)\right).$$

It follows that

$$\Pr[\exists j \in [k] \text{ s.t. } A(T,j)] \geq 1 - \left(1 - \Omega\left(\frac{1}{k} \cdot (\log^{2m} n)\right)\right)^k \geq 1 - e^{-\Omega(\log^{2m} n)}.$$

Finally, by a union bound, we have

$$\Pr[\forall T \subset [q], \exists j \in [k] \text{ s.t. } A(T,j)] \geq 1 - e^{-\Omega(\log^{2m} n)} \cdot \binom{q}{m} \geq 1 - \mathsf{ngl}(n),$$

as claimed.

As with most lower bounds for black-box zero-knowledge, a careful reading reveals that Theorem 2 also applies to more liberal definitions of zero-knowledge, such as ε-zero-knowledge and zero-knowledge with expected polynomial time simulators. Additionally, note that the proof of Lemma 9 never assume that S is a zero-knowledge simulator, and works just as well for any PPT oracle machine S.

Remark 10. *By examining the technical inner workings of the proof of Canetti, Kilian, Petrank and Rosen [CKPR01] (which also uses a random termination verifier), we discovered that part of their analysis implicitly presents a lower bound for the number of queries made by black-box simulators for parallel zero-knowledge protocols. Compared with Theorem 2 and our analysis, the result of [CKPR01] establishes a weaker bound (and is arguably more complicated); this is not surprising, since establishing a parallel lower bound was not their goal.*

Specifically, [CKPR01] implicitly establishes a $\log^{\omega(1)}(k)$ lower bound on the number of simulator queries, whereas we were able to establish a lower bound of $k^{1/m}/(\log^2 n)$. Nevertheless, we believe that by adapting our parameters (which may seem strange for their setting), their analysis could be strengthened to match our lower bounds (we have not verified all the details, however).

Acknowledgments. We thank to Iftach Haitner and Johan Håstad for useful discussion in the early stage of this research.

References

[Bar01]   Barak, B.: How to go beyond the black-box simulation barrier. In: FOCS 2001, pp. 106–115 (2001)

[BG02]   Barak, B., Goldreich, O.: Universal arguments and their applications. In: Computational Complexity, pp. 162–171 (2002)

[BL02]   Barak, B., Lindell, Y.: Strict polynomial-time in simulation and extraction. In: STOC 2002, pp. 484–493 (2002)

[Blu86]   Blum, M.: How to prove a theorem so no one else can claim it. In: Proc. of the International Congress of Mathematicians, pp. 1444–1451 (1986)

[CDS94]   Cramer, R., Damgård, I., Schoenmakers, B.: Proof of Partial Knowledge and Simplified Design of Witness Hiding Protocols. In: Desmedt, Y.G. (ed.) CRYPTO 1994. LNCS, vol. 839, pp. 174–187. Springer, Heidelberg (1994)

[CKPR01] Canetti, R., Kilian, J., Petrank, E., Rosen, A.: Black-box concurrent zero-knowledge requires $\tilde{\omega}(\log n)$ rounds. In: STOC 2001, pp. 570–579 (2001)

[DNS04]   Dwork, C., Naor, M., Sahai, A.: Concurrent zero-knowledge. J. ACM 51(6), 851–898 (2004)

[FS89]   Feige, U., Shamir, A.: Zero Knowledge Proofs of Knowledge in Two Rounds. In: Brassard, G. (ed.) CRYPTO 1989. LNCS, vol. 435, pp. 526–544. Springer, Heidelberg (1990)

[FS90]   Feige, U., Shamir, A.: Witness indistinguishable and witness hiding protocols. In: STOC, pp. 416–426 (1990)

[GK96a]   Goldreich, O., Kahan, A.: How to construct constant-round zero-knowledge proof systems for NP. Journal of Cryptology 9(3), 167–190 (1996)

[GK96b]   Goldreich, O., Krawczyk, H.: On the composition of zero-knowledge proof systems. SIAM Journal on Computing 25(1), 169–192 (1996)

[GMR89]   Goldwasser, S., Micali, S., Rackoff, C.: The knowledge complexity of interactive proof systems. SIAM Journal on Computing 18(1), 186–208 (1989)

[GMW91]   Goldreich, O., Micali, S., Wigderson, A.: Proofs that yield nothing but their validity for all languages in NP have zero-knowledge proof systems. J. ACM 38(3), 691–729 (1991)

[GO94]   Goldreich, O., Oren, Y.: Definitions and properties of zero-knowledge proof systems. Journal of Cryptology 7, 1–32 (1994)

[Gol01]   Goldreich, O.: Foundations of Cryptography — Basic Tools. Cambridge University Press (2001)

[Gol02]   Goldreich, O.: Concurrent zero-knowledge with timing, revisited. In: STOC 2002, pp. 332–340 (2002)

[Hai09]   Haitner, I.: A parallel repetition theorem for any interactive argument. In: FOCS 2009, pp. 241–250 (2009)

[HRS09]   Haitner, I., Rosen, A., Shaltiel, R.: On the (Im)Possibility of Arthur-Merlin Witness Hiding Protocols. In: Reingold, O. (ed.) TCC 2009. LNCS, vol. 5444, pp. 220–237. Springer, Heidelberg (2009)

[Kat08]   Katz, J.: Which Languages Have 4-Round Zero-Knowledge Proofs? In: Canetti, R. (ed.) TCC 2008. LNCS, vol. 4948, pp. 73–88. Springer, Heidelberg (2008)

[KPR98]   Kilian, J., Petrank, E., Rackoff, C.: Lower bounds for zero knowledge on the internet. In: FOCS 1998, pp. 484–492 (1998)

[MP06]   Micali, S., Pass, R.: Local zero knowledge. In: STOC 2006, pp. 306–315 (2006)

[PPS⁺08] Pandey, O., Pass, R., Sahai, A., Tseng, W.-L.D., Venkitasubramaniam, M.: Precise Concurrent Zero Knowledge. In: Smart, N.P. (ed.) EUROCRYPT 2008. LNCS, vol. 4965, pp. 397–414. Springer, Heidelberg (2008)

[PTV10]   Pass, R., Tseng, W.-L.D., Venkitasubramaniam, M.: Eye for an Eye: Efficient Concurrent Zero-Knowledge in the Timing Model. In: Micciancio, D. (ed.) TCC 2010. LNCS, vol. 5978, pp. 518–534. Springer, Heidelberg (2010)

[PV08]   Pass, R., Venkitasubramaniam, M.: On Constant-Round Concurrent Zero-Knowledge. In: Canetti, R. (ed.) TCC 2008. LNCS, vol. 4948, pp. 553–570. Springer, Heidelberg (2008)

[Ros00]   Rosen, A.: A Note on the Round-Complexity of Concurrent Zero-Knowledge. In: Bellare, M. (ed.) CRYPTO 2000. LNCS, vol. 1880, pp. 451–468. Springer, Heidelberg (2000)

Simultaneously Resettable
Arguments of Knowledge

Chongwon Cho[1], Rafail Ostrovsky[12],
Alessandra Scafuro[3], and Ivan Visconti[3]

[1] Department of Computer Science, UCLA
[2] Department of Mathematics, UCLA, USA
{ccho,rafail}@cs.ucla.edu
[3] Dipartimento di Informatica, University of Salerno, Italy
{scafuro,visconti}@dia.unisa.it

Abstract. In this work, we study simultaneously resettable arguments of knowledge. As our main result, we show a construction of a constant-round simultaneously resettable witness-indistinguishable argument of knowledge (simres$\mathcal{WIA}$oK, for short) for any **NP** language. We also show two applications of simres$\mathcal{WIA}$oK: the first constant-round simultaneously resettable zero-knowledge argument of knowledge in the Bare Public-Key Model; and the first simultaneously resettable identification scheme which follows the knowledge extraction paradigm.

1 Introduction

Interaction and private randomness are the two fundamental ingredients in Cryptography. They are especially important for achieving zero-knowledge proofs [15]. In [7] Canetti, Goldreich, Goldwasser and Micali showed that when private randomness is limited and re-used in multiple instances of a proof system, it is still possible to preserve the zero-knowledge requirement. The setting proposed by [7] is of a malicious verifier that resets the prover, therefore forcing the prover to run several protocol executions using the same randomness. This setting applies to protocols where the prover is implemented by a stateless device. Therefore, a prover can only count on the limited (hardwired) randomness while it can be adaptively reset any polynomial number of times. The resulting security notion against such powerful verifiers is referred to as *resettable zero knowledge* ($r\mathcal{ZK}$) and is provably harder to achieve than concurrent zero knowledge [11,18]. Feasibility results have been achieved in [7,17] in the standard model with the following round-complexity: polylogarithmic for $r\mathcal{ZK}$ and constant for resettable witness indistinguishability ($r\mathcal{WI}$, in short). Since then, it was also shown how to achieve resettable zero knowledge in the Bare Public-Key (BPK) model, introduced by Canetti et al. [7], where one can obtain better round complexity and assumptions [19,10,1,22,21]. Very recently, it has been shown [13] that resettable statistical zero knowledge for non-trivial languages is possible.

R. Cramer (Ed.): TCC 2012, LNCS 7194, pp. 530–547, 2012.

The "reverse" of the above question has been considered by Barak, Goldreich, Goldwasser and Lindell [4] where a malicious prover resets a verifier, called *resettable soundness*. In [4], it has been shown how to obtain resettable soundness along with $\mathcal{ZK}$ in a constant number of rounds.

Barak et al. [4] proposed the challenging *simultaneous resettability conjecture*, where one would like to prove that a protocol is secure against both a resetting malicious prover and a resetting malicious verifier. The existing machinery turned out to be insufficient, and a definitive answer required almost a decade. In the work of Deng, Goyal and Sahai [9] they showed a resettably sound $r\mathcal{ZK}$ argument for **NP** with polynomial round complexity. Very recently, results in the BPK model for simultaneous resettability have been obtained in [8,2] with a constant number of rounds.

Arguments of knowledge under simultaneous resettability. Argument systems are often used with a different goal than proving membership of an instance in a language. Indeed, it is commonly required to prove knowledge (possession) of a witness instead of the truthfulness of a statement. Since arguments of knowledge serve as major building blocks in Cryptography (e.g., in identification schemes[1]), it is an interesting question whether the previous results for arguments of membership extend to arguments of knowledge. Unfortunately, arguments of knowledge have been achieved so far only when one party can reset. That is, we have $r\mathcal{ZK}$ arguments of knowledge [7] and, separately, resettably sound $\mathcal{ZK}$ arguments of knowledge [4]. Instead, when reset attacks are possible in both directions, no result is known even when only $r\mathcal{WI}$ with resettable argument of knowledge is desired.

It is important to note that resettable security for ZAPs comes almost for free because of the minimal round complexity (1 or 2 rounds). However, it is not known how to accommodate for knowledge extraction, unless one relies on non-standard (e.g., non-falsifiable) assumptions. For the case of resettably sound $r\mathcal{ZK}$, all the above results [9,8,2] critically use an instance-dependent technique along with ZAPs: when the statement is true (i.e., when proving $r\mathcal{ZK}$), the prover/simulator can run ZAPs which allow the use of multiple witnesses. Such use of multiple witnesses gives some flexibility that turns out to be very useful to prove resettable zero knowledge. Instead, when the statement is false, the protocols are designed so that adversarial malicious prover must stick with some fixed messages during the execution of protocol. Therefore, rewinding capabilities do not help the resetting malicious prover since he can not change those fixed messages. This is critically used in the proofs of resettable soundness in order to reach a contradiction when a prover proves a false statement. It is easy to see that the above approach fails when arguments of knowledge are considered. Indeed, when the malicious resetting prover proves a true statement, the same freedom that allows one to prove $r\mathcal{ZK}/r\mathcal{WI}$, also gives extra power to the malicious prover. Consequently, designing an extractor appears problematic and new

---

[1] Bellare et al. in [5] gave various definitions for identification schemes when the adversary can also reset the proving device.

techniques seem to be needed so that the simultaneous resettability conjecture is resolved even when we consider knowledge extraction.

Our results. Our main result is the first construction of a constant-round simultaneously resettable witness-indistinguishable argument of knowledge[2] (in short, simres$\mathcal{WI}$AoK) for any **NP** language. Our protocol is based on the novel use of ZAPs and resettably sound zero-knowledge arguments, which improves over the techniques previously used in [9,8] as well as concurrent and independent work[3] of [16].

We show several applications of our main result. First, we show that by combining two executions of our protocol for simres$\mathcal{WI}$AoK, we obtain a constant-round simultaneously resettable zero-knowledge argument of knowledge in the BPK model. This improves the results of [8,2] which do not enjoy witness extraction with respect to adversarial resetting provers.

As another application of our main protocol, we also consider the question of secure identification under simultaneous resettability and show how to use the above simres$\mathcal{WI}$AoK to obtain the first simultaneously resettable identification scheme which follows the knowledge extraction paradigm. We describe it by extending the work of Bellare, et al. [5].

In addition, in the full version of this paper, we show how to obtain a constant-round resettably sound concurrent zero knowledge argument of knowledge in the BPK model by relying on collision-resistant hash functions only (CRHFs, for short) (i.e., we do not require ZAPs, and thus trapdoor permutations).

Notation. We denote by $n \in \mathbb{N}$ the security parameter and by PPT the property of an algorithm of running in probabilistic polynomial-time. A function ϵ is negligible in n (or just negligible) if for every polynomial $p(\cdot)$ there exists a value $n_0 \in \mathbb{N}$ such that for all $n > n_0$ it holds that $\epsilon(n) < 1/p(n)$. We denote by $x \leftarrow \mathcal{D}$ the sampling of an element x from the distribution $\mathcal{D}$. We also use $x \xleftarrow{\$} \mathsf{A}$ to indicate that the element x is sampled from set A according to the uniform distribution. Let $\mathcal{P}, \mathcal{V}$ be interactive Turing machines, we denote by $\langle \mathcal{P}(\cdot), \mathcal{V}(\cdot) \rangle(x)$ the random variable representing the local output of $\mathcal{V}$ when interacting with $\mathcal{P}$ where x is the common input and the randomness of each machine is uniformly and independently chosen.

Blum's protocol. We will use the 3-round $\mathcal{WI}$PoK protocol of Blum [6] for the **NP**-complete language Graph Hamiltonicity (HC) as main ingredient of our construction. We refer to Blum's protocol as BL and to BL1, BL2, BL3 its three rounds.

---

[2] In this work, we will never consider the case of resettable soundness along with non-resettable argument of knowledge. Therefore, each time we mention together resettable soundness and argument of knowledge, we mean that both soundness and witness extraction hold against a malicious resetting prover.

[3] In a very recent and independent work [16], Goyal and Maji achieved simultaneously resettable secure computation. Their work achieves (with simulation-based security) simultaneous resettability with polynomial round complexity assuming also the existence of lossy trapdoor encryption.

2   Resettably Sound r$\mathcal{WI}$ Arguments of Knowledge

Our goal is to obtain a construction that is resettably-sound resettable $\mathcal{WI}$ *and* a resettable argument of knowledge in a constant number of rounds. The only known constant-round simultaneously-resettable $\mathcal{WI}$ protocol is rZAP which is not an argument of knowledge and as discussed previously there is not much hope to transform it in an argument of knowledge (even without considering resettability).

A typical paradigm: determining message and consistency proof. Typically, protocols dealing with a resetting adversary ([7,4,9]) rely on the following paradigm: the resetting party is required to provide a special message (called *determining message*) that determines her own action for the rest of the protocol. Namely, for each protocol message the resetting party is required to prove that such message is consistent with the determining message (we call this proof a *consistency proof*). Moreover, the actual randomness used by the honest party in the protocol depends on the determining message (typically the honest party applies a pseudorandom function (PRF) on it). The combination of the randomness depending on the determining message and the consistency proof given by the resetting party, suppresses the resetting power of the adversary. Indeed, due to the consistency proof, the resetting party can not change a message previously played without first having changed the determining message (unless she is able to fake the consistency proof). However, if she changes the determining message, then the honest party plays the protocol with (computationally) fresh randomness (unless the pseudo-randomness of the PRF is violated). We will follow this paradigm to construct our simultaneously resettable witness indistinguishable argument of knowledge as well. Recall that as specified above, we do not know how to from rZAPs that are already simultaneously resettable and try to transform them in arguments of knowledge. Our starting point is Blum's proof of knowledge [6]. In the following discussion we show incrementally how to transform such protocol to enjoy resettable witness indistinguishability and resettable soundness (this transformation is already known in literature) to finally present our novel technique to obtain also *resettable* argument of knowledge.

Resettable $\mathcal{WI}$ and stand-alone argument of knowledge [4]. When the verifier can reset the prover, following the above paradigm, it is easy to construct a resettable $\mathcal{WI}$ system starting from Blum's protocol. In Blum's protocol the only message from $\mathcal{V}$ to $\mathcal{P}$ is the challenge. The modified resettable version requires that $\mathcal{V}$ sends a statistically binding commitment of the challenge as determining message. The only other protocol message of $\mathcal{V}$ is the opening of the commitment which, due to the binding property, is itself a proof that the message is consistent with the determining message. Note that such modified protocol is no longer an argument of knowledge since the extractor has the same power of the malicious verifier. In order to allow only the extractor to cheat, the next step is to avoid the opening as a proof of consistency. Instead of the actual opening of the commitment, $\mathcal{V}$ is required to send the challenge along with a ressound (non-black-box) $\mathcal{ZK}$ argument ([3]). The (non-black box) extractor can send an arbitrary challenge and prove consistency with the determining message by using the (stand-alone) non-black-box simulator (recall that only $\mathcal{V}$ might reset here). The resulting protocol is resettable $\mathcal{WI}$ and (stand-alone) argument of knowledge (r$\mathcal{WI}$AoK for short) and it is known from [4].

We use a modified version of such protocol. We require that the commitment sent by the verifier is statistically hiding (instead of statistically binding), and we use the statistical zero-knowledge argument of knowledge of [20].

Achieving Resettable Soundness and Resettable Argument of Knowledge: existent solutions do not work. We now deal with the case in which also the prover can reset. By the BGGL compiler [4], we know that any constant-round public-coin $\mathcal{WI}$ argument system can be upgraded to resettable soundness by simply requiring the honest verifier to apply a PRF on the first message received from the prover. However, since our aim is to obtain simultaneous resettability, we need to start from the r$\mathcal{WI}$AoK protocol shown before, which is not public coin. Thus, following the paradigm and the technique of [9], we require that as first message, $\mathcal{P}$ sends the commitment of the randomness that will be used in the protocol: this is the determining message. Then upon each protocol message $\mathcal{P}$ proves that the message is honestly computed using the randomness committed in the determining message: this is the consistency proof. Since we are now in the setting in which both parties can reset each other the consistency proof must be provided with a simultaneous resettable tool. For this purpose we use rZAPs that are constant-round simultaneously resettable $\mathcal{WI}$ proofs. We denote the theorem to be proved with rZAP as "consistency theorem", since $\mathcal{P}$ proves that a message is honestly computed and consistent with the randomness committed in the determining message.

The technical problem using rZAPs is that since guarantee $\mathcal{WI}$, the theorem being proved is required to have more than one witness (note that the simultaneously resettable protocol of [9] can not be used here since we aim to a constant-round construction). Recall that we want to use rZAP to provide the proof of consistency with the determining message. If the determining message is a statistically binding commitment of the randomness, then there exists a unique opening, which implies the existence of only one witness. On the other hand, if we use a statistically hiding commitment, then any opening is a legitimate witness, the theorem is always true and the benefit of the determining message vanishes. The solution to overcome this problem is to change the theorem to be proved with rZAP so that it admits more than one witness.

In [9] the consistency theorem is augmented with the theorem "$x \in L$" that we call "*trapdoor* theorem" recalling FLS paradigm [12] but with a different purpose. We call it trapdoor to stress out that it is an escape for the prover that can pass the consistency proof essentially having freedom to change messages among resets. Hence in [9,8], along with each protocol message, $\mathcal{P}$ is required to prove that either the protocol message is computed honestly with the randomness committed in the determining message, (i.e., the "consistency theorem") or $x \in L$ (i.e., the "trapdoor theorem").

This solution can be seen as an instance-dependent technique. Indeed, it is easy to see that a malicious prover can play messages inconsistently with the determining message and still pass the consistency check, therefore exploiting its resetting power, only when $x \in L$. Instead, when proving soundness, since $x \notin L$, the trapdoor theorem is false, hence due to soundness of rZAPs, the malicious prover is forced to play according to the determining message therefore honestly following the protocol specifications.

Unfortunately, such an instance-dependent solution suffices to prove resettable soundness but fails completely when one would like to prove witness extraction (i.e., the argument of knowledge property). The reason is that, when proving witness extraction, we have to construct an extractor that works against any malicious prover, even one who uses the witness of the trapdoor theorem when proving consistency of the protocol messages. This possible behavior harms the extractor in two ways (recall that the witness can be computed from two distinct transcripts of Blum's protocol that have the *same* first message): 1) upon seeing the challenge of the verifier/extractor, $\mathcal{P}$ resets it and changes the first message of Blum's protocol according to the challenge; 2) $\mathcal{P}$ acts as a resetting verifier in the non-black-box $\mathcal{ZK}$ protocol, therefore preventing the extractor to use the stand-alone non-black-box simulator. Even though this is not harmful for the soundness property (a malicious prover can perform this attack only when $x \in L$), this attack kills the existence of the extractor. Therefore the above construction is only resettable $\mathcal{WI}$ and resettable sound. Concluding, the instance-dependent technique of [9] inherently prevents the existence of any extractor. New ideas are required to solve the problem.

Achieving Resettable *Argument of Knowledge: the new technique.* We propose a new "trapdoor" theorem that forces the resetting prover to honestly follow the protocol regardless of whether $x \in L$ or not.

The idea is the following. We require $\mathcal{P}$ to run *two* parallel executions of the rWIAoK shown above, that we denote as subprotocols π_0, π_1. In the determining message, in addition to the commitment of the random tape that will be used to run each sub-protocol, we require that $\mathcal{P}$ commits to a single bit. Then, the trapdoor theorem in sub-protocol π_d will be the following: "d is the bit committed in the determining message". Since in the determining message there is only one bit committed (the other two are commitments of random tapes), due to the statistical binding property of the commitment, the trapdoor theorem is true in only one sub-protocol. Hence, in at least one of the sub-protocols the trapdoor theorem is false regardless of whether $x \in L$ or not, and in such sub-protocol $\mathcal{P}$ is forced to honestly follow the rWIAoK protocol, playing consistently with the determining message.

More specifically, the final protocol goes as follows. $\mathcal{P}$ first sends the determining message which consists of the statistically binding commitment of the random tapes that will be used in each sub-protocol and of a single bit. Each sub-protocol is augmented with rZAPs sent by $\mathcal{P}$ to $\mathcal{V}$ in which $\mathcal{P}$ proves consistency with the determining message. Therefore, in each sub-protocol π_d, along with each message of the rWIAoK protocol, $\mathcal{P}$ provides a rZAP for the following compound theorem: either the message is honestly computed and consistent with the determining message, or d is the bit committed in the determining message. Finally, the verifier will accept the proof if and only if *both* sub-protocol executions are accepting.

It is easy to see that any malicious prover can not escape from following the determining message in at least one of the subprotocols. Indeed, let b be the bit committed in the determining message. If on one hand, in sub-protocol π_b, a malicious $\mathcal{P}$ is not forced to be honest and can then use the resetting power to prove any false theorem (indeed among resets $\mathcal{P}$ can change the protocol messages without changing the determining message), on the other hand, in sub-protocol $\pi_{\bar{b}}$, the trapdoor theorem is false, thus the only way to provide an accepting rZAP is to follow the honest behavior playing messages derived from the determining message. Therefore, in sub-protocol $\pi_{\bar{b}}$, the extractor is guaranteed that 1) for sessions starting with the same determining message, the first round of Blum's protocol does not change, so that playing with two distinct challenges yields the extraction of the witness; 2) the extractor can run the stand-alone non-black-box $\mathcal{ZK}$ simulator without being detected. Hence we have the following: sub-protocol $\pi_{\bar{b}}$ is resettably-sound and resettable argument of knowledge, while sub-protocol π_b is not sound. Note that in both sub-protocols, the resettable $\mathcal{WI}$ property is still preserved.

2.1   Formal Construction of simresWIAoK

We formally describe how to build a constant-round simultaneously resettable $\mathcal{WI}$ AoK (simresWIAoK) starting from Blum's protocol (BL protocol). We denote by SHCom, a two-round statistically hiding commitment scheme. We denote by SBCom the commitment procedure of a non-interactive statistically binding commitment scheme. We denote by $c \leftarrow$ SBCom(v, s) (resp. SHCom) the output of the commitment of the value v computed with randomness s. We use the resettably-sound statistical (non-black-box) $\mathcal{ZK}$ AoK of [20] that we denote by resSZK. In our construction, we require that $\mathcal{P}$, at each round of the protocol (except the last that is the opening of commitments as required by BL protocol), provides a proof that either the messages are honestly computed according to the randomness committed in the first round, or the "trapdoor" condition is satisfied. Formally, $\mathcal{P}$ provides rZAPs for the following **NP** languages (except the language Λ_{SHCom} that is proved only by $\mathcal{V}$ using resSZK protocol).

Λ_{BL1}: correctness and consistency of the first round of Blum's protocol (BL1). A tuple $(x, m, c_{r_b}, c_b) \in \Lambda_{\text{BL1}}$ if there exist (r_b, s_b) such that $c_{r_b} = $ SBCom(r_b, s_b) and m is honestly computed according to BL1 for the graph x using randomness $f_{r_b}(c_b)$.

Λ_{V}: correctness and consistency of verifier's messages of the protocol resSZK. A tuple $(m_P, m_V, c_{r_b}, c_b) \in \Lambda_{\text{V}}$ if there exist (r_b, s_b) such that $c_{r_b} = $

$\mathsf{SBCom}(r_b, s_b)$ and m_V is honestly computed according to the verifier's procedure of the protocol resSZK having in input prover's message m_P (m_P corresponds to the concatenation of all messages played by the prover so far) using randomness $f_{r_b}(c_b)$.

Λ_{trap}: **trapdoor theorem** (true only for sub-protocol b). The pair $(c_s, b) \in \Lambda_{\mathsf{trap}}$ if there exists s such that $c_s = \mathsf{SBCom}(b, s)$.

Λ_{SHCom}: validity of the opening (proved by $\mathcal{V}$). The pair $(c_s, m) \in \Lambda_{\mathsf{SHCom}}$ if there exists s such that $c_s = \mathsf{SHCom}(m, s)$. Note that for a statistically hiding commitment scheme, any pair (c_s, m) is actually in Λ_{SHCom}. Nevertheless, $\mathcal{V}$ proves this theorem using the argument of knowledge resSZK.

Protocol simres$\mathcal{WI}$AoK consists of two phases (see Fig. 1). In the first phase, $\mathcal{P}$ and $\mathcal{V}$ generate the random tapes that they will use to run the sub-protocols. $\mathcal{P}$ sends $\mathcal{V}$ the commitments c_{r_0}, c_{r_1} of two random strings r_0, r_1 and the commitment c_s of a random bit b. This message is the *determining message* on which $\mathcal{V}$ applies a PRF to generate a pseudo-random tape (to be used to execute the sub-protocols). The second phase consists of a parallel execution of π_0 and π_1 (see Fig. 2). $\mathcal{P}$ runs each sub-protocol on theorem x, randomness r_0, r_1, and the witnesses for computing the rZAPs as inputs (i.e., the opening of the commitments of the determining message). $\mathcal{V}$ runs each sub-protocol using the pseudo-random tapes determined by the determining message received from $\mathcal{P}$. Each sub-protocol is resettable $\mathcal{WI}$, while only one of the two sub-protocols is resettably-sound and a resettable AoK. Since $\mathcal{V}$ accepts the proof only if *both* executions are accepting, the final protocol is also a resettably-sound resettable AoK.

Protocol simres$\mathcal{WI}$AoK

Inputs: common input $x \in \mathsf{HC}$.
$\mathcal{P}$'s input: witness y, randomness ω. $\mathcal{V}$'s input: randomness r.

$\mathcal{P}$: $b \xleftarrow{\$} \{0, 1\}$; $r_0, r_1, s_0, s_1 \xleftarrow{\$} \{0, 1\}^n$.
   Send $c_{r_0} \leftarrow \mathsf{SBCom}(r_0, s_0)$, $c_{r_1} \leftarrow \mathsf{SBCom}(r_1, s_1)$, $c_s \leftarrow \mathsf{SBCom}(b, s)$.
   Run in parallel $\pi_0^{\mathcal{P}}(x, y, r_0, s_0)$; $\pi_1^{\mathcal{P}}(x, y, r_1, s_1)$.
$\mathcal{V}$: upon receiving $\mathsf{dm} = (c_{r_0}, c_{r_1}, c_s)$ from $\mathcal{P}$.
   $\mathsf{R_{V0}} \leftarrow f_r(x||c_{r_0}||c_s)$; $\mathsf{R_{V1}} \leftarrow f_r(x||c_{r_1}||c_s)$;
   Run in parallel $\pi_0^{\mathcal{V}}(x, \mathsf{R_{V0}})$; $\pi_1^{\mathcal{V}}(x, \mathsf{R_{V1}})$.

Fig. 1. Simultaneously Resettable Argument of Knowledge

The sub-protocol π_d is described in Fig. 2. We omit the first round of the rZAP and the first round of the statistically hiding commitment scheme SHCom. rZAPs are computed with independent randomness. We stress out that the determining message for $\mathcal{V}$ is the first prover's message: $\mathsf{dm} = (c_{r_0}, c_{r_1}, c_s)$. The determining message for $\mathcal{P}$ is the first verifier's message: (c_0, c_1).

Sub-protocol: $\pi_d = \langle \pi_d^{\mathcal{P}}(x, y, r_d, s_d), \pi_d^{\mathcal{V}}(x, \mathsf{R}_{\mathsf{V}d}) \rangle$.

Inputs: common input: x ($\in$ HC). $\mathcal{P}$'s input: witness y for $\mathcal{R}_{\mathsf{HC}}$; witness (r_d, s_d) to prove rZAP's consistency theorem. $\mathcal{V}$'s input: randomness $\mathsf{R}_{\mathsf{V}d}$. Protocols BL [6] and resSZK [20] are used as sub-protocols.

- $\mathcal{V}$: Pick challenge for BL protocol: $ch_d \xleftarrow{\$} \{0,1\}^n$. Send $\mathsf{c}_d \leftarrow \mathsf{SHCom}(ch_d)$ to $\mathcal{P}$.
- $\mathcal{P}$: upon receiving c_d (this is the determining message for $\mathcal{P}$):
  1. Generate randomness $\mathsf{R}_{\mathsf{P}d} \leftarrow f_{r_d}(x||\mathsf{c}_d)$.
  2. Compute the step BL1 for the instance x using randomness $\mathsf{R}_{\mathsf{P}d}$. Let us denote the output as $\mathsf{mBL1}^d$.
  3. Send $\mathsf{mBL1}^d$ to $\mathcal{V}$ along with the rZAP for theorem: $((x, \mathsf{mBL1}^d, \mathsf{c}_{r_d}, \mathsf{c}_d) \in \Lambda_{\mathsf{BL1}} \vee (\mathsf{c}_s, d) \in \Lambda_{\mathsf{trap}})$.
- $\mathcal{V}$: if rZAP is accepting send ch_d to $\mathcal{P}$.

  Prove theorem $(\mathsf{c}_d, ch_d) \in \Lambda_{\mathsf{SHCom}}$ using resSZK protocol. Let $m_{\mathsf{P}_{\mathsf{rszk}}}^d$ be the prover's message of sub-protocol resSZK (sent by $\mathcal{V}$ to $\mathcal{P}$) and $m_{\mathsf{V}_{\mathsf{rszk}}}^d$ be the verifier's message of resSZK (sent by $\mathcal{P}$ to $\mathcal{V}$):
  1. $(\mathcal{P} \to \mathcal{V})$ at each round of the protocol resSZK, upon receiving $m_{\mathsf{P}_{\mathsf{rszk}}}^d$ from $\mathcal{V}$, $\mathcal{P}$ computes $m_{\mathsf{V}_{\mathsf{rszk}}}^d$ using randomness $\mathsf{R}_{\mathsf{P}d}$ and sends $m_{\mathsf{V}_{\mathsf{rszk}}}^d$ to $\mathcal{V}$ along with an rZAP for the theorem $((m_{\mathsf{P}_{\mathsf{rszk}}}^d, m_{\mathsf{V}_{\mathsf{rszk}}}^d, \mathsf{c}_{r_d}, \mathsf{c}_d) \in \Lambda_{\mathsf{V}} \vee (\mathsf{c}_s, d) \in \Lambda_{\mathsf{trap}})$.
  2. $(\mathcal{V} \to \mathcal{P})$ at each round of the protocol resSZK upon receiving $m_{\mathsf{V}_{\mathsf{rszk}}}^d$ from $\mathcal{P}$, if rZAP is accepting $\mathcal{V}$ computes the next resSZK's prover message and sends it to $\mathcal{P}$. Otherwise it aborts.
- $\mathcal{P}$: upon successfully completing the resSZK protocol compute step BL3 and send the message $\mathsf{mBL3}^d$ to $\mathcal{V}$.
- If $\mathsf{mBL3}^d$ is the correct third message of BL protocol $\mathcal{V}$ outputs accept, else outputs abort.

Fig. 2. Sub-protocol $\pi_d = (\pi_d^{\mathcal{P}}(\cdot), \pi_d^{\mathcal{V}}(\cdot))$

2.2 Security Proof

In this section we provide the high-level proof of the simultaneous resettable witness indistinguishability property and the resettable argument of knowledge property of the protocol depicted in Fig. 1.

Resettable-soundness. Towards showing resettable soundness we start with the following observations. Recall that by dm we denote the determining message sent by $\mathcal{P}^*$ in the first round consisting of the commitment of two random seeds and the commitment of a bit (let us call the bit committed b).

1. The randomness used by $\mathcal{V}$ depends on dm. In a resetting attack, malicious prover $\mathcal{P}^*$ activates $\mathcal{V}$ by selecting theorem and randomness, denoted by (x, j) which forces $\mathcal{V}$ to run with the same randomness r_j among several executions. However, the randomness actually used by $\mathcal{V}$ at each session is determined by the output of the PRF on seed r_j and input (x, dm). Thus, even if activated with the same random tape r_j, when receiving a new determining message, $\mathcal{V}$ executes the protocol with a fresh pseudo-random tape.

Note that, due to the computational indistinguishability of the PRF, soundness holds against a computationally bounded adversary.

2. In sub-protocol π_b, the resetting power of $\mathcal{P}^*$ is effective since $\mathcal{P}^*$ can honestly prove the trapdoor theorem of the rZAP. Therefore, $\mathcal{P}^*$ is not forced to use the randomness committed in the determining message among multiple resetting attacks. Specifically, $\mathcal{P}^*$ can mount the following attack. $\mathcal{P}^*$ initiates a session labelled by (x, j, dm). In the sub-protocol π_b, upon the reception of challenge ch_b from $\mathcal{V}$, $\mathcal{P}^*$ resets $\mathcal{V}$ (while keeping the *same* determining message) back to the second round (the point after $\mathcal{V}$ has sent the commitment of the challenge). Then, $\mathcal{P}^*$ changes the message $\mathsf{mBL1}^b$ according to the challenge ch_b previously seen. This is possible using the trapdoor theorem, therefore $\mathcal{P}^*$ does not need to stick with the randomness committed in the determining message. Since the determining message is the same as before the reset, $\mathcal{V}$ will use the same challenge in the sub-protocol π_b. Thus, in this sub-protocol, $\mathcal{P}^*$ can prove any theorem by obtaining the challenge in advance and thus π_b is not resettable sound.

3. In sub-protocol $\pi_{\bar{b}}$, the trapdoor theorem is always false, thus resetting $\mathcal{V}$ is ineffective. Indeed, in order to provide an accepting transcript, $\mathcal{P}^*$ must provide an rZAP that only exists when the "consistency" theorem is true, that is, each of $\mathcal{P}^*$'s message is honestly computed according to the randomness committed in the determining message. By the statistically binding property of SBCom (there exists only one opening for the commitments c_s and $c_{r_{\bar{b}}}$) and the soundness of rZAP (any unbounded $\mathcal{P}^*$ cannot prove a false theorem), $\mathcal{P}^*$ must be consistent with the randomness committed in the determining message. Therefore, $\pi_{\bar{b}}$ is resettably sound.

Assume that there exists a PPT malicious prover $\mathcal{P}^*$ and a pair (x, j) such that $\mathcal{V}$ accepts x with non-negligible probability for some $x \notin \mathsf{HC}$. By observation 1, such a transcript is indexed by determining message dm. Thus, the accepting transcript can be labelled by triple (x, j, dm). By observation 2, for the same determining message dm, there are polynomially many distinct transcripts for sub-protocol π_b ($\mathcal{P}^*$ can reset $\mathcal{V}$ polynomially many times and change the protocol messages). All these (partial) transcripts of π_b can be accepting for $x \notin \mathsf{HC}$ since soundness does not hold for π_b. However, by observation 3, for a fixed triple (x, r_j, dm), there exists only one possible accepting transcript for sub-protocol $\pi_{\bar{b}}$ since $\mathcal{P}^*$ is forced to honestly follow the BL protocol according to the randomness committed in the determining message. Therefore the soundness of BL is preserved when $\mathcal{P}^*$ resets $\mathcal{V}$ in $\pi_{\bar{b}}$. Since $\mathcal{V}$ accepts if and only if the executions of *both* sub-protocols are accepting, protocol simresWIAoK is resettably sound.

Resettable argument of knowledge. To prove resettable argument of knowledge we show an expected PPT extractor that extracts the witness from any malicious prover $\mathcal{P}^*$ with probability that is negligibly close to the probability that $\mathcal{P}^*$ convinces an honest verifier. Let (x, j, dm) be the label of the session in which $\mathcal{P}^*$ provides an accepting proof. The goal of the extractor is to obtain two accepting transcripts with the same BL1 message and two distinct challenges (for at least one sub-protocol) for the same label.

Our extractor consists of two phases. In the first phase it follows the honest verifier procedure. When $\mathcal{P}^*$ has completed its execution, if there exists an accepting session labeled by (x, j, dm) that we call "target session", the extractor proceeds to the second phase. In the second phase, the extractor obtains a distinct accepting transcript for the target session by cheating in the "opening" of the commitment by sending a challenge that is distinct from the one sent in the first phase and simulating the zero knowledge proof given by the verifier.

The crucial step of this phase is to detect the sub-protocol in which $\mathcal{P}^*$ is stuck with the randomness committed in dm and must follow the protocol honestly. Indeed, in such sub-protocol, the extractor can use the stand-alone simulator and open the statistically hiding commitment to any challenge. Note that the non-black-box simulator of the protocol resSZK takes as input the code of the malicious verifier. Thus, in order to use the simulator, the extractor must carefully prepare a machine which internally handles the interaction with $\mathcal{P}^*$ and forwards to the simulator only the messages belonging to the resSZK protocol played in one of the sub-protocol. One of the tasks of such machine is detecting the sub-protocol in which $\mathcal{P}$ is forced to be honest. Once the right sub-protocol has been detected, by the statistically-hiding property of SHCom, and by the statistical zero-knowledge property of protocol resSZK run by $\mathcal{V}$ instead of the opening, we are guaranteed that upon each rewind, $\mathcal{P}^*$ provides another accepting transcript for the target session with the same probability of the first phase. Finally, by the proof of knowledge property of Blum's protocol, collecting two distinct transcripts allows the extractor to compute the witness. The actual extractor requires an intermediate estimation step (as shown in [14]) in which the probability of having another accepting transcript for the label (x, j, dm) is estimated. More details on the formal description of the extractor, the augmented machine and the formal proof can be found in the full version of this work.

Resettable witness indistinguishability. Recall that the protocol mainly consists of a single message from $\mathcal{P}$ to $\mathcal{V}$, the determining message $(\mathsf{c}_{r_0}, \mathsf{c}_{r_1}, \mathsf{c}_s)$, and the parallel execution of π_0 and π_1. Such protocol can be seen as a parallel repetition of (Π_0, Π_1) where Π_b is the protocol π_b augmented with the message $(\mathsf{c}_s, \mathsf{c}_{r_b})$ sent from $\mathcal{P}$ to $\mathcal{V}$ and $b = 0, 1$.

Assume that there exists a resetting PPT distinguisher $\mathcal{V}^*$ for (Π_0, Π_1). That is, $\mathcal{V}^*$ distinguishes whether $\mathcal{P}$ runs *both* protocols using witnesses sampled from distribution $Y_0 = \{\bar{y}^0(\bar{x})\}_{\bar{x}}$ or from distribution $Y_1 = \{\bar{y}^1(\bar{x})\}_{\bar{x}}$. Let us denote by $H_{0,0}$ the experiment in which $\mathcal{P}$ uses witnesses sampled from Y_0 when running both protocols $(\Pi_b, \Pi_{\bar{b}})$, where b is the bit committed in c_s, and by $H_{1,1}$ the experiment in which $\mathcal{P}$ uses witnesses sampled from Y_1 in both $(\Pi_b, \Pi_{\bar{b}})$. We prove by hybrid arguments that experiments $H_{0,0}$ and $H_{1,1}$ are computationally indistinguishable. Let n denote the number of theorems and t the bound on the prover's random tapes. Consider the following hybrids.

$H_{1,0}$: In this hybrid, in each session, $\mathcal{P}$ uses witnesses sampled from Y_1 to run protocol Π_b and the bit b is committed in the determining message in such session. The only difference between experiment $H_{1,0}$ and $H_{0,0}$ is in the witness used in Π_b. Assume that there exists a distinguisher between hybrids $H_{0,0}$ and $H_{1,0}$ then it is possible to construct an adversary $\mathcal{V}^*_{\mathsf{BL}}$ for the $\mathcal{WI}$ property of sub-protocol BL of Π_b. Note that, when b is the bit committed in the determining message, the trapdoor theorem is true in Π_b. $\mathcal{V}^*_{\mathsf{BL}}$, on input $(\bar{x}, Y_0, Y_1)$, runs $\mathcal{V}^*$ as sub-routine and honestly executes the protocol $\Pi_{\bar{b}}$ using the witness belonging to Y_0. Instead for the execution of Π_b it forwards the messages received from $\mathcal{V}^*$ and belonging to BL protocol to the external prover, while it simulates the remaining messages belonging to Π_b. The first difficulty in such reduction seems to be the fact that $\mathcal{V}^*$ can mount a reset attack asking the prover of Π_b to run with the same randomness while changing the challenge of BL protocol. Instead, $\mathcal{V}^*_{\mathsf{BL}}$ can only mount a concurrent attack against the external BL's prover. Nevertheless, $\mathcal{V}^*_{\mathsf{BL}}$ can replicate the same attack of $\mathcal{V}^*$ for the following reasons. The randomness of the honest prover executing protocol Π_b is computed on the determining message (the commitment of BL's challenge) received from $\mathcal{V}^*$. Due to the pseudo-randomness of PRF, when $\mathcal{V}^*$ changes the determining message the prover of Π_b plays with fresh randomness. By the resettably-sound argument of knowledge property of the resSZK protocol and by the computational binding property of SHCom we have that $\mathcal{V}^*$ can not maintain the same determining message and query the prover with two distinct BL's challenges. Thus the resetting power is suppressed and $\mathcal{V}^*_{\mathsf{BL}}$ can replicate the same attack as $\mathcal{V}^*$. The second difficulty is that for each protocol message the honest prover of Π_b is required to send a rZAP proving that the messages are consistent with the randomness committed in the determining message. However, in the reduction $\mathcal{V}^*_{\mathsf{BL}}$ forwards the messages received by an external prover of BL's protocol, therefore it can not prove the consistency with the determine message. Nevertheless, since we are in the case in which the trapdoor theorem is true, $\mathcal{V}^*_{\mathsf{BL}}$ can forward the external messages and computes the rZAPs using the witness of the trapdoor theorem. Due to the resettable $\mathcal{WI}$ property of rZAP such deviation from the honest prover is not detected by any PPT $\mathcal{V}^*$. Then, by the $\mathcal{WI}$ of BL protocol hybrids $H_{0,0}$ and $H_{1,0}$ are computationally indistinguishable.

$H_{0,1}^{i,j}$ (with $1 \leq i \leq n$, $1 \leq j \leq t$): In hybrid $H_{0,1}^{i,j}$, in session (i,j), $\mathcal{P}$ runs protocol $\Pi_{\bar{b}}$ using the witness sampled from Y_1, while protocol Π_b is run by using a witness sampled from Y_0, and b is the bit committed in the determining message of such session. The only difference between experiment $H_{0,1}^{i,j}$ and $H_{0,1}^{i-1,j-1}$ is that in experiment $H_{0,1}^{i,j}$, in session (i,j), the witness is sampled from Y_1 in the sub-protocol where the trapdoor theorem is false. Note that $H_{0,1}^{0,0} = H_{1,0}$. Assume that there exists a distinguisher between $H_{0,1}^{i,j}$ and $H_{0,1}^{i-1,j-1}$ then it is possible to construct an adversary for the hiding of the commitment scheme SBCom. The reduction works as follows. $\mathcal{A}$ playing in the hiding experiment obtains the challenge commitment C. Then it runs $\mathcal{V}^*$ as sub-routine and simulates the honest prover $\mathcal{P}$ as in experiment $H_{0,1}^{i-1,j-1}$, except that in session (i,j) it proceeds as follows. It computes $\mathsf{c}_{r_0}, \mathsf{c}_{r_1}$ as the honest prover, while it sets $\mathsf{c}_s = C$, and sends the first round to $\mathcal{V}^*$. Then

$\mathcal{A}$ uniformly chooses a bit b and executes the protocol π_b using a witness sampled from distribution Y_1 and protocol $\pi_{\bar{b}}$ using the witness sampled from distribution Y_0. Note that $\mathcal{A}$ can run both sub-protocols without knowing the opening of C since also the honest $\mathcal{P}$ never uses such witness in the protocol execution. When $\mathcal{V}^*$ terminates its execution, $\mathcal{A}$ hands the output of $\mathcal{V}^*$ to the distinguisher and outputs whatever the distinguisher outputs. If C is a commitment of b then the experiment simulated by $\mathcal{A}$ is distributed identically to experiment $H_{0,1}^{i-1,j-1}$. Else if C is a commitment of $\bar{b}$ then the experiment is distributed as experiment $H_{0,1}^{i,j}$. By the computational hiding of SBCom we have that experiments $H_{0,1}^{i,j}$ and $H_{0,1}^{i-1,j-1}$ are computational indistinguishable.

$H_{1,1}$: In this hybrid, $\mathcal{P}$ uses a witness sampled from Y_1 to run protocol Π_b and the bit b is committed in the determining message. The only difference between experiment $H_{0,1}^{n,t}$ and experiment $H_{1,1}$ is in the witness used to run sub-protocol Π_b. By the same arguments put forth in proving the indistinguishability of hybrid $H_{1,0}$ and $H_{0,0}$, experiments $H_{0,1}^{n,t}$ and $H_{1,1}$ are computational indistinguishable. This completes the proof.

Theorem 1. *If trapdoor permutations and collision-resistant hash functions exist, then the protocol shown in Fig. 1 is a Simultaneously Resettable Witness Indistinguishable Argument of Knowledge.*

3   Application in the BPK Model

Here we show that by combining two instances of simres$\mathcal{WI}$AoK we obtain the first constant-round simultaneously resettable $\mathcal{ZK}$ AoK (simres$\mathcal{ZK}$AoK) in the BPK model.

High-level overview of protocol and proof. The construction is very simple since it takes advantage of the properties guaranteed by the protocol simres$\mathcal{WI}$AoK. We use it twice, once for a proof given by the verifier and once for a proof given by the prover. First, the verifier uses simres$\mathcal{WI}$AoK to prove knowledge of its secret key (one out of two possible sets of pre-images of a OWF), then the prover commits to its witness and finally uses simres$\mathcal{WI}$AoK to prove that the committed message is either a witness for the theorem $x \in L$ or a secret key. The intuition of why the protocol works is the following. First of all, the secret key of the verifier is protected by the one-wayness of the OWF, by the r$\mathcal{WI}$ property of the simres$\mathcal{WI}$AoK given by the verifier and by the resettable argument of knowledge of the simres$\mathcal{WI}$AoK given by the prover. Indeed, we will be able to prove that the witness extracted from the proof given by the prover can only be a witness for $x \in L$, otherwise we break either the hardness of the OWF or the r$\mathcal{WI}$ property of simres$\mathcal{WI}$AoK. Instead, the security for the prover comes from the existence of a simulator against any resetting verifier. Indeed, we can design a simulator as follows: the simulator starts a main thread that is always updated with new messages until the simulator is stuck. This event happens when the simulator is supposed to commit to a witness and to then play the second simres$\mathcal{WI}$AoK. At this point, the simulator suspends the main thread and starts some rewinding threads in order to extract the secret key used by the adversarial verifier in that session. Once this is done, the simulator continues the main thread since it is not stuck anymore (i.e., it can simply commit to the extracted secret and use it as witness in the second simres$\mathcal{WI}$AoK). Since the number of identities of possible verifiers in the BPK model is polynomially bounded, we have that the simulator has to start only an expected polynomial number of rewinding threads, and thus its expected running time is polynomial. The indistinguishability of the view comes from the hiding of the commitment scheme and the r$\mathcal{WI}$ property of the second simres$\mathcal{WI}$AoK. Instead the resettable argument of knowledge of the first simres$\mathcal{WI}$AoK (i.e., the one given by the verifier) is helpful for guaranteeing the expected running time of the simulator. The commitment played in between the two executions of the simres$\mathcal{WI}$AoK plays an important role in breaking a possible malleability attack of the malicious sender.

The formal description of the protocol is provided in Fig. 3. For underlying primitives, we use a non-interactive statistically binding commitment scheme, denoted by SBCom, and a one-way function $g : \{0,1\}^* \to \{0,1\}^*$. In the protocol we use the following two **NP** relations: 1) a pair $((y,g),x) \in \mathcal{R}_{\Lambda_{\text{ow}}}$ if x is such that $y = g(x)$; 2) a pair $((c,m),r) \in \mathcal{R}_{\text{SBCom}}$ if the string r is such that $c = \text{SBCom}(m,r)$.

Theorem 2. *If trapdoor permutations and collision-resistant hash functions exist, then protocol simres$\mathcal{ZK}$AoK is a constant-round simultaneously resettable zero-knowledge argument of knowledge in the BPK model.*

For lack of space, the formal proof can be found in the full version of this paper.

4   Simultaneously Resettable Identification Schemes

In this section, we present the second application of our main protocol, the first construction of a simultaneously resettable identification scheme. Identification schemes represent one of the most successful practical applications of cryptographic protocols. The basic goal of an identification scheme is to prevent an adversary $\mathcal{A}$ from impersonating a honest user $\mathcal{P}$ to another honest user $\mathcal{V}$. However, this is not sufficient for some applications. Indeed, consider the case in which $\mathcal{V}$ provides a service to $\mathcal{P}$, and the service is restricted only to a small community controlled by $\mathcal{V}$. Then, $\mathcal{P}$ could give to another party T that is not in the small community, some partial information about his secret that is sufficient for T to obtain the service from $\mathcal{V}$, while still T does not know $\mathcal{P}$'s secret. The proof of knowledge property allows us to do secure identification as well as preventing the attack described above. When the identification protocol is a proof of knowledge, the sole fact that T convinces $\mathcal{V}$ is sufficient to claim that one can extract the whole secret from T. This implies that T obtained $\mathcal{P}$'s secret key corresponding to his identity, and this is unlikely to happen in scenarios where the

---

Protocol simres$\mathcal{ZK}$AoK

Ingredients: One-way function g, statistically binding commitment scheme SBCom, sub-protocol simres$\mathcal{WI}$AoK.

Key-Registration Phase:
$\mathcal{V}$ chooses a pair of secrets (sk_0, sk_1) where $sk_b \in \{0,1\}^n$ and $b \in \{0,1\}$. Then $\mathcal{V}$ generates the corresponding public key (pk_0, pk_1) such that $pk_b = g(sk_b)$ for $b \in \{0,1\}$. $\mathcal{V}$ publishes (pk_0, pk_1) in public file F and stores sk_b as its secret trapdoor information with $b \xleftarrow{\$} \{0,1\}$. We assume that the i-th verifier $\mathcal{V}$ has public key (pk_0^i, pk_1^i) and secret key sk_b^i.

Main-Execution Phase:
Common input: NP-statement $x \in L$ and the verifier's identity i. Hence, prover $\mathcal{P}$ knows public key (pk_0^i, pk_1^i) in F, chosen by $\mathcal{V}$.
Input for $\mathcal{P}$: Witness w such that $(x, w) \in \mathcal{R}_L$ and randomness r_P.
Input for $\mathcal{V}$: Randomness r_V, secret key sk_b^i.

- $\mathcal{P}$: Obtain a sufficiently long pseudo-random tape $r_P' \leftarrow f_{r_P}(x||pk_0^i||pk_1^i)$. From now on, $\mathcal{P}$ uses r_P' for the execution in the rest of protocol. For convenience, we assume that r_P consists of four partitions, $r_P'(1)$, $r_P'(2)$, $r_P'(3)$ and $r_P'(4)$.
- $(\mathcal{V} \rightarrow \mathcal{P})$: $\mathcal{V}$ proves, by using simres$\mathcal{WI}$AoK, the following statement: There exists sk_b^i such that $((pk_0^i, g), sk_b^i) \in \mathcal{R}_{\Lambda_{\text{ow}}} \vee ((pk_1^i, g), sk_b^i) \in \mathcal{R}_{\Lambda_{\text{ow}}}$. For the execution of simres$\mathcal{WI}$AoK, $\mathcal{P}$ uses random tape $r_P'(1)$.
- $(\mathcal{P} \rightarrow \mathcal{V})$: If the above proof is rejecting, then $\mathcal{P}$ aborts. Otherwise, $\mathcal{P}$ commits to w and 0^n as $c_0 \leftarrow$ SBCom$(w, r_P'(2))$ and $c_1 \leftarrow$ SBCom$(0^n, r_P'(3))$. Then, $\mathcal{P}$ sends c_0 and c_1 to $\mathcal{V}$.
- $(\mathcal{P} \rightarrow \mathcal{V})$: $\mathcal{P}$ by using simres$\mathcal{WI}$AoK and random tape $r_P'(4)$ proves to $\mathcal{V}$ the following statements:
  1. $\exists\,(w, r)$ such that $(x, w) \in \mathcal{R}_L \wedge ((c_0, w), r) \in \mathcal{R}_{\text{SBCom}}$ **OR**
  2. $\exists\,(sk, r)$ such that $((pk_0^i, g), sk) \in \mathcal{R}_{\Lambda_{\text{ow}}} \wedge ((c_1, sk), r) \in \mathcal{R}_{\text{SBCom}}$ **OR**
  3. $\exists\,(sk, r)$ such that $((pk_1^i, g), sk) \in \mathcal{R}_{\Lambda_{\text{ow}}} \wedge ((c_1, sk), r) \in \mathcal{R}_{\text{SBCom}}$.
- $\mathcal{V}$: output "accept" if and only if the proof provided by $\mathcal{P}$ is accepting.

Fig. 3. Constant-Round Simultaneously Resettable $\mathcal{ZK}$AoK in the BPK Model

same secret key is used for other critical tasks such as digital signatures. As discussed in the introduction, our simultaneously resettable identification scheme follows the above proof of knowledge paradigm. This extends the previous work of Bellare et al. [5] to a setting in which every party can be reset. We emphasize that our simultaneously resettable identification scheme is easily obtained from our main protocol simres$\mathcal{WI}$AoK, so achieving a constant round complexity.

Identification protocols secure against reset attacks. We introduce the notion of Reset-Reset-1 security as a generalization of the Concurrent-Reset-1 CR1 notion introduced in [5]. CR1 considers an adversary I, called impersonator, that plays in two phases. In the first phase, it interacts with a prover as a resetting verifier (Reset phase). In the second phase, it has no access to the prover anymore, but it tries to impersonate such a prover to an honest verifier (Concurrent phase). In the second phase, I is not allowed to reset the verifier. In our new definition

Reset-Reset-1 (RR1) the impersonator is allowed to reset in both phases. The formal definition is a straightforward extension of the one given in [5] and can be found in the full version of this work.

The protocol $\mathcal{ID}$. Let $f : \{0,1\}^n \to \{0,1\}^*$ be a one-way function, let n be the security parameter. The public key of $\mathcal{P}$ is the pair $(\mathsf{pk_0}, \mathsf{pk_1})$, the secret key is x_d for a randomly chosen bit d, such that $\mathsf{pk_0} = f(x_d) \vee \mathsf{pk_1} = f(x_d)$. The protocol simply consists in $\mathcal{P}$ running the simres$\mathcal{WI}$AoK protocol with $\mathcal{V}$ to prove that it *knows* the preimage of either $\mathsf{pk_0}$ or $\mathsf{pk_1}$. Formally, let $\Lambda_{\mathcal{ID}}$ be the following language $\Lambda_{\mathcal{ID}} = \{(y_0, y_1)\colon$ there exists $x \in \{0,1\}^n$ s.t. $y_0 = f(x) \vee y_1 = f(x)\}$, then the identification scheme consists of $\mathcal{P}$ proving the statement $(\mathsf{pk_0}, \mathsf{pk_1}) \in \Lambda_{\mathcal{ID}}$ using simres$\mathcal{WI}$AoK.

Theorem 3. *If a constant-round simultaneously resettable $\mathcal{WI}$AoK protocol exists and one-way functions exist, then the above protocol is constant-round and secure in the RR1 setting.*

Proof. Let $pk = (\mathsf{pk_0}, \mathsf{pk_1})$ be the public key of a player $\mathcal{P}$. Assume that there exists a PPT adversary I playing the RR1 experiment, that succeeds in impersonating an honest $\mathcal{P}$ with non-negligible probability. This means that I is able to prove to an honest $\mathcal{V}$ that her identity is $pk = (\mathsf{pk_0}, \mathsf{pk_1})$. Then we show that I can be used to construct an adversary against the one-wayness of f, or a distinguisher for the resettable $\mathcal{WI}$ property of the simres$\mathcal{WI}$AoK protocol. The resettable argument of knowledge property of simres$\mathcal{WI}$AoK protocol is crucial to put forth both reductions.

Recall that, in the RR1 game, I plays the first phase interacting as a resetting verifier $\mathcal{V}^*$ with $\mathcal{P}$ and in the second phase interacts as a resetting prover $\mathcal{P}^*$ with $\mathcal{V}$ trying to impersonate $\mathcal{P}$.

First we show an adversary $\mathcal{A}$ that breaks the one-wayness of f. $\mathcal{A}$ has in input a challenge y that is the output of $f(x)$ for some unknown x. The reduction works as follows. $\mathcal{A}$ picks $d \in \{0,1\}$, $x_d \in \{0,1\}^n$ and computes $\mathsf{pk_d} = f(x_d)$ and $\mathsf{pk_{\bar d}} = y$. Then it runs I as subroutine, in the first phase $\mathcal{A}$ simulates the honest prover playing the simres$\mathcal{WI}$AoK protocol with witness x_d. In the second phase, $\mathcal{A}$ simulates the honest verifier to I. If I provides an accepting proof, then $\mathcal{A}$ runs the extractor of the simres$\mathcal{WI}$AoK protocol and, by the resettable argument of knowledge property, except with negligible probability, it obtains the witness used by I in the proof. In order to run the extractor, $\mathcal{A}$ prepares an augmented machine that internally contains all messages belonging to the first phase so that they can be internally played with I, while the messages sent by I in the second phase are forwarded to the extractor. Now note that during the extraction process the extractor rewinds the machine several times changing the protocol messages (of the second phase), therefore I could change her messages accordingly. Note that however, since there is a separation between the first phase and the second phase, this does not require to re-play messages of the first phase. Since, by assumption f is a one-way function, the probability that the witness extracted corresponds to a pre-image of y is negligible.

Now, assume that the witness extracted from I is x_d. Then we can construct a distinguisher $\mathcal{A}_{\mathcal{WI}}$ for the resettable witness indistinguishability property of simres$\mathcal{WI}$AoK. $\mathcal{A}_{\mathcal{WI}}$ works as follows. It computes $\mathsf{pk}_0 = f(x_0), \mathsf{pk}_1 = f(x_1)$ and activates an external prover for the simres$\mathcal{WI}$AoK protocol with inputs $((\mathsf{pk}_0, \mathsf{pk}_1), (x_0, x_1))$. In the first phase, when I runs as a verifier, $\mathcal{A}_{\mathcal{WI}}$ forwards all messages to the external prover of the simres$\mathcal{WI}$AoK. In the second phase, when I runs as a prover, $\mathcal{A}_{\mathcal{WI}}$ follows the procedure of the honest verifier. Then, if I provides an accepting proof, then $\mathcal{A}_{\mathcal{WI}}$ runs the extractor of the simres$\mathcal{WI}$AoK protocol. Finally by the resettable argument of knowledge property, except with negligible probability, it obtains the witness used by I in the proof, i.e. it obtains x_0 or x_1. Now notice that in the previous experiment, when we tried to invert the one-way function, the witness extracted corresponded to the one used in the first phase, while I was verifying the proof. Since this second experiment is identical to the previous one, it is again true that the extracted witness corresponds to the one used by the prover. Since the prover now is the external prover of simres$\mathcal{WI}$AoK, we have that the above adversary $\mathcal{A}_{\mathcal{WI}}$ breaks the r$\mathcal{WI}$ property of simres$\mathcal{WI}$AoK. By the r$\mathcal{WI}$ property of simres$\mathcal{WI}$AoK, this event happens with negligible probability only and thus I wins the RR1 game with negligible probability.

Acknowledgments. Research supported in part by NSF grants 0830803, 09165174, 1065276, 1118126 and 1136174, US-Israel BSF grant 2008411, B. John Garrick Foundation, IBM, OKAWA Foundation, Lockheed-Martin Corporation and the Defense Advanced Research Projects Agency through the U.S. Office of Naval Research under Contract N00014-11-1-0392. The views expressed are those of the authors and do not reflect the official policy or position of the Department of Defense or the U.S. Government. The work of the third and fourth authors has been done while visiting UCLA and is supported in part by the European Commission through the FP7 programme under contract 216676 ECRYPT II.

References

1. Alwen, J., Persiano, G., Visconti, I.: Impossibility and Feasibility Results for Zero Knowledge with Public Keys. In: Shoup, V. (ed.) CRYPTO 2005. LNCS, vol. 3621, pp. 135–151. Springer, Heidelberg (2005)
2. Arita, S.: A constant-round resettably-sound resettable zero-knowledge argument in the bpk model. Cryptology ePrint Archive, Report 2011/404 (2011), http://eprint.iacr.org/
3. Barak, B.: How to go beyond the black-box simulation barrier. In: FOCS, pp. 106–115 (2001)
4. Barak, B., Goldreich, O., Goldwasser, S., Lindell, Y.: Resettably-sound zero-knowledge and its applications. In: FOCS, pp. 116–125 (2001)
5. Bellare, M., Fischlin, M., Goldwasser, S., Micali, S.: Identification Protocols Secure against Reset Attacks. In: Pfitzmann, B. (ed.) EUROCRYPT 2001. LNCS, vol. 2045, pp. 495–511. Springer, Heidelberg (2001)
6. Blum, M.: How to Prove a Theorem So No One Else Can Claim It. In: Proceedings of the International Congress of Mathematicians, pp. 1444–1451 (1986)

7. Canetti, R., Goldreich, O., Goldwasser, S., Micali, S.: Resettable zero-knowledge (extended abstract). In: STOC, pp. 235–244 (2000)
8. Deng, Y., Feng, D., Goyal, V., Lin, D., Sahai, A., Yung, M.: Resettable Cryptography in Constant Rounds – The Case of Zero Knowledge. In: Lee, D.H. (ed.) ASIACRYPT 2011. LNCS, vol. 7073, pp. 390–406. Springer, Heidelberg (2011)
9. Deng, Y., Goyal, V., Sahai, A.: Resolving the simultaneous resettability conjecture and a new non-black-box simulation strategy. In: FOCS, pp. 251–260. IEEE Computer Society (2009)
10. Di Crescenzo, G., Persiano, G., Visconti, I.: Constant-Round Resettable Zero Knowledge with Concurrent Soundness in the Bare Public-Key Model. In: Franklin, M. (ed.) CRYPTO 2004. LNCS, vol. 3152, pp. 237–253. Springer, Heidelberg (2004)
11. Dwork, C., Naor, M., Sahai, A.: Concurrent zero-knowledge. In: Proceedings of the 20th Annual ACM Symposium on Theory of Computing, STOC 1998, pp. 409–418. ACM (1998)
12. Feige, U., Lapidot, D., Shamir, A.: Multiple noninteractive zero knowledge proofs under general assumptions. SIAM J. Comput. 29(1), 1–28 (1999)
13. Garg, S., Ostrovsky, R., Visconti, I., Wadia, A.: Resettable Statistical Zero Knowledge. In: Cramer, R. (ed.) TCC 2012. LNCS, vol. 7194, pp. 494–511. Springer, Heidelberg (2012)
14. Goldreich, O., Kahan, A.: How to construct constant-round zero-knowledge proof systems for np. J. Cryptology 9(3), 167–190 (1996)
15. Goldwasser, S., Micali, S., Rackoff, C.: The knowledge complexity of interactive proof-systems. In: Proceedings of the 17th Annual ACM Symposium on Theory of Computing, STOC 1985, pp. 291–304. ACM (1985)
16. Goyal, V., Maji, H.K.: Stateless cryptographic protocols. In: FOCS 2011 (2011)
17. Kilian, J., Petrank, E.: Concurrent and resettable zero-knowledge in polylogarithmic rounds. In: Proceedings of the 33rd Annual ACM Symposium on Theory of Computing, STOC 2001, pp. 560–569. ACM (2001)
18. Kilian, J., Petrank, E., Rackoff, C.: Lower bounds for zero knowledge on the internet. In: Proceedings of 39th IEEE Conference on the Foundations of Computer Science, FOCS 1998, pp. 484–492 (1998)
19. Micali, S., Reyzin, L.: Soundness in the Public-Key Model. In: Kilian, J. (ed.) CRYPTO 2001. LNCS, vol. 2139, pp. 542–565. Springer, Heidelberg (2001)
20. Pass, R., Rosen, A.: New and improved constructions of non-malleable cryptographic protocols. In: Proceedings of the 37th Annual ACM Symposium on Theory of Computing, STOC 2005, pp. 533–542. ACM (2005)
21. Scafuro, A., Visconti, I.: On round-optimal zero knowledge in the bare public-key model. In: EUROCRYPT. LNCS. Springer, Heidelberg (2012)
22. Yung, M., Zhao, Y.: Generic and Practical Resettable Zero-Knowledge in the Bare Public-Key Model. In: Naor, M. (ed.) EUROCRYPT 2007. LNCS, vol. 4515, pp. 129–147. Springer, Heidelberg (2007)

Subspace LWE

Krzysztof Pietrzak[*]

IST Austria

Abstract. The (decisional) learning with errors problem (LWE) asks to distinguish "noisy" inner products of a secret vector with random vectors from uniform. The learning parities with noise problem (LPN) is the special case where the elements of the vectors are bits. In recent years, the LWE and LPN problems have found many applications in cryptography.

In this paper we introduce a (seemingly) much stronger *adaptive* assumption, called "subspace LWE" (SLWE), where the adversary can learn the inner product of the secret and random vectors after they were projected into an adaptively and adversarially chosen subspace. We prove that, surprisingly, the SLWE problem mapping into subspaces of dimension d is almost as hard as LWE using secrets of length d (the other direction is trivial.)

This result immediately implies that several existing cryptosystems whose security is based on the hardness of the LWE/LPN problems are provably secure in a much stronger sense than anticipated. As an illustrative example we show that the standard way of using LPN for symmetric CPA secure encryption is even secure against a very powerful class of related key attacks.

1 Introduction

The (search version of the) learning with errors problem (LWE) is specified by parameters $\ell, q \in \mathbb{N}$ and an error distribution χ over $\mathbb{Z}_q$. It asks to find a secret vector $\mathbf{s} \in \mathbb{Z}_q^\ell$ given any number of "noisy" inner products of $\mathbf{s}$ with random vectors. Formally, these products are samples from a distribution $\Lambda_{\chi,\ell}(\mathbf{s})$ over $\mathbb{Z}_q^{\ell+1}$ which is defined by sampling a uniform $\mathbf{r} \overset{\$}{\leftarrow} \mathbb{Z}_q^\ell$ and an error $e \leftarrow \chi$, and outputting $(\mathbf{r}, \mathbf{r}^\mathsf{T}\mathbf{s} + e)$ (where multiplications and additions are all modulo q.)

An important special case of this problem is Regev's LWE problem [Reg05] where χ is a so called discrete Gaussian distribution and q is polynomial or exponential in a security parameter. Another important case is the learning parities with noise problem (LPN) where $q = 2$.

The decisional version of the LWE problem asks to *distinguish* samples of the form $\Lambda_{\chi,\ell}(\mathbf{s})$ from uniform (which might be easier than to actually output $\mathbf{s}$ as required by the computational version of the problem). The decisional LWE

---

[*] Supported by the European Research Council under the European Union's Seventh Framework Programme (FP7/2007-2013) / ERC Starting Grant (259668-PSPC).

R. Cramer (Ed.): TCC 2012, LNCS 7194, pp. 548–563, 2012.

problem has been proven polynomially equivalent to the computational version if q is prime [Reg05], and in particular for LPN [BFKL94, KS06]. In this paper we will always consider the decisional version of the problem, and we also only prove the main result for the case where q is prime.

Regev's LWE. The LWE problem has proven to be extremely useful to construct cryptographic schemes. One reason is its versatility, pretty much any cryptographic primitive known to date can be based on LWE. Another reason is its hardness. The best known algorithms against Regev's LWE (where χ is a discrete Gaussian and $q = poly(\ell)$) need time and space $2^{\Theta(\ell)}$ [BKW00] to recover $\mathbf{s} \in \mathbb{Z}_q^\ell$,[1] and unlike for most other assumptions on which public-key crypto can be based, no faster quantum algorithms for the problem are known. But most strikingly, Regev's LWE is as hard as *worst-case* (standard) lattice assumptions [Reg05, Pei09].

An incomplete list of cryptosystems whose security can be reduced to LWE is public-key encryption secure against chosen plaintext [Reg05, KTX08, PVW08] and chosen ciphertext attacks [PW08, Pei09], circular-secure encryption [ACPS09], identity-based encryption [GPV08, CHKP10, ABB10a, ABB10b], oblivious transfer [PVW08], collision-resistant hash functions [PR06, LMPR08] and public-key identification schemes [Lyu08, Lyu09].

LPN. The learning parity with noise (LPN) problem [BFKL94, BKW00, Kea93] is the special case of the LWE problem where $q = 2$ (i.e. we work over bits) and the error distribution is the Bernoulli distribution for some constant parameter $\tau, 0 < \tau < 1/2$, denoted Ber_τ, and defined as $\Pr[x = 1 ; x \leftarrow \mathrm{Ber}_\tau] = \tau$. The LPN problem is closely related to the problem of decoding random linear codes,[2] a well studied question in coding theory. The LPN problem seems less versatile than the general LWE problem, and so far only "minicrypt" primitives (i.e. primitives known to be equivalent to one-way functions) were constructed under the LPN assumption. Alekhnovich [Ale03] constructs a public-key encryption scheme from a relaxed LPN assumption where the error τ is not constant but upper bounded as a function of ℓ as $\tau = O(1/\sqrt{\ell})$.

The Appeal of the LPN problem comes from the fact that LPN based schemes can be extremely efficient, just requiring relatively few bit-level operations to compute an inner product of two bit-vectors. Constructions from LPN include PRGs [FS96] and encryption schemes [GRS08, ACPS09] and public-key authentication schemes [Ste94], but by far most work has been done on efficient LPN based authentication schemes which we'll discuss in more detail in Section 4.

Subspace LWE. The LWE problem has been shown to be very robust with respect to *leakage*. Distinguishing LWE samples remains hard even if we the adversary can learn a function $f(\mathbf{s})$ about the secret $\mathbf{s}$ as long as $f(.)$ is compressing

---

[1] This is slightly better than a trivial brute-force search which takes time $\approx 2^{\ell \log q} = 2^{\Theta(\ell \log \ell)}$ but only linear space.

[2] The only difference is that in the decoding problem one is given a fixed number of samples (typically a small multiple of the length of the secret), whereas in the LPN problem the adversary can ask for arbitrary many samples.

[AGV09] or hard to invert [DKL09, DGK+10, GKPV10]. In this paper we show that the LWE problem is also very robust to *tampering* with the secret vector $\mathbf{s}$ and the randomness vector $\mathbf{r}$ (albeit not with the noise e.)

We define a (seemingly) much stronger *adaptive* version of LWE which we call "Subspace LWE", or SLWE for short. In the SLWE problem the adversary is not restricted to just ask for samples $\mathbf{r}, \mathbf{r}^\mathsf{T}.\mathbf{s} + e$ from $\Lambda_{\chi,\ell}(\mathbf{s})$ as in LWE, but has access to a more powerful oracle which she can query adaptively. The oracle takes as input the description of two affine mappings $\phi_r, \phi_s : \mathbb{Z}_q^\ell \to \mathbb{Z}_q^\ell$ and outputs a sample

$$\mathbf{r}, \phi_r(\mathbf{r})^\mathsf{T}.\phi_s(\mathbf{s}) + e \quad \text{where} \quad \mathbf{r} \xleftarrow{\$} \mathbb{Z}_q^\ell \ , \ e \leftarrow \chi$$

An affine mapping $\phi_r : \mathbb{Z}_q^\ell \to \mathbb{Z}_q^\ell$ (similarly ϕ_s) is given by a matrix and a vector $\phi_r = [\mathbf{X}_r \in \mathbb{Z}_q^{\ell \times \ell}, \mathbf{x}_r \in \mathbb{Z}_q^\ell]$ and its evaluation is defined as

$$\phi_r(\mathbf{r}) = \mathbf{X}_r.\mathbf{r} + \mathbf{x}_r$$

Without additional restrictions, the SLWE problem as just defined is easy to break. By choosing the input to the oracle appropriately,[3] one can e.g. learn samples of the form $\mathbf{s}[i] + e$, $e \leftarrow \chi$ ($\mathbf{s}[i]$ denotes the ith element of $\mathbf{s}$.) For distributions χ as used in LPN or Regev's LWE one can efficiently learn $\mathbf{s}[i]$ (and thus the entire $\mathbf{s}$) from just a few such samples.

We prove that the SLWE problem (using secrets in $\mathbb{Z}_q^\ell$ and error distribution χ) is almost as hard as the standard (q, χ, d)-LWE problem with secrets of length $d \leq \ell$ if the adversary is restricted in the sense that she is only allowed to query ϕ_r, ϕ_s which "overlap" in an $d + \delta$ (or more) dimensional subspace where $\delta \in \mathbb{N}$ is a statistical security parameter. Formally this means $\mathbf{X}_r^\mathsf{T}.\mathbf{X}_s$ must have rank at least $d + \delta$. We call this the $(q, \chi, \ell, d + \delta)$-SLWE problem. Let us mention that the other direction – showing that (q, χ, ℓ, d)-SLWE is at most as hard as (q, χ, d)-LWE – is trivial.

The precise statement of our result asserts that for any $\ell, d, \delta \in \mathbb{N}, d + \delta \leq \ell$, the $(q, \chi, \ell, d + \delta)$-SLWE problem is at most an additive term $2/q^{\delta+1}$ easier than the standard (q, χ, d)-LWE problem. For large fields, where q is superpolynomial, $2/q^{\delta+1}$ is negligible already for $\delta = 0$. For small fields, in particular the important case $q = 2$ as used in LPN, we must choose some δ to be a statistical security parameter.

The above formulation of SLWE is somewhat redundant, in the sense that an adversary who is restricted to always choose ϕ_s to be the identity function, is as powerful (i.e. can learn exactly the same distribution from the SLWE oracle) as the adversary described above. We chose to explicitly allow the adversary to choose affine mappings for the randomness and the secret separately, as for the applications it is sometimes more convenient to think of the adversary being able to apply a mapping to the secret key (like in the setting of related key attacks we'll discuss), or to the randomness (e.g. to show that LWE is hard, even if the randomness comes from a bit-fixing source.)

---

[3] Set $\mathbf{x}_r = \mathbf{x}_s = 0^\ell$ and $\mathbf{X}_r, \mathbf{X}_s$ the zero matrix with a single one in the ith diagonal element. The oracle will output $\mathbf{r}, \mathbf{r}[i]\mathbf{s}[i] + e$, the last element is $\mathbf{s}[i] + e$ if $\mathbf{r}[i] = 1$.

When q is not Prime. The reduction from SLWE to LWE assumes that q is prime, as we use the fact that $\mathbb{Z}_q$ is a field and $\mathbb{Z}_q^m$ is a vector space.[4] We believe that the proof of the reduction can be adapted to the case where q is composite.

The case where q is prime covers the cryptographically most interesting cases of LPN and Regev's LWE. Also the reduction from the search to decision version of LWE [Reg05] only works for prime q (of polynomial size.) But the case where q is not prime has found cryptographic applications too. In particular, the case where $q = p^e$ for a prime p and $e > 1$ has been used in the construction of an encryption scheme with circular security [ACPS09]. The case where q is a product of distinct, small primes has been used in [Pei09].

Applications of SLWE. In Section 4 we'll discuss some applications of the SLWE problem. In particular, the fact that SLWE is equivalent to LWE implies stronger security notions – like security against related-key attacks – that one can give for existing schemes whose security is reduced to the LWE problem. In subsequent work, the hardness of SLPN has been used to construct efficient authentication schemes and even MACs from LPN. These schemes differ significantly from previous schemes which all were extensions of the Hopper-Blum protocol.

Outline. In Section 2 we first define the LWE and the new subspace LWE (SLWE) problem. In Section 3 we state and prove our main technical result (Theorem 1) which bounds the hardness of the SLWE problem in terms of the hardness of he standard LWE problem. In Section 4 we describe in more detail some applications of this result which were already mentioned in the introduction.

2   Hard Learning Problems

2.1   Notation

We denote the set of integers modulo an integer $q \geq 1$ by $\mathbb{Z}_q$. We will use normal, bold and capital bold letters like x, $\mathbf{x}$, $\mathbf{X}$ to denote single elements, vectors and matrices over $\mathbb{Z}_q$, respectively. For $\mathbf{x} \in \mathbb{Z}_q^\ell$, $|\mathbf{x}| = \ell$ denotes the length and $\mathbf{wt}(\mathbf{x})$ denotes the Hamming weight of the vector $\mathbf{x}$, i.e. the number of indices $i \in \{1, \ldots, |\mathbf{x}|\}$ where $\mathbf{x}[i] \neq 0$. For $\mathbf{v} \in \mathbb{Z}_2^m$ we denote with $\overline{\mathbf{v}}$ its inverse, i.e. $\overline{\mathbf{v}}[i] = 1 - \mathbf{v}[i]$ for all i. For a distribution χ, $x \leftarrow \chi$ denotes sampling a value x with distribution χ. For a set $\mathcal{S}$, $x \xleftarrow{\$} \mathcal{S}$ denotes sampling a value x with the uniform distribution over $\mathcal{S}$.

$\mathbf{x}_{\downarrow \mathbf{v}}, \mathbf{X}_{\downarrow \mathbf{v}}$: For two vectors $\mathbf{v} \in \mathbb{Z}_2^\ell$ and $\mathbf{x} \in \mathbb{Z}_q^\ell$, we denote with $\mathbf{x}_{\downarrow \mathbf{v}}$ the vector (of length $\mathbf{wt}(\mathbf{v})$) which is derived from $\mathbf{x}$ by deleting all the bits $\mathbf{x}[i]$ where $\mathbf{v}[i] = 0$. If $\mathbf{X} \in \mathbb{Z}_q^{\ell \times m}$ is a matrix, then $\mathbf{X}_{\downarrow \mathbf{v}} \in \mathbb{Z}_q^{\mathbf{wt}(\mathbf{v}) \times m}$ denotes the submatrix we get by deleting the ith row if $\mathbf{v}[i] = 0$.

---

[4] The fact that $\mathbb{Z}_q^m$ is a vector space is e.g. used in the proof of Lemma 1.

$\mathbf{x_v}, \mathbf{X_v}$: For $\mathbf{x}, \mathbf{X}, \mathbf{v}$ as in the previous item, $\mathbf{x_v}$ denotes the vector where the ith entry is $\mathbf{x}[i] \wedge \mathbf{v}[i]$. Think of $\mathbf{x_v}$ as $\mathbf{x}$ where all entries of $\mathbf{x}$ where $\mathbf{v}$ is 0 are set to 0. $\mathbf{X_v}$ denotes the matrix $\mathbf{X}$ where the ith row is set to all 0 if $\mathbf{v}[i] = 0$.

2.2   The (Subspace) LWE Problem

The (search version of the) learning with errors (LWE) problem is specified by parameters $\ell, q \in \mathbb{N}$ and an error distribution χ over $\mathbb{Z}_q$. It asks to find a secret vector $\mathbf{s} \in \mathbb{Z}_q^\ell$ given any number of "noisy" inner products of $\mathbf{s}$ with random vectors.

Formally, let $\Lambda_{\chi,\ell}(\mathbf{s})$ be the distribution over $\mathbb{Z}_q^{\ell+1}$ where a sample is given by

$$(\mathbf{r}, \mathbf{r}^\mathsf{T}.\mathbf{s} + e) \leftarrow \Lambda_{\chi,\ell}(\mathbf{s}) \qquad \text{where} \qquad \mathbf{r} \overset{\$}{\leftarrow} \mathbb{Z}_q^\ell \ , \ e \leftarrow \chi$$

Let U_q^m denote the uniform distribution over $\mathbb{Z}_q^m$ and $U_q = U_q^1$. The decisional LWE problem asks to distinguish samples from $\Lambda_{\chi,\ell}(\mathbf{s})$ with a uniform $\mathbf{s}$ from a random oracle (outputting $U_q^{\ell+1}$ samples.) For any $\mathbf{s}$, $\Lambda_{U_q,\ell}(\mathbf{s})$ is the same as the uniform distribution $U_q^{\ell+1}$. It will be convenient for the proof to think of the random oracle as outputting samples from $\Lambda_{U_q,\ell}(\mathbf{s})$ for some random $\mathbf{s}$ instead of $U_q^{\ell+1}$.

Definition 1 (Decisional Learning with Errors Problem (LWE)). *The (decisional) (q, χ, ℓ)-LWE problem is (t, Q, ε) hard if for every distinguisher D running in time t and making Q oracle queries,*

$$\left| \Pr\left[\mathbf{s} \overset{\$}{\leftarrow} \mathbb{Z}_q^\ell \ : \ D^{\Lambda_{\chi,\ell}(\mathbf{s})} = 1 \right] - \underbrace{\Pr\left[\mathbf{s} \overset{\$}{\leftarrow} \mathbb{Z}_q^\ell \ : \ D^{\Lambda_{U_q,\ell}(\mathbf{s})} = 1 \right]}_{\Pr[D^{U_q^{\ell+1}} = 1]} \right| \leq \varepsilon. \qquad (1)$$

Usually one defines the LWE problem by considering a distinguisher who gets a polynomial number of samples as input and not access to an oracle (which doesn't take inputs anyway.) We use this oracle based definition so it is more similar to the SLWE problem we define below, where the oracle does take adaptively chosen inputs.

An affine projection $\phi : \mathbb{Z}_q^\ell \to \mathbb{Z}_q^\ell$ is given by a matrix/vector tuple $\mathbf{X} \in \mathbb{Z}_q^{\ell \times \ell}, \mathbf{x} \in \mathbb{Z}_q^\ell$ and defined as $\phi(\mathbf{v}) \overset{\text{def}}{=} \mathbf{X}^\mathsf{T}\mathbf{v} + \mathbf{x}$.

For $\mathbf{s} \in \mathbb{Z}_q^{\ell \times \ell}$ and affine projections $\phi_r = [\mathbf{X}_r, \mathbf{x}_r]$, $\phi_s = [\mathbf{X}_s, \mathbf{x}_s]$ we define the distribution $\Gamma_{\chi,\ell,d}(\mathbf{s}, \phi_r, \phi_s)$ over $\mathbb{Z}_q^{\ell+1} \cup \perp$ as

$$\perp \ \leftarrow \ \Gamma_{\chi,\ell,d}(\mathbf{s}, \phi_r, \phi_s) \qquad \text{if} \qquad \mathsf{rank}(\mathbf{X}_r^\mathsf{T}\mathbf{X}_s) < d$$

and

$$[\ \mathbf{r} \ , \ \phi_r(\mathbf{r})^\mathsf{T}.\phi_s(\mathbf{s}) + e \ ] \leftarrow \Gamma_{\chi,\ell,d}(\mathbf{s}, \phi_r, \phi_s) \qquad \text{where} \qquad \mathbf{r} \overset{\$}{\leftarrow} \mathbb{Z}_q^\ell \ , \ e \leftarrow \chi$$

otherwise. With $\Gamma_{\chi,\ell,d}(\mathbf{s}, .)$ we denote the oracle which on input ϕ_r, ϕ_s outputs a sample of $\Gamma_{\chi,\ell,d}(\mathbf{s}, \phi_r, \phi_s)$.

Definition 2 (Subspace Learning with Errors Problem (SLWE)). *The (decisional) (q, χ, ℓ, d)-SLWE problem is (t, Q, ε) hard if for every distinguisher D running in time t and making Q oracle queries,*

$$\left| \Pr\left[\mathbf{s} \xleftarrow{\$} \mathbb{Z}_q^\ell \ : \ D^{\Gamma_{\chi,\ell,d}(\mathbf{s},.)} = 1 \right] - \Pr\left[\mathbf{s} \xleftarrow{\$} \mathbb{Z}_q^\ell \ : \ D^{\Gamma_{U_q,\ell,d}(\mathbf{s},.)} = 1 \right] \right| \leq \varepsilon. \quad (2)$$

Note that by definition the $\Gamma_{U_q,\ell,d}(\mathbf{s},)$ oracle outputs $\perp$ if the input satisfies $\mathrm{rank}(\mathbf{X}_r^\mathsf{T}\mathbf{X}_s) < d$ and a uniform sample $U_q^{\ell+1}$ otherwise. In particular, like $\Lambda_{U_q,\ell}(\mathbf{s})$, the output distribution of $\Gamma_{U_q,\ell,d}(\mathbf{s},)$ is independent of $\mathbf{s}$.

3   The Hardness of SLWE

Theorem 1 below is the main technical result stating that the SLWE problem mapping into subspaces of dimension d is almost as hard as the standard LWE problem with secrets of length d. But let's first look at the (easy) other direction as stated by Claim 1 below.

Claim 1 ((q, χ, ℓ, d)-SLWE at most as hard as (q, χ, d)-LWE). *If (q, χ, ℓ, d)-SLWE is (s, t, ϵ) hard then (q, χ, d)-LWE is (s', t, ϵ) hard where $s' = s - poly(t, \ell)$.*

Proof (of Claim). To prove this claim we will show how, for any error distribution χ', one can efficiently generate (q, χ', d)-LWE samples which have distribution $\Lambda_{\chi',d}(\mathbf{s}')$ (for some uniform $\mathbf{s}' \in \mathbb{Z}_q^d$) given access to a (q, χ', ℓ, d)-SLWE oracle $\Gamma_{\chi',\ell,d}(\mathbf{s},.)$ (for some uniform $\mathbf{s} \in \mathbb{Z}_q^\ell$.) We do so without known knowing the distribution χ' or $\mathbf{s}$.

Given such a transformation, we then can use any distinguisher D who breaks the (q, χ, d)-LWE assumption with advantage ϵ as defined in eq.(1), to break the (q, χ, ℓ, d)-SLWE assumption as in eq.(2) with the same advantage by simply transforming the SLWE samples (where the oracle uses either the error distribution $\chi' = \chi$ or $\chi' = U_q^{\ell+1}$, but we don't know which) to LWE samples (with the same unknown error distribution χ') before forwarding them do D.

Let $\mathbf{v} \stackrel{\text{def}}{=} 1^d \| 0^{\ell-d}$. To generate samples as described above, query $\Gamma_{\chi',\ell,d}(\mathbf{s},.)$ so it outputs samples $\Lambda_{\chi',d}(\mathbf{s}')$ where $\mathbf{s}' \in \mathbb{Z}_q^d$ consists of, say, the first d elements of $\mathbf{s} \in \mathbb{Z}_q^\ell$, i.e. $\mathbf{s}' := \mathbf{s}_{\downarrow \mathbf{v}}$. This can be done by making q queries $\mathbf{X}_s, \mathbf{X}_r, \mathbf{x}_s, \mathbf{x}_r$ to $\Gamma_{\chi',\ell,d}(\mathbf{s},.)$ where $\mathbf{x}_s = \mathbf{x}_r = 0^\ell$ and $\mathbf{X}_s = \mathbf{X}_r$ is 1 in the first d diagonal entries and 0 everywhere else. The output of the SLWE oracle on these queries are samples of the form

$$\mathbf{r}, \underbrace{(\mathbf{X}_r \mathbf{r} + \mathbf{x}_r)^\mathsf{T}}_{\mathbf{r}_{\downarrow \mathbf{v}}^\mathsf{T}} \underbrace{(\mathbf{X}_s \mathbf{s} + \mathbf{x}_s)}_{\mathbf{s}_{\downarrow \mathbf{v}}} + e \quad \text{where} \quad e \leftarrow \chi', \ \mathbf{r} \xleftarrow{\$} \mathbb{Z}_q^\ell$$

from which we then get an $\Lambda_{\chi',d}(\mathbf{s}_{\downarrow \mathbf{v}})$ sample $\mathbf{r}_{\downarrow \mathbf{v}}, \mathbf{r}_{\downarrow \mathbf{v}}^\mathsf{T}\mathbf{s}_{\downarrow \mathbf{v}} + e$ by replacing $\mathbf{r}$ with $\mathbf{r}_{\downarrow \mathbf{v}}$. Note that these samples have the right distribution, which means $\mathbf{s}_{\downarrow \mathbf{v}}$ and the q $\mathbf{r}_{\downarrow \mathbf{v}}$'s are uniformly random as required. This is easy to see recalling that $\mathbf{s}$ and the q $\mathbf{r}$'s are uniform. $\square$

In the proof of Theorem 1, we'll need the following simple technical Lemma:

Lemma 1. *For* $q, d, \delta \in \mathbb{N}$, *let* $\Delta(q, d, \delta)$ *denote the probability that a random matrix in* $\mathbb{Z}_q^{(d+\delta) \times d}$ *has rank less than* d, *then*

$$\Delta(q, d, \delta) \leq \frac{2}{q^{\delta+1}} .$$

Proof. Assume we sample the d columns of a matrix $M \in \mathbb{Z}_q^{(d+\delta) \times d}$ one by one. For $i = 1, \ldots, d$ let E_i denote the event that the first i columns are linearly independent, then

$$\Pr[\neg E_i | E_{i-1}] = \frac{q^{i-1}}{q^{d+\delta}} = q^{i-1-d-\delta}$$

as $\neg E_i$ happens iff the ith column (sampled uniformly from a space of size $q^{d+\delta}$) falls into the space (of size q^{i-1}) spanned by the first $i - 1$ columns. We get further

$$\Delta(q, d, \delta) = \Pr[\neg E_d] \leq \sum_{i=1}^{d} \Pr[\neg E_i | E_{i-1}] = \sum_{i=1}^{d} q^{i-1-d-\delta} \leq \frac{2}{q^{\delta+1}}$$

$\square$

Theorem 1 ((q, χ, ℓ, d)**-SLWE almost as hard as** (q, χ, d)**-LWE**)
For $q, d, \delta, \ell \in \mathbb{N}$. *If the* ($q, \chi, d$)-*LWE problem is* ($s, t, \epsilon$) *hard, then the* ($q, \chi, \ell, d + \delta$)-*SLWE problem is* ($s', t, \epsilon'$) *hard where*

$$s' = s - poly(\ell, t) \qquad \epsilon' = \epsilon + \frac{2t}{q^{\delta+1}}$$

Proof (of Theorem 1). To prove the theorem we will show how to sample outputs of an SLWE oracle $\Gamma_{\chi', \ell, d+\delta}(\hat{s}, .)$ for some uniformly random $\hat{s} \in \mathbb{Z}_q^\ell$ and adversarially chosen inputs, given only standard LWE samples $\Lambda_{\chi', d}(s)$ for some uniform $s \in \mathbb{Z}_q^d$. This sampling is done without knowing s or the error distribution χ'.

Given such a transformation, we then can use any distinguisher D who breaks the ($q, \chi, \ell, d + \delta$)-SLWE assumption with advantage ϵ to break the standard (q, χ, d)-LWE assumption with the same advantage, minus the probability that the transformation will fail (which, unlike in the previous claim, is non-zero.)

Recall that an LWE sample $\Lambda_{\chi', d}(s \in \mathbb{Z}_q^d)$ is of the form

$$\mathbf{r}, \mathbf{r}^\mathsf{T}.s + e \quad \text{where} \quad e \leftarrow \chi' \quad \mathbf{r} \xleftarrow{\$} \mathbb{Z}_q^d \tag{3}$$

For $\mathbf{X}_r, \mathbf{X}_s \in \mathbb{Z}_q^{\ell \times \ell}$, $\mathbf{x}_s, \mathbf{x}_r \in \mathbb{Z}_q^\ell$, we'll show how to transform such a sample into a an SLWE sample $\Gamma_{\chi', \ell, d+\delta}(\hat{s}, [\mathbf{X}_r, \mathbf{x}_r, \mathbf{X}_s, \mathbf{x}_s])$. If $\mathsf{rank}(\mathbf{X}_r^\mathsf{T}.\mathbf{X}_s) < d + \delta$ this sample is simply $\perp$, so from now one we assume that this rank is at least $d + \delta$, in this case the sample has the form

$$\hat{\mathbf{r}}, (\mathbf{X}_r.\hat{\mathbf{r}} + \mathbf{x}_r)^\mathsf{T}(\mathbf{X}_s.\hat{s} + \mathbf{x}_s) + e \quad \text{where} \quad e \leftarrow \chi' \quad \hat{\mathbf{r}} \xleftarrow{\$} \mathbb{Z}_q^\ell \tag{4}$$

In our transformation, the SLWE secret $\hat{\mathbf{s}} \in \mathbb{Z}_q^\ell$ is defined as a function of the LWE secret $\mathbf{s} \in \mathbb{Z}_q^d$ as follows

$$\mathbf{R} \xleftarrow{\$} \mathbb{Z}_q^{\ell \times d} \qquad \mathbf{b} \xleftarrow{\$} \mathbb{Z}_q^\ell \qquad \hat{\mathbf{s}} = \mathbf{R}.\mathbf{s} + \mathbf{b} \qquad (5)$$

Note that we only know $\mathbf{R}, \mathbf{b}$ (which we sampled), but will not get $\hat{\mathbf{s}}$ as we don't know $\mathbf{s}$. Also note that $\hat{\mathbf{s}}$ is uniformly random as it is blinded with a uniform $\mathbf{b}$. Define the set $\mathcal{L} \subseteq \mathbb{Z}_q^\ell$, which is the set of solutions to a system of linear equations, as

$$\mathcal{L} = \{\mathbf{y} \ : \ \mathbf{y}.\mathbf{X}_r^\mathsf{T}.\mathbf{X}_s.\mathbf{R} = \mathbf{r}^\mathsf{T} - \mathbf{x}_r^\mathsf{T}.\mathbf{X}_s.\mathbf{R}\}. \qquad (6)$$

If $\mathbf{X}_r^\mathsf{T}.\mathbf{X}_s.\mathbf{R}$ has rank at least d, then $\mathcal{L}$ is not empty as the linear equation considered in eq.(6) is (over)defined (we will bound the probability that the rank is d later.) In this case the LWE sample is transformed into an SLWE sample as

$$\underbrace{\mathbf{r}, \mathbf{r}^\mathsf{T}.\mathbf{s} + e}_{\text{LWE Sample (3)}} \ \to \ \underbrace{\hat{\mathbf{r}}, \mathbf{r}^\mathsf{T}.\mathbf{s} + z + e}_{\text{SLWE Sample (4)}} \quad \text{where} \ \ \hat{\mathbf{r}} \xleftarrow{\$} \mathcal{L} \qquad (7)$$

and the z is computed from known values as

$$z \stackrel{\text{def}}{=} (\hat{\mathbf{r}}^\mathsf{T}.\mathbf{X}_r^\mathsf{T} + \mathbf{x}_r^\mathsf{T}).\mathbf{X}_s.\mathbf{b} + (\mathbf{X}_r.\hat{\mathbf{r}} + \mathbf{x}_r)^\mathsf{T}.\mathbf{x}_s$$

It follows from the three claims below that this sampling gives the right distribution.

Claim. If $\mathbf{T} \stackrel{\text{def}}{=} \mathbf{X}_r^\mathsf{T}.\mathbf{X}_s.\mathbf{R}$ has rank $\geq d$ then $\hat{\mathbf{r}} \xleftarrow{\$} \mathcal{L}$ is uniformly random (given $\mathbf{x}_s, \mathbf{x}_r, \mathbf{X}_s^\mathsf{T}, \mathbf{X}_r^\mathsf{T}, \mathbf{R}, \mathbf{b}$.)

Proof (of Claim). Fix some $\mathbf{t} \in \mathbb{Z}_2^\ell$ of weight $\mathbf{wt}(\mathbf{t}) = d$ such that $\mathbf{T}_{\downarrow \mathbf{t}}$ has full rank. Such a $\mathbf{t}$ exists as $\mathbf{T}$ has rank d.

By eq.(6), $\hat{\mathbf{r}} \xleftarrow{\$} \mathcal{L}$ is a random solution to the equation

$$\hat{\mathbf{r}}.\mathbf{T} = \mathbf{r}^\mathsf{T} - \mathbf{x}_r^\mathsf{T}.\mathbf{X}_s.\mathbf{R}$$

or equivalently (using $\hat{\mathbf{r}}.\mathbf{T} = \hat{\mathbf{r}}_{\downarrow \mathbf{t}}.\mathbf{T}_{\downarrow \mathbf{t}} + \hat{\mathbf{r}}_{\downarrow \bar{\mathbf{t}}}.\mathbf{T}_{\downarrow \bar{\mathbf{t}}}$)

$$\hat{\mathbf{r}}_{\downarrow \mathbf{t}}.\mathbf{T}_{\downarrow \mathbf{t}} = \mathbf{r}^\mathsf{T} - \mathbf{x}_r^\mathsf{T}.\mathbf{X}_s.\mathbf{R} - \hat{\mathbf{r}}_{\downarrow \bar{\mathbf{t}}}.\mathbf{T}_{\downarrow \bar{\mathbf{t}}} \qquad (8)$$

Now sampling a random $\hat{\mathbf{r}}$ can be done as follows. First sample $\hat{\mathbf{r}}_{\downarrow \bar{\mathbf{t}}} \xleftarrow{\$} \mathbb{Z}_q^{\ell - d}$ uniformly. The remaining d positions $\hat{\mathbf{r}}_{\downarrow \mathbf{t}} \in \mathbb{Z}_q^d$ are then uniquely determined by $\mathbf{r}$ and given by the solution to the equation (8).

As $\mathbf{T}_{\downarrow \mathbf{t}}$ is a full rank square matrix eq.(8) defines a bijection between $\hat{\mathbf{r}}_{\downarrow \mathbf{t}}$ and $\mathbf{r}$. As $\mathbf{r}$ is chosen uniformly at random, also $\hat{\mathbf{r}}_{\downarrow \mathbf{t}}$ is uniformly random. Thus the entire $\hat{\mathbf{r}}$ is uniform as claimed. $\qquad \square$

Claim. The $\hat{\mathbf{r}}, \mathbf{r}^\mathsf{T}.\mathbf{s} + z + e$ as sampled in (7) is an SLWE sample for secret $\hat{\mathbf{s}}$, randomness $\hat{\mathbf{r}}$ and error e.

Proof (of Claim).

$$\hat{\mathbf{r}}, (\mathbf{X}_r.\hat{\mathbf{r}} + \mathbf{x}_r)^\mathsf{T}.(\mathbf{X}_s.\hat{\mathbf{s}} + \mathbf{x}_s) + e \qquad\qquad \text{(SLWE sample)}$$

$$= \hat{\mathbf{r}}, (\mathbf{X}_r.\hat{\mathbf{r}} + \mathbf{x}_r)^\mathsf{T}.(\mathbf{X}_s.\hat{\mathbf{s}}) + (\mathbf{X}_r.\hat{\mathbf{r}} + \mathbf{x}_r)^\mathsf{T}.\mathbf{x}_s + e$$

$$\stackrel{(5)}{=} \hat{\mathbf{r}}, (\mathbf{X}_r.\hat{\mathbf{r}} + \mathbf{x}_r)^\mathsf{T}.(\mathbf{X}_s.(\mathbf{R}.\mathbf{s} + \mathbf{b})) + (\mathbf{X}_r.\hat{\mathbf{r}} + \mathbf{x}_r)^\mathsf{T}.\mathbf{x}_s + e$$

$$= \hat{\mathbf{r}}, (\hat{\mathbf{r}}^\mathsf{T}.\mathbf{X}_r^\mathsf{T} + \mathbf{x}_r^\mathsf{T}).(\mathbf{X}_s.\mathbf{R}.\mathbf{s}) + \underbrace{(\hat{\mathbf{r}}^\mathsf{T}.\mathbf{X}_r^\mathsf{T} + \mathbf{x}_r^\mathsf{T}).\mathbf{X}_s.\mathbf{b} + (\mathbf{X}_r.\hat{\mathbf{r}} + \mathbf{x}_r)^\mathsf{T}.\mathbf{x}_s}_{z} + e$$

$$\stackrel{(6)}{=} \hat{\mathbf{r}}, \mathbf{r}^\mathsf{T}.\mathbf{s} + z + e$$

□

We have shown how to simulate an SLWE oracle $\Gamma_{\chi',\ell,d+\delta}(\hat{\mathbf{s}}, .)$ from standard LWE samples $\Lambda_{\chi',d}(\mathbf{s})$. This simulation goes well as long as we never get a query containing $\mathbf{X}_r, \mathbf{X}_s$ where $\mathsf{rank}(\mathbf{X}_r^\mathsf{T}.\mathbf{X}_s) \geq d + \delta$ (so the sample is not just $\perp$) but where $\mathsf{rank}(\mathbf{X}_r^\mathsf{T}.\mathbf{X}_s.\mathbf{R}) < d$ (in this case $\mathcal{L}$ can be empty.) The following claims bounds the probability of this happening.

Claim. Consider any $\mathbf{X} \in \mathbb{Z}_q^{\ell \times \ell}$ with $\mathsf{rank}(\mathbf{X}) \geq d + \delta$, then (with Δ as defined in Lemma 1)

$$\Pr[\mathsf{rank}(\mathbf{X}.\mathbf{R}) < d \; : \; \mathbf{R} \stackrel{\$}{\leftarrow} \mathbb{Z}_q^{\ell \times d}] \leq \Delta(q, d, \delta)$$

Proof (of Claim). Since the matrix $\mathbf{X}$ has rank at least $d + \delta$, without loss of generality, we can assume that the first $d + \delta$ rows of $\mathbf{X}$ are linearly independent. Since $\mathbf{R}$ is a random matrix, the upper $(d+\delta) \times d$ submatrix of $\mathbf{X}.\mathbf{R}$ is a random matrix in $\mathbb{Z}_q^{(d+\delta) \times d}$ and (by definition) such a matrix has rank strictly less than d with probability at most $\Delta(q, d, \delta)$. Thus $\mathbf{X}.\mathbf{R}$ has rank strictly less than d with at most the same probability. □

Using the union bound, we can upper bound the probability that for any of the t queries the matrix $\mathbf{X} = \mathbf{X}_r^\mathsf{T}.\mathbf{X}_s$ chosen by the distinguisher D will satisfy $\mathsf{rank}(\mathbf{X}.\mathbf{R}) < d$ by

$$t \cdot \Delta(q, n, d) \leq \frac{2 \cdot t}{q^{\delta+1}}$$

This error probability is thus an upper bound on the gap of the success probability ϵ' of D (in breaking SLWE) and the success probability ϵ we get in breaking LWE using the transformation.

Above we ignored the fact that D can choose its queries, and thus the matrix $\mathbf{X} = \mathbf{X}_r^\mathsf{T}.\mathbf{X}_s$, *adaptively*. To show that adaptivity does not help in picking an $\mathbf{X}$ where $\mathbf{X}.\mathbf{R}$ has rank $< d$ we must show that the view of D is *independent* of $\mathbf{R}$ (except for the fact that so far no query was made where $\mathsf{rank}(\mathbf{X}.\mathbf{R}) < d$.) To see this first note that $\hat{\mathbf{s}} = \mathbf{s}.\mathbf{R} + \mathbf{b}$ is independent of $\mathbf{R}$ as it is blinded with a uniform $\mathbf{b}$. In fact, the only reason we use this blinding is to enforce this independence. The $\hat{\mathbf{r}}$ are independent as they are uniform given $\mathbf{R}$ as shown in the first Claim in the proof of this theorem. □

4   Applications

In this section we discuss some consequences and applications which use the fact that the new subspace LWE problem is as hard as the classical LWE problem.

4.1   Security against Related Key Attacks

Theorem 1 implies that many existing schemes whose security is based on the standard LWE/LPN assumption are secure against attacks not anticipated by the designers of the schemes.

As an illustrative example below we discuss the simple construction of symmetric CPA secure secure encryption from LPN [GRS08]. We show that this simple scheme is not only CPA secure, but it's even secure against powerful related key-attacks. The scheme from [GRS08] is defined as follows

Public Parameters
  - Constants $0 < \tau < 0.5$, $\delta > 0$, $\ell \in \mathbb{N}$.
  - An error correcting code $\mathsf{E} : \mathbb{Z}_2^m \to \mathbb{Z}_2^n$, $\mathsf{D} : \mathbb{Z}_2^n \to \mathbb{Z}_2^m$, where D can correct up to $(\tau + \delta)\ell$ errors.

Key Generation: $\mathsf{KG}(\ell)$ samples and outputs $\mathbf{s} \xleftarrow{\$} \mathbb{Z}_2^\ell$.

Encryption: $\mathsf{Enc}(K, \mathbf{m})$ samples $\mathbf{R} \xleftarrow{\$} \mathbb{Z}_2^{\ell \times n}$, $\mathbf{e} \xleftarrow{\$} \mathsf{Ber}_\tau^n$ and outputs the ciphertext $(\mathbf{R}, \mathbf{R}^T.\mathbf{s} \oplus \mathbf{e} \oplus \mathsf{E}(\mathbf{m}))$.

Decryption: $\mathsf{Dec}(K, (\mathbf{R}, \mathbf{z}))$ outputs $\mathsf{D}(\mathbf{z} \oplus \mathbf{R}^T.\mathbf{s})$.

Correctness. To see that this scheme is correct, note that on input a correctly generated ciphertext $(\mathbf{R}, \mathbf{R}^T.\mathbf{s} \oplus \mathbf{e} \oplus \mathsf{E}(\mathbf{m}))$, the decryption algorithm outputs $\mathsf{D}(\mathbf{e} \oplus \mathsf{E}(\mathbf{m}))$, which is equal to $\mathbf{m}$ unless the error vector $\mathbf{e}$ has weight more than $(\tau + \delta)\ell$. As the bits of $\mathbf{e}$ are i.i.d. with each bit being one with probability τ, the probability of $\mathbf{e}$ having such high weight can be upper bounded (using the Chernoff bound) by an exponentially small probability $2^{-\gamma \cdot \ell}$ (for some $\gamma > 0$ which depends on τ, δ).

CPA Security. Recall that an encryption scheme is IND-CPA secure if no efficient adversary $\mathcal{A}$ can win the following game with probability noticeably better than $1/2$:

1. We sample a key $\mathbf{s} \xleftarrow{\$} \mathbb{Z}_2^\ell$ and a bit $b \xleftarrow{\$} \{0,1\}$.
2. $\mathcal{A}$ gets access to an oracle $\mathsf{Enc}_b(\mathbf{s}, .)$ where
   - $\mathsf{Enc}_0(\mathbf{s}, \mathbf{m}) = \mathsf{Enc}(\mathbf{s}, \mathbf{m})$ (encrypt $\mathbf{m}$)
   - $\mathsf{Enc}_1(\mathbf{s}, \mathbf{m}) = \mathsf{Enc}(\mathbf{s}, 0^{|\mathbf{m}|})$ (encrypt dummy message)
3. $\mathcal{A}$ outputs b' and wins if $b = b'$.

The IND-CPA security of the [GRS08] encryption scheme follows quite easily from the LPN assumption, i.e. the fact that samples $(\mathbf{R}, \mathbf{R}^T.\mathbf{s} + \mathbf{e})$ are pseudorandom.

RKA Security. Classical security notions, like IND-CPA security, model the encryption scheme as a "black-box", where an adversary can only observe the legitimate input-output behavior of the scheme. Unfortunately, in the last decade it became evident that such idealized models fail to capture many real-world attacks where an adversary can attack an actual physical implementation of the scheme. An important example is direct leakage from the secret state, typically by side-channel attacks or malware. To deal with this issue, in the last years many "intrusion-resilient" and "leakage-resilient" schemes have been proposed [ISW03, MR04, Dzi06, DP07, DP08, ADW09, Pie09, CDD+07, KV09, DW09, DKL09].

But the key can also leak indirectly, for example due to key-dependent messages [BRS03, HK07, BHHO08, HU08, ACPS09, BHHI10, BG10, ABBC10]. Here, as the name suggest, one considers a setting where the encrypted message can depend on the secret key. Another important setting are related-key attacks (RKA). In an RKA attack on a encryption scheme the adversary can not only ask for encryptions under the secret key $\mathbf{s}$, but also under "related" keys. RKA attacks were first considered by Biham [Bih94] and Knudsen [Knu92], and were extensively studied in the last decade [Luc04, BDK06, BDK08, FKL+00, JD04, ZZWF07, BC10]. Bellare and Kohno [BK03] initiated a formal study of RKA attacks. All this works consider RKA security of deterministic primitives, usually block-ciphers.

Very recently [AHI11] initiated a formal study of RKA security for probabilistic encryption [GM84]. As in [BK03], they define RKA with respect to related-key-deriving functions (RKD) Φ. Φ-RKA-IND-CPA security of an encryption scheme is then defined almost like standard IND-CPA security, but where the adversary can additionally apply any function $\phi \in \Phi$ to the secret key $\mathbf{s}$, i.e.

1. We sample a key $\mathbf{s} \xleftarrow{\$} \mathbb{Z}_2^\ell$ and a bit $b \xleftarrow{\$} \{0,1\}$.
2. $\mathcal{A}$ gets access to an oracle $\mathsf{Enc}_b^\Phi(\mathbf{s}, ., .)$ where
   - $\mathsf{Enc}_0^\Phi(\mathbf{s}, \mathbf{m}, \phi \in \Phi) = \mathsf{Enc}(\phi(\mathbf{s}), \mathbf{m})$ (encrypt $\mathbf{m}$)
   - $\mathsf{Enc}_1^\Phi(\mathbf{s}, \mathbf{m}, \phi \in \Phi) = \mathsf{Enc}(\phi(\mathbf{s}), 0^{|\mathbf{m}|})$ (encrypt dummy message)
3. $\mathcal{A}$ outputs b' and wins if $b = b'$.

In [AHI11] it is shown that [GRS08] is $\Phi^\oplus$-RKA-IND-CPA secure where $\Phi^\oplus$ is the class of XOR relations. This class contains, for every $\Delta \in \mathbb{Z}_2^\ell$, the function $\phi_\Delta(\mathbf{s}) \stackrel{\text{def}}{=} \mathbf{s} \oplus \Delta$.

This is an interesting class of relations as (1) it captures realistic RKA and (2) many existing schemes (mostly block-ciphers) have actually been shown to be insecure against $\Phi^\oplus$-RKA. Unfortunately $\Phi^\oplus$-RKA security does not imply any security in the realistic scenario where an adversary can not only flip, but set some of the bit of the secret key. Neither does it cover the case where the adversary can exchange the *position* of the key bits.

Using Theorem 1 we can show that the scheme is in fact secure against a much more powerful class of "affine relations", which as special cases contains the relations just mentioned. Let Φ_d^{aff} be the class which contains the functions

$$\phi_{\mathbf{X},\mathbf{x}}(\mathbf{s}) = \mathbf{X}^T.\mathbf{s} \oplus \mathbf{x}$$

for every $\mathbf{X} \in \mathbb{Z}_2^{\ell \times \ell}, \mathbf{x} \in \mathbb{Z}_2^\ell$ where $\mathsf{rank}(\mathbf{X}) \geq d$.

Proposition 1. *Under the (decisional) (τ, ℓ, d)-SLPN assumption[5] (which by Theorem 1 is equivalent to the standard LPN assumption), the encryption scheme from [GRS08] is Φ_d^{aff}-RKA-IND-CPA secure.*

Proof. For any $\phi \in \Phi_d^{\mathsf{aff}}$, samples of the from $\mathbf{R}, \mathbf{R}^{\mathsf{T}}.\phi(\mathbf{s}) + \mathbf{e}$ are pseudorandom by assumption. So the outputs of both $\mathsf{Enc}_0^{\Phi_d^{\mathsf{aff}}}(\mathbf{s}, ., .)$ and $\mathsf{Enc}_1^{\Phi_d^{\mathsf{aff}}}(\mathbf{s}, ., .)$ are pseudorandom and thus indistinguishable. $\square$

Φ_d^{aff} is a very powerful class of relations, and captures many realistic settings. It contains $\Phi^{\oplus}$, but also the class of relations $\Phi_d^{\mathsf{set}} \subset \Phi_d^{\mathsf{aff}}$ which allows to overwrite all but d bits of the input, and the class $\Phi^{\mathsf{perm}} \subset \Phi_0^{\mathsf{aff}}$ which allows to permute the key bits.[6] Previous to our work no scheme was known to be provably secure against Φ_d^{aff}, or even just for one of the special cases Φ_d^{set} (for $d > 0$) or Φ^{perm}. In fact, no *deterministic* encryption scheme can be secure against Φ^{perm}, and no "natural"[7] deterministic scheme can be secure against Φ_1^{set}.

4.2 Weak Randomness and New Constructions

The RKA security example from the previous section used the fact that an adversary can apply any affine function to the LWE secret. There are also natural implications from the fact that she can apply a mapping to the randomness $\mathbf{r}$. For example, it implies that LWE is hard, even if the randomness $\mathbf{r}$ used to generate the samples $\mathbf{r}^{\mathsf{T}}\mathbf{s} + e$ is not uniform, but comes from a bit-fixing source [CGH+85]. Let us stress that the (comparably small) amount of randomness necessary to sample the error e must be uniform.

Theorem 1 not only has implications for existing constructions, but in subsequent work has inspired completely new constructions, most notably the authentication schemes and message authentications codes proposed in [KPC+11].

References

[ABB10a] Agrawal, S., Boneh, D., Boyen, X.: Efficient Lattice (H)IBE in the Standard Model. In: Gilbert, H. (ed.) EUROCRYPT 2010. LNCS, vol. 6110, pp. 553–572. Springer, Heidelberg (2010)

[ABB10b] Agrawal, S., Boneh, D., Boyen, X.: Lattice Basis Delegation in Fixed Dimension and Shorter-Ciphertext Hierarchical IBE. In: Rabin, T. (ed.) CRYPTO 2010. LNCS, vol. 6223, pp. 98–115. Springer, Heidelberg (2010)

[ABBC10] Acar, T., Belenkiy, M., Bellare, M., Cash, D.: Cryptographic Agility and Its Relation to Circular Encryption. In: Gilbert, H. (ed.) EUROCRYPT 2010. LNCS, vol. 6110, pp. 403–422. Springer, Heidelberg (2010)

---

[5] This is the $(2, \mathsf{Ber}_\tau, \ell, d)$-SLWE problem as given in Definition 2.

[6] $\Phi^{\mathsf{perm}} \subset \Phi_0^{\mathsf{aff}}$ as it only contains $\phi_{\mathbf{x},\mathbf{x}}$ where $\mathbf{x} = 0^\ell$ and $\mathbf{X}$ is a (full rank) permutation matrix.

[7] We need that every bit of the secret key is relevant, i.e. $\mathsf{Enc}(\mathbf{s}, \mathbf{m}) \neq \mathsf{Enc}(\mathbf{s}', \mathbf{m})$ with good probability for $\mathbf{s} \neq \mathbf{s}'$.

[ACPS09]   Applebaum, B., Cash, D., Peikert, C., Sahai, A.: Fast Cryptographic Primitives and Circular-Secure Encryption Based on Hard Learning Problems. In: Halevi, S. (ed.) CRYPTO 2009. LNCS, vol. 5677, pp. 595–618. Springer, Heidelberg (2009)

[ADW09]   Alwen, J., Dodis, Y., Wichs, D.: Leakage-Resilient Public-Key Cryptography in the Bounded-Retrieval Model. In: Halevi, S. (ed.) CRYPTO 2009. LNCS, vol. 5677, pp. 36–54. Springer, Heidelberg (2009)

[AGV09]   Akavia, A., Goldwasser, S., Vaikuntanathan, V.: Simultaneous Hardcore Bits and Cryptography against Memory Attacks. In: Reingold, O. (ed.) TCC 2009. LNCS, vol. 5444, pp. 474–495. Springer, Heidelberg (2009)

[AHI11]   Applebaum, B., Harnik, D., Ishai, Y.: Semantic security under related-key attacks and applications. In: 2nd Innovations in Computer Science (ICS), pp. 45–60 (2011)

[Ale03]   Alekhnovich, M.: More on average case vs approximation complexity. In: 44th FOCS, pp. 298–307. IEEE Computer Society Press (October 2003)

[BC10]   Bellare, M., Cash, D.: Pseudorandom Functions and Permutations Provably Secure against Related-Key Attacks. In: Rabin, T. (ed.) CRYPTO 2010. LNCS, vol. 6223, pp. 666–684. Springer, Heidelberg (2010)

[BDK06]   Biham, E., Dunkelman, O., Keller, N.: New Cryptanalytic Results on IDEA. In: Lai, X., Chen, K. (eds.) ASIACRYPT 2006. LNCS, vol. 4284, pp. 412–427. Springer, Heidelberg (2006)

[BDK08]   Biham, E., Dunkelman, O., Keller, N.: A Unified Approach to Related-Key Attacks. In: Nyberg, K. (ed.) FSE 2008. LNCS, vol. 5086, pp. 73–96. Springer, Heidelberg (2008)

[BFKL94]   Blum, A., Furst, M.L., Kearns, M., Lipton, R.J.: Cryptographic Primitives Based on Hard Learning Problems. In: Stinson, D.R. (ed.) CRYPTO 1993. LNCS, vol. 773, pp. 278–291. Springer, Heidelberg (1994)

[BG10]   Brakerski, Z., Goldwasser, S.: Circular and Leakage Resilient Public-Key Encryption under Subgroup Indistinguishability (or: Quadratic Residuosity Strikes Back). In: Rabin, T. (ed.) CRYPTO 2010. LNCS, vol. 6223, pp. 1–20. Springer, Heidelberg (2010)

[BHHI10]   Barak, B., Haitner, I., Hofheinz, D., Ishai, Y.: Bounded Key-Dependent Message Security. In: Gilbert, H. (ed.) EUROCRYPT 2010. LNCS, vol. 6110, pp. 423–444. Springer, Heidelberg (2010)

[BHHO08]   Boneh, D., Halevi, S., Hamburg, M., Ostrovsky, R.: Circular-Secure Encryption from Decision Diffie-Hellman. In: Wagner, D. (ed.) CRYPTO 2008. LNCS, vol. 5157, pp. 108–125. Springer, Heidelberg (2008)

[Bih94]   Biham, E.: New types of cryptanalytic attacks using related keys. Journal of Cryptology 7(4), 229–246 (1994)

[BK03]   Bellare, M., Kohno, T.: A Theoretical Treatment of Related-key Attacks: RKA-PRPs, RKA-PRFs, and Applications. In: Biham, E. (ed.) EUROCRYPT 2003. LNCS, vol. 2656, pp. 491–506. Springer, Heidelberg (2003)

[BKW00]   Blum, A., Kalai, A., Wasserman, H.: Noise-tolerant learning, the parity problem, and the statistical query model. In: 32nd ACM STOC, pp. 435–440. ACM Press (May 2000)

[BRS03]   Black, J., Rogaway, P., Shrimpton, T.: Encryption-Scheme Security in the Presence of Key-Dependent Messages. In: Nyberg, K., Heys, H.M. (eds.) SAC 2002. LNCS, vol. 2595, pp. 62–75. Springer, Heidelberg (2003)

[CDD⁺07] Cash, D., Ding, Y.Z., Dodis, Y., Lee, W., Lipton, R.J., Walfish, S.: Intrusion-Resilient Key Exchange in the Bounded Retrieval Model. In: Vadhan, S.P. (ed.) TCC 2007. LNCS, vol. 4392, pp. 479–498. Springer, Heidelberg (2007)

[CGH⁺85] Chor, B., Goldreich, O., Hastad, J., Friedman, J., Rudich, S., Smolensky, R.: The bit extraction problem of t-resilient functions (preliminary version). In: FOCS, pp. 396–407 (1985)

[CHKP10] Cash, D., Hofheinz, D., Kiltz, E., Peikert, C.: Bonsai Trees, or How to Delegate a Lattice Basis. In: Gilbert, H. (ed.) EUROCRYPT 2010. LNCS, vol. 6110, pp. 523–552. Springer, Heidelberg (2010)

[DGK⁺10] Dodis, Y., Goldwasser, S., Tauman Kalai, Y., Peikert, C., Vaikuntanathan, V.: Public-Key Encryption Schemes with Auxiliary Inputs. In: Micciancio, D. (ed.) TCC 2010. LNCS, vol. 5978, pp. 361–381. Springer, Heidelberg (2010)

[DKL09] Dodis, Y., Kalai, Y.T., Lovett, S.: On cryptography with auxiliary input. In: Mitzenmacher, M. (ed.) 41st ACM STOC, pp. 621–630. ACM Press (May/June 2009)

[DP07] Dziembowski, S., Pietrzak, K.: Intrusion-resilient secret sharing. In: 48th FOCS, pp. 227–237. IEEE Computer Society Press (October 2007)

[DP08] Dziembowski, S., Pietrzak, K.: Leakage-resilient cryptography. In: 49th FOCS, pp. 293–302. IEEE Computer Society Press (October 2008)

[DW09] Dodis, Y., Wichs, D.: Non-malleable extractors and symmetric key cryptography from weak secrets. In: Mitzenmacher, M. (ed.) 41st ACM STOC, pp. 601–610. ACM Press (May/June 2009)

[Dzi06] Dziembowski, S.: Intrusion-Resilience via the Bounded-Storage Model. In: Halevi, S., Rabin, T. (eds.) TCC 2006. LNCS, vol. 3876, pp. 207–224. Springer, Heidelberg (2006)

[FKL⁺00] Ferguson, N., Kelsey, J., Lucks, S., Schneier, B., Stay, M., Wagner, D., Whiting, D.L.: Improved Cryptanalysis of Rijndael. In: Schneier, B. (ed.) FSE 2000. LNCS, vol. 1978, pp. 213–230. Springer, Heidelberg (2001)

[FS96] Fischer, J.-B., Stern, J.: An Efficient Pseudo-random Generator Provably as Secure as Syndrome Decoding. In: Maurer, U.M. (ed.) EUROCRYPT 1996. LNCS, vol. 1070, pp. 245–255. Springer, Heidelberg (1996)

[GKPV10] Goldwasser, S., Kalai, Y., Peikert, C., Vaikuntanathan, V.: Robustness of the learning with errors assumption. In: 1st Innovations in Computer Science (ICS) (2010)

[GM84] Goldwasser, S., Micali, S.: Probabilistic encryption. Journal of Computer and System Sciences 28(2), 270–299 (1984)

[GPV08] Gentry, C., Peikert, C., Vaikuntanathan, V.: Trapdoors for hard lattices and new cryptographic constructions. In: Ladner, R.E., Dwork, C. (eds.) 40th ACM STOC, pp. 197–206. ACM Press (May 2008)

[GRS08] Gilbert, H., Robshaw, M., Seurin, Y.: How to Encrypt with the LPN Problem. In: Aceto, L., Damgård, I., Goldberg, L.A., Halldórsson, M.M., Ingólfsdóttir, A., Walukiewicz, I. (eds.) ICALP 2008, Part II. LNCS, vol. 5126, pp. 679–690. Springer, Heidelberg (2008)

[HK07] Halevi, S., Krawczyk, H.: Security under key-dependent inputs. In: Ning, P., De Capitani di Vimercati, S., Syverson, P.F. (eds.) ACM CCS 2007, pp. 466–475. ACM Press (October 2007)

[HU08] Hofheinz, D., Unruh, D.: Towards Key-Dependent Message Security in the Standard Model. In: Smart, N.P. (ed.) EUROCRYPT 2008. LNCS, vol. 4965, pp. 108–126. Springer, Heidelberg (2008)

[ISW03]   Ishai, Y., Sahai, A., Wagner, D.: Private Circuits: Securing Hardware against Probing Attacks. In: Boneh, D. (ed.) CRYPTO 2003. LNCS, vol. 2729, pp. 463–481. Springer, Heidelberg (2003)

[JD04]   Jakimoski, G., Desmedt, Y.: Related-key Differential Cryptanalysis of 192-Bit Key AES Variants. In: Matsui, M., Zuccherato, R.J. (eds.) SAC 2003. LNCS, vol. 3006, pp. 208–221. Springer, Heidelberg (2004)

[Kea93]   Kearns, M.J.: Efficient noise-tolerant learning from statistical queries. In: 25th ACM STOC, pp. 392–401. ACM Press (May 1993)

[Knu92]   Knudsen, L.R.: Cryptanalysis of LOKI91. In: Zheng, Y., Seberry, J. (eds.) AUSCRYPT 1992. LNCS, vol. 718, pp. 196–208. Springer, Heidelberg (1993)

[KPC+11]   Kiltz, E., Pietrzak, K., Cash, D., Jain, A., Venturi, D.: Efficient Authentication from Hard Learning Problems. In: Paterson, K.G. (ed.) EUROCRYPT 2011. LNCS, vol. 6632, pp. 7–26. Springer, Heidelberg (2011)

[KS06]   Katz, J., Shin, J.S.: Parallel and Concurrent Security of the HB and HB$^+$ Protocols. In: Vaudenay, S. (ed.) EUROCRYPT 2006. LNCS, vol. 4004, pp. 73–87. Springer, Heidelberg (2006)

[KTX08]   Kawachi, A., Tanaka, K., Xagawa, K.: Concurrently Secure Identification Schemes Based on the Worst-Case Hardness of Lattice Problems. In: Pieprzyk, J. (ed.) ASIACRYPT 2008. LNCS, vol. 5350, pp. 372–389. Springer, Heidelberg (2008)

[KV09]   Katz, J., Vaikuntanathan, V.: Signature Schemes with Bounded Leakage Resilience. In: Matsui, M. (ed.) ASIACRYPT 2009. LNCS, vol. 5912, pp. 703–720. Springer, Heidelberg (2009)

[LMPR08]   Lyubashevsky, V., Micciancio, D., Peikert, C., Rosen, A.: SWIFFT: A Modest Proposal for FFT Hashing. In: Nyberg, K. (ed.) FSE 2008. LNCS, vol. 5086, pp. 54–72. Springer, Heidelberg (2008)

[Luc04]   Lucks, S.: Ciphers Secure against Related-Key Attacks. In: Roy, B., Meier, W. (eds.) FSE 2004. LNCS, vol. 3017, pp. 359–370. Springer, Heidelberg (2004)

[Lyu08]   Lyubashevsky, V.: Lattice-Based Identification Schemes Secure Under Active Attacks. In: Cramer, R. (ed.) PKC 2008. LNCS, vol. 4939, pp. 162–179. Springer, Heidelberg (2008)

[Lyu09]   Lyubashevsky, V.: Fiat-Shamir with Aborts: Applications to Lattice and Factoring-Based Signatures. In: Matsui, M. (ed.) ASIACRYPT 2009. LNCS, vol. 5912, pp. 598–616. Springer, Heidelberg (2009)

[MR04]   Micali, S., Reyzin, L.: Physically Observable Cryptography (Extended Abstract). In: Naor, M. (ed.) TCC 2004. LNCS, vol. 2951, pp. 278–296. Springer, Heidelberg (2004)

[Pei09]   Peikert, C.: Public-key cryptosystems from the worst-case shortest vector problem: extended abstract. In: Mitzenmacher, M. (ed.) 41st ACM STOC, pp. 333–342. ACM Press (May/June 2009)

[Pie09]   Pietrzak, K.: A Leakage-Resilient Mode of Operation. In: Joux, A. (ed.) EUROCRYPT 2009. LNCS, vol. 5479, pp. 462–482. Springer, Heidelberg (2009)

[PR06]   Peikert, C., Rosen, A.: Efficient Collision-Resistant Hashing from Worst-Case Assumptions on Cyclic Lattices. In: Halevi, S., Rabin, T. (eds.) TCC 2006. LNCS, vol. 3876, pp. 145–166. Springer, Heidelberg (2006)

[PVW08]   Peikert, C., Vaikuntanathan, V., Waters, B.: A Framework for Efficient and Composable Oblivious Transfer. In: Wagner, D. (ed.) CRYPTO 2008. LNCS, vol. 5157, pp. 554–571. Springer, Heidelberg (2008)

[PW08]   Peikert, C., Waters, B.: Lossy trapdoor functions and their applications. In: Ladner, R.E., Dwork, C. (eds.) 40th ACM STOC, pp. 187–196. ACM Press (May 2008)

[Reg05]   Regev, O.: On lattices, learning with errors, random linear codes, and cryptography. In: Gabow, H.N., Fagin, R. (eds.) 37th ACM STOC, pp. 84–93. ACM Press (May 2005)

[Ste94]   Stern, J.: A New Identification Scheme Based on Syndrome Decoding. In: Stinson, D.R. (ed.) CRYPTO 1993. LNCS, vol. 773, pp. 13–21. Springer, Heidelberg (1994)

[ZZWF07] Zhang, W., Zhang, L., Wu, W., Feng, D.: Related-Key Differential-Linear Attacks on Reduced AES-192. In: Srinathan, K., Pandu Rangan, C., Yung, M. (eds.) INDOCRYPT 2007. LNCS, vol. 4859, pp. 73–85. Springer, Heidelberg (2007)

Bounded-Collusion IBE from Key Homomorphism

Shafi Goldwasser[1,*], Allison Lewko[2,**], and David A. Wilson[3, ***]

[1] MIT CSAIL and Weizmann Institute
shafi@csail.mit.edu
[2] UT Austin
alewko@cs.utexas.edu
[3] MIT CSAIL
dwilson@mit.edu

Abstract. In this work, we show how to construct IBE schemes that are secure against a bounded number of collusions, starting with underlying PKE schemes which possess linear homomorphisms over their keys. In particular, this enables us to exhibit a new (bounded-collusion) IBE construction based on the quadratic residuosity assumption, without any need to assume the existence of random oracles. The new IBE's public parameters are of size $O(t\lambda \log I)$ where I is the total number of identities which can be supported by the system, t is the number of collusions which the system is secure against, and λ is a security parameter. While the number of collusions is bounded, we note that an exponential number of total identities can be supported.

More generally, we give a transformation that takes any PKE satisfying *Linear Key Homomorphism*, *Identity Map Compatibility*, and the *Linear Hash Proof Property* and translates it into an IBE secure against bounded collusions. We demonstrate that these properties are more general than our quadratic residuosity-based scheme by showing how a simple PKE based on the DDH assumption also satisfies these properties.

1 Introduction

The last decade in the lifetime of cryptography has been quite exciting. We are witnessing a paradigm shift, departing from the traditional goals of secure and authenticated communication and moving towards systems that are simultaneously highly secure, highly functional, and highly flexible in allowing selected access to encrypted data. As part of this development, different "types" of encryption systems have been conceived and constructed to allow greater ability to meaningfully manipulate and control access to encrypted data, such as bounded and fully homomorphic encryption (FHE), identity-based encryption (IBE), hierarchical identity-based encryption (HIBE), functional encryption (FE), attribute

* Supported by NSF CCF-0729011, NSF CCF-1018064, DARPA FA8750-11-2-0225.
** Supported by a Microsoft Research Ph.D. Fellowship.
*** Supported by NSF CCF-1018064, DARPA FA8750-11-2-0225.

R. Cramer (Ed.): TCC 2012, LNCS 7194, pp. 564–581, 2012.

based encryption (ABE), and others. As is typical at any time of rapid innovation, the field is today at a somewhat chaotic state. The different primitives of FHE, IBE, HIBE, FE, and ABE are being implemented based on different computational assumptions and as of yet we do not know of general constructions.

One way to put some order in the picture is to investigate reductions between the various primitives. A beautiful example of such a result was recently shown by Rothblum [29], who demonstrated a simple reduction between any semantically secure private key encryption scheme which possesses a simple homomorphic property over its ciphertexts to a full-fledged semantically secure public key encryption scheme. The homomorphic property requires that the product of a pair of ciphertexts c_1 and c_2, whose corresponding plaintexts are m_1 and m_2, yields a new ciphertext $c_1 \cdot c_2$ which decrypts to $m_1 + m_2 \mod 2$.

In this paper, we continue this line of investigation and show how public-key encryption schemes which posses a linear homomorphic property over their keys as well as hash proof system features with certain algebraic structure can be used to construct an efficient identity-based encryption (IBE) scheme that is secure against bounded collusions. The main idea is simple. In a nutshell, the homomorphism over the keys will give us a way to map a set of public keys published by the master authority in an IBE system into a new user-specific public key that is obtained by taking a linear combination of the published keys. By taking a linear combination instead of a subset, we are able to achieve smaller keys than a strictly combinatorial approach would allow. Our constructions allow the total number of potential identities to be exponential in the size of the public parameters of the IBE. The challenge will be to prove that the resulting cyptosystem is secure even in the presence of a specified number of colluding users. For this, we rely on an algebraic hash proof property.

To explain our results in the context of the known literature, let us quickly review some relevant highlights in the history of IBEs. The Identity-Based Encryption model was conceived by Shamir in the early 1980s [30]. The first constructions were proposed in 2001 by Boneh and Franklin [6] based on the hardness of the bilinear Diffie-Hellman problem and by Cocks [13] based on the hardness of the quadratic residuosity problem. Both works relied on the random oracle model. Whereas the quadratic residuosity problem has been used in the context of cryptography since the early eighties [22], computational problems employing bilinear pairings were at the time of [6] relative newcomers to the field. Indeed, inspired by their extensive usage within the context of IBEs, the richness of bilinear group problems has proved tremendously useful for solving other cryptographic challenges (e.g. in the area of leakage-resilient systems).

Removing the assumption that random oracles exist in the construction of IBEs and their variants was the next theoretical target. A long progression of results ensued. At first, partial success for IBE based on bilinear group assumptions was achieved by producing IBEs in the standard model provably satisfying a more relaxed security condition known as selective security [11,4], whereas the most desirable of security guarantees is that any polynomial-time attacker who can request secret keys for identities of its choice cannot launch a successful chosen-ciphertext attack (CCA) against a new adaptively-chosen challenge identity. Enlarging the arsenal of computational complexity bases for IBE, Gentry, Peikert, and Vaikuntanathan [21] proposed an IBE based on the intractability of the learning with errors (LWE) problem, still in the random oracle model. Ultimately, fully (unrelaxed) secure IBEs were constructed in the standard model (without assuming random oracles) under the decisional Bilinear Diffie-Hellman assumption by Boneh and Boyen [5] and Waters [34], and most recently under the LWE assumption by Cash, Hofheinz, Kiltz, and Peikert [12] and Agrawal, Boneh, and Boyen [1]. Constructing a fully secure (or even selectively secure) IBE without resorting to the random oracle model based on classical number theoretic assumptions such as DDH in non-bilinear groups or the hardness of quadratic residuosity assumptions remains open.

A different relaxation of IBE comes up in the work of Dodis, Katz, Xu, and Yung [16] in the context of their study of the problem of a bounded number of secret key exposures in public-key encryption. To remedy the latter problem, they introduced the notion of *key-insulated* PKE systems and show its equivalence to *IBEs semantically secure against a bounded number of colluding identities.* This equivalence coupled with constructions of key-insulated PKE's by [16] yields a generic combinatorial construction which converts any semantic secure PKE to a bounded-collusion semantic secure IBE, without needing a random oracle.

New Results. The goal of our work is to point to a new direction in the construction of IBE schemes: the utilization of homomorphic properties over keys of PKE schemes (when they exist) to obtain IBE constructions. This may provide a way to diversify the assumptions on which IBEs can be based. In particular, we are interested in obtaining IBE constructions based on quadratic residuosity in the standard model.

In recent years, several PKE schemes were proposed with interesting homomorphisms over the public keys and the underlying secret keys. These were constructed for the purpose of showing circular security and leakage resilience properties. In particular, for both the scheme of Boneh, Halevi, Hamburg, and Ostrovski [8] and the scheme of Brakerski and Goldwasser [9], it can be shown that starting with two valid (public-key, secret-key) pairs $(pk_1, sk_1), (pk_2, sk_2)$, one can obtain a third valid pair as $(pk_1 \cdot pk_2, sk_1 + sk_2)$.

We define properties of a PKE scheme allowing homomorphism over keys that suffice to transform the PKE into an IBE scheme with bounded collusion resistance. As examples of our general framework, we show how to turn the schemes of [8] and a modification of [9] into two IBE schemes in the standard model (that is, without random oracles), which are CPA secure against bounded collusions. Namely, security holds when the adversary is restricted to receive t secret keys for identities of its choice for a pre-specified t. We allow the adversary to choose its attack target adaptively. The security of the scheme we present here is based on the intractability of the quadratic residuosity problem. In the full version of this paper, we also present a second scheme with security based on the intractability of DDH. Letting the public parameters of the IBE be of size $O(n\lambda)$ where λ is a security parameter, the new DDH-based IBE will be secure as long as the adversary is restricted to receive t secret keys for adaptively chosen ID's where $t = n - 1$. The QR-based IBE will be secure as long as the adversary is restricted to receive t secret keys for $t = \frac{n}{\log I} - 1$, where I is the total number of users (or identities) that can be supported by the system. There is no upper bound on I, which can be exponential in the size of public parameters.

Let us compare what we achieve to the constructions obtained by [16]. In their generic combinatorial construction, they start with any PKE and obtain a bounded-collision IBE, requiring public parameters to be of size $O(t^2 \log I)$ times the size of public keys in the PKE scheme and secret keys to be of size $O(t \log I)$ times the size of secret keys in the PKE scheme for t collisions and I total identities supported. Their approach employs explicit constructions of sets $S_1, \ldots, S_I$ of some finite universe U such that the union of any t of these does not cover any additional set. There is a public key of the PKE scheme generated for each element of U, and each set S_i corresponds to a set of $|S_i|$ PKE secret keys. There are are intrinsic bounds on the values of $I, |U|, t$ for which this works, and [16] note that their values of $|U| = \Theta(t^2 \log I)$ and $|S_i| = \Theta(t \log I)$ for each i are essentially optimal. In contrast, by exploiting the algebraic homomorphisms over the keys, we require public parameters of size roughly $O(t \cdot \log I)$ times the size of public keys and secret keys which are $O(\lambda)$ (within a constant times the size of PKE secret keys) for our quadratic residuosity based scheme. (This is assuming a certain relationship between the security parameter λ and n. See the statement of Theorem 2 for details.)

In [16], they also provide a DDH-based key-insulated PKE scheme which is more efficient than their generic construction. It has $O(t\lambda)$ size public parameters and $O(\lambda)$ size secret keys. Viewing their scheme in the identity based context results in, perhaps surprisingly, the DDH based scheme we obtain by exploiting the homomorphism over the keys in BBHO [8]. In the full version of this paper, we describe this scheme and show it can be proved secure against t collisions using our framework.

1.1 Overview of the Techniques

The basic idea is to exploit homomorphism over the keys in a PKE system Π. The high-level overview is as follows.

Start with a PKE Π with the following properties:

1. The secret keys are vectors of elements in a ring R with operations $(+, \cdot)$ and the public keys consist of elements in a group G.
2. If (pk_1, sk_1) and (pk_2, sk_2) are valid keypairs of Π and $a, b \in R$, then $a sk_1 + b sk_2$ is also a valid secret key of Π, with a corresponding public key that can be efficiently computed from pk_1, pk_2, a, b. For the schemes we present, this public key is computed as $pk_1^a \cdot pk_2^b$.

We note that many existing cryptosystems have this property, or can be made to have this property with trivial modifications, including [8], [9], and [14].

The trusted master authority in an IBE will then choose n pairs of (pk_i, sk_i) $(i = 1, ..., n)$ using the key generation algorithm of Π, publish the n pk_i values, and keep secret the corresponding n sk_i's. Each identity is mapped to a vector $id_1...id_n$ in R^n (we abuse terminology slightly here since R is only required to be a ring and not a field, but we will still call these "vectors"). The secret key for the identity is computed as a coordinate-wise linear combination of the vectors $sk_1, \ldots, sk_n$, with coefficients $id_1, \ldots, id_n$ respectively, i.e.

$$\mathrm{SK}_{ID} := \sum_{i=1}^{n}(sk_i \cdot id_i)$$

where all additions take place in R.

Anyone can compute the matching public key PK_{ID} using the key homomorphism and the published pk_i values. Since by the key homomorphism (PK_{ID}, SK_{ID}) is still a valid key pair for the original PKE, encryption and decryption can function identically to before. The encryptor simply runs the encryption algorithm for Π using PK_{ID}, and the decryptor runs the decryption algorithm for Π using SK_{ID}.

We refer to the combination of a PKE scheme with this homomorphic property over keys and a mapping for identities as having the *linear key homomorphism* and *identity map compatibility* properties. To prove security for the resulting bounded-collusion IBE construction, one can intuitively see that we need the map taking identities to vectors to produce linearly independent outputs for any distinct $t + 1$ identities. This is required to ensure that any t colluding users will not be able to compute a secret key for another user as a known linear combination of their own secret keys. To obtain our full security proof, we define an algebraic property of the PKE scheme in combination with the identity map, called the *linear hash proof property*, which encompasses this requirement on any $t+1$ images of the map and more. The definition of this property is inspired by the paradigm of hash proof systems (introduced by Cramer and Shoup [14]), though it differs from this in many ways. We define valid and invalid ciphertexts for our systems, where valid ciphertexts decrypt properly and invalid ciphertexts should decrypt randomly over the set of many secret keys corresponding to a single public key. We require that valid and invalid ciphertexts are computationally indistinguishable. So far this is quite similar to the previous uses of hash proof systems. However, the identity-based setting introduces a further challenge in proving security by changing to an invalid ciphertext, since now the adversary's view additionally includes the secret keys that it may request for other identities. Hence, we must prove that an invalid ciphertext decrypts randomly over the subset of secret keys that are consistent not only with the public keys, but also with the received secret keys.

Controlling the behavior over this set of consistent keys in the QR-based setting is particularly challenging, where the mathematical analysis is quite subtle due to the fact that secret keys must be treated as integers in a bounded range while public keys are elements of a subgroup of $\mathbb{Z}_N$. To prove the linear hash proof property for our QR-based system, we employ technical bounds concerning the intersection of a shifted lattice in $\mathbb{Z}^n$ with a "bounding box" of elements of $\mathbb{Z}^n$ whose coordinates all lie within a specified finite range.

1.2 Other Related Work

In addition to those referenced above, constructions of IBE schemes in the standard model in the bilinear setting were also provided by Gentry [20] under the q-ABHDE assumption, and by Waters [35] under the bilinear Diffie-Hellman and decisional linear assumptions. Another construction based on quadratic residuosity in the random oracle model was provided by Boneh, Gentry, and Hamburg [7]. Leakage-resilient IBE schemes in various models have also been constructed, for example by Alwen, Dodis, Naor, Segev, Walfish, and Wichs [2], by Brakerski, Kalai, Katz, and Vaikuntanathan [10], and by Lewko, Rouselakis, and Waters [26].

The property we require for our PKE schemes in addition to key homomorphism is a variant of the structure of hash proof systems, which were first introduced by Cramer and Shoup as a paradigm for proving CCA security of PKE schemes [14]. Hash proof systems have recently been used in the context of leakage-resilience as well ([28], for example), extending to the identity-based setting in [2]. We note that the primitive of identity-based hash proof systems introduced in [2] takes a different direction than our work, and the instantiation they provide from the quadratic residuosity assumption relies on the random oracle model.

The relaxation to bounded collusion resistance has also been well-studied in the context of broadcast encryption and revocation schemes, dating back to the introduction of broadcast encryption by Fiat and Naor [17]. This work and several follow up works employed combinatorial techniques [31,32,33,18,25,19]. Another combinatorial approach, the subset cover framework, was introduced by Naor, Naor, and Lopspeich [27] to build a revocation scheme. In this framework, users are associated with subsets of keys. The trusted system designer can then broadcast an encrypted message by selecting a family of subsets which covers all the desired recipients and none of the undesired ones. An improvement to the NNL scheme was later given by Halevy and Shamir [24], and these techniques were then extended to the public key setting by Dodis and Fazio [15].

2 Preliminaries

2.1 IND-CPA Security for Bounded-Collusion IBE

We define IND-CPA security for bounded-collusion IBE in terms of the following game between a challenger and an attacker. We let t denote our threshold parameter for collusion resistance. The game proceeds in phases:

Setup Phase. The challenger runs the setup algorithm to produce the public parameters and master secret key. It gives the public parameters to the attacker.

Query Phase I. The challenger initializes a counter to be 0. The attacker may then submit key queries for various identities. In response to a key query, the challenger increments its counter. If the resulting counter value is $\leq t$, the challenger generates a secret key for the requested identity by running the key generation algorithm. It gives the secret key to the attacker. If the counter value is $> t$, it does not respond to the query.

Challenge Phase. The attacker specifies messages m_0, m_1 and an identity ID^* that was not queried in the preceding query phase. The challenger chooses a random bit $b \in \{0, 1\}$, encrypts m_b to identity ID^* using the encryption algorithm, and gives the ciphertext to the attacker.

Query Phase II. The attacker may again submit key queries for various identities *not equal to ID^**, and the challenger will respond as in the first query phase. We note that the same counter is employed, so that only t total queries in the game are answered with secret keys.

Guess. The attacker outputs a guess b' for b.

We define the advantage of an attacker $\mathcal{A}$ in the above game to be $Adv_{\mathcal{A}} = |Pr[b = b'] - \frac{1}{2}|$. We say a bounded-collusion IBE system with parameter t is *secure* if any PPT attacker $\mathcal{A}$ has only a negligible advantage in this game.

2.2 Complexity Assumption

We formally state the QR assumption. We let λ denote the security parameter.

Quadratic Residuosity Assumption. We let $N = pq$ where p, q are random λ-bit primes. We require $p, q \equiv 3 \pmod 4$, i.e. N is a Blum integer. We let $\mathbb{J}_N$ denote the elements of $\mathbb{Z}_N^*$ with Jacobi symbol equal to 1, and we let $\mathbb{QR}_N$ denote the set of quadratic residues modulo N. Both of these are multiplicative subgroups of $\mathbb{Z}_N^*$, with orders $\frac{\phi(N)}{2}$ and $\frac{\phi(N)}{4}$ respectively. We note that $\frac{\phi(N)}{4}$ is odd, and that -1 is an element of $\mathbb{J}_N$, but is not a square modulo N. As a consequence, $\mathbb{J}_N$ is isomorphic to $\{+1, -1\} \times \mathbb{QR}_N$. We let u denote an element of $\mathbb{QR}_N$ chosen uniformly at random, and h denote an element of $\mathbb{J}_N$ chosen uniformly at random. For any algorithm $\mathcal{A}$, we define the advantage of $\mathcal{A}$ against the QR problem to be:

$$Adv_N^{\mathcal{A}} \, |Pr\left[\mathcal{A}(N, u) = 1\right] - Pr\left[\mathcal{A}(N, h) = 1\right]|.$$

We further restrict our choice of N to values such that $\mathbb{QR}_N$ is cyclic. We note that this is satisfied when p, q are strong primes, meaning $p = 2p' + 1, q = 2q' + 1$, where p, q, p', q' are all distinct odd primes. This restriction was previously imposed in [14], where they note that this restricted version implies the usual formulation of the quadratic residuosity assumption if one additionally assumes that strong primes are sufficiently dense. We say that the QR assumption holds if for all PPT $\mathcal{A}$, $Adv_N^{\mathcal{A}}$ is negligible in λ.

Furthermore, we note that this definition is equivalent to one in which $\mathcal{A}$ receives a random element h of $\mathbb{J}_N \backslash \mathbb{QR}_N$ instead of $\mathbb{J}_N$.

2.3   Mapping Identities to Linearly Independent Vectors

To employ our strategy of transforming PKE schemes with homomorphic properties over keys into IBE schemes with polynomial collusion resistance, we first
need methods for efficiently mapping identities to linearly independent vectors
over various fields. This can be done using generating matrices for the Reed-
Solomon codes over $\mathbb{Z}_p$ and dual BCH codes over $\mathbb{Z}_2$. The proofs of the following
lemmas can be found in the full version.

Lemma 1. *For any prime p and any $t + 1 < p$, there exists an efficiently-
computable mapping $f : \mathbb{Z}_p \to \mathbb{Z}_p^{t+1}$ such that for any distinct $x_1, x_2, ... x_{t+1} \in \mathbb{Z}_p$,
the vectors $f(x_1), f(x_2), ... f(x_{t+1})$ are linearly independent.*

Lemma 2. *For any positive integer k and any $t + 1 < 2^k$, there exists an
efficiently-computable mapping $f : \{0, 1\}^k \to \{0, 1\}^{(t+1)k}$ such that for any distinct $x_1, x_2, ... x_{t+1} \in \{0, 1\}^k$, the vectors $f(x_1), f(x_2), ... f(x_{t+1})$ are linearly independent over $\mathbb{Z}_2$.*

3   From PKE to Bounded Collusion IBE: General Conditions and Construction

We start with a public key scheme and an efficiently computable mapping f
on identities that jointly have the following useful properties. We separate the
public keys of the PKE into public parameters (distributed independently of the
secret key) and user-specific data; the latter is referred to as the "public key".

3.1   Linear Key Homomorphism

We say a PKE has linear key homomorphism if the following requirements hold.
First, its secret keys are generated randomly from d-tuples of a ring R for some
positive integer d, with a distribution that is independent and uniform in each
coordinate over some subset R' of R. Second, starting with any two secret keys
sk_1, sk_2 each in R^d and any $r_1, r_2 \in R$, the component-wise R-linear combination
formed by $r_1 sk_1 + r_2 sk_2$ also functions as a secret key, with a corresponding public
key that can be computed efficiently from r_1, r_2 and the public keys pk_1 and pk_2
of sk_1 and sk_2 respectively, fixing the same public parameters. We note that
$r_1 sk_1 + r_2 sk_2$ may not have all entries in R', but it should still function properly
as a key.

3.2   Identity Map Compatibility

We say the identity mapping f is compatible with a PKE scheme with linear
key homomorphism if f maps identities into n-tuples of elements of R. Letting
I denote the number of identities, the action of f can be represented by a $I \times n$
matrix with entries in R. We denote this matrix by F and its rows by $f_1, \ldots, f_I$.

3.3   Linear Hash Proof Property

We now define the strongest property we require, which we call the linear hash proof property. This property is inspired by the paradigm of hash proof systems, but we deviate from that paradigm in several respects. In hash proof systems, a single public key corresponds to many possible secret keys. There are two encryption algorithms: a valid one and an invalid one. Valid ciphertexts decrypt properly when one uses any of the secret keys associated to the public key, while invalid ciphertexts decrypt differently when different secret keys are used. Our linear hash proof property will consider several public keys at once, each corresponding to a set of many possible secret keys. The adversary will be given these public keys, along with some linear combinations of fixed secret keys corresponding to the public keys. We will also have valid and invalid encryption algorithms. Our valid ciphertexts will behave properly. When an invalid ciphertext is formed for a public key corresponding to a linear combination of the secret keys that is *independent* of the revealed combinations, the invalid ciphertext will decrypt "randomly" when one chooses a random key from the set of secret keys that are consistent with the adversary's view.

To define this property more formally, we first need to define some additional notation. We consider a PKE scheme with linear key homomorphism which comes equipped with a compatible identity map f and an additional algorithm InvalidEncrypt which takes in a message and a *secret key sk* and outputs a ciphertext (note that the invalid encryption algorithm does not necessarily need to be efficient). The regular and invalid encryption algorithms produce two distributions of ciphertexts. We call these *valid* and *invalid* ciphertexts. Correctness of decryption must hold for valid ciphertexts.

We let $(sk_1, pk_1), (sk_2, pk_2), \ldots, (sk_n, pk_n)$ be n randomly generated key pairs, where all of $sk_1, \ldots, sk_n$ are d-tuples in a ring R (here we assume that the key generation algorithm chooses R, d and then generates a key pair. We fix R and then run the rest of the algorithm independently n times to produce the n key pairs). We define S to be the $n \times d$ matrix with entries in R whose i^{th} row contains sk_i.

Fix any $t + 1$ distinct rows of the matrix of identity vectors F, denoted by $f_{i_1}, \ldots, f_{i_{t+1}}$. We let $sk_{ID_{i_{t+1}}}$ denote the secret key $f_{i_{t+1}} \cdot S$ and $pk_{ID_{i_{t+1}}}$ denote the corresponding public key (computed via the key homomorphism). We let $Ker_R(f_{i_1}, \ldots, f_{i_t})$ denote the kernel of the $t \times n$ submatrix of F formed by these rows; that is, it consists of the vectors $v \in R^n$ such that $f_{i_j} \cdot v = 0$ for all j from 1 to t.

Now we consider the set of possible secret key matrices given the public and secret key information available to an adversary who has queried identities $i_1, \ldots, i_t$. We let W denote the set of matrices in $R^{n \times d}$ whose columns belong to $Ker_R(f_{i_1}, \ldots, f_{i_t})$ and whose rows w_i satisfy that $sk_i + w_i$ has the same public key as sk_i for all i. Since W's columns are orthogonal to the identity vectors $f_{i_1}, \ldots, f_{i_t}$, adding an element of W to S does not change any of the secret keys $f_{i_j}S$. Furthermore, by construction, adding an element of W to S does not change the public keys associated with the scheme.

We define the subset $\tilde{S}$ of $R^{n \times d}$ to be the set of all matrices in $S + W :=$ $\{S + W_0 | W_0 \in W\}$, intersected with the set of all matrices of n secret keys that can be generated by the key generation algorithm (i.e. those with components in R'). Intuitively, $\tilde{S}$ is the set of all possible $n \times d$ secret key matrices that are "consistent" with the n public keys $pk_1, \ldots, pk_n$ and the t secret keys $\boldsymbol{f}_{i_1} \cdot S, \ldots, \boldsymbol{f}_{i_t} \cdot S$. In other words, after seeing these values, even an information-theoretic adversary cannot determine S uniquely - only the set $\tilde{S}$ can be determined.

We say that a **PKE scheme with linear key homomorphism is a linear hash proof system with respect to the compatible map** f if the following two requirements are satisfied. We refer to these requirements as *uniform decryption of invalid ciphertexts* and *computational indistinguishability of valid/invalid ciphertexts*.

Uniform Decryption of Invalid Ciphertexts. With all but negligible probability over the choice of $sk_1, pk_1, \ldots, sk_n, pk_n$ and the random coins of the invalid encryption algorithm, for any choice of distinct rows $\boldsymbol{f}_{i_1}, \ldots, \boldsymbol{f}_{i_{t+1}}$ of F, an invalid ciphertext encrypted to $pk_{ID_{i_{t+1}}}$ must decrypt to a message distributed negligibly close to uniform over the message space when decrypted with a secret key chosen at random from $\boldsymbol{f}_{i_{t+1}} \cdot \tilde{S}$. More precisely, an element of $\tilde{S}$ is chosen uniformly at random, and the resulting matrix is multiplied on the left by $\boldsymbol{f}_{i_{t+1}}$ to produce the secret key.

Computational Indistinguishability of Valid/Invalid Ciphertexts. Second, we require valid and invalid ciphertexts are computationally indistinguishable in the following sense. For any fixed (distinct) $\boldsymbol{f}_{i_1}, \ldots, \boldsymbol{f}_{i_{t+1}}$, we consider the following game between a challenger and an attacker $\mathcal{A}$:

$Game_{hp}$: The challenger starts by sampling $(sk_1, pk_1), \ldots, (sk_n, pk_n)$ as above, and gives the attacker the public parameters and $pk_1, \ldots, pk_n$. The attacker may adaptively choose distinct rows $\boldsymbol{f}_{i_1}, \ldots, \boldsymbol{f}_{i_{t+1}}$ in F in any order it likes. (For convenience, we let $\boldsymbol{f}_{i_{t+1}}$ always denote the vector that will be encrypted under, but we note that this may be chosen before some of the other $\boldsymbol{f}_i$'s.) Upon setting an $\boldsymbol{f}_{i_j}$ for $j \neq t + 1$, the attacker receives $\boldsymbol{f}_{i_j} \cdot S$. When it sets $\boldsymbol{f}_{i_{t+1}}$, it also chooses a message m. At this point, the challenger flips a coin $\beta \in \{0, 1\}$, and encrypts m to the public key corresponding to $\boldsymbol{f}_{i_{t+1}} \cdot S$ as follows. We let pk_{ch} denote the public key corresponding to $\boldsymbol{f}_{i_{t+1}} \cdot S$. If $\beta = 0$, it calls Encrypt with m, pk_{ch}. If $\beta = 1$, it calls InvalidEncrypt with $m, \boldsymbol{f}_{i_{t+1}} \cdot S$. It gives the resulting ciphertext to the attacker, who produces a guess β' for β.

We denote the advantage of the attacker by $Adv_{\mathcal{A}}^{hp} = \left| \mathbb{P}[\beta = \beta'] - \frac{1}{2} \right|$. We require that $Adv_{\mathcal{A}}^{hp}$ be negligible for all PPT attackers $\mathcal{A}$.

3.4   Construction

Given a PKE scheme (KeyGen, Encrypt, Decrypt) and an identity mapping f having the properties defined above, we now construct a bounded-collision IBE scheme. We let t denote our collision parameter, and n will be the dimension of the image of f.

$Setup(\lambda) \rightarrow$ PP, MSK. The setup algorithm for the IBE scheme calls the key generation algorithm of the PKE scheme to generate n random $sk_1, pk_1, \ldots, sk_n, pk_n$ pairs, sharing the same public parameters. The public parameters PP of the IBE scheme are defined to be these shared public parameters as well as $pk_1, \ldots, pk_n$. The master secret key MSK is the collection of secret keys $sk_1, \ldots, sk_n$.

$KeyGen(ID, \text{MSK}) \rightarrow \text{SK}_{ID}$. The key generation algorithm takes an identity in the domain of f and first maps it into R^n as $f(ID) = (id_1, \ldots, id_n)$. It then computes SK_{ID} as an R-linear combination of $sk_1, \ldots, sk_n$, with coefficients $id_1, \ldots, id_n$: $\text{SK}_{ID} = \sum_{i=1}^{n} id_i sk_i$.

$Encrypt(m, \text{PP}, ID) \rightarrow$ CT. The encryption algorithm takes in a message in the message space of the PKE scheme. From the public parameters PP, it computes a public key corresponding to SK_{ID} using the linear key homomorphism property (we note that the mapping f is known and efficiently computable). It then runs the PKE encryption algorithm on m with this public key to produce CT.

$Decrypt(\text{CT}, \text{SK}_{ID}) \rightarrow m$. The decryption algorithm runs the decryption algorithm of the PKE, using SK_{ID} as the secret key.

3.5   Security

Theorem 1. *When a PKE scheme (KeyGen, Encrypt, Decrypt) with linear key homomorphism and a compatible identity mapping f satisfy the linear hash proof property, then the construction defined in Section 3.4 is a secure bounded-collusion IBE scheme with collusion parameter t.*

Proof. We first change from the real security game defined in Section 2.1 to a new Game′ in which the challenger calls the invalid encryption algorithm to form an invalid ciphertext. We argue that if the adversary's advantage changes by a non-negligible amount, this violates the computational indistinguishability of valid/invalid ciphertexts. To see this, we consider a PPT adversary $\mathcal{A}$ whose advantage changes non-negligibly. We will construct a PPT adversary $\mathcal{A}'$ against Game$_{hp}$. The challenger for Game$_{hp}$ gives $\mathcal{A}'$ the public parameters and $pk_1, \ldots, pk_n$, which $\mathcal{A}'$ forwards to $\mathcal{A}$. When $\mathcal{A}$ requests a secret key for an identity corresponding to $\boldsymbol{f}_{i_j}$, $\mathcal{A}'$ can forward $\boldsymbol{f}_{i_j}$ to its challenger and obtain the corresponding secret key. When $\mathcal{A}$ declares m_0, m_1 and some ID^* corresponding to $\boldsymbol{f}_{i_{t+1}}$, $\mathcal{A}'$ chooses a random bit $b \in \{0,1\}$ and sends $m_b, \boldsymbol{f}_{i_{t+1}}$ to its challenger. It receives a ciphertext encrypting m_b, which it forwards to $\mathcal{A}$. We note here that the $t+1$ distinct identities chosen by $\mathcal{A}$ correspond to distinct rows of F. If the challenger for $\mathcal{A}'$ is calling the regular encryption algorithm, then $\mathcal{A}'$ has properly simulated the real security game for $\mathcal{A}$. If it is calling the invalid encryption algorithm, then $\mathcal{A}'$ has properly simulated the new game, Game′. Hence, if $\mathcal{A}$ has a non-negligible change in advantage, $\mathcal{A}'$ can leverage this to obtain a non-negligible advantage in Game$_{hp}$.

In Game', we argue that information-theoretically, the attacker's advantage must be negligible. We observe that in our definition of the linear hash proof property, the subset $\tilde{S}$ of $R^{n \times d}$ is precisely the subset of possible MSK's that are consistent with the public parameters and requested secret keys that the attacker receives in the game, and each of these is equally likely. Since the invalid ciphertext decrypts to an essentially random message over this set (endowed with the uniform distribution), the attacker cannot have a non-negligible advantage in distinguishing the message.

4   QR-Based Construction

We now present a PKE scheme with linear key homomorphism and a compatible identity mapping f such that this is a linear hash proof system with respect to f under the quadratic residuosity assumption.

QR-based PKE Construction. We define the message space to be $\{-1, 1\}$. The public parameters of the scheme are a Blum integer $N = pq$, where primes $p, q \equiv 3 \bmod 4$ and $\mathbb{QR}_N$ is cyclic, and an element g that is a random quadratic residue modulo N. Our public keys will be elements of $\mathbb{Z}_N$, while our secret keys are elements of the ring $R := \mathbb{Z}$. We define the subset R' to be $[\rho(N)]$. We will later provide bounds for appropriate settings of $\rho(N)$.

- $Gen(1^\lambda)$: The generation algorithm chooses an element sk uniformly at random in $[\rho(N)]$. This is the secret key. It then calculates the public key as $pk = g^{sk}$.
- $Enc_{pk}(m)$: The encryption algorithm chooses an odd $r \in [N^2]$ uniformly at random, and calculates $Enc(m) = (g^r, m \cdot pk^r)$.
- $Dec_{sk}(c_1, c_2)$: The decryption algorithm computes $m = c_2 \cdot (c_1^{sk})^{-1}$.

We additionally define the invalid encryption algorithm:

- $InvalidEnc_{sk}(m)$: The invalid encryption algorithm chooses a random $h \in \mathbb{J}_N \setminus \mathbb{QR}_N$ (i.e. a random non-square). It produces the invalid ciphertext as $h, m \cdot h^{sk}$.

Key Homomorphism. Considering N, g as global parameters and only $pk = g^{sk}$ as the public key, we have homomorphism over keys through multiplication and exponentiation in G for public keys and arithmetic over the integers for secret keys.

For secret keys $sk_1, sk_2 \in \mathbb{Z}$ and integers $a, b \in \mathbb{Z}$, we can form the secret key $sk_3 := ask_1 + bsk_2$ and corresponding public key $pk_3 = pk_1^a \cdot pk_2^b$ in G.

4.1   Compatible Mapping and Resulting IBE Construction

Our compatible map f is obtained from Lemma 2 (Section 2.3). We may assume that our identities are hashed to $\{0, 1\}^k$ for some k using a collision-resistant hash function, so they are in the domain of f. The image of each identity under f is a vector with 0,1 entries of length $n = k(t + 1)$, where t is our collusion parameter. For every $t + 1$ distinct elements of $\{0, 1\}^k$, their images under f are linearly independent (over $\mathbb{Z}_2$ as well as $\mathbb{Q}$).

A formal description of our construction follows. This is an instance of the general construction in Section 3.4, but we state it explicitly here for the reader's convenience. We assume that messages to be encrypted are elements of $\{-1, +1\}$, and identities are elements of $\{0, 1\}^k$. For each identity ID, we let ID^{T} denote the row vector of length n over $\{0, 1\}$ obtained by our mapping from $\{0, 1\}^k$ to binary vectors of length n.

Setup. The setup algorithm chooses a Blum integer N such that $\mathbb{QR}_N$ is cyclic and a random element $g \in \mathbb{QR}_N$. It then generates n key pairs of the PKE $((pk_1, sk_1), (pk_2, sk_2), ...(pk_n, sk_n))$ using the common g, and publishes the public keys (along with N, g) as the public parameters. The master secret key consists of the corresponding secret keys, $sk_1, \ldots, sk_n$. These form an $n \times 1$ vector $\dot{S}$ with entries in $[\rho(N)]$ (the ith component of S is equal to sk_i for $i = 1 \ldots n$).

KeyGen(ID). The key generation algorithm receives an $ID \in \{0, 1\}^k$. By Lemma 2 (Section 2.3), we then have a mapping f that takes this ID to a vector $(id_1, id_2, ...id_n)$, such that the vectors corresponding to $t + 1$ different ID's are linearly independent. The secret key for ID will be an element of $\mathbb{Z}$, which is computed as a linear combination of the values $sk_1, \ldots, sk_n$, with coefficients $id_1, \ldots, id_n$ respectively. We express this as $\mathrm{SK}_{ID} := \sum_{i=1}^{n}(sk_i \cdot id_i)$, where the sum is taken over $\mathbb{Z}$. Since the mapping f provided in Section 2.3 produces vectors $(id_1, \ldots, id_n)$ with 0,1 entries, the value of SK_{ID} is at most $\rho(N)n$. Since n will be much less than $\rho(N)$, this will require roughly $\log \rho(N)$ bits to represent.

Encrypt(ID, m, PP). We let $\mathrm{PK}_{ID} := \prod_{i=1}^{n}(pk_i^{id_i})$. Anyone can compute this using the multiplicative key homomorphism and the published pk_i values. Since by the key homomorphism (PK_{ID}, SK_{ID}) is still a valid keypair for the original PKE, encryption and decryption can function as for the PKE. In other words, the encryptor runs the encryption algorithm for the PKE scheme with PK_{ID} as the public key to produce the ciphertext CT.

Note that for ciphertexts, we now have

$$Enc_{PK_{ID}}(m) = (g^r, m \cdot ((PK_{ID})^r))$$
$$= \left(g^r, m \cdot \prod_{i=1}^{n}(pk_i^{id_i \cdot r})\right) = \left(g^r, m \cdot \prod_{i=1}^{n} g^{id_i \cdot sk_i \cdot r}\right).$$

All arithmetic here takes place modulo N.

This can alternately be expressed as: $Enc_{PK_{ID}}(m) = \left(g^r, m \cdot g^{(ID)^{\mathrm{T}} Sr}\right)$ where $S = (sk_i)_{n \times 1}$ is a vector over $\mathbb{Z}$ containing the n PKE secret keys of the master secret key.

Decrypt(CT, SK_{ID}). The decryption algorithm runs the decryption algorithm of the PKE with SK_{ID} as the secret key.

4.2   Security of the IBE

We now prove security of IBE scheme up to t collusions. This will follow from Theorem 1 and the theorem below.

Theorem 2. *Under the QR assumption, the PKE construction in Section 4 is a linear hash proof system with respect to f when $\rho(N)$ is sufficiently large. When $\log(N) = \Omega(n^2 \log n)$, $\rho(N) = N^\ell$ for some constant ℓ suffices.*

We note that when $\rho(N) = N^\ell$, our secret keys are of size $O(\log N) = O(\lambda)$. We prove this theorem in two lemmas.

Lemma 3. *Under the QR assumption, computational indistinguishability of valid and invalid ciphertexts holds.*

Proof. We suppose there exists a PPT adversary $\mathcal{A}$ with non-negligible advantage in Game$_{hp}$. We will create a PPT algorithm $\mathcal{B}$ with non-negligible advantage against the QR assumption. We simplify/abuse notation a bit by letting $f_1, \ldots, f_{t+1}$ denote the distinct rows of f that are chosen adaptively by $\mathcal{A}$ during the course of the game (these were formerly called $f_{i_1}, \ldots, f_{i_{t+1}}$).

$\mathcal{B}$ is given (N, h), where N is a Blum integer such that $\mathbb{QR}_N$ is cyclic and h is either a random element of $\mathbb{J}_N \backslash \mathbb{QR}_N$ or a random element of $\mathbb{QR}_N$. Crucially, $\mathcal{B}$ does not know the factorization of N. $\mathcal{B}$ sets g to be a random element of $\mathbb{QR}_N$.

It chooses an $n \times 1$ vector $S = (sk_i)$, whose entries are chosen uniformly at random from $[\rho(N)]$. For each i from 1 to n, the ith entry of S is denoted by sk_i. It computes $pk_i = g^{sk_i} \bmod N$ and gives the public parameters PP $= (N, g, pk_1, \ldots, pk_n)$ to $\mathcal{A}$. We note that $\mathcal{B}$ knows the MSK $= S$, so it can compute $f_1 \cdot S, \ldots, f_t \cdot S$ and give these to $\mathcal{A}$ whenever $\mathcal{A}$ chooses the vectors $f_1, \ldots, f_t$.

At some point, $\mathcal{A}$ declares a message m and a vector f_{t+1} corresponding to identity ID^*. $\mathcal{B}$ encrypts m using the following ciphertext: $\left(h, m \cdot h^{(ID^{*\mathrm{T}})S} \right)$.

We consider two cases, depending on the distribution of h.

Case 1: h is random in $\mathbb{QR}_N$. When h is a random square modulo N, we claim that the ciphertext is properly distributed as a valid ciphertext. More precisely, we claim that the distribution of h and the distribution of g^r for a random odd $r \in [N^2]$ are negligibly close. This follows from the fact that $\mathbb{QR}_N$ is cyclic of order $\frac{\phi(N)}{4}$, and the reduction of a randomly chosen odd $r \in [N^2]$ modulo $\frac{\phi(N)}{4}$ will be distributed negligibly close to uniform.

Case 2: h is random in $\mathbb{J}_N \backslash \mathbb{QR}_N$. In this case, $\mathcal{B}$ has followed the specification of the invalid encryption algorithm.

Thus, if $\mathcal{A}$ has a non-negligible advantage in distinguishing between valid and invalid ciphertexts, then $\mathcal{B}$ can leverage $\mathcal{A}$ to obtain non-negligible advantage against the QR assumption.

Lemma 4. *Uniform decryption of invalid ciphertexts holds when $\rho(N)$ is sufficiently large. When $\log(N) = \Omega(n^2 \log n)$, $\rho(N) = N^\ell$ for some constant ℓ suffices.*

Proof. We choose S with uniformly random entries in $[\rho(N)]$. We then fix any $t+1$ distinct rows of F, denoted by $\boldsymbol{f}_1, \ldots, \boldsymbol{f}_{t+1}$. We must argue that the value of $\boldsymbol{f}_{t+1} \cdot S$ modulo 2 is negligibly close to uniform, conditioned on $\boldsymbol{f}_1 \cdot S, \ldots, \boldsymbol{f}_t \cdot S$ and S modulo $\frac{\phi(N)}{4}$. To see why this is an equivalent statement of the uniform decryption of invalid ciphertexts property for our construction, note that the decryption of an invalid ciphertext is computed as follows. We let sk denote the secret key the ciphertext was generated with, and sk^* denote another secret key for the same public key used for decryption: $Dec(sk^*, (h, mh^{sk})) = m(-1)^{sk-sk^*}$, since $sk \equiv sk^* \bmod \phi(N)/4$ in order to both have the same public key. If we think of S as fixed and $\tilde{S}$ as the set of vectors with entries in $[\rho(N)]$ that yield the same values of $\boldsymbol{f}_1 \cdot S, \ldots, \boldsymbol{f}_t \cdot S$ and S modulo $\frac{\phi(N)}{4}$, we can restate our goal as showing that the distribution of $\boldsymbol{f}_{t+1} \cdot S' \bmod 2$ is negligibly close to uniform, where S' is chosen uniformly at random from $\tilde{S}$.

We know by Lemma 2 that the vectors $\boldsymbol{f}_1, \ldots, \boldsymbol{f}_{t+1}$ are linearly independent as vectors over $\mathbb{Z}_2$. This implies that these vectors are linearly independent as vectors over $\mathbb{Q}$ as well. We let $Ker_\mathbb{Q}(\boldsymbol{f}_1, \ldots, \boldsymbol{f}_t)$ denote the $(n-t)$-dimensional kernel of these vectors as a subspace of $\mathbb{Q}^n$.

Our strategy is to prove that this space contains a vector $\boldsymbol{p}$ with integer entries that is *not* orthogonal to $\boldsymbol{f}_{t+1}$ modulo 2. Then, for every S' in $S+W$, $S' + \frac{\phi(N)}{4}\boldsymbol{p}$ is also in $S+W$. Here we are using the notation from Section 3 where we defined W. In this instance, $S+W$ is the set of vectors yielding the same values as S for $\boldsymbol{f}_1 \cdot S, \ldots, \boldsymbol{f}_t \cdot S$ and S modulo $\frac{\phi(N)}{4}$. $\tilde{S}$ is then the intersection of $S+W$ with the set of vectors having all of their entries in $[\rho(N)]$.

To complete the argument, we need to prove that for most elements of $S' \in \tilde{S}$ (all but a negligible proportion), $S' + \frac{\phi(N)}{4}\boldsymbol{p}$ will also be in $\tilde{S}$ (i.e. have entries in $[\rho(N)]$). This will follow from showing that there exists a $\boldsymbol{p}$ with reasonably bounded entries, and also that the set $\tilde{S}$ contains mostly vectors whose entries stay a bit away from the boundaries of the region $[\rho(N)]$.

We will use the following lemmas. The proof the second can be found in the full version.

Lemma 5. *Let A be a $t \times n$ matrix of rank t over $\mathbb{Q}$ with entries in $\{0, 1\}$. Then there exists a basis for the kernel of A consisting of vectors with integral entries all bounded by $n^{\frac{t}{2}} t^{\frac{t}{4}}$.*

Proof. This is an easy consequence of Theorem 2 in [3], which implies the existence of a basis with entries all bounded in absolute value by $\sqrt{det(AA^\mathrm{T})}$. We note that AA^T is a $t \times t$ matrix with integral entries between 0 and n. Dividing each row by n, we obtain a matrix with rational entries between 0 and 1, and can then apply Hadamard's bound [23] to conclude that the determinant of this rational matrix has absolute value at most $t^{\frac{t}{2}}$. Thus, the determinant of AA^T has absolute value at most $n^t t^{\frac{t}{2}}$. Applying Theorem 2 in [3], the lemma follows.

Lemma 6. *We suppose that M is $d \times n$ matrix with integral entries all of absolute value at most B and rank d over $\mathbb{Q}$. Then there exists another $d \times n$ matrix M' with integral entries of absolute value at most $2^{d-1}B$ that has the same rowspan as M over $\mathbb{Q}$ and furthermore remains rank d when its entries are reduced modulo 2.*

Combining these two lemmas, we may conclude that there exists a basis for $Ker_{\mathbb{Q}}(\boldsymbol{f}_1, \ldots, \boldsymbol{f}_t)$ with integral entries all having absolute value at most $C := 2^{n-t-1} n^{\frac{t}{2}} t^{\frac{t}{4}}$ that remains of rank $n - t$ when reduced modulo 2. Now, if all of these basis vectors are orthogonal to $\boldsymbol{f}_{t+1}$ modulo 2, then these form a $(n-t)$-dimensional space that is contained in the kernel of the $(t+1)$-dimensional space generated by $\boldsymbol{f}_1, \ldots, \boldsymbol{f}_t, \boldsymbol{f}_{t+1}$ in $\mathbb{Z}_2^n$. This is a contradiction. Thus, at least one of the basis vectors is not orthogonal to $\boldsymbol{f}_{t+1}$ modulo 2. Since it is orthogonal to $\boldsymbol{f}_1, \ldots, \boldsymbol{f}_t$ over $\mathbb{Q}$ and has integral entries of absolute value at most C, this is our desired $\boldsymbol{p}$.

Now, the set of vectors $\tilde{S}$ can be described as the intersection of the set

$$S + \frac{\phi(N)}{4} Ker_{\mathbb{Z}}(\boldsymbol{f}_1, \ldots, \boldsymbol{f}_t)$$

with the set of vectors with coordinates all in $[\rho(N)]$, where $Ker_{\mathbb{Z}}(\boldsymbol{f}_1, \ldots, \boldsymbol{f}_t)$ denotes the vectors in $Ker_{\mathbb{Q}}(\boldsymbol{f}_1, \ldots, \boldsymbol{f}_t)$ with integral entries. Since we have a bound C on the size of entries an integer basis for the kernel, we can argue that if the coordinates of S are sufficiently bounded away from 0 and $\rho(N)$, then there will be many vectors in $\tilde{S}$, negligibly few of which themselves have entries outside of $(\frac{\phi(N)}{4}C, \rho(N) - \frac{\phi(N)}{4}C)$. Both this bound and the probability that S is indeed sufficiently bounded away from 0 and $\rho(N)$ depend only on the relationship between n and $\rho(N)$. In the full version of this paper, we prove the following lemma:

Lemma 7. *With $\rho(N), n, \boldsymbol{p}, S,$ and $\tilde{S}$ defined as above, when $\log N = \Omega(n^2 \log n)$, we can set $\rho(N) = N^\ell$ for some constant ℓ so that the fraction of $S' \in \tilde{S}$ such that $S' + \frac{\phi(N)}{4}\boldsymbol{p}$ is not also in $\tilde{S}$ is negligible with all but negligible probability over the choice of S.*

Thus, ignoring negligible factors, we can consider $\tilde{S}$ as partitioned into pairs of the form S' and $S' + \frac{\phi(N)}{4}\boldsymbol{p}$. For each S', the values of $\boldsymbol{f}_{t+1} \cdot S'$ and $\boldsymbol{f}_{t+1} \cdot \left(S' + \frac{\phi(N)}{4}\boldsymbol{p}\right)$ modulo 2 are different. Thus, the distribution of $\boldsymbol{f}_{t+1} \cdot S' \mod 2$ over $S' \in \tilde{S}$ is sufficiently close to uniform.

5   Open Problems

It remains to find additional constructions within this framework based on other assumptions; in particular, lattice-based constructions may be possible. It would also be interesting to extend this framework to accommodate stronger security requirements, such as CCA-security. Finally, constructing a fully collusion-resistant IBE from the QR assumption in the standard model remains a challenging open problem.

Acknowledgements. We thank David Zuckerman for pointing us to the use of dual BCH codes for the mapping f and Mark Lewko for pointing us to [3].

References

1. Agrawal, S., Boneh, D., Boyen, X.: Efficient Lattice (H)IBE in the Standard Model. In: Gilbert, H. (ed.) EUROCRYPT 2010. LNCS, vol. 6110, pp. 553–572. Springer, Heidelberg (2010)
2. Alwen, J., Dodis, Y., Naor, M., Segev, G., Walfish, S., Wichs, D.: Public-Key Encryption in the Bounded-Retrieval Model. In: Gilbert, H. (ed.) EUROCRYPT 2010. LNCS, vol. 6110, pp. 113–134. Springer, Heidelberg (2010)
3. Bombieri, E., Vaaler, J.: On siegel's lemma. Inventiones Mathematicae 73, 11–32 (1983), doi:10.1007/BF01393823
4. Boneh, D., Boyen, X.: Efficient Selective-ID Secure Identity-Based Encryption Without Random Oracles. In: Cachin, C., Camenisch, J.L. (eds.) EUROCRYPT 2004. LNCS, vol. 3027, pp. 223–238. Springer, Heidelberg (2004)
5. Boneh, D., Boyen, X.: Secure Identity Based Encryption Without Random Oracles. In: Franklin, M. (ed.) CRYPTO 2004. LNCS, vol. 3152, pp. 443–459. Springer, Heidelberg (2004)
6. Boneh, D., Franklin, M.: Identity-Based Encryption from the Weil Pairing. In: Kilian, J. (ed.) CRYPTO 2001. LNCS, vol. 2139, pp. 213–229. Springer, Heidelberg (2001)
7. Boneh, D., Gentry, C., Hamburg, M.: Space-efficient identity based encryption without pairings. In: FOCS, pp. 647–657 (2007)
8. Boneh, D., Halevi, S., Hamburg, M., Ostrovsky, R.: Circular-Secure Encryption from Decision Diffie-Hellman. In: Wagner, D. (ed.) CRYPTO 2008. LNCS, vol. 5157, pp. 108–125. Springer, Heidelberg (2008)
9. Brakerski, Z., Goldwasser, S.: Circular and Leakage Resilient Public-Key Encryption under Subgroup Indistinguishability (or: Quadratic Residuosity Strikes Back). In: Rabin, T. (ed.) CRYPTO 2010. LNCS, vol. 6223, pp. 1–20. Springer, Heidelberg (2010)
10. Brakerski, Z., Tauman Kalai, Y., Katz, J., Vaikuntanathan, V.: Overcoming the hole in the bucket: Public-key cryptography resilient to continual memory leakage. In: FOCS, pp. 501–510 (2010)
11. Canetti, R., Halevi, S., Katz, J.: A Forward-secure Public-key Encryption Scheme. In: Biham, E. (ed.) EUROCRYPT 2003. LNCS, vol. 2656, pp. 255–271. Springer, Heidelberg (2003)
12. Cash, D., Hofheinz, D., Kiltz, E., Peikert, C.: Bonsai Trees, or How to Delegate a Lattice Basis. In: Gilbert, H. (ed.) EUROCRYPT 2010. LNCS, vol. 6110, pp. 523–552. Springer, Heidelberg (2010)
13. Cocks, C.: An Identity Based Encryption Scheme Based on Quadratic Residues. In: Honary, B. (ed.) Cryptography and Coding 2001. LNCS, vol. 2260, pp. 360–363. Springer, Heidelberg (2001)
14. Cramer, R., Shoup, V.: Universal Hash Proofs and a Paradigm for Adaptive Chosen Ciphertext Secure Public-Key Encryption. In: Knudsen, L.R. (ed.) EUROCRYPT 2002. LNCS, vol. 2332, pp. 45–64. Springer, Heidelberg (2002)
15. Dodis, Y., Fazio, N.: Public Key Broadcast Encryption for Stateless Receivers. In: Feigenbaum, J. (ed.) DRM 2002. LNCS, vol. 2696, pp. 61–80. Springer, Heidelberg (2003)

16. Dodis, Y., Katz, J., Xu, S., Yung, M.: Key-Insulated Public Key Cryptosystems. In: Knudsen, L.R. (ed.) EUROCRYPT 2002. LNCS, vol. 2332, pp. 65–82. Springer, Heidelberg (2002)
17. Fiat, A., Naor, M.: Broadcast Encryption. In: Stinson, D.R. (ed.) CRYPTO 1993. LNCS, vol. 773, pp. 480–491. Springer, Heidelberg (1994)
18. Gafni, E., Staddon, J., Yin, Y.L.: Efficient Methods for Integrating Traceability and Broadcast Encryption. In: Wiener, M. (ed.) CRYPTO 1999. LNCS, vol. 1666, pp. 372–387. Springer, Heidelberg (1999)
19. Garay, J.A., Staddon, J., Wool, A.: Long-Lived Broadcast Encryption. In: Bellare, M. (ed.) CRYPTO 2000. LNCS, vol. 1880, pp. 333–352. Springer, Heidelberg (2000)
20. Gentry, C.: Practical Identity-Based Encryption Without Random Oracles. In: Vaudenay, S. (ed.) EUROCRYPT 2006. LNCS, vol. 4004, pp. 445–464. Springer, Heidelberg (2006)
21. Gentry, C., Peikert, C., Vaikuntanathan, V.: Trapdoors for hard lattices and new cryptographic constructions. In: STOC, pp. 197–206 (2008)
22. Goldwasser, S., Micali, S.: Probabilistic encryption. J. Comput. Syst. Sci. 28(2), 270–299 (1984)
23. Hadamard, J.: Resolution d'une question relative aux determinants. Bull. Sci. Math. 17, 240–246 (1893)
24. Halevy, D., Shamir, A.: The LSD Broadcast Encryption Scheme. In: Yung, M. (ed.) CRYPTO 2002. LNCS, vol. 2442, pp. 47–60. Springer, Heidelberg (2002)
25. Kumar, R., Rajagopalan, S., Sahai, A.: Coding Constructions for Blacklisting Problems without Computational Assumptions. In: Wiener, M. (ed.) CRYPTO 1999. LNCS, vol. 1666, pp. 609–623. Springer, Heidelberg (1999)
26. Lewko, A., Rouselakis, Y., Waters, B.: Achieving Leakage Resilience through Dual System Encryption. In: Ishai, Y. (ed.) TCC 2011. LNCS, vol. 6597, pp. 70–88. Springer, Heidelberg (2011)
27. Naor, D., Naor, M., Lotspiech, J.: Revocation and Tracing Schemes for Stateless Receivers. In: Kilian, J. (ed.) CRYPTO 2001. LNCS, vol. 2139, pp. 41–62. Springer, Heidelberg (2001)
28. Naor, M., Segev, G.: Public-Key Cryptosystems Resilient to Key Leakage. In: Halevi, S. (ed.) CRYPTO 2009. LNCS, vol. 5677, pp. 18–35. Springer, Heidelberg (2009)
29. Rothblum, R.: Homomorphic Encryption: From Private-Key to Public-Key. In: Ishai, Y. (ed.) TCC 2011. LNCS, vol. 6597, pp. 219–234. Springer, Heidelberg (2011)
30. Shamir, A.: Identity-Based Cryptosystems and Signature Schemes. In: Blakely, G.R., Chaum, D. (eds.) CRYPTO 1984. LNCS, vol. 196, pp. 47–53. Springer, Heidelberg (1985)
31. Stinson, D.R.: On some methods for unconditionally secure key distribution and broadcast encryption. Des. Codes Cryptography 12(3), 215–243 (1997)
32. Stinson, D.R., van Trung, T.: Some new results on key distribution patterns and broadcast encryption. Des. Codes Cryptography 14(3), 261–279 (1998)
33. Stinson, D.R., Wei, R.: Combinatorial properties and constructions of traceability schemes and frameproof codes. SIAM J. Discret. Math. 11(1), 41–53 (1998)
34. Waters, B.: Efficient Identity-Based Encryption Without Random Oracles. In: Cramer, R. (ed.) EUROCRYPT 2005. LNCS, vol. 3494, pp. 114–127. Springer, Heidelberg (2005)
35. Waters, B.: Dual System Encryption: Realizing Fully Secure IBE and HIBE under Simple Assumptions. In: Halevi, S. (ed.) CRYPTO 2009. LNCS, vol. 5677, pp. 619–636. Springer, Heidelberg (2009)

A Unified Approach to Deterministic Encryption: New Constructions and a Connection to Computational Entropy

Benjamin Fuller[1], Adam O'Neill[2], and Leonid Reyzin[2]

[1] Boston University and MIT Lincoln Laboratory
[2] Boston University

Abstract. We propose a general construction of deterministic encryption schemes that unifies prior work and gives novel schemes. Specifically, its instantiations provide:

- A construction from any trapdoor function that has sufficiently many hardcore bits.
- A construction that provides "bounded" multi-message security from lossy trapdoor functions.

The security proofs for these schemes are enabled by three tools that are of broader interest:

- A weaker and more precise sufficient condition for semantic security on a high-entropy message distribution. Namely, we show that to establish semantic security on a distribution M of messages, it suffices to establish indistinguishability for all conditional distribution $M|E$, where E is an event of probability at least $1/4$. (Prior work required indistinguishability on *all* distributions of a given entropy.)
- A result about computational entropy of conditional distributions. Namely, we show that conditioning on an event E of probability p reduces the quality of computational entropy by a factor of p and its quantity by $\log_2 1/p$.
- A generalization of leftover hash lemma to correlated distributions.

We also extend our result about computational entropy to the average case, which is useful in reasoning about leakage-resilient cryptography: leaking λ bits of information reduces the quality of computational entropy by a factor of 2^λ and its quantity by λ.

1 Introduction

Public-key cryptosystems require randomness: indeed, if the encryption operation is deterministic, the adversary can simply use the public key to verify that the ciphertext c corresponds to its guess of the plaintext m by encrypting m. However, such an attack requires the adversary to have a reasonably likely guess for m in the first place. Recent results on deterministic public-key encryption (DE) (building on work in the information-theoretic symmetric-key setting [38,17,14]) have studied how to achieve security when the randomness

R. Cramer (Ed.): TCC 2012, LNCS 7194, pp. 582–599, 2012.

comes only from m itself [3,5,7,27,8,40]. DE has a number of practical applications, such as efficient search on encrypted data and securing legacy protocols (cf. [3]). It is also interesting from a foundational standpoint; indeed, its study has proven useful in other contexts: Bellare et al. [4] showed how it extends to a notion of "hedged" public-key encryption that reduces dependence on external randomness for probabilistic encryption more generally, and Dent et al. [13] adapted its notion of privacy to a notion of confidentiality for digital signatures.

However, our current understanding of DE is somewhat lacking. The constructions of [3,5,7,27], as well as their analysis techniques, are rather disparate, and some natural questions arise from them. Namely, does the scheme of [5] inherently require using the Goldreich-Levin hardcore bit? Can it be made to work with trapdoor functions rather than permutations? Is the single-message security achieved by [5,7,27] an inherent limitation of standard model (i.e., non-random-oracle) schemes? In this work our main goal is to provide a *unified framework* for the construction of DE and to shed light on these questions.

1.1   Our Results

A SCHEME BASED ON TRAPDOOR FUNCTIONS. We propose a general *Encrypt-with-Hardcore* (EwHCore) construction of DE from trapdoor functions (TDFs), which generalizes the basic idea behind the schemes of [3,5] and leads to a unified framework for the construction of DE. Let f be a TDF with a hardcore function hc, and let $\mathcal{E}$ be any probabilistic public-key encryption algorithm. Our scheme encrypts an input message x by computing $y = f(x)$ and then encrypting y using $\mathcal{E}$ with $hc(x)$ as the coins; that is, the encryption of x is $\mathcal{E}(f(x); hc(x))$.

Intuitively, this scheme requires that the output of hc be sufficiently long to provide enough random coins for $\mathcal{E}$ (in fact, it need only be sufficiently long to be used as a seed for a psuedorandom generator), and that it not reveal any partial information about x (because $\mathcal{E}$ does not necessarily protect the privacy of its random coins). There are two nontrivial technical steps needed to make intuition precise. First, we define a condition required of hc (which we call "robustness") and show that it is sufficient for security of the resulting DE. Second, through a computational entropy argument, we show how to make *any* sufficiently long hc robust by applying a randomness extractor.

This general scheme admits a number of instantiations depending of f and hc. For example, when f is any trapdoor function and hc is a random oracle (RO), we obtain the construction of [3][1]. When f is an iterated trapdoor permutation (TDP) and hc is a collection Goldreich-Levin (GL) [23] bits extracted at each iteration, we obtain the construction of [5]. When f is a lossy trapdoor function (LTDF) [35] and hc is a pairwise-independent hash, we get a variant of the construction of [7] (which is less efficient but has a more straightforward analysis). We also obtain a variant of the construction of Hemenway et al. [27]

---

[1] Technically, this construction does not even need a TDF because of the random oracle model; however, it may be prudent to use a TDF because then it seems more likely that the instantiation of the random oracle will be secure as it may be hardcore for the TDF.

under the same assumption as they use (see Section 5.2 for details). Note that in all but the last of these cases, the hardcore function is *already* robust (without requiring an extractor), which shows that in prior work this notion played an implicit role.

Moreover, this general scheme not only explains past constructions, but also gives us new ones. Specifically, if f is a trapdoor function with enough hardcore bits, we obtain:

- DE that works on the uniform distribution of messages;

- DE that works on any distribution of messages whose min-entropy is at most logarithmically smaller than maximum possible;

- assuming sufficient hardness distinguishing the output of hc from uniform (so in particular of inverting f), DE that works on even-lower entropy message distributions.

Prior results require more specific assumptions on the trapdoor function (such as assuming that it is a permutation or that it is lossy—both of which imply enough hardcore bits) in order to get constructions that work even just on the uniform distribution of messages. Furthermore, our results yield more efficient schemes (though sometimes under stronger assumptions) even in the permutation case, by avoiding iteration.

Notably, we obtain the *first* DE scheme without random oracles based on the hardness of syndrome decoding using the Niederreiter trapdoor function [32], which was shown to have linearly many hardcore bits by Freeman et al. [19] (and moreover to be "correlated input" secure) but is not known to be lossy. (A scheme in the random oracle model follows from [3].) Additionally, the RSA [37] and Paillier [34] trapdoor permutations have linearly many hardcore bits under certain computational assumptions (the "Small Solutions RSA" [39] and "Bounded Computational Composite Residuosity" [9] assumptions respectively). Therefore, we can use these TDPs to instantiate our scheme efficiently under the same computational assumptions. Before our work, DE schemes from RSA and Paillier either required many iterations [5] or decisional assumptions that imply lossiness of these TDPs [30,19,7].

SECURITY FOR MULTIPLE MESSAGES: DEFINITION AND CONSTRUCTION. An important caveat is that, as in [5,7], we can prove the above standard-model DE schemes secure only for the encryption of a *single* high-entropy plaintext, or, what was shown equivalent in [7], an unbounded number of messages drawn from a *block source* [10], where each subsequent message brings "fresh" entropy. On the other hand, the strongest and most practical security model for DE introduced by [3] considers the encryption of an unbounded number of plaintexts that have individual high entropy but may not have any conditional entropy. In order for EwHCore to achieve this, the hardcore function hc must also be robust on *correlated inputs*. (A general study of correlated-input security for the case of hash functions rather than hardcore functions was concurrently initiated in [25].) In particular, it follows from the techniques of [3] that a RO hash satisfies such a notion. This leads to a multi-message secure scheme in the RO model (as obtained in [3]). We thus have a large gap between what is (known to be) achievable with random oracles versus in the standard model.

To help bridge this gap, we propose a notion of "q-bounded" security for DE, where up to q high-entropy but arbitrarily correlated messages may be encrypted under the same public key (whose size may depend polynomially on q). We feel that if one is limited to the standard model, this notion is useful. Indeed, it seems that the requirement of previous results in the standard model—that messages come from a block source—may be difficult to guarantee: all that's needed to violate it is a single message that has low conditional entropy. Following [7], we also extend our security definition to unbounded multi-message security where messages are drawn from what we call a "q-block source" (essentially, a block source where each "block" consists of q messages which may be arbitrarily correlated but have individual high entropy); Theorem 4.2 of [7] extends to show that q-bounded multi-message security and unbounded multi-message security for q-block sources are equivalent for a given min-entropy.

Using our EwHCore construction and a generalization of the leftover hash lemma discussed below, we show q-bounded DE schemes (for long enough messages), for any polynomial q, based on LTDFs losing an $1 - O(1/q)$ fraction of the input. It is known how to build such LTDFs from the decisional Diffie-Hellman [35], d-linear [19], and decisional composite residuosity [7,19] assumptions.

1.2   Our Tools

Our results are enabled by three tools that may be of more general applicability.

A MORE PRECISE CONDITION FOR SECURITY OF DE. We revisit the definitional equivalences for DE proven by [5] and [7]. At a high level, they showed that the semantic security style definition for DE (called PRIV) introduced in the initial work of [3], which asks that a scheme hides all public-key independent[2] functions of messages drawn from some distribution is in some sense equivalent to an indistinguishability based notion for DE, which asks that it is hard to distinguish ciphertexts of messages drawn from one of two possible distributions. Notice that while PRIV can be meaningfully said to hold for a given message distribution, IND inherently talks of *pairs* of distributions. The works of [5,7] compensated for this by giving an equivalences in terms of *min-entropy*. That is, they showed that PRIV for all message distributions of min-entropy μ is implied by indistinguishability with respect to all pairs of plaintext distributions of min-entropy slightly less than μ.

We demonstrate a more precise equivalence that, for a *fixed* distribution $\mathbf{M}$, identifies a class of pairs of distributions such that if IND holds on those pairs, then PRIV holds on $\mathbf{M}$. By re-examining the equivalence proof of [5], we show that PRIV on $\mathbf{M}$ is implied by IND on all pairs of "slightly induced" distributions of $\mathbf{M} \mid \mathsf{E}$, where E is an arbitrary event of probability at least $1/4$.

---

[2] As shown in [3], the restriction to public-key independent functions is inherent here.

This first tool is needed to argue that "robustness" of hc is sufficient for security EwHCore (essentially, a robust hardcore function is one that remains hardcore on a slightly induced distribution[3]).

CONDITIONAL COMPUTATIONAL ENTROPY. We investigate how conditioning reduces computational entropy of a random variable X. Suppose you have a distribution that has *computational* entropy (such as the pair $f(r)$, hc(r) for a random r). Suppose you condition that distribution on an event E of probability p. How much computational entropy is left?

To make this question more precise, we should note that computational entropy is parameterized by quality (how distinguishable is X from a variable Z that has true entropy) and quantity (how much true entropy is there in Z).

We prove an intuitively natural result: conditioning on an event of probability p reduces the quality of metric entropy by a factor of p and the quantity of metric entropy by $\log_2 1/p$ (note that this means that the reduction in quantity and quality is the same, because the quantity of entropy is measured on log scale). Naturally, the answer becomes so simple only once the correct notion of entropy is in place. Our result holds for Metric* entropy (defined in [2,18]). This entropy is convertible (with some loss) to HILL entropy [26,2], which can then be used with randomness extractors to get pseudorandom bits.

Our result improves the bounds of Dziembowski and Pietrzak [18, Lemma 3], where the loss in the *quantity* of entropy was related to its original *quality*. The use of metric entropy simplifies the analogous result of Reingold et al. [36, Theorem 1.3] for HILL entropy. (See [20] for information on other related work [22, Lemma 3.1] and [11, Lemma 16].)

We use this result to show that randomness extractors can be used to convert a hardcore function into a robust one, through a computational entropy argument for slightly induced distributions. The result is also applicable to leakage-resilient cryptography, as demonstrated by [18]. To make the result useful in more contexts, we also provide an average-case entropy formulation, which can be helpful in situations in which not all leakage is equally informative. For the information-theoretic case, it is known that leakage of λ bits reduces the average entropy by at most λ ([15, Lemma 2.2]). We show essentially the same[4] for the computational case: if λ bits of information are leaked, then the amount of computational Metric* entropy decreases by at most λ and its quality decreases by at most 2^λ (again, this entropy can be converted to HILL entropy and be used in randomness extractors [15,28]).

(CROOKED) LEFTOVER HASH LEMMA FOR CORRELATED DISTRIBUTIONS. We show that the leftover hash lemma (LHL) [26, Lemma 4.8], as well its generalized form [15, Lemma 2.4] and the "crooked" LHL [16]) extend in a natural way to

---

[3] One could alternatively define robustness as one that remains hardcore on inputs of slightly lower entropy; however, in our proofs of robustness we would then need to go through an additional argument that distributions of lower entropy are induced by distributions of higher entropy.

[4] In case of randomized leakage, the information-theoretic result of [15, Lemma 2.2(b)] gives better bounds.

"correlated" distributions. That is, suppose we have t random variables (sources) $X_1, \ldots, X_t$, where each X_i individually has high min-entropy but may be fully determined by the outcome of some other X_j (though we assume $X_i \neq X_j$ for all $i \neq j$). We would like to apply a hash function H such that $H(X_1), \ldots, H(X_t)$ is indistinguishable from t independent copies of the uniform distribution on the range of H (also over the choice of the key for H, which is made public). We show that this is the case assuming H is $2t$-wise independent. (The standard LHL is thus $t = 1$; previously, Kiltz et al. [31] showed this for $t = 2$.) Naturally, this requires the output size of H to be about a $1/t$ fraction of its input size, so there is enough entropy to extract.

2   Preliminaries

We omit standard cryptographic definitions (see the full version for precise definitions [20]). The security parameter is denoted by k, and 1^k denotes the string of k ones. Vectors are denoted in boldface, for example $\mathbf{x}$. For convenience, we extend algorithmic notation to operate on each vector of inputs componentwise. For example, if A is an algorithm and $\mathbf{x}, \mathbf{y}$ are vectors then $\mathbf{z} \xleftarrow{\$} A(\mathbf{x}, \mathbf{y})$ denotes that $\mathbf{z}[i] \xleftarrow{\$} A(\mathbf{x}[i], \mathbf{y}[i])$ for all $1 \leq i \leq |\mathbf{x}|$. We write P_X for the distribution of random variable X and $P_X(x)$ for the probability that X puts on value $x \in \mathcal{X}$, i.e., $P_X(x) = \Pr[X = x]$. Denote by $|X|$ the size of the support of X, i.e., $|X| = |\{x \text{ s.t. } P_X(x) > 0\}|$. We often identify X with P_X when there is no danger of confusion. For a function $f : \mathcal{X} \to \mathbb{R}$, we denote the expectation of f over X by $\mathbb{E} f(X) \stackrel{\text{def}}{=} \mathbb{E}_{x \in X} f(x) \stackrel{\text{def}}{=} \sum_{x \in \mathcal{X}} P_X(x) f(x)$.

We will use the notions of min-entropy and average min-entropy (defined in [15]). For vector-valued $\mathbf{X}$ the min-entropy is the minimum of the components (see [3,5]). We use the standard notions of *collision probability* of X denoted $\mathrm{Col}(X)$ and *statistical distance* of X and Y denoted $\Delta(X, Y)$. We denote the *computational distance* between two random variables X, Y with respect to a distinguisher D as $\delta^D(X, Y)$.

Dodis et al. [15, Lemma 2.2] characterized the effect of auxiliary information on average min-entropy, namely, $\tilde{H}_\infty(A|(B, C)) \geq \tilde{H}_\infty((A, B)|C) - |B| \geq \tilde{H}_\infty(A|C) - |B|$.

We will use extractors (defined in [33]) and average-case extractors (defined in [15, Section 2.5]) and denote both by ext.

For a *(probabilistic) public-key encryption scheme*, which is a triple of algorithms $\Pi = (\mathcal{K}, \mathcal{E}, \mathcal{D})$ defined in the usual way, we will use the standard notion of *IND-CPA security* as defined in [24].

We use the standard definition of a *lossy trapdoor function (LTDF) generator* (defined in [35]) which we denote as a pair $\mathsf{LTDF} = (\mathcal{F}, \mathcal{F}')$ of algorithms.

COMPUTATIONAL ENTROPY. We use the standard notion of HILL entropy as defined in [26]. Additionally, we use a notion known as "metric-star" entropy (this notion was used in [18,21]):

Definition 1. *A distribution X has* Metric* *entropy at least k, denoted* $H_{\epsilon,s}^{\text{Metric}^*}(X) \geq k$ *if for all deterministic distinguishers D of size at most s, with outputs in $[0,1]$, there exists a distribution Y with $H_\infty(Y) \geq k$ and $\delta^D(X,Y) \leq \epsilon$.*

Equivalence (with a loss in quality) between Metric* and HILL entropy was shown in [2, Theorem 5.2]. Extractors can be applied to distributions with computational entropy to obtain pseudorandom outputs. This is well-known for HILL entropy, but the only known way to extract from Metric* entropy is first to convert Metric* to HILL entropy by using [2, Theorem 5.2]. Conditional entropy has been extended to the computational case (for both HILL [28] and Metric entropy [21]). Conditional Metric* can be defined similarly, by making the distinguisher deterministic with outputs in $[0,1]$. The Metric* to HILL conversion can be extended to the computational case as shown in [11, Lemma 18], [21, Theorem 2.7]. Average-case extractors can be used on distributions with conditional Metric* entropy by first using applying [21, Theorem 2.7].

2.1 Deterministic Encryption

An encryption scheme $\Pi = (\mathcal{K}, \mathcal{E}, \mathcal{D})$ is *deterministic* if $\mathcal{E}$ is deterministic.

SEMANTIC SECURITY OF DE. We recall the semantic-security style PRIV notion for DE from [3]. (More specifically, it is a "comparison-based" semantic-security style notion; this was shown equivalent to a "simulation-based" formulation in [5].) To encryption scheme $\Pi = (\mathcal{K}, \mathcal{E}, \mathcal{D})$, an adversary $A = (A_0, A_1, A_2)$, and $k \in \mathbb{N}$ we associate the left-most and middle experiments in Figure 1. We require that there are functions $v = v(k), \ell = \ell(k)$ such that (1) $|\mathbf{x}| = v$, (2) $|\mathbf{x}[i]| = \ell$ for all $1 \leq i \leq v$, and (3) the $\mathbf{x}[i]$ are all distinct with probability 1 over $(\mathbf{x}, t) \xleftarrow{\$} A_1(state)$ for any $state$ output by A_0. (Since in this work we only consider the definition relative to deterministic Π requirement (3) is without loss of generality.) In particular we say A *outputs vectors of size v* for v as above. Define the *PRIV advantage* of A against Π as

$$\mathbf{Adv}_{\Pi,A}^{\text{priv}}(k) = \Pr\left[\mathbf{Exp}_{\Pi,A}^{\text{priv-1}}(k) \Rightarrow 1\right] - \Pr\left[\mathbf{Exp}_{\Pi,A}^{\text{priv-0}}(k) \Rightarrow 1\right].$$

Let $\mathbb{M}$ be a class of distributions on message vectors. Define $\mathbb{A}_\mathbb{M}$ to be the class of adversaries $\{A = (A_0, A_1, A_2)\}$ such that for each $A \in \mathbb{A}_\mathbb{M}$ there is a $M \in \mathbb{M}$ for which $\mathbf{x}$ has distribution M over $(\mathbf{x}, t) \xleftarrow{\$} A_1(state)$ for any $state$ output by A_0. We say that Π is *PRIV secure for $\mathbb{M}$* if $\mathbf{Adv}_{\Pi,A}^{\text{priv}}(\cdot)$ is negligible for any PPT $A \in \mathbb{A}_\mathbb{M}$. Note that (allowing non-uniform adversaries as usual) we can without loss of generality consider only those A with "empty" A_0, since A_1 can always be hardwired with the "best" state. However, following [5] we explicitly allow state because it greatly facilitates some proofs.

INDISTINGUISHABILITY OF DE. Next we recall the indistinguishability-based formulation of security for DE [5,7]. To an encryption scheme $\Pi = (\mathcal{K}, \mathcal{E}, \mathcal{D})$, an adversary $D = (D_1, D_2)$, and $k \in \mathbb{N}$ we associate the right-most experiment in Figure 1. We make the analogous requirements on D_1 as on A_1 in the PRIV definition. Define the *IND advantage* of D against Π as $\mathbf{Adv}_{\Pi,D}^{\text{ind}}(k) =$

$2 \cdot \Pr\left[\mathbf{Exp}_{\Pi,D}^{\mathrm{ind}}(k) \Rightarrow 1\right] - 1$. Let $\mathbb{M}^*$ be a class of *pairs* of distributions on message vectors. Define $\mathbb{D}_{\mathbb{M}^*}$ to be the class of adversaries $\{D = (D_1, D_2)\}$ such that for each $D \in \mathbb{D}_{\mathbb{M}^*}$, there is a pair of distributions $(\boldsymbol{M}_0, \boldsymbol{M}_1) \in \mathbb{M}^*$ such that for each $b \in \{0,1\}$ the distribution of $\mathbf{x} \overset{\$}{\leftarrow} D_1(b)$ is $\boldsymbol{M}_b$. We say that Π is *IND secure for* $\mathbb{M}^*$ if $\mathbf{Adv}_{\Pi,D}^{\mathrm{ind}}(\cdot)$ is negligible for any PPT $D \in \mathbb{D}_{\mathbb{M}^*}$.

Expr $\mathbf{Exp}_{\Pi,A}^{\mathrm{priv}\text{-}1}(k)$:	**Expr $\mathbf{Exp}_{\Pi,A}^{\mathrm{priv}\text{-}0}(k)$:**	**Expr $\mathbf{Exp}_{\Pi,A}^{\mathrm{ind}}(k)$:**
$(pk, sk) \overset{\$}{\leftarrow} \mathcal{K}(1^k)$	$(pk, sk) \overset{\$}{\leftarrow} \mathcal{K}(1^k)$	$(pk, sk) \overset{\$}{\leftarrow} \mathcal{K}(1^k)$
$state \overset{\$}{\leftarrow} A_0(1^k)$	$state \overset{\$}{\leftarrow} A_0(1^k)$	$b \overset{\$}{\leftarrow} \{0,1\}\,;\, (\mathbf{x},t) \overset{\$}{\leftarrow} D_1(b)$
$(\mathbf{x}_1, t_1) \overset{\$}{\leftarrow} A_1(state)$	$(\mathbf{x}_1, t_1),(\mathbf{x}_0, t_0) \overset{\$}{\leftarrow} A_1(state)$	$\mathbf{c} \overset{\$}{\leftarrow} \mathcal{E}(pk, \mathbf{x})$
$\mathbf{c} \overset{\$}{\leftarrow} \mathcal{E}(pk, \mathbf{x}_1)$	$\mathbf{c} \overset{\$}{\leftarrow} \mathcal{E}(pk, \mathbf{x}_0)$	$d \overset{\$}{\leftarrow} D_2(pk, \mathbf{c})$
$g \overset{\$}{\leftarrow} A_2(pk, \mathbf{c}, state)$	$g \overset{\$}{\leftarrow} A_2(pk, \mathbf{c}, state)$	If $b = d$ ret 1 else ret 0
If $g = t_1$ ret 1 else ret 0	If $g = t_1$ ret 1 else ret 0	

Fig. 1. Security experiments for deterministic encryption

3   Our Tools

3.1   A Precise Definitional Equivalence for DE

While the PRIV definition is meaningful with respect a single message distribution $\boldsymbol{M}$, the IND definition must inherently talk of *pairs* of different message distributions. Thus, in proving an equivalence between the two notions, the best we can hope to show is that PRIV security for a message distribution $\boldsymbol{M}$ is equivalent to IND security for some *class of pairs* of message distributions (depending on $\boldsymbol{M}$). However, prior works [5,7] did not provide such a statement. Instead, they showed that PRIV security on *all* distributions of a given entropy μ is equivalent to IND security on all pairs of distributions of slightly less entropy.

INDUCED DISTRIBUTIONS. To state our result we first give some definitions relating to a notion of "induced distributions." Let X, X' be distributions (or random variables) on the same domain. For $\alpha \in \mathbb{N}$, we say that X' is an α-*induced distribution of* X if X' is a conditional distribution $X' = X \mid \mathsf{E}$ for an event E such that $\Pr[\mathsf{E}] \geq 2^{-\alpha}$. We call E the *corresponding event* to X'. We require that the pair (X, E) is efficiently samplable. Define $X[\alpha]$ to be the class of all α-induced distributions of X. Furthermore, let X_0, X_1 be two α-induced distributions of X with corresponding events $\mathsf{E}_0, \mathsf{E}_1$ respectively. Define $X^*[\alpha] = \{(X_0, X_1)\}$ to be the class of all pairs (X_0, X_1) for which there is a pair (X_0', X_1') of α-induced distributions of X such that X_0 (resp. X_1) is statistically close to X_0' (resp. X_1').[5]

THE EQUIVALENCE. We are now ready to state our result. The following theorem captures the "useful" direction that IND implies PRIV:

---

[5] We need to allow a negligible statistical distance for technical reasons. Since we will be interested in indistinguishability of functions of these distributions this will not make any appreciable difference, and hence we mostly ignore this issue in the remainder of the paper.

Theorem 1. *Let* $\Pi = (\mathcal{K}, \mathcal{E}, \mathcal{D})$ *be an encryption scheme. For any distribution* M *on message vectors, PRIV security of* Π *with respect to* M *is implied by IND security of* Π *with respect to* $M^*[2]$. *In particular, let* $A \in \mathbb{A}_M$ *be a PRIV adversary against* Π. *Then there is a IND adversary* $D \in \mathbb{D}_{M^*[2]}$ *such that for all* $k \in \mathbb{N}$

$$\mathbf{Adv}_{\Pi,A}^{\mathrm{priv}}(k) \leq 162 \cdot \mathbf{Adv}_{\Pi,D}^{\mathrm{ind}}(k) + \left(\frac{3}{4}\right)^k.$$

Furthermore, the running-time of D *is the time for at most that for* k *executions of* A *(but 4 in expectation).*

The theorem essentially follows from the techniques of [5]; details are given in [20]. Thus, our contribution here is not in providing any new technical tools used in proving this result but rather in extracting it from the techniques of [5]. In particular, our more precise statement allows us to use results about entropy of conditional distributions, which we explain next. Looking ahead, it also simplifies proofs for schemes based on one-wayness, because it is easy to argue that one-wayness is preserved on slightly induced distributions (the alternative would require an argument that distributions of lower entropy are induced by distributions of higher entropy).

To establish a definitional *equivalence*; that is, also show that PRIV implies IND, we need to further restrict the latter to pairs (that are statistically close to pairs) of *complementary* 2-induced distributions of M (which we did not do above for conceptual simplicity), where we call X_0, X_1 *complementary* if $\mathsf{E}_1 = \overline{\mathsf{E}_0}$. We stress that this further restriction is not needed for the "useful" implication above and for our security proofs.

3.2 Measuring Computational Entropy of Induced Distributions

We study how conditioning a distribution reduces its computational entropy. This result is used later in the work to show that randomness extractors can convert a hardcore function into a robust one; it also applicable to leakage-resilient cryptography. This result is simplest to understand when stated in terms of $\mathtt{Metric}^*$ computational entropy (defined in [18]) It is easy to see that conditioning on an event E with probability P_E reduces (information-theoretic) min-entropy by at most $\log P_\mathsf{E}$. We show that the same holds for the computational notion of $\mathtt{Metric}^*$ entropy if one considers reduction in both quantity and quality:

Lemma 1. *Let* X, Y *be discrete random variables. Then*

$$H_{\epsilon/P_Y(y),s'}^{\mathtt{Metric}^*}(X|Y = y) \geq H_{\epsilon,s}^{\mathtt{Metric}^*}(X) - \log 1/P_Y(y) \text{ where } s' \approx s.$$

The use of $\mathtt{Metric}^*$ entropy and an improved proof allow for a simpler and tighter formulation than results of [18, Lemma 3] and [36, Theorem 1.3] (see the full version for a comparison [20]). The proof is similar to [36] and can be found in the full version [20].

If we now consider averaging over all values of Y, we obtain the following simple formulation that expresses how much average entropy is left in X from the point of view of someone who knows Y. (This scenario naturally occurs in leakage-resilient cryptography, as exemplified in [18]).

Theorem 2. *Let X, Y be discrete random variables. Then*

$$H_{\epsilon|Y|,s'}^{\texttt{Metric}^*}(X|Y) \geq H_{\epsilon,s}^{\texttt{Metric}^*}(X) - \log |Y|, \text{ where } s' \approx s.$$

This statement is similar to the statement in the information-theoretic case (where the reduction is only in quantity) from [15, Lemma 2.2]. In the full version [20], we compare the theorem to [11, Lemma 16] and [22, Lemma 3.1].

To apply a randomness extractor, we must convert conditional $\texttt{Metric}^*$ to conditional $\texttt{HILL}$ entropy using [21, Theorem 2.7], this conversion loses some quality. Thus, the conversion should be applied only when necessary (for instance, repeated conditioning is best measured in $\texttt{Metric}^*$ entropy, and then converted to $\texttt{HILL}$ entropy once at the end). Here we provide a "HILL-to-HILL" formulation of Lemma 1.

Corollary 1. *Let X be a discrete random variable over χ and let Y be a discrete random variable. Then,*

$$H_{\epsilon',s'}^{\texttt{HILL}}(X|Y=y) \geq H_{\epsilon,s}^{\texttt{HILL}}(X) - \log 1/P_Y(y)$$

where $\epsilon' = \epsilon/P_Y(y) + \sqrt[3]{\frac{\log|\chi|}{s}}$, and $s' = \Omega(\sqrt[3]{s/\log|\chi|})$.

The Corollary follows by combining Lemma 1, [2, Theorem 5.2], and setting $\epsilon_{HILL} = \sqrt[3]{\log|\chi|/s}$ (see the full version for justification of parameters [20]).

3.3   A (Crooked) Leftover Hash Lemma for Correlated Distributions

The following generalization of the (Crooked) LHL to correlated input distributions will be very useful to us when considering bounded multi-message security in Section 6. Since our generalization of the classical LHL is a special case of our generalization of the Crooked LHL, we just state the latter here.

Lemma 2. (CLHL for Correlated Sources) *Let $\mathcal{H} \colon \mathcal{K} \times D \to R$ be a 2t-wise δ-dependent function for $t > 0$ with range R, and let $f : R \to S$ be a function. Let $\mathbf{X} = (X_1, \ldots, X_t)$ where the X_i are random variables over D such that $\mathrm{H}_\infty(X_i) \geq \mu$ for all $1 \leq i \leq n$ and moreover $\Pr[X_i = X_j] = 0$ for all $1 \leq i \neq j \leq t$. Then*

$$\Delta((K, f(\mathcal{H}(K, \mathbf{X}))), (K, f(\mathbf{U}))) \leq \frac{1}{2}\sqrt{|S|^t(t^2 2^{-\mu} + 3\delta)}$$

where $K \xleftarrow{\$} \mathcal{K}$ and $\mathbf{U} = (U_1, \ldots, U_t)$ where the U_i are all uniform and independent over R (recall that functions operate on vectors $\mathbf{X}$ and $\mathbf{U}$ component-wise).

One can further extend Lemma 2 to the case of average conditional min-entropy using the techniques of [15]. Note that the lemma implies the corresponding generalization of the classical LHL by taking $\mathcal{H}$ to have range S and f to be the identity function. The proof of the lemma, which extends the proof of the Crooked LHL in [7], is given in the full version [20].

4   Encrypt-with-Hardcore Scheme from Robust HCFs

We define a new notion of *robust* HCFs. Intuitively, robust HCFs are those that remain hardcore when the input is conditioned on any event that occurs with good probability.

Definition 2. *Let $\mathcal{F}$ be a TDF generator and let* hc *be a HCF such that* hc *is hardcore for $\mathcal{F}$ with respect to a distribution $\boldsymbol{X}$ on input vectors. For $\alpha = \alpha(k)$, we say* hc *is α-robust for $\mathcal{F}$ on $\boldsymbol{X}$ if* hc *is also hardcore for $\mathcal{F}$ with respect to the class $\boldsymbol{X}[\alpha]$ of α-induced distributions of $\boldsymbol{X}$.*

DISCUSSION. Robustness is interesting even for the classical definition of hardcore bits, where hc is boolean and a single uniform input x is generated in the security experiment. Here robustness means that hc remains hardcore even when x is conditioned on an event that occurs with good probability. It is clear that not every hardcore bit in the classical sense is robust — note, for example, that while every bit of the input to RSA is well-known to be hardcore assuming RSA is one-way [1], they are not even 1-robust since we may condition on a particular bit of the input being a fixed value.

THE SCHEME. Let $\Pi = (\mathcal{K}, \mathcal{E}, \mathcal{D})$ be a probabilistic encryption scheme, $\mathcal{F}$ be a TDF generator, and hc_f be a HCF. Assume that hc outputs binary strings of the same length as the random string r needed by $\mathcal{E}$. Define the associated "*Encrypt-with-Hardcore*" deterministic encryption scheme $\mathsf{EwHCore}[\Pi, \mathcal{F}, \mathrm{hc}] = (\mathcal{DK}, \mathcal{DE}, \mathcal{DD})$ with plaintext-space $\mathsf{PtSp} = \{0,1\}^k$ via

Alg $\mathcal{DK}(1^k)$:	**Alg $\mathcal{DE}((pk, f), x)$:**	**Alg $\mathcal{DD}((sk, f^{-1}), c)$:**
$(pk, sk) \xleftarrow{\$} \mathcal{K}(1^k)$	$r \leftarrow \mathrm{hc}_f(x)$	$y \leftarrow \mathcal{D}(sk, c)$
$(f, f^{-1}) \xleftarrow{\$} \mathcal{F}(1^k)$	$c \leftarrow \mathcal{E}(pk, f(x); r)$	$x \leftarrow f^{-1}(y)$
Return $((pk, f), (sk, f^{-1}))$	Return c	Return x

SECURITY ANALYSIS. To gain some intuition, suppose hc is hardcore for $\mathcal{F}$ on some distribution $\boldsymbol{X}$ on input vectors. One might think that PRIV security of $\mathsf{EwHCore} = \mathsf{EwHCore}[\Pi, \mathcal{F}, \mathrm{hc}]$ on $\boldsymbol{X}$ then follows by IND-CPA security of Π. However, this is not true. For example, hc may be a "natural" hardcore function (i.e., that outputs some bits of the input), and $\mathcal{E}$ may output some of its coins in the clear. This is how our notion of robustness comes into play, giving us the following theorem (for a proof and further discussion, see [20]):

Theorem 3. *Suppose Π is IND-CPA secure,* hc *is 2-robust for $\mathcal{F}$ on a distribution $\boldsymbol{M}$ on input vectors. Then $\mathsf{EwHCore}[\Pi, \mathcal{F}, \mathrm{hc}]$ is PRIV-secure on $\boldsymbol{M}$.*

5   Single-Message Instantiations of EwHCore

5.1   Getting Robust Hardcore Functions

AUGMENTED TRAPDOOR FUNCTIONS. In order to describe the conversion procedure, it is useful to introduce the notion of an "augmented" version of a TDF, which augments the description of the TDF with keying material for a HCF. More formally, let $\mathcal{F}$ be a trapdoor function generator and let H be a keyed function with keyspace $\mathcal{K}$. Define the H-*augmented version of* $\mathcal{F}$, denoted $\mathcal{F}[H]$, that on input 1^k returns $(f, K), (f^{-1}, K)$ where $(f, f^{-1}) \xleftarrow{\$} \mathcal{F}(1^k)$ and $K \xleftarrow{\$} \mathcal{K}$; evaluation is defined for $x \in \{0,1\}^k$ as $f(x)$ (i.e., evaluation just ignores K) and inversion is defined analogously.

MAKING ANY LARGE HARDCORE FUNCTION ROBUST. We show that by applying a randomness extractor in a natural way, one can convert *any* large hardcore function in the standard sense to one that is robust (with some loss in parameters). However, while the conversion procedure is natural, proving that it works turns out to be non-trivial.

Let $\mathcal{F}$ be a TDF generator, and let hc: $\{0,1\}^k \to \{0,1\}^\ell$ be an HCF for $\mathcal{F}$ on an input distribution X such that $H_\infty(X) \geq \mu$. Let ext : $\{0,1\}^\ell \times \{0,1\}^d \to \{0,1\}^m \times \{0,1\}^d$ be a strong average-case $(\ell - \alpha, \epsilon_{\text{ext}})$-extractor for $\alpha \in \mathbb{N}$. (Here we view ext as a keyed function with the *second* argument as the key.) Define a new "*extractor-augmented*" HCF hc[ext] for $\mathcal{F}[\text{ext}]$ such that hc[ext]$_s(x) = $ ext(hc$(x), s)$ for all $x \in \{0,1\}^k$ and $s \in \{0,1\}^d$. The following characterizes the α-robustness of hc[ext].

Lemma 3. *Fix* $X' \in X[\alpha]$, *and suppose there is a distinguisher* D' *against* hc[ext] *on* X'. *Then there is a distinguisher* D *against* hc *on* X *such that for all* $k \in \mathbb{N}$

$$\mathbf{Adv}^{\text{hcf}}_{\mathcal{F},X',\text{hc[ext]},D'}(k) \leq O\left(\sqrt[3]{\mathbf{Adv}^{\text{hcf}}_{\mathcal{F},X,\text{hc},D}(k)} + 2^\alpha \cdot \mathbf{Adv}^{\text{hcf}}_{\mathcal{F},X,\text{hc},D}(k) \right) + \epsilon_{\text{ext}} .$$

Furthermore, the running-time of D *is* $O((t_{D'}(k+\ell))^3)$, *where* $t_{D'}$ *is the running-time of* D.

Note that when $\alpha = \log(k)$ the security loss in the reduction is polynomial (in our application we just need $\alpha = 2$). The proof, which appears in the full version [20], relies crucially on Corollary 1.

The above conversion procedure notwithstanding, we give specific examples of hardcore functions that are already robust.

ROBUST GOLDREICH-LEVIN BITS FOR ANY TDF. In [20] we show that the Goldreich-Levin [23] (GL) hardcore function is robust. Specifically, if the function that extracts i-many independent GL bits is hardcore for $\mathcal{F}$, then it is also $O(\log k)$-robust for $\mathcal{F}$.

ROBUST BITS FOR ANY LTDF. Peikert and Waters [35] showed that LTDFs admit a simple, large hardcore function, namely a pairwise-independent hash function (the same argument applies also to universal hash functions or, more generally, randomness extractors). By using average conditional min-entropy, in [20] we show that this hardcore function is $O(\log k)$ robust.

5.2   Putting It Together

Equipped with the above results, we describe instantiations of the Encrypt-with-Hardcore scheme that both explain prior constructions and produce novel ones.

USING AN ITERATED TRAPDOOR PERMUTATION. The prior trapdoor permutation based DE scheme of Bellare et al. [5] readily provides an instantiation of EwHCore by using an iterated trapdoor permutation as the TDF. Let $\mathcal{F}$ be a TDP and hc be a hardcore bit for $\mathcal{F}$. For $i \in \mathbb{N}$ denote by $\mathcal{F}^i$ the TDP that iterates $\mathcal{F}$ i-many times. Define the Blum-Micali-Yao (BMY) [6,41] hardcore function for $\mathcal{F}^i$ via $\mathcal{BMY}^i[\mathsf{hc}](f, x) = \mathsf{hc}(x)\|\mathsf{hc}(f(x))\|\ldots\|\mathsf{hc}(f^{i-1})$. Bellare et al. [5] used the specific choice of $\mathsf{hc} = \mathcal{GL}$ (the Goldreich-Levin bit) in their scheme, which is explained by the fact that the GL bit is robust, and one can show that BMY iteration expands one robust hardcore bit to many (on a non-uniform distribution, the bit should be hardcore on all "permutation distributions" of the former).

However, due to our augmentation procedure to make any large hardcore function robust, we are no longer bound to any specific choice of hc. For example, we may choose hc to be a natural hardcore bit. In fact, it may often be the case that $\mathcal{F}$ has many simultaneously hardcore natural bits, and therefore our construction will require fewer iterations of the TDP than the construction of [5].

USING A LOSSY TDF. Using the fact that extractors are robust hardcore functions for LTDFs, we get an instantiation of the Encrypt-with-Hardcore scheme from LTDFs that is an alternative to the prior scheme of Boldyreva et al. [7] and the concurrent work of Wee [40]. Our scheme requires an LTDF with residual leakage $s \leq \mathrm{H}_\infty(X) - 2\log(1/\epsilon) - r$, where r is the number of random bits needed in $\mathcal{E}$ (or the length of a seed to a pseudorandom generator that can be used to obtain those bits).

USING 2-CORRELATED PRODUCT TDFs. Hemenway et al. [27] show a construction of DE from a *decisional 2-correlated product TDF*, namely where $\mathcal{F}$ has the property that $f_1(x), f_2(x)$ is indistinguishable from $f_1(x_1), f_2(x_2)$ where x_1, x_2 are sampled independently (in both cases for two independent public instances f_1, f_2 of $\mathcal{F}$). They show such a trapdoor function is a secure DE scheme for uniform messages. To obtain an instantiation of EwHCore under the same assumption, we can use $\mathcal{F}$ as the TDF, and an independent instance of the TDF as hc. When a randomness extractor is applied to the latter, robustness follows from Lemma 3, taking into account that the lemma holds even if the output of the hardcore function is not uniform, as long as it has high HILL entropy.

USING ANY TDF WITH A LARGE HCF. Our most novel instantiations in the single-message case come from considering TDFs that have a sufficiently large HCF but are not necessarily lossy or an iterated TDP. Let us first consider instantiations on the uniform message distribution Freeman et al. [19] shown that the Niederreiter TDF [32] has linearly many (simultaneous) hardcore bits under the "Syndrome Decoding Assumption (SDA)" and "Indistinguishability Assumption (IA)" (as defined in [19, Section 7.2]). Furthermore, the RSA [37] and Paillier [34] TDPs have linearly many hardcore bits under certain computational assumptions, namely the "Small Solutions RSA (SS-RSA) Assumption" [39] and the "Bounded Computational Composite Residuosity (BCCR) Assumption" [9] respectively. Because these hardcore functions are sufficiently long, they can be made robust via Lemma 3 and give us a linear number of *robust* hardcore bits—enough to use as randomness for $\mathcal{E}$ (expanded by a pseudorandom generator if necessary). Thus, by Theorem 3, we obtain:

Corollary 2. *Under SDA+IA for the Niederreiter TDF, DE for the uniform message distribution exists. Similarly, under SS-RSA the RSA TDP or BCCR for the Paillier TDP respectively, DE for the uniform message distribution exists.*

In particular, the first statement provides the first DE scheme without random oracles based on the hardness of syndrome decoding. (A scheme in the random oracle model follows from [3].) Moreover, the schemes provided by the second statement are nearly as efficient as the ones obtained from lossy TDFs (since they do not use iteration), and the latter typically requires decisional assumptions (in contrast to the computational assumptions used here).

If we do not wish to rely on specific assumptions, we can also get DE from strong but general assumptions, such as sub-exponential hardness. We can also obtain DE for nonuniform message distributions (the strength of the assumption needed will depend on how far the entropy of the message space is from the maximum). See [20] for details.

6   Bounded Multi-message Security and its Instantiations

6.1   The New Notion and Variations

THE NEW NOTION. Our notion of q-bounded multi-message security (or just q-bounded security) for DE is quite natural, and can be viewed as analogous to other forms of "bounded" security (see e.g. [12]). In a nutshell, it asks for security on up to q arbitrarily correlated but high-entropy messages (where we allow the public-key size to depend on q). Fix an encryption scheme $\Pi = (\mathcal{K}, \mathcal{E}, \mathcal{D})$. For $q = q(k)$ and $\mu = \mu(k)$, let $\mathbb{M}^{q,\mu}$ be the class of distributions on message vectors $M^{\mu,q} = (M_1^{\mu,q}, \ldots, M_q^{\mu,q})$ where $H_\infty(M_i^{\mu,q}) \geq \mu$ and for all $1 \leq i \leq q$ and $M_{1,q}^\mu, \ldots, M_{q,q}^\mu$ are distinct with probability 1. We say that Π is q-*bounded multi-message PRIV (resp. IND) secure for μ-sources* if it is PRIV (resp. IND) secure for $\mathbb{M}^{q,\mu}$. By Theorem 1, PRIV on $\mathbb{M}^{q,\mu}$ is equivalent to IND on $\mathbb{M}^{q,\mu-2}$.

UNBOUNDED MULTI-MESSAGE SECURITY FOR q-BLOCK SOURCES. We also consider unbounded multi-message security for what we call a q-*block source*, a generalization of a block-source [10] where every q-th message introduces some "fresh" entropy. Fix an encryption scheme $\Pi = (\mathcal{K}, \mathcal{E}, \mathcal{D})$. For $q = q(k)$, $n =$

$n(k)$, and $\mu = \mu(k)$, let $\mathbb{M}^{q,n,\mu}$ be the class of distributions on message vectors $M^{q,n,\mu} = (M_1^{q,n,\mu}, \ldots, M_{qn}^{q,n,\mu})$ such that $H_\infty(X_{qi+j} \mid X_1 = x_1, \ldots, X_{qi-1} = x_{qi-1}) \geq \mu$ for all $1 \leq i \leq n$, all $0 \leq j \leq q-1$, and all outcomes $x_1, \ldots, x_{qi-1}$ of $X_1, \ldots, X_{qi-1}$. We say that Π is q-bounded multi-message PRIV (resp. IND) secure for (μ, n)-block-sources if Π is PRIV (resp. IND) secure on $\mathbb{M}^{q,n,\mu}$. Using a similar argument to [7, Theorem 4.2], one can show equivalence of PRIV on $\mathbb{M}^{q,n,\mu}$ to IND on $\mathbb{M}^{q,n,\mu}$.

6.2   Our Basic Scheme

We cannot trivially achieve q-bounded security by running, say, q copies of a scheme secure for one message in parallel (and encrypting the i-th message under the i-th public key), since this approach would lead to a stateful scheme. The main technical tool we use to achieve the notion is Lemma 2. Combined with [15, Lemma 2.2], this tells us that a $2q$-wise independent hash function is robust on correlated input distributions of sufficient min-entropy:

Proposition 1. *For any q, let $\mathsf{LTDF} = (\mathcal{F}, \mathcal{F}')$ be an LTDF generator with input length n and residual leakage s, and let $\mathcal{H}: \mathcal{K} \times D \to R$ where $r = \log|R|$ be a $2q$-wise independent hash function. Then $\mathcal{H}$ is a 2-robust hardcore function for $\mathcal{F}$ on any input distribution $X = (X_1, \ldots, X_q)$ such that $H_\infty(X) \geq q(s + r) + 2\log q + 2\log(1/\epsilon) - 2$ for negligible ϵ.*

By Theorem 3, we obtain a q-bounded DE scheme based on lossy trapdoor functions that lose a $1 - O(1/q)$ fraction of its input. Specifically, we can use the DDH-based construction of Peikert and Waters [35], the Paillier-based one of [7,19], or the one from d-linear of [19] for any polynomial q.

6.3   Our Optimized Scheme

We show that by extending some ideas of [7], we obtain a more efficient DE scheme meeting q-bounded security that achieves better parameters.

INTUITION AND PRELIMINARIES. Intuitively, for the optimized scheme we modifying the scheme of [7] to first pre-process an input message using a $2q$-wise independent permutation (instead of pairwise as in [7]). However, there are two issues to deal with here. First, for $q > 1$ such a permutation is not known to exist (in an explicit and efficiently computable sense). Second, Lemma 2 applies to t-wise independent *functions* rather than permutations.

To solve the first problem, we turn to $2q$-wise "δ-dependent" permutations (as constructed in e.g. [29]). Namely, say that a permutation $H: \mathcal{K} \times D \to D$ is *t-wise δ-dependent* if for all distinct $x_1, \ldots, x_t \in D$

$$\Delta((H(K, x_1), \ldots, H(K, x_t)), (P_1, \ldots, P_t)) \leq \delta,$$

where $K \xleftarrow{\$} \mathcal{K}$ and $P_1, \ldots, P_t$ are defined iteratively by taking P_1 to be uniform on D and, for all $2 \leq i \leq t$, taking P_i to be uniform on $R \setminus \{p_1, \ldots, p_{i-1}\}$ where $p_1, \ldots, p_{i-1}$ are the outcomes of $P_1, \ldots, P_{i-1}$ respectively.

To solve the second problem, we show that a t-wise δ-dependent permutation is a t-wise δ'-dependent function where δ' is a bit bigger than δ (see [20] for details, where we also restate Lemma 2 in terms of δ-dependent permutations).

THE CONSTRUCTION. We now detail our construction. Let $\mathsf{LTDF} = (\mathcal{F}, \mathcal{F}')$ be an LTDF and let $\mathcal{P} \colon \mathcal{K} \times \{0,1\}^k \to \{0,1\}^k$ be an efficiently invertible family of permutations on k bits. Define the associated deterministic encryption scheme $\Pi[\mathsf{LTDF}, \mathcal{P}] = (\mathcal{DK}, \mathcal{DE}, \mathcal{DD})$ with plaintext-space $\mathrm{PtSp} = \{0,1\}^k$ via

Alg $\mathcal{DK}(1^k)$:	**Alg** $\mathcal{DE}((f, K), x)$:	**Alg** $\mathcal{DD}((sk, f^{-1}), c)$:
$(f, f^{-1}) \xleftarrow{\$} \mathcal{F}(1^k)$; $K \xleftarrow{\$} \mathcal{K}$	$c \leftarrow f(\mathcal{P}(K, x))$	$x \leftarrow f^{-1}(\mathcal{P}^{-1}(K, c))$
Return $((f, K), (f^{-1}, K))$	Return c	Return x

We have the following result:

Theorem 4. *Suppose* LTDF *is a lossy trapdoor function on* $\{0,1\}^n$ *with residual leakage* s, *and let* $q, \epsilon > 0$. *Suppose* $\mathcal{P}$ *is a* $2q$-wise δ-dependent permutation on $\{0,1\}^n$ for $\delta = t^2/2^n$. *Then for any* q-message IND *adversary* $B \in \mathbb{D}_{\mathbf{M}^{q,\mu}}$ *with min-entropy* $\mu \geq qs + 2\log q + \log(1/\epsilon) + 5$, *there is an LTDF distinguisher* D *such that for all* $k \in \mathbb{N}$

$$\mathbf{Adv}^{\mathrm{ind}}_{\Pi[\mathsf{LTDF}, \mathcal{P}], B}(k) \leq \mathbf{Adv}^{\mathrm{ltdf}}_{\mathsf{LTDF}, D}(k) + \epsilon .$$

Furthermore, the running-time of D *is the time to run* B.

An efficiently invertible $2q$-wise δ-dependent permutation on $\{0,1\}^n$ for $\delta = t^2/2^n$ can be obtained from [29] using key length $nt + \log(1/\delta) = n(t+1) - 2t$. Comparing the above to Proposition 1, we see that we have dropped the r in the entropy bound (indeed, there is no hardcore function here).

Acknowledgements. The authors are grateful to Mihir Bellare, Alexandra Boldyreva, Kai-Min Chung, Sebastian Faust, Marc Fischlin, Serge Fehr, Péter Gács, Bhavana Kanukurthi, Fenghao Liu, Payman Mohassel, Krzysztof Pietrzak, Adam Smith, Ramarathnam Venkatesan, and Hoeteck Wee for helpful discussions, improvements to our analysis, and useful references. The work was supported, in part, by NSF awards 0546614, 0831281, 1012910, 1012798, CNS-0915361, CNS-0952692, NSF CAREER award 0545659, and NSF Cyber Trust award 0831184.

References

1. Alexi, W., Chor, B., Goldreich, O., Schnorr, C.P.: RSA and Rabin functions: Certain parts are as hard as the whole. SIAM J. Comput. 17(2) (1988)
2. Barak, B., Shaltiel, R., Wigderson, A.: Computational analogues of entropy. In: 11th International Conference on Random Structures and Algorithms, pp. 200–215 (2003)
3. Bellare, M., Boldyreva, A., O'Neill, A.: Deterministic and Efficiently Searchable Encryption. In: Menezes, A. (ed.) CRYPTO 2007. LNCS, vol. 4622, pp. 535–552. Springer, Heidelberg (2007)

4. Bellare, M., Brakerski, Z., Naor, M., Ristenpart, T., Segev, G., Shacham, H., Yilek, S.: Hedged Public-Key Encryption: How to Protect against Bad Randomness. In: Matsui, M. (ed.) ASIACRYPT 2009. LNCS, vol. 5912, pp. 232–249. Springer, Heidelberg (2009)
5. Bellare, M., Fischlin, M., O'Neill, A., Ristenpart, T.: Deterministic Encryption: Definitional Equivalences and Constructions without Random Oracles. In: Wagner, D. (ed.) CRYPTO 2008. LNCS, vol. 5157, pp. 360–378. Springer, Heidelberg (2008)
6. Blum, M., Micali, S.: How to generate cryptographically strong sequences of pseudo-random bits. SIAM J. Comput. 13(4), 850–864 (1984)
7. Boldyreva, A., Fehr, S., O'Neill, A.: On Notions of Security for Deterministic Encryption, and Efficient Constructions without Random Oracles. In: Wagner, D. (ed.) CRYPTO 2008. LNCS, vol. 5157, pp. 335–359. Springer, Heidelberg (2008)
8. Brakerski, Z., Segev, G.: Better Security for Deterministic Public-Key Encryption: The Auxiliary-Input Setting. In: Rogaway, P. (ed.) CRYPTO 2011. LNCS, vol. 6841, pp. 543–560. Springer, Heidelberg (2011)
9. Catalano, D., Gennaro, R., Howgrave-Graham, N.: Paillier's trapdoor function hides up to $O(n)$ bits. J. Cryptology (2002)
10. Chor, B., Goldreich, O.: Unbiased bits from sources of weak randomness and probabilistic communication complexity. SIAM J. Comput. 17(2) (1988)
11. Chung, K.-M., Kalai, Y.T., Liu, F.-H., Raz, R.: Memory Delegation. In: Rogaway, P. (ed.) CRYPTO 2011. LNCS, vol. 6841, pp. 151–168. Springer, Heidelberg (2011)
12. Cramer, R., Hanaoka, G., Hofheinz, D., Imai, H., Kiltz, E., Pass, R., Shelat, A., Vaikuntanathan, V.: Bounded CCA2-Secure Encryption. In: Kurosawa, K. (ed.) ASIACRYPT 2007. LNCS, vol. 4833, pp. 502–518. Springer, Heidelberg (2007)
13. Dent, A.W., Fischlin, M., Manulis, M., Stam, M., Schröder, D.: Confidential Signatures and Deterministic Signcryption. In: Nguyen, P.Q., Pointcheval, D. (eds.) PKC 2010. LNCS, vol. 6056, pp. 462–479. Springer, Heidelberg (2010)
14. Desrosiers, S.P.: Entropic security in quantum cryptography. Quantum Information Processing 8(4), 331–345 (2009)
15. Dodis, Y., Ostrovsky, R., Reyzin, L., Smith, A.: Fuzzy extractors: How to generate strong keys from biometrics and other noisy data. SIAM J. Comput. 38(1), 97–139 (2008)
16. Dodis, Y., Smith, A.: Correcting errors without leaking partial information. In: STOC, pp. 654–663 (2005)
17. Dodis, Y., Smith, A.: Entropic Security and the Encryption of High Entropy Messages. In: Kilian, J. (ed.) TCC 2005. LNCS, vol. 3378, pp. 556–577. Springer, Heidelberg (2005)
18. Dziembowski, S., Pietrzak, K.: Leakage-resilient cryptography. In: FOCS, pp. 293–302 (2008)
19. Freeman, D.M., Goldreich, O., Kiltz, E., Rosen, A., Segev, G.: More Constructions of Lossy and Correlation-Secure Trapdoor Functions. In: Nguyen, P.Q., Pointcheval, D. (eds.) PKC 2010. LNCS, vol. 6056, pp. 279–295. Springer, Heidelberg (2010)
20. Fuller, B., O'Neill, A., Reyzin, L.: A unified approach to deterministic encryption: New constructions and a connection to computational entropy. Cryptology ePrint Archive (2012)
21. Fuller, B., Reyzin, L.: Computational entropy and information leakage. Tech. rep., Boston University (2011), http://cs-people.bu.edu/bfuller/metricEntropy.pdf
22. Gentry, C., Wichs, D.: Separating succinct non-interactive arguments from all falsifiable assumptions. In: STOC, pp. 99–108. ACM, New York (2011)

23. Goldreich, O., Levin, L.A.: A hard-core predicate for all one-way functions. In: STOC, pp. 25–32 (1989)
24. Goldwasser, S., Micali, S.: Probabilistic encryption. J. Comput. Syst. Sci. 28(2), 270–299 (1984)
25. Goyal, V., O'Neill, A., Rao, V.: Correlated-Input Secure Hash Functions. In: Ishai, Y. (ed.) TCC 2011. LNCS, vol. 6597, pp. 182–200. Springer, Heidelberg (2011)
26. Håstad, J., Impagliazzo, R., Levin, L.A., Luby, M.: A pseudorandom generator from any one-way function. SIAM J. Comput. 28(4), 1364–1396 (1999)
27. Hemenway, B., Lu, S., Ostrovsky, R.: Correlated product security from any one-way function and the new notion of decisional correlated product security. Cryptology ePrint Archive, Report 2010/100 (2010), http://eprint.iacr.org/
28. Hsiao, C.-Y., Lu, C.-J., Reyzin, L.: Conditional Computational Entropy, or Toward Separating Pseudoentropy from Compressibility. In: Naor, M. (ed.) EUROCRYPT 2007. LNCS, vol. 4515, pp. 169–186. Springer, Heidelberg (2007)
29. Kaplan, E., Naor, M., Reingold, O.: Derandomized constructions of k-wise (almost) independent permutations. Algorithmica 55(1), 113–133 (2009)
30. Kiltz, E., O'Neill, A., Smith, A.: Instantiability of RSA-OAEP under chosen-plaintext attack. IACR Cryptology ePrint Archive 2011, 559 (2011)
31. Kiltz, E., Pietrzak, K., Stam, M., Yung, M.: A New Randomness Extraction Paradigm for Hybrid Encryption. In: Joux, A. (ed.) EUROCRYPT 2009. LNCS, vol. 5479, pp. 590–609. Springer, Heidelberg (2009)
32. Niederreiter, H.: Knapsack-type cryptosystems and algebraic coding theory. Problems of Control and Information Theory 15, 367–391 (1986)
33. Nisan, N., Zuckerman, D.: Randomness is linear in space. Journal of Computer and System Sciences, 43–52 (1993)
34. Paillier, P.: Public-Key Cryptosystems Based on Composite Degree Residuosity Classes. In: Stern, J. (ed.) EUROCRYPT 1999. LNCS, vol. 1592, pp. 223–238. Springer, Heidelberg (1999)
35. Peikert, C., Waters, B.: Lossy trapdoor functions and their applications. In: STOC, pp. 187–196 (2008)
36. Reingold, O., Trevisan, L., Tulsiani, M., Vadhan, S.: Dense subsets of pseudorandom sets. In: 2008 49th Annual IEEE Symposium on Foundations of Computer Science, pp. 76–85. IEEE (2008)
37. Rivest, R.L., Shamir, A., Adleman, L.M.: A method for obtaining digital signatures and public-key cryptosystems. Commun. ACM 21(2), 120–126 (1978)
38. Russell, A., Wang, H.: How to fool an unbounded adversary with a short key. IEEE Transactions on Information Theory 52(3), 1130–1140 (2006)
39. Steinfeld, R., Pieprzyk, J., Wang, H.: On the Provable Security of an Efficient RSA-Based Pseudorandom Generator. In: Lai, X., Chen, K. (eds.) ASIACRYPT 2006. LNCS, vol. 4284, pp. 194–209. Springer, Heidelberg (2006)
40. Wee, H.: Dual projective hashing and its applications—lossy trapdoor functions and more. In: Eurocrypt (2012)
41. Yao, A.C.C.: Theory and applications of trapdoor functions (extended abstract). In: FOCS, pp. 80–91 (1982)

A Dichotomy for Local Small-Bias Generators

Benny Applebaum[1,*], Andrej Bogdanov[2,**], and Alon Rosen[3,***]

[1] School of Electrical Engineering, Tel-Aviv University
benny.applebaum@gmail.com
[2] Department of Computer Science and Engineering and Institute for Theoretical
Computer Science and Communications, Chinese University of Hong Kong
andrejb@cse.cuhk.edu.hk.
[3] Efi Arazi School of Computer Science, IDC Herzliya
alon.rosen@idc.ac.il

Abstract. We consider pseudorandom generators in which each output bit depends on a constant number of input bits. Such generators have appealingly simple structure: they can be described by a sparse input-output dependency graph G and a small predicate P that is applied at each output. Following the works of Cryan and Miltersen (MFCS '01) and by Mossel et al (FOCS '03), we ask: which graphs and predicates yield "small-bias" generators (that fool linear distinguishers)?

We identify an explicit class of degenerate predicates and prove the following. For most graphs, all *non-degenerate* predicates yield small-bias generators, $f \colon \{0,1\}^n \to \{0,1\}^m$, with output length $m = n^{1+\epsilon}$ for some constant $\epsilon > 0$. Conversely, we show that for most graphs, *degenerate* predicates are not secure against linear distinguishers, even when the output length is linear $m = n + \Omega(n)$. Taken together, these results expose a dichotomy: every predicate is either very hard or very easy, in the sense that it either yields a small-bias generator for almost all graphs or fails to do so for almost all graphs.

As a secondary contribution, we give evidence in support of the view that small bias is a good measure of pseudorandomness for local functions with large stretch. We do so by demonstrating that resilience to linear distinguishers implies resilience to a larger class of attacks for such functions.

Keywords: small-bias generator, dichotomy, local functions, NC0.

1 Introduction

In recent years there has been interest in the study of cryptographic primitives that are implemented by *local* functions, that is functions in which each output bit depends on a constant number of input bits. This study has been in large

* Supported by Alon Fellowship, by the Israel Science Foundation (grant No. 1155/11), and by the Check Point Institute for Information Security.
** Supported in part by Hong Kong RGC GRF grant CUHK 410309.
*** Supported by the Israel Science Foundation (grant No. 334/08).

R. Cramer (Ed.): TCC 2012, LNCS 7194, pp. 600–617, 2012.

part spurred by the discovery that, under widely accepted cryptographic assumptions, local functions can achieve rich forms of cryptographic functionality, ranging from one-wayness and pseudorandom generation to semantic security and existential unforgeability [6].

Local functions have simple structure: they can be described by a sparse input-output dependency graph and sequence of small predicates applied at each output. Besides allowing efficient parallel evaluation, this simple structure makes local functions amenable to analysis, and gives hope for understanding their computational properties. Given that the cryptographic functionalities that local functions can achieve are quite complex, it is very interesting and appealing to try to understand which properties of local functions (namely, graphs and predicates) are necessary and sufficient for them to implement such functionalities.

In this work we focus on the study of local pseudorandom generators with large stretch. We give evidence that for most graphs, all but a handful of "degenerate" predicate types yield pseudorandom generators with output length $m = n^{1+\varepsilon}$ for some constant $\varepsilon > 0$. Conversely, we show that for almost all graphs, degenerate predicates are not secure even against linear distinguishers. Taken together, these results expose a dichotomy: every predicate is either very hard or very easy, in the sense that it either yields a small-bias generator for almost all graphs or fails to do so for almost all graphs.

1.1   Easy, Sometimes Hard, and Almost Always Hard Predicates

Recall that a pseudorandom generator is a length-increasing function $f\colon \{0,1\}^n \to \{0,1\}^m$ such that no efficiently computable test can distinguish with noticeable advantage between the value $f(x)$ and a randomly chosen $y \in \{0,1\}^m$, when $x \in \{0,1\}^n$ is chosen at random. The additive *stretch* of f is defined to be the difference between its output length m and its input length n.

In the context of constructing local pseudorandom generators of superlinear stretch, we may assume without loss of generality that all outputs apply the same predicate $P\colon \{0,1\}^d \to \{0,1\}$.[1] We are interested in understanding which d-local functions $f_{G,P}\colon \{0,1\}^n \to \{0,1\}^m$, described by a graph G and a predicate P, are pseudorandom generators. For a predicate P, we will say

- P is *easy* if $f_{G,P}$ is *not* pseudorandom for every G (against a given class of adversaries),
- P is *sometimes hard* if $f_{G,P}$ is pseudorandom for some G, and
- P is *almost always hard* if $f_{G,P}$ is pseudorandom for a $1 - o(1)$ fraction of graphs G.[2]

---

[1] If this is not the case, project on the outputs labeled by the most frequent predicate. This decreases the stretch only by a constant factor as there are only 2^{2^d} different predicates.

[2] One cannot hope for *always* hard predicates, for which $f_{G,P}$ is pseudorandom for *all* graphs, as there are simple examples of "easy" graphs G for which $f_{G,P}$ fails to be pseudorandom regardless of P.

Cryan and Miltersen [17] and Mossel et al. [27] identified several classes of predicates that are easy for polynomial time algorithms when the stretch is a sufficiently large linear function. These include four types of predicates:

1. linear predicates, i.e., $P(w) = b + \Sigma_i w_i \pmod 2$ where $b \in \{0,1\}$,
2. unbalanced predicates, i.e., $\Pr_w[P(w) = 1] \neq \frac{1}{2}$,
3. predicates that are biased towards one input, i.e., $\Pr_w[P(w) = w_i] \neq \frac{1}{2}$,
4. predicates that are biased towards a pair of inputs, i.e., $\Pr_w[P(w) = w_i + w_j \pmod 2] \neq \frac{1}{2}$.

We call such predicates *degenerate*. It turns out that all predicates of locality at
most 4 are degenerate.

On the positive side, Mossel et al. [27] also gave examples of 5-bit predicates
that are sometimes (exponentially) hard against linear distinguishers. Applebaum et al. [5] show that when the locality is sufficiently large, almost always
hard predicates against linear distinguishers exist.

Pseudorandomness against linear distinguishers means that there is no subset
of output bits whose XOR has noticeable bias. This notion, due to Naor and
Naor [28], was advocated in the context of local pseudorandom generators by
Cryan and Miltersen [17]. A bit more formally, for a function $f : \{0,1\}^n \to \{0,1\}^m$, we let

$$\mathsf{bias}(f) = \max_L |\Pr[L(f(\mathcal{U}_n)) = 1] - \Pr[L(\mathcal{U}_m) = 1]|,$$

where the maximum is taken over all affine functions $L : \mathbb{F}_2^m \to \mathbb{F}_2$. A small-bias
generator is a function f for which $\mathsf{bias}(f)$ is small (preferrably negligible) as a
function of n.

1.2   Our Results

We fully classify predicates by showing that all predicates that are not known
to be *easy*, are *almost always hard*.

Theorem 1 (Non-degenerate predicates are hard). *Let* $P : \{0,1\}^d \to \{0,1\}$ *be any non-degenerate predicate. Then, for every* $\varepsilon < 1/4$ *and* $m = n^{1+\varepsilon}$:

$$\Pr_G[\mathsf{bias}(f_{G,P}) \leq \delta(n)] > 1 - o(1),$$

where $\delta(n) = \exp(-\Omega(n^{1/4-\varepsilon}))$ *and* G *is randomly chosen from all* d-regular hypergraphs with n nodes (representing the inputs) and m hyperedges (representing
the outputs).

The theorem shows that, even when locality is large, the only easy predicates
are degenerate ones, and there are no other "sources of easiness" other than ones
that already appear in predicates of locality 4 or less.

Conversely, we show that degenerate predicates are easy for *linear* distinguishers (as opposed to general polynomial-time distinguishers).

Theorem 2 (Linear tests break degenerate predicates). *For every* $m = n + \Omega(n)$, *and every degenerate predicate* $P : \{0,1\}^d \to \{0,1\}$

$$\Pr_G[\text{bias}(f_{G,P}) > \Omega(1/\log(n))] > 1 - o(1),$$

where G is randomly chosen from all d-regular hypergraphs with n nodes and m hyperedges.

The proof of Theorem 2 mainly deals with degenerate predicates that are correlated with a pair of their inputs; In this case, we show that the non-linear distinguisher which was previously used in [27] and was based on a semi-definite program for MAX-2-LIN [21] can be replaced with a simple linear distinguisher. (The proof for other degenerate predicates follows from previous works).

Taken together, Theorems 1 and 2 expose a dichotomy: a predicate can be either easy (fail for almost all graphs) or hard (succeeds for almost all graphs). One possible interpretation of our results is that, from a designer point of view, a strong emphasis should be put on the choice of the predicate, while the choice of the input-output dependency graph may be less crucial (since if the predicate is appropriately chosen then most graphs yield a small-bias generator). In some sense, this means that constructions of local pseudorandom generators with large stretch are robust: as long as the graph G is "typical," any non-degenerate predicate can be used (our proof classifies explicitly what is a typical family of graphs and in addition shows that even a mixture of different non-degenerate predicates would work).

1.3   Why Polynomial Stretch?

While Applebaum et al. [6] give strong evidence that local pseudorandom generators exist, the stretch their construction achieves is only sublinear, that is $m = n + n^{1-\varepsilon}$. (This stretch can be achieved even for 4-local predicates which are necessarily degenerate.) In contrast, the regime of large (polynomial or even linear) stretch is not as well understood, and the only known constructions are based on non-standard assumptions. (See Section 1.5.)

Local generators of large stretch have several applications in cryptography and complexity, such as secure computation with constant overhead [24] and strong (average-case) inapproximability results for constraint-satisfaction problems [7]. These results are not known to follow from other (natural) assumptions. It should be mentioned that it is possible to convert small polynomial stretch of $m = n^{1+\varepsilon}$ into arbitrary (fixed) polynomial stretch of $m = n^c$ at the expense of constant blow-up in the locality. (This follows from standard techniques, see [4] for details). Hence, it suffices to focus on the case of $m = n^{1+\varepsilon}$ for some fixed ε.

The proof of Theorem 1 yields exponentially small bias when $m = O(n)$, and sub-exponential bias for $m = n^{1+\varepsilon}$ where $\varepsilon < 1/4$. We do not know whether this is tight, but it can be shown that some non-degenerate predicates become easy (to break on a random graph) when the output length is $m = n^2$ or even $m = n^{3/2}$. In general, it seems that when m grows the number of hard predicates of locality d decreases, till the point $m^\star$ where all predicates become easy. (By [27], $m^\star \le n^{d/2}$.) It will be interesting to obtain a classification for larger output lengths, and to find out whether a similar dichotomy happens there as well.

1.4   Why Small-Bias?

Small-bias generators are a strict relaxation of cryptographic pseudorandom generators in that the tests $L : \mathbb{F}_2^m \to \mathbb{F}_2$ are restricted to be affine (as opposed to arbitrary efficiently computable functions). Even though affine functions are, in general, fairly weak distinguishers, handling them is a necessary first step towards achieving cryptographic pseudorandomness. In particular, affine functions are used extensively in cryptanalysis and security against them already rules out an extensive class of attacks.

For local pseudorandom generators with linear stretch, Cryan and Miltersen conjectured that affine distinguishers are as powerful as polynomial-time distinguishers [17]. In Section 5, we attempt to support this view by showing that resilience against small-bias, by itself, leads to robustness against other classes of attacks.

Small-bias generators are also motivated by their own right being used as building blocks in constructions that give stronger forms of pseudorandomness. This includes constructions of local cryptographic pseudorandom generators [7,4], as well as pseudorandom generators that fool low-degree polynomials [14], small-space computations [23], and read-once formulas[11].

1.5   Related Work

The function $f_{G,P}$ was introduced by Goldreich [22] who conjectured that when $m = n$, one-wayness should hold for a random graph and a random predicate. This view is supported by the results of [22,29,3,16,26,20,25] who show that a large class of algorithms (including ones that capture DPLL-based heuristics) fail to invert $f_{G,P}$ in polynomial-time.

At the linear regime, i.e., when $m = n + \Omega(n)$, it is shown in [12] that if the predicate is degenerate the function $f_{G,P}$ can be *inverted* in polynomial-time. (This strengthens the results of [17,27] who only give distinguishers.) Recently, a strong self-amplification theorem was proved in [13] showing that for $m = n + \Omega_d(n)$ if $f_{G,P}$ is hard-to-invert over tiny (sub-exponential small) fraction of the inputs with respect to sub-exponential time algorithm, then the same function is actually hard-to-invert over almost all inputs (with respect to sub-exponential time algorithms).

Pseudorandom generators with *sub-linear* stretch can be implemented by 4-local functions based on standard intractability assumptions (e.g., hardness of factoring, discrete-log, or lattice problems) [6], or even by 3-local functions based on the intractability of decoding random linear codes [8]. However, it is unknown how to extend this result to polynomial or even linear stretch since all known stretch amplification procedures introduce a large (polynomial) overhead in the locality. In fact, for the special case of 4-local functions (in which each output depends on at most 4 input bits), there is a provable separation: Although such functions can compute sub-linear pseudorandom generators [6] they *cannot* achieve polynomial-stretch [17,27].

Alekhnovich [1] conjectured that for $m = n + \Theta(n)$, the function $f_{G,P}$ is pseudorandom for a random graph and when P is a *randomized* predicate which computes $z_1 \oplus z_2 \oplus z_3$ and with some small probability $p < \frac{1}{2}$ flips the result. Although this construction does not lead directly to a local function (due to the use of noise), it was shown in [7] that it can be derandomized and transformed into a local construction with linear stretch. (The restriction to linear stretch holds even if one strengthen Alekhnovich's assumption to $m = \text{poly}(n)$.)

More recently, [4] showed that the pseudorandomness of $f_{G,P}$ with respect to a random graph and output length m, can be reduced to the one-wayness of $f_{H,P}$ with respect to a random graph H and related output length m' (for certain settings of the stretch and security parameters). The current paper complements this result as it provides a criteria for choosing the predicate P.

2 Techniques and Ideas

In this section we give an overview of the proof of Theorem 1. Let $f : \{0,1\}^n \to \{0,1\}^m$ be a d-local function where each output bit is computed by applying some d-local predicate $P : \{0,1\}^d \to \{0,1\}$ to a (ordered) subset of the inputs $S \subseteq [n]$. Any such function can be described by a list of m d-tuples $G = (S_1, \ldots, S_m)$ and the predicate P. Under this convention, we let $f_{G,P} : \{0,1\}^n \to \{0,1\}^m$ denote the corresponding d-local function.

We view G as a d-regular hypergraph with n nodes (representing inputs) and m hyperedges (representing outputs) each of size d. (We refer to such a graph as an (m, n, d)-graph.) Since we are mostly interested in polynomial stretch we think of m as $n^{1+\varepsilon}$ for some fixed $\varepsilon > 0$, e.g., $\varepsilon = 0.1$.

We would like to show that for almost all (m, n, d)-graphs G, the function $f_{G,P}$ fools all linear tests L, where P is non-degenerate. Following [27], we distinguish between *light* linear tests which depend on less than $k = \Omega(n^{1-2\varepsilon})$ outputs, and *heavy* tests which depend on more than k outputs.

From our definition of non-degenerate predicates, it immediately follows that such predicates P satisfy two forms of "non-linearity": (1) (2-*resilience*) P is uncorrelated with any linear function in two or fewer inputs; and (2) (*algebraic nonlinearity*) P is not linear as a polynomial over $\mathbb{F}_2$. Both properties are classical design criteria which are widely used in practical cryptanalysis (cf. [30]). We use the fist property to fool light linear tests (tests that depend on a small number of outputs) and the second one to fool heavy linear tests (tests that depend on a large number of outputs).

2.1 Fooling Light Tests

Our starting point is a result of [27] which shows that if the predicate is the parity predicate $\oplus$ and the graph is a good expander, the output of $f_{G,\oplus}(\mathcal{U}_n)$

perfectly fools all light linear tests. In terms of expectation, this can be written as

$$\mathop{\mathsf{E}}_{x}[L(f_{G,\oplus}(x)) = 0],$$

where we think of $\{0,1\}$ as $\{\pm 1\}$, and let $L : \{\pm 1\}^m \to \{\pm 1\}$ be a light linear test. Our key insight is that the case of a general predicate P can be reduced to the case of linear predicates.

More precisely, let ξ denote the outcome of the test $L(f_{G,P}(x))$. Then, by looking at the Fourier expansion of the predicate P, we can write ξ as a convex combination over the reals of exponentially many summands of the form $\xi_i = L(f_{G_i,\oplus}(x))$ where the G_i's are subgraphs of G in the sense that the j-th hyperedge of G_i is a subset of the j-th hyperedge of G. (The exact structure of G_i is determined by the Fourier representation of P.) When x is uniformly chosen, the random variable ξ is a weighted sum (over the reals) of many dependent random variables ξ_i's. However, if all the subgraphs are good expanders, the expectation of each summand ξ_i is zero, and so, by the linearity of expectation, the expectation of ξ is also zero.

It turns out that when the predicate is 2-resilient the size of each hyperedge of G_i is at least 3, and therefore if every 3-uniform subgraph of G is a good expander $f_{G,P}$ (perfectly) passes all light linear tests. Fortunately, it turns out that most graphs G satisfy this property. We emphasize that the argument crucially relies on the *perfect* bias of XOR predicates, as there are exponentially many summands. (See Section 3.1 for full details.)

2.2   Fooling Heavy Tests

Consider a heavy test which involves $t \geq k$ outputs. Switching back to zero-one notation, assume that the test outputs the value $\xi = P(x_{S_1}) + \ldots + P(x_{S_t})$ (mod 2) where $x \overset{R}{\leftarrow} U_n$. Our goal is to show that ξ is close to a fair coin. For this it suffices to show that the sum ξ can be rewritten as the sum (over $\mathbb{F}_2$) of ℓ random variables

$$\xi = \xi_1 + \ldots + \xi_\ell \quad (\text{mod } 2), \tag{1}$$

where each random variable ξ_i is an *independent* non-constant coin, i.e., $\Pr[\xi_i = 1] \in [2^{-d}, 1 - 2^{-d}]$. In this case, the statistical distance between ξ and a fair coin is exponentially small (in ℓ), and we are done as long as ℓ is large enough.

In order to partition ξ, let us look at the hyperedges $S_1, \ldots, S_t$ which are involved in the test. As a first attempt, let us collect ℓ distinct "independent" hyperedges that do not share a single common variable. Renaming the edges, we can write ξ as

$$(P(x_{T_1}) + \ldots + P(x_{T_\ell})) + (P(x_{S_{\ell+1}}) + \ldots + P(x_{S_t})) \quad (\text{mod } 2),$$

where the first ℓ random variables are indeed statistically independent. However, the last $t - \ell$ hyperedges violate statistical-independence as they may be correlated with more than one of the first ℓ hyperedges. This is the case, for example, if S_j has a non-empty intersection with both T_i and T_r.

This problem is fixed by collecting ℓ "strongly-independent" hyperedges T_1, $\ldots, T_\ell$ for which every S_j intersects at most a single T_i. (Such a large collection is likely to exist since t is sufficiently large.) In this case, for any fixing of the variables outside the T_i's, the random variable ξ can be partitioned into ℓ independent random variables of the form $\xi_i = P(x_{T_i}) + \sum P(x_{S_j})$, where the sum ranges over the S_j's which intersects T_i. This property (which is a relaxation of Eq. 1) still suffices to achieve our goal, as long as the ξ_i's are *non-constant*.

To prove the latter, we rely on the fact that P has algebraic degree 2. Specifically, let us assume that S_i and T_j have no more than a single common input node. (This condition can be typically met at the expense of throwing a small number of the T_i's.) In this case, the random variable $\xi_i = P(x_{T_i}) + \sum P(x_{S_j})$ cannot be constant, as the first summand is a degree 2 polynomial in x_{T_i} and each of the last summands contain at most a single variable from T_i. Hence, ξ_i is a non-trivial polynomial whose degree is lower-bounded by 2. This completes the argument. Interestingly, non-linearity is used only to prove that the ξ_i's are non-constant. Indeed, linear predicates fail exactly for large tests for which the ξ_i's become fixed due to local cancelations. (See Section 3.2 for details.)

2.3   Proving Theorem 2

When P is a degenerate predicate and G is random, the existence of a linear distinguisher follows by standard arguments. The cases of linear or biased P are trivial, and the case of bias towards one input was analyzed by Cryan and Miltersen. When P is biased towards a pair of inputs, say the first two, we think of P as an "approximation" of the parity $x_1 \oplus x_2$ of its first two inputs. If P happened to be the predicate $x_1 \oplus x_2$, one could find a short "cycle" of output bits that, when XORed together, causes the corresponding input bits to cancel out. In general, as long as the outputs along the cycle do not share any additional input bits, the output of the test will be biased, with bias exponential in the length of the cycle. In Section 4 we show that a random G is likely to have such short cycles, and so the corresponding linear test will be biased.

3   Non-degenerate Predicates Are Hard

In this section we prove Theorem 1. We follow the outline described in Section 2 and handle light linear tests and heavy linear tests separately.

3.1   Fooling Light Tests

In this section we show that if the predicate P is 2-resilient (see definition below) and the graph G is a good expander, the function $f_{G,P}$ is k-wise independent, and in particular fools linear tests of weight smaller than k. We will need the following definitions.

Super expansion. Let G be an (m, n, d)-graph. A graph H is (k, a) subgraph of G if it can be constructed by choosing $\ell \leq k$ distinct hyperedges of G and for each selected hyperedge S_j removing some of the nodes while leaving $b_j \geq a$ nodes. We say that G is (k, a) *super-expander* if the hyperedges $T = T_1, \ldots, T_\ell$ of every (k, a)-subgraph H of G touch more than $b\ell/2$ nodes where $b = \sum |T_j| / \ell$ is the average cardinality of the hyperedges of H. We say that G is (k, a)-*linear* if the hyperedges of every (k, a)-subgraph of G are linearly independent viewed as vectors in $\mathbb{F}_2^n$.

Fourier coefficients. For a set $T \subseteq [d]$, let $\chi_T : \{\pm 1\}^d \to \{\pm 1\}$ be the Parity function defined by $(x_1, \ldots, x_d) \mapsto (-1)^{\sum_{t \in T} x_t}$. It is well known that every predicate $P : \{\pm 1\}^d \to \{\pm 1\}$ can be expressed as a convex combination of parities, i.e., $P(x) = \sum_{T \subseteq [d]} \alpha_T \chi_T(x)$ where $\alpha_T \in \mathbb{R}$. The predicate is a-*resilient* if α_T is zero for every set T of size smaller or equal to a.

The following lemma shows that resiliency combined with (k, a)-linearity leads to k-wise independence.

Lemma 1. *If P is $(a-1)$-resilient and the (m, n, d)-graph G is (k, a)-linear then $f_{G,P}$ is k-wise independent generator, i.e., the m r.v.'s $(y_1, \ldots, y_m) = f_{G,P}(\mathcal{U}_n)$ are k-wise independent.*

Proof. Fix an $\ell \leq k$ outputs of $f_{G,P}$, and let $S_1, \ldots, S_\ell$ be the corresponding hyperedges. We should show that $\mathsf{E}_x[\prod_i P(x_{S_i})] = 0$. For every $x \in \{0, 1\}^n$ we have:

$$\prod_{i=1}^{\ell} P(x_{S_i}) = \prod_{i=1}^{\ell} \sum_{T \subseteq [d], |T| \geq a} \alpha_T \chi_T(x_{S_i}) = \sum_{T = (T_1, \ldots, T_\ell), |T_i| \geq a} \prod_i \alpha_{T_i} \chi_{S_{i,T_i}}(x),$$

where $S_{i, \{K_1, \ldots, K_b\}}$ denotes the set $\{S_{i, K_1}, \ldots, S_{i, K_b}\}$ and $S_{i,j}$ denotes the j-th entry of the tuple S_i. Hence, by the linearity of expectation, it suffices to show that

$$\mathsf{E}_x\left[\prod_i \chi_{S_{i,T_i}}(x) \right] = 0,$$

for every $(T_1, \ldots, T_\ell)$ where $T_i \subseteq [d], |T_i| \geq a$. (Recall that the α_{T_i}'s are constants and thus can be ignored.) Observe that $\prod_i \chi_{S_{i,T_i}}(x)$ is just a parity function, which, by (k, a)-linearity, is *non-constant*. Since every non-constant parity function is balanced (guaranteed to have zero expectation value), the claim follows. □

Next, we show that (k, a)-linearity is implied by super-expansion, and that a random graph is likely to be super-expanding.

Lemma 2. *Let $d \geq 3$ be a constant. Let $\Delta \leq \sqrt{n}/\log n$ and $3 \leq a \leq d$.*

1. *Every $(\Delta n, n, d)$-graph which is (k, a)-super-expander is also (k, a)-linear.*
2. *A random $(\Delta n, n, d)$-graph is whp an $(\alpha n/\Delta^2, a)$-super-expander where α is a constant that depends on a, d.[3]*

---

[3] With high probability (whp) means with probability $1 - o(1)$ as n gets large.

Proof. The proof of the first item parallels the standard relation between lossless-expansion and unique/odd-expansion. Let G be a (k, a)-super-expander. Observe that if G is not (k, a)-linear then there must be (k, a)-subgraph H whose edges sum-up to zero (over $\mathbb{F}_2^n$). We argue that G cannot have such a subgraph. Indeed, by counting edges, in each (k, a)-subgraph H the average degree of the participating nodes is smaller than 2, and so there exists at least one node which participates in a single hyperedge. Hence, the sum of the hyperedges (over $\mathbb{F}_2^n$) is non-zero.

To prove the second item, we calculate the probability that a random $(\Delta n, n, d)$-graph fails to be (k, a)-super-expander. First we bound the probability that there exists a subgraph H with ℓ hyperedges and average degree $b \geq a$ that violates expansion. This probability is bounded by

$$\binom{\Delta n}{\ell} \cdot 2^{d\ell} \cdot \binom{n}{b\ell/2} \cdot \left(\frac{b\ell}{2n}\right)^{b\ell} < \left(\frac{e\Delta n}{\ell} \cdot 2^d \cdot \left(\frac{2en}{b\ell}\right)^{b/2} \left(\frac{b\ell}{2n}\right)^b\right)^\ell$$

$$= \left(e2^d \left(\frac{be}{2}\right)^{b/2} \Delta \left(\frac{\ell}{n}\right)^{b/2-1}\right)^\ell$$

$$\leq \left(c_{d,a} \Delta \left(\frac{\ell}{n}\right)^{a/2-1}\right)^\ell$$

where $c_{d,a}$ is a constant which depends on d and a, and the second inequality is due to $a \leq b \leq d$. Let us denote the above quantity by $p_{\ell,n,\Delta,a,d}$. By a union-bound G fails to be (k, a)-super-expander with probability at most $\sum_{2\leq\ell\leq k} p_{\ell,n,\Delta,a,d}$.

Let us fix $a \geq 3$, and assume that $\Delta \leq n^{\frac{1}{2}}/\log n$ and $k = \alpha n/\Delta^2$ where $\alpha = 1/(2c_{d,a})^2$ is a constant. Indeed, in this case

$$p_\ell \leq \left(c_{d,a}\frac{\Delta\sqrt{\ell}}{\sqrt{n}}\right)^\ell \leq \left(c_{d,a}\frac{\sqrt{\ell}}{\log n}\right)^\ell.$$

Observe that for $\ell = 1, 2, 3$, the quantity p_ℓ is $o(1)$, for $4 \leq \ell \leq 10\log n$ the quantity $p_\ell \leq O(1/\log^2 n)$ and for $10\log n \leq \ell \leq \alpha n/\Delta^2$ the quantity p_ℓ is at most $O(1/n^{10})$. It follows that each of these three intervals contributes $o(1)$ to the overall failure probability. ∎

By combining the lemmas, we obtain the following corollary.

Corollary 1. *If P is 2-resilient and $m = \Delta n$ for constant Δ, then whp over the choice of an (m, n, d)-graph G, the function $f_{G,P}$ is k-wise independent for $k = \Omega(n)$. If $\Delta = n^\varepsilon$, the above holds with $k = \Omega(n^{1-2\varepsilon})$.*

By taking $\varepsilon < 1/4$, 2-resiliency suffices for $\omega(\sqrt{n})$-wise independence whp.

3.2   Fooling Heavy Tests

In this section we show that if the predicate P is non-linear and the graph G has large sets of "independent" hyperedges, the function $f_{G,P}$ fools linear tests of weight larger than k. Formally, we will need the following notion of independence.

(k, ℓ, b)-*independence*. Let S be a collection of k distinct hyperedges. A subset $T \subseteq S$ of ℓ distinct hyperedges is an (ℓ, b)-independent set of S if the following two properties hold: (1) Every pair of hyperedges $(T, T') \in T$ are of distance at least 2, namely, for every pair $T_i \neq T_j \in T$ and $S \in S$,

$$T_i \cap S = \emptyset \text{ or } T_j \cap S = \emptyset;$$

and (2) For every $T_i \in T$ and $S \neq T_i$ in S we have

$$|T_i \cap S| < b.$$

A graph is (k, ℓ, b)-independent if every set of hyperedges of size larger than k has an (ℓ, b)-independent set.

Our key lemma shows that good independence and large algebraic degree guarantee resistance against heavy linear tests.

Lemma 3. *If G is (k, ℓ, b)-independent and P has an algebraic degree of at least b, then every linear test of size at least k has bias of at most $\frac{1}{2}e^{-2\ell/2^d}$.*

Proof. Fix some test $S = (S_1, \ldots, S_k)$ of size k, and let $T = (T_1, \ldots, T_\ell)$ be an (ℓ, b)-independence set of S. Fix an arbitrary assignment σ for all the input variables which do not participate in any of the T_i's and choose the other variables uniformly at random. In this case, we can partition the output of the test y to ℓ summands over ℓ disjoint blocks of variables, namely

$$y = \sum_{i \in [k]} P(x_{S_i}) = \sum_{i \in [\ell]} z_i(x_{T_i}),$$

where the sum is over $\mathbb{F}_2$ and

$$z_i(x_{T_i}) = P(x_{T_i}) + \sum_{S:T_i \neq S, S \cap T_i \neq \emptyset} P(x_{S \cap T_i}, \sigma_{S \setminus T_i}).$$

We need two observations: (1) the random variables z_i's are statistically independent (as each of them depends on a disjoint block of inputs); and (2) the r.v. z_i is non-constant and, in fact, it takes each of the two possible values with probability at least 2^{-d}. To prove the latter fact it suffices to show that $z_i(x)$ is a non-zero polynomial (over $\mathbb{F}_2$) of degree at most d. Indeed, recall that z_i is the sum of the polynomial $P(x_{T_i})$ whose degree is in $[b, d]$, and polynomials of the form $P(x_{S \cap T_i}, \sigma_{S \setminus T_i})$ whose degree is smaller than b (as $|S \cap T_i| < b$). Therefore the degree of z_i is in $[b, d]$.

To conclude the proof, we note that the parity of ℓ independent coins, each with expectation in $(\delta, 1 - \delta)$, has bias of at most $\frac{1}{2}(1 - 2\delta)^\ell$. (See, e.g., [27].)   □

We want to show that a random graph is likely to be $(k, \ell, 2)$-independent.

Lemma 4. *For every positive ε and δ. A random $(n^{1+\varepsilon}, n, d)$-graph is, whp, $(n^{2\varepsilon+\delta}, n^{\delta/2}, 2)$ independent.*

Proof. We will need the following claim. Call a hyperedge S b-intersecting if there exists another hyperedge S' in the graph for which $|S' \cap S| \geq b$. We first bound the number of b-intersecting hyperedges.

Claim. Let b be a constant. Then, in a random $(m = n^{1+\varepsilon}, n, d)$-graph, whp, the number of b-intersecting hyperedges is at most $n^{2(1+\varepsilon)-b} \log n$.

Hence, whp, at most $O(n^{2\varepsilon} \log n)$ of the hyperedges are 2-intersecting, and for $\varepsilon < 1/4$ there are at most $o(\sqrt{n})$ such hyperedges.

Proof (of Claim). Let X be the random variable which counts the number of b-intersecting hyperedges. First, we bound the expectation of X by $m^2 d^{2b}/n^b = d^{2b} \cdot n^{2(1+\varepsilon)-b}$. To prove this, it suffices to bound the expected number of pairs S_i, S_j which b-intersects. Each such pair b-intersects with probability at most d^{2b}/n^b, and so, by linearity of expectation, the expected number of of intersecting pairs is at most $m^2 d^{2b}/n^b$. Now, by applying Markov's inequality, we have that $\Pr[X > \frac{\log n}{d^{2b}} \mathsf{E}[X]] < d^{2b}/\log n = o(1)$, and the claim follows. (A stronger concentration can be obtained via a martingale argument.) □

We can now prove Lemma 4. Assume, without loss of generality, that $\varepsilon > 1$ (as if the claim holds for some value of ε it also holds for smaller values). First observe that, whp, all the input nodes in G have degree at most $2n^\varepsilon$. As by a multiplicative Chernoff bound, the probability that a single node has larger degree is exponentially small in n^ε. We condition on this event and the event that there are no more than $r = n^{2\varepsilon} \log n$ 2-intersecting edges. Fix a set of $k = n^{2\varepsilon+\delta}$ hyperedges. We extract an $(\ell, 2)$-independent set by throwing away the 2-expanding edges, and then by iteratively inserting an hyperedge T into the independent set and removing all the hyperedges S that share with T a common node, and the hyperedges which share a node with an edge, that shares a node with T. At the beginning we removed at most r edges, and in each iteration we remove at most $(d2n^\varepsilon)^2$ edges, hence there are at least $\ell \geq \frac{k-r}{4d^2 n^{2\varepsilon}} > n^{\delta/2}$ hyperedges in the independent set. □

Combining the lemmas together we get:

Corollary 2. *Fix some positive ε and δ. If P has an algebraic degree of at least 2 and $m = n^{1+\varepsilon}$, then, whp over the choice of a random (m, n, d)-graph, the function $f_{G,P}$ has at most sub-exponential bias (i.e., $\exp(-\Omega(n^\delta))$) against linear tests of size at least $n^{2\varepsilon+2\delta}$.*

By combining Corollaries 1 and 2, we obtain Theorem 1.

4   Linear Tests Break Degenerate Predicates

In this section we prove Theorem 2; That is, we show that the assumptions that P is non-linear and 2-resilient are necessary for P to be a hard predicate. Clearly the assumption that P is non-linear is necessary even when $m = n + 1$.

When $m \geq Kn$ for a sufficiently large constant K (depending on d), it follows from work of Cryan and Miltersen [17] that if P is not 1-resilient, then for any $f : \{\pm 1\}^n \to \{\pm 1\}^m$, the output of f is distinguishable from uniform with constant advantage by some linear test. When P is 1-resilient but not 2-resilient, Mossel, Shpilka, and Trevisan show that f is distinguishable from uniform by a polynomial-time algorithm, but not by one that implements a linear test.

Here we show that if P is not 2-resilient, then the output of $f_{G,P}$ is distinguishable by linear tests with non-negligible advantage with high probability over the choice of G.

Claim. Assume P is unbiased and 1-resilient but $|E[P(z)z_1 z_2]| = \alpha > 0$. Then for every $\ell = o(\log n)$, with probability $1 - (2^{-\Omega(\ell)} + d\ell/n)$ over the choice of G, there exists a linear test that distinguishes the output of $f_{G,P}$ from random with advantage α^ℓ.

Proof. Let H be the directed graph with vertices $\{1, \ldots, n\}$ where every hyper-edge $(i_1, i_2, \ldots, i_d)$ in G induces the edge (i_1, i_2) in H.

Let ℓ be the length of the shortest directed cycle in H and without loss of generality assume that this cycle consists of the inputs $1, 2, \ldots, \ell$ in that order. Let z_i be the name of the output that involves inputs i and $i+1$ for i ranging from 1 to ℓ (where i is taken modulo ℓ) and S_i the corresponding hyperedge. With probability at least $1 - d\ell/n$, input i does not participate in any hyperedge besides S_i and S_{i+1} and all other inputs participate in at most one of the hyperedges $S_1, \ldots, S_\ell$.

We now calculate the bias of the linear test that computes $z_1 \oplus \ldots \oplus z_\ell$. For simplicity, we will assume that $d = 3$; larger values of d can be handled analogously but the notation is more cumbersome. We will denote the entries in S_i by i, $i+1$ and i'. Then the fourier expansion of $z_i(x_{S_i})$ has the form

$$z_i(x_{S_i}) = \alpha x_i x_{i+1} + \beta x_i x_{i'} + \gamma x_{i+1} x_{i'} + \delta x_i x_{i+1} x_{i'}$$

The Fourier expansion of the expression $E[z_1(x_{S_1}) \ldots z_\ell(x_{S_\ell})]$ can be written as a sum of 4^ℓ products of different monomials participating in the above terms. The only monomial that does not vanish is the one containing all the α-terms, namely

$$E\left[\prod_{i=1}^n \alpha x_i x_{i+1}\right] = \alpha^\ell.$$

All the other products of monomials contain at least one unique term of the form $x_{i'}$, and this causes the expectation to vanish.

It remains to argue that with high probability ℓ is not too large. We show that with probability $1 - O((4/K)^\ell)$, H has a directed cycle of length ℓ, as long as $\ell < \log_{2K}(n/4)$. Let X denote the number of directed cycles of length ℓ in H. The number of potential directed cycles of length in H is $n(n-1) \ldots (n-\ell+1) \geq (n-\ell)^\ell$. Each of these occurs uniquely in H with probability

$$(Kn)(Kn-1)\ldots(Kn-\ell+1)\left(\frac{1}{n(n-1)}\right)^\ell\left(1 - \frac{1}{n(n-1)}\right)^{Kn-\ell} \geq \left(\frac{Kn-\ell}{n^2}\right)^\ell.$$

Therefore $E[X] \geq (K/4)^\ell$. The variance can be upper bounded as follows. The number of *pairs* of cycles of length ℓ that intersect in i edges is at most $\binom{\ell}{i} n^{2\ell-i-1}$, and the covariance of the indicators for these cycles is at most $(K/n)^{2\ell-i}$. Adding all the covariances up as i ranges from 1 to ℓ, it follows that

$$\mathsf{Var}[X] \leq E[X] + \sum_{i=1}^{\ell} \binom{\ell}{i} n^{2\ell-i-1} \left(\frac{K}{n}\right)^{2\ell-i} \leq E[X] + \frac{2^\ell K^{2\ell}}{n}.$$

By Chebyshev's inequality,

$$\Pr[X = 0] \leq \frac{\mathsf{Var}[X]}{E[X]^2} < \frac{2}{E[X]}$$

as long as $\ell < \log_{2K}(n/4)$. □

5   Small Bias vs. Cryptographic Security for Local Functions

It is not difficult to come up with examples of generators that have (exponentially) small bias against linear distinguishers but are not cryptographically secure. However, we do not know of any such examples of generators that are local and have at least linear stretch: To the best of our knowledge, all local functions of linear stretch that are known to implement small-biased generators could be pseudorandom generators against all polynomial-time adversaries.

Therefore it may be plausible to conjecture that if P is almost always hard against linear adversaries, then P is almost always hard against polynomial-time adversaries. While this conjecture cannot be proven without resolving the existence of pseudorandom generators, we give evidence in support of it: We show that if P is almost always hard against linear adversaries, then $f_{G,P}$ is not only small-biased but (1) it is k-wise independent and (2) it cannot be inverted by myopic backtracking algorithms.

First, we observe that for local functions the small-bias property immediately implies k-wise independence. (This is in general false for non-local functions.)

Lemma 5. *Let $f : \{0,1\}^n \to \{0,1\}^m$ be a d-local function which is 2^{-kd}-biased. Then it is also k-wise independent.*

Proof. Assume towards a contradiction that f is not k-wise independent. Then, there exists a set of k outputs T and a linear distinguisher L for which $\varepsilon = |\Pr[L(y_T) = 1] - \Pr[L(u) = 1]| > 0$, where $y = f(x)$ for a uniformly random x and u is a uniformly random string of length k. Since f is d-local, y_T is sampled by using fewer than kd bits of randomness and therefore $\varepsilon \geq 2^{-kd}$. □

Recall that the proof of our main theorem, Theorem 1, establishes k-wise independence as an intermediate step (Section 3.1). However, the above lemma is stronger in the sense that it holds for every fixed graph and every output length including ones that are not covered by the main theorem.

By plugging in known results about k-wise independent distributions, it immediately follows that if a local function is sufficiently small-biased, then it is pseudorandom against $\mathbf{AC}^0$ circuits [15], linear threshold functions over the reals [18], and degree-2 threshold functions over the reals [19].

Attacks on local functions, which are actively studied at the context of algorithms for constraint-satisfaction problems, appear to be based mainly on "local" heuristics (DPLL, message-passing algorithms, random-walk based algorithms) or linearization [9]. Hence, it appears that in the context of local functions, the small-bias property already covers all "standard" attacks. We support this intuition by showing that if P is non-degenerate, then the outputs of $f_{G,P}$ are not merely min-wise independent, but have a stronger property: Even after reading an arbitrary set of t-outputs, the posterior distribution on *every* set of ℓ inputs, while not uniform, still has large min-entropy. We call this property *robustness*.

The notion of robustness was used by Cook et al. [16] to prove that myopic backtracking algorithms cannot invert $f_{G,P}$ in polynomial time when $m = n$. We now argue that for $f_{G,P}$, robustness is almost always a consequence of small bias, and conclude that $f_{G,P}$ cannot be inverted by myopic backtracking algorithms even when $m = n^{1+\varepsilon}$, $\varepsilon < 1/4$, as long as P is non-degenerate. (The analysis of [16] also applies to some degenerate predicates.)

5.1   Robustness and Myopic Backtracking Algorithms

Robustness. Let $f : \{0,1\}^n \to \{0,1\}^m$. Let $L \subset [n]$ be a set of inputs, and $t, h \in [m]$. We say that f is (t, L, h)-robust if for every set of outputs $T \subset [m]$ of size t and every string $z \in \{0,1\}^t$ the following holds. Let $x \in \{0,1\}^n$ be a uniformly chosen string conditioned on the event $f(x)_T = z$, i.e., the outputs which are indexed by T equal to z. Then the random variable $x_L = (x_i)_{i \in L}$ has min-entropy of h, namely, for every fixed $w \in \{0,1\}^{|L|}$, $\Pr[x_L = w] \leq 2^{-h}$. The function is (t, ℓ, h)-robust if it is (t, L, h)-robust for every ℓ-size input set L.

In the full version of this work, we prove that if $f_{G,P}$ is k-wise independent with respect to random graph, then it is also robust for shorter output length.

Lemma 6. *Suppose that P is a predicate for which $f_{G,P} : \{0,1\}^n \to \{0,1\}^m$ is k-wise independent, whp over the choice of a random (m, n, d) graph G. Then, whp over the choice of a random $(m - r, n, d)$ graph H, the function $f_{H,P} : \{0,1\}^n \to \{0,1\}^{m-r}$ is (t, ℓ, h)-robust, where $h = \min\left(\ell, r \cdot (\ell/n)^d/2, k - t\right)$.*

In the case of linear stretch, $m = n + O(n)$, where k is linear as well (Corollary 1), one can get (t, ℓ, h)-robustness with linear parameters at the expense of linear decrease in the output length (e.g., $r = m/2$). When the output is polynomial $m = n^{1+\varepsilon}$ (for $\varepsilon < 1/4$), we get (t, ℓ, h)-robustness for inverse-polynomial parameters, again at the expense of a linear decrease in the output length (e.g., $r = m/2$).

Robustness is especially useful if the actual number of preimages of $y = f_{G,P}(x)$ is relatively small compared to 2^h. In this case, an algorithm which attempts to guess ℓ bits of a preimage x based on t outputs is likely to be wrong (obtain a partial assignment that does not correspond to any preimage of y.) We show that in our setting of parameters (when the output length is large) most inputs have a small number of siblings under $f_{G,P}$ (where G is random). The proof of the following lemma is given in the full version.

Lemma 7. *Let P be any nonconstant predicate. For every $\eta > 0$ there exists a constant M such that when $m > 2^{Md}n \log n$,*

$$\Pr_{G,x}\left[|\{x' \mid x' \text{ is a preimage of } f_{G,P}(x)\}| < M\right] > 1 - \eta.$$

Myopic DPLL algorithms. We now show how the simple statistical properties proved in the above lemmas yield lower-bounds for DPLL algorithms who attack $f_{G,P}$. The high-level argument is similar to the one used in [3,16] and it is only sketched here. Consider the following myopic backtracking DPLL algorithm, whose input consists of $y = f_{G,P}(x)$ where x is uniformly chosen. The algorithm is allowed to read the entire graph G, but it reads the values of y in an incremental way. Specifically, in each iteration the algorithm adaptively chooses an input variable x_i and asks to reveal r new output bits of y. Then it guesses the value of x_i based on its current state and on the output bits that were already revealed (including the ones that were revealed in previous iterations). If the algorithm reaches a contradiction, i.e., its partial assignment to x is consistent with some output it backtracks.

Suppose that $f_{G,P}$ satisfies Lemmas 6 and 7. Since $f_{G,P}$ is k-wise independent the algorithm does not backtrack in the first k/r steps (as some patrial assignment is consistent with every value of k outputs). Since f is $(r \cdot \ell, \ell, h)$-robust and the number of siblings of a random x is at most M whp, the partial assignment chosen by the algorithm after $\ell < k$ steps is likely to be globally inconsistent (there are 2^h locally consistent assignments while there are only $M \ll 2^h$ globally consistent assignments). Hence, with all but negligible probability, the algorithm will err during the first ℓ steps, and therefore will backtrack at some point after more than k steps. It can be shown (by standard lower-bound on resolution [10,2]) that, for a random graph, the backtracking phase takes super-polynomial time. (By plugging in the exact parameters the lower-bound is exponential $2^{\Omega(n)}$ when $m = O(n)$ or sub-exponential $\exp(n^\delta)$ when $m = n^{1+\varepsilon}$.)

References

1. Alekhnovich, M.: More on average case vs approximation complexity. In: FOCS, pp. 298–307. IEEE Computer Society (2003)
2. Alekhnovich, M., Ben-Sasson, E., Razborov, A.A., Wigderson, A.: Pseudorandom generators in propositional proof complexity. SIAM Journal of Computation 34(1), 67–88 (2004)
3. Alekhnovich, M., Hirsch, E.A., Itsykson, D.: Exponential lower bounds for the running time of DPLL algorithms on satisfiable formulas. J. Autom. Reasoning 35(1-3), 51–72 (2005)

4. Applebaum, B.: Pseudorandom generators with long stretch and low locality from random local one-way functions. Electronic Colloquium on Computational Complexity (ECCC) 18 (2011)
5. Applebaum, B., Barak, B., Wigderson, A.: Public-key cryptography from different assumptions. In: 42nd ACM Symposium on Theory of Computing (STOC 2010), pp. 171–180 (2010)
6. Applebaum, B., Ishai, Y., Kushilevitz, E.: Cryptography in NC^0. SIAM Journal on Computing 36(4), 845–888 (2006)
7. Applebaum, B., Ishai, Y., Kushilevitz, E.: On pseudorandom generators with linear stretch in NC^0. Journal of Computational Complexity 17(1), 38–69 (2008)
8. Applebaum, B., Ishai, Y., Kushilevitz, E.: Cryptography with constant input locality. Journal of Cryptology 22(4), 429–469 (2009)
9. Arora, S., Ge, R.: New Algorithms for Learning in Presence of Errors. In: Aceto, L., Henzinger, M., Sgall, J. (eds.) ICALP 2011, Part I. LNCS, vol. 6755, pp. 403–415. Springer, Heidelberg (2011)
10. Ben-Sasson, E., Wigderson, A.: Short proofs are narrow - resolution made simple. In: STOC, pp. 517–526 (1999)
11. Bogdanov, A., Papakonstantinou, P., Wan, A.: Pseudorandomness for read-once formulas. In: Proceedings of the 52nd Annual Symposium on Foundations of Computer Science (2011) (to appear)
12. Bogdanov, A., Qiao, Y.: On the Security of Goldreich's One-Way Function. In: Dinur, I., Jansen, K., Naor, J., Rolim, J. (eds.) APPROX and RANDOM 2009. LNCS, vol. 5687, pp. 392–405. Springer, Heidelberg (2009)
13. Bogdanov, A., Rosen, A.: Input Locality and Hardness Amplification. In: Ishai, Y. (ed.) TCC 2011. LNCS, vol. 6597, pp. 1–18. Springer, Heidelberg (2011)
14. Bogdanov, A., Viola, E.: Pseudorandom bits for polynomials. SIAM J. Comput. 39(6), 2464–2486 (2010)
15. Braverman, M.: Poly-logarithmic independence fools AC^0 circuits. In: Annual IEEE Conference on Computational Complexity, pp. 3–8 (2009)
16. Cook, J., Etesami, O., Miller, R., Trevisan, L.: Goldreich's One-Way Function Candidate and Myopic Backtracking Algorithms. In: Reingold, O. (ed.) TCC 2009. LNCS, vol. 5444, pp. 521–538. Springer, Heidelberg (2009)
17. Cryan, M., Miltersen, P.B.: On Pseudorandom Generators in NC^0. In: Sgall, J., Pultr, A., Kolman, P. (eds.) MFCS 2001. LNCS, vol. 2136, pp. 272–284. Springer, Heidelberg (2001)
18. Diakonikolas, I., Gopalan, P., Jaiswal, R., Servedio, R.A., Viola, E.: Bounded independence fools halfspaces. SIAM Journal of Computation 39(8), 3441–3462 (2010)
19. Diakonikolas, I., Kane, D.M., Nelson, J.: Bounded independence fools degree-2 threshold functions. In: FOCS, pp. 11–20 (2010)
20. Etesami, S.O.: Pseudorandomness against depth-2 circuits and analysis of goldreich's candidate one-way function. Technical Report EECS-2010-180, UC Berkeley (2010)
21. Goemans, M., Williamson, D.: Improved approximation algorithms for maximum cut and satisfiability problems using semidefinite programming. JACM: Journal of the ACM 42 (1995)
22. Goldreich, O.: Candidate one-way functions based on expander graphs. Electronic Colloquium on Computational Complexity (ECCC) 7(090) (2000)
23. Impagliazzo, R., Nisan, N., Wigderson, A.: Pseudorandomness for network algorithms. In: Proceedings of the 26th Annual ACM Symposium on Theory of Computing, pp. 356–364 (1994)

24. Ishai, Y., Kushilevitz, E., Ostrovsky, R., Sahai, A.: Cryptography with constant computational overhead. In: Ladner, R.E., Dwork, C. (eds.) STOC, pp. 433–442. ACM (2008)
25. Itsykson, D.: Lower bound on average-case complexity of inversion of goldreich's function by drunken backtracking algorithms. In: Computer Science - Theory and Applications, 5th International Computer Science Symposium in Russia, pp. 204–215 (2010)
26. Miller, R.: Goldreich's one-way function candidate and drunken backtracking algorithms. Distinguished major thesis, University of Virginia (2009)
27. Mossel, E., Shpilka, A., Trevisan, L.: On ϵ-biased generators in NC^0. In: Proc. 44th FOCS, pp. 136–145 (2003)
28. Naor, J., Naor, M.: Small-bias probability spaces: Efficient constructions and applications. SIAM Journal on Computing 22(4), 838–856 (1993); Preliminary version in Proc. 22th STOC (1990)
29. Panjwani, S.K.: An experimental evaluation of goldreich's one-way function. Technical report, IIT, Bombay (2001)
30. Siegenthaler, T.: Correlation-immunity of nonlinear combining functions for cryptographic applications. IEEE Transactions on Information Theory 30(5), 776–778 (1984)

Randomness Condensers for Efficiently Samplable, Seed-Dependent Sources

Yevgeniy Dodis[1,*], Thomas Ristenpart[2,**], and Salil Vadhan[3,***]

[1] New York University
dodis@cs.nyu.edu
[2] University of Wisconsin–Madison
rist@cs.wisc.edu
[3] Harvard University
salil@seas.harvard.edu

Abstract. We initiate a study of randomness condensers for sources that are efficiently samplable but may depend on the seed of the condenser. That is, we seek functions $\mathsf{Cond} : \{0,1\}^n \times \{0,1\}^d \to \{0,1\}^m$ such that if we choose a random seed $S \leftarrow \{0,1\}^d$, and a source $X = \mathcal{A}(S)$ is generated by a randomized circuit $\mathcal{A}$ of size t such that X has min-entropy at least k given S, then $\mathsf{Cond}(X; S)$ should have min-entropy at least some k' given S. The distinction from the standard notion of randomness condensers is that the source X may be correlated with the seed S (but is restricted to be efficiently samplable). Randomness *extractors* of this type (corresponding to the special case where $k' = m$) have been implicitly studied in the past (by Trevisan and Vadhan, FOCS '00).

We show that:

- Unlike extractors, we can have randomness condensers for samplable, seed-dependent sources whose computational complexity is smaller than the size t of the adversarial sampling algorithm $\mathcal{A}$. Indeed, we show that sufficiently strong collision-resistant hash functions are seed-dependent condensers that produce outputs with min-entropy $k' = m - \mathcal{O}(\log t)$, i.e. logarithmic *entropy deficiency*.
- Randomness condensers suffice for key derivation in many cryptographic applications: when an adversary has negligible success probability (or negligible "squared advantage" [3]) for a uniformly random key, we can use instead a key generated by a condenser whose output has logarithmic entropy deficiency.
- Randomness condensers for seed-dependent samplable sources that are robust to side information generated by the sampling algorithm imply soundness of the Fiat-Shamir Heuristic when applied to any constant-round, public-coin interactive proof system.

* Partially supported by NSF Grants CNS-1065134, CNS-1065288, CNS-1017471, CNS-0831299 and Google Faculty Award. Work done in part while visiting Microsoft Research Redmond.
** Partially supported by NSF Grant CNS-1065134. Work done in part while visiting Microsoft Research Redmond.
*** Supported by NSF grant CCF-1116616. Work done in part while visiting Microsoft Research SVC and Stanford University.

R. Cramer (Ed.): TCC 2012, LNCS 7194, pp. 618–635, 2012.
© International Association for Cryptologic Research 2012

1   Introduction

Randomness extractors — functions that convert sources of biased and/or correlated bits into almost uniformly distributed bits — have a wide variety of applications in cryptography and other parts of theoretical computer science. However, to extract randomness from rich models of sources, e.g. sources for which we only have a lower bound on their min-entropy (or even sources where each bit is mildly unpredictable given the previous ones), deterministic functions cannot be randomness extractors [30]. Thus the general definition of randomness extractor by Nisan and Zuckerman [27] allows the extractor to be probabilistic — the extractor is given a uniformly random *seed* that it can use as a catalyst for extraction.

The need for a seed, however, is a problem in some applications of randomness extractors. First, if the reason for extraction is lack of access to high-quality random bits, then we may not have any way to generate the seed.[1] (In algorithmic applications of randomness extractors, it is often possible to try all possible seeds, and combine the results obtained for each extractor output. But this does not work in most cryptographic applications. Even one bad seed can compromise one's secrets, and thus eliminate security.) Second, even if we can generate a uniformly random seed, it is crucial that the weak random source from which we extract is independent from the seed. This means that it is problematic to generate the seed once and for all (perhaps using an expensive source of randomness) in hope that it can be used for all future randomness extractions. If there is any chance that the future weak sources can be influenced by the seed, then the extractor guarantees will be lost. For example, if the seed is stored in some hardware random number generator (RNG) that extracts from physical sources of randomness within the computer (e.g. timing of various events), these sources may be affected by the internal computations of the RNG itself and thus we have correlations between the seed and the sources.

Such considerations and others have motivated a revival in the study of *deterministic extractors* over the past decade, i.e. extractors that do not require a seed. Since deterministic extraction is impossible for general weak sources of randomness, this body of work has sought to identify the richest classes of sources for which deterministic extraction is possible, and construct explicit extractors for those sources. Most of the studied models of such "extractable sources" (e.g. bit-fixing sources [9], discrete control sources [26] or multiple independent sources [8]) implicitly or explicitly require independence between different portions of the source. To avoid this, Trevisan and Vadhan [34] suggested studying the class of *samplable sources*, sources generated by efficient algorithms, e.g. polynomial-sized circuits. They showed that for every t, there exist (non-explicit) deterministic extractors for sources generated by circuits of size t, provided that the min-entropy of the source is $\omega(\log t)$. Moreover, this result is based on a probabilistic argument, and can be viewed as giving an explicit *seeded* extractor that

---

[1] Actually, using *2-source extractors* [8,11], the seed can also be weakly random, but it still needs to be independent from the source.

works for *seed-dependent* sources in the following sense. We generate once and for all a random seed S for the extractor, then an adversary $\mathcal{A}$ of size t generates a source $X = \mathcal{A}(S)$ (using additional randomness) with the property that X has enough min-entropy given S, and our extractor $\mathsf{Ext}(X; S)$ produces an output that is statistically close to uniform given S. (We remark that [34] also gave an explicit and seedless extractor for samplable sources having min-entropy rate close to 1 based on some strong complexity assumptions, and subsequent works have given explicit and seedless extractors for sources sampled by weaker models of computation, such as small-space algorithms [24,25,23] and constant-depth circuits [35].)

A deficiency of the above extractors is that their computational complexity is $\mathrm{poly}(t)$ — larger than the complexity of the adversary generating the source. As observed in [34], this is inherent. If the adversary has more resources than the extractor, then it can randomly generate inputs on which the first few bits of the extractor's output is constant (and this will be a high min-entropy source). More precisely, if the adversary's running time is larger than the extractor's by a factor of t, it can fix roughly $\log t$ bits of the output (and generate a source on n bits of min-entropy approximately $n - \log t$).

The starting point for our paper is the observation that the above attack is not so bad. If the adversary can only reduce the min-entropy of the extractor's output by a logarithmic number of bits, we have still achieved something very nontrivial and useful. Indeed, we will have what is called a *randomness condenser* [28,29] — which takes an n-bit source with at least some k bits of min-entropy and outputs an m-bit source with at least some k' bits of min-entropy. Randomness condensers are nontrivial when the output entropy *deficiency* $m - k'$ is smaller than the input entropy deficiency $n - k$ (otherwise we could condense just by truncating the source). They have been extensively studied in the literature as a building block towards constructing randomness extractors (starting with [29], and continuing in some of the latest extractors [20]), as well as bipartite expander graphs [33,7].

Here we note that condensers are useful in their own right. If the entropy deficiency of the output is at most β, then any event that occurs with probability p under a uniformly random string can occur under the condenser's output with probability at most $p' = 2^{\beta} \cdot p$. For example, if p is negligible and β is logarithmic, then p' is also negligible.

Motivated by the above, we initiate a study of condensers for samplable sources.

DEFINING SEED-DEPENDENT RANDOMNESS CONDENSERS. We define a condenser for seed-dependent samplable sources to be a function $\mathsf{Cond} : \{0,1\}^n \times \{0,1\}^d \to \{0,1\}^m$ with the following property. If $S \leftarrow U_d$, and $X = \mathcal{A}(S)$ is a source with (min-)entropy at least k given S, generated by a randomized circuit $\mathcal{A}$ of size at most t, then we require that $\mathsf{Cond}(X; S)$ should be (close to) a source with min-entropy at least k' given S. We provide a number of variants of this definition, using different measures of conditional entropy, and also consider the case that $\mathcal{A}$ generates side information along with X (to be discussed more below).

CONDENSERS FROM CR HASHING. We show that sufficiently strong collision-resistant hash functions provide good seed-dependent condensers for samplable sources. Here the seed is simply a description of a hash function h from the family, and $\mathsf{Cond}(x; h) = h(x)$. We show that if efficient algorithms can find collisions in the hash functions with probability at most $2^\beta/2^m$, then the condenser output will have min-entropy $k' \approx m - \beta$ given the seed (for sources of min-entropy larger than m). Note that a birthday attack will find collisions with probability $O(t^2/2^m)$ in time t. If time t algorithms cannot do much better, e.g. the probability of finding collisions is at most $\mathrm{poly}(t)/2^m$, then we can achieve entropy deficiency $\beta = O(\log t)$, within a constant factor of the lower bound mentioned above.

CONDENSERS AND KEY DERIVATION. We formalize the applicability of seed-dependent condensers to key derivation. Specifically, we consider using the output of a condenser as a key in a cryptographic application, and show that for "unpredictability" applications (where an adversary can win in a security game with at most negligible probability), security is preserved if the output entropy deficiency β is small enough (e.g. logarithmic). For indistinguishability applications, we follow [3] and show that security is preserved if the "squared advantage" is negligible, which can be achieved for a number of applications. These results provide the first formal evidence that when seed-dependent sources arise in practice [21] security is not immediately compromised.

CONDENSERS AND FIAT–SHAMIR. We investigate seed-dependent condensers for adversaries $\mathcal{A}(S)$ that generate some *side information* Z in addition to X (with the requirement that X has min-entropy at least k given S and Z), analogously to the notion of average-case extractors introduced by [12]. We observe that the most natural generalization of our condenser definition to this setting, namely requiring that $\mathsf{Cond}(X; S)$ has min-entropy at least k' given S and Z, is impossible to achieve: the adversary $\mathcal{A}(S)$ can simply compute $Z = \mathsf{Cond}(X; S)$ as its side information. However, it seems plausible to have good condensers if we provide the side information also as input to the condenser. While this may not be feasible in some applications (because we do not know the side information), we show that condensers satisfying this definition can be used to obtain a sound implementation of the Fiat–Shamir Heuristic for all constant-round, public-coin interactive proof systems (ones with *statistical* soundness), and hence show that such protocols cannot be zero knowledge (by connections established by Dwork et al. [14]). This novel connection between the Fiat–Shamir Heuristic and randomness condensing is obtained by observing a close relation between seed-dependent condensers for samplable sources tolerating side information and some conjectures of Barak, Lindell, and Vadhan [4] (made in the study of zero knowledge and Fiat–Shamir). In fact, this connection only requires condensers for "leaky sources" — ones that are uniform prior to conditioning on the adversary's side information — and we show that such condensers are also *necessary* for soundness of the Fiat–Shamir Heuristic. It remains an intriguing open problem to give a construction of condensers for leaky sources based on some more well-studied complexity assumptions.

2   Definitions and Preliminaries

ENTROPY AND STATISTICAL DISTANCE. We start by defining the relevant notions of entropy that we use, which are min-entropy, collision (also known as Renyi) entropy and Shannon entropy. The *Shannon entropy* and *min-entropy* of a random variable X are defined as $\mathbf{H}_1(X) \overset{\text{def}}{=} \mathbb{E}_{x \leftarrow X}[-\log \Pr[X = x]]$ and $\mathbf{H}_\infty(X) \overset{\text{def}}{=} -\log(\max_x \Pr[X = x])$. We also define *average (aka conditional) Shannon entropy* and *average min-entropy* of a random variable X conditioned on another random variable Z by $\mathbf{H}_1(X|Z) \overset{\text{def}}{=} \mathbb{E}_{(x,z) \leftarrow (X,Z)}[-\log \Pr[X = x | Z = z]]$ and $\mathbf{H}_\infty(X|Z) \overset{\text{def}}{=} -\log(\mathbb{E}_{z \leftarrow Z}[\max_x \Pr[X = x | Z = z]])$ respectively, where $\mathbb{E}_{z \leftarrow Z}$ denotes the expected value over $z \leftarrow Z$.

The *collision probability* of a random variable X is defined as $\mathbf{Col}(X) \overset{\text{def}}{=} \sum_x \Pr[X = x]^2$, and the *collision entropy* of X is $\mathbf{H}_2(X) = \log(1/\mathbf{Col}(X))$. It is easy to see that for any X, $\mathbf{H}_\infty(X) \leq \mathbf{H}_2(X) \leq \mathbf{H}_1(X)$ and $\mathbf{H}_2(X) \leq 2\mathbf{H}_\infty(X)$. We can also define *average* collision probability and collision entropy of a random variable X conditioned on another random variable Z by $\mathbf{Col}(X|Z) = \mathbb{E}_{z \leftarrow Z}[\mathbf{Col}(X|Z = z)]$ and $\mathbf{H}_2(X|Z) = \log(1/\mathbf{Col}(X|Z))$. Once again, $\mathbf{H}_\infty(X|Z) \leq \mathbf{H}_2(X|Z) \leq \mathbf{H}_1(X|Z)$ and $\mathbf{H}_2(X|Z) \leq 2\mathbf{H}_\infty(X|Z)$.

We denote with $\text{dist}_D(X, Y)$ the advantage of a function D in distinguishing the random variables X, Y: $\text{dist}_D(X, Y) \overset{\text{def}}{=} |\Pr[D(X) = 1] - \Pr[D(Y) = 1]|$. The *statistical distance* between two random variables X, Y is defined by

$$\mathsf{SD}(X, Y) \overset{\text{def}}{=} \frac{1}{2} \sum_x |\Pr[X = x] - \Pr[Y = x]| = \max_D \text{dist}_D(X, Y)$$

We say that X and Y are ε-*close* if $\mathsf{SD}(X, Y) \leq \varepsilon$. We also note that any tuple (X, Z) is ε-close to (X', Z) such that $\mathbf{H}_\infty(X'|Z) \geq \mathbf{H}_2(X|Z) - \log(1/\varepsilon)$, which is often much better than bounding $\mathbf{H}_\infty(X|Z) \geq \frac{1}{2} \cdot \mathbf{H}_2(X|Z)$.

3   Seed-Dependent Condensers

We now generalize the notion of a condenser to the *seed-dependent* setting, in which the adversarial sampler $\mathcal{A}$ of size t can depend on the seed S. As we will see, seed-dependent condensers are useful for important applications such as cryptographic key derivation.

Definition 3.1 (Seed-Dependent Condenser). *Let $c, c' \in \{1, 2, \infty\}$. An efficient function* $\mathsf{Cond} : \{0,1\}^n \times \{0,1\}^d \to \{0,1\}^m$ *is a seed-dependent* $([\mathbf{H}_c \geq k] \to_\varepsilon [\mathbf{H}_{c'} \geq k'], t)$-*condenser if for all probabilistic adversaries $\mathcal{A}$ of size at most t who take a random seed $S \leftarrow \{0,1\}^d$ and output (using more coins) a sample $X \leftarrow \mathcal{A}(S)$ of entropy $\mathbf{H}_c(X|S) \geq k$, the joint distribution $(S, \mathsf{Cond}(X; S))$ is ε-close to some (S, R), where $\mathbf{H}_{c'}(R|S) \geq k'$.*

The quantity $\beta \overset{\text{def}}{=} m - k'$ is called the entropy deficit *of the condenser. When $c = c'$ is clear from the context, we say that Cond is a seed-dependent $(k \to_\varepsilon k', t)$-condenser. We omit the reference to ε and/or t when $\varepsilon = 0$ and/or $t = \infty$, respectively.*

A notion for traditional condensers arises by replacing $\mathcal{A}$ in the definition above with an unbounded circuit that does not take the seed S as input. Unlike with traditional condensers, seed-dependent condensers require that $\mathcal{A}$ be efficient. Otherwise, an inefficient $\mathcal{A}$ can, by repeatedly evaluating the condenser using the seed S, always find a high entropy distribution of inputs that map to a low entropy output distribution. Second, while a seed-dependent extractor can be defined as a special case of the definition above corresponding to $k' = m$, Proposition 3.3 below implies that it is impossible to build a (non-trivial) seed-dependent extractor.

The following lemma (see proof in [13]) will be useful in several of our later results.

Lemma 3.2. *Let* $c \in \{1, 2, \infty\}$. *Then,*

- **"Output $(\infty \to 2 \to 1)$":** *If* $c' \geq c''$ *and* Cond *is a seed-dependent* $(([\mathbf{H}_c \geq k] \to_\varepsilon [\mathbf{H}_{c'} \geq k']), t)$-*condenser, then* Cond *is also a seed-dependent* $(([\mathbf{H}_c \geq k] \to_\varepsilon [\mathbf{H}_{c''} \geq k']), t)$-*condenser.*

- **"Output $(2 \to \infty)$":** *For any* $\gamma > 0$, *if* Cond *is seed-dependent* $(([\mathbf{H}_c \geq k] \to_\varepsilon [\mathbf{H}_2 \geq k']), t)$-*condenser, then* Cond *is also a seed-dependent* $(([\mathbf{H}_c \geq k] \to_{\varepsilon+\gamma} [\mathbf{H}_\infty \geq k' - \log(1/\gamma)]), t)$-*condenser and also a seed-dependent* $(([\mathbf{H}_c \geq k] \to_\varepsilon [\mathbf{H}_\infty \geq k'/2]), t)$-*condenser.*

- **"Input $(1 \to 2 \to \infty)$":** *If* $c' \leq c''$ *and* Cond *is seed-dependent* $(([\mathbf{H}_{c'} \geq k] \to_\varepsilon [\mathbf{H}_c \geq k']), t)$-*condenser, then* Cond *is also a seed-dependent* $(([\mathbf{H}_{c''} \geq k] \to_\varepsilon [\mathbf{H}_c \geq k']), t)$-*condenser.*

Thus, it is somewhat preferable (but also the hardest) to build a seed-dependent $([\mathbf{H}_2 \geq k] \to_\varepsilon [\mathbf{H}_\infty \geq k'])$ condenser, since it implies $([\mathbf{H}_c \geq k] \to_\varepsilon [\mathbf{H}_{c'} \geq k'])$-condenser for any $c, c' \in \{2, \infty\}$. In contrast, it is preferable to base a security of a given application on a $([\mathbf{H}_\infty \geq k] \to_\varepsilon [\mathbf{H}_2 \geq k'])$-condenser, since such condensers are likely to have slightly better parameters k and k'.

The following negative result shows that the output entropy deficiency $\beta = m - k'$ must be at least roughly $\log t$ to work for samplers computable in time t, if the condenser is computable in time significantly less than t. In particular, we cannot hope for a seed-dependent *extractor* (i.e. $\beta = 0$) that is computable in time significantly less than t, generalizing an observation of Trevisan and Vadhan [34] about deterministic extractors for samplable sources.

Proposition 3.3. *Let* Cond $: \{0,1\}^n \times \{0,1\}^d \to \{0,1\}^m$ *be computable by a circuit of size* t', *and let* $\beta \in [0, m]$, $\varepsilon, \delta \in (0, 1/2)$. *Then for* Cond *to be a* $(([\mathbf{H}_\infty \geq n - \alpha] \to_\varepsilon [\mathbf{H}_1 \geq m - \beta]), t)$-*condenser for* $\alpha = \lceil (\beta + 1)/(1 - \varepsilon - \delta) \rceil$, *it must be that* $\alpha \geq \log t - \log t' - O(\log(1/\delta))$ *or* $\alpha \geq m$.

Note that as $\varepsilon, \delta \to 0$, the ratio between α and β approaches 1. Thus, the proposition says that if we want to decrease the entropy deficiency by any significant factor, we must settle for output entropy deficiency $\beta \approx \alpha$ that is at least roughly $\log t$.

HANDLING SIDE INFORMATION.   One can naturally generalize the notion of (regular) extractors and condensers to handle some *side information* Z about the source X, yielding the notion of *average-case* extractors/condensers [12]. Formally, the adversarial sampler $\mathcal{A}$ produces a pair (X, Z) such that $\mathbf{H}_c(X|Z) \geq k$, and one requires that the joint distribution $(Z, S, \mathsf{Ext}(X; S))$ is ε-close to (Z, S, U_m) (for condensers, that $(Z, S, \mathsf{Cond}(X; S))$ is ε-close to (Z, S, R) where $\mathbf{H}_{c'}(R|(S, Z)) \geq k'$).

However, things become a bit trickier in the seed-dependent case that we introduce in this work. Naturally, the sampler $\mathcal{A}$ now takes the seed S to produce the pair (X, Z). Unfortunately, this means that $\mathcal{A}$ can now run the condenser $\mathsf{Cond}(X; S)$ and simply record all or part of this output in the side information Z. This still leaves the entropy of X high enough (say, if k is noticeably larger than m), but now the output entropy k' drops to 0. Thus, to make a meaningful but *satisfiable* definition in the case of side information, we will relax the syntax of the condenser Cond to also take the side information Z as part of its input. While less convenient for some applications, now the previous attack no longer applies, since the sampler $\mathcal{A}(S)$ has to choose Z before $R = \mathsf{Cond}((X, Z); S)$ is derived, making it much harder to "correlate" R and Z. Therefore we say that a condenser is a average-case, seed-dependent ($[\mathbf{H}_c \geq k] \to_\varepsilon [\mathbf{H}_{c'} \geq k'], t$)-condenser if $(Z, S, \mathsf{Ext}((X, Z); S))$ is ε-close to (Z, S, R) where $S \leftarrow \{0, 1\}^d$, $(X, Z) \leftarrow \mathcal{A}(S)$ with $\mathbf{H}_c(X|(S, Z)) \geq k$, and $\mathbf{H}_{c'}(R|(S, Z) \geq k')$. A formal definition can be found in the full version [13].

We notice that Lemma 3.2 clearly extends to the average-case setting. Also, when Z is empty, this still generalizes the "worst-case" seed-dependent condenser from Definition 3.1. However, the introduction of side information makes the notion of seed-dependent condenser very non-trivial to satisfy even when the source X is perfectly uniform, but some side information $Z = f(X)$ is "leaked" to the attacker. Indeed, we show in Section 6 that this special case of average-case condensers (see Definition 6.1) is exactly what is needed to instantiate the Fiat-Shamir heuristic.

Finally, an equivalent way to think about average-case condensers is to interpret the output (X, Z) of the sampler as a single (variable-length) source X', so that the condenser is simply applied to X', but a subset of (known) *physical bits* Z of X' is leaked to the attacker/distinguisher.

4   Condensers from Collision Resistance

In this section we show that a sufficiently strong collision-resistant hash function (CRHF) gives a good seed-dependent (but *not average-case*) ($[\mathbf{H}_2 \geq k] \to_0 [\mathbf{H}_2 \geq k']$) condenser, which also implies non-trivial bounds for other input/output entropy settings when $c, c' \in \{2, \infty\}$, by Lemma 3.2.

Definition 4.1. *A family of hash function* $\mathcal{H} = \{h : \{0, 1\}^* \to \{0, 1\}^m\}$ *is* (t, δ)*-collision-resistant if for any (non-uniform) attacker* $\mathcal{B}$ *of size at most* t, $\Pr[H(X_1) = H(X_2) \wedge X_1 \neq X_2] \leq \delta$ *where* $H \leftarrow \mathcal{H}$ *and* $(X_1, X_2) \leftarrow \mathcal{B}(H)$.

The proof of the following theorem appears in the full version [13].

Theorem 4.2. *Fix any $\beta > 0$. If $\mathcal{H}$ is a $(2t, 2^{\beta-1}/2^m)$-collision-resistant hash function family, then $\mathsf{Cond}(X; H) \stackrel{\text{def}}{=} H(x)$ for $H \leftarrow \mathcal{H}$ is a seed-dependent $(([\mathbf{H}_2 \geq m - \beta + 1] \to [\mathbf{H}_2 \geq m - \beta]), t)$-condenser with entropy deficit β and no error.*

In particular, it is also a seed-dependent $(([\mathbf{H}_\infty \geq m - \beta + 1] \to [\mathbf{H}_2 \geq m - \beta]), t)$-condenser and $(([\mathbf{H}_\infty \geq m - \beta + 1] \to_\varepsilon [\mathbf{H}_\infty \geq m - \beta + \log \varepsilon]), t)$-condenser.

PARAMETERS. To obtain good entropy deficit β as a function on the sampler's complexity t, we need to understand the best possible $(2t, \delta)$-collision-resistant security of $\mathcal{H}$. Clearly, a birthday attack (essentially) implies that $\delta = \Omega(t^2/2^m)$, since the attacker can pick t random points, evaluate h on them, and hope for some collision. Conversely, this bound is tight in the random oracle model, and state-of-the-art hash functions more or less assume that the "birthday attack" is the only possible attack on a good CRHF design. For example, birthday attacks are currently the best known attacks on many popular hash functions, such as SHA-256, SHA-512, and the new SHA-3 functions, as well as discrete-log based CRHFs over many elliptic curve groups (c.f., [32]). Thus, under such (strong but reasonable) assumptions, all the above popular hash functions achieve $\delta = O(t^2/2^m)$, which means that we can set $2^{\beta-1} = O(t^2)$ resulting in $\beta = 2 \log t + O(1)$. More generally, if the best collision-finding attack has success probability $\delta = \text{poly}(t)/2^m$, then $\beta = O(\log t)$.

Corollary 4.3. *Assuming the existence of $(t, \frac{O(t^2)}{2^m})$-collision-resistant hash functions, there exists a seed-dependent $(([\mathbf{H}_2 \geq m - \beta + 1] \to [\mathbf{H}_2 \geq m - \beta]), t)$-condenser with entropy deficit $\beta = 2 \log t + O(1)$ and no error.*

In particular, it is also a seed-dependent $(([\mathbf{H}_\infty \geq m - \beta + 1] \to [\mathbf{H}_2 \geq m - \beta]), t)$-condenser with entropy deficit $\beta = 2 \log t + O(1)$ and no error, and $(([\mathbf{H}_\infty \geq m - \beta + 1] \to_\varepsilon [\mathbf{H}_\infty \geq m - \beta - \log(1/\varepsilon)]), t)$-condenser with entropy deficit $\beta' = (2 \log t + \log(1/\varepsilon) + O(1))$ and error ε.

AVERAGE-CASE SETTING? Unfortunately, the proof of Theorem 4.2 does not extend to average-case seed-dependent condensers. The problem is that when estimating the value $\mathbf{Col}(H(X, Z)|(H, Z))$, one already needs to sample two sources X_1 and X_2 corresponding to the *same side information* Z, which seems to be hard. A bit more formally, a natural attempt to define a collision-finding adversary $\mathcal{B}$ would be to first let $\mathcal{B}(H)$ run $\mathcal{A}(H)$ to produce a tuple (X_1, Z_1), and then run $\mathcal{A}(H)$ several more times to try to produce a second tuple (X_2, Z_2) with the hope that $Z_2 = Z_1$. But this will not be guaranteed to be efficient unless Z is very short (e.g., just a few bits). In some sense, the difficulty of handling side information might be expected, since we show that average-case seed-dependent condensers are enough to instantiate the random oracle in the Fiat-Shamir heuristic (see Section 6), which is a long-standing open problem.

5   Application to Key Derivation

Consider any cryptographic primitive P (e.g., digital signatures, encryption, etc.), which uses randomness $R \in \{0,1\}^m$ to derive its secret (and, public, if needed) key(s). Without loss of generality, we can assume that R itself is the secret key. In the "ideal" setting, $R \leftarrow \{0,1\}^m$ is chosen uniformly at random, and the attacker $\mathcal{B}$ against P obtains no knowledge about the choice of R, except for what is revealed by P. In practice, however, R is not perfectly uniform. For example, it may be the output of a system random number generator (RNG) that attempts to extract uniform bits from a source of entropy. To guarantee security for the widest range of settings, we ask for the key-derivation to be secure even against seed-dependent[2], adversarially-manipulated sources. However, Proposition 3.3 shows that, at least in general, no extractors exist that work for such a strong adversarial model. We therefore turn to seed-dependent condensers, showing that these yield strong positive results about the security of key-derivation.

Towards this, we model the "real" seed-dependent setting as follows. Let $S \leftarrow \{0,1\}^d$ be a random seed that is chosen and $X \leftarrow \mathcal{A}(S)$ is sampled by an adversarial sampler $\mathcal{A}$. Finally, the cryptographic primitive P uses $R \leftarrow \mathsf{Cond}(X; S)$ as the key. While the above model is the one of greatest most direct practical interest, we will actually consider the more general case of *average-case* condensing, in which an attacker $\mathcal{B}$ against P obtains part of the input to the condenser, the side-information Z. The resulting real/ideal settings for deriving the key for P are formalized by the procedures $\mathsf{Real}(\mathcal{A})$ and $\mathsf{Ideal}(\mathcal{A})$:

$\mathsf{Real}(\mathcal{A})$:	$\mathsf{Ideal}(\mathcal{A})$:
$S \leftarrow \{0,1\}^d$	$S \leftarrow \{0,1\}^d$
$(X, Z) \leftarrow \mathcal{A}(S)$	$(X, Z) \leftarrow \mathcal{A}(S)$
$R \leftarrow \mathsf{Cond}((X, Z); S)$	$R \leftarrow \{0,1\}^m$
Return (R, S, Z)	Return (R, S, Z)

The two procedures are parameterized by a sampler $\mathcal{A}$ that on input the seed S outputs a pair (X, Z). We assume that the sampler $\mathcal{A}$ has size at most t and produces a source X of (conditional) min-entropy $\mathbf{H}_\infty(X|(S, Z)) \geq k$, for some parameters t and k. We call such samplers (t, k)-*bounded*. Sometimes, to emphasize the dependence on the sampler complexity t and source min-entropy k, we will refer to the above two settings as the (t, k)-*real* and (t, k)-*ideal* models, respectively.

The side information Z naturally models information about the random source X that may be leaked to an adversary via a side channel. However, in most or all practical scenarios, our assumption that the value of Z is known and available to the condenser is unrealistic. Thus, we will also state our results for the analogous models *without side information*, meaning we omit Z in both the real and ideal models.

---

[2] For example the Linux RNG folds back into its entropy pool prior outputs [21].

DEFINING REAL/IDEAL SECURITY. We assume that the security of the cryptographic primitive P is defined via an interactive game between a probabilistic attacker $\mathcal{B}(s, z)$ and a probabilistic challenger $\mathcal{C}(r)$. Here one should think of s and z as particular values of the seed and the side information, respectively, and r as a particular value used by the challenger in the key generation algorithm of P. We note that $\mathcal{C}$ only uses the secret key r and does not directly depend on s and z. In particular, in the ideal model, the values s and z are not really useful to the actual attacker $\mathcal{B}$, since the key r used by the challenger $\mathcal{C}$ is chosen completely independently from these values. Still, we include them for consistency.

At the end of the game, $\mathcal{C}(r)$ outputs a bit b, where $b = 1$ indicates that the attacker "won the game". Since $\mathcal{C}$ is fixed by the definition of P (e.g., $\mathcal{C}$ runs the unforgeability game for signature or the semantic security game for encryption, etc.), we denote by $\mathcal{D}_\mathcal{B}(r, s, z)$ the (abstract) distinguisher which simulates the entire game between $\mathcal{B}(s, z)$ and $\mathcal{C}(r)$ and outputs the bit b. We also let

$$\mathbf{Adv}_\mathcal{B}(r, s, z) \stackrel{\text{def}}{=} \Pr[\mathcal{D}_\mathcal{B}(r, s, z) = 1] - c$$

be the advantage of $\mathcal{B}(s, z)$ to win the game against $\mathcal{C}(r)$, where $c = 0$ for unpredictability applications (one-way functions, signatures, etc.) and $c = 1/2$ for indistinguishability applications (encryption, pseudorandom functions, etc.). Thus, $\mathbf{Adv}_\mathcal{B}(\cdot) \in [0, 1]$ for unpredictability applications and $\mathbf{Adv}_\mathcal{B}(\cdot) \in [-\frac{1}{2}, \frac{1}{2}]$ for indistinguishability applications. When $\mathcal{B}$ is clear from the context, we simply write $\mathbf{Adv}(r, s, z)$.

In the following security definition for P, we will use the letter T to denote the maximum allowable resources of $\mathcal{B}$, which include all the efficiency measures we might care about in the corresponding application, such as the circuit size, number of oracle queries, etc. We say that such a $\mathcal{B}$ is T-limited.

Definition 5.1. Given a sampler $\mathcal{A}$ and an attacker $\mathcal{B}$, we define their ideal advantage $\mathbf{\Delta}(\mathcal{A}, \mathcal{B}) \stackrel{\text{def}}{=} | \mathbb{E}[\mathbf{Adv}_\mathcal{B}(\mathsf{Ideal}(\mathcal{A}))] |$. We say that P is (T, δ)-secure in the (t, k)-ideal model if for any (t, k)-bounded sampler $\mathcal{A}$ and any T-limited attacker $\mathcal{B}$, $\mathbf{\Delta}(\mathcal{A}, \mathcal{B}) \leq \delta$. Similarly, given $\mathcal{A}$ and $\mathcal{B}$, we define their real advantage $\widetilde{\mathbf{\Delta}}(\mathcal{A}, \mathcal{B}) \stackrel{\text{def}}{=} | \mathbb{E}[\mathbf{Adv}_\mathcal{B}(\mathsf{Real}(\mathcal{A}))] |$. We say that P is (T', δ')-secure in the (t, k)-real model if for any (t, k)-bounded sampler $\mathcal{A}$ and any T'-limited attacker $\mathcal{B}$, $\widetilde{\mathbf{\Delta}}(\mathcal{A}, \mathcal{B}) \leq \delta'$.

5.1   Simple Bound for Unpredictability Applications

As our first attempt, we would like to argue that if P is (T, δ)-secure in the ideal setting, then P is also (T', δ')-secure in the real setting, where T' is not much lower than T, and, more importantly, δ' is not much larger than δ. With traditional extractors, this is done by arguing that the derived real key R is (statistically) ε-close to U_m, even conditioned on S and Z. This means that $\delta' \leq \delta + \varepsilon$. Unfortunately, in the seed-dependent settings it is impossible to achieve statistical extraction, as shown by Proposition 3.3. In this section, we observe that is not strictly necessary to argue statistical extraction: if the original ideal security δ is low enough, a good enough condenser (achievable even in the seed-dependent setting) might result in "real" security δ' not much larger than the "ideal" security δ. At least, we show that this intuition is true for unpredictability applications (where, recall, $\mathbf{Adv}(\cdot) \geq 0$) in the following lemma.

Lemma 5.2. *Assume P is some unpredictability application which is (T, δ)-secure in the (t, k)-ideal model, and Cond is an average-case seed-dependent $(([\mathbf{H}_\infty \geq k] \to_\varepsilon [\mathbf{H}_\infty \geq k']), t)$-condenser with entropy deficit $\beta = m - k'$. Then P is (T, δ')-secure in the (t, k)-real model, where $.\delta' \leq \varepsilon + \delta \cdot 2^\beta..$ If instead Cond is an (non-average-case) seed-dependent $(([\mathbf{H}_\infty \geq k] \to_\varepsilon [\mathbf{H}_\infty \geq k']), t)$-condenser, then P is (T, δ')-secure in the (t, k)-real model without side information.*

PARAMETERS. In essence, Lemma 5.2 states that the security δ degrades exponentially with the entropy deficit β of our seed-dependent condenser. Recall that $\beta = O(\log t)$ is the best we can hope for (by Proposition 3.3); this would give a meaningful security guarantee $\delta' \approx \delta \cdot \text{poly}(t)$, as long as $\delta \ll 1/\text{poly}(t)$.

For example, for the non-average-case setting, we can combine the bound in Lemma 5.2 with the construction from Corollary 4.3 to show that a $\mathcal{O}(t^2)/2^m$-collision-resistant hash function suffices for real model security.

5.2   General Bound through Squared Advantage

The bound of Lemma 5.2 only holds for unpredictability applications, and also requires seed-dependent condensers guaranteeing the *min-entropy* of the extracted key R. In this section we show a more general bound which also holds for indistinguishability applications, has better dependence on the entropy deficit of the condenser, and needs a slightly weaker type of seed-dependent condenser for *collision* entropy. However, the small price we pay for such improvements is that we can no longer directly relate the real-security δ' of our application to its ideal security δ. Rather, we use the notion of the *squared advantage* $\mathbf{\Delta}_2(\mathcal{A}, \mathcal{B})$, and will relate $\widetilde{\mathbf{\Delta}}(\mathcal{A}, \mathcal{B})$ to $\mathbf{\Delta}_2(\mathcal{A}, \mathcal{B})$, which will in turn relate δ' to the "square-security" σ which we define below. This notion of squared advantage/security was implicitly introduced by Barak et al. [3] in the "seed-independent" setting (to improve the entropy loss of the Leftover Hash Lemma), who also showed that for many important applications the value σ is not "too much worse" than δ (see the full version for more details [13]).

Definition 5.3. *Given a sampler $\mathcal{A}$ and an attacker $\mathcal{B}$, we define their (ideal) square advantage $\mathbf{\Delta}_2(\mathcal{A}, \mathcal{B}) \stackrel{\text{def}}{=} \mathbb{E}[\mathbf{Adv}_\mathcal{B}(\mathsf{Ideal}(\mathcal{A}))^2]$. We say that P is (T, σ)-square-secure in the (t, k)-ideal model if for any (t, k)-bounded sampler $\mathcal{A}$ and any T-limited attacker $\mathcal{B}$, $\mathbf{\Delta}_2(\mathcal{A}, \mathcal{B}) \leq \sigma$.*

We can now state our improved bound, and then compare it to our previous bound from Lemma 5.2. The proof appears in the full version [13].

Lemma 5.4. *Assume P any application which is (T, σ)-square-secure in the (t, k)-ideal model, and Cond is an average-case seed-dependent $(([\mathbf{H}_\infty \geq k] \to_\varepsilon$*

$[\mathbf{H}_2 \geq k'])$, t)-*condenser with entropy deficit* $\beta = m - k'$. *Then* P *is* (T, δ')-*secure in the* (t, k)-*real model, where* $.\delta' \leq \varepsilon + \sqrt{\sigma \cdot 2^\beta}$.. *If instead* Cond *is an (non-average-case) seed-dependent* $(([\mathbf{H}_\infty \geq k] \to_\varepsilon [\mathbf{H}_\infty \geq k']), t)$-*condenser, then* P *is* (T, δ')-*secure in the* (t, k)-*real model without side information.*

Using Corollary 4.3, we obtain a nearly optimal security degradation in the real model with no side information:

Corollary 5.5. *Assuming the existence of* $(t, \frac{O(t^2)}{2^m})$-*collision-resistant hash functions, if* P *is* (T, σ)-*square-secure in the* $(t, m - 2\log t + O(1))$-*ideal model with no side information, then using a collision-resistant function as a condenser makes* P *to be* (T, δ')-*secure in the* $(t, m - 2\log t + O(1))$-*real model with no side information, where* $\delta' \leq O(t \cdot \sqrt{\sigma})$.

6   Side-Information and Fiat-Shamir

One of the earliest and most influential applications of the Random Oracle Model in cryptography (predating its formalization by Bellare and Rogaway [5]) was to analyze the Fiat–Shamir Heuristic [15]. In the Fiat–Shamir Heuristic, a hash function is used to eliminate interaction in constant-round public-coin protocols, replacing the verifier's random challenges with hashes of the transcript so far. If the hash function is modeled as a random oracle, then this heuristic is known to preserve soundness of the underlying protocol (up to a factor polynomial in the number of queries made by the adversary to the random oracle). However, there are no natural examples of protocols for which the Fiat–Shamir Heuristic has been proven sound when the hash function is implemented by an efficiently computable family of functions.

The original motivation for the Fiat–Shamir Heuristic was as a method to convert identification schemes into digital signature schemes, and the method gave rise to many efficient digital signature schemes in practice [15,31,19] (albeit with only a proof in the Random Oracle Model). Another compelling motivation for understanding the soundness of the Fiat–Shamir Heuristic is its close connection to the zero-knowledge property of the underlying protocols, as pointed out by Dwork, Naor, Reingold, and Stockmeyer [14]. Dwork et al. showed that the soundness of the Fiat–Shamir Heuristic on a given protocol is essentially equivalent to that protocol *not* being (auxiliary-input) zero knowledge unless the underlying language is in BPP.[3] There are many constant-round public-coin protocols whose zero knowledge status is a long-standing open problem (e.g. ones obtained by starting some underlying basic zero-knowledge protocol and

---

[3] The forward direction is shown as follows: if there is an efficiently computable family of hash functions for which the Fiat–Shamir heuristic is sound, then it is infeasible to simulate a verifier that has a random hash function from the family as auxiliary input, and obtains its challenges by applying the hash function to the transcript so far. Indeed, an efficient simulator would constitute a prover strategy that generates accepting proofs for the Fiat-Shamir-collapsed protocol, which would only be possible for inputs in the language.

applying parallel repetition to make the soundness error negligible). While these protocols cannot be *black-box* zero knowledge (for nontrivial languages) [16], they may still be non-black-box (auxiliary-input) zero knowledge.

Indeed, Barak [2] constructed a constant-round, public-coin (non-black-box) zero-knowledge argument system for NP (assuming the existence of collision-resistant hash functions), thereby yielding a natural protocol on which the Fiat–Shamir heuristic is unsound (for any efficiently computable family of hash functions). Goldwasser and Kalai [17] extended Barak's techniques to construct 3-message public-coin identification schemes on which the Fiat–Shamir Heuristic is unsound. In both of these counterexamples to the Fiat–Shamir Heuristic, the initial interactive protocol is only *computationally* sound, and the results seem to use this in an essential way.

Thus, Barak, Lindell, and Vadhan [4] conjectured that there *is* a sound implementation of the Fiat–Shamir Heuristic for any *statistically sound* interactive proof of language membership (and thus that there can be no constant-round public-coin zero-knowledge *proof* system with negligible soundness for a language outside BPP). Indeed, they provided a plausible property for a family of hash functions that suffices for it to provide a sound implementation of Fiat–Shamir on proof systems. While they conjectured that such hash families exist, it remains open to construct one based on a standard complexity assumption.

The significance of statistical soundness for reducing interaction was further highlighted by the recent work of Kalai and Raz [22], who showed that a method proposed by Aiello et al. [1] (based on Private Information Retrieval) can be used to convert (statistically sound) interactive proofs into 2-message argument systems. However, this construction does not subsume Fiat–Shamir, because the 2-message argument system it produces is private coin (so the verifier's first message cannot be published as a CRS and shared by all verifiers, as needed for the application to digital signatures) and it does not have the connection to zero knowledge mentioned above.

Here we show that condensers for seed-dependent samplable sources that can handle side information (i.e. *average-case* condensers) imply hash functions for which the Fiat–Shamir Heuristic is sound for proof systems. In fact, we only require condensers for the case that the initial source X is uniform and the adversary's side-information Z consists of a bounded-length "leakage" $f(X, S)$ on the source and seed, for an efficiently computable leakage function f. We also show a partial converse — some form of such condensers are also *necessary* for the Fiat–Shamir heuristic to be sound for all proof systems.

Our results are inspired by a similarity between the definition of condensers for samplable sources and the aforementioned conjectures of Barak et al. [4]. While the existence of such condensers and hash functions remains an open problem, the connection between randomness condensing and the Fiat–Shamir Heuristic, along with our construction of condensers without side information (Theorem 4.2), seem to yield a clearer picture of what is needed for the Fiat–Shamir Heuristic to work. (In particular, we find the definition of a seed-dependent average-case condenser more natural than the conjectures in [4].)

We begin by defining the restricted form of average-case condensers that we relate to the Fiat–Shamir heuristic:

Definition 6.1 (Condensers for Leaky Sources). *Let $c, c' \in \{1, 2, \infty\}$. An efficient function* Cond $: \{0,1\}^n \times \{0,1\}^\alpha \to \{0,1\}^m$ *is an $(\varepsilon, [\mathbf{H}_{c'} \geq k'], t)$-condenser for leaky sources if for all probabilistic adversaries $\mathcal{A}$ of size at most t who take a random source $X \leftarrow \{0,1\}^n$ and output a string $Z := \mathcal{A}(X)$ of length α, the joint distribution $(Z, \mathsf{Cond}(X, Z))$ is ε-close to (Z, R), where $\mathbf{H}_{c'}(R|Z) \geq k'$.*

When $\varepsilon = 0$, we will refer to Cond *as an $([\mathbf{H}_{c'} \geq k'], t)$-condenser for leaky sources. The quantity $\beta \stackrel{\text{def}}{=} m - k'$ is called the* entropy deficit *of the condenser.*

Thus, instead of allowing an arbitrary efficiently samplable source X that has high entropy given the adversary's side information Z, we restrict to $X \leftarrow \{0,1\}^n$ and Z of bounded length α. For natural measures of conditional entropy, this implies that $\mathbf{H}(X|Z) \geq n - \alpha$, so an average-case condenser for entropy $k = n - \alpha$ is also condenser for leaky sources according to Definition 6.1. Note that in the case of leaky sources, we do not provide the condenser with a seed; that is because any seed can be viewed as part of the uniformly random source X. Indeed, average-case condensers with seeds imply seedless condensers for leaky sources; further discussion and formal results are in the full version [13].

Now we define the Fiat–Shamir heuristic more precisely. Let (P, V) be a public-coin interactive protocol, where the parties receive no inputs (except a security parameter κ), there are $2r + 1$ messages exchanged starting with P. We denote the lengths of P's messages by $\ell = \ell(\kappa)$ and the lengths of V's messages by $m = m(\kappa)$.

Definition 6.2. *For a language $L = L(\kappa) \subseteq \{0,1\}^\ell$, we say that (P, V) is a (t, ε)-sound interactive argument for L iff there is no prover strategy P^* of circuit size at most t that convinces V to accept on a transcript whose first message is not in L with probability greater than ε.*

We say that (P, V) is an ε-sound interactive proof for L iff it is an (∞, ε) interactive argument for L (i.e. it holds for computationally unbounded prover strategies P^*).

Ordinarily, interactive proofs are formulated with the input x (whose membership in L is being determined) being provided separately as a common input to P and V. However, incorporating x into the first message of the protocol is notationally more convenient for us.

Fiat and Shamir [15] suggested a way to remove the interaction from protocols as above, by replacing the verifier's messages with hashes of the transcript:

Definition 6.3. *For an interactive protocol (P, V) as above, $\alpha = r \cdot \ell + (r-1) \cdot m$, and a family of hash functions $\mathcal{H} = \mathcal{H}(\kappa) = \{h : \{0,1\}^\alpha \to \{0,1\}^m\}$, the Fiat–Shamir collapse of (P, V) using $\mathcal{H}$ is the 2-message public-coin protocol (P', V') defined as follows:*

(1) *V' sends P' a random hash function $H \leftarrow \mathcal{H}$,*

(2) P' sends V' a tuple $(M_1, M_2, \ldots, M_{r+1}) \in (\{0,1\}^\ell)^{r+1}$,

(3) V' accepts iff V accepts on the transcript $(M_1, R_1, M_2, R_2, \ldots, M_r, R_r, M_{r+1})$ where $R_i \stackrel{\text{def}}{=} H(M_1, R_1, \ldots, M_{i-1}, R_{i-1}, M_i)$ for each $i \in [r]$.

We say that the Fiat-Shamir heuristic using $\mathcal{H}$ is (t, ε')-sound on (P, V) iff (P', V') is a (t, ε')-sound interactive argument for the language $L' = \{(M_1, \ldots, M_{r+1}) : M_1 \in L\}$.

Now we prove that we can use condensers for leaky sources to construct hash functions for which the Fiat–Shamir heuristic is secure:

Theorem 6.4. Let (P, V) be an interactive protocol as above, and let $\alpha = r \cdot \ell + (r-1) \cdot m$. Given $\mathsf{Cond} : \{0,1\}^n \times \{0,1\}^\alpha \to \{0,1\}^m$, define $\mathcal{H} = \{h_x : \{0,1\}^\alpha \to \{0,1\}^m\}_{x \in \{0,1\}^n}$ by $h_x(z) = \mathsf{Cond}(x, z)$.

Then if (P, V) is an ε_1-sound interactive proof for some language L and Cond is an $(\varepsilon_2, [\mathbf{H}_\infty \geq m - \beta], t)$-condenser for leaky sources, then the Fiat-Shamir heuristic is (t', ε')-sound on (P, V), for $t' = t - (r-1) \cdot t_{\mathsf{Cond}} - O(n)$ and

$$\varepsilon' = 2^{r\beta} \cdot \varepsilon_1 + \frac{2^{r\beta} - 1}{2^\beta - 1} \cdot \varepsilon_2 \leq 2^{r\beta} \cdot (\varepsilon_1 + \varepsilon_2).$$

For intuition about the parameters, consider the standard, polynomial-time asymptotic setting. Here all length parameters of the proof system (ℓ, m) are some fixed polynomial in the security parameter κ, and we are interested in protocols whose soundness error ε_1 is negligible, i.e. $\varepsilon_1 = \kappa^{-\omega(1)}$. We focus on constant-round proof systems, so $r = O(1)$. We take the length $n = \mathrm{poly}(\kappa)$ of the condenser source to be significantly larger than $m + \alpha = r \cdot (\ell + m)$. This means that the condenser should work for sources with entropy at least $k = n - \alpha$, which is significantly larger than m. By analogy with Theorem 4.2, we can hope for the output to have min-entropy deficiency $\beta = O(\log t)$, which is $O(\log \kappa)$ for any polynomial $t = t(\kappa)$, possibly with some negligible statistical difference $\varepsilon_2 = \kappa^{-\omega(1)}$. Thus the new soundness error satisfies

$$\varepsilon' \leq 2^{r\beta} \cdot (\varepsilon_1 + \varepsilon_2) = 2^{O(\log \kappa)} \cdot (\kappa^{-\omega(1)} + \kappa^{-\omega(1)}) = \kappa^{-\omega(1)},$$

which is still negligible.

For intuition about the proof, consider a cheating prover strategy, that given the description X of a random hash function from the family, tries to construct a transcript $(M_1, R_1, \ldots, M_r, R_r, M_{r+1})$ such that $M_1 \notin L$, the original verifier accepts, and each R_i is the hash of the prefix preceding it, i.e.

$$R_i = h_X(M_1, R_1, \ldots, M_i) = \mathsf{Cond}(X, (M_1, R_1, \ldots, M_i)).$$

Viewing $Z_i = (M_1, R_1, \ldots, M_i)$ as the adversary's side information (which is of length at most $r \cdot \ell + (r-1) \cdot m$), the condenser property says that R_i is ε_2-close to having min-entropy deficiency at most β given the prefix $M_1, R_1, \ldots, M_i$. Compared to R_i being uniform and independent of the prefix, this should increase the soundness error by an additive ε_2 and a multiplicative 2^β. Incurring this blow up for each of the rounds i yields the bound in the theorem. The formal proof is given in the full version [13].

Many interactive proofs of interest have only three messages (i.e. $r = 1$ above) and have optimal soundness $\varepsilon_1 = 1/2^m$, meaning that for every initial prover message not in L, there is at most 1 verifier challenge that can lead to an accepting transcript. Examples include parallel repetitions of Blum's Hamiltonicity protocol [6], the Goldwasser-Micali-Rackoff Quadratic Residuosity Protocol (to which Fiat-Shamir was originally applied) [18], and any Σ protocol [10]. Setting $r = 1$ and $\varepsilon_1 = 1/2^m$, we see that the resulting soundness error is $\varepsilon' = 2^\beta/2^m + \varepsilon_2$, which is small even for entropy deficiency β that is quite close to m, i.e. the output entropy of the condenser need only be $k' = m - \beta = \log(1/\varepsilon_3)$ to achieve soundness error $\varepsilon_2 + \varepsilon_3$:

Corollary 6.5. *Let* Cond, $\mathcal{H}$, *and* (P, V) *be as in Theorem 6.4. Suppose further that* (P, V) *has 3 messages (i.e.* $r = 1$), *and has soundness* $\varepsilon_1 = 1/2^m$, *where* m *is the length of the verifier's challenge.*

Then if Cond *is a* $(\varepsilon_2, [\mathbf{H}_\infty \geq \log(1/\varepsilon_3)], t)$-*condenser for leaky sources computable in time* t_{Cond}, *it follows that the Fiat-Shamir heuristic is* (t', ε')-*sound on* (P, V), *for* $t' = t - O(n)$ *and* $\varepsilon' = \varepsilon_2 + \varepsilon_3$.

Theorem 6.4 and Corollary 6.5 are stated using average min-entropy as the entropy measure for the output of the condenser. We now discuss their extensions to other entropy measures.

If the condenser output is only guaranteed to have high collision entropy given the seed and the adversary's side information, we can deduce that it is statistically close to having high average-min-entropy. Indeed, if $\mathbf{H}_2(A|B) \geq k$, then for every $\gamma > 0$, (A, B) is γ-close to some (A', B) such that $\mathbf{H}_2(A'|B) \geq k - \log(1/\gamma)$. Thus we can switch from min-entropy to collision entropy at a price of increasing the entropy deficiency by at most $\log(1/\gamma)$ and increasing ε by at most γ.

If the condenser output is only guaranteed to have high Shannon entropy, we can only deduce that the Fiat–Shamir Heuristic has soundness error bounded by a constant. This is still quite nontrivial, and indeed the soundness error can be made negligible without adding interaction by repeating the heuristic with several independent hash functions. This case (obtaining constant error using condensers for Shannon entropy) actually follows from the results in [4] and the connection between condensers for leaky sources and the conjectures in [4]. Moreover, in the full version [13], we give a converse, that soundness of the Fiat-Shamir transform implies the existence of condensers for leaky sources.

References

1. Aiello, W., Bhatt, S.N., Ostrovsky, R., Rajagopalan, S.: Fast Verification of Any Remote Procedure Call: Short Witness-Indistinguishable One-Round Proofs for NP. In: Montanari, U., Rolim, J.D.P., Welzl, E. (eds.) ICALP 2000. LNCS, vol. 1853, pp. 463–474. Springer, Heidelberg (2000)
2. Barak, B.: How to go beyond the black-box simulation barrier. In: 42nd Annual Symposium on Foundations of Computer Science, pp. 106–115. IEEE, Las Vegas (2001), preliminary full version http://www.wisdom.weizmann.ac.il/~boaz

3. Barak, B., Dodis, Y., Krawczyk, H., Pereira, O., Pietrzak, K., Standaert, F.-X., Yu, Y.: Leftover Hash Lemma, Revisited. In: Rogaway, P. (ed.) CRYPTO 2011. LNCS, vol. 6841, pp. 1–20. Springer, Heidelberg (2011)
4. Barak, B., Lindell, Y., Vadhan, S.: Lower bounds for non-black-box zero knowledge. Journal of Computer and System Sciences 72(2), 321–391 (2006), special Issue on FOCS 2003
5. Bellare, M., Rogaway, P.: Random oracles are practical: A paradigm for designing efficient protocols. In: Denning, D., Pyle, R., Ganesan, R., Sandhu, R., Ashby, V. (eds.) First ACM Conference on Computer and Communication Security, November 3-5, pp. 62–73. ACM (1993)
6. Blum, M.: Coin flipping by telephone. In: Proc. 1982 IEEE COMPCON, High Technology in the Information Age, pp. 133–137 (1982)
7. Capalbo, M., Reingold, O., Vadhan, S., Wigderson, A.: Randomness conductors and constant-degree lossless expanders. In: 34th Annual ACM Symposium on Theory of Computing (STOC 2002), pp. 659–668. ACM, Montréal (2002); joint session with CCC 2002
8. Chor, B., Goldreich, O.: Unbiased bits from sources of weak randomness and probabilistic communication complexity. SIAM Journal on Computing 17(2), 230–261 (1988)
9. Chor, B., Goldreich, O., Håstad, J., Friedman, J., Rudich, S., Smolensky, R.: The bit extraction problem or t-resilient functions. In: Proceedings of the 26th IEEE Symposium on Foundation of Computer Science, pp. 396–407 (1985)
10. Cramer, R., Damgård, I., Schoenmakers, B.: Proof of Partial Knowledge and Simplified Design of Witness Hiding Protocols. In: Desmedt, Y.G. (ed.) CRYPTO 1994. LNCS, vol. 839, pp. 174–187. Springer, Heidelberg (1994)
11. Dodis, Y., Elbaz, A., Oliveira, R., Raz, R.: Improved Randomness Extraction from Two Independent Sources. In: Jansen, K., Khanna, S., Rolim, J.D.P., Ron, D. (eds.) APPROX 2004 and RANDOM 2004. LNCS, vol. 3122, pp. 334–344. Springer, Heidelberg (2004)
12. Dodis, Y., Ostrovsky, R., Reyzin, L., Smith, A.: Fuzzy extractors: How to generate strong keys from biometrics and other noisy data. SIAM Journal on Computing 38(1), 97–139 (2008)
13. Dodis, Y., Ristenpart, T., Vadhan, S.: Randomness condensers for efficiently samplable, seed-dependent sources, full version of this paper. Available from authors' websites
14. Dwork, C., Naor, M., Reingold, O., Stockmeyer, L.: Magic functions. In: 40th Annual Symposium on Foundations of Computer Science, pp. 523–534. IEEE, New York (1999)
15. Fiat, A., Shamir, A.: How to Prove Yourself: Practical Solutions to Identification and Signature Problems. In: Odlyzko, A.M. (ed.) CRYPTO 1986. LNCS, vol. 263, pp. 186–194. Springer, Heidelberg (1987)
16. Goldreich, O., Krawczyk, H.: On the composition of zero-knowledge proof systems. SIAM Journal on Computing 25(1), 169–192 (1996)
17. Goldwasser, S., Kalai, Y.T.: On the (in)security of the Fiat-Shamir paradigm. In: 44th Annual Symposium on Foundations of Computer Science, pp. 102–113. IEEE, Cambridge (2003)
18. Goldwasser, S., Micali, S., Rackoff, C.: The knowledge complexity of interactive proof systems. SIAM Journal on Computing 18, 186–208 (1989)
19. Guillou, L.C., Quisquater, J.-J.: A "Paradoxical" Identity-Based Signature Scheme Resulting from Zero-Knowledge. In: Goldwasser, S. (ed.) CRYPTO 1988. LNCS, vol. 403, pp. 216–231. Springer, Heidelberg (1990)

20. Guruswami, V., Umans, C., Vadhan, S.P.: Unbalanced expanders and randomness extractors from parvaresh–vardy codes. J. ACM 56(4) (2009)
21. Gutterman, Z., Pinkas, B., Reinman, T.: Analysis of the linux random number generator. In: 27th IEEE Symposium on Security and Privacy, pp. 371–385. IEEE Computer Society (2006)
22. Kalai, Y.T., Raz, R.: Probabilistically Checkable Arguments. In: Halevi, S. (ed.) CRYPTO 2009. LNCS, vol. 5677, pp. 143–159. Springer, Heidelberg (2009)
23. Kamp, J., Rao, A., Vadhan, S.P., Zuckerman, D.: Deterministic extractors for small-space sources. In: Kleinberg, J.M. (ed.) STOC, May 21-23, pp. 691–700. ACM, Seattle (2006)
24. Koenig, R., Maurer, U.: Extracting randomness from generalized symbol-fixing and Markov sources. In: Proceedings of 2004 IEEE International Symposium on Information Theory, p. 232 (June 2004)
25. Koenig, R., Maurer, U.: Generalized Strong Extractors and Deterministic Privacy Amplification. In: Smart, N.P. (ed.) Cryptography and Coding 2005. LNCS, vol. 3796, pp. 322–339. Springer, Heidelberg (2005)
26. Lichtenstein, D., Linial, N., Saks, M.: Some extremal problems arising from discrete control processes. Combinatorica 9(3), 269–287 (1989)
27. Nisan, N., Zuckerman, D.: Randomness is linear in space. Journal of Computer and System Sciences 52(1), 43–53 (1996)
28. Raz, R., Reingold, O.: On recycling the randomness of states in space bounded computation. In: Annual ACM Symposium on Theory of Computing, Atlanta, GA, pp. 159–168 (electronic). ACM, New York (1999), http://dx.doi.org/10.1145/301250.301294
29. Reingold, O., Shaltiel, R., Wigderson, A.: Extracting randomness via repeated condensing. SIAM Journal on Computing 35(5), 1185–1209 (electronic) (2006), http://dx.doi.org/10.1137/S0097539703431032
30. Santha, M., Vazirani, U.V.: Generating quasi-random sequences from semi-random sources. J. Comput. Syst. Sci. 33(1), 75–87 (1986)
31. Schnorr, C.P.: Efficient signature generation by smart cards. Journal of Cryptology 4(3), 161–174 (1991)
32. Shamir, A., Tauman, Y.: Improved Online/Offline Signature Schemes. In: Kilian, J. (ed.) CRYPTO 2001. LNCS, vol. 2139, pp. 355–367. Springer, Heidelberg (2001)
33. Ta-Shma, A., Umans, C., Zuckerman, D.: Lossless condensers, unbalanced expanders, and extractors. Combinatorica 27(2), 213–240 (2007), http://dx.doi.org/10.1007/s00493-007-0053-2
34. Trevisan, L., Vadhan, S.: Extracting randomness from samplable distributions. In: 41st Annual Symposium on Foundations of Computer Science, pp. 32–42. IEEE, Redondo Beach (2000)
35. Viola, E.: Extractors for circuit sources. In: IEEE Symposium on Foundations of Computer Science, FOCS (2011)

Uniqueness Is a Different Story: Impossibility of Verifiable Random Functions from Trapdoor Permutations

Dario Fiore[1],[*] and Dominique Schröder[2],[**]

[1] Department of Computer Science, New York University
[2] Department of Computer Science, University of Maryland

Abstract. Verifiable random functions (VRFs) are pseudorandom functions with the additional property that the owner of the seed SK can issue publicly-verifiable proofs for the statements "$f(SK,x) = y$", for any input x. Moreover, the output of VRFs is guaranteed to be unique, which means that $y = f(SK,x)$ is the only image that can be proven to map to x. Despite their popularity, constructing VRFs seems to be a challenging task and only a few constructions based on specific number-theoretic problems are known. Basing a scheme on general assumptions is still an open problem. Towards this direction, Brakerski *et al.* showed that verifiable random functions cannot be constructed from one-way permutations in a black-box way.

In this paper we continue the study of the relationship between VRFs and well-established cryptographic primitives. Our main result is a separation of VRFs and adaptive trapdoor permutations (ATDPs) in a black-box manner. This result sheds light on the nature of VRFs and is interesting for at least three reasons:

- First, the separation result of Brakerski *et al.* gives the impression that VRFs belong to the "public-key world", and thus their relationship with other public-key primitives is interesting. Our result, however, shows that VRFs are strictly stronger and cannot be constructed (in a black-box way) form primitives like e.g., public-key encryption (even CCA-secure), oblivious transfer, and key-agreement.
- Second, the notion of VRFs is closely related to weak verifiable random functions and verifiable pseudorandom generators which are both implied by TDPs. Dwork and Naor (FOCS 2000) asked whether there are transformation between the verifiable primitives similar to the case of "regular" PRFs and PRGs. Here, we give a negative answer to this problem showing that the case of verifiable random functions is essentially different.
- Finally, our result also shows that *unique* signatures cannot be instantiated from ATDPs. While it is well known that standard signature schemes are equivalent to OWFs, we essentially show that the uniqueness property is crucial to change the relations between primitives.

* Work done while at ENS Paris.
** Postdoctoral fellow of the DAAD.

R. Cramer (Ed.): TCC 2012, LNCS 7194, pp. 636–653, 2012.

1   Introduction

Verifiable random functions (VRF) were introduced by Micali, Rabin, and Vadhan [1]. VRFs are random functions with the additional property that they provide a proof verifying the input-output relationships. Formally, a VRF is defined by a key pair (SK, PK) such that: the secret seed SK allows the evaluation of the function $y \leftarrow F(SK, x)$ on any input x and the generation of a proof π. This proof is publicly verifiable i.e., given the public key PK one can efficiently verify (using π) that the statement "$F(SK, x) = y$" holds. For security, VRFs must satisfy two properties: pseudorandomness and uniqueness. Roughly speaking, *pseudorandomness* states that the function looks random at any input x for which no proof has been issued. *Uniqueness* guarantees that for any x, there exists only one image y for which a valid proof can be produced (even for maliciously chosen public keys).

In some sense a VRF can be seen as the public-key equivalent of a pseudorandom function. This fascinating primitive has many applications, both theoretical and practical: 3-rounds resettable zero-knowledge [2], non-interactive lottery systems and micropayment schemes [3], a verifiable transaction escrow scheme [4], and updatable zero-knowledge sets [5]. However, despite their popularity, constructing VRFs seems to be challenging. In particular, only a few schemes are known so far, e.g., [1,6,7,8,9,10] (see Section 1.3 for a brief description of these works). Furthermore, all known schemes are based on specific number-theoretic problems such as RSA or different assumptions relying on bilinear maps. Constructing a VRF based on general assumptions is still an open problem.

In modern cryptography, almost all cryptographic primitives base their security on unproven computational assumptions that are considered reasonable by the community[1]. In particular, the existence of one-way functions (OWF) is one of the major open problems in cryptography. A common methodology for proving the security of a cryptographic primitive, and for better understanding its relation to other primitives, are black-box reduction techniques that can be described as follows. Let P and Q be two primitives. A *construction* of P from Q is black-box if the primitive P has only oracle access to Q (i.e., P does not have access to the code of this primitive, but can evaluate it). A *security reduction* of P to Q is black-box if for any (efficient) adversary $\mathcal{A}$ that breaks P there exists an (efficient) algorithm $\mathcal{S}$ that has black-box access to $\mathcal{A}$ and breaks Q. This approach has been extensively formalized by Reingold *et al.* who gave different "flavors" of black-box reductions depending on the "degree" of black-box access [11].

Black-box constructions and black-box proofs give clearly a limited view on the relation between the different primitives as no conclusions beyond the black-box access can be made. Nevertheless, the approach is well established as most of the cryptographic proofs are black-box and it is strong enough to show that many cryptographic primitives, such as pseudorandom functions, digital signatures, private-key encryption, are equivalent to the existence of one-way functions

---

[1] If one makes exception of a few cases that are proven secure in an information-theoretic sense.

(OWFs), which is considered to be one of the most basic assumptions. On the other hand, other primitives (e.g., public-key encryption) are believed to exist only under stronger assumptions (e.g., the existence of trapdoor permutations). Though such primitives and/or assumptions look different, it might be possible that many of them are related or even equivalent. Therefore, identifying the minimal assumptions on which one can base the security of a primitive is considered one of the most important goals for a better and deeper understanding of the cryptography world.

On the negative side, Impagliazzo and Rudich introduced a methodology for proving separations between primitives in the sense of black-box constructions, e.g., proving that *Q does not imply P in a black-box way* [12]. In their work they ruled out any black-box construction of key-agreement protocols (KA) from one-way functions. Gertner *et al.* show that the breakthrough result of Impagliazzo and Rudich can be seen as defining two separated worlds in which the cryptographic primitives can be divided: the "private cryptography" world that contains all those primitives that are equivalent to OWFs, and private-key encryption; the "public cryptography" world that contains harder primitives such as trapdoor permutations, public-key encryption (PKE), KA and oblivious transfer (OT) [13].

It is worth to mention that another methodology, called *meta-reductions*, for separating primitives in a black-box sense is known. Since we do not follow this approach, we refer the reader to e.g., [14,15,16].

1.1 Our Results

We investigate the relationship between verifiable random functions and well-studied cryptographic primitives. The first step towards this goal was recently given by Brakerski, Goldwasser, Rothblum, and Vaikuntanathan who separated VRFs from one-way permutations [17]. The authors introduce the notion of *weak verifiable random functions (wVRFs)* that can be seen as the public key analogue to weak-PRFs: pseudorandomness only holds with respect to randomly chosen inputs. Moreover, they construct wVRFs from (enhanced) trapdoor permutations and show that wVRFs are essentially equivalent to non-interactive zero knowledge proof (NIZK) systems in the common reference string model. In the private key setting, it is well known that "regular" PRFs can be constructed from weak PRFs in a black-box way [18,19]. Thus, a natural direction to study the relation between the primitives is to build a VRF out of any wVRF.

Another work that is closely related to this topic is the study of *verifiable pseudorandom generators* (VPRGs) due to Dwork and Naor [20]. Roughly speaking, a VPRG is a pseudorandom generator that allows the owner of the seed to prove the correctness of subsets of the generated bits while the other bits remain indistinguishable from random. Dwork and Naor constructed VPRGs from trapdoor permutations. Again, in the case of "regular" PRFs we know how to turn a PRG into a PRF in a black-box way [21]. Dwork and Naor left open the question if a similar transformation can be found in the public key setting [20], namely:

Is it possible to construct a VRF from VPRGs and/or weak-VRFs in a black-box way?

In this paper, we give a negative answer to this question and, more generally, we show that no black-box constructions of VRFs from (enhanced) trapdoor permutations exist.

Theorem 1 (informal). *There exists no black-box reduction of verifiable random functions to trapdoor permutations.*

Our result is actually more general than the above indicates; it separates the weaker primitive of verifiable unpredictable functions (VUFs) from the stronger primitive of *adaptive* trapdoor functions. The difference between VRFs and VUFs is that in the latter the output should be unpredictable instead of pseudorandom. Therefore, VUFs can also be seen as *"unique signatures"*, where, for every public key, each message can have at most one valid signature[2].

Adaptive trapdoor functions (ATDFs), recently introduced by Kiltz, Mohassel, and O'Neill in [22], are essentially strictly stronger than trapdoor functions as the adversary is given access to an inversion oracle.

Implications of Our Result. Our result sheds light on the nature of VRFs and explains why this primitive seems so hard to construct. First, given the separation result of Brakerski *et al.*, one can naturally think of VRFs as though they belong to the "public cryptography" world. Then, if we consider the relationship between VRFs and the other public-key primitives, our result highlights that VRFs are much stronger as they cannot be implied by most of the primitives in this world: basically everything which is implied by TDPs, e.g. semantically-secure public-key encryption, oblivious transfer, key-agreement. Moreover, since ATDPs imply CCA-secure PKE [22], then VRFs are separated even from it. On the positive side we observe that we can obtain a construction of VRFs from identity-based encryption with *unique key derivation* following the idea of Abdalla *et al.* [9][3]. Combining this positive result with our impossibility result confirms the impossibility result of IBE from TDPs [23].

Second, our result points out the hardness of achieving the uniqueness property in the context of digital signatures: While signature schemes are equivalent to OWFs, *unique* signatures cannot be instantiated from (adaptive) TDPs in a black-box way.

Finally, since both weak-VRFs and VPRGs are implied by TDPs, our result rules out the possibility of constructing VRFs from weak-VRFs and/or VPRGs

---

[2] At this stage, it is interesting to observe unique and deterministic signatures are two distinct primitives. Consider for example the signature $\sigma = \sigma' \| 0$ where σ' is deterministic and the verification algorithm ignores the last bit. Then it is obvious that uniqueness could be easily violated by flipping the last bit.

[3] Precisely, the unique key derivation algorithm immediately implies a VUF, which can then be turned into a VRF using the original idea of Micali, Rabin and Vadhan.

(in a black-box way). Thus, it seems that there is no hope that the approaches used in the private key world to build PRFs from weak-PRFs and PRGs can be adopted to the case of the public verifiable primitives. This shows that the verifiable analogous of these primitives are essentially different.

1.2   Overview of the Techniques

Our starting point is the so-called "two oracles" technique of Hsiao and Reyzin [24]. The main idea of this technique is to construct two oracles, say $\mathcal{O}$ and $\mathcal{B}$, such that $\mathcal{O}$ is used in the constructions, whereas both oracles $\mathcal{O}$ and $\mathcal{B}$ can be accessed by the adversaries. This approach is slightly weaker than the single oracle technique because it "only" rules out fully-black-box reductions (instead of any black-box reduction).

Our Oracles. In our case the oracle $\mathcal{O}$ is an ideal random trapdoor permutation oracle that is modeled as a triple of random functions (g, e, d) such that: $g(\cdot)$ maps trapdoors to public keys; $e(ek, \cdot)$ is a random permutation for every public key ek and $d(td, \cdot)$ is the inverse of $e(ek, \cdot)$ when $g(td) = ek$. Due to the fact that $\mathcal{O}$ is truly random, $\mathcal{O}$ is secure even in the sense of adaptive trapdoor permutations. The oracle $\mathcal{B}$ is designed to break any black-box construction of VUF based on $\mathcal{O}$.

Therefore, the core of our separation theorem is the definition of the weakening oracle $\mathcal{B}$. The proof then consists of two main parts:

(i) showing an efficient adversary that can break the unpredictability of the VUF by making a polynomial number of queries to $\mathcal{B}$;
(ii) showing an ATDP construction that is secure against any adversary that makes at most polynomially-many oracle queries.

The design of $\mathcal{B}$ is rather technical. In particular, the main difficulty is to prevent an attacker from exploiting $\mathcal{B}$ to break the one-wayness of the ATDP. A naïve construction would be an oracle that takes as input a VUF public key and returns $y^* \leftarrow F(SK, x^*)$, i.e., the evaluation of the function on a random point x^*. This oracle would clearly break the unpredictability of the VUF, but it would also be too strong. Consider, for instance, an adversary $\mathcal{A}$ that is given as input a public key ek^* of a trapdoor permutation and that is challenged to invert it on a random point b^*. Now, $\mathcal{A}$ might encode (ek^*, b^*) into PK in such a way that the evaluation of $F(SK, x^*)$ requires to invert b^*. But then the attacker would learn all informations about b^*'s inverse. To prevent these "dangerous" queries we modify $\mathcal{B}$ such that it takes as input a certain number of triples (x_i, y_i, π_i), where π_i is a valid proof for "$F(SK, x_i) = y_i$". The idea follows from the intuition that the attacker can encode b^* (and ek^*) into PK in only two ways:

(i) $F(SK, \cdot)$ needs to invert b^* on a large fraction of the inputs,
(ii) $F(SK, \cdot)$ needs to invert b^* only on a negligible fraction of the inputs.

Now, suppose that $\mathcal{A}$ encodes b^* into PK as defined in the first case. In order to query the oracle, $\mathcal{A}$ has to provide valid proofs. But if $\mathcal{A}$ can compute all proofs, then the attacker must already know b^*'s inverse. Otherwise, if b^* is encoded into PK as described in the second case, then the probability that evaluating $F(SK, x^*)$ on a random input x^* requires to invert b^* is negligible. Hence, returning y^* does not reveal any useful informations to $\mathcal{A}$. Although this idea seems very promising, it raises another issue. In fact $\mathcal{A}$ might overcome this limitation by choosing all the x_i's from the small fraction that does not require to invert b^*. We solve this issue by defining a two-steps oracle $\mathcal{B} = (\mathcal{B}_1, \mathcal{B}_2)$ such that $\mathcal{B}_1$ chooses the values x_i's and $\mathcal{B}_2$ is the actual oracle as described above, such that it works properly only if the inputs x_i's are chosen by $\mathcal{B}_1$.

Finally, an important detail towards the definition of $\mathcal{B}$ is that it simulates the run of $F^{\mathcal{O}}(SK, x^*)$ using a *different* oracle $\mathcal{O}'$ and a *different* secret key SK' such that SK' still corresponds to PK under $\mathcal{O}'$. The idea is that, if $\mathcal{O}'$ is close enough to $\mathcal{O}$ (as it should be the case while trying to break the VUF), then evaluating $F^{\mathcal{O}'}(SK', x^*)$ produces the same output as $F^{\mathcal{O}}(SK, x^*)$. On the other hand, with high probability $\mathcal{O}$ and $\mathcal{O}'$ are *not* close when an ATDP adversary invokes $\mathcal{B}$.

1.3 Other Related Work

Verifiable Random Functions. Goldwasser and Ostrovsky introduce the notion of unique signatures (calling them *invariant signatures*) and they show that in the common random string model they are equivalent to non-interactive zero-knowledge proofs [25]. Later, Micali, Rabin and Vadhan formally define VRFs and propose a construction (in the plain model) [1]. The authors follow two main steps: (1) they construct a verifiable unpredictable function (VUF) based on the RSA problem and then (2) they show a generic transformation to convert a VUF into a VRF using the Goldreich-Levin theorem [26] (that extracts one random bit from polynomially-many unpredictable bits). The hope of this two-steps approach is that a VUF should be easier to realize than a VRF, but the second step is very inefficient. Finally, Lysyanskaya proposes a VUF relying on a strong version of the Diffie-Hellman assumption [6].

The subsequent works suggest direct and (more) efficient constructions of VRFs without relying on the Goldreich-Levin transformation. Dodis suggests an instantiation on the sum-free generalized DDH assumption [7], and Dodis and Yampolskiy give a construction based on the bilinear Diffie-Hellman inversion assumption [8]. Abdalla, Catalano, and Fiore show the relationship between VRFs and a certain class of identity-based encryption schemes [9]. Moreover, the authors propose a construction based on the weak bilinear Diffie-Hellman inversion assumption. All the schemes mentioned so far share the limitation of supporting only a small domain (i.e., of superpolynomial size). The only exception is the recent scheme by Hohenberger and Waters, who give the first construction having a large input space [10]. Another closely related work is one of Dodis and Puniya who construct NIZK from verifiable random permutations (VRPs), that are the verifiable analog of pseudorandom permutations [27]. The author also show how to convert a VRF into a VRP.

Black-Box Separations. After the seminal result of Impagliazzo and Rudich many follow up works studied the relation between different primitives, such as, e.g., [13,28,29,30,23,23,31,32]. We discuss these works in the full version [14].

2   Preliminaries

Adaptive Trapdoor Permutations. Adaptive trapdoor permutations (AT-DPs) are defined similar to a trapdoor permutation, but in the security definition the adversary is provided with an oracle that inverts the function on arbitrary images (except on the challenge value). A formal definition is given in [22,14].

Verifiable Random Functions. Verifiable random functions (VRF) are similar to pseudorandom functions, but differ in two main aspects: Firstly, the output of the function is publicly verifiable, i.e., there exists an algorithm Π that returns a proof π which shows that y is the output of the function on input x. Secondly, the output of the function is unique, i.e., no two images (and proofs) exist that verify under the same preimage.

Definition 1 (Verifiable Random Functions). *A family of functions* $\mathcal{F} = \{f_s : \{0,1\}^{n(\lambda)} \to \{0,1\}^{m(\lambda)}\}_{s \in \{0,1\}^{seed(\lambda)}}$ *is a family of* **Verifiable Random Functions** *if there exists a tuple of algorithms* (KG, F, Π, V) *with the following functionalities:*

$KG(1^\lambda)$ *outputs a pair of keys* (PK, SK).
$F(SK, x)$ *is a deterministic algorithm that evaluates* $f_s(x)$.
$\Pi(SK, x)$ *is an algorithm that outputs a proof* π *related to* x.
$V(PK, x, y, \pi)$ *outputs 1 if* π *is a valid proof for "$f_s(x) = y$", else it outputs 0.*

A tuple (KG, F, Π, V) *is said to be a VRF if it satisfies the following properties:*

Domain Range Correctness *For all values* $x \in \{0,1\}^{n(\lambda)}$, *over the choices of* (PK, SK), *we have that* $F(SK, x) \in \{0,1\}^{m(\lambda)}$ *holds with all but negligible probability.*

Completeness *For all* $x \in \{0,1\}^{n(\lambda)}$ *if* $\Pi(SK, x) = \pi$ *and* $F(SK, x) = y$ *then* $V(PK, x, y, \pi)$ *outputs 1 with overwhelming probability (over the choices of* (PK, SK) *and the coin tosses of* V).

Uniqueness *There exist no values* $(PK, x, y_1, y_2, \pi_1, \pi_2)$, *unless with negligible probability over the coin tosses of* V, *such that for distinct* y_1 *and* y_2 *it holds that* $V(PK, x, y_1, \pi_1) = V(PK, x, y_2, \pi_2) = 1$.

Pseudorandomness *For all PPT adversaries* $\mathcal{A} = (\mathcal{A}_1, \mathcal{A}_2)$ *we require that the probability* $\mathcal{A}$ *succeeds in the experiment* $\text{pseudo}_{\mathcal{A}}^f$ *is at most* $\frac{1}{2} + \text{negl}(\lambda)$, *where the experiment is defined in Figure 1.*

Verifiable unpredictable functions (VUF) are similar to VRFs, except that unpredictability must hold instead of pseudorandomness:

Definition 2 (Verifiable Unpredictable Functions). *A tuple* (KG, F, Π, V) *is a* **verifiable unpredictable function** *if the probability that any PPT adversary* $\mathcal{A}$ *succeeds in the experiment* $\text{predict}_{\mathcal{A}}^f$, *defined in Figure 1, is at most negligible.*

Experiment pseudo$_{\mathcal{A}}^f$
$(PK, SK) \leftarrow KG(1^\lambda);$
$(x^*, \text{state}) \leftarrow \mathcal{A}_1^{Func(SK,\cdot)}(PK)$
$b \xleftarrow{\$} \{0,1\};$
$y_0 \leftarrow F(SK, x); \ y_1 \xleftarrow{\$} \{0,1\}^{m(\lambda)}$
$b' \leftarrow \mathcal{A}_2^{Func(SK,\cdot)}(\text{state}, y_b)$
Output 1 iff $b' = b$
   and x^* was not asked
   to the $Func(SK, \cdot)$ oracle.

Experiment predict$_{\mathcal{A}}^f$
$(PK, SK) \leftarrow KG(1^\lambda);$
$(x^*, y^*) \leftarrow \mathcal{A}^{Func(SK,\cdot)}(PK)$
Output 1 iff $y^* = F(SK, x^*)$ and
   x^* was not asked
   to the $Func(SK, \cdot)$ oracle.

Fig. 1. This Figure show the experiment of pseudorandomness and unpredictability. In both experiments the oracle $Func(SK, \cdot)$ computes $F(SK, \cdot)$ and $\Pi(SK, \cdot)$ and returns their output.

3 The Black-Box Separation

We first give a high-level overview of the main ideas of our proof before going into the details afterwards. Our starting point is the "two oracles" separation technique of Hsiao and Reyzin [24]. In the context of VRFs, we have to construct two oracles $\mathcal{O}$ and $\mathcal{B}$ relative to which ATDPs exist while VUFs do not. In particular, the constructions are restricted to have black-box access only to $\mathcal{O}$, while the adversary may access both $\mathcal{O}$ and $\mathcal{B}$.

The core of our separation are the two oracles, $\mathcal{O}$ and $\mathcal{B}$. The oracle $\mathcal{O} = (g, e, d)$ realizes a random trapdoor permutation (we give a formal definition in Section 3.2). The second oracle is a weakening oracle such that relative to $\langle \mathcal{O}, \mathcal{B} \rangle$ a secure construction of adaptive trapdoor permutations exists while any given candidate (and correct) VUF construction $(KG^\mathcal{O}, F^\mathcal{O}, \Pi^\mathcal{O}, V^\mathcal{O})$ is insecure[4]. To prove this result, we build an adversary that wins the unpredictability game with non-negligible probability. Since the description of the oracle $\mathcal{B}$ is rather technical, we first describe the high-level intuitions that guides us to the design of $\mathcal{B}$.

3.1 Towards the Definition of $\mathcal{B}$

Towards the definition of such $\mathcal{B}$, the main difficulty is to design an oracle that is strong enough to help predicting a value of the VUF while simultaneously being too weak to invert the ATDP.

A naïve approach for $\mathcal{B}$ would be the one that immediately breaks the VUF, by taking the VUF's public key PK and a value x as input; it then would return $F^\mathcal{O}(SK, x)$. Of course, any VUF construction breaks down in the presence of such oracle. So, it would remain to show that an ATDP is still secure in the presence of such $\langle \mathcal{O}, \mathcal{B} \rangle$, which unfortunately is not the case. To see this, consider the following VUF defined through $KG^\mathcal{O}, F^\mathcal{O}, \Pi^\mathcal{O}, V^\mathcal{O}$ (where $\Pi^\mathcal{O}(SK, \cdot) = F^\mathcal{O}(SK, \cdot)$): The $KG^\mathcal{O}$ algorithm queries $ek \leftarrow g(td)$ on a random $td \in \{0,1\}^\lambda$ and sets $PK = ek$ and $SK = td$. The function evaluation algorithm on input x

---

[4] By $\langle \mathcal{O}, \mathcal{B} \rangle$ we mean that the algorithm $A^{\langle \mathcal{O}, \mathcal{B} \rangle}$ gets access to both oracles.

obtains $y \leftarrow d(td, x)$ and outputs y. $V(PK, x, y)$ simply checks that $e(ek, y) = x$. Observe that this construction is sound and unique (but trivially insecure). Now, we construct an adversary $\mathcal{A}$ against the ATDP that exploits the above defined $\mathcal{B}$ to invert the challenge (ek^*, b^*). This attacker inverts the challenge by simply submitting $(PK = ek^*, x = y^*)$ to $\mathcal{B}$! This means that the oracle $\mathcal{B}$ that we sketched before is too strong and reveals too much information.

As one can guess, the problem are those queries to $\mathcal{B}$ that are "dangerous" in the sense that they extract too much useful information to invert the ATDP. Starting from this (toy) example we modify $\mathcal{B}$ to prevent such "dangerous queries". The first important observation is that our adversary against the unpredictability only needs to predict *some* value, rather than a specific one. This means, the attacker only needs to find y^* for a fresh $x^* \in \{0, 1\}^n$. Therefore, our first modification consists of changing the input that is provided to $\mathcal{B}$. Basically, we let $\mathcal{B}$ choose x^* on which it evaluates $y^* \leftarrow F^{\mathcal{O}}(SK, x^*)$. This new definition of $\mathcal{B}$ still allows us to break the security of the VUF and it also avoids direct inversion queries as the attack can no longer query x directly to $\mathcal{B}$.

However, this modification is not sufficient to avoid that an ATDP adversary exploits the access to $\mathcal{B}$. The problem is that an attacker $\mathcal{A}$ might encode its challenge (ek^*, b^*) into the public key PK. For instance, $\mathcal{A}$ could create and submit a public key such that any function evaluation will require to invert b^* according to the permutation $e(ek^*, \cdot)$. We show how to prevent such queries starting from the following basic intuition.

Assume that a value $b \in \{0, 1\}^\lambda$ is (somehow) encoded into the public key PK and recall that we denote by x the input of $F^{\mathcal{O}}(SK, \cdot)$. Then we have two mutually exclusive cases:

1. $F^{\mathcal{O}}(SK, \cdot)$ inverts b on a large fraction of the x's;
2. $F^{\mathcal{O}}(SK, \cdot)$ inverts b only on a negligible fraction of the x's (even on no x in the most extreme case).

Now, recall that a VUF attacker is allowed to query the function (and see the corresponding proofs) for inputs of her choice. Therefore, if $\mathcal{A}$ queries the function oracles on a sufficiently large number of the x's, then $\mathcal{A}$ will learn the inverses of all the "frequent" b's of type 1 with high probability. On the other hand, for any b of type 2, the probability that running $F^{\mathcal{O}}(SK, x)$ on a random x asks to invert b is negligible.

Ensuring $\mathcal{A}$ Has Access to the Function Oracles. The above intuition suggests that any algorithm querying $\mathcal{B}$ must provide as additional input sufficiently many triples (x_i, y_i, π_i) such that π_i is a valid proof for "$F^{\mathcal{O}}(SK, x_i) = y_i$". This way, if a ATDP adversary embeds a "type 1" b into PK, then it must know its inverse in order to provide the above triples. Or, if a "type 2" b is encoded into PK, then with high probability the attacker $\mathcal{A}$ will not gain any further information on its inverse from seeing the evaluation of $F^{\mathcal{O}}(SK, x^*)$ for a random x^*.

Although such restriction seems to capture the right intuition, we observe that it is not sufficient to prevent the adversary from exploiting $\mathcal{B}$. To see this, assume that $\mathcal{A}$ encodes its challenge (ek^*, b^*) into PK such that b^* is of type

1, namely $F^{\mathcal{O}}(SK, x)$ queries $d(td^*, b^*)$ on a large fraction of the x's. Then, if the attacker $\mathcal{A}$ is allowed to choose the inputs $x_1, \ldots, x_\ell$ provided to $\mathcal{B}$, then it might take all of them from the small fraction that does not require to invert b^*. In this case our previous argument would fail.

Therefore, in order to prevent these dangerous queries, we deny $\mathcal{A}$ choosing the inputs $x_1, \ldots, x_\ell$. That is, we define a two-steps oracle $\mathcal{B} = (\mathcal{B}_1, \mathcal{B}_2)$ where $\mathcal{B}_1$ chooses ℓ random inputs, and $\mathcal{B}_2$ evaluates the VUF *only* if it gets as input values and proofs for x's that were chosen by $\mathcal{B}_1$. For this we will require that $\mathcal{B}_1$ is essentially a random function that, given as input a VUF public key and a collection of oracle circuits implementing a VUF, outputs ℓ random strings.

Furthermore, observe that this restriction is not a problem for the attacker that we build against the VUF, because it has access to the function oracles, $F(SK, \cdot)$ and $\Pi(SK, \cdot)$, that compute these values and proofs for her. On the other hand, an ATDP adversary now has restricted power as it does not know b^*'s inverse.

Avoiding Malicious Keys. Finally, the last type of dangerous queries that we have to handle are those where the attacker $\mathcal{A}$ queries $\mathcal{B}$ on an "invalid" public key PK. By "invalid" we mean that PK is not the output of an honest execution of the key generation algorithm $KG^{\mathcal{O}}(SK)$. The problem is again that an evaluation of $F^{\mathcal{O}}(SK, x)$ can reveal "sensitive" informations about the trapdoor permutation. Indeed, observe that an execution of $F^{\mathcal{O}}$ must use the $d(\cdot, \cdot)$ oracle in a *significant* way or the VUF cannot be secure.[5] Thus, one may think about designing $\mathcal{B}$ in such a way that it rejects any queries that involve invalid public keys. However, this solution is still dangerous as $\mathcal{B}$ might be used to test the validity of public keys. We solve the issue by defining $\mathcal{B}$ such that it computes the answer using a different key SK' and a different oracle $\mathcal{O}''$ but that the new function $F^{\mathcal{O}''}(SK', \cdot)$ behaves in almost all cases as the original one $F^{\mathcal{O}}(SK, \cdot)$. More precisely, the oracle $\mathcal{B}$ evaluates $F^{\mathcal{O}''}(SK', \cdot)$ using a key SK' (that is most likely different from SK) and an oracle $\mathcal{O}''$ which is also different from the real oracle $\mathcal{O}$. The key SK' is computed such that it corresponds to the "real" key PK under $\mathcal{O}''$ (i.e., $PK \leftarrow KG^{\mathcal{O}''}(SK')$). The idea is to construct $\mathcal{O}''$ such that is close to $\mathcal{O}$. Then we can show that evaluating $F^{\mathcal{O}''}(SK', x)$ is basically the same as evaluating $F^{\mathcal{O}}(SK, x)$.

The hope is that $\mathcal{O}''$ differs from $\mathcal{O}$ in the points that may represent dangerous queries. If this is the case, then we are done as computing $F^{\mathcal{O}''}(SK', x)$ will not reveal sensitive informations on the real ATDP. More precisely, our oracle $\mathcal{B}$ selects uniformly at random a secret key SK' and an oracle $\mathcal{O}''$ such that $PK = KG^{\mathcal{O}''}(SK')$ and $\mathcal{O}''$ agrees with $\mathcal{O}$ on those points that are already known to the adversary.

Discovering All ATDP Public Keys. In order to correctly simulate a run of $F^{\mathcal{O}''}$ it is important that our oracle has discovered all the ATDP public keys ek that may be needed while running $F^{\mathcal{O}''}$. More precisely it needs to know all

---

[5] For instance, if $F^{\mathcal{O}}$ does not use the oracles, then an exponentially-strong adversary could always evaluate the circuit associated to F.

the public keys that were generated during the honest execution of $KG^{\mathcal{O}}(SK)$. So, to discover these public keys we define $\mathcal{B}$ such that it runs $V^{\mathcal{O}}$ on all the received triples (x_i, y_i, π_i) and collect all the queries made by the algorithm. Since by Assumption 1, the algorithm KG generates at most q of such ek's, it is sufficient to repeat the above step on sufficiently many triples, say q^c for some constant c that we will specify later. This allows us to discover all the public keys with high probability.

3.2   The Formal Separation Theorem

In this section we formalize the techniques that we use to prove our result. The core of our proof is the description of two oracles $\mathcal{O}$ and $\mathcal{B}$. The first oracle $\mathcal{O} = (g, e, d)$ implements a perfectly random trapdoor permutation and it is obvious that a secure ATDP exists relative to $\mathcal{O}$ (where the security follows from the randomness of the function). Therefore, we follow the strategy of defining a "weakening" oracle $\mathcal{B}$ whose main task is to break the security of a given VUF construction. This approach is formalized in the following theorem:

Theorem 1 (formally restated). *Let $\mathcal{O} = (g, e, d)$ be a random trapdoor permutation oracle. Then, there exists an oracle $\mathcal{B}$ such that for every VUF construction $(KG^{\mathcal{O}}, F^{\mathcal{O}}, \Pi^{\mathcal{O}}, V^{\mathcal{O}})$ which is correct and unique we have:*

(i) there is an adversary $\mathcal{A}$ such that $\mathcal{A}^{\langle \mathcal{O}, \mathcal{B} \rangle}$ breaks the security of the VUF with non-negligible probability;

(ii) there exists an ATDP construction $(G^{\mathcal{O}}, E^{\mathcal{O}}, D^{\mathcal{O}})$ relative to $\mathcal{O}$ such that no adversary $\mathcal{A}^{\langle \mathcal{O}, \mathcal{B} \rangle}$ can break its security with non-negligible probability.

We formally prove this theorem defining the oracles $\mathcal{O}$ and $\mathcal{B}$ in the following paragraphs. Afterwards, we prove the theorem by stating two separate lemmata. The first one, given in Section 4, shows the insecurity of the VUF, whereas the second lemma (Section 5) proves the existence of a secure ATDP.

The Oracle $\mathcal{O}$. We prove our separation in a relativized model where each algorithm has access to a random trapdoor permutation oracle $\mathcal{O} = (g, e, d)$ where g, e and d are sampled uniformly at random from the set of all functions with the following conditions:

- $g : \{0,1\}^{\lambda} \rightarrow \{0,1\}^{\lambda}$ takes a trapdoor key td and outputs a public key ek.
- $e : \{0,1\}^{\lambda} \times \{0,1\}^{\lambda} \rightarrow \{0,1\}^{\lambda}$ is a function that takes in input a public key ek and a value a and outputs b. For every $ek \in \{0,1\}^{\lambda}$, $e(ek, \cdot)$ is required to be a permutation over $\{0,1\}^{\lambda}$.
- $d : \{0,1\}^{\lambda} \times \{0,1\}^{\lambda} \rightarrow \{0,1\}^{\lambda}$ is a function that on input a pair (td, b) outputs the unique $a \in \{0,1\}^{\lambda}$ such that $e(g(td), a) = b$.

Since the permutation is defined over $\{0,1\}^{\lambda}$, it is easy to see that the oracle is also an enhanced TDP.

Notation. We write $\mathcal{A}^{\mathcal{O}}$ to denote that an algorithm $\mathcal{A}$ is given access to an oracle $\mathcal{O}$. We will use square brackets to denote queries and mappings. For instance, we write $[e(ek, a)]$ to denote a query to e with input ek and a. Otherwise $e(ek, a)$ refers the actual value of the function e on the given input. We write $[e(ek, a) = b]$ to denote that there is a mapping between a and b in the function $e(ek, \cdot)$. Also, for ease of presentation, we will sometimes abuse the notation and write $\mathcal{O}(\alpha)$ to denote the answer of $\mathcal{O}$ on a query α which depends on the type of α. For example if $\alpha = [e(ek, a)]$, then $\mathcal{O}(\alpha) = e(ek, a)$.

Let $\mathcal{O}_k$ (with $k \in \{1, 2\}$) be a partial (aka suboracle) oracle. We define the set of all public keys that are contained into the queries of $\mathcal{O}_k$ as

$$Z(\mathcal{O}_k) = \{ek : [g(\cdot) = ek] \in \mathcal{O}_k \text{ or } [e(ek, \cdot) = \cdot] \in \mathcal{O}_k\}.$$

Suboracles. Let $\mathcal{O}_1$ and $\mathcal{O}_2$ be two (possibly partial) trapdoor permutation oracles. We write $\mathcal{O}_1 \diamond_c \mathcal{O}_2$ to denote the oracle that answers with $\mathcal{O}_1$ only if $\mathcal{O}_2$ is not defined. Otherwise, it answers with $\mathcal{O}_2$. If $\mathcal{O}_1 = (g_1, e_1, d_1)$ and $\mathcal{O}_2 = (g_2, e_2, d_2)$ are two trapdoor permutation oracles as defined above, then its composition is defined by composing each algorithm, namely:

$$\mathcal{O}_1 \diamond_c \mathcal{O}_2 = (g_1 \diamond_c g_2, e_1 \diamond_c e_2, d_1 \diamond_c d_2)$$

This definition needs some more explanation. We want that the oracle obtained from the composition of two oracles preserves the properties of the two individual oracles. In particular, we require that $(e_1 \diamond_c e_2)(ek, \cdot)$ is a permutation for any valid ek. The problem is that the permutations e_1 and e_2 may contain collisions, namely there exist ek and two distinct values $a, a' \in \{0, 1\}^{\lambda}$ such that $e_2(ek, a) = e_1(ek, a')$. To handle such collisions we use the same technique suggested in [33]. We define $e = e_1 \diamond_c e_2$ as follows: let ek, a, b be values such that $[e_2(ek, a) = b] \in \mathcal{O}_2$. We set $e(ek, a) = b$. If there exists a value $a' \neq a$ such that $[e_1(ek, a') = b] \in \mathcal{O}_1$, then let $b' = e_1(ek, a)$ and set $e(ek, a') = b'$. The composition $d = d_1 \diamond_c d_2$ is defined to be consistent with g and e.

VUF in the Presence of Our Oracle. For a simpler exposition we make some general assumptions on any VUF construction with access to the oracle $\mathcal{O} = (g, e, d)$. First, we consider a slightly relaxed definition of the VUF algorithms (KG, F, Π, V) as follows. The algorithm $KG(SK)$ takes as input a secret key $SK \in \{0, 1\}^n$ and outputs $PK \in \{0, 1\}^n$. The input of F and Π are the secret key SK and a value $x \in \{0, 1\}^n$. The output of F is the function value $y \in \{0, 1\}^n$, whereas the output from Π is the corresponding π, respectively. Finally, V is given in input the public key PK, an input x, an output y and a proof π and outputs 1 if it accepts the proof, or 0 otherwise. In the above description n is a function of the security parameter λ.

Recall that we assume towards contradiction that there exists a black-box reduction of VUFs to ATDPs. Then we denote by $(KG^{\mathcal{O}}, F^{\mathcal{O}}, \Pi^{\mathcal{O}}, V^{\mathcal{O}})$ the corresponding VUF construction. According to our notation, each algorithm has access to the (g, e, d) oracles and they have to use them in a "significant" way to implement a secure primitive. Also, by definition of black-box reduction, this construction is a correct VUF implementation, that satisfies completeness and uniqueness according to Definition 1.

Assumption 1. *For a simpler exposition, in our proofs we use the following assumptions:*

- *each algorithm is unbounded, but makes at most $q = poly(\lambda)$ oracle queries during its execution;*
- *every query $d(td, \cdot)$ is followed by a query $g(td)$;*
- *the proof algorithm is deterministic;*
- *the verification algorithm is deterministic;*
- *the completeness of the VUF holds in a perfect sense.*

Before proceeding with the description of the breaking oracle, we briefly justify these assumptions. The first condition is reasonable because we consider only efficient constructions and moreover, it allows us to easily quantify the advantage of our adversaries. The second one avoids queries of the adversary to $d(\cdot, \cdot)$ using a trapdoor key without knowing the corresponding public key. This assumption is also common and has been previously used in e.g., [23]. Assuming that the proof algorithm is deterministic is not a restriction as we can turn any VRF with a probabilistic proof algorithm into one having a deterministic algorithm by applying a PRF to the input and the private seed of the VRF to derive the randomness. Completeness and uniqueness follow easily from the VRF (note that uniqueness only holds w.r.t. to the output of the function and not w.r.t. the proof). The rest follows easily applying a standard hybrid argument. The assumptions on deterministic verification and perfect completeness have already been addressed in [17], hence we omit the discussion here.

A Formal Definition of $\mathcal{B}$. Here, we provide a formal description of our oracle $\mathcal{B}$, which is composed by the following two algorithms $(\mathcal{B}_1, \mathcal{B}_2)$:

Algorithm $\mathcal{B}_1$:
    INPUT: A collection of oracle circuits $VUF^{\mathcal{O}} = (KG^{\mathcal{O}}, F^{\mathcal{O}}, \Pi^{\mathcal{O}}, V^{\mathcal{O}})$ implementing a VUF, and a VUF public key PK
    OUTPUT: $x_1, \dots, x_\ell \in \{0,1\}^n$.
    COMPUTATION: To each input $(VUF^{\mathcal{O}}, PK)$, the algorithm $\mathcal{B}_1$ associates a random function $f : \{0,1\}^n \to \{0,1\}^n$. For $i = 1$ to ℓ, it computes $x_i = f(i)$, and finally it returns $x_1, \dots, x_\ell$.

Algorithm $\mathcal{B}_2$:
    INPUT: A collection of oracle circuits $VUF^{\mathcal{O}} = (KG^{\mathcal{O}}, F^{\mathcal{O}}, \Pi^{\mathcal{O}}, V^{\mathcal{O}})$ implementing a VUF, a VUF public key PK and a set $\{(x_i, y_i, \pi_i)\}_{i=1}^\ell$ such that $x_i \in \{0,1\}^n$, $y_i \in \{0,1\}^m$, and π_i is in the range of $\Pi(\cdot, \cdot)$.
    OUTPUT: $x^* \in \{0,1\}^n$, $y^* \in \{0,1\}^m$.
    COMPUTATION: The oracle performs the following computation:
        - **Step 1:** Invoke $(x_1', \dots, x_\ell') \leftarrow \mathcal{B}_1(VUF^{\mathcal{O}}, PK)$ and check that the values $x_1, \dots, x_\ell$ received as input are equal to $(x_1', \dots, x_\ell')$ returned by $\mathcal{B}_1$. Otherwise, output $\perp$.

- **Step 2:** For all $i = 1$ to ℓ run the algorithm $V^{\mathcal{O}}(PK, x_i, y_i, \pi_i)$ and collect into a partial oracle $\mathcal{O}_Q$ all the queries that are made during each run. If there is some j such that the verification algorithm does not accept, stop and output $\perp$.
- **Step 3:** Find a secret key SK' and a partial oracle $\mathcal{O}'$ such that:
  1. $KG^{\mathcal{O}'}(SK') = PK$, $F^{\mathcal{O}'}(SK', x_i) = y_i$ and $\Pi^{\mathcal{O}'}(SK', x_i) = \pi_i$.
  2. $\mathcal{O}' \supseteq \mathcal{O}_Q$ and $|\mathcal{O}'| \leq |\mathcal{O}_Q| + q$ where q is the same value defined in Assumption 1.
- **Step 4:** Define $\mathcal{O}'' = \mathcal{O} \diamond_c \mathcal{O}'$
- **Step 5:** Choose x^* uniformly at random in $\{0,1\}^n$ such that $x^* \neq x_i$ for all $i = 1$ to ℓ. Run $y^* \leftarrow F^{\mathcal{O}''}(SK', x^*)$ and $\pi^* \leftarrow \Pi^{\mathcal{O}''}(SK', x^*)$.
- **Step 6:** Run $V^{\mathcal{O}''}(PK, x^*, y^*, \pi^*)$. If $V^{\mathcal{O}''}$ asks a query α such that $\mathcal{O}''(\alpha) \neq \mathcal{O}(\alpha)$, then return $\perp$. Otherwise output y^*.

Complexity of $\mathcal{B}$. Based on Assumption 1, we evaluate the cost of each query to $\mathcal{B}$ in terms of queries to the oracle $\mathcal{O}$. Since the function f chosen by $\mathcal{B}_1$ is completely independent of $\mathcal{O}$, we do not count its cost. Instead a query to $\mathcal{B}_2$ counts $\ell q + 3q + |\mathcal{O}'|$ queries to $\mathcal{O}$ in total. This cost is obtained as follows: Step 2 makes ℓq queries as it evaluates V ℓ times, Step 3 is made offline, Step 4 counts $|\mathcal{O}'|$ queries that are needed to perform the $\diamond_c$ operation and finally Step 5 and Step 6 require $2q$ and q queries respectively.

4   Insecurity of VUFs Relative to Our Oracles

In this section we formally show that for every candidate black-box construction $(KG^{\mathcal{O}}, F^{\mathcal{O}}, \Pi^{\mathcal{O}}, V^{\mathcal{O}})$ of a VUF from ATDP there is an efficient adversary $\mathcal{A}$ that breaks the unpredictability of the VUF with non-negligible probability $1 - \delta$ by making a polynomial number of oracle queries to $\langle \mathcal{O}, \mathcal{B} \rangle$.

Let q be the maximum number of oracle queries that can be made by the VUF algorithms (according to Assumption 1) and $c \in \mathbb{N}$ be a sufficiently large constant specified below. Without loss of generality, in the following proof we assume $q \geq 2$ and we fix c such that $\delta \leq \frac{3}{eq^{c-1}}$ and our adversary has non-negligible advantage at least $1 - \delta$. Also we set $\ell = q^c$.

Our adversary $\mathcal{A}$ works as follows:

INPUT: A public key PK and access to the function oracles $F(SK, \cdot), \Pi(SK, \cdot)$.
OUTPUT: $x^*, y^* \in \{0,1\}^n$.
ALGORITHM: Our algorithm performs the following steps:
1. Query $\mathcal{B}_1$ on input $(KG^{\mathcal{O}}, F^{\mathcal{O}}, \Pi^{\mathcal{O}}, V^{\mathcal{O}}), PK$ and obtain $x_1, \dots, x_\ell$.
2. Query the VUF oracles $F(SK, \cdot), \Pi(SK, \cdot)$ on x_i for all $i = 1$ to ℓ. Let $\{y_1, \pi_1, \dots, y_\ell, \pi_\ell\}$ be the values obtained from such queries.
3. Query $\mathcal{B}_2$ on input $(KG^{\mathcal{O}}, F^{\mathcal{O}}, \Pi^{\mathcal{O}}, V^{\mathcal{O}}), PK, \{x_1, y_1, \pi_1, \dots, x_\ell, y_\ell, \pi_\ell\}$.
4. If $\mathcal{B}_2$ returns $\perp$, then halt and fail. Otherwise, if $\mathcal{B}_2$ returns (x^*, y^*), then output (x^*, y^*).

Then we are able to state the following lemma:

Lemma 1. *The adversary $\mathcal{A}$ defined above with input PK and oracle access to $\langle \mathcal{O}, \mathcal{B} \rangle$ wins the unpredictability experiment with probability at least $1 - \frac{3}{eq^{c-1}}$ and makes at most $2q^{c+1} + 4q$ oracle queries.*

The proof is given in the full version [14].

5   Security of ATDPs Relative to Our Oracles

In this section we show the existence of a trapdoor permutation $(G^{\mathcal{O}}, E^{\mathcal{O}}, D^{\mathcal{O}})$ that is adaptively one-way even against adversaries that have access to $\mathcal{B}$. The construction is straightforward as each algorithm forwards its input to the corresponding oracle, namely: $G^{\mathcal{O}}(td) = g(td), E^{\mathcal{O}}(ek, a) = e(ek, a)$ and $D^{\mathcal{O}}(td, b) = d(td, b)$.

By the randomness of the oracle $\mathcal{O}$, it is easy to see that the above construction is a secure ATDP when the adversary is given access only to $\mathcal{O}$. Therefore, in order to prove its security relative to the oracle $\mathcal{B}$, we will show that $\mathcal{B}$ does not help to break the one-wayness of $(G^{\mathcal{O}}, E^{\mathcal{O}}, D^{\mathcal{O}})$, namely that $\mathcal{B}$ can be simulated to the adversary $\mathcal{A}$. Now we can state the following lemma:

Lemma 2. *Let $(G^{\mathcal{O}}, E^{\mathcal{O}}, D^{\mathcal{O}})$ be an adaptive trapdoor permutation where each algorithm forwards its input to $g, e,$ and d respectively. Then, for every adversary $\mathcal{A}$ that has access to $\langle \mathcal{O}, \mathcal{B} \rangle$ and makes at most q oracle queries, there is a sufficiently large λ such that the probability that $\mathcal{A}$ succeeds in the adaptive one-wayness experiment against the above construction is at most negligible.*

5.1   Defining the Simulator

Recall that the main idea is to show that $\mathcal{A}$ can simulate the oracle $\mathcal{B}$ locally. To do so, we show that for every $\mathcal{A}$, there exists a simulator $\mathcal{S}$ that gets the same input as $\mathcal{A}$, but which does not have access to $\mathcal{B}$. We then show that the success probability of $\mathcal{S}$ is close to that of $\mathcal{A}$.

Intuition for the Simulator. In the first step, the simulator generates a random trapdoor permutation oracle $\mathcal{O}_{\mathcal{S}}$ locally, except for the portion concerning the permutation $e(ek^*, \cdot)$. In particular $\mathcal{O}_{\mathcal{S}}$ is defined progressively by choosing its answers uniformly at random. Moreover, we construct $\mathcal{S}$ such that it collects into a partial oracle $\mathcal{O}^*$ all the queries of the form $[e(ek^*, \cdot)]$ that $\mathcal{A}$ makes during the simulation. This way, $\mathcal{S}$ knows all the trapdoors of all the public keys (but ek^*) and is therefore able to evaluate all inversion queries $d(td, \cdot)$ where $g(td) \neq ek^*$.

The first three steps of the algorithm $\mathcal{B}_2$ can easily be simulated as in the real case. The first difference comes up into Step 4 where $\mathcal{S}$ has to define the oracle $\mathcal{O}''$. The difficulty here is that the simulator does not know the entire $\mathcal{O}$ and thus it cannot compute the composition $\mathcal{O} \diamond_c \mathcal{O}'$. We solve this problem using an idea similar to the one used in [33]. Namely, we define $\mathcal{O}''$ such that it is consistent with the partial oracles that are known to S so far (i.e., $\mathcal{O}_S, \mathcal{O}^*$ and $\mathcal{O}'$) and we forward all other queries to $\mathcal{O}$. This solves most of the problematic cases due to the fact that the adversary $\mathcal{A}$ only knows *queried* mappings (which are also known to S since it has stored all of them).

One remaining issue are those queries $[d(td', b)]$ such that td' is the trapdoor that is "virtually" associated to ek^* (i.e., $[g(td') = ek^*] \in \mathcal{O}'$) and there is no known mapping $[e(ek^*, \cdot) = b]$ in $\mathcal{O}^*$. Indeed, recall that the simulator does not know the real trapdoor td^* such that $[g(td^*) = ek^*] \in \mathcal{O}$, and also notice that forwarding these unknown queries to $\mathcal{O}$ would inevitably lead to an inconsistent mapping. Assume for example that $\alpha = [d(td', b)]$ is answered with $\mathcal{O}(\alpha) = a$. Then we have a mapping $[e(ek^*, a) = b] \in \mathcal{O}''$, but it is very unlikely that $[e(ek^*, a) = b]$ is in $\mathcal{O}$. Such inconsistencies could potentially be discovered in Step 6 which would cause the simulation to output $\perp$ while it should not.

Fortunately, we show how to handle such queries by using the external inversion oracle $I(ek^*, \cdot)$. Finally, the last remaining problem is the query $\alpha = [d(td', b^*)]$. We cannot answer this query correctly (at least as long as the inverse of b^* has not been discovered before), however we will show that this case only happens with negligible probability. The main idea is that either $\mathcal{A}$ cannot provide an accepting input to $\mathcal{B}_2$ or (in the case that we have passed all the checks and have reached Step 5) the probability that this query cannot be answered is very small.

The full description of the simulator and the proof are provided in the full version [14].

Acknowledgments. We thank the anonymous referees for their valuable comments. We would like to thank Yevgeniy Vahlis for helpful clarifications about black-box separation techniques and Michel Abdalla for helpful discussions on this work. We also thank Jonathan Katz and Arkady Yerukhimovich for helpful discussions about their augmented black-box model. The work described in this paper has been supported in part by the European Commission through the ICT programme under contract ICT-2007-216676 ECRYPT II and by the Emmy Noether Program Fi 940/2-1 of the German Research Foundation (DFG).

References

1. Micali, S., Rabin, M.O., Vadhan, S.P.: Verifiable random functions. In: 40th Annual Symposium on Foundations of Computer Science, pp. 120–130. IEEE Computer Society Press (1999)
2. Micali, S., Reyzin, L.: Soundness in the Public-key Model. In: Kilian, J. (ed.) CRYPTO 2001. LNCS, vol. 2139, pp. 542–565. Springer, Heidelberg (2001)
3. Micali, S., Rivest, R.L.: Transitive Signature Schemes. In: Preneel, B. (ed.) CT-RSA 2002. LNCS, vol. 2271, pp. 236–243. Springer, Heidelberg (2002)
4. Jarecki, S., Shmatikov, V.: Handcuffing Big Brother: an Abuse-Resilient Transaction Escrow Scheme. In: Cachin, C., Camenisch, J. (eds.) EUROCRYPT 2004. LNCS, vol. 3027, pp. 590–608. Springer, Heidelberg (2004)
5. Liskov, M.: Updatable Zero-Knowledge Databases. In: Roy, B. (ed.) ASIACRYPT 2005. LNCS, vol. 3788, pp. 174–198. Springer, Heidelberg (2005)

6. Lysyanskaya, A.: Unique Signatures and Verifiable Random Functions from the DH-DDH Separation. In: Yung, M. (ed.) CRYPTO 2002. LNCS, vol. 2442, pp. 597–612. Springer, Heidelberg (2002)

7. Dodis, Y.: Efficient Construction of (Distributed) Verifiable Random Functions. In: Desmedt, Y.G. (ed.) PKC 2003. LNCS, vol. 2567, pp. 1–17. Springer, Heidelberg (2002)

8. Dodis, Y., Yampolskiy, A.: A Verifiable Random Function with Short Proofs and Keys. In: Vaudenay, S. (ed.) PKC 2005. LNCS, vol. 3386, pp. 416–431. Springer, Heidelberg (2005)

9. Abdalla, M., Catalano, D., Fiore, D.: Verifiable Random Functions from Identity-Based Key Encapsulation. In: Joux, A. (ed.) EUROCRYPT 2009. LNCS, vol. 5479, pp. 554–571. Springer, Heidelberg (2009)

10. Hohenberger, S., Waters, B.: Constructing Verifiable Random Functions with Large Input Spaces. In: Gilbert, H. (ed.) EUROCRYPT 2010. LNCS, vol. 6110, pp. 656–672. Springer, Heidelberg (2010)

11. Reingold, O., Trevisan, L., Vadhan, S.P.: Notions of Reducibility between Cryptographic Primitives. In: Naor, M. (ed.) TCC 2004. LNCS, vol. 2951, pp. 1–20. Springer, Heidelberg (2004)

12. Impagliazzo, R., Rudich, S.: Limits on the provable consequences of one-way permutations. In: 21st Annual ACM Symposium on Theory of Computing, pp. 44–61. ACM Press (1989)

13. Gertner, Y., Kannan, S., Malkin, T., Reingold, O., Viswanathan, M.: The relationship between public key encryption and oblivious transfer. In: 41st Annual Symposium on Foundations of Computer Science, pp. 325–335. IEEE Computer Society Press (2000)

14. Fiore, D., Schröder, D.: Uniqueness is a different story: Impossibility of verifiable random functions from trapdoor permutations. Cryptology ePrint Archive, Report 2010/648 (2010), http://eprint.iacr.org/

15. Bresson, E., Monnerat, J., Vergnaud, D.: Separation Results on the "One-More" Computational Problems. In: Malkin, T. (ed.) CT-RSA 2008. LNCS, vol. 4964, pp. 71–87. Springer, Heidelberg (2008)

16. Fischlin, M., Schröder, D.: On the Impossibility of Three-Move Blind Signature Schemes. In: Gilbert, H. (ed.) EUROCRYPT 2010. LNCS, vol. 6110, pp. 197–215. Springer, Heidelberg (2010)

17. Brakerski, Z., Goldwasser, S., Rothblum, G.N., Vaikuntanathan, V.: Weak Verifiable Random Functions. In: Reingold, O. (ed.) TCC 2009. LNCS, vol. 5444, pp. 558–576. Springer, Heidelberg (2009)

18. Naor, M., Reingold, O.: Synthesizer and their applications to the parallel construction of pseudo-random functions. Journal of Computer and System Sciences 58 (1999)

19. Maurer, U.M., Sjödin, J.: A Fast and Key-Efficient Reduction of Chosen-Ciphertext to Known-Plaintext Security. In: Naor, M. (ed.) EUROCRYPT 2007. LNCS, vol. 4515, pp. 498–516. Springer, Heidelberg (2007)

20. Dwork, C., Naor, M.: Zaps and their applications. SIAM Journal on Computing 36, 1513–1543 (2007)

21. Goldreich, O., Goldwasser, S., Micali, S.: How to construct random functions. Journal of the ACM 33, 792–807 (1986)

22. Kiltz, E., Mohassel, P., O'Neill, A.: Adaptive Trapdoor Functions and Chosen-Ciphertext Security. In: Gilbert, H. (ed.) EUROCRYPT 2010. LNCS, vol. 6110, pp. 673–692. Springer, Heidelberg (2010)

23. Boneh, D., Papakonstantinou, P.A., Rackoff, C., Vahlis, Y., Waters, B.: On the impossibility of basing identity based encryption on trapdoor permutations. In: 49th Annual Symposium on Foundations of Computer Science, pp. 283–292. IEEE Computer Society Press (2008)
24. Hsiao, C.-Y., Reyzin, L.: Finding Collisions on a Public Road, or Do Secure Hash Functions Need Secret Coins? In: Franklin, M. (ed.) CRYPTO 2004. LNCS, vol. 3152, pp. 92–105. Springer, Heidelberg (2004)
25. Goldwasser, S., Ostrovsky, R.: Invariant Signatures and Non-interactive Zero-Knowledge Proofs Are Equivalent. In: Brickell, E.F. (ed.) CRYPTO 1992. LNCS, vol. 740, pp. 228–245. Springer, Heidelberg (1993)
26. Goldreich, O., Levin, L.A.: A hard-core predicate for all one-way functions. In: 21st Annual ACM Symposium on Theory of Computing, pp. 25–32. ACM Press (1989)
27. Dodis, Y., Puniya, P.: Feistel Networks Made Public, and Applications. In: Naor, M. (ed.) EUROCRYPT 2007. LNCS, vol. 4515, pp. 534–554. Springer, Heidelberg (2007)
28. Dodis, Y., Oliveira, R., Pietrzak, K.: On the Generic Insecurity of the Full Domain Hash. In: Shoup, V. (ed.) CRYPTO 2005. LNCS, vol. 3621, pp. 449–466. Springer, Heidelberg (2005)
29. Boneh, D., Canetti, R., Halevi, S., Katz, J.: Chosen-ciphertext security from identity-based encryption. SIAM Journal on Computing 36, 915–942 (2006)
30. Gertner, Y., Malkin, T., Myers, S.: Towards a Separation of Semantic and CCA Security for Public Key Encryption. In: Vadhan, S.P. (ed.) TCC 2007. LNCS, vol. 4392, pp. 434–455. Springer, Heidelberg (2007)
31. Rosen, A., Segev, G.: Chosen-Ciphertext Security via Correlated Products. In: Reingold, O. (ed.) TCC 2009. LNCS, vol. 5444, pp. 419–436. Springer, Heidelberg (2009)
32. Katz, J., Schröder, D., Yerukhimovich, A.: Impossibility of Blind Signatures from One-Way Permutations. In: Ishai, Y. (ed.) TCC 2011. LNCS, vol. 6597, pp. 615–629. Springer, Heidelberg (2011)
33. Vahlis, Y.: Two Is a Crowd? A Black-Box Separation of One-Wayness and Security under Correlated Inputs. In: Micciancio, D. (ed.) TCC 2010. LNCS, vol. 5978, pp. 165–182. Springer, Heidelberg (2010)

Author Index